Chemistry in Action

Chemistry in Action

Michael Freemantle

MACMILLAN

First published 1987 by
THE MACMILLAN PRESS LTD
Houndmills, Basingstoke, Hampshire RG21 2XS
and London
Companies and representatives
throughout the world

ISBN 0–333–44497–3 (hardcover)
ISBN 0–333–37310–3 (paperback)

A catalogue record for this book is available
from the British Library

Printed in Hong Kong

Reprinted (with corrections) 1987, 1989. 1990, 1991, 1992

CONTENTS

PREFACE

Chemistry in Action aims to provide a modern, comprehensive and systematic treatment of the core chemistry required to advanced level standard. The content is based on a rigorous analysis of the most recently available syllabuses of the following examining boards: AEB, UCLES, JMB, L, NISEC, OLE, O & C and SUJB. It also covers all the inter-board common core in chemistry published by the GCE Boards in a booklet entitled *Common Cores at Advanced Level* (1983).

The book is structured and written in a style that gives easy access to the theory and knowledge of chemistry needed by 'A' level students: learning objectives, summaries and examination questions are included to help guide the student through his or her studies. The book is also designed to stimulate and sustain the interest of the student in chemistry. Numerous examples of 'chemistry in action' in both developed countries and the developing world are included to demonstrate the importance and relevance of chemistry in industry, society, the environment, history and literature.

Each chapter begins with a short article or extract from a magazine, journal or book. The aim of this is not only to show the broad relevance of chemistry but also to draw the student into the chapter. The learning objectives which follow relate to the entire chapter. The chapter is then split into several sections, each of which is devoted to a major topic. Summaries are provided at the end of each section and examination questions at the end of each chapter. The answers to these questions (given at the end of the book) are my sole responsibility. Boxed items and photographs have two main functions. First, they provide examples of 'chemistry in action'. Secondly, they are included to stimulate the interest of the student.

The content of the book develops in a conventional manner. The first 10 chapters are devoted to physical chemistry. The chapters on inorganic chemistry follow the order of the groups in the Periodic Table — moving from left to right. The final four chapters are devoted to organic chemistry.

The chemical nomenclature, symbols and terminology used in this book are based on the recommendations of the Association of Science Education published in *Chemical Nomenclature, Symbols and Terminology for Use in School Science* (Third Edition, 1985).

I would like to thank Denise Johnston-Burt for providing the initial inspiration for this book and also Mary Waltham and her colleagues at Macmillan Education for their continuous support and hard work throughout the various stages of the book. Finally, I would like to thank my family for their encouragement and understanding whilst I was writing it.

November, 1986 Michael Freemantle

ACKNOWLEDGEMENTS

The author and publishers wish to thank the following who have kindly given permission for the use of copyright material:

The Associated Examining Board, Joint Matriculation Board, Northern Ireland Schools Examination Council, Oxford and Cambridge Schools Examination Board, Scottish Examination Board, Southern Universities' Joint Board, University of Cambridge Local Examinations Syndicate, University of London School Examinations Board, University of Oxford Delegacy of Local Examinations and the Welsh Joint Education Committee for questions from past examination papers.

The Association for Science Education for an extract and tables from *Chemical Nomenclature, Symbols and Terminology for Use in School Science*, Third edition, 1985.

Bell & Hyman Ltd for a diagram from *Fifth Form Chemistry* by K. H. Farrah, 1966 and from *Modern Physical Chemistry* by G. F. Liptrot, J. J. Thompson and G. R. Walker, 1982.

The Bodley Head for an extract from *Ulysses* by James Joyce.

Chemistry International for 'Chemistry's Cornucopia' by Sir George Porter, 1986, Vol. 8, No. 1; 'Periodicity' by Norman E. Holden, 1984, No. 6; and an extract from 'What's in a Name', 1985, Vol. 7, No. 5.

Chatto & Windus Ltd for '*Dulce et Decorum Est*' from *The Collected Poems of Wilfred Owen* edited by C. Day Lewis.

Longman Group Ltd for figures from *Chemistry: Facts, Patterns and Principles* by Kneen, 1972, Addison Wesley.

Methuen & Co. for adapted figure from *Chain Reactions: An Introduction* by F. S. Dainton, 1956.

New Internationalist Publications Ltd for material from October (1984) and January (1980) editions of *New Internationalist*.

New Scientist for 'Understanding the Electron' by L. Milgrom and I. Anderson (27.10.83) and 'The Poor World Needs Chemists' by M. Freemantle (28.4.83); and for 'The Dead Sea' by D. Neev and K. O. Emery, *Science Journal*, December, 1966.

Newsweek, Inc for 'The Physics of Chemistry' by S. Begley (2.11.81) and an extract from 'Snow's Flaky Little Secrets' by S. Begley and J. Carey (16.1.84).

The Observer for 'Superbugs Start to Buzz' by J. McLoughlin (13.9.81) and an extract from 'Chaos = Gas' by J. Wain (10.10.82).

The Royal Society for a figure from British Library R&D Report No. 5626, *A Study of the Scientific Information System in the United Kingdom*, 1981.

The Royal Society of Chemistry for extracts from: 'Neither Le Chatelier's nor a Principle?' by J. and V. Gold, Sept. 1984; 'Hydrocarbons' by J. Brooks, May, 1983; 'The Firework Industry' by R. Lancaster; and 'Bhopal's Poison Gas Tragedy — Could it Happen in Britain?', Feb., 1985, all from *Chemistry in Britain*. Also for extracts from: 'Two British Women Chemists' by G. W. Rayner-Canham, July, 1983; and 'Trimethylamine — A Pungent Experience' by G. R. Willey, Nov., 1985, all from *Education in Chemistry*.

Times Newspapers Ltd for 'The Universe Strikes Back' (27.5.84) by B. Silcock and 'Crystal but not so Clear' (18.12.83) by B. Silcock from *The Sunday Times*.

World Bank for data from *World Development Report* 1985(6), Oxford University Press, New York, 1985(6).

Academic Press for the figure from Brooks, J.(ed.), *Organic Maturation Studies and Fossil Fuel Exploration* (1981).

The author and publishers wish to acknowledge, with thanks, the following photographic sources:

Aerofilms; Albright and Wilson, Australia; Author/IUPAC photographs; Basingstoke and District Hospital; BBC Hulton Picture Library; Courtesy of BDH Chemicals Ltd, Dorset; Biomedicinska Centrum, Uppsala, Sweden; Biophoto Associates; Niels Bohr Institutet, Copenhagen; Courtesy of British Airways; British Nuclear Fuels Ltd; Jim Brownbill; Camera Press, London; J. Allan Cash; CERN; CIBA GEIGY Ltd; FAO photographs J. Van Acker; G. de Sabatino; E. Kennedy; F. Mattioli; F. Botts; Michael Freemantle; Fife News; International Labour Office; ISC Chemicals Ltd; IUPAC; Kinetico/Indusfoto; Lilly Industries Ltd; Museum of Science and Industry, Manchester; NASA; Nobelstiftelsen; OXFAM; Popperfoto; RTZ Photographic Library; Science Photo Library, London; Shell Photographic Service; Snamprogetti Ltd; Courtesy of SRI-Asia; Alan Thomas; Courtesy of the Tourist Office, Lyon, France; UAC International; UNESCO photographs W. Behreuat; K. Chernush; Paul Almasy; J. H. A. Kleijo; D. Roger; Leon Herschtritt; Sunil Jonah; M. Serraillier; F. Boissonnet; UNICEF photographs Sennett; Jacques Danois; Arild Vollan; United Nations; WHO photographs J. Mohr; A. S. Kochar; R. de Silva; K. E. Mott; T. Kelly; P. Larsen; P. Pittet; W. Wiese.

The publishers have made every effort to trace the copyright holders, but if they have inadvertently overlooked any, they will be pleased to make the necessary arrangements at the first opportunity.

1 ATOMIC STRUCTURE

The Universe Strikes Back

Everything in the world of sub-atomic physics seemed to be falling into place last year [1983], when scientists at Cern, the European nuclear research centre in Geneva, discovered the so-called W and Z particles, exactly as the theorists had predicted. The 'fundamental' particles—the basic building blocks of matter—had apparently been marshalled into some sort of basic order.

Now it turns out that the celebrations may have been premature. Further results from the very experiments that first revealed the W and the Z particles have called into question the 'standard model'—the comprehensive theory laboriously built up over the past 10 years, from which the existence of the W and Z was predicted in the first place.

An experiment at CERN to investigate fundamental particles

By definition, a fundamental particle, such as the electron or the quark, is one that cannot be broken down into smaller components. Protons and neutrons, for example, can each be built up from three quarks. But the latest Cern results suggest that the inferred neat set of basic building blocks, rounded off so nicely by the discovery of the W and Z particles, might not be as complete—or as basic—as was thought.

The Cern experiments may have found new fundamental particles heavier than the W and Z, which is clearly contradicted by the standard model. This extra mass could be explained as simply due to high-energy, or excited, versions of some already known fundamental particles—which would mean that these latter particles are not fundamental after all. Either way, back to the drawing board.

Another possibility is that they *are* new fundamental particles, and are indeed the first experimental support for a theory called super-symmetry. This theory, which is an attempt to find a place for gravity in comprehensive explanations of the four basic forces of nature, would require that the standard model be considerably expanded. More work lies in wait.

Physicists have long yearned to unify the forces of gravity, electromagnetism and the two—the 'strong' and the 'weak'—that act only at the nuclear level. Electromagnetism and the 'weak' force have already been shown to be different manifestations of a single force. Gravity is proving most difficult, but super-symmetry shows promise.

The third possibility, which would leave the standard model intact, is that Cern has stumbled on an ill-defined entity known to physicists as a Higgs particle.

So far experimental traces of only a handful of the new particles have been identified, but physicists still think they are on to something. Further results are eagerly awaited.

Bryan Silcock

LEARNING OBJECTIVES

After you have studied this chapter you should be able to

1. Describe the main features of the structure of the atom.
2. Outline the historical developments that have led to our modern knowledge of the atom.
3. Explain the terms: **mass number, relative atomic mass, isotope** and **atomic number.**
4. Calculate the relative atomic mass of an element.
5. Sketch and explain the main features of a **mass spectrometer.**
6. Interpret a simple mass spectrograph of an element.
7. Indicate how **atomic spectra** and **ionisation energies** provide evidence for the **electronic structure** of atoms.
8. Work out the electronic structure of a given element.
9. Comment on the **dual nature of the electron.**
10. Describe the nature and properties of the three principal **types of radiation.**
11. Briefly describe the methods of detecting radiation.
12. Write equations for **nuclear transformations.**
13. Plot **decay curves** and calculate **half-lives** from experimental data.
14. Discuss the occurrence and uses of natural radioactivity.
15. Describe how nuclear transformations may be brought about artificially.
16. Distinguish between **nuclear fusion** and **nuclear fission.**
17. Outline some of the uses of radioactivity and **nuclear energy.**
18. Indicate the hazards associated with radioactivity and nuclear energy.

1.1
**Sub-atomic
Particles**

ELECTRONS, PROTONS AND NEUTRONS

The first modern atomic theory was proposed by John Dalton. He proposed that an element was composed of atoms of identical size and mass. These particles were indivisible and remained unchanged during a chemical reaction. He worked out the relative weight of the atoms of elements such as hydrogen, oxygen, nitrogen and suphur. He gave a symbol to each element.

A number of discoveries in the late nineteenth century showed that the atom was not indivisible but consisted of sub-atomic particles. The first evidence came from the observations of rays emanating from a negative electrode. Crooke (1879) and Goldstein (1886) both demonstrated these **cathode rays** in a number of experiments. In Crooke's paddle-wheel experiment, for example, a small paddle-wheel was propelled along glass rails by cathode rays. In 1890 **X-rays** were discovered by William Roentgen. The following year Henri Becquerel showed that uranium naturally emitted invisible radiation comparable to X-rays. Both Roentgen and Becquerel were awarded Nobel Prizes for their work.

The Electron

The electron was the first sub-atomic particle to be identified. In 1874 G. J. Stoney proposed that an electric current consisted of negative particles called electrons. However, the credit for the discovery and identification of the electron is almost universally given to Sir Joseph John Thomson.

Thomson discovered the electron through his work on cathode rays. Relevant features of the discharge tube he used to produce cathode rays are shown in figure 1.1. At low pressures and high voltages (15 000 volts and above), he was able to produce cathode rays which produced a well defined spot on a fluorescent screen. The spot could be deflected by an electric field produced by secondary electrodes. The spot was also deflected by a magnetic field perpendicular to the electric field (this is not shown in the diagram). This led him to conclude that cathode rays consisted of streams of negatively charged particles called electrons. By using his measurements of the magnetic field strength, electric field and extent of deflection of the spot, he was able to determine the charge-to-mass ratio (e/m) of these particles. He found that, whatever gas was used to fill the discharge tube, the value for e/m remained the same. He concluded that all atoms contained electrons.

In 1909 Millikan, through his famous oil-drop experiment, determined the charge e on an electron. Combined with Thomson's value for e/m it was then possible to calculate the mass m of an electron. Nowadays the accepted values for these are:

$$e = 1.602 \times 10^{-19} \text{ C}$$
$$m_e = 9.110 \times 10^{-28} \text{ g}$$

Sir Joseph John Thomson, who discovered the electron in 1897. He won the Nobel Prize for Physics in 1906. His son, Sir George Paget, through his work on electron diffraction through gold foil, confirmed de Broglie's theory that free electrons behave both like waves and like particles. Sir George Paget shared the Nobel Prize for Physics in 1937 with C. J. Davisson

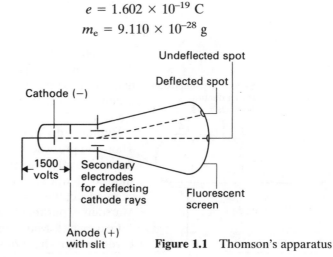

Figure 1.1 Thomson's apparatus

Millikan and Mulliken

Robert A. Millikan   Robert S. Mulliken

There is sometimes confusion between Robert Andrews Millikan and Robert Sanderson Mulliken, both of whom won Nobel Prizes.

Robert Andrews Millikan was the American physicist who determined the charge on an electron by means of his *oil-drop experiment*. In this experiment he used X-rays to charge electrically very small oil drops. The drops were allowed to settle between two horizontal metal plates. The mass of a single drop was determined by measuring its rate of fall. The plates were then charged and as a result the rate of fall of the charged drops changed. The measurement of the velocity of the drops enabled Millikan to calculate the charge on the drops. Although the charges were not identical for each drop they were found to be simple multiples of the same value—the value of the charge on an electron. Millikan won the Nobel Prize for Physics in 1923.

Robert Sanderson Mulliken, an American chemist and physicist, won the Nobel Prize for Chemistry in 1966 for his theoretical studies of the nature of the chemical bond and molecular structure. In the 1920s he applied the theory of quantum mechanics to the study of chemical bonding and the interpretation of molecular spectra. In particular, he formulated the concept of molecular orbitals and showed how electrons could be delocalised in such bonding (*see* chapter 2).

The Proton

The proton was the second sub-atomic particle to be discovered. In 1886 Goldstein had observed positively charged rays emanating from a perforated cathode. He called these canal rays, although they later became known as positive rays (*see* figure 1.2).

Rutherford, in 1899, discovered α- and β-radiation. About the same time, Thomson proposed a plum-pudding model of the atom (*see* below) to account for the negatively and positively charged parts of the atom. In 1902 Rutherford showed that the α-radiation he had identified earlier was due to positively charged helium. However, it was not until 1914 after the famous Geiger and Marsden experiment that positive particles were identified.

Geiger and Marsden were two of Rutherford's students. In 1910 they bombarded a thin sheet of gold foil with a beam of α-particles (*see* figure 1.3). Some

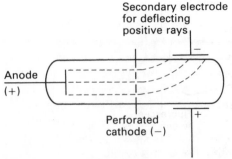

Secondary electrode for deflecting positive rays

Anode (+)

Perforated cathode (−)

Figure 1.2   Positive rays

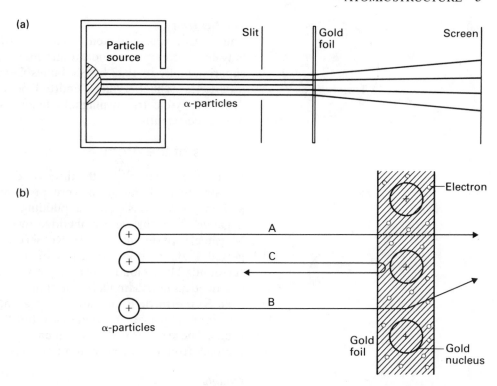

Figure 1.3 The Geiger and Marsden experiment. (a) A stream of α-particles were directed at a thin sheet of gold leaf. Most passed straight through, but a few rebounded back to the source. (b) Rutherford proposed that the rebounding particles had collided with the core of the atom—the nucleus. This evidence led Rutherford to put forward his nuclear model of the atom

passed straight through the foil (course A). Others were deflected off-course (course B). To everyone's surprise about 1 in 20 000 were deflected backwards (course C). 'It was almost as incredible', said Rutherford in a famous statement, 'as if you fired a fifteen-inch shell at a piece of tissue paper and it came back and hit you'. The experiment provided evidence that the atom consisted of a very small positively charged nucleus surrounded by a large space containing light negatively charged electrons.

Rutherford then went on to propose the existence of the proton and showed that its mass was over 1800 times that of the electron.

Ernest Rutherford

Ernest Rutherford was born near Nelson, New Zealand, on 30 August 1871. At the age of 27 he became Professor of Physics at McGill University, Montreal. He soon became one of the leading figures in the rapidly advancing field of radioactivity. He discovered several radioactive elements and identified two types of radiation: α- and β-radiation. With Frederick Soddy he discovered the phenomenon of radioactive half-life. In 1907 he moved to the University of Manchester where, with Hans Geiger, he showed that α-particles were doubly charged helium ions. In 1908 he was awarded the Nobel Prize for Chemistry for his work on radioactivity. In 1910, with Geiger and Marsden, he discovered that α-particles could be deflected by thin metal foils. This led to his model for the structure of the atom (1911). In 1914, the year in which he was knighted, he proposed the existence of the proton and in 1920 he predicted the existence of the neutron. The following year he was awarded the Order of Merit. He was President of the Royal Society from 1915 to 1930 and raised to a peerage in 1931. He died on 19 October, 1937. He was perhaps one of the most distinguished scientists of this century.

(a)

(b)

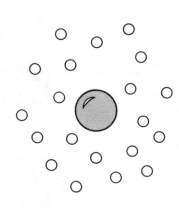

(c)

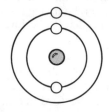

Figure 1.4   Early models of the atom.
(a) Thomson's plum-pudding model
(1904): a ball of positive charge with
electrons embedded in it. (b) Rutherford's
nuclear atom (1911): a positive nucleus
surrounded by a cloud of electrons.
(c) Bohr's model (1913): a nucleus
surrounded by electrons in circular orbits

The Neutron

The neutron was predicted by Rutherford in 1920 to account for the differences between atomic mass and atomic number (*see* below). It was finally detected experimentally in 1932 by Sir James Chadwick. He bombarded beryllium with α-particles. The beryllium emitted penetrating particles which could not be deflected by electric or magnetic fields. Since these particles were neutral, they were called neutrons.

Models of the Atom

During the discoveries of the three fundamental particles, a number of models of the structure of the atom were proposed (*see* figure 1.4). Thomson's plum-pudding model consisted of a 'pudding' or sphere of positive electricity. Rings of negative electrons were embedded in this pudding. Following the Geiger and Marsden experiment in 1910, Rutherford proposed a different model. He suggested that an atom consisted of a small heavy positively charged **nucleus** surrounded by a cloud of light negatively charged electrons. Later, in 1913, Bohr described his famous model of the atom which is still used nowadays (*see* section 1.2). Like Rutherford, Bohr viewed the atom as a positive nucleus surrounded by electrons. However, he suggested that the electrons travelled in stable circular orbits. These orbits had different energies. Electrons could gain or lose energy by jumping from one orbit to another.

Quanta

In 1900 Max Planck suggested that energy could only be absorbed or emitted in discrete quantities called quanta. The magnitude E of these bundles of energy were related to the frequency v of the energy by a constant h, known as Planck's constant:

$$E = hv$$

Thus, the higher the energy of a quantum the higher its frequency.

In 1905, Einstein postulated that all light was composed of particles called **photons**. He used these photons to explain why metal surfaces emitted electrons when, under certain conditions, light was shone on them. The phenomenon was known as the **photoelectric effect**.

Light was thus seen to have a **dual nature**. It could behave as waves—by forming interference and diffraction patterns, for example—or it could behave as a beam of particles—as in the photoelectric effect.

In 1924, de Broglie extended this concept of a dual nature to electrons. In later experiments a beam of electrons was shown to exhibit interference and diffraction patterns like a beam of light. This discovery heralded the beginning of **quantum mechanics**. This was based on the Bohr model of the atom and also on Wolfgang **Pauli's exclusion principle** (1925). This principle determined exactly how electrons were arranged in the discrete energy levels around the nucleus.

In 1926 Schrödinger proposed his famous **wave equation** for particles. This was quickly followed in 1927 by **Heisenberg's uncertainty principle.** This stated that the position and velocity of a particle could not be measured with absolute precision simultaneously. At any time only one or the other could be measured with precision.

Quantum mechanics culminated in the work of Dirac. In 1927 Heisenberg had used matrices as a mathematical tool in his work on quantum mechanics. Dirac developed the **matrix formulations** of quantum mechanics and by this means was able to resolve theoretically the wave–particle paradox. He showed that in some situations the intensity of a wave was equivalent to the density of particles in it. At other times particle density was so low that the wave nature could be ignored altogether. Through this sort of work he was able to predict the existence of **electron spin** and **anti-particles** such as the positron. His work thus heralded a new era in the theory of the structure of the atom.

More and More Elementary Particles

In recent decades the term elementary particle is often used to refer to the indivisible parts of an atom. Until the work of Dirac, only three elementary particles were thought to exist—the electron, the proton and the neutron. Soon after the prediction of Dirac that **anti-particles** existed, other physicists proposed the existence of other elementary particles. The most famous was Hideki Yukawa who, in 1935, proposed the existence of the meson. This is the elementary particle needed to hold protons and neutrons together in the nucleus. In 1938 he also predicted the existence of what he called an 'intermediate vector boson'. Since then hundreds of elementary particles have been predicted. Some have been identified by bombarding matter with particles of enormous velocities. High-energy particle accelerators are used for this purpose.

Anti-particles have the same mass (if any) but the opposite electrical charge to particles. For example, the positron is the electron's anti-particle. It has a positive charge. The anti-proton has a negative electric charge.

Nowadays, the truly fundamental or elementary particles are considered to be **quarks and leptons**. Quarks were discovered by Gell-mann and independently by Zweig in 1964. There are now thought to be at least 18 types of quark. These include the bottom quark, the bottom anti-quark, the charmed quark and a quark endowed with a 'flavour' known as beauty. A beauty-flavoured meson is made up of two quarks one of which has the property beauty. Some quarks have naked beauty whilst for others the beauty is hidden! The 'flavours' of quarks distinguish their quantum properties (*see* next section). As well as quarks, six leptons and a dozen other particles that act as carriers of forces are thought to exist as fundamental particles. The leptons are a class of indivisible particles which includes the electron. Until recently, there was no direct experimental evidence that fundamental particles such as the quark and lepton existed. Their existence had been predicted on the basis of hypothetical models of the atom proposed by theoretical physicists.

A proton is thought to consist of a set of three quarks, and a neutron of another set of three quarks. The quarks are held together by a strong force known as the colour force—although it has nothing to do with colours as we know them. The colour force is due to gluons. Different types of gluons have different colours. When quarks bind together in a proton or neutron, gluons are exchanged. This theory of the colour force is called quantum chromodynamics or QCD for short. This theory is comparable in many ways to the fundamental theory of electromagnetic interactions which Dirac introduced in the late 1920s and which was developed over the next 20 years or so. In this theory the force between two electrically charged particles is explained by the exchange of photons. This theory is called quantum electrodynamics or QED for short.

The discovery of the so-called W and Z particles by scientists in Geneva in 1983, however, has put both theories into question. A schematic summary of the present state of knowledge concerning atomic structure is shown in figure 1.5.

The study of elementary particles has, until recently, fallen almost exclusively into the realm of theoretical and high-energy physics. In the last few years, however, chemists have been examining the effects of bombarding molecules with muons and other elementary particles. Even so, most modern chemists still rely heavily on the three fundamental particles—the electron, proton and neutron—to help explain the nature of chemical reactions and chemical systems.

The word *quark* originates from a line in James Joyce's *Finnegan's Wake*: 'Three quarks for Muster Mark!/Sure he hasn't got much of a bark.'

THE MASS OF AN ATOM

Mass Number

This is the number of protons and neutrons in a nucleus. It is given the symbol A.

A nucleus is often called a **nuclide** and the nuclear particles—protons and neutrons—sometimes called **nucleons**.

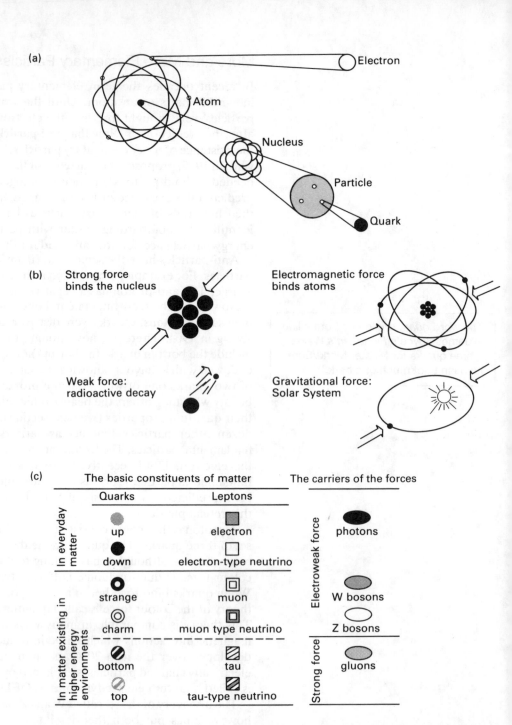

Figure 1.5 A summary of the present state of knowledge concerning atomic structure, field forces and the constituent particles of matter. (a) Structure of the atom. (b) The four types of force. (c) Particles and force carriers

Atomic Number

The atomic number of an element is the number of protons in the nucleus. It is given the symbol Z. Atomic number is related to mass number by the following equation:

$$A = Z + N$$

where N is the number of neutrons in the nucleus.

Each element has its own atomic number. In other words, no two elements can have the same atomic number. Atomic number is not only equal to the number of protons in a nucleus but also equal to the number of electrons surrounding the nucleus. This is because there is no net charge on an atom. The number of protons thus equals the number of electrons. This does not apply to ions which are, of course, charged particles.

The first evidence of atomic numbers was produced in 1913 by Henry Moseley whilst working at Oxford. He bombarded solid metal targets with cathode rays. In 1909 Barkla and Kaye had already shown that a solid element emitted X-rays characteristic of that element when bombarded by a fast stream of cathode rays. Moseley analysed the X-rays using photographic techniques. He found that the wavelengths of the X-rays increased with increasing atomic weight of the metal. He then showed that the square root of the frequency of the X-rays was directly proportional to an integral number which is labelled Z:

$$\sqrt{\text{Frequency}} \propto Z$$

This number, he found, was approximately half the value of the atomic weight. He concluded that this number—the atomic number—was a fundamental property of the atom. It was the number of protons in an atom of the element.

Moseley, through his work, was able to predict the existence of three missing elements—those with atomic numbers 43, 61 and 75. These were later identified as technetium, promethium and rhenium respectively.

Moseley was killed in action in World War I.

Symbols of Nuclides

The mass number of a nuclide is often indicated as a superscript and the atomic number as a subscript on the left of the symbol for an element. For example, $^{12}_{6}C$ means that carbon—as always—has an atomic number of 6. This particular nuclide has a mass number of 12. Another nuclide of carbon is $^{14}_{6}C$. Since all carbon nuclides have an atomic number of 6, this nuclide is often written as ^{14}C or carbon-14.

Isotopes

These are different atomic species of the same element. They differ in the number of neutrons in their nuclei. They thus have the same atomic number but different mass numbers. The values for A, Z and N for three isotopes of carbon are shown in table 1.1.

Table 1.1 Isotopes of carbon

Isotope	Atomic number (no. of protons) Z	No.of neutrons N	Mass number A
$^{12}_{6}C$	6	6	12
$^{13}_{6}C$	6	7	13
$^{14}_{6}C$	6	8	14

Isotopic abundance

Most elements consist of mixtures of isotopes. The abundance of each in the mixture is called its isotopic abundance. For example, silicon occurs in naturally occurring compounds as 92.28% ^{28}Si; 4.67% ^{29}Si; and 3.05% ^{30}Si. Note that these percentages add up to 100%. The **fractional abundances** of these isotopes are 0.9228, 0.0467 and 0.0305 respectively. These add up to 1.0000.

Atomic Mass Unit (amu)

Nowadays the standard for this unit is taken to be the mass of the $^{12}_{6}C$ nuclide. This has a mass of 12.0000 amu. An atomic mass unit is thus one-twelfth of the mass of this nuclide. The actual mass of this unit is 1.661×10^{-27} kg. The masses of the three fundamental particles are,

$$\text{Mass of the proton} = 1.007\ 277 \text{ amu}$$
$$\text{Mass of the neutron} = 1.008\ 665 \text{ amu}$$
$$\text{Mass of the electron} = 0.000\ 548\ 6 \text{ amu}$$

The **isotopic mass** of any nuclide can be calculated using these values. For example, the isotopic mass of $^{35}_{17}Cl$ is the sum of the masses of 17 protons, 18 neutrons and 17 electrons:

$$17(1.007\ 277 \text{ amu}) + 18(1.008\ 665 \text{ amu}) + 17(0.000\ 548\ 6 \text{ amu})$$
$$= 35.289\ 005 \text{ amu}$$

However, accurate experimental determinations of the isotopic mass of $^{35}_{17}Cl$ have shown the value to be 34.968 85 amu. The difference between the calculated and experimental values is 0.320 16 amu. This is known as the **mass defect** or mass deficit (*see* section 1.3).

On the **relative atomic mass scale**, isotopic masses are divided by one-twelfth of the mass of the $^{12}_{6}C$ nuclide. Thus, the **relative isotopic mass** of $^{35}_{17}Cl$ is

$$\frac{34.968\ 85\ \text{amu}}{(1/12) \times 12.0000\ \text{amu}} = 34.968\ 85$$

Note that on the relative atomic mass scale the units have disappeared.

The **relative atomic mass**, A_r of an element is the average of the relative isotopic masses weighted to take account of the isotopic abundances. It is calculated by multiplying the relative isotopic mass of each isotope by its fractional abundance and adding all these values together.

EXAMPLE

Calculate the relative atomic mass of chlorine from the following data:

Isotope	Relative isotopic mass	Fractional abundance
$^{35}_{17}Cl$	34.97	0.7553
$^{37}_{17}Cl$	36.95	0.2447

SOLUTION

$$A_r = (34.97 \times 0.7553) + (36.95 \times 0.2447) = 35.45$$

MASS SPECTROMETRY

This technique has two important uses:
- The determination of relative isotopic masses and abundances of isotopes.
- The determination of the relative molecular masses and structures of organic compounds.

A mass spectrometer

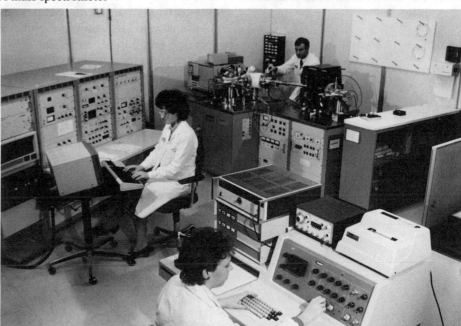

How a Mass Spectrometer Works

The essential features of a mass spectrometer are shown in figure 1.6. A gaseous sample of the material under investigation is allowed to enter the ionisation chamber. Here, an electron gun bombards the sample with electrons, forming positively charged ions. These are mainly singly charged ions. The ions are then accelerated by an electric field and deflected along a circular path by a magnetic field. The lighter the particles the greater the deflection. The intensity of the ion beam is detected electrically, amplified and finally recorded.

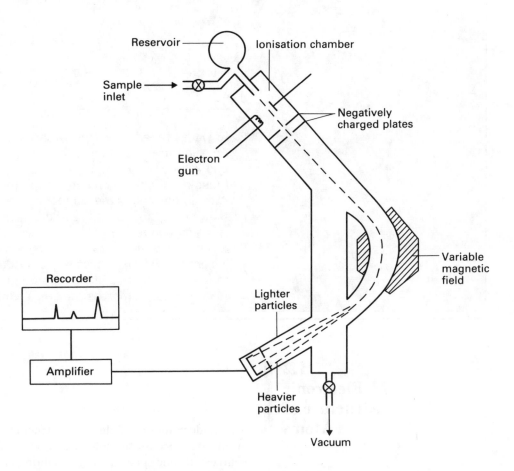

Figure 1.6   The essential features of a mass spectrometer

The **mass spectrum** is produced by increasing the magnetic field so that ions of higher mass/charge ratios are gradually brought into line with the detector. The mass spectrum is usually calibrated using ^{12}C so that relative isotopic masses can be read directly. The relative heights of the peaks give the relative isotopic abundances.

The mass spectrum of a chlorine sample is represented in figure 1.7. It shows that the abundance of ^{35}Cl is 75% and that of ^{37}Cl 25%.

The **mass spectrograph** operates on the same principles as a mass spectrometer. The main difference is that a mass spectrograph uses a photographic plate to detect ions instead of an electrical device.

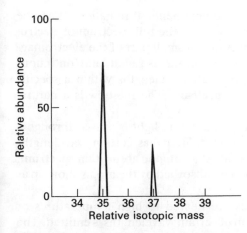

Figure 1.7   The mass spectrum of chlorine

SUMMARY

1. Until the late 1920s, only three **fundamental particles** were thought to exist.

Table 1.2   Three fundamental particles

Particle	Mass/amu	Charge	Discovered by	Date	Experiment
electron	0	−	Thomson	1897	cathode ray tube
proton	1	+	Rutherford	1914	Geiger and Marsden experiment with gold foil
neutron	1	0	Chadwick	1932	bombardment of beryllium with α-particles

2. Nowadays quarks and leptons are thought to be the fundamental particles. An electron is a lepton, and a proton and a neutron each consist of three quarks bound by gluons.
3. **Mass number** is the number of protons and neutrons in a nucleus.
4. **Atomic number** is the number of protons in a nucleus.
5. **Isotopes** of an element have the same atomic number but different mass numbers.
6. **Relative atomic mass** is the average of the relative isotopic masses weighted to allow for **isotopic abundance**. **Relative isotopic mass** is the mass of the **nuclide** compared to one-twelfth of the $^{12}_{6}C$ nuclide.
7. Relative isotopic masses can be determined using a **mass spectrometer.**

1.2
The Electronic Structure of Atoms

The modern theory of electronic structure originates from the Bohr model of the atom. Evidence for this and later models of the atom derives principally from two sources: atomic spectra and ionisation energies.

ATOMIC SPECTRA

A **spectrum** is a display or dispersion of the components of radiation. It can be obtained by using a **spectrometer**. Figure 1.8 shows the full spectrum of **electromagnetic radiation**. Note that visible light is only a small part of the electromagnetic spectrum. Visible light is an example of **continuous radiation**. Continuous radiation is radiation containing radiation of all wavelengths within a specific range. Its spectrum is called a **continuous spectrum**. The rainbow is a natural example of this type of spectrum.

When a beam of continuous radiation such as white light is passed through a gaseous sample of an element, the radiation emitted has certain wavelengths missing. The spectrum of this radiation is called an **atomic absorption spectrum**. The wavelengths of the radiation that has been absorbed by the atoms show up as dark lines on the continuous spectrum.

When elements in their gaseous states are heated to high temperatures or subjected to electrical discharges, radiation of certain wavelengths is emitted. The spectrum of this radiation is called an **atomic emission spectrum** (*see* figure 1.9).

An absorption or emission spectrum which consists of lines is called a **discontinuous** or **line spectrum.** Any part of the spectrum where the lines converge is called a **continuum**.

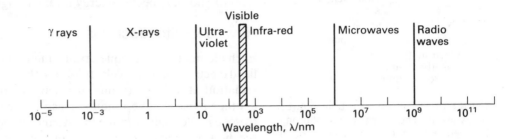

Figure 1.8 The electromagnetic spectrum

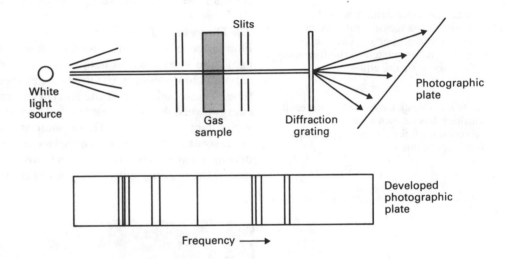

Figure 1.9 Emission spectroscopy

Bohr Relates the Lines to the Electrons

Why do gaseous elements absorb or emit radiation of fixed wavelengths and not a continuous range of radiation? One of Bohr's great achievements was to answer this question. He related the lines in the atomic spectra to changes in the energies of electrons in atoms. He maintained that an electron in an atom could not have a continuous range of energy values. It could only have certain fixed values. These were called **discrete** or **quantised** levels. To each energy value he gave a number which he called a **quantum number**. An electron could jump from one energy level to another by emitting or absorbing a fixed amount of energy. This was called a quantum of energy.

An electron in its lowest energy level was said to be in its **ground state**. Electrons in higher energy levels were said to be in **excited states**. The **transition** of an electron to a higher energy level was called an **excitation**.

What exactly were these quanta of energy which were absorbed or emitted during transitions of electrons from one energy level to another? Bohr was already aware of the work of Planck and Einstein. He was also aware that Einstein had postulated that light consists of photons. Bohr suggested that an electron could only jump from one energy level to a higher one by absorbing a photon of exactly the right energy. Since the energy of a photon was related to wavelengths (or frequency), this meant that an electron has to absorb a photon of exactly the right

Niels Bohr (1885–1962), who won the Nobel Prize for Physics in 1922

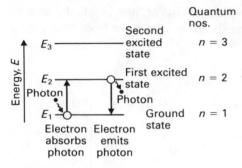

Figure 1.10 Electron transitions between energy levels. The energy of the photon absorbed or emitted by an electron equals the difference between the two energy levels

$$\Delta E = E_2 - E_1$$

(The Greek letter delta Δ is used to indicate the change or difference in the value of some property.) The energy difference ΔE is related to the frequency ν of the photon by the equation

$$\Delta E = h\nu$$

where h is Planck's constant. Thus each transition has its own specific frequency. This corresponds to a specific line in an atomic spectrum

frequency for absorption to occur. Conversely, when an electron descended from a higher to a lower energy level, it emitted a photon with a specific frequency. This frequency corresponded to a specific line in the spectrum. Thus each line in an atomic spectrum corresponded to electrons jumping from one specific energy level to another specific energy level (*see* figure 1.10).

How to Excite Electrons

Each element has a unique arrangement of electrons and thus a unique range of fixed electron energy levels. It follows that the wavelengths and frequencies of the radiation absorbed or emitted when electrons jump from one energy level to another must also be unique. This uniqueness forms the basis of atomic spectroscopy. We shall examine the atomic emission spectrum of hydrogen below and see how it relates to the electron energy levels in the hydrogen atom. The uniqueness of the arrangement of electrons and their energy levels in atoms of a specific element also has a number of important practical applications. We shall consider two: the fluorescent light and the laser.

The Fluorescent Light

Fluorescence is a form of **luminescence**. The essential features of a fluorescent light tube are shown in figure 1.11. A fluorescent light tube contains mercury vapour at low pressure. Electrons are ejected from a hot filament at one end of the tube. These collide with the mercury atoms, exciting their electrons to higher energy levels. As these electrons fall back to lower energy levels, photons of ultraviolet light are emitted. These photons collide with the atoms which form the fluorescent coating inside the surface of the tube. The electrons in these atoms become excited. They then return to lower energy levels emitting light characteristic of the coating. Different types of coating give different coloured light.

Fluorescent lighting

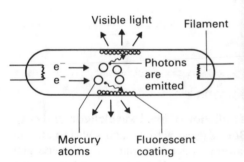

Figure 1.11 The fluorescent light tube

Luminescence

Certain substances, when stimulated by ultra-violet or other forms of radiation, emit visible light. This phenomenon is called luminescence. The incident radiation promotes electrons of atoms, ions or molecules into their excited states. As the electrons return to their ground states, they emit the visible light.

If the return to the ground state occurs promptly after excitation, the phenomenon is called **fluorescence**. However, if the electrons remain in their excited states and emit light over a period of time, the phenomenon is called **phosphorescence**.

cont'd.

Chemical reactions of certain compounds such as luminol (5-amino-2,3-dihydro-1,4-phthalazinedione) result in **chemiluminescence**.

Luminol

The amount and nature of the chemiluminescence depend on factors such as concentration of reagents, pH and the presence of a catalyst.

Naturally occurring forms of luminescence include lightning. This is an example of **electroluminescence**. Certain species of bacteria, crustaceans, fish, fungi, jellyfish, molluscs, protozoa, sponges and worm emit light. Perhaps the most famous examples of animals which emit light are the firefly and the glowworm. The phenomenon is known as **bioluminescence** or 'living' light. It is produced by the reaction of oxygen with a substance in the organism called luciferin. The reaction is catalysed by an enzyme called luciferase. Mammals, birds, reptiles, amphibians and leafy plants do not exhibit luminescence.

Samples of some minerals—such as fluorite, CaF_2—fluoresce when subjected to ultra-violet light. Whether the sample fluoresces or not depends on the exact composition of the mineral.

Barite, a mineral consisting mainly of $BaSO_4$, exhibits **thermoluminescence**. This is the ability to glow in the dark after heating to a temperature below that of red heat.

Luminescence is distinct from **incandescence**. The latter is the emission of light when a substance is heated to a high temperature.

The Laser

We have seen that when electrons in an excited state fall back to lower energy levels, they release photons of characteristic wavelength. This is called **stimulated emission**.

The word *laser* is an acronym for *l*ight *a*mplification by *s*timulated *e*mission of *r*adiation. A typical laser consists of three essential components: a ruby rod; a flash lamp; and a pair of mirrors placed at either end of the rod (*see* figure 1.12). One of these mirrors is partially transparent.

A flash of ultra-violet light from the flash lamp excites electrons in the rod. A few of the excited electrons immediately and spontaneously fall back to lower energy levels. In so doing they emit photons. These photons are reflected back into the rod by the mirrors at either end. This stimulates a chain reaction whereby all the remaining excited electrons simultaneously fall back to their lower energy states. This produces a highly intense pulse of light with a specific direction and frequency. Since one of the mirrors is partially transparent it allows the laser pulse to pass through.

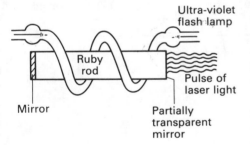

Figure 1.12   The ruby laser

The Atomic Emission Spectrum of Hydrogen

The atomic emission spectrum of hydrogen shows three distinct series of lines (*see* figure 1.13). The series in the visible region of the spectrum is called the **Balmer series**. In 1885, Balmer fitted these lines to the following equation:

$$\frac{1}{\lambda} = R_\infty \left(\frac{1}{2^2} - \frac{1}{n^2} \right)$$

where λ is the wavelength, R_∞ a constant known as the 'Rydberg constant' and n a whole number.

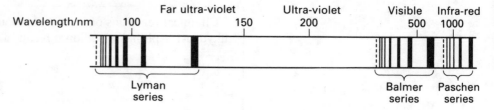

Figure 1.13   The atomic emission spectrum of hydrogen

The series in the infra-red region is called the **Paschen series**. The equation for this series is

$$\frac{1}{\lambda} = R_\infty \left(\frac{1}{3^2} - \frac{1}{n^2} \right)$$

The series in the ultra-violet region is called the **Lyman series**. The equation is

$$\frac{1}{\lambda} = R_\infty \left(\frac{1}{1^2} - \frac{1}{n^2} \right)$$

Bohr related the values for n in these equations to the quantum numbers of the energy levels for the electron of a hydrogen atom (*see* figure 1.14). When this electron exists in its ground state, its quantum number n is 1. Each line in the Lyman series corresponds to an electron falling back from a higher energy level to its ground state. The Balmer series corresponds to electrons falling from higher energy levels to the first excited state (quantum number $n = 2$). The Paschen series corresponds to electrons falling to the second excited state.

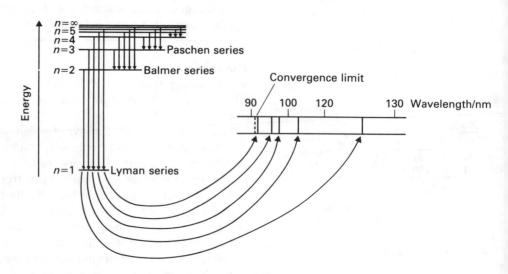

Figure 1.14   Relating transitions to spectral lines

Notice how the lines of each series converge as wavelength decreases (*see* figure 1.13 and 1.14). The wavelengths represented by the broken lines are called **convergence limits**. As the quantum numbers of the energy levels increase, the energy levels become closer and closer together until they converge. The convergence limits correspond to transitions of electrons in these highest energy levels.

What happens if an electron has an even higher energy? In this case the electron becomes detached from the atom. The atom is thus ionised. It has become a positive ion. The energy required to lift an electron from its ground state and detach it from its atom is called the **ionisation energy**. Ionisation energies provide valuable information about the electronic structure of atoms.

IONISATION ENERGIES

The energy required to remove one mole of electrons from one mole of atoms of an element is known as the **first ionisation energy**, $\Delta H_{1,m}^{\ominus}$ (*see* chapter 5 for an exact definition of $\Delta H^{\ominus}$). The energy required to remove one mole of electrons from one mole of singly charged positive ions is known as the **second ionisation energy**, $\Delta H_{2,m}^{\ominus}$. For example, the first and second ionisation energies of sodium are given by

$$Na(g) \rightarrow Na^+(g) + e^- \qquad \Delta H_{1,m}^{\ominus} = +495 \text{ kJ mol}^{-1}$$
$$Na^+(g) \rightarrow Na^{2+}(g) + e^- \qquad \Delta H_{2,m}^{\ominus} = +4563 \text{ kJ mol}^{-1}$$

Note that ionisation energies refer to atoms and ions in their gaseous states. Notice also that the amount of energy required to remove a second electron from the sodium atom is almost 10 times that required for the first electron. An element can have several ionisation energies—the exact number corresponding to its atomic number.

Experimental Determination of Ionisation Energies

Ionisation energies can be determined experimentally from atomic spectra and by the electron impact method.

Atomic spectra. As we have seen, the convergence limit of a series of lines on an atomic spectrum corresponds to the ionisation of an atom. The frequency of radiation at the convergence limit can be converted to ionisation energy by using the equation

$$\Delta E = h\nu$$

We have seen that for hydrogen there are several series and thus several convergence limits. So which convergence limit do we take? First of all we need to remember that ionisation energy is the energy needed to lift an electron from its ground state and remove it completely from its atom. Reference to figure 1.14 shows that the Lyman series corresponds to electrons falling back to their ground state. We thus use the convergence limit of this series.

The electron impact method. A discharge tube is filled with a gaseous sample of the element under investigation. The sample is subjected to bombardment of electrons which accelerate from the cathode of the discharge tube. The potential between the cathode and a grid is gradually increased. A sudden increase in current corresponds to the ionisation of the sample. The ionisation energy can be calculated from the potential difference when ionisation occurs.

Relating Ionisation Energies to the Bohr Model of the Atom

Bohr proposed that the electron in the hydrogen atom could only exist in certain fixed or quantised energy levels. The same held for other elements. Electrons rotated in circular orbits (sometimes called shells) outside the nucleus. Each shell had a fixed energy level and a quantum number assigned to it. The electrons rotated in these shells with a fixed angular momentum. The value of the angular momentum depended on the quantum number. The shells were given letters K, L, M and so on. For each shell there was a maximum number of electrons which it could contain (*see* figure 1.15).

As we shall see in more detail later in the chapter and in chapter 11, the arrangement of electrons in these shells corresponds to the arrangement of elements in the Periodic Table. The electron arrangement also corresponds to certain patterns which emerge in the ionisation energies of the elements. For example, the first ionisation energies of elements generally decrease down a group (*see* table 1.3). This is because the outer-shell electrons become further and further away from the nucleus on descending a group. The nucleus thus exerts less

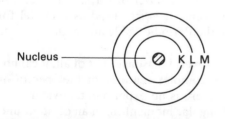

Shell	Quantum number	Maximum number of electrons permitted
K	1	2
L	2	8
M	3	18

Figure 1.15 The Bohr model of the atom

attraction on the electrons and they become easier to remove. On the other hand, as we cross a period the ionisation energies tend to increase. This corresponds to an increase in the atomic number and the number of electrons in a specific shell.

Table 1.3   First ionisation energies and electronic configurations

element	atomic number	quantum no., n = shell	1 K	2 L	3 M	first ionisation energy/ kJ mol^{-1}
group 0						
helium	2		2			2372
neon	10		2	8		2081
argon	18		2	8	8	1520
period 3						
sodium	11		2	8	1	495
magnesium	12		2	8	2	738
aluminium	13		2	8	3	577
silicon	14		2	8	4	787
phosphorus	15		2	8	5	1060
sulphur	16		2	8	6	1000
chlorine	17		2	8	7	1255
argon	18		2	8	8	1520

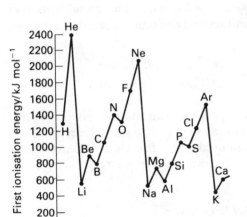

Figure 1.16   First ionisation energies of the elements with atomic numbers 1 to 20

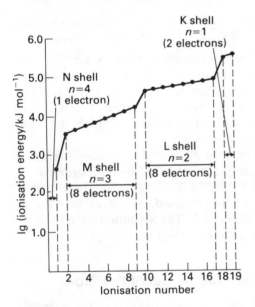

Figure 1.17   Successive ionisation energies of the potassium atom

When the first ionisation energies of elements are plotted against atomic number, certain patterns emerge (*see* figure 1.16). The highest values are those for the noble gases. These all have outer shells full of electrons. They are thus very stable. The alkali metals have the lowest values. They each have one electron in their outer shells and are very reactive.

Evidence for the electronic arrangement of electrons in atoms is also provided by comparing the first, second and successive ionisation energies of a particular element with **ionisation number**. Ionisation number is the number of electrons removed. Thus, the first ionisation has an ionisation number of 1, the second 2 and so on. When the logarithms to base 10 (these have the symbol 'lg') of the ionisation energies are plotted against ionisation number, we see a sudden jump each time an electron is removed from a full shell (*see* figure 1.17). Values increase as we move from electrons in the higher energy shells to electrons in the lowest energy shell, that is the K shell. This we would expect since more energy is required to lift an electron from a low energy level than from a higher energy level (*see* figure 1.14).

WAVE NATURE OF THE ELECTRON

Defects of the Bohr Model

The Bohr model of the atom is still useful in a number of ways. It is useful in explaining the lines in the hydrogen spectrum. It is still used as a model for explaining the arrangement of elements in the Periodic Table and patterns in the ionisation energies of elements.

However, the Bohr model is deficient for several reasons. First of all, it cannot be used to account for some of the more complex features of the spectra of elements other than hydrogen. Secondly, there is no experimental evidence to show that an electron rotates with a fixed angular momentum in an orbit around the nucleus. And even if it did, the electron would gradually lose energy and slow down. It would then be drawn in towards the nucleus. If this were the case the atom would effectively collapse. Indeed all matter would collapse!

The Dual Nature of the Electron

A breakthrough in the development of the quantum theory of the atom came in 1925. In that year Louis de Broglie suggested that a wavelength could be assigned to an electron. It was already known that electromagnetic radiation could have both a wave and particle nature—the particle being the photon. The energy E of the photon can be related to its wavelength λ or its frequency ν by the following equations:

$$E = h\nu = hc/\lambda$$

(where c is the speed of light). By using **Einstein's equation**

$$E = mc^2$$

and substituting this value for E into the above equation we obtain

$$\lambda = \frac{h}{mc}$$

Note that this equation relates wavelength to the momentum of the photon.

De Broglie suggested than an almost identical equation could be written **for the electron**

$$\lambda = \frac{h}{mv} \qquad \text{the de Broglie equation}$$

where λ is the wavelength of the electron, m its mass and v its velocity. This equation formed the basis of the **new quantum theory**.

Momentum = Mass × Velocity

$= m \times c$

Those Greek Letters Again!

We have already met a number of Greek letters in this chapter, for example:

α in α-particles
β in β-particles
ν in the equation $E = h\nu$
λ the symbol for wavelength
Δ in the equation $\Delta E = E_2 - E_1$

Greek letters are used extensively in mathematics and science. The reason is that there are not enough in our own alphabet (the Latin alphabet) to go around. The table below gives the full Greek alphabet.

Greek letter		Greek name	English equivalent	Greek letter		Greek name	English equivalent
A	α	alpha	a	N	ν	nu	n
B	β	beta	b	Ξ	ξ	xi	x
Γ	γ	gamma	g	O	o	omicron	o
Δ	δ	delta	d	Π	π	pi	p
E	ϵ	epsilon	e	P	ρ	rho	r
Z	ζ	zeta	z	Σ	σ	sigma	s
H	η	eta	e	T	τ	tau	t
Θ	θ	theta	th	Y	υ	upsilon	u
I	ι	iota	i	Φ	ϕ	phi	ph
K	κ	kappa	k	X	χ	chi	ch
Λ	λ	lambda	l	Ψ	ψ	psi	ps
M	μ	mu	m	Ω	ω	omega	o

De Broglie viewed the electron as a standing wave which could be fitted into an orbit. Only an integral number of wavelengths were permitted in an orbit. The number of wavelengths in an orbit correspond to the quantum of the electron.

In 1927 the wave-like properties of the electron were confirmed experimentally by Davisson and Germer and by G. P. Thomson. They found that a beam of electrons, like a beam of light, could be diffracted through a crystal and through metal foil. Other evidence supported both the wave and corpuscular (particle) nature of the electron. This evidence is summarised in table 1.4.

Table 1.4 Evidence for the wave and corpuscular nature of the electron

Evidence	Wave theory	Corpuscular theory
diffraction	consistent	inconsistent
reflection	consistent	consistent
refraction	consistent	consistent
interference	consistent	inconsistent
photoelectric effect	inconsistent	consistent

Orbitals

The position and momentum of an electron cannot both be known with absolute certainty simultaneously. This is the **Heisenberg uncertainty principle**. However, although the position of an electron cannot be determined exactly, the **probability** of finding an electron in a certain position at any time can be found. A region or volume of space within which there is a high probability of finding an electron is known as an **orbital**. The term orbital should not be confused with the term orbit used in the Bohr theory. An **orbit** in his theory was the path of an electron around a nucleus.

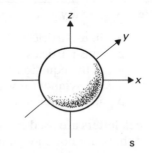

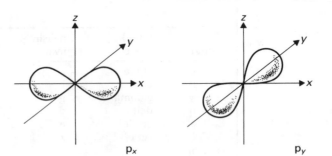

Figure 1.18 Shapes of s and p orbitals

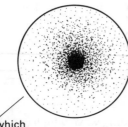

Boundary within which there is a 95% chance of finding an electron

Figure 1.19 Cross section of charged cloud

Electrons can occupy four types of orbital. These are called s, p, d and f orbitals. These orbitals can be represented by three-dimensional surface boundaries. The regions or volumes enclosed by these boundaries are usually taken to be those within which there is a 95% chance of finding a particular electron. Figure 1.18 shows the shapes of the s and p orbitals. Note that the s orbital has the shape of a sphere and the p orbitals dumb-bell shapes.

Since an electron has a negative charge, its orbital can be regarded as a spread of charge. This is often called a **charged cloud** (*see* figure 1.19).

QUANTUM NUMBERS

The Quantum Characteristics of an Electron

The Bohr theory gave the four electron **shells** K, L, M and N the quantum number n where n equals 1, 2, 3 and 4 respectively. These numbers represented the increasing energy levels of the shells.

However, close examination of atomic spectra shows that the lines representing transitions between these quantised energy levels are in fact split into finer lines. This indicates that the electron shells are, in fact, split into **sub-shells** each with its own quantised energy level. These sub-shells have been labelled after the types of lines in the atomic spectra to which they correspond. They are

      s   for sharp
      p   for principal
      d   for diffuse
      f   for fundamental

An s sub-shell consists of an s orbital. A p sub-shell consists of up to three p orbitals (*see* figure 1.18) and a d sub-shell of up to five d orbitals.

The line corresponding to transitions between these sub-shells are split even further when elements are subjected to a magnetic field. This is called the **Zeeman effect**. The evidence shows that only p, d and f sub-shells can be split into further energy levels. Each energy level corresponds to an orbital in the sub-shell. A magnetic field has no effect on an s sub-shell because s orbitals are spherical.

In the absence of a magnetic field all the orbitals in a sub-shell have the same energy. Orbitals having the same energy are said to be **degenerate**.

The atomic spectra of elements, under certain conditions, can be split due to the spin of an electron. All electrons spin about their own axis. The spin can be either clockwise or anti-clockwise (*see* figure 1.20).

We see, then, that the energy level of an electron is determined by four characteristics:

- its shell
- its sub-shell
- its orbital
- its spin

Each of these characteristics is assigned a quantum number.

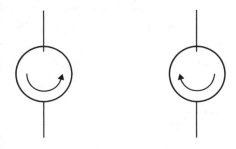

Figure 1.20 Electron spin: the opposite spins of a pair of electrons can be represented by a pair of arrows thus: ↑ ↓

The Four Quantum Numbers

The principal quantum number (n)**.** This is assigned to the **shell** in which the electron is found. This can have values

$$n = 1, 2, 3, \ldots$$

For example

Shell	n
K	1
L	2
M	3

The higher the value of n, the higher the energy level of the shell.

The subsidiary quantum number (*l*). This is assigned to the **sub-shell** in which an electron is found. It can have values

$$l = 0, 1,..., (n - 1)$$

where *n* is the principal quantum number. The number of orbitals in a sub-shell equals $2l + 1$. For example

Sub-shell	*l*	No. of orbitals
s	0	1
p	1	3
d	2	5

The relative energy level of all the s, p and d sub-shells of the first four shells are shown in figure 1.21. Note that the 4s sub-shell is exceptional in that it has a lower energy than the 3d sub-shell.

The magnetic quantum number (*m*). As we have seen, the energy values of the orbitals in a sub-shell are normally degenerate. However, under the influence of a magnetic field they become discrete or quantised. The magnetic quantum number of these discrete energy levels can have values

$$m = -l \text{ to } +l$$

For example

Orbital	*m*
	−1
p	0
	+1

The spin quantum number (*s*). The spin quantum number of an electron can have one of two values: $+\frac{1}{2}$ or $-\frac{1}{2}$.

The four quantum numbers of all the electrons in the first three shells are summarised in table 1.5. Notice that each electron has its own set of quantum numbers and this set is different for each electron.

Table 1.5   Quantum numbers

Shell *n*	Sub-shell $l = 0,..., n-1$		Orbital $m = -l,..., +l$	Spin $\pm\frac{1}{2}$		Maximum no. of electrons	
K 1	s	0	0	$+\frac{1}{2}$	$-\frac{1}{2}$	2	2
L 2	s	0	0	$+\frac{1}{2}$	$-\frac{1}{2}$	2	8
	p	1	−1	$+\frac{1}{2}$	$-\frac{1}{2}$	6	
			0	$+\frac{1}{2}$	$-\frac{1}{2}$		
			+1	$+\frac{1}{2}$	$-\frac{1}{2}$		
M 3	s	0	0	$+\frac{1}{2}$	$-\frac{1}{2}$	2	18
	p	1	−1	$+\frac{1}{2}$	$-\frac{1}{2}$	6	
			0	$+\frac{1}{2}$	$-\frac{1}{2}$		
			+1	$+\frac{1}{2}$	$-\frac{1}{2}$		
	d	2	−2	$+\frac{1}{2}$	$-\frac{1}{2}$	10	
			−1	$+\frac{1}{2}$	$-\frac{1}{2}$		
			0	$+\frac{1}{2}$	$-\frac{1}{2}$		
			+1	$+\frac{1}{2}$	$-\frac{1}{2}$		
			+2	$+\frac{1}{2}$	$-\frac{1}{2}$		

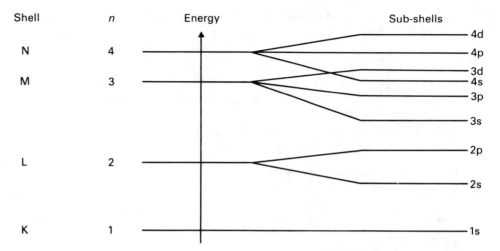

Figure 1.21   Energy levels

ELECTRONIC CONFIGURATION

The **electronic configuration** of an element describes how the electrons of its atoms are arranged in their shells, sub-shells and orbitals. The term normally applies to atoms in their **ground state**. The electronic configuration of an atom with one or more electrons in an excited state is called an **excited-state configuration**. There are three rules for determining the exact ground-state electronic configuration of an element.

Rule 1: the aufbau principle. The word *aufbau* means 'building up'. This principle states that the electrons in their ground states occupy orbitals in order of the orbital energy levels. The lowest energy orbitals are always filled first.

EXAMPLE

Hydrogen; atomic number = 1; no. of electrons = 1

This electron must occupy the s orbital of the K shell since, of all the available orbitals, this has the lowest energy (*see* figure 1.21). An electron in this orbital is called a 1s electron. The electron configuration of hydrogen is written $1s^1$.

Rule 2: the Pauli exclusion principle. This states that an orbital cannot contain more than two electrons and then only if they have opposite spins.

EXAMPLE

Lithium; atomic number = 3; no. of electrons = 3

The orbital with the lowest energy is the 1s orbital. This can only take two electrons. These must have opposite spins:

$$\boxed{\uparrow \;\; \downarrow} \qquad \textit{allowed}$$

An orbital occupied by two electrons with parallel spins is not permitted:

$$\boxed{\uparrow \;\; \uparrow} \;\; \text{or} \;\; \boxed{\downarrow \;\; \downarrow} \qquad \textit{not allowed}$$

The remaining electron of the lithium atom must occupy the orbital with the next lowest energy. This is the 2s orbital. The electronic configuration of lithium is thus $1s^2 2s^1$.

Rule 3: Hund's rule. This states that the orbitals of a sub-shell must be occupied singly and with parallel spins before they can be occupied in pairs.

EXAMPLE

Nitrogen; atomic number = 7; no. of electrons = 7

The following configuration of electrons in the nitrogen atom is *allowed*:

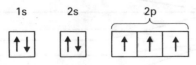

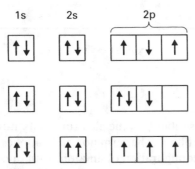

This configuration is written as

either 1s² 2s² 2p³

or 1s² 2s² 2p$_x^1$ 2p$_y^1$ 2p$_z^1$

The following three configurations for the electrons in the nitrogen atom are *not allowed*:

Figure 1.22 Ground-state electronic configuration of the nitrogen atom

Nitrogen has the electronic configuration $1s^2 2s^2 2p^3$. The three electrons occupying the 2p sub-shell must occupy the three separate 2p orbitals singly. They must also have parallel spins (*see* figure 1.22).

The electronic configurations of elements with atomic numbers 1 to 20 is shown in table 1.6.

Table 1.6 Ground-state electronic configurations of elements with atomic numbers 1 to 20

Atomic number	Element	Configuration
1	hydrogen	$1s^1$
2	helium	$1s^2$
3	lithium	$1s^2 2s^1$
4	beryllium	$1s^2 2s^2$
5	boron	$1s^2 2s^2 2p^1$
6	carbon	$1s^2 2s^2 2p^2$
7	nitrogen	$1s^2 2s^2 2p^3$
8	oxygen	$1s^2 2s^2 2p^4$
9	fluorine	$1s^2 2s^2 2p^5$
10	neon	$1s^2 2s^2 2p^6$
11	sodium	$1s^2 2s^2 2p^6 3s^1$
12	magnesium	$1s^2 2s^2 2p^6 3s^2$
13	aluminium	$1s^2 2s^2 2p^6 3s^2 3p^1$
14	silicon	$1s^2 2s^2 2p^6 3s^2 3p^2$
15	phosphorus	$1s^2 2s^2 2p^6 3s^2 3p^3$
16	sulphur	$1s^2 2s^2 2p^6 3s^2 3p^4$
17	chlorine	$1s^2 2s^2 2p^6 3s^2 3p^5$
18	argon	$1s^2 2s^2 2p^6 3s^2 3p^6$
19	potassium	$1s^2 2s^2 2p^6 3s^2 3p^6 4s^1$
20	calcium	$1s^2 2s^2 2p^6 3s^2 3p^6 4s^2$

SUMMARY

1. Each line in an **atomic spectrum** corresponds to electrons jumping from one energy level to another.
2. During electron transitions **photons** are emitted or absorbed.
3. The energy of **transition** ΔE is related to the frequency ν of the photon by $\Delta E = h\nu$.
4. The **convergence limit** of the **Lyman series** can be used to obtain the **ionisation energy** of hydrogen.
5. The **Bohr model** of the atom assumes that electrons rotate with a fixed angular momentum in circular **orbits** around the nucleus. Each orbit has a fixed or quantised energy level.
6. Experimental evidence suggests that the electron has a **dual nature**, that is, it has the properties of both waves and particles.
7. The **Heisenberg uncertainty principle** states that the position and momentum of an electron cannot both be known with absolute certainty simultaneously.
8. An electron has four **quantum numbers**:

   n the principal quantum number   for its shell
   l the subsidiary quantum number for its sub-shell
   m the magnetic quantum number   for its orbital
   s the spin quantum number   for its spin

9. The **ground-state electronic configuration** of an element can be determined by the application of three rules:

   Rule 1: the **aufbau principle**
   Rule 2: the **Pauli exclusion principle**
   Rule 3: **Hund's rule**

1.3
Radioactivity

THE DISCOVERY OF RADIOACTIVITY

Antoine Henri Becquerel (1852–1908). He discovered radioactivity in 1896 and won the Nobel Prize for Physics for his work in 1903

Radioactivity is the spontaneous emission of radiation by an element due to the splitting of atomic nuclei. The phenomenon was discovered by a Frenchman, Henri Becquerel, in 1896. He found that uranium salts caused photographic plates to 'fog' even when covered by a layer of black paper. He also found that these salts could ionise gases and thereby discharge an electroscope.

In 1898 Mme Curie used this ionisation method to examine the radioactivity of pitchblende. Pitchblende is a uranium ore consisting mainly of U_3O_8. She found that the radioactivity of a given mass of uranium was greater than that of a pure uranium salt. By chemical methods of separation, she then separated two radioactive elements from the ore. She called these polonium and radium. The radium was about a million times more radioactive than the uranium. Soon after, her collaborator, Debierne, discovered actinium—another radioactive element.

Mme Curie suggested that the radioactivity of radium was due to the **disintegration** of its atoms. She identified two types of radiation. These were called α-radiation and β-radiation. The α-radiation could be deflected by a positively charged plate. In 1900 her husband, Pierre, discovered a third type of radiation. This was called γ-radiation.

During the next few years, two British physicists, Ernest Rutherford and Frederick Soddy, performed a number of experiments on radium, radon and uranium. They showed that α-radiation consisted of positive particles. These were later shown to be helium nuclei. They also demonstrated that radioactivity led to the formation of other types of elements.

Marie Curie was born Manya Sklodowska in Warsaw, Poland, in 1867. She shared the Nobel Prize for Physics with her husband Pierre in 1903. In 1911 she received the Nobel Prize for Chemistry for the isolation of radium. She died from leukaemia in 1934. Her death was undoubtedly caused by prolonged exposure to radiation

IONISING RADIATION

The Curies showed that the radiation emitted by radioactive materials causes the ionisation of gases and other substances. They identified three types of ionising radiation: α-, β- and γ- (alpha, beta and gamma respectively).

α-*Radiation*

This consists of streams of high-velocity helium nuclei. The nuclei have a mass of 4 and a charge of +2. They can be written as $_2^4\text{He}$ or He^{2+}.

The loss of an α-particle thus results in the loss of two protons and two neutrons from a nucleus. Since two protons are lost, the atomic number decreases by 2 (*see* table 1.7) and a new element is consequently formed. For example, when the nuclide radium-226 loses an α-particle, the nuclide radon-222 is formed. This is written as

$$_{88}^{226}\text{Ra} \rightarrow {}_{86}^{222}\text{Rn} + {}_2^4\text{He}$$

Since helium nuclei are positively charged, they attract electrons and are thus highly ionising. The ionisation results in the formation of helium atoms:

$$\text{He}^{2+} + 2e^- \rightarrow \text{He}$$

Small amounts of helium are indeed found in some radioactive minerals.

The positive charge of the helium nucleus also results in α-radiation being deflected towards a negative electric field (*see* figure 1.23). However, the size of the helium nuclei limits the range and penetrating power of this type of radiation compared to β- and γ-radiation.

Table 1.7 Types of radiation

Radiation	Charge	Mass	Change in atomic number	Change in mass number
α	+2	+4	−2	−4
β	−1	0	+1	none
β⁺	+1	0	−1	none
γ	0	0	none	none

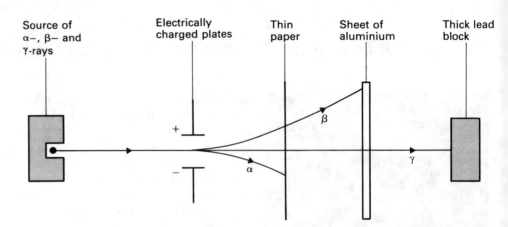

Figure 1.23   The penetrating power of α-,β- and γ-radiation

β-*Radiation*

There are two types of β-radiation. These are β- (or β⁻) radiation and β⁺-radiation. **β-radiation** consists of streams of electrons moving at speeds comparable to the speed of light. The electrons ($_{-1}^{0}e$) are emitted from unstable nuclei as a result of the splitting up of a neutron ($_{0}^{1}n$)

$$_{0}^{1}n \rightarrow {}_{1}^{1}p + {}_{-1}^{0}e$$

Since an additional proton ($_{1}^{1}p$) is formed in the nucleus, the atomic number increases by one. A new element is thus formed. For example, when the nuclide thorium-234 loses an electron, the nuclide protactinium-234 is formed. Note that the mass numbers of both nuclides are the same:

$$_{90}^{234}Th \rightarrow {}_{91}^{234}Pa + {}_{-1}^{0}e$$

β⁺-radiation is due to the emission of positrons from a nucleus. Positrons are similar to electrons except that they have a positive charge. They are formed by the conversion of a proton to a neutron:

$$_{1}^{1}p \rightarrow {}_{0}^{1}n + {}_{+1}^{0}e$$

Since the new nuclide has one less proton but one more neutron, there is no net change in mass number. However, the atomic number decreases by 1. For example

$$_{19}^{38}K \rightarrow {}_{18}^{38}Ar + {}_{+1}^{0}e$$

γ-*Radiation*

This is high-energy electromagnetic radiation. γ-Rays are comparable to X-rays although they have shorter wavelengths. Since they have a high energy and short wavelength, they have a long range and a high penetrating power. They are less ionising than α- and β-radiation. γ-Rays are emitted when a nuclide emits α- or β-particles. They are not deflected by electric or magnetic fields.

γ-Rays are also emitted in a process called **electron capture** (ec). In this process a nucleus captures an electron from a K or L shell. This results in a proton being converted to a neutron. The mass number of the nuclide remains the same but its atomic number decreases by one. For example

$$_{18}^{37}Ar + {}_{-1}^{0}e \rightarrow {}_{17}^{37}Cl$$

The Detection and Measurement of Radioactivity

Radioactivity is measured in terms of either absolute activity or specific activity. The **absolute activity** of a radioactive substance is measured in curies. The **specific activity** of a radioactive material is the radioactivity per unit mass of the material.

The Curie

The unit of radioactivity is the curie (Ci). One curie is the amount of radioactivity which has the same rate of disintegration as one gram of radium-226. One gram of radium-226 undergoes 3.7×10^{10} disintegrations per second. The millicurie (mCi) is 3.7×10^7 disintegrations per second and the microcurie (μCi) 3.7×10^4 disintegrations per second.

One gram of radium-226, incidentally, produces 0.0001 cm³ of radon per day.

Since most radioisotopes occur in mixtures with stable isotopes, specific activity is a measure of the relative abundance of the isotopes. Specific activity is measured in counts per minute (cpm) or counts per second (cps). In practice, only a given fraction of the actual disintegrations occurring are counted. This is called **relative counting**.

One of the earliest devices for detecting radiation was a cloud chamber invented by C. T. R. Wilson in 1911. The **Wilson cloud chamber** contained dust-free air saturated with water vapour. The air was allowed to expand rapidly causing it to become supersaturated with water vapour. The passage of ionising radiation through the chamber produced ions which acted as nuclei onto which the water condensed as tiny droplets. The track of these ions could thus be seen and photographed.

Nowadays various methods are used for the detection and measurement of radioactivity. Three of the most important are the gas ionisation, scintillation and photochemical methods.

Gas ionisation

The **Geiger–Müller tube** and counter relies on gas ionisation. The tube consists of a chamber filled with argon. The radiation passes through a thin mica window and ionises the argon. The positive argon ions move to a cathode and the electrons to an anode. This creates an electrical pulse, which is amplified. The pulse registers as clicks, light flashes or meter readings on the counter (*see* figure 1.24).

Scintillation Methods

Some materials absorb the energy of radioactive emissions and convert it into light. Zinc sulphide is an example of such a material. The impact of a radioactive particle causes a small flash of light. A **scintillation counter** consists of a photoelectric tube with a window coated with zinc sulphide. The flashes of light cause pulses of electric current to pass through the photoelectric tube. These are amplified and counted.

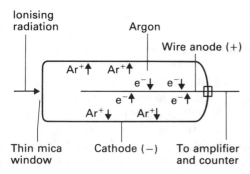

Figure 1.24 The Geiger–Müller tube

Measurement of radioactivity

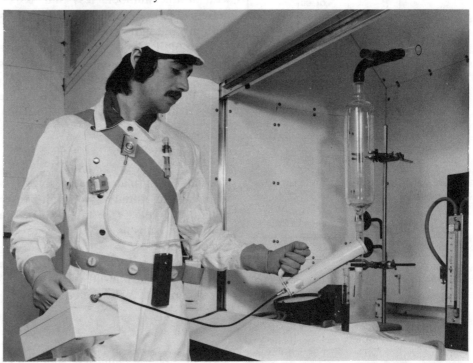

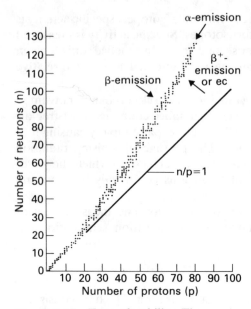

Figure 1.25 Zone of stability. The arrows show the types of emission which result in the decay of unstable isotopes towards the zone of stability

Table 1.8 Occurrence of stable isotopes with odd and/or even numbers of protons and neutrons

Number of protons	neutrons	Number of stable isotopes
even	even	166
even	odd	53
odd	even	57
odd	odd	8

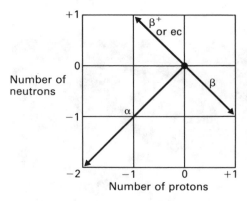

Figure 1.26 Decay of unstable isotopes towards zone of stability

Table 1.9 Half-lives

Isotope		Half-life
polonium-212	$^{212}_{84}$Po	3×10^{-7} seconds
bismuth-214	$^{214}_{83}$Bi	19.7 minutes
radium-224	$^{224}_{88}$Ra	3.64 days
lead-210	$^{210}_{82}$Pb	19.7 years
carbon-14	$^{14}_{6}$C	5.7×10^3 years
uranium-238	$^{238}_{92}$U	4.5×10^9 years
thorium-232	$^{232}_{90}$Th	1.39×10^{10} years

Photochemical Methods

This type of method is sometimes called autoradiography. The radioactive sample is placed on a film of photographic emulsion containing silver halide. The radioactivity of the sample is estimated by developing the film and viewing it under high magnification.

STABLE AND UNSTABLE ISOTOPES

Isotopes may be classified as stable or unstable. **Stable isotopes** do not undergo radioactive decay. They therefore persist in nature. Oxygen-16 and carbon-12 are examples of stable isotopes. There are about 280 naturally occurring stable isotopes on Earth. Most naturally occurring elements consist of one stable isotope in high abundance along with small amounts of other stable and unstable isotopes.

Unstable isotopes are radioactive. They are known as radioactive isotopes, **radioisotopes** or radionuclides. Radioisotopes may be classified as natural or artificial (*see* below). Both types decay by emitting α- or β-particles. The decay continues until a stable isotope is formed.

Zone of Stability

Figure 1.25 shows a plot of the number of neutrons against the number of protons for stable isotopes. The stable isotopes form a band or zone of stability. The zone of stability has some notable features. First of all, the stable isotopes of elements with an atomic number of less than 20 have a neutron-to-proton ratio (n/p) equal to 1. $^{12}_{6}$C and $^{16}_{8}$O are examples of stable isotopes falling in this part of the band. As atomic number (number of protons) increases, so does the ratio n/p for stable isotopes. Examples of stable isotopes with n/p ratios greater than 1 are $^{208}_{82}$Pb and $^{197}_{79}$Au. The stability band ends with $^{209}_{83}$Bi. Naturally occurring isotopes with higher atomic numbers do occur but they are unstable and thus undergo radioactive decay. Stable isotopes are predominantly isotopes with even numbers of neutrons and/or protons (*see* table 1.8). The eight stable isotopes with odd numbers of protons and neutrons include $^{2}_{1}$H (deuterium), $^{6}_{3}$Li and $^{14}_{7}$N. Only two elements with atomic numbers less than 83 do not occur naturally. These are technetium $^{98}_{43}$Tc and promethium $^{147}_{61}$Pm. Note that technetium has odd numbers of both protons and neutrons.

Unstable isotopes which occur above the stability band decay towards stable isotopes by β-emission. Radioisotopes below the band decay towards stable isotopes by β⁺-emission or electron capture. α-Emission occurs predominantly when the atomic number of the unstable isotope is above 82. These processes are depicted graphically in figure 1.26. They correspond to the changes shown in table 1.7.

Half-Life

The **half-life** $t_{1/2}$ of an unstable isotope is the time taken for its radioactivity to drop to half of its initial value. Half-lives can vary from millionths of a second to millions of years (*see* table 1.9). They are independent of the mass of the radioactive substance and they are not influenced by catalysts or changes in temperature.

The radioactive decay of an isotope is invariably a first-order rate process (*see* section 9.1). The half-life of an isotope is related to its rate constant, known as the decay constant, λ, by the equation

$$t_{1/2} = \frac{0.693}{\lambda}$$

Both $t_{1/2}$ and λ are characteristic of a particular isotope.

Magic Numbers

Isotopes with the following numbers of protons or neutrons are exceptionally stable:

2, 8, 20, 28, 50, 82 and 126

These numbers have been called 'magic numbers'. They are thought to indicate closed nuclear shells comparable to the closed electron shells of the noble gases. For comparison, the atomic numbers of the noble gases are

2, 10, 18, 36, 54 and 86

Stable isotopes with these magic numbers of protons and neutrons include

$^{4}_{2}$He, $^{16}_{8}$O and $^{208}_{82}$Pb

Decay Curves

Once the half-life of an isotope is known, it is possible to plot a decay curve for the isotope. This shows the decrease in mass or activity against time.

EXAMPLE

Phosphorus-32, $^{32}_{15}$P, has a half-life of 14.3 days. Plot its decay curve.

SOLUTION

We may choose *any* value for the initial mass or *any* value for the initial activity. Let us assume, for convenience, that the initial mass is 100 g and the initial activity is 200 cpm. Using these values we can obtain the data shown in table 1.10.

A plot of mass against activity gives the decay curve (*see* figure 1.27).

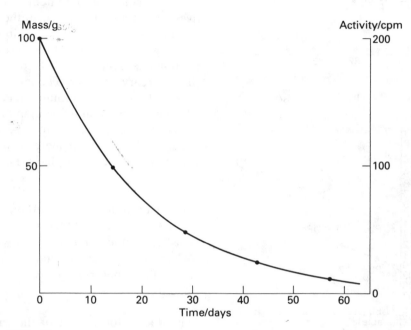

Figure 1.27 Decay curve

Table 1.10 Data for decay curve

Time/days	Mass/g	Activity/cpm
0	100	200
14.3	50	100
28.6	25	50
42.9	12.5	25
57.2	6.25	12.5

Decay Series

Thorium-232 decays by emitting an α-particle ($^{4}_{2}$He):

$$^{232}_{90}\text{Th} \rightarrow \,^{228}_{88}\text{Ra} + \,^{4}_{2}\text{He}$$

$^{232}_{90}$Th is called the **parent** and $^{228}_{88}$Ra the **daughter**. Radium-228 is itself unstable and decays by emitting a β-particle to form actinium-228:

$$^{228}_{88}\text{Ra} \rightarrow \,^{228}_{89}\text{Ac} + \,^{0}_{-1}\text{e}$$

The daughter thus becomes a parent. This parent–daughter process continues until a stable isotope is formed, in this case the lead-208 isotope ($^{208}_{82}$Pb). The complete process is known as a decay series or disintegration series. The complete decay series for thorium-232 is shown in figure 1.28.

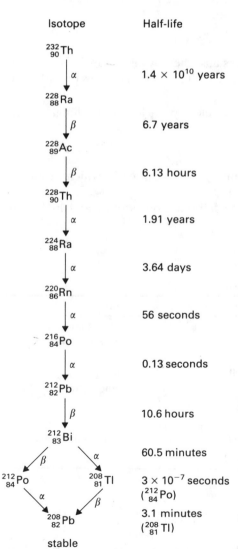

Isotope	Half-life
$^{232}_{90}$Th	
α	1.4×10^{10} years
$^{228}_{88}$Ra	
β	6.7 years
$^{228}_{89}$Ac	
β	6.13 hours
$^{228}_{90}$Th	
α	1.91 years
$^{224}_{88}$Ra	
α	3.64 days
$^{220}_{86}$Rn	
α	56 seconds
$^{216}_{84}$Po	
α	0.13 seconds
$^{212}_{82}$Pb	
β	10.6 hours
$^{212}_{83}$Bi	
β / α	60.5 minutes
$^{212}_{84}$Po / $^{208}_{81}$Tl	3×10^{-7} seconds ($^{212}_{84}$Po)
α / β	3.1 minutes ($^{208}_{81}$Tl)
$^{208}_{82}$Pb	
stable	

Figure 1.28 The thorium series

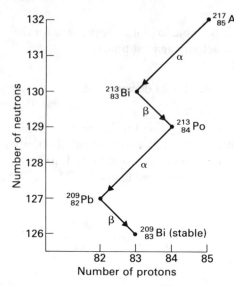

Figure 1.29   The decay of $^{217}_{85}$At

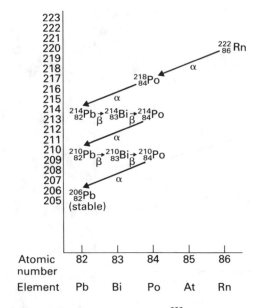

Figure 1.30   The decay of $^{222}_{86}$Rn

A decay series or part of a decay series may be represented graphically by plotting number of neutrons against number of protons or by plotting mass number against atomic number. Figure 1.29 shows the decay of astatine-217. This is part of the neptunium decay series. Traces of astatine-217 have been detected with naturally occurring isotopes of neptunium in uranium minerals.

Figure 1.30 shows the change in mass number and atomic number during the decay of radon-222. This is part of the uranium-238 series.

Natural Radioisotopes

During the creation of the Earth, 25 radioisotopes were formed. Since then numerous other radioisotopes have been formed both by radioactive decay and also by cosmic bombardment of the Earth's atmosphere. Those with long half-lives persist and thus occur in the greatest abundance, whereas those with very short half-lives may only occur in trace amounts.

The greatest source of radioisotopes on Earth are uranium minerals such as pitchblende. About 0.0004% of the Earth's crust is uranium. It is more plentiful than mercury or silver. The most abundant uranium isotope is ^{238}U. It accounts for over 99% of the uranium on Earth. Some of this finds its way into the oceans, which contain about 0.002 ppm uranium.

The radioisotopes hydrogen-3 (or tritium) and carbon-14 are continuously produced in the atmosphere by bombardment by neutrons produced by cosmic rays entering the atmosphere. For example

$$^{14}_{7}N + ^{1}_{0}n \rightarrow ^{14}_{6}C + ^{1}_{1}H$$

Carbon-14 in the form of carbon dioxide is absorbed by all plants during photosynthesis. Consequently, carbon-14 is found in all living things. This provides the basis for carbon dating (*see* below).

Carbon-14 emits β-radiation and a chargeless particle of vanishingly small mass called a neutrino (ν):

$$^{14}_{6}C \rightarrow ^{14}_{7}N + ^{0}_{-1}e + \nu$$

This process forms part of the carbon–nitrogen–oxygen cycle which produces energy in main-sequence stars (*see* below).

Certain radioisotopes also occur in the fall-out from atomic and hydrogen bombs (*see* table 1.11).

Table 1.11   Some naturally occurring radioisotopes

Radioisotope		$t_{1/2}$/years	Source
hydrogen-3	$^{3}_{1}$H	12.3	cosmic bombardment
carbon-14	$^{14}_{6}$C	5700	of atmospheric $^{14}_{7}$N
strontium-90	$^{90}_{38}$Sr	80 000	fall-out from atomic
caesium-137	$^{137}_{55}$Cs	27	and hydrogen bombs
radium-226	$^{226}_{86}$Ra	1620	
thorium-230	$^{230}_{90}$Th	80 000	uranium minerals
uranium-238	$^{238}_{92}$U	4.5×10^9	

ARTIFICIAL NUCLEAR TRANSFORMATIONS

Both stable and unstable (that is, radioactive) isotopes can be produced by bombarding nuclei with high-energy particles. The first artificial nuclear transformation was achieved by Rutherford in 1915. He passed α-particles obtained from $^{214}_{84}$Po through nitrogen. This produced the stable isotope $^{17}_{8}$O (*see* reaction 1, table 1.12). This was the first time one element had been artificially converted to another. The conversion of one element to another is known as the **transmutation** of an element. The first radioisotope produced artificially was $^{30}_{15}$P. This was achieved in 1933 by bombarding aluminium nuclei with α-particles (reaction 2, table 1.12).

Table 1.12 Types of nuclear reaction

	Type	Example
1.	(α,p)	$^{14}_{7}$N + $^{4}_{2}$He → $^{17}_{8}$O + $^{1}_{1}$H
2.	(α,n)	$^{27}_{13}$Al + $^{4}_{2}$He → $^{30}_{15}$P + $^{1}_{0}$n
3.	(p,n)	$^{23}_{11}$Na + $^{1}_{1}$H → $^{23}_{12}$Mg + $^{1}_{0}$n
4.	(p,α)	$^{9}_{4}$Be + $^{1}_{1}$H → $^{6}_{3}$Li + $^{4}_{2}$He
5.	(p,γ)	$^{14}_{7}$N + $^{1}_{1}$H → $^{15}_{8}$O + γ
6.	(d,p)	$^{31}_{15}$P + $^{2}_{1}$H → $^{32}_{15}$P + $^{1}_{1}$H
7.	(d,n)	$^{27}_{13}$Al + $^{2}_{1}$H → $^{28}_{14}$Si + $^{1}_{0}$n
8.	(n,p)	$^{14}_{7}$N + $^{1}_{0}$n → $^{14}_{6}$C + $^{1}_{1}$H
9.	(n,γ)	$^{59}_{27}$Co + $^{1}_{0}$n → $^{60}_{27}$Co + γ
10.	(n,α)	$^{27}_{13}$Al + $^{1}_{0}$n → $^{24}_{11}$Na + $^{4}_{2}$He

Key:	*Particle*	*Symbol*	*Particle*	*Symbol*
	helium nucleus	α or $^{4}_{2}$He	deuterium nucleus	d or $^{2}_{1}$H
	proton	p or $^{1}_{1}$H	gamma-ray	γ
	neutron	n or $^{1}_{0}$n		

Artificial nuclear transformations are also known as artificial nuclear reactions or simply as **nuclear reactions**. They can be classified in terms of the bombarding particles and the type of particle or radiation emitted. In a (p,n) nuclear reaction, for example, nuclei are bombarded with protons (p) and neutrons (n) are emitted (*see* reaction 3, table 1.12).

Note that, in nuclear reactions, the sums of the atomic and mass numbers must always be balanced on the left- and right-hand sides of the equation. In reaction 4, for example:

$$\text{mass nos.} \quad \overbrace{9 + 1}^{10} \quad \to \quad \overbrace{6 + 4}^{10}$$
$$^{9}_{4}\text{Be} + ^{1}_{1}\text{H} \to ^{6}_{3}\text{Li} + ^{4}_{2}\text{He}$$
$$\text{atomic nos.} \quad \underbrace{4 + 1}_{5} \quad \to \quad \underbrace{3 + 2}_{5}$$

The (n,γ) type of nuclear reaction (reaction 9) is also known as **neutron capture**. This is because a nucleus captures a neutron without emitting another particle. Only γ-radiation is emitted. Neutron capture is used to obtain radioactive cobalt-60 from stable cobalt-59. Cobalt-60 is used in cancer therapy (*see* below).

Particle Accelerators

For nuclear reactions to occur, the bombarding particles must have a high kinetic energy. This is required to overcome the electrostatic forces of repulsion exerted by the *target* nuclei. The particles thus have to be accelerated to a high velocity. The two most important types of particle accelerator are the cyclotron and the linear accelerator (*see* figure 1.31).

The Cyclotron
This has two D-shaped metallic boxes called dees. The dees are oppositely charged and separated by a gap. Positively charged particles are pushed out of a

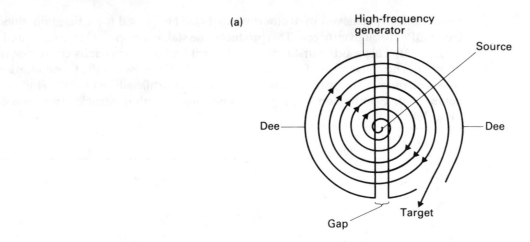

(a)

High-frequency generator

Source

Dee

Dee

Gap

Target

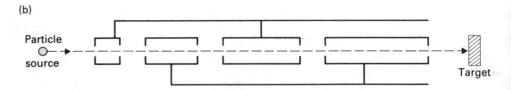

(b)

Particle source

Target

Figure 1.31 Particle accelerators: (a) the cyclotron; (b) the linear accelerator

source in the middle of the gap. The particles are immediately attracted to the negative dee. On reaching this dee, the polarity is reversed. The particles then accelerate round towards the other dee, which is now negatively charged. On reaching the gap, the polarity is once again reversed. The polarity is reversed with high frequency, causing the particles to spiral around between the dees with increasing velocity.

The Linear Accelerator

This works on a similar principle to the cyclotron. It consists of a series of metal tubes of increasing length. Alternate tubes are connected and have similar charges. Positive particles accelerate to the first tube, which has a negative charge. Immediately on leaving this tube the polarity is reversed. The tube thus repels the

A linear accelerator (or 'linac' for short) at CERN

particles. The particles are immediately attracted to the next tube. The two series of tubes are connected to a high-frequency potential source.

Fast Neutrons

Neutrons with a high velocity are called **fast neutrons**. They can be produced by bombarding a target with high-energy particles from a particle accelerator. A target containing beryllium-9 is often used for this purpose. The target is bombarded with deuterium nuclei

$$\mathrm{^9_4Be + {}^2_1H \rightarrow {}^{10}_5B + {}^1_0n}$$

This is an example of a (d,n) reaction (*see* table 1.12).

Transuranium Elements

Elements with atomic numbers of 93 or more do not occur naturally. They have to be made artificially by nuclear reactions. They are called the transuranium elements. Elements with atomic numbers from 93 to 105 have all been produced artificially. The first two were made in 1940 by bombarding uranium-238 with neutrons:

$$\mathrm{^{238}_{92}U + {}^1_0n \rightarrow {}^{239}_{92}U}$$

Uranium-239 is radioactive. It decays by β-emission:

$$\mathrm{^{239}_{92}U \rightarrow {}^{239}_{93}Np + {}^0_{-1}e}$$

The element formed by this process is called neptunium (Np), after the planet Neptune. Neptunium also decays by β-emission, forming plutonium (Pu). This is named after the planet Pluto, which lies just beyond Neptune.

$$\mathrm{^{239}_{93}Np \rightarrow {}^{239}_{94}Pu + {}^0_{-1}e}$$

Most of the elements of atomic numbers 99 or above have been made by using **heavy bombarding projectiles** such as $\mathrm{^{12}_6C}$ or $\mathrm{^{14}_7N}$. For example, einsteinium, $\mathrm{^{248}_{99}Es}$, was made by bombarding uranium-238 with $\mathrm{^{14}_7N}$:

$$\mathrm{^{238}_{92}U + {}^{14}_7N \rightarrow {}^{248}_{99}Es + 4{}^1_0n}$$

USES OF ISOTOPES

The uses of isotopes, particularly radioisotopes, are numerous. Table 1.13 shows examples of just some of the **industrial uses** of radioisotopes. Any one of the techniques mentioned in the table is also used in other industries. For example, radioisotopes are used for leak detection in:

- the beverage industry, for detecting leaks in storage tanks;
- civil engineering, for detecting leaks in buried water mains;
- electricity generation, for detecting leaks in heat exchangers;
- the oil industry, for detecting leaks in buried oil pipelines;
- sewage and effluent control, for detecting leaks of sewage and effluent.

Isotopes are also used extensively in scientific research. For instance, they are used to determine the mechanisms of chemical reactions. An example is the use of water labelled with the stable oxygen isotope ^{18}O to examine the hydrolysis of esters such as ethyl ethanoate (*see* also section 19.3). By using mass spectrometry to detect the ^{18}O isotope, it has been found that the oxygen atom in the water molecule is transferred to ethanoic acid and not ethanol

$$\mathrm{H-{}^{18}OH + CH_3\overset{\displaystyle O}{\overset{\|}{C}}OC_2H_5 \longrightarrow CH_3\overset{\displaystyle O}{\overset{\|}{C}}-{}^{18}OH + C_2H_5-OH}$$

Table 1.13 Radioisotopes at work

Industry	Use	Example of application
agriculture	tracers	to study root growth and function
aircraft	wear testing	to study engine wear
chemical	radioactivation analysis	analysis of elements by measuring radiation emitted
civil engineering	density gauges	measurement of soil density
gas	leak location	to detect leaks in gas mains
iron and steel	level gauges	to measure level of molten metals in furnaces
laundering	sterilisation	to sterilise contaminated hospital blankets
mechanical engineering	thickness gauges	to measure thickness of sheets, tubes and rods
plastics	polymerisation	to initiate polymerisation using γ-radiation
textiles	static elimination	to eliminate static on cotton and synthetic woven yarns

A sterilised male tsetse fly, having been exposed to low-level radiation, is marked for identification before being released into an agricultural area in Upper Volta. This is part of an experiment to study the habits of the fly and find effective controls against the disease trypanosomiasis or 'sleeping sickness'. The disease is carried by the fly and attacks both humans and domestic animals and herds. It is endemic in certain parts of Africa

Radioisotopes have been used to trace siltation mechanisms in docks, harbours and estuaries

Using radioisotopes to produce a photographic image of the combustion cans of a jet engine at the Non Destructive Test Centre, Heathrow Airport. Radioisotopes are used widely in industry for non-destructive testing

Metabolic Pathways

Metabolism is the sum of all the chemical reactions in the cells of a living organism. These metabolic reactions convert nutrients into useful energy or cell constituents. They tend to occur in series of small steps. A series of metabolic reactions is called a **metabolic pathway**.

Radioisotopes are used widely as tracers in biological research. Carbon-14, tritium, phosphorus-32 and sulphur-35 are all used to trace metabolic pathways in living systems. For example, the uptake of phosphorus from fertilised soil by plants can be traced by using fertilisers containing phosphorus-32.

Radiation Therapy

Ionising radiation can destroy living tissues. Malignant tissues are more sensitive to exposure than normal tissue. Cancer can be treated by the use of γ-rays obtained from a cobalt-60 source. The radiation is directed at the tumour site for a few minutes each day for between two to six weeks. During such treatment, all other parts of the patient's body must be carefully shielded to prevent damage to normal tissue.

Carbon Dating

A small percentage of carbon dioxide in the atmosphere contains the radioactive isotope $^{14}_{6}C$. This isotope is absorbed by plants during the process of photosynthesis. The tissues of all living plants and animals thus contain the isotope. The level of radioactivity in living tissues is constant because the loss by radioactive decay is compensated by addition of the isotope from the atmosphere. However, as soon as a plant or animal dies the isotope is no longer absorbed. The level of radioactivity thus decreases.

The radioactivity of the $^{14}_{6}C$ isotope is due to β-decay

$$^{14}_{6}C \rightarrow {}^{14}_{7}N + {}^{0}_{-1}e$$

The method of carbon dating was developed by W. F. Libby who won the Nobel Prize for Chemistry in 1960. The method is now widely used by archaeologists, anthropologists and geologists to date samples up to 35 000 years old. The accuracy of the method is about ±300 years. The method is primarily applied to wood, seeds, mollusc shells and bones. To determine the age of a sample, the β-emission count per minute per gram of carbon is measured. The age of the sample can then be found directly from the radioactive decay curve for $^{14}_{6}C$.

Samples of charcoal taken from Stonehenge have been shown to be about 4000 years old by carbon dating

EXAMPLE

The half-life of $^{14}_{6}C$ is 5700 years. Living tissue in active contact with the atmosphere has an activity of 15.3 counts per minute per gram of carbon. Use this information to

(a) determine the decay constant for the decay of $^{14}_{6}C$;
(b) plot a decay curve for $^{14}_{6}C$;
(c) calculate the age of Crater Lake in Oregon, USA. A tree overturned in the eruption which created the lake was found to have β-activity of 6.5 counts per minute per gram of carbon.

SOLUTION

(a) The decay constant can be found from the equation

$$t_{1/2} = \frac{0.693}{\lambda}$$

Thus

$$\lambda = \frac{0.693}{5700}$$

$$= 0.00012 \text{ years}^{-1}$$

(b) A decay curve is a plot of activity against time. The data required to plot the curve can be calculated from the half-life and initial count (see table 1.14). The decay curve is shown in figure 1.32.
(c) The age of the lake can be obtained directly from the decay curve as shown by the broken line in figure 1.32. The age is 7000 years.

Table 1.14 Data required for the carbon dating decay curve

Number of half-lives	Time/years	Activity/counts per minute per gram of carbon
0	0	15.3
1	5 700	7.6
2	11 400	3.8
3	17 100	1.9
4	22 800	1.0
5	28 500	0.5

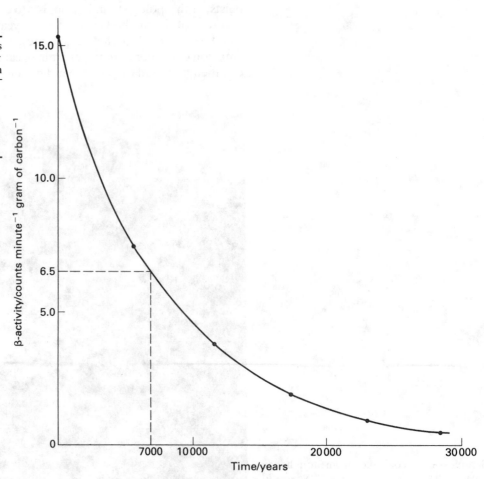

Figure 1.32 Radioactive decay curve for the $^{14}_{6}C$ isotope

How Old are the Earth and the Moon?

The Moon.

Many rocks of the Earth and the Moon contain radioisotopes with half-lives of the order 10^9 to 10^{10} years. By measuring and comparing the abundance of these radioisotopes to the abundances of their decay products, it is possible to estimate the ages of these rocks. Three of the most important geological dating methods are based on the abundances of $^{40}_{19}K$ (half-life, 1.4×10^9 years), $^{87}_{37}Rb$ (half-life 6×10^{10} years) and $^{238}_{92}U$ (half-life 4.50×10^9 years).

The Potassium–Argon Method
Rocks such as mica and some kinds of feldspar contain small amounts of the radioisotope potassium-40. This decays by electron capture to argon-40:

$$^{40}_{19}K + {}^{0}_{-1}e \rightarrow {}^{40}_{18}Ar$$

Calculations based on the ratio of the concentration of potassium-40 to argon-40 in a sample make it possible to estimate the age of the sample.

The Rubidium–Strontium Method
Some of the oldest known terrestrial rocks such as the granite from the west coast of Greenland contain rubidium. About one-third of all rubidium atoms are rubidium-87. This decays to stable strontium-87. Calculations using the ratio of the abundance of rubidium and strontium isotopes give the age of these rocks.

The Uranium–Lead Method
Uranium isotopes decay to lead isotopes. The age of rocks, such as apatite, which contain uranium as a trace element can be found by comparing the abundances of specific isotopes of uranium and lead in a sample.

All three methods have been used to date terrestrial rocks. The data suggest that the age of the Earth is 4.6×10^9 years. These methods have also been used to estimate the age of lunar rocks brought back by the Apollo 11–17 expeditions. The ages of these rocks range from 3.2 to 4.2×10^9 years.

NUCLEAR FISSION AND FUSION

We have already seen that the experimental values for isotopic masses are lower than the calculated values. This loss of mass is called the **mass defect** (or mass deficit). The mass defect corresponds to the amount of energy required to overcome the repulsive forces of the like charged particles in the nucleus and bind them together. For this reason it is called the **binding energy**. Without it, the particles in a nucleus would repel each other and the nucleus would fly apart. Binding energy is related to mass defect by Einstein's equation

$$E = mc^2$$

where E is energy, m the mass and c the velocity of light.

Binding energy is usually expressed in terms of MeV per nucleon. MeV stands for million electronvolts. One electronvolt is the energy gained or lost when a particle with a unit charge passes through a potential difference of one volt. One MeV is approximately equal to 9.6×10^{10} J mol^{-1}.

The binding energy for a nucleon in a helium nucleus is about 7 MeV, for example. For the chlorine-35 nucleus, it is 8.5 MeV per nucleon.

The higher the binding energy per nucleon, the greater the stability of the nucleus. Figure 1.33 shows the variation of binding energy with mass number. Note that the elements with the highest stability are those with mass numbers around 60. These include $^{56}_{26}$Fe, $^{59}_{28}$Co, $^{59}_{28}$Ni and $^{64}_{29}$Cu. Elements with lower mass numbers can, in theory at least, increase their stability by increasing their mass number. In practice, it has only been possible to increase the mass numbers of those elements with the very lightest nuclei such as hydrogen. Helium is exceptionally stable and does not fit onto the curve. The mass number of these elements is increased by a process called nuclear fusion (see below).

Elements with high mass numbers can become more stable by lowering their mass numbers and thus becoming lighter. The splitting of a nucleus is known as nuclear fission (see below).

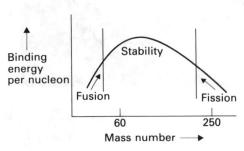

Figure 1.33 Variation of binding energy with mass number

Nuclear Fission

This is the breaking down of heavy nuclei into fragments. It can be induced by bombarding heavy nuclei with neutrons. The fission is usually accompanied by the emission of neutrons. The pattern of fragmentation is variable even for the same type of nucleus. For example, three possible fissions of the uranium-235 isotope are

$$^{235}_{92}U + ^{1}_{0}n \longrightarrow \begin{cases} ^{90}_{38}Sr + ^{144}_{54}Xe + 2^{1}_{0}n \\ ^{93}_{36}Kr + ^{140}_{56}Ba + 3^{1}_{0}n \\ ^{95}_{42}Mo + ^{139}_{57}La + 2^{1}_{0}n \end{cases}$$

The average number of neutrons lost per fission of a uranium-235 nucleus is 2.5. Each fission results in loss of mass due to the mass defect (see above). The energy released due to this loss of mass is approximately 200 MeV per fission of each uranium-235 nucleus. Since two or more neutrons are emitted from every single uranium-235 nucleus, a **nuclear chain reaction** is initiated (see figure 1.34). However, this does not occur for small amounts of fissionable material. This is because the neutrons are lost from the surface of the mass. However, above a **critical mass** the neutrons are captured before they can leave the fissionable material. The result is an explosive chain reaction with the release of massive amounts of energy. The complete chain reaction and loss of energy take place in less than a microsecond (1×10^{-6} seconds).

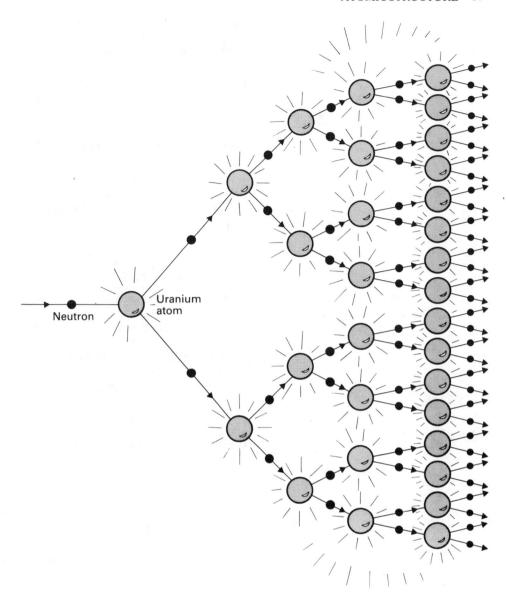

Figure 1.34 Nuclear chain reaction

Nuclear Fusion

This is the fusing together of nuclei of light atoms to form heavier nuclei. The process releases an enormous amount of energy. Figure 1.35 shows the fusion between two isotopes of hydrogen to form helium.

Fusion reactions are far more difficult to produce than chemical reactions. This is because the nuclei repel one another due to their positive charges. This prevents the close contact required if nuclear fusion is to occur. However, if the nuclei collide with one another with a sufficiently high speed, the repulsive barrier is overcome. This can be achieved by heating the mixture of light isotopes to temperatures of a 100 million degrees Celsius or so.

When a gas is heated to a very high temperature, the electrons achieve sufficient energy to escape from the nuclei. As a result, the gas becomes ionised. This ionised gas is called **plasma**. It has different properties to the ordinary gas at lower temperatures. The stars and the Sun exist as plasma. Indeed, plasma is the most common form of matter in the Universe.

Figure 1.35 The deuterium–tritium fusion reaction

$$^2_1H \ + \ ^3_1H \rightarrow \ ^4_2He + \ ^1_0n \ + \ energy$$

deuterium   tritium

Nuclear Fusion in the Stars

Elements are synthesised in the stars by the process of nuclear fusion. For example, oxygen-16 is synthesised by the fusion of carbon-12 and helium-4. γ-Rays are emitted in the process.

$$^{12}_{6}C + \,^{4}_{2}He \rightarrow \,^{16}_{8}O + \gamma$$

The synthesis of oxygen-16 is only one step in a series of fusion processes. Other isotopes formed in the series include neon-20, fluorine-18, magnesium-24 and silicon-30.

These fusion processes occur in two distinct phases. Most of the heavy nuclei are formed by a slow 'quiescent' fusion during which progressively heavier nuclei are synthesised. Quiescent fusion releases thermonuclear energy over millions of years.

Main-Sequence Stars

Most observed stars including the Sun are main-sequence stars. Main-sequence stars are stable spheres of gas. They change little over many millions of years. They depend primarily on what is called the **carbon–nitrogen–oxygen cycle** for their energy. This is also known as the carbon cycle. The cycle slowly converts hydrogen into helium by a series of nuclear reactions including nuclear fusion reactions.

This cycle is also known as the **proton–proton reaction**. It is dominant for low-mass main-sequence stars like the Sun.

When a star has exhausted its store of hydrogen, it collapses into a white dwarf, neutron star or black hole. The Sun is composed of about 95% hydrogen and 5% helium and heavier elements. The Sun has an estimated lifetime of about 10×10^{10} years—half of which is still in the future!

IN (hydrogen)

$$^{12}_{6}C + \,^{1}_{1}H \longrightarrow \,^{13}_{7}N + \gamma$$
$$^{13}_{7}N \longrightarrow \,^{13}_{6}C + \beta^{+} + \nu$$
$$^{13}_{6}C + \,^{1}_{1}H \longrightarrow \,^{14}_{7}N + \gamma$$
$$^{14}_{7}N + \,^{1}_{1}H \longrightarrow \,^{15}_{8}O + \gamma$$
$$^{15}_{8}O \longrightarrow \,^{15}_{7}N + \beta^{+} + \nu$$
$$^{15}_{7}N + \,^{1}_{1}H \longrightarrow \,^{12}_{6}C + \,^{4}_{2}He + \gamma$$

OUT (helium)

When stars above a certain mass have exhausted their supply of nuclei which can fuse together, the second stage takes place. This is called explosive nucleosynthesis. Since the star no longer has an energy source from fusion, it collapses and then explodes as a **supernova**. The explosion heats the star to several billion degrees Celsius for a few tenths of a second, resulting in further nuclear fusion. The explosion ejects most of the stellar mass into the interstellar medium, leaving behind a dense, compact object such as a neutron star or a **black hole**.

NUCLEAR ENERGY

The use of nuclear power stations to generate electricity has increased dramatically since the 1950s when they were first used. Britain started its commercial nuclear power programme in 1956. The first nuclear power stations, Calder Hall and Chapelcross, started up in 1959. Statistics from the International Atomic Energy Agency (IAEA) show that, by 1982, there were 673 nuclear power reactors in operation, under construction or being planned in 36 countries.

How Does a Reactor Work?

Nuclear reactors tap the energy released when uranium-235 undergoes the types of fission described above. The energy released in these reactions is absorbed by a coolant, which then transfers the heat to a secondary coolant. The heat from the secondary coolant is used in a turbine or engine to generate electrical power.

A typical reactor consists of five essential components.

The fuel. Most reactors use uranium-235. In its naturally occurring forms, the concentration of this uranium isotope is only 0.7%. Some reactors use enriched uranium which has a concentration of 1 to 2%.

The moderator. This is an inert material such as water or graphite which surrounds the fuel. The neutrons released by the fission process are slowed down by the moderator. This gives the neutrons an opportunity to collide with other uranium-235 nuclei and thus enhances the reaction.

The coolant. The coolant removes the heat from the fission reaction. Coolants used include water, liquid sodium, air and carbon dioxide.

Control rods. Control rods of cadmium or boron are used to ensure that the reaction proceeds at a controlled rate. They function by absorbing excess neutrons. The rate of the fission reaction and thus the output of the reactor can be controlled by raising or lowering the control rods.

Shielding. Since the reactor is highly radioactive, it has to be shielded to prevent radiation escaping. The reactor is thus housed in thick concrete which acts as the shield.

Types of Reactor

The most common type of reactor is the pressurised water reactor (PWR). This uses pressurised water as both the moderator and the coolant (*see* figure 1.36). The pressurised water reactor is a type of light water reactor (LWR). Another LWR is the boiling water reactor (BWR) which uses steam rather than pressurised water as a moderator and coolant. The first type of reactor used in Britain was the Magnox reactor. In this reactor the natural uranium fuel is contained in magnesium

Counter-current concentration of solutions of uranium ore

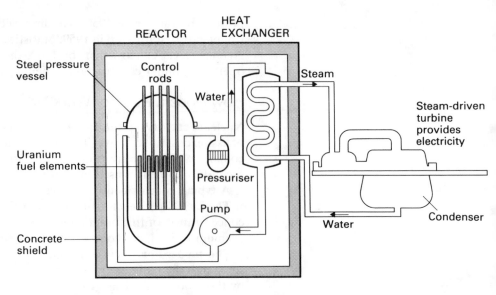

Figure 1.36   The pressurised water reactor

oxide. Graphite is used as a moderator and carbon dioxide as a coolant gas. The advanced gas-cooled reactor (AGR) uses helium gas as a coolant.

All the above reactors are types of **thermal reactor**. They all depend on ^{235}U as a fuel. The estimated reserves of naturally occurring ^{235}U are about 30 years. However, uranium ores are abundant in ^{238}U. This can be converted to plutonium-239 by the following reaction:

$$^{238}_{92}U + ^{1}_{0}n \rightarrow ^{239}_{92}U \rightarrow ^{239}_{93}Np + \beta$$
$$\searrow$$
$$^{239}_{94}Pu + \beta$$

A fast or breeder reactor uses $^{239}_{94}Pu$ as a fuel. It maintains a controlled chain reaction by using fast neutrons (*see* above). It thus does not need a moderator to slow down neutrons. Since the **fast reactor** can use depleted uranium and plutonium obtained as a by-product from the processing of uranium ores, it can be linked into the thermal reactor fuel cycle (*see* figure 1.37).

A single 1 kW bar of an electric fire can run on:
- one kilogram of coal for two hours,
- one kilogram of oil for three hours,
- one kilogram of uranium in thermal reactors for 5 years;
- one kilogram of uranium in fast reactors for 300 years.

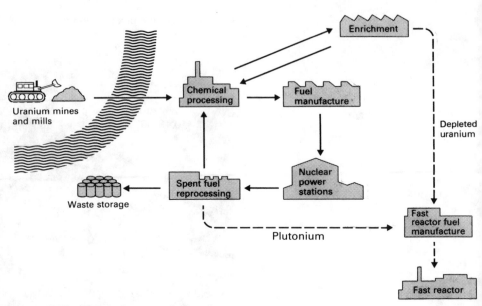

Figure 1.37   Thermal reactor fuel cycle

The Prototype Fast Reactor at Dounreay, Scotland, which has been working since 1975. The sphere belongs to the original Dounreay Fast Reactor which was closed down in 1977

A drainage channel at the Drigg National Disposal Centre, near Sellafield, in Cumbria. The Sellafield Works is the UK centre for the reprocessing of irradiated fuel from which by-product plutonium and unused uranium are extracted for future use. Irradiated fuel from both British and overseas nuclear reactors is reprocessed. Separation of uranium, plutonium and fission products is by counter-current solvent extraction

It has been estimated that by using fast reactors Britain's existing stocks of uranium could provide 250 years of electricity at current rates of consumption. The number of fast reactors at present operating, under construction or planned is low compared to the number of thermal reactors.

Nuclear Wastes and Their Treatment

Nuclear waste is usually divided into three categories: low-, intermediate- and high-level wastes.

Low-Level Wastes

These are usually discharged into the environment. Air containing radioactive dust is filtered and then discharged to the atmosphere. Effluent water from nuclear power stations may contain a low level of radioactivity. It is pumped into the sea or other waterways.

Intermediate-Level Wastes

These include fuel cans, solid scrap, structural materials, sludges and some liquid wastes. They are usually stored at the nuclear plant. Some low-activity scrap is enclosed in concrete and sunk in the deep ocean.

High-Level Wastes

These are the highly active products of the fission reactions. The level of radioactivity in the waste will take about 500 years to fall below that of the mined uranium ore from which the nuclear fuel is produced. The waste is stored as a concentrated liquid at the plant. It is kept in double-walled stainless steel tanks contained in stainless-steel lined concrete vaults. They will be eventually converted to steel-clad blocks of glass and after some 50 years or so buried deep in stable rock or under the deep ocean bed. The total amount of highly active waste produced in Britain from 1956 to 1983 was about 2000 cubic metres. The bulk of this is stored at a reprocessing plant at Sellafield.

Nuclear Fusion Reactors

At present, many countries are working on projects to harness energy from nuclear fusion to generate electric power. The generation of electricity from controlled nuclear fusion requires three basic steps:

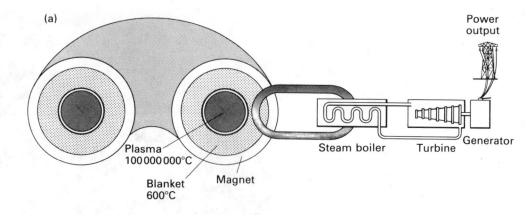

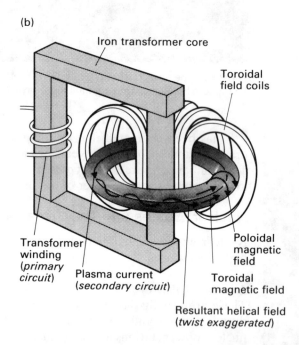

Figure 1.38 (a) A fusion reactor. The JET project at Culham in Oxfordshire uses the tokamak plasma confinement system. This was first developed in the USSR in the late 1960s. It consists of a toroidal vacuum chamber surrounded by a ring of magnetic field coils. Over these is a large transformer core which produces a current through the plasma. The current not only produces the second (poloidal) field but also heats the plasma. (b) The tokamak configuration

- The creation of plasma and raising its temperature to over 100 million °C.
- Holding the plasma away from the container wall long enough to allow abundant reactions to occur.
- The design of a fusion reactor which will provide electricity economically.

The Joint European Torus (JET) experiment which is being carried out at Culham in Oxfordshire uses magnetic fields to hold the plasma in place whilst electric currents heat them to the temperatures required for fusion to take place (*see* figure 1.38).

Arguments Against the Use of Nuclear Energy

Arguments against the use of nuclear energy focus on four main areas: the risk of an accident; terrorist attacks; disposal of wastes; and alternative sources of energy.

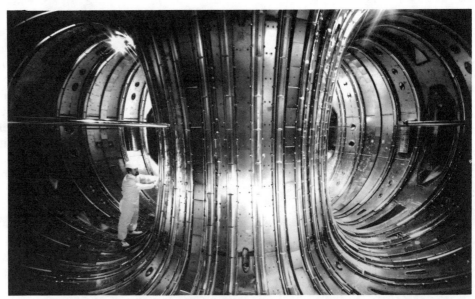

The interior of the vacuum vessel of JET (Joint European Torus), the world's largest toroidal fusion experiment. JET, based at Culham, is financed by, and has staff from, all EEC countries together with Switzerland and Sweden. JET commenced operation in June 1983. It aims, during its seven-year experimental programme, to reach plasma conditions approaching those required in a fusion reactor

The Risk of an Accident

The worst ever nuclear accident occurred at Chernobyl, Ukraine, USSR on Saturday 26 April 1986. A series of operator errors resulted in the build up of a head of steam in the reactor. The steam reacted with hot zirconium in the reactor to produce hydrogen. The gas pressure inside the core increased until the top of the core blew off. On coming into contact with air the gaseous mixture exploded causing a fire which ignited the graphite moderator and caused it to burn for several days. Radioactive material from the core was released into the atmosphere forming a radioactive cloud estimated to be 30 miles across and 100 miles long. The cloud spread across much of Europe causing widespread con-

An aerial view of the Chernobyl nuclear power plant, photographed on 9th May 1986, nearly two weeks after the disaster in which Unit 4 (arrowed) was crippled by an explosion, releasing a cloud of radioactive material into the air

tamination and alarm. The accident caused a rising number of deaths and injuries in the vicinity of Chernobyl. Over 80 000 people within a six-mile radius of the reactor were evacuated soon after the accident. Some experts estimate that between 1000 and 10 000 people will die from cancer over the next decade as a result of being irradiated by the accident.

Soon after the accident the Chairman of the Central Electricity Generating Board (CEGB) in the UK said that such a disaster 'cannot happen here'. The Chernobyl reactor would not have been licensed as safe for use today in any western country. Even so, the accident raised a number of serious questions concerning safety in the nuclear energy industry.

The worst ever nuclear accident in the West occurred in a pressurised-water reactor (PWR) at Three Mile Island, Pennsylvania, USA on 28 March 1979. A series of equipment failures and human mistakes led to a loss of coolant and partial core meltdown. A series of protective barriers prevented a serious release of radioactivity and no one was either injured or killed as a result of the accident.

Terrorist Attacks

The plutonium used in fast reactors is also used in nuclear weapons. It might therefore prove an attractive target for a terrorist hijack. However, plutonium is essentially self-protecting. It is highly toxic and because it is also highly radio-active it is surrounded by heavy shielding. Nevertheless, there is a risk that plutonium could be diverted from peaceful to military uses. The aim of the Non-Proliferation Treaty which has been signed by 100 countries is to prevent proliferation. It requires surveillance including physical checks of stocks and movements of fissile materials.

Disposal of Wastes

The problems of storing and finally disposing of highly active wastes which may take hundreds if not thousands of years before they become inactive present another argument against the use of nuclear energy. At present, we have no experience of containing materials for such long periods. It is claimed that even low-level wastes can result in the build-up of unacceptably high levels of radio-activity in the local environment of nuclear power stations.

Alternative Sources of Energy

As we shall see in chapter 5, alternative sources of energy, including coal, are sufficient to provide energy for another 300 years or so. Other energy sources including hydrogen extracted from water provide a much cleaner source of energy, although at present they are not economic. Those who argue against the use of nuclear energy point to the availability and lower risks of other sources of energy.

NUCLEAR WEAPONS

It is estimated that there are probably more than 40 000 nuclear warheads in the world today. Over 95% of these are in the hands of the two superpowers—the USA and the USSR. The total explosive strength of these is equivalent to about one million Hiroshima bombs. This is equivalent to more than 3 tons of TNT for every man, woman and child on Earth! This is enough to kill each one of us not just once but many times over!

The Atomic Bomb

The atomic bomb is a device which relies on nuclear fission for its release of energy. The energy released is about one million times greater than an equal weight of chemical high explosive.

The atomic bomb which was dropped on Hiroshima on 6 August 1945 used

The energy released by an atomic bomb is about one million times greater than that released by an equal weight of chemical high explosive

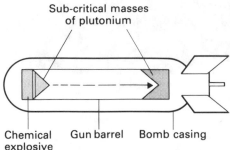

Sub-critical masses of plutonium

Chemical explosive   Gun barrel   Bomb casing

Figure 1.39   The atomic bomb. The atomic bomb contains two sub-critical masses of a fissionable isotope such as plutonium-239. On detonation of a chemical explosive, one of these masses is propelled towards the other. When they collide the total mass becomes super-critical, resulting in a nuclear explosion

(a)

Uranium-238 jacket

Lithium deuteride

Fission bomb trigger

(b)

Lithium deuteride

Fission bomb trigger

Figure 1.40   (a) The hydrogen (fusion) bomb. (b) The neutron bomb

uranium-235 enriched to about 90% (*see* figure 1.39). The bomb was produced as a result of The Manhattan Project. This was the code name for the US effort during World War II to produce the atomic bomb. The bomb that was dropped on Nagasaki on 9 August 1945 used plutonium. Most modern nuclear atomic bombs use plutonium.

The Hydrogen Bomb

The hydrogen bomb (or fusion bomb) is a nuclear weapon which relies primarily on nuclear fusion for its release of energy. The hydrogen bomb is about a thousand times more powerful than the atomic bomb. There are two types of hydrogen bomb. The tritium bomb uses the fusion of tritium and deuterium, whereas the deuterium bomb uses deuterium only. The deuterium used is in the form of lithium deuteride, LiD, where D is $^2_1$H. The fusion is triggered by a fission bomb containing uranium-235 or plutonium-239 (*see* figure 1.40).

The equation for the fusion in the deuterium bomb is

$$^2_1H + {}^2_1H \rightarrow {}^3_2He + {}^1_0n + energy$$

The neutrons produced in this reaction have sufficiently high energy to set off further fission reactions in the uranium jacket. The hydrogen bomb is thus a fission–fusion–fission device.

The Neutron Bomb

Neutron bombs are similar to fusion bombs except that they do not have an outer uranium jacket to capture the neutrons. The neutrons thus escape into the atmosphere on explosion. The neutrons carry about 30% of the energy released when a neutron bomb explodes.

A Campaign for Nuclear Disarmament (CND) rally

Women ring the US Air Force Base at Greenham Common in Berkshire to protest against the installation of nuclear missiles on British soil

> *Articles 47 and 56 of the Final Document of the Special Session of the United Nations General Assembly on Disarmament, 1978*
>
> **47.** Nuclear weapons pose the greatest danger to mankind and to the survival of civilization. It is essential to halt and reverse the nuclear arms race in all its aspects in order to avert the danger of war involving nuclear weapons. The ultimate goal in this context is the complete elimination of nuclear weapons.
> **56.** The most effective guarantee against the danger of nuclear war and the use of nuclear weapons is nuclear disarmament and the complete elimination of nuclear weapons.

Nuclear Disarmament

There are few people today who are not in favour of **nuclear disarmament**. The key question is whether such disarmament should proceed mutilaterally or unilaterally. Multilateralists maintain that disarmament should follow from negotiations and agreements between the superpowers. Meanwhile, they consider, a defence policy based on nuclear weapons is the best way of maintaining peace. Unilateralists question the effectiveness and point to the risks of a defence policy based on nuclear weapons.

There is also immense concern about the vast expenditure on not only nuclear weapons but armaments in general. For example, in sheer tonnage, there is more explosive material on Earth than there is food (*see* figure 1.41).

ORIGINS OF RADIATION AND ITS EFFECTS

We are all exposed to radiation. Some of this is natural and some man-made (*see* figure 1.42).

Exposure to radiation is measured as a 'radiation dose equivalent'. The SI unit of radiation dose equivalent is the Sievert (Sv). The internationally recommended upper limit of dosage to the general public is 5 milli-Sieverts to the whole body per year. The average radiation dose for people in the UK is 2.39 milli-Sieverts.

Accidental exposure to high levels of radiation can damage or kill healthy living cells in the body, causing immediate illness or death. Exposure to lower levels over a long term may result in cancer—both solid tumours and leukaemia. It may also result in genetic physical defects to children born to parents exposed to radiation.

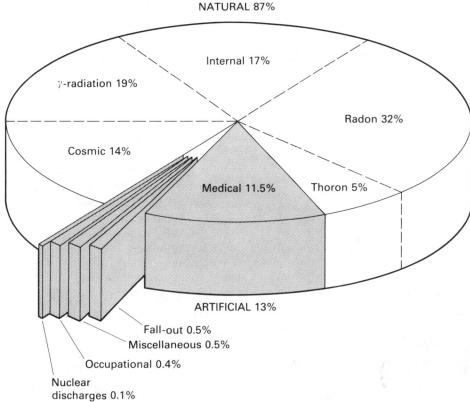

Figure 1.42 Composition of total radiation exposure of UK population

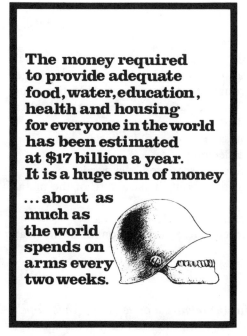

The money required to provide adequate food, water, education, health and housing for everyone in the world has been estimated at $17 billion a year. It is a huge sum of money ... about as much as the world spends on arms every two weeks.

Figure 1.41 This illustration first appeared in the *New Internationalist*, January, 1980

SUMMARY

1. **Radioactivity** is the spontaneous emission of radiation by an element due to the splitting of atomic nuclei.
2. The three principal types of **radiation** are:

   α-radiation (helium nuclei, $^4_2$He)
   β-radiation (electrons, $^0_{-1}$e)
   γ-radiation
3. **Radioisotopes** are unstable isotopes. They decay towards a **zone of stability.**
4. The **half-life** of an unstable isotope is the time taken for its radioactivity to drop to half its initial value. Its value is given by

$$t_{1/2} = \frac{0.693}{\lambda}$$

   where λ is the decay constant.
5. A **decay series** is a sequence of radioactive decay reactions leading to the formation of a **stable isotope.**
6. The conversion of one element to another is known as **transmutation**. It can be achieved artificially by **nuclear reactions**. These involve the bombardment of nuclei with particles such as protons and α-particles.
7. **Nuclear fission** is the breaking down of heavy nuclei into lighter nuclei.
8. **Nuclear fusion** is the fusion of lighter nuclei to form heavier nuclei.
9. Nuclear fission and nuclear fusion reactions generate **vast amounts of energy.**

Examination Questions

1. (a) Describe briefly the experimental evidence that led to the discovery of the neutron.
   (b) Give *one* piece of experimental evidence for the wave nature of the electron.

   (OLE)

2. In 1909, Geiger and Marsden reported the results of experiments in which a beam of α-particles was directed at very thin metallic foils. The large majority of the α-particles went through the foils without deflection, but some were deflected from their straight-line paths and a very small number were deflected or scattered backwards.
   (a) What is an α-particle? State its mass number and its charge.
   (b) Why were most α-particles not deflected?
   (c) Why were some α-particles deflected or scattered backwards?
   (d) What do these experiments illustrate about the structure of the metal atoms?
   (e) State one other method of deflecting α-particles.
   (f) $^{230}_{90}$Th decays by emitting an α-particle. Give the equation for the nuclear reaction involved.

   (JMB)

3. What do you understand by the terms (a) *relative atomic mass*, (b) *isotope*?
   Outline, with the aid of a labelled diagram, the use of the mass spectrometer in the determination of relative atomic mass.
   The mass spectrum of ethanol contains peaks at m/e values of 15, 28, 31, 45 and 46. Account for the appearance of these peaks as fully as you can.

   (UCLES)

4. (a) Draw a carefully labelled diagram of a mass spectrometer, and describe the principles of its operation.
   (b) The highly toxic environmental pollutant tetrachlorodibenzodioxin (TCDD, $C_{12}H_4O_2Cl_4$) is detected in soil samples at the 1 part in 10^{12} level by mass spectrometry on the molecular ion $(TCDD)^+$ at mass number 322.

   TCDD   DDE

   The pesticide residue DDE ($C_{14}H_8Cl_4$) which contaminates soil samples also gives a molecular ion $(DDE)^+$ at mass number 322.
   (i) Assuming that C, H and O are essentially isotopically pure, calculate the mass numbers of all the molecular ions $(TCDD)^+$ and $(DDE)^+$.
   (ii) Accurate isotopic masses are, ^{12}C 12,0000; ^{1}H 1.0078; ^{16}O 15.9949; ^{35}Cl 34.9688; ^{37}Cl 36.9651. Explain briefly why these are not all whole numbers.
   (iii) What mass difference must the mass spectrometer be able to resolve in order to distinguish between TCDD and DDE on the basis of their peaks at mass number 322?

   (O & C)

5. (a) State the meaning of **each** of the following: (i) atomic number; (ii) mass number; (iii) isotopes.

(b) The relative atomic mass of an element may be defined as the weighted mean of the masses of all the atoms in a normal isotopic sample of the element on the scale where the mass of one atom of carbon-12 has the value of twelve exactly.

The element copper has a relative atomic mass of 63.55 and contains atoms with mass numbers of 63 and 65.

Calculate the percentage composition of a normal isotopic sample of copper.

(c) Chlorine contains $^{35}_{17}Cl$ and $^{37}_{17}Cl$ in a 3:1 ratio and bromine contains $^{79}_{35}Br$ and $^{81}_{35}Br$ in a 1:1 ratio. Assuming that carbon contains atoms of mass number 12 only and hydrogen atoms of mass number 1 only, calculate the possible values for the molecular ion peaks in the mass spectrum of a compound of formula $C_2H_3Br_2Cl$ and the relative heights of these peaks.

(d) Explain what is meant by the *mass deficit* of a nucleus.

(AEB, 1985)

6. (a) The spectrum of atomic hydrogen includes a series of lines in the ultra-violet region. The separation of these lines decreases with increasing frequency (i.e. decreasing wavelength), and eventually the lines coalesce. There is a similar series which starts in the visible region of the spectrum. Explain these facts.

(b) Describe **one** example of the use of isotopes which **either** throws light on the mechanism of an organic reaction, **or** helps to solve a problem of industrial or technological importance.

(c) The element boron has two isotopes of relative atomic masses 10.01 and 11.01. The relative atomic mass of naturally occurring boron differs slightly according to its origin. The value (as found by a chemical method) for boron obtained from a Californian mineral was 10.84, while that for boron from an Italian mineral was 10.82. Calculate the percentage of the lighter isotope in the two samples of boron. Indicate (without giving any detailed explanation) what experiment might be performed to decide whether or not this difference in the relative atomic mass of boron from the two sources was real.

(OLE)

7. (a) Describe the atomic spectrum of hydrogen and indicate how the ionisation energy of hydrogen can be obtained from the spectrum.

(b) Sketch a graph to show how the successive ionisation energies of carbon vary with the number of electrons removed, and explain the shape of the graph.

(c) Explain how the variation in first ionisation energies along Period 2(Li–Ne) provides evidence for electron sub-shells.

(d) State and explain the variation in first ionisation energy down Group I in the Periodic Table.

(JMB)

8. (a) Write an essay on radioactive decay. In your answer reference should be made to the origin and nature of the emission, and to isotopes, neutron–proton ratios and half-lives.

(b) Describe **two** applications of radionuclides.

(NISEC)

9. (a) Identify the three main types of radioactive emission and describe their main properties.
   (b) Give and explain the principles of *two* examples of the uses of radioactivity or of radioactive isotopes.
   (c) The following is a part of the uranium decay series:

   $$^{238}_{92}U \rightarrow {}^{234}_{90}Th \rightarrow {}^{234}_{91}Pa \rightarrow X$$

   (i)   Which particles are emitted in each of the first two stages?
   (ii) If a β-particle is emitted in stage three, identify the isotope X.
   (iii) If the activitiy of $^{234}_{90}Th$ is reduced to 25% in 48 days, what is the half-life of $^{234}_{90}Th$?
   (iv) On graph paper draw, using appropriate scales, the rate of decay curve for $^{234}_{91}Pa \rightarrow X$, which has a half-life of one minute.

   (L)

10. (a) Outline the principles involved in determining the atomic mass of an isotope using a mass spectrometer.
    (b) The following decay series of $^{235}_{92}U$ has been somewhat abbreviated

    $$^{235}_{92}U \xrightarrow{2\alpha,\, 2\beta} X \xrightarrow{P} {}^{211}_{82}Pb \xrightarrow{Q} {}^{207}_{82}Pb$$

    (i)   Identify the isotope X by giving its mass and atomic number.
    (ii) Identify both the number and types of radiation referred to as P and Q respectively.
    (iii) If the average atomic mass of the two isotopes of lead $^{211}_{82}Pb$, $^{207}_{82}Pb$ was 209.32, calculate the percentage composition by mass of the mixture.
    (c) What are the exact natures of α- and β-radiations and how does each type of radiation behave when passed through
    (i)   a magnetic field.
    (ii) paper.
    (iii) a gas?
    (d) Identify the types of radiation J, K and L in the following nuclear transformations:

    $$^{9}_{4}Be + {}^{1}_{1}H \rightarrow {}^{10}_{5}B + J$$
    $$^{9}_{4}Be + {}^{4}_{2}He \rightarrow {}^{12}_{6}C + K$$
    $$^{14}_{7}N + {}^{2}_{1}H \rightarrow {}^{16}_{9}F + L$$

    Classify these types of radiation as particulate or non-particulate in nature.
    (e) The ratio of tritium $^{3}_{1}H$ to hydrogen atoms $^{1}_{1}H$ in a sample of water was $1:1 \times 10^{19}$. If the half-life of the tritium atoms is 12.25 years, calculate the actual number of tritium atoms remaining in 10 g of this water after a period of 49 years.

    (AEB, 1984)

11. (a) P is a radioisotope which undergoes transitions as follows:

    $$P \xrightarrow{\beta \text{ emission}} Q \xrightarrow{\beta \text{ emission}} R \xrightarrow{\alpha \text{ emission}} S$$

    If the atomic number of P is 88, and its mass number 228, what are the atomic number and mass number of isotope S?
    (b) In each of the following pairs, state whether or not both species have the same half-life:
    (i)   1 gram ^{212}Pb and 100 gram ^{212}Pb
    (ii) 1 gram ^{212}Pb and 1 gram $^{212}Pb^{2+}$

(iii) 1 mole ^{210}Pb and 1 mole ^{212}Pb

(iv) 1 mole ^{210}Pb and 1 mole ^{210}PbO

(SEB)

12. (a) What are x, Y and z in each of the following nuclear equations?

   (i) $^{35}_{17}\text{Cl} + ^{1}_{0}\text{n} \rightarrow ^{x}_{z}\text{Y} + ^{1}_{1}\text{H}$

   (ii) $^{7}_{3}\text{Li} + ^{2}_{1}\text{H} \rightarrow 2^{x}_{z}\text{Y} + ^{1}_{0}\text{n}$

(b) The half-life of the isotope of lead of mass number 210 is 21 years. How long after the isolation of a sample of this isotope will only one-eighth of the original mass be left?

(c) A radioactive isotope of the element thorium, of mass number 232 and atomic number 90, decays by α-particle emission to give an element X, which then loses a β-particle to give an element Y, which also loses a β-particle to give an element Z. What are the mass number and the atomic number of the element Z?

(OLE)

2 CHEMICAL BONDS

Kenichi Fukui

Roald Hoffmann

The Physics of Chemistry

A funny thing happened on the way to this year's [1981] Nobel Prizes in Science. The chemistry award honoured two men who used quantum mechanics—the bedrock of modern physics—to explain how molecules form. And the physics award went to three researchers who discovered how to read the chemical signatures of such complex mixtures as pollutants and rust. But if the $180 000 prizes announced last week showed that physics and chemistry meet in the subatomic world, they also suggested that Eastern and Western science recently have been poles apart. The work of Japanese scientist Kenichi Fukui, who shared the chemistry award with Cornell University chemist Roald Hoffmann, went largely unappreciated in his own country. 'The Japanese are very conservative when it comes to new theory,' he said last week. 'But once you get appreciated in the US or Europe, then the appreciation spreads back to Japan.'

Hoffmann and Fukui, of Kyoto University, based their research on an updated image of the atom. The old model looked like a tiny solar system, with a nuclear 'sun' and electron 'planets' in discrete circular orbits. Quantum mechanics—which looks at particles as waves—draws a radically different picture: the nucleus is surrounded by clouds of electrons that might be shaped like dumbbells or clover leaves, among other possibilities. Fukui, 63, figured out that the outermost clouds play the key role in chemical reactions. His 'frontier orbital theory' allows chemists to all but ignore the inner electrons and, by calculating the shape and density of the outer electron clouds, roughly predict how molecules will combine.

Rules: Hoffmann, 44, independently extended the theory that Fukui had formulated 25 years ago to devise his own rules that tell whether a chemical reaction is possible. 'If we have the orbitals of the starting molecules and the final products, we can make a go/no-go decision about the reaction,' he explains. Those rules now allow chemists to choose the right pathway to new drugs.

Fukui's own Japanese colleagues ignored his theory—possibly, he says, because 'there aren't enough scientists in pure chemistry here to apply it.' Indeed, Japan's technological excellence sometimes obscures the country's relative weakness in pure scientific research. 'We Japanese are very strong in learning and understanding when we have a clear target in view, but perhaps not quite so strong when it comes to finding a new direction,' recently wrote Makoto Kikuchi, of the Sony Research Center, in *Physics Today*.

For physics laureate Kai Siegbahn of Sweden's Uppsala University, the honour carried a special gratification—his father won the prize in 1925. 'It's a decided advantage if you start discussing physics every day at the breakfast table,' he says. Siegbahn, 63, developed electron spectroscopy, which lets scientists infer the chemical structure of unknown samples by studying electrons knocked out of atoms by light or X-rays. The technique is so sensitive that it can tell whether even 1 per cent of a single atomic layer on the surface has rusted.

Two men who literally shined a light on the innermost secrets of matter shared the other half of the physics prize. Arthur Schawlow of Stanford and Nicolaas Bloembergen of Harvard pioneered laser spectroscopy by exciting atoms with the intense, pure beam of a laser. The atoms absorb, scatter and reradiate the light in telltale ways. Schawlow, 60, was one of the inventors of the laser; Bloembergen, 61, blended three laser beams into a fourth to produce infra-red and ultra-violet laser light, which enable scientists to explore everything from how gases explode in a combustion engine to the movement of molecules in living tissues. And that may foreshadow Nobel Prizes yet to come: by looking at life on an atomic level, scientists are slowly drawing biology, too, into the realm of atoms and electrons.

Sharon Begley

LEARNING OBJECTIVES

After you have studied this chapter you should be able to

1. Outline the main features of the **electronic theory of valency**.
2. (a) Describe briefly the nature and characteristics of **ionic bonds, covalent bonds, coordinate bonds, metallic bonding, hydrogen bonding** and **van der Waals forces**.
   (b) Give examples of elements or compounds in which these types of bond exist.
3. Use the **electron transfer model** to show how ionic bonds are formed.
4. Show how **ionisation energy** and **electron affinity** influence the formation of ionic bonds.
5. Use **dot and cross models** to represent ionic, covalent and coordinate bonding in simple elements and compounds.
6. Distinguish between the principal types of **molecular orbital**.
7. Show how **electronegativity** influences the formation of ionic and covalent bonds.
8. Indicate how bonding influences
   (a) the **size** and **shape** of atoms, ions and molecules,
   (b) the physical **structure** of elements and compounds,
   (c) the physical and chemical **properties** of elements and compounds.

2.1 The Nature of Chemical Bonds

INTRODUCTION

A **chemical bond** is a force which holds together two or more atoms, ions, molecules or any combination of these. The force is the electrostatic force of attraction between negatively charged electrons and positively charged nuclei. The extent of this force of attraction depends primarily on the outer-shell electronic configurations of the atoms (*see* chapter 1). For example, noble gases do not readily form chemical bonds. This is because they have stable outer electron shells. On the other hand, elements whose atoms have a single electron in their outer shells readily form bonds. Hydrogen is an example.

When two hydrogen atoms come close together, they exert a force of attraction on one another. If they get too close, however, they repel each other. The optimum distance is when the forces of attraction equal the forces of repulsion. At this distance the potential energy is at a minimum. The distance is called the **bond length**. This is discussed in more detail later in this chapter. The variation of potential energy with distance between nuclei is shown in figure 2.1. This type of curve is known as the **Morse curve**. The energy needed to separate two atoms bonded together and place them at a distance where they do not exert a force of attraction on one another is called the **bond energy** or bond dissociation energy. These are determined experimentally as bond enthalpies—a term we shall define in chapter 5.

The ability of an atom to form chemical bonds is called its **valency**. However, this term is now largely outdated since it is now more common to consider chemical bonding in terms of specific types of bond rather than bonding in general. The electrons which form chemical bonds are called **valence electrons**. These occur in the highest-energy orbitals of an atom (*see* chapter 1). The outside shell of an atom which contains these orbitals is called the **valence shell**.

The Electronic Theory of Valency

Our modern ideas of the nature of the chemical bond are based on the electronic theory of valency. This theory was developed independently by G. N. Lewis and W. Kossel. Both published their theories in 1916. According to this theory, when atoms form bonds they try to achieve the most stable (that is the lowest-energy) electronic configuration. They can do this in two ways.

1. They can **gain or lose electrons** to form ions. If they gain electrons they become **anions**. If they lose electrons they become **cations**. The ions have full and thus stable outer-shell electron configurations. The bond is the electrostatic force of attraction between the anion and the cation. This type of bond was called an **electrovalent bond**. Nowadays it is more common to call it an **ionic bond**.

2. Atoms can also achieve stable outer electron configurations by **sharing electrons**. This is called **covalent bonding**. A covalent bond consists of a shared pair of electrons. However, in some molecules and ions, both the shared electrons come from one atom. This is called **coordinate bonding**. It is a type of covalent bond. Coordinate bonding is also known as **dative covalency**.

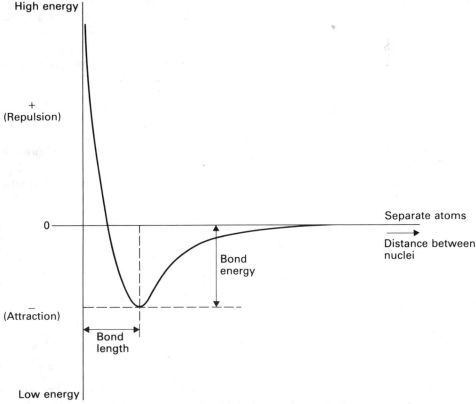

Figure 2.1   Potential energy curve for the hydrogen molecule (Morse curve)

The Octet Rule

When an atom forms a chemical bond by gaining, losing or sharing valency electrons, its electron configuration becomes the same as that of a noble gas either at the end of the same period or at the end of the previous period in the Periodic Table. Noble gases, with the exception of helium, have a stable octet of electrons (that is, eight electrons) in the outer shells of their atoms. The formation of chemical bonds to achieve noble-gas configurations is thus called the octet rule. It applies to both ionic and covalent bonds.

Other Types of Bonding

Bonding in metals is a special case. It is neither ionic nor covalent. Solid metals consist of positive metal ions closely packed together. They are held together by free electrons which 'float' in a 'sea' around the ions. This type of bonding is called **metallic bonding**.

There are two further types of chemical bonding which we shall also consider in this chapter. They are **hydrogen bonding** and **van der Waals forces**. They are weak compared to other types of bonding.

IONIC BONDING

An ionic bond is the electrostatic force of attraction between two oppositely charged ions. Kossel (1916) suggested that an ionic bond (or electrovalent bond as he called it) could be formed as a result of the transfer of one or more electrons from one atom to another. For example, the ionic bond in sodium chloride is formed by the transfer of an electron from a sodium atom to a chlorine atom.

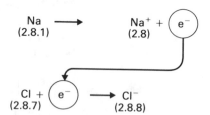

The electron configurations are written in brackets. This transfer of an electron can be represented by a *dot and cross model*.

$$Na\bullet \;+\; {}^{x}_{x}\!\overset{x\,x}{\underset{x\,x}{Cl}}{}^{x} \longrightarrow Na^{+} \quad {}^{x}_{\bullet}\!\overset{x\,x}{\underset{x\,x}{Cl}}{}^{x\,\;-}_{x}$$

The most typical ionic compounds consist of metallic cations from Groups I and II and non-metal anions from Groups VI and VII. Kossel's **electron transfer model** can be used to represent the formation of these compounds. Here are three examples:

1.   Li   +   F   →   Li$^+$ +   F$^-$ Lithium fluoride
           (2.1)   (2.7)   (2)   (2.8)
    Group   I   VII

2.   2Li   +   O   →   2Li$^+$ +   O^{2-} Lithium oxide
           (2.1)   (2.6)   (2)   (2.8)
    Group   I   VI

3.   Ca   +   O   →   Ca^{2+} +   O^{2-} Calcium oxide
       (2.8.8.2)   (2.6)   (2.8.8)   (2.8)
    Group   II   VI

Notice that the ions in these compounds have the same electron configurations as noble gases. When an element and a group of ions have the same electron

Table 2.1 Members of the neon and argon isoelectronic series

neon $1s^22s^22p^6$ (2.8)	argon $1s^22s^22p^63s^23p^6$ (2.8.8)
Ne	Ar
O^{2-}	S^{2-}
F^-	Cl^-
Na^+	K^+
Mg^{2+}	Ca^{2+}

configurations, they are said to form an **isoelectronic series**. Members of the neon and argon isoelectronic series are shown in table 1.1

Stability of Ionic Compounds

The ease of formation of an ionic compound depends on the ease of formation of its cations and anions. For this to be energetically favourable, the electron-donating atom must have a low ionisation energy and the electron-accepting atom must have a high electron affinity. **Electron affinity** is an energy term. It is a measure of the attraction (or affinity) of an atom for an electron. It can be defined as the energy change which occurs when one mole of singly charged anions is formed from one mole of atoms. For example

$$Cl(g) + e^- \rightarrow Cl^-(g) \qquad \text{Electron affinity} = -364 \text{ kJ mol}^{-1}$$

Strictly speaking this is called the first electron affinity. This is to distinguish it from the second and subsequent electron affinities. The second electron affinity is the energy change which occurs when one mole of doubly charged anions is formed from one mole of singly charged anions.

Notice that, like ionisation energy, electron affinity refers to atoms and ions in their gaseous states. Notice also that the ionisation of chlorine is accompanied by a loss of energy. The process is thus exothermic and the value for the electron affinity negative. Most first electron affinities are negative. The more negative the value, the greater the electron affinity, and thus the greater the ease of formation of anions.

The second electron affinity of any element is always endothermic. This is because energy must be absorbed to overcome the repulsive forces between the two negative particles. For example

$$O^-(g) + e^- \rightarrow O^{2-}(g) \qquad \text{Second electron affinity} = +791 \text{ kJ mol}^{-1}$$

We have seen that the formation of an ionic compound is energetically favourable when one element has a low ionisation energy and the other a high electron affinity. We have also seen that both ionisation energy and electron affinity apply to atoms and ions in their gaseous states. We might thus infer that an ionic compound is in its most stable and energetically favourable state when it is in its gaseous state. This is clearly not so. Most ionic compounds are most stable in their solid state. This is because in this state they exist as **lattices**. Figure 2.2 shows the familiar model of the sodium chloride lattice. The lattice structure accounts for many of the physical properties of ionic compounds. We shall look at some of these properties later in the chapter. We shall also examine the various types of lattice structures in the following chapter.

Why is it, then, that ionic compounds are more stable as lattices than as gases? The answer lies in the **lattice energy**. This is the energy needed to bring two gaseous cations and gaseous anions together to form a solid lattice. For example

$$Na^+(g) + Cl^-(g) \rightarrow Na^+Cl^-(s) \qquad \text{Lattice energy} = -787 \text{ kJ mol}^{-1}$$

The lattice energy is exothermic and thus results in a lowering of the energy. Lower energy means greater stability. In chapter 5 we shall see how lattice energy (or more specifically, lattice enthalpy) is related to ionisation energy, electron affinity and other energy terms through a cycle known as the Born–Haber cycle.

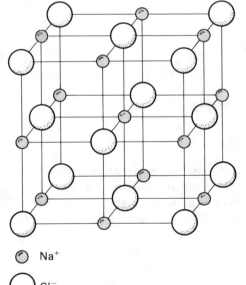

$\otimes$ Na^+

$\bigcirc$ Cl^-

Figure 2.2 Model of the sodium chloride lattice

COVALENT BONDING

A covalent bond consists of two electrons shared between two adjacent atoms. Each atom contributes one electron. Covalent compounds exist as either molecules or giant structures (see below). However, covalent bonds also exist in certain types of ion. A covalent bond, whether in a molecule, giant structure or ion, may be represented by a *dot and cross* (*see* figure 2.3). Each dot and cross represents an electron in the valence shell of an atom. A double covalent bond is represented by two pairs of dots and crosses and a triple bond by three pairs.

Lewis formulae are also used to show covalent bonds in molecules or ions. In these formulae, each electron is represented by a dot. A single covalent bond is thus represented by a pair of dots *between* two atoms. The Lewis formulae of hydrogen chloride, oxygen and ethyne are shown in figure 2.3.

Notice that each atom in a covalent molecule has a full shell of electrons—the hydrogen atoms have two electrons (in the K shell) and the chlorine, oxygen and carbon atoms are all surrounded by eight electrons (in the L or M shells). The octet rule is thus satisfied.

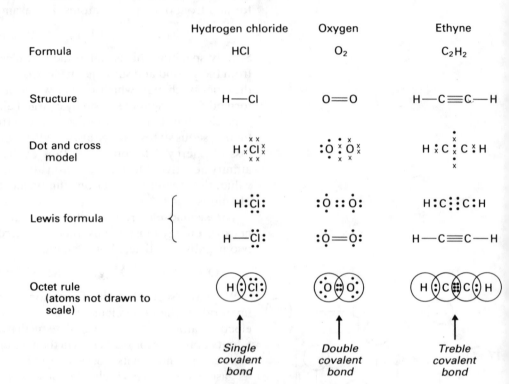

Figure 2.3   Dot and cross models and Lewis formulae

Molecular Orbitals

We saw in the previous chapter that electrons exist in orbitals. These orbitals have characteristic shapes and are defined by the subsidiary quantum number.

The electrons in covalent compounds exist in molecular orbitals. Each orbital can take one or two electrons of opposite spin. There are three types of molecular orbital:

. bonding orbitals,
   non-bonding orbitals,
   anti-bonding orbitals.

Bonding Orbitals

These are the orbitals represented by the dots and crosses in figure 2.3. They are formed by two atomic orbitals merging or overlapping. A σ(sigma) bond is formed by the *single* overlap of

- two s atomic orbitals (*see* figure 2.4a), or
- two p atomic orbitals (*see* figure 2.4a), or
- an s and a p atomic orbital (not shown in figure 2.4).

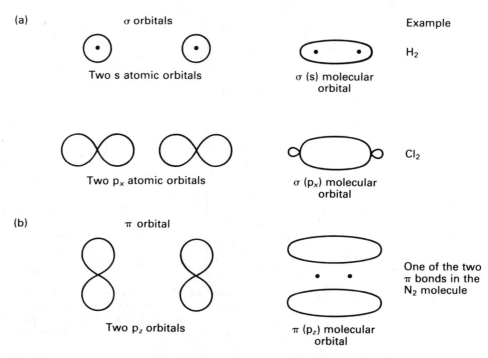

Figure 2.4 Bonding orbitals: (a) σ (sigma) orbitals; (b) π (pi) orbital

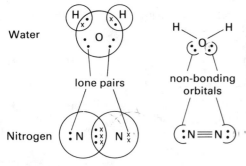

Figure 2.5 Lone pairs and non-bonding orbitals

A π (pi) bond is formed by the *double* overlap of two p orbitals (*see* figure 2.4b). This only occurs where the two atoms are already bonded by a sigma bond. For example, the triple bond of the N_2 molecule consists of two π bonds and one σ bond.

Non-Bonding Orbitals

In order to achieve a stable octet of electrons in its outer shell, an oxygen atom readily shares two of its six outer-shell electrons (*see* figure 2.5). This leaves four electrons which do not participate in bonding. These electrons are called **lone-pairs**.

The nitrogen molecule also contains a lone-pair of electrons. Each nitrogen atom in the molecule has five outer-shell electrons. Two of these exist as a lone-pair in a non-bonding orbital. The other three participate in bonding (*see* figure 2.5).

Anti-Bonding Orbitals

We have seen that a σ or π bonding orbital consists of two overlapping atomic orbitals. The electrons in these orbitals bond the molecule together. Electrons can also exist in orbitals which pull the molecule apart. These orbitals are called anti-bonding orbitals. They are labelled with an asterisk. Thus a σ anti-bonding orbital is denoted σ* (called sigma starred). An anti-bonding orbital has a higher energy than its corresponding bonding orbital (*see* figure 2.6). Anti-bonding orbitals can thus only be occupied by electrons which have been excited. Since the bonding electrons in molecules are normally in their ground state, anti-bonding orbitals are not normally occupied by electrons.

Hybrid Atomic Orbitals

In many molecules, particularly organic molecules, some of the atomic orbitals involved in covalent bonding combine together forming hybrid atomic orbitals. Figure 2.7 shows the shapes of the three types of hybrid atomic orbitals formed from s and p orbitals. An s and a p orbital mix to form two hybrid sp orbitals.

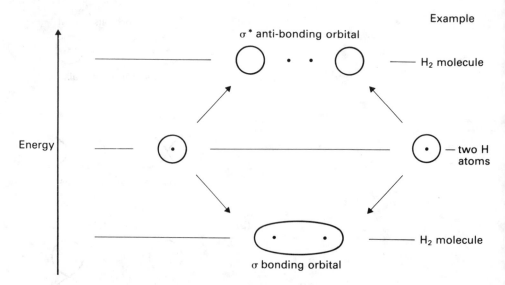

Figure 2.6 Molecular orbitals. The combination of atomic orbitals leads to the same number of molecular orbitals. For example, the s orbitals in two hydrogen atoms combine to form two molecular orbitals—one with a high energy and the other with a low energy. The former is known as a σ* (sigma starred) anti-bonding orbital and the latter as a σ bonding orbital

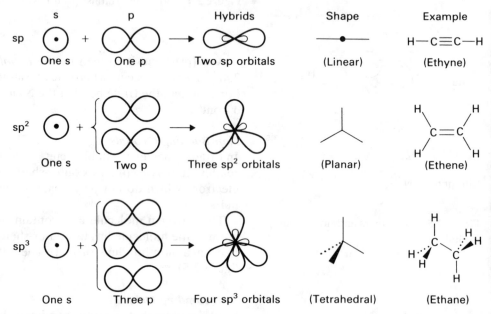

Figure 2.7 Hybrid atomic orbitals

Orbitals formed by mixing one s and two p orbitals are called sp^2 orbitals. There are three orbitals in a set of sp^2 orbitals. The hybrid orbitals in a set are equivalent. They cannot be distinguished from one another in shape, location or in energy. A set of sp^3 atomic orbitals consists of four equivalent orbitals formed by mixing one s and three p orbitals.

Delocalised Orbitals

So far, we have only considered bonding orbitals which bond two specific atoms together. Such bonding orbitals are called **localised orbitals**. However, in some molecules it is not possible to be specific about the location of the bonding orbitals.

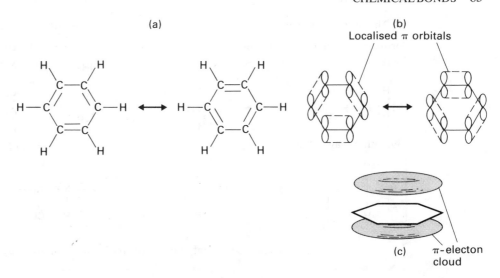

Figure 2.8 (a) and (b) Resonance hybrid structure. (c) Delocalised π orbitals (π-electron cloud). The six 2p orbitals of the carbon atoms give rise to six π molecular orbitals, three of which are low-energy bonding orbitals and three higher-energy anti-bonding orbitals. The three bonding orbitals overlap to form a π-electron cloud. This is shown as 'ring doughnuts' in the drawing for benzene (c)

A classic case is benzene. Benzene may be considered as a **resonance hybrid** of two structures. Figure 2.8a and b shows the two extremes of this hybrid. Each extreme is due to the overlap of adjacent pairs of p orbitals. These are, of course, π bonds. The σ bonds also bond the carbon atoms and the hydrogen atoms to the carbon atoms. The π bonding orbitals in the benzene molecule all overlap and are indistinguishable from one another. It is thus impossible to say which π electrons or orbitals bond which pair of carbon atoms. The orbitals are thus called delocalised π orbitals (*see* figure 2.8c). They can be thought of as clouds of electrons above and below the ring of carbon atoms. For this reason, the benzene molecule is commonly represented as

Coordinate Bonding

In some molecules and ions a covalent bond is formed by one atom donating and sharing a pair of electrons with another atom. The pair of electrons donated is always a lone-pair. The atom which donates the lone-pair is called the **electron-pair donor**. The atom accepting the pair of electrons is called the **electron-pair acceptor**. This type of bonding is also called dative bonding.

The nitrogen atom in an ammonia molecule has a lone-pair of electrons and thus readily forms a coordinate bond. An example is the bond formed between the nitrogen atom and the aluminium atom in the compound NH_3AlCl_3. Three different ways of representing this bonding in the compound are shown in figure 2.9. Note that the coordinate bond results in the formation of a positive charge on the electron-pair donor and a negative charge on the electron-pair acceptor.

Polyatomic ions such as the ammonium ion, NH_4^+, and the oxonium ion, H_3O^+ (this is also known as the hydroxonium or hydronium ion), contain co-ordinate bonds as well as the usual covalent bonds. The coordinate bond is represented by an arrow (*see* figure 2.9). In fact, it is impossible to distinguish the

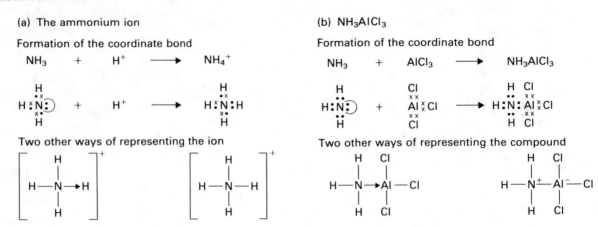

Figure 2.9 Coordinate bonds: (a) the ammonium ion; (b) NH_3AlCl_3

four bonds in an ammonium ion or the three bonds in an oxonium ion. The bonds are thus delocalised and the charge on the ion is distributed over the whole ion. The ion has a **resonance structure**. This is a hybrid of the various possible distributions of electrons. The different distributions are known as **limiting forms** (or canonical forms). The resonance structures and limiting forms of the oxoanions CO_3^{2-} and NO_3^- are shown in figure 2.10. The sulphate ion SO_4^{2-} also exists as a resonance structure with limiting forms.

These simple polyatomic ions are sometimes called **covalent ions**. They are part of a much larger group of ions known as **complex ions**. A complex ion consists of a central atom surrounded by other atoms, ions or groups of atoms. These are called **ligands**. They are bonded to the central atom by coordinate bonds. By far the most important examples of complex ions are the d-block metal ions such as $[Cu(NH_3)_4]^{2+}$ and $[CoCl_4]^{2-}$. Such ions are considered in detail in chapter 14.

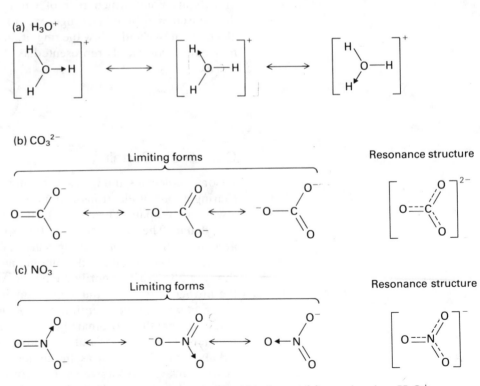

Figure 2.10 Resonance structures and limiting forms: (a) oxonium ion, H_3O^+; (b) carbonate ion, CO_3^{2-}; (c) nitrate ion, NO_3^-

IONIC AND COVALENT CHARACTER

Ionic and covalent bonding are two extreme cases. Most chemical bonds lie between these two extremes. Figure 2.11 shows the two extremes and the partial covalent and ionic nature of bonds in between. **Electron density maps** of sodium chloride and the hydrogen molecule show the two extremes clearly (figure 2.12). But even sodium chloride shows slight covalent character. This is indicated by the squaring of the outer contours. The electron density map of the hydrogen molecule shows no distortion. The bond in the hydrogen molecule is thus purely covalent (*see* figure 2.12).

Ionic bond

Ionic bond with partial covalent character

Polar covalent bond

'Pure' covalent bond

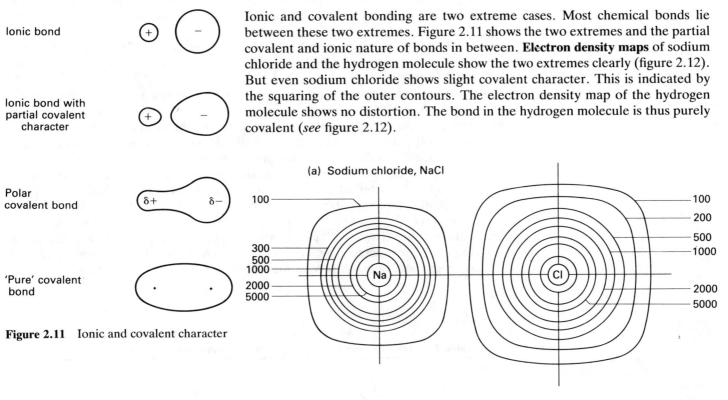

Figure 2.11 Ionic and covalent character

(a) Sodium chloride, NaCl

(b) Hydrogen molecule, H_2

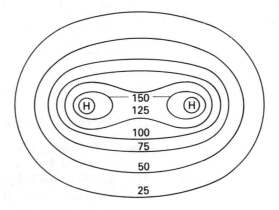

Figure 2.12 Electron density maps. The numbers represent the electron densities

Ionic Character of Covalent Bonds

Molecules consisting of two identical atoms such as H_2 and Cl_2 are always purely covalent. However, when two different atoms are joined by a covalent bond, the bonding electrons are not shared equally. This is due to the different electron attracting abilities of the two atoms. The result is a **polar covalent bond**. This type of bond has a small positive and a small negative pole. This is called a **dipole**. The charges are indicated by $\delta+$ and $\delta-$. The Greek letter delta δ is used to indicate that the charges are small. The product of the distance between the two charges and the magnitude of the charges is called the **electric dipole moment**.

The magnitude of the charges in a polar covalent bond depends on the difference in the electron attracting ability of the two atoms joined by the covalent bond. The electron attracting ability of an atom is sometimes called its **electronegativity**.

There have been several attempts to define an electronegativity scale. The scale most commonly used is that of Pauling. It is based on enthalpy values derived experimentally.

Electronegativity increases from left to right across a period in the Periodic Table. With the exception of part of Group III, it also decreases down every group. Table 2.2 shows some electronegativity values calculated by Pauling. The scale arbitrarily assigns an electronegativity value of 4.0 to the most electronegative of all elements—fluorine. Note that electronegativity values have no units.

Table 2.2 also shows how these electronegativity values can be used to estimate the ionic character of covalent bonds. The higher the difference in values, the greater the percentage ionic character. A difference of 2.1 corresponds to about 50% ionic character. Bonds with differences of 2.1 or more are almost entirely ionic. Those with lower values are polar covalent. The lower the value, the lower the percentage ionic character. A zero value indicates zero ionic character and thus a purely covalent bond.

Examples of the use of these electronegativity values to determine bond type:

Bond	Electronegativity difference	Type of bond
F—Li	4.0 − 1.0 = 3.0	ionic
Cl—C	3.0 − 2.5 = 0.5	polar covalent
Cl—Cl	3.0 − 3.0 = 0.0	covalent

Table 2.2 Electronegativity values for some elements

			H 2.1			
Li 1.0	Be 1.5	B 2.0	C 2.5	N 3.0	O 3.5	F 4.0
Na 0.9	Mg 1.2	Al 1.5	Si 1.8	P 2.1	S 2.5	Cl 3.0
K 0.8	Ca 1.0		Ge 1.9	As 2.0	Se 2.4	Br 2.8

The Degree of Covalent Character of Ionic Bonds: Fajans' Rules

The positive charge on the cation in an ionic compound can attract electrons towards it from the anion. This results in distortion of the anion (*see* figure 2.11). This distortion is called **polarisation**. The ability of a cation to attract electrons and distort an anion is called its polarising power. The degree of polarisation of an ionic bond is a measure of the **degree of covalent character**. Electron density maps of purely ionic compounds such as lithium fluoride show no distortion at all.

In 1923 Fajans suggested that all compounds possess a degree of covalency. Furthermore he proposed that the polarisation of an ionic bond and thus *the degree of covalency is high if*:

1. *The charges on the ions are high*
    For example
      C^{4+} favours covalent character
      Na^+ favours ionic character

2. *The cation is small*
    For example

Ion	Radius/nm	
C^{4+}	0.015	favours covalent character
Na^+	0.095	favours ionic character

3. *The anion is large*
   For example

Ion	Radius/nm	
F^-	0.136	favours ionic character
I^-	0.216	favours covalent character

These three rules are called **Fajans' rules**.

Ionic and Covalent Compounds

On crossing a period, the ionic nature of compounds decreases and their covalent nature increases. This means that we cannot say that a compound is completely ionic or completely covalent. What we can say, however, is that some compounds are predominantly ionic and others predominantly covalent.

Good examples of this are provided by the chlorides and oxides. Chlorides and oxides on the left-hand side of the Periodic Table tend to be predominantly ionic, whereas those on the right-hand side are predominantly covalent. Those in between exhibit both ionic and covalent character. Tables 2.3 and 2.4 provide examples of compounds that are normally regarded as ionic and covalent, respectively.

Table 2.3 Examples of ionic compounds

Salts
silver bromide	$AgBr$
barium sulphate	$BaSO_4$
calcium carbonate	$CaCO_3$
potassium nitrate	KNO_3
sodium chloride	$NaCl$

Mineral acids
hydrochloric acid	HCl
nitric acid	HNO_3
sulphuric acid	H_2SO_4

Bases
calcium oxide	CaO
sodium hydroxide	$NaOH$

Table 2.4 Examples of covalent compounds

Inorganic
water	H_2O
carbon dioxide	CO_2
ammonia	NH_3
nitrogen dioxide	NO_2

Organic
methane	CH_4
ethanol	C_2H_5OH
glucose	$C_6H_{12}O_6$
phenol	

Granulated sugar is packed into bags at a sugar processing plant in the Philippines. All sugars are covalent compounds

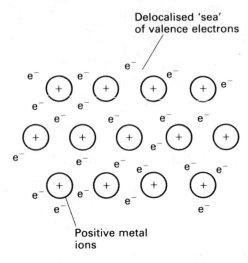

Figure 2.13 Metallic bonding

METALLIC BONDING

X-ray studies have shown that metals in their solid state exist as crystals. The crystals consist of positive ions bound together in fixed lattice positions by a freely moving 'sea' of electrons (*see* figure 2.13). The binding together of the ions in the lattice by these electrons is called metallic bonding.

The electrons involved in metallic bonding are the outer-shell or valence electrons of the metal atoms. The valence electrons are not controlled by specific metal nuclei. They are delocalised. Furthermore, these delocalised electrons do not occupy bonding orbitals in pairs. Instead they occupy giant orbitals which stretch throughout the crystal lattice. The orbitals are grouped together in **bands**. Orbitals within a band have very similar energy levels. The valence electrons of metals in Group I of the Periodic Table partly fill the lowest energy band. This is called the s-band. Metals in Group II completely fill the **s-band**. When a band is filled, the electrons then begin to fill the next lowest energy band (*see* figure 2.14).

Two Types of Band

The valence orbitals of the atoms in a crystal lattice combine to form two sets of energy levels. These are called the valence band and the conduction band. The valence band has the lower energy level. Electrons in a partially filled conduction band can move readily throughout the crystal.

Metals
In metals, the valence band is filled with electrons and the conduction band only partially filled. There is no gap between the two bands. *See* figure 2.14a.

Semiconductors
In semiconductors, the valence band is full but the conduction band is empty. There is only a small gap between the two bands. *See* figure 2.14b.

Insulators
In insulators, the valence band is full and, like semiconductors, the conduction band is empty. However, the gap between the two bands is large enough to prevent electrons transferring from the valence band to the conduction band. It is sometimes called the 'forbidden' energy gap. *See* figure 2.14c.

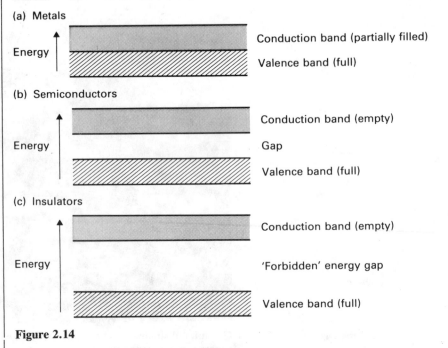

Figure 2.14

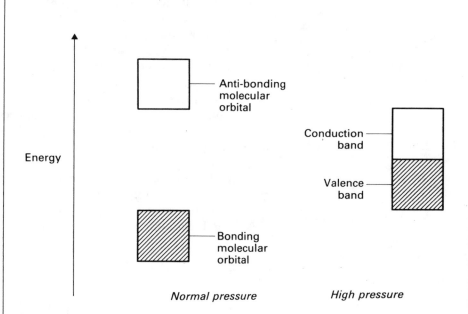

Figure 2.15   Metallic hydrogen

If hydrogen is subjected to very high pressures, it becomes metallic. Compression of the gas forces the molecules closer to each other. The normally empty anti-bonding molecular orbitals form a conduction band with an energy level much closer to the level of the bonding molecular orbitals which form the valence band (*see* figure 2.15).

Metals in their solid state consist of positive ions bound together in rigid lattices

Weak and Strong Hydrogen Bonds

There are now thought to be two types of hydrogen bonds—weak and strong. Weak and strong hydrogen bonds differ in both bond length and bond energy.

	Weak	Strong
bond energy/kJ mol^{-1}	10 to 30	400
bond length/nm	approx. 30	23 to 24

The hydrogen bonds shown in figures 2.16, 2.17 and 2.18 are all examples of weak or normal hydrogen bonds. There is experimental evidence that strong hydrogen bonds exist in

the bifluoride ion	$[F{-}H{-}F]^-$
the hydrated hydroxide ion	$[HO{-}H{-}OH]^-$
the hydrated oxonium ion	$[H_2O{-}H{-}OH_2]^+$

and in several organic and inorganic compounds.

HYDROGEN BONDING

The most renowned example of hydrogen bonding occurs in water. Since the oxygen atom is more electronegative than the two hydrogen atoms (*see* table 2.2), it attracts the bonding electrons away from the hydrogen atoms. As we have seen, this is called a polar covalent bond. The result is a dipole. The oxygen atom has a small negative charge $\delta-$ and each hydrogen atom a small positive charge $\delta+$.

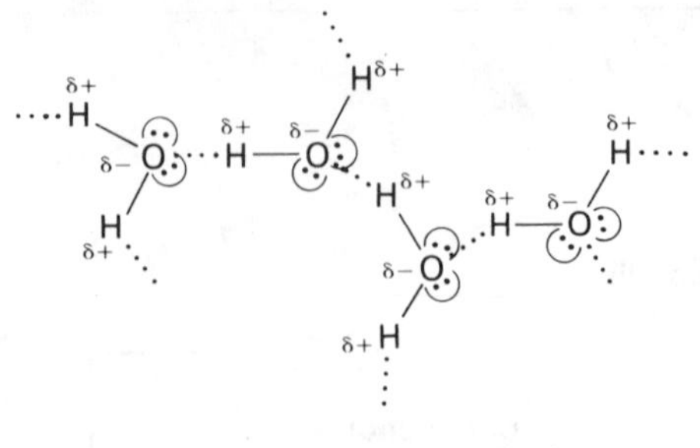

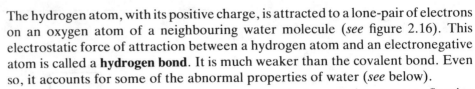

Figure 2.16 Hydrogen bonding in water

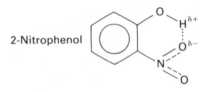

Ammonia

Methanol

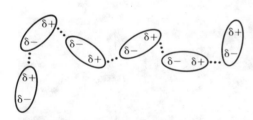

Hydrogen fluoride

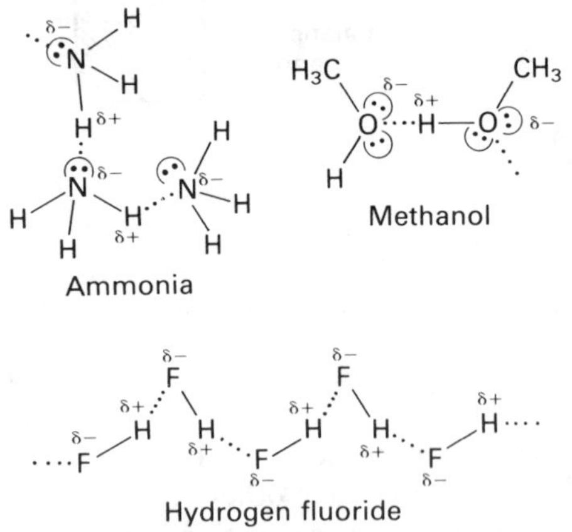

Figure 2.17 Intermolecular hydrogen bonding

The hydrogen atom, with its positive charge, is attracted to a lone-pair of electrons on an oxygen atom of a neighbouring water molecule (*see* figure 2.16). This electrostatic force of attraction between a hydrogen atom and an electronegative atom is called a **hydrogen bond**. It is much weaker than the covalent bond. Even so, it accounts for some of the abnormal properties of water (*see* below).

Hydrogen bonding occurs most commonly with atoms such as oxygen, fluorine and nitrogen. These atoms not only have high electronegativities but are also small. Hydrogen bonding does not occur significantly with larger atoms such as chlorine and sulphur even though the electronegativities of these elements are comparable to nitrogen (*see* table 2.2).

The hydrogen bonds shown in figure 2.17 are examples of **intermolecular hydrogen bonds**. An intermolecular bond is a bond between two molecules. Hydrogen bonding also occurs within molecules (*see* figure 2.18). This is called **intramolecular hydrogen bonding**. It is less common than intermolecular hydrogen bonding.

2-Nitrophenol

Figure 2.18 Intramolecular hydrogen bonding

VAN DER WAALS FORCES

Dipole–Dipole Attraction

Hydrogen bonding is an extreme example of what is called dipole–dipole attraction (*see* figure 2.19). It is extreme because the dipole–dipole forces of attraction in hydrogen bonding are much stronger than the dipole–dipole forces of attraction between other types of molecule. Dipole–dipole attraction is often called permanent dipole–permanent dipole attraction since it only occurs between molecules with permanent dipole moments. The (permanent) dipole moments of some molecules are shown in table 2.5. Note that only molecules with polar covalent bonds have dipole moments. Note also that the two dipoles of the linear CO_2 molecule cancel each other out. SO_2 does have a dipole moment, however, because the molecule is non-linear. We shall discuss the shapes of molecules in detail in the next section of this chapter. We shall also see later in the chapter how dipole–dipole attraction accounts for the properties of molecules.

Figure 2.19 Dipole–dipole attractions

(a) (b)

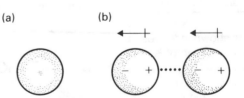

Figure 2.20 Dipole–induced dipole attraction. (a) A noble-gas molecule consists of a single atom. It is spherical. (b) A fluctuation of the electron cloud causes a **temporary dipole**. The temporary dipole **induces** a dipole in a neighbouring molecule. This results in a **weak** and temporary **force of attraction** between the two atoms

Table 2.5 Dipole moments

Molecule		Dipole moment/ debye	Comment	
argon	Ar	0	non-polar	
carbon dioxide	CO_2	0	$\overset{\delta+}{O}=\overset{\delta-}{C}=\overset{\delta+}{O}$	(the two dipoles cancel)
hydrogen chloride	HCl	1.05	$\overset{\delta+}{H}-\overset{\delta-}{Cl}$	
ammonia	NH_3	1.48		
sulphur dioxide	SO_2	1.63		
chlorobenzene	C_6H_5Cl	1.67		
water	H_2O	1.84		

(One debye is 3.338×10^{-30} coulomb metre)

NOTE: The dipole moment of a bond is often represented as an arrow $\mapsto$ pointing along the bond from $\delta+$ to $\delta-$. For example

Dipole–Induced Dipole Attraction

Dipole attraction also occurs between non-polar molecules. This is caused by the temporary fluctuations in electron density in a molecule. Even the electron clouds of noble-gas molecules (that is, of single atoms) become distorted from their normal symmetrical shapes (*see* figure 2.20). The result is a temporary dipole. The temporary dipole can induce a dipole in a neighbouring molecule. The forces of attraction between these fluctuating dipoles are often called **dispersion forces** or **London forces**. They are much weaker than permanent dipole–permanent dipole interactions.

So What Are van der Waals Forces?

There is often confusion over the term 'van der Waals forces'. In 1873 Johannes van der Waals suggested the existence of intermolecular attractive forces between molecules. The explanation of the forces in non-polar molecules was first given by Fritz London in 1930. Van der Waals forces are sometimes taken to include London forces only. Nowadays, however, there is a tendency to use the term 'van der Waals forces' to include all weak intermolecular forces, that is both permanent dipole–permanent dipole and temporary dipole–induced dipole attractions. Hydrogen bonding, which is much stronger and also both intermolecular and intramolecular in nature, is normally excluded from the term van der Waals forces.

SUMMARY

1. According to the **electronic theory of valency**, atoms try to achieve the most stable (that is, the lowest-energy) electron configuration when they form bonds.
2. The **octet rule** states that atoms try to achieve the electron configuration of noble gases when they form bonds.
3. An **ionic bond** is an electrostatic force of attraction between two oppositely charged ions.
4. **Electron affinity** is the energy change which occurs when one mole of singly charged anions is formed from one mole of atoms.
5. A **covalent bond** consists of two electrons shared between two adjacent atoms. Each atom contributes one electron.
6. There are three types of **molecular orbital: bonding orbital, non-bonding orbital** and **anti-bonding orbital**. σ and π bonds are two types of bonding orbital.
7. **Hybrid orbitals** are atomic orbitals consisting of a mixture of two types of atomic orbital.
8. The π bonds in a benzene molecule form a **delocalised molecular orbital**.
9. A **coordinate bond** is a type of covalent bond formed by the donation of a **lone-pair** of electrons by one atom.
10. The **ionic or covalent character** of a bond depends on the **electronegativities** of the two atoms forming the bond.
11. A **hydrogen bond** is an electrostatic force of attraction between a hydrogen bond and an electronegative atom. It is a dipole–dipole attraction.
12. Weak **dipole–dipole** and **dipole–induced dipole** attractions between molecules are called **van der Waals forces**.

2.2
The Influence of Bonding on Size, Shape, Structure and Properties

THE SIZE OF ATOMS

Atomic Volume

This is the volume of one mole of the atoms of an element. It can be calculated by dividing the mass of one mole of the atoms by the density of the element. Since there are 6.022×10^{23} atoms per mole of atoms, it is possible to calculate the volume and thus the radius of a single atom.

However, there are two problems associated with such calculations. First of all, what density do we take? Do we use gas or liquid density—and under what conditions of temperature and pressure? Secondly, such calculations assume that an element exists as unbonded atoms. But as we have seen, only the noble gases exist as unbonded atoms. Some elements, such as chlorine and sulphur, exist as molecules and others, such as the metals, exist as lattices. We therefore cannot consider the volume of an atom in isolation but only the volume of an atom as part of the structure of the element.

This second difficulty is reflected by the varying densities of the allotropes of carbon and sulphur (*see* table 2.6). The differences in the densities between the allotropes are due to the different structural arrangements of the atoms in the allotropes. The structural arrangements themselves depend on how the atoms bond together. It follows that the volume and also the radius of an atom depend on how it is bonded to other atoms.

Table 2.6 Densities of allotropes of carbon and sulphur

Element	Allotrope	Density/ g cm^{-3}
carbon	graphite	2.09 to 2.23
	diamond	3.50 to 3.53
sulphur	monoclinic	1.94
	orthorhombic	2.07
	rhombohedral	2.21

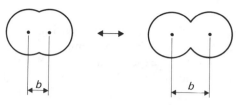

b = Bond length

Figure 2.21 Bond length. Owing to the vibration of molecules, the internuclear distance or **bond length** is not fixed. The figure shows the linear vibration of a simple diatomic molecule. Bond length thus cannot simply be defined as the distance separating the centres of two atoms bonded together. A more exact definition is: 'A bond length is the distance which separates two atoms, measured between the centres of mass of the two atoms, and which corresponds to a minimum of the energy of the bond.' The minimum energy is shown by the Morse curve (*see* figure 2.1)

Table 2.7 Carbon–carbon bond lengths

Bond	Bond length/nm
C—C	0.1541
C=C	0.1337
C≡C	0.1204

Atomic Radius

The determination of atomic radius also presents problems. First of all, the atom is not a sphere with a well defined surface and radius. It is, rather, a nucleus surrounded by a cloud of electrons. The probability of finding an electron increases with distance from the nucleus to a maximum. The probability then tails off to zero at infinity. Secondly, even if we did choose a specific radius, it would be impossible to measure it.

What can be determined experimentally, however, are internuclear distances— or more specifically bond lengths (however, *see* figure 2.21). These can be determined using X-ray and electron diffraction techniques. The radius of an atom is taken to be half the closest internuclear distance.

Van der Waals Radius
For non-bonded atoms, the closest internuclear distance is called either the non-bonded radius or the van der Waals radius. This is shown in figure 2.22a.

Covalent Radius
Covalent radius is half the bond length (or bond distance as it is sometimes called) of two identical atoms covalently bonded together (*see* figure 2.22b). The bond length in the chlorine molecule, Cl_2, is 0.1988 nm, for example. The covalent radius of chlorine is thus 0.0994 nm.

The covalent radius of one type of atom can be used to calculate the covalent radius of another. For instance, the experimentally determined value of the C—Cl bond length in CH_3Cl is 0.1767 nm. By subtracting the value for chlorine (0.0994 nm), the covalent radius of carbon is found to be 0.0773 nm. This method is based on the **principle of additivity**. The principle states that atomic radii are additive. The bond length of the C—Cl bond is thus the sum of the covalent radii of carbon and chlorine. The principle can only be applied to single covalent bonds. Double and triple covalent bonds have shorter bond lengths (*see* table 2.7).

The length of a single covalent bond also depends on its environment in the molecule. For example, the C—H bond length ranges from 0.1070 nm in a trisubstituted carbon atom to 0.115 nm in CH_3CN.

Figure 2.22 Atomic radii: (a) van der Waals radius; (b) covalent radius; (c) metallic radius

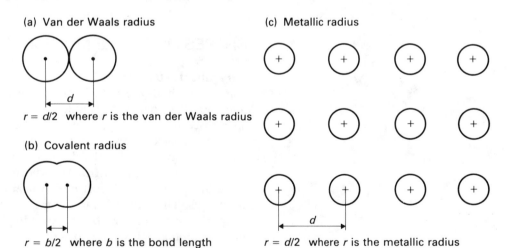

(a) Van der Waals radius

$r = d/2$ where r is the van der Waals radius

(b) Covalent radius

$r = b/2$ where b is the bond length and r is the covalent radius

(c) Metallic radius

$r = d/2$ where r is the metallic radius

Metallic Radius
Metallic radius is usually taken as half the internuclear distance between two neighbouring ions in the crystalline metal (*see* figure 2.22c). The term **atomic radius** is usually taken to be the covalent radius for non-metallic atoms and the metallic radius for metallic atoms.

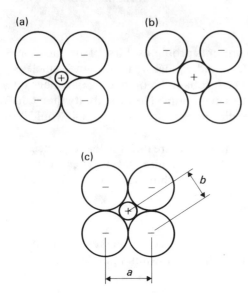

Figure 2.23 Ionic radii. (a) Anions touch but cation does not touch anions.
(b) Cation touches anions but anions do not touch. (c) The *assumption* usually made is that the cation touches the anions and the anions touch. The distance *a* is determined experimentally. This is twice the **anionic radius**. This enables the interionic distance *b* to be calculated: *b* is the sum of the anionic and cationic radii. The **cationic radius** can thus be calculated

Ionic radius

Ionic radius is half the internuclear distance between two neighbouring monatomic ions in a crystalline salt. Its determination once again presents a problem since it is the interionic distances that are determined experimentally and not the ionic radii. The interionic distances depend on the packing of the ions in a crystal lattice. Figure 2.23 shows three possibilities for packing ions in a face-centred cubic lattice. Unfortunately, the experimentally determined interionic distances do not distinguish between these three possibilities. The problem is thus how to apportion the interionic distance between the two ionic radii. In other words, where does one ion end and the other begin. This is certainly not clear from electron density maps of salts (*see* for example, figure 2.12). To overcome this problem, it is usually assumed that the interionic distance is the sum of the two ionic radii. It is also assumed that the ions are spheres and that any two neighbouring spheres touch each other. This is possibility (c) in figure 2.23. Once one ionic radius is known, others can be calculated by applying the principle of additivity.

Comparing the Radii

Ionic radii are invariably the smallest of the various types of radii. Table 2.8 compares values for the various types of radii of three elements in Period 3 of the Periodic Table. Note that both anionic and van der Waals radii are relatively large. A comparison of ionic size with atomic size for all Period 3 elements except argon is shown in figure 11.9. The atomic size is based on the covalent radius. Note how the cations are small and the anions large compared with atoms.

Table 2.8 Comparing the values of different types of atomic radii

| | Radius/nm | | | | | |
	Na	ion	P	ion	Cl	ion
cationic	0.095	Na^+	0.034	P^{5+}	0.026	Cl^{7+}
covalent	0.156	–	0.110	–	0.099	–
metallic	0.190	–	0.128	–	–	–
anionic	–	–	0.212	P^{3-}	0.181	Cl^-
van der Waals	–	–	0.180	–	0.185	–

SHAPES OF SIMPLE MOLECULES AND IONS

Experimental Determination

Various methods are available for determining the bond lengths and bond angles and thus the shapes of simple molecules and ions. They include microwave spectroscopy and X-ray, neutron and electron diffraction techniques. We shall see, in the next chapter, how X-ray diffraction is used to determine crystal structure. However, for the determination of the shapes of simple molecules in the gas phase, **electron diffraction** is normally used. The technique depends on the wave-like properties of electrons. A beam of electrons is passed through a sample of the gas. The molecules in the gas scatter the electrons, forming a *diffraction pattern*. The pattern can be analysed to give bond lengths and bond angles. The method of analysis is analogous to that used for the analysis of X-ray diffraction patterns.

Valence-Shell Electron-Pair Repulsion (VSEPR) Theory

This theory is a development of the Sidgwick–Powell theory of electron-pair repulsion. The theories assume that the spatial arrangement of molecules and ions is determined by the repulsion between electron pairs. Each electron pair tries to separate itself as much as possible from other electron pairs. The shapes of molecules and ions are thus determined by the electron pairs rather than by the atoms. As we have already seen, pairs of electrons in their ground state form either bonding orbitals or exist in non-bonding orbitals as lone-pairs. The shapes and bond angles of molecules and ions are primarily determined by the number of electron pairs—whether they are bonding or non-bonding (see table 2.9). However, the number of electron pairs alone does not account completely for the shapes and bond angles. For example, it does not account for the differences between the shapes and bond angles of the methane, water and ammonia molecules (see table 2.10). All these molecules contain four electron pairs. In the modern VSEPR theory, the lone-pair of electrons repels other pairs more

Table 2.9 Electron-pair repulsion theory

Number of electron pairs	Maximum separation/deg	Shape
2	180	linear
3	120	trigonal planar
4	109.5	tetrahedral
6	90	octahedral

Table 2.10 Shapes of simple molecules and ions

Shape	Example	Bond angle/deg	Other examples
linear	Cl——Hg——Cl	180	$BeCl_2$, CO_2, H—C≡C—H
trigonal planar	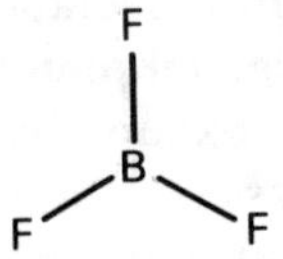	120	BCl_3, C_2H_4, CO_3^{2-}, NO_3^-
tetrahedral	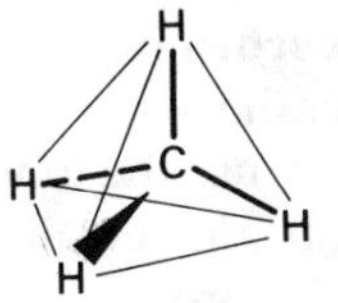	109.5	SiF_4, NH_4^+, PO_4^{3-}
tetrahedral (bent)	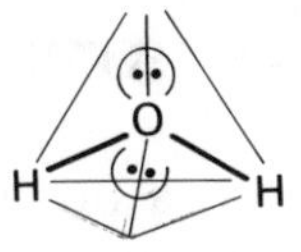	104.5	H_2S
tetrahedral (pyramidal)	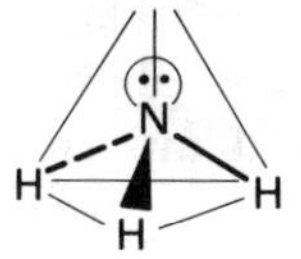	106.7	PF_3, H_3O^+
trigonal bipyramidal	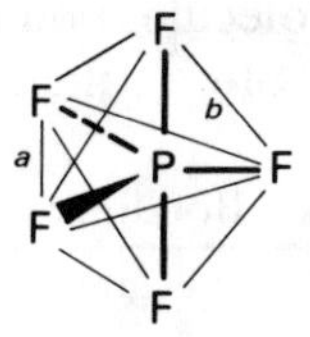	$a=120$ $b=90$	PCl_5
octahedral		90	$[PF_6]^-$

Bonds:
——— in plane of this page
--- going behind the page
▶ coming out of the page

strongly than a bonding pair of electrons. Consequently, the strength of repulsion between electron pairs decreases as follows:

strongest   lone pair–lone pair
↓   lone pair–bonding pair
least strong   bonding pair–bonding pair

Examples of covalent compounds and ions with these and other shapes are shown in table 2.10. Note the convention for indicating the shape of molecules in three dimensions. A heavy line represents a bond in the plane of the paper, a broken line represents a bond going behind the plane of the paper and a thin solid triangle represents a bond coming out of the plane of the paper. Each of these represents a shared or bonding electron pair, that is a σ covalent bond. Lone-pairs are represented thus (..).

Elements consisting of **linear** molecules with single covalent bonds are relatively common. Chlorine, Cl_2, and hydrogen, H_2, are two examples. Compounds consisting of linear molecules with single covalent bonds are less common. The two most well known examples are mercury(II) chloride, $HgCl_2$, and beryllium chloride, $BeCl_2$. More common are linear molecules with multiple covalent bonds. These include carbon dioxide, CO_2, and ethyne, C_2H_2.

The **trigonal planar** shape is found in molecules of the boron halides, ethene and those oxoanions with three oxygen atoms (*see* figure 2.10).

The classic example of a **tetrahedral** structure is the methane molecule, CH_4. Remember that the four bonding orbitals are in fact formed from four sp^3 hybrid orbitals. This tetrahedral structure also occurs in diamond and forms the basis of the **stereochemistry** of most organic compounds. Both ammonia and water molecules have tetrahedral shapes. However, neither is perfectly tetrahedral. This is because the lone-pairs take up more space than the bonding electron pairs. As a result, the tetrahedral bond angle of 109.5° is reduced to 106.7° for ammonia and 104.5° for water.

BONDING, STRUCTURE AND PHYSICAL PROPERTIES

Bonding has a pronounced influence on the structure and physical structure of both elements and compounds. In elements, bonding can be either metallic or covalent. For compounds, it is either ionic or covalent. However, as we saw above (figure 2.11), these are extreme cases. Generalisations concerning bonding, structure and physical properties are therefore risky. Even so, there are some obvious and important underlying trends which are worth noting.

For a start, as we see in table 2.11, metals and ionic compounds both exist as giant crystal lattices. However, we shall leave the detailed treatment of solid structure until the next chapter. Covalent elements and compounds exist either as simple molecules bound by weak intermolecular forces or as giant structures. The latter are often called macromolecules.

The molecules you can see—diamonds! A diamond is a macromolecule. It consists of a giant three-dimensional structure of carbon atoms bound together by covalent bonds

Table 2.11 Bonding, structure and physical properties

Bond type	Structure	Examples	Melting point/°C	Boiling point/°C	$\Delta H^{\ominus}_{fus,m}/$ kJ mol^{-1}	$\Delta H^{\ominus}_{vap,m}/$ kJ mol^{-1}
metallic	giant lattice	Mg	649	1090	9.07	132.3
		Cu	1083	2567	13.06	305.8
ionic	giant lattice	NaCl	801	1413	30.32	–
		CaF_2	1360	~2500	17.22	–
covalent	simple molecular	S_8	119	445	1.24	10.5
		H_2O	0	100	6.03	41.2
	giant molecular	C (diamond)	3550	4827	–	–
		SiO_2 (quartz)	1610	2230	14.28	–

Note: For a definition of the terms, $\Delta H^{\ominus}_{fus,m}$ and $\Delta H^{\ominus}_{vap,m}$ *see* chapter 5

Not all structures of elements and compounds fall into the simple categories shown in table 2.11. Graphite, for example, exists as layers of carbon atoms. The carbon atoms are linked by covalent bonds whilst the layers are bound by weak van der Waals forces. Cadmium iodide is another example. It also has a layer structure, although the bonds in each layer are intermediate between ionic and covalent.

Let us now examine in turn various physical properties which are influenced by bonding.

States of Matter

All metallic elements with the exception of mercury exist as solids at room temperature. Non-metals exist as solids or gases. The only exception is bromine, which is a liquid.

Ionic compounds also exist as solids at room temperature. Exceptions are the mineral acids, which only exist as ionic compounds in polar solvents (*see* below).

Covalent elements and compounds consisting of simple molecules exist as gases, liquids and solids with low melting points. Covalent elements and compounds which exist as giant structures are solids with very high melting points.

Volatility

Elements and compounds with giant structures have high melting and boiling points and high enthalpies of fusion and vaporisation (*see* table 2.11). They are said to be **non-volatile**. They normally exist as solids. Simple molecular compounds are **volatile**. They have low melting and boiling points and low enthalpies of fusion and vaporisation. As we have seen, they may exist as gases, liquids or solids.

Hardness

Elements and compounds with giant structures all tend to be hard. Metals are **malleable** and **ductile**. This means they can be hammered into sheets or drawn into wire. Giant ionic and molecular compounds tend to be brittle.

Colour

Metals are grey-black or brown/yellow in appearance. Most ions are colourless, although some have distinct colours. For example, the manganate(VII) anion, MnO_4^-, is purple. The cations of d-block metals are often coloured. The colour of a salt depends on the electronic interaction of the cation and the anion. The salt may be coloured even though the anion is colourless. A good example of this is provided by copper(II) salts. They have a variety of colours even though the anion in each case is colourless (*see* table 2.12).

Covalent elements and compounds can be coloured or colourless. For example, water and ammonia are both colourless, whereas nitrogen dioxide is brown. All the halogens are coloured. Organic compounds tend to be colourless unless a **chromophore** is present.

Table 2.12 Colours of copper salts

Copper(II) salt	Formula	Colour
copper(II) ethanoate	$Cu(CH_3COO)_2$	green
copper(II) sulphate	$CuSO_4$	white
copper(II) sulphate-5-water	$CuSO_4 \cdot 5H_2O$	blue
copper(II) chloride	$CuCl_2$	yellow
copper(II) chloride-2-water	$CuCl_2 \cdot 2H_2O$	blue-green
copper(II) nitrate-6-water	$Cu(NO_3)_2 \cdot 6H_2O$	blue
copper(II) sulphide	CuS	black

A **chromophore** is the part of a molecule responsible for the colour of the compound. Chromophores frequently have one or more double bonds. For example

Electrical Conductivity and Resistance

All metals are good conductors of electricity in both their solid and liquid states. Solid ionic compounds are poor conductors of electricity. However, when molten or in aqueous solution, the crystal lattice is broken up and the ions are free to move. They thus become good conductors of electricity. They are called **electrolytes** since they undergo chemical decomposition when electricity is passed through them (*see* chapter 10). Covalent compounds are non-conductors, although some react with water to form electrolytes. Hydrogen chloride, HCl, and ammonia, NH_3, are examples. As usual, there are exceptions. Graphite, for example, is a good conductor of electricity even though it is a covalent compound. This is because it has a layer structure (*see* next chapter).

The resistance of a metal depends on three factors—its nature, its temperature and its length and cross section.

Nature of the metal. The resistance to an electric current is probably due to the vibration of the metal ions in their lattice positions. These vibrations restrict the movement of electrons and thus resist the current.

Temperature of the metal. The metal ions vibrate increasingly as temperature is increased. The resistance of metals thus increases with temperature. It follows that metals are better conductors at lower temperatures.

Length and cross-sectional area of the metal conductor. We shall examine the relationship between this and electrical resistance in chapter 10.

Solubility

Metals are insoluble in both polar and non-polar solvents. They are soluble in liquid metals, however. Ionic compounds are soluble in polar solvents such as water but insoluble in non-polar solvents such as tetrachloromethane, CCl_4. When ionic compounds dissolve in polar solvents such as water, the lattice breaks up and the ions become solvated by the solvent molecules (*see* figure 2.24). This means that the polar solvent molecules become oriented around the ions. Giant molecular compounds are insoluble in all solvents. Simple molecular compounds tend to be insoluble in polar solvents but soluble in non-polar solvents.

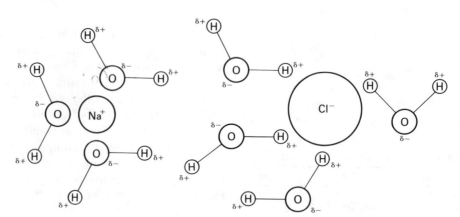

Figure 2.24   Solvation

BOND TYPE AND CHEMICAL PROPERTIES

The characteristic chemical properties of metals arise from their low ionisation potentials and low electronegativities. The chemical reactivity of a specific metal relates to its position in the electrochemical series. This is discussed in chapter 10.

Reactions involving ionic compounds are usually rapid, especially if they involve the combination of ions of opposite charge. For example, the neutralisation of strong acids with strong bases occurs virtually instantaneously. A notable feature of ionic compounds is that their ions can act independently when in solution. For example, when silver nitrate is added to any solution containing chloride ions, silver chloride is precipitated regardless of the cation. For example, in the reaction

$$AgNO_3(aq) + NaCl(aq) \rightarrow AgCl(s) + NaNO_3(aq)$$

only the Ag^+ and Cl^- react:

$$Ag^+(aq) + Cl^-(s) \rightarrow AgCl(s)$$

Na^+ and NO_3^- are called **spectator ions** since they do not participate in the reaction.

Reactions involving covalent compounds tend to be slower than reactions involving ions—although not always. TNT is a covalent compound! By far the most important branch of chemistry involving covalent compounds is organic chemistry. Organic chemistry is characterised by the wide diversity of covalent compounds formed by one element—carbon.

THE INFLUENCE OF INTERMOLECULAR FORCES ON STRUCTURE AND PROPERTIES

Intermolecular dipole–dipole attractions such as hydrogen bonding play an important part in determining the structure and physical properties of many elements and compounds. For example, dipole–dipole attractions result in a net pull inwards on water molecules at the surface of the liquid. This accounts for the concave meniscus of water in glass and the spherical nature of a water droplet.

Hydrogen Bonding

Hydrogen bonding in certain compounds accounts for the abnormally high values of their melting and boiling points, surface tension and enthalpy of vaporisation. Figure 2.25 shows graphically how values for melting point and boiling point and enthalpy of vaporisation vary down Group VI of the Periodic Table. Water, the only hydride in this group which exhibits hydrogen bonding, has abnormally high

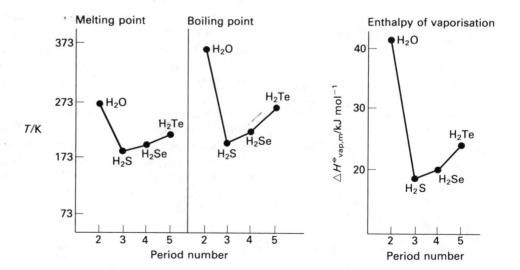

Figure 2.25 Melting and boiling points and enthalpies of vaporisation of Group VI hydrides

The abnormally high enthalpy of vaporisation of water—due to hydrogen bonding—enables these cyclists to cool off rapidly!

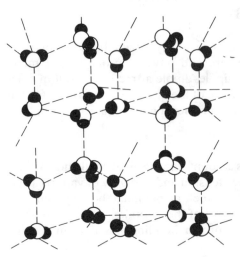

Figure 2.26   Structure of ice

values for all these properties. A similar pattern is found for Group V and Group VII hydrides. Ammonia (Group V) and hydrogen fluoride (Group VII) both have abnormally high melting and boiling points.

Hydrogen fluoride and water can both be regarded as **polymerised** in their liquid and solid states due to intermolecular hydrogen bonding. The model of the three-dimensional structure of ice is shown in figure 2.26. Each water molecule is hydrogen bonded to four other water molecules. Two of the hydrogen bonds are due to its own hydrogen atoms and two are due to the hydrogen atoms of other water molecules. Ice thus has a tetrahedral structure similar to diamond (*see* section 3.2) but not exactly the same. This hydrogen bonding accounts not only for the high freezing point of water but also for the low density of ice. On melting, the hydrogen bonds break and the water molecules become more closely packed.

Experimental determinations of the relative molecular mass of carboxylic acids such as ethanoic acid and benzoic acid have shown that they exist as **dimers**—that is, two molecules bonded together. The bonds responsible are thought to be hydrogen bonds. As a result the boiling points of carboxylic acids are higher than compounds with similar relative molecular masses. For example, ethanoic acid, CH_3COOH ($M_r = 60$), has a boiling point of 391 K whereas propanone, CH_3COCH_3 ($M_r = 58$), has a boiling point of 329 K.

The layer structures of acids such as nitric acid and sulphuric acid are also due to hydrogen bonding. Figure 2.27 shows the hydrogen bonds between molecules of sulphuric acid.

The examples of hydrogen bonding we have been considering so far have all been intermolecular. Intramolecular hydrogen bonding also influences the structure and properties of some compounds. An example is provided by the 2- and 4-nitrophenol isomers. 2-Nitrophenol boils at 489 K compared with 532 K for 4-nitrophenol. The 2-nitrophenol molecule has an intramolecular hydrogen bond (*see* figure 2.18) whereas molecules of 4-nitrophenol form intermolecular hydrogen bonds.

(a) Ethanoic acid (b) Sulphuric acid

Figure 2.27 (a) Ethanoic acid; (b) sulphuric acid

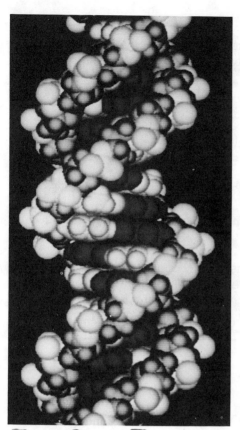

A model of the double helix

Perhaps one of the most important and certainly one of the most famous examples of the role of intramolecular hydrogen bonding in determining structure is provided by deoxyribonucleic acid (DNA). DNA consists of a **double helix**. The two strands of the double helix are hydrogen bonded together (*see* chapter 20).

Van der Waals Forces

Non-polar compounds and elements such as the noble gases can be liquefied and solidified. This suggests that there are forces of attraction between atoms or molecules in non-polar elements and compounds. Energy is needed to melt a solid and to boil a liquid. The energy is needed to break down these forces of attraction. These forces of attraction are the van der Waals forces referred to earlier in the chapter. They exist between all simple molecules whether they are polar or non-polar. Substances with molecules or atoms that are attracted to one another only by these forces are characterised by low melting and boiling points compared with other substances of about the same relative molecular mass.

The forces of attraction are due to dipole–induced dipole attraction. This results, as we have seen, from the fluctuations in the distribution of electrons in the atoms or molecules. Molecules with large and diffuse electron distributions are attracted to one another more strongly than smaller molecules with more tightly held electrons. If we compare methane with butane, for example, we see that the electron distribution in butane is more diffuse and there are thus more opportunities for attractive interactions. The boiling point of methane is $-164°C$. Butane has a much higher boiling point, $-0.5°C$.

Methane

Butane

Similarly, iodine, which has 106 electrons in its molecule I_2, has a much higher boiling point (165°C) than chlorine, Cl_2 (boiling point $-35°C$), which has only 34 electrons in each molecule. The ordered arrangement of non-polar molecules such as iodine or carbon dioxide in their crystals is due to these weak forces of interaction.

Van der Waals forces are also responsible for holding together the individual layers in layer-structured solids such as graphite and cadmium iodide. Since the forces are weak compared to the strong covalent bonds which hold the atoms together in a layer, they are easily broken. Consequently, both graphite and cadmium iodide have a slippery or greasy texture.

SUMMARY

1. **Bond length** is the distance which separates two atoms, measured between the centres of mass of the two atoms, and which corresponds to a minimum energy of the bond.
2. **Covalent radius** is half the bond length of two identical atoms covalently bonded together.
3. **Metallic radius** is half the internuclear distance between two neighbouring cations in the crystalline metal.
4. **Ionic radius** is half the internuclear distance between two neighbouring monatomic ions in a crystalline salt.
5. **Van der Waals radius** is half the closest internuclear distance between two non-bonded atoms.
6. The **valence-shell electron-pair repulsion (VSEPR) theory** assumes that the spatial arrangement of molecules and ions is determined by the repulsion between electron pairs.
7. Metals and ionic compounds exist as giant crystal lattices.
8. Covalent compounds exist as simple molecules or giant structures.
9. **Hydrogen bonding** accounts for some of the abnormal properties of covalent compounds such as water, ammonia and hydrogen fluoride.

Examination Questions

1. (a) Explain briefly what is meant by (i) *ionic bonding* and (ii) *covalent bonding*. For each, give **one** example of a compound containing such a bond.
   (b) Draw a diagram to illustrate the shape of each of the following species: ammonium ion; ammonia; amide ion (NH_2^-).
   (c) Explain the trend in the normal boiling points of the compounds given in the table below.

Compound	CH_4	NH_3	H_2O
Normal boiling point/°C	-164	-33	100

   (JMB)

2. Draw diagrams to illustrate the shape and symmetry of s and p atomic orbitals, labelling them appropriately.

   Show how the idea of a hybridised orbital is necessary to describe the shape of molecules such as ethane, ethene and ethyne, and give a simple description of bonding in these molecules.

   The shapes of many simple inorganic molecules can be explained in terms of the electrostatic repulsions of bonding and non-bonding pairs of electrons. Show how these ideas predict the shapes of the XeF_4 and SF_4 molecules.

   (OLE)

3. (a) Explain why the concept of hybridised carbon orbitals is useful when discussing the bonding in hydrocarbons.
   (b) All the carbon–carbon bonds in benzene have a length of 0.140 nm, whereas the carbon–carbon bond lengths of ethane and ethene are 0.154 nm and 0.133 nm respectively. Show how these facts can be accounted for in terms of the electron distributions in these three hydrocarbons.
   (c) The average standard enthalpy of hydrogenation for an alkene is -122 kJ mol^{-1}. The standard enthalpy of hydrogenation of benzene is -207 kJ

mol^{-1}. Estimate the delocalisation energy of benzene and explain its significance.

   (d) Show that the reactions of benzene demonstrate the stability of an aromatic system.

(OLE)

4. (a) The electron affinity of bromine is 330 kJ mol^{-1}. What does this mean?

   (b) What is the *electronegativity* of an element regarded as measuring (or representing)? State which is the most electronegative element.

(OLE)

5. Write an essay on *intermolecular forces*. Your answer should include an account of how the different types of intermolecular forces arise, their relative strengths and how they influence the physical properties of various substances.

(UCLES)

6. (a) Explain the meaning of the term *hydrogen bond*, illustrating your answer with reference to the reasons why this bond is formed between water molecules. Outline its influence on the properties of water and on the properties of hydrogen fluoride.

   (b) Give a description, in terms of orbitals, of the interaction between hydrogen atoms in forming the bond in H_2. Include a description of the atomic orbitals involved.

   (c) Give a simple treatment of van der Waals forces between molecules (such as H_2) leading to the liquefaction of gases. Mention **briefly** the forces between CO_2 molecules in the solid.

   (d) Describe the deviations from ideal behaviour shown by gases due to forces such as those in (a) and (c) above.

(WJEC)

7. (a) Explain what is meant by (i) *hydrogen bonding*, using hydrogen fluoride as an example, and (ii) *bond polarity*, using chloroethane as an example.

   (b) Use the concepts of hydrogen bonding and of bond polarity to account for the observed differences in the boiling points within the following group of compounds.

	Bp./°C
$CH_3CH_2CH_2CH_3$	0
$CH_3CH_2OCH_3$	11
CH_3CH_2CHO	49
$CH_3CH_2CH_2OH$	97

   (c) By reference to solubility, melting and boiling points and colligative properties, describe how hydrogen bonding influences the physical properties of carboxylic acids.

(JMB)

8. State which of the following substances are polar (i.e. the molecule has a dipole moment) and which are non-polar (i.e. the molecule does not have a dipole moment): (a) CH_3Cl; (b) $C(CH_3)_4$; (c) CO_2; (d) SO_2.

(OLE)

9. (a) State the principles of the Sidgwick–Powell theory of electron-pair repulsion and molecular shape.
   (b) Illustrate the application of this theory by considering how it accounts for the shapes of the following species: $BeCl_2$; $SnCl_2$; BCl_3; PCl_3. (For each species your answer should include a clear sketch, appropriately labelled, showing the spatial arrangements of the atoms.)
   (c) Which of the following molecules have dipole moments: SO_2; CO_2; CH_3Cl; CH_4?

(OLE)

10. (a) What are the shapes of the following species: SO_3^{2-}; SO_4^{2-}; H_2S; N_2O; H_2O_2? For each species your answer should consist of a clear sketch, appropriately labelled, and a brief description. Details of bonding or explanations are *not* required.
    (b) Describe the crystal structures of the following: graphite; sodium chloride.
    (c) Discuss the bonding in the complex $NH_3 \cdot BF_3$.
    Your answers to (b) and (c) should include appropriate diagrams.

(OLE)

11. Discuss the assumptions which are made when predicting the structures of molecules formed from non-transition [non d-block] elements by considering the repulsions between electron pairs in the molecule. Illustrate your answer
    (a) by reference to the structures of BF_3, NF_3 and PF_5,
    (b) by explaining why the bond angle in NH_3 is greater than that in H_2O.
    Comment on the structure of, and bonding in, XeF_4, PF_6^- and NH_4^+.
    Deduce the shapes of PF_3, H_3O^+ and ICl_2^-.

(JMB)

12. Give an account of the bonding present in **each** of the following substances and show how their physical properties are related to the type of bonding:
    (a) copper, (b) diamond, (c) iodine, (d) ice.

(UCLES)

13. For each of the following solids relate the physical properties to the structure and the types of bonding present: (a) sodium chloride; (b) diamond; (c) paraffin wax; (d) ice; (e) copper.

(O & C)

3 GASES, LIQUIDS AND SOLIDS

To a Snowflake

What heart could have thought you?—
Past our devisal
(O filigree petal)
Fashioned so purely,
Fragilely, surely,
From what Paradisal
Imagineless metal
Too costly for cost?
Who hammered you, wrought you,
From argentine vapour?—
God was my shaper
Passing surmisal,
He hammered, He wrought me,
From curled silver vapour.
To lust of His mind:—
Thou coulds't not have thought me!
So purely, so palely,
Mightily, frailly,
Insculped and embossed,
With His hammer of wind,
And His graver of frost

Francis Thompson

Snow's Flaky Little Secrets

A microscopic speck of dust, trapped in a molecule of water vapour, rides the buffeting winds inside a winter storm cloud. As the particle is frosted with droplets of supercooled water, it begins the six-mile plunge to Earth. The falling ice crystal is then sculpted by the varying temperature and humidity—lengthening here, a spiky branch pushing out there—until it grows into a shape as individual as a fingerprint. The resulting snowflake may look like Cleopatra's needle or a fern or a chunky hexagon, but it will be different from every other snowflake around. And although such particulars are understandably meaningless to people digging out from billions of flakes, they are becoming crucial to scientists who are trying to predict everything from avalanches to metal failures. Only now, after centuries of befuddlement, are scientists beginning to make sense of snow.

Sharon Begley and John Carey

Only now, after centuries of befuddlement, are scientists beginning to make sense of snow

Crystal But Not So Clear

Why do the ice crystals that fall from the sky as snowflakes take on intricate six branched star shapes. . . ? . . . The best informed team of crystallographers, meteorologists, physicists and mathematicians in the world could not predict the shape of snowflakes from first principles . . . .

. . . The problem is so difficult because a whole range of interrelated factors are involved: the rate at which heat is lost from different parts of the growing crystal, for example, depends on their curvature, but curvature also affects the melting temperature—and so on. Solving all the necessary mathematical equations on paper is next to impossible. Even with computers it is extremely difficult, as the slow progress shows.

As for the six arms of a typical snowflake, this arrangement is related to the shape and electrical charges on the water molecule. . . .

Bryan Silcock

LEARNING OBJECTIVES   *After you have studied this chapter you should be able to*
1. Compare the properties of gases, liquids and solids.
2. Define and apply
   (a) **Boyle's law,**
   (b) **Charles' law,**
   (c) **Avogadro's principle,**
   (d) **Dalton's law of partial pressures,**
   (e) **Graham's law of diffusion.**
3. Use the **ideal gas equation** to calculate relative molecular mass.
4. Describe experiments for
   (a) determining relative molecular mass,
   (b) measuring the **distribution of molecular speeds.**
5. Outline the essential features of the **kinetic theory of gases.**

6. Sketch a graph to show the distribution of molecular speeds.
7. Indicate how **real gases** deviate from ideal gas behaviour.
8. Relate **critical constants** to the **liquefaction of gases.**
9. Show how **X-ray diffraction** can be used to determine crystal structure.
10. (a) Recognise models and diagrams and draw diagrams of the principal types of **lattice structures** of metallic crystals, ionic crystals and covalent crystals.
    (b) For each type, give an example of a solid which has this structure.
11. Relate crystal structure to physical properties of a solid.

3.1 Gases and Liquids

STATES OF MATTER

In chapter 1 we saw how all matter is made up of atoms. We also examined the electronic and nuclear structure of atoms. In chapter 2 we turned our attention to the various types of bonding between atoms, ions and molecules and saw how the bonding related to the structure and properties of matter.

In this chapter we now look at the behaviour and properties of the three states of matter. These are

- gases
- liquids
- solids

Some of the characteristics that distinguish these three states of matter are shown in table 3.1.

Table 3.1 Properties of gases, liquids and solids

	Gases	Liquids	Solids
volume	takes volume of container; greatly influenced by temperature and pressure	fixed	fixed
shape	fills container	not fixed—fills or partially fills container	fixed
compressibility	high	almost zero	zero
density	low	moderate to high	high

Gases
Most gases are covalent compounds, although, of course, the noble gases are exceptions. The most notable characteristics of a gas are its compressibility and its ability to expand. Gases do not have their own shape but rather expand to fill the shapes of their containers uniformly. Since a gas assumes the shape of its container, a gas cannot have a fixed volume. Its volume is the volume of the container. A confined gas exerts a constant pressure on the wall of its container uniformly in all directions. One final point worth noting about gases—gases may be mixed with each other in any proportion. In other words, 'there's always room for more!'

Liquids
Liquids may be metallic, ionic or covalent. Examples are mercury, dilute hydrochloric acid and benzene, respectively. Like a gas, a liquid does not have a fixed

---

Chaos = Gas

The Dutch scientist J. B. Van Helmont (1577–1644), needing a word for something previous scientists had seldom (never?) had to take notice of, by his own account went to the Greek word *chaos*, which describes the unformed mass of the elements before the Divine Mind worked on them at the Creation.

The Greek letter we represent by 'chi' is of course the one that looks like an overgrown X; it was natural for a Dutchman to transliterate this as 'g', since the Dutch sound of a 'g' is more gutteral and breathy, more what the OED meticulously calls a 'spirant'. If an English scientist had made Van Helmont's discoveries, we might now be speaking of chass, doubtless with a hard ch as in 'chronic' or 'chiropody'. The spelling wasn't, in fact, settled for a long time; 'gaz' was an accepted alternative until well into the nineteenth century, and in France, as we know, it won the day.

John Wain

A liquid takes the shape of its container

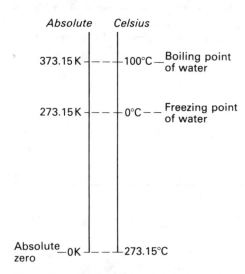

Figure 3.1   Scales of temperature

Evangelista Torricelli

The torr is named after Evangelista Torricelli, the Italian mathematician and physicist. He served as Galileo's secretary and succeeded him as the court mathematician and philosopher to Grand Duke Ferdinando II of Tuscany. In 1643 he proposed that atmospheric pressure determines the height to which a fluid will rise in a tube inverted over a saucer of the same liquid. This led to the development of the barometer. He also proved that the rate of flow of a liquid through an opening is proportional to the square root of the height of the liquid. This is called Torricelli's theorem.

---

shape. A liquid takes the shape of its container as it finds it own level under the influence of gravity. However, unlike a gas, a liquid does have a fixed volume. Liquids can only be compressed a very small amount and then only under high pressures.

Solids

Like liquids, solids may also be metallic, ionic or covalent. We shall examine the various metallic, ionic and covalent structures of solids in the second part of this chapter. Solids distinguish themselves from liquids and gases in that they have fixed shapes and fixed volumes. They also cannot be compressed even under quite high pressures.

THE GAS LAWS

The gas laws mathematically relate the temperatures, pressures and volumes of gases. Their correct application depends on using the correct units. So before we proceed further it is important that we look at the units and measurement of temperature, pressure and volume.

Temperature

There are two scales of temperature used in science. The **absolute scale of temperature** was proposed by Lord Kelvin in 1848. The unit of this scale, the kelvin (K) is named after him. In any absolute scale of measurement, only positive values are permitted. Zero on an absolute scale of measurement indicates the complete absence of the quantity being measured. On the absolute scale of temperature , 0 K is called **absolute zero**.

The **Celsius scale of temperature** is not absolute since negative values are possible. Figure 3.1 compares the two scales of temperature. The SI unit of temperature is the kelvin (*see* appendix A). It should always be used for calculations involving temperature. Note that the unit does *not* have the degree symbol, °.

Temperatures can be converted from the Celsius scale to the absolute scale by adding 273.15. It is worth noting that a difference in temperature is the same for both scales. For example, a rise in temperature of 10°C is the same as a rise in temperature of 10 K.

Pressure

An atmosphere (atm) is often used as a unit of pressure. It is the pressure of the atmosphere, equivalent to a height of 760 mm of mercury at 0°C when measured at sea level. Pressure is also expressed in millimetres of mercury (mmHg). For low pressures, the torr is often used. It is equivalent to 1 mmHg. Thus one atmosphere equals 760 torr.

Although the atmosphere, mmHg and torr are all used in scientific work, they should strictly be used only for rough comparisons or as ratios. The SI unit of pressure is the pascal (Pa). One pascal equals one newton per metre square (1 N m^{-2}). One atmosphere equals 101 325 Pa.

Volume

The SI units of volume are the cubic metre (m^3) or the cm^3 or dm^3. One litre (l) equals 1 dm^3. Note that

$$1 \text{ m}^3 \ = 1 \times 10^3 \text{ dm}^3$$
$$1 \text{ dm}^3 = 1 \times 10^3 \text{ cm}^3$$

Standard Temperature and Pressure (s.t.p.)

As we shall see in the section below, gas volume varies with temperature and pressure. Thus, when comparing volumes of gases, the temperature and pressure must be specified. It is usual to quote gas volume and other physical properties under conditions of standard temperature and pressure. By convention, standard temperature is 273 K (0°C) and standard pressure was, until recently, 101 325 Pa (that is, 1 atm or 760 mmHg). However, it is now 100 kPa.

Standard-State Pressure (SSP)

In 1984, the International Union of Pure and Applied Chemistry (IUPAC) formally approved and published a recommendation, initiated by the Commission on Thermodynamics, that the conventional standard-state pressure for thermodynamic data be changed from the traditional 1 atm (101.325 kPa) to 100 kPa (1 bar). The change in SSP causes no change in standard enthalpy changes and the entropy values of solids and liquids. However, it does result in small changes in the tabulated values of the molar entropies of gases. As a result of this change, it has been suggested that the term 'normal boiling point' (which refers to the boiling point of a liquid at 101 325 Pa be replaced with the term 'atmospheric boiling point' or 'a.b.n'. The boiling point of a liquid at the new standard pressure (100 kPa) would then be referred to as the 'baric boiling point' or 'b.b.n.'.

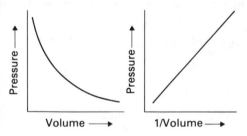

Figure 3.2   Boyle's law

Boyle's Law

The quantitative relationship between the volume and pressure of a gas was first stated by Robert Boyle in 1662. His law states that gas volume varies inversely with pressure at constant temperature. The law applies to any fixed amount of gas. It can be represented graphically in various ways (*see* figure 3.2). The left-hand graph shows that at low pressure the volume of a fixed amount of gas is high. The volume decreases as the pressure increases. This is expressed mathematically as

$$V \propto 1/p$$

where V = volume and p = pressure. Thus

$$pV = k$$

where k is a proportionality constant.

The usual way of expressing Boyle's law, however, is

$$p_1V_1 = p_2V_2 \tag{1}$$

Thus, when a gas is allowed to expand (or is compressed) at constant temperature from an initial volume of V_1 to a final volume of V_2, the final pressure p_2 can be calculated so long as the initial pressure p_1 is known.

Charles' Law

In 1787 Charles showed that gas volume varies directly with temperature at constant pressure. This is represented graphically in figure 3.3. The graph shows that gas volume increases linearly with temperature. This can be expressed mathematically as

$$V \propto T$$

where T = absolute temperature. However, Charles' law is more usually expressed in the form

$$V_1/T_1 = V_2/T_2 \tag{2}$$

Charles' law was developed by Joseph **Gay-Lussac** who, in 1802, stated that the volume of a gas changes by 1/273 of its volume at 0°C for every 1°C change in temperature. Thus, if we take any volume of any gas at 0°C and reduce its temperature by 273°C at constant pressure, the final volume will be zero. This

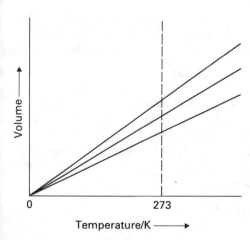

Figure 3.3   Charles' law

Absolute Zero

Absolute zero cannot be achieved. However, temperatures to within 0.001 K have been achieved in the laboratory. At such temperatures, random motions of molecules virtually vanish. This leads to some remarkable properties. For example, metals chilled near absolute zero lose almost all of their electrical resistance and become superconducting. Another example is provided by liquid helium. Near absolute zero temperature, its viscosity disappears and it becomes superfluid.

corresponds to a temperature of $-273°C$ or 0 K. This is the temperature called absolute zero. In reality it can never be achieved. Figure 3.3 shows how the volume–temperature plots of a gas all extrapolate back to zero volume at 0 K.

The General Gas Law

By combining Boyle's law and Charles' law (equations (1) and (2)), the following equation is obtained;

$$\frac{p_1 V_1}{T_1} = \frac{p_2 V_2}{T_2} \qquad (3)$$

This is called the general gas law or sometimes the combined gas law. This law enables the volume of a gas to be calculated at a specific temperature and pressure so long as the volume of the gas is known at another temperature and pressure.

The general gas law can also be written as

$$\frac{pV}{T} = \text{constant}$$

The exact value of the constant depends on the amount of gas. When the amount is one mole (*see* chapter 4) the constant is designated R. It is known as the **molar gas constant** or simply the gas constant. When the pressure is expressed in atmospheres, the value of R is given by

$$R = 0.082 \text{ atm dm}^3 \text{ K}^{-1} \text{ mol}^{-1}$$

In SI units it becomes

$$R = 8.314 \text{ J K}^{-1} \text{ mol}^{-1}$$

The general gas law for one mole of gas can now be expressed as

$$\frac{pV_m}{T} = R$$

where V_m is the volume of one mole of the gas. For n moles of a gas the equation is

$$\frac{pV}{T} = nR$$

or

$$pV = nRT \qquad (4)$$

This is called the **ideal gas equation.**

The ideal gas equation is an **equation of state**. For a gas, an equation of state relates pressure, volume and temperature.

A gas that obeys the ideal gas equation exactly is called an **ideal gas** or perfect gas. An ideal gas does not exist in reality. Real gases obey the ideal gas equation closely at low pressures and higher temperatures. We shall examine the deviation of real gases from the ideal gas equation in more detail below.

Using the Ideal Gas Equation to Calculate Relative Molecular Mass

The ideal gas equation can be used to calculate the **relative molecular mass**, M_r, of a gas directly. So far we have not defined relative molecular mass, although we have defined relative atomic mass, A_r (*see* section 1.1). For a simple gas molecule, the relative molecular mass is simply the sum of the relative atomic masses for all the atoms in the molecule. Thus, for carbon dioxide,

$$\begin{aligned} M_r(CO_2) &= A_r(C) + 2A_r(O) \\ &= 12 + (2 \times 16) \\ &= 44 \end{aligned}$$

The relative molecular mass expressed as grams per mole is called the *molar mass* (*see* chapter 4). Thus the molar mass of CO_2 is 44 grams per mole. Two moles of CO_2 have a mass of 88 grams and n moles a mass of $n \times 44$ grams. In general we can write

$$n = \frac{m}{M}$$

where n is the amount of substance in moles (or more simply the number of moles), m is the mass of the substance in grams and M is the molar mass. The units of n are grams per mole.

We can substitute this expression directly into the ideal gas equation (4) to obtain

$$pV = \frac{m}{M}RT \tag{5}$$

Thus, if we know the mass and volume of a gas at a given temperature and pressure, we can calculate the molar mass, M. Since

$$M = M_r \text{ g mol}^{-1}$$

we can obtain M_r directly.

EXAMPLE
A 512 cm^3 sample of a gas weighed 1.236 g at 20°C and a pressure of one atmosphere. Calculate the relative molecular mass of the gas. ($R = 8.314$ J K^{-1} mol^{-1}.)

SOLUTION
We are required to find the value of M_r given

$$\begin{aligned}
&\qquad\qquad\qquad\qquad\quad \textit{SI units (see appendix A)}\\
V &= 512 \text{ cm}^3 = 0.000\,512 \text{ m}^3\\
m &\qquad\qquad = 1.236 \text{ g}\\
T &= 20°C \quad = 293.15 \text{ K}\\
p &= 1 \text{ atm} \quad = 101\,325 \text{ Pa} = 101\,325 \text{ Jm}^{-3}\\
R &\qquad\qquad = 8.314 \text{ J K}^{-1} \text{ mol}^{-1}
\end{aligned}$$

By rearranging equation (5) we obtain

$$M = \frac{m \times R \times T}{p \times V}$$

$$= \frac{(1.236 \text{ g}) \times (8.314 \text{ J K}^{-1} \text{ mol}^{-1}) \times (293.15 \text{ K})}{(101\,325 \text{ J m}^{-3}) \times (0.000\,512 \text{ m}^3)}$$

$$= 58.07 \text{ g mol}^{-1}$$

Thus

$$M_r = 58.07$$

Avogadro's Law

In 1811, Amedeo Avogadro put forward his famous law which states that equal volumes of all gases at the same temperature and pressure contain equal numbers of molecules. The law is also known as Avogadro's principle, Avogadro's theory or Avogadro's hypothesis.

The number of molecules in one mole of a gas is always 6.022×10^{23}. This quantity 6.022×10^{23} is known as Avogadro's constant, L. It used to be known (and still is in some quarters) as Avogadro's number. However, strictly speaking it is not a number but a physical quantity with units of mol^{-1}. The constant can apply

Amedeo Avogadro (1776–1856)

Amedeo Avogadro was an Italian physicist and chemist. His great achievements included

- establishing that the formula of water was H_2O and not HO as had previously been thought;
- distinguishing between atoms and molecules—indeed, he coined the term 'molecule';
- distinguishing between atomic weight and molecular weight;
- and, of course, his famous law.

equally to atoms, molecules, ions, electrons and even chemical bonds and equations.

Since one mole of any gas always contains the same number of molecules, it follows from Avogadro's law that one mole of any gas always occupies the same volume. This volume, at s.t.p., can be calculated from the ideal gas equation (4) by putting $n = 1$ and substituting values for R and standard temperature and pressure in SI units. It works out at 22.4 dm^3. This is known as the **molar volume**.

Gas Density

Since one mole of any gas occupies 22.4 dm^3 at s.t.p., the density of a gas is easily calculated. For example, one mole (44 grams) of CO_2 occupies 22.4 dm^3. The density of CO_2 at s.t.p. is thus given by

$$d(CO_2) = \frac{44 \text{ g mol}^{-1}}{22.4 \text{ dm}^3 \text{ mol}^{-1}}$$

$$= 1.96 \text{ g dm}^{-3}$$

It should be noted that the calculation is based on the assumptions that
(a) CO_2 obeys Avogadro's law at s.t.p. and
(b) CO_2 is an ideal gas and thus obeys the ideal gas equation.
We shall see later that real gases—and CO_2 is one—deviate from ideal gas behaviour markedly under certain conditions.

The density of gas can be obtained at temperatures and pressures other than s.t.p. by using a rearranged form of the ideal gas equation (5). Thus

$$d = \frac{m}{V} = \frac{p \times M}{R \times T} \tag{6}$$

The density of a gas or vapour is often compared to that of hydrogen and expressed as the **relative vapour density**

$$\text{Relative vapour density} = \frac{\text{Density of gas}}{\text{Density of hydrogen}}$$

The experimental determination of gas densities and the comparison of these with the density of hydrogen formed the basis of the early determination of molecular weights of many gases and liquids. Hydrogen was always assigned an 'atomic weight' of one.

> The terms **atomic weight** and **molecular weight** correspond approximately to the modern terms relative atomic mass and relative molecular mass, respectively.

Experimental Determination of Relative Molecular Mass

The relative molecular masses of non-volatile substances can be determined experimentally by **colligative methods**. These methods are described in chapter 6. The relative molecular masses of gases and volatile liquids can be determined accurately by use of a mass spectrometer (*see* section 1.1). Relative molecular masses of gases and volatile liquids can also be found by **vaporisation methods**. These include syringe methods and the famous Victor Meyer's method.

The Syringe Method

There are several variations of this method. For gases the method originally devised by **Regnault** in 1845 is often used. A large glass syringe of known volume is weighed with and without the gas whose relative molecular mass is to be determined. Substitution of the values for the mass and volume of the gas together with those for temperature, pressure and the molar gas constant into equation (6) enables the molar mass and thus the relative molecular mass to be calculated.

For **liquids** a hypodermic syringe is used to inject the liquid through a rubber cap into a gas syringe heated so as to vaporise the liquid. The mass of the hypodermic syringe before and after injection gives the mass of the liquid. This equals the mass of the gas. The volume of the gas is measured with the gas syringe.

Once again equation (6) is used to calculate the molar mass and thus the relative molecular mass. Both methods rely on the accurate determination of the temperature and pressure of the gas.

The measurement of temperature and pressure can be avoided by using a comparative method whereby the gas density of the gas of unknown molar mass is compared with the gas density of a gas of known molar mass under identical conditions of temperature and pressure. Trichloromethane (chloroform) and propanone (acetone) are often used for this purpose. A schematic diagram of the method is shown in figure 3.4. The unknown molar mass is calculated from equation (6) as follows:

$$d_1 = \frac{p \times M_1}{R \times T}$$

where d_1 is the density of the gas of known molar mass, M_1, and

$$d_2 = \frac{p \times M_2}{R \times T}$$

where d_2 is the density of the gas of unknown molar mass, M_2. By eliminating the common factor p/RT from these two equations we obtain

$$M_2 = \frac{d_2 \times M_1}{d_1}$$

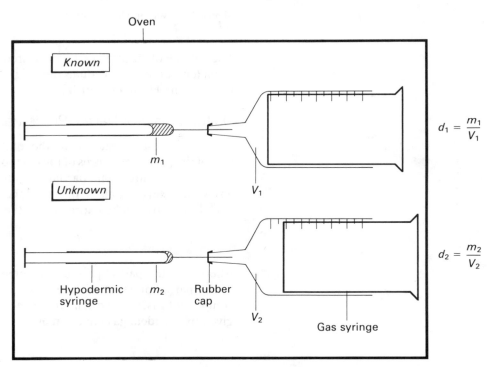

Figure 3.4   Syringe method for determining relative molecular mass

Victor Meyer's Method

This method is used to determine the molar mass of volatile liquids. It employs the same principle as the syringe method. That is, the gas density is found by measuring the volume of gas obtained from a known mass of the liquid. Equation (6) is then used to obtain the molar mass and thus the relative molecular mass.

A diagram of Victor Meyer's apparatus is shown in figure 3.5. The liquid under investigation is contained in a Hofmann bottle. The bottle is allowed to rest on a rod. The rod is then withdrawn and the bottle falls onto a bed of sand. The heat from the steam jacket causes the liquid to vaporise and thus remove the stopper.

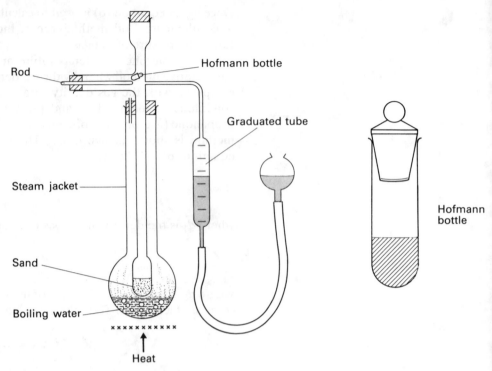

Figure 3.5   Victor Meyer's method

The volume of air displaced is measured by a graduated tube. This is measured at room temperature and equals the volume of gas obtained from the known mass of liquid in the Hofmann bottle.

Dalton's Law of Partial Pressures

In 1801 John Dalton stated that the total pressure of a mixture of gases equals the sum of the partial pressures of the component gases. This is known as Dalton's law of partial pressures. The partial pressure of a component gas is the pressure the gas would exert if it alone occupied the total volume at the same temperature.

Dalton's law can be expressed mathematically

$$p_{total} = p_a + p_b + p_c + \ldots \tag{7}$$

where p_a is the partial pressure of component a and so on.

The partial pressure of a component gas can be calculated if the masses of the component gases in a container are known. The partial pressure of component a is given by the ideal gas equation as

$$p_a = \frac{n_a RT}{V} \tag{8}$$

If we repeat this for all the components and substitute these expressions into equation (7) we obtain

$$p_{total} = (n_a + n_b + n_c + \ldots)\frac{RT}{V} \tag{9}$$

By combining equations (8) and (9) we obtain

$$p_a = \left(\frac{n_a}{n_a + n_b + n_c + \ldots} \right) p_{total} \tag{10}$$

The quantity

$$\frac{n_a}{n_a + n_b + n_c + \ldots}$$

is known as the **mole fraction** of component a. It is the ratio of the number of moles of gas a to the total number of moles of gas present.

Partial pressures of gases required to produce anaesthesia in mice

Gas	Anaesthetic pressure/atm
N_2	33
Ar	15
SF_6	6.1
N_2O (laughing gas)	1.5
Xe	0.95
ether	0.030
$CHCl_3$ (chloroform)	0.008

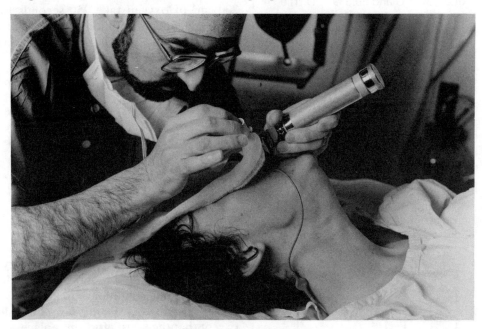

The partial pressure of an anaesthetic gas needed to produce anaesthesia is called the anaesthetic pressure. It varies widely from gas to gas as the table shows. The anaesthetic pressures for other mammals and humans are very comparable

Graham's Laws

We have seen that a gas or vapour does not have a definite volume or shape. When it is placed in a container it spreads uniformly throughout the whole volume of the container. This process is called **diffusion**. The **rates of diffusion** of two gases may be compared by measuring the time taken for a given volume of each gas to diffuse through a porous surface under identical conditions. Graham found that light gases diffuse through porous surfaces more rapidly than heavy gases. In 1833, he summarised his results in the following law: the rate of diffusion of a gas is inversely proportional to the square root of its density. This is **Graham's law of diffusion.**

This can be expressed mathematically as

$$r \propto \sqrt{\frac{1}{d}}$$

where r is the rate of diffusion and d the gas density. The ratio of the rates of diffusion of two gases a and b diffusing under identical conditions is given by

$$\frac{r_a}{r_b} = \sqrt{\frac{d_b}{d_a}}$$

From equation (6) we see that, under the same conditions of temperature and pressure, density is proportional to molar mass. Thus

$$\frac{r_a}{r_b} = \sqrt{\frac{M_b}{M_a}}$$

Atmolysis

This is the separation of two gases of different densities by diffusion. It is used to separate uranium isotopes. Naturally occurring uranium consists of 0.7% ^{235}U and 99.3% ^{238}U. The former isotope is needed to make the atomic bomb and also for nuclear power stations (*see* chapter 1). To separate the two isotopes, their volatile hexafluorides, UF_6, are allowed to diffuse through a porous barrier. The ratio of the densities of these two hexafluorides is 349 to 352. The lighter one, $^{235}UF_6$, diffuses 1.004 times faster than the other. Thus the mixture of gases emerging from the porous barrier is richer in $^{235}UF_6$. The procedure must be repeated thousands of times to obtain significant enrichment of the gaseous mixture.

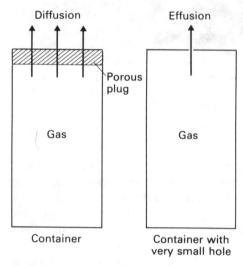

Figure 3.6 There is sometimes confusion between diffusion and effusion

Since the rate of diffusion is inversely proportional to the time taken for it to diffuse, we obtain the following expression:

$$\frac{t_b}{t_a} = \sqrt{\frac{M_b}{M_a}} \tag{11}$$

EXAMPLE

200 cm^3 of oxygen take 250 seconds to diffuse through a porous diaphragm. Under identical conditions, 200 cm^3 of an unkown gas X take 177 seconds to diffuse. What is the relative molecular mass of the unknown gas?

SOLUTION

The molar mass of oxygen, O_2, is 32 g mol^{-1}. Using equation (11) we thus obtain

$$M(X) = \left(\frac{177 \text{ s}}{250 \text{ s}} \right)^2 \times 32 \text{ g mol}^{-1}$$
$$= 16 \text{ g mol}^{-1}$$

Thus

$$M_r(X) = 16$$

Effusion

The process of effusion is similar to that of diffusion (*see* figure 3.6). Effusion is the passage of a gas through a single very small hole. Graham studied the rates of effusion of gases and found that the relative rates of effusion of different gases under the same conditions are inversely proportional to the square roots of their densities. This is **Graham's law of effusion**. It is the same as his law of diffusion except that the word effusion replaces diffusion.

The relative molecular mass of ozone, O_3, which cannot be obtained pure, can be determined by application of this law.

THE KINETIC THEORY OF GASES

The kinetic theory of gases is a mathematical model of an ideal gas. The theory can be used to account for the gas laws described above.

Assumptions

The kinetic theory of gases is based on five assumptions:
1. The gas consists of particles of negligible volume.
2. The particles are in continuous random motion.
3. The particles exert no attractive forces on each other.
4. The particles are perfectly elastic. Thus, no kinetic energy is lost on collision.
5. The average kinetic energy of the particles is proportional to the absolute temperature.

The Fundamental Equation of the Kinetic Theory

The following equation forms the basis of the kinetic theory of gases. It can be derived using the assumptions given above.

$$\bullet \qquad pV = \tfrac{1}{3}Nm\overline{c^2} \tag{12}$$

where N is the number of gas particles, m is the mass of a gas particle and $\overline{c^2}$ is the **mean square speed** of a gas particle, that is, the average value of c^2.

Accounting for the Gas Laws

The fundamental equation (12) of the kinetic theory can be used to account for

- the ideal gas equation
- Boyle's law
- Charles' law
- Avogadro's law
- Dalton's law of partial pressures
- Graham's law of diffusion

We shall content ourselves with just two examples.

The ideal gas equation. Assumption 5 of the kinetic theory can be expressed mathematically as

$$\tfrac{1}{2}mN\overline{c^2} \propto T \tag{13}$$

where $\tfrac{1}{2}m\overline{c^2}$ is the average **kinetic energy** of a gas particle and $\tfrac{1}{2}Nm\overline{c^2}$ is the average kinetic energy for N particles.

By comparing equations (12) and (13) we obtain

$$pV \propto T$$

and thus

$$\frac{pV}{T} = \text{constant} \tag{14}$$

We have already seen how the ideal gas equation (4) can be derived from equation (14).

Boyle's law. For a definite mass of gas N is constant. For a given temperature $\tfrac{1}{2}m\overline{c^2}$ is also constant. Thus from equation (12)

$$pV = \text{constant}$$

As we have seen, this is a form of Boyle's law.

Molecular Speeds

The kinetic theory of gases assumes that a gas consists of particles of negligible volume. The smallest particle of a gas is a molecule. A molecule may be monatomic—as in the case of noble gases—or it may consist of two or more atoms.

The speed of a gas molecule can be calculated using the following equation:

$$c_{rms} = \sqrt{\frac{3RT}{M}} \tag{15}$$

where c_{rms} is called the **root mean square speed** (or r.m.s. speed). This is the square root of $\overline{c^2}$. Equation (15) can be derived from equation (12). It follows from equation (15) that the smaller the molecule the faster it moves. The hydrogen molecule, H_2, which is the smallest of all molecules travels at approximately 4300 miles per hour at 25°C. The root mean square speeds of some simple molecules are shown in table 3.2.

Maxwell–Boltzmann Distribution

The speed of any particular molecule in a gas is always changing. This is due to collisions and the resultant change in energy. However, at any given instant the distribution of molecular speeds is always constant under the same conditions. This is because of the large number of molecules involved.

The speeds of molecules in a gas are spread over a wide range. This spread is called the Maxwell–Boltzmann distribution. Figure 3.7 shows graphically the Maxwell–Boltzmann distributions of molecular speeds at two different temperatures. Note that at the higher temperature there is a higher spread of speed. Notice also that the distribution is shifted to higher speeds at a higher temperature.

The Maxwell–Boltzmann distribution applies not only to molecular speeds but also to molecular energies. Figure 3.8 shows a typical curve for the distribution of molecular energies.

Table 3.2 4300 miles per hour!

Gas	Formula	c_{rms}/m s^{-1}	
hydrogen	H_2	1930	(4300 mph)
helium	He	1365	
methane	CH_4	680	
ammonia	NH_3	660	
water	H_2O	640	
nitrogen	N_2	515	
oxygen	O_2	480	
carbon dioxide	CO_2	410	

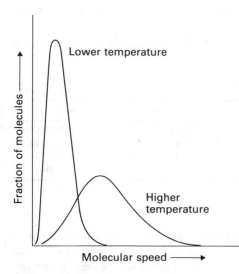

Figure 3.7 Distribution of molecular speeds

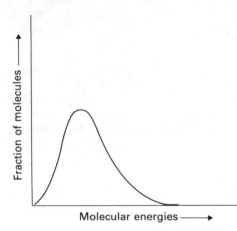

Figure 3.8   Distribution of molecular energies

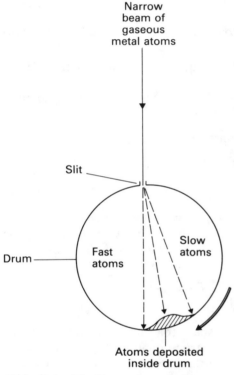

Figure 3.9   The Zartmann experiment

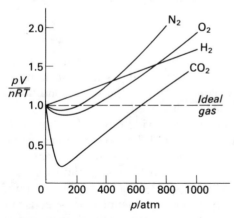

Figure 3.10   The deviation of four real gases from ideal gas behaviour at 273 K

Measuring the Distribution of Molecular Speeds

The Maxwell–Boltzmann distribution of molecular speeds was first verified by Zartmann in 1931. In the **Zartmann experiment** a narrow beam of vaporised metal atoms is directed at a rotating cylindrical drum (*see* figure 3.9). The drum contains a fine slit. Once, during each revolution of the drum, the beam of atoms enters the slit in the drum. They deposit on the far side of the drum. The fastest atoms deposit first and the slowest ones last. All atoms with a particular speed thus deposit in the same place. The greater the fraction of atoms with that speed the greater the intensity of metal deposited at that place. The distribution of intensities thus corresponds to the distribution of speeds.

REAL GASES

Real gases do not obey the ideal gas equation, $pV = nRT$. The deviation from ideal gas behaviour can be shown by plotting pV/nRT against p. For an ideal gas this is a straight line (*see* figure 3.10) parallel to the pressure axis. The extent and nature of the deviation depends on

- the gas
- pressure
- temperature.

Figure 3.10 shows the deviation for four different gases. Notice that the deviation of a real gas from ideal gas behaviour is more pronounced at high pressures. Real gases also deviate more from ideal gas behaviour at lower temperatures.

The deviations of real gases from ideal gas behaviour can be attributed to

- the attractive forces between gas molecules;
- the volume of a gas molecule (the kinetic theory assumes this is negligible).

Van der Waals Equation

In 1873 Johannes van der Waals modified the ideal gas equation to take account of the deviations of real gases. The van der Waals equation is

$$\left(p + \frac{an^2}{V^2}\right)(V - nb) = nRT$$

where a is a constant which corrects for the attractive forces between gas molecules and b is a constant which corrects the volume of the gas by allowing for the part of the volume occupied by the gas molecules. Van der Waals equation is an equation of state.

Values for van der Waals constants have been determined for many gases (*see* table 3.3).

Table 3.3   Van der Waals constants

Gas	Formula	a/atm dm^6 mol^{-2}	b/dm^3 mol^{-1}
hydrogen	H_2	0.244	0.0266
oxygen	O_2	1.36	0.0318
nitrogen	N_2	1.39	0.0391
carbon dioxide	CO_2	3.59	0.0427

Liquefaction of Gases

A gas liquefies when the attractive forces between the molecules are sufficient to bind them together in liquid form. Attractive forces are greater when molecules are closer together. This occurs at high pressure. Attractive forces are opposed by the motion of molecules. The motion (and kinetic energy) of molecules is higher at higher temperatues. Liquefaction is thus favoured by lower temperatures.

Table 3.4 Pressures needed to liquefy CO_2 at different temperatures

Temperature/K	Pressure/atm
223	6.7
263	26.1
293	56.5
303	71.2

A gas becomes more difficult to liquefy the higher its temperature. As its temperature increases, a higher pressure must be employed to liquefy it (*see* table 3.4). Above a certain temperature it is impossible to liquefy a gas. This is called the **critical temperature**, T_c. The minimum pressure required to liquefy a gas at its critical temperature is called its **critical pressure**, p_c. The volume occupied by one mole of the gas at its critical temperature and pressure is called its **critical volume**, V_c. The values for T_c, p_c and V_c are constants for gases. **Critical constants** for four gases are shown in table 3.5.

Table 3.5 Critical constants

Gas	Formula	Critical pressure/atm	Critical volume/dm^3 mol^{-1}	Critical temp./K
hydrogen	H_2	12.8	0.068	33.2
nitrogen	N_2	33.5	0.090	126.0
oxygen	O_2	49.7	0.074	154.8
carbon dioxide	CO_2	73.0	0.095	304.2

In 1863, Andrews studied the relationship between the pressure and volume of a fixed mass of carbon dioxide at different temperatures. He obtained a series of isotherms known as **Andrews' curves** (*see* figure 3.11). An **isotherm** is a pressure–volume plot at constant temperature. The isotherm for CO_2 at 321 K shows that the gas does not liquefy whatever the pressure or volume at this temperature. This is because 321 K is above the critical temperature of CO_2. The critical temperature of CO_2 is 304 K. The isotherm at this temperature is called the **critical isotherm**. The point P represents the gas at its critical temperature, pressure and volume. At this point it is in its **critical state**. Figure 3.11 shows two isotherms below the critical temperature of CO_2. Let us consider the one at 286 K. Along the curve from A to B the gas contracts as the pressure increases. Between B and C there is a large change of volume but no pressure change. This corresponds to the liquefaction of the gas at this temperature. From C to D there is little change in volume as pressure increases. The **compressibility** of liquids is very low compared to that of gases.

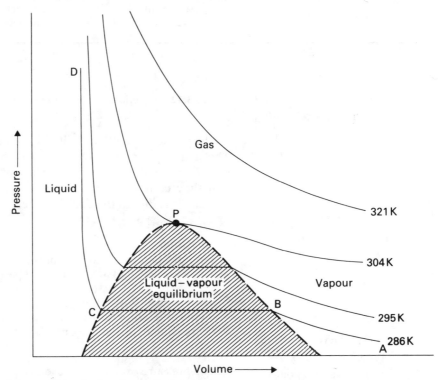

Figure 3.11 Andrew's curves for CO_2

LIQUIDS

Kinetic Concept of the Liquid State

Liquids are intermediate in character between solids and gases (*see* figure 3.12). For example, the attractive forces between particles (atoms, ions or molecules) in a liquid are intermediate in strength between those of solids and those of gases. As a result, the particles are held together in a definite volume. However, unlike solids, the attractive forces are not sufficient to hold the particles together in a regular structure. Liquids thus do not have shape. There is experimental evidence, however, that small groups of particles in liquids do arrange themselves in small, ordered and short-lived **clusters**. This is more pronounced in polar liquids than in non-polar liquids.

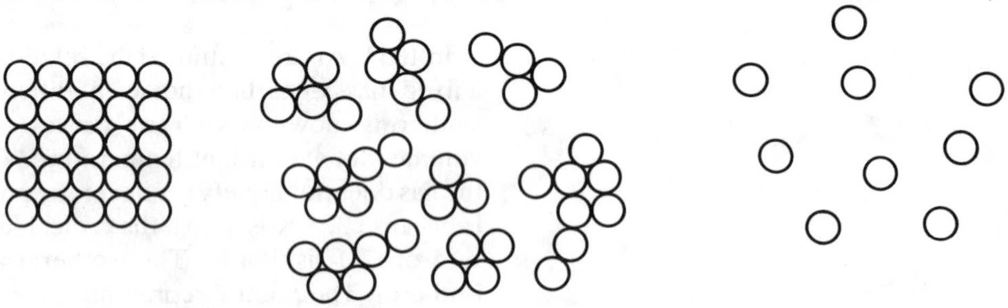

	Solid	Liquid	Gas
attractive forces between particles	high	intermediate	weak
motion of particles	zero	intermediate	high
space between particles	negligible	small	high
order	regular packing in crystals	loose clusters of particles	none

Figure 3.12   Kinetic concept of the states of matter

The particles in a liquid, like those in a gas, are in a state of constant motion. The first experimental evidence for this was provided by Brown in 1827. He examined small grains on the surface of water through a microscope. He noticed that they continually moved in a zigzag fashion. This is known as **Brownian motion**. It is due to the collisions between liquid particles and the grains.

Evaporation
When left open to the atmosphere, some particles of a liquid escape into the gas phase. This is called evaporation. The rate of evaporation increases with

- increasing surface area
- increasing temperature
- decreasing external pressure

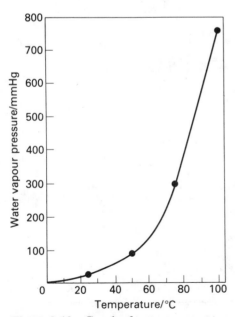

Figure 3.13   Graph of water vapour pressure against temperature

The pressure exerted by these escaping particles is called the **vapour pressure** of the liquid. Figure 3.13 shows how the vapour pressure of water increases with

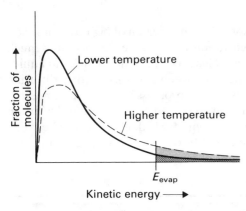

Figure 3.14 Distribution of kinetic energies of molecules at two different temperatures

temperature. Similar curves are obtained for other liquids. The increase in vapour pressure with temperature can be explained in terms of the kinetic theory. The kinetic energies of particles in a liquid follow a Maxwell–Boltzmann distribution like that for gas particles (*see* figure 3.7). Only those particles with kinetic energies sufficiently high to overcome the attractive forces between the particles in the liquid can escape. This energy is indicated as E_{evap} in figure 3.14. Since energy is lost from the liquid, it cools. However, if liquids are exposed to the open atmosphere at room temperature, this energy flows back into the liquid from the surroundings. The liquid thus continues to evaporate. If the temperature of the liquid is increased, more particles achieve the energy E_{evap} or higher and the rate of evaporation thus increases.

Boiling

Boiling occurs when the vapour pressure of the liquid equals the external pressure. The temperature at which this occurs is called the **boiling point** of the liquid. During boiling, bubbles of vapour are formed inside the liquid and it is these that cause the turbulence. The boiling point of a liquid can be reduced by lowering the external pressure. On the other hand, boiling can be suppressed by increasing pressure. At constant pressure, the boiling point remains constant until all the liquid has boiled away.

Freezing

In a liquid the motion and thus the kinetic energies of particles are sufficiently high to prevent the attractive forces holding the particles together in a crystal lattice. However, as the liquid cools, the attractive forces overcome the motions of those particles with low kinetic energies. As a result, these particles are held in fixed positions in a crystal lattice. The temperature at which the solid and liquid are in equilibrium is called the **freezing point**. A full discussion of equilibrium between liquids and solids and liquids and gases is found in chapter 6.

Properties of Liquids

We have seen that liquids have a **fixed volume** and take the **shape** of the container. We also saw that **densities** of liquids are much higher than those of gases. In general, liquid densities are similar in value to the densities of solids. The **compressibility** of liquids is very low since there is only little free space between liquid particles.

There are three other important properties which we have yet to consider. They can all be explained in terms of the kinetic concept of liquids.

Fluidity and Viscosity

Like gases, liquids can flow. This property is known as fluidity. Resistance to flow is called viscosity. Fluidity and viscosity are influenced by a number of factors. The most important are the attractive forces between molecules and the shape, structure and relative molecular mass of these molecules. The fluidity of liquids consisting of large molecules is lower than those consisting of small molecules. Viscosity of liquids is about 100 times higher than that of solids.

Surface Tension

A molecule in the bulk of a liquid is pulled in all directions equally by the intermolecular forces. However, at the surface of a liquid these forces are unbalanced, with the result that there is a net pull inwards. The surface of a liquid is thus in a state of tension—it is always trying to contract. The surface tension of a liquid is the minimum force needed to overcome this inward pull and thus expand the surface area. Surface tension accounts for the spherical shapes of free-falling droplets of liquid.

A free-falling drop of water. Surface tension accounts for its spherical shape

Diffusion

This is the process whereby a substance spreads from a region of high concentration or pressure to a region of lower concentration or pressure. Diffusion is much slower through a liquid than through a gas. This is because the particles in liquids are much more closely packed than particles between gases. A particle diffusing through a liquid is subject to frequent collisions. Its progress is therefore hampered. In gases, there is plenty of space for particles to spread at a rapid rate. Diffusion occurs between mutually soluble or miscible liquids. It does not occur between immiscible liquids. All gases are miscible and thus they can all diffuse through each other.

SUMMARY

1. **Standard temperature and pressure** (s.t.p.) is 0°C (273.15 K) and 100 kPa.

2. **Boyle's law:**

$$p_1 V_1 = p_2 V_2 \qquad \text{(constant temperature)}$$

3. **Charles' law:**

$$V_1/T_1 = V_2/T_2 \qquad \text{(constant pressure)}$$

4. **The general gas law:**

$$\frac{p_1 V_1}{T_1} = \frac{p_2 V_2}{T_2}$$

5. **The ideal gas equation:**

$$pV = nRT$$

6. **Avogadro's law** states that equal volumes of all gases at the same temperature and pressure contain equal numbers of molecules.

7. The **molar mass** of a gas can be determined from the following relationship:

$$d = \frac{m}{V} = \frac{p \times M}{R \times T}$$

8. **Dalton's law of partial pressures:**

$$p_{\text{total}} = p_a + p_b + p_c + \dots \qquad \text{(constant temperature)}$$

9. **Graham's law of diffusion:**

$$\frac{r_a}{r_b} = \sqrt{\frac{d_b}{d_a}}$$

10. **The fundamental equation of the kinetic theory:**

$$pV = \tfrac{1}{3}Nm\overline{c^2}$$

11. **Root mean square speed** is given by

$$c_{\text{rms}} = \sqrt{\frac{3RT}{M}}$$

12. The **Maxwell–Boltzmann distribution** gives the proportion of particles in a gas with a given speed or energy. The higher the temperature, the greater the speeds or energies.

13. **Van der Waals equation for real gases**:

$$\left(p + \frac{an^2}{V^2} \right)(V - nb) = nRT$$

14. The **critical temperature** is the temperature above which it is impossible to liquefy a gas. The **critical pressure** is the minimum pressure needed to liquefy a gas at its critical temperature. The **critical volume** is the volume of one mole at its critical temperature and pressure.

3.2 Solids

Solids consist of closely packed particles. These particles may be atoms, molecules or ions. Most solids are **crystalline**. This means that the particles are highly ordered in a regular three-dimensional arrangement.

The particles of some solids do not possess sufficient order to define a regular crystalline structure. These solids are said to be **amorphous**. Glass is amorphous. Its atoms have a random arrangement. Most **polymers** are amorphous. Polymers have macromolecular **chain structures** formed from small molecules called monomers (*see* chapter 20). Polymeric macromolecules have variable dimensions and thus cannot pack closely together in a regular arrangement.

At one time charcoal, coke and soot—all forms of carbon—were thought to be amorphous. However, X-ray analysis has shown that all these forms of carbon consist of small graphite-like crystals.

X-RAY CRYSTALLOGRAPHY

Diffraction

The structures of crystals can be analysed by a technique variously known as X-ray analysis, X-ray crystallography, X-ray diffraction or X-ray spectrometry.

When an advanced front of water waves hits a barrier with a narrow slit, the waves emerge as circular waves (*see* figure 3.15). This is called **diffraction**. Diffraction occurs with all types of radiation, including radio-waves, light waves and X-rays. Where there are a series of slits, each slit acts as a source of spherical or circular waves. These **interfere** with each other, cancelling each other out or reinforcing each other to produce bigger waves. The result is bent or diffracted waves.

Dorothy Crowfoot Hodgkin (*b*. 1910) at an IUPAC Congress in Manchester in 1985. Professor Hodgkin won the 1964 Nobel Prize for Chemistry for her determination of the structure of vitamin B_{12} using X-ray crystallography

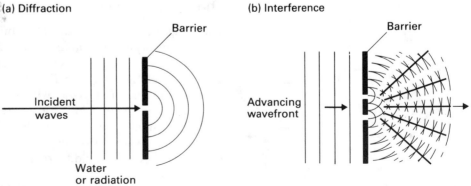

Figure 3.15 (a) Diffraction; (b) interference

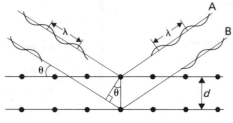

Figure 3.16 X-ray diffraction

When X-rays strike a crystal some are absorbed whilst others are not absorbed and pass straight through the crystal. The absorbed X-rays cause excitation of the electrons in the atoms of a crystal. As electrons return to the lower-energy states, X-rays are re-emitted. Those that are in phase reinforce each other and can be detected. Those that are out of phase destroy each other and cannot be detected (*see* figure 3.16). The re-emitted X-rays thus produce diffraction patterns.

Sir William Henry Bragg   Sir William Lawrence Bragg

In 1912 the German physicist Max von Laue suggested that crystals might serve as diffraction gratings for X-rays. He produced a diffraction pattern from a copper(II) sulphate-5-water crystal.

Following this discovery, Sir William Henry Bragg (1862–1942) and his son Sir William Lawrence Bragg (1890–1971) developed and used X-ray diffraction techniques to determine crystal structure. The first structure they determined was that of zinc sulphide—which has a cubic form. Together, the two Braggs derived the famous relationship known as the Bragg equation. They were jointly awarded the Nobel Prize for Physics in 1915. In later years, they determined the crystal structures of numerous inorganic compounds, minerals, metals and proteins.

The Bragg Equation

The Braggs showed that the absorption and emission of X-rays by crystals is mathematically equivalent to the reflection of light from parallel planes. X-rays of wavelength λ impinge on a crystal at an angle of incidence θ. The path length of an X-ray (path A in figure 3.17) which strikes the top layer of atoms in a crystal is shorter than that of an X-ray which strikes the second layer (path B). If the two emitted waves are to be in phase and reinforce each other, their path lengths must differ by a number of wavelengths. This difference is $n\lambda$ where n is the whole number and λ the wavelength. The Braggs thus showed that the angle of reflection of X-rays could be related to the distance d between the two layers of atoms:

$$2d \sin \theta = n\lambda$$

Figure 3.17 The Bragg equation: $2d \sin \theta = n\lambda$

This is the **Bragg equation**.

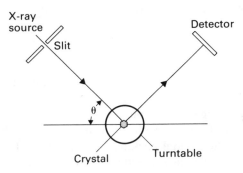

Figure 3.18 X-ray spectrometer

X-Ray Spectrometry

A crystal may have several planes from which reflection can take place. The X-ray spectrometer thus directs a beam of X-rays at a crystal mounted on a turntable. A photographic plate, ionisation chamber or Geiger counter is positioned to detect the reflected rays (*see* figure 3.18). The crystal is rotated. As angles are passed that satisfy the Bragg equation, signals flash out. The intensity of the flashes and the angle of rotation enable values for n and θ to be determined.

Analysis of the diffraction pattern can thus provide detailed information about the spacing between layers in a crystal. However, such analysis is often complicated and time-consuming. Nowadays computers are used extensively in X-ray analysis.

Powdered specimens can also be used for X-ray analysis. The diffracted X-ray beams are cones. They show up as curved lines on a photographic film.

CRYSTAL STRUCTURE

The arrangement of atoms, molecules or ions in a crystal can be depicted as a **crystal lattice**. This is sometimes called a space lattice. The particles in the lattice are joined by imaginary lines called **lattice lines**. The crystal lattice is composed of a basic unit called a unit cell. The **unit cell** is repeated throughout the crystal.

There are only seven types of unit cell. These are

- simple cubic
- simple tetragonal
- simple orthorhombic
- simple monoclinic
- hexagonal
- rhombohedral
- triclinic

Table 3.6 Types of simple cubic crystal lattice

Unit cell (primitive)	Crystal lattice (primitive or multi-primitive)
	simple cubic (primitive)
	face-centred cubic (multi-primitive)
	body-centred cubic (multi-primitive)

These unit cells are sometimes called primitive unit cells or **crystal systems**. Each of these crystal systems is represented by a primitive lattice. In addition, there are seven multi-primitive lattices. There are thus 14 types of crystal lattice in all. The multi-primitive lattices all fall within the seven crystal systems.

Simple cubic, for example, is a primitive unit cell. It forms three types of crystal lattice (*see* table 3.6).

The simple cubic unit cell and simple cubic primitive lattice are both shown in figure 3.19. The **face-centred cubic lattice** consists of a system of intersecting simple cubic unit cells. The corners of each one coincide with the face centres of others (*see* figure 3.20). There are similar intersections in all faces. The **body-centred cubic lattice** consists of interpenetrating simple cubic unit cells. Each corner of each cube coincides with the centre of the body of another.

Crystalline solids may be divided into three **classes**:

- metallic
- ionic
- covalent

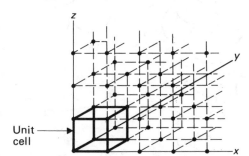

Figure 3.19 Simple cubic primitive unit cell and simple cubic primitive lattice

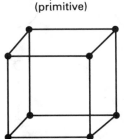
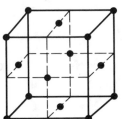
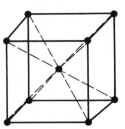

Simple cubic (primitive) Face-centred cubic (multi-primitive) Body-centred cubic (multi-primitive)

Figure 3.20 Cubic lattices

Crystals in each class have one or other of the 14 crystal lattices referred to above. We shall now examine each class in turn.

Metallic Crystals

We shall start with metallic crystals, since these provide examples of the simplest crystal structures.

The metal ions in a metallic lattice may be regarded as spheres. In solid metals, these spheres are packed together as closely as possible. The most efficient arrangement in a single layer of spheres is **hexagonal packing** (*see* figure 3.21). Each sphere is surrounded by six other spheres (in the same plane). The centres of any three adjacent spheres form a triangle. Less efficient is square packing where the centres of four adjacent spheres form a square.

Hexagonal layers of spheres can be closely stacked on top of each other in two ways: **hexagonal close packing** (hcp) or **face-centred cubic packing** (fcp). Fcp is also called cubic close packing. Both types are called close packing since they are very efficient in the use of space. In both cases 74% of the available volume is occupied by the spheres.

In hexagonal close packing, a second layer sits on top of the first layer so that each sphere in the second layer is in contact with three spheres of the first layer (*see* figure 3.22). The third layer is directly above the first, the fourth directly above the second, and so on. The hexagonal layers are thus arranged in the

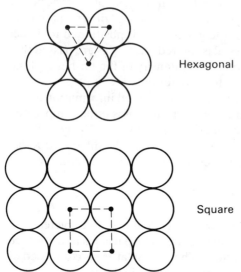

Hexagonal

Square

Figure 3.21 Types of packing in a single layer

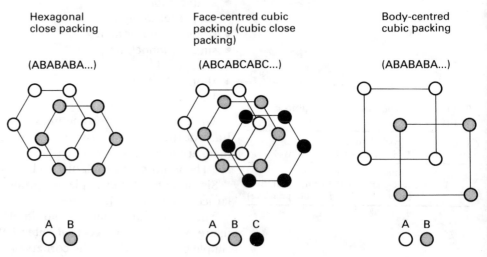

Figure 3.22 Bird's eye views of different packing types

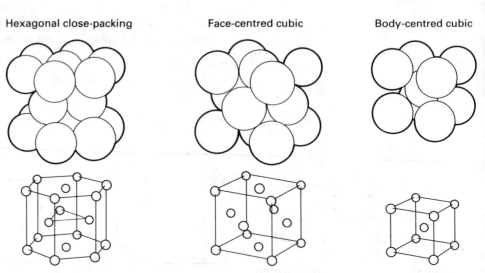

Figure 3.23 Three-dimensional views of different packing types

Table 3.7

Hexagonal close-packed	Face-centred cubic (cubic close packed)	Body-centred cubic
magnesium	aluminium	alkali metals
titanium	calcium	barium
cobalt	nickel	vanadium
zinc	copper	chromium
cadmium	silver	iron
	gold	
	lead	

sequence, ABABABABA... . The sequence of layers in face-centred cubic packing is more complicated, the sequence being ABCABCABC ... .

Layers of spheres in the square formation can build up to form the **body-centred cubic** structure. This structure is not so closely packed as the other two. Only 68% of the available volume is occupied by the spheres.

Three-dimensional views of all three structures are shown in figure 3.23. Almost all metals have one or other of these three crystal structures. Examples are shown in table 3.7.

Coordination Number

Each sphere in a body-centred cubic structure is at the centre of a cube with spheres at each corner (*see* figures 3.20 and 3.23). Each sphere is thus in contact with eight other spheres. The number 8 is called the coordination number of the lattice. The coordination numbers of the two close packing lattices are both 12.

Ionic Crystals

Ions in an ionic crystal are held together by electrostatic forces. The structure of the crystal lattice must therefore preserve electrical neutrality.

Details and diagrams of four of the most important types of ionic crystal are shown in figures 3.24 to 3.27. Each type of ion in an ionic lattice has its own coordination number. Thus, in the caesium chloride lattice (*see* figure 3.24) each Cs^+ is in contact with eight Cl^- ions. Its coordination number is thus 8. Similarly, each Cl^- ion is in contact with eight Cs^+ ions. Its coordination number is also 8. The caesium chloride lattice is thus said to have 8:8 coordination. The coordination of the sodium chloride lattice is 6:6 (*see* figure 3.25). Notice that in each case electrical neutrality is preserved.

The coordination and crystal structure of ionic lattices is largely determined by two factors:

- the ratio of cations to anions
- the radius ratio of the ions

The ratios of cations to anions in the caesium chloride (CsCl), sodium chloride (NaCl) and zinc blende crystal lattices are all 1:1. They are said to have an AB-

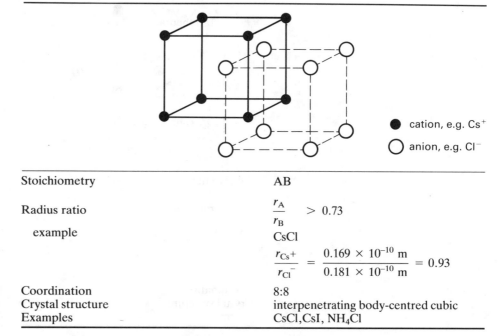

	● cation, e.g. Cs^+
	○ anion, e.g. Cl^-

Stoichiometry	AB
Radius ratio	$\dfrac{r_A}{r_B} > 0.73$
example	CsCl
	$\dfrac{r_{Cs^+}}{r_{Cl^-}} = \dfrac{0.169 \times 10^{-10}\ m}{0.181 \times 10^{-10}\ m} = 0.93$
Coordination	8:8
Crystal structure	interpenetrating body-centred cubic
Examples	CsCl, CsI, NH_4Cl

Figure 3.24 Caesium chloride structure

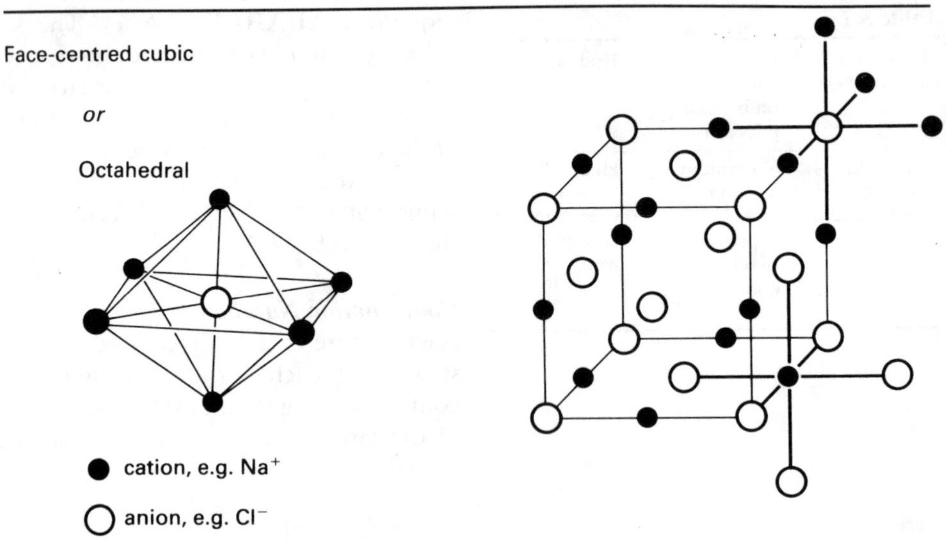

Face-centred cubic

or

Octahedral

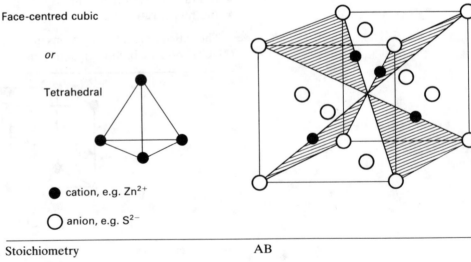

● cation, e.g. Na^+

○ anion, e.g. Cl^-

Stoichiometry	AB
Radius ratio	$0.73 > \dfrac{r_A}{r_B} > 0.41$
example	NaCl
	$\dfrac{r_{Na^+}}{r_{Cl^-}} = \dfrac{0.095 \times 10^{-10}\ m}{0.181 \times 10^{-10}\ m} = 0.52$
Coordination	6:6
Crystal structure	face-centred cubic (and octahedral)
Examples	alkali metal halides

Figure 3.25   Sodium chloride (rock salt) structure

Face-centred cubic

or

Tetrahedral

● cation, e.g. Zn^{2+}

○ anion, e.g. S^{2-}

Stoichiometry	AB
Radius ratio	$0.41 > \dfrac{r_A}{r_B} > 0.22$
example	ZnS
	$\dfrac{r_{Zn^{2+}}}{r_{S^{2-}}} = \dfrac{0.074 \times 10^{-10}\ m}{0.184 \times 10^{-10}\ m} = 0.40$
Coordination	4:4
Crystal structure	face-centred cubic (and tetrahedral)
Examples	ZnS, most copper halides, HgS

Figure 3.26   Zinc blende structure

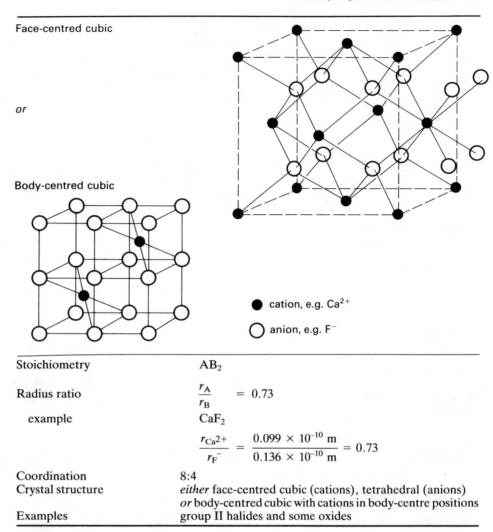

Face-centred cubic

or

Body-centred cubic

● cation, e.g. Ca^{2+}

○ anion, e.g. F^-

Stoichiometry	AB_2
Radius ratio	$\dfrac{r_A}{r_B} = 0.73$
example	CaF_2
	$\dfrac{r_{Ca^{2+}}}{r_{F^-}} = \dfrac{0.099 \times 10^{-10} \text{ m}}{0.136 \times 10^{-10} \text{ m}} = 0.73$
Coordination	8:4
Crystal structure	*either* face-centred cubic (cations), tetrahedral (anions) *or* body-centred cubic with cations in body-centre positions
Examples	group II halides and some oxides

Figure 3.27   Fluorite structure

Table 3.8

Radius ratio	Coordination
>0.73	8:8
0.41–0.73	6:6
0.22–0.41	4:4

In this section all the compounds mentioned are assumed to be purely ionic in character. The ions are considered to be hard spheres with well defined radii. However, as we have already seen (section 2.1) many compounds have both ionic and covalent character. Thus, ionic compounds which have appreciable covalent character will deviate from the general rules described in this section.

type **stoichiometry**. Fluorite (calcium fluoride, CaF_2) has an AB_2-type stoichiometry. Stoichiometry is dealt with in detail in the following chapter.

The ratio of the ionic radius of the cation (A) to the ionic radius of the anion (B) is called the **radius ratio** r_A/r_B. In general, the higher the value of the radius ratio, the higher the coordination of the lattice (*see* table 3.8).

It is often easier to view the structure of ionic crystals as two parts—one for the anions and one for the cations. The caesium chloride structure, for example, consists of a cubic structure of cations and a cubic structure of anions. Together they form two interpenetrating body-centred cubic structures (*see* figure 3.24). The sodium chloride or rock salt type of structure also consists of two cubic structures—one for the cations and one for the anions. Together they form two intersecting face-centred cubic structures. Each type of ion in this structure also has an **octahedral** arrangement (*see* figure 3.25).

The zinc blende type of structure has a face-centred cubic structure (*see* figure 3.26). If we view this as a cubic structure of cations, then the anions have a tetrahedral structure inside the cube. Alternatively, if we view the anions as the cubic structure then the cations have the **tetrahedral** arrangement.

The fluorite structure (*see* figure 3.27) differs from those described above in that its stoichiometry is AB_2 and also in that it has two different coordination numbers—8 and 4. Each Ca^{2+} ion is in contact with eight F^- ions and each F^- ion is in contact with four Ca^{2+} ions. The fluorite structure can be viewed as a face-centred cubic arrangement of cations within which the anions are arranged

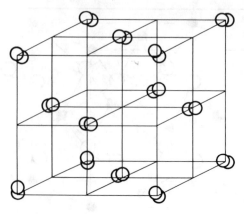

Figure 3.28 The iodine crystal: iodine has a face-centred cubic structure

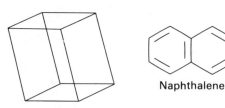

Monoclinic unit cell

Naphthalene

Figure 3.29 Naphthalene has a monoclinic structure

Asbestos

Asbestos is a polysilicate mineral. Its fibrous nature is due to silicate ions covalently bonded together to form polymeric chains.

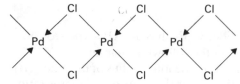

Figure 3.30 Chain structure of palladium chloride

tetrahedrally. Alternatively, the structure can be thought of as a body-centred cubic arrangement with the cations occupying the body-centre positions.

Covalent Crystals

Covalent crystals may be broadly classified into two types: molecular and macro-molecular.

Molecular Crystals

Molecular crystals consist of molecules held in lattice sites by weak intermolecular forces such as van der Waals forces or hydrogen bonds. We have already met examples of such crystals in the previous chapter. Ice, for example, consists of a lattice of water molecules held in position by hydrogen bonds (*see* figure 2.26).

The iodine crystal is another example. It has a face-centred cubic structure (*see* figure 3.28). The lattice positions are occupied by diatomic iodine molecules, I_2. The lattice is held together by weak van der Waals forces. Solid carbon dioxide has a similar structure with the CO_2 molecules occupying the lattice sites. Naphthalene also forms molecular crystals. Its flat molecules are arranged in a **monoclinic** lattice (*see* figure 3.29).

Macromolecular Crystals

These are also known as giant molecular crystals. In contrast to ionic and metallic crystals, which consist of ions, and to molecular crystals, which consist of molecules, macromolecular crystals consist of lattices of atoms. They are also known as atomic or network crystals. Macromolecular crystals may be grouped into three types: chain structures, layer structures and giant three-dimensional structures.

Chain structures. We have already seen that amorphous materials such as polymers have chain structures. We shall deal with organic polymers in chapter 20. Many silicate minerals also have polymeric chain structures. Asbestos is an example. Some inorganic compounds with covalent characteristics also have chain structures. Palladium chloride (*see* figure 3.30) and copper(II) chloride are notable examples. These chain structures may be regarded as one-dimensional macromolecules.

Layer structures. These may be regarded as two-dimensional macromolecules. Graphite is the classic example of a substance with a layer structure. Each layer consists of carbon atoms covalently bonded together into hexagons (*see* figure 3.31a). Adjacent layers are held together by weak van der Waals forces. The layers readily slide over one another. This accounts for its greasy texture.

Graphite conducts electricity in the plane of the layers but not at right angles to them. This is due to delocalised π orbitals extending between the layers. De-

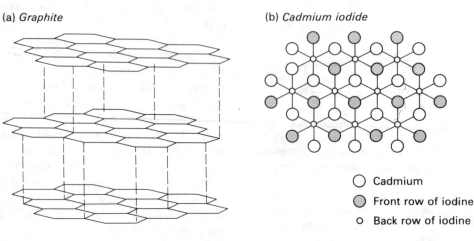

(a) *Graphite*

(b) *Cadmium iodide*

○ Cadmium
◐ Front row of iodine
∘ Back row of iodine

Figure 3.31 Layer structures: (a) graphite; (b) cadmium iodide

Talc and Mica

Talc and mica are both silicate minerals with the overall silicate composition $Si_2O_5{}^{2-}$. Talc has the formula $Mg_3(OH)_2(Si_2O_5)_2$. The composition of micas are variable. Lepidolite, for example, has the formula $KLi_2Al(OH)_2(Si_2O_5)_2$.

Both talc and mica have layer structures. Each layer consists of two-dimensional polymers. The layer structure accounts for the slippery texture of talc and also for the ease with which mica separates into sheets.

The layer structure of talc accounts for its slippery texture

localised electrons are thus free to move through the layers. Other examples of substances with layer structures are cadmium iodide, CdI_2 (*see* figure 3.31b), and aluminium chloride, $AlCl_3$. For example, the bonding of the cadmium and iodine atoms within each layer of CdI_2 is intermediate between ionic and covalent. The layers are held together by weak van der Waals forces. Like graphite, cadmium iodide has a greasy texture.

Three-dimensional covalent structures. Diamond has a giant three-dimensional structure (*see* figure 3.32). Each carbon atom is covalently bonded to four other carbon atoms in a tetrahedral arrangement. It has the same structure as zinc blende but with each zinc and sulphur ion replaced by a carbon atom. The whole diamond structure is, in fact, one giant molecule.

ISOMORPHISM, POLYMORPHISM AND ALLOTROPY

Substances which have the same type of chemical formula and which crystallise with the same crystal lattice are said to be **isomorphous**. For example, sodium nitrate, $NaNO_3$, and the calcite form of calcium carbonate, $CaCO_3$, exhibit isomorphism. They both crystallise as rhombohedra.

A **compound** which can exist in two or more crystalline forms is said to exhibit **polymorphism**. Silica (silicon(IV) oxide, SiO_2) is a polymorphous substance. It has a giant molecular structure with the silicon atoms covalently bonded to four oxygen atoms in a tetrahedral arrangement (*see* figure 3.33). Quartz is a crystalline form of silica. It has a hexagonal structure. Silica also crystallises in the ortho-rhombic and cubic forms at higher temperatures:

$$\underset{\text{(hexagonal)}}{\text{Quartz}} \underset{}{\overset{870°C}{\rightleftharpoons}} \underset{\substack{\text{(ortho-}\\\text{rhombic)}}}{\text{Tridymite}} \overset{1470°C}{\rightleftharpoons} \underset{\text{(cubic)}}{\text{Cristobalite}} \overset{1710°C}{\rightleftharpoons} \text{Liquid}$$

The temperature at which one form converts to another is called the **transition temperature**.

An **element** which can exist in two or more solid forms (crystalline or amorphous) is said to exhibit allotropy. The different forms are called **allotropes**. About one-half of all elements are allotropes.

Carbon, for example, exists as diamond or graphite. Sulphur exists in two crystalline forms—orthorhombic or monoclinic—depending on the temperature

Figure 3.32   Diamond lattice

Quartz is a crystalline form of SiO_2

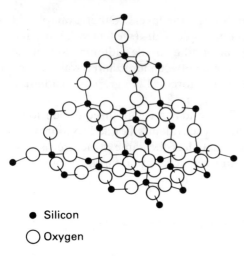

● Silicon

○ Oxygen

Figure 3.33 Silicon(IV) oxide lattice

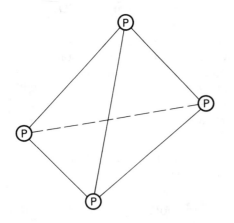

Figure 3.35 White phosphorus

> The particles (atoms, ions and molecules) in a solid lattice all vibrate—except at absolute zero. Occasionally a particle at the surface of the solid achieves sufficient kinetic energy to overcome the attractive forces that bind the particle in the solid. It thus escapes. It is the escaping particles which give rise to the vapour pressure.

(*see* also chapter 6). Both crystalline forms are examples of molecular crystals. The molecules are puckered rings containing eight covalently bonded sulphur atoms (*see* figure 3.34). Solid sulphur can also exist in a third allotropic form—plastic sulphur. This form is unstable. It consists of long chains of sulphur atoms which, at room temperature, break up and reform to produce S_8 molecules crystallising in the orthorhombic lattice.

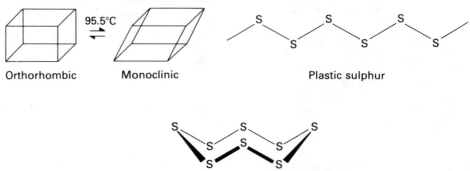

Orthorhombic   Monoclinic   Plastic sulphur

A molecule of sulphur, S_8

Figure 3.34 Sulphur

Phosphorus can exist in three allotropic forms. The most stable form is red phosphorus (*see* also chapter 6). This exists as a giant molecule with each phosphorus atom covalently bonded to three other phosphorus atoms. White phosphorus exists as molecular crystals. Each molecule contains four phosphorus atoms covalently bonded in a tetrahedral arrangement (*see* figure 3.35). A third allotrope, black phosphorus, is only formed under high pressures. It exists as a macromolecular layer lattice.

PROPERTIES OF SOLIDS

Compared with the other two states of matter—gases and liquids—solids have the highest degree of order. It is this high degree of order which accounts for many of the physical properties of solids.

The **compressibility** of solids is zero. All solids thus have a fixed volume. One of the most familiar properties of solids is their ability to retain their shapes. Compared to gases and liquids, solids can withstand considerable external stress. For a particular substance, the density of the solid is much higher than that of the gas and slightly higher than that of the liquid. Water is an exception since the density of ice is lower than that of liquid water. Most solids have sharp **melting points** although some, such as glasses, soften over a range of temperature.

All solids have **vapour pressures** although they tend to be very low. This is particularly the case with ionic solids.

Enthalpies of fusion are much lower than the enthalpies of vaporisation of the corresponding liquids.

Solids exhibit a wide variation in **rigidity**. Some solids—ionic solids, for example—tend to shatter under stress. This is called brittleness. Others—such as rubbers—are **elastic**. They return to their original shape once the stress is removed. Many metals are malleable and ductile. A malleable metal can be beaten into a sheet whereas a ductile metal can be drawn out into a wire.

The conducting properties of solids vary widely. All metals conduct heat and electricity. Non-metals tend to be insulators although, as we have seen, graphite is a conductor of electricity. Graphite is an **anisotropic** material. This means that its properties depend on direction. For example, graphite only conducts electricity in the direction along its layers. Crystals may also be anisotropic. Their refractive index, for example, may vary depending on direction.

When the properties of a material are uniform in all directions, the material is said to be **isotropic**. Cubic crystals are isotropic since they are fully symmetrical.

The physical properties of solids depend, to a large extent, on the type of bonding and the structure of the solid. The relationship between bonding and properties was discussed in chapter 2. Table 3.9 summarises the trends for solids.

Table 3.9 Structure and properties of solids

	Metallic	Ionic	Covalent	
			Molecular	Macromolecular
examples	Cu, Mg	NaCl, KNO$_3$	I$_2$, naphthalene	diamond, SiO$_2$
particles	positive ions and mobile electrons	cations and anions	molecules	atoms
bonding	metallic	ionic	1. covalent within molecules 2. van der Waals or hydrogen bonds between molecules	1. covalent between atoms 2. van der Waals forces (chain and layer structures only)
properties				
melting point	high	high	low	very high
boiling point	high	high	low	very high
hardness	hard, malleable and ductile	hard, brittle	soft	very hard
electrical	good conductors (solids and liquids)	non-conductors (electrolytes when molten or in aqueous solution)	non-coductors	non-conductors (except graphite)
solubility				
(a) in water	insoluble	soluble	insoluble	insoluble
(b) in non-polar solvents	insoluble	insoluble	soluble	insoluble

SUMMARY

1. Crystal structure may be determined by **X-ray diffraction** techniques.
2. The **Bragg equation** is 2d sin θ = $n\lambda$.
3. The arrangement of particles in a crystal can be depicted as a **crystal lattice**. The **unit cell** is repeated through the lattice.
4. There are three classes of crystalline solids: metallic, ionic and covalent.
5. The most efficient arrangement of cations in a single layer of a metallic crystal is hexagonal packing. The most efficient arrangement of cations in a metallic lattice is **hexagonal close packing** or **face-centred cubic packing**.
6. The **coordination** and crystal structure of ionic lattices is largely determined by
   (a) the ratio of cations to anions,
   (b) the **radius ratio** of the ions.
7. **Body-centred cubic** and **face-centred cubic** are two common types of structure for ionic crystals.
8. There are two classes of covalent crystal: molecular and macromolecular.
9. **Macromolecular crystals** include crystals with chain structures, layer structures and three-dimensional network structures.
10. An element which exists in more than one crystalline form is said to exhibit **allotropy**.
11. A compound which exists in more than one crystalline form is said to exhibit **polymorphism**.

Examination Questions

1. (a) State Avogadro's law.
   (b) What can you deduce from the fact that at s.t.p. 22.4 dm^3 of carbon dioxide contain rather more than 6.02×10^{23} molecules?
   (c) (i) State Graham's law of diffusion.
      (ii) The ratio of the rate of diffusion of a gas Y to that of nitrogen was found to be 0.366. Calculate the relative molecular mass of Y.

   (OLE)

2. Describe an experiment by which you would determine the relative molecular mass of a gas or vapour.

   State Dalton's law of partial pressures and show how it may be derived from the ideal gas equation.

   A mixture of oxygen and nitrogen, in which the mole fraction of oxygen is 0.25, is at a pressure of 1.0 atm. Calculate the partial pressure of each gas in the mixture (assumed to be ideal).

   (L)

3. (a) Define the terms *relative atomic mass* and *relative molecular mass*. Does *relative molecular mass* have the same significance when applied to potassium chloride as it has when applied to benzene?
   (b) What two variables determine the average velocity of the molecules of an ideal gas? State concisely how this velocity depends on these variables. (The derivation of formulae is *not* expected.)
   (c) On reacting potassium chlorate(VII) (potassium perchlorate), $KClO_4$, with fluorosulphonic acid, FSO_3H, a gas X is evolved.
      (i) 0.245 g of X are found to occupy 112 cm^3 at 293 K and at a pressure of 5.20×10^4 Pa. Calculate to three significant figures the relative molecular mass of X.
      (ii) By mass spectrometry, X is found to contain only the elements Cl, O and F. Suggest a molecular formula for X.

   (OLE)

4. (a) State Graham's law of diffusion.
   (b) Calculate to four significant figures the ratio of the rate of diffusion through a porous membrane of hydrogen chloride gas composed wholly of H^{35}Cl molecules to the rate of diffusion of a gas composed wholly of H^{37}Cl molecules, both gases being at the same temperature and pressure. (For this calculation use the following values for the relative atomic masses of the isotopes involved: H = 1.01; ^{35}Cl = 34.97; ^{37}Cl = 36.95.)

   (OLE)

5. (a) (i) Use the basic assumptions of the kinetic theory of gases to derive the expression $pV = \frac{1}{3}nmc^2$ for an ideal gas.
      (ii) What do you understand by the terms real gas and critical temperature?
   (b) (i) Sketch a plot of $\dfrac{pV}{nRT}$ against p for a gas Y (critical temperature approximately 155 K) at 273 K as the pressure increases from 0 to 1.013×10^8 Pa, showing *ideal* and *real* behaviour.

(ii) Explain the differences between the plot for the real gas and that for the ideal gas.

(iii) What changes, if any, would you expect to see in the real plot at 223 K and 323 K?

(c) 0.100 00 g of a gas Y at 273 K had the following pV values at the stated pressure p:

pV/J	7.076	7.066	7.050	7.035	7.025
p/Pa	20 000	40 000	60 000	80 000	100 000

(i) Estimate graphically the value of pV at zero pressure.

(ii) By means of the expression $pV = \dfrac{m}{M} RT$ find the relative molecular mass of Y.

(SUJB)

6. (a) Using the kinetic concept of matter, explain the following:
   (i) the melting of a solid;
   (ii) the evaporation of a liquid;
   (iii) the saturated vapour pressure of a liquid.

   (b) (i) Calculate a value for the gas constant, R, stating the units, given that the density of argon at s.t.p. is 1.784 g dm^{-3}.
   (ii) Use your value from (i) to estimate the average kinetic energy of an argon atom at 273 K, given that for 1 mole of a gas the total kinetic energy of the gas molecules equals $^3/_2 pV$.
   (iii) Draw curves to show how the distribution of kinetic energies of the molecules in a sample of argon gas changes when the temperature of the gas is raised. State the relationship between the areas under the curves you have drawn.

(AEB, 1984)

7. (a) 5 g of a substance R (boiling point 40°C) occupies a volume of 0.957 dm^3 at 50°C and 102.0 kPa.
   (i) Describe briefly an experiment that could be carried out to obtain these results in the laboratory.
   (ii) Calculate the relative molecular mass of compound R.

   (b) The *critical temperature* and *critical pressure* for butane are 425 K and 3.78×10^3 kPa respectively.
   What is meant by these terms and explain why butane can be conveniently stored as a *liquid* in a can, whereas methane cannot.

   (c) A compound shows significant deviations from the ideal gas law at normal conditions of temperature and pressure.
   (i) Write out the ideal gas equation indicating the significance of each term used in the equation.
   (ii) What changes in conditions of temperature and pressure would result in further deviations from the ideal gas law? Explain the significance of these changes.
   (iii) Explain how van der Waals changed the ideal gas equation to allow for such deviations.
   (iv) What can be deduced about the molecular and intramolecular bonds in this compound at normal temperature and pressure to account for these deviations?

(AEB, 1984)

8. (a) State concisely the essential properties of X-rays and crystals which make it possible to use X-rays to determine crystal structures.
(b) Describe with the aid of a sketch the structure of any **one** crystal which has an ionic lattice and which is appreciably soluble in water. Explain briefly (i) how the force holding this lattice together differs from that in the lattice of diamond, (ii) why the ionic solid you have chosen dissolves in water, whereas diamond does not.
(c) Crystals of metallic copper have a cubic unit cell which contains four copper atoms. By X-ray diffraction, the length of the side of this unit cell is found to be 3.61×10^{-10}m ($= 3.61 \times 10^{-8}$cm). The density of copper is 8920 kg m^{-3} ($= 8.920$ g cm^{-3}), and its relative atomic mass is 63.5. Use these data to calculate the Avogadro constant.

(OLE)

9. Write an essay on the *solid state*. Include in your account an explanation of the various types of bonds which occur in solids and, wherever possible, draw clear annotated diagrams of the arrangement of the structural units present. You may choose to refer to some, or all, of the following solids:

one or more *named* metallic halides, the allotropes of carbon, a typical *named* metallic element, a typical *named* non-metallic element (other than carbon), ice.

Indicate briefly how the physical properties of your examples are dependent on their bonding and structure.

With the help of clear, neat diagrams briefly indicate the structural difference between the two close-packed arrangements—*cubic* and *hexagonal*.

(NISEC)

10. Give an account of the bonding in, and the structure of, the following solids: (a) diamond; (b) iodine; (c) phosphorus(v) oxide; (d) ice; (e) magnesium (a metal with hexagonal close packing).

(AEB, 1984)

11. (a) What is meant by the term *coordination number* when applied to a description of crystal structure?
(b) name a metal halide which has:
    (i) 6:6 coordination;
    (ii) 8:8 coordination.
    (iii) Say why one of the metal halides in (i) and (ii) has higher coordination than the other.
(c) This section will show you how to draw a diagram of the fluorite structure which is the structure of calcium fluoride.
    (i) Draw a cube with a calcium ion (×) at the centre of each face and a calcium ion (×) at each corner.
    (ii) Draw a second cube which is the same size as the cube in (i). Divide this second cube into eight smaller cubes of equal size. Place a fluoride ion (○) at the centre of each of these eight smaller cubes. An internal cube with these eight fluoride ions, one at each corner, will be formed. The unit cell of fluorite is formed by the superimposition of diagrams (i) and (ii) on each other. The entire crystal of calcium fluoride is simply a three-dimensional repeating pattern of the unit cell.
    (iii) Superimpose the two diagrams on each other and so draw a new diagram which represents the unit cell.
    (iv) Are the fluoride ions shared by any other cube of the entire crystal other than the unit cell cube?

(v) How many cubes of the entire crystal are shared by the calcium ions at each corner of the unit cell?

(vi) Repeat (v) for the calcium ions at the centre of each face.

(vii) Hence deduce, with some explanation, the empirical formula of calcium fluoride.

(viii) How many moles of calcium fluoride are in each unit cell?

(ix) Explain why the deduced empirical formula is compatible with the positions of calcium and fluorine in the Periodic Table.

(SUJB)

12. (a) Explain the term *allotropy*.

(b) Draw diagrams to illustrate the structures of the two allotropes of carbon and relate each structure to the physical properties of the allotrope.

(c) Cite the evidence which supports the view that the allotropes mentioned in (b) are of the same element.

(d) Briefly explain why carbon forms quadricovalent rather than electro-valent bonds in its compounds.

(e) Draw, and explain, the molecular shapes of the three hydrocarbons: (i) ethane, (ii) ethene, and (iii) ethyne.

(f) Say why the bonds in tetrachloromethane (carbon tetrachloride) are polar but the molecule has zero dipole moment.

(SUJB)

4 STOICHIOMETRY

Seven Million—and Growing

Whilst the world's energy and material resources are declining, there is one resource which is growing rapidly, if not explosively. That resource is information. This is particularly the case with chemical information. In 1984 over seven million compounds were documented in the chemical literature. The Ninth Collective Index to *Chemical Abstracts* (a journal which provides synopses of chemical publications throughout the world—including publications in German, Chinese, Japanese and Russian) lists about 100 000 entries under 'aluminium' (and its compounds).

Each year almost half a million new compounds are documented. This is about one a minute. The compounds include about 1500 pesticide chemicals, 4000 drugs, 38 000 potentially toxic substances and 50 000 commercial or potentially useful chemicals. This presents formidable problems. For a start, there is the problem of determining and monitoring the environmental safety of these chemicals. There is also the related question of the storage and retrieval of information on these compounds. This in turn depends on an accurate system of naming compounds.

About 500 000 new chemical compounds are documented each year

What is IUPAC?

The International Union of Pure and Applied Chemistry (IUPAC) is the international representative body for chemistry recognised by all the other sciences and by the academies of science that are responsible for science at national levels. Indeed, it is these national academies of science or their national committees for chemistry which are the members of IUPAC and it is their subscriptions which finance it.

IUPAC is recognised throughout the world as the authority on chemical nomenclature, terminology, symbols, atomic weights and related topics. Its reports and recommendations on these matters are generally accepted as authoritative and definitive. In many countries the reports form the basis for drafting legislation and regulations relating to chemical manufacturing, international commerce and matters relevant to food, health and the environment. Most chemistry publications including research journals adhere to IUPAC terminology and nomenclature.

IUPAC also provides advice on chemical matters to international authorities such as the World Health Organization (WHO), the Food and Agriculture Organisation (FAO) and the United Nations Educational, Scientific and Cultural Organisation (UNESCO).

Professor C. N. R. Rao, President of IUPAC, 1985–1987.

The most widely used system of nomenclature of chemical compounds is that recommended by the International Union of Pure and Applied Chemistry (IUPAC). It is a comprehensive system—and still developing.

In the UK the Chemical Nomenclature Advisory Service (CNAS) of the Laboratory of the Government Chemist has produced a Euro-list of about 20 000 chemical names showing their correct IUPAC nomenclature and their classification under the European Common Tariff Rules. This is used by HM Customs and Excise. This list, which was published in 1981, is available in the six official languages of the European Community (Danish, Dutch, English, French, German and Italian).

In recent years, computer technology has played an increasingly important role in storing and retrieving chemical information (*see* figure 4.1). Much of the information is stored in databases. Each database may contain hundreds of thousands of references to the scientific and patent literature (*see* table 4.1).

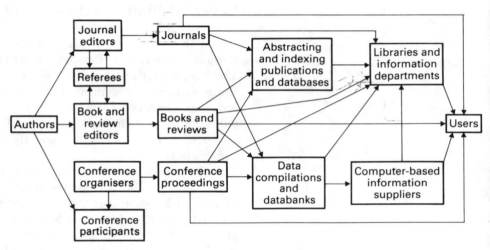

Figure 4.1 Major flows of information through the British scientific information system

Table 4.1 Some major systems of data retrieval

CIS (The Chemical Information System). This contains one of the most extensive collections of computer-readable chemical data now available to the general public

BRS (Bibliographic Retrieval Services). This system contains databases which include Chemical Abstracts Condensates (CAC), Chemical Abstracts Search (CASEARCH) and MEDLARS

MEDLARS (Computerised literature retrieval services of the (US) National Library of Medicine). It contains a number of databases on an on-line network. These include Chemline which has over 400 000 references

LIS (The Lockheed Information System). This contains approximately 125 databases

INPADOC (The International Patent Documentation Centre, Vienna). This covers patent literature in 14 different languages from more than 50 countries

Patents

There are now well over 20 million patent documents in existence. About 800 000 new ones are published each year (*see* figure 4.2). That means an average of 15 000 each week or one every 10 seconds. INPADOC has over 10 million of these patent documents stored on computer. To classify them, it uses the International Patent Classification (IPC) system. This subdivides the whole field of science and technology into about 60 000 subdivisions. To produce information, INPADOC uses COM, Computer Output on Microfiche, as one output medium. One microfiche contains 207 pages of information and one index page.

(12) **UK Patent Application** (19) **GB** (11) **2 120 226   A**

(21) Application No **8311310**
(22) Date of filing **26 Apr 1983**
(30) Priority data
(31) **21132**
(32) **7 May 1982**
(33) **Italy (IT)**
(43) Application published
30 Nov 1983
(51) INT CL³
C01B 33/20 35/00
(52) Domestic classification
C1A 410 411 425 518 519
528 CA D60 G50 G50D60
G58 G58D60
U1S 1345 1356 1388 C1A
(56) Documents cited
GB A 2062603
GB A 2024790
GB 1587921
(58) Field of search
C1A
(71) Applicants
Snamprogetti SpA,
(Italy),
Corso Venezia 16,
Milan,
Italy.
(72) Inventors
Giovanni Perego,
Vittorio Fattore,
Marco Taramasso.

(54) **Synthetic crystalline porous materials comprising silicon and boron oxides**

(57) Synthetic crystalline porous materials of a zeolitic nature, consisting of silicon and boron oxides, are prepared by reacting, under hydrothermal conditions, a silicon derivative, a boron derivative, a mixture of tetramethylammonium and tetrabutylammonium hydroxides or, as an alternative, a mixed quaternary base of (methyl-butyl) ammonium hydroxide, optionally with an alkali metal hydroxide or ammonium hydroxide being present. The materials have the empirical formula:

$$a\ R_2O.(1-a)Me_2O.B_2O_3.xSiO_2$$

wherein a is from 0 to 1, x is from 30 to 120, Me is a H^+ cation, an NH_4^+ cation or an alkali metal cation, and R is (i) a mixture of a tetramethylammonium cation and tetrabutylammonium cation, and/or (ii) a quaternary (methyl-butyl) ammonium cation.

Figure 4.2

Women Inventors

Each Tuesday, at noon, about 1200 American patents are issued. But only about 20 (1.7%) bear the name of a woman. Since the US Patent Office opened in 1790 over four million patents have been issued—yet women have received only about 60 000 (or 1.5%).

LEARNING OBJECTIVES

After you have studied this chapter you should be able to
1. Discuss the importance of **stoichiometry** in
   (a) chemistry in general,
   (b) the chemical industry in particular.
2. State the **laws of chemical combination**.
3. Write the **trivial and systematic names** of common inorganic compounds—given their formulae.
4. Write the **formula** of common inorganic compounds—given their systematic names.
5. Use **oxidation numbers** in writing the names and determining the formulae of common inorganic compounds.
6. Use **valencies** to determine the formulae of both inorganic and organic compounds.
7. Calculate
   (a) **molar masses**,
   (b) the number of particles in a given amount of a substance,
   (c) **percentage composition** of elements present in a compound,
   (d) **empirical formulae**,
   (e) concentrations of solutions.
8. Write **balanced chemical equations**.
9. Interpret chemical equations.
10. Perform calculations using chemical equations.

4.1
Stoichiometry
and Nomenclature

INTRODUCTION

Stoichiometry is the study of the quantitative composition of chemical substances and also the quantitative changes that take place during chemical reactions. The word 'stoichiometry' was coined in 1792. It derives from two Greek words: *stoicheion* meaning 'elementary constituents' and *metrein* meaning 'to measure.' Much of the chemistry of the eighteenth and nineteenth centuries was concerned with determining the weight ratio of elements in compounds. This work led to our modern knowledge of atomic masses and also led to the discovery of new elements.

The Stoichiometry of Slimming

The largest single source of energy in most plants and animals is the tricarboxylic acid cycle (**TCA cycle**). This is also known as the citric acid cycle or Krebs cycle. The cycle operates in the mitochondria of animal cells.

During the cycle, 'energy-rich' phosphate esters are produced. One of the most important of these is adenosine triphosphate (or ATP for short). The ATP is converted to adenosine diphosphate (ADP), releasing energy. The ADP is then recharged by adding another phosphate to form ATP. Most of the energy needed by the body for functions such as muscle movement come from these processes.

The complete oxidation of one molecule of glucose to form carbon dioxide and water provides sufficient energy to recharge 38 molecules of ADP. In mammals, about 50% of the carbohydrates consumed are normally oxidised to carbon dioxide and water to produce energy and about 30 to 40% are converted to fat.

Stoichiometry is of fundamental importance in modern chemistry. It is the basis of quantitative chemical analysis, for example. In the chemical industry, a knowledge of stoichiometry is essential for the calculation of the yields of chemical products and the efficiency of chemical processes. In the aerospace and transport industries, scientists and engineers use stoichiometric procedures to calculate fuel needs. Stoichiometric procedures also help the biochemist to follow the metabolic processes that take place in organisms. The process of energy production in living cells is an example.

How much fuel is needed to fly this aircraft? A knowledge of the stoichiometry of fuel combustion helps the aerospace engineer to answer this question

Chemical Substances

All matter is composed of chemical substances. Chemical substances may be divided into two types: pure substances and mixtures (*see* figure 4.3).

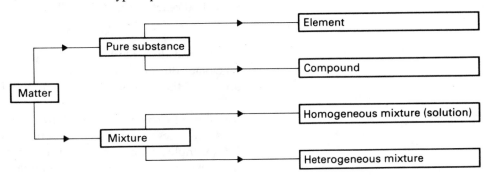

Figure 4.3   Classification of matter

Pure Substances

Pure substances have a fixed composition and well defined chemical and physical properties. They are always homogeneous (*see* below). There are two types of pure substance—elements and compounds.

An **element** is a pure substance which cannot be split up into any simpler pure substance. Elements are usually classified as metals or non-metals (*see* chapter 11).

A **compound** is a pure substance composed of two or more elements combined in fixed and definite proportions. For example, the compound carbon dioxide, CO_2, is composed of the two elements, carbon and oxygen. Carbon dioxide invariably contains 27.37% carbon and 72.73% oxygen, by mass. This is equally true whether we take our sample of carbon dioxide from the North Pole, the South Pole, the Sahara Desert or from the Moon. Thus carbon and oxygen are invariably combined in a fixed and definite proportion in carbon dioxide.

Mixtures

Mixtures are substances consisting of two or more pure substances. They have variable compositions and may consist of a single phase in which case they are called **homogeneous**. A **solution** is a homogeneous mixture. They may also exist in two or more phases. They are then called **heterogeneous**. Soil is an example of a heterogeneous mixture.

Types of Particle

All chemical substances, whether elements, compounds or mixtures, are made up of three types of particles. We have already met these in the previous chapters. They are:

- **atoms**—an atom consists of electrons, neutrons and protons (*see* chapter 1); an atom is characterised by the number of protons in its nucleus, and this number is called its atomic number
- **molecules**—a molecule consists of two or more atoms combined in a whole number ratio
- **ions**—an ion is an electrically charged atom or group of atoms; the charge is due to the gain or loss of electrons

Entities

An **entity** is any chemical or isotopically distinct atom, molecule, ion, radical, complex, etc., capable of identification as a separately distinguishable entity. A set of chemical identical entities is known as a **chemical species**. Chemical names and formulae may refer to either chemical species or entities depending on the context. The term **chemical substance** used above refers to matter capable of preparation in sufficient quantity to show distinct chemical behaviour.

Laws of Chemical Combination

During chemical reactions, one or more substances change their chemical composition to form one or more new substances. The substances undergoing the change are called **reactants** and the substances formed are called **products**. Thus

$$\text{Reactants} \quad \rightarrow \quad \text{Products}$$

The composition of chemical compounds and the changes in composition that take place during chemical reactions are governed by four important laws. These are called the laws of chemical combination.

The Law of Conservation of Matter (Lavoisier, 1774)

This law states that matter cannot be created or destroyed in a chemical reaction. For example, if 4 grams of hydrogen burns in oxygen to form 36 grams of water, the mass of oxygen involved in the reaction must be 32 grams.

Chemical equation: $2H_2(g) + O_2(g) \rightarrow 2H_2O(g)$
Combining weights: 4 g + 32 g → 36 g

The Law of Constant Composition (Proust, 1797).

This states that a pure compound always consists of the same elements combined in the same proportion by weight. This law was implicit in the definition of a compound given above.

The Law of Multiple Proportions (Dalton, 1803)

This states that when two elements A and B combine to form more than one compound, the weights of A which combine with a fixed weight of B are in the proportion of small whole numbers. For example, since $A_r(H) = 1$ and $A_r(O) = 16$, the weight ratio H:O in water, H_2O, is 2:16, and the weight ratio H:O in hydrogen peroxide, H_2O_2, is 2:32 or 1:16. Thus, the weight ratio of hydrogen combined with a fixed weight of oxygen (say 16 grams) in water and in hydrogen peroxide is 2:1.

The Law of Reciprocal Proportions (Richter, 1792)

This states that the weights of two or more substances which chemically react separately with a fixed weight of a third substance are also the weights which react with each other or simple multiples of them. For example, consider the following two reactions.

Chemical equation: $2H_2(g) + O_2(g) \rightarrow 2H_2O(g)$
Combining weights: 4 g + 32 g → 36 g

Chemical equation: $C(s) + O_2(g) \rightarrow CO_2(g)$
Combining weights: 12 g + 32 g → 44 g

On the basis of this information, the law of reciprocal proportions tells us to expect that carbon and hydrogen combine in the weight ratio, 12:4. This is indeed the weight ratio of carbon and hydrogen in methane, CH_4.

Weight and Mass

The four laws of chemical combination form the basis of stoichiometry. They were first stated in the late eighteenth and early nineteenth centuries in terms of combining weights. Nowadays it is common to use the term mass instead of weight. This is because weight is variable. Weight is the gravitational force of attraction exerted on a body. Mass on the other hand is invariable. Mass is a measure of quantity of matter. Mass is measured in kilograms and weight in newtons.

NAMES AND FORMULAE

The formulae (or formulas) and in many cases the names of pure substances (elements and compounds) indicate the stoichiometry of the substance. The naming or **nomenclature** of chemical substances falls broadly into two categories: inorganic nomenclature and organic nomenclature. In this section we shall confine ourselves mainly to inorganic nomenclature and leave a more systematic treatment of organic nomenclature until chapter 17. Both treatments will provide no more than outlines and in no way can be regarded as exhaustive.

Symbols

Every element has a symbol. A full list is given in appendix C. Some symbols derive from the Latin names of the elements (*see* table 4.2). Notice that these are all metals. The first parts of these Latin names used to be used in naming cations. For example, Cu^+ was called the *cuprous* ion and Cu^{2+} the *cupric* ion. We shall look at further examples of this older system below. Latin names are still used for naming metals in some complex anions (*see* below also).

Formulae

The formula of an element or compound which exists as molecules consists of symbols of the element(s) in one molecule and shows the number of atoms of each element in one molecule. This is also called the molecular formula. For example, one molecule of oxygen contains two atoms of oxygen. Its formula is O_2. One molecule of ammonia contains one atom of nitrogen and three atoms of hydrogen. Its formula is NH_3.

The formulae of metallic, ionic or covalent network substances show the simplest ratios of the atoms or ions in the lattice. A formula of an ion shows the ratio of atoms of each element present in the ion. In the fluorite (calcium fluoride) lattice, for example, there are two fluoride ions to every calcium ion. The formula is thus CaF_2.

The **structural formula** of a covalent element or compound shows how the bonds in one molecule are arranged (*see* table 4.3). When all the bonds are fully displayed, the formula is called a **displayed formula**.

Table 4.2 Symbols derived from Latin names

English name	Symbol	Latin name
iron	Fe	ferrum
copper	Cu	cuprum
tin	Sn	stannum
gold	Au	aurum
sodium	Na	natrium
lead	Pb	plumbum
silver	Ag	argentum

Table 4.3 Types of formula

Name	Formula	Empirical formula	Typical structural formula
methane	CH_4	CH_4	
ethene	C_2H_4	CH_2	
benzene	C_6H_6	CH	

The **empirical formula** shows the simplest ratio of the number of atoms or ions in a substance. For network substances, there is no difference between formula and empirical formula. However, for covalent substances consisting of simple molecules, the formula is often a simple multiple of the empirical formula (*see* table 4.3).

INORGANIC NOMENCLATURE

There are two important systems of nomenclature as far as we are concerned. One is that of the International Union of Pure and Applied Chemistry (IUPAC) and the other is that of the Association for Science Education (ASE). Both systems have much in common. However, the IUPAC system relies more heavily on the use of **trivial names**. Many substances have trivial names. These names often are traditional and well-established but do not comply with any system of nomenclature. More importantly, they do not provide information about the chemical composition of the substance. Water (H_2O) and ozone (O_3) are examples of trivial names.

The ASE system relies much more on **systematic names**. These names are based on well defined rules. In this book we shall use ASE nomenclature, although for many substances this will be the same as the IUPAC nomenclature.

The most important rule of nomenclature is that the name of a substance should be unambiguous. For example, magnesium oxide is an unambiguous name since magnesium only forms one oxide, MgO. However, carbon oxide is an ambiguous name since carbon forms two oxides, CO and CO_2. These are thus named carbon monoxide and carbon dioxide respectively.

Systematic names—particularly in IUPAC nomenclature—rely heavily on the use of **numerical prefixes** to indicate the number of atoms or number of groups in a substance (*see* table 4.4). The prefix *mono* is often dropped. For example, nitrogen monoxide, NO, is also called nitrogen oxide.

A crystal of
bluestone
blue vitriol
cupric sulphate
copper sulphate
hydrated copper sulphate
copper sulphate pentahydrate
copper(II) sulphate-5-water

All these names are or have been used for the compound with the formula $CuSO_4 \cdot 5H_2O$.

Table 4.4 Numerical prefixes

Number of atoms	Prefix
1	mono
2	di
3	tri
4	tetra
5	penta
6	hexa
7	hepta
8	octa
9	nona
10	deca

Oxidation Numbers

These are used extensively in both ASE and IUPAC nomenclature. The oxidation number of an element is equal to its combining power with oxygen. Elements in both ionic and covalent compounds can have oxidation numbers. There are a number of rules for calculating oxidation numbers. The three most important are:

The terms *oxidation number* and *oxidation state* have virtually the same meaning.

1. The oxidation number of
   uncombined elements $= 0$
   one-atom ions $=$ charge
   each hydrogen atom $= +1$
   each oxygen atom $= -2$
2. The sum of the oxidation numbers of the elements in a compound is zero.
3. For oxoanions (also known as oxyanions) the oxidation number of the element in the ion is given by
   $(2 \times$ no. of oxygen atoms$)$ − numerical value of charge

As a general rule, when an element has more than one oxidation number the numerical value is written as a Roman numeral after the name. Examples are shown in table 4.5. Oxoanions are given the numerical value of the oxidation number of the element which combines with oxygen in the ion.

For algebraic manipulations, oxidation numbers are normally written using Arabic numerals. In nomenclature, Roman numerals are used. For example

Ion	MnO_4^-
Oxidation number of Mn	$(2 \times 4) - 1 = +7$
Name of ion	manganate(VII)

Table 4.5 Oxidation numbers

	Formula	Oxidation number	
Elements			
hydrogen	H_2	0	
carbon	C	0	
chlorine	Cl_2	0	
nitrogen	N_2	0	
copper	Cu	0	
Compounds			
methane	CH_4	C = −4	H = +1
carbon dioxide	CO_2	C = +4	O = −2
hydrogen chloride	HCl	H = +1	Cl = −1
hydrogen sulphide	H_2S	H = +1	S = −2
sulphur dioxide	SO_2	S = +4	O = −2
sulphur trioxide	SO_3	S = +6	O = −2
Cations			
sodium	Na^+	+1	
copper(i)	Cu^+	+1	
copper(ii)	Cu^{2+}	+2	
iron(ii)	Fe^{2+}	+2	
iron(iii)	Fe^{3+}	+3	
Anions			
oxide	O^{2-}	−2	
chloride	Cl^-	−1	
carbonate	CO_3^{2-}	C = +4	
chlorate(v)	ClO_3^-	Cl = +5	
chlorate(vii)	ClO_4^-	Cl = +7	
sulphate	SO_4^{2-}	S = +6	
manganate(vii)	MnO_4^-	Mn = +7	

Names of Elements

Elements are not named in any systematic way. They are thus trivial and provide no chemical information about the element (*see* table 4.6).

The number of atoms in one molecule of a simple covalent element is often called the **atomicity** of the element. For example, all noble gases are monatomic. Their molecules consist of one atom only. Hydrogen, oxygen, nitrogen and chlorine are all diatomic. Their molecules consist of two atoms.

All the **isotopes** (*see* section 1.1) of an element have the same name. The only exception is hydrogen which has a name for each isotope:

Isotope:	protium	deuterium	tritium
Symbol:	^1_1H	^2_1H (or D)	^3_1H (or T)

Many **allotropes** (*see* section 3.2) have trivial names (*see* table 4.7).

Table 4.6 Symbols and formulae of five elements

Name	Symbol	Formula
neon	Ne	Ne
carbon	C	C
copper	Cu	Cu
oxygen	O	O_2
sulphur	S	S_8

Table 4.7 Names of allotropes

Element	Allotropes
carbon	diamond graphite
phosphorus	white phosphorus, P_4 red phosphorus black phosphorus
oxygen	oxygen, O_2 ozone, O_3

Naming Binary Compounds

A binary compound contains two elements. Binary compounds can be named systematically as follows. The name of the element closer to the bottom or to the left-hand side of the Periodic Table is usually given first. The suffix '-ide' is added to the name of the second element. For covalent compounds, a prefix is often added to the first and/or second name to indicate the ratio of atoms in the compound. For example

sulphur trioxide	SO_3
dinitrogen tetroxide	N_2O_4
diiodine hexachloride	I_2Cl_6

Table 4.8 Many hydrides have trivial names

water	H_2O
ammonia	NH_3
methane	CH_4
phosphine	PH_3
hydrazine	N_2H_4

Table 4.9 Modern and old names for some d-block cations

Ion	Modern name	Old name
Fe^{2+}	iron(II)	ferrous
Fe^{3+}	iron(III)	ferric
Cu^+	copper(I)	cuprous
Cu^{2+}	copper(II)	cupric
Sn^{2+}	tin(II)	stannous
Sn^{4+}	tin(IV)	stannic
Cr^{2+}	chromium(II)	chromous
Cr^{3+}	chromium(III)	chromic

Table 4.10 Names of polyatomic cations

H_3O^+	oxonium
NH_4^+	ammonium
$CH_3NH_3^+$	methylammonium
$C_6H_5NH_3^+$	phenylammonium

Table 4.11 Names of anions

Cl^-	chloride
O^{2-}	oxide
O_2^{2-}	peroxide
I_3^-	triiodide
OH^-	hydroxide
S^{2-}	sulphide
CN^-	cyanide

Table 4.12 Sulphur analogues of oxoanions

Oxoanion	Sulphur analogue
sulphate, SO_4^{2-}	thiosulphate, $S_2O_3^{2-}$
cyanate, NCO^-	thiocyanate, NCS^-

Alternatively, where a binary compound contains an element which may exist in more than one **oxidation state**, the oxidation number is indicated. For example

iron(II) chloride	$FeCl_2$
iron(III) chloride	$FeCl_3$

The hydrides, particularly those of non-metals, tend to have trivial names (*see* table 4.8).

Naming Cations

Cations of elements which only form one stable ion are given the same name as the element. All elements in Groups I and II fall into this category. For example

Na^+	sodium
Mg^{2+}	magnesium

d-Block elements often form more than one stable cation. Such ions take the name of the element followed by a Roman numeral in parentheses indicating the oxidation number of the ion.

The **older system** of naming such ions is still used nowadays. This distinguishes between the two most common oxidation states. The ion with the higher oxidation number is given the suffix *-ic* and the one with the lower oxidation number *-ous*. These are often but not always added to a segment of the Latin name of the element (*see* table 4.9).

Polyatomic cations often have the suffix *-ium* (*see* table 4.10).

Naming Anions

Monatomic and some polyatomic anions end in *-ide* (*see* table 4.11).

The most common oxonanions end in *-ate*. Those with an element in a low oxidation state end in *-ite*. The prefix *hypo-* is used for an ion with an element in a very low oxidation state and the prefix *per-* for an ion with an element in a high oxidation state. The prefix *thio-* is also used for sulphur analogues of oxonanions (*see* table 4.12).

In the ASE system of nomenclature, the oxidation number of an element in an oxoanion is sometimes used instead of or together with a prefix or suffix. The list shown in table 4.13 is not exhaustive. Other examples of polyatomic anions will be encountered in this book.

Table 4.13 Polyatomic anions

Ion	Oxidation state	Name	ASE name
NO_2^-	+3	nitrite	nitrite
NO_3^-	+5	nitrate	nitrate
ClO^-	+1	hypochlorite	chlorate(I)
ClO_2^-	+3	chlorite	chlorate(III)
ClO_3^-	+5	chlorate	chlorate(V)
ClO_4^-	+7	perchlorate	chlorate(VII)
MnO_4^-	+7	permanganate	manganate(VII)
CrO_4^{2-}	+6	chromate	chromate(VI)
$Cr_2O_7^{2-}$	+6	dichromate	dichromate(VI)
SO_3^{2-}	+4	sulphite	sulphite
SO_4^{2-}	+6	sulphate	sulphate

Names and Formulae of Complex ions

A complex ion consists of a central atom to which are bonded atoms, ions or groups of atoms. These are called **ligands**.

The formula of the complex ion is enclosed in square brackets. The charge on the ion is put outside the closing bracket. Inside the bracket the symbol of the central atom is placed first. This is followed by the formulae of anionic ligands and then neutral ligands in the alphabetical order of the donor atom (*see* chapter 14). Polyatomic ligands are placed in curved brackets.

In naming complex ions, the names of the ligands are cited first. They are placed in alphabetical order—ignoring the numerical prefixes. The name of the complex ion terminates with the name of the metal together with the oxidation state of the metal in parentheses. For complex cations, the English name of the metal is used (*see* table 4.14). For complex anions, a segment of the English or Latin name of the metal is used together with the suffix -*ate*.

Table 4.14 Names of complex ions

$[Cu(NH_3)_4]^{2+}$	tetraamminecopper(II)
$[Fe(SCN)(H_2O)_5]^{2+}$	pentaaquathiocyanatoiron(III)
$[Fe(CN)_6]^{4-}$	hexacyanoferrate(II)
$[FeF_4(H_2O)_2]^-$	diaquatetrafluoroferrate(III)

The names and formulae of some of the more common ligands are shown in table 4.15. Table 4.16 shows the names of metals used in anionic complexes.

Table 4.15 Ligands

	Formula	Name
neutral	H_2O	aqua
	NH_3	ammine
	CO	carbonyl
	NO	nitrosyl
anionic	F^-	fluoro
	Cl^-	chloro
	Br^-	bromo
	I^-	iodo
	S^{2-}	sulpho
	OH^-	hydroxo
	CN^-	cyano
	SCN	thiocyanato
	NO_2^-	nitro
	$S_2O_3^{2-}$	thiosulphato

Table 4.16 Metals in anionic complexes

Element	Name in complex (based on English name)
Ti	titanate
V	vanadate
Cr	chromate
Mn	manganate
Co	cobaltate
Zn	zincate
Hg	mercurate
B	borate
Al	aluminate
Ge	germanate

Element	Name in complex (based on Latin name)
Fe	ferrate
Ni	niccolate (ASE: nickelate)
Cu	cuprate
Ag	argentate
Au	aurate
Sn	stannate
Pb	plumbate

Naming Salts

The name of a salt consists of the name of the cation followed by the name of the anion. The prefix 'hydrogen' is attached to the name of the anion for acid salts. Numerical prefixes are not used in naming salts except for some acid salts. For double salts, the cations are given in alphabetical order. For basic salts, the anions are given in alphabetical order (*see* table 4.17).

Salt hydrates can be named in two ways. When one or more water molecules are known to be coordinated to a central atom in a complex ion, the system for naming

Table 4.17 Names of salts

Salt	Type of salt	Name
KI	simple	potassium iodide
$Mg(NO_3)_2$	simple	magnesium nitrate
$NaHSO_4$	acid	sodium hydrogensulphate
NaH_2PO_4	acid	sodium dihydrogenphosphate(V)
$KNaCO_3$	double	potassium sodium carbonate
$MgCl(OH)$	basic	magnesium chloride hydroxide

complex ions described above can be used. For the more common salt hydrates, the extent of hydration is indicated by the word 'hydrate' preceded by a numerical prefix. However, the ASE system of nomenclature prefers the use of an Arabic numeral and the word 'water'. Thus, $CuSO_4 \cdot 5H_2O$ is copper(II) sulphate pentahydrate or, in the ASE system, copper(II) sulphate-5-water.

Naming Acids

Acids are named after their anions. The acids of anions with the suffix *-ide* become *hydro -ic*; those with suffix *-ate* become *-ic* and those with the suffix *-ite* become *-ous*. The ASE system of nomenclature also uses oxidation numbers for the less common oxoacids (*see* table 4.18).

Thioacids are sulphur analogues of oxoacids. Thus

$$H_2SO_4 \qquad \text{sulphuric acid}$$
$$H_2S_2O_3 \qquad \text{thiosulphuric acid}$$

Table 4.18 Names of acids

HCl	hydrochloric acid
HNO_2	nitrous acid
HNO_3	nitric acid
H_2SO_3	sulphurous acid
H_2SO_4	sulphuric acid
H_3PO_4	phosphoric(v) acid
$HClO_3$	chloric(v) acid

Formula of Ionic Compounds

The formula of an ionic compound shows the simplest ratio of ions in the ionic lattice. Since the lattice is electrically neutral, the formula must show this neutrality (*see* table 4.19).

Table 4.19 Formulae of ionic compounds

Ionic compound	Ratio of ions	Formula
sodium chloride	$Na^+ : Cl^-$ 1:1	NaCl
magnesium chloride	$Mg^{2+} : Cl^-$ 1:2	$MgCl_2$
iron(III) nitrate	$Fe^{3+} : NO_3^-$ 1:3	$Fe(NO_3)_3$
sodium carbonate	$Na^+ : CO_3^{2-}$ 2:1	Na_2CO_3
sodium phosphate(v)	$Na^+ : PO_3^{3-}$ 3:1	Na_3PO_4
magnesium oxide	$Mg^{2+} : O^{2-}$ 1:1	MgO
aluminium sulphate	$Al^{3+} : SO_4^{2-}$ 2:3	$Al_2(SO_4)_3$
iron(II) phosphate	$Fe^{2+} : PO_4^{3-}$ 3:2	$Fe_3(PO_4)_2$

Valency

When writing the formulae of both covalent molecules and ionic compounds, it is often useful to use the valency of an atom, ion or group of atoms. Valency is the number of electrons used by an atom in bonding. For a cation, the valency is the number of electrons lost in forming the ion. For an anion, the valency is the number of electrons gained in forming the ion. The valency of an atom in a covalent molecule is the number of electrons shared by the atom in forming the bond.

Valencies are always small whole numbers. Some common valencies are shown in table 4.20. Inert gases such as helium have zero valencies. They do not normally form compounds. Some elements, particularly d-block metals, have multiple valencies (*see* table 4.21). Table 4.22 gives examples of how valencies can be used to find the formulae of simple compounds.

Table 4.20 Some common valencies

Metals			Non-metals		
Name	Symbol	Valency	Name	Symbol	Valency
lithium	Li	1	helium	He	0
sodium	Na	1	neon	Ne	0
potassium	K	1	hydrogen	H	1
silver	Ag	1	chlorine	Cl	1
calcium	Ca	2	bromine	Br	1
magnesium	Mg	2	iodine	I	1
zinc	Zn	2	oxygen	O	2
aluminium	Al	3	nitrogen	N	3
			carbon	C	4

Table 4.21 Multiple valencies

Element	Symbol	Valencies
copper	Cu	1 or 2
iron	Fe	2 or 3
lead	Pb	2 or 4
phosphorus	P	3 or 5
sulphur	S	2, 4 or 6
tin	Sn	2 or 4

Table 4.22 Using valencies to find formulae

Valencies	Name	Formula
Fe = 3, I = 1	iron(III) iodide	FeI_3
Zn = 2, Br = 1	zinc bromide	$ZnBr_2$
Cu = 1, O = 2	copper(I) oxide	Cu_2O

Valencies sometimes have the same numerical values as oxidation numbers—but not always. Carbon, for example, always has a valency of 4 whereas its oxidation number varies from −4 to +4. In organic chemistry it is thus more convenient to describe carbon in terms of its valency than its oxidation numbers. Table 4.23 compares the valency and oxidation numbers of carbon in five compounds.

Table 4.23 Valency and oxidation numbers of carbon

Name of compound	Formula	Valency of carbon	Oxidation number of carbon
methane	CH_4	4	−4
chloromethane	CH_3Cl	4	−2
dichloromethane	CH_2Cl_2	4	0
trichloromethane	$CHCl_3$	4	+2
tetrachloromethane	CCl_4	4	+4

Notes: (a) The valencies of both chlorine and hydrogen are 1.
(b) The oxidation numbers of chlorine and hydrogen in these compounds are −1 and +1 respectively.

Non-Stoichiometric Compounds

Most compounds are stoichiometric. This means that the ratio of atoms or ions in the compound is a simple whole number ratio. However, some compounds do not have constant integral ratios. The exact ratio may vary depending on the method of preparation. The following are examples.

iron(II) sulphide	$Fe_{1.1}S$ to $FeS_{1.1}$
iron(II) oxide	$Fe_{1.06-1.19}O$
copper(I) sulphide	$Cu_{\sim1.8}S$

These compounds are often called non-stoichiometric compounds.

The non-stoichiometry of these compounds arises from defects in their crystal lattices. In some cases atoms take up interstitial positions between the ions in a crystal. In other cases atoms take the place of ions in regular lattice positions. This results in ionic vacancies in the lattice, thus preserving electrical neutrality.

Strictly speaking non-stoichiometric compounds are not compounds since they do not have a constant composition. For this reason, the term non-stoichiometric solid is often preferred.

Berthollides

Non-stoichiometric compounds are sometimes called berthollides after Claude Louis Berthollet. Berthollet thought that the composition of compounds varies continuously within limits. This was finally refuted by Proust who, in 1799, put forward his famous law of constant composition.

SUMMARY

1. There are two types of pure substance: **elements** and **compounds**.
2. **Mixtures** may be **homogeneous** or **heterogeneous**.
3. Matter is made up of three types of particles: **atoms, molecules** and **ions**.
4. The four **laws of chemical combination** are
   (a) the **law of conservation of matter**,
   (b) the **law of constant composition**,
   (c) the **law of multiple proportions**,
   (d) the **law of reciprocal proportions**.
5. Common compounds often have both **trivial** and **systematic** names.
6. The name of a substance should be unambiguous.
7. A **formula** shows the simplest ratio of atoms present in a molecule, ion or network substance.
8. The **oxidation number** of an element is equal to its combining power with oxygen.
9. **Valency** is the number of electrons used by an atom or ion in bonding.

4.2
Stoichiometric
Calculations

THE MOLE

Amount of Substance

The amount of substance is a physical quantity. It has the symbol n and its base unit is the mole. The mole requires careful definition. The following definition and passage is taken from *Chemical Nomenclature, Symbols and Terminology* published by the Association for Science Education (ASE) in 1985:

> The mole is the amount of substance of a system which contains as many elementary entities as there are carbon atoms in 0.012 kilogram of carbon-12. The elementary entities must be specified and may be atoms, molecules, ions, electrons, other particles, or specified groups of such particles.
>
> The amount of substance is proportional to the number of specified elementary entities of that substance.
>
> The proportionality factor is the same for all substances; its reciprocal is L, the Avogadro constant $[(6.022\ 045 \pm 0.000\ 031) \times 10^{23}\ mol^{-1}]$. Note that the Avogadro constant is a physical quantity with the dimension [amount-of-substance]$^{-1}$, and is not a pure number. The term 'Avogadro number' should not be used.

For example (taking the Avogadro constant to be $6.02 \times 10^{23}\ mol^{-1}$):

one mole of hydrogen atoms	is 6.02×10^{23} hydrogen atoms
one mole of oxygen molecules, O_2	is 6.02×10^{23} oxygen molecules, O_2
one mole of chloride ions, Cl^-	is 6.02×10^{23} chloride ions, Cl^-
one mole of electrons, e^-	is 6.02×10^{23} electrons, e^-

By turning these statements around we can show that the Avogadro constant is a physical quantity and not a pure number. For example, there are 6.02×10^{23} oxygen molecules, O_2, per mole of oxygen molecules, O_2.

Notice that in each of these examples we are careful to specify exactly the entities or particles to which the amount of substance refers. This is essential if the statement is to be unambiguous. For example, 'one mole of hydrogen' is ambiguous since it could mean either 'one mole of hydrogen atoms, H' or 'one mole of hydrogen molecules, H_2'.

Molar Mass

Mass is a physical quantity. Its base unit is the kilogram although in chemical calculations it is often more convenient to use grams.

Molar mass is the mass per unit amount of substance. It is normally expressed as the number of grams per mole.

The molar mass of atoms of an element is related to the relative atomic mass (*see* section 1.1) of the element by the equation

$$M = A_r \text{ g mol}^{-1}$$

where M denotes the molar mass of atoms of the element and A_r is the relative atomic mass of the element. This expression can also be used to relate the molar masses of monatomic ions to the relative atomic masses of the elements. For example, since $A_r(O) = 16.0$ and $A_r(Ca) = 40.1$

for oxygen atoms, O	$M(O) = 16.0$ g mol^{-1}
and for oxygen ions, O^{2-}	$M(O^{2-}) = 16.0$ g mol^{-1}
for calcium atoms, Ca	$M(Ca) = 40.1$ g mol^{-1}
and for calcium ions, Ca^{2+}	$M(Ca^{2+}) = 40.1$ g mol^{-1}

The molar mass of a molecular substance has the same numerical value as its relative molecular mass (*see* section 4.1). Thus

$$M = M_r \text{ g mol}^{-1}$$

where M_r is the relative molecular mass. This expression can also be used to calculate the molar masses of polyatomic ions and ionic compounds. The values of M_r for polyatomic ions and ionic compounds are calculated in the same way as for molecular compounds.

EXAMPLE

What is the molar mass of carbonate ions, CO_3^{2-}?

SOLUTION

$$
\begin{aligned}
M_r(CO_3^{2-}) &= A_r(C) + 3A_r(O) \\
&= 12.0 + (3 \times 16.0) \\
&= 60.0
\end{aligned}
$$

Therefore $M(CO_3^{2-}) = 60.0$ g mol^{-1}

Note that neither A_r nor M_r have units.

The mass of several moles of a substance can be calculated using the following relationship:

$$m = n \times M$$

where m denotes the mass of the substance in units of grams, n denotes the amount of substance in units of moles and M denotes the molar mass of the substance in units of grams per mole.

Similarly, the mass of a substance can be converted to the amount of substance using the following relationship:

$$n = \frac{m}{M} \qquad (1)$$

EXAMPLE

Calculate the amount of methane molecules, CH_4, in 8.0 grams of methane, CH_4.

SOLUTION

$$
\begin{aligned}
m(CH_4) &= 8.0 \text{ g} \\
M_r(CH_4) &= A_r(C) + 4A_r(H) \\
&= 12.0 + (4 \times 1.0) \\
&= 16.0
\end{aligned}
$$

Therefore $M(CH_4) = 16.0$ g mol^{-1}

We substitute these values into equation (1) above. Thus

$$n(CH_4) \quad = \frac{8.0 \text{ g}}{16.0 \text{ g mol}^{-1}}$$

$$= 0.5 \text{ mol}$$

Number of Particles

The number of particles of a substance can be calculated from either the amount of the substance or the mass of the substance using the following equations:

$$N = n \times L$$

$$= \frac{m}{M} \times L$$

where N denotes the number of particles and L is the Avogadro constant.

EXAMPLE

Calculate the number of water molecules, H_2O, in 100 g of water.

SOLUTION

$$N(H_2O) = \frac{m(H_2O)}{M(H_2O)} \times L$$

$$= \frac{100 \text{ g}}{18.0 \text{ g mol}^{-1}} \times (6.02 \times 10^{23} \text{ mol}^{-1})$$

$$= 33.4 \times 10^{23}$$

Percentage Composition

We saw in the previous section that the formula of a molecular compound shows the number of atoms of each element in one molecule of the compound. The formula also shows the number of moles of atoms of each element in one mole of molecules. For an ionic compound, the formula shows not only the ratio of the different ions present in the lattice but also the ratio of moles of the different ions present in the compound. The moles in these ratios can be converted to masses as shown above. In this way the formula of a compound can be used to calculate the percentage composition by mass of a compound.

EXAMPLE

Calculate the percentage composition of benzene, C_6H_6.

SOLUTION

The formula C_6H_6 shows that one mole of benzene molecules contains 6 moles of carbon atoms and 6 moles of hydrogen atoms. Thus

$$\% \text{C in } C_6H_6 = \frac{\text{mass of 6 moles of C atoms}}{\text{mass of 1 mole of } C_6H_6 \text{ molecules}} \times 100$$

$$= \frac{6 \times 12.0 \text{ g}}{1 \times 78.0 \text{ g}} \times 100$$

$$= 92.3$$

$$\% \text{H in } C_6H_6 = \frac{\text{mass of 6 moles of H atoms}}{\text{mass of 1 mole of } C_6H_6 \text{ molecules}} \times 100$$

$$= \frac{6 \times 1.0 \text{ g}}{1 \times 78.0 \text{ g}} \times 100$$

$$= 7.7$$

Empirical Formula

This shows the simplest ratio of atoms of each element or the simplest ratio of the different ions present in a compound.

The empirical formula can be determined from the composition of a compound. There are two stages in this determination:

Stage 1: the number of moles of each element present in a known mass of the compound is found;

Stage 2: the simplest whole number ratio of moles is found; this gives the empirical formula.

EXAMPLE
A sugar was found to contain 40% by mass of carbon, 6.7% by mass of hydrogen and 53.3% by mass of oxygen. What is its empirical formula?

SOLUTION
The percentages show that 100 grams of the sugar contain 40 grams of carbon, 6.7 grams of hydrogen and 53.3 grams of oxygen. We can thus find the empirical formula as follows:

Stage 1:

		carbon	hydrogen	oxygen
mass (m)	=	40 g	6.7 g	53.3 g

since $n = \dfrac{m}{M}$,

$$\text{no. of moles } (n) = \frac{40 \text{ g}}{12.0 \text{ g mol}^{-1}} \quad \frac{6.7 \text{ g}}{1.0 \text{ g mol}^{-1}} \quad \frac{53.3 \text{ g}}{16.0 \text{ g mol}^{-1}}$$

$$= 3.33 \text{ mol} \qquad 6.7 \text{ mol} \qquad 3.33 \text{ mol}$$

Stage 2

$$\begin{array}{l}\text{simplest whole} \\ \text{number ratio}\end{array} = \frac{3.33 \text{ mol}}{3.33 \text{ mol}} \qquad \frac{6.7 \text{ mol}}{3.33 \text{ mol}} \qquad \frac{3.33 \text{ mol}}{3.33 \text{ mol}}$$

$$= \qquad 1 \qquad\qquad 2 \qquad\qquad 1$$

The empirical formula of the sugar is thus CH_2O.

Molar Constants and Molar Quantities

The Avogadro constant is a **molar constant**. We have already met one other molar constant. That was the molar gas constant (*see* section 3.1). This is equal to 8.314 J K^{-1} mol^{-1}. There is one other molar constant we shall need to use. This is the Faraday constant. This is the electric charge associated with one mole of electrons or one mole of singly charged ions (*see* chapter 10). It has a value of 96 480 coulombs per mole, that is 9.648×10^4 C mol^{-1}.

Molar mass is a **molar quantity**. Another example is the molar volume of a gas, which at s.t.p. is 22.4 dm^3 mol^{-1}. We shall meet some more molar quantities in the next chapter when we come to consider thermodynamic quantities. For example, molar heat capacity is an example of a thermodynamic molar quantity. It has the units J K^{-1} mol^{-1}.

Note that molar constants and molar quantities almost invariably include 'mol^{-1}' in their units. This is because the word molar means 'divided by the amount of substance' or 'per unit amount of substance'. In chemical calculations this unit is almost invariably the mole.

Composition of Homogeneous Mixtures

Concentration is the amount of substance per unit volume of solution. It is usually expressed as the number of moles of solute in one litre (that is, one cubic decimetre) of solution:

$$[X] = \frac{n(X) \text{ mol}}{V \text{ dm}^3} \qquad (2)$$

where $n(X)$ denotes the amount of solute X, V denotes the volume of solution and [X] denotes the concentration of solute X in solution. [X] is sometimes called the 'molar concentration' of X. The term molarity is not recommended by IUPAC or ASE.

> A solution with a concentration of 0.5 mol dm^{-3} is often referred to as a 0.5 M solution. M stands for the word **molar**.

EXAMPLE

Calculate the concentration of a solution obtained by dissolving 0.53 g of pure anhydrous sodium carbonate, Na_2CO_3, in water to make 250 cm^3 of solution.

SOLUTION

$$M(Na_2CO_3) = 106 \text{ g mol}^{-1}$$

Using equation(1)

$$n(Na_2CO_3) = \frac{0.53 \text{ g}}{106 \text{ g mol}^{-1}}$$

$$= 5.0 \times 10^{-3} \text{ mol}$$

Since

$$250 \text{ cm}^3 = 0.250 \text{ dm}^3$$

and using equation(2)

$$[Na_2CO_3(aq)] = \frac{5.0 \times 10^{-3} \text{ mol}}{0.250 \text{ dm}^3}$$

$$= 2.0 \times 10^{-2} \text{ mol dm}^{-3} \text{ (or } 2.0 \times 10^{-2} \text{M)}$$

The term **mass concentration** is also used in chemical calculations. Using grams as units of mass and dm^3 as volume units, it is given by

$$[X] = \frac{m \text{ g}}{V \text{ dm}^3}$$

In the above example, the mass concentration of the sodium carbonate solution is 2.12 g dm^{-3}.

The term **molality** is also used for the composition of solutions (*see* chapter 6). It is not a concentration term since volume is not involved. Molality can be expressed as the number of moles of solute in one kilogram of solvent. Its units are thus mol kg^{-1}.

The composition of a homogeneous mixture can also be expressed in terms of the **mole fraction** of the components in the mixture (*see* section 3.1). The mole fraction of component A in a mixture of A and B is given by

$$x_A = \frac{n_A}{n_A + n_B}$$

where n_A denotes the number of moles of A, n_B denotes the number of moles of B and x_A denotes the mole fraction of A.

CHEMICAL EQUATIONS

A chemical equation is a representation of a reaction in terms of the symbols and formulae of the elements and compounds involved. The relative amounts of reactants and products, measured in moles, are indicated by the coefficients in the balanced equation. These coefficients are sometimes called the **stoichiometric**

coefficients. Nowadays, there is a growing tendency to include the physical states of the reactants and products in equations. These are

(g)	gas
(l)	liquid
(s)	solid
(aq)	aqueous, meaning dissolved in water

Chemical equations describing reactions can be deduced experimentally by identifying the reactants and products and measuring the relative amounts of each reactant and product involved in the reaction.

Writing a Chemical Equation

This involves four stages:

Stage 1:   The word equation is written. For example

$$Magnesium + Carbon\ dioxide \rightarrow Magnesium\ oxide + Carbon$$

Stage 2:   The words are replaced by the formulae of the reactants and products

$$Mg + CO_2 \rightarrow MgO + C$$

Stage 3:   The equation is balanced

$$2Mg + CO_2 \rightarrow 2MgO + C$$

This is called the balanced equation or stoichiometric equation. Balancing is required by the law of conservation of matter. Thus, in our example, no magnesium, carbon or oxygen atoms can be created or destroyed. The number of atoms of each element on either side of the equation must therefore be equal.

Stage 4:   The physical states are finally put in

$$2Mg(s) + CO_2(g) \rightarrow 2MgO(s) + C(s)$$

Types of Chemical Equation

Let us consider the following balanced equation:

$$Fe(s) + CuSO_4(aq) \rightarrow FeSO_4(aq) + Cu(s)$$

This equation represents the reaction system as a whole. The reaction can also be represented by an **ionic equation**:

$$Fe(s) + Cu^{2+}(aq) \rightarrow Fe^{2+}(aq) + Cu(s)$$

In this equation the sulphate ions, SO_4^{2-}, have been removed since they do not take part in the reaction. They are called **spectator ions**.

The reaction between iron and copper(II) is an example of a redox reaction (*see* chapter 10). It can be split up to show separately the reduction and the oxidation taking place in the overall reaction:

Reduction	$Cu^{2+}(aq) + 2e^- \rightarrow Cu(s)$
Oxidation	$Fe(s) \rightarrow Fe^{2+}(aq) + 2e^-$

These two equations are called **half-equations**. They are particularly useful in electrochemistry in showing the processes that take place at electrodes (*see* chapter 10).

Interpreting Chemical Equations

The following simple stoichiometric equation can be interpreted in two ways:

$$H_2(g) + Br_2(g) \rightarrow 2HBr(g)$$

First of all we can take this equation to mean that, overall, one mole of hydrogen molecules, H_2, reacts with one mole of bromine molecules, Br_2, to produce two moles of hydrogen bromide molecules, HBr. This is sometimes called the **molar interpretation**.

We could also say that, **overall**, one molecule of hydrogen, H_2, reacts with one molecule of bromine, Br_2, to produce two molecules of hydrogen bromide, HBr. This is sometimes called the **molecular interpretation**.

Both the molar and molecular interpretations are valid. However, we could not conclude from the equation that one molecule of hydrogen, H_2, collides with one molecule of bromine, Br_2, to produce two molecules of HBr. This is because the reaction, like most other reactions, takes place in a number of steps. These steps are called **the reaction mechanism** (*see* chapter 9). In the example above, the steps include

$$Br_2 \rightarrow Br^\bullet + Br^\bullet$$
$$Br^\bullet + H_2 \rightarrow HBr + H^\bullet$$
$$H^\bullet + Br_2 \rightarrow HBr + Br^\bullet$$

This is, in fact, an example of a **chain reaction** involving **reaction intermediates** called **radicals** (*see* chapter 9). Other reaction steps and side reactions are also involved in this reaction mechanism. The stoichiometric equation thus only shows the overall reaction. It does not provide information about the reaction mechanism.

Calculations Using Chemical Equations

Chemical equations are the starting points for many types of calculations in chemistry. In this section we shall consider a few examples of these calculations. We shall meet further examples later in the book.

Calculating the Masses of Reactants and Products

We have seen that a balanced chemical equation shows the relative amounts of reactants and products, measured in moles, involved in a reaction. Once we know the amounts involved we can calculate the masses of reactants and products.

EXAMPLE
Calculate the mass of silver chloride formed when excess sodium chloride solution is added to a solution containing 0.1 moles of silver, Ag^+.

SOLUTION
The first step in any such calculation is to write the equation for the reaction:

$$Ag^+(aq) + Cl^-(aq) \rightarrow AgCl(s)$$

Since excess chloride ions are used, we can assume that all the Ag^+ ions are converted to AgCl. The equation shows that one mole of Ag^+ ions produces one mole of AgCl(s). We can now calculate the mass of AgCl(s) produced as follows:

$$\text{amount of } Ag^+ \text{ used} = 0.1 \text{ mol}$$

Therefore

$$\text{amount of AgCl produced} = 0.1 \text{ mol}$$

Since

$$M(AgCl) = 143.3 \text{ g mol}^{-1}$$
$$\text{mass of AgCl produced} = (0.1 \text{ mol}) \times (143.3 \text{ g mol}^{-1})$$
$$= 14.33 \text{ g}$$

Finding the Concentrations of Solutions

Calculations based on stoichiometric equations form the basis of quantitative chemical analysis. The following is an example of how the concentration of solution can be determined from the mass of a product formed in a reaction. This form of quantitative chemical analysis is known as **gravimetric analysis**.

EXAMPLE

Sufficient potassium iodide solution was added to a lead(II) nitrate solution to precipitate all the lead as lead (II) iodide. The mass of lead(II) iodide produced was 2.305 g. The volume of lead(II) nitrate solution was 250 cm³. What was the concentration of lead(II) nitrate?

SOLUTION

We have already met the equation for the reaction:

$$Pb(NO_3)_2(aq) + 2KI(aq) \rightarrow PbI_2(s) + 2KNO_3(aq)$$

The equation shows that one mole of lead(II) iodide can be produced from one mole of lead(II) nitrate. Thus

$$\text{mass of } PbI_2 \text{ produced} = 2.305 \text{ g}$$

Since

$$M(PbI_2) = 460.99 \text{ g mol}^{-1}$$

$$\text{amount of } PbI_2 \text{ produced} = \frac{2.305 \text{ g}}{460.99 \text{ g mol}^{-1}}$$

$$= 0.005 \text{ mol}$$

Therefore

$$\text{amount of } Pb(NO_3)_2 \text{ in solution} = 0.005 \text{ mol}$$

Using equation (2) we obtain $[Pb(NO_3)_2(aq)]$

$$= \frac{0.005 \text{ mol}}{0.250 \text{ dm}^3}$$

$$= 0.02 \text{ mol dm}^{-3} (\text{or } 0.02 \text{ M})$$

Calculating the Volumes of Gaseous Reactants and Products

Stoichiometric equations can be used to calculate the volumes of gaseous reactants or products of a reaction.

EXAMPLE

Calculate the maximum volume of CO_2 at s.t.p. that can be obtained when dilute hydrochloric acid is added to 10 grams of calcium carbonate.

SOLUTION

The equation is

$$CaCO_3(s) + 2HCl(aq) \rightarrow CaCl_2(aq) + CO_2(g) + H_2O(l)$$

The equation shows that one mole of CO_2 is produced from one mole of $CaCO_3$. Thus

$$\text{mass of } CaCO_3 = 10 \text{ g}$$

Since

$$M(CaCO_3) = 100 \text{ g mol}^{-1}$$

$$\text{amount of } CaCO_3 = 0.1 \text{ mol}$$

Therefore

$$\text{amount of } CO_2 \text{ produced} = 0.1 \text{ mol}$$

At s.t.p.

$$\text{molar volume of } CO_2 = 22.4 \text{ dm}^3 \text{ mol}^{-1}$$

Therefore

$$\text{volume of } CO_2 \text{ produced at s.t.p.} = (0.1 \text{ mol}) \times (22.4 \text{ dm}^3 \text{ mol}^{-1})$$

$$= 2.24 \text{ dm}^3$$

Electrolysis Calculations

The reactions that take place at electrodes can be represented by half-equations. For example, the following reaction shows that one mole of lead(II) ions are discharged at a cathode by two moles of electrons:

$$Pb^{2+} + 2e^- \rightarrow Pb$$

In chapter 10 we shall see how such half-reactions form the basis of calculating the mass of substances discharged at electrodes during electrolysis.

Yield

The mass of product formed in a chemical reaction is called the **yield**. The yield of a product calculated from the chemical equation is called the **theoretical yield**. The yield actually obtained in an experiment or in an industrial process is called the **actual yield**.

When the actual yield equals the theoretical yield, the yield is said to be **quantitative**. A chemical reaction which produces a quantitative yield is often called a **stoichiometric process**. Many ionic reactions produce quantitative yields. Neutralisations and ionic precipitations are examples. However, many yields of chemical reactions are not quantitative. This is particularly the case with organic chemical reactions. It is then often necessary to calculate the **percentage yield**. This is given by the following relationship:

$$\text{Percentage yield} = \frac{\text{Actual yield}}{\text{Theoretical yield}} \times 100$$

The percentage yield is a measure of the **efficiency** of the reaction. A chemical reaction which produces a quantitative yield is thus 100% efficient. Chemical reactions which do not produce quantitative yields are sometimes called **non-stoichiometric processes**. However, this is really a misnomer since the laws of chemical combination require that all chemical reactions are stoichiometric, that is, that combinations must involve exact whole numbers of particles (atoms, ions or molecules) in fixed, characteristic ratios. The non-stoichiometry of a reaction can be due to a number of factors such as impurities in the reactants, **side reactions** and practical factors such as loss on crystallisation.

Percentage yields of non-stoichiometric processes can often be increased by using an excess of one or more of the reactants. In this case the theoretical yield is calculated from the amount of reactant which is *not* used in excess. This reactant is called the **limiting reactant**.

EXAMPLE

Bromobenzene can be prepared by the reaction of benzene with excess bromine in the presence of iron(III) chloride. In one experiment, 23.0 grams of bromobenzene were obtained from a 20.0 gram sample of benzene. Calculate the percentage yield.

SOLUTION

The equation for this reaction is

$$C_6H_6 + Br_2 \rightarrow C_6H_5Br + HBr$$

Since   mass of benzene = 20 g
and $M(C_6H_6)$ = 78 g mol^{-1}

$$\text{amount of } C_6H_6 = \frac{20 \text{ g}}{78 \text{ g mol}^{-1}}$$

$$= 0.256 \text{ mol}$$

The equation above shows that, assuming 100% conversion, one mole of C_6H_6 produces one mole of C_6H_5Br. Thus

theoretical yield of C_6H_5Br = 0.256 mol
Since $M(C_6H_5Br)$ = 157 g mol^{-1}
theoretical yield of C_6H_5Br = (0.256 mol) × (157 g mol^{-1})
= 40.2 g

Thus $\text{percentage yield} = \frac{23.0 \text{ g}}{40.2 \text{ g}} \times 100$

$$= 57.2\%$$

INDUSTRIAL STOICHIOMETRY

The overall economic objective of a production department in the chemical industry is to produce chemicals of sufficient quality and in sufficient quantity that they can be sold at a profit. This requires that all resources should be used as efficiently as possible. This cannot be achieved unless the chemical process itself is efficient.

The starting point for considering the efficiency of a chemical process is the stoichiometry of the process. In theory, but not usually in practice, the total mass of products of a chemical process should equal the total mass of reactants used in the process. This is called the **mass balance**. It follows directly from the law of conservation of matter.

In practice, an excess of one or more reactants is used in a process. In addition, side reactions may produce by-products. The term 'mass balance' is thus often taken to mean that the total mass of materials used at the beginning of (and during) the process must equal the total mass of products, by-products, unused reactants and solvents at the end of the process.

A distinction is usually made between the theoretical mass balance and the actual mass balance. The theoretical mass balance includes the predicted yield of desired product. This is based on calculations performed by chemical engineers and also on experimental trials carried out in a laboratory or pilot plant.

In many processes the actual masses do not balance due to unavoidable and sometimes avoidable wastage of material. This wastage can be shown up by comparing the theoretical mass balance with the actual mass balance.

In industry the terms starting materials, raw materials or feedstock are used instead of reactants. For a process to be economically viable it is essential that the optimum yield of desired product is obtained from the raw materials. The optimum yield is not necessarily the theoretical yield or even the maximum possible yield. Production of the maximum possible yield might, for example,

The efficient operation of a modern chemical plant depends on a thorough knowledge of the stoichiometry of the chemical processes involved

require too high a consumption of an expensive starting material or the process might take too long and thus prove uneconomic. A number of factors have to be considered when determining the overall efficiency of a process and expected yield is only one of them. In many industrial processes the optimum yield of a desired product is only a fraction of the mass of starting materials.

The actual yield of a chemical process can be influenced by a number of factors. These include

- temperature
- pressure
- rate of agitation (stirring)
- presence of a catalyst
- purity of starting materials
- effectiveness of isolation procedures.

We shall examine the influence of such factors on chemical reactions when we come to consider chemical equilibria (chapter 7) and chemical kinetics (chapter 9).

Flow Diagrams

These show the flow of materials during an industrial chemical process. Many flow diagrams also include the quantities of materials flowing in and out at each stage of the process. The masses of materials used at each stage must equal the masses of material produced. These materials include wash water, solvents and effluent (discharges). The overall mass balance for the process is shown as a total of the materials used (in) and a total of the materials produced (out). The two totals must balance.

Figure 4.4 Flow diagram for the manufacture of sodium hydroxide from sodium carbonate

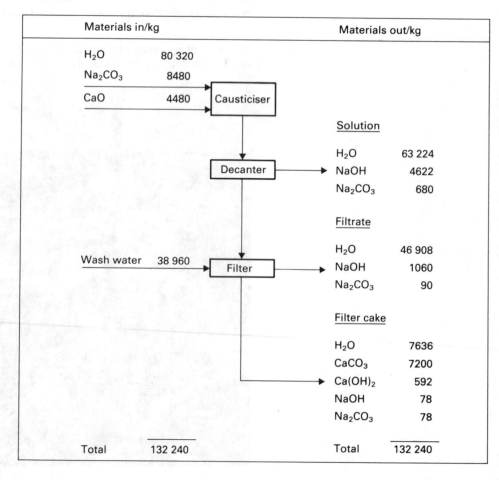

Figure 4.4 shows a simplified flow diagram for the manufacture of batches of caustic soda, NaOH, from sodium carbonate by the bossage process. The flow diagram is based on the following process.

Manufacture of Sodium Hydroxide

Most sodium hydroxide produced in Britain is made electrolytically (*see* chapter 13). However, in countries where electricity is expensive, it is made almost exclusively from sodium carbonate.

Manufacture of Caustic Soda, NaOH, from Sodium Carbonate, Na₂CO₃

1. An 88 800 kg batch of sodium carbonate solution containing 8480 kg of Na_2CO_3 is added to an agitated vessel called a causticiser.
2. 4480 kg of quicklime, CaO, are then added. This is the stoichiometric amount. The mixture is then heated in the agitated vessel. The result is a mixture of solids and liquid called a **slurry**.
3. The slurry is first decanted to remove most of the liquid. The resultant sludge is washed with twice its weight of water and then filtered. The wet solids left in the filter are called the filter cake.

The equations for the process are

$$CaO(s) + H_2O(l) \rightarrow Ca(OH)_2(aq)$$
$$Na_2CO_3(aq) + Ca(OH)_2(aq) \rightarrow CaCO_3(s) + 2NaOH(aq)$$

Laboratory tests show that under the conditions in the reaction vessel, a 90% yield is obtained after 2 hours.

SUMMARY

1. The **amount of substance** is a physical quantity. Its unit is the **mole**.
2. **Molar mass** is the mass per unit amount of substance. It is related to **relative molecular mass** by

$$M = M_r \text{ g mol}^{-1}$$

3. The mass of several moles of a substance is given by

$$m = n \times M$$

4. The number of particles in a given amount of substance can be calculated from the following relationship:

$$N = n \times L$$

   where L is the **Avogadro constant**.
5. An **empirical formula** shows the simplest ratio of atoms of each element or the simplest ratio of the different ions present in a compound.
6. The **concentration** of a solution is given by

$$[X] = \frac{n(X) \text{ mol}}{V \text{ dm}^3}$$

7. The **mole fraction** of A in a mixture of A and B is given by

$$x_A = \frac{n_A}{n_A + n_B}$$

8. A **chemical equation** may be interpreted on a molecular level or on a molar level. The overall chemical equation for a reaction does not provide information about the **reaction mechanism**.
9. The yield of a product calculated from the chemical equation is called the **theoretical yield**.
10. The term **mass balance** means that the total mass of materials that flow into an industrial chemical process or a stage of the process must equal the total mass of materials that flow out.

Examination Questions

1. Identify, by name or formula, each of the species A to I in the reaction sequences described below.
    (a) Silver nitrate solution is treated with dilute aqueous sodium hydroxide to give a brown precipitate A which dissolves in an excess of aqueous ammonia to give a colourless solution of B.
    (b) Aluminium sulphate solution is treated with sodium carbonate solution to give a white precipitate C and a gas D.
    (c) Copper(I) oxide is treated with dilute sulphuric acid to give a red-brown solid E and a blue solution of F. However, if copper(I) oxide is treated with warm concentrated hydrochloric acid, a solution is formed which, when poured into a large excess of water, gives a white precipitate G.
    (d) Iron(III) chloride solution is treated with aqueous ammonia to give a precipitate H which dissolves in an excess of concentrated hydrochloric acid to give a yellow solution of I.

    (JMB)

2. Give the oxidation number of nitrogen in each of the following species: NO; NO_2; NH_4^+; NO_3^-; NO_2^-.

    (OLE)

3. State and explain what happens in *each* of the following experiments. Write balanced questions for the reactions that occur where possible.
    (a) Aqueous sodium hydroxide is gradually added to aqueous aluminium sulphate until it is present in excess.
    (b) A few drops of silicon tetrachloride are added to water containing universal indicator solution.
    (c) Ammonium sulphate is warmed with aqueous sodium hydroxide and a strip of filter paper which has been immersed in aqueous copper(II) sulphate is held just above the mouth of the test tube.
    (d) Aqueous chlorine is shaken with aqueous potassium iodide to which a few drops of an organic solvent such as tetrachloromethane have been added.

    (UCLES)

4. Dry hydrogen chloride is passed over 23.8 g of heated tin until it is completely converted to a chloride of tin weighing 38.0 g.
    Calculate the empirical (simplest) formula of the tin chloride.

    (SEB)

5. 50 cm^3 nitrogen monoxide (nitric oxide) were mixed with 100 cm^3 oxygen. When the reaction was complete, the resulting gases were passed into water.
    (a) Write a balanced equation for the reaction between nitrogen monoxide and oxygen.
    (b) Calculate the total volume of the gaseous mixture after the reaction was complete, but *before* adding to water. (Assume all volume measurements at the same temperature and pressure.)
    (c) Calculate the final volume of the remaining gas *after* adding to water.

    (SEB)

6. The white allotrope of phosphorus when boiled with aqueous sodium hydroxide solution produces a mixture of neutral sodium salts, A and B, and a mixture of hydrogen gas and phosphine gas, PH_3.

    The salt A has the percentage composition by mass: Na, 26%; P, 35%; H, 2.2%; O, 36%. One mole of phosphine molecules is oxidised by four moles of iodine atoms in the presence of water to produce the free acid of A and hydriodic acid, HI. Calculate the empirical formula of the salt A and write a balanced equation for the reaction between phosphine and aqueous iodine.

    The salt B has a percentage composition by mass: Na, 36%; P, 25%; H, 0.8%; O, 38%. The free acid of B can be obtained by the aqueous hydrolysis of phosphorus trichloride, PCl_3. Calculate the empirical formula of the salt B and write a balanced equation for the reaction between phosphorus trichloride and water.

    In a titration 10.0 cm^3 of a 0.10 M solution of sodium salt A was oxidised to sodium phosphate(v), Na_3PO_4, by 33.4 cm^3 of 0.02 M sodium dichromate(vi) in acid solution. Calculate the oxidation number change for the salt A and explain your result in relation to its formula.

    In another titration 10.0 cm^3 of 0.10 M solution of sodium salt B was oxidised to sodium phosphate(v), Na_3PO_4, by 16.7 cm^3 of 0.02 M sodium dichromate(vi) in acid solution. Calculate the oxidation number change for the salt B and explain your result in relation to its formula.

    (L)

7. This question concerns a hydrated double salt A, $Cu_w(NH_4)_x(SO_4)_y \cdot zH_2O$, whose formula may be determined from the following experimental data.
    (a) 2 g of salt A was boiled with excess sodium hydroxide and the ammonia expelled collected by absorption in 40 cm^3 of 0.5 M hydrochloric acid in a cooled flask. Subsequently this solution required 20 cm^3 of 0.5 M sodium hydroxide for neutralisation.
        (i) Calculate the number of moles of ammonium ion in 2 g of salt A.
        (ii) Calculate the mass of ammonium ion present in 2 g of salt A.
    (b) A second sample of 2 g of salt A was dissolved in water and treated with an excess of barium chloride solution. The mass of the precipitate formed, after drying, was found to be 2.33 g.
        (i) Calculate the number of moles of sulphate ion in 2 g of salt A.
        (ii) Calculate the mass of sulphate ion in 2 g of salt A.
    (c) A third sample of 10 g of salt A was dissolved to give 250 cm^3 of solution. 25 cm^3 of this solution, treated with an excess of potassium iodide, gave iodine equivalent to 25 cm^3 of 0.1 M sodium thiosulphate solution.

    $$2Cu^{2+} + 4I^- \rightarrow 2CuI + I_2$$
    $$I_2 + 2S_2O_3^{2-} \rightarrow 2I^- + S_4O_6^{2-}$$

        (i) Calculate the number of moles of copper(ii) ion in 2 g of salt A.
        (ii) Calculate the mass of copper(ii) ion in 2 g of salt A.
    (d) (i) Calculate the mass of water of crystallisation in 2 g of salt A.
        (ii) Calculate the number of moles of water of crystallisation in 2 g of salt A.
    (e) Determine the formula of the hydrated salt A.

    (L)

5 ENERGETICS

Solar Fuels

The Hydrogen Tree

A radical new approach to solar energy may come from biotechnology—a broad class of systems that have at their core photosynthetic and biological energy conversion processes.

Plants have long 'known' how to use the energy of sunlight to split water, but they do not evolve hydrogen explicitly, since it is needed only for internal energetic processes within the plant itself as a means for reducing carbon dioxide. However, it may be possible to develop new biological structures that in fact evolve hydrogen. At IIASA [The International Institute for Applied Systems Analysis, Laxenburg, Austria], Cesare Marchetti considered the concept of

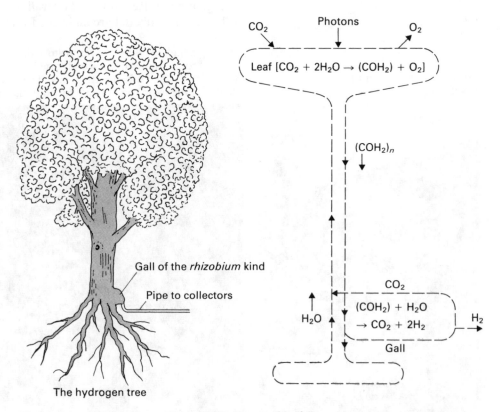

The hydrogen tree

Figure 5.1 Graphic presentation of the proposed hydrogen tree with a very schematic chemistry

hydrogen-producing trees. The concept is essentially one of replacing expensive solar collectors and solar cells with tree leaves. Swollen plant tissues, so-called galls, located at the tree trunk would be genetically programmed to use the solar energy captured in the leaves for generating hydrogen gas as a by-product of photosynthesis. The hydrogen gas would be collected within the galls and piped to a central storage system. The essential features of such a system already exist in nature. Many insects and bacteria induce the formation of galls in different types of plants. These various kinds of galls, which number in the tens of thousands, then provide the shelter or nutrients needed by the organism that caused them. In at least one case, that of *Rhizobium* bacteria in symbiosis with leguminous plants, substantial hydrogen is produced in the galls, though currently it simply escapes to the atmosphere. It has been estimated that in this way US soybean plantations leak about 30 billion cubic metres of hydrogen annually. Adapting this potential so that the plants can be easily integrated with some sort of collection system will depend on advances in the techniques of genetic engineering. The gall actuates a reverse of photosynthesis and makes hydrogen (or methane) available in an enclosed cavity that can be tapped by a collector pipe.

Wood

Eighty-six per cent of all the wood consumed annually in the developing countries is used for fuel, and of this total at least half is used for cooking. Nearly everywhere, reliance on charcoal as a source of fuel is increasing. In Tanzania, for example, the charcoal share of the wood fuel burned, which was 3% in 1970, is expected to rise to 25% by the year 2000. In principle this is discouraging, because in preparing the charcoal more than half the wood's energy is wastefully burned away. But charcoal makes wood energy easier and cheaper to transport, and the growing reliance on it is a result of the increasing distance from harvest site to the user. Also, charcoal is preferred because of its steady and concentrated heat, its smokeless burning, and because it can easily be extinguished when the fire is no longer needed. Charcoal can also substitute for fossil fuels, which in some places is an urgent need. Regardless of overall inefficiency, it seems clear that more meals will be cooked over charcoal in the future.

There is a high incidence of respiratory diseases and pulmonary heart disease in women exposed to the fumes of firewood, dried animal dung, or even kerosene inside their homes in crowded urban settlements. The smoke generated by these fuels contains almost all the toxic components found in the smoke emitted when fossil fuels are burnt

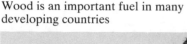

Wood is an important fuel in many developing countries

Ethanol

The most popular but not necessarily the most economic biomass conversion process is undoubtedly the production of ethyl alcohol (ethanol, C_2H_5OH) from sugar-cane and maize. The world's leading producer is Brazil, producing 3.2 billion litres of alcohol from sugar-cane, sorghum and cassava. The alcohol can be blended with petrol to a proportion of up to 20% alcohol and run in a conventional car engine without adjustment. However, in case of need, cars can run on pure alcohol after adjustments.

The cost of alcohol in Brazil is higher than most gasoline sold in Europe in 1980. Nevertheless, the indirect benefits are judged very beneficial to Brazil. They include the saving of foreign exchange, the creation of new employment, the encouragement of domestic technology and industry.

The United States is also deeply involved in the promotion of ethanol production and has set a target of approximately 3477 million litres a year of alcohol for fuel use by 1982. Most of the distilleries use corn as feedstock. Other countries known to be interested in bioconversion to ethanol include Australia (cassava) and New Zealand (beet).

Some economists feel deep disquiet at the use of food grains to produce motor spirit. They feel that it will lead to the rich getting transport and the poor getting starvation with land currently used for food production being used for fuel. Several situations can be envisaged in which production is economic. The developing countries with a surplus agricultural production but an energy deficit, such as Brazil, Sudan, and Thailand, are likely to have the strongest incentives to develop large biomass energy programmes in order to reduce their dependence on imported energy. Most of the countries with viable alcohol programmes are likely to belong to this group. Many of the large developing countries, such as Bangladesh and Pakistan, however, are net importers of both agricultural products and energy. In most of these countries ethanol production is likely to be attractive only if based on surplus low-cost biomass materials such as molasses and agricultural crop residues (or sugar-cane during periods of world sugar surpluses). In countries with surplus energy, such as Mexico, Nigeria and Venezuela, there is little incentive to launch major biomass energy programmes.

LEARNING OBJECTIVES

After you have studied this chapter you should be able to

1. Explain the terms

isolated system	**standard states**
state function	**Trouton's rule**
equation of state	**delocalised energy**
potential energy	**reversible process**
internal energy	**feasibility of a reaction**
heat	**useful work**
enthalpy	

2. State the **first law of thermodynamics**.
3. Distinguish between **exothermic** and **endothermic** reactions.
4. Outline how the **enthalpy of combustion**, the **enthalpy of neutralisation** and bond enthalpies can be determined experimentally.
5. Calculate the **molar enthalpy of combustion** and the **molar enthalpy of formation** from experimental data.
6. State **Hess's law** and apply it to simple calculations of enthalpy changes.
7. Devise, interpret and use **enthalpy diagrams**.
8. Interpret and use **bond and lattice enthalpies**.
9. Use the **Born–Haber cycle** to calculate lattice enthalpies.

10. Distinguish between the **enthalpy of solution** and the **enthalpy of hydration**.
11. Discuss the significance of the term **entropy**.
12. Explain the importance of **Gibbs free energy**.
13. Calculate entropy and Gibbs free energy changes given the appropriate thermodynamic data.
14. Sketch a typical **Ellingham diagram** and use it to show how Gibbs free energies apply to the extraction of metals from their oxides.
15. Describe the various **sources of energy** available to man.
16. Distinguish between **renewable** and **non-renewable** forms of energy.
17. Outline the various means of **energy storage**.
18. Discuss the past, present and future **energy needs** of man and point to the problems associated with present and future energy sources.

5.1
Thermodynamics

The word thermodynamics derives from the Greek words *thermos*, meaning heat, and *dynamis*, meaning power.

Thermodynamics is the study of energy. It is concerned with the forms energy can take, how efficiently it can be used and to what extent energy can be made available for useful work.

Classical thermodynamics is the study of energy and work in macroscopic systems. This means it is concerned with the bulk properties of a system. These include pressure, volume and temperature. It is not concerned with the motions, forces and interactions of individual particles.

The study of the energy relationships and the statistical behaviour of large groups of particles is called **statistical thermodynamics**. This branch of thermodynamics applies the laws of statistics to component microscopic particles. It relies heavily on the mathematical implications of quantum theory.

Chemical thermodynamics focuses on energy transfer during chemical reactions and on the work done by chemical systems. Chemical thermodynamics is an important branch of chemistry. For example, it can be used:

- to predict whether or not a chemical reaction is likely to occur when two different substances are mixed;
- to enable the amount of energy theoretically required by or released during reactions to be calculated;
- to predict the extent to which a reaction will proceed before reaching a condition of equilibrium.

SYSTEMS AND SURROUNDINGS

A **system** is that part of the world which we are interested in or which we are investigating. We might be investigating it with an experiment or we might be doing some calculations on it. A system might be a beaker containing a known amount of water or it might be a heat exchanger used in a chemical plant.

The rest of the world—outside the system—is called the **surroundings.** The surroundings include all other objects which might act on the system under investigation.

An **isolated system** is completely insulated from its surroundings. There is thus no energy transfer between an isolated system and its surroundings. An isolated system is an ideal system. It cannot be achieved in practice. However, although it cannot be achieved in practice, the concept of an isolated system *is* important in estimating the maximum theoretical energy differences between a system and surroundings.

STATE FUNCTIONS

The macroscopic quantities that describe the state of a thermodynamic system are called state variables or **state functions**. State functions are connected by mathematical relationships called **equations of state** (*see* section 3.1). For example, the pressure, volume and temperature of n moles of an ideal gas are related by the following equation of state:

$$pV = nRT$$

This is, of course, the ideal gas equation (*see* section 3.1). Pressure, volume and temperature are all state functions. When the state of a system is altered, the change in any state function depends only on the initial and final states of the system. It does not depend on how the change is accomplished. For example, if the temperature of a beaker of water is raised from an initial temperature T_1 to a final temperature T_2, the change in temperature is given by

$$\Delta T = T_2 - T_1$$

The Greek letter Δ (called delta) is used to refer to a change in a quantity. The change is always found by subtracting the initial value of the quantity from its final value.

THE ENERGY OF A SYSTEM

The word energy comes from the Greek word *energia*, meaning 'in-work'.

Energy is the capacity to do work. The unit of energy is the same as that for work. It is the joule and it has the symbol J.

There are numerous different forms of energy. For example

- chemical energy
- electrical energy
- mechanical energy
- nuclear energy
- solar energy

Each of these forms of energy relates to a specific system or type of system. For example, chemical energy relates to chemical systems and solar energy relates to the energy of the Sun. All these forms of energy consist essentially of two types of energy. These are kinetic energy and potential energy.

Kinetic energy is the energy of a body due to its motion. The kinetic energy is related to the mass m of a body and its velocity v by the relationship:

$$\text{Kinetic energy} = \tfrac{1}{2}mv^2$$

Potential energy is the energy stored by a body.

INTERNAL ENERGY

A chemical system may consist of atoms, molecules and ions or any combination of these. These particles all have kinetic energy and potential energy. The combined kinetic energy and potential energy of all the particles in the system is called the **internal energy** of the system.

The kinetic energy is due to the motions of the particles. These motions may be translational, rotational or vibrational (*see* figure 5.2).

The potential energy of the particles is due to the electrostatic forces of attraction between the particles and within the particles. For example, the electrons in an atom have potential energy with respect to the other electrons in the atom as well as with respect to the positive charges on the nucleus. The

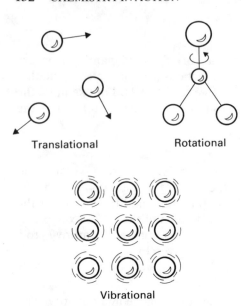

Translational   Rotational

Vibrational

Figure 5.2   Types of motion

potential energy of a particular electron with an electrical charge q_1 in the vicinity of other electrical charges q_2 is given by:

$$\text{Potential energy} = \frac{q_1 q_2}{r}$$

where r is the distance between the electron and the centre of the other charges. Bond energy is also a form of potential energy (*see* figure 2.1).

Temperature

Temperature is, as we saw above, a state function. It is a measure of the average kinetic energy of all the particles in a system. If we transfer energy to a system, the kinetic energy of the particles in the system increases. The temperature therefore rises. This, in fact, is the fifth assumption of the kinetic theory of gases (*see* section 3.1).

TRANSFER OF ENERGY

Internal energy is a function of the state of a system. It is given the symbol U. The absolute value of the internal energy of a system cannot be determined experimentally. What can be measured, however, is the change in internal energy of a system. This is given by

$$\Delta U = U_{\text{final}} - U_{\text{initial}}$$

ΔU has a negative value if U_{final} is smaller than U_{initial}. This occurs if the system loses energy, that is, if energy is transferred from the system to the surroundings.

There are two fundamental ways of transferring energy to or from a system. These are heat and work.

Heat

The transfer of energy caused by a difference in temperature between a system and its surroundings or between a system and another system is called **heat**. The amount of energy transferred in this way is written q. Heat is proportional to the mass m of the system and the temperature change ΔT associated with this transfer of energy:

$$q \propto m\Delta T$$

For a system consisting of a particular substance, this becomes

$$q = mc\Delta T$$

where c is the specific heat capacity of the substance.

The **specific heat capacity** of a substance is the energy required to raise the temperature of one kilogram of the substance by one kelvin. For example, the specific heat capacity of water is 4184 J kg^{-1} K^{-1}.

The **molar heat capacity** of a substance is the energy required to raise the temperature of one mole of the substance by one kelvin. It is denoted C_{m}.

Heat is not a property of a system. It is therefore not a state function. For example, a beaker of water may have a temperature of 50°C but it does not have heat. To get to 50°C energy may have been transferred to the system—assuming that it was previously at a lower temperature. Alternatively, if the beaker of water was previously at a higher temperature energy would have been transferred out of it in order to reach a temperature of 50°C. Thus, the energy transferred to or from the system—that is, the heat—does not describe the system. It only describes what happens to the system before it reaches its final state, which in our example was 50°C.

Work

Work is also a form of transfer of energy from one system to another or to its surroundings. Work is done by a system when the system exerts a force to overcome resistance. Work can be expressed in terms of force and distance:

$$\text{Work} = \text{Force} \times \text{Distance moved in direction of force}$$

The type of work we most commonly meet in chemistry is expansion. For example, expansion can occur when a gas is evolved during a chemical reaction (*see* figure 5.3). In this case, the work w done by a system is given by

$$w = -P\Delta V \tag{1}$$

where P is the external pressure and ΔV the change in volume. For many chemical reactions performed in a laboratory, the external pressure is simply the atmospheric pressure. The negative sign in the equation is necessary because work is done by the system. The system therefore *loses* energy.

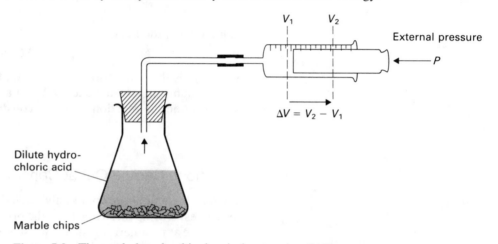

Figure 5.3 The work done by this chemical system is $-P\Delta V$

THE FIRST LAW OF THERMODYNAMICS

We have seen that a chemical system has an internal energy, U. This energy can change by an amount ΔU if energy is transferred in to or out of the system. The transfer of energy in to or out of the system can take the form of heat q or the form of work w. The quantities ΔU, q and w are related by the equation

$$\Delta U = q + w \tag{2}$$

This mathematical relationship is called the **first law of thermodynamics**.

If q has a positive value, the system has gained energy.

If w has a *positive* value, the system has also gained energy. This means that work has been done *on* the system. If w has a *negative* value, the work has been done *by* the system. The system has thus lost energy.

ΔU has a positive value if, overall—that is, taking account of both q and w—the system has gained energy.

ΔU has a negative value, if overall, the system has lost energy.

ΔU, q and w all have energy units—that is, joules.

The first law of thermodynamics is no more than one expression of the **law of conservation of energy**. This states that energy cannot be created or destroyed, although it may be converted from one form into another. A system thus cannot create or destroy energy. The changes in energy of system (ΔU) is due to the transfer of energy in to or out of the system. This can only be done by heat or work.

Reactions Carried Out at Constant Volume

We saw above that the work done by a chemical system frequently results in expansion. We can thus substitute equation (1) into equation (2) and obtain

$$\Delta U = q - P\Delta V \qquad (3)$$

For a chemical system which involves only solid and/or liquid reactants and products, the expansion ΔV of the system is often negligible or zero. If the expansion is zero, the work done by the system is also zero. We can express this mathematically as follows: since

$$\Delta V = 0$$

then

$$P\Delta V = 0$$

and thus, from equation (1),

$$w = 0$$

Equation (2) then becomes

● $$\Delta U = q_V$$

where q_V is the heat absorbed at constant volume.

The change in internal energy ΔU of a system is thus the heat absorbed by the system when the reaction occurs at constant volume.

Reactions Carried Out at Constant Pressure

Chemical reactions are commonly carried out in open vessels such as beakers and test-tubes. During these reactions the pressure of the system is constant. It usually equals the atmospheric pressure. We saw above that the heat absorbed at constant volume equals the change in the thermodynamic state function U or internal energy. So does the heat absorbed (or lost) by a system at constant pressure equal a change in any thermodynamic state function? The answer is yes. The thermodynamic state function is **enthalpy**. It is denoted by the capital letter H and defined by the following mathematical relationship:

$$H = U + pV$$

where U is the internal energy of the system and pV is the energy of the system due to the space it occupies. This term is sometimes called the external energy.

We should note that the small letter p is used for pressure and not the capital P. We use P for external pressure—that is, the pressure acting on a system—whereas p is used for internal pressure. Internal pressure is the pressure of the system acting on its surroundings.

Since p, V and U are all properties of a system, that is they are all state functions, it follows from the above equation that H must also be a state function.

The change in enthalpy that occurs when a process is carried out at constant pressure is given by

$$\Delta H = \Delta U + p\Delta V \qquad (4)$$

We can substitute equation (3) into this equation to get:

$$\Delta H = q_p - P\Delta V + p\Delta V$$

where q_p is the heat absorbed by the system when the pressure of the system is constant. In open vessels the internal pressure of the system equals the external pressure, which is, as we have seen, usually the atmospheric pressure. We can thus assume that

$$p = P$$

and thus

●
$$\Delta H = q_p$$

The enthalpy change, ΔH, of a system is thus the heat absorbed by the system when the reaction occurs at constant pressure. It is this quantity which is normally determined during calorimetric experiments (*see* below).

We should finally note that for reactions involving solids and liquids only the term $p\Delta V$ in equation (4) is negligible or zero. Thus, for such reactions,

$$\Delta H \approx \Delta U$$

SUMMARY

1. An **isolated system** is completely insulated from its **surroundings**.
2. **Energy** is the capacity to do work.
3. For any thermodynamic **state function** X

$$\Delta X = X_{final} - X_{initial}$$

   The change ΔX does not depend on the path taken during the change but only on the initial and final states of the system.
4. The **work** done by a system is given by

$$w = -P\Delta V$$

5. The **molar heat capacity**, C_m, of a substance is the energy required to raise the temperature of one mole of the substance by one kelvin.
6. The **first law of thermodynamics** is

$$\Delta U = q + w$$

7. The **law of conservation of energy** states that energy cannot be created or destroyed.
8. The heat absorbed by a system at constant volume equals the change in **internal energy**

$$\Delta U = q_V$$

9. The heat absorbed by a system at constant pressure equals the change in **enthalpy**

$$\Delta H = q_p$$

5.2
Enthalpies

The branch of chemistry concerned with the enthalpy changes that occur during chemical reactions and phase changes is called **thermochemistry**. The experimental technique used for the determination of enthalpy is **calorimetry**.

ENTHALPY CHANGE OF REACTION

Many chemical reactions are accompanied by a change in enthalpy. This means that the total enthalpy of the products of the reaction is different from the total enthalpy of the reactants. The enthalpy change is given by

$$\Delta H = H_2 - H_1$$

Exothermic reaction

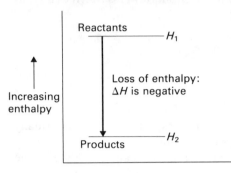

Endothermic reaction

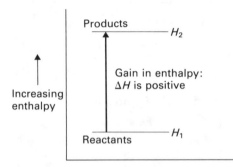

Figure 5.4 Exothermic and endothermic reactions

where H_1 is the enthalpy of the reactants and H_2 is the enthalpy of the products. ΔH is sometimes called the heat of reaction.

If the enthalpy of the products is lower than the enthalpy of the reactants, the reaction is said to be **exothermic** (*see* figure 5.4). Since there is a drop in enthalpy, ΔH has a negative value. The energy that is lost during an exothermic reaction is first transferred to the reaction mixture. The temperature of the reaction mixture therefore increases. This energy is then transferred to the surroundings. Thus, once the reaction is complete, the temperature falls.

If the enthalpy of the products is higher than the enthalpy of the reactants, the reaction is said to be **endothermic**. Since there is an increase in enthalpy, ΔH has a positive value. The energy absorbed during the reaction is first taken from the reaction mixture. The temperature of the reaction mixture therefore decreases. Since the temperature of the surroundings is now higher than that of the reaction mixture, energy flows into the reaction mixture from the surroundings. Thus, once the reaction is complete, the temperature increases.

STANDARD ENTHALPY CHANGES

The enthalpy changes that occur during a chemical reaction or phase change vary depending on temperature, pressure and the physical states of the substances involved. The enthalpy changes also depend on the amount of substances involved in the process. It is therefore normal to quote the standard molar enthalpy change for a process. The word 'change' is usually dropped from this term.

The **standard molar enthalpy** of a reaction is the enthalpy change under standard conditions per mole of the reaction as specified by a balanced chemical equation. The standard conditions usually chosen are 298 K and one atmosphere pressure. Under these conditions, the reactants and products are said to be in their standard states. For example, the standard state of hydrogen is a gas whereas the standard state of water is a liquid.

The standard molar enthalpy of a reaction is written $\Delta H_m^{\ominus}(298 \text{ K})$. It is essential that $\Delta H_m^{\ominus}(298 \text{ K})$ should refer unambiguously to a specific equation or formula. For example,

(a) $2H_2(g) + O_2(g) \rightarrow 2H_2O(l)$ $\Delta H_m^{\ominus}(298 \text{ K}) = -571.6 \text{ kJ mol}^{-1}$

This shows that when two moles of hydrogen gas, $H_2(g)$, combine with one mole of oxygen gas, $O_2(g)$, to form two moles of liquid water, $H_2O(l)$, at a pressure of one atmosphere and a temperature of 298 K, the enthalpy change is -571.6 kJ. The negative sign means the reaction is exothermic.

If the same reaction is represented by the following equation, then the standard molar enthalpy of reaction is half the above value:

(b) $H_2(g) + \tfrac{1}{2}O_2(g) \rightarrow H_2O(l)$ $\Delta H_m^{\ominus}(298K) = -285.8 \text{ kJ mol}^{-1}$

It is important to specify the states after the formulae in the equation. For example, compare the standard molar enthalpy for the following reaction with that for the reaction represented by equation (b) above:

(c) $H_2(g) + \tfrac{1}{2}O_2(g) \rightarrow H_2O(g)$ $\Delta H_m^{\ominus}(298 \text{ K}) = -241.8 \text{ kJ mol}^{-1}$

In reaction (b) above the water formed in the liquid state but in reaction (c) it is a gas (steam). The difference of 44.0 kJ mol^{-1} in the standard molar enthalpies between the two reactions is the standard molar enthalpy (change) of vaporisation of water.

We should thus avoid using phrases such as 'the standard molar enthalpy of oxidation of hydrogen' as they are ambiguous. The standard molar enthalpy of oxidation of hydrogen could refer to any of the three balanced chemical equations shown above and thus have any one of the three widely differing values shown above.

Standard Molar Enthalpy of Formation, $\Delta H_{f,m}^{\ominus}(298\ K)$

The standard molar enthalpy of formation of a substance is the enthalpy change per mole of the substance when the substance is formed from its constituent elements in their standard states at 298 K and one atmosphere pressure.

Let us relate this statement to the example above. We saw that the oxidation of hydrogen can be represented by at least three different balanced chemical equations. We thus have at least three different standard molar enthalpies:

(a) $\qquad\qquad \Delta H_m^{\ominus}\{298\ K,\ 2H_2O(l)\} = -571.6\ kJ\ mol^{-1}$

(b) $\qquad\qquad \Delta H_m^{\ominus}\{298\ K,\ H_2O(l)\} = -285.8\ kJ\ mol^{-1}$

(c) $\qquad\qquad \Delta H_m^{\ominus}\{298\ K,\ H_2O(g)\} = -241.8\ kJ\ mol^{-1}$

Which of these is the standard molar enthalpy of formation? The value in (c) can be ruled out since the standard state of water at 298 K is not steam, $H_2O(g)$. The standard molar enthalpy of formation could have two different values depending on how we write the equation. However, since the definition given above specifies the enthalpy change per mole we must choose (b). Thus, the standard molar enthalpy of formation $\Delta H_{f,m}^{\ominus}(298\ K)$ for water is $-285.8\ kJ\ mol^{-1}$.

Table 5.1 gives the standard molar enthalpies of formation of some common substances in their standard states. Note that these include not only compounds but also ions. The ions and many of the compounds cannot be made by direct combination of their constituent elements. However, by application of Hess's law (which we shall deal with in section 5.3 below) it is possible to calculate them.

Most compounds have negative enthalpies of formation. They are said to be **exothermic compounds**. A few have positive enthalpies of formation. They are called **endothermic compounds**. Benzene is an example of an endothermic compound.

Notice that the enthalpies of formation of solid compounds tend to be more negative than the enthalpies of formation of gases. The standard molar enthalpy of formation of a compound or ion is a measure of its stability relative to its constituent elements. The more negative the value, the more stable the compound or ion.

By definition, the standard molar enthalpy change of formation of an element in its standard state at 298 K is always zero. For example,

$$\Delta H_{f,m}^{\ominus}\{298\ K,\ O_2(g)\} = 0$$

Table 5.1 Standard molar enthalpies of formation

	$\Delta H_{f,m}^{\ominus}(298\ K)/kJ\ mol^{-1}$
$AgCl(s)$	−127.1
$Al_2O_3(s)$	−1676
$CaCO_3(s)$	−1206
$CO_2(g)$	−393.5
$CuSO_4(s)$	−770
$HCl(g)$	−92.3
$HCl(aq)$	−167.2
$H_2O(l)$	−285.8
$H_2O(g)$	−241.8
$MgO(s)$	−601.8
$Na^+(aq)$	−240.1
$NH_3(g)$	−46.1
$SO_2(g)$	−296.8
$SO_4^{2-}(aq)$	−909.3
$Zn^{2+}(aq)$	−152.4
$CH_4(g)$	−74.8
$C_2H_6(g)$	−84.7
$CHCl_3(l)$	−135.4
$C_2H_5OH(l)$	−277.6
$CH_3COOH(l)$	−484.2
$C_6H_{12}O_6(s)$	−1274
$C_{12}H_{22}O_{11}(s)$	−2222

Standard Molar Enthalpy of Combustion, $\Delta H_{c,m}^{\ominus}(298\ K)$

The standard molar enthalpy of combustion of a substance is the enthalpy change per mole of a substance when the substance in its standard state is completely burnt in oxygen at 298 K and a pressure of one atmosphere.

For example, the standard molar enthalpy of combustion of propane is $-2219.7\ kJ$ per mole of propane. We write this as

$$C_3H_6(g) + 4\tfrac{1}{2}O_2(g) \rightarrow 3CO_2(g) + 3H_2O(l) \qquad \Delta H_{c,m}^{\ominus}(298\ K) = -2219.7\ kJ\ mol^{-1}$$

The standard molar enthalpies of combustion of a range of compounds are shown in table 5.2.

Table 5.2 Standard molar enthalpies of combustion

Substance	Formula	$\Delta H_{c,m}^{\ominus}(298\ K)/$ $kJ\ mol^{-1}$
hydrogen	$H_2(g)$	−285.8
carbon (graphite)	$C(s)$	−393.5
carbon (diamond)	$C(s)$	−395.4
methane	$CH_4(g)$	−890.2
ethene	$C_2H_4(g)$	−1410.9
ethane	$C_2H_6(g)$	−1559.7
propane	$C_3H_6(g)$	−2219.7
butane	$C_4H_{10}(g)$	−2878.6
benzene	$C_6H_6(l)$	−3267.4
methanol	$CH_3OH(l)$	−726.3
ethanol	$C_2H_5OH(l)$	−1366.9
glucose	$C_6H_{12}O_6(s)$	−2816.0

Standard Molar Enthalpy of Neutralisation, $\Delta H_{n,m}^{\ominus}(298\ K)$

The standard molar enthalpy of neutralisation is the enthalpy change per mole of water formed in the neutralisation between an acid and an alkali at 298 K and one atmosphere pressure.

For the neutralisation of a strong acid such as hydrochloric acid, $HCl(aq)$, and a strong alkali such as sodium hydroxide, $NaOH(aq)$, the standard molar enthalpy of neutralisation is almost invariably $-57.1\ kJ\ mol^{-1}$. The reason is that the following ionic reaction is common to all neutralisations of strong acids with strong alkalis:

$$H^+(aq) + OH^-(aq) \rightarrow H_2O(l) \qquad \Delta H^{\ominus}_{n,m}(298 \text{ K}) = -57.1 \text{ kJ mol}^{-1}$$

An exception is the neutralisation of chloric(VII) acid, $HClO_4(aq)$, with potassium hydroxide. This neutralisation is more exothermic than the others due to the precipitation of potassium chlorate(VII), $KClO_4$. The precipitation is an exothermic process.

OTHER ENTHALPY CHANGES

So far we have only considered enthalpies of chemical reactions. Enthalpy changes also occur with other types of change. We have just met one example immediately above. That was the enthalpy of precipitation. Enthalpy changes also occur when a substance is dissolved and a solution is diluted. We shall consider the enthalpy of solution in section 5.3. Enthalpy changes also accompany the transformation of one allotrope to another (*see* section 3.2) and the process of isomerisation (*see* section 17.1).

Phase Changes

A change in phase—melting, vaporisation or sublimation, for example—always results in an enthalpy change. The temperature at which a phase change occurs is called the transition temperature (*see* section 3.2).

The **standard molar enthalpy of fusion**, $\Delta H^{\ominus}_{\text{fus,m}}$, is the enthalpy change that accompanies the fusion of one mole of a substance at its melting point at one atmosphere pressure.

The **standard molar enthalpy of vaporisation**, $\Delta H^{\ominus}_{\text{vap,m}}$, is the enthalpy change that accompanies the vaporisation of one mole of a substance at its boiling point at one atmosphere pressure.

These two terms are sometimes called the molar heats of fusion and vaporisation respectively.

Some standard molar enthalpies of fusion and vaporisation are shown in table 5.3.

Table 5.3   Standard molar enthalpies of fusion and vaporisation

	T_f/K	$\Delta H^{\ominus}_{\text{fus,m}}/\text{kJ mol}^{-1}$	T_b/K	$\Delta H^{\ominus}_{\text{vap,m}}/\text{kJ mol}^{-1}$
HCl	159.0	1.99	188.1	16.15
H_2O	273.2	6.01	373.2	41.09
NH_3	195.5	5.65	239.8	23.35
CH_4	90.7	0.94	111.7	8.20
$CHCl_3$	209.6	9.20	334.4	29.4
CCl_4	250.2	2.51	349.8	30.0
C_2H_5OH	156.2	5.02	351.7	38.58
C_6H_6	278.6	9.83	353.23	30.8

T_f = freezing point
T_b = boiling point
The enthalpy values relate to the transition termperature

We have already noted (in section 2.2) that the value for the molar enthalpy of vaporisation of water is anomalously high. This is due to hydrogen bonding. The standard molar enthalpy of fusion is a measure of the energy needed to overcome the forces of attraction that exist between particles in the solid state. Similarly, the standard molar enthalpy of vaporisation is a measure of the energy needed to overcome the forces of attraction that exist between particles in the liquid state.

Latent heat

The term 'latent heat' means 'hidden heat'. It was introduced in the eighteenth century by a Scottish chemist called Joseph Black to describe the 'heat' supplied when a solid changes to a liquid without a change in temperature. The specific latent heat of fusion is the 'heat' required to change one kilogram of a solid at its melting point to a liquid. The specific latent heat of vaporisation is the 'heat' required to change one kilogram of a liquid at its boiling point to a vapour. The term 'latent heat' is now more or less redundant in chemistry although it is still used elsewhere.

Trouton's Rule

In 1884 Frederick Trouton established that the molar enthalpy of vaporisation of a liquid at its normal boiling point divided by the normal boiling point on the Kelvin scale is constant. This is known as Trouton's rule. In SI units the constant is 88 J K^{-1} mol^{-1}. The rule only holds approximately and then only for certain liquids.

Examples of Trouton's rule

$\Delta H^{\ominus}_{\text{vap,m}} \times T^{-1}/\text{J K}^{-1} \text{ mol}^{-1}$	
$CHCl_3$	87.9
CCl_4	85.8
C_6H_6	87.2

CALORIMETRY

A calorimeter is an insulated vessel used for measuring the quantity of energy released or absorbed during a chemical or physical change. It is useful for determining changes in internal energy, changes in enthalpy and heat capacities, for example.

In a typical calorimetry experiment the energy evolved during a chemical change is transferred to water or to the reaction mixture itself and the temperature increase determined. The energy evolved in the reaction is given by the following relationship:

$$q = [C_{vessel} + (m \times c_{contents})]\Delta T \qquad (5)$$

where q is the energy transferred to the vessel and contents as heat; C_{vessel} is the heat capacity of the vessel; m is the mass of the contents (water or reaction mixture); $c_{contents}$ is the specific heat capacity of the contents; and ΔT is the temperature change of the contents.

Approximate measurements of the enthalpy change for reactions in solution can be made using a simple calorimeter such as a polystyrene beaker. The heat capacity of this is negligible and so the term C_{vessel} in equation (5) can be ignored. Thus

$$q = m \times c_{contents} \times \Delta T \qquad (6)$$

EXAMPLE

40 cm^3 of hydrochloric acid with a concentration of 2.0 mol dm^{-3} were poured into a polystyrene beaker. The same volume of sodium hydroxide with the same concentration was then added to the acid. The temperature rise was found to be 13.7 K. Calculate the approximate enthalpy of neutralistion for the reaction given that the specific heat capacity of water is 4.184×10^{-3} kJ g^{-1} K^{-1}.

SOLUTION

The enthalpy change for the neutralisation is given by

$$\Delta H_{n,m} = - \frac{q_p}{n(H_2O)} \qquad (7)$$

where q_p is the energy evolved in the experiment; and n is the number of moles of water formed in the experiment. The negative sign is necessary because energy is evolved in the process. The reaction system thus loses energy.

There are three stages in this calculation:

Stage 1: Determination of q_p

The reaction is carried out at constant pressure and so we can put $q = q_p$. Since the heat capacity of the calorimeter is negligible we can use equation (6). This is an approximate calculation so we can assume that the density and specific heat capacity of the reaction mixture are the same as that for water. Thus

$$m \quad\ = 80 \text{ g}$$
$$c_{contents} = 4.184 \times 10^{-3} \text{ kJ g}^{-1} \text{ K}^{-1}$$
$$\Delta T \quad\ = 13.7 \text{ K}$$

Substituting these values into equation (6) we obtain

$$q_p = (80 \text{ g}) \times (4.184 \text{ kJ g}^{-1} \text{ K}^{-1}) \times (13.7 \text{ K})$$
$$= 4.586 \text{ kJ}$$

Stage 2: Determination of $n(H_2O)$

The reaction is

$$HCl(aq) + NaOH(aq) \rightarrow NaCl(aq) + H_2O(l)$$

Thus, one mole of HCl(aq) produces one mole of H$_2$O(l). The amount of H$_2$O(l) produced in the reaction thus equals the amount of HCl(aq) consumed in the reaction. The amount of HCl(aq) consumed in the reaction is given by

$$
\begin{aligned}
n(\text{HCl}) &= \text{volume} \times \text{concentration} \\
&= (40 \times 10^{-3} \text{ dm}^3) \times (2.0 \text{ mol dm}^{-3}) \\
&= 0.08 \text{ mol}
\end{aligned}
$$

Thus
$$
n(\text{H}_2\text{O}) = 0.08 \text{ mol}
$$

Stage 3: Determination of $\Delta H_{n,m}$

The molar enthalpy change for this exothermic reaction is found by substituting values for q_p and $n(\text{H}_2\text{O})$ into equation (7):

$$
\begin{aligned}
\Delta H_{n,m} &= -\frac{4.586 \text{ kJ}}{0.08 \text{ mol}} \\
&= -57.3 \text{ kJ mol}^{-1}
\end{aligned}
$$

In more accurate calorimetric determinations the value for ΔT has to be corrected for cooling losses (*see* figure 5.5).

For the calorimetric determination of enthalpy changes or internal energy changes (but not heat capacities), the term $[C_{\text{vessel}} + (m \times c_{\text{contents}})]$ in equation (5) can be eliminated altogether. This is done by carrying out two experiments. First of all the reaction is carried out as normal to find the corrected temperature rise ΔT_1 due to the energy q_1 evolved in the reaction. In a second experiment the calorimeter is calibrated by heating it and its contents electrically. The amount of electricity q_2 required to raise the temperature of the contents by ΔT_2 is given by

$$
q_2 = I \times V \times t \tag{8}
$$

where I is the current in amperes, V is the potential difference in volts, and t is the time in seconds. This value for q_2 is used in the following equation to obtain q_1:

$$
q_1 = q_2 \times \frac{\Delta T_1}{\Delta T_2} \tag{9}
$$

This equation can be derived directly from equation (5).

A Bomb Calorimeter

This type of calorimeter is used to determine the energy changes that accompany chemical reactions such as combustion. The calorimeter consists of a strong sealed vessel enclosed by an insulated water jacket (*see* figure 5.6). A known mass of the sample is held in an atmosphere of pure oxygen at about 25 atmospheres. The sample is then ignited electrically. The energy released during the combustion is transferred to the water jacket and the temperature rise is measured. The temperature rise is often no more than a degree or two. Some modern bomb calorimeters are so sensitive that they require only very small amounts of sample. This is particularly important where the sample is expensive and only very small amounts are available.

Since the combustion in a bomb calorimeter is carried out at constant volume, the energy released is the internal energy change, ΔU.

A separate control experiment is then carried out to calibrate the calorimeter. This can be done electrically as described above. In this case it is necessary to correct for the electrical energy supplied to ignite the sample. Alternatively a second material can be used to calibrate the calorimeter. The molar change of internal energy under the conditions of the experiment of this second material must be shown accurately. Benzoic acid is often used for this purpose.

A Flame Calorimeter

This is used to determine the enthalpies of combustion of volatile liquids such as alcohols and liquid hydrocarbons (*see* figure 5.7). The liquid is burnt in a plentiful

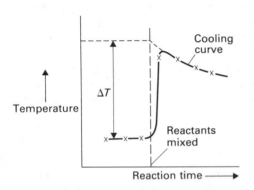

Figure 5.5 An accurate value for ΔT is obtained by plotting temperature against time every 30 seconds. The cooling curve is extrapolated to a reaction time of 0 seconds

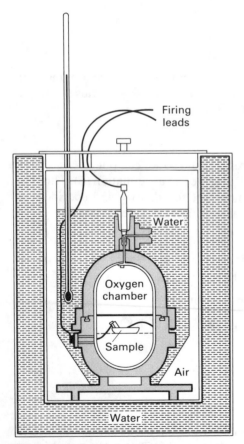

Figure 5.6 Bomb calorimeter

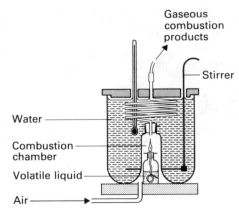

Gaseous combustion products

Stirrer

Water

Combustion chamber

Volatile liquid

Air

Figure 5.7   Flame calorimeter

supply of air and the temperature rise in the surrounding water bath is measured. The mass of liquid burnt can be found by weighing the bottle containing the liquid before and after the experiment. A second experiment is then performed to calibrate the calorimeter.

EXAMPLE

The combustion of a 1.6 g sample of methanol in a flame calorimeter resulted in a temperature rise of 7.8 K. When a 12.0 V supply of electricity with a current of 2.0 A was passed through the contents of the calorimeter for 20 minutes, the temperature rose by 6.2 K. Calculate the molar enthalpy of combustion of methanol under the conditions of the experiment.

SOLUTION

The molar enthalpy of combustion of methanol is given by

$$\Delta H_{c,m} = - \frac{q_p}{n(CH_3OH)} \qquad (10)$$

The negative sign is required since combustion is exothermic. As before, the calculation breaks down into three parts:

Stage 1: Determination of q_p

We use equation (9) to calculate q_p. Since the combustion is carried out at constant pressure, $q_1 = q_p$. Thus,

$$q_p = q_2 \times \frac{\Delta T_1}{\Delta T_2}$$

We are given

$$\Delta T_1 = 7.8 \text{ K}$$
$$\Delta T_2 = 6.2 \text{ K}$$

q_2 is calculated using equation (8) with

$$I = 2.0 \text{ A}$$
$$V = 12.0 \text{ V}$$
$$= 12.0 \text{ m}^2 \text{ kg s}^{-3} \text{ A}^{-1}$$
$$t = 20 \times 60 \text{ s}$$
$$= 1200 \text{ s}$$

Putting these values into equation (8) we obtain

$$q_2 = (2.0 \text{ A}) \times (1200 \text{ s}) \times (12.0 \text{ m}^2 \text{ kg s}^{-3} \text{ A}^{-1})$$

$$= 28\ 800 \text{ m}^2 \text{ kg s}^{-3}$$
$$= 28.8 \text{ kJ}$$

We can now substitute values for q_2, and ΔT_1 and ΔT_2 into the above equation:

$$q_p = (28.8 \text{ kJ}) \times \left(\frac{7.8 \text{ K}}{6.2 \text{ K}} \right)$$

$$= 36.2 \text{ kJ}$$

Stage 2: Determination of $n(CH_3OH)$

Since

$$M_r(CH_3OH) = 32$$
$$M(CH_3OH) = 32 \text{ g mol}^{-1}$$

Thus

$$n(CH_3OH) = \frac{1.6 \text{ g}}{32 \text{ g mol}^{-1}}$$

$$= 0.05 \text{ mol}$$

The use of relative molecular mass (M_r) and molar mass (M) to calculate the amount of a substance in units of moles is explained in section 4.2.

Stage 3: Determination of $\Delta H_{c,m}$
We can now substitute values for q_p and $n(CH_3OH)$ directly into equation (10):

$$\Delta H_{c,m} = -\frac{36.2 \text{ kJ}}{0.05 \text{ mol}}$$
$$= -724 \text{ kJ mol}^{-1}$$

THE ENERGY VALUE OF FOODS

We all consume energy in the form of food and drink. Energy transferred to our bodies as heat—by sitting by a fire or by drinking a hot beverage for example—is negligible in comparison. Figure 5.8 shows the approximate energy value of some common foods. There are three components of food that provide this energy. These are fats, proteins and carbohydrates. Ethanol in alcoholic beverages also provides energy. The energy values for these are shown in table 5.4.

Table 5.4   Energy values of foods

	Energy value/	
	kJ g^{-1}	kcal g^{-1}
carbohydrates	16	3.8
proteins	17	4.1
fats	38	9.1
alcohol	29	6.9

The kilocalorie (kcal) is sometimes called the large calorie, the Calorie or even the calorie by nutritionists. Kilocalories can be converted into joules using the following conversion factor:

$$1 \text{ kcal} = 4.184 \text{ kJ}$$

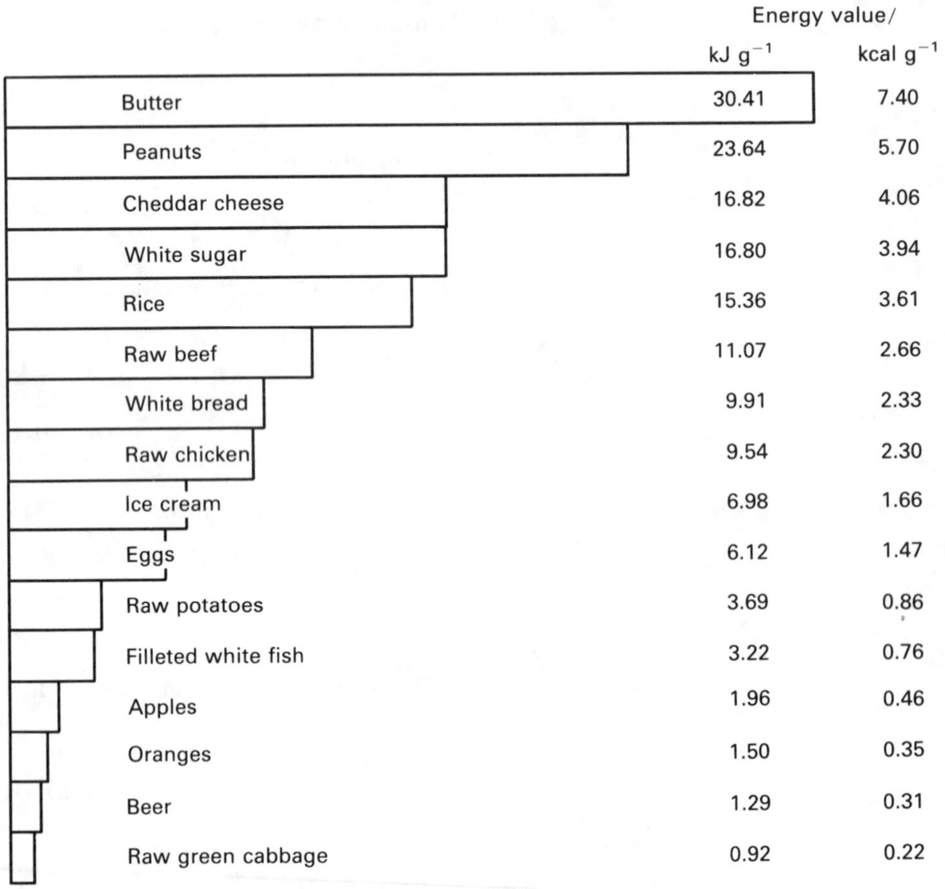

Food	kJ g^{-1}	kcal g^{-1}
Butter	30.41	7.40
Peanuts	23.64	5.70
Cheddar cheese	16.82	4.06
White sugar	16.80	3.94
Rice	15.36	3.61
Raw beef	11.07	2.66
White bread	9.91	2.33
Raw chicken	9.54	2.30
Ice cream	6.98	1.66
Eggs	6.12	1.47
Raw potatoes	3.69	0.86
Filleted white fish	3.22	0.76
Apples	1.96	0.46
Oranges	1.50	0.35
Beer	1.29	0.31
Raw green cabbage	0.92	0.22

Figure 5.8   The energy values of some common foods

The energy value of a particular food can be determined experimentally by burning a measured mass of the food in a bomb calorimeter (*see* figure 5.6) and determining the temperature rise as described above.

Essentially three things can happen to the energy we consume:
1. Some of it is **wasted**—through our faeces. In a healthy adult the amount lost in this way is negligible.

100 g of white bread contains about 50 g carbohydrates, 8 g protein, 2 g fat and 39 g water. It has an energy value of about 10 kJ g^{-1} (2.4 kcal g^{-1}). Brown bread and wholemeal bread contain less carbohydrates but slightly more protein and water than white bread. Their energy values are also slightly lower than white bread

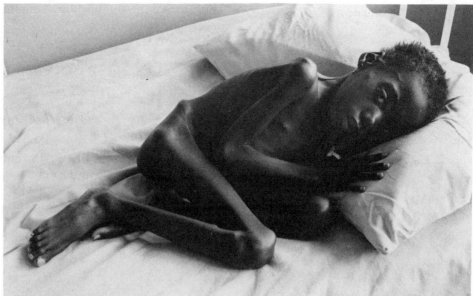

This infant is suffering from marasmus. Infants with this condition are severely wasted—their body weight being below 75% of that expected for their age. The condition, which is widespread in infants in developing countries, is caused by an energy-deficient diet and particularly a lack of proteins

Jamaican-born Olympic athlete Tessa Sanderson, who set a new British javelin record in 1980. Well over 50 kJ min^{-1} of energy are required for javelin throwing and other vigorous athletic activities

Table 5.5 Approximate energy requirements for common activities

	approximate energy requirement/	
	kJ min^{-1}	kcal min^{-1}
sitting	6	1.5
standing	10	2.5
walking	16	3.8
running	40	9.6

2. If we consume more energy than we expend, the surplus is **stored** as fat. We put on roughly a pound of fat for every 15 000 kilojoules we consume in excess. This is equivalent to about 3500 kilocalories.
3. Most of the energy we consume is **used up** by our bodies. The exact amount we use depends on our age, sex, height, weight and the amount of activity we do. For men, the range is usually between 9200 kJ and 12 100 kJ per day although for manual workers it can be more. For women, the range is 6700 to 8800 kJ per day.

Energy is essentially expended in three ways:

- Energy is needed for the chemical changes that go on inside our bodies. These changes come under the name 'metabolism'. For example, energy is needed to metabolise the food we consume.
- Energy is needed to keep us warm in cold weather.
- Energy is needed for muscular activity. Table 5.5 shows approximate energy requirements for four common activities.

HESS'S LAW

Many reaction enthalpies cannot be determined experimentally because the reactions cannot be brought about in the laboratory. For example, it is not possible to determine the enthalpy of formation of ethanol in the laboratory because it is not possible to synthesise ethanol from carbon, hydrogen and oxygen. Such reaction enthalpies can be calculated from the enthalpies of other reactions by an application of the first law of thermodynamics. The application is called **Hess's law of constant heat summation**. It states that if a given change can be brought about in more than one way, the overall enthalpy change is the same for each way.

This means that if a reaction, in theory, can be divided up into a number of steps the enthalpy change for the overall reaction equals the sum of the enthalpy changes for each step (*see* figure 5.9).

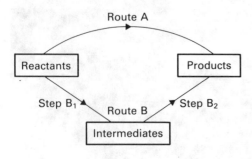

Figure 5.9   From Hess's law

$$\Delta H^{\ominus}_{\text{route A}} = \Delta H^{\ominus}_{\text{route B}}$$
$$= \Delta H^{\ominus}_{\text{step B}_1} + \Delta H^{\ominus}_{\text{step B}_2}$$

The Greek letter Σ means 'the sum of'. The sum must allow for the number of moles of each reactant and product as shown in the balanced equation representing the reaction.

Let us take the combustion of graphite as an example. The following are both conceivable routes:

	Equation	Standard molar enthalpy change
Route A:	$C(\text{graphite}) + O_2(g) \rightarrow CO_2(g)$	$\Delta H^{\ominus}_{\text{A,m}}$
Route B:		
step B_1	$C(\text{graphite}) + \frac{1}{2}O_2(g) \rightarrow CO(g)$	$\Delta H^{\ominus}_{\text{B}_1,\text{m}}$
step B_2	$CO(g) + \frac{1}{2}O_2(g) \rightarrow CO_2(g)$	$\Delta H^{\ominus}_{\text{B}_2,\text{m}}$

The net result is the same by either route. That is, one mole of graphite combines with one mole of oxygen to form one mole of carbon dioxide. By Hess's law we have

$$\Delta H^{\ominus}_{\text{A,m}} = \Delta H^{\ominus}_{\text{B}_1,\text{m}} + \Delta H^{\ominus}_{\text{B}_2,\text{m}} \tag{11}$$

Route A is, in fact, the formation of carbon dioxide from its elements in their standard states. Thus

$$\Delta H^{\ominus}_{\text{A,m}} = \Delta H^{\ominus}_{\text{f,m}}(CO_2)$$

Step B_1 is the formation of carbon monoxide from its elements in their standard states. Thus

$$\Delta H^{\ominus}_{\text{B}_1,\text{m}} = \Delta H^{\ominus}_{\text{f,m}}(CO)$$

When these standard molar enthalpies of formation are put into equation (11) we obtain

$$\Delta H^{\ominus}_{\text{f,m}}(CO_2) = \Delta H^{\ominus}_{\text{f,m}}(CO) + \Delta H^{\ominus}_{\text{B}_2,\text{m}}$$

On rearrangement this equation becomes

$$\Delta H^{\ominus}_{\text{B}_2,\text{m}} = \Delta H^{\ominus}_{\text{f,m}}(CO_2) - \Delta H^{\ominus}_{\text{f,m}}(CO)$$

Thus, the standard molar enthalpy change for step B_2 can be calculated from the standard molar enthalpies of formation of the reactants and products. The standard molar enthalpy of formation of an element is zero and so the standard molar enthalpy of formation of oxygen need not be included in the above expression.

It is possible to derive a similar relationship for any chemical reaction. Expressed in general terms this relationship is:

● $$\Delta H^{\ominus}_{\text{r,m}} = \Sigma[\Delta H^{\ominus}_{\text{f,m}}(\text{products})] - \Sigma[\Delta H^{\ominus}_{\text{f,m}}(\text{reactants})] \tag{12}$$

where $\Delta H^{\ominus}_{\text{r,m}}$ is the standard molar enthalpy of reaction.

Equation (12) is the most useful form of Hess's law.

EXAMPLE

Calculate the standard molar enthalpy of formation of carbon monoxide from the following information:

$C(\text{graphite}) + O_2(g) \rightarrow CO_2(g)$ $\Delta H^{\ominus}_{\text{f,m}}(CO_2, 298\text{ K}) = -393.5\text{ kJ mol}^{-1}$

$CO(g) + \frac{1}{2}O_2(g) \rightarrow CO_2(g)$ $\Delta H^{\ominus}_{\text{r,m}}(298\text{ K}) = -283.0\text{ kJ mol}^{-1}$

SOLUTION

We can apply Hess's law in the form of equation (12) to the second reaction, and obtain

$$\Delta H^{\ominus}_{\text{r,m}}(298\text{ K}) = \underset{\text{product}}{[\Delta H^{\ominus}_{\text{f,m}}(CO_2, 298\text{ K})]} - \underset{\text{reactants}}{[\Delta H^{\ominus}_{\text{f,m}}(CO, 298\text{ K}) + \frac{1}{2}\Delta H^{\ominus}_{\text{f,m}}(O_2, 298\text{ K})]}$$

We can put

$$\Delta H^{\ominus}_{\text{f,m}}(O_2, 298\text{ K}) = 0$$

Thus, by rearranging the equation above we obtain

$$\Delta H_{f,m}^{\ominus}(CO,298\ K)\ =\ \Delta H_{f,m}^{\ominus}(CO_2,298\ K) - \Delta H_{r,m}^{\ominus}(298\ K)$$
$$=\ (-393.5\ kJ\ mol^{-1}) - (-283.0\ kJ\ mol^{-1})$$
$$=\ -110.5\ kJ\ mol^{-1}$$

Enthalpy Diagrams

The information in the example above can be summarised in the form of an enthalpy diagram (*see* figure 5.10). The top line in this diagram is called the datum line or datum level. It corresponds to $H^{\ominus} = 0$ and represents elements in their standard states. For an endothermic reaction, the datum line would be at the bottom.

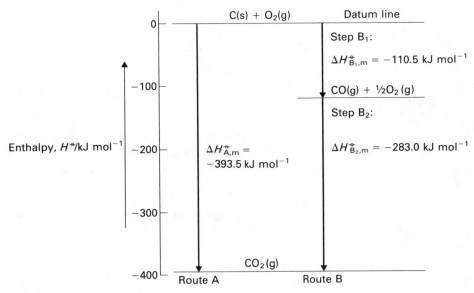

Figure 5.10 Enthalpy diagram for the formation of CO_2

EXAMPLE

(a) Calculate the standard molar enthalpy of formation of ethanol from the following *experimental* data:

$$\Delta H_{c,m}^{\ominus}(C_2H_5OH,\ 298\ K)\ =\ -1367\ kJ\ mol^{-1}$$
$$\Delta H_{f,m}^{\ominus}(CO_2,\ 298\ K)\qquad =\ -393.5\ kJ\ mol^{-1}$$
$$\Delta H_{f,m}^{\ominus}(H_2O,\ 298\ K)\qquad =\ -285.8\ kJ\ mol^{-1}$$

(b) Draw the enthalpy diagram for the combustion of ethanol.

SOLUTION

(a) We must first write the equation for the complete combustion of ethanol:

$$C_2H_5OH(l)\ +\ 3O_2(g) \rightarrow 2CO_2(g) + 3H_2O(l)\quad \Delta H_{c,m}^{\ominus}(298\ K) = -1367\ kJ\ mol^{-1}$$

Applying equation (12) to this reaction and remembering that the standard molar enthalpy of formation of elements in their standard states is zero we obtain

$$\Delta H_{c,m}^{\ominus}(C_2H_5OH,298\ K) = 2\Delta H_{f,m}^{\ominus}(CO_2,298\ K) + 3\Delta H_{f,m}^{\ominus}(H_2O,298\ K) - \Delta H_{f,m}^{\ominus}(C_2H_5OH,298\ K)$$

By rearranging this equation and substituting the above values for the standard molar enthalpies, we obtain

$$\Delta H_{f,m}^{\ominus}(C_2H_5OH,298\ K)\ =\ 2(-393.5\ kJ\ mol^{-1}) + 3(-285.8\ kJ\ mol^{-1}) - (-1367\ kJ\ mol^{-1})$$
$$=\ -277.4\ kJ\ mol^{-1}$$

(b) The enthalpy diagram will show two routes to the formation of the combustion products of ethanol from the elements in their standard states: route A is the overall reaction; route B is divided into two steps, B$_1$—the formation of C$_2$H$_5$OH(l) from its elements in their standard states, and B$_2$—the combustion of C$_2$H$_5$OH(l). Thus, we can proceed as follows.

Route A:

$$2C(graphite) + 3H_2(g) + 3\tfrac{1}{2}O_2(g) \rightarrow 2CO_2(g) + 3H_2O(l) \quad \Delta H^{\ominus}_{A,m}$$

where

$$\begin{aligned}\Delta H^{\ominus}_{A,m} &= 2\Delta H^{\ominus}_{f,m}(CO_2, 298\ K) + 3\Delta H^{\ominus}_{f,m}(H_2O, 298\ K) \\ &= -1644.4\ kJ\ mol^{-1}\end{aligned}$$

Note that the term mol^{-1} refers to the reaction as specified by the equation, that is, two moles of CO$_2$, three moles of H$_2$O. The term mol^{-1} thus means per mole of the specified equation.

Route B:

step B$_1$ $$2C(graphite) + 3H_2(g) + \tfrac{1}{2}O_2(g) \rightarrow C_2H_5OH(l) \qquad \Delta H^{\ominus}_{B_1,m}$$

where

$$\begin{aligned}\Delta H^{\ominus}_{B_1,m} &= \Delta H^{\ominus}_{f,m}(C_2H_5OH, 298\ K) \\ &= -277.4\ kJ\ mol^{-1}\end{aligned}$$

Note once again that $\Delta H^{\ominus}_{B_1,m}$ refers to 'per mole of the specified balanced chemical equation'. C$_2$H$_5$OH thus need not be included in the symbol. However, if the equation is not specified, the entity to which the enthalpy change refers must

Figure 5.11   Enthalpy diagram for the combustion of ethanol

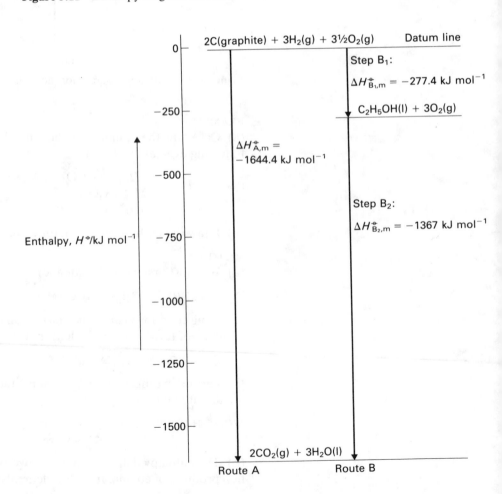

be included in the symbol. For example, $\Delta H^{\ominus}_{f,m}(C_2H_5OH, 298\ K)$ refers to 'per mole of C_2H_5OH formed from its elements in their standard states'.

step B_2 $$C_2H_5OH(l) + 3O_2(g) \rightarrow 2CO_2(g) + 3H_2O(l) \qquad \Delta H^{\ominus}_{B_2,m}$$

where

$$\Delta H^{\ominus}_{B_2,m} = \Delta H^{\ominus}_{c,m}(C_2H_5OH, 298\ K)$$
$$= -1367\ kJ\ mol^{-1}$$

The standard molar enthalpies for route A and steps B_1 and B_2 can now be included in an enthalpy diagram (*see* figure 5.11). This is the enthalpy diagram for the combustion of ethanol.

SUMMARY

1. $\Delta H = H_2 - H_1$, where ΔH is the **enthalpy change**.
   (a) If ΔH is negative, the reaction is **exothermic**.
   (b) If ΔH is positive, the reaction is **endothermic**.
2. The **standard molar enthalpy** is the enthalpy change per mole under standard conditions. **Standard conditions** are 298 K and one atmosphere pressure. Standard molar enthalpy at 298 K has the symbol $\Delta H^{\ominus}_m(298\ K)$.
3. $\Delta H^{\ominus}_{f,m}\{298\ K, \text{an element}\} = 0$.
4. For most neutralisations of a strong acid with a strong alkali

$$\Delta H^{\ominus}_{n,m}(298\ K) = -57.1\ kJ\ mol^{-1}$$

5. **Trouton's rule**:

$$\frac{\Delta H^{\ominus}_{vap,m}}{T} = 88\ J\ K^{-1}\ mol^{-1}$$

6. Enthalpy changes, internal energy changes and heat capacities can be determined experimentally using calorimeters.
   (a) A **bomb calorimeter** is used for determining the internal energy change that accompanies combustion.
   (b) A **flame calorimeter** is used to determine the enthalpies of combustion of volatile organic liquids.
7. The energy evolved in a typical calorimetry experiment is given by

$$q = [C_{vessel} + (m \times c_{contents})]\Delta T$$

   (a) The term C_{vessel} can be ignored if the heat capacity of the calorimeter is negligible.
   (b) For accurate calorimetric determinations ΔT must be corrected for cooling losses.
   (c) A second ('control') experiment is often carried out to calibrate a calorimeter.
8. Fats, proteins and carbohydrates in food and drink provide the energy we need to keep our bodies functioning.
9. **Hess's law** of constant heat summation states that, if a given change can be brought about in more than one way, the overall enthalpy change is the same for each way.
10. The most useful form of Hess's law is

$$\Delta H^{\ominus}_{r,m} = \Sigma[\Delta H^{\ominus}_{f,m}(\text{products})] - \Sigma[\Delta H^{\ominus}_{f,m}(\text{reactants})]$$

11. Information concerning the enthalpy changes relating to a specific chemical reaction can be summarised in the form of an **enthalpy diagram**.

5.3
Energy and
Structure

BOND ENTHALPIES

The **standard bond dissociation enthalpy** is the enthalpy change that accompanies the breaking of one mole of bonds with the molecules and resulting fragments being in their standard states at 298 K and a pressure of one atmosphere.

A bond dissociation enthalpy refers to a specific bond in a molecule. Its value depends on the local environment. For example, in the methane molecule there are four C—H bonds and thus four bond dissociation enthalpies. These are referred to as the first, second, third and fourth bond dissociation enthalpies respectively.

First: $CH_4(g) \rightarrow CH_3(g) + H(g)$ $\Delta H_{d_1,m}^{\ominus}(298 \text{ K}) = +425 \text{ kJ mol}^{-1}$

Second: $CH_3(g) \rightarrow CH_2(g) + H(g)$ $\Delta H_{d_2,m}^{\ominus}(298 \text{ K}) = +470 \text{ kJ mol}^{-1}$

Third: $CH_2(g) \rightarrow CH(g) + H(g)$ $\Delta H_{d_3,m}^{\ominus}(298 \text{ K}) = +416 \text{ kJ mol}^{-1}$

Fourth: $CH(g) \rightarrow C(g) + H(g)$ $\Delta H_{d_4,m}^{\ominus}(298 \text{ K}) = +335 \text{ kJ mol}^{-1}$

Similarly, for water, there are two O—H bond dissociation enthalpies.

First: $HO-H(g) \rightarrow HO(g) + H(g)$ $\Delta H_{d_1,m}^{\ominus}(298 \text{ K}) = +494 \text{ kJ mol}^{-1}$

Second: $H-O(g) \rightarrow H(g) + O(g)$ $\Delta H_{d_2,m}^{\ominus}(298 \text{ K}) = +430 \text{ kJ mol}^{-1}$

Notice that bond dissociation enthalpies are endothermic. Energy must be absorbed to break a bond.

If we average the four standard bond dissociation enthalpies for the methane molecule, we obtain a value of +412 kJ mol^{-1}. The average C—H standard bond dissociation enthalpy for all types of molecule including the methane molecule is +413 kJ mol^{-1}. This is called the average or mean bond enthalpy, the bond enthalpy term or simply the **bond energy**. A list of some mean bond enthalpies is shown in table 5.6.

The mean bond enthalpy of an element or compound which exists as diatomic molecules is exactly the same as its bond dissociation enthalpy. For such elements and compounds, the bond dissociation enthalpy has twice the value of the enthalpy of atomisation.

The **standard enthalpy of atomisation** of an element is the enthalpy change which accompanies the production of one mole of isolated gaseous atoms from the element in its standard state at 298 K and one atmosphere pressure. For example, the standard enthalpy of atomisation of hydrogen is given by

$$\tfrac{1}{2}H_2(g) \rightarrow H(g) \qquad \Delta H_{a,m}^{\ominus}(298 \text{ K}) = +218 \text{ kJ mol}^{-1}$$

whereas the standard bond dissociation enthalpy for hydrogen is given by

$$H_2(g) \rightarrow 2H(g) \qquad \Delta H_{d,m}^{\ominus}(298 \text{ K}) = +436 \text{ kJ mol}^{-1}$$

For solid elements the enthalpy of atomisation is equal to the enthalpy of sublimation of the solid. **Sublimation** is the change from solid to gas on heating and from gas to solid on cooling without passing through the liquid phase.

Determination of Bond Enthalpies

Bond enthalpies can be determined by spectroscopic methods (*see* section 17.1), by electron impact methods (*see* section 1.2) or by thermochemical methods. Thermochemical methods involve the calorimetric determination of the enthalpies of combustion, formation and other processes. These data are then used to calculate bond enthalpies. Bond enthalpies cannot be determined directly by thermochemical methods.

Experimental evidence for the consistency of bond enthalpies over a number of compounds is provided by the standard molar enthalpies of combustion of alkanes

Table 5.6   Mean bond enthalpies

Bond	$\Delta H_{d,m}^{\ominus}(298 \text{ K})/\text{kJ mol}^{-1}$
H—H	435
H—Cl	431
C—C	347
C—H	413
C—O	335
C—Cl	326
C—N	293
N—N	159
N—H	389
O—O	138
O—H	464
Cl—Cl	243

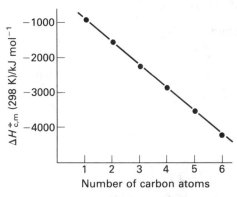

Figure 5.12 Standard molar enthalpies of combustion for the first six members of the homologous series of alkanes

in their homologous series (*see* section 17.1). When these are plotted against number of carbon atoms, a straight line is obtained (*see* figure 5.12). The difference in the standard molar enthalpy of combustion between each successive member of the series is about 660 kJ mol^{-1} (*see* table 5.7). The difference in the structure of each successive member of a homologous series is —CH$_2$—. The value of 660 kJ mol^{-1} thus corresponds to the dissociation and formation of bonds in the combustion of one mole of the entity —CH$_2$—. The combustion of each —CH$_2$— involves

- the dissociation of two C—H bonds;
- the dissociation of one C—C bond;
- the formation of two C=O bonds to produce a molecule of CO_2;
- the formation of two H—O bonds to produce a molecule of H_2O.

Table 5.7   Standard molar enthalpies of combustion of alkanes

Number of C atoms	Alkane	Formula	$\Delta H_{c,m}^{\ominus}$(298 K)/kJ mol^{-1}
1	methane	CH_4	−890
2	ethane	C_2H_6	−1560
3	propane	C_3H_8	−2220
4	butane	C_4H_{10}	−2877
5	pentane	C_5H_{12}	−3520
6	hexane	C_6H_{14}	−4195

The consistent value of 660 kJ mol^{-1} suggests that the bond dissociation enthalpies for the C—H and C—C bonds are fairly consistent for all alkanes.

Determination of the Mean Bond Enthalpy of the C—H Bond

The mean bond enthalpy of the C—H bond can be determined from experimental data by first finding the standard enthalpy of atomisation of methane:

$$CH_4(g) \rightarrow C(g) + 4H(g) \qquad \Delta H_{a,m}^{\ominus}(CH_4, 298 \text{ K})$$

The standard enthalpy of atomisation is four times the mean C—H bond enthalpy:

$$\Delta H_{a,m}^{\ominus}(CH_4, 298 \text{ K}) = 4 \, \Delta H_{d,m}^{\ominus}(C—H, 298 \text{ K}) \qquad (13)$$

The standard enthalpy of atomisation is four times the mean C—H bond enthalpy: mentally. However, it can be calculated by application of Hess's law to three standard enthalpies which can be determined experimentally. These are:

1. The standard enthalpy of atomisation of graphite

$$C(\text{graphite}) \rightarrow C(g) \qquad \Delta H_{a,m}^{\ominus}(298 \text{ K}) = +715 \text{ kJ mol}^{-1}$$

This enthalpy is found by measuring the temperature dependence of the vapour pressure of graphite.

2. The standard enthalpy of atomisation of hydrogen

$$\tfrac{1}{2}H_2(g) \rightarrow H(g) \qquad \Delta H_{a,m}^{\ominus}(298 \text{ K}) = +218 \text{ kJ mol}^{-1}$$

This enthalpy can be determined spectroscopically.

3. The standard enthalpy of formation of methane

$$C(\text{graphite}) + 2H_2(g) \rightarrow CH_4(g) \qquad \Delta H_{f,m}^{\ominus}(298 \text{ K}) = -75 \text{ kJ mol}^{-1}$$

This value can be calculated from the standard enthalpy of combustion of methane and the standard enthalpies of formation of carbon dioxide and water. All three of these enthalpies can be determined experimentally by calorimetric methods.

We can now apply Hess's law (equation (12)) to the atomisation of methane:

$$\begin{aligned}
\Delta H_{a,m}^{\ominus}(CH_4, 298\ K) &= \Delta H_{a,m}^{\ominus}\{C(graphite), 298\ K\} \\
\text{(atomisation)} &\quad + 4\Delta H_{a,m}^{\ominus}(\tfrac{1}{2}H_2, 298\ K) \qquad \text{(products)} \\
&\quad - \Delta H_{f,m}^{\ominus}(CH_4, 298\ K) \qquad\qquad \text{(reactant)} \\
&= [(+715) + 4(+218) - (-75)]\ kJ\ mol^{-1} \\
&= +1662\ kJ\ mol^{-1}
\end{aligned}$$

Applying equation (13) we obtain

$$\Delta H_{d,m}^{\ominus}(C\!-\!H,\ 298\ K) = +\frac{1662}{4}\ kJ\ mol^{-1}$$

$$= +415.5\ kJ\ mol^{-1}$$

The information required to calculate the mean bond enthalpy for C—H can be summarised in an enthalpy diagram (*see* figure 5.13).

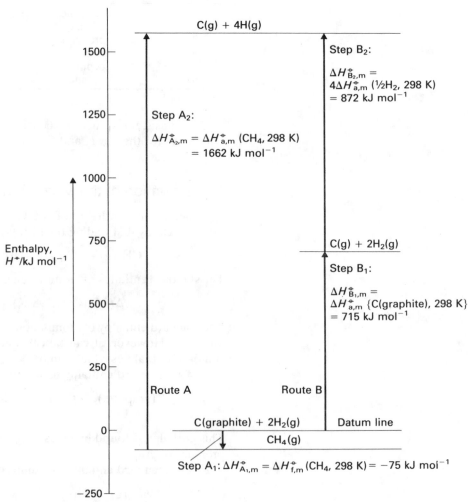

Figure 5.13 Enthalpy diagram summarising the data used for determining the mean C–H bond enthalpy in methane

Interpretation and Uses of Bond Enthalpies

A knowledge of bond enthalpies can help us to understand the structure of covalent compounds and also the mechanisms (*see* chapter 9) of their chemical reactions.

In section 2.1 we saw how the Pauling scale of electronegativities of elements was based on bond enthalpies. In general, as the difference in electronegativity between two elements forming a bond decreases, so does the bond enthalpy. This is illustrated by comparing the electronegativity differences between hydrogen and the halogens with the bond enthalpies of the hydrogen halides (*see* table 5.8).

Table 5.8 Bond enthalpies of the hydrogen halides

Bond	Electronegativity difference between hydrogen and halogen	$\Delta H_{d,m}^{\ominus}(298 \text{ K})/\text{kJ mol}^{-1}$
H—F	1.9	+565
H—Cl	0.9	+431
H—Br	0.7	+364
H—I	0.3	+297

The data in table 5.9 show that bond enthalpies increase as the number of bonds between two atoms decreases. This follows since the strength of a bond depends on the number of bonding electrons. As the number of bonding electrons increases, the bond length decreases and the bond becomes stronger.

Table 5.9 Bond enthalpies decrease with increasing bond length

Bond	Number of bonding electrons	Bond length/nm	$\Delta H_{d,m}^{\ominus}(298 \text{ K})/\text{kJ mol}^{-1}$
C—C	2	0.154	+347
C=C	4	0.134	+619
C≡C	6	0.120	+837
Cl—Cl	2	0.199	+242
Br—Br	2	0.228	+193
I—I	2	0.266	+151

The size of atoms also influences bond enthalpies and bond lengths. Table 5.9 also compares values for the bonds in the diatomic halogen molecules as the group is descended. On descending the group, the atoms become larger, with the result that bond lengths are increased due to increasing repulsion between the atoms. The bond thus becomes weaker and the bond enthalpies decrease.

Bond enthalpies can also be used to calculate the enthalpies of reactions involving covalent molecules. As an approximate guide, the following relationship can be used:

$$\Delta H_{r,m}^{\ominus} = \Sigma \Delta H_{d,m}^{\ominus}(\text{bonds broken}) - \Sigma \Delta H_{d,m}^{\ominus}(\text{bonds formed}) \qquad (14)$$

EXAMPLE
Use the data in tables 5.6 and 5.9 to calculate the standard molar enthalpy change of the hydrogenation of ethene.

SOLUTION
Bonds broken: C=C, H—H
Bonds formed: C—C, 2(C—H)

By applying equation (14) we obtain

$$\Delta H_{r,m}^{\ominus} = \Delta H_{d,m}^{\ominus}(\text{C=C}) + \Delta H_{d,m}^{\ominus}(\text{H—H}) - \Delta H_{d,m}^{\ominus}(\text{C—C}) - 2\Delta H_{d,m}^{\ominus}(\text{C—H})$$

The required data are found in tables 5.6 and 5.9:

$$\Delta H_{r,m}^{\ominus}(298\ K) = [(619) + (435) - (347) - 2(413)]\ kJ\ mol^{-1}$$
$$= -119\ kJ\ mol^{-1}$$

The Stability of Benzene

In the above example the experimental value for the standard molar enthalpy of hydrogenation of ethene is $-137\ kJ\ mol^{-1}$. This is slightly larger than the calculated value of $-119\ kJ\ mol^{-1}$. For the hydrogenation of benzene, however, there is a much larger discrepancy between the experimental value and the value calculated from bond enthalpies.

We can see this by looking at the hydrogenation of cyclohexene. The calculated standard molar enthalpy of hydrogenation of this compound is virtually the same as that calculated for ethene. This is because exactly the same bonds are broken and formed. Thus

Cyclohexene   Cyclohexane $\Delta H_{r,m}^{\ominus}(298\ K) = -121\ kJ\ mol^{-1}$

Benzene contains three double bonds and so we can calculate the standard molar enthalpy of hydrogenation of benzene by just multiplying the value for cyclohexene by 3. Thus

$+\quad 3H_2(g) \longrightarrow$ $\Delta H_{r,m}^{\ominus}(298\ K) = -363\ kJ\ mol^{-1}$

The experimental value for the standard molar enthalpy of hydrogenation of benzene is $-208\ kJ\ mol^{-1}$. The difference is thus $155\ kJ\ mol^{-1}$. This is a measure of the stability of benzene due to the delocalisation of its π electrons (*see* section 2.1). For this reason it is sometimes called the **delocalisation enthalpy**, **resonance energy** or **stabilisation energy**.

LATTICE ENTHALPIES

We have seen how bond enthalpies provide a measure of the strength of covalent bonds. For ionic compounds the comparable enthalpy is the lattice enthalpy. This is also called **lattice energy**.

The **standard molar lattice enthalpy** is the enthalpy change which accompanies the formation of one mole of the solid ionic compound from its gaseous ions under standard conditions.

Lattice enthalpies cannot be determined directly. However, they can be calculated from experimentally determined enthalpies by application of Hess's law. Let us take the sodium chloride lattice, for example. The formation of the lattice from its elements in their standard states can be broken down into five steps. The overall reaction we called route A.

Route A: $Na(s) + \frac{1}{2}Cl_2(g) \rightarrow Na^+Cl^-(s)$ $\Delta H_{A,m}^{\ominus} = -411\ kJ\ mol^{-1}$

where $\Delta H_{A,m}^{\ominus}$ is the standard molar enthalpy of formation of sodium chloride.

Route B: This is broken down into five steps, one of which is the formation of the lattice from its gaseous ions.

step B_1 $Na(s) \rightarrow Na(g)$ $\Delta H_{B_1,m}^{\ominus} = +108\ kJ\ mol^{-1}$

This process is the atomisation or sublimation of sodium.

step B_2 $Na(g) \rightarrow Na^+(g) + e^-$ $\Delta H^{\ominus}_{B_2,m} = +495$ kJ mol^{-1}

This is the ionisation of gaseous sodium. The enthalpy change is the first ionisation energy of sodium (*see* section 2.2).

step B_3 $\frac{1}{2}Cl_2(g) \rightarrow Cl(g)$ $\Delta H^{\ominus}_{B_3,m} = +122$ kJ mol^{-1}

This is the atomisation of chlorine. The enthalpy of atomisation of chlorine is half its bond dissociation enthalpy.

step B_4 $Cl(g) + e^- \rightarrow Cl^-(g)$ $\Delta H^{\ominus}_{B_4,m} = -360$ kJ mol^{-1}

This is the ionisation of chlorine. The enthalpy corresponds to the first electron affinity of chlorine (*see* section 3.1).

step B_5 $Na^+(g) + Cl^-(g) \rightarrow Na^+Cl^-(s)$ $\Delta H^{\ominus}_{B_5,m} = ?$

This is the formation of the lattice from its gaseous ions.

$\Delta H^{\ominus}_{B_5,m}$ is the standard molar lattice enthalpy of sodium chloride. We can now calculate its value by application of Hess's law. The following equation is analogous to equation (11):

$$\Delta H^{\ominus}_{A,m} = \Delta H^{\ominus}_{B_1,m} + \Delta H^{\ominus}_{B_2,m} + \Delta H^{\ominus}_{B_3,m} + \Delta H^{\ominus}_{B_4,m} + \Delta H^{\ominus}_{B_5,m}$$

By rearranging this equation and substituting the values shown above we obtain

$$\Delta H^{\ominus}_{B_5,m} = [(-411) - (108 + 495 + 122 - 360)] \text{ kJ mol}^{-1}$$
$$= -776 \text{ kJ mol}^{-1}$$

This is the standard molar lattice enthalpy of sodium chloride.

The Born–Haber Cycle

The information used in the above calculation of lattice enthalpy can be summarised in an enthalpy diagram known as the Born–Haber cycle. The Born–Haber cycle for sodium chloride consists of route A and the five steps of route B (*see* figure 5.14). The cycle can be used to determine the lattice enthalpy of sodium chloride. Experimental values for the enthalpies of route A and steps B_1 to B_4 are put into the cycle. The enthalpy needed to complete the cycle is the lattice enthalpy.

Interpreting Lattice Enthalpies

We have seen that bond dissociation enthalpies are endothermic since energy must be absorbed to break bonds. Lattice enthalpies, on the other hand, are exothermic. This is because lattice enthalpies relate to the formation of bonds and energy is released when bonds are formed.

Table 5.10 shows the lattice enthalpies of a number of compounds. Notice that, as the electronegativity of the halogens decreases down the group (*see* section 2.2), so does the lattice enthalpy.

Lattice enthalpies determined through the Born–Haber cycle can be regarded as experimental values since the lattice enthalpies are calculated from experimental values. Lattice enthalpies can also be calculated using a theoretical model. The model is based on the geometry of the crystal lattice. It assumes that the ions are point charges. The potential energy V required to bring two ions with charges q_1 and q_2 from an infinite distance to a distance r from each other is given by

$$V = \frac{q_1 q_2}{r}$$

The exact calculation of lattice enthalpy from this potential energy expression is quite complicated. However, for compounds such as the alkali halides, it does

Table 5.10 Lattice enthalpies

	Lattice enthalpy/kJ mol^{-1}	
	Theoretical	Experimental
NaCl	−766	−776
NaBr	−731	−719
NaI	−686	−670
AgCl	−769	−921
AgBr	−759	−876
AgI	−736	−862
ZnS	−3430	−3739

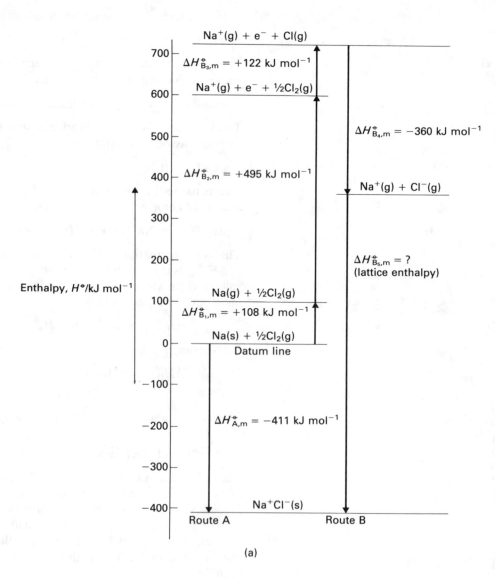

(a)

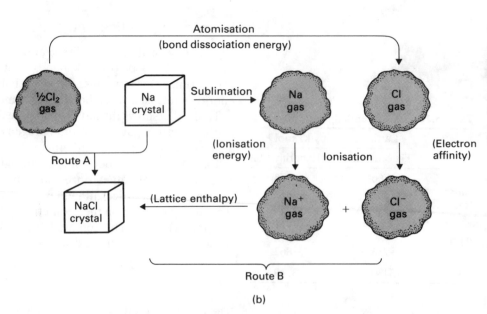

(b)

Figure 5.14　(a) and (b) Born–Haber cycle for sodium chloride

give theoretical values of lattice enthalpies which are in good agreement with experimental values. For example, the theoretical value for the lattice enthalpy of sodium chloride is -766 kJ mol^{-1}, whereas the experimental value determined through the Born–Haber cycle is -776 kJ mol^{-1}.

There is a far greater discrepancy for other compounds, however. For example, the theoretical value for silver chloride is -769 kJ mol^{-1}, whereas the experimental value is -921 kJ mol^{-1}. This shows that the bonding in silver chloride is stronger than in sodium chloride. This is because the bonding is partially covalent in character. Agreement tends to be far better when the electronegativity difference between the two ions is high. A comparison of the theoretical and experimental values of lattice enthalpies thus enables us to assess the degree of ionic character of a compound.

Enthalpies of Solution and Hydration

The enthalpy change that occurs when one mole of solute is dissolved in a solvent to form an infinitely dilute solution is called the **enthalpy of solution**—or quite often the heat of solution. It is easily measured experimentally.

The enthalpy change that occurs can be attributed to two factors. First of all the bonds that hold the particle together in the solid have to be broken. Secondly, the particles become solvated.

Let us take the case of an ionic solid. First there is an enthalpy change when the ionic lattice in the solid breaks down. Secondly, the ions become solvated in the water, that is, they become hydrated. The overall process can be exothermic or endothermic. For example, when sodium chloride dissolves in water, there is a small net gain in energy. The process is thus endothermic.

$$Na^+Cl^-(s) + (aq) \rightarrow Na^+(aq) + Cl^-(aq) \qquad \Delta H^{\ominus}_{solution,m} = +4 \text{ kJ mol}^{-1}$$

The breaking down of the lattice is obviously the reverse of its formation. Thus, the energy required to break down one mole of the lattice into isolated gaseous ions is equal to the reverse of the energy change which accompanies the formation of one mole of the lattice from its gaseous ions. The latter energy change is the lattice enthalpy. For sodium chloride its value is -776 kJ mol^{-1}. We can therefore write

$$NaCl(s) \rightarrow Na^+(g) + Cl^-(g) \qquad \begin{aligned} \Delta H^{\ominus}_m &= -\Delta H^{\ominus}_{lattice,m}\{NaCl(s)\} \\ &= +776 \text{ kJ mol}^{-1} \end{aligned}$$

The enthalpy change that accompanies the hydration of one mole of both these gaseous ions is called the **enthalpy of hydration**.

$$Na^+(g) + Cl^-(g) + (aq) \rightarrow Na^+(aq) + Cl^-(aq) \qquad \Delta H^{\ominus}_{hydration,m} = -772 \text{ kJ mol}^{-1}$$

Values for hydration enthalpies cannot be found experimentally. However, they can be found by application of Hess's law:

$$\Delta H^{\ominus}_{solution,m} = -\Delta H^{\ominus}_{lattice,m} + \Delta H^{\ominus}_{hydration,m}$$

and thus

$$\Delta H^{\ominus}_{hydration,m} = \Delta H^{\ominus}_{solution,m} + \Delta H^{\ominus}_{lattice,m}$$

Substituting values for NaCl(s) we obtain

$$\begin{aligned} \Delta H^{\ominus}_{hydration,m}\{NaCl(s)\} &= [(+4) + (-776)] \text{ kJ mol}^{-1} \\ &= -772 \text{ kJ mol}^{-1} \end{aligned}$$

As usual, the relationship between these three enthalpies can be displayed in the form of an enthalpy diagram (*see* figure 5.15). The datum line is omitted from the enthalpy diagram since the enthalpy of formation from elements in their standard states is not involved in the calculation. Furthermore, since we are only considering enthalpy changes, we can omit the absolute scale of enthalpy, $H^{\ominus}$.

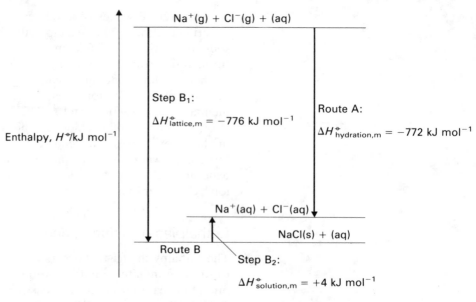

$$Na^+(g) + Cl^-(g) + (aq)$$

Enthalpy, $H^\ominus$/kJ mol^{-1}

Step B$_1$:

$\Delta H^\ominus_{lattice,m} = -776$ kJ mol^{-1}

Route A:

$\Delta H^\ominus_{hydration,m} = -772$ kJ mol^{-1}

$$Na^+(aq) + Cl^-(aq)$$

$$NaCl(s) + (aq)$$

Route B

Step B$_2$:

$\Delta H^\ominus_{solution,m} = +4$ kJ mol^{-1}

Figure 5.15   Enthalpy diagram for aqueous sodium chloride

The enthalpy of hydration for sodium chloride is the sum of the individual hydration enthalpies of the sodium and chloride ions:

$$\Delta H^\ominus_{hydration,m}\{NaCl(s)\} = \Delta H^\ominus_{hydration,m}\{Na^+(s)\} + \Delta H^\ominus_{hydration,m}\{Cl^-(s)\}$$

The hydration enthalpies of individual ions can be estimated by comparing the hydration enthalpies of a range of ionic compounds with a common ion. For such estimations the hydration enthalpy of the hydration ion, H^+, is usually taken as a standard. Its value is -1075 kJ mol^{-1}. The hydration enthalpies of some alkali metal ions and halide ions are shown in table 5.11. Note how the values decrease as each group is descended.

ENTROPY

The first law of thermodynamics states that whilst energy may be transferred between a system and its surroundings, energy is never created or destroyed. It therefore imposes the conservation of energy as a restriction on chemical and physical changes. At one time it was thought that all chemical reactions were exothermic. In other words, a chemical reaction could only proceed if the system lost energy. However, many chemical and physical changes are now known which are endothermic. Thus, energy or enthalpy change alone cannot help us to predict whether a reaction will occur or not. To predict whether a reaction will occur spontaneously or not it is necessary to introduce another thermodynamic state function called **entropy**. Entropy is given the symbol S.

Entropy can be thought of as a measure of the chaos, randomness or disorder of a system. Gases, for example, are relatively disordered compared to solids. The entropy of gases is much higher than the entropy of solids.

But how does entropy help us to predict whether a change will take place or not? To answer this, let us consider a system consisting of two bulbs joined by a stopcock (see figure 5.16). The two bulbs contain different gases. When the stopcock is opened, the gases will mix spontaneously by a process called diffusion (*see* section 3.1). After mixing the gases have a higher state of disorder than before mixing. They therefore have a higher entropy after mixing. During this spontaneous process, the entropy of the system therefore increases. There is no energy change in this process. The overall enthalpy of the gases before and after mixing is exactly the same. However, mixing does result in the more random dispersal of energy.

Table 5.11   Hydration enthalpies

Ion	$\Delta H^\ominus_{hydration,m}$/kJ mol^{-1}
Group I	
Li$^+$	-499
Na$^+$	-390
K$^+$	-305
Group VII	
F$^-$	-457
Cl$^-$	-382
Br$^-$	-351
I$^-$	-307

Before

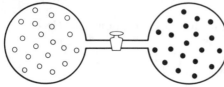

After

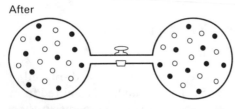

Figure 5.16   The spontaneous mixing of two gases results in an increase in entropy but no overall energy change

In many chemical reactions, energy is also dispersed. For example, combustion reactions are exothermic processes. During combustion, energy is released and dispersed to the surroundings. Entropy can thus be viewed as a measure of the dispersal of energy. During chemical reactions, there is always a dispersal of energy either from the chemical system to its surroundings or from the surroundings to the chemical system. Thus during a chemical reaction there is always an entropy change. It is this entropy change which, along with the enthalpy change of the reaction, we must consider when attempting to predict whether a chemical reaction will proceed spontaneously or not. However, before we consider the relationship of entropy change and enthalpy change to the feasibility of a reaction, we must first look at the second law of thermodynamics.

The Second Law of Thermodynamics

This states that the total entropy of a system and its surroundings increases during all spontaneous processes. The second law is, perhaps, one of the most comprehensive generalisations in the whole of science. There are many statements of the second law. The essential theme of them all is that disorder is all the time increasing.

Some Statements of the Second Law of Thermodynamics

1. Every system which is left to itself will, on the average, change toward a condition of maximum probability. (G. N. Lewis)
2. The state of maximum entropy is the most stable state for an isolated system. (Enrico Fermi)
3. When any actual process occurs it is impossible to invent a means of restoring *every* system concerned to its original condition. (G. N. Lewis)
4. Every physical or chemical process in nature takes place in such a way as to increase the sum of the entropies of all the bodies taking part in the process. (Max Planck)
5. Heat cannot flow spontaneously from a colder to a warmer body.
6. Gain in information is loss in entropy. (G. N. Lewis)
7. Entropy is time's arrow. (A. Eddington)

It follows from the second law of thermodynamics that, for spontaneous processes.

$$\Delta S_{total} > 0$$

where the change in entropy, ΔS_{total}, during a chemical or physical change is given by

$$\bullet \qquad \Delta S_{total} = \Delta S_{system} + \Delta S_{surroundings} \qquad (15)$$

Entropy Changes of Chemical Reactions

The entropy of one mole of a substance in its standard state at a specified temperature is called the **standard molar entropy**. It has the symbol, $S_m^{\ominus}$. Its units are J K^{-1} mol^{-1}. The standard molar entropies for a number of elements and compounds at 25°C are given in table 5.12. Notice that the entropies of gases tend to be high whereas those of solids are generally much lower. The entropy of a fixed amount of matter increases in the order

$$\text{Solid} \rightarrow \text{Liquid} \rightarrow \text{Gas}$$

Standard molar entropies are sometimes called absolute entropies. They are not the entropy changes when a compound is formed from its constituent elements. It should be noted that standard molar entropies of elements are not zero.

Table 5.12 Standard molar entropies

	S_m (298 K)/J K^{-1} mol^{-1}
Solids	
C(diamond)	2.4
Cu	33.1
SiO$_2$	41.8
CuO	43.5
I$_2$	116.8
CuSO$_4$·5H$_2$O	360.2
Liquids	
H$_2$O	69.9
Hg	76.0
Br$_2$	151.6
C$_2$H$_5$OH	160.7
C$_6$H$_6$	173.3
CHCl$_3$	201.8
Gases	
H$_2$	131.0
CH$_4$	186.2
H$_2$O	189.0
O$_2$	205.0
CO$_2$	213.6
Cl$_2$	233.0
NO$_2$	239.9

The third law of thermodynamics states that, the entropy of a perfect ionic crystal at absolute zero, 0 K, is zero.

The standard molar entropy change of a reaction, $\Delta S_m^{\ominus}$, is given by

$$\Delta S_m^{\ominus} = \Sigma S_m^{\ominus}(\text{products}) - \Sigma S_m^{\ominus}(\text{reactants}) \qquad (16)$$

EXAMPLE

Calculate the standard molar entropy change for the complete combustion of one mole of hydrogen gas at 25°C. Use the data in table 5.12 for your answer.

SOLUTION

The equation for this reaction is

$$H_2(g) + \tfrac{1}{2}O_2(g) \rightarrow H_2O(g)$$

Applying equation (16) we obtain

$$\Delta S_m^{\ominus} = S_m^{\ominus}\{H_2O(g)\} - S_m^{\ominus}\{H_2(g)\} - \tfrac{1}{2}S_m^{\ominus}\{O_2(g)\}$$

Substituting values at 298 K from table 5.12 gives us

$$\begin{aligned} \Delta S_m^{\ominus} &= \{(189.0) - (131.0) - \tfrac{1}{2}(205.0)\} \text{ J K}^{-1} \text{ mol}^{-1} \\ &= -44.5 \text{ J K}^{-1} \text{ mol}^{-1} \end{aligned}$$

You will notice that the entropy change for this reaction is negative. This might have been expected since the total amount of reactant gases shown by the equation is 1.5 moles whereas only one mole of gaseous product is formed. The amount of gas therefore decreases during the reaction. On the other hand, we know that combustion reactions are exothermic. Energy is thus dispersed and so we might expect an increase in entropy and not a decrease. Furthermore, once initiated, the combustion of hydrogen gas at 25°C is very spontaneous. Should not the entropy change for the reaction therefore be positive—as required by the second law of thermodynamics? The answer is no—or not necessarily so. The second law requires that the *total* entropy of a system and its surroundings increases during a spontaneous process. The entropy we have calculated is for the system only—the system being the reactants and products involved in the combustion of hydrogen gas at 25°C. So how do we calculate the entropy change of the surroundings?

Entropy Changes in the Surroundings

By means of thermodynamic arguments it is possible to show that entropy change is equal to energy transferred as heat, q, divided by the temperature, T, thus

$$\Delta S = \frac{q}{T} \qquad (17)$$

The change in entropy can relate to either the system or the surroundings. However, there is one condition. The energy q must be transferred reversibly. In thermodynamics a **reversible process** is one that is carried out infinitely slowly and carefully so that it is virtually in a state of balance at all times.

In an exothermic process the energy lost by a reaction system is equal to the energy gained by its surroundings. Similarly, in an endothermic process, the energy absorbed by the system is equal to the energy lost by the surroundings. We can thus write

$$q_p(\text{system}) = -q_p(\text{surroundings}) \qquad (18)$$

We saw above that at constant pressure the energy transferred as heat during a chemical reaction is equal to the enthalpy change, ΔH. Thus

$$\Delta H = q_p(\text{system}) \qquad (19)$$

We can now use equation (18) to rewrite equation (17) as

$$\Delta S_{\text{surroundings}} = -\frac{\Delta H}{T} \qquad (20)$$

The Total Entropy Change of a Chemical Reaction

We saw earlier that the total entropy change of a spontaneous process equals the entropy change of the system plus the entropy change of the surroundings (*see* equation (15)). The entropy change of a chemical reaction system is given by equation (16) and the entropy change of the surroundings is given by equation (20) above. We are thus able to calculate the total entropy change which accompanies a chemical change and see whether the result obeys the second law of thermodynamics.

EXAMPLE
Calculate the total entropy change that accompanies the combustion of one mole of hydrogen gas at 25°C. Does the result obey the second law of thermodynamics?

SOLUTION
The total entropy change accompanying a process is given by equation (15). If we insert equation (20) into this we obtain

$$\Delta S^{\ominus}_{total,m} = \Delta S^{\ominus}_{system,m} - \frac{\Delta H^{\ominus}_{c,m}(298 \text{ K})}{T} \tag{21}$$

This equation applies to the combustion of one mole of hydrogen gas under standard conditions as specified by the equation in the previous example.
We are given,

$$T = 298 \text{ K}$$

The value for $\Delta S^{\ominus}_{system,m}$ was calculated in the previous example as

$$\Delta S^{\ominus}_{system,m} = -44.5 \text{ J K}^{-1} \text{ mol}^{-1}$$

The standard enthalpy of combustion of hydrogen is given in table 5.2 as

$$\Delta H^{\ominus}_{c,m}(298 \text{ K}) = -285.8 \text{ kJ mol}^{-1}$$
$$= -285.8 \times 10^3 \text{ J mol}^{-1}$$

Thus

$$\Delta S^{\ominus}_{total,m} = (-44.5 \text{ J K}^{-1} \text{ mol}^{-1}) - \left(\frac{-285.8 \times 10^3 \text{ J mol}^{-1}}{298 \text{ K}} \right)$$
$$= +914.6 \text{ J K}^{-1} \text{ mol}^{-1}$$

In conclusion, although the enthalpy change for the reaction system is negative, the total entropy change which accompanies the reaction is positive. The result thus obeys the second law of thermodynamics.

GIBBS FREE ENERGY

We have seen that there are two factors which determine whether a reaction will proceed spontaneously or not. These are:

1. **Energy**. Systems tend to a minimum in potential energy. For example, when a ball is placed at the top of a hill, it will roll downwards. It loses potential energy until it reaches a minimum at the bottom of the hill. In reaction systems at constant pressure the change in energy is given by the **enthalpy change**. In exothermic reactions the system loses energy until the total enthalpy of the system reaches a minimum.

2. **Entropy**. Systems tend to maximum disorder. If we look back at the previous example of the combustion of methane, we see that the enthalpy term was far larger than the entropy term. The **driving force** of the reaction was thus the enthalpy change. In endothermic reactions, however, the entropy term dominates.

In these reactions the reaction system can gain energy and still proceed spontaneously.

How then, do we balance these two factors? The answer lies in a thermodynamic state function known as **free energy** or the **Gibbs function,** G. This function is related to equation (21) which, expressed in general terms, is

$$\Delta S_{total} = \Delta S_{system} - \frac{\Delta H}{T} \qquad (22)$$

If we multiply this equation through by $-T$ we obtain

$$-T\Delta S_{total} = \Delta H - T\Delta S_{system}$$

The term $-T\Delta S_{total}$ is equal to the free energy change of a reaction *system*:

● $$\Delta G = -T\Delta S_{total} \qquad (23)$$

Thus, on rearrangement, equation (22) becomes

● $$\Delta G = \Delta H - T\Delta S \qquad (24)$$

where ΔG, ΔH and ΔS all refer to the reaction system.

The Gibbs free energy change therefore takes into account the enthalpy change of a reaction system and its entropy change. Note that the entropy term includes temperature as a factor. This is necessary since at low temperatures systems are more ordered than at high temperatures.

We have seen that for a reaction to be spontaneous ΔS_{total} must be positive. Since temperature is always positive on the absolute scale of temperature, it thus follows from equation (23) that ΔG must be negative for a spontaneous reaction. In other words, when a spontaneous reaction occurs at constant temperature and pressure, the free energy of the system decreases. We express this mathematically as

$$\Delta G < 0$$

The **standard molar Gibbs free energy of formation** is the free energy change that accompanies the formation of one mole of the substance from its elements in their standard states. Gibbs free energy has the usual energy units, kJ mol^{-1}.

Some standard molar Gibbs free energies of formation are shown in table 5.13. Notice that the magnitudes of the values for solids are much greater than those for gases. We saw earlier in this chapter that the enthalpy change of formation was a measure of the stability of a compound. However, a much truer indication of its stability is the free energy of formation. Its value for a compound represents the work that has to be done to return it to its elements in their standard states. And

Table 5.13   Standard molar Gibbs free energies of formation

Compound or ion	Formula	$\Delta G^{\ominus}_{f,m}$/kJ mol^{-1}
ammonia	$NH_3(g)$	−16
calcium carbonate (calcite)	$CaCO_3(s)$	−1129
calcium ion	$Ca^{2+}(aq)$	−553
carbonate ion	$CO_3^{2-}(aq)$	−528
carbon dioxide	$CO_2(g)$	−394
copper(II) sulphate-5-water	$CuSO_4 \cdot 5H_2O$	−1880
water	$H_2O(l)$	−244
	$H_2O(g)$	−229
hydrogen chloride	$HCl(g)$	−95
hydrochloric acid	$HCl(aq)$	−131
methane	$CH_4(g)$	−51
ethane	$C_2H_6(g)$	−33
benzene	$C_6H_6(l)$	+124
trichloromethane	$CHCl_3(l)$	−72
ethanol	$C_2H_5OH(l)$	−174

like enthalpies of formation, the standard free energies of formation of elements in their standard states are zero.

The standard free energy change of a reaction can be calculated in two ways. It can be calculated from the free energies of formation of products and reactants using the following equation:

● $$\Delta G_m^{\ominus} = \Sigma \Delta G_{f,m}^{\ominus}(\text{products}) - \Sigma \Delta G_{f,m}^{\ominus}(\text{reactants}) \qquad (25)$$

Alternatively it can be calculated from standard molar changes in enthalpy and entropy for the reaction using equation (24).

EXAMPLE

(a) Calculate the standard Gibbs free energy change at 25°C for the thermal decomposition of calcium carbonate given that, for this reaction,

$$\Delta H_m^{\ominus} = +178 \text{ kJ mol}^{-1}$$
$$\Delta S_m^{\ominus} = +161 \text{ J K}^{-1} \text{ mol}^{-1}$$

(b) Is the reaction spontaneous at 25°C?

(c) At what temperature would the reaction become spontaneous—assuming that $\Delta H_m^{\ominus}$ and $\Delta S_m^{\ominus}$ are independent of temperature?

SOLUTION

The balanced chemical equation for this reaction is

$$CaCO_3(s) \rightarrow CaO(s) + CO_2(g)$$

(a) To find $\Delta G_m^{\ominus}$ we use equation (24). Thus

$$\Delta G_m^{\ominus} = (178 \times 10^3 \text{ J mol}^{-1}) - [(298 \text{ K}) \times (161 \text{ J K}^{-1} \text{ mol}^{-1})]$$
$$= +130 \text{ kJ mol}^{-1}$$

(b) Since $\Delta G_m^{\ominus}$ is positive at 25°C (298 K), the reaction is *not* spontaneous at this temperature.

(c) The reaction is spontaneous when

$$\Delta G_m^{\ominus} < 0$$

that is when

$$\Delta H_m^{\ominus} - T\Delta S_m^{\ominus} < 0$$

Rearranging, we obtain

$$T\Delta S_m^{\ominus} > \Delta H_m^{\ominus}$$

or

$$T > \frac{\Delta H_m^{\ominus}}{\Delta S_m^{\ominus}}$$

$$> \frac{178 \times 10^3 \text{ J mol}^{-1}}{161 \text{ J K}^{-1} \text{ mol}^{-1}}$$

$$> 1106 \text{ K}$$

The reaction is thus spontaneous at temperatures above 1106 K.

The standard Gibbs free energies of four reactions are shown in table 5.14. All of these except reaction (c) are spontaneous since they have negative ΔG values. Reaction (c) does not occur spontaneously since its value is positive.

Table 5.14   Standard Gibbs free energies for some reactions

Reaction	$\Delta G_m^{\ominus}$/kJ mol^{-1}	$\Delta H_m^{\ominus}$/kJ mol^{-1}	$T\Delta S_m^{\ominus}$/kJ mol^{-1}
(a) $H_2(g) + Br_2(l) \rightarrow 2HBr(g)$	−106	−72	+34
(b) $2H_2(g) + O_2(g) \rightarrow 2H_2O(l)$	−474	−572	−98
(c) $2Ag_2O(s) \rightarrow 4Ag(s) + O_2(g)$	+22	+61	+39
(d) $Ca^{2+}(aq) + CO_3^{2-}(aq) \rightarrow CaCO_3(s)$	−48	+13	+61

IMPORTANCE OF ΔG

We have already seen that Gibbs free energy is useful in two ways. First of all, the Gibbs free energy of formation of a compound is a measure of the stability of the compound. Secondly, the free energy change of a reaction is a measure of the **feasibility** of a reaction. Only reactions with negative ΔG values can take place. Gibbs free energy is also important for a number of other reasons.

Equilibrium

When a system is in dynamic equilibrium,

$$\Delta G^{\ominus} = 0$$

If a reaction has a large and negative value for ΔG, then the equilibrium will lie strongly to the right—that is in favour of the products. If the value is large and positive the equilibrium lies in favour of the reactants.

Useful Work

We have already seen that when a system undergoes a change at constant pressure the energy transferred as heat is called the enthalpy change ΔH. If we rearrange equation (24) we can see that this enthalpy change can be split up into two parts, ΔG and $T\Delta S$:

$$\Delta H = \Delta G + T\Delta S$$

ΔG is part of the enthalpy change which is available for doing useful work (*see* figure 5.17). The other part cannot be used to do work. This corresponds to the entropy term $T\Delta S$.

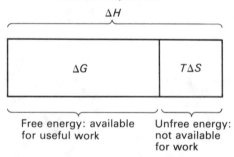

Total energy change at constant temperature

ΔH

| ΔG | $T\Delta S$ |

Free energy: available for useful work   Unfree energy: not available for work

Figure 5.17 Enthalpy change can be split into two parts

$$\Delta H = \Delta G + T\Delta S$$

Extraction of Metals from Oxides

Metals can be extracted from their oxides by reduction (*see* section 10.5). Reducing agents used for this purpose include carbon or other metals. How can we tell whether carbon or another metal will reduce a metal oxide? Let us consider the following two cases.

CASE 1

$$\text{Metal} + \text{Oxide of carbon} \rightarrow \text{Metal oxide} + \text{Carbon}$$

For example

$$2Mg(s) + CO_2(g) \rightarrow 2MgO(s) + C(s)$$

In this case the metal acts as a reducing agent. It reduces the carbon in carbon dioxide.

CASE 2

$$\text{Metal oxide} + \text{Carbon} \rightarrow \text{Metal} + \text{Oxide of Carbon}$$

For example

$$2NiO(s) + 2C(s) \rightarrow 2Ni(s) + 2CO(g)$$

In this case the carbon acts as a reducing agent. It reduces the metal oxide.

We see immediately that case 2 is the reverse of case 1. For some metals case 1 applies, and for others case 2 applies. So how can we predict for a given metal which case will apply? In other words, which is feasible, case 1 or case 2?

The answer lies in the stability of the oxides. We saw above that the free energy of formation of a compound is a measure of its stability. We also saw that, for a reaction to occur spontaneously, ΔG must be less than zero,

$$\Delta G < 0$$

We can now apply this condition to equation (25) and obtain the following condition if a reaction is to occur spontaneously:

$$\Delta G^{\ominus}_{f,m}(\text{products}) < \Delta G^{\ominus}_{f,m}(\text{reactants})$$

In both of the cases above, one of the products and one of the reactants are elements. Since the standard molar free energies of formation of these are zero, the following condition must apply if these reactions are to occur spontaneously:

$$\Delta G^{\ominus}_{f,m}(\text{product oxide}) < \Delta G^{\ominus}_{f,m}(\text{reactant oxide})$$

The more stable oxide—that is, the oxide with the more negative $\Delta G^{\ominus}_{f,m}$ value—must therefore be the product and not the reactant in both cases. The standard molar free energies of formation of the oxides in our examples are shown in table 5.15. These values show us that, in the first example above, magnesium oxide is more stable than carbon dioxide. Magnesium oxide is therefore a product and not a reactant. The forward reaction as shown above occurs spontaneously and not the reverse. In other words, case 1 applies.

In our second example we have a problem. According to the values listed above for $\Delta G^{\ominus}_{f,m}$ the product of the reaction in our example for case 2 should be NiO(s) and not CO(g). Why then, does case 2 apply? The answer is temperature. The reaction does not occur at 298 K.

Table 5.15

Oxide	$\Delta G^{\ominus}_{f,m}/\text{kJ mol}^{-1}$
CO_2(g)	−394
MgO(s)	−569
NiO(s)	−213
CO(g)	−137

Ellingham Diagrams

The standard molar free energies of formation of metal oxides increase with temperature whereas that for carbon monoxide decreases. The variation of standard molar free energy of formation with temperature can be shown on an Ellingham diagram (*see* figure 5.18). In this diagram the higher the lines the less stable the oxide. Thus carbon monoxide is less stable than nickel(II) oxide at temperatures below about 680 K whereas above this temperature it is more stable. And it is at temperatures above 680 K that this reaction is normally carried out.

Figure 5.18 Simplified Ellingham diagram

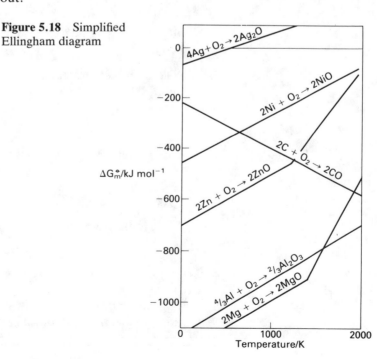

Ellingham diagrams show at a glance which oxide will be reduced. At a given temperature, the oxide with the higher line will be the one that is reduced. Thus, at 1000 K magnesium will reduce aluminium oxide, whereas at 2000 K aluminium reduces magnesium oxide. On the other hand, the diagram shows that magnesium will reduce zinc oxide at any temperature below 2000 K. The 'elbows' in the lines indicate phase changes—melting or boiling—of the metals. These result in increases in entropy and thus a more negative value for ΔG. Notice also that the free energy of formation of silver oxide is negative at lower temperatures but becomes positive as the temperature increases. The crossover point (where $\Delta G_m^{\ominus} = 0$) is at 440 K.

SUMMARY

1. The **standard bond dissociation enthalpy** is the enthalpy change that accompanies the breaking of one mole of bonds with the molecules and resulting fragments being in their standard states at 298 K and a pressure of one atmosphere.

2. The **standard molar lattice enthalpy** is the enthalpy change which accompanies the formation of one mole of the solid ionic compound from its gaseous ions under standard conditions.

3. The **Born–Haber cycle** is an enthalpy diagram from which the lattice enthalpy of a salt can be determined.

4. The **standard molar enthalpy of solution** is given by

$$\Delta H_{solution,m}^{\ominus} = \Delta H_{hydration,m}^{\ominus} - \Delta H_{lattice,m}^{\ominus}$$

5. The **second law of thermodynamics** states that the total entropy of a system and its surroundings increases during all spontaneous processes.

6. For all **spontaneous processes** the total change in **entropy** must be greater than zero:

$$\Delta S_{total} > 0$$

where

$$\Delta S_{total} = \Delta S_{system} + \Delta S_{surroundings}$$

7. $\Delta S_m^{\ominus} = \Sigma S_m^{\ominus}(\text{products}) - \Sigma S_m^{\ominus}(\text{reactants})$

8. **Gibbs free energy** change is related to the total entropy change by

$$\Delta G = -T\Delta S_{total}$$

9. The Gibbs free energy change is a measure of the **feasibility** of a reaction.

10. For a **reaction system**

$$\Delta G = \Delta H - T\Delta S$$

11. A reaction will only occur spontaneously if

$$\Delta G < 0$$

12. $\Delta G_m^{\ominus} = \Sigma \Delta G_{f,m}^{\ominus}(\text{products}) - \Sigma \Delta G_{f,m}^{\ominus}(\text{reactants})$

13. For a system in dynamic equilibrium

$$\Delta G^{\ominus} = 0$$

14. ΔG is that part of an enthalpy change which is available for doing useful work.

15. **Ellingham diagrams** show the variation of standard molar free energy of formation with temperature.

5.4
Energy sources

SUN, EARTH AND MOON

We have already discussed or referred to various forms of energy in this chapter and earlier chapters. The most important forms are

- chemical energy (enthalpy, internal energy),
- electrical energy,
- electromagnetic energy (light),
- thermal energy (energy transferred as heat),
- mechanical energy (motion),
- nuclear energy.

In addition we have seen that the energy of a body due to its motion is called kinetic energy and that energy which is stored is called potential energy. In effect, since energy is the capacity for doing work, all energy may be considered to be stored energy. The first law of thermodynamics tells us that energy may be converted from one form to another without any of it being destroyed. The store remains complete. However, when energy is converted from one form to another, some of it is wasted. In other words the amount of useful energy is always less after a change. Thus, as the energy of the Universe changes, the amount of useful energy decreases, and the amount of useless energy increases. A measure of the uselessness is entropy. It always increases after a change.

There is little danger over the next few million years that we shall exhaust the store of useful energy. The problem is that some stores of energy are easily tapped whereas others are exceedingly difficult. And as we 'progress' towards the twenty-first century we are rapidly consuming all the stores of energy that are easily consumed and thus becoming more and more reliant on the less accessible stores of energy.

Figure 5.19 shows the most important sources of energy and their origins—the Sun, the Earth and the Moon. Of these, the Sun is by far and away the most important source of energy. About 30% of the Sun's energy hitting the Earth is reflected back into space by dust particles and cloud. Another 47% is used to heat up the Earth's surface and 22% is consumed in the evaporation cycle which makes rain. No more than 0.1% drives the wind, waves and ocean currents and a minute 0.03% is consumed in photosynthesis. Yet it is through photosynthesis that we have our principal forms of stored energy—fossil fuels and biomass (*see* below).

Renewable and Non-Renewable Forms of Energy

Renewable energy sources replenish themselves naturally in a relatively short space of time. They will thus always be available. They include

- biomass,
- geothermal energy,
- hydroelectric power,
- solar energy,
- tidal energy.

Non-renewable sources of energy are those which cannot be replaced once used up. They are

- fossil fuels: coal, oil and natural gas,
- nuclear fuels: uranium ores.

An energy source which could potentially supply up to ten times our energy needs for several thousand years is called an **indefinitely sustainable energy source**. Such energy sources include nuclear fission, nuclear fusion, solar energy together with biomass, controlled thermonuclear fusion and, finally, geothermal energy.

In one week . . .

Each year the Earth's surface receives 3×10^{24} J of energy from the Sun. This compares with an estimated 2.5×10^{22} J of energy stored in proven reserves of natural gas, coal, oil and uranium. This means that in one week the Earth receives an amount of energy from the Sun which is more than twice the total known energy reserves of the Earth.

Harnessing the energy of the wind on Lake Victoria, Kenya

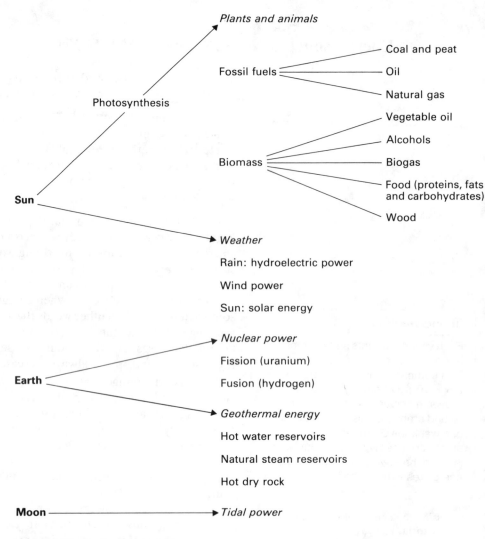

Figure 5.19   Sources of energy

Conventional Energy Sources

These are the non-renewable sources of energy that have been used predominantly in the past. They are the fossil fuels—coal, gas and oil and more recently uranium. We discussed the nuclear fission of uranium in chapter 1. In chapter 18 we shall look at oil and natural gas in some detail. In this section, therefore, we shall confine our attention to **coal** (*see* figure 5.20).

Coal is one of mankind's primary energy sources. It is a rock formed from vegetable matter by the process of **metamorphism**. Metamorphic rocks are those which have undergone changes in composition under high pressure and/or high temperature (*see* section 13.3). The first stage in this process is the formation of **peat**. This is decomposed organic matter. Coal is formed from peat which has been covered with sediments. These sediments are called overburden. The overburden reduces the moisture content of the peat.

Coal is classified by three criteria:

- Purity—this relates to the percentage carbon content.
- Type—this depends on the composition of the original vegetation.
- Rank—this depends on the degree of metamorphism.

Figure 5.20 A proposed molecular *model* of low-rank coal. Coal consists of a complicated mixture of chemical substances containing carbon, hydrogen and oxygen along with small amounts of nitrogen, sulphur and some trace elements. Coal also contains various amounts of moisture and minerals

A typical mining village in the coal-mining area in Bihar, India. Coal is one of mankind's primary energy sources

Anthracite, which is used as a domestic fuel, has a relatively high energy value

Table 5.16   Carbon content and energy values of fuels

Fuel	Percentage carbon content — by mass	Energy value/ kJ kg⁻¹
wood	50.0	19 800
peat	59.9	18 700
lignite	61.8	20 900–25 600
brown coal	69.5	27 200
bituminous coal	78.7	32 100
anthracite	91.0	32 600

The lowest rank of coal is **brown coal** or **lignite** (*see* table 5.16). It is closely related to peat and has a relatively low carbon content and high moisture content. **Bituminous coal** has less moisture and is used in industry. The driest and hardest coal is **anthracite**. This is used as a domestic fuel.

In recent years **coal gasification** has become increasingly economically viable due to technological developments. The products of coal gasification include carbon monoxide, carbon dioxide, hydrogen, methane and nitrogen. They are used as fuels or as raw materials for the manufacture of chemicals and fertilisers.

ALTERNATIVE ENERGY SOURCES

These are sources of energy that are currently being developed to replace conventional energy sources. The most important are nuclear fission, wind, wave and tidal energy sources and solar energy by direct conversion and biomass. We dealt with nuclear fission in chapter 1. In the two sections below we shall consider the direct conversion of solar energy and biomass.

The Direct Conversion of Solar Energy

We saw in figure 5.19 that the various forms of solar energy include fossil fuels, biomass, wind and water power. All these provide indirect means of converting solar energy to forms of energy which are useful to us. How can solar radiation be directly trapped and used?

The most important method of trapping solar energy is by means of **solar panels**. These are made of metals since they are good conductors of heat. Copper is most commonly used. It is covered with a black surface to absorb the sunlight. Solar panels are usually mounted on the roof of a house and used for domestic heating.

A solar energy experiment at Dakar University, Senegal

The **photovoltaic cell** also provides a means of collecting the Sun's energy. The problem with photocells is their low efficiency. They are no more than about 20% efficient. At present, much research is being carried out to improve the efficiency of photocells. One possibility is the use of a glass panel treated with uranium oxide and the rare earth element neodymium in order to increase the absorption of light by the cell.

The use of solar energy to produce **hydrogen** as a fuel **from water** has aroused much scientific interest over recent years. Hydrogen is the most abundant element in the Universe although it is not so on Earth (*see* chapter 12). Most terrestrial hydrogen occurs in the form of water. Once the hydrogen is produced it can be converted to other fuel forms such as methanol. The easiest method of producing hydrogen from water is by means of **electrolysis**. The efficiency of this process is about 83%. However, at present, the economics of using solar-produced electricity to produce hydrogen from water by electrolysis are poor.

A more attractive long-term possibility is the direct solar production of hydrogen from water by **photochemical conversion**. This process is known as **photolysis**. Two approaches are currently being investigated. These are the biological approach and the biochemical approach.

In the biological approach, living organisms are used to decompose water into hydrogen and oxygen (*see* 'the hydrogen tree' at the beginning of this chapter). This process is called **biophotolysis**. In the biochemical approach, enzymes obtained from biological organisms are used. Hydrogen has already been produced from water using this approach. But the rate and duration of production is low.

There is also much experimental interest in the use of artificial chloroplast and synthetic **chlorophyll** for the photochemical conversion of water.

The use of solar collectors which can concentrate solar radiation sufficiently to generate the high temperatures needed for the **thermochemical conversion** of water is also a long-term possibility. The equation for this process is

$$H_2O(g) \rightarrow H_2(g) + \tfrac{1}{2}O_2(g)$$

Biomass

Biomass includes all animal and plant materials—both dead and alive. Food, wood, the organic wastes of animals and plants are all forms of biomass. Biomass is an important form of stored energy. The carbohydrates, fats and proteins in food all provide us with energy, for example (*see* section 5.2). Wood, of course, has long been used as fuel by mankind (*see* the beginning of this chapter). In recent decades biomass has become increasingly important as a source of fuels such as biogas and alcohol.

Biogas—like natural gas—is predominantly methane. Biological processes are already used extensively at sewage works and on farms to convert domestic and agricultural sewage into biogas. A biogas plant uses animal wastes and plants as raw material. These are allowed to rot in a biogas generator or digester. The digester excludes oxygen. Under these conditions, certain species of bacteria break down the waste to form methane gas. The gas is collected and can be burned directly for domestic heating and cooking or used to produce electricity in a generator (*see* figure 5.21).

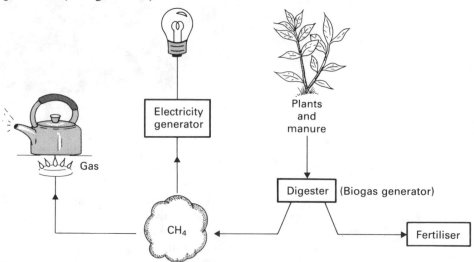

Figure 5.21 Biogas can be produced from plants and animal waste. In some Third World countries it is used for domestic purposes. The residue, which has a high nitrogen content, can be used as a fertiliser

The biogas produced from this sewage treatment plant is used as a fuel

One Cow and a Mud Jar

The conventional biogas plant running on animal waste requires dung from at least five head of cattle. Such a plant can produce about 3 m³ of gas per day and enough residue to fertilise about 15 000 m² of farmland each year.

But there is a problem. Millions of poor farmers have only one cow. The Centre of Science for Villages in Wardha, India, has developed the world's cheapest and probably the smallest biogas plant. It is a mud jar which can be made by any village potter. To prevent the methane gas from seeping out of the porous jars, the Centre has experimented with an indigenous paste made from lime and the sap of local trees.

Biogas production is growing rapidly in Third World countries. China has about 4.3 million biogas production plants installed over the country and about 100 000 personnel trained in biogas production. India has over 60 000 and South Korea 30 000 biogas production plants in operation.

The use of 'green waste' in developing countries is attractive for several reasons. First of all, it can be produced in rural areas. Secondly, waste can be used as fertiliser—the residue from such processing has a high nitrogen content. Land which is unfit for food crops can be used to grow crops for biogas production. Aquatic weeds such as the wild hyacinths which choke canals and reservoirs in some parts of the world can also be used to produce biogas. Finally, the increased use of biogas serves to reduce the use of wood as a fuel.

Another **bio-fuel** is **alcohol**. Alcohol has long been used as a fuel—in spirit lamps and as methylated spirits, for example. Alcohol can be produced from plants such as sugar cane and cassava by fermentation and distillation. Brazil, in its Proalcohol programme, extracts ethanol from crops and has designed and manufactured cars to run on the fuel (*see* the beginning of this chapter).

There is also considerable potential and interest in the extraction and use of **vegetable oils** as diesel fuels. Oils extracted from soya beans, sunflowers, peanuts, rapeseed, palms, castor beans, eucalyptus, squash and coconut have all been tested for use as fuels and lubricating oils over recent years.

An oil palm plantation in Ghana

ENERGY STORAGE

One of the problems of supplying energy is matching supply with demand. The body, for example, consumes energy in the form of carbohydrate, fats and proteins. If we supply ourselves with too much of this energy we store the surplus as fat. If the demand for energy outstrips the supply, we use up the fat and become slimmer. Carbohydrates, proteins and fats are called **fuel nutrients**.

A **fuel** is a source of energy which can be readily converted into thermal energy when required and stored when not required. Conventional fuels such as coal, wood and oil are all readily stored when supply exceeds demand. However, other forms of energy cannot be stored directly.

Electricity is an example. Electricity is one of the most useful forms of energy. It can be generated from conventional fuels. But when supply exceeds demand, it cannot be stored directly. It has to be converted into other forms. Three methods of doing this are:

Pump storage. At times of low demand, electricity can be used to pump water from a low reservoir to a high reservoir. Electricity is regenerated by returning the water to a low reservoir through a turbo-generator.

Compression of air. Excess electricity can also be used to compress air in an underground store. The electricity can then be regenerated by allowing the air to expand in a turbine.

Production of hydrogen. In the future it may be possible to convert water to hydrogen on a large scale by electrolysis, for example. At present such storage is not economically viable.

Batteries and Fuel Cells

These allow a store of chemical energy to be converted to electrical energy as required.

A **battery** is a portable source of electricity. It consists of an electrochemical cell or several electrical cells connected in series (*see* section 10.5). A distinction is usually made between primary cells and secondary cells. Primary cells cannot be recharged whereas secondary ones can be recharged. The lead–acid accumulator used in cars consists of secondary cells.

Unlike in a battery, the chemicals in a **fuel cell** are continuously replaced as they are used up. The most common type of fuel cell is the hydrogen–oxygen fuel cell (*see* section 10.5). This has been used in space and conveniently produces not only electricity but also water.

A battery is a portable source of electricity

PAST, PRESENT AND FUTURE

It has been estimated that the amount of fossil fuels that will be consumed between AD 2000 and AD 2050 will be over three times the amount consumed between AD 1 and AD 2000. Even in the second half of this century almost twice as much energy will have been consumed as in the preceding 1950 years (*see* table 5.17). The major portion of the consumption of energy over the past 100 years has been due to the industrialisation of what is now called the Developed World. Figure 5.22 shows that developing countries such as Nigeria, Indonesia and Egypt consume a mere fraction of the energy consumed by countries such as the UK and Australia. Amongst the largest consumer of energy per capita is the United States. It consumes more than twice as much as the United Kingdom.

The primary energy source of industrialised countries has always been fossil fuels. However, as we have seen, these are non-renewable and rapidly running out. The so-called energy crisis is really a fossil fuel crisis. Estimates suggest that there may be enough coal left to last a few more hundred years, enough oil to last

Table 5.17   Consumption of fossil fuels

Period of time AD	Relative amounts of fossil fuels consumed
1–1860	8
1860–1950	4
1950–2000	20
2000–2050	100

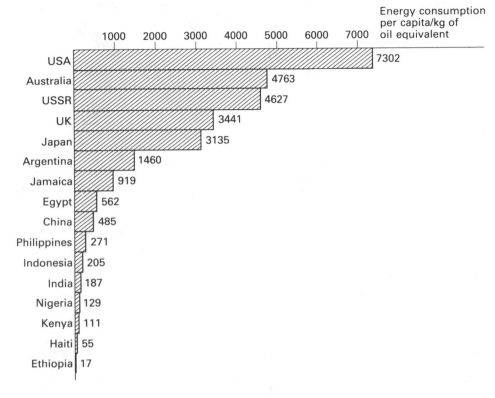

Figure 5.22 Energy consumption per person in some industrialised and Third World countries in 1984

perhaps 70 or so years and natural gas could run out within 50 years. The need to tackle and solve the energy problem is becoming increasingly urgent. The urgency of the problem is compounded by three factors:

- the increasing world population;
- social, economic and political pressures for economic expansion in industrialised countries;
- increasing awareness in Third World countries that their economic situation could be improved by increasing their energy consumption.

Up and out of the chimney. A typical domestic household in the UK wastes about half of its energy

In pursuing a solution to the energy problem, a number of factors have to be taken into consideration. The most important are the following.

Demand
To what extent can or will industrialised countries moderate their demand for energy through appropriate social, economic and political policies?

Efficiency and Wastage
The efficiency of energy conversion is variable. The electrical generator is about 98% efficient. Yet the overall efficiency of converting fossil fuels to electrical energy is about 35 to 40%. Much energy is lost in its transmission and distribution for example. In Europe only 42% of the energy produced is converted into useful energy. The ordinary domestic household in the UK wastes about 50% of its energy (*see* figure 5.23).

Typical conversion efficiencies for various converters are shown in table 5.18. Note that the solar cell has an efficiency of only about 10%. The question, then, is to what extent can these efficiencies be improved and to what extent can the wastage be reduced?

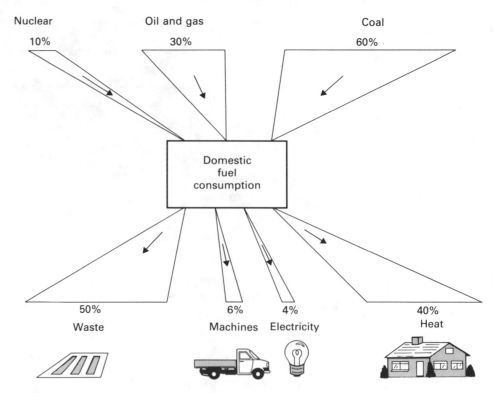

Figure 5.23   Domestic fuel consumption

Table 5.18   Typical conversion efficiencies

Converter	Conversion		Percentage efficiency of conversion
electrical generator	mechanical	→ electrical	98
dry battery	chemical	→ electrical	90
car battery	chemical	→ electrical	74
fuel cell	chemical	→ electrical	70
domestic oil burner	chemical	→ thermal	65
thermal power plant	chemical	→ thermal	41
diesel engine	chemical	→ thermal	38
car engine	chemical	→ thermal	25
nuclear power plant	nuclear	→ thermal	30
solar cell	light	→ electrical	10

Alternative Energy Sources

Figures produced at a World Energy Conference showed that oil and gas production should reach a peak between 1985 and 2000. The production of energy

Table 5.19   Energy production

Resource	Energy production/10^{18} J			
	Actual	Predicted		
	1972	1985	2000	2020
coal	66	115	170	269
oil	115	216	195	106
gas	46	77	143	125
nuclear	2	23	88	314
solar, geothermal and photosynthesis	26	33	56	100
other	0	0	4	40
TOTAL	255	464	656	954

A future means of using the Sun's energy? An artist's impression of how an array of solar cells with a transmitting antenna might direct solar energy to Earth

from coal, nuclear fuels and solar and geothermal sources on the other hand should steadily increase over the next 30 years or so (*see* table 5.19). In the very long term it is probable that nuclear fusion and nuclear fission together with solar energy will become the predominant sources of our energy. Meanwhile the major problem is to what extent can suitable, safe and cost-effective technologies be developed to produce energy from these sources.

Environmental Considerations

One of the major problems of fossil fuels and particularly coal is **pollution**. We are all familiar with news items concerning marine pollution due to oil spills. The combustion of fossil fuels also results in the pollution of the atmosphere. Every minute of every day, hundreds of millions of tonnes of coal are burned. The products—apart from useful energy—are smoke, grit, soot, carbon dioxide and

Each year about 3000 million tonnes of carbon dioxide are released into the atmosphere from the combustion of fossil fuels

sulphur oxides. In the modern coal-fired power station, dust and grit are removed by filtering devices called 'arresters'. Chimneys then take the remaining gases to the upper atmosphere. The consequent increase in concentration of carbon dioxide in the upper atmosphere is thought to result in what is known as the **greenhouse effect**. This limits the amount of thermal energy escaping from the Earth (*see* figure 5.24a). On the other hand, the accumulation of dust in the upper atmosphere results in cooling since the solar radiation is reflected away from the Earth (*see* figure 5.24b).

Nuclear fission also has its own environmental problems (*see* also section 1.3). There are two major dangers.

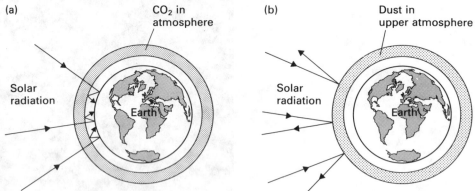

Figure 5.24 (a) The greenhouse effect: thermal energy is trapped in the Earth's atmosphere by atmospheric CO_2. This warms the Earth. (b) Dust particles in the upper atmosphere reflect solar radiation away from the Earth. This cools the Earth

Accidents. No major accidents have occurred in the West although the incidents at Brown's Ferry, Harrisburg and Three Mile Island in the United States are well known. These were near misses. The Chernobyl accident in the USSR has already been described in chapter 1.

Leakage and disposal of radioactive waste. This issue has become increasingly controversial in recent years. The general public is constantly exposed to radiation. About 78% of this comes from natural sources such as granite rocks and cosmic rays. Less than one-half of one per cent derives from nuclear power stations and atmospheric testing of nuclear weapons. The remainder—about 21%—arises from medical X-rays (*see* section 1.3).

The problem of providing enough energy for our increasing needs is not only how to conserve the environment and minimise damage to it but also how to use our environment most effectively. In some Third World countries there is an increasing danger that fuel crops are replacing traditional food crops. To put it crudely, the poor are going without food so that the rich can drive cars. Although developing countries are generally increasing their use of oil, many of the poorest and non-oil-producing countries of the Third World are, of necessity, still using wood as their main fuel (*see* the beginning of the chapter). For some countries, wood accounts for 90% of energy production, with consequent detriment of the environment.

Food versus fuel. This is a problem in some Third World countries. This eight-year-old street urchin in Bogota, Colombia, normally begs or steals, like a wild scavenger, to find his food. He is seen here taking a free meal from a centre for waifs. His legs show the wheelmarks of a bus

The Forests of Nepal

Wood provides about 85% of Nepal's total energy. It is also used there for building houses. Every person in Nepal uses an average of 600 kg of wood each year. But the forests in Nepal grow at only about 80 kg per person each year. Thus the forests of Nepal—one of the country's greatest natural assets—are disappearing at a potentially disastrous rate.

SUMMARY

1. **Renewable energy sources** replenish themselves naturally in a relatively short space of time. They include biomass and solar energy.
2. **Non-renewable sources of energy** cannot be replaced once used up. They are the **fossil fuels** (coal, oil and natural gas) and also nuclear fuels.
3. **Alternative energy sources** such as nuclear fission, biomass and solar energy are currently being developed to replace **conventional energy sources** such as coal, oil and natural gas.
4. **Solar energy** can be directly converted into useful energy by solar panels and photocells. It is converted naturally by photosynthesis.
5. **Biomass** can be used to produce bio-fuels such as biogas (methane) and alcohol.
6. A **fuel** is a source of energy which can be stored.
7. The **efficiency** of energy conversion varies immensely depending on the type of converter. Much of the energy produced in the world is wasted.
8. The provision of energy for our present and future needs is having a pronounced impact on our **environment**.

Examination Questions

1. (a) (i) Define the term 'internal energy change of a reaction (ΔU)'.
     (ii) Define the term 'enthalpy change of a reaction (ΔH)'.
     (iii) Why is there a difference between ΔU and ΔH?
   (b) Given the following data:

$$C_2H_4(g) + H_2(g) \rightarrow C_2H_6(g) \qquad \Delta H^{\ominus} = -136 \text{ kJ mol}^{-1}$$
$$C_6H_6(g) + 3H_2(g) \rightarrow C_6H_{12}(g) \qquad \Delta H^{\ominus} = -208 \text{ kJ mol}^{-1}$$

   what can you conclude about the bonding in C_6H_6?
   (c) Calculate the *enthalpy of formation* of methane using the information given in the table below.

Substance	Enthalpy of combustion/kJ mol^{-1}
methane (gas)	−890
hydrogen (gas)	−286
carbon (graphite)	−394

(JMB)

2. (a) Define standard enthalpy of combustion of ethane (C_2H_6) and standard enthalpy of formation of propane (C_3H_8).
   (b) (1) Give a **brief** account of the determination of the standard enthalpy of combustion of a solid or liquid of known relative molecular mass.
     (ii) 5.6 dm³ (measured at s.t.p.) of a mixture of propane and butane on complete combustion evolved 654 kJ of heat. Calculate the percentage composition of the mixture by volume.
   (The enthalpies of combustion of propane and butane are 2220 and 2877 kJ mol^{-1} respectively and the molar volume is 22.4 dm³ at s.t.p.)

(c) Using the following data which give the energy liberated when 1 g of each substance is burned in excess oxygen, calculate the standard enthalpy of formation of ethene:

Substance	Energy liberated/kJ
graphite	32
hydrogen	143
ethene	50

(H = 1; C = 12)

(d) Explain **briefly** *either* why calcium chloride has a high melting point *or* why it is very soluble in water.

(SUJB)

3. (a) (i) Define 'standard enthalpy of combustion, $\Delta H_c^{\ominus}$'.
      (ii) Given the following standard enthalpies of combustion:

$$\Delta H_c^{\ominus}(C_2H_5OH) = -1371 \text{ kJ mol}^{-1}$$
$$\Delta H_c^{\ominus}(CH_3CHO) = -1167 \text{ kJ mol}^{-1}$$

calculate the standard enthalpy change for the oxidation of ethanol to ethanal.

  (b) (i) By reference to methane, CH_4, distinguish between *bond dissociation enthalpy* and *bond enthalpy term*.
      (ii) Given the standard enthalpies of atomisation of methane and ethane.

$$\Delta H_{at}^{\ominus}(CH_4) = +1662 \text{ kJ mol}^{-1}$$
$$\Delta H_{at}^{\ominus}(C_2H_6) = +2924 \text{ kJ mol}^{-1}$$

calculate the bond enthalpy terms for C—C and C—H bonds.

   (iii) If the first, second and third bond dissociation enthalpies of methane are +420, +475 and +421 kJ mol^{-1}, respectively, calculate the fourth.

  (c) How do you account for the fact that
      (i) ammonium nitrate is readily soluble in water even though the standard enthalpy of solution has a positive value;
      (ii) although the enthalpy of combustion of cane sugar is about -6000 kJ mol^{-1}, cane sugar is not observed to oxidise in air at ordinary temperatures?

(L)

4. (a) State Hess's law.
   (b) The standard enthalpy of formation $\Delta H_f^{\ominus}$ of benzene (i.e. ΔH for reaction (i) below, at 25°C and one atmosphere pressure), and $\Delta H_f^{\ominus}$ of sodium hydroxide (i.e. ΔH for reaction (ii)) cannot be determined directly.

   (i) $6C(graphite) + 3H_2(g) = C_6H_6(l)$

   (ii) $Na(s) + \frac{1}{2}H_2(g) + \frac{1}{2}O_2(g) = NaOH(s)$

Each of these two $\Delta H_f^{\ominus}$ values can, however, be obtained from the ΔH values, all measurable by suitable calorimetry, for **three** processes or reactions. Write down these three processes or reactions (in words, or as equations) for (i) benzene, (ii) sodium hydroxide. (You are not required to show how the ΔH values are used to give $\Delta H_f^{\ominus}$).

(OLE)

5. (a) State Hess's law.
   (b) When 2.76 g (0.02 mol) of potassium carbonate was added to 30.0 cm³ of approximately 2 mol dm⁻³ hydrochloric acid the temperature rose by 5.2°C.
      (i) Write an equation for this reaction.
      (ii) Calculate the enthalpy change of this reaction per mole of potassium carbonate. Assume that the specific heat capacities of all solutions are 4.2 J g⁻¹ K⁻¹, and that all solutions have a density of 1.0 g cm⁻³.
      (iii) Explain why the hydrochloric acid need only be *approximately* 2 mol dm⁻³.
   (c) When 2.00 g (0.020 mol) of potassium hydrogencarbonate was added to 30.0 cm³ of the same hydrochloric acid the temperature fell by 3.7°C.
      (i) Write an equation for this reaction.
      (ii) Calculate the enthalpy change of this reaction per mole of potassium hydrogencarbonate.
   (d) When potassium hydrogencarbonate is heated it decomposes into potassium carbonate, water and carbon dioxide. By applying Hess's law and using your results in (b) and (c), calculate the enthalpy change for the thermal decomposition of potassium hydrogencarbonate. Give all working and explanation, carefully noting whether enthalpy changes are exothermic or endothermic.

   (UCLES)

6. (a) State Hess's law and define (i) *standard state* of a substance and (ii) *standard enthalpy change* for a reaction.
   (b) Describe briefly how you would determine experimentally the enthalpy of neutralisation of a dilute solution of hydrochloric acid by a dilute solution of sodium hydroxide.
   (c) The standard enthalpy of combustion of propane (C_3H_8) at 298 K is -2220.0 kJ mol⁻¹. Calculate the standard enthalpy of formation ($\Delta H^{\ominus}_f$) of propane at 298 K given that at this temperature $\Delta H^{\ominus}_f = -285.9$ kJ mol⁻¹ for $H_2O(l)$ and $\Delta H^{\ominus}_f = -393.5$ kJ mol⁻¹ for $CO_2(g)$.

   (OLE)

7. Define *enthalpy change of formation* and *lattice energy*.
   Discuss the factors that determine the magnitude of a lattice energy.
   Draw Born–Haber cycles for the formation of (a) crystalline potassium chloride, (b) an aqueous solution of potassium chloride. Use these cycles to calculate (i) the enthalpy change of formation of potassium chloride, (ii) the enthalpy change of solution of potassium chloride.

   Data

Enthalpy term	ΔH/kJ mol⁻¹
enthalpy change of atomisation of potassium	+90
enthalpy change of atomisation of chlorine	+121
first ionisation energy of potassium	+418
electron affinity of chlorine*	−364
enthalpy change of hydration of potassium ion	−322
enthalpy change of hydration of chloride ion	−364
lattice energy of potassium chloride	−701

   *i.e. the enthalpy change for the process $Cl(g) + e^- \rightarrow Cl^-(g)$.

   Comment briefly on your answer to (ii) in relation to the solubility of potassium chloride in water.

   (UCLES)

8. The following is a list of bond enthalpies at 298 K.

Bond	$\Delta H/\text{kJ mol}^{-1}$
C—C	348
C═C	612
H—H	436
C—H	412

(a) By reference to methane, distinguish between 'bond enthalpy term' and 'bond dissociation enthalpy'.

(b) Using the bond enthalpies given above, calculate the enthalpy change of the reaction:

$$C_2H_4(g) + H_2(g) \rightarrow C_2H_6(g)$$

(c) In the reaction in (b), does the entropy decrease, increase or stay the same? Explain your answer.

(d) (i) Using the result of your calculation in (b), calculate the enthalpy of hydrogenation of butadiene, assuming the formula to be CH_2═CHCH═CH_2.

(ii) The experimental value for this enthalpy change is -239 kJ mol^{-1}. Suggest an explanation for the difference.

(L)

9. (a) Draw a thermodynamic cycle to show how the enthalpy of formation of a metal halide MX (M = metal, X = halide) is related to:

(i) the enthalpy of atomisation of M and of X;

(ii) the ionisation enthalpy of M;

(iii) the electron affinity of X; and

(iv) the lattice enthalpy of MX.

Indicate clearly on your diagrams which steps in the cycle are exothermic and which steps are endothermic.

(b) Explain in terms of the cycle why calcium does not form a chloride CaCl analogous to KCl.

(c) If it is assumed that a substance of known structure is completely ionic it is possible to calculate its lattice enthalpy electrostatically. When the calculation is done for the potassium halides KF and KI and for the silver halides AgF and AgI the results obtained may be compared with lattice enthalpies calculated from a thermodynamic cycle as follows.

Halide	KF	KI	AgF	AgI
Lattice enthalpy/kJ mol^{-1}				
calculated from thermodynamic cycle	801	629	955	876
calculated electrostatically	795	632	870	736

Comment on the differences between the values of lattice enthalpies calculated from a thermodynamic cycle and those calculated electrostatically for the two sets of halides.

(O & C)

10. Discuss the difference between the *enthalpy change* of a chemical reaction and the *free energy change* of a chemical reaction.

If the variation of the equilibrium constant of a reaction relative to temperature is studied, what relationship can be found to (a) the enthalpy change of that reaction, and (b) the free energy change of the reaction? Illustrate your answer by reference to suitable reactions.

(L)

⑥ PHASE EQUILIBRIA

Whisky

Whisky, from the Scottish Gaelic *uisge beatha*, 'water of life', is one of the distilled alcoholic beverages. Other distilled alcoholic beverages include brandies, rums and vodkas. Percentages of alcohol commonly range from 40% to 50%, expressed as proof. In the United States a spirit with 50% alcohol by volume is termed *100 proof*. Scotch, Canadian and American whiskies combined account for about one-half of the distilled spirits sold in the United States. The percentage of whisky sold has been decreasing as the sale of vodkas and light rums has been rising.

Production
Whisky, like the other distilled spirits, is generally distilled from a fermented, or alcohol-containing, mash of grains, which may include barley, rye, oats, wheat, or corn. Because distillation requires an alcohol-containing liquid, it is necessary to ferment the grain. Part of the process of whisky production includes converting the starch in the grain to sugar so that fermentation and ultimately distillation can take place. The grain is first milled, or ground to a meal. It is then cooked until it becomes gelatinous, releasing the starch from its tough coating. Barley malt is added because it is rich in amylase, an enzyme that enables starch to be converted to sugar. The mash is now ready for fermentation, and laboratory yeast is added. It ferments for 72 hours, creating an alcoholic liquid known as beer. The beer undergoes selective distillation to become whisky. Certain desirable secondary products called cogeners, or flavourings, are collected with the alcohol.

  Distillation may be carried out in either a batch process, using a pot still, or a continuous process, using a patent or Coffey still. The pot still is uniquely suited for full-flavoured spirits, such as malt whisky and Irish whiskey. The continuous process is best suited for light whiskies. The product of either process is colourless and of varied flavour. Fuller-flavoured whiskies mellow with wood maturation. This ageing will vary with the spirit (higher-distillation-proof spirits need less age because they have less flavour), the material of the cask (whether it is new or used and how deeply the interior is charred), the size of the cask (which determines the wood-to-liquid ratio), and the storage conditions (which are affected by temperature and humidity). The common amber colour and some of the flavour of the matured whisky are acquired from the storage cask.

Types
Straight whiskies are flavourful. They are made from at least 51% of a particular grain. Water may be added to reduce the proof to a bottling strength of 80 proof minimum. Straight whiskies are aged in new, charred white-oak barrels for at least two years.

The predominant ingredient in light whisky is corn; the higher the amount of corn, the lighter the flavour. Bourbon, the most popular whisky made in the United States, must have at least 51% corn in the mash; it may have as much as 79%. Other grains present are rye and barley malt, which also contribute flavour. Light whisky is stored in seasoned charred oak casks, which impart little colour or flavour.

Blended whisky, erroneously known as rye whisky, is a combination of straight (at least 20%) and light whisky. The final product may have as many as 40 to 50 different components. Tennessee whisky is usually made from corn, but any grain may be used. A bourbon-type whisky, it is very full because it is treated with maple-wood charcoal to remove the lighter flavours. Canadian whisky is always distilled in the patent still and is always a blend. Most Canadian whisky is at least six years old when sold. Delicate and light bodied, it is often confused with American blended whisky and thus called rye. Nearly all Scotch whiskies are blends. They are usually distilled from barley malt cured with peat, giving the spirit a smoky flavour.

Harriet Lembeck

Alcohol (Ethanol)

Status
Legal for adults.

Nature
A drink produced by the fermentation of fruits, vegetables or grain.

Effects
Alcohol is absorbed into the blood stream via the stomach and takes effect within 5–10 minutes. Effects vary according to individual health, weight and sex but, as a rough measure, three single whiskies drunk in one hour might result in 0.05% alcohol content in the blood—this would lift spirits and lessen inhibitions (a single whisky is equivalent to one glass of wine or half a pint of beer). Six double whiskies in an hour might produce uncontrollable behaviour and impair the functioning of the central nervous system; and 12 double whiskies in an hour might produce profound anaesthesia, near coma and death. Women get drunk more easily than men because they have less water per body weight. They also stay drunk longer if they are on the pill but get drunk slower during menstruation. Hangovers are actually the body's response of shock at being subjected to a substantial dose of a poisonous substance.

Dangers
Frequent intoxication damages the mouth, oesophagus, stomach and especially the liver, where overloading of the metabolising process can cause hepatitis and cirrhosis. Heavy drinking affects the heart and is linked to brain disorders. Alcoholic drink provides calories, giving energy and thus reducing appetite. But it contains no vitamins, minerals, amino acids or other essential nutrients. So if food consumption decreases there is a danger of malnutrition and if it continues one of obesity (with attendant heart problems). Alcohol is particularly dangerous when taken with barbiturates or tranquillisers, causing deep sedation, a drop in blood pressure and possible breathing failure.

Addictiveness
Highly addictive. Severe withdrawal symptoms (delirium tremens) are acute panic, delusions, exhaustion and trembling to the point of seizure. This lasts 3–10 days. Severe delirium tremens has a fatality rate of 20%, higher than any drug except the barbiturates.

After you have studied this chapter you should be able to

1. Distinguish between the terms **phase equilibrium** and **chemical equilibrium**.
2. Describe the main features of **dynamic equilibria**.
3. Interpret **phase diagrams** of **one- and two-component systems**.
4. Draw, label and compare the phase diagrams of
   (a) water,
   (b) carbon dioxide,
   (c) sulphur,
   (d) phosphorus.
5. Discuss the **solubility** of **gases in liquids**.
6. State and apply **Henry's law**.
7. State **Raoult's law** and apply it to a system of two **completely miscible liquids**.
8. Explain the basis of **fractional distillation**.
9. Discuss and give examples of the **deviation** of liquid mixtures from **ideal behaviour**.
10. Sketch and interpret the three types of phase diagrams for **partially miscible liquids**.
11. Explain the basis of **steam distillation**.
12. Use data obtained from steam distillation to calculate molar masses.
13. Define the **distribution law** and discuss its limitations.
14. Describe the main features of **solvent extraction**.
15. Sketch and interpret
    (a) typical **cooling curves**,
    (b) typical phase diagrams of the **solid–liquid equilibria** of two-component systems.
16. Explain the four principal **colligative properties** of solutions of non-volatile solutions and show how these can be used to determine the relative molecular masses of non-volatile solutes.
17. Explain the term **colloid** and give examples of the different types of colloid.
18. Outline the main features and applications of the principal types of **chromatography**.

6.1
Equilibria and Phases

Static Equilibria

The *Concise Oxford Dictionary* defines 'equilibrium' as 'state of balance'. It also defines 'a body in stable equilibrium' as 'tending to recover equilibrium after a disturbance'. A ball at the bottom of a ditch is in **stable equilibrium** (*see* figure 6.1). It soon recovers its stable position if it is pushed to one side or the other. On the other hand, a ball lying on the edge of a ditch is in a position of **unstable equilibrium**. Even a slight push means that it falls and cannot recover its original position of equilibrium. Both these examples of stable and unstable equilibria are also examples of **static equilibria**. In other words, the system at equilibrium is not in motion. It is static.

There are many other types of equilibrium. **Mechanical equilibrium**, for example, is a form of static equilibrium. It exists when the pressure is constant throughout all parts of a system.

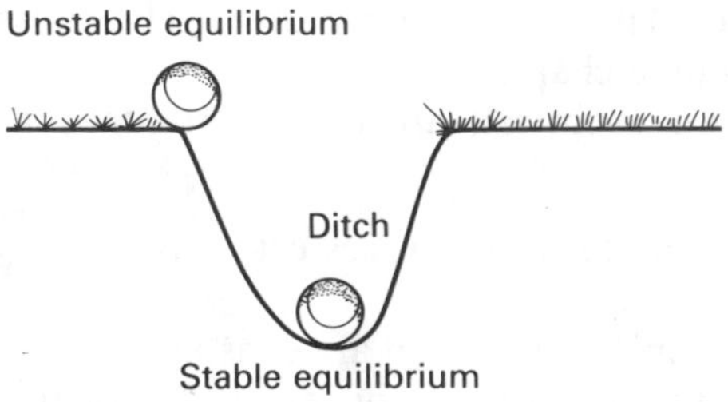

Unstable equilibrium

Ditch

Stable equilibrium

Figure 6.1 Static equilibria

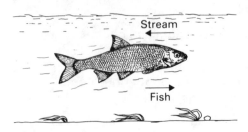

Figure 6.2 Dynamic equilibrium: the fish appears to be still. It is swimming upstream at the same speed as the stream

Dynamic Equilibria

In chemistry, we are not so much concerned with static equilibria as dynamic equilibria. A **dynamic equilibrium** exists when two reversible or opposite processes are balanced. An example of this is walking up an escalator at the same speed that the escalator is moving downwards. The overall change in position is zero. The two opposing motions are balanced. Another example is a fish swimming upstream at the same speed as the stream (*see* figure 6.2). The fish appears to be static. It is in dynamic equilibrium with the stream.

Dynamic equilibria can be divided into two categories:

- physical equilibria,
- chemical equilibria.

From our point of view the most important types of physical equilibria are **phase equilibria**—that is, the equilibria that exist between the phases of a system. The dynamic nature of phase equilibria can be understood by considering them in the light of the kinetic theory. Consider the equilibrium between a liquid and its vapour for example (*see* figure 6.3). The two phases are in dynamic equilibrium when the rate of evaporation equals the rate of condensation. In other words, in a given time, the number of particles leaving the liquid equals the number of particles entering it.

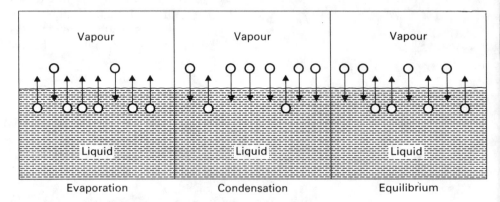

Figure 6.3 Water–vapour equilibrium

A system is in **chemical equilibrium** when the rate of the forward reaction is equal to the rate of the reverse reaction. For example, when the rate of the following reaction,

$$N_2(g) + 3H_2(g) \rightarrow 2NH_3(g)$$

equals the rate of the reverse of this reaction,

$$2NH_3(g) \rightarrow N_2(g) + 3H_2(g)$$

the system is in dynamic equilibrium. We call this a reversible reaction and write it with the sign $\rightleftharpoons$. Thus

$$N_2(g) + 3H_2(g) \rightleftharpoons 2NH_3(g)$$

We shall look at chemical equilibria in detail in the following chapter. The rates of chemical reactions are dealt with in detail in chapter 9.

Dynamic equilibria, whether physical or chemical, are characterised by a number of features:

- the dynamic equilibria consist of forward and reverse processes;
- the forward and reverse processes are balanced;
- the overall properties of a system in dynamic equilibrium are constant;
- the equilibrium can be achieved from either the forward or reverse direction;
- dynamic equilibrium can only be achieved in a closed system.

A **closed system** is a system in which there is no net gain or loss of matter in the system. Thus, in the ammonia example above, if ammonia, NH_3, is allowed to escape from the reaction system, the reaction cannot achieve equilibrium.

PHASES AND COMPONENTS

States of Matter

We saw in chapter 3 that there are three states of matter:

- gas,
- liquid,
- solid.

However, the states of some matter do not fall conveniently into these categories.

Glass has the properties of a solid but the structure of a liquid

Liquid crystal displays

What's the Matter?

Plasma

Plasma's properties are sufficiently different from those of gases, liquids and solids for it to be considered the fourth state of matter. Plasma consists of positive ions and unbound electrons. It occurs in the Earth's lower atmosphere. Over 99% of the matter in the Universe appears to exist as plasma. The Sun and stars are highly ionised plasmas formed at very high temperatures.

Glass

This is also a substance which does not fall conveniently into any of the three categories of gas, liquid or solid. Glass is an amorphous substance which has properties of a solid and the structure of a liquid. It is in fact a liquid cooled below its freezing point without crystallising. It is thus a supercooled liquid and may be considered as intermediate between liquid and solid.

Liquid Crystals

These are more or less the opposite to glass. They have the properties of liquids—they flow for example—yet their structure has some order like that of crystals. Liquid crystals are used in calculators and watches. Like glass, they may also be considered to be intermediate between liquid and solid.

Phases

A **phase** is defined as a portion of matter which is homogeneous. Subdivision of a phase will only produce smaller portions indistinguishable from one another. A phase may consist of two or more substances. A solution is a single-phase substance for example. Gases always mix freely with one another in any proportion to give a fully homogeneous mixture—that is a single phase. Both solids and liquids may exist in more than one phase.

Phases are separated from one another by physical boundaries known as **phase boundaries** (*see* figure 6.4).

Equilibria which occur within a single phase are called **homogeneous equilibria**. Ionic equilibria in solutions (*see* chapter 8) are examples of homogeneous equilibria. Equilibria which occur between physically distinct regions, that is between phases, are called **heterogeneous equilibria** or phase equilibria.

Components

A phase system may be classified as one-component, two-component (or binary) or three-component (or tertiary). The number of components in a phase system is

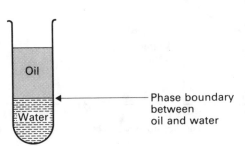

Phase boundary between oil and water

Figure 6.4 Phase boundary

the smallest number of chemical species needed to define all the phases of the system. The ice–water–water vapour phase system is a one-component system. The single component is water, H_2O.

Upsetting the Equilibria

We saw above that the equilibrium between a liquid and its vapour is a dynamic equilibrium. The equilibrium can be presented as

$$Liquid \rightleftharpoons Vapour$$

This equilibrium is a balance between two **phase transitions**, that is, the transition from liquid to vapour and the transition from vapour to liquid.

Similarly the equilibrium between solid and liquid can be represented by

$$Solid \rightleftharpoons Liquid$$

The equilibrium between solid and its vapour is represented by

$$Solid \rightleftharpoons Vapour$$

In each case the two phase transitions are the reverse of one another.

All these equilibria can be disturbed by changes in temperature and pressure. For example, the equilibrium between liquid and vapour is upset if the temperature is increased. An increase in temperature results in particles in both the liquid and vapour phases gaining kinetic energy (*see* section 3.1). More particles in the liquid phase will thus have sufficient energy to escape into the vapour phase. As a result, the equilibrium shown above shifts to the right.

Experimental data on the influence of temperature and pressure on phase equilibria can be represented by **phase diagrams**. Some of the most important features of phase diagrams are shown in the phase diagram of water.

THE PHASE DIAGRAM OF WATER

This is shown in figure 6.5. The **regions** bounded by the curves represent conditions where a single phase is stable. For example, at any temperature and

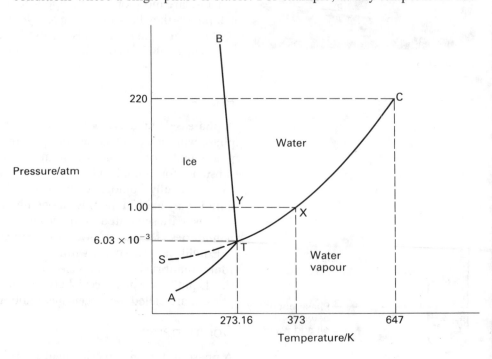

Figure 6.5 Phase diagram for water

pressure enclosed by the curves BT and TC, water exists in the liquid state. At any temperature and pressure below the curves AT and TC water exists in the vapour state.

The **curves** represent conditions under which two phases are in equilibrium. For example, along the curve TC water and water vapour are in equilibrium. This is the vapour pressure curve of water (*see* figure 3.13). At point X on this curve, liquid water and water vapour are in equilibrium at a temperature of 373 K (100°C) and a vapour pressure of one atmosphere (101.325 kPa). This is the boiling point of water at one atmosphere.

Curve AT is the vapour pressure curve for ice. This is often called the **sublimation curve**.

Curve BT is the **melting temperature curve**. This shows the effect of pressure on the melting point of ice. It shows that as pressure increases the melting point decreases slightly. This is unusual. Increase of pressure usually favours the formation of the solid—as we shall see in the carbon dioxide phase diagram below. In the case of water, higher pressures result in breaking down of the hydrogen bonds which bind the water molecules in ice together in an open structure. As the hydrogen bonds break down the denser liquid phase is thus formed (*see* section 2.2).

At point Y on curve BT ice is in equilibrium with water at a temperature of 273 K (0°C) and a pressure of one atmosphere. This is the freezing point of water at one atmosphere.

The curve ST represents the vapour pressure of water below its freezing point. Since water does not normally exist as a liquid below its freezing point, any point on this curve represents water in a **metastable** condition. This means that water is not in its most stable state at this temperature and pressure. The phenomenon represented by points on this curve is known as **supercooling**.

There are two **points** on the phase diagram of particular interest. First of all note that the vapour pressure curve terminates at point C. This is called the **critical point** of water. At temperatures and pressures above this point water vapour cannot be converted into liquid water no matter how great the applied pressure (*see also* section 3.1). In other words the vapour and liquid forms of water are indistinguishable above this point. The critical temperature of water is 647 K and the critical pressure 220 atmospheres.

Point T is known as the **triple point**. At this point, ice, water and water vapour are in equilibrium. It occurs at 273.16 K and 6.03×10^{-3} atmospheres. These are the only temperature and pressure values at which all three phases can exist together in equilibrium.

Frost

Frost can be formed in two ways: either from dew or directly from damp air.

From Dew

Dew is water formed from damp air when the temperature drops across the curve TC in figure 6.5. The frost is formed when the dew freezes, that is, when the temperature drops across curve BT.

Directly from Damp Air

Frost is only formed from dew when the vapour pressure of water is above the pressure of the triple point T, that is, above 6.03×10^{-3} atmospheres. If the vapour pressure is below this pressure, the frost is formed directly from damp air without involving dew. In this case the temperature drops across curve AT in figure 6.5. Frost formed in this way is known as **hoar frost**.

THE PHASE DIAGRAM OF CARBON DIOXIDE

This is shown in figure 6.6. It is similar to the phase diagram of water although there are two important differences.

First of all, the triple point occurs well above one atmosphere, at 5.11 atmospheres. Thus, at all pressures below this point, carbon dioxide cannot exist as a liquid. If solid carbon dioxide (dry ice) is warmed at one atmosphere it **sublimes** at 195 K ($-78°C$). This means it passes directly into the gas phase without becoming a liquid.

The other notable difference is the curve BT. This slopes to the right. Molecules of carbon dioxide in the solid phase are packed more closely than in the liquid phase. Thus, unlike water, the solid carbon dioxide is denser than the liquid. This is typical for most substances. An increase of applied pressure thus favours formation of solid carbon dioxide. Consequently, as pressure increases the melting point also increases.

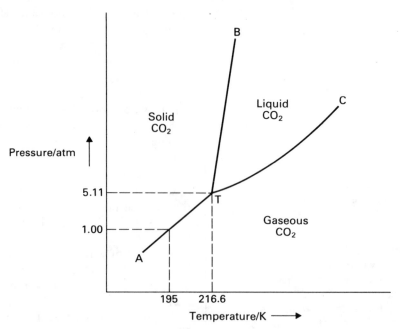

Figure 6.6 Phase diagram for carbon dioxide

THE PHASE DIAGRAM OF SULPHUR

In section 3.2 we saw that compounds which exist in more than one crystalline form are said to exhibit **polymorphism**. Elements which exist in more than one crystalline form are said to exhibit **allotropy**. Sulphur, for example, can exist in two allotropic forms: α-sulphur which has a rhombic crystalline form and β-sulphur which has a monoclinic crystalline form. The S_8 molecules in α-sulphur are more densely packed than in β-sulphur.

The variation of free energy (*see* chapter 5) of the two allotropes of sulphur and also its liquid form with temperature is shown in figure 6.7. The free energy of any substance decreases with increasing temperature. For sulphur, the α-allotrope has the lowest free energy at temperatures less than 368.5 K and is thus most stable at these temperatures. At temperatures between 368.5 K (95.5°C) and 393 K (120°C) the β-allotrope is most stable. Above 393 K, the liquid is most stable.

When an element can exist in two or more allotropic forms each of which is stable over a range of conditions, it is said to exhibit **enantiotropy**. The temperature at which two enantiotropes are in equilibrium is called the **transition temperature**. For sulphur, the transition temperature at one atmosphere is 368.5 K.

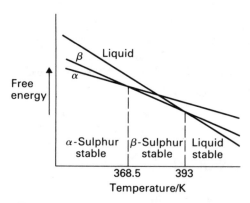

Figure 6.7 Variation of the free energy of sulphur with temperature at one atmosphere

$$\alpha\text{-Sulphur} \underset{\text{(stable} < 368.5 \text{ K)}}{\overset{368.5 \text{ K (95.5°C)}}{\rightleftharpoons}} \beta\text{-Sulphur} \atop \text{(stable} > 368.5 \text{ K)}$$

The effect of pressure on transition temperature is shown by the curve AB in the phase diagram of sulphur in figure 6.8. As pressure increases, the transition temperature increases.

Sulphur has three triple points—A, B and C. At A, for example, two solid phases and the vapour phase are in equilibrium. The two solid phases are the two enantiotropes of sulphur. The dotted curves represent metastable conditions. The dotted curve AD, for example, is the vapour pressure curve of α-sulphur above its transition temperature.

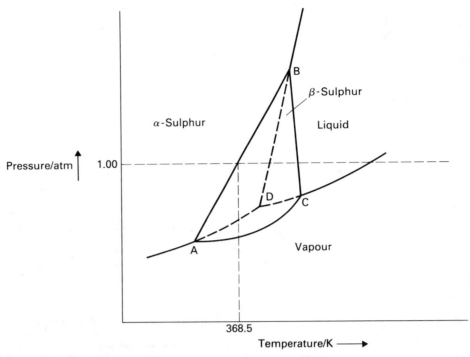

Figure 6.8   Phase diagram for sulphur

Enantiotropy of Other Elements

Sulphur is not the only element to exhibit enantiotropy. Tin, for example, has two enantiotropes—grey tin and white tin. The transition temperature at one atmosphere is 286.2 K (13.2°C).

$$\text{Grey tin} \overset{268.2\ K}{\rightleftharpoons} \text{White tin}$$

Iron is another example. It exists in three enantiotropic forms, α-iron, γ-iron and δ-iron.

$$\underset{\substack{\text{(body-centred}\\\text{cubic)}}}{\alpha\text{-Iron}} \overset{1181\ K}{\rightleftharpoons} \underset{\substack{\text{(face-centred}\\\text{cubic)}}}{\gamma\text{-Iron}} \overset{1661\ K}{\rightleftharpoons} \underset{\substack{\text{(body-centred}\\\text{cubic)}}}{\delta\text{-Iron}}$$

THE PHASE DIAGRAM OF PHOSPHORUS

When an element exists in more than one crystalline form only one of which is stable it is said to exhibit **monotropy**.

Phosphorus exhibits monotropy. We saw in section 3.2 that phosphorus has three forms. The stable monotrope is red phosphorus. At one atmosphere this is stable up to 690 K (*see* figure 6.9). White phosphorus and black phosphorus are both metastable monotropes. Black phosphorus only occurs at high pressures—this is not shown in figure 6.9. Phosphorus has a triple point at 862.5 K (589.5°C) and 43.1 atmospheres. At this point, red phosphorus, liquid phosphorus and its vapour are in equilibrium.

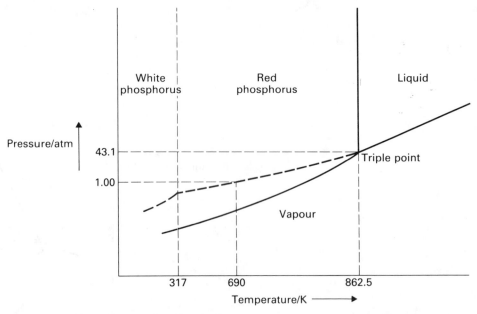

Figure 6.9 Phase diagram for phosphorus

SUMMARY

1. A **static equilibrium** is one which is not in motion.
2. A **dynamic equilibrium** exists when two reversible or opposite processes are balanced.
3. A **phase** is a homogeneous portion of matter.
4. The **number of components** in a phase system is the smallest number of chemical species needed to define all the phases of the system.
5. In a **phase diagram**
    (a) a **region** bounded by curves represents conditions where a single phase is stable;
    (b) a **curve** represents conditions under which **two phases** are in equilibrium;
    (c) a **triple point** represents the conditions at which **three phases** are in equilibrium.
6. An element that exists in two or more allotropic forms which are stable over a range of conditions is said to exhibit **enantiotropy.**
7. An element that exists in more than one crystalline form only one of which is stable is said to exhibit **monotropy.**

6.2
Multicomponent
Systems

We saw in chapter 4 that **mixtures** may be homogeneous, in which case they exist as a single phase, or they may be heterogeneous. A heterogeneous mixture consists of two or more phases. In this section we shall be examining phase equilibria of some heterogeneous mixtures. However, we shall be principally interested in phase equilibria involving homogeneous mixtures—that is solutions. In particular we shall be looking at three types of solution:

- gases in liquids,
- liquids in liquids,
- solids in liquids.

A solution consists of at least two components. The component in the greater quantity is called the **solvent**. The minor component is called the **solute**. A solution which cannot dissolve more solute at a given temperature is called a saturated solution.

A **saturated solution** is in dynamic equilibrium with undissolved solute. If we take the case of a saturated solution of sodium chloride for example, solid sodium chloride is in dynamic equilibrium with dissolved sodium chloride:

$$NaCl(s) \rightleftharpoons NaCl(aq)$$

The composition of a solution can be expressed in various ways including molality and mole fraction. These were discussed in section 4.2.

GASES IN LIQUIDS

Table 6.1 Absorption coefficients of some gases in water at s.t.p.

Gas	Absorption coefficient
NH_3	1300
HCl	500
SO_2	80
CO_2	1.7
O_2	0.05
N_2	0.024
He	0.009

The solubility of a gas in a liquid can be expressed in terms of its **absorption coefficient**. This is the volume of the gas, converted to s.t.p. (*see* section 3.1), which dissolves in unit volume of the liquid at the temperature and pressure specified.

The absorption coefficient depends on four factors:

- the gas,
- the liquid,
- temperature,
- pressure.

Values for the absorption coefficients of several different gases in water at s.t.p. are shown in table 6.1. The values range from very low for the virtually insoluble helium to very high for the highly soluble ammonia.

Examples of the influence of the liquid—or solvent—on the solubility of a particular gas are clearly shown by the values for the solubility of oxygen in table 6.2.

Table 6.2 Absorption coefficients of oxygen in various solvents at 293 K

Solvent	Absorption coefficient
water	0.028
ethanol	0.144
propanone	0.208

Influence of Temperature

In general, the solubility of a gas decreases with increasing temperature. The solubility of ammonia in water is a clear example of this (*see* table 6.3). Let us take the following equilibrium for example:

$$O_2(g) \rightleftharpoons O_2(aq)$$

At 273 K the absorption coefficient of oxygen in water is 0.05 whereas at 293 K it is 0.03. This is because the reverse reaction in the equilibrium is endothermic. Thus, as temperature increases, heat energy is absorbed shifting the equilibrium to the left. Conversely, the process of dissolving a gas in a solvent to form a saturated solution is an exothermic process. There are exceptions to this. For example, the solubility of noble gases in liquid hydrocarbons increases with increasing temperature. The process of forming a saturated solution in this case is endothermic.

Table 6.3 Solubility of ammonia in water at different temperatures

Temperature/K	Absorption coefficient
273	1300
293	710
363	240

Influence of Pressure

The relationship between gas pressure and solubility is given by Henry's law. This states that the mass (m) of a gas dissolved in a given volume of liquid with which it is in equilibrium is proportional to the pressure (p) of the gas at a particular temperature. Thus

$$m \propto p$$

and

$$m = Kp$$

K is a constant of proportionality. It is an equilibrium constant for dissolving the gas.

$$K = \frac{\text{Concentration of gas in saturated solution at equilibrium}}{\text{Concentration of gas in gas phase at equilibrium}}$$

Thus, the solubility of a gas increases with the pressure of a gas. It is for this reason that divers must avoid rapid decompression when they rise to the surface. The phenomenon is also important in industry. In the industrial manufacture of hydrogen by the Bosch process (*see* section 12.1), carbon dioxide is removed from hydrogen by bubbling the gaseous mixture through water at high pressure. At about 50 atmospheres the carbon dioxide readily dissolves in water thus removing it almost completely from the hydrogen.

TWO-COMPONENT LIQUID MIXTURES

We shall consider three categories of two-component liquid mixtures:

	Examples
• completely miscible liquids	water and ethanol
• partially miscible liquids	water and butan-1-ol
• immiscible liquids	water and oil

COMPLETELY MISCIBLE LIQUIDS

These are liquids which completely dissolve in one another in all proportions, forming a homogeneous mixture or solution. As a starting point in our treatment of such systems we shall develop a model of an ideal solution and then show how real or non-ideal solutions deviate from this model. Although no solution is completely ideal, some approach ideality. Examples are:

• a solution of benzene and methylbenzene (toluene);
• a liquid mixture of nitrogen and oxygen;
• a mixture of alkanes, in crude oil for example.

An ideal solution has the following properties:

1. Intermolecular attractions between like molecules and unlike molecules are equal. There is thus little or no tendency for molecules of one component to hinder or help the escape into the vapour phase of molecules of the other component.
2. There is no enthalpy change on mixing of the two components.
3. There is no volume change on mixing the two components.

Ideal solutions obey **Raoult's law**. This states that the partial vapour pressure (p_A) of component A in a solution equals the vapour pressure of the pure component (p_A°) multiplied by the mole fraction (x_A) of the component in the solution.

Compressed Air Illness

Compressed air illness is also known as caisson disease. The disease can affect divers who have been working under high pressure and return to normal atmospheric pressure too quickly. As they rise, the pressure decreases and thus the solubility of the nitrogen in the bloodstream also decreases. This can cause bubbles of nitrogen to form in the bloodstream. These bubbles cause pain known as 'the bends'. They also block the circulation of blood in the small blood vessels of the brain and parts of the body. This is known as decompression sickness.

Compressed air illness is avoided by returning slowly to normal atmospheric pressure or by using an atmosphere containing 80% helium and 20% oxygen. Helium does not cause compressed air illness since it is far less soluble than nitrogen (*see* table 6.1).

For an ideal solution consisting of two components A and B, Raoult's law may be expressed mathematically as

$$p_A = x_A p_A^\circ$$

and

$$p_B = x_B p_B^\circ$$

The vapour of an ideal solution may be considered to be an ideal gas, and Dalton's law (*see* section 3.1) therefore applies. Thus the total vapour pressure (p) of the solution is given by

$$p = p_A + p_B$$

Hence

$$p = x_A p_A^\circ + x_B p_B^\circ$$

The total pressure of the two components and the partial pressure of the more volatile component increase as the mole fraction of the more volatile component increases. The partial vapour pressure of the less volatile component decreases at the same time. This is shown in figure 6.10.

The more volatile component is the one whose molecules have a greater tendency to escape into the vapour phase. The more volatile component is thus the one with the higher vapour pressure. Note that the mole fraction of either A or B can be used to express the composition of the solution. This is because

$$x_A + x_B = 1$$

Note also that when

$$x_B = 0 \qquad p_A = p_A^\circ$$

and when

$$x_A = 0 \qquad p_B = p_B^\circ$$

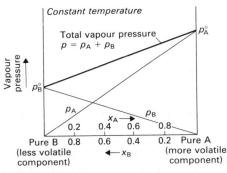

Figure 6.10   Raoult's law

Phase Diagrams of Completely Miscible Liquids

The top line in figure 6.10 is the total pressure curve for an ideal solution consisting of two components A and B. At all vapour pressures and compositions above the line, the mixture is a liquid. At all vapour pressures and compositions below it, the mixture is a vapour. At all points *on the line*, the liquid and vapour are in equilibrium

$$\text{Liquid} \rightleftharpoons \text{Vapour}$$

Let us now consider the conditions under which three mixtures in this system boil.

Mixture C is a liquid at one atmosphere (*see* figure 6.11). To make it boil without increasing its temperature, its pressure would have to be reduced from one atmosphere (point C) to point C' on the line. This corresponds to the boiling pressure of C at this temperature. At this point the liquid and vapour are in equilibrum.

Mixture D is a vapour at one atmosphere. This is represented by point D. To condense it without reducing its temperature, its pressure would have to be increased to point D' on the line. This corresponds to the boiling pressure of D at this temperature.

Mixture E boils at one atmosphere at the temperature of the system. Point E represents the vapour pressure and the composition of the *liquid* when the vapour and liquid are in equilibrium. But what is the composition of the *vapour* at this point? The vapour is a mixture of the more volatile component A and the less volatile component B. The particles of the more volatile component have a greater tendency to escape from the liquid than the particles of the less volatile

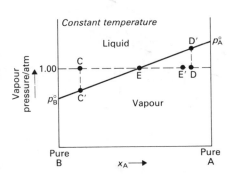

Figure 6.11

component. The vapour mixture is thus proportionately richer in component A. The mole fraction of A in the vapour is thus greater than the mole fraction of A in the liquid. The composition of the vapour is represented by point E′ in figure 6.11.

If we plot the vapour pressure of the liquid mixture (that is the solution) against *vapour composition* over the complete range of compositions, we obtain the lower curve in figure 6.12. The top line is a plot of vapour pressure of the liquid mixture against *liquid composition*. It corresponds to the line in figure 6.11 and the top line in figure 6.10.

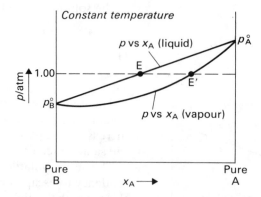

Figure 6.12 Constant-temperature phase diagram showing composition of both liquid and vapour

When temperature is increased, the partial pressures of both components and the total vapour pressure of the mixture increase. This follows from Charles' law (*see* section 3.1). Figure 6.13 represents the vapour pressure–liquid composition lines for three different temperatures. As the temperature increases, the slope increases. A mixture with composition $x_A(F)$ boils at one atmosphere pressure when the temperature equals T_3. Similarly, a mixture with composition $x_A(G)$ boils at T_2 and so on.

If we plot boiling point against composition we obtain the type of **boiling point curve** shown in figure 6.14. Point C represents a liquid with composition $x_A(C)$ at its boiling point $T(C)$. At this point, the vapour pressure of the mixture equals the

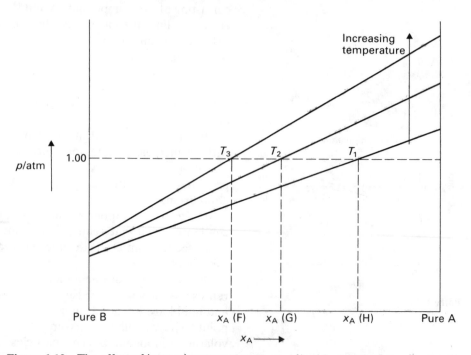

Figure 6.13 The effect of increasing temperature on a liquid–vapour phase diagram

external pressure and is in equilibrium with the liquid. For the reasons given above, the vapour will be richer in the more volatile component. This is the component with the lower boiling point. The vapour at this boiling temperature is represented by point D. If we plot all such points for compositions over the complete range we obtain the upper curve in figure 6.15.

Liquid–vapour phase diagrams at constant pressure are of crucial importance in the study of distillation. If, for example, we distil a mixture with composition $x_A(C)$, the distillate will have the composition $x_A(D)$. If we then redistil the distillate, the new distillate will have the composition $x_A(E)$ and so on. By repeated distillation, it is theoretically possible to separate components A and B completely. This is the basis of fractional distillation.

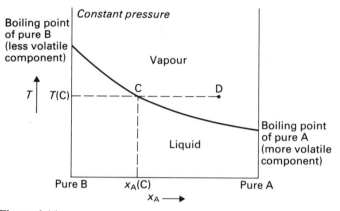

Figure 6.14

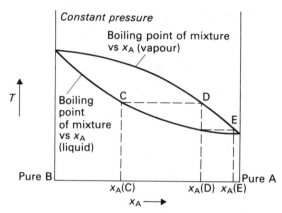

Figure 6.15 Constant-pressure phase diagram showing composition of both liquid and vapour

Fractional Distillation

Fractional distillation has a number of important applications, for example

- the manufacture of oxygen, nitrogen and the noble gases from liquid air,
- the refining of petroleum,
- the production of whisky and other alcoholic drinks (*see* introductory reading at start of this chapter and also below).

A diagram of a typical laboratory apparatus used for fractional distillation is shown in figure 6.16. The vertical column is filled with glass beads or randomly orientated short pieces of glass tubing. Alternatively, the column may have bulbous surfaces. The column allows the ascending vapour to come into contact with the descending liquid.

Let us consider what happens when we fractionally distil a two-component mixture with composition $x_A(C)$ (*see* figure 6.17). As we heat up this mixture, its temperature increases to point C. It then boils. The vapour produced is richer in the more volatile component A. At the boiling point, this vapour and the liquid are in equilibrium. This is represented by the **tie-line** CD in the phase diagram. The vapour rises in the fractionating column. As it does, its temperature decreases until it condenses. This temperature drop is represented by the vertical line DD' in the phase diagram. At point D' a new equilibrium is established between the condensate, which has the composition $x_A(D)$, and its vapour, which has the composition $x_A(E)$. The liquid condensate descends the column whilst the vapour ascends. Thus, at various stages in the column the descending liquid and the ascending vapour are in equilibrium. These equilibria are represented by the tie-lines. As the vapours rise in the column through each successive equilibrium, they become richer in the more volatile component. Eventually, they pass out of the column at the top, are condensed and collected. At the same time, the liquid in the flask becomes richer in the less volatile component and consequently its boiling point increases.

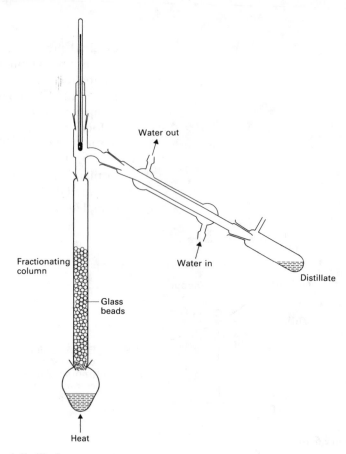

Figure 6.16   Fractional distillation apparatus

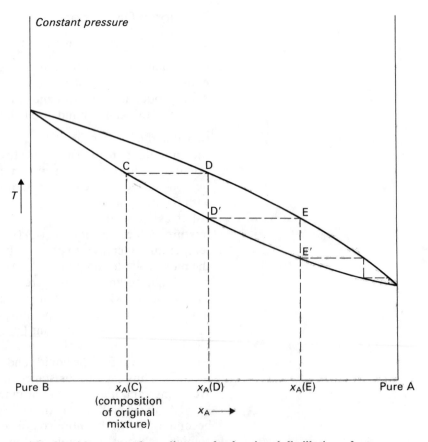

Figure 6.17   Liquid–vapour phase diagram for fractional distillation of a two-component mixture. The horizontal tie-lines represent equilibria in the fractionating column

Since vapour is removed from the top of the column, the equilibria in the column are continuously changing. Good separation is only achieved if the flask is heated slowly to allow time for the equilibria to become established.

Enguli

In Uganda, a spirit called 'enguli' is made by the fractional distillation of beer in home-made stills. The beer is obtained by the fermentation of molasses and banana juice. Three fractions are collected during the distillation of the beer.

The *first fraction* contains toxic lower-boiling aldehydes, ketones and alcohols. For example:

Substance	Boiling point/°C	Toxicity
propanal	48	toxic
propanone	56	toxic
methanol	64	very toxic: can cause blindness

This fraction is discarded.

The *second fraction* is the 'enguli'. It contains water and ethanol. Ethanol (also known as ethyl alcohol) has a boiling point of 78°C and is harmless if consumed in small quantities (*see* introductory reading at start of this chapter).

The *third fraction* contains alcohols with boiling points ranging from 120 to 130°C. This fraction is also discarded.

In Uganda, licensed 'enguli' producers sell their product to industrial distilleries which produce a spirit known as 'waragi'. Home-made enguli and similar home-made drinks in other East African countries can be dangerous since the second fraction is often contaminated with the toxic first and third fractions. For this reason, most East African countries discourage or ban the production and consumption of such drinks.

In practice, fractional distillation is normally used to separate multicomponent liquid mixtures into several fractions.

Non-Ideal Solutions

All liquid mixtures deviate from ideal behaviour to a greater or lesser extent. However, only those which deviate to a lesser extent can be separated by fractional distillation. Such mixtures are called **zeotropic**. Benzene and methylbenzene form a zeotropic mixture.

Deviations from ideal behaviour may be positive or negative. A mixture which exerts a vapour pressure greater than that predicted by Raoult's law is said to show a positive deviation. Figure 6.18 is a typical vapour pressure–composition diagram for a non-ideal solution showing positive deviation. The curves should be compared with the broken lines. These represent ideal behaviour.

Figure 6.19 is the corresponding diagram for a non-ideal solution which exhibits **negative deviation** from Raoult's law. Examples of two-component liquid mixtures which show positive and negative deviations from Raoult's law are listed in table 6.4.

Positive deviations occur for two reasons:

- First of all, attractions or bonding between molecules of components A and B are weak. Bonding between molecules of components A and B can be represented by A...B.
- Secondly, bonding of the A...A and B...B types are broken on mixing the two components. These bonds are often hydrogen bonds.

Beer making in Africa

Table 6.4 Liquid mixtures which deviate from Raoult's law

Positive deviation	Negative deviation
benzene/ethanol	methanol/water
water/ethanol	nitric acid/water
ethyl ethanoate/ethanol	trichloromethane/propanone
ethanol/carbon disulphide	

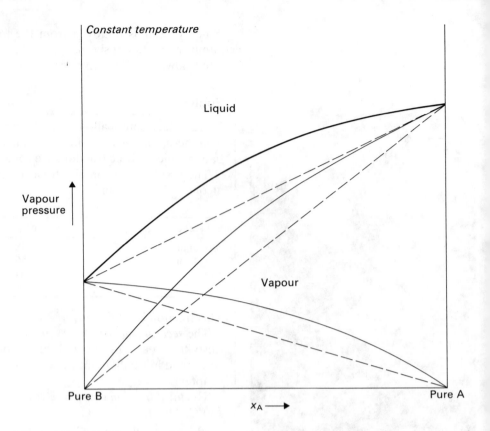

Figure 6.18 Liquid–vapour phase diagram showing positive deviation from Raoult's law

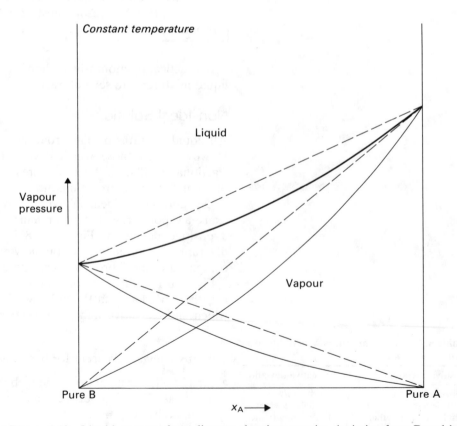

Figure 6.19 Liquid–vapour phase diagram showing negative deviation from Raoult's law

As a result of these two factors, molecules of both A and B have a greater tendency to escape from the liquid phase into the vapour phase. The liquid mixture thus has a higher vapour pressure.

Negative deviations occur because A...B attractions or bonds are stronger than A...A and B...B attractions or bonds. The molecules are consequently held back in the liquid phase and the vapour pressure is lower.

Liquid mixtures which deviate widely from ideal behaviour cannot be separated by fractional distillation. They are called **azeotropic**. The following are examples of azeotropic mixtures: hydrochloric acid/water, ethanol/water, trichloromethane/propanone and benzene/ethanol.

The reason azeotropic mixtures cannot be separated by fractional distillation is that they have maximum or minimum boiling points. Those which exhibit positive deviation from Raoult's law have **minimum boiling points**. Those which exhibit negative deviations have **maximum boiling points** (*see* figures 6.20 and 6.21).

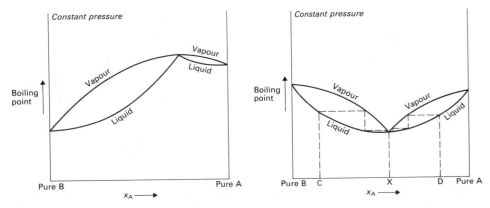

Figure 6.20 Liquid–vapour phase diagram showing maximum boiling point

Figure 6.21 Liquid–vapour phase diagram showing minimum boiling point

Examples of Azeotropes
• Ethanol/benzene minimum boiling point: 67.8°C mole fraction of ethanol in azeotrope: 0.54
• Trichloromethane/propanone maximum boiling point: 64.7°C mole fraction of trichloromethane in azeotrope: 0.64

Let us consider a liquid mixture of initial composition C in figure 6.21. Fractional distillation results in pure B and a mixture having the composition X. This mixture is called the **azeotrope**, the azeotropic mixture or the **constant boiling mixture**. If the initial composition of the mixture is D, then fractional distillation would result in pure A and the constant boiling mixture.

Although such mixtures cannot be separated by fractional distillation, they can be separated by other means. For example, ethanol can be separated from a liquid mixture of ethanol and water by shaking the mixture with silica gel. The silica gel removes the water.

PARTIALLY MISCIBLE LIQUIDS

If the deviation from Raoult's law is sufficiently great, the mixture may no longer remain homogeneous over the complete range of compositions. In this case the liquids are said to be **partially miscible**. Figure 6.22 shows the three possible types of phase diagram for partially miscible liquids.

Let us examine type (a) in a little more detail. If we add component A to component B at a constant temperature T_d, A will dissolve in B to form a solution until point X is reached. At this point B is saturated with A and a second phase appears. This new phase is A saturated with B. On adding more A the proportion of this second phase increases until point Y is reached. At this point the phase consisting of B saturated with A disappears resulting in a single phase consisting of A saturated with B.

Points X and Y thus represent the compositions of saturated solutions of A in B and B in A respectively at temperature T_d. At points on the tie-line between X and

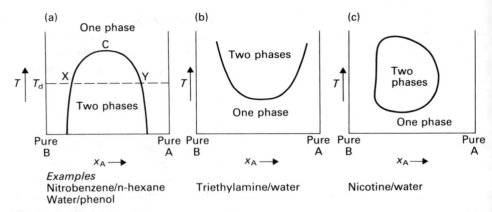

Figure 6.22 Phase diagrams for partially miscible liquids

Y the saturated solutions co-exist in equilibrium. Their compositions remain constant but the relative proportions of each change. The mole fraction x_A represents the overall composition of the two solutions.

The temperature at point C is the maximum temperature at which two phases can form. This is called the **critical solution temperature**. The composition at this point is called the **critical composition**.

IMMISCIBLE LIQUIDS

Some pairs of liquids are insoluble in one another. Thus, when mixed in any proportion, they form two separate layers. Such liquids are said to be **immiscible**. Examples include: mercury/water, carbon disulphide/water, chlorobenzene/water, nitrobenzene/water and phenylamine/water.

The total vapour pressure of a pair of immiscible liquids is found experimentally to equal the sum of the vapour pressures of the two pure components. Thus

$$p_{total} = p_A^\circ + p_B^\circ$$

Steam Distillation

Since the total vapour pressure is higher than that of either component, the temperature at which the mixture boils is lower than that of either component. This is the principle of steam distillation. Water boils at 100°C at one atmosphere pressure. However, when steam is bubbled through a liquid with which water is immiscible, the vapour pressure is increased. As a consequence the mixture boils at a lower temperature (*see* figure 6.23).

A diagram of an apparatus used for steam distillation is shown in figure 6.24. The required product is collected in a flask and separated from water using a separating funnel.

Steam distillation is used

- to purify liquids such as nitrobenzene and phenylamine which have high boiling points. Distillation at a lower temperature avoids the risk of thermal decomposition. Nowadays vacuum distillation is usually preferred to steam distillation for this purpose.
- to extract oils from plants, for example in perfumery.

Calculation of Molar Mass Using Steam Distillation

The molar mass of a liquid which is immiscible with water can be calculated from steam distillation data. From Raoult's law

$$p_A = x_A p_{total}$$

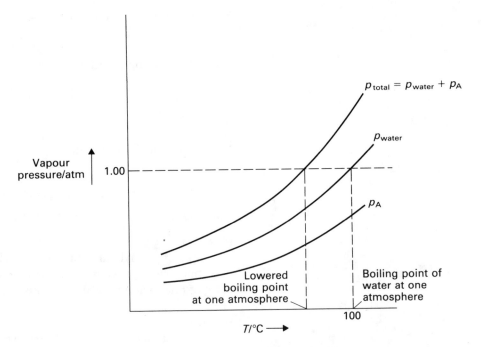

Figure 6.23   Vapour pressure curves for two immiscible liquids

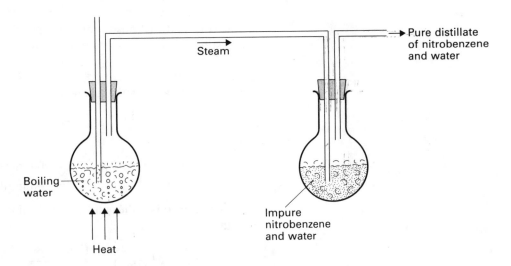

Figure 6.24   Steam distillation of nitrobenzene and water

where p_A and x_A are the partial vapour pressure and mole fraction of liquid A whose molar mass is to be determined. Similarly, for water

$$p_{water} = x_{water} p_{total}$$

By eliminating p_{total} from these two equations we obtain

$$p_A/p_{water} = x_A/x_{water} \qquad (1)$$

Since

$$x_A = \frac{n_A}{n_A + n_{water}}$$

where n_A and n_{water} are the numbers of moles of liquid A and water and

$$x_{water} = \frac{n_{water}}{n_A + n_{water}}$$

by substituting these equations into equation (1), we obtain

$$p_A/p_{water} = n_A/n_{water} \qquad (2)$$

We saw in section 4.1 that

$$n = \frac{m}{M}$$

where m is the mass in grams and M is the molar mass in grams per mole. Thus, from equation (2) we obtain

$$\frac{p_A}{p_{water}} = \frac{m_A/M_A}{m_{water}/M_{water}} \qquad (3)$$

This equation can be used to calculate the molar mass (M_A) of liquid A.

EXAMPLE

A sample of phenylamine (aniline), $C_6H_5NH_2$, was steam distilled at 98.6°C and one atmosphere pressure. The distillate was found to contain 25.5 g of water and 7.4 g of phenylamine. Calculate the molar mass of phenylamine.

SOLUTION

To find the molar mass of $C_6H_5NH_2$ we use equation (3). Thus

$$M_{C_6H_5NH_2} = \frac{m_{C_6H_5NH_2} \times M_{water} \times p_{H_2O}}{m_{water} \times p_{C_6H_5NH_2}}$$

Since the mixture boils at one atmosphere (760 mmHg), the sum of the two partial pressures must equal one atmosphere:

$$p_{water} + p_{C_6H_5NH_2} = 760 \text{ mmHg}$$

Tables of data show that, at 98.6°C,

$$p_{water} = 720 \text{ mmHg}$$

Thus

$$\begin{aligned} p_{C_6H_5NH_2} &= 760 \text{ mmHg} - 720 \text{ mmHg} \\ &= 40 \text{ mmHg} \end{aligned}$$

Substituting this value and the experimental data into the above equation we obtain

$$M_{C_6H_5NH_2} = \frac{(7.4 \text{ g}) \times (18 \text{ g mol}^{-1}) \times (720 \text{ mmHg})}{(25.5 \text{ g}) \times (40 \text{ mmHg})}$$

$$= 94.02 \text{ g mol}^{-1}$$

Note that the molar mass calculated from the formula $C_6H_5NH_2$ is 93.13 g mol^{-1}.

The Distribution Law

When a solute is soluble in two immiscible liquids, the solute distributes itself between the two liquids. The ratio in which it is distributed is governed by the **distribution law**. This states that a solute distributes itself between two immiscible liquids in a constant ratio of concentrations irrespective of the amount of solute added.

This is sometimes called the partition law. It is based on experimental evidence. Consider, for example, the distribution of iodine between the immiscible solvents, water and tetrachloromethane (*see* figure 6.25). If we shake up iodine with these two solvents, some will dissolve in the water and some in the tetrachloromethane, CCl_4. Eventually a dynamic equilibrium is established. At this point the rate at

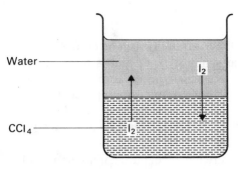

Figure 6.25 The distribution of iodine between two immiscible solvents

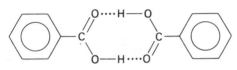

Figure 6.26 Association of benzoic acid

Figure 6.27 Separating funnel

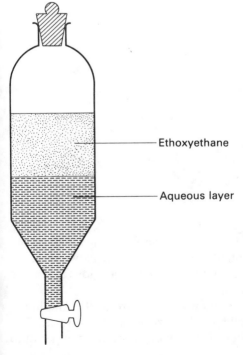

which iodine passes from CCl_4 to water equals the rate at which it passes from water to CCl_4.

$$I_2(aq) \rightleftharpoons I_2(CCl_4)$$

No matter how much iodine we use the ratio of the final concentration is constant. The constant is called the **distribution** (or partition) **coefficient**, K. It is given by

$$K = \frac{[I_2(CCl_4)]}{[I_2(aq)]} \qquad (4)$$

At 25°C the value of K for this equilibrium is 85. This means that the concentration of iodine in CCl_4 is 85 times higher than in water. This is because iodine is a non-polar solute. It is thus much more soluble in non-polar solvents such as CCl_4 than in polar solvents such as water. The corresponding value for the distribution of iodine in benzene and water is about 400.

The law only holds under certain conditions:

- The temperature must be constant.
- The two solutions must be reasonably dilute.
- The solute must not react, associate or dissociate in the solvents.

For example, the distribution coefficient for benzoic acid in benzene and water increases with increasing concentrations in the two layers. This is due to the formation of benzoic acid dimers in the benzene layer. These dimers are formed as a result of hydrogen bonding (*see* figure 6.26 and also chapters 2 and 3).

Applications of the Distribution Law

The distribution law has a number of important applications. Two of the most important are liquid-phase chromatography and solvent extraction. We shall deal with chromatography in general in section 6.3 of this chapter.

Solvent Extraction

Two immiscible liquids are sometimes used as selective solvents for components in a mixture. The mixture is shaken with the two immiscible liquids. The liquids are then allowed to separate. Each layer is then subjected to a number of extractions with the other solvent.

This technique is used to separate uranium salts from nuclear fission products. For example, the uranyl salt, $UO_2(NO_3)_2$, may be separated from NaCl by using butanol and water. The uranyl salt is more soluble in butanol and NaCl more soluble in water. The technique of **counter-current extraction** can be used for this purpose. One of the solvents is passed over the solid mixture which is located in a column or tower. The other solvent is then passed over the mixture in the opposite direction. The process is repeated several times.

Perhaps the most common laboratory example of solvent extraction is **ether extraction**. This is used to separate the products of an organic synthesis from water. The aqueous solution is shaken up with ethoxyethane (ether) in a separating funnel and allowed to separate (*see* figure 6.27). The product is thus removed from inorganic impurities which are soluble in water. The solution may be subjected to several separations. Finally the exothyethane is evaporated off, leaving the organic product.

The technique is particularly useful when the product is volatile or thermally unstable. In this case the solvent has to be evaporated or boiled off at a low temperature. Ethoxyethane with its boiling point of 34.5°C is thus a suitable solvent for this purpose—so long as there are no naked flames in the laboratory! Repeated extractions using small portions of solvent are more efficient than using a single but larger volume of solvent.

SOLID–LIQUID EQUILIBRIA

Phase diagrams for two-component solid–liquid equilibria can be quite complex. In this section we shall consider only the simplest type, that is temperature–composition phase diagrams of liquid mixtures which form a simple eutectic. We shall explain the term eutectic below. Temperature–composition phase diagrams for solid–liquid systems can be constructed from **cooling curves**. To do so, the cooling curves are plotted for the two pure components and for mixtures with various compositions of the two components.

Figure 6.28 shows the cooling curves for pure benzene, pure naphthalene and a mixture of the two. The first part of each curve represents the liquid cooling. For the two pure components, the curve then flattens out. This occurs at the melting point. At this point the solid and liquid are in equilibrium. The temperature remains constant due to the enthalpy decrease which accompanies freezing. The small blip is due to supercooling. When all the liquid has solidified, the temperature drops once again. The curve in the middle shows what happens when a mixture of benzene and naphthalene is cooled. Solid begins to form at point A. Depending on the composition of the mixture, the solid will be either benzene or naphthalene. As the solid of one component forms the liquid becomes richer in the other component. At point B, a mixture of both solids form. The solid mixture is in equilibrium with the liquid. Finally, the solid mixture cools.

> **CAUTION**
>
> Benzene, naphthalene and other aromatic compounds are highly poisonous. They should be handled with care.

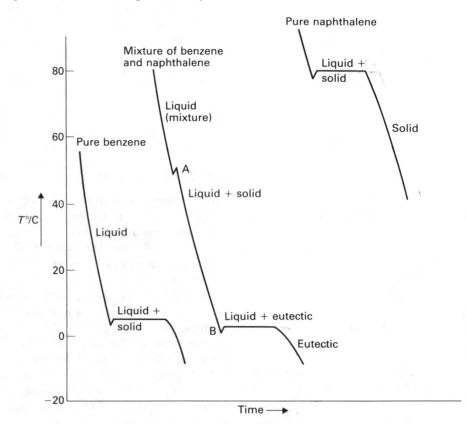

Figure 6.28 Cooling curves

If we plot points A and B for mixtures of benzene and naphthalene ranging from pure benzene to pure naphthalene we obtain the phase diagram shown in figure 6.29. The cooling curves corresponding to mixtures with compositions a, b and c are shown in figure 6.30.

Let us consider the mixture with composition a (*see* figure 6.29). The liquid mixture cools until point a'. At this point solid benzene begins to form. The liquid thus becomes richer in naphthalene. Consequently, as the temperature

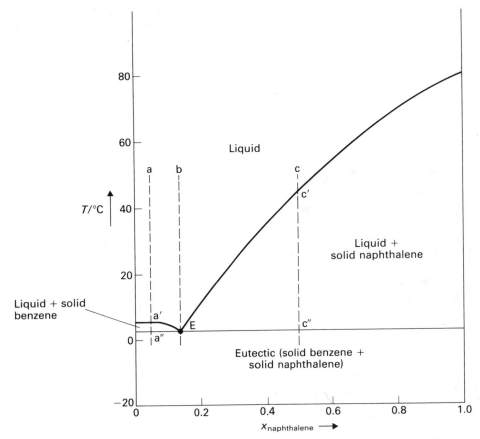

Figure 6.29 Temperature–composition phase diagram for the benzene/naphthalene system at constant pressure

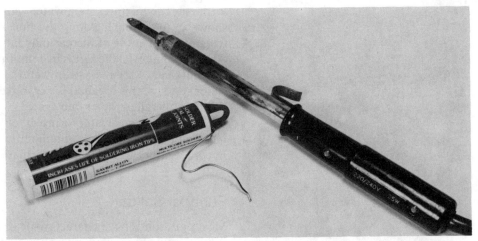

Figure 6.30 Cooling curves for figure 6.29

Eutectic

This word derives from the Greek word *eutektos* meaning 'easily melted'. A eutectic is a mixture of two or more substances which have a minimum melting point. Although a eutectic is a mixture, it behaves like a pure compound in some ways.

An **alloy** is a mixture of a metal and one or more other elements.

---

drops the composition of the liquid mixture travels along the curve a′–E. All the time solid benzene is forming. At point E a solid mixture of benzene and naphthalene is formed. This mixture is called the eutectic mixture or simply the eutectic. The point is called the eutectic point.

Let us now consider a liquid mixture of composition c. This starts to form solid naphthalene at point c′. The liquid composition then becomes richer in benzene and travels along the curve c′–E. At point E, the eutectic is formed.

Alloys

Solid–liquid phase diagrams are particularly important in the study of alloys and are thus of much importance to metallurgists. An ordinary electrical solder is a

Solder is an alloy of tin and lead

mixture of tin and lead. The eutectic mixture melts at the relatively low temperature of 183°C. The solder can be strengthened by using silver.

Figure 6.31 shows the phase diagram for zinc and cadmium. Note that the

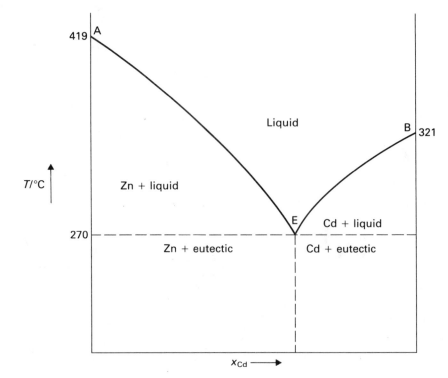

Figure 6.31 Temperature–composition phase diagram for zinc and cadmium

eutectic mixture at point E always has the same composition. Thus, to the left of E at temperatures equal to or lower than 270°C the solid consists of zinc and the eutectic mixture. Microscopic examinations show that crystals of zinc are embedded in the eutectic mixture. The zinc is formed whilst cooling along the line from A to E. To the right of E at temperatures equal to or lower than 270°C the solid consists of a mixture of cadmium and the eutectic mixture. Small crystals of cadmium are interspersed in the finely divided eutectic mixture. It should be stressed that the eutectic is a mixture and not a compound.

Cryohydrates

Inorganic salts such as sodium chloride, potassium chloride and potassium iodide form eutectic mixtures with water. A typical phase diagram for such a system is shown in figure 6.33. Along the line AE, ice is formed. It thus represents the **freezing-point curve** of increasingly more concentrated solutions of salt in water. Along the line BE, the salt is in equilibrium with the solution. The line is thus the **solubility curve** for the salt in water. Point E is the eutectic point. For salt and water systems, this is called the **cryohydric point**. The eutectic mixture formed at this point is called the **cryohydrate**.

Common salt and water form a cryohydrate which freezes at −21.2°C. Since salt and water mixtures freeze at temperatures lower than 0°C, salt is spread on roads in cold weather to prevent the formation of ice.

Purity of Solids

Most pure solids have characteristic sharp melting points. A glance at any of the solid–liquid phase diagrams above shows that the presence of even a small

---

Dendritic Crystals

Some metals crystallise in a dendritic manner. This means they branch as they grow. Crystallisation starts at a nucleus and outward growth continues in three directions. Later secondary and tertiary arms grow, forming a skeleton-type crystal. The outward growth ceases when the dendrite arms meet a neighbouring crystal (see figure 6.32).

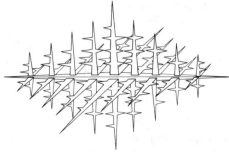

Figure 6.32 A dendritic crystal

In a typical winter, over two million tonnes of salt and grit are used to keep Britain's roads free from ice—and injury.

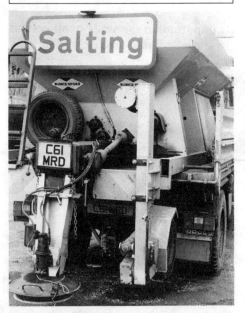

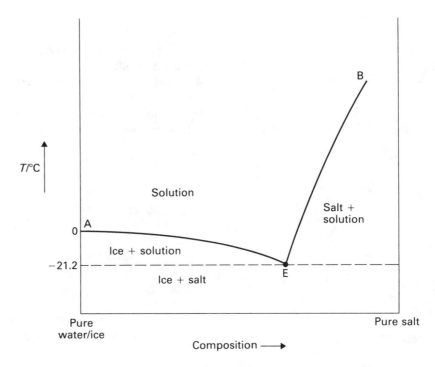

Figure 6.33 Temperature–composition phase diagram for salt and water at constant pressure. Note that the curve BE does not start at pure salt but rather at about 90% salt. This is because solutions containing more than 90% vaporise at temperatures higher than B

amount of a second component can lower the freezing point or melting point. This fact can be used to determine the identity of unknown organic compounds by the method of **mixed melting points**. The unknown compound is matched with a known compound with the same melting point. Approximately equal amounts of each are powdered and mixed together thoroughly. The melting point is determined. If it is not sharp and is lower than that of the two separate samples, then the samples are not identical. If, on the other hand, the melting point is sharp and is not lowered then the two samples are identical.

SUMMARY

1. The solubility of a gas in a liquid can be expressed in terms of its **absorption coefficient**.
2. **Henry's law** states that the mass of a gas dissolved in a given volume of liquid with which it is in equilibrium is proportional to the pressure of the gas at a particular temperature.
3. **Ideal solutions** obey Raoult's law.
4. For a two-component system, **Raoult's law** can be expressed mathematically as

$$p_A = x_A p_A^\circ$$
$$p_B = x_B p_B^\circ$$

5. A **boiling-point curve** forms a liquid–vapour phase diagram at constant pressure.
6. A **tie-line** in a liquid–vapour phase diagram relates the composition of the liquid to the composition of the vapour at equilibrium.
7. **Zeotropic mixtures** can be separated by **fractional distillation**.
8. **Non-ideal solutions** show either **positive deviation** or **negative deviaton** from Raoult's law.

(continued)

9. An **azeotrope** is a **constant boiling mixture**. It cannot be separated by fractional distillation. They have either **maximum boiling points** or **minimum boiling points**.
10. **Partially miscible liquids** form two liquid phases over a limited range of compositions and temperatures.
11. **Immiscible liquids** are insoluble in one another.
12. The technique of **steam distillation** can be used to purify high-boiling organic liquids and also to determine molar mass.
13. The **distribution law** states that a solute distributes itself between two immiscible liquids in a constant ratio of concentrations irrespective of the amount of solute added.
14. Repeated **solvent extractions** using small portions of solvent are more efficient than a single extraction with a larger volume of solvent.
15. Temperature–composition phase diagrams for **solid–liquid systems** can be constructed from **cooling curves**.
16. A **eutectic** is a mixture of two or more substances with a minimum melting point.
17. A eutectic mixture formed by a salt and water is called a **cryohydrate**.
18. The **method of mixed melting points** relies on the lowering of the melting point of a pure substance by a second component.

6.3 Colligative Properties, Colloids and Chromatography

COLLIGATIVE PROPERTIES OF SOLUTIONS

A **colligative property** is one which depends upon the number of particles of the solute but not on the nature of the solute. In this section we shall be concerned with the following four colligative properties:

- lowering of vapour pressure
- elevation of boiling point
- depression of freezing point
- osmotic pressure

All four of these properties apply to solutions containing non-volatile solutes. A **non-volatile solute** is one which exerts a negligible vapour pressure.

Lowering of Vapour Pressure

In a pure solvent particles can escape all over the surface of the liquid (*see* figure 6.34). However, if the solvent contains dissolved solute particles the escape of the solvent particles from the surface is hindered. If the solute has a lower pressure than the solvent, then the vapour pressure is reduced. This follows from Raoult's law (*see* previous section). The extreme case occurs with a solution containing a non-volatile solute. In this case the vapour pressure of the solution is almost entirely due to the solvent particles. The vapour pressure–composition diagram for this case is shown in figure 6.35. This is hypothetical since in reality all solids exert a vapour pressure although it may be very small.

It is evident from figure 6.35 that, as the mole fraction of non-volatile solute B increases, the vapour pressure of the solution is lowered. An expression for the

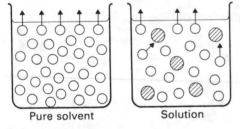

Pure solvent   Solution

○ Solvent particle
⊘ Solute particle

Figure 6.34 Non-volatile solute particles inhibit solvent particles from escaping from the surface and thus lower vapour pressure

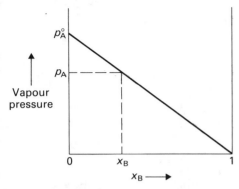

Figure 6.35 The vapour pressure–composition phase diagram for a solution containing a non-volatile solute

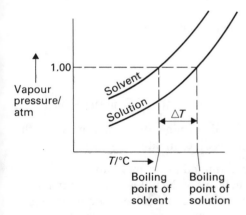

Figure 6.36 Elevation of boiling point at one atmosphere pressure

Table 6.5 Ebullioscopic constants

Solvent	K_b/K mol^{-1} kg
benzene	2.53
ethanoic acid	3.07
water	0.52
camphor	3.07

lowering of vapour pressure for dilute solutions can be derived from Raoult's law. For solvent A Raoult's law is

$$p_A = x_A p_A^\circ \qquad (5)$$

For a two-component system,

$$x_A + x_B = 1$$

and thus

$$x_A = 1 - x_B$$

If we substitute this into equation (5) we obtain

$$p_A = p_A^\circ(1 - x_B)$$
$$= p_A^\circ - p_A^\circ x_B$$

Rearranging

$$\frac{p_A^\circ - p_A}{p_A^\circ} = x_B$$

This shows that the lowering of the vapour pressure, given by $p_A^\circ - p_A$, is proportional to the mole fraction of the non-volatile solute, x_B.

Elevation of Boiling Point

A solution containing a non-volatile solute is less volatile than the pure solvent. It therefore boils at a higher temperature. Figure 6.36 shows the variation of vapour pressure with temperature for a solvent and solution containing a non-volatile solute. The increase in boiling point is proportional to the lowering of the vapour pressure:

$$\Delta T \propto p_A^\circ - p_A$$

where ΔT is the increase in boiling point. And thus

$$\Delta T = K(p_A^\circ - p_A)$$

By application of Raoult's law to this expression, it is possible to derive another expression relating ΔT to the composition of the solution, namely

$$\Delta T = K_b m_B \qquad (6)$$

In this equation m_B is the **molality** of the solute. Molality is a term which we met in section 4.2. It is the number of moles of solute in one kilogram of solvent.

K_b is called the **ebullioscopic constant**. It is also known as the molal boiling-point elevation constant or simply as the boiling-point constant. Some values for K_b are given in table 6.5.

Experimental Determination of the Elevation of Boiling Point

The elevation of boiling point can be determined experimentally by **Landsberger's method**. Solvent vapour is passed through the solvent in a graduated tube (*see* figure 6.37). The enthalpy change of vaporisation resulting from the condensation of the vapour causes the solvent to boil. Superheating is thus avoided. **Superheating** occurs if the temperature of a liquid rises above its boiling point.

After the temperature of the boiling solvent has been measured, the solvent is cooled and a weighed amount of solute is dissolved in it. Solvent vapour is then passed through the solution until it boils. The temperature of the solution is then measured. Since the temperature rise is usually small the temperature must be measured accurately. A **Beckmann thermometer** is used for this purpose. The concentration of the solution can be found from the mass of the solute and the final volume of the solution measured in the graduated tube.

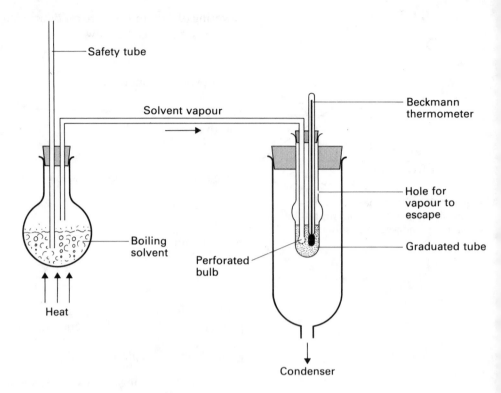

Figure 6.37   Apparatus for determining elevation of boiling point

Determination of Molar Mass Using Elevation of Boiling Point

The Landsberger method can be used to determine the molar mass of the dissolved solute. There are three parts to this determination:

1. The elevation of boiling point ΔT is determined experimentally.
2. The molality of the solute is expressed in terms of the unknown molar mass and the experimentally determined masses of solute and solvent required to give this elevation of boiling point.
3. The ebullioscopic constant K_b can either be found from tables or be determined experimentally. In the latter case the experiment is repeated using the same solvent but a solute of known molar mass.

The required expression for molality is derived from a mathematical definition of molality:

$$m_B = \frac{n_B}{m'_A \times 10^{-3}} \tag{7}$$

where n_B is the number of moles of solute and m'_A the mass of solvent in grams. It follows that $m'_A \times 10^{-3}$ is the mass of solute in kilograms. Since

$$n_B = \frac{m'_B}{M_B}$$

where m'_B is the mass of solute in grams and M_B is the unknown molar mass of the solute, we can reformulate equation (7) as

$$m_B = \frac{m'_B}{M_B \times m'_A \times 10^{-3}}$$

This is the expression for molality required for the second part of the determination.

We can now put it into equation (6) together with ΔT determined in part 1 and K_b obtained in part 3. Thus

$$\underbrace{\Delta T}_{\text{part 1}} = \underbrace{K_b}_{\text{part 3}} \times \underbrace{\frac{m_B'}{M_B \times m_A' \times 10^{-3}}}_{\text{part 2}}$$

We can now rearrange this equation to give the unknown molar mass in terms of experimentally determined values, thus

$$M_B = \frac{K_b \times m_B'}{\Delta T \times m_A' \times 10^{-3}} \qquad (8)$$

EXAMPLE

Propanone has a boiling point of 56.2°C. 1.00 g of a non-volatile solute was dissolved in 10 g of propanone. The solution boiled at 57.4°C. Calculate the molar mass and thus the relative molecular mass of the solute given that the ebullioscopic constant of propanone is 1.71 K mol^{-1} kg.

SOLUTION

When 1.00 g of the solute was dissolved in 10 g of solvent the elevation of boiling point was 1.2°C. We can put this directly into equation (8):

$$M_B = \frac{(1.71 \text{ K mol}^{-1} \text{ kg}) \times (1.00 \text{ g})}{(1.2 \text{ K}) \times (10 \times 10^{-3} \text{ kg})}$$
$$= 142.5 \text{ g mol}^{-1}$$

The molar mass of the non-volatile solute is thus 142.5 g mol^{-1}. Its relative molecular mass is 142.5.

Depression of Freezing Point

We noted in the previous section that the addition of a second component to a liquid can lower the freezing point (see figure 6.33). The depression in freezing point ΔT when a solute is added to a solvent is related to the molality m of the solute. The relationship is similar to that for the elevation of boiling point:

$$\Delta T = K_f m \qquad (9)$$

where K_f is the **cryoscopic constant**. This is also known as the molal freezing-point depression constant or simply the freezing-point constant.

The cryoscopic constant is specific for a given solvent. Some values are shown in table 6.6.

The depression of freezing point can be determined experimentally using the apparatus shown in figure 6.38. The inner glass tube contains a known mass of solvent. The solvent is cooled slowly with continuous stirring to minimise supercooling (see previous section). A Beckmann thermometer is used to record the temperature every half minute. A cooling curve is plotted and the freezing point determined. The solvent is then warmed to melt it and a known mass of solute is added. The solvent is stirred until the solute dissolves. The solution is cooled, the temperature recorded every half minute and the new freezing point determined as before.

Table 6.6 Cryoscopic constants

Solvent	K_f/K mol^{-1} kg
benzene	5.12
ethanoic acid	3.90
water	1.86
camphor	40.0

Determination of Relative Molecular Mass Using Depression of Freezing Point

The depression of freezing point can be used to determine the relative molecular mass of a non-volatile solute.

In **Rast's method** camphor is used as a solvent since it has a large cryoscopic constant (see table 6.6). In other methods, ethanoic acid or other solvents are used.

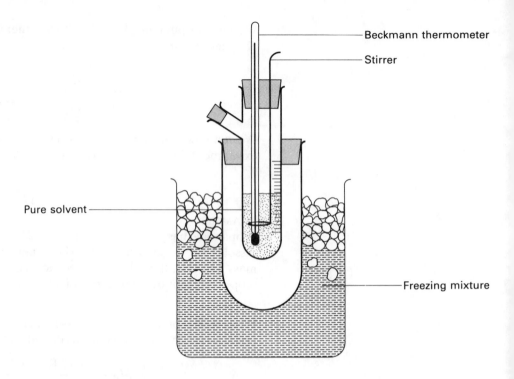

Beckmann thermometer

Stirrer

Pure solvent

Freezing mixture

Figure 6.38   Apparatus for determining depression of freezing point

EXAMPLE
Ethanoic acid has a freezing point of 16.63°C. On adding 2.5 g of an organic solute to 40 g of the acid, the freezing point was lowered to 15.48°C. Calculate the relative molecular mass of the solute.

SOLUTION
To solve this problem we need an expression comparable to equation (8). The following equation can be derived from equations (9) and (7) in the same way that equation (8) was derived from equations (6) and (7):

$$M_B = \frac{K_f \times m_B'}{\Delta T \times m_A' \times 10^{-3}} \qquad (10)$$

We have

$$m_A' = 40 \text{ g}$$
$$m_B' = 2.5 \text{ g}$$
$$\Delta T = 16.63°C - 15.48°C$$
$$= 1.15°C$$
$$= 1.15 \text{ K}$$

Finally, from table 6.6

$$K_f = 3.90 \text{ K mol}^{-1} \text{ kg}$$

Substituting these values into equation (10) we obtain

$$M_B = \frac{(3.90 \text{ K mol}^{-1} \text{ kg}) \times (2.5 \text{ g})}{(1.15 \text{ K}) \times (40 \times 10^{-3} \text{ kg})}$$
$$= 212 \text{ g mol}^{-1}$$

The relative molecular mass of the organic solute is thus 212.

Osmotic Pressure

Osmosis is the passage of a solvent through a membrane from a dilute solution or the solvent itself into a concentrated solution. The phenomenon can be demonstrated by using the apparatus shown in figure 6.39. One end of a tube is closed with an animal membrane and the tube filled with sugar solution. The tube is then placed in a beaker of water. The water passes through the membrane into the sugar solution.

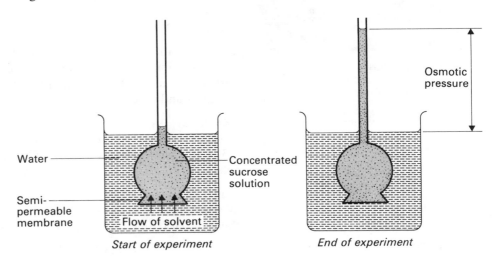

Start of experiment   *End of experiment*

Figure 6.39   Osmotic pressure

A membrane which allows solvent particles but not solute particles to pass through it is called a **semi-permeable membrane**. A semi-permeable membrane allows solvent particles to pass in both directions. However, since the concentration of solvent is lower when the concentration of solution is higher, there is a net flow towards the concentrated solution. This results in a pressure difference across the membrane. The pressure which must be applied to a solution to prevent this flow across the membrane is called the **osmotic pressure**. It is given the Greek symbol π.

Osmotic pressure is a colligative property since it depends on the number but not the nature of the solute particles.

Osmotic pressure is important in biological processes. For example, some types of animal cells such as red blood cells are filled with salt solution. The walls of these cells are lined with plasma membranes. When red blood cells are put into water, they swell and burst due to osmosis. However, if they are put into more concentrated salt solutions they shrink.

Plant cells contain salt solutions in spaces known as vacuoles. A vacuole is surrounded by a thin layer of cytoplasm which acts as a semi-permeable membrane. It controls the uptake of water by the plant cell.

When the pressure applied to a concentrated solution is greater than the osmotic pressure, solvent flows from the concentrated solution through the membrane into the dilute solution. This process is known as **reverse osmosis**. The process is used commercially to obtain drinkable water from sea water.

Experiments using sucrose solutions show that,

- at a constant temperature, osmotic pressure is directly proportional to the concentration difference across the membrane;
- at a fixed concentration, the osmotic pressure is directly proportional to absolute temperature.

The relationship between osmotic pressure and temperature is analogous to the ideal gas equation (*see* section 3.1). It is given by the **van't Hoff equation**

$$\pi V = nRT \qquad (11)$$

The amoeba is a single cell. Water continually enters its body by osmosis.

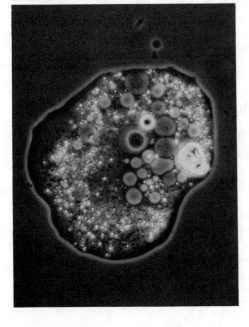

These reverse osmosis modules produce ultra-pure water for use in the semiconductor industry

where π is the osmotic pressure, V is the volume of solution, n is the amount of solute in moles, T is the absolute temperature and R is the molar gas constant. This equation can also be expressed as

$$\pi = cRT$$

where c is the concentration ($c = n/V$).

The van't Hoff equation is only approximate and holds only at low concentrations

Determination of Relative Molecular Mass Using Osmotic Pressure

The van't Hoff equation can be used to determine the relative molecular mass from experimentally determined values of osmotic pressure. The method is particularly useful for determining the average relative molecular masses of polymers and other macromolecular substances.

EXAMPLE
A sugar solution, with concentration 2.5 g dm^{-3}, gave an osmotic pressure of 8.3×10^{-4} atmospheres at 25°C. Calculate the relative molecular mass of the solute.

SOLUTION
We can find a value for the relative molecular mass M_r of the solute directly from the van't Hoff equation (11). We have the following data:

$$\pi = 8.3 \times 10^{-4} \text{ atm}$$
$$= 8.3 \times 10^{-4} \times 101\ 325 \text{ N m}^{-2}$$
$$V = 1 \text{ dm}^3$$
$$= 10^{-3} \text{ m}^3$$
$$n = \frac{2.5 \text{ g}}{M_r \text{ g mol}^{-1}}$$
$$R = 8.314 \text{ J K}^{-1} \text{ mol}^{-1}$$
$$T = 25°C$$
$$= 298 \text{ K}$$

Substituting these values into the van't Hoff equation and rearranging we obtain

$$M_r = \frac{(2.5 \text{ g}) \times (8.314 \text{ J K}^{-1} \text{ mol}^{-1}) \times (298 \text{ K})}{(8.3 \times 10^{-4} \times 101\ 325 \text{ N m}^{-2}) \times (10^{-3} \text{ m}^3)}$$
$$= 7.36 \times 10^4 \text{ g mol}^{-1} \qquad \text{(noting that 1 J = 1 N m)}$$

Thus

$$M_r = 7.36 \times 10^4$$

Anomalous Values of Relative Molecular Mass

We have seen that three types of colligative property can be used to determine relative molecular mass. These are

- elevation of boiling point,
- depression of freezing point,
- osmotic pressure.

Anomalous values are obtained if the solute associates or dissociates in solution. Carboxylic acids, for example, **associate** to form dimers in organic solvents (*see* figure 6.26). This is due to hydrogen bonding.

Electrolytes such as sodium chloride **dissociate** in aqueous solutions

$$NaCl(s) + H_2O(l) \rightarrow Na^+(aq) + Cl^-(aq)$$

The colligative properties of electrolyte solutions thus depend on the number but not the nature of the solute ions present in solution. In the case of sodium chloride, there are two moles of ions for every mole of NaCl(s). The relative molecular mass of NaCl determined from the elevation of boiling point is thus approximately half that calculated from its formula.

Comparison of values of relative molecular mass determined experimentally from colligative properties with those calculated from the formulae of the solutes allow the degree of association or dissociation to be determined.

COLLOIDS

We saw in the previous section that a solution is a homogeneous mixture of at least two components. Particles in solution are generally very small and cannot be seen. These particles may be atoms, molecules or ions and their diameters are generally less than 5 nm (that is, 5×10^{-9} m).

A **suspension** is a heterogeneous mixture of two components. It consists of particles suspended in a second medium. Over a period of time the particles settle. The suspended particles usually have a diameter of 1000 nm (that is 10^{-6} m) or more.

Intermediate between solutions and suspensions are **colloidal dispersions** or colloids. Colloids consist of a discontinuous phase, that is a **disperse phase**, in a **continuous phase**, that is a dispersion medium.

Colloids are distinguished from suspensions by the size of the dispersed particles. Colloidal particles are approximately 1 to 500 nm in diameter.

Colloidal particles do not settle and cannot be separated from the dispersion medium by ordinary techniques such as filtration and centrifugation.

Colloids are classified according to the original phases of their constituents. The classification is shown in table 6.7.

Smoke is a colloidal dispersion of solid particles dispersed in air. The disperse phase is the solid particles. The continuous phase is the air.

Table 6.7 Colloids

Disperse phase	Continuous phase	Name	Examples
liquid	gas	aerosol	mist, clouds, fog, paint and insecticide sprays
solid	gas	aerosol	smoke, dust
gas	liquid	foam	whipped cream, froth
solid	liquid	sol	paints, milk of magnesia
liquid	liquid	emulsion	milk, mayonnaise
gas	solid	(solid foam)	polyurethane, cork
solid	solid	(solid) sol	alloys
liquid	solid	gel	jellies, gelatin, agar

The Tyndall Effect

When a beam of light is passed through a colloidal dispersion, some of the light is scattered. This is called the **Tyndall effect**. This effect is noticeable when a stream of sunlight passes into a darkened and dusty room. It is also produced by the projector beam in a smoky cinema. On close examination, tiny points of light can be seen bouncing in a random zig-zag motion. This is known as **Brownian motion**. It is observed with all transparent or translucent colloids consisting of a gas or liquid dispersion medium.

CHROMATOGRAPHY

Chromatography is a technique used to separate the components of a mixture. There are two types of chromatography. Both depend on the partition of the

Consuming a colloid: jelly

A substance is **adsorbed** when it is taken up at the surface of another substance. Adsorption should not be confused with absorption. In absorption the **absorbed** substance diffuses throughout the bulk of another substance (*see* figure 6.40).

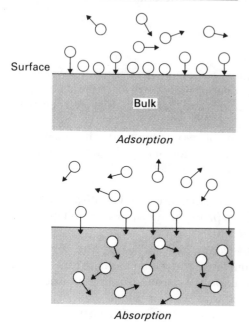

Figure 6.40   Adsorption and absorption

The name 'chromatography' bears little relation to the technique itself. The word originates from the Greek word for 'colour writing'. This is because the earliest chromatographic methods were confined to the separation of coloured substances from mixtures.

components of the mixture between a **stationary phase** and a **mobile phase**.

In **adsorption chromatography** the stationary phase is a *solid*. The solid adsorbs a portion of each component in the mixture.

In **partition chromatography** the stationary phase is a *liquid*. The components are partitioned between the liquid and the mobile phase.

In both types of chromatography, portions of each component remain dissolved in the mobile phase. The mobile phase flows continuously over the stationary phase and as it does so separates the components on the stationary phase.

Both chromatographic methods thus consist essentially of two steps:

1. Distribution of the components between the two phases.
2. Separation of the components on or in the stationary phase by a continuous flow of the mobile phase.

The distribution of a component between the two phases is given by the partition coefficient D (*see* also previous section)

$$D = \frac{\text{Concentration of component in the moving phase}}{\text{Concentration of component in the stationary phase}}$$

A component of a mixture with a high value for D remains largely dissolved in the mobile phase and thus passes over the stationary phase rapidly. A component with a small value for D remains largely adsorbed on the stationary phase. As the moving phase passes over it, this component moves slowly along the stationary phase.

Chromatography is particularly useful in organic synthesis in separating and recovering the components of a mixture. It is used in both quantitative and qualitative analysis to identify separated components of a mixture and also to determine the purity of a substance.

Column Chromatography

This is the simplest form of adsorption chromatography. It is used to separate mixtures of solids.

The stationary phase is a solid adsorbent such as alumina. The moving phase is a solvent. A glass column is first packed with a slurry of the two phases. A pad of glass wool is used at the bottom of the column to prevent the slurry from running out (*see* figure 6.41). The solid is allowed to settle and then the solvent is run off until its level is just above the level of the solid. The solid mixture is then dissolved in a small amount of the solvent and added to the top of the column. The solution is then allowed to run through the column. Successive quantities of solvent are added to ensure the solvent level remains above the stationary phase. The process of passing the solvent through the column is called **elution**. The solvent is called the **eluent**. The various components of the mixture separate as they gradually move down the column. If the components are coloured—as in the case of pigments and dyes—the components can be separated visually. Figure 6.42 shows the separation of two components. Component 1 has the higher partition coefficient and thus moves through the column faster. The time taken for a component to pass through the column is called its **elution time**. Successive portions of the eluted components are collected in flasks or tubes. The solvent can be removed by distillation to leave the pure component. Figure 6.42 is an idealised picture of the separation of the two components.

Thin-Layer Chromatography

Thin-layer chromatography (TLC) is a form of adsorption chromatography. It is used extensively in organic chemistry to identify compounds and establish their purity. The solid adsorbent phase is usually silica gel, alumina or cellulose containing a binder such as starch. A slurry of the solid is spread thinly and

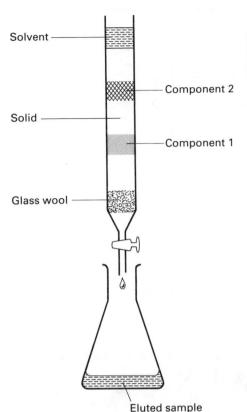

Figure 6.41 Column chromatography

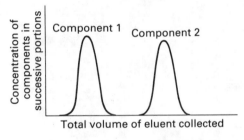

Figure 6.42 Elution peaks (idealised)

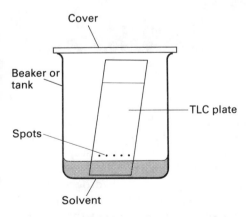

Figure 6.43 Thin-layer chromatography (TLC)

uniformly over a glass plate and allowed to dry. A spot of the sample solution is placed near one edge of the plate. To facilitate identification of the components, spots of known compounds may also be placed alongside. When the spots have dried the plate is dipped in a small amount of the solvent in a beaker or tank (*see* figure 6.43). The beaker or tank is covered to prevent evaporation. The solvent slowly rises up the plate by capillary action. As the **solvent front** passes the spots, the solutes begin to move. The rate at which they move depends on their partition coefficients. When the solvent front is near the top of the plate, the plate is removed and allowed to dry. The various components of a coloured mixture can be visually identified. If they are colourless, the plate has to be developed. This can be done by spraying with a reagent, by ultra-violet radiation, by heat or by exposure to a gas such as iodine vapour. Each component has a characteristic R_f **value**. The R_f value is related to its partition coefficient and is given by

$$R_f = \frac{\text{Distance travelled by component from original spot}}{\text{Distance travelled by solvent from original spot}}$$

Figure 6.44 shows a developed plate with five spots:

Unknown, X
Pure A
Pure B
Two samples of C

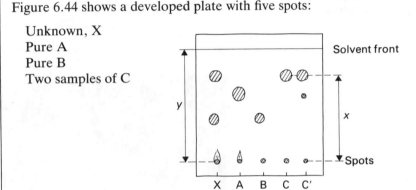

Figure 6.44 TLC plate

The plate shows that X contains both B and C. Furthermore C is a pure sample whereas C′ is impure since another spot shows that a small amount of A is also present. The R_f value for C is given by

$$R_f(\text{C}) = \frac{x}{y}$$

Paper Chromatography

This technique is a form of partition chromatography since the stationary phase is a liquid. The stationary phase is water adsorbed on paper. The mobile phase is usually an organic liquid. Paper chromatography is similar in many ways to TLC.

There are two types of paper chromatography. In **ascending** paper chromatography, a paper strip is suspended from a glass rod in a tank (*see* figure 6.45). The strip dips into a small amount of solvent at the bottom of the tank. In **descending** paper chromatography the solvent is contained in a trough at the top of the tank (*see* figure 6.46). The paper chromatograms are dried and developed in the same manner as for TLC. The advantage of paper chromatograms is that they can be easily stored for future reference and study.

Gas Chromatography

There are two types of gas chromatography.

In **gas–solid chromatography** the stationary phase is a solid—usually silica gel or alumina. It is thus a form of adsorption chromatography. However, the method is rarely used nowadays.

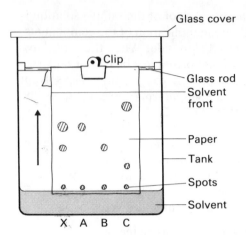

Figure 6.45 Ascending paper chromatography. The left-hand spot X is the unknown. A, B and C are reference spots. The chromatogram shows that X contains A and B

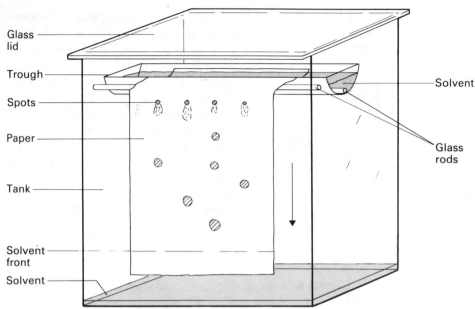

Figure 6.46 Descending paper chromatography

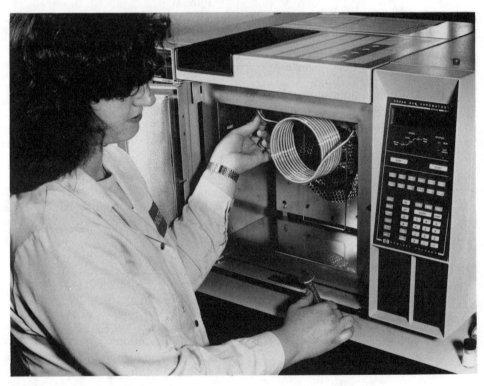

Gas–liquid chromatography column

Much more common is **gas–liquid chromatography**. This is a form of partition chromatography. The stationary phase is a high-boiling liquid—a long-chain alkane for example. The liquid is supported by a porous inert solid such as charcoal, alumina or silicon(IV) oxide. This is packed into a narrow coiled column (*see* figure 6.47). The mobile phase is a gas—usually nitrogen, hydrogen, helium or argon. This is called the carrier gas. A hypodermic syringe is used to inject the sample into the chamber. The sample vaporises and is carried through the column by the carrier gas. The various components of the mixture flow through the column at different rates due to their different partition coefficients. Each component finally passes through a detector which gives a signal to a recorder. The result is a gas chromatogram. This contains a series of peaks (*see* figure 6.48).

One of the most common detectors measures the difference between the **thermal conductivity** of the pure carrier gas and the carrier gas containing a separated component. This type of detector is called a **katharometer**. It is used for both organic and inorganic compounds. The **flame-ionisation detector** is used when organic compounds are to be separated. The carrier gas containing the separated components is burnt in a hydrogen flame. The number of ions produced in the flame is measured electrically. The **electron captor detector** is used to analyse compounds containing halogens, sulphur, phosphorus and nitrogen. β-Particles (that is electrons: *see* section 1.3) ionise the carrier gas, causing it to

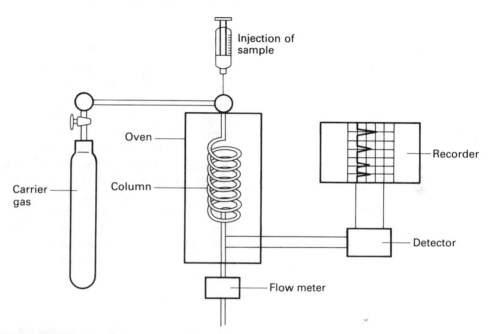

Figure 6.47 Schematic representation of gas–liquid chromatography

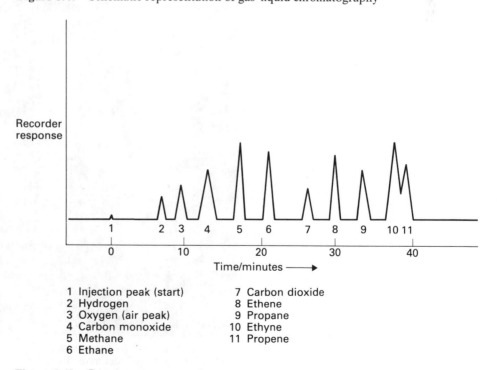

1 Injection peak (start)
2 Hydrogen
3 Oxygen (air peak)
4 Carbon monoxide
5 Methane
6 Ethane
7 Carbon dioxide
8 Ethene
9 Propane
10 Ethyne
11 Propene

Figure 6.48 Gas chromatogram of a mixture of gases

conduct. The components of the sample are reduced by the electrons and thus absorb them to varying extents. The reduction of conductivity caused by this is related to the amount of component. This method of detection is very sensitive. It is used to measure the amounts of insecticides containing halogens found in foods and other products.

Figure 6.48 shows a gas–liquid chromatogram of a mixture of gases. The injection peak corresponds to the time of injection. Each subsequent peak corresponds to a component of the sample mixture. The distance from the injection peak to each component peak is called the retention time. The retention time is characteristic for a compound and the conditions of the experiment. The conditions include temperature, rate of flow of gas and the nature and concentration of the stationary phase.

Liquid Chromatography

Liquid chromatography embraces a group of separation techniques in which compounds are partitioned between a mobile liquid phase which flows over a stationary phase of large surface area. Paper chromatography, thin-layer chromatography, column chromatography and **high-performance liquid chromatography** (HPLC) are all forms of liquid chromatography.

HPLC has become increasingly widely used over recent years. It is complementary to gas chromatography and is especially of value in the determination of non-volatile compounds. For example, it is used for the rapid estimation of additives, contaminants and natural components of foodstuffs. Large surface areas and thus highly efficient column chromatographic separations are obtained by using support materials which are small and uniform in size. However, columns packed with such **microparticulate** support materials are highly resistant to liquid flow. High pressures are therefore required to force mobile phases through them. For this reason HPLC is also used as an abbreviation for high-pressure liquid chromatography.

High performance liquid chromatography (HPLC) has become an increasingly important separation technique in recent years—particularly in pharmacy and food analysis.

SUMMARY

1. A **colligative property** depends on the number but not the nature of solute particles.
2. The **elevation of boiling point** by a **non-volatile solute** is given by

$$\Delta T = K_b m_B$$

3. The **depression of freezing point** by a solute is given by

$$\Delta T = K_f m$$

4. **Osmotic pressure** is the pressure required to prevent a solvent flowing across a **semi-permeable membrane** from a region of high solute concentration to one of low solute concentration.
5. The **van't Hoff equation** is

$$\pi V = nRT$$

6. A **colloid** consists of a **disperse phase** and a **continuous phase**. Colloidal particles are about 1 to 500 nm in diameter.
7. In **adsorption chromatography** the stationary phase is a solid.
8. In **partition chromatography** the stationary phase is a liquid.
9. **Column chromatography** and **thin-layer chromatography** are forms of adsorption chromatography.
10. **Paper chromatography** and **gas–liquid chromatography** are forms of partition chromatography.
11. In **high-performance liquid chromatography** (HPLC) a **mobile liquid phase** is forced by pressure through a column containing a **microparticulate stationary phase**.

Examination Questions

1. (a) Draw on labelled axes, at least 10 cm in length, the phase diagram (pressure against temperature) for water. (Only one solid phase need be included.) Label the triple and critical points on your diagram with T and C respectively.
   (b) Mark on your diagram the phase or phases which can exist in equilibrium under conditions represented by (i) each area, (ii) each line.
   (c) Explain what the triple and critical points represent.
   (d) (i) How does the melting point of ice change with increasing pressure? Give an explanation in terms of the densities of liquid water and ice.
      (ii) Account for the difference in densities in terms of the structures of the two phases.
      (iii) Describe briefly a situation in which this difference in densities is important.

(O & C)

2. (a) What do you understand by the term 'vapour pressure of a liquid (or solid)'? Sketch a graph to show how the vapour pressure of liquid water/kPa varies with temperature over the range 0–200°C. Label the axes. (1 atmosphere = 101.3 kPa)
      Indicate any important temperature/pressure points.
   (b) What do you understand by the statement that the triple point of water occurs at 610 Pa and 273.1 K?

Sketch a graph to show the phase diagram of water, indicate important temperatures and explain the significance of the various lines on the diagram. Label the axes.

Under what conditions is *hoar frost* formed?

(c) Sketch the phase diagram for carbon dioxide and explain why solid carbon dioxide sublimes if left in the open air at temperatures above $-78°C$, given that the triple point is (518 kPa, 217 K). How does this phase diagram differ from that of water?

(d) Sketch the phase diagram for sulphur, drawing attention to important features and explaining why the solid has two melting points, according to the rate at which it is heated. What physical phenomenon is illustrated by this diagram?

(SUJB)

3. (a) (i) State Henry's law.
   (ii) At 293 K, 1 cm³ of water dissolves 0.12 cm³ of the rare gas xenon (measured at s.t.p.) when the pressure of this gas is one atmosphere. Assuming that in air one molecule in every 1.7×10^7 molecules is a xenon atom, calculate to two significant figures the number of xenon atoms in one cm³ of water saturated with air at atmospheric pressure at 293 K.

   (b) A compound A (m.p. 100°C) and a compound B (m.p. 200°C) are completely miscible in the liquid state. They are completely immiscible in the solid state, and do not combine to form any new substance. In the liquid mixture with the lowest freezing point (50°C) the mole fraction of A is 0.30.
   (i) Sketch the temperature–composition diagram for this system, and label each area on your sketch.
   (ii) Describe what will happen on cooling a liquid mixture in which the mole fraction of A is 0.50.

   (c) Purified water for use in a chemistry laboratory can be obtained by using ion exchange. Explain the **principles** involved in this method of purifying water.

(OLE)

4. State Raoult's law.

Show, with the help of diagrams, how (i) the boiling point and (ii) the freezing point of a solvent are affected by the presence of a non-volatile solute. Describe in detail an experiment based on Raoult's law that you have carried out in the laboratory for finding the molar mass of a *named* substance. Draw a labelled diagram of the apparatus you used. From what limitations do these methods suffer?

Ethanol, boiling point 78.5°C at 101 kPa and methanol, b.p. 64.5°C at 101 kPa, are *completely miscible* and form an *ideal* solution (or mixture). Explain the meaning of the words in *italics*.

Sketch and label (a) a vapour pressure–composition curve (at constant temperature) and (b) a boiling point–composition curve (at constant pressure) for this sytem. Mark on (a) the separate vapour pressure contributions made by the two components and, using a more complete form of diagram (b), explain fully what is meant by *fractional distillation*.

(NISEC)

5. (a) How does the vapour pressure over a mixture of two volatile liquids at constant temperature vary with the composition of the mixture when (i) they are miscible, producing an ideal mixture; (ii) they are immiscible?

(b) Describe, with practical details, *one* laboratory use of the distillation of immiscible liquids.

Methanol (b.p. 64.7°C) and trichloromethane (b.p. 61.2°C) form a non-ideal mixture. In the table below are given the mole fractions, x, of $CHCl_3$ present in the liquid and vapour phases at various temperatures under a confining pressure of 1 atm.

$T/°C$	Liquid x_{CHCl_3}	Vapour x_{CHCl_3}	Liquid x_{CHCl_3}	Vapour x_{CHCl_3}
64.0	0.015	0.040	—	—
62.0	0.065	0.155	—	—
60.0	0.125	0.260	0.990	0.950
58.0	0.190	0.365	0.970	0.880
56.0	0.280	0.470	0.930	0.800
54.0	0.460	0.560	0.810	0.710
53.5	0.650	0.650	0.650	0.650

(c) On graph paper, draw accurately the labelled boiling-point diagram for this liquid mixture. Draw lines on your graph to indicate what happens when a small amount is distilled from a mixture containing 80 mol% of CH_3OH, is condensed, redistilled and recondensed. Estimate the composition of the liquid thus produced and its boiling point.

(O & C)

6. (a) Propan-1-ol (boiling point 97°C) and water form an azeotropic mixture with a boiling point of 88°C and a composition of 72% propan-1-ol/28% water by mass. Sketch a fully labelled boiling point/composition diagram for a propan-1-ol/water mixture and use it to explain what is meant by the term *azeotropic mixture*.

Explain carefully what happens when a mixture of propan-1-ol and water, containing 50% propan-1-ol by mass, is fractionally distilled.

A propan-1-ol/water mixture shows a positive deviation from Raoult's law. Explain what this statement means and account for the deviation.

(b) Use the data in the table below to calculate the percentage by mass of phenylamine in the distillate when a mixture of phenylamine and water is steam distilled at standard atmospheric pressure (101.3 kPa). What is the temperature of this steam distillation?

Temperature/°C	Vapour pressure/kPa	
	Water	Phenylamine
95	84.5	4.4
96	87.7	4.5
97	90.9	4.7
98	94.3	4.9
99	97.7	5.1
100	101.3	5.3

(UCLES)

7. (a) Describe, with the aid of a diagram, the experimental process of steam distillation.
(b) Explain the theory underlying this process and show how the composition of the distillate may be calculated in terms of vapour pressure.
(c) When an organic compound (A) was steam distilled at a pressure of 101.3 kPa (760 mmHg), the temperature of distillation was 99°C. If 80%

(by mass) of the distillate was found to be water, calculate the relative molecular mass of A.

(The saturated vapour pressure of water at 99°C is 98.4 kPa (733 mmHg).)

(JMB)

8. (a) What do you understand by the term *partition coefficient*?

An aqueous solution of butanedioic acid, $HO_2CCH_2CH_2CO_2H$, was shaken with ethoxyethane ('ether') at 25°C until equilibrium was attained. 25.0 cm^3 of the aqueous layer required 30.0 cm^3 of 0.0500 mol dm^{-3} aqueous sodium hydroxide and 50.0 cm^3 of the ether layer required 20.0 cm^3 of 0.0250 mol dm^{-3} aqueous sodium hydroxide for neutralisation. Calculate the partition coefficient at 25°C.

State and explain how you would expect the partition coefficient of butanedioic acid between (i) water and hexane, (ii) water and 2-methyl-propan-1-ol to compare with that between water and ethoxyethane.

(b) Explain why ethoxyethane is particularly suitable for the extraction of an organic compound from an aqueous solution.

Calculate the mass of an organic compound A that can be extracted from 100 cm^3 of an aqueous solution containing 5.0 g of A by shaking with

(i) 100 cm^3 of ethoxyethane in one portion.

(ii) two successive 50 cm^3 portions of ethoxyethane.

Comment briefly on the results obtained in (i) and (ii).

(Partition coefficient of A between ethoxyethane and water = 3.0)

(UCLES)

9. Gold melts at 1063°C and thallium at 302°C. Cooling gold/thallium alloys at various compositions gave the following results:

Percentage of gold by mass	80	60	40	25	20	10
Beginning of freezing, °C	835	610	315	140	160	232
End of freezing, °C	131	131	131	131	131	131

Draw, on graph paper, a phase diagram of the gold/thallium system, labelling each area and stating the significance of each line and intersection of lines.

Sketch and explain the temperature against time curve obtained when cooling slowly an alloy containing 72% gold. What would be the physical state of this alloy at 500°C?

Sketch and explain the temperature against time curve obtained when cooling slowly an alloy containing 28% gold. What is the special significance of an alloy of this composition?

Give **one** practical application of **any** alloy which has a phase diagram similar to that of gold/thallium.

(O & C)

10. (a) The depression in freezing point of a solvent by a solute is known as a *colligative property*. Explain what is meant by this term and give **two** other examples of colligative properties.

(b) Describe how you would determine the relative molecular mass of a compound dissolved in a suitable solvent by an experiment involving **one** of these properties.

(c) Explain, with reasons, whether your method would be satisfactory for determining the relative molecular mass of (i) ethanoic (*acetic*) acid in a concentrated solution of the acid in a suitable solvent and (ii) a protein molecule.

(d) Calculate the freezing point of (i) a solution of 22.5 g of cane sugar ($C_{12}H_{22}O_{11}$) in 450 g of water and (ii) a solution of 3.0 g of potassium chloride in 400 g of water. From which solution will ice first separate? (The freezing point (cryoscopic) constant of water is 1.86 K mol^{-1} kg.)

(JMB)

11. (a) What is meant by the terms *vapour pressure* and *mole fraction*?
   (b) For a solution of a single non-volatile solute in a solvent, state the relationship between the lowering of the vapour pressure of the solvent by the solute relative to the vapour pressure of the pure solvent and (i) the mole fraction of the solvent and (ii) the mole fraction of the solute, at a fixed temperature.
   (c) (i) If the vapour pressure of pure water (relative molecular mass 18.015) at 300.0 K is 3.5649 kPa, what will be the vapour pressure of a solution of 0.1000 mol of glucose in 1.000 kg of water?
      (ii) What will be the quantitative effect, if any, on the vapour pressure of dissolving 0.2000 mol of urea in such a glucose solution?
   (d) Draw labelled graphs to show:
      (i) how the vapour pressure of water varies with temperature in the range 273–373 K, and
      (ii) the relative positions of the vapour pressure curve of a pure solvent and that of a solution of an involatile solute in that solvent near the boiling point of the pure solvent.
   (e) Use the second graph to explain the basis of a method for the determination of relative molecular masses based on boiling point measurements, stating clearly any approximations and assumptions which have to be made.

(OLE)

12. (a) At a certain temperature, the osmotic pressure of a particular solution of cane sugar in water is found to be one atmosphere. What does this statement mean?
   (b) At 290 K, a solution containing 20.0 g of a polymer per dm^3 of water is found to have an osmotic pressure of 1600 N m^{-2}. Calculate to three significant figures the relative molecular mass (molecular weight) of the polymer.

(OLE)

13. Write an essay on the applications of chromatography in analysis and separation.

(O & C)

14. Describe the basic *principles* that apply to all forms of chromatography.
   Give experimental details of **four** different types of chromatography, showing how these principles are applied in practice.
   For each of the following pairs of substances, suggest a suitable chromatographic method for separation.
   (a) propane and butane,
   (b) $LaCl_3$ and $CeCl_3$,
   (c) 2-nitrophenylamine and 4-nitrophenylamine (*o*- and *p*-nitroaniline).
In each case give reasons for your choice of technique.

(O & C)

7 CHEMICAL EQUILIBRIA

Chemistry's Cornucopia

Chemistry, Health and Wealth

Throughout history, the lot of most of mankind has been rather miserable. Until a few centuries ago it was real slavery and into the last century, although legally free, most men and women had to labour so hard to earn a living that they were effectively slaves to their work. Different professions have had different approaches to this problem. As Max Perutz has put it 'The Priest persuades the humble people to endure their hard lot, the politician urges them to rebel against it and the scientist thinks of a method that does away with the hard lot altogether'.

No branch of science has done more, or promises more in this respect, than chemistry. It has provided a cornucopia of good things, both of necessities and luxuries, which have improved our health, and our wealth and also I believe our happiness. Man is himself a biochemical system living in a chemical world. His *health* has been improved out of all recognition by better nutrition, better hygiene and by the drugs which have doubled his lifespan, relieved pain and made it possible for many handicapped people to lead a more normal life.

Sir George Porter, who won the Nobel Prize for Chemistry in 1967, presents a lecture on 'Chemistry's Cornucopia' at an IUPAC Congress in Manchester

His *wealth*, judged in terms of the general availability of the necessities of life is many times what it was even a few decades ago. This is particularly true of the wealth of food now available in countries which, only a few years ago, were poor to the point of starvation.

Most people today would also class as necessities the plastics, fibres and paints which have made it possible for everybody to be well clothed (though not all may want it!) and to live in a bright, clean environment without having to employ the labour of others to keep it so.

The luxuries, which are also fast becoming universal, often owe as much to chemistry as to the other technologies. This is obvious of such things as motor fuels, cosmetics and dyestuffs but it is also true of electronic devices of every kind, and mechanical, labour-saving appliances. Some of the largest manufacturers of heavy electrical equipment employ more chemists than physicists and, on the newer and lighter side, the silicon *chip* is a highly purified chemical element treated with other elements in a very sophisticated, chemically pure environment.

But perhaps the most successful of all, over the last two or three decades, has been the contribution of chemistry to agriculture. The 'green revolution' did exactly what the King of Brobdingnag of Gulliver's travels had asked for and 'made two ears of corn or two blades of grass grow where only one grew before' and those who brought it about, according to the King, 'would deserve better of mankind and do a more essential service for the country than a whole race of politicians put together'. This was a proud achievement of chemistry, depending heavily on fertilisers and new insecticides, plant growth substances and the like. It is a huge industry: in the United States crop production amounts to US $74 billion a year and the cost of the pesticides which make this possible US $4.2 billion. The world production of plant foods in the form of fertilisers has increased threefold over the last twenty years, and eighty times since the beginning of the century.

Since the war, the improvement of food productivity has been quite dramatic. In Britain, the same area of land yields twice as much wheat or potatoes, and the cow gives twice as much milk using less land. Fewer farm workers on less land supply a larger and better fed population with less imports and £1500 million worth of exports of farm products.

The cornucopia is overflowing, and the world, as a whole, is now awash with surplus food. Since 1964, *world-wide* production of wheat and food grains has doubled and kept ahead of population growth. 60% of this can be attributed to improved agricultural technology and the other 40% springs from price guarantees for producers and improved marketing conditions. Remaining areas of endemic food insufficiency include elements of the Middle East and South Saharan Africa. In 1985 China, for example, was expected to produce the largest wheat crop in the world. Wheat production in the EEC has doubled in seven years. It is the same story round the world in those countries which used to be held up as unable to feed themselves. India is effectively self-sufficient, Pakistan nearly.

Over-reaction Against 'Chemicals'

How did mankind welcome these achievements? Did everybody thank the chemist and congratulate him on what had been done for the benefit of mankind? Well, not really! A decade or two ago, a period of depression began about science and technology. The 'post industrial revolution' was discovered. Rachel Carson and many others questioned not only some genuinely worrying aspects of chemical pollution but often the whole ethic of modern technology. All science became suspect and none more than chemistry.

Quite quickly the generally accepted idea of 'Better living through Chemistry' was transformed into an association of chemistry, in the public mind, with pollution, world shortages and iniquities of the multinational corporations. Many of the essentials to the green revolution, such as the pesticides, became primarily, in the layman's mind, destroyers of life. Since then we have had several chemical disasters; the 41 drums of dioxin-contaminated waste from Seveso in Italy (which caused no injury to anybody during transportation but concentrated many minds on chemical waste disposal) and the terrible leak of methyl isocyanate in Bhopal, India which killed 2500 people and injured over 100 000. The fact that the plant was making a 'safer' pesticide, Sevin (which, unlike DDT, does not accumulate in the body), does little to compensate for the tragedy. Add to these the dreadful thalidomide story, the atmospheric pollution by chlorofluorocarbons, the dangers of hazardous chemical waste disposal, and 'chemistry' is no longer, to many people, the beautiful word that it was.

But high crop yields depend on pesticides and weed killers, as much as upon fertilisers, because food plants have to compete with weeds and pests, viruses and fungi. Without them, as Max Perutz has pointed out, the production of grain would fall by nearly a half in three years and we would have a famine of catastrophic proportions, like the Irish potato famine, which was caused by fungal infection. That is the prospect for so-called 'organic farming'. DDT is no more toxic to man than is aspirin and less so than the organic phosphates and carbonates which superseded It. It eliminated plague and typhus and less completely, malaria, from most parts of the world. Not quite completely however; I was recently in Manila [in the Philippines] where I heard that, only fifty miles north, one half of the population still suffered from malaria of a particularly lethal kind.

Man-made insecticides can create important problems by accumulating in the food chain, and those known to be dangerous in this way have been largely replaced. New pesticides are being synthesised all the time; they have to be because pests become clever and develop a resistance to them. It is a big and

Chemistry can benefit agricultural production in many ways, for example through fertilisers, pesticides and plastics. Tunnel greenhouses constructed from polythene sheeting protect sensitive crops during the cold weeks of the year. They also produce larger numbers of transplants ready for field planting early in the season and thus ensure the longest period of production in the open. This photograph, taken near Basrah, Southern Iraq, shows the large-scale use of low tunnels of plastic film to protect tomatoes from frost

expensive business; thousands of compounds have to be synthesised and tested on animals, and in the fields, for every one that is commercially used. But only in this way has it become possible to feed an ever-growing world population.

The memory of how the people of the world suffered without science is soon lost, and younger people never knew that world. They see only the remaining problems and human errors, some of which we can solve and prevent but some of which we shall always have to live with. It is ethically no more justified to take life by intentionally doing nothing than by some positive action which unintentionally causes death. We can introduce safer motor cars up to a point, but eventually the only way to further safety is to give up your car, and most people decide that the risk is worth it, even though it is the second most important cause of untimely death in the developed countries.

An over-reaction against 'chemicals' can do more harm than the disasters themselves. One can congratulate the food and drug laws of the United States for being so strict that that country avoided the thalidomide disaster. But other delays caused by these controls also condemned thousands of cardiac and other patients to suffering and even death. Such is the public ignorance of the most elementary facts about the natural world, about atoms and molecules and chemical substances, that reasoned explanation has little place in such things. A public which has become accustomed to scientific answers in black and white is not prepared to discuss technological risks in various shades of grey. And what on earth can we say to somebody who objects to having chemicals in his food?

Accepting that it is our destiny to live in a chemical world we had better learn some chemistry. How can we help people to understand that we are, ourselves, chemically constituted bodies, living wholly in a world of chemical substances, eating them, burning them, smelling them (for better or worse), and spending every second of our lives looking at them and manipulating them in some way or other: that every little bit of us is a chemical entity; all our living, illnesses and dying are chemical processes? One might as well try to do without chemistry as attempt to stop the world and get off. Since it appears that the media, like television, which began with so much promise, are going to contribute little enlightenment, we must go look again to the teaching in our schools, particularly at the most elementary level. This will be one of the most important tasks for chemists in the next few decades.

An equally important task will be to get our politicians and policy makers (including, I have to add, some *scientists* who make policy) to understand the importance of chemistry.

Sir George Porter

LEARNING OBJECTIVES

After you have studied this chapter you should be able to

1. State the three characteristics of a system in **chemical equilibrium**.
2. Give examples of systems in chemical equilibrium.
3. Define the **equilibrium law**.
4. Calculate **equilibrium constants**.
5. Predict the influence of the following on the **position of equilibrium** and on equilibrium constant:
   (a) a change in concentration of reactants or products;
   (b) a change in applied pressure;
   (c) the presence of a catalyst;
   (d) a change in temperature.
6. Outline the main stages in the **manufacture of ammonia**.
7. Explain the influence of pressure, temperature and catalysts on the **yield** and **throughput** of ammonia.
8. Point to the economic importance of ammonia and its compounds.
9. Outline the main stages in the modern Contact process for the **manufacture of sulphuric acid**.
10. Explain the influence of pressure, temperature and catalysts on the **yield** and **throughput** of sulphuric acid.
11. Point to the economic importance of sulphuric acid.

7.1 The Equilibrium Constant

REVERSIBLE REACTIONS

Many chemical reactions **go to completion**. This means that the reaction continues until one of the reactants is completely used up in the reaction. When this happens the reaction stops—it has gone to completion. An example of this type of reaction is the reaction between hydrochloric acid and calcium carbonate:

$$CaCO_3(s) + 2HCl(aq) \rightarrow CaCl_2(aq) + CO_2(g) + H_2O(l)$$

Many other reactions are **reversible**. Consider the reaction between silver ions, Ag^+, and iron(II) ions, Fe^{2+}, for example. When a solution containing Ag^+ ions is mixed with a solution containing Fe^{2+} ions, the following reaction occurs:

$$Ag^+(aq) + Fe^{2+}(aq) \rightarrow Ag(s) + Fe^{3+}(aq)$$

This is called the **forward reaction**. During this reaction, the concentrations of the reactants decrease.

As soon as the products $Ag(s)$ and Fe^{3+} are formed by the forward reaction, they react:

$$Ag(s) + Fe^{3+}(aq) \rightarrow Ag^+(aq) + Fe^{2+}(aq)$$

This is called the **reverse reaction**. The forward and reverse reactions together form a **reversible reaction**. The sign $\rightleftharpoons$ is used to indicate this type of reaction.

In a reversible reaction both forward and reverse reactions continue indefinitely. However, there comes a point when there is no further observable change. At this point the rate of the forward reaction equals the rate of the reverse reaction. The reaction has reached **chemical equilibrium**. This is a dynamic equilibrium since the two opposing processes are proceeding at equal rates.

At equilibrium the concentrations of all the species in the reaction mixture, that is Ag^+. Fe^{2+}, $Ag(s)$ and Fe^{3+}, remain constant. A reaction mixture in equilibrium is called an **equilibrium mixture**.

A position of chemical equilibrium can be attained from either direction. This can be demonstrated using the above reaction.

It should be noted that the reaction system remains in dynamic equilibrium only so long as the system remains isolated. An **isolated system** is one in which there is no exchange of matter or energy between the system and its surroundings.

In summary, we can see that there are three characteristics of a system in chemical equilibrium:

1. the rates of the forward and reverse reactions are equal;
2. the system does not undergo any observable change;
3. the system is closed.

THE EQUILIBRIUM LAW

Let us consider a general reversible reaction,

$$aA + bB \rightleftharpoons cC + dD$$

It is possible to show experimentally that at equilibrium

$$K_c = \left(\frac{[C]^c [D]^d}{[A]^a [B]^b} \right)_{eq}$$

where [ ] is concentration. This is a mathematical expression of the law of chemical equilibrium or simply the **equilibrium law**. The law states that, for any equilibrium at a given temperature, the product of the concentrations of the products, each raised to a power equal to its coefficient in the equation, divided by the product of the concentrations of the reactions, each raised to a power equal to its coefficient in the equation, is constant. Quite a mouthful! The mathematical form is much easier. The constant is K_c. It is called the **equilibrium constant** and applies to reactions in solution.

The right-hand side in the above expression is the **equilibrium quotient**. Note that it only contains concentration terms. It should not include terms for

- any pure solid,
- any pure liquid,
- any solvent.

For reactions involving gases, the equilibrium constant is expressed in terms of the partial pressures of the gases rather than their concentrations. The equilibrium constant is then given the symbol K_p.

EXAMPLE
Express the equilibrium constant for the reaction between hydrogen and iodine in terms of the partial pressures of the gases.

SOLUTION
The reaction is

$$H_2(g) + I_2(g) \rightleftharpoons 2HI(g)$$

The equilibrium constant is thus

$$K_p = \left(\frac{p_{HI}^2}{p_{H_2} p_{I_2}} \right)_{eq}$$

Note that the concentrations or partial pressures of the products—that is the species on the right-hand side of the chemical equation—always form the numerator and the concentrations or partial pressures of the reactants—that is the species on the left-hand side of the chemical equation—always form the denominator.

The concentration of a gas can be related to the pressure of the gas by the ideal gas equation (*see* section 3.1):

$$pV = nRT$$

Rearranging

$$\frac{n}{V} = \frac{p}{RT} = \left(\frac{1}{RT} \right) p$$

where n/V is the concentration of the gas, i.e. [gas]. Since R is a constant,

$$[gas] \propto p$$

at a given temperature

Units of the Equilibrium Constant

The equilibrium constant may or may not have units depending on the form of the equilibrium expression. In the examples above, the equilibrium constant has no units since those of the numerator cancel out those of the denominator. When they do not cancel out, the equilibrium constant has units related to concentration or pressure.

EXAMPLE 1
What are the units of the equilibrium constant of the following reaction?

$$Fe^{3+}(aq) + SCN^-(aq) \rightleftharpoons [Fe(SCN)]^{2+}(aq)$$

SOLUTION 1
The equilibrium constant is given by

$$K_c = \left(\frac{[[Fe(SCN)]^{2+}]}{[Fe^{3+}][SCN^-]} \right)_{eq}$$

The units are thus

$$\frac{(mol \ dm^{-3})}{(mol \ dm^{-3})(mol \ dm^{-3})}$$

This gives

$$(mol \ dm^{-3})^{-1} \quad or \quad dm^3 \ mol^{-1}$$

EXAMPLE 2
What are the units of the equilibrium constant of the following reaction?

$$N_2O_4(g) \rightleftharpoons 2NO_2(g)$$

SOLUTION 2
The equilibrium constant is given by

$$K_p = \left(\frac{p_{NO_2}^2}{p_{N_2O_4}} \right)_{eq}$$

The units are thus

$$\frac{atm^2}{atm} \quad or \quad \frac{Pa^2}{Pa}$$

This gives

$$atm \quad or \quad Pa$$

Heterogeneous Equilibria

All the equilibria we have considered so far have been examples of homogeneous equilibria. In the synthesis of hydrogen iodide, for example, both the product and the two reactants were in the gaseous phase.

An example of a reaction which leads to a heterogeneous equilibrium is the thermal dissociation of calcium carbonate:

$$CaCO_3(s) \rightleftharpoons CaO(s) + CO_2(g)$$

The equilibrium constant for this reaction is given by

$$K_p = (p_{CO_2})_{eq}$$

Note that terms for the two solids are not included in the expression. In this example K_p is the **dissociation pressure** of calcium carbonate. It shows that if calcium carbonate is heated in a closed vessel, its dissociation pressure at a fixed temperature is independent of the amount of calcium carbonate. We shall see in the next section how equilibrium constant varies with temperature. In our example, the dissociation pressure only exceeds one atmosphere at temperatures above 900°C. Calcium carbonate must therefore be heated to this temperature before carbon dioxide readily escapes to the atmosphere.

Salt Hydrates

Salt hydrates such as $CuSO_4 \cdot 5H_2O$ thermally dissociate into lower hydrates and the anhydrous salt in stages. At each stage the water vapour establishes a vapour pressure above the solid phase. The equilibrium constant for the dissociation at each stage is given by

$$K_p = p^x$$

where x is the number of moles of water vapour produced by the dissociation of the hydrate. For example, copper(II) sulphate-5-water dissociates in three stages:

1. $CuSO_4 \cdot 5H_2O(s) \rightleftharpoons CuSO_4 \cdot 3H_2O(s) + 2H_2O(g)$ $K_p = (p_{H_2O}^2)_{eq}$
2. $CuSO_4 \cdot 3H_2O(s) \rightleftharpoons CuSO_4 \cdot H_2O(s) + 2H_2O(g)$ $K_p = (p_{H_2O}^2)_{eq}$
3. $CuSO_4 \cdot H_2O(s) \rightleftharpoons \underset{\text{anhydrous}}{CuSO_4(s)} + H_2O(g)$ $K_p = (p_{H_2O})_{eq}$

In stage 1, $CuSO_4 \cdot 5H_2O$ and $CuSO_4 \cdot 3H_2O$ are called a **hydrate pair**. The hydrate pair in stage 2 is $CuSO_4 \cdot 3H_2O$ and $CuSO_4 \cdot H_2O$.

THE EFFECT OF CONDITIONS

The Influence of Concentration

The magnitude of an equilibrium constant is an indication of the extent of the chemical reaction. If it is large, then the concentrations of the products are high in relation to the concentrations of the reactants. In this case the **position of equilibrium** *lies to the right*. The position of equilibrium *lies to the left* if the equilibrium constant is low. In this case the concentrations of the products are low in relation to the concentrations of the reactants.

When a chemical equilibrium is subjected to a change in concentration, the position of equilibrium changes. What does not change is the equilibrium constant. At a constant temperature the value of the equilibrium constant remains the same regardless of any changes in the concentration of reactants or products.

How does the equilibrium position change with a change in concentration? Let us consider the reaction between the iron(III) ion and the thiocyanate ion to form the complex ion, thiocyanatoiron(III). At equilibrium all three ions are present:

$$Fe^{3+}(aq) + SCN^-(aq) \rightleftharpoons [Fe(SCN)]^{2+}(aq)$$

$$K_c = \left(\frac{[[Fe(SCN)]^{2+}]}{[Fe^{3+}][SCN^-]} \right)_{eq}$$

K_c is constant at a given temperature. If extra Fe^{3+} is now added to the equilibrium mixture, the equilibrium constant *must remain constant*. Since the concentration of Fe^{3+} has increased, the quotient is now smaller than the equilibrium quotient

$$\left(\frac{[[Fe(SCN)]^{2+}]}{[Fe^{3+}][SCN^-]} \right) \quad < \quad \left(\frac{[[Fe(SCN)]^{2+}]}{[Fe^{3+}][SCN^-]} \right)_{eq}$$

As a result, a reaction occurs so that the concentration of Fe^{3+} decreases and the quotient on the left-hand side approaches the equilibrium quotient. In other words, the forward reaction occurs.

Le Chatelier's Principle

A hundred years ago, in a brief paper presented to the Académie des Sciences of Paris, Henry Le Chatelier enunciated his famous principle, a fairly literal translation of which runs as follows:

Any system in stable chemical equilibrium, when subjected to the influence of an external cause which tends to change either its temperature or condensation (pressure, concentration, number of molecules in unit volume), throughout or in only some of its parts, can undergo only such internal modifications which, if they occurred on their own, would bring about a change of temperature or of condensation of a sign contrary to that resulting from the external cause.

Perhaps because of its length or its involved style, the statement has subsequently been cast in many different forms* but, as Le Chatelier reiterated in 1933, nearly 50 years later, this early version was in his view the only correct and exact one.

Throughout the 100 years since its inception, the principle, which Le Chatelier regarded as '*très simple*', has been at the centre of much controversy. The main scientific questions in this debate concern the validity of the general statement. Should its scope be restricted or, at least, should its terminology be made more precise? Is it capable of being given quantitative expression?

Whilst these discussions deal with somewhat abstract principles, it is perhaps important also to note that, as usually stated, Le Chatelier's principle fails to provide the correct answer in certain cases even for the reaction $N_2 + 3H_2 \rightleftharpoons 2NH_3$, the 'simple' example often selected for discussion in textbooks. If we consider the perturbation of this equilibrium by the addition of a small amount of nitrogen (at constant temperature and pressure), the direction of the effect depends on the mole fraction of nitrogen already present. If that mole fraction is less than 0.5, the addition results in the formation of a little more ammonia, but the opposite effect is found if the mole fraction of nitrogen is greater than 0.5. This means that the system sometimes obeys Le Chatelier's principle and sometimes violates it.

Le Chatelier's principle, despite the beguiling superficial simplicity of the basic idea, thus turns out to be a chimaera. The search for the accurate definition of a *general and simple* principle has mesmerised, tantalised, and defeated some of the best minds (amongst them the Nobel laureates Lippmann, Braun, Rayleigh, Ostwald, Planck, Bijvoet and Prigogine).

Jean Gold and Victor Gold

* A typical simplified statement of Le Chatelier's principle is 'when a system in equilibrium is subjected to a change, the system will alter in such a way as to lessen the effect of that change'.

Concentrations of reactants not only influence the position of equilibrium but also the rate at which chemical equilibrium is attained. We shall see in chapter 9 that, in most reactions, the higher the reactant concentration, the faster the rate at which equilibrium is attained.

Influence of Pressure

The effect of pressure on equilibrium systems which do not involve gases is negligible. We shall thus only consider systems involving gases.

A change in pressure of a gaseous equilibrium mixture does not affect the equilibrium constant. Nor does a change in applied pressure. However, a change in pressure can affect the position of equilibrium.

Let us now consider the effect of changing the total pressure on the position of the following equilibrium:

$$N_2O_4(g) \rightleftharpoons 2NO_2(g)$$

The equilibrium constant is given by

$$K_p = \left(\frac{p^2_{NO_2}}{p_{N_2O_4}} \right)_{eq}$$

K_p is constant at a given temperature. p_{NO_2} and $p_{N_2O_4}$ are the partial pressures of NO_2 and N_2O_4 respectively. They are related to the total pressure P of the system by their mole fractions x_{NO_2} and $x_{N_2O_4}$ as follows:

$$p_{NO_2} = x_{NO_2}P$$
$$p_{N_2O_4} = x_{N_2O_4}P$$

We thus obtain the following equilibrium expression:

$$K_p = \left(\frac{x^2_{NO_2}P^2}{x_{N_2O_4}P} \right)_{eq}$$
$$= \left(\frac{x^2_{NO_2}P}{x_{N_2O_4}} \right)_{eq}$$

At constant temperature the quotient on the right-hand side must be constant. Thus, if the total pressure P increases, then $x^2_{NO_2}/x_{N_2O_4}$ must decrease. This means that x_{NO_2} must decrease and $x_{N_2O_4}$ must increase. In other words, the reverse reaction occurs.

Finally, we should note that pressure, like concentration, influences the rate at which chemical equilibria are attained. As the pressure of reactants increases, the number of collisions of the reactant particles also increases. Equilibrium is thus attained faster.

Catalysts and Equilibria

A catalyst is a substance which increases the rate of a chemical reaction (*see* chapter 9). In a reversible reaction a catalyst speeds up the rates of both the forward and reverse reactions equally. Furthermore, a catalyst has no influence on the concentrations of reactants or products. For these reasons a catalyst has no influence on the equilibrium constant or the position of equilibrium. It only affects the rate at which equilibrium is attained.

Influence of Temperature

Unlike a change in concentration or pressure, a change in temperature results in a change in the value of the equilibrium constant. We should also note that an increase in temperature increases the rate at which equilibrium is attained. We shall consider the affect of temperature on rate of reaction in detail in chapter 9.

The equilibrium constants for the synthesis of ammonia at four different temperatures are shown in table 7.1a. The equilibrium constant is given by

$$K_p = \left(\frac{p^2_{NH_3}}{p_{N_2} p^3_{H_2}} \right)_{eq}$$

Note that the units of K_p are atmospheres^{-2}. The values in the table show that the equilibrium constant of an exothermic reaction decreases with increasing temperature.

Table 7.1 Influence of temperature on equilibrium constant

(a) $N_2(g) + 3H_2(g) \rightleftharpoons 2NH_3(g)$,
$\Delta H^{\ominus}_{r,m}(298 \text{ K}) = -92 \text{ kJ mol}^{-1}$

T/K	K_p/atm^{-2}
400	1.0×10^2
500	1.6×10^{-1}
600	3.1×10^{-3}
700	6.3×10^{-5}

(b) $N_2O_4(g) \rightleftharpoons 2NO_2(g)$,
$\Delta H^{\ominus}_{r,m}(298 \text{ K}) = +57 \text{ kJ mol}^{-1}$

T/K	K_p/atm
275	2.2×10^{-2}
350	4.5
500	1.5×10^3

Values for the equilibrium constant for the thermal dissociation of dinitrogen tetraoxide at different temperatures are shown in table 7.1b. The equilibrium constant is given by

$$K_p = \left(\frac{p^2_{NO_2}}{p_{N_2O_4}} \right)_{eq}$$

The units for K_p in this case are atmospheres. The results in the table show that the equilibrium constant of an endothermic reaction increases with increasing temperature.

If we thus take an equilibrium mixture of these two nitrogen oxides at the equilibrium temperature and then increase the temperature of the mixture, a new equilibrium mixture will be formed. This will correspond to the higher value of the equilibrium constant. The equilibrium quotient of partial pressures must therefore increase until it reaches the equilibrium value corresponding to the higher temperature. This means that the partial pressure of NO_2 must increase and the partial pressure of N_2O_4 must decrease. The forward reaction thus occurs in this case.

SUMMARY

1. There are three **characteristics** of a system in **chemical equilibrium**:
   (a) the rates of the forward and reverse reactions are equal:
   (b) the system does not undergo any observable change;
   (c) the system is closed.
2. For the reaction

$$aA + bB \rightleftharpoons cC + dD$$

   the **equilibrium law** is

$$K_c = \left(\frac{[C]^c [D]^d}{[A]^a [B]^b} \right)_{eq}$$

   K_c is the **equilibrium constant**. The right-hand side is called the **equilibrium quotient**.
3. Terms for pure solids, pure liquids and any solvent are not included in the equilibrium quotient.
4. For the reaction

$$N_2O_4(g) \rightleftharpoons 2NO_2(g)$$

   the equilibrium law can be expressed in terms of the partial pressures at equilibrium

$$K_p = \left(\frac{p^2_{NO_2}}{p_{N_2O_4}} \right)_{eq}$$

5. Factors affecting equilibrium

Influence	Equilibrium constant	Equilibrium position	Rate at which equilibrium is attained
change in concentration	no change	changes	changes
change in pressure	no change	changes	changes
addition of catalyst	no change	no change	changes
change in temperature	changes	changes	changes

6. The influence of temperature on **equilibrium position**

Forward reaction	Increase in temperature	Decrease in temperature
exothermic	*equilibrium*: moves to left (reverse reaction occurs)	*equilibrium*: moves to right (forward reaction occurs)
endothermic	moves to right	moves to left

7.2
Manufacture of Ammonia and Sulphuric Acid

The influence of temperature, pressure and catalysts on rates of reaction and chemical equilibria have a pronounced economic impact on the manufacture of many chemical products. In this section we shall first consider the manufacture of ammonia and look at the importance of these factors on its manufacture. We shall then turn our attention to the manufacture of sulphuric acid.

THE INDUSTRIAL PRODUCTION OF AMMONIA

In Britain there are eight ammonia plants. Together they have a capacity of producing over 2 million metric tonnes of ammonia each year. The world production is now around 5 million metric tonnes each year. Figure 7.1 shows how the growth of ammonia production is related to world population growth. So why is all this ammonia needed?

The answer lies in nitrogenous fertilisers. About 80% of all ammonia produced is used to make fertilisers. Nitrogenous fertilisers provide a convenient form of the soluble nitrogen needed by most plants. The remaining 20% of the ammonia produced is used to make polymers, explosives and other products. These uses are summarised in table 7.2.

Table 7.2 Uses of ammonia and related products

Use	Examples of product
fertilisers	ammonium sulphate, urea, ammonium nitrate
synthetic fibres and other polymers	nylon, rayon, polyurethane
cleaning	ammonium hydroxide
refrigeration	liquid ammonia
dry cells, flux	ammonium chloride
explosives	concentrated nitric acid
dyestuff	azo-compounds

Distribution of fertilisers in Nigeria. Nitrogenous fertilisers provide nitrogen in a soluble form for plants

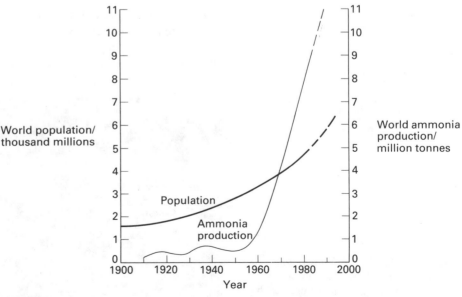

Figure 7.1 World population growth and growth of ammonia production

The Manufacture of Ammonia

The first commercial process used for making ammonia was the **cyanamide process**. Lime (CaO) and carbon were heated to produce calcium carbide, CaC_2. This was then heated under nitrogen to produce calcium cyanamide, $CaCN_2$. The ammonia was produced by hydrolysis of the calcium cyanamide:

$$CaCN_2(s) + \underset{\text{steam}}{3H_2O(g)} \rightarrow 2NH_3(g) + CaCO_3(s)$$

The process consumed a lot of energy and was expensive.

In 1911 Haber discovered that ammonia could be synthesised directly from nitrogen and hydrogen by using an iron catalyst. The first plant to manufacture ammonia by this method used hydrogen obtained by the electrolysis of water. Later, the hydrogen was produced from water by reduction with coke. This was a much more economic way of producing hydrogen.

Fritz Haber, 1868–1934

In 1908, the German chemist Fritz Haber discovered that ammonia could be produced from hydrogen and atmospheric nitrogen using an iron catalyst. High pressures and a moderately high temperature were necessary for the process. This discovery enabled the Germans to continue manufacturing explosives during World War I. During the war the Allied blockade had prevented German access to the nitrate deposits which they had previously used as raw materials for the manufacture of explosives.

The year after Haber discovered the process for synthesising ammonia he developed a glass electrode to measure the pH of solutions (*see* chapter 10).

Haber received the Nobel Prize for Chemistry in 1918. Following Hitler's rise to power in 1933 Haber was forced to emigrate.

Note: The manufacture of nitric acid and nitrates from ammonia is described in section 15.3

The Modern Manufacturing Process

In the modern manufacturing process ammonia is synthesised from nitrogen and hydrogen at 380 to 450°C and 250 atmospheres using an iron catalyst:

$$N_2(g) + 3H_2(g) \rightleftharpoons 2NH_3(g)$$

The *nitrogen* is obtained from the air. The *hydrogen* is obtained by reduction of water (steam) using either methane from natural gas or **naphtha**. Naphtha is a liquid mixture of aliphatic hydrocarbons obtained when refining crude oil (*see* chapter 18).

This ammonia–urea complex in Pakistan can produce up to 1000 tonnes of ammonia and 1725 tonnes of urea each day

The operation of a modern ammonia plant is a complicated affair. A sim scheme for an ammonia plant based on natural gas is shown in figure 7.2. are eight stages in this manufacture.

Stage 1. Sulphur is removed from natural gas. This is necessary since sulphur can poison a catalyst (*see* section 9.2).

Stage 2. Steam is reduced to hydrogen at 750°C and 30 atmospheres using a nickel catalyst:

$$CH_4(g) + H_2O(g) \rightleftharpoons CO(g) + 3H_2(g)$$

Stage 3. Air is introduced and the hydrogen burnt:

$$2H_2(g) + O_2(g) \rightleftharpoons 2H_2O(g)$$

A mixture of water vapour, carbon monoxide and nitrogen is produced. The water vapour is reduced to hydrogen as in stage 2.

Stage 4. The carbon monoxide produced in stages 2 and 3 is oxidised to carbon dioxide in the following 'shift' reaction:

$$CO(g) + H_2O(g) \rightleftharpoons CO_2(g) + H_2(g)$$

The process takes place in two 'shift reactors'. The first uses an iron oxide catalyst at about 400°C. The second uses copper as a catalyst at 220°C.

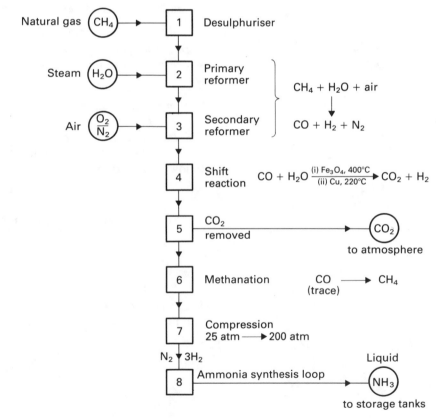

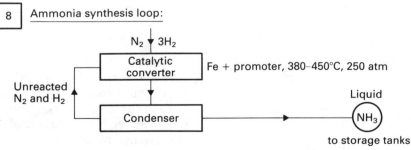

Figure 7.2 Stages in the operation of an ammonia plant

Stage 5. Carbon dioxide is 'scrubbed' from the gaseous mixture by an alkaline buffer solution of potassium carbonate or by a solution of an amine such as monoethanolamine, $NH_2CH_2CH_2OH$. The carbon dioxide is eventually liquefied, used directly in the manufacture of urea or vented to the atmosphere.

Stage 6. Stage 4 leaves about 0.3% carbon monoxide. This could poison the iron catalyst during ammonia synthesis (stage 8). The carbon monoxide is thus removed by methanation using a nickel catalyst at 325°C.

Stage 7. The gaseous mixture, which now contains about 74% hydrogen and 25% nitrogen is compressed from 25 or 30 atmospheres to 200 atmospheres. Since this raises the temperature, the mixture is cooled immediately afterwards.

Stage 8. Gas from the compressor now enters an 'ammonia synthesis loop'. The scheme in figure 7.2 is a very simplified representation. First the gases enter a catalyst converter. This uses an iron catalyst and operates at 380–450°C. The gas leaving the converter contains no more than 15% ammonia. This is condensed and the liquid is stored. The unreacted gases are recycled to the converter.

Choosing the Best Conditions for the Process

Conditions for the synthesis of ammonia have to be carefully chosen if the process is to be efficient and thus economically viable. The three most important considerations in this context are:

- *yield*—how much ammonia is produced?
- *rate*—how fast is it produced?
- *energy*—how much energy is lost during the process?

Let us look at stage 8 of the process—the ammonia synthesis—and examine how pressure, temperature and catalysts influence the efficiency of the process.

The Influence of Pressure

As we have seen, the manufacture of ammonia can be represented by the following equation:

$$N_2(g) + 3H_2(g) \rightleftharpoons 2NH_3(g)$$

The equilibrium constant is given by

$$K_p = \left(\frac{p^2_{NH_3}}{p_{N_2} \, p^3_{H_2}} \right)_{eq}$$

We can convert this into an expression involving the mole fractions x of the gases and the total pressure P of the system as follows:

$$K_p = \left(\frac{x^2_{NH_3} P^2}{x_{N_2} P x^3_{H_2} P^3} \right)_{eq}$$

This simplifies to

$$K_p = \left(\frac{x^2_{NH_3}}{x_{N_2} x^3_{H_2}} \times \frac{1}{P^2} \right)_{eq}$$

K_p must remain constant at a given temperature. If the total pressure P of the system increases, the term $1/P^2$ in this expression decreases. It follows that $x^2_{NH_3}/x_{N_2} x^3_{H_2}$ must increase to keep K_p constant. An increase in total pressure thus leads to an increase in x_{NH_3} and a decrease in x_{N_2} and x_{H_2}. The forward reaction is thus favoured by an increase of pressure. In other words, high yields of ammonia are favoured by high pressures.

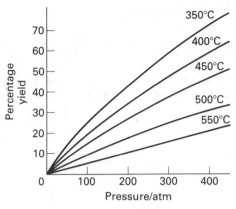

Figure 7.3 Effect of temperature and pressure on yield (the term 'percentage yield' is defined in section 4.2)

The Influence of Temperature and Catalysts

The synthesis of ammonia is an exothermic process (*see* table 7.1a). Thus, as temperature rises, the reverse reaction is favoured (*see* previous section). This means that the best yields are obtained at lower temperatures (*see* figure 7.3). Unfortunately, at low temperatures the rate of this reaction is very slow. A slow rate of reaction means a slow rate of production. In other words the **throughput** is low and this is costly. To obtain the optimum throughput, a balance has to be struck between

- high yield and low rate of reaction (at low temperatures)
or
- low yield and high rate of reaction (at high temperatures).

The rate of reaction is, of course, increased by the use of a catalyst. A catalyst can thus enable the process to be carried out more efficiently at a lower temperature. The efficiency of the iron catalyst used in the ammonia synthesis is enhanced by the addition of small amounts of what are called **promoters**. The ones used to promote the efficiency of the iron catalyst are potassium oxide and aluminium oxide.

A careful consideration of the economics of the ammonia synthesis show that the optimum yield and throughput are obtained at a temperature of around 400°C and a pressure of 250 atmospheres.

Energy Balance

A typical ammonia plant produces about 1000 tonnes of ammonia each day. This requires about 6000 tonnes of steam to be raised each day. This is needed to drive the steam turbines which drive the compressors. Fortunately, the chemical processes involved in producing ammonia are exothermic. All the energy released from the processes in the early stages of the manufacture is used to produce high-pressure steam. The energy released in the ammonia synthesis itself (stage 8) is used to maintain the temperature of the catalyst converter around 400°C. The overall thermal efficiency of the plant is about 60%. In other words, 40% of the energy input provided by the natural gas is lost.

Construction and Siting of an Ammonia Plant

The design, commissioning and operation of a modern ammonia plant requires the services of expert engineers and relies on advanced engineering techniques. For example, the compressors used in stage 3 to compress air and in stage 7 to compress the **synthesis gas** (nitrogen and hydrogen) have to be designed to withstand very high pressures—in some cases up to 350 atmospheres. These compressors are driven by steam turbines which take in steam at 100 atmospheres and over 400°C. The turbines run at several thousand revolutions per minute.

Reactors also have to be designed to very high specifications. At the high pressures and temperatures under which the reactors operate, hydrogen can attack steel by diffusing into the metal. It then reacts with the carbon in the steel to form methane. This forms blisters and makes the steel brittle. To resist this, special alloys containing chromium, molybdenum and nickel are used.

The siting of an ammonia plant is also an important economic consideration. Ideally the plant should be located close to

- sources of energy;
- water which can be supplied in large quantities;
- transport: road, river, sea or rail.

Four of the British ammonia plants are located at Billingham on the River Tees. The site was originally chosen because it is near the Durham coalfields. It is also convenient nowadays because of its proximity to North Sea oil and gas.

THE INDUSTRIAL PRODUCTION OF SULPHURIC ACID

> The commercial prosperity of a nation can be measured by the amount of sulphuric acid it consumes.
> **Baron von Justus Liebig, 1803–1873**

Over 100 million metric tonnes of sulphuric acid are produced each year. The United Kingdom accounts for just under 3% of this production—about 3 to 3½ million metric tonnes annually. About 28% of this is used for agricultural purposes—including the manufacture of fertilisers. The pie-chart in figure 7.4 provides a breakdown of the major uses in the United Kingdom in 1981. Some specific uses are also described in the box.

The Modern Contact Process

Nowadays almost all the sulphuric acid produced in the world is made by the Contact process. The process involves three stages.

Stage 1. In this stage sulphur dioxide is produced. Liquid sulphur is sprayed into a combustion furnace where it burns in air at about 1000°C:

$$S(l) + O_2(g) \rightarrow SO_2(g) \qquad \Delta H_{r,m}^{\ominus} = -297 \text{ kJ mol}^{-1}$$

The air must be dry to avoid the formation of mist. The sulphur is obtained by the Frasch process from deposits in Poland, Mexico and the USA (*see* section 15.4). Elemental sulphur is also obtained as a by-product in the refining of oil and purification of natural gas. French or Canadian natural gases, for example, contain up to 25% hydrogen sulphide.

Sulphur dioxide is also produced by roasting a sulphide mineral such as zinc sulphide or iron pyrites:

$$2ZnS(s) + 3O_2(g) \rightarrow 2ZnO(s) + 2SO_2(g) \qquad \Delta H_{r,m}^{\ominus} = -441 \text{ kJ mol}^{-1}$$

It should be noted that both these oxidation processes are **irreversible** and **exothermic**.

Stage 2. In this stage sulphur trioxide is produced:

$$2SO_2(g) + O_2(g) \rightleftharpoons 2SO_3(g) \qquad \Delta H_{r,m}^{\ominus} = -192 \text{ kJ mol}^{-1}$$

The reaction is **reversible** and **exothermic**. High yields of sulphur trioxide are favoured by low temperatures and high pressures. In practice, the process is carried out at little more than atmospheric pressure. This is to ensure a good flow of gas. The extra yield obtained from higher pressure would not justify the additional cost.

The **feed gas** (sulphur dioxide and oxygen) is passed through a converter. This contains a series of beds containing the vanadium(v) oxide catalyst and promoters. Since the conversion to sulphur trioxide is exothermic, the temperature increases. The gases are thus passed through **heat exchangers** to reduce their temperature. Heat exchangers work on the same principles as the Liebig condensers used in chemical laboratories. At each reaction bed, the percentage conversion to sulphur trioxide increases. The temperature of the catalyst beds is not allowed to drop below about 400°C as the catalyst becomes inactive at lower temperatures. The final yield in a converter containing four catalyst beds is 98%.

Stage 3. During this stage the following process occurs:

$$SO_3(g) + H_2O(l) \rightleftharpoons H_2SO_4(aq) \qquad \Delta H_{r,m}^{\ominus} = -130 \text{ kJ mol}^{-1}$$

However, the sulphur trioxide cannot be directly absorbed by water. This is because the water vapour above the water forms a stable mist of minute droplets of sulphuric acid. Thus 98% sulphuric acid is used to absorb the sulphur trioxide.

A sample of SO_2 from a sulphuric acid plant being analysed in a product development laboratory using the Reisch test. In this test SO_2 is bubbled through an aqueous solution containing potassium iodide and iodine. The SO_2 solution reduces the iodine to iodide. Excess iodine remaining in the solution is titrated with sodium thiosulphate solution using a starch indicator

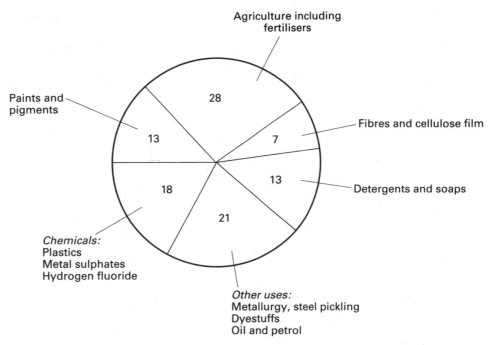

Figure 7.4   Uses of sulphuric acid in the UK (1981)

Uses of Sulphuric Acid

Fertilisers

About 26% of all sulphuric acid is used to make superphosphate fertilisers (*see* section 15.3). Another 2% is used to make ammonium sulphate.

Detergents

The sodium salts of the sulphonates of linear alkylbenzenes are used as the main active constituents of domestic washing powders. Sulphuric acid or oleum is used to manufacture these sulphonates.

Pigments

Sulphuric acid is used in the first stage of the sulphate process for the manufacture of titanium(IV) oxide, TiO_2. This is used as a white paint pigment.

Fibres

Sulphuric acid is used to manufacture caprolactam from cyclohexanone. Caprolactam is the monomer of nylon-6.

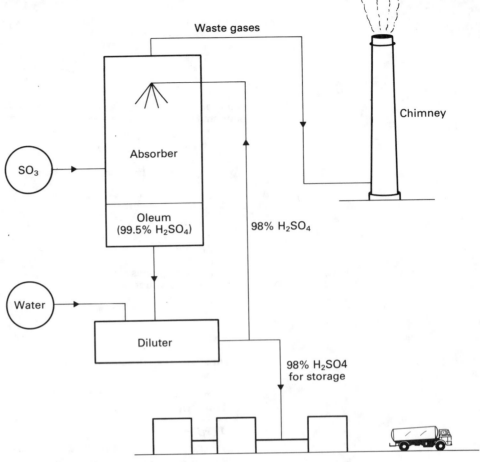

Figure 7.5 Stage 3: the absorption process

The concentration is allowed to increase to 99.5% (*see* figure 7.5). It is then diluted with water to form 98% acid. Some of this is returned to the absorber. The rest is pumped away to storage. If the concentration is allowed to exceed 99.5% the vapour pressure of sulphur trioxide becomes too high to allow complete absorption. As a result, a visible mist appears. The 99.5% sulphuric acid is sometimes called oleum and given the formula $H_2S_2O_7$; thus

$$H_2SO_4(l) + SO_3(g) \rightleftharpoons H_2S_2O_7(l)$$

Double Absorption Sulphuric Acid Plant

In a **single absorption** sulphuric acid plant, the processes described above occur in three separate stages

- Stage 1: production of SO_2.
- Stage 2: 98% conversion of SO_2 to SO_3.
- Stage 3: absorption of the SO_3 in 98% H_2SO_4 to form oleum.

In a **double absorption** sulphuric acid plant, absorption occurs in stages *2 and 3*.

- Stage 1: production of SO_2
- Stage 2: (a) conversion of SO_2 to SO_3;
    (b) absorption of SO_3 in 98% H_2SO_4 to form oleum—this is known as *intermediate absorption*;
    (c) conversion of remaining SO_2 to SO_3 giving a final conversion of 99.5%;
- Stage 3: *final absorption* of SO_3 in 98% H_2SO_4 to form oleum.

The double absorption plant enables a 99.5% conversion of sulphur dioxide to sulphur trioxide to be achieved. Wastage of sulphur dioxide to the atmosphere

Part of a double absorption Contact process plant for sulphuric acid production

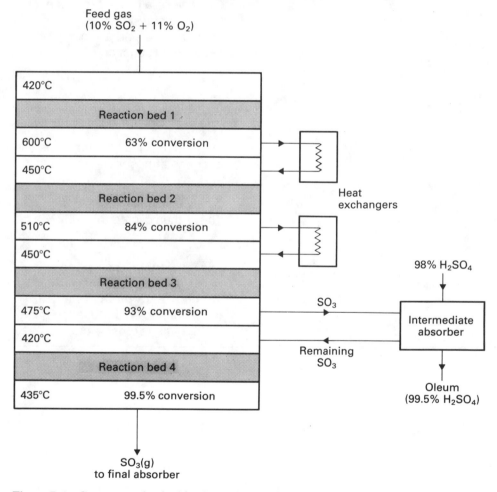

Figure 7.6 Converter of a double absorption sulphuric acid plant

and therefore pollution is minimised. A minimum conversion efficiency of 99.5% is now a mandatory requirement for all new sulphur-burning plants built in the United Kingdom.

Figure 7.6 shows how the intermediate absorber is incorporated into the converter in stage 2 of the manufacture of sulphuric acid in a double absorption plant. The feed gas, which contains about 10% sulphur dioxide and 11% oxygen, is passed into the converter. This contains four catalyst beds. The first bed produces a 66% conversion to sulphur trioxide. After bed 3 all the sulphur trioxide is removed and absorbed in an intermediate absorber. Unabsorbed gases return to join the remaining sulphur dioxide and oxygen for final conversion. The sulphur trioxide from bed 4 then passes on to the final absorber.

SUMMARY

1. (a) Ammonia is made by the **Haber process**:

$$N_2(g) + 3H_2(g) \rightleftharpoons 2NH_3(g)$$

The nitrogen is obtained from air. The hydrogen is obtained from water or methane.

(b) The first seven stages of the process produce the **synthesis gas** (N_2 and $3H_2$). The synthesis of the gas takes place in stage 8. An iron catalyst is used. The **optimum yield** and **throughput** are obtained at 400°C and 250 atmospheres.

(c) Ammonia is used to make fertilisers, nitric acid and many other products.

2. (a) Sulphuric acid is made by the **Contact process**:

Stage 1. Production of SO_2

$$S(l) + O_2(g) \rightarrow SO_2(g)$$

Stage 2. Conversion of sulphur dioxide to sulphur trioxide

$$2SO_2(g) + O_2(g) \rightleftharpoons 2SO_3(g)$$

A vanadium(v) oxide catalyst is used. The process is carried out at just over one atmosphere and 400°C.

Stage 3. This is the **absorption stage**. The sulphur trioxide is absorbed by 98% sulphuric acid to produce **oleum**. This is 99.5% sulphuric acid. This is then diluted to the 98% acid. The process can be represented by

$$SO_3(g) + H_2O(l) \rightleftharpoons H_2SO_4(aq)$$

(b) In a **double absorption** sulphuric acid plant, absorption takes place in two stages. **Intermediate absorption** takes place in stage 2 and **final absorption** in stage 3.

(c) Sulphuric acid is used to make fertilisers, paints, detergents, fibres and many other products.

Examination Questions

1. (a) State Le Chatelier's principle.
   (b) For the reaction

   $$H_2(g) + I_2(g) \rightleftharpoons 2HI(g) \qquad \Delta H^{\ominus} = 52\ kJ\ mol^{-1}$$

   give the expression for the equilibrium constant in terms of partial pressures (K_p).
   (c) What would be the effect, if any, on the equilibrium above of
      (i) increasing the total pressure,
      (ii) increasing the temperature?

(d) The following table gives the concentrations of the components in the above equilibrium at 600 K.

Concentration of H_2	Concentration of I_2	Concentration of HI
1.71×10^{-3} mol l^{-1}	2.91×10^{-3} mol l^{-1}	1.65×10^{-2} mol l^{-1}

Calculate the equilibrium constant in terms of concentrations (K_c) at this temperature.

(JMB)

2. (a) What is meant by the terms *homogeneous equilibrium* and *equilibrium constant*?
   (b) Consider the reaction

$$PCl_5(g) \rightleftharpoons PCl_3(g) + Cl_2(g).$$

For such a system, of total volume dm^3 and a total pressure of 202.6 kPa, at equilibrium the percentage dissociation of $PCl_5(g)$ at temperatures of 200°C and 300°C are 48.5 and 97.0 respectively.
   (i) Write an expression to represent the equilibrium constant, K_p.
   (ii) Calculate the value for K_p at 200°C.
   (iii) What percentage dissociation of PCl_5 would result if the pressure of the system was reduced to 135.1 kPa? (Temperature remaining constant.)
   (iv) How is the value obtained in (iii) explained in terms of Le Chatelier's principle?
   (v) What can be inferred from the above information regarding the enthalpy change of reaction?
   (c) How would the position of the above equilibrium be affected by
   (i) a decrease in total pressure.
   (ii) the presence of a catalyst.
   (iii) an increase in concentration of chlorine.
   (iv) a decrease in temperature?
   Which of these factors would affect the rate at which equilibrium would be established?

(AEB, 1982)

3. (a) Explain what is meant by *dynamic equilibrium*.
   (b) The hydrogenation of coal can be used to manufacture substitute natural gas. The major reaction involved is

$$C(s) + 2H_2(g) \rightleftharpoons CH_4(g)$$

For this reaction, give expressions for the equilibrium constants (i) K_c, and (ii) K_p.
   (c) For the same reaction, describe and explain the effects of the following changes on both the position of equilibrium and the value of K_p.
   (i) increase in temperature at constant pressure;
   (ii) increase in the partial pressure of methane at constant volume;
   (iii) increase in total pressure at constant temperature;
   (iv) addition of a catalyst.

(O & C)

4. (a) 1.113×10^{-3} mol of iodine are introduced into a vessel of volume 249.0×10^{-3} dm^3. The vessel is closed and evacuated, and then

heated to 1175 K, at which temperature the pressure in the vessel is found to be 50.25 kPa.
  (i) Calculate K_p at 1175 K for the equilibrium $I_2(g) \rightleftharpoons 2I(g)$.
  (ii) At what pressure would 90% of the iodine molecules be dissociated into atoms?
(b) By what method may chlorine atoms be produced at 300 K?

(OLE)

5. Write a short account of the factors affecting the position of equilibrium of a balanced homogeneous gaseous reaction, the rate at which equilibrium is attained and the value of the equilibrium constant.
(a) Explain why there is a constant pressure of carbon dioxide at a particular temperature over calcium carbonate undergoing thermal dissociation

$$CaCO_3(s) \rightleftharpoons CaO(s) + CO_2(g)$$

and why a current of air is blown through a lime kiln during the production of quicklime, CaO.
(b) Using partial pressures, show that for gaseous reactions of the type

$$XY(g) \rightleftharpoons X(g) + Y(g)$$

at a given temperature, the pressure at which XY is exactly one-quarter dissociated is numerically equal to *fifteen* times the value of the equilibrium constant at that temperature.
(c) When one mole of ethanoic acid (acetic acid) is maintained at 25°C with one mole of ethanol, one-third of the ethanoic acid remains when equilibrium is attained. How much would have remained if *three-quarters of one mole of ethanol* had been used instead of one mole at the same temperature?

(SUJB)

6. (a) For the reaction $A + B \rightleftharpoons 2C + 2D$, give an expression for and state the units of the appropriate equilibrium constant (K_c or K_p) for *each* of the following situations when:
  (i) reactants and products are liquids;
  (ii) reactants and products are gases;
  (iii) A and D are gases but B and C are liquids.
(b) For the reaction:

$$N_2(g) + 3H_2(g) \rightleftharpoons 2NH_3(g) \qquad \Delta H = -92 \text{ kJ mol}^{-1}$$

discuss the effect on the equilibrium of:
  (i) increasing the temperature at constant pressure;
  (ii) decreasing the pressure at constant temperature;
  (iii) adding a catalyst.
(c) 1 mole of nitrogen and 3 moles of hydrogen were mixed at 593 K and 2×10^7 Pa. At equilibrium the mixture contained 1.5 moles of ammonia.
  (i) Write the equilibrium expression for K_p.
  (ii) Find the mole fraction of each component and thus their partial pressures.
  (iii) Use these values, from (ii), to calculate K_p.
  (iv) Explain why these physical conditions are *not* used in practice.
(d) Sketch a graph to show how the amounts of the three components of the equilibrium in (b) vary with time, starting with the initial mixing of the reactants.

(SUJB)

7. (a) State what is meant by:
        (i) *dynamic equilibrium*;
        (ii) *the equilibrium law*.
    (b) For the reaction:

$$N_2(g) + 3H_2(g) \rightleftharpoons 2NH_3(g)$$

        (i) calculate the mole percentage of ammonia in the equilibrium mixture formed at 400°C and 3×10^7 Pa pressure, when gaseous hydrogen and nitrogen are mixed in a 3:1 mole ratio, and there is 61% conversion of nitrogen to ammonia;
        (ii) write an expression for K_p in terms of the partial pressures of the three gases in equilibrium;
        (iii) given that the value of K_p at 400°C is 2.0×10^{-14} Pa^{-2}, calculate the pressure at which ammonia is 95% dissociated into its elements at 400°C.

(AEB, 1985)

8. This question concerns the following reversible process.

$$2SO_2(g) + O_2(g) \rightleftharpoons 2SO_3(g) \qquad \Delta H^{\ominus} = -98 \text{ kJ mol}^{-1}$$

    (a) State Le Chatelier's principle.
    (b) Give and explain the effect on the position of equilibrium of
        (i) increasing the pressure at constant temperature,
        (ii) increasing the temperature at constant pressure.
    (c) Bearing in mind your answer to (b), how do you account for the fact that industrially, (i) a temperature of about 800 K, (ii) a pressure of 1 or 2 atm, and (iii) a catalyst are used?
    (d) At 800 K, the equilibrium partial pressures in atmospheres are: p_{SO_2} = 0.10; p_{O_2} = 0.70; p_{SO_3} = 0.80.
        (i) Calculate K_p.
        (ii) If the equilibrium mixture had been obtained starting from oxygen and sulphur dioxide only, in a sealed container at 800 K, what were the initial partial pressures of oxygen and sulphur dioxide?

    (e) In industry, this reaction is part of a series of reactions in which sulphur is converted into sulphuric acid. In the 1960s a double absorption/double conversion process was developed.
        (i) What is the main feature of a double absorption/double conversion process?
        (ii) Suggest a reason, other than an economic one, why this type of process might have been introduced.

(L)

8 IONS

The Dead Sea

All the data indicate that the water column of the Dead Sea is sharply stratified into two water masses separated by a stable boundary at a depth of 40 metres. Seasonal temperature fluctuations, from 19°C to 37°C, occur down to this depth, and the density and salinity of waters at this level vary with the seasons as well, of course, as with long-term changes in climate. However, below 40 metres these parameters remain unchanged throughout the year. Between 40 metres and 60 metres, temperature, density and salinity increase sharply; from 60 to 100 metres they increase more gradually. At greater depths most parameters are uniform, the temperature at the greatest sampling depth of about 350 metres being approximately 22°C and the salinity there being about 332 grammes per litre ($g\ l^{-1}$).

The chemical compositions of the two water masses are so different that it is highly unlikely that the lower one evolved from the upper. During 1959–60 the average surface salinity of the Dead Sea was almost 300 $g\ l^{-1}$, the chief anion being chloride. This was followed by bromide at 4.6 $g\ l^{-1}$, sulphate at 0.58 $g\ l^{-1}$ and bicarbonate at 0.23 $g\ l^{-1}$. The principal cations were sodium, magnesium, calcium and potassium.

However, below a depth of 40 metres these ions have quite different concentrations. Magnesium, potassium, chloride and bromide all exhibit a sharp increase in concentration at a depth of 40 metres and they vary with depth in a way similar to density and total salinity. At depths greater than 100 metres the average concentrations of these ions are all constant at values considerably higher than at the surface. Sodium ion concentration shows a small minimum just below the 40 metre boundary but it increases again below this depth and at depths greater than 100 metres it is 39.7 $g\ l^{-1}$, slightly higher than its surface value. Sulphate and bicarbonate concentrations decrease below 40 metres.

These variations were investigated by using the ratio of the concentration of each ion to that of magnesium, which was chosen as the reference because of its high solubility and high concentration in the Dead Sea. The ratios show that sodium chloride, calcium carbonate and calcium sulphate must largely have been removed from the water deeper than 40 metres but that magnesium, potassium and bromide have remained in solution.

Precipitation of a salt takes place from a fully saturated solution. Thus removal of sodium chloride, for example, from water below 40 metres is inferred because waters at depths less than 40 metres are undersaturated with sodium chloride, those at depths between 40 and 100 metres are only slightly undersaturated (some upward diffusion presumably having taken place) and those at depths greater than 100 metres are fully saturated. That sodium chloride has been precipitated from

The Dead Sea is sharply stratified into two water masses separated by a stable boundary at a depth of 40 metres. The chemical compositions of the two water masses are very different

the entire water body deeper than the 40 metre level is further indicated by the presence of a massive bed composed mainly of sodium chloride on the floor of the Dead Sea throughout the area deeper than 40 metres.

The lower water mass is, therefore, probably a 'fossil' water body from which sodium chloride has been precipitated for many thousands of years. At a late stage in the history of the Dead Sea this brine must have been sealed from the atmosphere by the accumulation of the more dilute upper water which apparently owes its origin to changes in the water balance of the region; at some stage inflow must have increased until it exceeded the rate of evaporation. This surface inflow would have contained dissolved salts but these must have been supplemented by highly saline spring waters and by upward diffusion of ions from the lower water mass. Only in this way can we account for the high salinity of the surface water of the present-day sea.

The gradual upward decrease in both temperature and salinity of the uppermost 60 metres of the lower water mass is believed to be caused by heat flow and ion diffusion. However, diffusion is such a slow process that, although the boundary will gradually sink further below the surface, a complete mixing of the entire water body of the Dead Sea will probably not occur for a long time to come.

The sediments now accumulating in the Dead Sea are largely chemical precipitates. These consist almost entirely of gypsum, a form of calcium sulphate, and aragonite and calcite, both forms of calcium carbonate. Gypsum and aragonite are precipitated from the surface water throughout the lake; in fact, gypsum is precipitated faster than the annual supply of sulphate ions from the Jordan River and other streams would imply.

A clue to this mystery is provided by the fact that gypsum is less abundant in deeper bottom sediments than calcite and aragonite, although it forms crusts along the wave-agitated zone of the shore. Geochemical investigation suggests that the precipitated gypsum that reaches the deep water is reduced by bacteria to hydrogen sulphide and calcium. The calcium then combines with carbonate ions to form calcite and the hydrogen sulphide diffuses upwards to the upper water mass where it reacts with dissolved oxygen to yield the sulphate ion again. This in turn combines with more calcium to renew the gypsum cycle. In this way the gross precipitation of gypsum exceeds the net annual precipitation; as the gypsum is precipitated, sulphate ions cycle to the sea bottom and return to be re-precipitated at the surface.

D. Neev and K. O. Emery

LEARNING OBJECTIVES

After you have studied this chapter you should be able to

1. Distinguish between **strong electrolytes, weak electrolytes** and **non-electrolytes** and give examples of each.
2. Outline the evidence for **electrolytic dissociation**.
3. State and apply **Ostwald's dilution law**.
4. Summarise the **Brønsted–Lowry theory** of **proton transfer** of acids and bases.
5. Distinguish between **strong and weak acids**.
6. Distinguish between **strong and weak bases**.
7. Give examples of
   (a) acids and bases which form **conjugate pairs**;
   (b) **amphoteric** substances.
8. Explain the terms **Lewis acid** and **Lewis base** and give examples of each.
9. Calculate **dissociation constants** and **pK values** given the appropriate data.
10. Explain the significance of
    (a) pK values;
    (b) the **ionic product** of water.
11. Calculate **pH values**
12. Explain the following terms:
    (a) **indicator dissociation constant**;
    (b) **indicator range**.
13. Sketch and interpret the four types of acid–base **titration curves**.
14. Give examples of the acid–base reaction of salts in water.
15. Explain how **buffer solutions** work and outline their uses.
16. Use the **Henderson equation** to calculate pH and concentrations.
17. Calculate solubilities and **solubility products**.
18. Explain and give an example of the **common ion effect**.
19. Write expressions for the **stability constants** for the formation of complex ions in aqueous solution.
20. Distinguish between **cation exchangers** and **anion exchangers** and give examples of each.
21. Outline the uses of **ionic exchange**.

8.1
Acids and Bases

INTRODUCTION

An **ion** is an electrically charged atom or group of atoms. Ions which are positively charged are called **cations**:

Na^+	sodium ion
Ca^{2+}	calcium ion
H_3O^+	oxonium ion
NH_4^+	ammonium ion

Ions which are negatively charged are called **anions**:

Cl^-	chloride ion
O^{2-}	oxide ion
SO_4^{2-}	sulphate ion
CH_3COO^-	ethanoate ion

We saw in chapter 2 that the cations and anions in ionic solids are bound together by electrostatic forces of attraction. These forces are called **ionic** or electrovalent **bonds**. It is these bonds which account for the shape and structure of ionic crystals (*see* chapter 3). In ionic crystals the ions are held together in a rigid **lattice**. They are thus not free to move.

When an ionic solid such as potassium iodide is melted or dissolved in a polar solvent, the cations and anions become **fully ionised**. The ions are thus free to move and conduct electricity. **Covalent** compounds may also dissociate to form ions in solution. For example, hydrogen chloride dissolves in water to form oxonium and chloride ions. This solution is known as hydrochloric acid.

ELECTROLYTES

Molten potassium iodide and hydrochloric acid are examples of electrolytes. An **electrolyte** conducts electricity due to the flow of charge carried by its ions. The passage of an electric current through an electrolyte is thus accompanied by the transfer of matter.

Electrolytes are compounds—normally acids, bases or salts—in their fused or aqueous states. Electrolytes are sometimes known as **electrolytic conductors**. They should be distinguished from **electronic conductors**. In electronic conductors the flow of charge is due to the flow of electrons and not the movement of ions. The passage of an electric current through an electronic conductor is thus not accompanied by the transfer of matter. All metals are electronic conductors.

Electrolytes may be divided into two categories: strong and weak electrolytes. A **strong electrolyte** is a compound which is fully ionised when molten or in solution. For example, hydrochloric acid is a strong electrolyte. When hydrogen chloride dissolves in water, it becomes fully ionised:

$$HCl(g) + H_2O(l) \rightarrow H_3O^+(aq) + Cl^-(aq)$$

Notice the arrow, which indicates that the process goes to completion.

A **weak electrolyte** is a compound which only partially dissociates into ions. Thus, in solution, an equilibrium is established between the undissociated molecules and the dissociated ions. Ethanoic acid is an example:

$$CH_3COOH(l) + H_2O(l) \rightleftharpoons CH_3COO^-(aq) + H_3O^+(aq)$$

Organic acids and organic bases are generally weak electrolytes.

A **non-electrolyte** does not dissociate into ions at all and thus does not conduct electricity. Most organic compounds are non-electrolytes.

Experimental Evidence for the Existence of Ions in Solution

We saw in chapter 3 that X-ray crystallography provides evidence for the existence of ions in solids. How do we know ions exist in solution? The evidence comes from a number of sources:

Colligative Properties
Colligative properties such as the depression of freezing point depend on the number but not the nature of solute ions present in solution (*see* chapter 6).

Conductivity
The passage of an electric current through an electrolyte can, as we saw above, be accounted for by the movement of ions carrying the charge. Strong electrolytes are better conductors of electricity than weak electrolytes. We shall look at the topic of conductivity in detail in chapter 10.

Electrolysis
When an electric current is passed through an electrolyte, chemical decomposition of the electrolyte occurs. This is called electrolysis. It provides further evidence of the existence of ions in solution. This topic will be considered again in chapter 10.

Ostwald's Dilution Law

In 1888 Ostwald applied the equilibrium law (*see* chapter 7) to the dissociation of electrolytes. He derived a relationship between the equilibrium constant and the degree of dissociation of an electrolyte. This relationship is known as Ostwald's dilution law. We can derive this law ourselves by considering the dissociation of a weak electrolyte. The electrolyte BA is dissolved in water to form a solution of concentration c. A fraction α of the BA then dissociates, forming the ions B^+ and A^-. The quantity α is called the **degree of dissociation**. The concentration of un-ionised BA at equilibrium is then $c(1 - \alpha)$. We thus have

	BA(aq)	$\rightleftharpoons$ B$^+$(aq)	+ A$^-$(aq)
initial concentration	c	0	0
equilibrium concentration	$c(1 - \alpha)$	αc	αc

The equilibrium law for this equilibrium is

$$K_c = \left(\frac{[B^+][A^-]}{[BA]} \right)_{eq}$$

Friedrich Wilhelm Ostwald won the Nobel Prize for Chemistry in 1909

Friedrich Wilhelm Ostwald (1853–1932)

Ostwald was a German physical chemist and philosopher. He was one of the co-founders of physical chemistry. He is famous for his dilution law. He also showed that indicators are weak acids or weak bases which have different colours in the undissociated and dissociated states (*see* below). In 1909 he won the Nobel Prize for Chemistry for his work on catalysis. It was Ostwald who first inferred that catalysts speed up chemical reactions by lowering the energy of activation of the reactions (*see* chapter 9). Ostwald was also the first person to develop a technical process for the oxidation of ammonia using a platinum catalyst. The process, which is named after him, is one stage of the manufacture of nitric acid (*see* chapter 15). Curiously enough, although Ostwald was in the forefront in several fields of chemistry, he was one of the last to oppose the rapidly developing atomic theory—a theory which had been first formulated by Dalton in 1808. It was not until after 1906 that he finally accepted the existence of atoms.

where K_c is the equilibrium constant. Substituting the concentration terms at equilibrium, we obtain

$$K_c = \frac{(\alpha c)(\alpha c)}{c(1 - \alpha)}$$

Thus

$$K_c = \frac{\alpha^2 c}{(1 - \alpha)}$$

This is known as **Ostwald's dilution law**. The constant K_c is the **dissocation constant** of the electrolyte. The higher the value of K_c the stronger the electrolyte.

For very weak electrolytes, α is very small. Thus

$$1 - \alpha \approx 1$$

($\approx$ means approximately equal to) and so

$$K_c \approx \alpha^2 c$$

Ostwald's dilution law is of considerable importance. In this and the following sections we shall see how it applies to the dissociation of acids, bases, water and salts.

ACIDS AND BASES

Acids typically have the following properties:

1. They are non-electrolytes when pure but become electrolytes when dissolved in water.
2. Aqueous solutions of acids change the colours of indicators.
3. Aqueous solutions of acids undergo the following characteristic reactions:
   (a) they react with metal oxides to produce salts and water;
   (b) they react with the more reactive metals to produce salts and hydrogen;
   (c) they react with carbonates to produce salts, carbon dioxide and water.

Compounds which neutralise acids are called **bases**. Not all bases are soluble in water. A base which is soluble in water is called an **alkali**. Aqueous solutions of alkalis are electrolytes. Alkaline solutions also change the colours of indicators.

In 1884 Arrhenius suggested that an acid was a substance which dissociated to give hydrogen ions, H^+. He also suggested that bases were substances which dissociated in water to produce hydroxide ions, OH^-.

Although his theory was and still is adequate for some purposes, it had its limitations. For example, it did not account for the basic properties of some organic substances such as triethylamine, $N(C_2H_5)_3$, which did not contain the group OH.

The Brønsted–Lowry Theory

In 1923 Brønsted and Lowry proposed the **proton transfer** theory of acids and bases. According to their theory, an *acid* is a substance consisting of molecules or ions which donates protons, and a *base* is a substance consisting of molecules or ions which accepts protons.

To examine this theory in detail, let us first take the case of a **strong acid** such as hydrochloric acid. When hydrogen chloride dissolves in water, its molecules dissociate fully, forming hydrogen ions, H^+, and chloride ions, Cl^-. The hydrogen ions are also known as **protons** since a hydrogen ion consists of a single proton only. These protons are donated to the water molecules, forming oxonium ions, H_3O^+:

$$HCl(g) + H_2O(l) \rightarrow H_3O^+(aq) + Cl^-(aq)$$

Svante August Arrhenius (1859–1927)

The Swedish physical chemist Arrhenius received the Nobel Prize for Chemistry in 1903 for his theory of electrolytic dissociation. In his doctoral dissertation at the University of Uppsala he argued that 'molecules' such as sodium chloride break apart spontaneously in solution forming ions which become the agents of electrolysis. He is also known for his book *Worlds in the Making* which was published in 1908. In this he supports the suggestion that the origin of life on Earth, and in the Universe generally, were living spores travelling through outer space.

In the case of a **weak acid** such as ethanoic acid, the acid molecules are only partially dissociated in water, and so an equilibrium is established:

$$CH_3COOH(aq) + H_2O(l) \rightleftharpoons H_3O^+(aq) + CH_3COO^-(aq) \qquad (1)$$

Even so, the acid molecules still donate protons to water molecules.

Hydrochloric acid and ethanoic acid can only donate a single proton each. They are known as **monoprotic** acids. They are also known as **monobasic** acids since they only have a single conjugate base each (*see* below).

An acid which donates more than one proton is called **polyprotic**. For example, sulphuric acid is **diprotic** since it can donate two protons. It is also called **dibasic** since it has two conjugate bases, HSO_4^- and SO_4^{2-}:

$$H_2SO_4(l) + H_2O(l) \rightarrow H_3O^+(aq) + HSO_4^-(aq)$$

or

$$H_2SO_4(l) + 2H_2O(l) \rightarrow 2H_3O^+(aq) + SO_4^{2-}(aq)$$

Strong bases such as the alkali metal hydroxides are fully ionised in water:

$$NaOH(s) + aq \rightarrow Na^+(aq) + OH^-(aq)$$

They are bases because the hydroxide ions can accept hydrogen ions:

$$OH^-(aq) + H_3O^+(aq) \rightarrow 2H_2O(l)$$

This reaction is common to all **neutralisation** reactions between acids and bases in aqueous solution. Note that in aqueous solutions the hydrogen ions exist in their hydrated forms, that is, as oxonium ions.

Let us now consider the reaction between the weak base ammonia, NH_3, and hydrogen chloride in the gas phase. In this reaction solid ammonium chloride is formed:

$$NH_3(g) + HCl(g) \rightleftharpoons NH_4^+(s) + Cl^-(s)$$

The reverse of this reaction is the **thermal dissociation** of ammonium chloride. Ammonia is a base because it accepts a proton from the hydrogen chloride. But notice that no water or hydroxide ions are involved in the equilibrium. The Brønsted–Lowry theory thus extends to acid–base reactions which occur in the absence of a solvent or even in the presence of a non-aqueous solvent.

If we examine any of the above equilibria a little more closely, we will see that there are four species present in the equilibrium mixture. Using the Brønsted–Lowry theory we also see that two of these species are acids and two are bases. For example

$$\underset{\text{base(1)}}{NH_3(g)} + \underset{\text{acid(2)}}{HCl(g)} \rightleftharpoons \underset{\text{acid(1)}}{NH_4^+(s)} + \underset{\text{base(2)}}{Cl^-(s)}$$

The equilibrium mixture is said to consist of two **conjugate pairs** of acids and bases:

NH_3 and NH_4^+ are a conjugate pair
HCl and Cl^- are also a conjugate pair

Notice that, in each conjugate pair, the acid and base differ from one another by a proton. Each acid has its own **conjugate base**. A strong acid always has a weak conjugate base. A weak acid always has a strong conjugate base.

Amphoteric Substances

A substance which can react as both an acid and a base is said to be **amphoteric**. Zinc hydroxide is an example. It reacts as a base with hydrochloric acid,

$$Zn(OH)_2(s) + 2HCl(aq) \rightarrow ZnCl_2(aq) + 2H_2O(l)$$

With sodium hydroxide, however, zinc hydroxide reacts as an acid, forming a compound called sodium tetrahydroxozincate(II),

$$Zn(OH)_2(s) + 2NaOH(aq) \rightarrow Na_2Zn(OH)_4(aq)$$

Under the Brønsted–Lowry theory, water can be an acid or a base. In the ethanoic acid equilibrium (see equation (1) above) water is clearly a base. However, in the following equilibrium it is an acid:

$$NH_3(aq) + H_2O(l) \rightleftharpoons NH_4^+(aq) + OH^-(aq)$$

Water can thus form two conjugate pairs:

$$H_2O(acid) \text{ and } OH^- \text{ (conjugate base)}$$

$$H_3O^+(acid) \text{ and } H_2O \text{ (conjugate base)}$$

Since water can either donate or accept a proton, it is sometimes called **amphiprotic**. Whether it acts as an acid or a base depends on the other species present.

Lewis Acids and Bases

In 1938, the American chemist G. N. Lewis produced a theory which extends the concept of acids and bases even further than that of Brønsted and Lowry. According to Lewis, an *acid* is a substance which can form a covalent bond by accepting an electron pair from a base, and a *base* is a substance that has an unshared electron pair which can form a covalent bond with an atom, molecule or ion.

A covalent bond formed by the donation of an electron pair is often called a dative or coordinate bond and represented by an arrow, thus, → (*see* section 2.1).

The reaction between boron trifluoride and ammonia is a classic example of a reaction between a Lewis acid and a Lewis base:

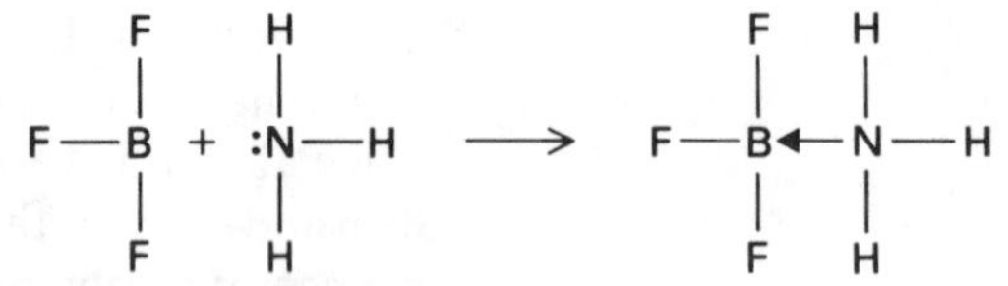

Lewis acid   Lewis base

The Lewis theory thus extends the range of acid–base reactions to include those which do not involve protons. A proton itself is an example of a simple Lewis acid since it can accept an electron pair:

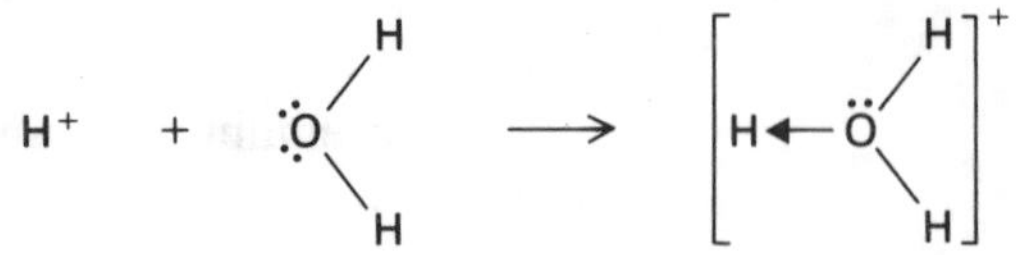

Lewis acid   Lewis base

Lewis acids include cations and Lewis bases include anions as may be seen from the following examples:

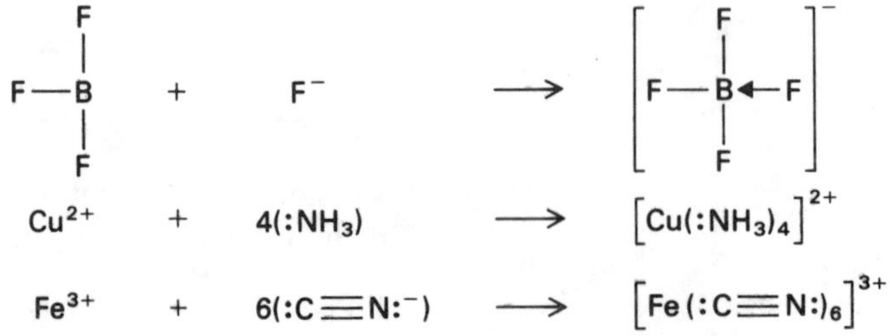

It is possible to titrate many Lewis acids and Lewis bases against one another using suitable indicators in the same way as conventional acids and bases are titrated.

The examples above show that Lewis acids and Lewis bases are important in the formation of complex ions (*see* chapter 14). They are also important in nucleophilic reactions in organic chemistry (*see* chapter 17).

DISSOCIATION CONSTANTS

The equilibrium law can be applied to aqueous solutions of acids. For example, the following equilibrium is established in an aqueous solution of ethanoic acid:

$$CH_3COOH(aq) + H_2O(l) \rightleftharpoons CH_3COO^-(aq) + H_3O^+(aq)$$

The equilibrium constant is given by

$$K_a = \left(\frac{[H_3O^+][CH_3COO^-]}{[CH_3COOH]} \right)_{eq}$$

K_a is called the **acid dissociation constant**. It has concentration units of mol dm^{-3}. It is also given by Ostwald's dilution law

$$K_a = \frac{\alpha^2 c}{1 - \alpha}$$

where c is the initial concentration of the acid and α is the degree of dissociation of the acid. The degree of dissociation can be found by determining the molar conductivities of the solution at a known concentration and also at infinition dilution (for details *see* chapter 10).

The acid dissociation constant is a measure of the strength of an acid. For an acid such as hydrochloric acid which is virtually fully dissociated in aqueous solution, its value is extremely large. On the other hand, for weak acids, values for K_a can be extremely small. It is often more convenient to compare the strengths of acids using pK_a values, where pK_a is given by

$$pK_a = -\lg K_a$$

Table 8.1 Dissociation constants of acids and bases in water at 25°C

Acids		K_a/mol dm^{-3}	pK_a
hydrochloric acid	HCl	1×10^7	-7
sulphuric acid	H$_2$SO$_4$	1.2×10^{-2}	1.92 (pK_{a_1})
phosphoric acid	H$_3$PO$_4$	7.08×10^{-3}	2.15 (pK_{a_1})
hydrofluoric acid	HF	5.62×10^{-4}	3.25
ethanoic acid	CH$_3$COOH	1.74×10^{-5}	4.75
phenol	C$_6$H$_5$OH	1.28×10^{-10}	9.31
Bases		K_b/mol dm^{-3}	pK_b
diethylamine	(C$_2$H$_5$)$_2$NH	9.55×10^{-4}	3.02
ethylamine	C$_2$H$_5$NH$_2$	5.62×10^{-4}	3.25
methylamine	CH$_3$NH$_2$	4.54×10^{-4}	3.34
ammonia	NH$_3$	1.74×10^{-5}	4.76
pyridine	C$_5$H$_5$N	5.62×10^{-9}	8.25
phenylamine	C$_6$H$_5$NH$_2$	4.27×10^{-10}	9.37

For most acids this gives a range of values from 1 to 14. Strong acids have low pK_a values and weak acids have large values. A list of values for K_a and pK_a for some acids is shown in table 8.1. Polyprotic acids such as H$_2$SO$_4$ and H$_3$PO$_4$ have more than one dissociation constant. For example, H$_3$PO$_4$, which is a triprotic acid, has three dissociation constants (*see* table 8.2). These are called the first, second and third dissociation constants, K_{a_1}, K_{a_2} and K_{a_3} respectively.

Table 8.2 Dissociation constants of H_3PO_4

Dissociation	Equilibrium	K_a/mol dm^{-3}	pK_a
first	$H_3PO_4(aq) + H_2O(l) \rightleftharpoons H_3O^+(aq) + H_2PO_4^-(aq)$	7.52×10^{-3}	2.12
second	$H_2PO_4^-(aq) + H_2O(l) \rightleftharpoons H_3O^+(aq) + HPO_4^{2-}(aq)$	6.23×10^{-8}	7.21
third	$HPO_4^{2-}(aq) + H_2O(l) \rightleftharpoons H_3O^+(aq) + PO_4^{3-}(aq)$	2.2×10^{-13}	12.67

EXAMPLE

Conductivity experiments on aqueous ethanoic acid with a concentration of 0.005 mol dm^{-3} showed that the degree of dissociation at this concentration and a temperature of 25°C was 0.057. Calculate values for K_a and pK_a.

SOLUTION

K_a is given by

$$K_a = \frac{\alpha^2 c}{1 - \alpha}$$

Substituting

$$\alpha = 0.057$$

and

$$c = 0.005 \text{ mol dm}^{-3}$$

we obtain

$$K_a = \frac{(0.057)^2 \times (0.005 \text{ mol dm}^{-3})}{(1 - 0.057)}$$

$$= 1.72 \times 10^{-5} \text{ mol dm}^{-3}$$

Thus

$$\text{p}K_a = -\lg K_a$$
$$= -\lg (1.72 \times 10^{-5})$$
$$= 5 - 0.233$$
$$= 4.767$$

The equilibrium law can also be applied to the dissociation of bases in water. For example, when ammonia is dissolved in water the equilibrium is

$$NH_3(aq) + H_2O(l) \rightleftharpoons NH_4^+(aq) + OH^-(aq)$$

The **base dissociation constant** K_b for this equilibrium is given by

$$K_b = \left(\frac{[NH_4^+][OH^-]}{[NH_3]} \right)_{eq}$$

The strength of bases can also be compared on the pK_b scale, where

$$\text{p}K_b = -\lg K_b$$

Strong bases have low pK_b values and weak bases have larger values. Values for K_b and pK_b for a range of bases are given in table 8.1.

The Ionic Product of Water

The electrical conductivity of even the purest water never falls to exactly zero. This is due to the **self-ionisation** of water. This can be represented by

$$2H_2O(l) \rightleftharpoons H_3O^+(aq) + OH^-(aq)$$

Applying the equilibrium law to this equilibrium we obtain

$$K_w = ([H_3O^+][OH^-])_{eq}$$

K_w is called the **ionic product of water**. It has units of (concentration)^2, that is mol^2 dm^{-6}. The exact value depends on temperature (*see* table 8.3). At 25°C its

Table 8.3 Variation of ionic product of water with temperature

Temperature/°C	K_w/mol^2 dm^{-6}
10	0.29×10^{-14}
20	0.68×10^{-14}
25	1.00×10^{-14}
30	1.47×10^{-14}
40	2.92×10^{-14}

value is 1.0×10^{-14}. This gives a value of $pK_w = 14$, where

$$pK_w = -\lg K_w$$

There is a simple relationship between the ionic product of water and the dissociation constants of an acid and its conjugate base. We can derive this relationship by considering the dissociation of ethanoic acid in water once again.

For ethanoic acid we have

$$CH_3COOH(aq) + H_2O(l) \rightleftharpoons CH_3COO^-(aq) + H_3O^+(aq)$$

$$K_a = \left(\frac{[H_3O^+][CH_3COO^-]}{[CH_3COOH]} \right)_{eq}$$

For ethanoate, which is the conjugate base of ethanoic acid, we have

$$CH_3COO^-(aq) + H_2O(l) \rightleftharpoons CH_3COOH(aq) + OH^-(aq)$$

$$K_b = \left(\frac{[CH_3COOH][OH^-]}{[CH_3COO^-]} \right)_{eq}$$

By multiplying these two expressions together we obtain

$$K_a \times K_b = ([H_3O^+][OH^-])_{eq}$$
$$= K_w$$

Taking logarithms

$$pK_a + pK_b = pK_w$$
$$= 14 \qquad \text{(at 25°C)}$$

Thus, if the pK_a value of an acid is known, the pK_b value of its conjugate base can be found.

Note that if an acid is strong, that is if it has a low pK_a, the pK_b of its conjugate base must be large. The conjugate base of a strong acid is thus weak—a point we noted above.

THE pH SCALE

The concentration of oxonium ions in solution can be expressed in terms of the pH scale. The pH of a solution is the logarithm to base 10 of the reciprocal of the numerical value of the oxonium ion concentration:

$$pH = \lg \left(\frac{1}{[H_3O^+]} \right)$$
$$= -\lg [H_3O^+]$$

The pH of a neutral solution can be calculated directly from the ionic product of water:

$$K_w = [H_3O^+][OH^-] = 10^{-14} \text{ mol}^2 \text{ dm}^{-6}$$

Since

$$[H_3O^+] = [OH^-]$$
$$[H_3O^+] = 10^{-7} \text{ mol dm}^{-3}$$

and

$$pH = 7 \qquad \text{for neutral solutions}$$

Acidic soluions always have a pH less than 7 (*see* table 8.4).

EXAMPLE
Calculate the pH of a solution of ethanoic acid with a concentration of 0.01 mol dm^{-3} given that the degree of dissociation, α, is 0.057.

Origin of pH

The term pH was originally introduced by the Danish biochemist Soren Peter Lauritz Sorensen (1868–1939) whilst working on methods to improve the quality control of beer. Sorensen is also famous for his research on the physical properties of proteins and for devising analytical methods for their determination. Sorensen and his wife were the first to crystallise the protein egg albumin.

Table 8.4 The pH scale

pH	H_3O^+	OH^-	
0	10^0	10^{-14}	
1	10^{-1}	10^{-13}	
2	10^{-2}	10^{-12}	
3	10^{-3}	10^{-11}	increasing
4	10^{-4}	10^{-10}	acidity
5	10^{-5}	10^{-9}	
6	10^{-6}	10^{-8}	
7	10^{-7}	10^{-7}	neutrality
8	10^{-8}	10^{-6}	
9	10^{-9}	10^{-5}	
10	10^{-10}	10^{-4}	increasing
11	10^{-11}	10^{-3}	alkalinity
12	10^{-12}	10^{-2}	
13	10^{-13}	10^{-1}	
14	10^{-14}	10^0	

SOLUTION

At equilibrium

$$CH_3COOH(aq) + H_2O(l) \rightleftharpoons H_3O^+(aq) + CH_3COO^-(aq)$$

concentration: $c(1 - \alpha)$ $c\alpha$ $c\alpha$

Thus

$$
\begin{aligned}
pH &= -lg\,(c\alpha) \\
&= -lg\,(0.01 \times 0.057) \\
&= -lg\,(5.7 \times 10^{-4}) \\
&= 3.24
\end{aligned}
$$

pH Values of Bases

The concentration of hydroxide ions in a solution can be expressed in terms of pOH. This is given by

$$pOH = -lg\,[OH^-] \tag{2}$$

It is possible to derive a simple expression relating pH and pOH as follows. Since

$$K_w = [H_3O^+][OH^-]$$
$$lg\,K_w = lg\,[H_3O^+] + lg\,[OH^-]$$
$$-pK_w = -pH - pOH$$

Thus

$$pH + pOH = pK_w$$

and

$$pH + pOH = 14 \tag{3}$$

This expression enables the pH values of bases to be calculated directly. For this reason, pOH values are not usually quoted.

EXAMPLE

Calculate the pH value of a 0.001 mol dm^{-3} solution of NaOH.

SOLUTION

$$[OH^-] = 10^{-3} \text{ mol dm}^{-3}$$

Thus, from equation (2)

$$pOH = 3$$

Substituting into equation (3)

$$pH + 3 = 14$$

Thus

$$pH = 11$$

The pH scale is shown in table 8.4.

Measurement of pH

The pH of a solution is usually measured potentiometrically. An electrode, such as a **glass electrode**, which is sensitive to H_3O^+ concentration, is put into the solution together with a **reference electrode** and the electromotive force (e.m.f.) between the two electrodes is measured. The e.m.f. can be related directly to the pH of the solution and monitored on a pH meter. For accurate measurements a **hydrogen electrode** is used instead of a glass electrode. We shall look at the measurement of pH in detail in chapter 10.

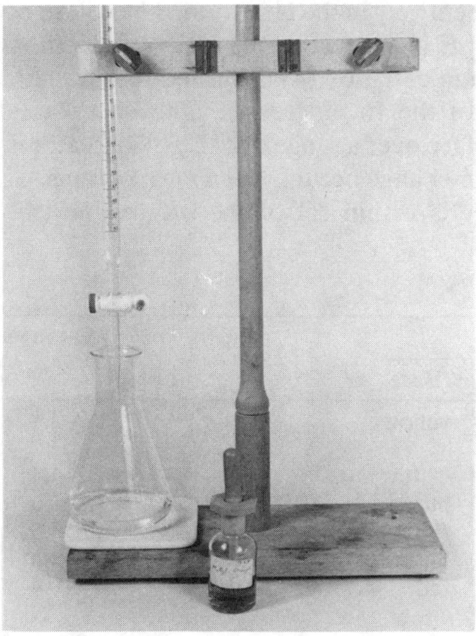

ACID–BASE INDICATORS AND TITRATIONS

Acid–base indicators (also known as pH indicators) are substances which change colour with pH. They are usually weak acids or weak bases. When dissolved in water they dissociate slightly forming ions. Let us consider an indicator which is a weak acid with the general formula HIn. At equilibrium, the following equilibrium is established with its conjugate base:

$$HIn(aq) + H_2O(l) \rightleftharpoons H_3O^+(aq) + In^-(aq) \qquad (4)$$

acid   conjugate base
(colour A)   (colour B)

The acid and its conjugate base have different colours. At low pH values the concentration of H_3O^+ is high and thus the equilibrium position lies to the left. The equilibrium solution thus has colour A. At high pH values the concentration of H_3O^+ is low. The equilibrium position thus lies to the right and the equilibrium solution has colour B.

Phenolphthalein is an example of an indicator which establishes this type of equilibrium in aqueous solution (*see* figure 8.1). Phenolphthalein is a colourless weak acid which dissociates in water forming pink anions. Under acidic conditions the equilibrium is pushed to the left. The concentration of the anions is too low for the pink colour to be observable. However, under alkaline conditions the equilibrium is pulled to the right and the concentration of the anion becomes sufficient for the pink colour to be observed.

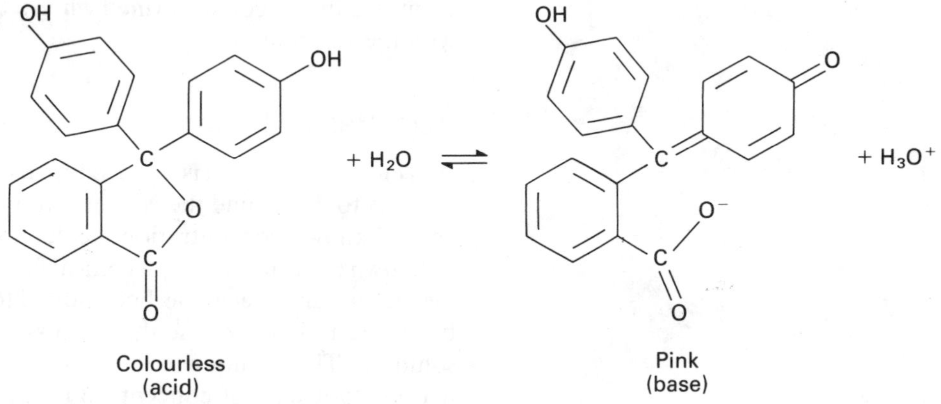

An acid–base indicator changes colour with pH. For example, methyl orange changes from red to yellow over the pH range 3.2 to 4.4

Colourless
(acid)

Pink
(base)

Figure 8.1   Phenolphthalein

We can apply the equilibrium law to indicator equilibria. In general, for a weak acid indicator

$$K_{In} = \left(\frac{[H_3O^+][In^-]}{[HIn]} \right)_{eq} \qquad (5)$$

K_{In} is known as the **indicator dissociation constant**.

The colour of the indicator changes from colour A to colour B or vice versa at its turning point. At this point

$$[HIn] = [In^-]$$

Thus, from equation (5)

$$K_{In} = [H_3O^+]_{eq}$$

The pH of the solution at the turning point is called the pK_{In}. Thus pK_{In} is the pH at which half the indicator is in the acid form and half in the form of its conjugate base.

Indicator Range

At a low pH a weak acid indicator is almost entirely in the HIn form. The colour of this form therefore predominates. As the pH increases, the intensity of colour A of the HIn diminishes and the equilibrium in equation (4) is pushed to the right. Consequently the intensity of colour B of the In⁻ increases. The observable colour change from A to B actually takes place over a range of pH. An indicator is most effective if a distinct observable colour change occurs over a narrow range of pH values. For most indicators the range is within ±1 of the pK_{In} value (*see* table 8.5).

Table 8.5 Indicators

Indicator	Colour		pK_{In}	pH range
	Acid	Base		
thymol blue (first change)	red	yellow	1.5	1.2–2.8
methyl orange	red	yellow	3.7	3.2–4.4
bromocresol green	yellow	blue	4.7	3.8–5.4
methyl red	yellow	red	5.1	4.8–6.0
bromothymol blue	yellow	blue	7.0	6.0–7.6
phenol red	yellow	red	7.9	6.8–8.4
thymol blue (second change)	yellow	blue	8.9	8.0–9.6
phenolphthalein	colourless	pink	9.4	8.2–10.0

A **universal indicator** is a mixture of indicators which give a gradual change in colour over a wide range of pH. The pH of a solution can be approximately identified by the colour formed when a few drops of universal indicator are mixed with the solution.

Acid–Base Titrations

An acid–base titration is a procedure used predominantly in quantitative chemical analysis to determine the concentration of either an acid or a base. Typically, an acid of known concentration is added from a burette to an alkaline solution of unknown concentration in a conical flask. The **end-point** occurs when the stoichiometric amount of acid has been added to the base. At this point, all the alkali has been neutralised and neither excess acid nor excess alkali are present in the solution. The solution consists of salt and water only. For example, when hydrochloric acid of concentration 0.1 mol dm⁻³ is added to 25 cm³ of sodium hydroxide of concentration 0.1 mol dm⁻³ the end-point occurs when 25.0 cm³ of

Students at a school in Sri Lanka carry out an acid–base titration

the acid have been added. This follows from the stoichiometric equation

$$HCl(aq) + NaOH(aq) \rightarrow NaCl(aq) + H_2O(l)$$

Acid–base titrations are often carried out using a visual indicator to determine end-point. However the end-point can also be determined potentiometrically using a pH meter or by a conductimetric method (*see* chapter 10).

Let us now assume that a titration is carried out by adding a base to an acid. If the pH is plotted against volume of base added, four types of curve are obtained depending on whether the acid and base are strong or weak. The four typical titration curves are shown in figure 8.2. Note that at the end-point there is a sharp increase in pH. The exception is the titration of a weak acid with a weak base. If a visual indicator is to be used to determine the end-point of an acid–base titration, its pH range should fall on the vertical portion of the titration. This gives a sharp colour change.

Strong Acid–Strong Base Titration

EXAMPLE
$$H_2SO_4(aq) + 2NaOH(aq) \rightarrow Na_2SO_4(aq) + 2H_2O(l)$$

The vertical portion of curve lies between pH 4 and pH 10. Thus, at the end-point the addition of one drop of base to the acid causes an increase of 6 pH units. Suitable indicators thus have a pH range lying between 4 and 10. Examples are methyl red and phenolphthalein. Note that if methyl orange is used the colour change is not so sharp.

Strong Acid–Weak Base Titration

EXAMPLE
$$HCl(aq) + NH_4OH(aq) \rightarrow NH_4Cl(aq) + H_2O(l)$$

The vertical portion of the curve lies between pH 4 and pH 8. A suitable indicator is methyl red or bromothymol blue. Phenolphthalein is unsuitable since its pH range lies on a flat portion of the curve.

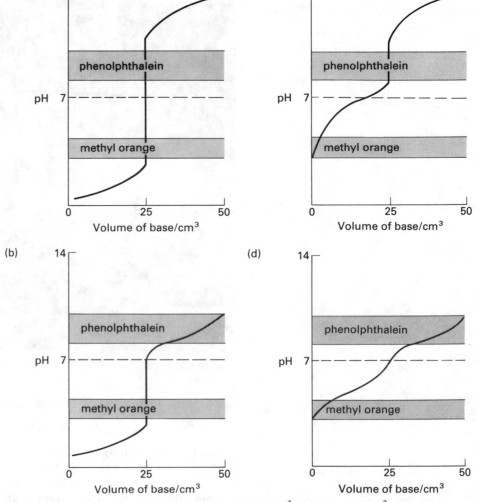

Figure 8.2 Titration curves for titrations of 25.00 cm³ of 0.10 mol dm⁻³ acid with 0.10 mol dm⁻³ base: (a) strong base with strong acid; (b) weak base with strong acid; (c) strong base with weak acid; (d) weak base with weak acid

Weak Acid–Strong Base Titration

EXAMPLE

$$CH_3COOH(aq) + NaOH(aq) \rightarrow CH_3COONa(aq) + H_2O(l)$$

The vertical portion of the curve lies between pH 6.5 and pH 11. A suitable indicator is thus phenol red or phenolphthalein. Indicators with a pH range of less than 6, such as methyl orange, are unsuitable since their pH ranges lie on the flat portion of the curve and thus do not give a sharp end-point.

Weak Acid–Weak Base Titration

EXAMPLE

$$CH_3COOH(aq) + NH_4OH(aq) \rightarrow CH_3COONH_4(aq) + H_2O(l)$$

For this type of titration there is no sharp increase of pH at the end-point. The pH changes gradually over a range of pH values. Thus, no indicator is suitable for this type of titration.

SUMMARY

1. A **strong electrolyte** is fully ionised when molten or in solution.
2. A **weak electrolyte** is only partially dissociated into ions when molten or in solution.
3. **Ostwald's dilution law** is

$$K_c = \frac{\alpha^2 c}{(1 - \alpha)}$$

4. According to the **Brønsted–Lowry theory**, an **acid** is a substance which donates protons, and a **base** is a substance which accepts protons.
5. A **strong acid** has a **weak conjugate base**.
6. A **weak acid** has a **strong conjugate base**.
7. An **amphoteric** substance can react as both an acid and a base.
8. A **Lewis acid** is a substance which can accept an electron pair from a base.
9. A **Lewis base** is a substance with an unshared electron pair.
10. $pK_a = -\lg K_a$, where K_a is the **acid dissociation constant**.
11. $pK_b = -\lg K_b$, where K_b is the **base dissociation constant**.
12. $pK_w = -\lg K_w$, where K_w is the **ionic product of water**.
13. $pK_a + pK_b = pK_w = 14$.
14. $pH = -\lg [H_3O^+]$.
15. $pH + pOH = 14$.
16. The equilibrium of a weak acid **indicator** is given by

$$HIn(aq) + H_2O(l) \rightleftharpoons H_3O^+(aq) + In^-(aq)$$
acid   conjugate base
(colour A)   (colour B)

17. The **end-point** of an **acid–base titration** occurs when the stoichiometric amount of acid has been added to the base.
18. The **pH range** of a visual indicator must fall on the vertical portion of a **titration curve**.

8.2
Salts, Solubility and Stability

A **salt** may be defined as a compound, other than water, formed by the reaction of an acid and a base. In this section we consider those properties of salts which relate to ionic equilibria.

REACTIONS OF SALTS IN WATER

We shall see below that solubility is only a relative term. However, for the purposes of this discussion we may classify salts as either soluble or insoluble in water.

Some salts when dissolved in water form neutral solutions. Others form acidic or alkaline solutions. This is due to a reversible reaction between the salt ions and water during which conjugate acids or bases are formed. Whether the solution is neutral, acidic or alkaline depends on the type of salt. There are four types in this context.

Salt of a Strong Acid and Weak Base

This type of salt forms an acidic solution when added to water. An example is ammonium chloride, NH_4Cl. When added to water, the ammonium ion acts as an acid by donating a proton to water:

$$NH_4^+(aq) + H_2O(l) \rightleftharpoons NH_3(aq) + H_3O^+(aq)$$

The excess H_3O^+ ions generated in the process cause the solution to be acidic.

Salt of a Weak Acid and Strong Base

This type of salt forms an alkaline solution when added to water. An example is sodium ethanoate, CH_3COONa. The ethanoate ion acts as a base by accepting a proton from water acting as an acid:

$$CH_3COO^-(aq) + H_2O(l) \rightleftharpoons CH_3COOH(aq) + OH^-(aq)$$

The excess OH^- ions generated in the process cause the solution to be alkaline.

Salt of a Strong Acid and Strong Base

This type of salt dissolves in water to form a neutral solution. An example is sodium chloride, $NaCl$. When dissolved in water the salt is fully ionised and thus the concentration of Na^+ ions equals the concentration of Cl^-. Since neither of these ions undergo acid–base reactions with water, excess H_3O^+ or OH^- ions are not generated. The solution is thus neutral.

Salt of a Weak Acid and Weak Base

Ammonium ethanoate is an example of this type of salt. When dissolved in water the ammonium ion undergoes an acid reaction with water and the ethanoate ion undergoes a base reaction. Both these reactions are described above. An aqueous solution of a salt of a weak acid and a weak base may be slightly acidic, slightly alkaline or neutral depending on the relative concentrations of H_3O^+ and OH^- ions formed by the reactions of the cations and anions. This depends on the relative values of the dissociation constants of the cation and anion.

BUFFER SOLUTIONS

A buffer solution is a solution the pH of which does not change significantly when a small amount of acid or base is added to it.

We can divide buffer solutions into four categories.

Strong Acid Buffers

A strong acid such as nitric acid can act as a buffer with a low pH. Strong acids are fully dissociated in aqueous solution and thus the concentration of oxonium ions is high. The addition of a small amount of acid or base to the acid will thus have a negligible effect on the pH of the acid.

For example, if 1 cm^3 of hydrochloric acid with a concentration of 0.1 mol dm^{-3} is added to 100 cm^3 of 0.01 mol dm^{-3} nitric acid, the pH decreases from 2.00 to 1.96. This is a negligible decrease of 0.04. It can be checked by using the following equation to calculate the pH of the solution before and after the hydrochloric acid is added:

$$pH = -\lg \left(\frac{1}{[H_3O^+]} \right) \tag{6}$$

This negligible decrease should be compared to the addition of 1 cm^3 of 0.1 mol dm^{-3} HCl to 100 cm^3 of pure water. In this case the pH drops from 7.00 to 4.00. Clearly, pure water does not act as a buffer solution as it does not resist changes in pH. The concentrations of buffer solutions corresponds to the flat portions of the titration curves in figure 8.2. These portions of the curves are called **buffer regions**. In a buffer region pH is insensitive to small changes in the concentration of acid or base.

Strong Base Buffers

A strong base can be used as a buffer solution with a high pH. The addition of a small amount of acid or base has a negligible effect on the pH. For example, when 1 cm^3 of 0.1 mol dm^{-3} HCl is added to 100 cm^3 of 0.01 mol dm^{-3} NaOH solution, the pH drops from 12.00 to 11.96. Once again this is a drop of 0.04. This can be checked using equation (6) and also

$$pH + pOH = 14$$

Weak Acid Buffers

Buffer solutions with constant pH values of between 4 and 7 can be prepared from a weak acid and one of its salts. Ethanoic acid and sodium ethanoate are often used for this purpose. Sodium ethanoate in water is fully ionised:

$$CH_3COONa(s) + aq \rightarrow CH_3COO^-(aq) + Na^+(aq)$$

On the other hand, ethanoic acid is only partially ionised:

$$CH_3COOH(aq) + H_2O(l) \rightleftharpoons CH_3COO^-(aq) + H_3O^+(aq) \qquad (7)$$

If acid is added this equilibrium shifts to the left. The additional H$_3$O$^+$ ions are thus removed and the pH remains constant. The presence of sodium ethanoate in the buffer solution ensures that there is a large reservoir of CH$_3$COO$^-$ ions to cope with additions of acid.

If base is added it is neutralised by the oxonium ions:

$$H_3O^+(aq) + OH^-(aq) \rightarrow 2H_2O(l)$$

Removal of H$_3$O$^+$ ions by this reaction results in equilibrium (7) shifting to the right. The concentration of H$_3$O$^+$ ions and thus the pH of the solution remains constant. The presence of ethanoic acid in the buffer solution ensures that there is a large reservoir of undissociated CH$_3$COOH molecules ready to dissociate in order to cope with additions of base.

We can put the action of buffer solutions onto a quantitative basis by application of the equilibrium law. As we saw in the previous section, when this law is applied to the equilibrium for ethanoic acid we obtain the acid dissociation constant

$$K_a = \left(\frac{[H_3O^+][CH_3COO^-]}{[CH_3COOH]} \right)_{eq}$$

Taking logarithms to base 10 and rearranging we obtain

$$pH = pK_a + \lg \left(\frac{[CH_3COO^-]}{[CH_3COOH]} \right) \qquad (8)$$

where [CH$_3$COO$^-$] and [CH$_3$COOH] are the total concentrations of each species in the buffer solution. The acid dissociation constant of ethanoic acid is 1.74×10^{-5} (*see* table 8.1). This means that the dissociation equilibrium for ethanoic acid (*see* equation (7)) lies predominantly to the left. For this reason the ethanoic acid in the buffer solution contributes few of the CH$_3$COO$^-$ ions. The value of [CH$_3$COO$^-$] in equation (8) is almost entirely due to the contribution of the salt—sodium ethanoate. This is fully dissociated into Na$^+$ and CH$_3$COO$^-$ ions. Thus

$$[CH_3COO^-] = [CH_3COONa] = [salt]$$

Since the ethanoic acid is largely undissociated in the buffer solution, the concentration of the acid [CH$_3$COOH] in the equilibrium mixture (7) is approximately the same as the initial concentration of acid in the buffer solution. Thus

$$[CH_3COOH] = [acid]$$

We can now put these generalisations into equation (8), to get

$$pH = pK_a + lg \left(\frac{[salt]}{[acid]} \right) \qquad (9)$$

This is called the **Henderson equation** for a buffer solution consisting of a weak acid and its salt. It can be used to calculate

- the pH of a buffer solution;
- how much acid or salt is needed to make a buffer solution of required pH;
- the effect on the pH of a buffer solution when a small amount of acid or base is added.

EXAMPLE

(a) How much sodium ethanoate must be dissolved in 1 dm^3 of ethanoic acid with a concentration of 0.01 mol dm^{-3} to produce a buffer solution of pH 5.
(b) How would the pH of this buffer solution change if 1 cm^3 of NaOH solution with a concentration of 1.0 mol dm^{-3} were added to 1 dm^3 of the buffer.

SOLUTION

(a) Rearranging equation (9) we obtain

$$lg [salt] = pH - pK_a + lg [acid]$$

We are given

$$pH = 5.00$$

and

$$lg [acid] = lg (0.01) = -2$$

From table 8.1

$$pK_a = 4.75$$

Substituting these values into the rearranged equation, we obtain

$$lg [salt] = 5.00 - 4.75 - 2$$
$$= -1.75$$
$$= +0.25 - 2$$

Thus

$$[salt] = 1.78 \times 10^{-2} \text{ mol dm}^{-3}$$

This means that 1.78×10^{-2} moles of sodium ethanoate must be dissolved in 1 dm^3 of ethanoic acid to give a pH of 5.0.

The mass of one mole of sodium ethanoate is given by

$$M_r(CH_3COONa) = 82 \text{ g mol}^{-1}$$

The mass of 1.78×10^{-2} moles is thus

$$m = (1.78 \times 10^{-2} \text{ mol}) \times (82 \text{ g mol}^{-1})$$
$$= 1.46 \text{ g}$$

It is thus necessary to dissolve 1.46 g of sodium ethanoate in 1 dm^3 of the acid to make a buffer solution with pH 5.

(b) 1 cm^3 of 1.0 mol dm^{-3} NaOH solution contains 0.001 mol of NaOH. This reacts with CH_3COOH to form CH_3COONa. The concentration of CH_3COOH thus decreases by 0.001 mol dm^{-3} and that of CH_3COONa increases by 0.001 mol dm^{-3} (the small increase in volume can be ignored). Thus,

$$[salt] = (0.0178 + 0.001) \text{ mol dm}^{-3}$$
$$= 0.0188 \text{ mol dm}^{-3}$$
$$[acid] = (0.01 - 0.001) \text{ mol dm}^{-3}$$
$$= 0.009 \text{ mol dm}^{-3}$$

Thus

$$pH = 4.75 + \lg \left(\frac{0.0188}{0.009} \right)$$

$$= 4.75 + \lg (2.089)$$

$$= 4.75 + 0.32$$

$$= 5.07$$

The pH thus increases by a negligible 0.07 when 1 cm^3 of the alkali is added to 1 dm^3 of the buffer.

There is one *special case* we need to examine when considering weak acid buffers. The Henderson equation shows us that when the concentration of the salt equals the concentration of the acid, the pH of the buffer solution equals the pK_a of the acid. Thus when

$$[salt] = [acid]$$

$$pH = pK_a$$

For example, if we add 100 cm^3 of 0.1 mol dm^{-3} CH$_3$COOH (pK_a = 4.75) to 100 cm^3 of 0.1 mol dm^{-3} CH$_3$COONa, the pH of the buffer solution will be 4.75 at 25°C.

Weak Base Buffers

Buffer solutions with constant pH values of between 7 and 10 can be prepared from a weak base and one of its salts. A solution of ammonia and ammonium chloride is typically used for this purpose. In aqueous solution the ammonium chloride is fully dissociated:

$$NH_4Cl(aq) \rightarrow NH_4^+(aq) + Cl^-(aq)$$

The ammonia is only partially dissociated:

$$NH_3(aq) + H_2O(l) \rightleftharpoons NH_4^+(aq) + OH^-(aq) \qquad (10)$$

If acid is added it is neutralised by the OH$^-$ ions. As a consequence, equilibrium (10) shifts to the right. This shift maintains a constant concentration of OH$^-$ and thus a constant pH.

If base is added this equilibrium shifts to the left thus maintaining the concentration of OH$^-$ ions constant. The presence of ammonium chloride in the buffer solution ensures that there is a large reservoir of NH$_4^+$ ions to cope with additions of base.

The Henderson equation for a buffer solution made up of a weak base and one of its salts is

$$pH = pK_w - pK_b + \lg \left(\frac{[base]}{[salt]} \right)$$

Table 8.6 pH of biological and other systems

	pH
stomach juices	1.6–1.8
orange juice	2.6–4.4
vinegar	3.0
tomatoes	4.3
urine	4.8–7.5
saliva	6.35–6.85
milk	6.6–6.9
human blood	7.35–7.45
tears	7.4
egg white	8.0
sea water	8.0

Uses of Buffer Solutions

Buffer solutions play an important part in many industrial processes. They are used, for example, in electroplating and in the manufacture of dyes, photographic materials and leather. Buffer solutions are also used widely in chemical analysis and for the calibration of pH meters (*see* chapter 10).

Many biological and other systems depend on buffer action to preserve a constant pH. The normal pH of some of these systems is shown in table 8.6. The pH of human blood, for example, is maintained between 7.35 and 7.45 even though the concentration of carbon dioxide and thus carbonic acid in the blood varies greatly. The buffer in blood consists of a mixture of phosphate, hydrogen-carbonate and proteins. The pH of tears is also maintained at 7.4 by use of buffers

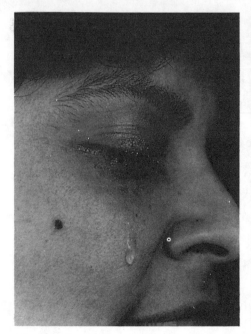

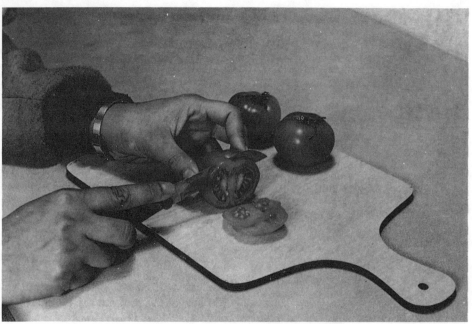

The pH of tears is maintained at 7.4 by protein buffers

Tomatoes depend on buffer action to preserve a constant pH of 4.3

consisting of proteins. Another example is the use of buffers in bacteriological research to maintain the pH of culture media used for the growth of bacteria.

SOLUBILITY

In chapter 2 we saw that ionic compounds tend to dissolve in polar solvents such as water. The solubility is due to the solvation—or in the case of water, hydration—of the ions by the polar solvent molecules. The enthalpy changes that accompany dissolving and hydration were discussed in chapter 5. Then in chapter 6 we saw that the solubility curve of a salt in water was part of a phase diagram for the two-component mixture.

We now turn our attention to the ionic equilibria that are established when ionic compounds dissolve in water.

Solubility Product

A solution is saturated when no more of the solute will dissolve in it. If the solute is an ionic compound, then the ions in the saturated solution are in dynamic equilibrium with the excess solid. In the case of silver chloride, for example, the following equilibrium is established:

$$AgCl(s) \rightleftharpoons Ag^+(aq) + Cl^-(aq)$$

By applying the equilibrium law to this we obtain

$$K_{sp} = ([Ag^+][Cl^-])_{eq}$$

K_{sp} is called the **solubility product** of the silver chloride.

The solubility of silver chloride in water can be expressed in terms of the concentrations c of *dissolved* AgCl in water. This concentration equals the concentrations of both the ions in solution:

$$[Ag^+] = [Cl^-] = c$$

Thus, the solubility of silver chloride can be related to its solubility product by

$$K_{sp} = c^2$$

or

$$c = \sqrt{K_{sp}}$$

Solubility of Metal Salts

Solubility is usually expressed as the mass of solute required to saturate 100 g of solvent at a given temperature. It is also sometimes expressed as the amount of solute in moles required to saturate one kilogram of water. Neither of these terms for solubility are concentration terms since they do not involve volume.

It should be noted here that **soluble** and **insoluble** are relative terms only. For example, if 50 g of magnesium carbonate, $MgCO_3$, are added to 100 g of water at room temperature, only 0.06 g of the salt dissolves. Magnesium carbonate is thus normally regarded as an insoluble salt. On the other hand, when 50 g of Epsom salts, $MgSO_4 \cdot 7H_2O$, are added to 100 g of water at room temperature, 13.6 g remain undissolved. Since 36.4 g do dissolve, however, magnesium sulphate is regarded as soluble. No metallic salts are completely insoluble in water. For example, lead(II) carbonate, $PbCO_3$, is normally regarded as highly insoluble in water. Even so, 0.000 017 g do dissolve in 100 g of water at room temperature.

The equilibrium established in a saturated solution of a salt with the general formula A_xB_y is

$$A_xB_y(s) \rightleftharpoons xA^{y+}(aq) + yB^{x-}(aq)$$

The solubility product is given by

$$K_{sp} = ([A^{y+}]^x [B^{x-}]^y)_{eq}$$

The units of K_{sp} are (concentration)$^{x+y}$. For example, the following equilibrium is established for a saturated solution of $Ca_3(PO_4)_2$ in water:

$$Ca_3(PO_4)_2(s) \rightleftharpoons 3Ca^{2+}(aq) + 2PO_4^{3-}(aq)$$

Thus

$$K_{sp} = ([Ca^{2+}]^3[PO_4^{3-}]^2)_{eq}$$

The units are thus (concentration)^5 or $mol^5 \ dm^{-15}$.

Values for K_{sp} are normally quoted only for electrolytes which are **sparingly soluble** in water. An electrolyte may be considered sparingly soluble if its solubility is less than 10^{-2} mol kg^{-1} water. The solubility products at 25°C of a number of sparingly soluble electrolytes are shown in table 8.7. We should note that K_{sp} values are only constant for a given solute at a given temperature.

Solubility products may be determined experimentally by a number of methods including titration, ion exchange (*see* below) and electrochemical methods (*see* chapter 10). They may also be determined by evaporating the saturated solution to dryness and weighing the residue.

Table 8.7 Solubility products in water at 25°C

Solute	$K_{sp}/(mol \ dm^{-3})^{x+y}$
AgCl	1.8×10^{-10}
Al(OH)$_3$	1.0×10^{-33}
BaSO$_4$	1.3×10^{-10}
CaCO$_3$	4.8×10^{-9}
Ca$_3$(PO$_4$)$_2$	1.0×10^{-29}
CuS	6.3×10^{-36}
Fe(OH)$_2$	7.9×10^{-16}
Fe$_2$S$_3$	1.0×10^{-88}
MgCO$_3$	1.0×10^{-5}
Mg(OH)$_2$	1.1×10^{-11}
SnCO$_3$	1.0×10^{-9}
Zn(OH)$_2$	2.0×10^{-17}
ZnS	1.6×10^{-23}

EXAMPLE

A saturated aqueous solution of calcium fluoride, CaF_2, was found to contain 0.0168 g dm^{-3} of the solute at 25°C. Calculate the K_{sp} for CaF_2.

SOLUTION

The equilibrium for this saturated solution is

$$CaF_2(s) \rightleftharpoons Ca^{2+}(aq) + 2F^-(aq)$$

Thus

$$K_{sp} = ([Ca^{2+}][F^-]^2)_{eq}$$

We first calculate the concentration c of the dissolved CaF_2. Since

$$M_r(CaF_2) = 78 \text{ g mol}^{-1}$$

$$c = \frac{0.0168 \text{ g dm}^{-3}}{78 \text{ g mol}^{-1}}$$

$$= 2.15 \times 10^{-4} \text{ mol dm}^{-3}$$

The equilibrium equation shows us that

$$[Ca^{2+}] = c = 2.15 \times 10^{-4} \text{ mol dm}^{-3}$$

$$[F^-] = 2c = 4.30 \times 10^{-4} \text{ mol dm}^{-3}$$

Substituting these values into the expression for K_{sp} we obtain

$$K_{sp} = (2.15 \times 10^{-4} \text{ mol dm}^{-3}) \times (4.30 \times 10^{-4} \text{ mol dm}^{-3})^2$$

$$= 4.0 \times 10^{-11} \text{ mol}^3 \text{ dm}^{-9}$$

Predicting Precipitation

The solubility product can be used to predict whether precipitation will occur or not. Suppose, for example, that we mix a dilute solution of barium chloride with a very dilute solution of sodium sulphate. How do we know whether precipitation of barium sulphate will occur or not? First of all we need to consider the equilibrium

$$BaSO_4(s) \rightleftharpoons Ba^{2+}(aq) + SO_4^{2-}(aq) \qquad (11)$$

The solubility product is given by

$$K_{sp} = ([Ba^{2+}][SO_4^{2-}])_{eq}$$

At 25°C K_{sp} has a constant value of 1.3×10^{-10}. If the product of the concentrations of the Ba^{2+} and SO_4^{2-} ions in the two solutions which are mixed together exceed the solubility product, then precipitation occurs until the product of the concentrations has been reduced to the solubility product. On the other hand, if the product is less than the solubiliity product, precipitation does not occur. In this case the equilibrium lies completely to the right. We can express this mathematically as follows:

If $[Ba^{2+}][SO_4^{2-}] > K_{sp}$ precipitation occurs

until $[Ba^{2+}][SO_4^{2-}] = K_{sp}$

But if $[Ba^{2+}][SO_4^{2-}] < K_{sp}$ precipitation does not occur

The term $[Ba^{2+}][SO_4^{2-}]$ is called the **ionic product**.

The Common Ion Effect

Let us consider the equilibrium established when silver chloride dissolves in water to form a saturated solution:

$$AgCl(s) \rightleftharpoons Ag^+(aq) + Cl^-(aq)$$

If we now add more chloride ions—by adding sodium chloride solution for example—the equilibrium shifts to the left. This effect is known as the **common ion effect**. In this case, chloride is the common ion. It is common to both the silver chloride and the sodium chloride.

Since the addition of a common ion results in the equilibrium shifting to the left, silver chloride is precipitated. The addition of a common ion has thus reduced its solubility. The same effect is produced if silver nitrate solution is added. In this case the common ion is silver, Ag^+.

It follows from the common ion effect that the solubility of an electrolyte in an aqueous solution containing a common ion is less than its solubility in water. For example, at 25°C the solubility of $BaSO_4$ in water is about 1×10^{-5} mol dm^{-3}. However, in 0.1 mol dm^{-3} H_2SO_4 the solubility is only 1×10^{-9} mol dm^{-3}. The latter value can be checked by using equation (11). Indeed the solubility product can be used to calculate how much solute is precipitated when a common ion is added.

The common ion effect also reduces the degree of dissociation of a weak electrolyte. Let us once again consider the equilibrium established in an aqueous solution of ethanoic acid:

$$CH_3COOH(aq) + H_2O(l) \rightleftharpoons CH_3COO^-(aq) + H_3O^+(aq)$$

If sodium ethanoate is added to this solution the equilibrium position moves to the left. In other words, the dissociation of ethanoic acid is decreased by the addition of the common ion, CH_3COO^-. The same effect is achieved if a small amount of strong acid is added. In this case the common ion is H_3O^+.

Applications of the Common Ion Effect

Gravimetric Analysis

The concentration of sulphate in aqueous solution may be determined gravimetrically by adding barium chloride solution to precipitate barium sulphate. The precipitate is filtered and then washed with dilute sulphuric acid rather than water. The presence of SO_4^{2-} ions in the wash ensures that the maximum amount of barium sulphate remains precipitated.

Qualitative Analysis

In classical schemes of qualitative inorganic analysis the identification of metal ions present in aqueous solution involves the application of the common ion effect. Certain metal ions are precipitated as sulphides when hydrogen sulphide is bubbled through the solution. For example, in order that a metal ion M^{2+} be precipitated as a sulphide, the following condition must be fulfilled:

$$[M^{2+}][S^{2-}] > K_{sp}$$

Thus, if $[S^{2-}]$ is low, only those metals whose sulphides have low K_{sp} values will be precipitated as sulphides. The concentration of S^{2-} ions in solution depends on the pH of the solution. This is because hydrogen sulphide is a weak acid in aqueous solution:

$$H_2S(aq) + 2H_2O(l) \rightleftharpoons 2H_3O^+(aq) + S^{2-}(aq)$$

When acid is added to the solution, this equilibrium shifts to the left. As a result the concentration of sulphide ions is lowered. This means that only those metals with the lowest K_{sp} values for their sulphides will be precipitated under these conditions. Metals which fall into this group are copper(II), mercury(II) and lead(II).

When the solution is made alkaline, the equilibrium shifts to the right. As a result the concentration of sulphide ions increases. Metals with higher K_{sp} values for their sulphides are thus precipitated as sulphides. Metals which fall into this group are zinc, nickel(II) and cobalt(II).

Salting Out Effect

Salting out is a term often used in industry for removing a salt from solution. An example occurs in the manufacture of soap. The chief constituent of soap is sodium stearate, $C_{17}H_{35}CO_2Na$—otherwise known as sodium octadecanoate. It is salted out by adding a concentrated solution of sodium chloride. This causes the product of the stearate and sodium ion concentrations to exceed the solubility product of sodium stearate. In this case the salting out is due to the common ion effect. The common ion is sodium, Na^+.

Sodium chloride—a strong electrolyte—can be salted out of solution by adding concentrated hydrochloric acid or by bubbling hydrogen chloride gas through the solution. The salting out is due to both the common ion effect and also the high degree of hydration of the ions formed from the acid. The latter reduces the amount of water available as solvent for the sodium chloride.

EQUILIBRIA INVOLVING COMPLEX IONS

We have already dealt with the naming and formulae of complex ions in chapter 4. In chapter 14 we shall look at the chemistry of complex ions involving d-block metals in some detail. In this section we shall confine our attention to equilibria involving complex ions.

The chemistry of complex ions in aqueous solution is the chemistry of replacing one **ligand** in the **coordination sphere** of a metal with another ligand. The formation of the hexaamminecobalt(II) ion, for example, involves the replacement of six water molecules in the coordination sphere around the cobalt(II) ion with six ammonia molecules. This can be written

$$[Co(H_2O)_6]^{2+}(aq) + 6NH_3(aq) \rightleftharpoons [Co(NH_3)_6]^{2+}(aq) + 6H_2O(l)$$

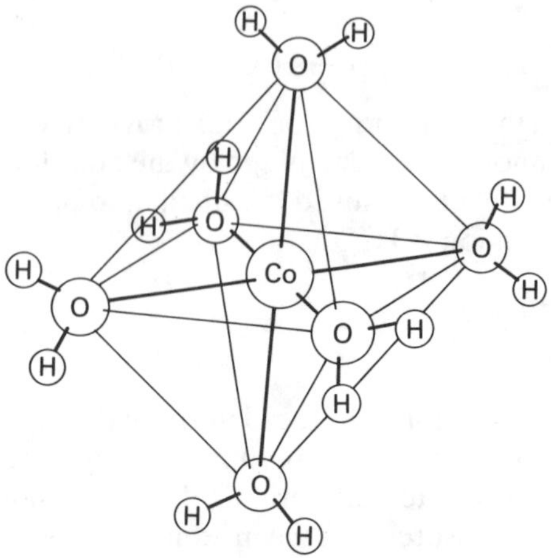

In the complex ion $[Co(H_2O)_6]^{2+}$ the water molecules are the ligands. Since there are six, the **coordination number** of the ion is 6. This ion is called the hydrated cobalt(II) ion or hexaaquacobalt(II) ion. Its structure is shown in figure 8.3.

For the sake of simplicity the above equilibrium is usually written

$$Co^{2+}(aq) + 6NH_3(aq) \rightleftharpoons [Co(NH_3)_6]^{2+}(aq) \tag{12}$$

When the symbol Co^{2+} is used to represent the cobalt(II) ion in aqueous solution, it is normally understood that it does not exist as the bare ion. It is hydrated.

The equilibrium constant for the formation of the $[Co(NH_3)_6]^{2+}$ ion is given by

$$K_{stab} = \left(\frac{[[Co(NH_3)_6]^{2+}]}{[Co^{2+}][NH_3]^6} \right)_{eq}$$

This constant is known as the **stability constant**. It is also known as the formation constant of the complex ion. The stability constants of a number of complex ions are shown in table 8.8. Note that they are generally large and thus the ions are generally very stable. However, some complex ions which have halide ions as ligands do have relatively low stability constants.

The term **instability constant** is also used. This is merely the reciprocal of the stability constant. For example

$$K_{inst} = \left(\frac{[Co^{2+}][NH_3]^6}{[[Co(NH_3)_6]^{2+}]} \right)_{eq}$$

Thus, whereas the stability constant is a formation constant of the complex ion, the instability constant is the dissociation constant of the ion

$$Co^{2+}(aq) + 6NH_3(aq) \underset{dissociation}{\overset{formation}{\rightleftharpoons}} [Co(NH_3)_6]^{2+}(aq)$$

The substitution of one ligand by another occurs stepwise. The number of steps equals the number of ligands. Each step has its own **stepwise stability constant**. Thus, $[Co(NH_3)_6]^{2+}$ has six stepwise stability constants. The stability constants quoted in table 8.8 are the overall stability constants. The overall stability constant of an ion is the product of its stepwise stability constants. Let us take a simple case to illustrate this. The formation of the $[Ag(NH_3)_2]^+$ ion takes place in two steps and thus has two stepwise stability constants, $K_{stab,1}$ and $K_{stab,2}$.

Figure 8.3 The hydrated Co^{2+} ion, $[Co(H_2O)_6]^{2+}$

Table 8.8 Stability constants

Complex ion	$K_{stab}/(mol\ dm^{-3})^{-n}$
$[Ag(NH_3)_2]^+$	1.6×10^7
$[Co(NH_3)_6]^{2+}$	1×10^5
$[Cu(NH_3)_4]^{2+}$	1×10^{12}
$[Ni(NH_3)_6]^{2+}$	6×10^8
$[AlF_6]^{3-}$	7×10^{-19}
$[SnF_6]^{2-}$	10^{25}
$[CuF]^+$	10
$[ZnF]^+$	5.0
$[AgCl_2]^-$	1×10^2
$[HgCl_4]^{2-}$	1.6×10^{16}
$[FeCl]^{2+}$	3.0
$[Al(OH)]^{2+}$	2×10^{28}
$[Fe(OH)]^{2+}$	1×10^{11}
$[Fe(CN)_6]^{3-}$	10^{31}
$[Ag(CN)_2]^-$	10^{21}

n = the coordination number of the ion

Step 1

$$Ag^+(aq) + NH_3(aq) \rightleftharpoons [Ag(NH_3)]^+(aq) \qquad K_{stab,1} = \frac{[[Ag(NH_3)]^+]}{[Ag^+][NH_3]}$$

$$= 2.3 \times 10^3 \text{ mol}^{-1} \text{ dm}^3$$

Step 2

$$[Ag(NH_3)]^+(aq) + NH_3(aq) \rightleftharpoons [Ag(NH_3)_2]^+(aq) \qquad K_{stab,2} = \frac{[[Ag(NH_3)_2]^+]}{[[Ag(NH_3)]^+[NH_3]}$$

$$= 7.1 \times 10^3 \text{ mol}^{-1} \text{ dm}^3$$

The overall stability constant is the product of $K_{stab,1}$ and $K_{stab,2}$:

$$Ag^+(aq) + 2NH_3(aq) \rightleftharpoons [Ag(NH_3)_2]^+(aq)$$

$$K_{stab} = \frac{[[Ag(NH_3)_2]^+]}{[Ag^+][NH_3]^2} = \frac{[[Ag(NH_3)]^+]}{[Ag^+][NH_3]} \times \frac{[[Ag(NH_3)_2]^+]}{[[Ag(NH_3)]^+][NH_3]}$$

$$= K_{stab,1} \times K_{stab,2}$$

$$= 1.6 \times 10^7 \text{ mol}^{-1} \text{ dm}^3$$

ION EXCHANGE

Ion exchange is the process of exchanging ions from a solid phase with ions in solution. The solid, which is insoluble, may be a natural material or a synthetic resin. Natural materials used for ion exchange include zeolites (these are complex sodium aluminium silicates) and green sands. Synthetic resins are complex polymeric materials. They are usually manufactured as small insoluble beads.

The surfaces of these solids contain electrically charged sites at more or less regular intervals. The sites are paired with simple ions of opposite charge. It is these ions which are exchanged with ions from solution.

Cation Exchangers

Cation exchange materials consist of three parts:

- The main body or backbone of the material. This is normally labelled R–.
- The active site. These are groups such as $-SO_3^-$ or $-CO_2^-$.
- The cation which is to be exchanged. This is usually H^+ (or H_3O^+).

When the solid comes into contact with a solution containing ions, an equilibrium is established. For example

$$RSO_3^-H^+(s) + Na^+(aq) \rightleftharpoons RSO_3^-Na^+(s) + H^+(aq)$$

Ion exchange can be carried out in the laboratory by packing the solid into a column (*see* column chromatography, chapter 6). The solution containing the ions to be exchanged is poured on top of the column and allowed to pass down the column. Water is used as an eluting agent. If the solution contains sodium chloride, for example, the sodium ions exchange with hydrogen ions. A dilute solution of hydrochloric acid emerges from the bottom of the column.

The ion exchange material can be **regenerated** by washing the column through with dilute hydrochloric acid. This displaces the equilibrium to the left and as a result the sodium ions are replaced by the hydrogen ions.

Anion Exchangers

An anion exchanger removes anions from solution. The following exchange is typical:

$$RNH_3^+OH^-(s) + Cl^-(aq) \rightleftharpoons RNH_3^+Cl^-(s) + OH^-(aq)$$

In a typical household in a hard water area, sufficient scale is deposited in one year to fill two dustbins. Most of this, of course, does not end up in dustbins, or the drains, but in kettles, in the boiler, in hot water pipes and elsewhere in the house—unless the water is softened

An anion exchanger can be regenerated using a base such as sodium hydroxide solution. The equilibrium then shifts to the left.

Many anion exchangers are synthetic resins containing quaternary ammonium groups, R_4N^+.

Uses of Ion Exchange

Volumetric Analysis

Ion exchange can be used to determine volumetrically the concentrations of ions which are otherwise difficult to determine. An example is the sulphate ion, SO_4^{2-}. If, for example, sodium sulphate solution is passed through a cation exchange column, the sodium ions will, as we saw above, exchange for hydrogen (or oxonium) ions. The sodium sulphate is thus converted to sulphuric acid which can then be titrated with, say, sodium hydroxide to determine the concentration of sulphate ions.

Separation of Ions

The ion exchange process involves the distribution of the ionic solute between the liquid or solution phase and the solid phase. Each ion has its own distribution coefficient. It follows that ions can be separated by **ion exchange chromatography**. Each type of ion moves down the column at a different rate. For example, a solution containing similar ions, such as sodium and potassium ions, can be washed through a cation exchange column with dilute acid. By this means the sodium and potassium ions can be separated.

Water Softening

A water softener removes calcium, magnesium and other ions from water. Water which contains more than 1.20×10^{-3} g dm^{-3} of dissolved calcium, magnesium and other salts is called hard water. **Hard water** forms a deposit of carbonate salts, called **scale**, on the inner surfaces of boilers, cooking utensils and pipes that carry hot water or steam. The salts precipitate the fatty acids from soap in the form of scum.

Various types of water softener have been developed industrially to cope with the problem of hard water. Polystyrene ion exchange filters, for example,

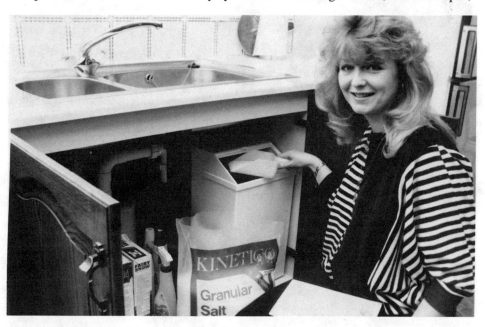

Water softening in a kitchen. The water is softened by passing it through a bed of ion exchange resin. The resin is regenerated periodically with brine (sodium chloride solution). After regeneration, the resin is rinsed with fresh water to ensure that no traces of salt are left

exchange sodium ions from the filter for calcium and magnesium ions:

$$2RSO_3^-Na^+(s) + Ca^{2+}(aq) \rightleftharpoons (RSO_3^-)_2Ca^{2+}(s) + 2Na^+(aq)$$

The sodium ions produced in the exchange do not precipitate soap. The ion exchange resin can be regenerated by washing the filter with a solution of sodium chloride.

Sodium aluminium silicate (zeolite), which is manufactured under the trade name of Permutit, also replaces the calcium and magnesium ions in water with sodium ions. The exchange can be represented by the following equation:

$$2NaAlSi_2O_6(s) + Ca^{2+}(aq) \rightleftharpoons Ca(AlSi_2O_6)_2(s) + 2Na^+(aq)$$
zeolite

De-ionised Water

Water can be de-ionised by passing it through a cation exchange column to exchange the cations with hydrogen ions. It is then passed through an anion exchange resin to exchange the anions with hydroxide ions. The hydrogen and hydroxide ions then combine to form water:

$$H_3O^+(aq) + OH^-(aq) \rightarrow 2H_2O(l)$$

Very pure water, suitable for conductivity experiments, can be obtained in this way.

SUMMARY

1. A **salt** is formed by the reaction of an acid and a base.
2. Some salts undergo acid–base reactions in water to form acidic or alkaline solutions.
3. A **buffer solution** is a solution the pH of which does not change significantly when a small amount of acid or base is added to it.
4. The **Henderson equation** for a weak acid buffer is

$$pH = pK_a + lg\left(\frac{[salt]}{[acid]}\right)$$

5. The equilibrium established in a saturated solution of a salt with the general formula A_xB_y is

$$A_x r B_y(s) \rightleftharpoons xA^{y+}(aq) + yB^{x-}(aq)$$

The **solubility product** of the salt is given by

$$K_{sp} = [A^{y+}]^x[B^{x-}]^y$$

6. The reduction of the solubility of a salt in a solution due to the addition of another substance with an ion in common with the salt is known as the **common ion effect**.
7. The removal of a salt from a solution by addition of another substance is known as **salting out**.
8. For the following **complex ion** equilibrium:

$$Co^{2+} + 6NH_3 \rightleftharpoons [Co(NH_3)_6]^{2+}$$

the **stability constant** is given by

$$k_{stab} = \left(\frac{[[Co[NH_3)_6]^{2+}]}{[Co^{2+}][NH_3]^6}\right)_{eq}$$

9. A typical **cation exchanger** has the formula $RSO_3^-H^+$. The H^+ ions are exchanged for other cations.
10. The following reaction is typical of an **anion exchanger**:

$$RNH_3^+OH^-(s) + Cl^-(aq) \rightleftharpoons RNH_3^+Cl^-(s) + OH^-(aq)$$

Examination Questions

1. (a) Define in terms of proton transfer: (i) an acid; (ii) a base.
   (b) For each of the three following species state whether in aqueous solution it is an acid or a base or both, and write an equation or equations to show the relationship of the acid and/or base to its conjugate species. (In each case label the conjugate species clearly.) (i) $C_6H_5NH_3^+$; (ii) PO_4^{3-}; (iii) HCO_3^-.

   (OLE)

2. (a) Write equations for each of the following reactions and, where appropriate, label the conjugate acid–base pairs.
   (i) The reaction between ammonia gas and water.
   (ii) The reaction between hydrogen chloride gas and water.
   (iii) The thermal dissociation of ammonium chloride.
   (b) Aqueous acids may be placed in order of acid strength according to their pK_a values. In the following, the stronger acids are nearer the top.

   $$HClO_4(aq) \rightleftharpoons H^+(aq) + ClO_4^-(aq)$$

   $$HCl(aq) \rightleftharpoons H^+(aq) + Cl^-(aq)$$

   $$CH_3CO_2H(aq) \rightleftharpoons CH_3CO_2^-(aq) + H^+(aq)$$

   $$H_2O(l) \rightleftharpoons H^+(aq) + OH^-(aq)$$

   (i) Give one piece of experimental evidence to justify the above order.
   (ii) Define K_a and pK_a.
   (iii) Write equations for the reactions between: H_2O and $HClO_4$; and CH_3CO_2H and HCl.
   (c) Explain concisely why
   (i) aqueous sodium chloride is neutral, whereas aqueous sodium sulphide is alkaline;
   (ii) aqueous copper(II) sulphate is acidic.

   (L)

3. (a) What is the *Brønsted–Lowry theory* of acids and bases?
   (b) Explain carefully the meaning of the terms *concentration* and *strength* as applied to acids or bases.
      Comment on the relative merits of pH and dissociation constant as measures of the strength of an acid or base.
   (c) What do you understand by the term *buffer solution*?
      Calculate the approximate pH of
   (i) 0.010 mol dm^{-3} hydrochloric acid,
   (ii) 0.010 mol dm^{-3} ethanoic acid,
   (iii) a buffer solution which contains 0.010 mol dm^{-3} of ethanoic acid and 0.10 mol dm^{-3} of sodium ethanoate.
      (Dissociation constant of ethanoic acid = 1.6×10^{-5} mol dm^{-3}.)

   (UCLES)

4. (a) What do you understand by the ionic product (K_w) of water?
   (b) The ionic product of water changes with temperature as shown:

K_w/mol^2 dm^{-6}	1.19×10^{-15}	2.93×10^{-15}	1.00×10^{-14}	1.47×10^{-14}
Temperature/K	278	283	298	303

   Discuss critically *each* of the following statements:
   (i) The ionisation of water is accompanied by the *absorption* of energy.

(ii) The degree of ionisation of water *increases* as the temperature *falls*.

(iii) The acidity of water *increases* with temperature *increase*.

(c) Calculate the pH of the following aqueous solutions at 298 K:

(i) 1.0×10^{-2} mol dm^{-3} nitric(v) acid;

(ii) 1.0×10^{-2} mol dm^{-3} calcium hydroxide;

(iii) 1.0×10^{-7} mol dm^{-3} hydrochloric acid;

(iv) A mixture of 25 cm^3 of 1 mol dm^{-3} sulphuric(vi) acid and 49.5 cm^3 of 1 mol dm^{-3} sodium hydroxide.

(d) Explain how a buffer solution works. Calculate the mass of sodium ethanoate (CH_3COONa) which would have to be added to 1 dm^3 of 1×10^{-1} mol dm^{-3} ethanoic acid ($K_a = 2 \times 10^{-5}$ mol dm^{-3}) to give a solution of pH = 4.70. (C = 12; O = 16; Na = 23)

(SUJB)

5. The ionic product of water at 25°C is 1.0×10^{-14} mol^2 dm^{-6}.

(a) Calculate the pH of 1.0×10^{-2} mol dm^{-3} nitric acid.

(b) Why is the pH of 1.0×10^{-8} mol dm^{-3} nitric acid 6.96 and not 8.00?

(c) The pH of saturated aqueous calcium hydroxide is 12.3 at 25°C. What is the hydroxide ion concentration of this solution?

(d) Write an expression for the solubility product of calcium hydroxide. Use the value of the hydroxide ion concentration you obtained in (c) to calculate the solubility product of calcium hydroxide at 25°C. Explain your reasoning and give the units.

(e) Explain why calcium hydroxide is not precipitated when aqueous ammonia is added to aqueous calcium nitrate, whereas it is precipitated if aqueous sodium hydroxide is used.

(UCLES)

6. (a) Define pH.

(b) At 25°C, K_a for the dissociation of ethanoic (acetic) acid in water is 1.8×10^{-5} mol l^{-1} and the ionic product (K_w) of water is 1×10^{-14} mol^2 l^{-2}.

(i) Calculate the basicity constant (K_b) of the ethanoate (acetate) ion in water at 25°C.

(ii) Calculate the pH of a 0.05 M aqueous solution of ethanoic acid at 25°C.

(c) The following data refer to aqueous solutions at 25°C.

Substance	pK_b
ammonia	4.75
methylamine	3.38
phenylamine (aniline)	9.40

Account for the base strengths of methylamine and phenylamine relative to that of ammonia.

(JMB)

7. (a) Define the term pH.

(b) How will litmus paper behave towards aqueous solutions of the following salts: sodium carbonate, potassium sulphate, aluminium chloride? Explain your answers.

(c) Classify the following species as Brønsted acids and/or bases. Explain by suitable equations the reason for the selection made.

$$S^{2-}; \quad HCO_3^-; \quad H_2O; \quad NH_3.$$

(d) Calculate the pH values of aqueous solutions (all at 25°C) containing
    (i)   hydrochloric acid of concentration 0.1 mol dm^{-3},
    (ii) sodium hydroxide of concentration 0.2 mol dm^{-3}, (given $K_w = 1 \times 10^{-14}$ mol^2 dm^{-6} at 25°C),
    (iii) ethanoic acid of concentration 0.5 mol dm^{-3} (given $K_a = 1.74 \times 10^{-5}$ mol dm^{-3} at 25°C),
    (iv) 40 cm^3 of hydrochloric acid of concentration 0.1 mol dm^{-3} mixed with 60 cm^3 of aqueous sodium hydroxide of concentration 0.05 mol dm^{-3}.

(e) (i)   Why is it necessary to quote the temperature of solutions in part (d)?
     (ii) What is the effect on pH of raising the temperature in (d) (iii) above?

(f) The dissociation of phenolphthalein can be written as follows

$$HX(aq) \rightleftharpoons H^+(aq) + X^-(aq) \qquad \text{where } K_a = 7 \times 10^{-10} \text{ mol dm}^{-3}$$

    (i)   What colours, if any, are associated with HX(aq), H$^+$(aq) and X$^-$(aq) respectively?
    (ii) What should the pH of a solution be at the end-point using this indicator assuming a strong base is added to a weak base?

(AEB, 1984)

8.   A solution contains 0.150 mol dm^{-3} of Na$_2$CO$_3$ and an unknown concentration of NaHCO$_3$. 25.0 cm^3 of this solution were titrated with a solution of HCl of 0.100 mol dm^{-3}. Using methyl orange as indicator, the titre was 87.5 cm^3 of the HCl. Calculate (i) the concentration of NaHCO$_3$ in the solution and (ii) the titre if phenolphthalein had been used as indicator.

(OLE)

9.   What is a *buffer solution*? Explain how an aqueous buffer solution containing a weak acid and a salt of the same weak acid functions on the addition of a small amount of (i) acid, (ii) alkali and on (iii) dilution with water.
    The acid dissociation (or ionisation) constant, K_a, of ethanoic acid in water at 298 K is 1.75×10^{-5} mol dm^{-3}. Explain this statement. The pH of a buffer solution may be calculated from the relation

$$pH = pK_a + \log \left(\frac{[\text{salt}]}{[\text{acid}]} \right) \qquad (1)$$

where p( ) = $-\log$ ( ).
    Using equation (1) calculate the pH of an 'acid-phosphate' buffer solution containing 0.10 mol dm^{-3} phosphoric(v) acid and 0.20 mol dm^{-3} sodium dihydrogenphosphate(v), NaH$_2$PO$_4$. The relevant dissociation constant, K_a, for the acid (H$_3$PO$_4$(aq) $\rightleftharpoons$ H$^+$(aq) + H$_2$PO$_4^-$(aq)) is 7.1×10^{-3} mol dm^{-3} and the salt may be considered to be fully ionised into H$_2$PO$_4^-$(aq) and Na$^+$(aq) ions.
    From equation (1) show how the pH of an aqueous solution of a weak monobasic acid, HA, at 'half-neutralisation', i.e. when half the original amount of acid has been neutralised by, say, sodium hydroxide solution, is numerically equal to the pK_a of the acid.
    Sketch the pH versus volume curve you would expect to obtain on the addition of aqueous sodium hydroxide (approximately 0.10 mol dm^{-3}) to 25.0 cm^3 of a weak monobasic acid ($K_a = 1 \times 10^{-5}$ mol dm^{-3}) of concentration 0.10 mol dm^{-3}. Mark clearly on your diagram
    (i)   the value of the initial pH of the acid solution,
    (ii) the pH at half-neutralisation and
    (iii) the end-point of the titration.

Show how you arrive at your values for (i) and (ii). What indicator, if any, would you use for this titration? Explain your answer.

(NISEC)

10. (a) Explain why a solution of the salt ammonium chloride in water is not neutral; a solution of concentration 0.10 mol dm^{-3} has pH = 6.0 at 298 K.

(b) Ammonia is a weak base, $pK_b = 4.76$.

Explain what is meant by this statement and deduce the value of the pH of a solution of ammonia at 298 K which contains 0.20 mol dm^{-3}, given $K_w = 1.0 \times 10^{-14} \text{ mol}^2 \text{ dm}^{-6}$ at 298 K.

(c) Sketch graphs (pH against volume of acid added) to compare and to illustrate what happens during the titrations of hydrochloric acid containing 0.20 mol dm^{-3} separately with ammonia and sodium hydroxide solutions of the same molar concentrations. (Take care to calibrate axes and show calculable values.)

(d) Use the graphs to illustrate the concept of *buffer action* and define the term *buffer solution*.

What is the approximate range of the buffer action and how is it deduced from the pK_b?

Give *two* examples of chemical or biochemical studies requiring the use of buffer solutions and add a brief note of explanation.

(e) Calculate the pH of a solution made by mixing 30.0 cm^3 of ethanoic acid (acetic acid) solution, which contains $0.100 \text{ mol dm}^{-3}$ of the acid, with 20.0 cm^3 of sodium hydroxide solution of the same molar concentration, given that $K_a = 1.8 \times 10^{-5} \text{ mol dm}^{-3}$ at 298 K.

Compare this result with that obtained by using hydrochloric acid of the same molar concentration instead of ethanoic acid, and comment.

(SUJB)

11. In working out each of the following calculations, state clearly the physico-chemical principles upon which they depend, and show all working clearly so that marks can be allocated even if errors occur.

(a) Distinguish carefully between the terms 'solubility' and 'solubility product' as applied to a sparingly soluble electrolyte. If the *numerical* values of the solubility products of silver chloride, AgCl, and silver chromate(VI), Ag_2CrO_4, are respectively 1.6×10^{-10} and 1.0×10^{-12}, calculate their respective molar solubilities in water at this temperature. (AgCl = 143.5, Ag_2CrO_4 = 332)

(b) The numerical value for the solubility product of silver ethanoate (acetate), CH_3COOAg, is 2.0×10^{-3}.

If 4.1 g of sodium ethanoate is added to 100 cm^3 of a saturated solution of silver ethanoate in water, calculate the molarity of silver ions remaining in solution and the mass of silver ethanoate precipitated. (CH_3COONa = 82, CH_3COOAg = 167)

(c) If the numerical value of the solubility product of silver iodide, AgI, is 1.0×10^{-16} and the instability constant for the ion

$$Ag(NH_3)_2^+(aq) \rightleftharpoons 2NH_3(aq) + Ag^+(aq)$$

has the numerical value 6.0×10^{-8}, calculate the mass of silver iodide which will dissolve in 1 dm^3 (1 litre) of 1.0 M ammonia solution,

$$AgI(s) + 2NH_3(aq) \rightleftharpoons Ag(NH_3)_2^+(aq) + I^-(aq)$$

(AgI = 235)

(SUJB)

12. (a) Explain the term solubility product and write expressions for, and the units of, the solubility products of calcium sulphate, aluminium hydroxide and lead bromide.

(b) Discuss *each* of the following:

(i) the solubility of silver chloride in water decreases when dilute hydrochloric acid is added but increases when concentrated hydrochloric acid or aqueous ammonia is added;

(ii) aqueous ammonia can precipitate certain metals as their hydroxides but the presence of ammonium chloride often prevents the precipitation.

(c) The solubility of strontium hydroxide ($Sr(OH)_2$) is 0.524 g in 100 cm^3 water. Calculate:

(i) The solubility of strontium hydroxide in water, mol dm^{-3};

(ii) the hydroxide ion concentration (mol dm^{-3}) in a saturated solution of strontium hydroxide;

(iii) the solubility product of strontium hydroxide;

(iv) the approximate solubility of strontium hydroxide (g dm^{-3}) in 1 dm^3 of 2×10^{-1} mol dm^{-3} strontium chloride solution;

(v) the volume of 1×10^{-2} mol dm^{-3} potassium chromate solution which must be added to 1 dm^3 saturated strontium hydroxide solution to precipitate strontium chromate ($SrCrO_4$).

(H = 1; O = 16; Sr = 87.6. Solubility product of strontium chromate is 3.6×10^{-5} mol^2 dm^{-6})

(SUJB)

13. (a) The solubility product of $BaSO_4$ in water at 25°C is 1.00×10^{-10} mol^2 dm^{-6}.

(i) In what volume of water will 2.33×10^{-2} g of $BaSO_4$ just dissolve at 25°C?

(ii) 20 cm^3 of 0.010 M $BaCl_2$ solution and 20 cm^3 of 0.005 M Na_2SO_4 solution are mixed at 25°C. Calculate whether the solubility product of $BaSO_4$ will be exceeded in the mixture.

(b) The solubility of $Na_2SO_4 \cdot 10H_2O$ increases rapidly with increasing temperature over the range 0°C to 32.4°C. Predict whether the dissolution of $Na_2SO_4 \cdot 10H_2O$ is exothermic or endothermic in this temperature range, and explain briefly the basis for your prediction.

(OLE)

⑨ CHEMICAL KINETICS

Explosives

An explosive is a stable material that, upon proper stimulation, rapidly changes from a solid or liquid into a hot, expanding gas. The pressure exerted on the surrounding materials by the expanding gas constitutes an explosion.

Confining an explosive greatly increases its propensity to detonate by increasing the speed of the reaction. For example, gunpowder, or black powder, confined within the paper wrapping of a firecracker explodes when ignited, but the same powder sprinkled in the open simply burns when ignited. During the relatively slow reaction of burning, the pressure of the hot gases does not increase fast enough to have an explosive effect.

Although many substances in various forms can be exploded (gasoline vapours, hydrogen, and finely dispersed coal or grain dust are all explosive substances, for example), only those substances specifically intended to produce an explosion are generally called explosives. They can be divided into two types: propellant explosives, such as gunpowder, and detonating explosives, such as TNT and dynamite.

Propellant Explosives

Gunpowder, which was developed by the Chinese in the tenth century and by the Arabs, independently, soon afterwards, was the first explosive to be used in firearms. In 1845 a nitrocellulose explosive called guncotton was introduced, but it proved too potent for its intended use as a firearms propellant.

Breakthroughs in the 1880s, however, resulted in satisfactory smokeless powders that, within 20 years, virtually supplanted black powder as a propellant. With the advent of the space age, many specialised explosives were developed for use as rocket propellants.

Detonating Explosives

Detonating explosives may be subdivided into initiating explosives and high explosives. Initiating explosives, which must be handled with extreme care, are the more sensitive of the two. Materials such as fulminate of mercury explode instantly when burned or ignited, which makes them desirable for use in blasting caps.

High explosives are less sensitive and can burn without producing an explosion. They can only be detonated by a severe shock, which is delivered by another explosive (usually a blasting cap) placed in or near the high explosive. Since they are relatively stable, large amounts of high explosives can be moved and handled safely.

In 1846 the Italian Ascanio Soberro invented nitroglycerine, an explosive so sensitive that it was virtually unusable. It became important later, however, when in 1867 the Swedish inventor Alfred Nobel combined it first with siliceous earth

A shell explodes as French troops advance through barbed wire entanglements on the Western front in World War I

and later with wood pulp to produce dynamite.

TNT (trinitrotoluene), which was first used in the early 1900s, has become the standard by which all other explosives are measured. It is used either by itself or mixed with other ingredients and explosives to produce many subtypes with differing performance characteristics.

Modern Explosives

Modern detonating explosives include PETN (pentaerythrite tetranitrate), which is used in blasting caps and detonating cord; RDX (also known as cyclonite), which is combined with other explosives and waxes to produce what are popularly known as plastic explosives; ammonium nitrate, an explosive of low detonating velocity that is used when a slow push or heave is more desirable than a shattering effect; and amatol, a mixture of ammonium nitrite and TNT that is used as a bursting charge.

Explosives are used in a wide variety of civilian and military tasks. Tunnel construction, obstacle clearing, and open pit and underground mining employ large quantities of explosives. They are also used as propellants for firearms and rockets; as bursting charges for bombs, mines, artillery projectiles, torpedos, and hand grenades; and for general engineering and demolition work.

LEARNING OBJECTIVES

After you have studied this chapter you should be able to

1. Explain the term **chemical kinetics** and give specific examples to illustrate its importance in (a) industry and (b) living systems.
2. Explain the terms (a) **reaction rate**, (b) **rate constant** and (c) **order of reaction**.
3. Outline techniques used to measure reaction rates and give specific examples of each.
4. List the five most important factors affecting reaction rate. Describe and give examples of how each factor affects the rate of reaction.
5. Briefly describe an experiment to show the effect of concentration on reaction rate.
6. Determine reaction orders and calculate rate constants from experimental data.
7. Define the term **half-life** and calculate first-order half-lives from experimental data.
8. Use the Arrhenius equation to calculate the effect of temperature on rate constant.
9. Explain what is meant by **rate-determining step** and give a specific example to illustrate this.
10. Use a rate equation to suggest a possible **reaction mechanism** for a reaction.
11. Explain the terms (a) **activation energy**, (b) **activated complex** and (c) **collision frequency**.
12. Briefly describe how (a) **collision theory**, (b) **transition state theory** can be used to explain changes in reaction rate.
13. Sketch and label the **energy profile** of a reaction and show how a **catalyst** affects the profile.
14. Briefly describe an experiment to show the effect of a catalyst on the rate of a reaction.
15. Distinguish between **heterogeneous catalysis** and **homogeneous catalysis** and give one example of each.
16. Give two specific examples of **enzymes** and indicate the importance of each.
17. Give two specific examples of **chain reactions** and indicate the importance of each.

9.1
Rates of Chemical Reactions

INTRODUCTION

The **rate of a reaction** is the rate at which products are formed or the rate at which reactants are used up in the reaction. Rates vary immensely from reaction to reaction. In some reactions, the products are formed and the reactants used up so fast that the reaction is virtually instantaneous. Explosions are examples. As soon as the reaction starts it is finished. Many ionic reactions take place instantaneously. For example, the following neutralisation reaction is instantaneous:

$$H^+(aq) + OH^-(aq) \rightarrow H_2O(l)$$

In other cases, reactions keep on going for months or even years. The fermentation of grape juice to form wine can take many months to complete. And then, when the wine is stored, the complex chemical processes which take place to bring out its full flavour may take years.

The rate of any particular reaction may vary depending on a number of factors. The most important of these are:

- concentration or pressure of reactants,
- temperature,
- presence of catalysts,
- availability of reactants.

The reaction between hydrogen and oxygen provides a startling example of how the rate of a specific reaction can vary. Hydrogen burns rapidly in a plentiful

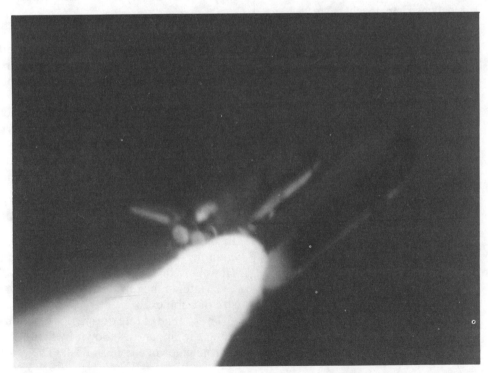

The Space Shuttle *Challenger*, travelling at 2000 miles per hour, 58.32 seconds after launch on its tenth flight. The Shuttle is fuelled by two solid-fuel rocket boosters and 378 000 gallons (1.72×10^6 dm^3) of liquid hydrogen and 140 000 gallons (0.64×10^6 dm^3) of liquid oxygen. The two liquids are contained in separate compartments of an external tank the size of a ten-storey building. The fuels are pumped into the shuttle's three engines where they mix and burn, generating enormous thrust and steam super-heated to over 3000°C. Moments after this picture was taken, the external tank exploded with the force of a small nuclear bomb

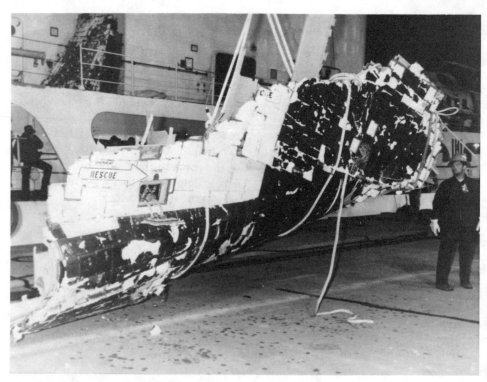

Wreckage from the *Challenger* is retrieved from the Atlantic Ocean and loaded onto a United States Coast Guard vessel

Most chemistry teachers are well aware of the danger of a hydrogen explosion during the preparation of hydrogen from zinc and hydrochloric acid. The flying glass resulting from this type of explosion can cause serious injuries—particularly to eyes. Yet, on a smaller scale, these self-same hydrogen explosions are commonly used in the laboratory. They are known as the 'pop' test for hydrogen.

supply of air or oxygen to form water:

$$2H_2(g) + O_2(g) \rightarrow 2H_2O(g)$$

However, if the supply of air is restricted, the reaction can be explosive.

Curiously enough, if a mixture of hydrogen and oxygen is *not* ignited, the reaction is so slow that there appears to be no reaction at all. Thus, the same reaction can proceed explosively or infinitesimally slowly depending on conditions.

Reaction rate also varies during a reaction. In most reactions the rate is fastest at the beginning and then gradually tails off. This can be demonstrated by adding dilute hydrochloric acid to calcium carbonate. Immediately the acid is added the mixture froths vigorously. This is due to the rapid production of carbon dioxide. As the reaction slows down, the bubbling subsides.

How Can the Rate of Reaction be Determined?

In the reaction between an acid and carbonate, the rate could be found by counting the number of bubbles of carbon dioxide given off in one minute. However, this would be impossible at the start when the reaction is vigorous. Even later in the reaction, the method would be inaccurate due to variations in the size of bubbles.

The change of rate during this reaction can be demonstrated more accurately by using the apparatus shown in figure 9.1. Dilute hydrochloric acid is added to marble chips in a conical flask. The reaction is

$$CaCO_3(s) + 2HCl(aq) \rightarrow CaCl_2(aq) + CO_2(g) + H_2O(l)$$

The carbon dioxide is collected in a gas syringe. The change of rate is followed by measuring the volume of carbon dioxide at regular time intervals during the reaction. The results shown in table 9.1 are typical. When these are plotted as a graph, a **rate curve** is obtained (*see* figure 9.2). A rate curve shows the change in amount of either the product formed or the reactant used up with time.

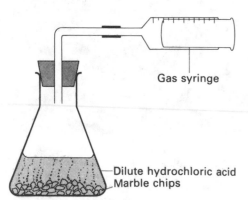

Gas syringe

Dilute hydrochloric acid
Marble chips

Figure 9.1 Apparatus used to investigate the rate of reaction between dilute hydrochloric acid and marble chips

Table 9.1 The volume of carbon dioxide collected at 15 second intervals during the reaction between marble chips and dilute acid

Time/s	Volume/cm³
0	0
15	27
30	47
45	61
60	69
75	75
90	80
105	80

The slope or **gradient** of the rate curve indicates how fast the reaction is going. The steeper the slope, the faster the reaction. In figure 9.2, the slope for the **initial rate** is drawn. Since the reaction is fastest at the beginning of the reaction, the slope for the initial rate is steepest. Gradually, as the reaction proceeds, the slope becomes less steep. Finally, when there is no further reaction, the curve flattens out. The tangent is a horizontal line and the gradient is zero.

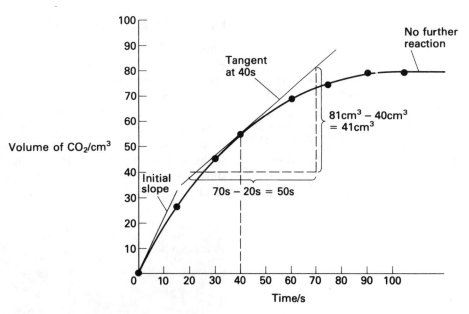

Figure 9.2   Rate curve for the reaction between dilute hydrochloric acid and marble chips

Using the Rate Curve to Obtain a Value for the Rate of Reaction

In general, for the rate of formation of a product in a reaction we can write

$$\text{Rate} = \frac{\text{Increase in amount of product}}{\text{Time taken for this increase}}$$

In our example, the amount measured is the volume of carbon dioxide. Thus

$$\text{Rate} = \frac{\text{Increase in volume of carbon dioxide}}{\text{Time taken for this increase}}$$

This can be written as

$$\text{Rate} = \frac{\Delta V}{\Delta t}$$

This is the **average rate** of reaction over the period of time Δt. At a specific time the **instantaneous rate** is given by

$$\text{Rate} = \frac{\mathrm{d}V}{\mathrm{d}t}$$

The instantaneous rate is the slope of the tangent to the curve at that time.

EXAMPLE

Using the rate curve shown in figure 9.2, calculate the rate of reaction at 40 seconds.

SOLUTION

The tangent to the curve at 40 seconds is shown on the graph.

$$\text{Slope of tangent} = \frac{41 \text{ cm}^3}{50 \text{ s}}$$

Thus, rate of production of CO_2 at 40 seconds is 0.82 cm^3 s^{-1}.

How Does Surface Area Influence Rate of Reaction?

The reaction between acid and marble chips can also be used to investigate the effect of surface area on reaction rate. The results shown in table 9.2 were obtained in one such investigation.

In this experiment, the only factor varied is surface area. The large chips have a small surface area compared to a similar mass of small chips. The results in the table show that the chips with a large surface area react faster than the chips with a small surface area. This is because reactions between solids and liquids take place at the surface of the solid. Increasing the surface area of a solid by increasing its **state of subdivision** thus increases its availability. This is why chemicals are often used in powdered form rather than large lumps.

Table 9.2 Experimental results obtained for the reaction between marble chips and dilute acid

Size of chips	Time to collect the initial 20 cm^3 of CO_2
large	55 seconds
medium	21 seconds
small	8 seconds

Granulated sugar dissolves more quickly in tea than does a sugar lump. This is because granulated sugar has a greater surface area than the same amount of lump sugar

METHODS OF MEASUREMENT

We saw above that the rates of reaction for the reaction between dilute hydrochloric acid and marble chips could be obtained by plotting volume of CO_2 produced against time. Experimental data required for the study of reaction rates can also be obtained by measuring other quantities. These include:

- concentration,
- pressure,
- optical rotation,
- absorption of radiation,
- electrical conductivity.

The following are some of the experimental methods used to obtain data on rates of chemical reactions.

Physical Methods

In gas reactions, gas pressure or volume is often measured. There is a single requirement. The number of moles of gaseous reactants must be different from the number of moles of gaseous products. If the number of moles of reactants equals the number of moles of products, there can be no change in pressure. This follows from Avogadro's principle. In the following reaction, there are three moles of gaseous reactants and two moles of gaseous products. The rate can thus be followed by measuring pressure or volume:

$$2NO(g) + O_2(g) \rightarrow 2NO_2(g)$$

But in the reaction between hydrogen and iodine, there is no change in volume. The rate can therefore not be followed by this method:

$$H_2(g) + I_2(g) \rightarrow 2HI(g)$$

Chemical Analysis

Experimental data on reaction rates can often be obtained by taking small samples of the reaction mixture and analysing them. This method can be used to determine the rate of saponification (*see* section 20.3) of ethyl ethanoate:

$$C_2H_5CO_2C_2H_5(aq) + NaOH(aq) \rightarrow C_2H_5CO_2Na(aq) + C_2H_5OH(aq)$$

At regular intervals during the reaction, a sample of reaction mixture is taken and titrated against acid. This gives the concentrations of sodium hydroxide in the reaction mixture during the reaction.

Optical Methods

If a reactant or product absorbs electromagnetic radiation, the reaction can be followed by measuring the amount of radiation absorbed. The absorption of ultra-violet or infra-red radiation (*see* section 17.1) is frequently used in the study of the rates of organic reactions. For reactions involving coloured compounds, a colorimeter can be used to measure colour intensity. The colour intensity of a compound or the amount of radiation absorbed by it is proportional to its concentration.

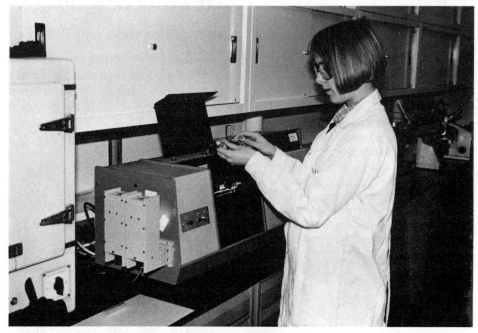

A tube containing a sample is placed into a polarimeter. Polarimeters can be used to determine the rates of reaction of optically active compounds

The rates of reactions involving optically active compounds can be determined using a polarimeter (*see* section 17.2). The polarimeter measures optical rotation. This method is particularly useful in investigating the rates of sugar reactions.

Electrochemical Methods

In many inorganic and organic reactions, rates can be determined from changes in electrical conductivity during the reaction. This method may be used to measure the rate of saponification of ethyl ethanoate (*see* above). Two inert electrodes connected to a conductivity meter are immersed in the reaction mixture (*see* figure 9.3). The conductivity is almost entirely due to the sodium hydroxide since it is the only strong electrolyte present. As the reaction proceeds, the sodium hydroxide is used up and so the conductivity decreases. The advantage of this method over the titration method is that it is not necessary to remove samples from the reaction mixture.

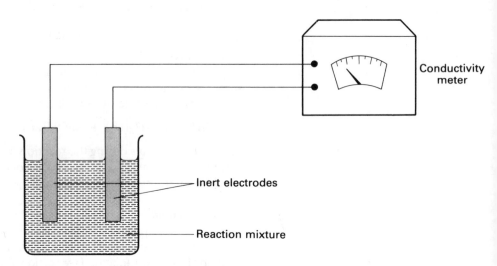

Figure 9.3 The rate of some reactions can be followed by measuring the electrical conductivity at regular intervals

RATE AND CONCENTRATION

Order of Reaction and Rate Constants

A glowing splint will relight if put into a gas jar or test tube of oxygen. This shows that if we suddenly increase the concentration of oxygen the rate of combustion also suddenly increases. On the other hand, if we decrease concentration, rate of reaction also decreases. In the last section we saw that the initial rate of reaction is usually fastest. Then, as the reaction proceeds, the rate of reaction decreases.

Experiments show that the rate of a reaction depends on concentration c of reactants:

$$\text{Rate} \propto c^n$$

where n is a constant called the **order** of the reaction. The order of a reaction tells us the exact dependence of rate on concentration. It is a constant which can only be found experimentally.

From the above relationship, we obtain the **rate equation**:

$$\text{Rate} = kc^n$$

where k is a constant of proportionality called the **rate constant**. This is also known as the **velocity constant**.

As an example, experimental studies of the reaction

$$2NO(g) + O_2(g) \rightarrow 2NO_2(g)$$

Increasing the concentration of oxygen increases the rate of combustion

Table 9.3   Units of rate constants

Order	Units
first	s^{-1}
second	$mol^{-1}\ dm^3\ s^{-1}$

Table 9.4   The variation of initial reaction rate with $[N_2O_5]$ for the reaction

$$2N_2O_5(g) \rightarrow 4NO_2(g) + O_2(g)$$

Rate/mol dm^{-3} h^{-1}	$[N_2O_5]$/mol dm^{-3}
0.016	0.010
0.032	0.020
0.064	0.040

Note: h = hour

have shown that the rate of this reaction is proportional to $[NO]^2$ and $[O_2]$. The symbol [ ] indicates concentration. The rate equation is thus

$$Rate = k[NO]^2[O_2]$$

The reaction is said to be **first order** in oxygen, **second order** in nitrogen monoxide and third order overall. The **order of a reaction** with respect to a reactant is the power to which the concentration of that reactant is raised in the rate equation. The **overall order** is the sum of these powers.

In general we can write

$$Rate = k[A]^m\ [B]^n$$

The reaction order is

m	with respect to reactant A
n	with respect to reactant B
$m + n$	overall

A knowledge of the order of a reaction can help us to predict the course of the reaction.

Gas Reactions

The rate of a gas reaction depends on the partial pressures of reactants. As the partial pressures decrease, the rate of reaction decreases. The partial pressure of a gas is equivalent to its concentration. Thus, in the study of chemical kinetics, the terms concentration and partial pressure are often used interchangeably.

Units of Rate Constants

These are derived from the rate equation. They thus depend on the order of reaction. Table 9.3 shows the units of the rate constants for first- and second-order reactions.

Experimental Determination of Order of Reaction and Rate Constants

A number of methods can be employed to determine the order of reaction and its rate constant. We shall start by considering how these can be obtained by simply inspecting the experimental data.

The Inspection Method

Some experimentally determined rate data for the following reaction are shown in table 9.4:

$$2N_2O_5(g) \rightarrow 4NO_2(g) + O_2(g)$$

The data clearly show that, on doubling the concentration, the rate of reaction also doubles. Thus

$$Rate \propto [N_2O_5]^1$$

The reaction is thus first order

$$Rate = k[N_2O_5]^1$$

The rate constant for the reaction is obtained by substitution of any pair of values from the table into the rate equation:

$$k = \frac{Rate}{[N_2O_5]}$$

$$= \frac{0.032\ mol\ dm^{-3}\ h^{-1}}{0.020\ mol\ dm^{-3}}$$

$$= 1.6\ h^{-1}$$

The Rate of Growth of Cancer Cells

A number of factors influence cell growth in animals. These include pH, nutrient levels and temperature. In laboratories, cells can be artificially grown in liquid media containing glucose. Glucose is an important nutrient. It provides energy by means of the process called respiration. During this process the glucose is converted to lactic acid and finally to carbon dioxide and water.

There have been numerous investigations into how cancer cells grow. In one study to determine how the rate of growth of cancer cells is related to nutrient levels, tumour cells were taken from rats and inoculated into a liquid growth medium. The medium contained salts, glucose, amino acids and blood serum. During the experiment, pH, temperature and oxygen and carbon dioxide levels were closely controlled. Over a period of 60 hours the cell density and the levels of glucose and lactic acid in the medium were determined every ten hours. The results were plotted on graphs (see figure 9.4). The curves show that the rate of cell growth corresponds to the rate of glucose depletion and the rate of lactic acid build-up in the medium.

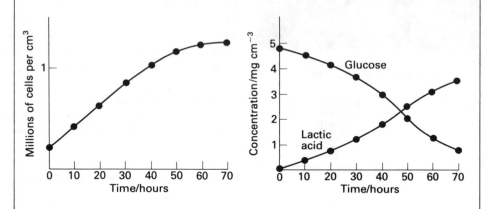

Figure 9.4   The 'rat rates'. The left-hand curve shows the increase in density of rat tumour cells over a period of 70 hours. The curves on the right show the corresponding decrease in glucose concentration and increase in lactic acid concentration

Table 9.5   The variation of initial reaction rate with $[NO_2]$ for the reaction
$$NO_2(g) + CO(g) \rightarrow NO(g) + CO_2(g)$$

Expt	Rate/mol dm^{-3} s^{-1}	$[NO_2]$/mol dm^{-3}
1	0.010	0.15
2	0.040	0.30
3	0.16	0.60
4	0.32	0.90

Table 9.5 shows some experimental data for the following reaction at 325°C:

$$NO_2(g) + CO(g) \rightarrow NO(g) + CO_2(g)$$

Now since

$$\text{Rate (expt 1)} \propto k[NO_2, \text{expt 1}]^n$$

and

$$\text{Rate (expt 2)} \propto k[NO_2, \text{expt 2}]^n$$

It follows that

$$\frac{\text{Rate (expt 2)}}{\text{Rate (expt 1)}} = \left(\frac{[NO_2, \text{expt 2}]}{[NO_2, \text{expt 1}]}\right)^n$$

Substituting values for these experiments into this equation, we obtain

$$\frac{0.040 \text{ mol dm}^{-3} \text{ s}^{-1}}{0.010 \text{ mol dm}^{-3} \text{ s}^{-1}} = \left(\frac{0.30 \text{ mol dm}^{-3}}{0.15 \text{ mol dm}^{-3}}\right)^n$$

$$4 = 2^n$$

Thus n must be 2. The reaction is therefore second order.

The rate constant is obtained by substituting any pair of values from the table into the rate equation:

$$k = \frac{\text{Rate}}{[NO_2]^2}$$
$$= \frac{0.040 \text{ mol dm}^{-3} \text{ s}^{-1}}{(0.30 \text{ mol dm}^{-3})^2}$$
$$= 0.44 \text{ mol dm}^{-3} \text{ s}^{-1}$$

Table 9.6 Rate data for the reaction between magnesium and dilute hydrochloric acid

Rate/mol dm^{-3} s^{-1}	HCl/mol dm^{-3}
0.004	0.50
0.011	0.80
0.022	1.10
0.034	1.40

The Rate Curve Method

It is not always possible to determine the order of a reaction and thus the rate constant simply by inspection of the experimental data. Table 9.6 shows some data obtained from a rate curve for the reaction between magnesium and hydrochloric acid (*see* figure 9.5).

$$Mg(s) + 2HCl(aq) \rightarrow MgCl_2(aq) + H_2(g)$$

The rate at a given concentration is the slope of the tangent to the curve at that concentration. To illustrate this, the tangent to the curve at [HCl] = 0.8 mol dm^{-3} is drawn in figure 9.5.

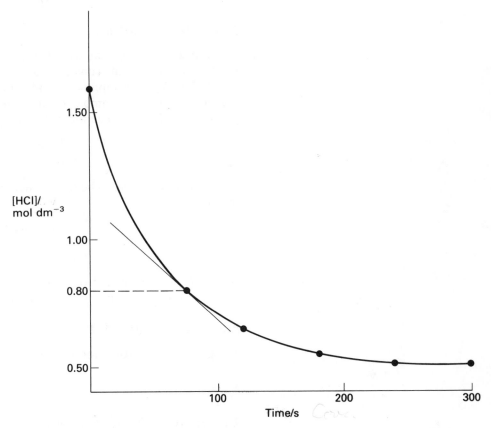

Figure 9.5 Rate curve for the reaction between magnesium and dilute hydrochloric acid

The order of the reaction is not evident from inspection of the data in table 9.6. The order can be found by plotting rate against [HCl]^n as follows:

Rate against [HCl]^1	(*see* figure 9.6)
Rate against [HCl]^2	(*see* figure 9.7)
Rate against [HCl]^3	(*see* figure 9.8)

A linear plot indicates the order of the reaction. Clearly the plot in figure 9.7 is

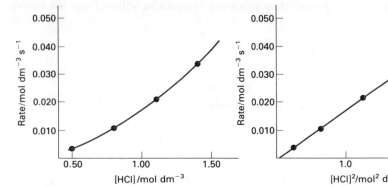

Figure 9.6 Variation of reaction rate with [HCl]

Figure 9.7 Variation of reaction rate with [HCl]^2

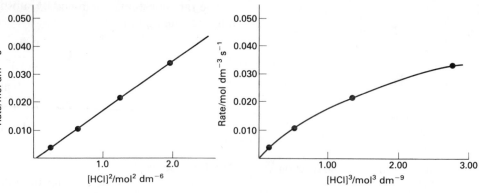

Figure 9.8 Variation of reaction rate with [HCl]^3

linear whereas those in figures 9.6 and 9.8 are not. The reaction is thus second order.

The rate constant for the reaction can, as usual, be obtained by substituting a pair of values into the rate equation:

$$\text{Rate} = k[\text{HCl}]^2$$

Non-Integral Orders

The inspection method and the rate curve method are only suitable for determining reaction orders and rate constants for reactions with integral reactions, that is with $n = 1$, 2 or 3 for example. Some reactions, however, have non-integral or zero reaction orders. For example, the decomposition of ethanal, CH_3CHO, in the gas phase at 720 K has a reaction order of 1.5:

$$CH_3CHO(g) \rightarrow CH_4(g) + CO(g)$$
$$\text{Rate} = k[CH_3CHO]^{1.5}$$

Such reaction orders can be determined by plotting log(rate) against log(concentration) as follows:

$$\text{Rate} = kc^n$$

Thus

$$\log(\text{rate}) = \log k + n \log c$$

A plot of log(rate) against log c should give a straight line the slope of which equals n. The intercept (when log $c = 0$) gives log k.

Integrated Rates Method

The general rate expression may be expressed as

$$-\frac{dc}{dt} = kc^n$$

The term dc/dt is the slope of the tangent to the rate curve. Since reactant concentrations decrease during a reaction, the gradient of the tangent is negative. However, by convention reaction rates are always positive. The negative sign is thus included in the rate equation to make the rate positive.

Let us now consider the general rate equation for a first-order reaction:

$$-\frac{dc}{dt} = kc$$

Integration of this gives

$$\ln c = \ln c_0 - kt$$

where ln c is the natural logarithm of concentration and c_0 is the concentration at $t = 0$. If the reaction is first order, a plot of ln c against t will be a straight line with slope $= -k$. The intercept equals ln c_0.

The general rate equation for a second-order reaction is

$$-\frac{dc}{dt} = kc^2$$

Integration of this equation gives

$$\frac{1}{c} = \frac{1}{c_0} + kt$$

If the reaction is second order, a plot of $1/c$ against t will be a straight line with slope $= k$. In this case the intercept equals $1/c_0$.

By plotting both graphs and finding out which is linear, it is possible to distinguish between a first- and second-order reaction and also determine the rate constant of the reaction.

Half-Life

We have already come across this term in the section on radioactivity (section 1.3). Half-life is the time taken for the initial concentration of a reactant to fall to half its value. It is a particularly useful measurement when investigating the order of a reaction. For example, in first-order reactions, half-life $t_{1/2}$ is related to the rate constant k by the simple equation

$$t_{1/2} = \frac{0.693}{k}$$

Rate of radioactive decay is independent of temperature. However, radioactive decay does follow a first-order pattern. The half-life of a radioactive process can be obtained by measuring the level of activity at regular intervals and plotting a decay curve. This approach forms the basis of the technique known as carbon dating.

THE EFFECT OF TEMPERATURE AND THE ARRHENIUS EQUATION

As a general rule of thumb, the rate of reaction doubles for every 10 K rise in temperature. This would seem to suggest that there is an exponential relationship between rate and temperature. The exact relationship was first proposed by a Swedish chemist called Arrhenius in 1889. It is known as the **Arrhenius equation**:

$$k = Ae^{-E_a/RT}$$

where k is the rate constant for the reaction; A is a constant for the reaction, sometimes known as the Arrhenius constant; E_a is another constant for the reaction, known as the activation energy; R is the gas constant; and T is the temperature, in kelvin. Note that the equation relates rate constant and *not* rate of reaction to temperature.

The logarithmic form of the equation is

$$\ln k = \ln A - E_a/RT$$

In terms of log to the base 10, this is

$$\log k = \log A - E_a/2.303RT$$

This is the most useful form of the equation.

The constants A and E_a for a given reaction can be obtained by plotting log k against $1/T$. The temperatures T must be in kelvin. The slope of the graph gives

The Swedish physical chemist Svante Arrhenius received the Nobel Prize for Chemistry in 1903 for his theory of electrolytic dissociation. In his doctoral dissertation at the University of Uppsala, he argued that 'molecules' such as sodium chloride break apart spontaneously in solution, forming ions which become the agents of electrolysis. However, he is most famous for the equation named after him

Table 9.7 Experimental results for the decomposition of nitrogen(v) oxide, N_2O_5

Temperature/°C	Rate constant/s^{-1}
10	3.83×10^{-6}
20	1.71×10^{-5}
30	6.94×10^{-5}
40	2.57×10^{-4}
50	8.78×10^{-4}

Table 9.8 Data needed for the Arrhenius plot for the decomposition of nitrogen(v) oxide, N_2O_5

T^{-1}/K^{-1}	$\text{Log}(k/s^{-1})$
0.00353	-5.42
0.00341	-4.77
0.00330	-4.16
0.00319	-3.59
0.00309	-3.06

$E_a/2.303R$. The Arrhenius constant can then be obtained by substituting values into the Arrhenius equation.

EXAMPLE

Table 9.7 gives rate constants obtained at different temperatures for the reaction

$$2N_2O_5(g) \rightarrow 2N_2O_4(g) + O_2(g)$$

Determine (a) the activation energy and (b) the Arrhenius constant for the reaction.

SOLUTION

(a) The activation energy can be obtained by plotting $\log(k/s^{-1})$ against T^{-1}/K^{-1}. The first task is to convert the data in table 9.7 into the form required for plotting. This is shown in table 9.8. The graph obtained from plotting these points is shown in figure 9.9. From the graph we see that

$$\text{Slope} = -5667 \text{ K}$$

The log form of the Arrhenius equation shows that

$$\text{Slope} = -E_a/2.303R$$

The value for R is 0.008 314 kJ K^{-1} mol^{-1}. Thus

$$E_a = 2.303 \times (0.008\ 314 \text{ kJ K}^{-1} \text{ mol}^{-1}) \times (5667 \text{ K})$$
$$= 108.5 \text{ kJ mol}^{-1}$$

(b) The Arrhenius constant can be obtained directly by rearranging the log form of the Arrhenius equation:

$$\log A = \log k + E_a/2.303RT$$

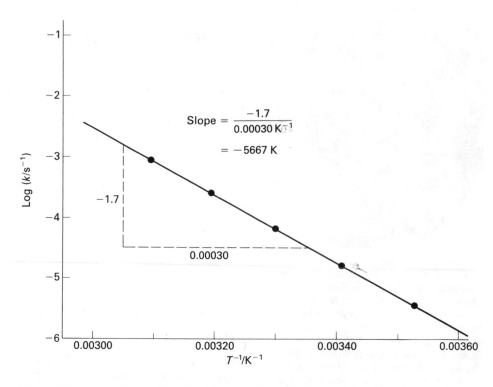

Figure 9.9 Arrhenius plot for the decomposition of nitrogen(v) oxide, N_2O_5

The value of k at any temperature may be used in the equation. For example, using the value of k at 30°C we obtain

$$\log A = -4.16 + \frac{108.5 \text{ kJ mol}^{-1}}{2.303 \times (0.008\ 314 \text{ kJ K}^{-1} \text{ mol}^{-1}) \times (303.2 \text{ K})}$$

$$= 14.53$$

$$A = 3.39 \times 10^{14} \text{ s}^{-1}$$

Why is the Study of the Effect of Temperature on Rates Important?

Temperature has a pronounced effect on the complex biological and chemical processes that take place in plants and animals. For example, the influence of temperature on rate of respiration is important in medicine. During some surgical operations, patients are cooled to slow down the metabolic rate, the rate of blood circulation and the rate of respiration.

Temperature also has a pronounced effect on the biochemical processes that take place in food when it is stored or cooked. The rate of food decomposition can be slowed down by various methods, including freezing (*see* figure 9.10), the use of preservatives such as sorbic acid, sulphur dioxide and nitrite and by suitable

In some heart and brain surgical operations, patients are cooled to slow their metabolic rates

Figure 9.10   Food stored at 5°C (the normal temperature of a refrigerator) may keep for several days. Frozen food, stored at −5°C to −2°C keeps for several weeks. Deep frozen food is stored at about −18°C. It keeps for several months

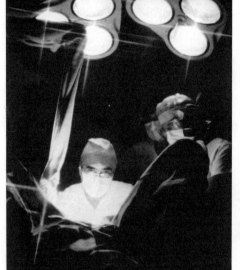

The rate of food decomposition is slowed down by cooling and freezing

Temperature not only affects the rate of chemical reactions; it can also influence the course of a reaction. For example, when ammonium nitrate is heated to temperatures of up to 200°C, the following reaction takes place at a moderate rate:

$$NH_4NO_3(s) \rightarrow N_2O(g) + 2H_2O(g)$$

This is the standard process for the laboratory preparation of dinitrogen oxide. However, if ammonium nitrate is heated to a high temperature under close confinement, the following reaction takes place explosively:

$$2NH_4NO_3(s) \rightarrow 2N_2(g) + 4H_2O(g) + O_2(g)$$

The thermal decomposition of ammonium nitrate can thus take place in two completely different ways depending on the temperature. In the next section we shall see how the experimental determination of reaction rates can lead us to predict the course of a chemical reaction.

The flavour and freshness of tea depend on the rates of a number of biochemical processes

storage. For example, the freshness and flavour of tea depends on a number of chemicals including a group called the theaflavins. On storage, the theaflavin content steadily decreases. The rate of loss of theaflavins is affected about equally by moisture and temperature.

The optimum efficiency of many industrial chemical processes depends on temperature. The second stage of the Contact process for the manufacture of sulphuric acid (*see* section 7.2) is one example:

$$2SO_2(g) + O_2(g) \rightarrow 2SO_3(g) \qquad \Delta H_{r,m}^{\ominus} = -192 \text{ kJ mol}^{-1}$$

Since the reaction is exothermic, the yield of sulphur trioxide can be increased by lowering the temperature. However, lowering temperature slows up the process. It is thus necessary to find the optimum balance between yield and rate of reaction. This is achieved by running this stage of the process at a temperature of 450 to 500°C using vanadium(v) oxide as a catalyst.

SUMMARY

1. The **rate of a reaction** is the rate at which products are formed or the rate at which reactants are used up in the reaction.
2. The rate of reaction may depend on
   (a) concentration or pressure of reactants,
   (b) temperature,
   (c) presence of catalysts,
   (d) availability of reactants—this includes state of subdivision of solids and surface area.
3. The **initial rate** of reaction is usually fastest. The rate then decreases during the reaction.
4. Reaction rate increases with the **state of subdivision** of a solid reactant.
5. The general **rate equation** is

$$\text{Rate} = kc^n$$

   where k is the **rate constant**, n is the **order of the reaction**, and c is the concentration of a reactant.
6. The order of a reaction and rate constant can be obtained from experimental data by
   (a) the **inspection method**,
   (b) the **rate curve method**,
   (c) the **integrated rates method**.
7. **Half-life** is the time taken for the initial concentration of reactant to fall to half its value. Half-life is related to rate constant by

$$t_{1/2} = \frac{0.693}{k}$$

8. The rates of most chemical reactions increase with temperature.
9. The **Arrhenius equation** relates rate constant k to temperature T. The most useful form of the equation is

$$\log k = \log A - E_a/2.303RT$$

   where A and E_a are constants for the reaction, and E_a is known as the **activation energy**.

**9.2
Reaction
Mechanisms and
Catalysis**

REACTION MECHANISMS

A chemical reaction may involve a sequence of individual reaction steps. This sequence of reaction steps is called the **reaction mechanism**. The mechanism for a chemical reaction is usually proposed on the basis of experimentally determined reaction rate data and also on the experimental detection of short-lived species which are not represented in the stoichiometric equations. These short-lived species are called **reactive intermediates**.

Let us consider the iodination of propanone in acid solution for example. The overall stoichiometric equation is given by

$$CH_3COCH_3(aq) + I_2(aq) \rightarrow CH_2ICOCH_3(aq) + HI(aq)$$

The experimentally derived rate equation for this reaction is

$$\text{Rate} = k[CH_3COCH_3][H_3O^+]$$

Note that the rate equation does *not* depend on the concentration of iodine. The rate equation thus does not correspond to the overall stoichiometric equation. This can be explained by proposing the following three-step reaction mechanism:

Step 1 CH_3COCH_3 + H_3O^+ $\xrightarrow{\text{slow}}$ $CH_3\overset{\overset{\displaystyle +OH}{\displaystyle \|}}{C}CH_3$ + H_2O

Step 2 $CH_3\overset{\overset{\displaystyle +OH}{\displaystyle \|}}{C}CH_3$ + H_2O $\xrightarrow{\text{fast}}$ $CH_2{=}\overset{\overset{\displaystyle OH}{\displaystyle |}}{C}CH_3$ + H_3O^+

Step 3 $CH_2{=}\overset{\overset{\displaystyle OH}{\displaystyle |}}{C}CH_3$ + I_2 $\xrightarrow{\text{fast}}$ CH_2ICOCH_3 + HI

Since the first step is the slowest step, it is the bottleneck. It thus controls and determines the overall rate of the reaction. This step is called the **rate-determining step**. The rate-determining step is always the slowest step of a reaction mechanism. The reactive intermediates in this three-step mechanism are

$$CH_3\overset{\overset{\displaystyle +\ OH}{\displaystyle \|}}{C}CH_3 \quad \text{and} \quad CH_2{=}\overset{\overset{\displaystyle OH}{\displaystyle |}}{C}CH_3$$

Molecularity

The number of species taking part in a step of the reaction mechanism is called the **molecularity** of that step.

> A **unimolecular** step involves one reactant species only in that step.
> A **bimolecular** step involves two reactant species in that step.
> A **termolecular** step involves three reactant species in that step.

In the above example, all three steps are bimolecular. Overall reaction order should not be confused with molecularity. A reaction which is first order overall may well involve a number of unimolecular and bimolecular steps.

THE COLLISION THEORY

The collision theory is a theory developed from the kinetic theory of gases (*see* section 3.1) to account for the influence of concentration and temperature on reaction rates. The theory is based on three major postulates:

- Reactions occur due to the collision of reactant particles.
- A collision only results in a reaction if a certain minimum energy is exceeded.
- A collision will not result in a reaction unless the colliding particles are correctly orientated to one another.

The collision theory does not necessarily apply to the reactant species represented by the stoichiometric equation. As we have seen, the stoichiometric equation represents the overall reaction. The mechanism of the reaction may involve a number of steps. The collision theory only applies to a specific step.

Concentration and the Collision Theory

We saw above that the rate of a chemical reaction generally increases with the concentration of one or more of the reactant species. According to the collision theory, as concentration increases so does the frequency of collisions. And as frequency of collisions increases, the probability of a collision having sufficient energy for a reaction to occur must also increase.

Proposed Mechanism for the Decomposition of N_2O_5

Overall Stoichiometric Equation

$$2N_2O_5(g) \rightarrow 4NO_2(g) + O_2(g)$$

Experimental Evidence
- Rate = $k[N_2O_5]$
- The species NO and NO_3 have been detected in the reaction mixture.

Proposed Mechanism

Step 1: $N_2O_5 \xrightarrow{\text{fast}} NO_2 + NO_3$

Step 2: $NO_2 + NO_3 \xrightarrow{\text{slow}} NO_2 + NO + O_2$

Step 3: $NO + NO_3 \xrightarrow{\text{fast}} 2NO_2$

Step 1 is unimolecular whilst steps 2 and 3 are bimolecular. Step 2 is the slowest step and thus the rate-determining step. By some fairly complicated mathematics, it can be shown that this reaction mechanism leads to the rate expression:

$$\text{Rate} = k[N_2O_5]$$

The mathematics depend on the steady-state approximation. This assumes that, after a while,

Rate of formation of reaction intermediates =
Rate of consumption of reaction intermediates

In this example, the reaction intermediates are NO and NO_3

Temperature and the Collision Theory

We saw in section 3.1 that the kinetic energy of molecules is spread over a wide range. The spread is called the **Maxwell–Boltzmann distribution**. As temperature increases, this spread shifts to higher energies. A Maxwell–Boltzmann distribution curve can be drawn showing the distribution of collision energies with respect to **collision frequency**. The collision frequency shown in figure 9.11 is the mean

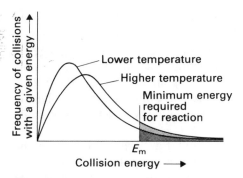

Figure 9.11 Maxwell–Boltzmann distribution of kinetic energies at two temperatures

number of collisions per unit time per unit volume with a given kinetic energy of relative translational motion along the line of centres of the colliding molecules. Collision frequency can be calculated from simple kinetic theory.

We can draw a vertical line through the distribution curve to represent the minimum collision energy required for a reaction to occur. The area enclosed by this line, the distribution curve and the horizontal axis represents the probability that the energy of collision will exceed this minimum energy. This area is shaded in figure 9.11. The curves show clearly that as temperature increases, this area also increases. It follows that, as temperature increases, the frequency of collisions with energy exceeding this minimum energy also increases. The rate of reaction therefore also increases.

The Arrhenius Equation (Again!)

According to the collision theory, the rate of reaction can be related mathematically to the collision frequency and the probability that the collision energy E exceeds the minimum energy E_m for the reaction as follows:

$$\text{Rate} = \text{Collision frequency} \times \text{Probability that } E > E_m$$

From this it is possible to derive the following relationship:

$$k = PZe^{-E_a/RT}$$

where k is the rate constant; P is the steric or probability factor with values between 0 and 1, and represents the fraction of colliding particles with correct orientation; Z is the collision number, which is related to the collision frequency; E_a is the activation energy, representing the minimum collision energy required by reacting molecules; R is the gas constant; and T is the absolute temperature.

The two factors P and Z can be combined into a single constant A known as the pre-exponential factor or the Arrhenius constant. This gives the famous Arrhenius equation, which we met in the previous section:

$$k = Ae^{-E_a/RT}$$

TRANSITION STATE THEORY

Transition state theory treats reacting molecules as a single entity. It deals with changes in the geometrical arrangements of the atoms involved as the system changes from reactants to products. A geometrical arrangement of atoms is called a **configuration**. As the reactant configurations change towards product configurations, there is an increase in potential energy until a maximum is reached. At this maximum, molecules have a critical configuration known as the **transition state** or **activated complex**. Only molecules with adequate total energy can attain

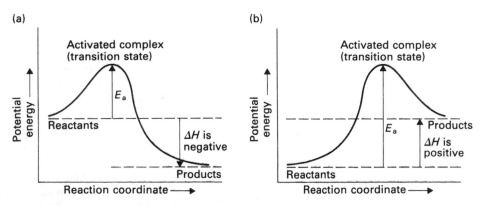

Figure 9.12 Potential energy–reaction coordinate graph for (a) exothermic reaction, and (b) endothermic reaction

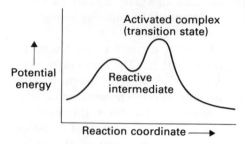

Figure 9.13 Potential energy–reaction coordinate graph showing a minimum for a reactive intermediate

this critical configuration. As the configuration changes from this transition state to the product configurations, there is a decrease in potential energy (*see* figure 9.12). The **reaction coordinate** in these two diagrams represents the changes in geometrical arrangement of the atoms of the reacting molecules taken as a whole entity as the system moves from the reactant arrangements through the critical arrangement to the product arrangements. Any reactive intermediate involved in the reaction appears at a minimum on the potential energy–reaction coordinate graph (*see* figure 9.13).

Transition state theory can be used to account for both A and E_a in the Arrhenius equation. By use of the theory and modern computer techniques, it is now possible to obtain an accurate picture of what is going on at the molecular level in a reaction.

CHAIN REACTIONS

A chain reaction is a self-sustaining chemical reaction in which the initial products participate in the formation of additional products. They are usually very rapid and often explosive.

Chain reactions involve three steps: an initiation step, a propagation step and a termination step.

Initiation Step

In this step an intermediate product is formed. The intermediate product can be an atom, an ion or a neutral molecule. The step can be initiated by light, nuclear radiation, thermal energy, anions or catalysts.

Propagation Step

During this step the intermediate products react with the original reactants to form further intermediate products and the final products. Propagation steps are repeated in a chain of reactions yielding more final and intermediate products.

Termination Step

In this step the intermediate products are finally consumed or destroyed. The reaction thus stops. A chain reaction may stop naturally or be stopped by an inhibitor.

Chain reactions play an important part in a number of branches of chemistry, particularly

- photochemistry,
- the chemistry of combustion,
- nuclear fission and fusion (*see* chapter 1),
- organic chemistry (*see* chapters 17–20).

Photochemistry

This is the branch of chemistry concerned with the interaction of light with matter. **Photosynthesis** is an example of a photochemical process.

Many chain ractions are initiated by light. The initiator is a photon, *hv* (*see* section 1.2). A classic example is the reaction between hydrogen and chlorine in the presence of light:

$$H_2(g) + Cl_2(g) \xrightarrow{hv} 2HCl(g)$$

This reaction is explosive. The three steps are as follows.

Initiation. In this step the covalent bond of the chlorine molecule is broken, producing two chlorine atoms, each with an unpaired electron:

$$Cl\text{–}Cl \rightarrow Cl^\bullet + Cl^\bullet$$

Photosynthesis is a complicated process which is still being investigated by photochemists and other scientists

This type of process is known as homolysis or **homolytic fission** (*see* section 17.3). It is also an example of **photolysis**. Photolysis means photochemical decomposition. The two chlorine atoms are the intermediate products. They are **radicals**. A radical is an atom or group of atoms possessing at least one unpaired electron. It should be noted that, although the initiation step is the slowest step of a chain reaction, it does not control the rate of the chain reaction.

Propagation. During this step the chlorine atoms react with hydrogen molecules, forming the final product hydrogen chloride and hydrogen radicals. The hydrogen radicals react with chlorine molecules, forming more product and more chlorine radicals:

$$Cl^\bullet + H_2 \rightarrow HCl + H^\bullet$$
$$H^\bullet + Cl_2 \rightarrow HCl + Cl^\bullet$$

These two reactions, which comprise the propagation step, are repeated millions of times.

Termination. The chain reaction is finally terminated by reactions such as

$$H^\bullet + H^\bullet \rightarrow H_2$$
$$Cl^\bullet + Cl^\bullet \rightarrow Cl_2$$

or

$$H^\bullet + Cl^\bullet \rightarrow HCl$$

A third body is necessary to absorb the energy of these termination reactions. The third body is usually the walls of the reaction vessel.

Quantum Yield

The absorption of one photon of radiation by a chlorine molecule in the above chain reaction may result in the formation of over a million molecules of hydrogen chloride. This is called the quantum yield. The quantum yields of photochemical reactions range from less than one to several million. A high quantum yield indicates that a chain reaction is involved.

Flash Photolysis

This is a technique used to produce radicals at sufficiently high concentrations for their identification. A simplified diagram of the apparatus used is shown in figure 9.14. The reaction mixture is subjected to a powerful flash of light from the photolysis flash tube. The light has energy of up to 10^5 J and a duration of about 10^{-4} s or less. Modern techniques use pulse lasers with a flash duration of about a nanosecond (10^{-9} s). The reaction can then be followed by recording a series of absorption spectra. The first flash is followed by a series of flashes from a low-intensity flash tube. These flashes are separated by milliseconds or microseconds and used to record the absorption spectra at these intervals.

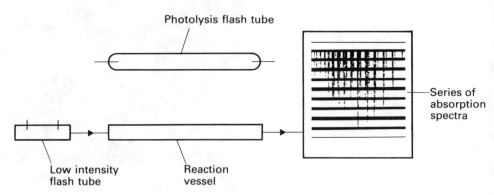

Figure 9.14   Flash photolysis

Combustion

A reaction with oxygen which produces thermal energy and light is called combustion. Combustion usually involves a complex sequence of radical reactions.

   An example is the combustion of hydrogen. Under certain conditions the reaction is explosive. Figure 9.15 shows experimental data for the reaction of a stoichiometric mixture of hydrogen and oxygen in a Pyrex reaction vessel. The shaded region represents the explosion region. For the combustion of hydrogen, the region is in the form of an **explosion peninsula**. The explosion region is bounded by an explosion boundary.

CATALYSIS

Hydrogen peroxide decomposes very slowly by itself to form hydrogen and oxygen:

$$2H_2O_2(aq) \rightarrow 2H_2O(l) + O_2(g)$$

However, in the presence of manganese(IV) oxide, the reaction is very rapid. Manganese(IV) oxide is a catalyst for the reaction. A **catalyst** is a substance which increases the rate of a chemical reaction without itself being consumed in the process.

Characteristics of Catalysts

Catalysts have a number of characteristics. The most important of these can be summarised as follows.

Stoichiometry

A catalyst does not affect the overall stoichiometry of a reaction.

Equilibrium and Yield

A catalyst speeds up the rates of both forward and backward reactions. It thus

Figure 9.15   Explosion diagram for the combustion of hydrogen:

$$2H_2(g) + O_2(g) \rightarrow 2H_2O(g)$$

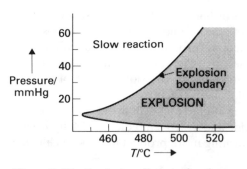

Combustion involves a complex sequence of radical reactions

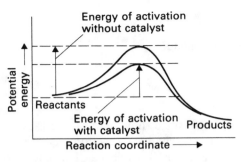

Figure 9.16 A catalyst reduces the energy of activation

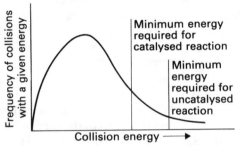

Figure 9.17 A catalyst increases the frequency of those collisions resulting in a reaction

speeds up the rate at which equilibrium is attained. However, it does not alter the position of equilibrium. This means that it does not alter the equilibrium concentrations. A catalyst thus does not alter the yield of a reaction.

Reaction Mechanism

A catalyst provides an alternative reaction pathway. The activation energy of this alternative route is lower (*see* figure 9.16). As a result, more reactant molecules possess the energy required for a successful collision. The total number of collisions resulting in a reaction per unit time increases accordingly and the rate of the reaction thus increases (*see* figure 9.17).

Catalysts are Often Specific

A catalyst may increase the rate of one reaction but not increase the rate of a similar reaction.

Chemical Involvement

A catalyst is chemically involved in a reaction. It is consumed in one reaction step and **regenerated** in a subsequent step. It can thus be used repeatedly without undergoing any permanent change. However, although a catalyst does not undergo a net chemical change, it may well change its physical form.

Autocatalysis

This is the catalysis of a reaction by one of the products of the reaction. The reaction is initially slow but as products are formed the reaction speeds up. One of the most common examples of autocatalysis is the oxidation of ethanedioic acid (oxalic acid) by potassium manganate(VII):

$$2MnO_4^-(aq) + 5C_2O_4^{2-}(aq) + 16H^+(aq) \rightarrow 2Mn^{2+}(aq) + 10CO_2(g) + 8H_2O(l)$$

The reaction is catalysed by Mn^{2+} ions. When the reaction is carried out at room temperature, initially it is very slow. However, as Mn^{2+} ions are formed, the reaction speeds up. This can easily be tested in the laboratory by adding some Mn^{2+} in the form of manganese(II) sulphate solution at the start of the reaction. The reaction then proceeds at a reasonable rate right from the start.

Intermediate Compound Theory of Catalysis

In 1889 Arrhenius suggested that a catalyst works by forming an intermediate compound. The catalyst (C) reacts with the reactant, known as the **substrate** (S), to form the intermediate compound CS. The intermediate compound then decomposes forming product P and unchanged catalyst C:

Step 1   C + S   →   CS
                                        intermediate
                                        compound

Step 2   CS   → C + P

The catalyst is thus consumed in the first step and regenerated in the next.

The intermediate compound theory can also be applied to reactions involving two reactants:

Step 1   A + C   →   AC
                                        intermediate
                                        compound

Step 2   AC + B   → AB + C

Overall reaction   A + B   → AB

TYPES OF CATALYSTS

It is convenient to discuss catalysts under three headings:

- homogeneous catalysts,
- heterogeneous catalysts,
- biological catalysts (enzymes).

These all might be regarded as positive catalysts since they increase the rate of reaction. A substance which decreases the rate of a chemical reaction is called an **inhibitor**. Inhibitors are sometimes known as **negative catalysts**. They do *not* function in the opposite way to catalysts, that is by raising the activation energy. Rather, they just inhibit the normal course of a chemical reaction by reacting with and removing reactive intermediates from the reaction sequence. An example of an inhibitor is glycerine (propane-1,2,3-triol), which slows down the decomposition of hydrogen peroxide. Dilute acid also inhibits this reaction.

HOMOGENEOUS CATALYSIS

Homogeneous catalysis occurs when the catalyst and the reacting system are in the same phase. This type of catalysis can often be explained in terms of the formation of an intermediate compound or intermediate radical or ion—as the following two examples show.

Homogeneous Catalysis in the Gas Phase

The catalytic decomposition of dinitrogen oxide in the gas phase is an example of this type of catalysis. Dinitrogen oxide is an anaesthetic known as 'laughing gas'. At room temperature it is comparatively inert. It only decomposes at temperatures in excess of 1000 K:

$$2N_2O(g) \rightarrow 2N_2(g) + O_2(g)$$

However, the decomposition is catalysed by traces of chlorine gas, particularly in the presence of light. The catalyst is thought to be chlorine radicals produced by the photolysis of the chlorine gas (*see* above). The chlorine radical reacts with N_2O forming the intermediate radical $ClO^\bullet$. The proposed mechanism is

Step 1	$N_2O(g) + Cl^\bullet(g) \rightarrow N_2(g) + ClO^\bullet(g)$
Step 2	$2ClO^\bullet(g) \rightarrow Cl_2(g) + O_2(g)$

Homogeneous Catalysis in the Aqueous Phase

Examples of this type of catalysis are numerous. The decomposition of hydrogen peroxide, for example, is catalysed by iodide ions. The intermediate ion in this case is IO^-. The steps are

Step 1	$H_2O_2(aq) + I^-(aq) \rightarrow H_2O(l) + IO^-(aq)$
Step 2	$H_2O_2(aq) + IO^-(aq) \rightarrow H_2O(l) + O_2(g) + I^-(aq)$

Acid catalysis and **base catalysis** of organic reactions are important examples of homogeneous catalysis. During acid catalysis a proton is transferred to a reactant, forming a positively charged intermediate compound. During base catalysis a reactant transfers a proton to the base. The reactant thus acquires a negative charge.

Acid catalysis and base catalysis are particularly important in organic chemistry. The iodination of propanone in acid solution (*see* above) is one example of this. Another is the acid hydrolysis of esters. The mechanism of these hydrolyses is similar to that for the iodination of propanone. The acid hydrolysis of ethyl ethanoate, for example, has many protonation and deprotonation steps. A simplified three-step version of this mechanism is shown in figure 9.18. The base-catalysed condensation of propanone to form 4-hydroxy-4-methylpentan-2-one is also thought to take place in three steps (*see* figure 9.19).

Some organic reactions such as the mutarotation of glucose (*see* section 17.2) are catalysed by both acids and bases.

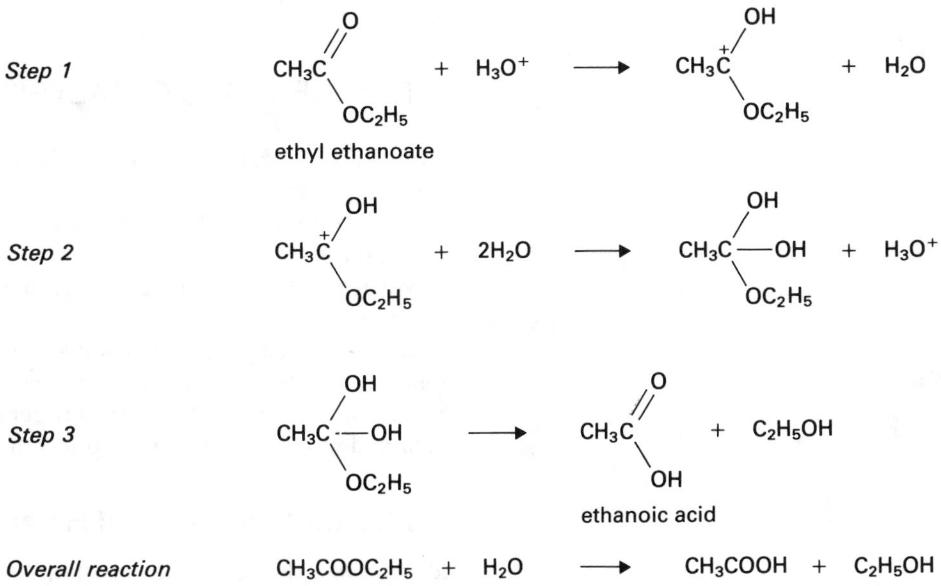

Step 1

$$CH_3C\overset{O}{\underset{OC_2H_5}{\big|}} + H_3O^+ \longrightarrow CH_3\overset{+}{C}\overset{OH}{\underset{OC_2H_5}{\big|}} + H_2O$$

ethyl ethanoate

Step 2

$$CH_3\overset{+}{C}\overset{OH}{\underset{OC_2H_5}{\big|}} + 2H_2O \longrightarrow CH_3C\overset{OH}{\underset{OC_2H_5}{\big|}}\!\!-OH + H_3O^+$$

Step 3

$$CH_3C\overset{OH}{\underset{OC_2H_5}{\big|}}\!\!-OH \longrightarrow CH_3C\overset{O}{\underset{OH}{\big|}} + C_2H_5OH$$

ethanoic acid

Overall reaction $\quad CH_3COOC_2H_5 + H_2O \longrightarrow CH_3COOH + C_2H_5OH$

Figure 9.18 A mechanism for the acid hydrolysis of ethyl ethanoate

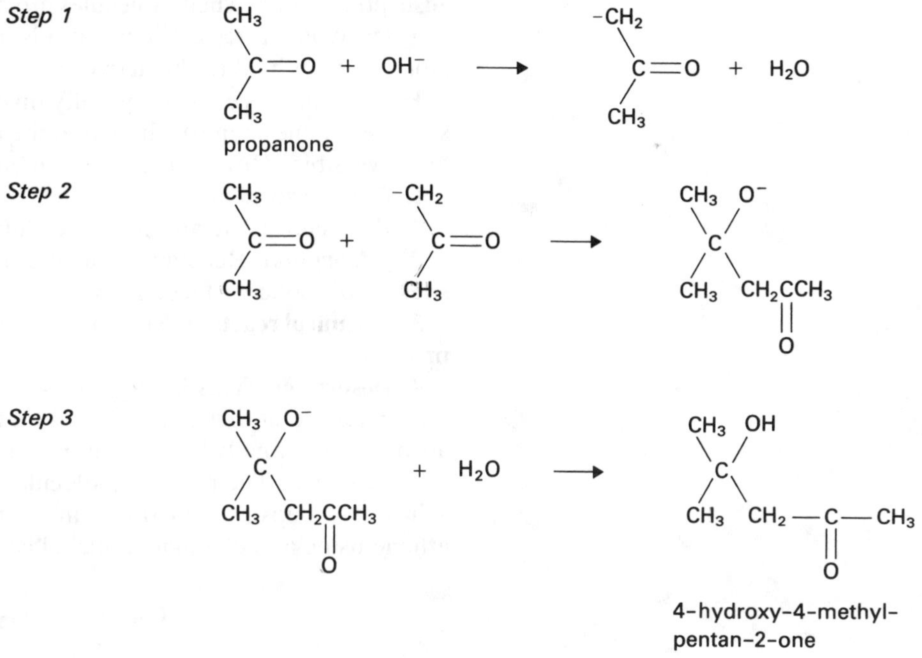

Step 1

$$\underset{CH_3}{\overset{CH_3}{\big\backslash}}C{=}O + OH^- \longrightarrow \underset{CH_3}{\overset{-CH_2}{\big\backslash}}C{=}O + H_2O$$

propanone

Step 2

$$\underset{CH_3}{\overset{CH_3}{\big\backslash}}C{=}O + \underset{CH_3}{\overset{-CH_2}{\big\backslash}}C{=}O \longrightarrow \underset{CH_3\ CH_2CCH_3}{\overset{CH_3\ O^-}{\big|}}C\underset{O}{\|}$$

Step 3

$$\underset{CH_3\ CH_2CCH_3}{\overset{CH_3\ O^-}{\big|}}C\underset{\|O}{} + H_2O \longrightarrow \underset{CH_3\ CH_2-C-CH_3}{\overset{CH_3\ OH}{\big|}}C\underset{\|O}{}$$

4-hydroxy-4-methyl-
pentan-2-one

Figure 9.19 Mechanism for the base-catalysed condensation of propanone

d-Block Metal Ions

Many redox reactions in solution are catalysed by d-block metal ions (*see* chapter 14). For example, a few drops of copper(II) sulphate solution catalyse the reaction between zinc and hydrochloric acid:

$$Zn(s) + 2HCl(aq) \rightarrow ZnCl_2(aq) + H_2(g)$$

Organometallic Complexes

These often act as intermediates in industrial processes. For example, the oxidation of ethene, C_2H_4, to ethanal, CH_3CHO (acetaldehyde) is catalysed by passing gaseous ethene into an aqueous solution of palladium(II) chloride and copper(II) chloride. The reaction mechanism is complicated and involves several intermediates. One of these is $[HOCH_2CH_2PdCl_2(OH)_2]^-$.

HETEROGENEOUS CATALYSIS

Heterogeneous catalysis occurs when the catalyst and reacting system are in different phases.

Many examples of heterogeneous catalysis can be explained in terms of the intermediate compound theory. For example, during the catalytic decomposition of hydrogen peroxide, the catalyst, manganese(IV) oxide, is consumed and then regenerated.

All heterogeneous processes take place at a **phase boundary**. The most commonly studied heterogeneous processes are those which involve gaseous molecules reacting at a solid surface. Heterogeneous catalysis at a solid surface can be explained in terms of the adsorption theory.

The Adsorption Theory of Heterogeneous Catalysis

Adsorption is the accumulation of molecules at a surface. Care should be taken to distinguish between adsorption and **absorption**. The latter is the penetration of molecules into the bulk. Two types of adsorption are recognised. **Physical adsorption** occurs when molecules are bound to 'active sites' on a solid surface by van der Waals forces. Chemical adsorption or **chemisorption** occurs when the molecules are held to the active sites on the surface by chemical bonds.

Heterogeneous catalysis usually involves both physical adsorption and chemisorption. Some chemists think that the mechanism involves five steps, all of which are reversible. However, no agreement has been reached on the mechanism as yet. The five steps are:

1. **Diffusion**. Reactant molecules diffuse to the solid surface.
2. **Adsorption**. Reactant molecules first become physically adsorbed on active sites on the solid surface. They then become chemisorbed.
3. **Chemical reaction**. Reactant molecules adjacent to one another react to form products.
4. **Desorption**. This is the reverse of adsorption. After the reaction, product molecules are at first chemisorbed on the surface. They then become physically adsorbed and finally break free from the surface.
5. **Diffusion**. The product molecules diffuse away from the surface.

The five steps are illustrated in figure 9.20 for the catalytic hydrogenation of ethene using a finely divided nickel catalyst:

$$C_2H_4(g) + H_2(g) \xrightarrow[400 \text{ K}]{\text{Ni}} C_2H_6(g)$$

Note that the hydrogen is bonded as single atoms to the nickel atoms at the surface. Note also that by being finely divided the catalyst exposes a greater surface area and is thus more effective as a catalyst.

Classes of Heterogeneous Catalyst

Heterogeneous catalysts may be divided into four classes. These are shown in table 9.9. The examples in the table also illustrate the importance of catalysts in industry.

Catalysed processes are responsible for generating some 20% of the gross national product (GNP) in the United States.

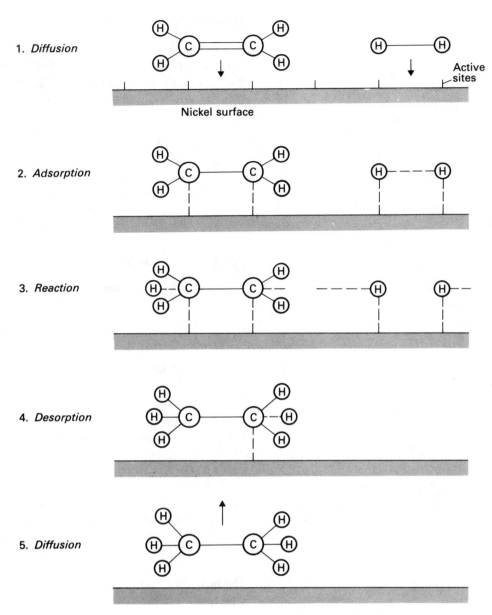

1. *Diffusion*

Active sites

Nickel surface

2. *Adsorption*

3. *Reaction*

4. *Desorption*

5. *Diffusion*

Figure 9.20 Schematic representation of the five steps of heterogeneous catalysis

Table 9.9 Classes of heterogeneous catalyst

Class	Example	Type of reaction catalysed	Example
d-block metal	Fe, Ni, Pd, Pt	hydrogenation, dehydrogenation, oxidation	manufacture of margarine, Ostwald process, catalytic converter, Haber process
semiconducting oxides	V_2O_5, NiO, CuO	oxidation and reduction	Contact process
insulator	Al_2O_3, SiO_2	hydration, dehydration	dehydration of ethanol to form ethene
acid catalysts	natural clays (aluminium silicates), H_2SO_4	cracking, isomerisation, polymerisation	cracking of long-chain hydrocarbons

Catalytic Converters

A catalytic converter is used in some exhaust systems to convert gases which are environmentally harmful into harmless gases. A diagram of a typical catalytic converter is shown in figure 9.21. The exhaust gases containing hydrocarbons and carbon monoxide are passed through a 'honeycomb' of small beads coated with platinum and palladium catalysts. The convertor is heated and extra air pumped into it. The hydrocarbons and carbon monoxide are converted to carbon dioxide and water, both of which are harmless.

Cars which include catalytic converters in their exhaust systems must use lead-free petrol to prevent the catalyst from being coated with lead.

Catalytic converters are also used to reduce nitrogen oxides to nitrogen.

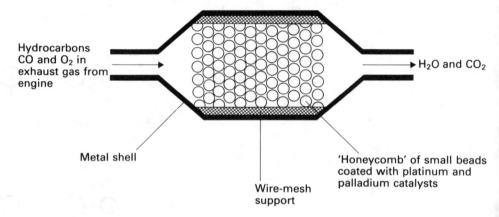

Figure 9.21   The catalytic converter

Promoters and Poisons

A **promoter** is a substance which enhances the performance of a catalyst. For example, small amounts of potassium and aluminium oxides are used as promoters to improve the performance of the iron catalyst in the Haber synthesis of ammonia (*see* section 7.2).

A catalyst **poison** inhibits the effectiveness of a catalyst. The hydrogenation of alkenes using a finely divided nickel catalyst is poisoned by carbon monoxide for example.

ENZYMES

Enzymes are proteins which catalyse the chemical reactions in living systems. They are often called biological catalysts or biocatalysts. Without them most biochemical reactions would be too slow to sustain life.

Properties of Enzymes

Enzymes have a number of distinct properties and characteristics. The following are the most important.

Size
The relative molecular masses of enzymes range from 10^5 to 10^7. Since the dimensions of enzymes are in the colloidal range (*see* section 6.3) they cannot be classified as either homogeneous or heterogeneous catalysts. They fall somewhere in between.

Selectivity
Enzymes vary in their degree of **specificity**. Enzymes with a low degree of

specificity catalyse a relatively broad range of biochemical reactions. Those with a high degree of specificity only catalyse a limited range of reactions.

Enzymes in the lipase group, for example, have a relatively low degree of specificity. They catalyse the hydrolysis of most esters. The enzyme β-glucosidase, on the other hand, is very specific. It catalyses the hydrolysis of β-glucosides but not α-glucosides (*see* chapter 20). Another example of a highly specific enzyme is urease. This enzyme is found in soya beans and jack beans. It catalyses the hydrolysis of urea (carbamide):

$$CO(NH_2)_2 + H_2O \rightarrow CO_2 + 2NH_3$$

However, it does not catalyse the hydrolysis of substituted ureas.

In general, enzymes are highly selective and only catalyse a specific reaction or type of reaction.

Efficiency

Even small amounts of certain enzymes can be highly efficient. This is because the enzyme molecules are regenerated during their catalytic activity. A typical enzyme molecule may be regenerated a million times in one minute.

An example is the enzyme rennin which is found in rennet. Rennet is used in cheese making. It coagulates over a million times its own weight of milk protein.

Another example is catalase. In one second, one molecule of this enzyme decomposes some 50 000 molecules of hydrogen peroxide at 0°C:

$$2H_2O_2(l) \rightarrow 2H_2O(l) + O_2(g)$$

In doing so it lowers the activation energy of the reaction from about 75 kJ mol^{-1} to about 21 kJ mol^{-1}. When colloidal platinum is used as a catalyst for this reaction, the activation energy is lowered to about 50 kJ mol^{-1} only.

Temperature

Enzymes operate most effectively at body temperature, that is about 37°C. When the temperature rises above 50 to 60°C, they are destroyed and thus become inactive. Figure 9.22 shows a plot of reaction rate against temperature for a typical enzyme-catalysed reaction.

Poisons

Enzymes are sensitive to poisons. In fermentation processes, for example, ethanol poisons the enzymes in yeast if the ethanol concentration is more than 15.5%. The fermentation thus stops. It is therefore normally impossible to produce wine or beer with an ethanol concentration higher than 15.5% by the fermentation process alone.

How Do Enzymes Work?

In 1902 Henri proposed that enzymes operate by forming a complex with a substrate molecule in a reversible process. The enzyme–substrate complex corresponds to the intermediate compound or transition state in the intermediate compound theory. The complex then breaks down to form products and regenerate the enzyme:

$$E + S \rightleftharpoons ES \rightarrow E + P$$

where E is the enzyme, S is the substrate, ES the complex, and P the products. This equation was first put forward by Michaelis and Menton in 1913 and is often known as the **Michaelis–Menton equation**.

The substrate molecule is thought to bind to a region on the surface of the enzyme called the active site. The activity at this site is helped by vitamins and minerals. **Trace elements** and particularly **d-block metals** such as copper, manganese, iron and nickel are essential for this activity.

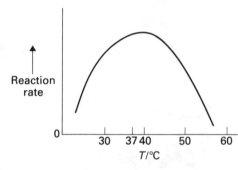

Figure 9.22 The variation of reaction rate against temperature for a typical enzyme-catalysed reaction

Certain enzymes require the assistance of a **coenzyme** for their activity. Coenzymes are relatively small organic molecules which bind to the active sites of an enzyme. B-complex vitamins are often components of these coenzymes. A schematic representation of the operation of a coenzyme is shown in figure 9.23.

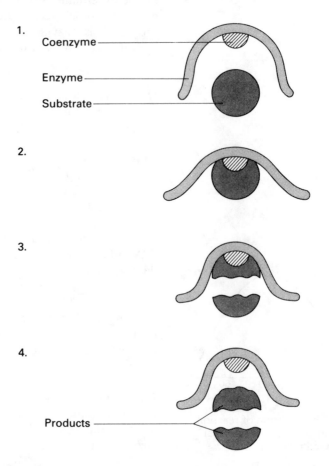

1.

Coenzyme

Enzyme

Substrate

2.

3.

4.

Products

Figure 9.23 A schematic representation of how a substrate molecule binds to an enzyme and coenzyme. After reaction, the products are released from the enzyme, which is then free to bind to another substrate molecule

Importance of Enzymes

Enzymes play an important part in controlling the chemical reactions which take place in all living systems. In the human body, for example, thousands of chemical reactions take place every second. Enzymes are also important in a number of industrial processes. They are used, for example, in the preparation and manufacture of foods, wine and beer, pharmaceuticals, detergents, textiles, leather and paper.

The Human Digestive System

The digestive system converts nutrients such as proteins, carbohydrates and fats to products which can be readily absorbed for use by the cells in the body.

The enzyme amylase, which is found in both saliva and the small intestine, helps convert starch to maltose. The maltose is converted to glucose in the small intestine with the aid of another enzyme called maltase.

Gastric enzymes such as pepsin and trypsin convert proteins to peptides in the stomach and small intestine. The peptides are then converted to amino acids in the small intestine by the action of enzymes called peptidases. Lipase is another enzyme found in the small intestine. It hydrolyses fats (lipids) forming fatty acids.

Metabolic Processes

These are the chemical processes that take place in living cells. A typical animal cell consists of a **nucleus** surrounded by a material called **cytoplasm**. This is bounded by the **cell membrane**. The cytoplasm contributes to the processing of food by extracting energy from it and producing the compounds needed by other cells. One of the enzymes found in cytoplasm is cytochrome oxidase. Cytochromes are a group of proteins that contain iron in the form of haem. Together with enzymes such as cytochrome oxidase they play a major role in **respiration**. They do so by oxidising food to produce energy. Cytochrome oxidase contains the trace element copper. Without this element, animals and plants cannot metabolise food effectively. In plants, cytochromes are found in chloroplasts. These are essential for photosynthesis.

Enzymes and Disease

Certain diseases are caused by enzyme deficiencies or excess levels of enzymes. The congenital disease phenylketonuria, for example, is due to a deficiency of the enzyme phenylalanine hydroxylase. The disease results in the accumulation of compounds in the body, causing brain damage and mental retardation. The brain damage can be prevented by a diet with a low content of the amino acid phenylalanine. Albinism is another disease caused by the lack of an enzyme, in this case the enzyme tyrosinase. In both of these cases the enzyme deficiency is caused by genetic mutation.

Heart attacks can result in high levels of enzymes in the blood. This is due to leakage of enzymes from the cells destroyed when heart tissue is destroyed.

Enzymes are also used to treat diseases—such as heart disease. Many heart attacks are caused by a blood clot forming in a coronary artery. A recent method of treatment uses the enzyme streptokinase to dissolve the clot. The enzyme is injected into the heart through a catheter. The catheter is inserted in the patient's groin and passed up the aorta into the blocked vessel in the heart.

Proenzymes

These are also known as zymogens. They are non-catalytic substances formed by plants and animals and only converted to enzymes when needed. For example, blood contains a substance called prothrombin. When needed, this can be converted to thrombin, an enzyme used to clot blood. It catalyses the conversion of the soluble protein, fibrinogen, in the blood to the insoluble fibrin.

Fermentation

This is a chemical process in which micro-organisms such as yeast and certain bacteria act on sugars to produce ethanol and other products. The fermentation of glucose to form ethanol may be represented by the equation:

$$C_6H_{12}O_6 \rightarrow 2C_2H_5OH + 2CO_2$$

The mechanism of this process is complicated. One of the major intermediates is 2-oxopropanoic acid (pyruvic acid):

$$CH_3-\underset{\underset{O}{\|}}{C}-COOH$$

There are over 12 steps involved in the process and each step requires a specific enzyme. Zymase is one of the 12 or so enzymes found in a typical yeast.

Ethanol can be manufactured industrially by the fermentation of cane sugar followed by distillation. Cane sugar contains about 50% sucrose, $C_{12}H_{22}O_{11}$. A

A fermentation yeast such as the wine yeast *Saccharomyces ellipsoideus* has the ability to ferment sugars up to 15.5% by volume of ethanol. Wild yeasts can only ferment up to about 5.5%. Film yeasts—such as those used to make sherry—can ferment up to 20% by volume ethanol.

A vineyard in France. More than 100 million acres spread over at least 50 different countries are used to cultivate vines. The total production of wine in the world is over 30 000 million litres (3×10^{10} dm^3) each year. European countries account for about 80% of this

syrup of the sugar called molasses is diluted before fermentation. During the fermentation process the sucrose is converted to glucose and fructose. The process is catalysed by the enzyme invertase.

Beer is made by fermentation of a mixture of malt and hops. The hops are added to give the beer its flavour. The malt is obtained from a cereal—usually barley. The starch in malt is converted to maltose by the action of an enzyme called diastase. Another yeast enzyme, maltase, then catalyses the conversion of maltose to glucose. Haze in beer can be removed by an enzyme called papain. This hydrolyses the proteins causing the haze. Papain is also used to tenderise beef.

In wine making the natural yeasts present on grape skins act on the sugars in grape juice to form ethanol. Sometimes a cultured yeast is also added. If the fermentation is allowed to proceed to completion, a dry wine is produced. If it is stopped early, a sweeter wine is produced.

Since solutions containing more than 15.5% ethanol poison the enzymes in yeast, distillation must be used to produce beverages with a higher alcohol content. Alternatively wines can be 'fortified' by adding a spirit such as vodka or brandy.

SUMMARY

1. A **reaction mechanism** is a sequence of reaction steps usually proposed on the basis of experimentally determined reaction rate data and on the experimental detection of **reactive intermediates**.
2. The **rate-determining step** in a reaction mechanism is the slowest step.
3. **Molecularity** of a step in a reaction mechanism is the number of species taking part in that step.
4. The influence of concentration and temperature on reaction rates can be explained in terms of the **collision theory**.
5. **Transition state theory** deals with changes in the geometrical arrangements of atoms in reacting molecules treated as a single entity.
6. The **transition state**, or activated complex, is a **critical configuration** which coincides with a maximum in potential energy along the **reaction**

coordinate. Reacting molecules which attain this critical configuration form product molecules.

7. **Chain reactions** involve three steps:
   (a) the **initiation step**,
   (b) the **propagation step**,
   (c) the **termination step**.

8. A **catalyst** is a substance which increases the rate of a chemical reaction without itself being consumed in the process.

9. The **intermediate compound theory** of catalysis suggests that a catalyst works by forming an intermediate compound with the **substrate.**

10. A substance which decreases the rate of a chemical reaction is called an **inhibitor**.

11. A **homogeneous catalyst** operates in the same phase as the reacting system.

12. A **heterogeneous catalyst** is in a different phase to the reacting system. It operates at a phase boundary.

13. **Physical adsorption** occurs when molecules are bound to active sites on a solid surface by van der Waals forces.

14. **Chemical adsorption** or chemisorption occurs when the molecules are held to the active sites on the surface by chemical bonds.

15. **Enzymes** are proteins which catalyse the chemical reactions in living systems.

16. **Coenzymes** assist enzymes. They are bound to the active sites of enzymes.

Examination Questions

1.   Reaction rates can be investigated by various physical and chemical techniques. For each of **three** reactions outline how the progress of the reaction can be followed and the rate of the reaction determined, specifying what measurements should be taken. A *different* technique should be used for each example. In **one** case indicate how the rate constant is obtained from the experimental measurements.

(NISEC)

2. (a) Explain how the rate of a reaction may be affected by:
   (i) changing the concentration of the reactant;
   (ii) changing the temperature of the reaction;
   (iii) the presence of a catalyst.
   (iv) Sketch a graph of concentration of reactant against time and indicate how you would find the initial rate of reaction from your graph.

   (b) (i) What do you understand by order of reaction, rate (velocity) constant and rate-determining step? Why cannot the rate equation be deduced from the overall chemical equation?
   (ii) For the reaction $T + U \rightarrow V$, the rate of the reaction was found to depend on the concentration of T and on the concentration of U *squared*.
       Write the rate equation for the reaction, state the overall order of the reaction, and derive the units of the rate constant.
   (iii) Most reactions are exothermic. Why are so few exothermic reactions spontaneous?

(c) The disintegration of a radioactive isotope follows first-order kinetics. Show diagrammatically how the activity of an isotope changes with time and illustrate the meaning of half-life $(t_{1/2})$. How does $t_{1/2}$ depend on the initial concentration (activity) of the isotope?

(d) The gas-phase reaction between A and B is zero order with respect to A and second order with respect to B. When the concentrations of A and B are 1.0 and 1.5 mol dm^{-3} respectively, the rate is 7.75×10^{-4} mol dm^{-3} s^{-1}.

   (i) Calculate the initial rate when the concentrations of A and B are 2.0 and 3.0 mol dm^{-3} respectively. State what would happen to the rate if:

   (ii) the volume of the reaction vessel was trebled, and

   (iii) the pressure of B was doubled.

(SUJB)

3. (a) Explain the terms *velocity constant* and *order of reaction*.

   (b) In an investigation of the recombination of X atoms to give X_2 molecules in the gas phase in the presence of argon, that is the reaction

$$X + X + Ar = X_2 + Ar$$

the following data were obtained.

With the concentration of argon fixed at 1.0×10^{-3} mol dm^{-3}

Initial concentration, [X] /mol dm^{-3}	Initial rate, $\dfrac{d[X_2]}{dt}$ /mol dm^{-3} s^{-1}
1.0×10^{-5}	8.70×10^{-4}
2.0×10^{-5}	3.48×10^{-3}
4.0×10^{-5}	1.39×10^{-2}

With the concentration of X atoms fixed at 1.0×10^{-5} mol dm^{-3}

Initial concentration, [Ar] /mol dm^{-3}	Initial rate, $\dfrac{d[X_2]}{dt}$ /mol dm^{-3} s^{-1}
1.0×10^{-3}	8.70×10^{-4}
5.0×10^{-3}	4.35×10^{-3}
1.0×10^{-2}	8.69×10^{-3}

Find the order of reaction with respect to (i) [X], (ii) [Ar] and hence determine the overall velocity [rate] constant for the formation of X_2 molecules.

   (c) Write a short account of the photochemical reaction between methane and chlorine.

(OLE)

4. Write down a general equation for the rate of a reaction and define, or explain, the terms used.

An investigation of the rate of hydrolysis of an ester in an excess of aqueous acid at 298 K gave the following information:

Time/min	0	10	20	30	40	50	60	70	80
Fraction of ester remaining	1	0.62	0.38	0.24	0.14	0.08	0.05	0.03	0.02

(a) Using graph paper, plot the fraction of ester remaining against time.
(b) Measure the gradient of the curve at the following values of the fraction of ester remaining, state the units in which the gradient is measured and explain briefly what you have done.
  Fraction of ester remaining: 0.8   0.6   0.4   0.3   0.2.
What information do you now possess?
(c) By plotting a graph of the measured gradients against other appropriate data, deduce the first-order rate constant for the hydrolysis under these conditions. State any assumption made. Give the units in which the rate constant is measured.
(*It is not necessary to use the integrated form of the rate equation.*)
(d) From the first graph, measure the time intervals for the fraction of ester remaining to be successively halved and average your result. Comment on what you have found, and give *one* example of where a similar phenomenon is found in chemistry.

(SUJB)

5. (a) State *five* factors which may affect the rate of a chemical reaction.
   (b) The reaction between iodide ions and persulphate (peroxodisulphate(VI)) ions in aqueous solution may be represented by the equation:

$$2I^-(aq) + S_2O_8^{2-}(aq) \rightarrow I_2(aq) + 2SO_4^{2-}(aq)$$

Outline a method of investigating the rate of this reaction so that the order of the reaction with respect to persulphate ions may be determined.
   (c) The hydrolysis of methyl ethanoate may be represented by the equation:

$$CH_3COOCH_3(l) + H_2O(l) \rightleftharpoons CH_3COOH(aq) + CH_3OH(aq)$$

When the hydrolysis was carried out in the presence of aqueous hydrochloric acid in a constant-temperature bath, the following results were obtained.

Time/s $\times 10^4$	Concentration of ester/mol dm^{-3}
0	0.24
0.36	0.156
0.72	0.104
1.08	0.068
1.44	0.045

(i) State the reasons for using the hydrochloric acid and the constant-temperature bath.
(ii) Plot a graph of ln[ester] against time.
(iii) By reference to the graph, show that the reaction is first order with respect to the ester.
(iv) Given the equation $\dfrac{d[ester]}{dt} = -k[ester]$, use your graph to determine a value for k and state its units.

(AEB, 1985)

6. (a) Mercury(II) chloride, $HgCl_2$, reacts with oxalate (ethanedioate) ions, $C_2O_4^{2-}$, in solution with the precipitation of Hg_2Cl_2.

$$2HgCl_2(aq) + C_2O_4^{2-}(aq) \rightarrow 2Cl^-(aq) + 2CO_2(g) + Hg_2Cl_2(s)$$

Initial rates at 373 K for solutions of the concentration shown and expressed as mol dm^{-3} min^{-1} of HgCl$_2$ reacted are given below.

Experiment	HgCl$_2$/mol dm^{-3}	K$_2$C$_2$O$_4$/mol dm^{-3}	Rate $\times$ 10^4/ mol dm^{-3} min^{-1}
1	0.0836	0.202	0.52
2	0.0836	0.404	2.08
3	0.0418	0.404	1.06

(i) Write down the rate equation for the reaction, expressing the rate in differential form.
(ii) Deduce the order of reaction with respect to each of the reactants.
(iii) Calculate the value of the rate constant and state its units, if any.
(iv) Compare the information expressed by the equation for the reaction with what has been found and comment.

(b) Sketch curves of reactant concentration against time for the decomposition of a gas A to show the relative rates for both first- and second-order decompositions. Label the graphs with explanatory notes.

(c) What do you understand by the term *half-life* of a reaction? How would the half-life of the decomposition of the gas A in (b) depend on the initial concentration of A according to the different orders?

(d) The rate at which hydrogen peroxide decomposes may be followed by titration against potassium manganate(VII) solution (potassium permanganate) under acid conditions, equal volumes being withdrawn at specified times and titrated:

Time/min	0	10	20	30
Volume of KMnO$_4$(aq)/cm^3	45.6	27.6	16.5	10.0

Determine the half-life of the decomposition, deduce the rate equation and explain how you might calculate the rate constant.

(SUJB)

7. (a) Explain what is meant by the *activation energy* of a reaction, and show how the concept provides a qualitative explanation of the effect of heat on the velocity of a chemical reaction.

(b) The table below gives values for the velocity constant, k, of the reaction between potassium hydroxide and bromoethane in ethanol at a series of temperatures, T.

k/dm^3 mol^{-1} s^{-1}	T/K
0.182	305.0
0.466	313.0
1.35	323.1
3.31	332.7
10.2	343.6
22.6	353.0

(i) Use these data to obtain, by a graphical method, a value for the activation energy of the reaction. ($R = 8.31$ J K^{-1} mol^{-1}.)
(ii) Calculate the initial rate of reaction in a solution at 318.0 K when the concentration of both potassium hydroxide and bromoethane are 0.100 mol dm^{-3}.

(OLE)

8. This question concerns the following reaction:

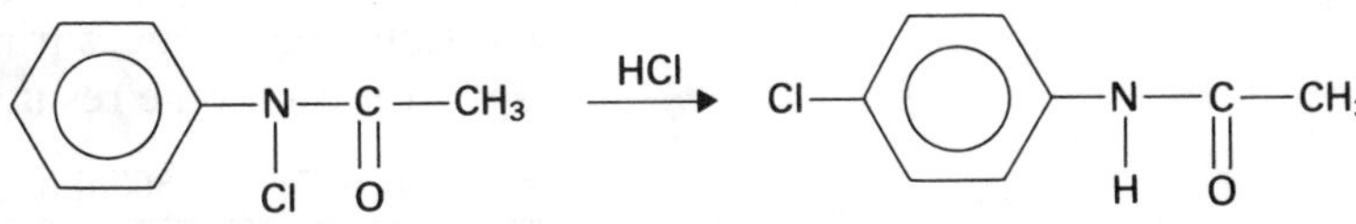

Compound A   Compound B

The progress of this reaction can be followed by using the fact that compound A reacts with acidified potassium iodide, liberating iodine, whereas compound B does not.

A series of experiments was carried out to determine the initial rate of reaction for various initial concentrations of compound A. The following data were obtained:

Concentration of compound A/mol dm^{-3}	Initial rate/ mol dm^{-3} s^{-1}
0.060	3.12 × 10^{-4}
0.120	6.23 × 10^{-4}
0.180	9.38 × 10^{-4}
0.240	12.5 × 10^{-4}

One possible pathway for the reaction is:

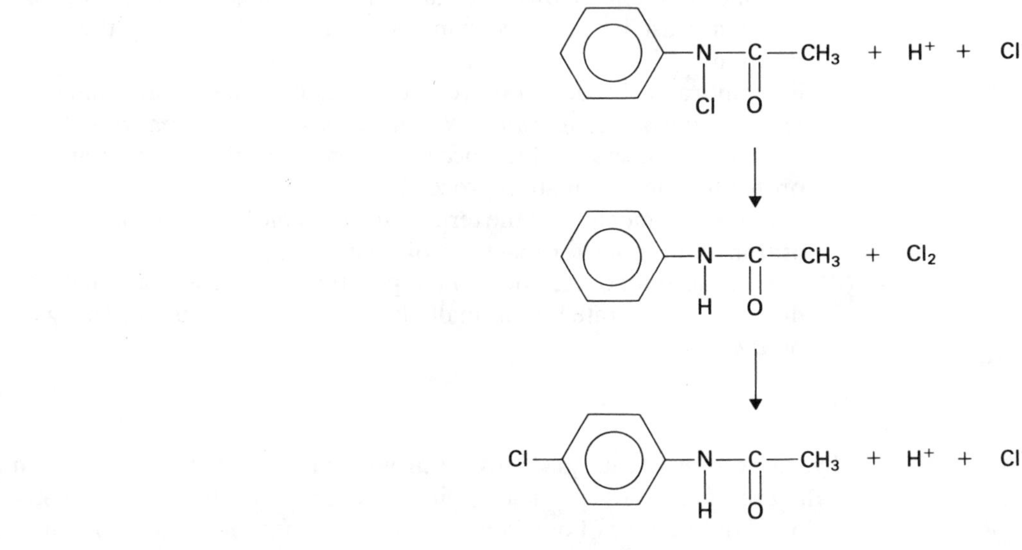

A sample of hydrochloric acid labelled with radioactive chloride ions was mixed with a sample of compound A, and the reaction allowed to proceed to completion. The chloride ions present at the end were precipitated as silver chloride, and the radioactivity of this material was compared to that of an equivalent sample of silver chloride prepared from another portion of labelled hydrochloric acid.

(a) Outline an experimental procedure for the determination of the concentration of the reactant (compound A) at intervals as the reaction proceeds.

(b) From the data in the table, deduce the order of reaction with respect to compound A. Explain your reasoning.

(c) Would you expect the silver chloride from the reaction mixture to be more or less radioactive than the silver chloride from the hydrochloric acid?

Explain how the suggested mechanism would lead to such a result.

(L Nuffield)

9. Explain the terms *half-life, rate constant* and *order of reaction*.

The base-catalysed decomposition of a substance A in aqueous solution was followed by measuring the volume increase of the solution at constant temperature. Some results were:

Run 1. $[OH^-] = 0.500$ mol dm^{-3}

Time/min	0	5	10	15	20	∞
ΔV_t/cm^3	0	0.5689	0.7782	0.8552	0.8835	0.9000

Run 2. $[OH^-] = 0.100$ mol dm^{-3}

Time/min	0				20	∞
ΔV_t/cm^3	0				0.4956	0.9000

Given that the concentration of unreacted A is proportional to $\Delta V_\infty - \Delta V_t$, plot a suitable graph for Run 1 to show that the reaction is first order with respect to A.

Calculate the first-order rate constants for both Runs 1 and 2.

Hence find the order of reaction with respect to OH^-.

What is the overall order of the reaction and the value of the overall rate constant for the reaction (give units)?

(O & C)

10. (a) Draw a fully labelled diagram to give the energy profile for an endothermic reaction, showing clearly that the activation energy of the reaction must be not less than the enthalpy change for the forward reaction.
    (b) For any radical reaction of your choice, illustrate the meaning of the terms *initiation, propagation, termination* and *chain reaction*. Mention **two** other processes which occur by chain reactions. Draw an energy profile for the termination process.
    (c) Illustrate the meaning of the term *transition state* by giving **three** examples of transition states for reactions of your choice.
    (d) Give an account of the effect of temperature on reaction rates including a diagram to illustrate how a small increase in temperature has a big effect on the rate.

(WJEC)

11. Give an account of catalysis, illustrating your answer with a range of examples of the use of catalysts, particularly in the laboratory, but also in industry.

You should discuss homogeneous and heterogeneous catalysis, and the ways in which catalysts provide an alternative reaction pathway.

(L)

12. Enzyme washing powders have only appeared in the shops during recent years, but enzymes themselves are not a new material.
    (a) What is an enzyme?
    (b) Give reasons why enzyme washing powders have appeared only recently.
    (c) The enzymes used in enzyme washing powders are digestive enzymes. Why is this?
    (d) Why is it important to soak clothes overnight with an enzyme washing powder rather than simply boil the clothes with it?

(UCLES)

10 ELECTROCHEMISTRY

Electrochemistry from Ancient Mesopotamia to Treating Brain Diseases

In 1800 Alessandro Volta, experimenting in Italy, developed the first modern electric battery. A few months after, two British scientists, Nicholson and Carlisle, succeeded in breaking matter down into its components by putting in electricity. In this experiment, the electricity was put into water which separated into hydrogen and oxygen.

Several years later the British scientist Davy proposed to the Royal Navy that fastening a small patch of iron to its copper-sheathed wooden ships would protect them against corrosion. The method became known as cathodic protection, and is used worldwide today to protect pipelines. Yet the problem of metal corrosion persists. In 1975 it cost the United States some $75 000 million or a quarter of the country's gross national product for that year.

Present-day applications of electrochemistry range from the tiny cells that regulate heartbeats in cardiac patients to the hydrogen fuel cells that power spacecraft; from electroplating and etching of micro-electronic parts to electrochemical machining of metals. Potential applications include the production of fuels, such as hydrogen from sunlight, by means of a 'photoelectrochemical system', and the purification of water by similar methods; the monitoring and analysis of minute amounts of pollutants in air and water; and the regulation of substances in the brain that are involved in afflictions such as Parkinson's disease.

Dry batteries developed by Leclanché and rechargeable lead accumulators developed by Planté were the only sources of electricity available for the first half of the nineteenth century. Without such sources of electricity probably there would be no portable radios, no electrically powered vehicles, no cardiac pacemakers, no electric torches, no electric watches and clocks, no portable tape recorders—and modern warfare, with its battery-powered weapons, would be impossible.

Electrochemistry has also opened up a new line of important industrial processes It is called 'electrophoretic deposition' and provides improved adherence of paints and plastic coatings, plus other technological advantages over classical spraying processes.

Another new technology—electrochemical machining—has been developed which makes possible operations that cannot be achieved by conventional machining. The mould of the form to be achieved serves as the cathode and is impressed on the material to be machined, which serves as the anode. A fast stream of electrolyte flows between cathode and anode and when an electric current is passed through it, the material is machined by dissolution of the anode. The electrolyte not only carries the current through but carries away the dissol-

Where would we be without dry batteries?

ution products and cools the system by drawing away heat.

Electrochemical processes are especially suitable for quantitative analysis of chemicals, because the signals produced can easily be fed into electronic instruments and processed by computer. Jan Heyrovsky won the Nobel Prize in 1964 for development of such a system. Improvements on this system have made it possible to measure as little as one-hundredth of one layer of atoms of one metal on the surface of another, while at the same time showing how the metals react with one another.

Such a level of detection has not been achieved in analytical chemistry by any other method known so far. The ease of automation, the selectivity and sensitivity of electrochemical systems have led to a wide use of electrochemical sensors for the majority of pollutants in the air and in natural waters, as well as in other media affecting human life and its environment.

The ancient 'battery' was discovered in Mesopotamia in the ruins of a Parthian village and is 2200 years old. It is technically a chemical power source of the primary cell type and consisted of an iron rod and a copper sheet bent into a cylinder around it, both placed in a ceramic jar that might have been filled with grape juice. Its likely use was that of a mysterious driving power source for the process of plating metal objects with gold or silver.

Dr R. M. Wightman of the Chemistry Department of Indiana University in the United States has been measuring the brain chemicals of rats by means of electrochemical electrodes. Through the information received from these probes, electrophysiologists are learning how the chemicals of the brain are involved in its function. This knowledge could lead eventually to new methods of controlling or preventing some diseases involving the brain.

For example, Parkinson's disease, characterised by marked tremors of the limbs, results from a shortage of dopamine in the brain of humans. Dopamine is what is known as a neurotransmitter—a chemical secreted by the nerve cells of the brain by means of which brain signals are transmitted. Dr Wightman revealed that, following several years of experiments, it is now possible to measure neurotransmitters directly by electrochemical methods. This might lead to a successful treatment for Parkinson's disease through replacement of the dopamine deficiency.

David Spurgeon

LEARNING OBJECTIVES   *After you have studied this chapter you should be able to*

1. Describe how **electrolytic conductivity** is measured.
2. Indicate how **molar conductivity** varies with concentration.
3. Calculate the **ionic mobility** of an electrolyte using **Kohlrausch's law of independent ionic mobilities**.
4. Compare the **conductimetric titrations** of a strong acid and a strong base with that of a weak acid and a strong base.
5. Balance **redox equations**.
6. Explain the terms
   (a) **oxidation number**,
   (b) **disproportionation**.
7. Give examples of **oxidising agents** and **reducing agents**.
8. Outline the essential features of the **ionic theory of electrolysis**.
9. State and apply **Faraday's laws of electrolysis**.
10. Give examples of industrial and other **applications** of electrolysis.
11. Define **oxidation** and **reduction** and give examples of both types of reaction.
12. Write **cell diagrams** for simple **electrochemical cells**.
13. Explain and interpret **standard redox potentials**.
14. Calculate the **e.m.f.** of an electrochemical cell.

15. Describe how cell e.m.f.s and **electrode potentials** are measured.
16. Outline the main features and importance of the **electrochemical series**.
17. Calculate equilibrium constants using the **Nernst equation**.
18. Explain how **pH** may be determined using a **glass electrode**.
19. Describe a typical **potentiometric titration**.
20. Determine the **feasibility** of a redox reaction by inspecting the electrode potentials of the **half-reactions**.
21. Explain the operation of typical **batteries** and **commercial cells**.
22. Indicate the importance of **electrochemistry** in **metallurgy**.
23. Outline the electrochemical features of **corrosion** and its prevention.

10.1 Conduction

INTRODUCTION

Electrochemistry is concerned with the **transfer of electrons** from one chemical species to another and the relationships between the electron transfer and the electrical currents that are generated or used during these processes.

We shall see in this chapter that electrochemistry is a topic which touches on many other areas of chemistry. We shall see, for example, how it touches upon ions in solution and pH (chapter 8), on energy and thermodynamics (chapter 5) and also on periodicity (chapter 11).

It is convenient, for the purposes of this chapter, to divide the topic of electrochemistry into five sections:

Conduction. In this section we shall be concerned with the properties and processes that occur in the bulk of an electrolyte when an electric current is passed through it.

Redox. This is concerned with the gain or loss of electrons from a substance.

Electrolysis. Electrolysis is concerned with the processes that occur at the surfaces of electrodes when an electric current is passed through an electrolyte.

Electrochemical cells. These are sometimes called voltaic or galvanic cells. In this section we shall be concerned with how processes of electron transfer generate electricity and how the tendency or potential of a chemical species to transfer electrons is related to its chemical reactivity.

Applications of electrochemistry. These are wide and varied.

Note that in the conductivity and electrolysis sections we are concerned with the passage through an electrolyte of an electric current driven by an external electric potential. The section on electrochemical cells is concerned with the reverse process—that is, the *generation* of an electric current.

CONDUCTORS AND CONDUCTIVITY

An electric current is a flow of electric charge. The flow of charge is carried by a **conductor**. In section 8.1 we defined two types of conductor:

Electronic Conductor
The flow of charge in an electronic conductor is due to a flow of electrons. There is therefore no transfer of material. Metals are electronic conductors.

Electrolytic Conductor
The flow of charge in an electrolytic conductor is due to the movement of ions. There is thus a transfer of material. Electrolytic conductors are called electrolytes.

They may be pure substances, such as molten salts, or solutions of salts, acids or bases in water.

Electrical Units

The SI base unit of **electric current** is the **ampere**. Electric current has the symbol I and the ampere the symbol A.

A **quantity of electric charge** is measured in coulombs (symbol C). A **coulomb** is the quantity of electric charge carried by one ampere in one second (symbol s). Thus

$$1 \text{ C} = 1 \text{ A s}$$

Electric charge is driven through a circuit by an electrical **potential difference**. This has the symbol V. The SI unit of potential difference is the **volt** which also has the symbol V. One joule of work is performed when one coulomb of charge is transferred through a potential difference of one volt. Thus

$$1 \text{ V} = 1 \text{ J A}^{-1} \text{s}^{-1}$$

Potential difference is related to electric current by **Ohm's law**:

$$V = IR$$

R is a proportionality constant called **resistance**. The unit of resistance is the ohm. This has the symbol Ω. One volt is required to drive an electric current of one ampere through a resistance of one ohm. Thus

$$1 \, \Omega = 1 \text{ V A}^{-1}$$

The reciprocal of electrical resistance is electrical **conductance**. This has the symbol G. The unit of electrical conductance is the siemens, S. Thus

$$1 \text{ S} = 1 \, \Omega^{-1}$$

Resistivity and Conductivity

The length and cross-sectional area of a conductor are related to its resistance by a property known as **resistivity**. The relationship is

$$\rho = \frac{AR}{l} \tag{1}$$

where ρ is the resistivity, A is the cross-sectional area with units m^2 and l is the length of the conductor with units m. The units of resistivity are ohm metres, Ω m.

The reciprocal of resistivity is **conductivity**, κ, or specific conductivity

$$\kappa = \frac{1}{\rho} = \frac{l}{AR} \tag{2}$$

The units of conductivity are $\Omega^{-1} \text{ m}^{-1}$ or $\Omega^{-1} \text{ cm}^{-1}$. The units S m^{-1} or S cm^{-1} are also used for conductivity. Thus

$$1 \, \Omega^{-1} \text{ cm}^{-1} = 1 \text{ S cm}^{-1}$$

The conductivity of an electrolyte solution is called the electrolytic conductivity.

Measuring Electrolytic Conductivity

The electrolytic conductivity of a solution can be determined using a **conductivity cell**. This consists of a glass cell containing two electrodes of known area rigidly fixed at a set distance apart. The electrodes are made of platinum black. The cell is dipped into the electrolyte solution as shown in figure 10.1. The solution is made up using **conductivity water**. This is water that has been passed through an ion exchange column (*see* section 8.2). Ordinary distilled water is not suitable since it

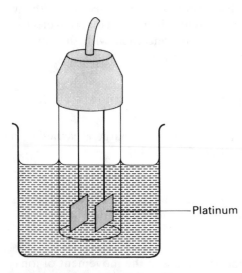

Figure 10.1 Conductivity cell

Platinum

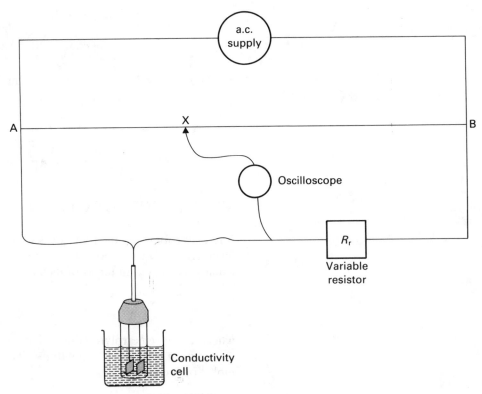

Figure 10.2 Wheatstone bridge circuit

contains traces of the ions H^+ and HCO_3^- from dissolved CO_2. These ions give the water a conductivity of the order 10^{-5} Ω^{-1} cm^{-1}.

The resistance of the cell is measured using a Wheatstone bridge circuit (*see* figure 10.2). The sliding contact is adjusted until a minimum response is obtained on the oscilloscope. The resistance of the cell, R_{cell}, is given by

$$R_{cell} = \frac{BX \times R_r}{AX}$$

Where R_r is the resistance of the variable resistor. The electrolytic conductivity of the solution can then be calculated using equation (2). A high-frequency source of alternating current is necessary for conductivity measurement. A direct current cannot be used because electrolysis of the solution would occur.

In equation (2), l/A is a constant for the cell. It is thus known as the **cell constant**. It can be found by using the cell to measure the resistance of a solution of known conductivity. A solution of potassium chloride with concentration 0.1 mol dm^{-3} is often used for this purpose.

EXAMPLE
The resistance of a 0.1 mol dm^{-3} solution of potassium chloride was found to be 35.2 Ω. Using the same cell, the resistance of a 0.1 mol dm^{-3} solution of silver nitrate was found to be 42.4 Ω. Given that the electrolytic conductivity of potassium chloride is 1.29×10^{-1} S cm^{-1}, calculate (a) the cell constant and (b) the electrolytic conductivity of the silver solution.

SOLUTION
(a) The cell constant is given by

$$\frac{l}{A} = R_{cell} \times \kappa$$

$$= (35.2\ \Omega) \times (1.29 \times 10^{-2}\ \text{S cm}^{-1})$$

Since $\qquad 1\,\text{S} = 1\,\Omega^{-1}$

$$\frac{l}{A} = 45.4 \times 10^{-2}\,\text{cm}^{-1}$$

(b) The conductivity of the $0.1\,\text{mol dm}^{-3}$ $AgNO_3$ solution is given by

$$\kappa = \frac{l}{A} \times \frac{1}{R_{\text{cell}}}$$
$$= (45.4 \times 10^{-2}\,\text{cm}^{-1}) \times (42.4\,\Omega^{-1})$$
$$= 1.07 \times 10^{-2}\,\text{S cm}^{-1}$$

MOLAR CONDUCTIVITY

The electrolytic conductivity of a solution depends on the concentration of the electrolyte in the solution. The conductivity of an electrolyte solution is therefore usually expressed in terms of its molar conductivity, Λ, given by

$$\Lambda = \frac{\kappa}{c} \qquad (3)$$

where κ is the electrolytic conductivity of the solution and c is the concentration of the solution expressed in mol dm^{-3}. The units of molar conductivity are Ω^{-1} cm^2 mol^{-1} or S cm^2 mol^{-1}.

EXAMPLE
A $0.20\,\text{mol dm}^{-3}$ solution of potassium chloride has an electrolytic conductivity of $2.48 \times 10^{-2}\,\text{S cm}^{-1}$ at 298 K. Calculate its molar conductivity.

SOLUTION
Using equation (3) we obtain

$$\Lambda = \frac{2.48 \times 10^{-2}\,\text{S cm}^{-1}}{0.20\,\text{mol dm}^{-3}}$$
$$= 12.4 \times 10^{-1}\,\text{S cm}^{-1}\,\text{dm}^3\,\text{mol}^{-1}$$

Since
$$1\,\text{dm}^3 = 10^3\,\text{cm}^3$$
$$\Lambda = 12.4 \times 10^2\,\text{S cm}^2\,\text{mol}^{-1}$$

Variation of Molar Conductivity with Concentration

In general, molar conductivity decreases with increasing concentration of the electrolyte. The exact nature of the relationship depends on whether the electrolyte is strong or weak.

Strong Electrolytes
At low concentrations the molar conductivity of a strong electrolyte is found to be proportional to the square root of concentration

$$\Lambda \propto \sqrt{c}$$

The exact nature of the relationship is given by an empirical equation first formulated by Kohlrausch:

$$\Lambda = \Lambda^\infty - b\sqrt{c}$$

where b is a constant and Λ^∞ is called the **molar conductivity at infinite dilution** or sometimes the **limiting molar conductivity**. Thus

$$\Lambda = \Lambda^\infty \qquad \text{when } c \to 0$$

A value for Λ^∞ for a strong electrolyte can be found by plotting molar conductivity against $\sqrt{c}$ and extrapolating back to $\sqrt{c} = 0$. Figure 10.3 shows such a graph for potassium chloride solution. Note that, when the value of $\sqrt{c}$ is greater than about 0.5 $(\text{mol dm}^{-3})^{1/2}$, the graph deviates from a straight line.

Weak Electrolytes

The molar conductivities of solutions of weak electrolytes are low compared to those of strong electrolyte solutions. This is because molar conductivity depends on the proportion of ions present in solution. Even at low concentrations, the degree of dissociation of weak electrolytes is low. However, when the concentration of the weak electrolyte becomes very low indeed, the degree of ionisation does rise sharply. The molar conductivity thus increases correspondingly. Figure 10.3 shows the graph of molar conductivity against the square root of concentration for ethanoic acid. The graph is not linear since weak electrolytes do not obey the Kohlrausch equation.

Ionic Mobilities

In 1875 Kohlrausch compared the molar conductivities at infinite dilution for pairs of strong electrolytes with common ions. He showed that the difference in molar conductivities between members of a pair was constant (*see* table 10.1). This led him to formulate his famous **law of independent ionic mobilities**. This states that the molar conductivity of an electrolyte at infinite dilution is the sum of the ionic mobilities of the ions forming the electrolyte and that these mobilities are independent of other ions.

Ionic mobility is related to the molar conductivity of an ion at infinite dilution. For a 1:1 electrolyte such as KCl, Kohlrausch's law can be expressed mathematically as

$$\Lambda^\infty = \Lambda^\infty_+ + \Lambda^\infty_-$$

Where Λ^∞_+ is the ionic mobility of the cation and Λ^∞_- is the ionic mobility of the anion.

Ionic mobility is a measure of the speed at which an ion travels through solution. In general, ions with a small ionic radius travel more slowly than ions with a large ionic radius. This is because a small ion has a higher charge density and is thus more highly solvated (*see* section 2.2). Its hydration sphere and thus its effective size is larger than that of an ion with a greater ionic radius. Its larger effective size hinders its mobility in the solution.

Ions with double or triple charges generally have higher ionic mobilities than singly charged ions. Exceptions are the oxonium and hydroxide ions, H_3O^+ and

Table 10.1 Molar conductivities at infinite dilution for pairs of strong electrolytes with a common ion

Pairs with a common anion	$\Lambda^\infty/\text{S cm}^2\text{ mol}^{-1}$
KCl	149.9
NaCl	126.5
difference	23.4
KNO$_3$	145.0
NaNO$_3$	121.6
difference	23.4

Pairs with a common cation	$\Lambda^\infty/\text{S cm}^2\text{ mol}^{-1}$
NaCl	126.5
NaNO$_3$	121.6
difference	4.9
KCl	149.9
KNO$_3$	145.0
difference	4.9

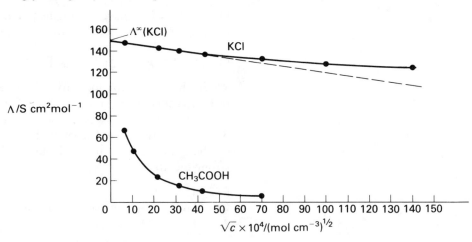

Figure 10.3 Variation of molar conductivity with concentration for potassium chloride solution and ethanoic acid

A CTD (conductivity/temperature/depth) profiler with a rosette of water samplers is deployed during a study of the Gulf Stream

OH^- ions, which have the highest of all ionic mobilities. This is due to the rapid transfer of protons between the ions.

Calculating Λ^∞ for Weak Electrolytes

When the molar conductivity of a weak electrolyte is plotted against concentration, a curve of the type shown in figure 10.3 for CH_3COOH is obtained. Since the curve does not intersect the vertical axis, it is impossible to obtain Λ^∞ for a weak electrolyte by extrapolating back to zero concentration. However, a value can be obtained by applying Kohlrausch's law of independent ionic mobilities to strong electrolytes.

EXAMPLE
Calculate Λ^∞ for ethanoic acid given

$$
\begin{aligned}
\Lambda^\infty(\text{HCl}) &= 426\ \Omega^{-1}\ cm^2\ mol^{-1} \\
\Lambda^\infty(\text{NaCl}) &= 126\ \Omega^{-1}\ cm^2\ mol^{-1} \\
\Lambda^\infty(\text{CH}_3\text{COONa}) &= 91\ \Omega^{-1}\ cm^2\ mol^{-1}
\end{aligned}
$$

SOLUTION

Ethanoic acid is a weak electrolyte. Hydrochloric acid, sodium chloride and sodium ethanoate are all strong electrolytes. Applying Kohlrausch's law to the strong electrolytes

$$\Lambda^{\infty}(HCl) = \Lambda^{\infty}(H^+) + \Lambda^{\infty}(Cl^-)$$
$$\Lambda^{\infty}(NaCl) = \Lambda^{\infty}(Na^+) + \Lambda^{\infty}(Cl^-)$$
$$\Lambda^{\infty}(CH_3COONa) = \Lambda^{\infty}(CH_3COO^-) + \Lambda^{\infty}(Na^+)$$

And the for the weak electrolyte

$$\begin{aligned}
\Lambda^{\infty}(CH_3COOH) &= \Lambda^{\infty}(CH_3COO^-) + \Lambda^{\infty}(H^+) \\
&= [\Lambda^{\infty}(CH_3COO^-) + \Lambda^{\infty}(Na^+)] - [\Lambda^{\infty}(Na^+) + \Lambda^{\infty}(Cl^-)] \\
&\quad + [\Lambda^{\infty}(H^+) + \Lambda^{\infty}(Cl^-)] \\
&= [(91) - (126) + (426)] \ \Omega^{-1} \ cm^2 \ mol^{-1} \\
&= 391 \ \Omega^{-1} \ cm^2 \ mol^{-1}
\end{aligned}$$

Degree of Dissociation

We saw above that the molar conductivity of an electrolyte depends on the proportions of ions present in solution and thus the degree of dissociation of the electrolyte. Arrhenius proposed that the molar conductivity of an electrolyte is related to its degree of dissociation by the following relationship:

$$\alpha = \frac{\Lambda}{\Lambda^{\infty}} \qquad (4)$$

At almost zero concentration or infinite dilution, we have

$$\Lambda = \Lambda^{\infty} \qquad \text{and thus} \qquad \alpha = 1$$

Equation (4) together with Ostwald's dilution law (see section 8.1) can be used to calculate the dissociation constant of a weak electrolyte from molar conductivities determined experimentally:

$$K_c = \frac{c\alpha^2}{1 - \alpha}$$

Substituting equation (4) into this, we obtain

$$K_c = \frac{c(\Lambda/\Lambda^{\infty})^2}{1 - (\Lambda/\Lambda^{\infty})}$$

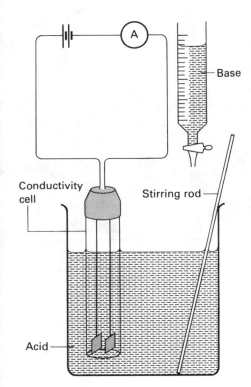

Figure 10.4 Conductimetric titration

Conductivity cell

Stirring rod

Base

Acid

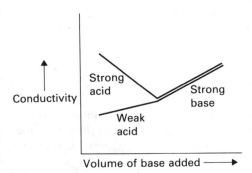

Figure 10.5 Conductimetric titration curves for titration of acid with strong base

Conductivity

Strong acid

Strong base

Weak acid

Volume of base added →

CONDUCTIMETRIC TITRATIONS

The end-point of a titration between an acid and a base can be determined by measuring the conductivity of the titrated solution at regular intervals (see figure 10.4). The end-point occurs when there is a sharp break in the titration curve. The exact shape of the titration curve depends on the dissociation constants of the acid and base. Figure 10.5 shows the approximate shapes of titration curves for the titration of a strong acid with a strong base and a weak acid with a strong base.

The titration of any acid with any base may be represented by the equation

$$H_3O^+(aq) + OH^-(aq) \rightarrow 2H_2O(l)$$

At the beginning of a titration of a strong acid with a strong base, the conductivity is high. This is because the strong acid is fully dissociated and thus the concentration of H_3O^+ ions is high. As base is added, the H_3O^+ ions are removed from solution by the OH^- ions and the conductivity of the solution decreases accordingly. At the neutralisation point, neither excess H_3O^+ nor excess OH^- ions are present. The conductivity is thus at its lowest. The end-point on the graph corresponds to the neutralisation point. After the end-point, base is added. The base contributes OH^- ions to the solution. The conductivity thus increases.

The titration of a weak acid with a strong base can be explained by employing similar arguments.

SUMMARY

1. **Conductivity** κ and **resistivity** ρ are related as

$$\kappa = \frac{1}{\rho} = \frac{l}{A} R$$

where l/A is the **cell constant**.
2. **Molar conductivity** is given by

$$\Lambda = \frac{\kappa}{c}$$

At infinite dilution

$$\Lambda = \Lambda^\infty$$

where Λ^∞ is the **limiting molar conductivity**.
3. **Kohlrausch's law of independent ionic mobilities** may be expressed as

$$\Lambda^\infty = \Lambda_+^\infty + \Lambda_-^\infty$$

4. **Ostwald's dilution law** may be expressed as

$$K_c = \frac{c(\Lambda/\Lambda^\infty)^2}{1 - (\Lambda/\Lambda^\infty)}$$

10.2
Redox ▌ OXIDATION AND REDUCTION

Oxidation may be defined in three ways:

1. It is the addition of oxygen to a substance.
2. It is the removal of hydrogen from a substance.
3. It is the loss of electrons from a substance.

Reduction may also be defined in three ways:

1. It is the removal of oxygen from a substance.
2. It is the addition of hydrogen to a substance.
3. It is the gain of electrons by a substance.

In this chapter we are principally concerned with the processes of **electron transfer**. We shall thus take oxidation to mean the loss of electrons and reduction to be the addition of electrons.

When oxidation takes place, reduction invariably takes place simultaneously. Similarly, when reduction takes place, oxidation also takes place at the same time. These reactions are called redox reactions—**redox** standing for *red*uction and *ox*idation.

A redox reaction therefore consists of two half-reactions, one of which is reduction and the other of which is oxidation.

Let us consider an example to illustrate these definitions. When a piece of iron is immersed in copper(II) sulphate solution, it soon becomes coated with copper. The reaction may be represented by the equation

$$Fe(s) + CuSO_4(aq) \rightarrow Cu(s) + FeSO_4(aq)$$

However, it is more convenient to represent the reaction by an **ionic equation**

$$Fe(s) + Cu^{2+}(aq) \rightarrow Fe^{2+}(aq) + Cu(s)$$

(reduction — above; oxidation — below)

This shows that the **iron is oxidised** to its hydrated cation $Fe^{2+}(aq)$ and the hydrated copper (II) ions $Cu^{2+}(aq)$ are reduced to copper atoms. The **oxidation half-reaction** is thus

$$Fe(s) \rightarrow Fe^{2+}(aq) + 2e^-$$

The two electrons lost by the iron atom are gained by the hydrated copper(II) ion:

$$Fe(s) \xrightarrow{2e^-} Cu^{2+}(aq)$$

The **reduction half-reaction** is thus

$$Cu^{2+}(aq) + 2e^- \rightarrow Cu(s)$$

REDOX REACTIONS

The following types of chemical reactions may all be classified as redox reactions.

Displacement Reactions

The reaction between iron and copper(II) ions is an example of this type of reaction. The iron displaces the copper from solution. Other examples are:

(a) *Overall reaction* $Zn(s) + Pb^{2+}(aq) \rightarrow Zn^{2+}(aq) + Pb(s)$
    Oxidation half-reaction $Zn(s) \rightarrow Zn^{2+}(aq) + 2e^-$
    Reduction half-reaction $Pb^{2+}(aq) + 2e^- \rightarrow Pb(s)$

(b) *Overall reaction* $Cl_2(g) + 2Br^-(aq) \rightarrow 2Cl^-(aq) + Br_2(aq)$
    Oxidation half-reaction $2Br^-(aq) \rightarrow Br_2(aq) + 2e^-$
    Reduction half-reaction $Cl_2(g) + 2e^- \rightarrow 2Cl^-(aq)$

In this reaction chlorine displaces bromine from a solution of bromide ions.

Reactions of Metals with Acids

This is also a type of displacement reaction. An example is the reaction between magnesium and hydrochloric acid. The magnesium displaces hydrogen from the· acid. The magnesium is oxidised to its hydrated cation and the hydrated proton is reduced to hydrogen gas.

    Overall reaction $Mg(s) + 2H^+(aq) \rightarrow Mg^{2+}(aq) + H_2(g)$
    Oxidation half-reaction $Mg(s) \rightarrow Mg^{2+}(aq) + 2e^-$
    Reduction half-reaction $2H^+(aq) + 2e^- \rightarrow H_2(g)$

Reactions of Metals with Water

These are also types of displacement reaction. Hydrogen is displaced from water in the form of gas. An example is the reaction between sodium metal and water.

Overall reaction	$2Na(s) + 2H_2O(l) \rightarrow 2NaOH(aq) + H_2(g)$
Oxidation half-reaction	$2Na(s) \rightarrow 2Na^+(aq) + 2e^-$
Reduction half-reaction	$2H_2O(l) + 2e^- \rightarrow 2OH^-(aq) + H_2(g)$

Reactions of Metals with Non-Metals

These may also be classified as *synthesis* reactions. An example is the formation of sodium chloride by burning sodium in chlorine.

Overall reaction	$2Na(s) + Cl_2(g) \rightarrow 2NaCl(s)$
Oxidation half-reaction	$2Na(s) \rightarrow 2Na^+(s) + 2e^-$
Reduction half-reaction	$Cl_2(g) + 2e^- \rightarrow 2Cl^-(s)$

Electrolysis is also a redox process (*see* following section).

BALANCING REDOX EQUATIONS

There are two important rules for writing redox equations:

Rule 1: In any ionic equation the charges must balance. This means that the sum of the charges on the left-hand side (LHS) must equal the sum of the charges on the right-hand side (RHS). This rule applies to ionic equations for both overall reactions and half-reactions.

EXAMPLES

	Charges on	
	LHS	RHS
(a) $Mg(s) + 2H^+ \rightarrow Mg^{2+}(aq) + H_2(g)$	2+	2+
(b) $2Br^-(aq) \rightarrow Br_2(aq) + 2e^-$	2−	2−
(c) $2Na(s) \rightarrow 2Na^+(aq) + 2e^-$	0	$(2+) + (2-) = 0$

Rule 2: The number of electrons lost in the oxidation half-reaction must equal the number of electrons gained in the reduction half-reaction. For instance, in the first example at the beginning of this section (the reaction between iron and hydrated copper(II) ions), the number of electrons lost in the oxidation half-reaction is two:

$$Fe(s) \rightarrow Fe^{2+}(aq) + 2e^-$$

Therefore the number of electrons gained in the reduction half-reaction must be two:

$$Cu^{2+}(aq) + 2e^- \rightarrow Cu(s)$$

To write a redox equation from two half-reactions, the following procedure can be employed:

1. The two half-equations are balanced separately by adding electrons to either the LHS *or* the RHS (this is rule 1 above).
2. The two half-reactions are balanced with respect to each other so that the number of electrons lost equals the number of electrons gained (this is rule 2 above).
3. The two half-equations are combined to give the redox equation. For example, adding the two half-reactions above and cancelling the electrons we obtain

$$Fe(s) + Cu^{2+}(aq) \rightarrow Fe^{2+}(aq) + Cu(s)$$

EXAMPLE

Balance the half-equations below and construct a redox equation for the oxidation of an aqueous solution of an iron(II) salt to an iron(III) salt by means of potassium manganate(VII) in acidic solution.

$$MnO_4^-(aq) + 8H^+(aq) \rightarrow Mn^{2+}(aq) + 4H_2O(l) \tag{5}$$

$$Fe^{2+}(aq) \rightarrow Fe^{3+}(aq) \tag{6}$$

SOLUTION

Step 1: Balance the two half-equations separately. For equation (5) we have

LHS	RHS
(1−) + (8+)	2+

To balance the two sides, five electrons must either be added to the LHS or subtracted from the RHS. Thus

LHS	RHS
(1−) + (8+) + (5−)	2+

The balanced equation is thus

$$MnO_4^-(aq) + 8H^+(aq) + 5e^- \rightarrow Mn^{2+}(aq) + 4H_2O(l) \tag{7}$$

Since electrons are added this is the reduction half-equation.

For equation (6) we have

LHS	RHS
2+	3+

To balance this we can add an electron to the RHS. Thus

LHS	RHS
2+	(3+) + (1−)

The balanced equation is thus

$$Fe^{2+}(aq) \rightarrow Fe^{3+}(aq) + e^- \tag{8}$$

Since an electron is removed from Fe^{2+} this is the oxidation half-equation.

Step 2: Balance the two half-equations so that the number of electrons are equal. We can do this by leaving equation (7) alone and by multiplying equation (8) by 5:

$$MnO_4^-(aq) + 8H^+(aq) + 5e^- \rightarrow Mn^{2+}(aq) + 4H_2O(l) \tag{7}$$

$$5Fe^{2+}(aq) \rightarrow 5Fe^{3+}(aq) + 5e^- \tag{9}$$

Step 3: Combine the two half-equations (7) and (9):

$$5Fe^{2+}(aq) + MnO_4^-(aq) + 8H^+(aq) \rightarrow 5Fe^{3+}(aq) + Mn^{2+}(aq) + 4H_2O(l)$$

This is the final balanced redox equation.

OXIDATION NUMBER

In section 4.1 we saw that the oxidation number of an element is equal to its combining power with oxygen. The three most important rules were given and examples of the oxidation numbers of some elements and ions were given in table 4.5.

Oxidation numbers are particularly important when considering redox reactions. There are two important rules concerning oxidation numbers and redox process.

Rule 1: Oxidation number increases during oxidation

EXAMPLES

(a)
$$Fe^{2+} \rightarrow Fe^{3+} + e^-$$
oxidation number   +2   +3

(b)
$$2Cl^- \rightarrow Cl_2 + 2e^-$$
oxidation number   2(−1)   0

(c)
$$C + O_2 \rightarrow CO_2$$
oxidation number of carbon
$$0 \qquad +4$$

Rule 2: Oxidation number decreases during reduction

EXAMPLES

(a)
$$Cu^{2+} + 2e^- \rightarrow Cu$$
oxidation number
$$+2 \qquad\qquad 0$$

(b)
$$MnO_4^- + 8H^+ + 5e^- \rightarrow Mn^{2+} + 4H_2O$$
oxidation number of manganese
$$+7 \qquad\qquad\qquad +2$$

(c)
$$PbO + H_2 \rightarrow Pb + H_2O$$
oxidation number of lead
$$+2 \qquad\qquad 0$$

Disproportionation

This is a redox reaction in which both reduction and oxidation of the same element occurs.

EXAMPLE

Nitrous acid disproportionates to form nitric acid and nitrogen monoxide:

$$3HNO_2(aq) \rightarrow HNO_3(aq) + 2NO(g) + H_2O(l)$$
oxidation number of nitrogen
$$+3 \qquad\qquad +5 \qquad\qquad +2$$

OXIDISING AGENTS

An oxidising agent is a substance which brings about oxidation. In bringing about oxidation, an oxidising agent is itself reduced. Common oxidising agents fall into three categories.

Non-Metallic Elements

These accept electrons to form anions. Chlorine is an example of this type of oxidising agent. It oxidises bromide ions, for instance. The ionic equation for the overall reaction is

$$Cl_2(g) + 2Br^-(aq) \rightarrow 2Cl^-(aq) + Br_2(g)$$

The bromide is thus oxidised:

$$2Br^-(aq) \rightarrow Br_2(aq) + 2e^-$$

In oxidising the bromide, the chlorine is itself reduced:

$$Cl_2(g) + 2e^- \rightarrow 2Cl^-(aq)$$

Oxygen and bromine are two other examples in this category. The reduction half-reactions are:

$$O_2 + 4e^- \rightarrow 2O^{2-}$$
$$Br_2 + 2e^- \rightarrow 2Br^-$$

Cations

These are usually metal ions. They accept electrons to form neutral atoms or molecules. Examples are

$$Cu^{2+} + 2e^- \rightarrow Cu$$
$$2H^+ + 2e^- \rightarrow H_2$$

Ions with an Element in a High Oxidation State

In bringing about the oxidation, the oxidation number of the ion is lowered. For example

$$MnO_4^- + 8H^+ + 5e^- \rightarrow Mn^{2+} + 4H_2O$$

oxidation number
of manganese $+7$ $+2$

$$Fe^{3+} + e^- \rightarrow Fe^{2+}$$

oxidation number $+3$ $+2$

Test for Oxidising Agents

Oxidising agents change the colour of moist starch iodide paper from white to blue. The paper is impregnated with starch and potassium iodide. The oxidising agent oxidises the iodide ion to form iodine:

$$2I^- \rightarrow I_2 + 2e^-$$

The iodine reacts with the starch to produce the blue colour.

REDUCING AGENTS

A reducing agent is a substance which brings about reduction. In bringing about reduction, a reducing agent is itself oxidised. Common reducing agents fall into three categories.

Metals

Metals which donate electrons and form ions in the process are reducing agents. Iron is a example of this type of reducing agent. It reduces copper(II) ions in the following reaction, for example:

$$Fe(s) + Cu^{2+}(aq) \rightarrow Fe^{2+}(aq) + Cu(s)$$

The half-reaction for the reduction of copper(II) ions is

$$Cu^{2+} + 2e^- \rightarrow Cu(s)$$

In reducing the copper(II) ions the iron is itself oxidised:

$$Fe(s) \rightarrow Fe^{2+}(aq) + 2e^-$$

All metals yielding their common ions are reducing agents. Examples are sodium and magnesium:

$$Na \rightarrow Na^+ + e^-$$
$$Mg \rightarrow Mg^{2+} + 2e^-$$

Non-Metals

Non-metals which remove elements with negative oxidation numbers from compounds with metals are also reducing agents. The most important elements with negative oxidation numbers in this context are oxygen and the halogens. Reducing agents in this category are carbon and carbon monoxide. For example

$$C(s) + 2CuO(s) \rightarrow 2Cu(s) + CO_2(g)$$

oxidation number
of carbon 0 $+4$

$$3CO(g) + Fe_2O_3(s) \rightarrow 2Fe(s) + 3CO_2(g)$$

oxidation number   +2   +4
of carbon

Notice that in both these examples the oxidation number of the carbon increases. The carbon is thus oxidised in both cases.

Ions with an Element in a Low Oxidation State

Ions with an element in a low oxidation state can bring about reduction. In doing so, the oxidation number of the ion increases. It is thus oxidised. Examples include iron(II) and sulphite ions:

$$Fe^{2+} \rightarrow Fe^{3+} + e^-$$

oxidation number   +2   +3

$$SO_3^{2-} + H_2O \rightarrow SO_4^{2-} + 2H^+ + 2e^-$$

oxidation number   +4   +6
of sulphur

Tests for Reducing Agents

Reducing agents decolourise acidified potassium manganate(VII) solution:

$$MnO_4^-(aq) + 8H^+(aq) + 5e^- \rightarrow Mn^{2+}(aq) + 4H_2O(l)$$
purple   colourless

They also change the colour of acidified potassium dichromate(VI) solution from orange to green:

$$Cr_2O_7^{2-}(aq) + 14H^+(aq) + 6e^- \rightarrow 2Cr^{3+}(aq) + 7H_2O(l)$$
orange   green

OXIDISING AND REDUCING AGENTS

Some substances can act as either oxidising or reducing agents depending on conditions.

Hydrogen peroxide is an example. In the absence of another oxidising agent, it acts as an oxidising agent:

It oxidises iodide ions, for example: $H_2O_2(aq) + 2H^+(aq) + 2e^- \rightarrow 2H_2O(l)$

$$H_2O_2(aq) + 2H^+(aq) + 2I^-(aq) \rightarrow 2H_2O(l) + I_2(s)$$

However, in the presence of a strong oxidising agent, it acts as a reducing agent:

$$H_2O_2(aq) \rightarrow O_2(g) + 2H^+(aq) + 2e^-$$

For example, it reduces manganate(VII) ions in the following reaction:

$$5H_2O_2(aq) + 2MnO_4^-(aq) + 6H^+(aq) \rightarrow 5O_2(g) + 2Mn^{2+}(aq) + 8H_2O(l)$$

SUMMARY

1. **Oxidation** is the loss of electrons from a substance.
2. **Reduction** is the gain of electrons by a substance.
3. A **redox** reaction consists of two **half-reactions: red**uction and **ox**idation.
4. In any ionic equation, the charges must balance.
5. The number of electrons lost in an oxidation half-reaction must equal the number of electrons gained in the reduction half-reaction.
6. **Oxidation number** increases during oxidation.
7. Oxidation number decreases during reduction.

8. **Disproportionation** is a redox reaction in which both reduction and oxidation of the same element occur.
9. An **oxidising agent** brings about oxidation and in so doing is itself reduced.
10. A **reducing agent** brings about reduction and in so doing is itself oxidised.

10.3 Electrolysis

When a direct current of electricity is passed through an electrolyte, chemical reactions take place at the electrodes. This process is called **electrolysis**. It literally means breaking up or decomposition by electricity.

We saw in section 8.1 that an **electrolyte** is a liquid which reacts chemically when electricity is passed through it. An electrolyte may be a molten salt such as molten lead(II) bromide or it may be an aqueous solution of an acid, base or salt.

Electrodes are the wires, rods or plates which make electrical contact with the electrolyte. The negative electrode is called the **cathode** and the positive electrode the **anode**. Electrodes which do not react chemically when in contact with electrolyes and when electricity is passed through them are called **inert electrodes**. Examples are graphite and platinum.

IONIC THEORY OF ELECTROLYSIS

In this theory, the direct current of electricity is carried through the electrolyte by ions. At the electrodes, electrons are transferred to or from the ions. The processes which take place at the electrodes are thus either reduction or oxidation half-reactions. Electrolysis is thus a reduction–oxidation (redox) process.

At the anode an oxidation half-reaction takes place. In this reaction anions lose electrons and become discharged. The anode thus acts as a sink for the electrons from the anions.

At the cathode a reduction half-reaction takes place. The cations gain electrons and become discharged. The cathode thus acts as a source of electrons for the cations.

EXAMPLE
The electrolysis of molten lead(II) bromide consists of two half-reactions:
   At the anode, bromide ions are discharged. The half-equation is

$$2Br^-(l) \rightarrow Br_2(g) + 2e^-$$

This is oxidation.
   At the cathode, lead ions are discharged. The half-equation is

$$Pb^{2+}(s) + 2e^- \rightarrow Pb(l)$$

This is reduction.
   We should note that the reactions which occur at the anode and the cathode are imposed upon the system by the polarity of the external circuit. The negative pole of the external battery supplies electrons to one electrode of the **electrolytic cell**. This forces the electrode to be negative. This is the cathode. Since the electrode is negative, it in turn forces the electrode reaction to be one which takes up electrons. Thus a reduction process occurs. At the other electrode, electrons flow from the electrolytic cell back to the external circuit, forcing this electrode to become a source of electrons for the external circuit and thus positive. It therefore acts as an anode. The positive sign forces the reaction which occurs at this electrode to be one which gives up electrons, that is, oxidation.

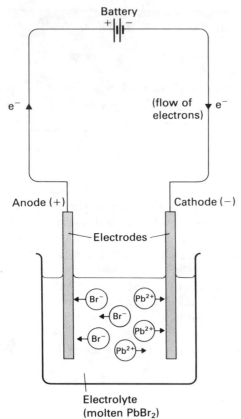

Figure 10.6 Electrolysis

A schematic diagram of the overall process is shown in figure 10.6.

Selective Discharge of Ions

An aqueous solution of an electrolyte contains more than one anion and more than one cation. For example, an aqueous solution of potassium chloride contains two types of anion, Cl^- and OH^-, and two types of cation, K^+ and H^+ (or H_3O^+).

The order in which ions are discharged at electrodes is determined by a number of factors including the nature of the electrode, the state of the electrolyte and the electrode potential of the ion.

The Nature of the Electrode

Electrodes may be inert or reactive. For example, the electrode reactions which take place during the electrolysis of an aqueous solution of copper(II) sulphate solution depend on whether reactive electrodes—such as copper—or inert electrodes—such as graphite—are used.

If **copper electrodes** are used, the following processes take place:

at the anode	$Cu(s) \rightarrow Cu^{2+}(aq) + 2e^-$	(oxidation)
at the cathode	$Cu^{2+}(aq) + 2e^- \rightarrow Cu(s)$	(reduction)

If **graphite electrodes** are used, the reactions are:

at the anode	$4OH^-(aq) \rightarrow 2H_2O(l) + O_2(g) + 4e^-$	(oxidation)
at the cathode	$Cu^{2+}(aq) + 2e^- \rightarrow Cu(s)$	(reduction)

The State of the Electrolyte

The half-reactions taking place at the electrode depend on whether the electrolyte is in a molten state or in an aqueous state and if in the aqueous state on the concentration. For example, the electrode reactions that take place during the electrolysis of molten potassium iodide are:

at the anode	$2I^-(l) \rightarrow I_2(g) + 2e^-$	(oxidation)
at the cathode	$K^+(l) + e^- \rightarrow K(l)$	(reduction)

However, if aqueous potassium iodide is used, the following electrode reactions take place:

at the anode	$2I^-(aq) \rightarrow I_2(aq) + 2e^-$	(oxidation)
at the cathode	$2H^+(aq) + 2e^- \rightarrow H_2(g)$	(reduction)

The Electrode Potential of the Ion

We shall define and discuss electrode potentials in the following section.

For cations (H^+ and metallic ions) the ease of discharge at the cathode (*see* table 10.2) follows the electrochemical series/electrode potential series (*see* later in chapter). Thus, if a solution contains both K^+ and H^+ ions, the H^+ ions will be discharged in preference to the K^+ ions. On the other hand, if a solution contains both Cu^{2+} and H^+ ions, the Cu^{2+} ions will be discharged in preference.

The ease of discharge of anions at the anode follows the order shown in table 10.3. Large ions such as sulphate are never discharged normally since hydroxide ions are discharged in preference.

FARADAY'S LAWS OF ELECTROLYSIS

The relationship between the mass of product formed at an electrode and the quantity of electricity passed through an electrolyte is given by Faraday's laws of electrolysis.

The **first law** states that the mass of a substance produced at an electrode during electrolysis is proportional to the quantity of electricity passed. The quantity of electricity is often expressed in terms of the Faraday. The Faraday is the charge

Table 10.2 Ease of discharge of cations

Cations in aqueous solution

K^+	
Na^+	
Mg^{2+}	
Al^{3+}	
Zn^{2+}	increasing ease
Fe^{2+}	of discharge
Pb^{2+}	
H^+	
Cu^{2+}	
Ag^+	

Table 10.3 Ease of discharge of anions

Anions in aqueous solution

SO_4^{2-}	
NO_4^-	increasing ease
Cl^-	of discharge
OH^-	
I^-	

carried by one mole of electrons or one mole of singly charged ions. It has a constant value of 96 500 C mol^{-1}. Thus

$$1 \text{ Faraday} = 96\,500 \text{ C mol}^{-1}$$
$$= 1 \text{ mole of electrons}$$
$$= 6.022 \times 10^{23} \text{ electrons}$$

where 6.022×10^{23} mol^{-1} is the Avogadro constant (*see* section 4.2).

EXAMPLE
The discharge of silver ions at the cathode during the electrolysis of silver nitrate solution is given by the half-reaction

$$\underset{\text{1 mol}}{Ag^+} + \underset{\text{1 mol}}{e^-} \rightarrow \underset{\text{1 mol}}{Ag}$$

Thus 1 Faraday of electricity (one mole of electrons) discharges 1 mole of silver ions and produces 1 mole of silver atoms. It follows that 2 Faradays produces 2 moles of silver atoms; and 3 Faradays produces 3 moles of silver atoms; and so on.

The **second law** states that the number of Faradays required to discharge one mole of an ion at an electrode equals the number of charges on the ion.

EXAMPLES

(a)
$$\underset{\text{1 mol}}{Pb^{2+}} + \underset{\text{2 mol}}{2e^-} \rightarrow \underset{\text{1 mol}}{Pb}$$

One mole of lead(II) ions requires 2 Faradays (2 moles of electrons) to be discharged at the cathode.

(b)
$$\underset{\text{1 mol}}{Al^{3+}} + \underset{\text{3 mol}}{3e^-} \rightarrow \underset{\text{1 mol}}{Al}$$

One mole of aluminium ions requires 3 Faradays (3 moles of electrons) to discharge at the cathode.

(c)
$$\underset{\text{2 mol}}{2Br^-} \rightarrow \underset{\text{1 mol}}{Br_2} + \underset{\text{2 mol}}{2e^-}$$

Two moles of bromide ions require 2 Faradays to discharge as one mole of bromine molecules at the anode. Thus, one mole of bromide ions require one Faraday.

EXAMPLE
Calculate the mass of lead produced at a cathode by a current of 2 amps flowing through molten lead(II) bromide for 30 minutes (Pb = 207).

SOLUTION
The half-reaction is

$$\underset{\text{1 mol}}{Pb^{2+}} + \underset{\text{2 mol}}{2e^-} \rightarrow \underset{\text{1 mol}}{Pb}$$

2 Faradays produce 1 mole of Pb atoms.
Thus, 2 Faradays produce 207 g of Pb atoms.
Hence, $2 \times 96\,500$ C produce 207 g of Pb atoms,

and so

$$1 \text{ C produces } \frac{207}{2 \times 96\,500} \text{ g of Pb}$$

Now 2 amps flowing for 30 minutes $= 2 \times 30 \times 60$ C

so $2 \times 30 \times 60$ C produce $\dfrac{207}{2 \times 96\,500} \times 2 \times 30 \times 60$ g of Pb

$$= 3.86 \text{ g}$$

Michael Faraday

The English chemist and physicist Michael Faraday was born in Newington, Surrey, in 1791. He is known for his pioneering experiments in electricity and magnetism. Some people consider Faraday to be the greatest experimentalist ever.

He received relatively little education. At the age of 14 he became an apprentice bookbinder. He soon became interested in science and after hearing a lecture by the famous chemist Humphrey Davy he wrote to him sending him the notes he made during the lecture. Davy appointed him as a laboratory assistant at the Royal Institution in London. Faraday was then 21.

During the years that followed, Faraday discovered two new chlorides of carbon. He also succeeded in liquefying chlorine and other gases. In 1825 he isolated benzene and in the same year was appointed Director of the laboratory. Following years of experimentation on electrolysis, he finally formulated his famous laws of electrolysis in 1834. By that time he had already discovered the phenomenon of electromagnetic induction.

Faraday became President of the Royal Society and wrote several books including *Experimental researches in chemistry and physics* (1858). In 1855, after suffering loss of memory, he ceased his research. He died in 1867.

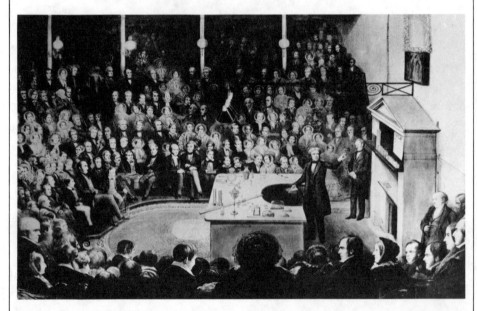

Christmas, 1855. Michael Faraday presents a lecture at the Royal Institution, London. Facing him in the front row are: the Prince Consort, on his left the Prince of Wales (afterwards Edward VII) and on his right the Duke of Edinburgh

INDUSTRIAL APPLICATIONS OF ELECTROLYSIS

Electrolysis has a number of important industrial applications. These include the extraction and purification of metals, electroplating and anodising and the manufacture of other chemicals.

Extraction of Metals

Metals in Groups I and II of the Periodic Table are extracted by electrolysis of fused halides. Sodium, for example, is obtained by electrolysis of molten sodium chloride in the Downs cell. Magnesium is obtained by the electrolysis of magnesium chloride which is obtained from dolomite and sea-water. Details of these and other extractions are given elsewhere in the book (*see* table 10.4).

Table 10.4 Extraction of metals

Metal	Source	Electrolyte	Ion discharged at cathode	*see* section
aluminium	bauxite	Al_2O_3 dissolved in molten cryolite	Al^{3+}	15.1
calcium	calcium chloride waste from Solvay process (Na_2CO_3 manufacture)	$CaCl_2$	Ca^{2+}	13.3
magnesium	dolomite and sea-water	$MgCl_2$	Mg^{2+}	13.3
sodium	rock salt	$NaCl$	Na^+	13.3
zinc	ZnS ores	$ZnSO_4$	Zn^{2+}	10.5

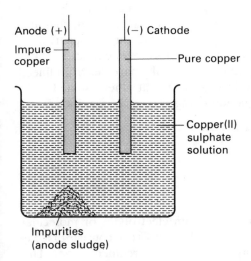

Anode (+)
(−) Cathode
Impure copper
Pure copper
Copper(II) sulphate solution
Impurities (anode sludge)

Figure 10.7 Purification of copper

Purification of Metals

Metals such as copper and zinc can be purified by electrolysis. The purification of metals is known as refining (*see* section 10.5). A schematic diagram of the purification of copper is shown in figure 10.7. The impure copper is the anode and the cathode is pure copper. Copper(II) sulphate solution can be used as an electrolyte.

At the anode the half-reaction is

$$Cu(s) \rightarrow Cu^{2+}(aq) + 2e^-$$

The impurities drop to the bottom of the vessel as anode sludge. This sludge may contain valuable metals such as gold and silver. Copper ions are discharged and deposited on the pure copper *cathode*. The half-reaction is

$$Cu^{2+}(aq) \rightarrow Cu(s) + 2e^-$$

Electroplating

In this process the object to be plated is the cathode. This is immersed in an electrolyte containing the ions of the plating metal. The anode is the pure plating metal. Successful electroplating requires that the electric current, concentration of electrolyte and temperature are exactly right. The cathode must also be clean.

Electroplating may involve a number of stages. For example, chromium plating of iron involves *four stages*:

1. The iron object, which is the cathode, is cleaned with sulphuric acid and then washed with de-ionised water.
2. The iron cathode is plated with copper.
3. It is then plated with nickel. This prevents corrosion.
4. Finally, the object is chromium plated.

Anodising

This is the process of coating aluminium objects with aluminium oxide. The object forms the anode. The electrolyte is dilute sulphuric acid. The coating of aluminium oxide resists corrosion.

Manufacture of Other Chemicals

The most important example of this application is the manufacture of sodium hydroxide, hydrogen and chlorine by means of the flowing mercury cathode cell. Details of this process are found in section 13.3.

(a)

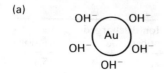

(b)

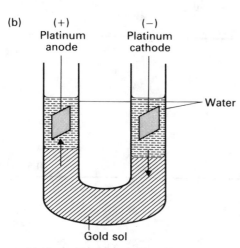

Figure 10.8 (a) Surface adsorption of ions; (b) electrophoresis

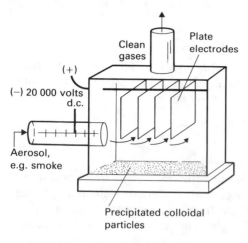

Figure 10.9 The Cottrell precipitator

Electrophoresis

The surfaces of colloidal particles suspended in solution are often electrically charged due to the adsorption of ions or electrons. For example, the gold particles in an aqueous gold sol adsorb hydroxide ions. When an electric current is passed through the sol, the negatively charged sol particles migrate to the anode (*see* figure 10.8). The migration of charged colloidal particles to an electrode is known as **electrophoresis**. Electrophoresis is also known as, cataphoresis.

Electrophoresis is particularly important in industry, medicine and biochemistry. It can be used to separate, identify and measure the amounts of proteins and lipids (*see* section 20.3) in the blood.

The Cottrell Precipitator

Electrically charged particles in aerosols such as mist, fog or smoke can be removed by application of the principle of electrophoresis. A device used for this purpose is the Cottrell precipitator (*see* figure 10.9). The aerosol is first passed through a negatively charged field. This ensures that sufficient charge is adsorbed on the particles. The aerosol is then passed through a positively charged field. The negatively charged colloidal particles are precipitated. The Cottrell precipitator is used in industry to remove smoke particles from flue gases and to recover valuable waste products which might otherwise be lost. The precipitator is also used in the air-conditioning systems of factories and office buildings.

Electrodialysis

Dialysis is the separation of colloidal particles from electrolytes or molecular solutes dissolved in the dispersion medium. A simple dialysis cell consists of a bag made of a semi-permeable material such as cellophane or parchment paper. This is dipped into the dispersion medium. The ionic and molecular species diffuse out of the bag into the dispersion medium. The colloidal particles remain in the bag.

Dialysis can be used to remove excess H^+ ions from an iron(III) oxide sol which has been stabilised by these ions (*see* figure 10.10).

Dialysis is a slow process. The dispersion medium has to be changed at frequent intervals if a high degree of removal of non-colloidal matter is to be achieved. If the non-colloidal matter is an electrolyte, dialysis can be accelerated by a

(a)

(b)

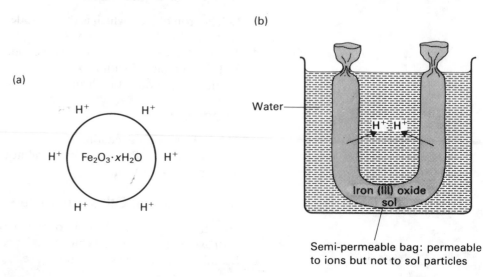

Figure 10.10 (a) Hydrated Fe_2O_3 sol particles are stabilised by the adsorption of H^+ ions. (b) Excess H^+ can be removed by dialysis

technique known as **electrodialysis** (*see* figure 10.11). Electrodialysis has been used on a large scale to produce potable water from brackish water or sea-water. However, alternative methods of removing salt are usually more economical.

The Artificial Kidney

Dialysis is used in medicine when kidneys fail to function. The dialysis is used to remove toxic waste products from blood. The process is known as haemodialysis. The process takes place in an artificial kidney or dialyser. A stream of blood taken from an artery circulates through the artificial kidney on one side of a semi-permeable membrane. On the other side a solution with an electrolyte composition similar to that of the patient's blood circulates. The waste products diffuse through the membrane into the solution. The pores of the membrane are too small to allow blood cells and proteins to pass through. The purified blood is returned to the patient's body through a vein.

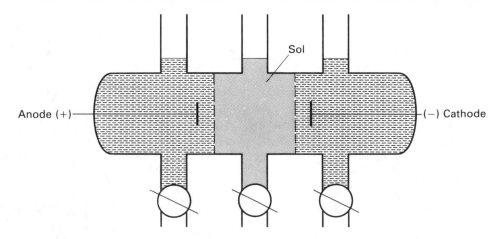

Figure 10.11   Electrodialysis

SUMMARY

1. **Electrolysis** occurs when electricity is passed through an **electrolyte** and chemical reactions take place at the **electrodes**.
2. The **anode** is the electrode at which oxidation occurs.
3. The **cathode** is the electrode at which reduction occurs.
4. The order in which ions are discharged at electrodes depends on
   (a) the nature of the electrode,
   (b) the state of the electrolyte,
   (c) the **electrode potential** of the ion.
5. **Faraday's first law of electrolysis** states that the mass of a substance produced at an electrode during electrolysis is proportional to the quantity of electricity passed.
6. **Faraday's second law of electrolysis** states that the number of **Faradays** required to discharge one mole of an ion at an electrode equals the number of charges on the ion.
7. Electrolysis is used to
   (a) extract metals,
   (b) purify metals,
   (c) electroplate metals.
8. **Electrophoresis** is the migration of charged colloidal particles to an electrode.
9. **Dialysis** is the separation of colloidal particles from electrolytes or molecular solutes dissolved in the dispersion medium.
10. Acceleration of dialysis by application of an electric field between two electrodes is known as **electrodialysis**.

10.4 Electrochemical Cells

An **electrochemical cell** is a device which produces an **electromotive force (e.m.f.)** as a result of chemical reactions taking place at the electrodes. An electrochemical cell is thus a device which converts chemical energy into electrical energy.

Each cell consists of two **half-cells**. In one half-cell an oxidation half-reaction takes place. In the other a reduction half-reaction takes place.

The Daniell Cell

This is a classic example of a simple electrochemical cell. One half-cell consists of a zinc rod dipping into a porous pot containing zinc sulphate solution (*see* figure 10.12). Oxidation takes place in this half-cell:

$$Zn(s) \rightarrow Zn^{2+}(aq) + 2e^-$$

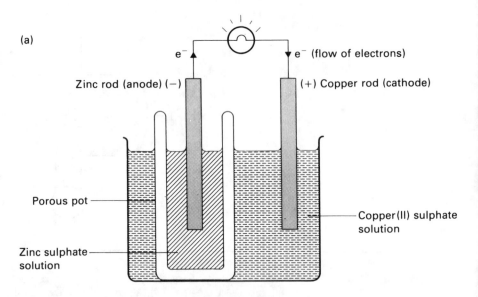

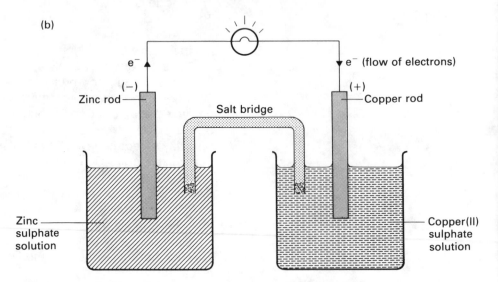

Figure 10.12 Electrochemical cells: (a) Daniell cell, (b) salt bridge

The zinc is the *anode* since oxidation occurs at this electrode. The other half-cell consists of a copper rod dipping into copper(II) sulphate solution. Reduction takes place in this half-cell:

$$Cu^{2+}(aq) + 2e^- \rightarrow Cu(s)$$

The copper is the *cathode* since reduction occurs at this electrode.

The porous pot is necessary to separate the two half-cells. It prevents the mixing of the two solutions but allows the passage of electricity.

Electrode Signs

The signs of the anode and cathode in an electrochemical cell are opposite to those in an electrolytic cell (*see* figure 10.13).

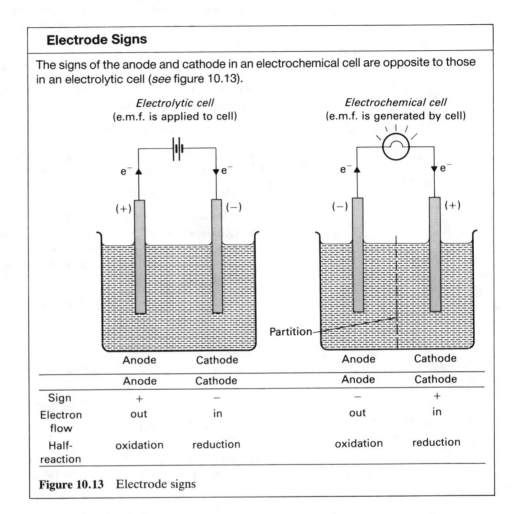

	Anode	Cathode	Anode	Cathode
Sign	+	−	−	+
Electron flow	out	in	out	in
Half-reaction	oxidation	reduction	oxidation	reduction

Figure 10.13   Electrode signs

The Salt Bridge

Simple electrochemical cells used in a laboratory often consist of two half-cells separated by a salt bridge (*see* figure 10.12). The salt bridge has the same function as the porous pot in a Daniell cell. It may consist of an inverted U-tube containing salt solution. It is plugged at both ends by cotton wool. Alternatively it may consist of a strip of filter paper soaked in salt solution. Salts used for this purpose include ammonium nitrate, potassium nitrate and potassium chloride.

CELL DIAGRAMS

A cell is represented in an electrical circuit by

The flow of electrons is thus,

A battery consisting of two cells in series is shown as

An electrochemical cell can be represented by a **cell diagram**. For example, the electrochemical cell with the salt bridge shown in figure 10.12 can be represented thus

$$Zn \mid Zn^{2+}(aq) \vdots Cu^{2+}(aq) \mid Cu$$

By convention, the half-cell with the cathode (the positive electrode) is placed on the right. This is the half-cell in which reduction takes place.

Thus $Cu^{2+}(aq) \mid Cu$ represents $Cu^{2+}(aq) + 2e^- \rightarrow Cu(s)$

The half-cell with the anode (negative electrode) is placed on the left. This is the half-cell in which oxidation takes place.

Thus $Zn \mid Zn^{2+}(aq)$ represents $Zn(s) \rightarrow Zn^{2+}(aq) + 2e^-$

A half-cell may be placed on the right in some cells and on the left in others. Which depends on its electrode potential and also that of the other half-cell. We shall discuss electrode potentials below. The copper–copper(II) ion half-cell, for example, is placed on the right in the Daniell cell (*see* above) but is on the left in the following cell:

$$Cu(s) \mid Cu^{2+}(aq) \vdots Ag^+(aq) \mid Ag(s)$$

This means that the copper is the anode and that oxidation takes place in the half-cell

$$Cu(s) \rightarrow Cu^{2+}(aq) + 2e^-$$

TYPES OF HALF-CELL

Half-cells are sometimes known as **redox electrodes** or redox couples. The three most common types are metal–metal ion, non-metal–ion and ion–ion half cells.

The Metal–Metal Ion Half-Cell
The zinc–zinc ion and copper–copper(II) ion electrodes used in the cells described above are typical examples.

The Non-Metal–Ion Half-Cell
The hydrogen electrode (strictly the hydrogen half-cell) is a good example of this (*see* figure 10.14). Hydrogen gas is allowed to bubble through a solution containing oxonium (hydrated hydrogen) ions. The electrode consists of platinum foil coated with platinum black. The electrode is inert although it allows the hydrogen to adsorb on its surface. The following equilibrium is established between the adsorbed layer of hydrogen molecules and the oxonium ions:

$$2H_3O^+(aq) + 2e^- \rightleftharpoons 2H_2O(l) + H_2(g)$$

or more simply:

$$2H^+(aq) + 2e^- \rightleftharpoons H_2(g)$$

In a cell diagram a hydrogen electrode is represented as

$Pt \mid H_2(g) \mid H^+(aq)$ if platinum is the anode, or
$H^+(aq) \mid H_2(g) \mid Pt$ if platinum is the cathode.

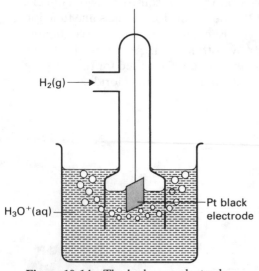

$H_2(g) \longrightarrow$

$H_3O^+(aq)$ —

—Pt black electrode

Figure 10.14   The hydrogen electrode

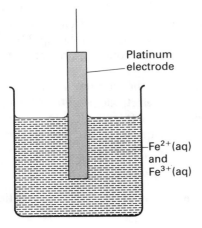

Figure 10.15   Ion–ion half-cell

Platinum
electrode

Fe²⁺(aq)
and
Fe³⁺(aq)

The Ion–Ion Half-Cell

This type of half-cell consists of an inert electrode such as a platinum electrode dipping into a solution containing ions of the same metal in two different oxidation states (*see* figure 10.15). A typical example is the iron(III)–iron(II) half-cell. In this the following equilibrium is established:

$$Fe^{3+}(aq) + e^- \rightleftharpoons Fe^{2+}(aq)$$

In a cell diagram this half-cell is represented as

$$Pt \mid Fe^{2+}(aq), Fe^{3+}(aq) \qquad \text{if platinum is the anode, or}$$
$$Fe^{3+}(aq), Fe^{2+}(aq) \mid Pt \qquad \text{if platinum is the cathode.}$$

ELECTRODE POTENTIALS

When a metal electrode is dipped into a solution containing ions of the same metal, atoms from the metal lattice pass into solution and form hydrated metal ions. At the same time hydrated metal ions gain electrons at the electrode and form metal atoms. They thus become part of the metal lattice. Eventually an equilibrium is established:

$$M^{z+}(aq) + ze^- \rightleftharpoons M(s)$$

In the forward process, that is the reduction process, the hydrated metal ions in solution consume electrons from the electrode (*see* figure 10.16). This results in a net deficit of electrons in the electrode and thus a positive charge on the electrode. On the other hand, in the reverse process, that is the oxidation process, atoms from the lattice pass into solution to form hydrated cations. This leaves a surplus of electrons on the electrode and results in a negative charge on the electrode.

The charge on the electrode depends on which of the two processes occurs more readily and thus on the position of equilibrium. The position of equilibrium

(a) $\qquad\qquad M^{z+}(aq) + ze^- \longrightarrow M(s)$

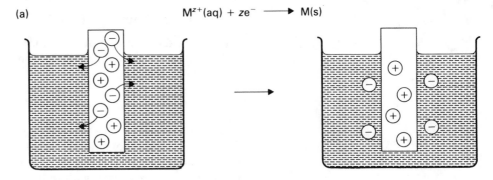

(b) $\qquad\qquad M(s) \longrightarrow M^{z+}(aq) + ze^-$

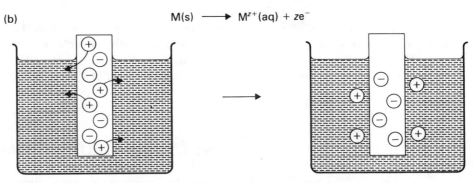

Figure 10.16   (a) Reduction: metal ions in solution gain electrons from the electrode leaving a positive charge on the electrode. (b) Oxidation: metal ions pass from the electrode into solution leaving an excess of electrons and thus a negative charge on the electrode

depends on a number of factors including the nature of the metal, the concentration of ions and the temperature. If the equilibrium lies to the right, reduction predominates and thus the electrode acquires a positive charge. If the equilibrium lies to the left, oxidation predominates and thus the electrode acquires a negative charge.

In either case there is a separation of charge and thus a potential difference between the electrode and the ions in solution. The **electrode potential** of a half-cell is the e.m.f. of a cell in which the electrode on the left is a hydrogen electrode and the electrode on the right is the electrode in question. For the electrode $M^{z+}(aq)/M$ the cell in question is

$$\text{Pt} \mid H_2(g) \mid H^+(aq) \mid\mid M^{z+}(aq) \mid M(s)$$

The electrode potential is written as $E_{M^{z+}/M}$ and refers to the reduction reaction taking place at the electrode

$$M^{z+}(aq) + ze^- \rightarrow M(s)$$

If the temperature of the half-cell increases, there will be an increasing tendency for the metal to dissolve in the solution and form hydrated metal ions. The reverse reaction will thus be favoured and the equilibrium will shift to the left. As a result, the potential difference between the electrode and the solution increases.

If, however, the concentration of the hydrated metal ions is increased, the equilibrium will shift to the right. In this case the potential difference between the electrode and the solution will be lowered.

Since electrode potentials depend on temperature, concentration and also pressure, it is necessary to standardise them if they are to be compared. Standard conditions are 298 K, one atmosphere pressure and a concentration of 1 mol dm^{-3}. The **standard electrode potential** of an electrode (or half-cell) is thus defined as the e.m.f. of a cell in which the electrode on the left is a standard hydrogen electrode and that on the right is the standard electrode in question. For the electrode M^{z+}/M the cell in question is

$$\text{Pt} \mid H_2(g), 1 \text{ atm} \mid H^+(aq), 1 \text{ mol } dm^{-3} \mid\mid M^{z+}(aq), 1 \text{ mol } dm^{-3} \mid M(s)$$

The standard electrode potential for this electrode is written as $E^{\ominus}_{M^{z+}/M}$.

As a consequence of this definition, the standard electrode potential of the (standard) hydrogen electrode is zero:

$$E^{\ominus}_{H^+/H_2} = 0$$

Redox Potentials

Electrode potentials of half-cells are often called redox potentials. By convention, redox potentials are always quoted as reduction potentials. The equilibrium half-reactions are thus written with reduction as the forward reaction:

$$\text{Oxidised species} + ze^- \rightleftharpoons \text{Reduced species}$$

A redox potential is a measure of the tendency for reduction to occur.

The **standard redox potentials** for a number of half-reactions are listed in table 10.5. The values range from about -3 volts to $+3$ volts. The more positive the redox potential the more likely reduction is to occur. The less positive, or more negative, the reduction potential the more likely oxidation is to occur.

THE e.m.f. OF ELECTROCHEMICAL CELLS

As we have seen, an electrochemical cell consists of two half-cells. Each half-cell has its own electrode potential. When these two half-cells are connected into an external circuit, two important things happen.

First of all, the cell redox reaction begins to occur. This reaction is the sum of

Table 10.5   Standard redox potentials

	Half-reaction			
	Oxidised species	$+ \, ze^-$ $\rightleftharpoons$	Reduced species	$E^{\ominus}/V$
	$Li^+(aq)$	$+ \, e^-$ $\rightleftharpoons$	$Li(s)$	-3.04
	$K^+(aq)$	$+ \, e^-$ $\rightleftharpoons$	$K(s)$	-2.92
	$Ca^{2+}(aq)$	$+ \, 2e^-$ $\rightleftharpoons$	$Ca(s)$	-2.87
	$Na^+(aq)$	$+ \, e^-$ $\rightleftharpoons$	$Na(s)$	-2.71
	$Mg^{2+}(aq)$	$+ \, 2e^-$ $\rightleftharpoons$	$Mg(s)$	-2.38
	$Al^{3+}(aq)$	$+ \, 3e^-$ $\rightleftharpoons$	$Al(s)$	-1.66
	$Zn^{2+}(aq)$	$+ \, 2e^-$ $\rightleftharpoons$	$Zn(s)$	-0.76
	$Sn^{2+}(aq)$	$+ \, 2e^-$ $\rightleftharpoons$	$Sn(s)$	-0.14
	$Pb^{2+}(aq)$	$+ \, 2e^-$ $\rightleftharpoons$	$Pb(s)$	-0.13
	$2H^+(aq)$	$+ \, 2e^-$ $\rightleftharpoons$	$H_2(g)$	0.00
	$Cu^{2+}(aq)$	$+ \, 2e^-$ $\rightleftharpoons$	$Cu(s)$	$+0.34$
	$I_2(g)$	$+ \, 2e^-$ $\rightleftharpoons$	$2I^-(aq)$	$+0.54$
	$Fe^{2+}(aq)$	$+ \, 3e^-$ $\rightleftharpoons$	$Fe(s)$	$+0.77$
	$Ag^+(aq)$	$+ \, e^-$ $\rightleftharpoons$	$Ag(s)$	$+0.80$
	$Br_2(l)$	$+ \, 2e^-$ $\rightleftharpoons$	$2Br^-(aq)$	$+1.07$
	$Cl_2(l)$	$+ \, 2e^-$ $\rightleftharpoons$	$2Cl^-(aq)$	$+1.36$
	$F_2(g)$	$+ \, 2e^-$ $\rightleftharpoons$	$2F^-(aq)$	$+2.87$

increasing:
(a) tendency for reverse reaction (oxidation) to occur
(b) tendency to lose electrons
(c) power as reducing agent

increasing:
(a) tendency for forward reaction (reduction) to occur
(b) tendency to gain electrons
(c) power as oxidising agent

the two half-reactions. As a result, the equilibrium in the two half-cells is disturbed. As the cell redox reaction proceeds, the concentration of the oxidised species in the half-cell containing the anode increases. On the other hand, the concentration of the oxidised species in the half-cell containing the cathode decreases. Eventually the redox reaction reaches equilibrium. When this occurs the battery has run down or is 'flat'.

Secondly, when the two half-cells are connected into an external circuit, current flows from the negative electrode, that is the anode, to the positive electrode, that is the cathode. The current is driven by the potential difference. In the external circuit the current is a flow of electrons. In the two half-cells the current is carried by the ions. As current is taken from the cell, the concentration of oxidised species increases and the concentrations of reduced species decreases. Consequently, the potential difference between the two electrodes decreases. At equilibrium, the potential difference between the two electrodes is zero.

The maximum value of the potential difference between the two electrodes is known as the electromotive force or e.m.f. of the cell. It occurs when no current is taken from the cell. We shall see below how the e.m.f. of a cell may be determined. The standard e.m.f. of a cell $E^{\ominus}$ is the e.m.f. of the cell under the standard conditions quoted above.

Calculating the e.m.f. of a Cell

There are two rules which can be applied to calculating the e.m.f. of a cell:

1. The half-cell with the more positive electrode potential is always placed on the right.

We saw above that this is the half-cell in which reduction occurs. The electrode of this half-cell is always positive.

2. The e.m.f. of an electrochemical cell is given by

$$E_{cell} = E_R - E_L \qquad (10)$$

where E_R is the electrode potential of the right-hand cell and E_L is the electrode potential of the left-hand cell.

EXAMPLE
From the following data determine (a) the standard cell e.m.f., (b) the cell reaction, (c) the cell diagram.

Data: $Zn^{2+}(aq) + 2e^- \rightleftharpoons Zn(s)$ $E^{\ominus}_{Zn^{2+}/Zn} = -0.76 \text{ V}$

$Ni^{2+}(aq) + 2e^- \rightleftharpoons Ni(s)$ $E^{\ominus}_{Ni^{2+}/Ni} = -0.25 \text{ V}$

SOLUTION

(a) The nickel–nickel(II) ion electrode is more positive. Thus

$$E^{\ominus}_{cell} = E^{\ominus}_{Ni^{2+}/Ni} - E^{\ominus}_{Zn^{2+}/Zn}$$
$$= (-0.25 \text{ V}) - (-0.76 \text{ V})$$
$$= +0.51 \text{ V}$$

(b) Reduction always occurs in the right-hand cell:

$$Ni^{2+}(aq) + 2e^- \rightarrow Ni(s)$$

Oxidation always occurs in the left-hand cell:

$$Zn(s) \rightarrow Zn^{2+}(aq) + 2e^-$$

The balanced redox equation is obtained by adding these two half-reactions. Thus

$$Zn(s) + Ni^{2+}(aq) \rightarrow Zn^{2+}(aq) + Ni(s)$$

(c) The cell diagram is written

$$Zn(s) \mid Zn^{2+}(aq) \mid\mid Ni^{2+}(aq) \mid Ni(s)$$

MEASURING CELL e.m.f.s AND ELECTRODE POTENTIALS

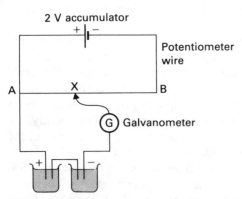

Figure 10.17 Determination of cell e.m.f.

The e.m.f. of a cell may be measured using a high-resistance voltmeter. The high resistance is necessary to ensure that only a negligible current is taken from the cell. We saw above that when current is taken from a cell the potential difference between the two electrodes is lowered. Alternatively, a bridge circuit may be used. In this case no current is taken from the cell. The contact wire is moved along the potentiometer wire until the galvanometer shows that no current is flowing (*see* figure 10.17). If the e.m.f. of the accumulator is 2 volts, the e.m.f. of the cell under standard conditions is given by

$$E^{\ominus}_{cell} = \frac{AX}{AB} \times 2 \text{ V}$$

The absolute potential difference between an electrode and solution cannot be measured. If a voltmeter is used, for example, it is necessary to dip a wire connecting the voltmeter into the solution. This wire would then constitute another electrode and thus a second half-cell. It is thus only possible to measure the potential difference between two electrodes. If the electrode potential of one half-cell is known, however, the electrode potential of the other can be calculated using equation (10). Since it is impossible to determine experimentally the absolute electrode potential of any electrode it is necessary to assign arbitrarily a value to one electrode and relate all other values to it. As we saw above, by convention, the standard electrode potential of the hydrogen electrode (or more strictly the hydrogen half-cell) is assigned a zero value. The electrode potential of another half-cell can then be measured by setting up an electrochemical cell with the hydrogen electrode as the other half-cell (*see* figure 10.18).

If the hydrogen electrode is the negative electrode, the cell diagram is

$$Pt \mid H_2(g) \mid H^+(aq) \mid\mid M^{z+}(aq) \mid M(s)$$

Thus

$$E^{\ominus}_{cell} = E^{\ominus}_{M^{z+}/M} - E^{\ominus}_{H^+/H_2}$$
$$= E^{\ominus}_{M^{z+}/M} - 0$$
$$= E^{\ominus}_{M^{z+}/M}$$

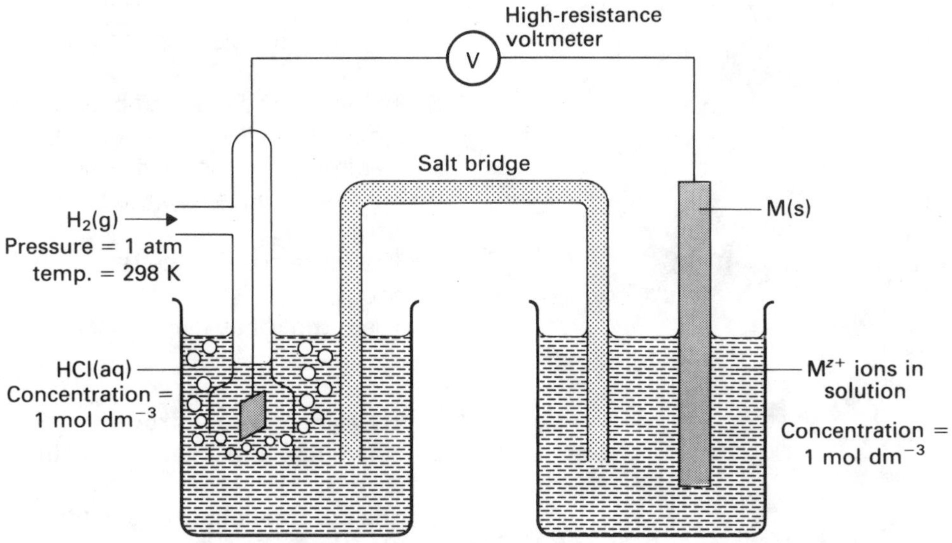

High-resistance voltmeter

V

Salt bridge

H$_2$(g)
Pressure = 1 atm
temp. = 298 K

M(s)

HCl(aq)
Concentration = 1 mol dm^{-3}

M^{z+} ions in solution
Concentration = 1 mol dm^{-3}

Figure 10.18   Measuring standard electrode potentials

The standard electrode potential of the half-cell in question thus equals the standard cell e.m.f.

If the hydrogen electrode is the positive electrode then

$$E^{\ominus}_{\text{cell}} = -E^{\ominus}_{\text{M}^{z+}/\text{M}}$$

The hydrogen electrode is known as the **primary reference electrode**.

The Calomel Electrode

The hydrogen electrode is relatively difficult to set up and operate under standard conditions. It is far easier to use the calomel electrode as a **secondary reference electrode** to **calibrate** other electrodes. Calomel is mercury(I) chloride, Hg$_2$Cl$_2$. A diagram of this electrode is shown in figure 10.19. The half-reaction for the electrode is

$$\text{Hg}_2\text{Cl}_2(\text{s}) + 2\text{e}^- \rightarrow 2\text{Hg(l)} + 2\text{Cl}^-(\text{aq})$$

A saturated solution of KCl is frequently used for the calomel electrode. The cell diagram is then represented as

$$\text{Hg}_2\text{Cl}_2(\text{s}),\text{Hg(l)} \mid \text{Cl}^-(\text{sat'd})$$

The electrode potential of the calomel electrode is known accurately over a range of concentrations and temperatures. When the concentration of the potassium chloride solution is 1 mol dm^{-3}, the electrode potential at 298 K is +0.2812 V. If a saturated solution of KCl is used, the electrode potential is +0.2415 V at 298 K.

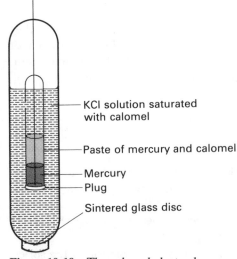

KCl solution saturated with calomel

Paste of mercury and calomel

Mercury

Plug

Sintered glass disc

Figure 10.19   The calomel electrode

SUMMARY

1. An **electrochemical cell** consists of two **half-cells** and is a device for converting chemical energy into electrical energy.
2. The **cell diagram** of the **Daniell cell** is

$$\text{Zn} \mid \text{Zn}^{2+}(\text{aq}) \mid\mid \text{Cu}^{2+}(\text{aq}) \mid \text{Cu(s)}$$

3. A **salt bridge** allows the passage of electricity between two electrodes but physically separates the two half-cells.

(continued)

4. The negative electrode in an electrochemical cell is the one in which oxidation takes place.
5. The **standard electrode potential** of an electrode (or half-cell) is the e.m.f. of a cell in which the electrode on the left is a standard hydrogen electrode and that on the right is the standard electrode in question.
6. The **redox potential** is a measure of the tendency for reduction to occur.
7. The more positive the redox potential, the more likely reduction is to occur.
8. The **e.m.f.** of an electrochemical cell is given by

$$E_{cell} = E_R - E_L$$

9. The **hydrogen electrode** is used as a **primary reference electrode**. By convention the standard electrode potential of the hydrogen electrode is zero.
10. The **calomel electrode** is often used as a **secondary reference electrode**.

10.5 Applications of Electrochemistry

We have noted above the wide-ranging influence of electrochemistry on other branches of chemistry. In this section we shall consider the importance of electrode potentials and electrochemical cells in some aspects of inorganic chemistry, thermodynamics, analytical chemistry and also some commercial and industrial applications.

THE ELECTROCHEMICAL SERIES

When elements are placed in order of their standard electrode potentials the electrochemical series is obtained. The series shown in table 10.6 can be obtained directly from the data shown in table 10.5. That part of the series which includes metals only is called the **electrochemical series of metals** This series closely approximates to the order of reactivity of metals.

The most electropositive and most reactive metals are at the top of the series. These metals are readily oxidised and thus ionise easily. As a consequence they are strong reducing agents. Metals at the bottom of the series do not oxidise readily. They are thus stable in their reduced form. Gold and mercury are examples.

Displacement Reactions

Metals high in the series reduce the oxidised forms of metals lower in the series. In effect they displace the metals from their oxides or from solutions of their salts. For example, zinc displaces copper from a solution of copper(II) sulphate. The redox equation for this reaction is

$$Zn(s) + Cu^{2+}(aq) \rightarrow Zn^{2+}(aq) + Cu(s)$$

On the other hand, magnesium, which is higher in the series, displaces zinc from its oxide:

$$Mg(s) + ZnO(s) \rightarrow MgO(s) + Zn(s)$$

Metals above hydrogen in the electrochemical series of elements reduce

Table 10.6 Electrochemical series of elements

	Metals	Non-metals	
	most electropositive, most reducing		
	lithium		
	potassium		
	calcium		
	sodium		
	magnesium		
	aluminium		
	zinc		
increasing reactivity	iron		
	tin		
	lead		
		hydrogen	
	copper		
		iodine	
	mercury		*increasing reactivity*
	silver		
		bromine	
		chlorine	
	gold		
	least electropositive, least reducing		

hydrogen ions to form hydrogen gas. For example

$$Mg(s) + 2H^+(aq) \rightarrow Mg^{2+}(aq) + H_2(g)$$

Note that this is a displacement reaction.

Whereas the reactivity of metals increases as the electrode potentials become more negative, the reactivity of non-metals increases as the electrode potentials become more positive. Chlorine, for example, displaces iodine from solution:

$$Cl_2(g) + 2I^-(aq) \rightarrow 2Cl^-(aq) + I_2(s)$$

Non-metals lower in the electrochemical series are thus more strongly oxidising.

Electrolysis

In section 10.2 we noted that the electrochemical series had an important bearing on the order of discharge of ions during electrolysis. As the electrode potentials of metals become less positive, that is, as the metals become more electropositive, they become increasingly difficult to discharge. On the other hand, non-metallic ions discharge more easily as the electrode potentials become more negative.

Metal Extraction and Corrosion

We shall discuss metal extraction and corrosion in more detail below. However, it is worth noting here that the stability of a metal in its reduced form is related to its position in the electrochemical series of metals. Metals low in the series are very stable in their reduced form whereas those high in the series, such as sodium and potassium, are relatively unstable. These metals oxidise easily. Indeed (almost) all metals above gold in the series occur naturally in oxidised form. Sodium, for example, occurs as sodium ions. Iron occurs as oxide ores. To obtain these as pure metals, it is therefore necessary to reduce the oxidised forms. Methods for doing this are described below.

Corrosion is an oxidation process which occurs at the surface of a metal. The corrosion may be due to reaction with oxygen, acids or other compounds. Gold, which is at the bottom of the series, does not corrode. Metals above hydrogen in the series corrode increasingly easily as the series is ascended. We shall discuss the mechanism of these in the section on metallurgy below.

CALCULATION OF EQUILIBRIUM CONSTANTS FOR REDOX REACTIONS

We saw above that the electrode potential of a half-cell varies with concentration of ions in solution. For the general half-reaction:

$$\text{Oxidised species} + ze^- \rightarrow \text{Reduced species}$$

the relationship between concentration and electrode potential is given by the **Nernst equation**:

● $$E = E^{\ominus} + \frac{RT}{zF} \ln \left(\frac{[\text{oxidised species}]}{[\text{reduced species}]} \right)$$

where R is the gas constant, T is the absolute temperature, z is the number of electrons transferred from the reduced species to the oxidised species, F is the charge of a mole of electrons (the Faraday), $E^{\ominus}$ is the standard electrode potential and ln is $\log_e$.

Let us see how the Nernst equation applies to the Daniell cell. In the left-hand half-cell we saw that the following half-reaction occurs:

$$Zn(s) \rightarrow Zn^{2+}(aq) + 2e^-$$

The oxidised species is $Zn^{2+}(aq)$ and the reduced species $Zn(s)$. For this half-cell,

$$E_{Zn^{2+}/Zn} = E^{\ominus}_{Zn^{2+}/Zn} + \frac{RT}{2F} \ln [Zn^{2+}]$$

The half-reaction in the right-hand half-cell is

$$Cu^{2+}(aq) + 2e^- \rightarrow Cu(s)$$

The Nernst equation is

$$E_{Cu^{2+}/Cu} = E^{\ominus}_{Cu^{2+}/Cu} + \frac{RT}{2F} \ln [Cu^{2+}]$$

These two Nernst equations can be combined to give the relationship of the cell e.m.f. to the concentrations of ions in the half-cells. Since

$$E_{cell} = E_{Cu^{2+}/Cu} - E_{Zn^{2+}/Zn}$$

it follows that

$$E_{cell} = E^{\ominus}_{cell} + \frac{RT}{2F} \ln \left(\frac{[Cu^{2+}]}{[Zn^{2+}]} \right)$$

$$= E^{\ominus}_{cell} + \frac{2.303\ RT}{2F} \lg \left(\frac{[Cu^{2+}]}{[Zn^{2+}]} \right)$$

Remember that

$$\ln x = 2.303 \lg x$$

When the redox reaction in the Daniell cell is at equilibrium, the e.m.f. of the cell is zero:

$$0 = E^{\ominus}_{cell} + \frac{2.303\ RT}{2F} \lg \left(\frac{[Cu^{2+}]}{[Zn^{2+}]} \right)$$

Thus

$$E^{\ominus}_{cell} = -\frac{2.303\ RT}{2F} \lg \left(\frac{[Cu^{2+}]}{[Zn^{2+}]} \right)$$

$$= \frac{2.303\ RT}{2F} \lg \left(\frac{[Zn^{2+}]}{[Cu^{2+}]} \right)$$

The equilibrium for the Daniell cell is

$$Zn(s) + Cu^{2+}(aq) \rightleftharpoons Zn^{2+}(aq) + Cu(s)$$

The equilibrium constant for this is given by

$$K_c = \frac{[Zn^{2+}]}{[Cu^{2+}]}$$

We can substitute this into the above equation as follows:

$$E_{cell}^{\ominus} = \frac{2.303\ RT}{2F}\ \lg K_c.$$

$$\frac{2.303\ RT}{F}$$

$$= \frac{2.303\ (8.314\ J\ K^{-1}\ mol^{-1}) \times (298\ K)}{9.648 \times 10^4\ C\ mol^{-1}}$$

$$= 0.059\ J\ C^{-1}$$

$$= 0.059\ V \qquad (\text{since } 1\ C\ V = 1\ J)$$

Under standard conditions the factor $2.303RT/F$ works to be 0.059 V. Thus

$$E_{cell}^{\ominus} = \frac{0.059}{2}\ \lg K_c$$

We thus have an expression relating the standard e.m.f. of the Daniell cell to the equilibrium constant of the redox reaction. Similar equations can be derived for other redox reactions. They provide a simple means of determining equilibrium constants from experimentally determined standard electrode potentials.

For the Daniell cell, the standard cell e.m.f. is 1.10 V. Thus

$$1.10\ V = \left(\frac{0.059\ V}{2}\right) \times (\lg K_c)$$

This gives

$$K_c = 10^{37.3}$$

The high value indicates that the reaction proceeds virtually to completion.

DETERMINATION OF pH

The half-reaction of the hydrogen electrode can be written

$$H^+(aq) + e^- \rightleftharpoons \tfrac{1}{2}H_2(g)$$

The concentration of the hydrogen ions in the half-cell can be related to the electrode potential of the hydrogen electrode by the Nernst equation:

$$E_{H^+/H_2} = E_{H^+/H_2}^{\ominus} + \frac{RT}{F}\ \ln\left(\frac{[H^+]}{p_{H_2}^{1/2}}\right)$$

where p_{H_2} is the partial pressure of the hydrogen.

If we assume that the partial pressure of the hydrogen gas is one atmosphere and since $E_{H^+/H_2}^{\ominus} = 0$, then

$$E_{H^+/H_2} = +(0.059\ V)\ \lg [H^+]$$

Since

$$pH = -\lg [H^+]$$

$$E_{H^+/H_2} = -(0.059\ V)\ pH$$

The electrode potential of the hydrogen electrode is thus directly proportional to pH. If a cell is set up consisting of a hydrogen electrode and another electrode, such as the calomel electrode, whose electrode potential is insensitive to hydrogen-ion concentration, then it is possible to show that

$$E_{cell} \propto pH$$

An electrochemical cell consisting of an electrode sensitive to hydrogen-ion concentration and a reference electrode thus provides an effective means of determining pH.

For routine pH determinations it is impracticable to use a hydrogen electrode. In practice a **glass electrode** is used. This consists of a thin-walled glass bulb, a silver wire and 0.1 mol dm^{-3} hydrochloric acid saturated with silver chloride. The electrode potential is proportional to hydrogen-ion concentration. It is used with a secondary reference electrode such as the calomel electrode.

The e.m.f. of a cell consisting of a glass electrode and a reference electrode can be measured using a high-resistance voltmeter. A **pH meter** is a high-resistance voltmeter calibrated with the pH scale (*see* figure 10.20). The glass electrode assembly can be calibrated using a buffer solution (*see* section 8.2) whose pH is accurately known.

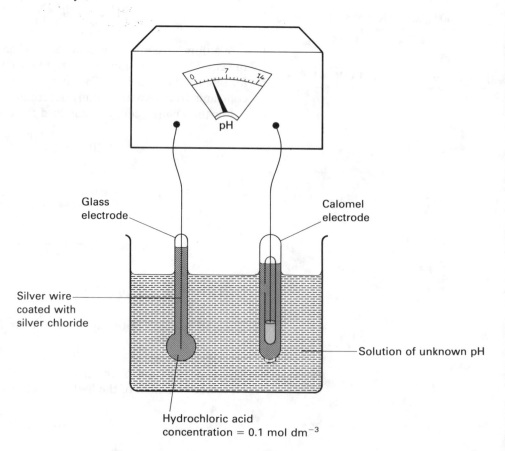

Figure 10.20   The pH meter

POTENTIOMETRIC TITRATIONS

Electrochemical cells may be set up as titration cells and used to determine the concentration of oxidised or reduced species. One species is added from a burette to the other in a beaker. The e.m.f. of the cell is measured by a voltmeter after each increment of titrant. In the case of an acid–base titration a pH meter can be used. Redox, precipitation and neutralisation (acid–base) titrations may all be performed potentiometrically.

1. For a **redox titration** an inert metal electrode is used as a detecting electrode. A calomel electrode is used as a reference electrode. Figure 10.21 shows the potentiometric titration of Fe^{2+} with cerium(IV), Ce^{4+}

(a)

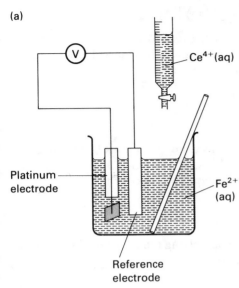

Platinum electrode

Reference electrode

(b)

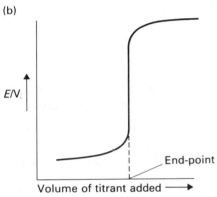

E/V

End-point

Volume of titrant added →

Figure 10.21 Potentiometric titration.
(a) Titration of Fe^{2+} with Ce^{4+}
$Fe^{2+}(aq) + Ce^{4+}(aq) \rightleftharpoons$
$\qquad Fe^{3+}(aq) + Ce^{3+}(aq)$
(b) Typical potentiometric titration curve

2. For **precipitation titrations** an **ion-selective electrode** and a reference electrode are used. For example, for the titration of potassium chloride with silver nitrate,

$$Ag^+(aq) + Cl^-(aq) \rightleftharpoons AgCl(s)$$

a silver electrode is used.

3. For **neutralisation titrations** (*see* section 8.1) a glass electrode is used together with a calomel electrode. The instrument readings can either be in volts or, if a pH meter is used, in pH units.

The end-point of a potentiometric titration corresponds to the position of the very steep—almost vertical—slope of the curve (*see* figure 10.21).

Both redox and neutralisation titrations may also be carried out using colour indicators.

FEASIBILITY OF REDOX REACTIONS

Thermodynamics predicts that certain reactions are capable of occurring whereas others are not. Reactions capable of occurring are said to be **feasible**. However, although a reaction may be feasible it may not necessarily occur spontaneously. Combustion reactions are examples. Most combustion reactions do not occur spontaneously although they are thermodynamically feasible. This is because the energy barrier has to be overcome before the reaction can start. So, under given conditions a reaction may be thermodynamically feasible but not kinetically feasible. The thermodynamic feasibility of a redox reaction can quite simply be determined by inspecting the electrode potentials of the two half-reactions.

A reduction half-reaction is feasible if its electrode potential is more positive than the electrode potential of the other half-reaction. Reduction is not feasible if the electrode potential is less positive than that of the other half-reaction. The half-reaction with the less positive electrode potential must be the oxidation half-reaction.

EXAMPLE
Given the following:

$$Cu^{2+}(aq) + 2e^- \rightleftharpoons Cu(s) \qquad E^\ominus = 0.34 \text{ V}$$
$$Ag^+(aq) + e^- \rightleftharpoons Ag(s) \qquad E^\ominus = 0.80 \text{ V}$$

(a) which is thermodynamically feasible:
the reduction of silver ions by copper *or*
the reduction of copper(ii) ions by silver?
(b) Write the redox equation and the cell diagram for the thermodynamically feasible reaction.

SOLUTION
(a) The silver ion–silver half-reaction has the more positive electrode potential. This must thus be the reduction half-reaction. Reduction of silver ions by copper is therefore feasible. Reduction of copper(ii) ions by silver is not feasible.
(b) The two half-reactions are thus
Oxidation $\qquad Cu(s) \rightarrow Cu^{2+}(aq) + 2e^-$
Reduction $\qquad Ag^+(aq) + e^- \rightarrow Ag(s)$
Multiplying the reduction half-reaction by 2 and then adding the two half-reactions we obtain

$$2Ag^+(aq) + Cu(s) \rightarrow Cu^{2+}(aq) + 2Ag(s)$$

Since the silver ion–silver half-cell has the more positive electrode potential this must be the right-hand half-cell. The cell diagram is thus

$$Cu \mid Cu^{2+}(aq) \mid\mid Ag^+(aq) \mid Ag$$

The feasibility of this reaction can also be predicted by inspecting the electrochemical series of metals. Since copper is higher in the series than silver, it is more strongly reducing. It thus reduces silver and not vice versa.

DETERMINING ΔG

We saw in section 5.3 that the Gibbs free energy change ΔG must be negative if a reaction is to proceed spontaneously. We also saw that ΔG is a measure of the feasibility of a reaction. $\Delta G_m^{\ominus}$ is related to the equilibrium constant of a reaction by the van't Hoff isotherm:

$$\Delta G_m^{\ominus} = -RT \ln K$$

The equilibrium constant of a redox reaction involving the transfer of z electrons is related to cell e.m.f. by the Nernst equation:

$$E_{cell}^{\ominus} = \frac{RT}{zF} \ln K$$

Combining these two equations we obtain

$$\Delta G_m^{\ominus} = -zFE_{cell}^{\ominus} \tag{11}$$

We thus have a means of using experimentally determined standard electrode potentials for calculating values of $\Delta G_m^{\ominus}$ and thus the maximum amount of electrical work a cell can supply.

EXAMPLE
Calculate the maximum amount of work obtainable from a Daniell cell under standard conditions given that $E_{cell}^{\ominus} = 1.10$ V.

SOLUTION
To solve this problem we calculate $\Delta G_m^{\ominus}$ using equation (11). The cell reaction is

$$Zn(s) + Cu^{2+}(aq) \rightleftharpoons Zn^{2+}(aq) + Cu(s)$$

Thus, $z = 2$
Given that $F = 9.648 \times 10^4$ C mol^{-1} we obtain

$$\Delta G_m^{\ominus} = -2 \times (9.648 \times 10^4 \text{ C mol}^{-1}) \times (1.10 \text{ V})$$
$$= -2.12 \times 10^5 \text{ C V mol}^{-1}$$

Since 10^3 C V = 1 kJ

$$\Delta G_m^{\ominus} = -2.12 \times 10^2 \text{ kJ mol}^{-1}$$

Thus, the maximum amount of work that can be obtained when one mole of zinc transfers two moles of electrons to copper(ii) ions is 212 kJ.

BATTERIES AND COMMERCIAL CELLS

We saw in section 5.4 that a battery is a portable source of electricity. It consists of one or more electrochemical cells. We also saw that a cell which does not regenerate reactants is called a primary cell. A secondary cell or storage cell is one which can be recharged by regenerating the cell reactants. We shall now examine three types of cell and find out which redox reactions take place in them.

The Dry Cell

The dry cell (see figure 10.22) is a primary battery or cell which is used in portable radios, clocks and torches. Once flat it cannot be re-used and has to be disposed

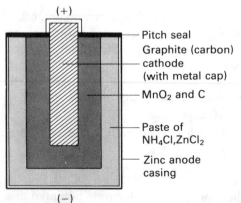

(+)

Pitch seal
Graphite (carbon) cathode (with metal cap)
MnO$_2$ and C
Paste of NH$_4$Cl,ZnCl$_2$
Zinc anode casing

(−)

Figure 10.22 Dry cell

of. However, it is a convenient portable source of electricity because all its components are solids or pastes which are tightly sealed from the environment. The anode is the zinc container which encases the dry cell. The cathode is a graphite rod surrounded by a layer of mangangese(IV) oxide and carbon. The electrolyte is a paste consisting of zinc chloride, ammonium chloride and water. The half-reactions at the electrodes are

at the anode $\quad\quad Zn(s) \rightarrow Zn^{2+}(aq) + 2e^-$
at the cathode $\quad 2MnO_2(s) + 2NH_4^+(aq) + 2e^- \rightarrow Mn_2O_3(s) + 2NH_3(aq) + H_2O(l)$

The overall equation is

$$Zn(s) + 2MnO_2(s) + 2NH_4^+(aq) \rightarrow Zn^{2+}(aq) + Mn_2O_3(s) + 2NH_3(aq) + H_2O(l)$$

A dry cell generates between 1.25 and 1.50 V.

Lead Storage Cell

This is also known as a lead accumulator. It is a secondary cell—the half-reactions at the electrodes are readily reversible. It consists of a lead anode and a grid of lead packed with lead(IV) oxide as the cathode. The electrolyte is sulphuric acid. The half-reactions at the electrodes are

at the anode $\quad\quad Pb(s) + SO_4^{2-}(aq) \rightleftharpoons PbSO_4(s) + 2e^-$
at the cathode $\quad PbO_2(s) + 4H^+(aq) + SO_4^{2-}(aq) + 2e^- \rightleftharpoons PbSO_4(s) + 2H_2O(l)$

The overall reaction is

$$Pb(s) + PbO_2(s) + 4H^+(aq) + 2SO_4^{2-}(aq) \underset{charging}{\overset{discharging}{\rightleftharpoons}} 2PbSO_4(s) + 2H_2O(l)$$

The current which can be obtained from the cell is increased by constructing the cathode of a number of plates which alternate with several anode plates (*see* figure 10.23). One such cell provides about 2 V. The storage battery used in cars normally consists of six of these cells arranged in series to provide 12 V.

The battery is recharged by applying a current from an external source. This reverses the electrode reactions. When the lead storage cell is fully charged, the

> The term *specific gravity* is often used instead of relative density when referring to battery acid. *Relative density* is the ratio of the mass of a given volume of a substance to the mass of an equal volume of water at a temperature of 4°C. Relative density has no units.

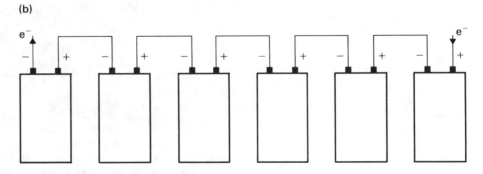

Figure 10.23 (a) Cell of lead storage battery; (b) car battery

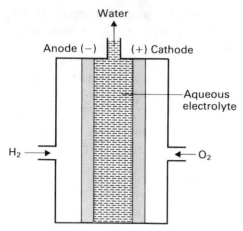

Figure 10.24 Simple fuel cell

sulphuric acid has a relative density of about 1.275. Upon discharging, lead(II) sulphate forms at both electrodes, thus reducing the concentration and relative density of the sulphuric acid. The relative density thus indicates the state of charge in the battery. The original relative density is obtained by recharging.

Fuel Cells

We have already referred to this type of cell in section 5.4. A fuel cell is a type of primary cell in which the reactants are continuously replaced as they are consumed and the products are continuously removed. A diagrammatic representation of a simple fuel cell is shown in figure 10.24. Hydrogen and oxygen are bubbled

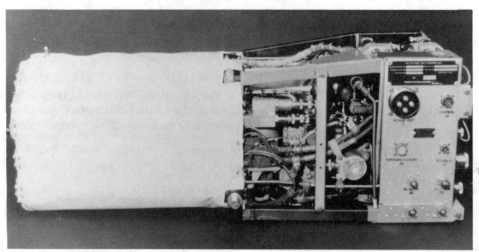

Electrical power for the Space Shuttle orbiter is supplied by a fuel cell system that produces power by the electrochemical conversion of hydrogen and oxygen. This is one of the three fuel cells that make up the generating system. Each fuel cell is capable of generating 12 kilowatts at peak and 7 kilowatts average power

Post-landing processing of the Space Shuttle Colombia is seen double in this scene in southern California. Uncommon rain water has given a mirror effect to the normally dry lake bed. Earlier the same day, Colombia, with its first four-member crew aboard, had touched down after successfully completing a 5 day 2 hour mission

through porous carbon electrodes into a concentrated solution of an alkali. The carbon electrodes contain a platinum catalyst. The half-reactions at the electrodes are

at the anode $2H_2(g) + 4OH^-(aq) \rightarrow 4H_2O(l) + 4e^-$
at the cathode $O_2(g) + 2H_2O(l) + 4e^- \rightarrow 4OH^-$

The overall reaction is

$$2H_2(g) + O_2(g) \rightarrow 2H_2O(l)$$

The water is removed and in spacecraft is consumed by the astronauts.

In some fuel cells an acidic electrolyte is used in which case the electrode reactions are

at the anode $H_2(g) \rightarrow 2H^+(aq) + 2e^-$
at the cathode $O_2(g) + 4H^+(aq) + 4e^- \rightarrow 2H_2O(l)$

the overall reaction is the same as above, however.

METALLURGY

Metals occur naturally as ores and minerals. A **mineral** is any naturally occurring inorganic solid with a crystalline structure. A mineral from which a metal can be extracted is called an **ore**. The science of extracting metals from ores and preparing them for use is called metallurgy. As we shall see below, the extraction of metals from ores relies heavily on the electrochemical or redox properties of the metals.

Examples of some important ores are shown in table 10.7. The metals are arranged in the same order as in the electrochemical series—that is the most reactive and electropositive metals at the top. It is perhaps dangerous to make sweeping generalisations about the relationship between the type of ore and the position of the metal in the series since there are many exceptions. However, it is noticeable that metals at the top tend to occur more commonly as chlorides, carbonates and sulphates. Metals in the middle of the series frequently occur as oxides and those lower as sulphides. Only metals towards the bottom of the series—such as copper, silver and gold—exist in their native state.

Extraction of Metals

We shall see in some detail in later chapters how specific metals are extracted from their ores. In this section we shall consider in general terms the three stages of metal extraction. These are concentration, reduction and refining.

Concentration

Many ores contain unwanted materials such as clay and granite. These unwanted materials are called **gangue**. The first stage in metal extraction is thus the removal of gangue. This process is known as concentration. Physical methods of separating gangue include flotation and magnetic separation. In the **flotation** method the ore is finely crushed and mixed with oil and water in a large tank. The mixture is agitated by blowing in air. A froth of oil and the desired mineral floats to the top of the tank where it is skimmed off.

Magnetic separation is used to separate magnetite, Fe_3O_4, from gangue. An electromagnet is used.

Chemical methods of separation include the **leaching** of ores. This involves the extraction of the metal as a soluble salt using a suitable aqueous solution. For example, copper(II) oxide ores can be leached using dilute sulphuric acid:

$$CuO(s) + 2H^+(aq) \rightarrow Cu^{2+}(aq) + H_2O(l)$$

Table 10.7 Some common ores

Element	Ore	Composition
K	sylvite	KCl
Ca	limestone	$CaCO_3$
	gypsum	$CaSO_4 \cdot 2H_2O$
Na	rock salt	NaCl
Mg	magnesite	$MgCO_3$
Al	bauxite	$Al_2O_3 \cdot nH_2O$
Ti	rutile	TiO_2
Zn	zinc blende	ZnS
Fe	haematite	$Fe_2O_3 \cdot nH_2O$
	pyrite	FeS_2
Sn	cassiterite	SnO_2
Pb	galena	PbS
Cu	chalcopyrite	$CuFeS_2$
Ag	argentite	Ag_2S
Au	native metal	Au

Bubbles of copper ore concentrate on the surface of a flotation tank in a plant in Papua New Guinea

Reduction

Reference to table 10.7 shows that most metals exist in an oxidised form. Sodium exists as Na^+ in compounds like sodium chloride and tin as SnO_2 for example. The second stage of metal extraction is to reduce these ores to the metal. Several methods are used. Metals which exist naturally as oxide ores may be reduced directly using carbon or carbon monoxide.

For example, in the extraction of iron from haematite in the blast furnace (*see* section 14.3) the overall reaction is

$$Fe_2O_3(s) + 3CO(g) \rightleftharpoons 2Fe(l) + 3CO_2(g)$$

Note that this is a redox reaction.

Other examples of this type of reduction were discussed in section 5.3. We saw that Ellingham diagrams provide a useful means of showing which metal oxides can be reduced at a given temperature.

Carbonate and sulphide ores are first converted to their oxides by heating. Zinc, for example, is obtained from zinc blende by first **roasting** the ore:

$$2ZnS(s) + 3O_2(g) \rightarrow 2SO_2(g) + 2ZnO(s)$$

The oxide is then reduced with carbon or carbon monoxide:

$$ZnO(s) + CO(g) \rightleftharpoons Zn(s) + CO_2(g)$$

Metals high in the electrochemical series are often reduced by electrolysis of the molten ore. Metals reduced in this way include aluminium, magnesium and sodium (*see* section 10.3 above and also sections 13.3 and 15.1). Inert electrodes such as graphite electrodes are used in the electrolytic cell. The metals are reduced at the cathode and collected at the bottom of the cell where they can be run off. The reduction half-reaction for the extraction of calcium from calcium chloride is, for example

$$Ca^{2+}(l) + 2e^- \rightarrow Ca(s)$$

Refining

This is the final stage in the extraction of metals. Various methods are used including distillation and electrolysis. Metals which are only weakly electropositive and thus do not react with water can be refined by electrolysis. These are the metals which fall below hydrogen in the electrochemical series. Details of the purification of copper were given in section 10.3 above.

CORROSION

Corrosion may be regarded as the natural tendency of metals to return to their oxidised states. Corrosion is a redox process. The main sources or causes of corrosion are:

- the atmosphere,
- submersion in water or aqueous solution,
- underground soil attack,
- corrosive gases,
- immersion in chemicals.

The most important of these is the atmosphere. The tendency to corrosion is largely determined by atmospheric humidity and particularly atmospheric pollution.

The most common example of corrosion is **rusting**. When iron comes into contact with acid, oxygen and other species found in the environment, electrochemical reactions take place. Since iron is higher in the electrochemical series than either hydrogen or oxygen, it tends to act as a reducing agent as it becomes oxidised:

$$Fe \rightarrow Fe^{2+} + 2e^-$$

Rusting is a form of corrosion. It takes place when iron comes into contact with acid, oxygen and other species found in the environment

The iron(II) ions are slowly converted to oxides of iron and deposit as flakes of rust with the formula $Fe_2O_3 \cdot nH_2O$. This hydrated oxide acts as an autocatalyst (*see* section 9.5) in the corrosion process.

The reduction processes which take place in rusting include

$$2H^+ + 2e^- \rightarrow H_2$$
$$2H_2O + O_2 + 4e^- \rightarrow 4OH^-$$

Various methods are used to inhibit corrosion. Coating the metal with paint or tin, for example, prevents attack by atmospheric moisture and oxygen. However, if the coating is flawed or broken, corrosion occurs. With *tin plating*, for example, iron is oxidised in preference to tin since the iron is higher in the electrochemical series. The iron and tin act as electrodes, the iron being the anode and the tin the cathode.

Corrosion of iron may be prevented by coating it with zinc. Iron coated with zinc is called *galvanised iron*. In this case, since zinc is higher in the electrochemical series than iron, it acts as the anode and the iron acts as the cathode. The zinc is thus oxidised in preference to the iron:

$$Zn \rightarrow Zn^{2+} + 2e^-$$

However, zinc is not used in food canning since it is higher than tin in the electrochemical series. It is thus more susceptible to attack by acids—in fruit juices, for example.

'Hot dip' galvanising

The use of zinc to protect iron from corrosion is called **cathodic protection** (*see* figure 10.25). A *sacrificial anode* is also used as a means of cathodic protection. This method has been used successfully to protect steel pipes under sea-water from corrosion. Magnesium anodes are attached to the pipe at regular intervals. Since magnesium is higher than iron in the electrochemical series, it is oxidised in preference to the iron.

Aluminium, although high in the electrochemical series, does not corrode readily since it oxidises to form aluminium oxide, Al_2O_3, which does not flake. It thus forms a strong protective coating.

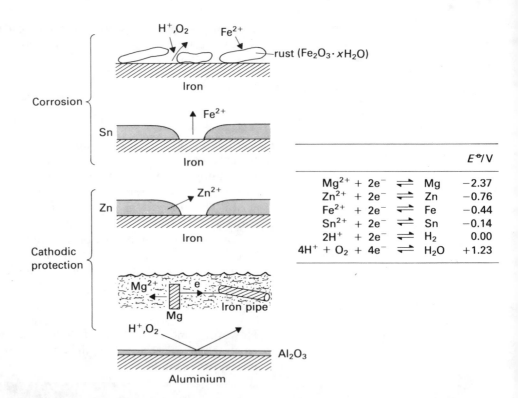

	$E^{\ominus}$/V
$Mg^{2+} + 2e^- \rightleftharpoons Mg$	−2.37
$Zn^{2+} + 2e^- \rightleftharpoons Zn$	−0.76
$Fe^{2+} + 2e^- \rightleftharpoons Fe$	−0.44
$Sn^{2+} + 2e^- \rightleftharpoons Sn$	−0.14
$2H^+ + 2e^- \rightleftharpoons H_2$	0.00
$4H^+ + O_2 + 4e^- \rightleftharpoons H_2O$	+1.23

Figure 10.25 Corrosion

SUMMARY

1. The **electrochemical series** can be used to
   (a) predict the **feasibility** of displacement reactions;
   (b) predict the **order of discharge** of ions during electrolysis.
2. The **electrochemical series of metals** is a list of metals arranged in order of their standard electrode potentials. The series closely approximates to the **order of reactivity** of metals.
3. The **Nernst equation** is

$$E = E^{\ominus} + \frac{RT}{zF} \ln\left(\frac{[\text{oxidised species}]}{[\text{reduced species}]}\right)$$

4. **pH** is related to the electrode potential of the hydrogen electrode by

$$E_{H^+/H_2} = -(0.059\ V)\ pH$$

5. Redox, precipitation and acid–base titrations may all be peformed **potentiometrically** using a reference electrode and a **detecting electrode**.
6. Values for $\Delta G_m^{\ominus}$ can be determined from experimentally determined standard electrode potentials using the following relationship:

$$\Delta G_m^{\ominus} = -zFE_{cell}^{\ominus}$$

7. A **primary cell** does not regenerate reactants.
8. A **secondary cell** can be recharged.
9. Three processes are involved in the **extraction of metals** from ores:
   (a) **concentration**,
   (b) **reduction**,
   (c) **refining**.
10. **Corrosion** is the natural tendency of metals to return to their oxidised states.
11. Corrosion can be prevented by **coating** the surface or by **cathodic protection**.

Examination Questions

1. (a) Explain the terms 'conductance (conductivity)' and 'molar conductance (conductivity)', and explain with the aid of suitable graphs how the variation of molar conductance with concentration of a solution gives information about the relative degree of ionisation of weak and strong electrolytes.

   (b) (i) Outline briefly using a suitable graph how the titration of a coloured solution against sodium hydroxide solution might be monitored conductimetrically where the use of an acid–alkali indicator is impossible.

   (ii) Discuss briefly the shape of the potentiometric titration curve (of E against cm^3 added) for the titration of ethanoic acid (acetic acid) against sodium hydroxide solution. Sketch the curve for deci-molar solutions and use it to explain how buffer solutions work.

   (c) If the acidity constant of ethanoic (acetic) acid is 1.8×10^{-5} mol dm^{-3} at 25°C, calculate the pH of the solution before and after adding 1.00 g of sodium ethanoate to 150 cm^3 of 0.05 M) ethanoic acid. ($CH_3COOH = 60$; $CH_3COONa = 82$)

   (d) derive Ostwald's dilution law and indicate how it provided a crucial test of the ionic theory propounded by Arrhenius.

   (SUJB)

2. (a) Explain the meaning of the following terms: *electrolytic conductivity (specific conductance)*; *cell constant*; *molar conductivity*.

   (b) (i) In a particular conductivity cell a 0.0200 mol dm^{-3} aqueous KCl solution had a resistance of 74.58 Ω. Given that the molar conductivity of this KCl solution is 138.3 Ω^{-1} cm^2 mol^{-1} determine the cell constant.

   (ii) In a study of the dependence of molar conductivity, Λ, on concentration, c, the following data were obtained for aqueous NaCl solutions:

c/mol dm^{-3}	0.0005	0.0010	0.0050	0.0100
Λ/Ω^{-1} cm^2 mol^{-1}	124.5	123.7	120.6	118.3

   Show that the data follow the relation

   $$\Lambda = \Lambda^\infty - a\sqrt{c}$$

   where Λ^∞ is the molar conductivity at infinite dilution and a is a constant. Hence find Λ^∞ for NaCl.

   (iii) For potassium ethanoate solution $\Lambda^\infty = 114.4$ Ω^{-1} cm^2 mol^{-1}; for hydrochloric acid solution $\Lambda^\infty = 425.0$ Ω^{-1} cm^2 mol^{-1}; and for potassium chloride solution $\Lambda^\infty = 149.9$ Ω^{-1} cm^2 mol^{-1}. Calculate Λ^∞ for ethanoic acid and explain the basis of your calculation.

   (OLE)

3. (a) Explain, *in terms of electron transfer*, what is meant by oxidation. Give the oxidation number (or oxidation state) of chlorine in
   (i) the chloride ion,
   (ii) the chlorate ion, ClO_3^-

   (b) Write balanced *ionic* equations for the following reactions:
   (i) chlorine gas displacing iodine from an aqueous solution containing iodide ions;
   (ii) solid iodine displacing chlorine gas from an aqueous solution con-

taining chlorate (ClO_3^-) ions, the iodine being converted to ions of formula IO_3^-.

(c) When chlorine gas is bubbled into hot aqueous sodium hydroxide, a mixture of chloride and chlorate (ClO_3^-) ions is formed. Write a balanced *ionic* equation for this reaction.

In terms of oxidation and reduction, comment on the role played by chlorine in this reaction.

(d) The ions of formula WO_4^{2-} (W, tungsten) contained in 20.0 cm^3 of 0.100 mol dm^{-3} aqueous solution were reduced chemically. The tungsten species was then re-converted at an electrode of an electrolytic cell to its original form by the passage of a steady current of 0.100 A flowing for 1 hour 36.5 minutes.

 (i) At which electrode did the re-oxidation take place, positive or negative?

 (ii) How many moles of tungsten are present in 20.00 cm^3 of 0.100 mol dm^{-3} $WO_4^{2-}(aq)$?

 (iii) How many moles of electrons were involved in the re-oxidation process? (Faraday constant, F, = 9.65×10^4 C mol^{-1})

 (iv) By comparing your answers to (ii) and (iii) find the oxidation state of tungsten in the reduced form. (You may assume that the oxidation state of oxygen remains unchanged throughout the process.)

 (v) Write a balanced ion–electron equation for *either* the oxidation *or* reduction reactions.

(NISEC)

4. (a) Discuss, with the aid of examples, the interpretation of 'oxidation' and 'reduction' in terms of electron transfer.

What do you understand by the term 'redox reaction'?

(b) What do you understand by 'disproportionation'? Give an example of such a reaction.

(c) Discuss the extent to which a table of standard electrode potentials is useful in predicting the course of chemical change.

(d) An aqueous solution of ammonium vanadate is reduced by boiling with zinc powder. The resulting solution may be re-oxidised to the vanadate with a solution of potassium manganate(VII). Describe experiments you would carry out, and the subsequent use of experimental data, to determine the change in oxidation state of the vanadium.

(L)

5. (a) When potassium nitrate is dissolved in pure sulphuric acid, a solution having a high conductance is formed and physical measurements show that six moles of solute particles are formed for each mole of potassium nitrate.

 (i) How may these observations be accounted for?

 (ii) What physical method could be used for demonstrating the observation on the number of particles formed when potassium nitrate is the solute?

 (iii) How might the solution be used in organic chemistry?

(b) When the solution referred to in (a) is diluted with water, placed in the cathode compartment of an electrolytic cell and a current passed for some time, reduction of the nitrogen-containing species occurs and from the reduced electrolyte a salt $[Z^+][HSO_4^-]$ may be isolated.

After treatment of one mole of the salt with 2 moles of sodium hydroxide a compound may be isolated containing only N, H and O; its RMM [relative molecular mass] is 33 and it contains 42.4% N and 9.09% H.

What is the formula of Z^+? What is the structure of Z^+?

(c) When 5 mmoles of the salt are titrated with potassium manganate(VII) solution in dilute sulphuric acid, 6 mmoles of the manganate(VII) are required for oxidation, manganese(II) being formed as the end-product of the manganate(VII) reduction. (1 mmole = 10^{-3} mole.)

What is the oxidation state of nitrogen in the product of the reaction with manganate(VII)?

(O & C)

6. (a) Balance the following redox equations using the principles of *either* electron transfer *or* change in oxidation state (number):
   (i) $Ag(s) + NO_3^-(aq) + H^+(aq) \rightarrow Ag^+(aq) + NO(g) + H_2O(l)$
   (ii) $Fe(CN)_6^{4-}(aq) + Cl_2(g) \rightarrow Fe(CN)_6^{3-}(aq) + 2Cl^-(aq)$

   (b) Discuss **briefly** the electrolysis of each of the following solutions:

Electrolyte	Cathode	Anode
sodium chloride	carbon	carbon
sodium hydroxide	carbon	carbon
sulphuric acid	platinum	silver
copper sulphate	copper	copper

   (c) (i) A current of 3.21 A was passed through fused aluminium oxide for 10 minutes. The volume of oxygen collected at the anode was 112 cm³ measured at s.t.p. Calculate the mass of aluminium obtained at the cathode and the charge of 1 mole of electrons (the Faraday).
   (O = 16; Al = 27. Molar volume = 22.4 dm³ at s.t.p.)
   (ii) When the same quantity of electricity was passed through the fused chloride of a metal M (relative atomic mass = 137.3), the mass of M obtained was 1.373 g. Calculate the charge on the cation M^{x+}
   (iii) The charge on the electron is 1.602×10^{-19} coulombs. Calculate a value for the Avogadro number (L).
   (iv) The standard electrode potentials of three metals X, Y and Fe are −0.14 V, −0.76 V and −0.44 V respectively. Explain which one of X or Y would be a more effective protection against the corrosion of iron.

(SUJB)

7. Calcium is manufactured by the electrolysis of molten calcium chloride.
   (a) Write an ion–electron equation for the production of calcium.
   (b) What mass of calcium is produced if a current of 20 A flows for 32 minutes 10 seconds?

(SEB)

8. (a) Sol particles may be prepared by a process of dispersion or condensation. What is the difference between these *two* processes? Using **one** such process explain the preparation of *either* iron(III) hydroxide *or* arsenic sulphide sol.
   (b) Such sols have to be freed from the presence of any electrolytes formed in their preparation by a process known as *dialysis*.
   (i) Explain why such removal is necessary and describe this process.
   (ii) What are the two rules relevant to the coagulation of colloids, sometimes referred to as the 'Hardy–Schulze' rules?
   (c) Two typical properties exhibited by sols are *electrophoresis* and the *Tyndall effect*.

Describe and explain these phenomena using the sol whose preparation was described in part (a) where appropriate.

(AEB, 1984)

9. (a) Outline briefly how the concept of *oxidation and reduction* has been developed into that of electron transfer reactions.
   (b) Explain what you understand by the following conventional representation of an electrochemical cell:

$$Zn(s) \mid ZnSO_4(aq) \mathbin{\|} CuSO_4(aq) \mid Cu(s)$$

   Sketch and label such a cell to show the essential features.
   Define the terms *anode* and *cathode*, and on your sketch show the direction of electron flow, the anode and the cathode.
   Illustrate how the electromotive force of the cell is calculated using the following values of the respective standard electrode potentials/V at 298 K.

$$Zn^{2+}/Zn \quad -0.763 \qquad Cu^{2+}/Cu \quad +0.337$$

   What other information is wanted normally?
   (c) When the following ions (as salts) in their standard states are introduced into a solution which also contains an excess of dilute sulphuric acid, deduce what happens given values of the standard redox potentials/V at 298 K:

$$Fe^{2+}, Fe^{3+}, Mn^{2+}, MnO_4^-$$
$$Fe^{2+}, Fe^{3+}/Pt \quad +0.76 \qquad MnO_4^-, Mn^{2+}, H^+/Pt \quad +1.52.$$

   Write separate ion–electron equations for the reactants and deduce the final equation for the reaction.
   (d) If 2.8 g of pure iron wire is dissolved, in the absence of air, in excess of dilute sulphuric acid to which 2.00 g of potassium manganate(VII) (potassium permanganate) is added, which one of the solid reagents would be in excess and by how much?

$$(Fe = 56, KMnO_4 = 158)$$

   Describe with essential practical detail only how potassium manganate(VII) solution is used in titrimetric (volumetric) analysis, and indicate any important limitation on its use.

(SUJB)

10. The standard electrode potential of the Cu^{2+}/Cu electrode is measured directly by connecting two half-cells by means of a salt bridge.
    (a) What chemical substances may be used in the Cu^{2+}/Cu half-cell and what conditions are necessary?
    (b) What chemical substances may be used in the other half-cell and what conditions are necessary?
    (c) What is the purpose of the salt bridge and what might it contain?
    (d) What is, by definition, the value of the potential of the standard half-cell in (b)?
    (e) Give the equations for the reactions in the two half-cells, including any state symbols.

(JMB)

11. (a) Define the term *standard electrode potential* and explain why a potential difference arises between a metal and a solution of one of its salts.
    (b) The reaction taking place in the electrochemical cell

$$Pt \mid Fe^{2+}, Fe^{3+} \mid\mid Ag^+ \mid Ag$$

is

$$Fe^{2+}(aq) + Ag^+(aq) \rightarrow Fe^{3+}(aq) + Ag(s).$$

Predict the effect (i.e. increase, decrease or no change) of the following changes upon the cell voltage, giving reasons for your predictions.

(i) Increase in the concentration of Fe^{3+}.

(ii) Increase in the concentration of both Fe^{3+} and Fe^{2+} by the same factor.

(iii) Addition of more solid silver.

(iv) Addition of some sodium chloride to the solution containing the Ag^+.

(v) Addition of an excess of potassium cyanide to the solution containing the Fe^{3+} and Fe^{2+}.

(O & C)

12. Outline the methods, both direct and indirect, available for the determination of equilibrium constants.

For the reaction:

$$4H_2O(l) + 3I_2(aq) + 2As(s) \rightleftharpoons 2HAsO_2(aq) + 6HI(aq)$$

(a) write a standard cell expression for the reaction, and hence determine the standard electrode potential of the cell,

(b) calculate the free energy change for the reaction under standard conditions,

(c) calculate K_c for the reaction.

(L, Nuffield)

13. (a) Outline the methods used for the extraction of:

(i) sodium, from sodium chloride;

(ii) zinc, from zinc(II) sulphide;

(iii) mercury, from mercury(II) sulphide.

(Details of the plant are not required).

(b) Discuss, in terms of ease of reduction, the *reasons* for employing essentially different methods of extraction in (a) above.

(c) Explain why you would not extract magnesium by displacing the metal from magnesium chloride with sodium.

(d) Draw a diagram to explain the essential features of metallic bonding. By referring to *two physical* features of metals, explain how the concept of metallic bonding explains this physical behaviour.

(e) Explain why galvanising, in contrast to tinning, is a more permanent method of protecting iron from corrosion.

(SUJB)

14. (a) Give one mechanism for the rusting of iron and explain one method of preventing rusting.

(b) Discuss the reactions between iron and

(i) chlorine gas,

(ii) aqueous hydrochloric acid.

(c) Iron reacts with molten potassium hydroxide in the presence of air to form a compound Z which dissolves in water to give a purple-coloured solution.

(i) A sample of an aqueous solution containing 19.8 g of Z and about 200 g of sulphuric acid per dm^3 was reduced with an excess of zinc. A 10.0 cm^3 sample of the reduced solution decolourised 20.0 cm^3 of

0.0100 M potassium manganate(VII) solution. Write an equation for the oxidation of iron(II) by manganate(VII) in acidic solution and calculate the number of moles of iron in 1.00 dm^3 of the original solution of Z.

(ii) A 10.0 cm^3 sample of the original solution of Z was itself decolourised by 30.0 cm^3 of 0.100 M iron(II) sulphate solution. The resulting solution was yellow in colour and gave a reddish brown precipitate on addition of aqueous ammonia.

What is the oxidation state of the iron in compound Z?

Suggest a formula for Z, given that it contains approximately 39% of potassium by mass.

(M signifies a concentration of 1.00 mol dm^{-3}.)

(O & C)

11 PERIODICITY

Mendeleyev and the Periodic Classification of the Elements

Introduction

Although most sources list Mendeleyev as the fourteenth and last child, a few sources, e.g. the Great Soviet Encyclopedia (pre-1917) list him as the seventeenth and last child.

Mendeleyev's birthdate was 27 January in the old style Julian calendar followed in Russia at that time. Russia converted to the present Gregorian calendar in February 1918, after the October Revolution (1917) so that the Revolution is now actually celebrated in November.

Mentioning 1984 to the man on the street brings to mind 'big brother', 'double-speak' and other artifacts of the world of George Orwell. What does it mean to scientists? 1984 marks the sesquicentennial of the birth of the Russian scientist Dmitri Ivanovich Mendeleyev, the chief architect of the Periodic Table. Mendeleyev, who was the youngest of fourteen children, was born in Tobolsk, Western Siberia on 8 February 1834. But wasn't Mendeleyev merely one of at least half a dozen nineteenth-century scientists who worked on the Periodic Table? What gave him such a prominent position, when he wasn't even one of the first to think of the concept? During this time period in the middle of the last century, the leading scientific centres were universities in England, France and Germany. How did a Russian become involved anyway? What was so significant about Mendeleyev's work that he became one of only seven scientists to have a chemical element named in their honour?

The original table was constructed using the ascending order of atomic weights of the elements. Why then did it take almost 70 years from the inception of the atomic weight concept to the Periodic Table? The table illustrates the Periodic Law, i.e. if chemical elements are arranged in the order of their atomic weights, then at certain regular intervals (periods) elements occur having similar chemical and physical properties. At the beginning of the nineteenth century, a chemical element was defined as a substance that could not be broken up into simpler parts (with different properties) by any known means. The table was first suggested in an era when no one had any clear idea of how many elements might exist. It gave great impetus to chemical work because regularities in element properties implied that the elements must have more in common than merely their resistance to all efforts at further separation.

In the past half-century, the only changes in the table have generally involved adding the name of a newly discovered element, but it wasn't always that way. There have been basic problems with the table as originally proposed. There were a number of pairs of elements which could only be accommodated into the table by ignoring the basic rule for the table's construction. How can you justify that for a seemingly universal law? There were over a dozen elements with somewhat similar properties for which only one location was available in the table and another half-dozen for which there was no space at all. In the first decade of this century, there were new elements being discovered almost daily and bewildered scientists could not figure out where to place them in the table.

John Dalton (1766–1844)

The equivalent weight of an element in any compound is equal to the atomic weight of the element divided by the valency of the element in that compound.

When Czar Alexander I died in December 1825, he left no direct male heir. His brother Constantine was next in line, but he renounced the throne and swore allegiance to his younger brother Nicholas. Some Army officers led a rebellion against Nicholas and in favour of Constantine called the Decembrist Rebellion. The leaders were later executed and the remaining dissidents sent to Siberia, a place of banishment for political exiles, e.g. following his abdication in 1917, Czar Nicholas II and his family were originally sent to Tobolsk (Mendeleyev's birthplace).

Atomic Weight Versus Equivalent Weight

The key to early periodic systems was the ordering of the elements by their atomic weight. The concept of atomic weight originated with the English scientist and teacher, John Dalton, in 1803. Dalton was attempting to explain why different amounts of various gases dissolve in a liquid by comparing the relative weights of these substances. His scale assigned a weight of 1 to hydrogen. When Dalton published his atomic theory, he included tables of atomic weight values. However, when two elements combine in a compound, it is insufficient merely to determine the percentage of each element present to obtain correct atomic weights. One must also determine the valency of each element in the compound. Valency is a measure of how many atoms of one element combine with an atom of the other element, e.g. is water HO, or H_2O? Unfortunately, this is a 'Catch-22' situation since it cannot be determined until the atomic weight is known. Early experimenters assumed, in the absence of evidence to the contrary, that a compound was made up of molecules containing one atom of each element. This monatomic hypothesis created difficulties in understanding reactions. The rule worked in many situations, but there was a large number of cases in early atomic weight tables where the assumed valency was not correct. Thus, equivalent weights (or proportional numbers as they were called, since they were ratios of elements) were quoted rather than the atomic weight. This problem of listing atomic weights for some elements and fractions of atomic weights for other elements obscured the true periodicity of the elements. It delayed development of the Periodic Table for some 60 years.

Actually an Italian lawyer and physicist, Amedeo Avogadro, solved this problem in 1811. He used the results of the French scientist Joseph Louis Gay-Lussac on combining volumes. Gay-Lussac determined that gaseous elements combined in chemical reactions in simple (numerical) volume ratios. Avogadro, then, hypothesised that equal volumes of different gases under the same conditions contain equal numbers of molecules, a possibility that Dalton had considered and rejected, although without the benefit of Gay-Lussac's data. Avogadro also assumed that common gaseous elements consisted of two-atom molecules. Unfortunately, Avogadro used the term molecule throughout the paper, either alone or with a series of qualifying adjectives such as integral, constituent and elementary. Each of these had a different meaning. In those days, the terms atom and molecule were often used interchangeably so that some scientists reading the paper understood Avogadro to imply that there could be half-atoms. Regrettably as a result of this confusion, his paper was ignored by chemists for the next half-century until it was rediscovered by the Italian chemist, Stanislao Cannizzaro, who presented Avogadro's solution to the world's leading chemists at the first International Scientific Conference in Karlsruhe, Germany (3–5 September 1860). This conference had been called because of the confused situation with regard to the terminology of molecule and atom and the fact that two sets of atomic weight values were in existence, one for inorganic compounds and another for organic. In spite of this aim, no conclusions were agreed upon at the Conference but Cannizzaro's presentation influenced the thinking of young chemists who attended, including Mendeleyev and the German chemist (Julius) Lothar Meyer.

Mendeleyev's early interest in natural science had been fostered by an older brother-in-law, who was a Decembrist. He was 26 years old and studying abroad at the University of Heidelberg, prior to obtaining his doctorate at the University of St Petersburg (Leningrad), when he attended the Conference. The 30-year-old Meyer was in his second year as a private teacher (privat-dozent for physics and chemistry) at the University of Breslau, when he heard and read Cannizzaro's paper at Karlsruhe. Both subsequently wrote textbooks, which spread the solution throughout the scientific community in the following years. Scientists began correcting equivalent weight values to atomic weights. As a result, elements with similar properties now fell one under another when they were arranged in rows by ascending order of atomic weight, and the periodicity of the elements became clearer.

Norman E. Holden

LEARNING OBJECTIVES

After you have studied this chapter you should be able to
1. Indicate the main stages of development which have led to the **modern Periodic Table of elements.**
2. Describe the principal features of the modern Periodic Table.
3. Show how the Periodic Table is related to the **electron structure** of elements.
4. Give examples of **s-block, p-block** and **d-block** elements and show where these elements are located in the Periodic Table.
5. Compare the properties of **metals, non-metals** and **metalloids** and give examples of each type of element.
6. Outline the **periodicity** of elements with respect to
   (a) structure and bonding,
   (b) physical properties, including (i) melting and boiling points, (ii) enthalpies, (iii) densities,
   (c) atomic and ionic radii,
   (d) ionisation energies, electron affinity and electronegativities,
   (e) chemical properties, including (i) compound formation, (ii) valency, (iii) oxidation state, (iv) redox properties
7. Discuss the periodicity of the chemical and physical properties of
   (a) oxides,
   (b) hydrides,
   (c) hydroxides,
   (d) halides.
8. Give examples of **diagonal relationships** in the Periodic Table.
9. Point to some anomalies in the Periodic Table.

11.1 Periodicity

THE PERIODIC TABLE

If we were asked 'What evidence is there for the periodicity of elements?', we might think of a number of examples. There is a good chance we would cite the example of the alkali metals. We all know that lithium, sodium and potassium react violently with water releasing a great amount of energy,

$$2K(s) + 2H_2O(l) \rightarrow 2K^+(aq) + 2OH^-(aq) + H_2(g)$$

We might, however, think first of the halogens. They are all oxidising agents, for example. They combine with metals to form halides. Many of us are familiar with the reactions of chlorine, bromine and iodine with red-hot iron to form iron(III) halides,

$$2Fe(s) + 3Br_2(g) \rightarrow 2FeBr_3(g)$$

After a little thought, we might also quote the example of the noble gases. They are all relatively inert and form few stable compounds.

In each case, we have arranged elements into groups. In each group, elements have similar chemical and physical properties.

One of the first attempts to arrange elements into groups was made by Lavoisier. In 1789 he published a book in which he presented a list of 'simple substances not decomposed by any known process of analysis'. The list included many of the common elements and also heat and light. He divided these 'substances' into four groups.

The *first group* included heat, light and gases such as oxygen and nitrogen.
The *second group* included elements such as sulphur and phosphorus. All elements in this group formed acidic oxides.
The *third group* included metals such as copper, tin, lead and zinc.
The *fourth group* contained 'simple earthy salt-forming substances'. These included lime, baryta, magnesium, alumina and silica—that is, the oxides of calcium, barium, magnesium, aluminium and silicon, respectively.

The Law of Triads

In 1817, the German chemist Johann Wolfgang Dobereiner (1780–1849) noticed that the relative atomic mass (or atomic weight as it was then known) of strontium was approximately the mean of the relative atomic masses of calcium and barium. In 1829, he reported that several groups of elements—each containing three elements—had similar physical and chemical properties. These groups were called **triads** (*see* table 11.1).

In the 1850s a number of attempts were made to arrange all the elements into triads and also to find numerical patterns among the atomic weights.

Spiral Arrangements

Over the next 50 years various spiral arrangements were produced (see figure 11.1). One of these was devised by De Chancourtois in 1863. He arranged the elements in order of their relative atomic masses in a spiral around a cylinder divided into

Table 11.1   Dobereiner's triads

Triads			Relative atomic masses		
1. Li	Na	K	7	23	39
2. S	Se	Te	32	79	128
3. Cl	Br	I	35.5	80	127
4. Ca	Sr	Ba	40	88	137

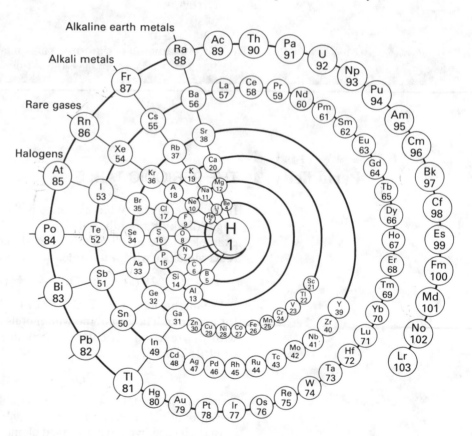

Figure 11.1   A spiral arrangement of the Periodic Table

Table 11.2 Part of Newlands' arrangement of elements into 'octaves'

H	Li	Be	B	C	N	O
1	2	3	4	5	6	7
F	Na	Mg	Al	Si	P	S
8	9	10	11	12	13	14
Cl	K	Ca	Cr	Ti	Mn	Fe
15	16	17	18	19	20	21

Newlands' 'law of octaves' was not well received and was ignored for years. When he presented his paper to the Chemical Society in London, he was asked sarcastically if he had ever tried arranging the elements in the order of the initial letters of their names. From 1868 to 1886 Newlands was chief chemist in a sugar refinery.

vertical strips. Elements with similar chemical and physical properties fell on the same vertical. There was little interest in his work.

The Law of Octaves

In 1864 the English analytical chemist John Newlands (1837–1898) noticed that, if elements were arranged in order of increasing atomic weights, the eighth element, starting from a given one, is a kind of repetition of the first—like the eighth note in an octave of music. He called this regularity the **law of octaves**. In 1865, he produced a table in which the elements were arranged in order of 'atomic numbers'. These numbers were the serial order of the atomic weights. Table 11.2 shows the first three rows of Newlands' table.

Although Newlands' arrangement of elements worked well for the first 17 elements, it broke down thereafter. He was forced to put two elements in some places and clearly there were few similarities between some elements in the columns. For example, phosphorus has little in common with manganese and iron little in common with sulphur.

Meyer's Atomic Volume Curve

In 1870, the German chemist Lothar Meyer (1830–1895) plotted the atomic volume of elements against their relative atomic masses. Atomic volume is given by

$$\text{Atomic volume} = \frac{\text{Relative atomic mass}}{\text{Density}}$$

From his graph (*see* figure 11.2), Meyer was able to produce a table showing a periodic arrangement of the elements.

Lothar Meyer

The German chemist demonstrated independently of Mendeleyev the principles underlying the Periodic Table of elements. His first version of the table published in his book *Modern Chemical Theory* (1864) contained 28 elements. In 1868 he prepared an expanded version of the table containing 57 elements. He demonstrated the periodicity by plotting the atomic volumes against atomic weight (nowadays the term relative atomic mass is used). This periodicity is better shown by a graph of atomic volume against atomic number (*see* figure 11.2).

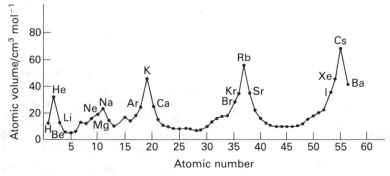

Figure 11.2 Variation of atomic volume with atomic number

Mendeleyev's Periodic Law

About the same time that Meyer produced his atomic volume curve, the Russian chemist Dmitri Mendeleyev (1834–1907) produced a form of the Periodic Table

Dmitri Ivanovitch Mendeleyev (1834–1907)

from which the modern Periodic Table has developed. In Mendeleyev's table, elements were arranged in order of increasing relative atomic masses in a manner similar to that of Newlands' table. The elements were also arranged into horizontal rows called **periods** and eight columns called **groups**. This arrangement was based on the properties of the elements and also the formulae of the compounds they formed (*see* table 11.3). For example, all the metals in Group I formed chlorides with the formula MCl.

Dmitri Mendeleyev

Dmitri Mendeleyev was born in Tobolsk in Siberia. Before entering university he trained and worked as a chemistry teacher. From 1859 to 1861 he worked abroad in Heidelberg, Germany. During this period he attended a conference at which the famous Italian chemist Cannizzaro circulated copies of his table of atomic weights. Mendeleyev worked on these and in 1860 produced his famous Periodic Table. By this time, he was Professor of General Chemistry at the University of St Petersburg (now Leningrad).

Table 11.3   The first five periods of Mendeleyev's Periodic Table

Period	I	II	III	IV	V	VI	VII	VIII
1	H							
2	Li	Be	B	C	N	O	F	
3	Na	Mg	Al	Si	P	S	Cl	
4	K	Ca	*	Ti	V	Cr	Mn	Fe Co Ni
	Cu	Zn	*	*	As	Se	Br	
5	Rb	Sr	Y	Zr	Nb	Mo	*	Ru Rh Pd
	Ag	Cd	In	Sn	Sb	Te	I	

The asterisks * marked spaces for elements which had yet to be discovered. They were scandium, gallium, germanium and technetium.

Unlike Newlands' table, however, Mendeleyev's contained a separate group for elements that did not fit into his arrangement. For example, whereas Newlands had forced cobalt and nickel into the same place in his table, Mendeleyev listed them separately. Mendeleyev also left gaps if no element was known to fit. He then made predictions about the properties of the elements that should fit into these spaces. Three of these elements—germanium, scandium and gallium—when discovered, were found to have properties remarkably similar to those predicted by Mendeleyev. For example, the formulae of germanium(IV) oxide and germanium(IV) chloride—GeO_2 and $GeCl_4$ respectively—were predicted by Mendeleyev.

On the basis of his Periodic Table, Mendeleyev put forward his 'Periodic Law'. This states that 'the properties of elements vary periodically with their atomic weights'.

However, there were problems. First of all, like Newlands, Mendeleyev had placed iodine *after* tellurium even though its relative atomic mass of 126.9 was lower than that of tellurium. Beryllium was also out of place in the table. With its relative atomic mass of 13.5 it should have been placed between carbon and nitrogen. Secondly, no place could be found in the table for the noble gases, which were first identified in the 1890s. Argon had been identified in 1893 by the British scientists John Rayleigh and William Ramsay and then, five years later, helium, neon, krypton and xenon were discovered.

These problems were not fully resolved until 1914 when Henry Moseley (*see* section 1.1) showed that the elements could be arranged in a periodic pattern according to their atomic numbers.

THE MODERN PERIODIC TABLE

The modern Periodic Table derives directly from Moseley's work on atomic numbers. The modern version of the Periodic Law states that 'the properties of elements vary periodically with their atomic numbers'.

In all modern Periodic Tables elements are thus arranged in order of their atomic numbers. As the atomic numbers of the elements increase, the electron structure of the atoms of the elements are 'built up' in accordance with the aufbau principle (*see* section 1.2). The structure of modern Periodic Tables thus relates to the electron structure of the elements. It is based on the electronic configuration of those electrons in the outermost shells and in the highest-energy sub-shells. The reason is that only these electrons determine the chemical properties of the elements and their compounds.

The modern Periodic Table has many forms. In 1984, the International Union of Pure and Applied Chemistry (IUPAC) provisionally recommended the format shown in table 11.4. Note that hydrogen is in Group 1 and that the halogens and

Figure 11.3   The Periodic Table

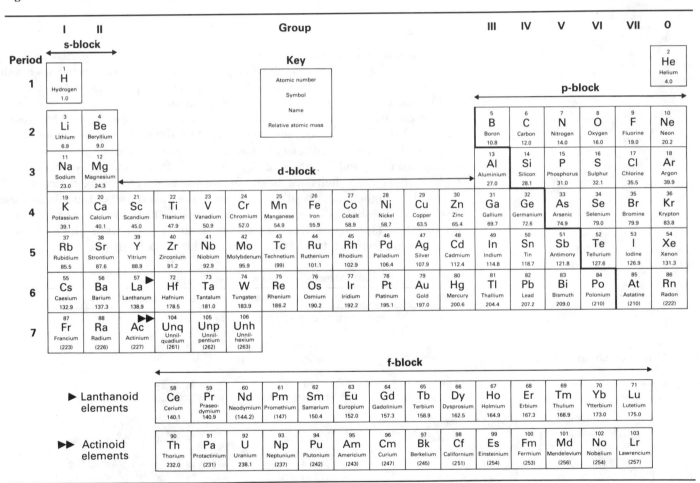

Table 11.4   Format recommended by IUPAC for the Periodic Table of elements

Group	1	2	3	4	5	6	7	8	9	10	11	12	13	14	15	16	17	18
	H																	He
	Li	Be											B	C	N	O	F	Ne
	Na	Mg											Al	Si	P	S	Cl	Ar
	K	Ca	Sc	Ti	V	Cr	Mn	Fe	Co	Ni	Cu	Zn	Ga	Ge	As	Se	Br	Kr
	Rb	Sr	Y	Zr	Nb	Mo	Tc	Ru	Rh	Pd	Ag	Cd	In	Sn	Sb	Te	I	Xe
	Cs	Ba	La*	Hf	Ta	W	Re	Os	Ir	Pt	Au	Hg	Tl	Pb	Bi	Po	At	Rn
	Fr	Ra	Ac**															
			*Ce	Pr	Nd	Pm	Sm	Eu	Gd	Tb	Dy	Ho	Er	Tm	Yb	Lu		
			**Th	Pa	U	Np	Pu	Am	Cm	Bk	Cf	Es	Fm	Md	No	Lr		

noble gases are in Groups 17 and 18 respectively. Figure 11.3 shows a form of the table recommended by the Association for Science Education (ASE) in 1985. This is the form we shall use in this book.

What's in a name—Unnilquadium

This is the IUPAC name for the element with atomic number 104. It is actually named after the atomic number: *un* = 1, *nil* = 0 and *quad* = 4. The element was reported first in 1964 by Soviet scientists. They called it kurchatovium after a Russian nuclear physicist and gave it the symbol Ku. In 1969 the element was produced in the USA and named rutherfordium (Rf) after the New Zealand physicist Lord Rutherford. The symbol for unnilquadium is Unq. Unnilquadium is the second element in the 6d row of elements. The third element in the row, which has an atomic number 105, is called unnilpentium (Unp) and element 106 has the name unnilhexium (Unh).

Periods

Horizontal rows of elements in the Periodic Table are called **periods**. All the elements in a period have a similar core of electrons. The core of electrons has the same structure as that of the noble gas at the end of the previous period (*see* table 11.5). The chemical properties of an element depend on the electrons in the highest-energy sub-shells. These are outside the core of electrons. The highest-energy sub-shell of the two elements in Period 1 is the 1s sub-shell (*see* table 11.6). The highest-energy sub-shells of elements in Period 2 are the 2s sub-shell (for Groups I and II) and 2p sub-shell (for Groups III to VII and 0). For Period 3 the highest-energy sub-shells are 3s and 3p for the same groups as in Period 2. In Period 4 the highest-energy sub-shells include not only the 4s and 4p sub-shells but also the 3d sub-shell. This is because the 3d sub-shell has a lower energy than the 4p sub-shell (*see* figure 1.21). Elements in this period whose highest-energy electrons are found in the 3d sub-shell are **d-block elements** (*see* chapter 14).

Table 11.5 Electron structure of the periods

Period			Electron structure
1		this consists of hydrogen, $1s^1$, and helium, $1s^2$, only	
2	core	helium	$1s^2$
	example	carbon	$1s^2 2s^2 2p^2$
3	core	neon	$1s^2 2s^2 2p^6$
	example	phosphorus	$1s^2 2s^2 2p^6 3s^2 3p^3$
4	core	argon	$1s^2 2s^2 2p^6 3s^2 3p^6$
	example	bromine	$1s^2 2s^2 2p^6 3s^2 3p^6 4s^2 3d^{10} 4p^5$
5	core	krypton	$1s^2 2s^2 2p^6 3s^2 3p^6 4s^2 3d^{10} 4p^6$
	example	rubidium	$1s^2 2s^2 2p^6 3s^2 3p^6 4s^2 3d^{10} 4p^6 5s^1$
6	core	xenon	$1s^2 2s^2 2p^6 3s^2 3p^6 4s^2 3d^{10} 4p^6 5s^2 4d^{10} 5p^6$
	example	barium	$1s^2 2s^2 2p^6 3s^2 3p^6 4s^2 3d^{10} 4p^6 5s^2 4d^{10} 5p^6 6s^2$

Table 11.6 Periods and highest-energy sub-shells

Groups

The vertical columns of elements in the Periodic Table are called **groups**. All the elements in a particular group have a characteristic electron configuration for the highest-energy sub-shell (*see* table 11.7). For example, for Group I this is s^1, for Group II, s^2, for Group III, $s^2 p^1$, and so on up to Group 0, the elements of which all have the configuration s^2p^6. Table 11.8 shows the electronic configuration of the highest-energy sub-shells and the cores for the five elements in Group IV.

Table 11.7 Groups and electronic configurations of highest-energy orbitals

			Groups					
	I	II	III	IV	V	VI	VII	0
Characteristic	s^1	s^2	s^2p^1	s^2p^2	s^2p^3	s^2p^4	s^2p^5	s^2p^6
Period 2	$2s^1$	$2s^2$	$2s^22p^1$	$2s^22p^2$	$2s^22p^3$	$2s^22p^4$	$2s^22p^5$	$2s^22p^6$
Period 3	$3s^1$	$3s^2$	$3s^23p^1$	$3s^23p^2$	$3s^23p^3$	$3s^23p^4$	$3s^23p^5$	$3s^23p^6$
Period 4	$4s^1$	$4s^2 3d^1$ to $3d^{10}$	$4s^24p^1$	$4s^24p^2$	$4s^24p^3$	$4s^24p^4$	$4s^24p^5$	$4s^24p^6$
		d-block elements						

Table 11.8 Electronic configurations of Group IV elements

Period	Element	Atomic number	Configuration
			Highest-energy sub-shell
2	carbon	6	$1s^2$ $\lceil 2s^22p^2$
3	silicon	14	$1s^2$ $2s^22p^6$ $\lceil 3s^23p^2$
4	germanium	32	$1s^2$ $2s^22p^6$ $3s^23p^63d^{10}$ $\lceil 4s^24p^2$
5	tin	50	$1s^2$ $2s^22p^6$ $3s^23p^63d^{10}$ $4s^24p^64d^{10}$ $\lceil 5s^25p^2$
6	lead	82	$1s^2$ $2s^22p^6$ $3s^23p^63d^{10}$ $4s^24p^64d^{10}$ $5s^25p^65d^{10}$ $\lceil 6s^26p^2$
			Core

Some of the groups have common names (*see* table 11.9). Two exceptions should be noted. Hydrogen (outer electronic configuration s^1) is in Group I but is not an alkali metal. Helium, in Group 0, is a noble gas but unlike the others in the group has the outer electronic configuration s^2. Elements in Group I (except

Table 11.9 Names and outer electronic configurations of the main groups

Group	Name	Outer electronic configuration	
I	alkali metals	s^1	s-block
II	alkaline earth metals	s^2	
III	no common name	s^2p^1	
IV	no common name	s^2p^2	
V	no common name	s^2p^3	
VI	sometimes known as the chalcogens	s^2p^4	p-block
VII	halogens	s^2p^5	
0	noble gases (also known as rare or inert gases)	s^2p^6	

hydrogen) and Group II are known as the **s-block** metals. Elements in Groups III to VII and in Group 0 (except helium) are known as the **p-block** elements. Elements in these two blocks are known variously as the major, representative, typical or **main group** elements.

The **d-block** elements are shown in figure 11.3. They consist of three series. The first series is in Period 4, the second in Period 5 and the third in Period 6 (*see* chapter 14). The **f-block** elements consist of two series. The lanthanoids or rare earth elements are in Period 6 and the actinoids in Period 7. Note that, in the version of the Periodic Table shown in figure 11.3, vertical columns of elements in the d- and f-blocks are not assigned group numbers.

Metals, Non-Metals and Metalloids

It is possible, on the basis of their electrical conductivities, to classify all the elements in the Periodic Table as metals, non-metals or metalloids (semi-metals).

Metals are all good conductors of electricity. Their conductivity decreases slowly as their temperature increases. With the exception of hydrogen and helium, the s-, d- and f-block elements are all metals. The s-block metals are sometimes known as reactive metals since they are the most reactive of all metals. The lower left triangle of p-block elements are also metals (*see* figure 11.4). They are sometimes known as the poor metals since their reactivities are low and they also exhibit some non-metallic character.

Non-metals are all insulators (except carbon in the form of graphite). They form the top right-hand triangle of p-block elements.

Metalloids have low electrical conductivities. The conductivities increase with increasing temperature. The metalloids are also known as semi-metals or semi-conductors. They form a diagonal band running from the top left-hand corner of the p-block elements to the bottom right-hand corner (*see* figure 11.4).

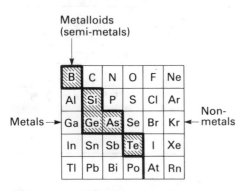

Figure 11.4   The p-block elements

PERIODICITY

We see from table 11.9 that the outer electronic configuration of the elements is a periodic function. For example, all the metals in Group I have one electron in their outer shell. The number of outer-shell electrons then increases across the two short periods (Periods 2 and 3) until a maximum of eight is reached. It then drops back to one. Since many physical and chemical properties of the elements depend on the outer electronic configuration of the elements, we would expect these properties also to vary periodically in line with the outer electronic configuration.

Examples of periodicity are dealt with elsewhere in this book. For example, the periodicity of ionisation energies is discussed in section 1.2 (*see* figure 1.16). The trends in the physical and chemical properties up or down the groups are considered in detail in the following chapters. In this chapter, therefore, we shall just summarise the major trends.

Structure and Physical Properties

The relationships between electronic configuration, bonding, structure and physical properties of elements were described in detail in chapter 2. Table 11.10 shows the variation of structure and bonding of the elements in Periods 2 and 3.

Table 11.10   Structures of elements in Periods 2 and 3

	Group							
	I	II	III	IV	V	VI	VII	0
Period 2	Li	Be	B	C	N	O	F	Ne
	metallic		B_{12}	diamond, graphite	N_2	O_2	F_2	atoms
			covalent (giant molecules)		covalent (simple molecules)			
Period 3	Na	Mg	Al	Si	P	S	Cl	Ar
	metallic			covalent (giant molecules)		S_8	Cl_2	atoms
						covalent (simple molecules)		

On crossing a period from left to right, the elements become less metallic in nature. On the other hand, on descending a group, the elements tend to become more metallic. Elements in the middle of the **short periods** (Periods 2 and 3) tend to have giant covalent structures, whereas those on the right tend to exist as simple covalent molecules.

These variations in the structures of the elements are reflected to some extent in the variations of melting point, boiling point, enthalpy of fusion, enthalpy of vaporisation and density (*see* figures 11.5 to 11.7). In general, we can conclude from these graphs that all these physical properties increase along a period until a maximum is reached. This maximum occurs with elements that have giant covalent structures. The noble gases—which exist as unbound atoms—and elements that exist as simple molecules all have low values for these properties. Thus, at room temperature and pressure, they all exist as gases or solids with low melting points.

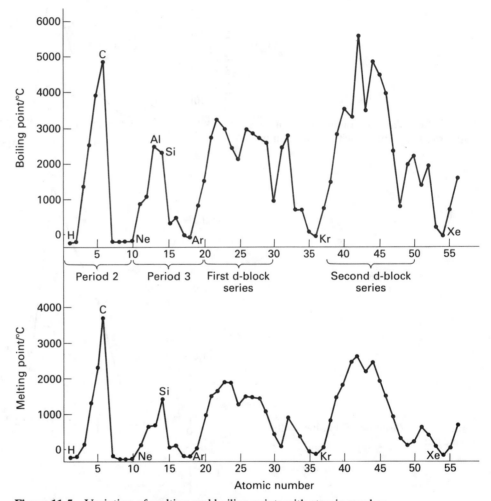

Figure 11.5 Variation of melting and boiling points with atomic number

Atomic and Ionic Radii

We have already noted above how Meyer's atomic volume curves provided evidence for the periodicity of elements (*see* figure 11.2). The atomic radii (that is covalent radii) of elements also exhibit periodicity. Figure 11.8 shows us that atomic radius

- decreases across a period,
- increases down a group.

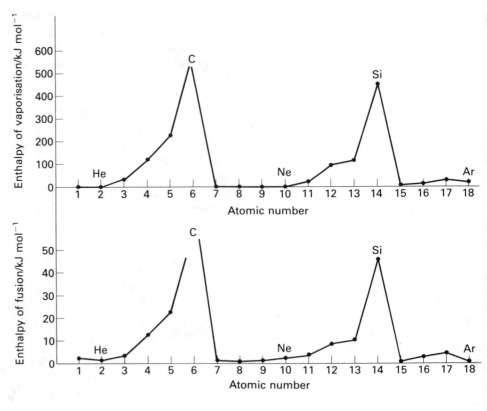

Figure 11.6 Variation of enthalpies of vaporisation and fusion with atomic number

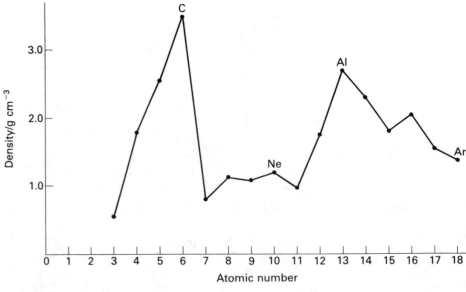

Figure 11.7 Variation of densities with atomic number

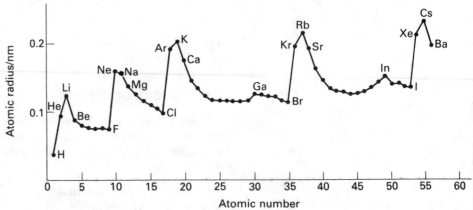

Figure 11.8 Graph of atomic radius against atomic number

Cationic radii are smaller and anionic radii larger than atomic radii (*see* figure 11.9). Both decrease across a period and increase down a group. Examples of this trend are provided by the alkali metals (*see* section 13.1 and table 13.2) and the halogens (*see* section 16.1 and table 16.1).

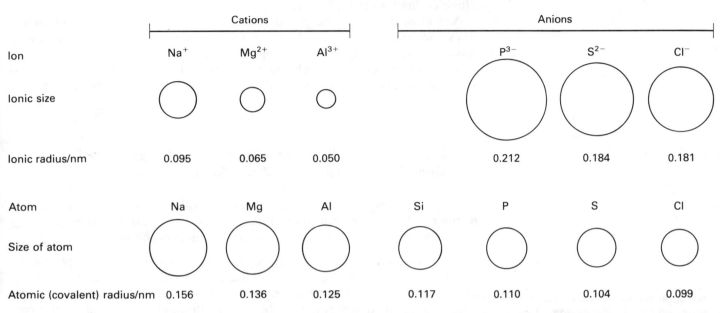

Figure 11.9 Comparison of ionic and covalent radii across Period 3

Ionisation Energy, Electron Affinity and Electronegativity

The first ionisation energies of the elements in Period 3 and Group 0 are shown in table 1.3. A plot of the first ionisation energies of elements with atomic numbers 1 to 20 is also shown in figure 1.16. The graph in figure 1.16 shows clearly that the first ionisation energy is lowest for Group I metals and then increases across a period until a maximum is reached with the noble gases. The graph and data in table 1.3 show that the first ionisation energies decrease down a group.

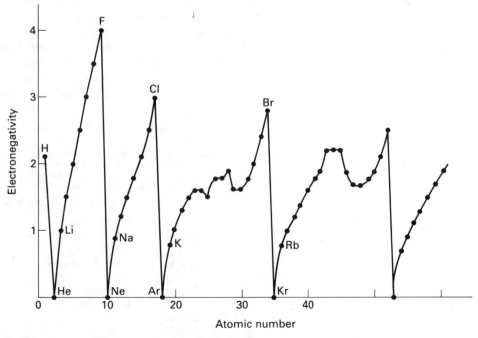

Figure 11.10 Periodic variation of electronegativities

The changes in electron affinities (*see* section 2.1) are less well defined than those for ionisation energies. This is because fewer experimental data are available. In general, however, electron affinities increase across a period, reaching a maximum with the halogens. There is no clear pattern in the changes in electron affinities down a group.

A clear pattern does emerge, however, for the electronegativities of elements. These are shown in table 2.2. When these values are plotted against atomic number, they clearly show that electronegativity increases across a period, reaching a maximum with the halogens (*see* figure 11.10). They then drop to zero for the noble gases. The graph also shows that electronegativities decrease down a group.

PERIODICITY IN THE CHEMICAL PROPERTIES OF THE ELEMENTS AND THEIR COMPOUNDS

The arrangement of elements in the Periodic Table according to their atomic number and outer electron configuration results in two major patterns emerging in the chemical properties of the main group elements and their compounds:

1. Elements with similar chemical properties fall in groups. For example, all the alkali metals are in Group I and all the halogens in Group VII.
2. The most electropositive elements and thus the most reactive metals fall in the bottom left-hand corner of the table. The electropositive nature of the elements then decreases on ascending the groups and on crossing the periods from left to right. On the other hand, the most electronegative elements and thus the most reactive non-metals fall in the top right-hand corner of the table. The electronegative nature of the elements thus increases on crossing the periods from Group I to Group VII but decreases on descending a group.

> Although quantitative values are often assigned for the electronegativities of elements (*see* table 2.2), the term 'electropositive' is normally used in a qualitative way only.

The electropositive or electronegative nature of the elements relates directly to the types of chemical reactions that the elements undergo and thus to the types of compounds formed by the elements. The s-block metals tend to form cations readily and thus form ionic compounds (*see* table 11.11). The p-block elements in the centre of the table tend to form covalent compounds only. The more electronegative p-block elements on the right side of the table can form either covalent or ionic compounds. The noble gases, of course, with their stable electronic configurations, form relatively few compounds.

As we have seen, the d-block elements fall between Groups II and III. They are metals but less electropositive and thus more electronegative than the s-block metals. As a result, their compounds—such as the oxides and chlorides—tend to be either ionic with a high degree of covalent character or covalent. In common with p-block elements in the centre of the table, they often form compounds that are macromolecular or have layer or chain structures.

Table 11.11   Trends in compound formation in Periods 2 and 3

		electronegative nature increases →
electropositive nature increases ↓	electronegative nature decreases ↓	electropositive nature decreases

Li	Be	B	C	N	O	F	Ne
Na	Mg	Al	Si	P	S	Cl	Ar
form cations and thus ionic compounds			form covalent compounds		form either covalent or ionic compounds		do not form stable compounds

Table 11.12 Examples of p-block ligands in complex ions

d-Block element	p-Block ligand	Complex ion
Cu	NH_3	$[Cu(NH_3)_4]^{2+}$
Fe	CN^-	$[Fe(CN)_6]^{3-}$

The d-block elements also tend to form both cationic and anionic complex ions, whereas the s-block metals tend not to. The p-block elements are often involved as ligands in both cationic and anionic complexes (*see* table 11.12).

The valencies (*see* chapter 4) of the main group elements also tend to exhibit a periodic variation. Table 11.13 shows how all the elements in Period 3 exhibit valencies of the same numerical value as their group numbers. Elements in Groups IV, V, VI and VII also all exhibit common valencies equal to the numerical value of their group numbers subtracted from eight.

The maximum attainable oxidation states of elements also exhibit a periodic variation (*see* figures 11.11 and 11.12). They tend to increase on crossing a period until a maximum is reached in Groups V, VI and VII. Note also how the elements with highest oxidation states also exhibit multiple oxidation states. Chlorine, for example, can exist in oxidation states ranging from -1 to $+7$.

Table 11.13 Common valencies of elements in Period 3

Group	I	II	III	IV	V	VI	VII	0
element	Na	Mg	Al	C	P	S	Cl	Ar
valency	1	2	3	4	5	6	7	0
				4	3	2	1	
examples	NaCl	$MgCl_2$	$AlCl_3$	CH_4	PCl_5	SO_3	Cl_2O_7	
				CH_4	PCl_3	H_2S	HCl	

For the three d-block series the maximum oxidation state occurs in the middle of the series (*see* figure 11.12). The d-block elements with the highest oxidation states also exhibit the greatest number of oxidation states. In the first series, for

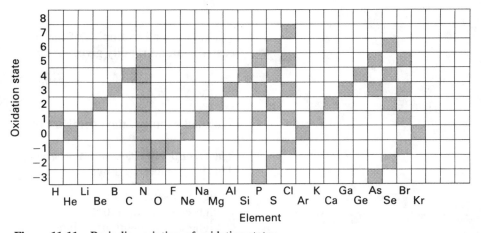

Figure 11.11 Periodic variation of oxidation states

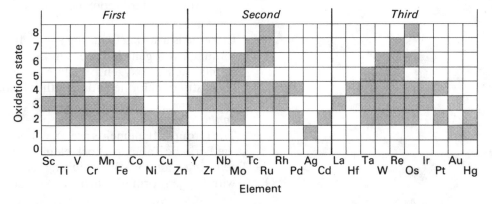

Figure 11.12 Periodic variation of oxidation states of the first, second and third d-block series

example, manganese exhibits five positive oxidation states ranging from +2 to +7.

Periodicity in Redox Properties

The redox properties of the elements also exhibit periodicity. In general, elements on the left of the table—that is the s-block metals—are strongly reducing. Then, on crossing a period from left to right the elements become weakly reducing and/or weakly oxidising. Finally, in Group VII, the elements become strongly oxidising. Let us examine this trend in a little more detail.

The s-Block Metals Are Strongly Reducing
This is characterised by:

    low ionisation energies,
    low electron affinities,
    low electronegativity values,
    high 'electropositivity' (a qualitative term—*see* comment above),
    negative standard redox potentials.

EXAMPLES
1. Reaction with air or oxygen,

$$2Mg(s) + O_2(g) \rightarrow 2MgO(s)$$

2. Reaction with chlorine,

$$2Na(s) + Cl_2(g) \rightarrow 2NaCl(s)$$

3. Reaction with dilute acids,

$$Mg(s) + 2H^+(aq) \rightarrow Mg^{2+}(aq) + H_2(g)$$

These are all examples of the reducing action of s-block metals, since in each case the metal readily gives up electrons,

$$Mg \rightarrow Mg^{2+} + 2e^-$$

$$Na \rightarrow Na^+ + e^-$$

We shall examine the chemistry of the s-block metals in detail in chapter 13.

The Group VII Elements Are Strongly Oxidising
This is characterised by:

    high ionisation energies,
    high electron affinities,
    high electronegativity values,
    low 'electropositivity',
    positive standard redox potentials.

EXAMPLE
Chlorine is a strong oxidising agent. It reacts violently with hydrogen in sunlight forming hydrogen chloride. On the other hand, it does not react with oxidising agents such as oxygen or dilute or concentrated acids. We shall examine the chemistry of chlorine and the other halogens in chapter 16.

In the Middle
The Group VII elements are, of course, p-block elements which lie on the right of the Periodic Table. p-Block elements in the middle of a period tend to be weakly reducing and/or weakly oxidising. For example, in Group IV, silicon reacts slowly with oxygen forming silicon(IV) oxide,

$$Si(s) + O_2(g) \rightarrow SiO_2(s)$$

In Group V, nitrogen can act either as a weak reducing agent or as a weak oxidising agent. For example, it is a weak reducing agent in its reaction with oxygen,

$$N_2(g) + O_2(g) \rightarrow 2NO(g)$$

In its reaction with hydrogen, on the other hand, it is a weak oxidising agent,

$$N_2(g) + 3H_2(g) \rightarrow 2NH_3(g)$$

The d-block elements are weak reducing agents. Iron, for example, when red hot, reacts with steam to form hydrogen,

$$Fe(s) + H_2O(g) \rightarrow FeO(s) + H_2(g)$$

Periodicity in the Properties of Compounds

The formation, structure and physical and chemical properties of compounds also exhibit periodicity. The oxides, hydrides, hydroxides and halides all provide examples of this.

Oxides

In general the reactivity of elements with oxygen decreases across the Periodic Table. In Period 3, for example, the two s-block metals, sodium and magnesium, and the two p-block elements, aluminium and phosphorus, all react vigorously with oxygen, forming oxides. In the same period, silicon and sulphur only react slowly with oxygen. Chlorine and argon on the right-hand side of the period do not react at all with oxygen.

The electropositive s-block metals form ionic oxides. Sodium oxide, Na_2O, and magnesium oxide, MgO, are examples. Oxides of elements in the middle of a period and on the right are predominantly covalent. The oxides of carbon, nitrogen and sulphur are examples.

The acid–base character of the oxides also changes from being basic on the left-hand side through amphoteric to acidic on the right-hand side. The s-block metals, for example, tend to form oxides that dissolve in water to form alkaline solutions,

$$Na_2O(s) + H_2O(l) \rightarrow 2Na^+(aq) + 2OH^-(aq)$$

The molecular oxides of the p-block elements such as carbon dioxide and sulphur trioxide all tend to be acidic. The trend from basic to acidic character is exhibited clearly by the oxides of elements in Period 3,

$$\underbrace{Na_2O \quad MgO}_{basic} \quad \underbrace{Al_2O_3}_{amphoteric} \quad \underbrace{SiO_2 \quad P_4O_{10} \quad SO_3 \quad Cl_2O_7}_{acidic}$$

The oxides of the d-block elements are generally insoluble in water and basic, although one or two—such as zinc—do exhibit amphoteric behaviour (see chapter 14).

We shall examine the chemistry of the oxides further in section 15.4.

Hydrides

The formation, structure and properties of the hydrides follow a similar although not identical pattern to that of the oxides.

The s-block metals, such as sodium and magnesium, tend to react vigorously when heated with dry hydrogen, forming ionic hydrides. The ionic hydrides are basic in character. The more electronegative p-block elements on the right of a period, such as sulphur and chlorine, react with hydrogen, forming covalent hydrides. These are acidic. Exceptions are methane, CH_4, which is neutral and ammonia, NH_3, which is basic.

The less electronegative p-block elements, such as aluminium, silicon and phosphorus, do not react with hydrogen when heated.

The d-block metals react with hydrogen on heating to form non-stoichiometric metallic hydrides.

The preparation, structure and properties of the hydrides are described in detail in chapter 12.

Hydroxides

The hydroxides of the most electropositive elements, such as sodium and calcium, are ionic and strongly basic. On the other hand, chlorine, which is highly electronegative, forms an acidic hydroxide. This is chloric(I) acid, HOCl. The bond between the chlorine and oxygen atom is covalent. The hydroxides of less electronegative elements are sometimes amphoteric. They tend to be unstable and form oxides.

Halides

The halides exhibit periodic variations similar to those of the oxides, hydrides and hydroxides. On traversing a period from the most electropositive elements to the most electronegative elements, both boiling points and melting points decrease (*see* table 11.14). Thus, the first three chlorides in Period 3 are all solids, the next three liquids and chlorine a gas at room temperature and pressure.

The ionic character of the chlorides decreases on crossing a period whilst the covalent character increases.

The s-block halides are generally salts of strong acids and strong bases. They thus dissolve in water to form neutral solutions. The chlorides of the p-block and d-block elements tend to react with water, forming acidic solutions. For example

$$SiCl_4(l) + 8H_2O(l) \rightarrow Si(OH)_4(s) + 4Cl^-(aq) + 4H_3O^+(aq)$$

The reactions of the chlorides of d-block elements in water are described in chapter 14, and the chemistry of the halides is described in more detail in chapter 16.

Table 11.14   Properties of the Period 3 chlorides

Group	I	II	III	IV	V	VI	VII
chloride	NaCl	$MgCl_2$	Al_2Cl_6	$SiCl_4$	PCl_3	SCl_2	Cl_2
boiling point/°C	1465	1498	180	57	76	59	−34
melting point/°C	808	714	192	−68	−92	−80	−101
state at r.t.p.	solid	solid	solid	liquid	liquid	liquid	gas
structure	3-D lattice	layer lattice	layer lattice	mole-cules	mole-cules	mole-cules	mole-cules
bond	ionic	ionic/ covalent	ionic/ covalent	covalent	covalent	covalent	covalent

Diagonal Relationships

We have already noted that, in general, the electropositive nature of an element decreases on crossing a period but increases on descending a group. This results in what are known as diagonal relationships. In each diagonal relationship, a pair of elements have similar chemical properties. The most important diagonal pairs are lithium and magnesium, beryllium and aluminium, and boron and silicon.

	Group		
	I	II	III	IV
Period 2	Li	Be	B	
Period 3		Mg	Al	Si

The diagonal relationships arise because the loss in electropositivity in going one step to the right across a period is compensated by the gain in electropositivity in going down a group one step. We shall examine these diagonal relationships in detail in chapter 13.

Anomalies

The head elements of the main group elements. Period 2 elements at the top of Groups I to VII are sometimes called the **head elements**. They are interesting because some of the properties of these elements and their compounds are out of character with other elements in their respective groups. These distinctive properties can be attributed to the smaller size of their atoms and their higher electronegativities and ionisation energies. For example, lithium and beryllium halides show more covalent character than the corresponding halides of metals lower in their groups. Lithium, unlike the other alkali metals, does not form a solid hydrogencarbonate. Furthermore, whereas the other alkali metal nitrates decompose on heating to form the nitrate and oxygen, lithium nitrate decomposes to give the oxide, oxygen and nitrogen dioxide. Finally, unlike other alkali metal hydroxides, lithium hydroxide is thermally unstable. The distinctive properties of lithium and the other head elements are discussed further in chapters 13, 14 and 16.

SUMMARY

1. Elements in the **modern Periodic Table** are arranged in order of their **atomic numbers**.
2. Elements in a **period** have a similar arrangement of electrons.
3. Elements in a **group** have the same outer electronic configuration.
4. All **s-block** elements (except hydrogen and helium) and all **d-** and **f-block** elements are **metals**.
5. Hydrogen and helium are **non-metals**. All other non-metals are **p-block** elements.
6. Elements become less metallic on crossing a period but more metallic on descending a group.
7. Physical properties of elements tend to increase along a period until a maximum is reached.
8. **Atomic and ionic radii** decrease across a period but increase down a group.
9. **First ionisation energies** of elements increase across a period but decrease down a group.
10. **Electronegativities** increase across a period to a maximum with the halogens. They decrease down a group.
11. The **most electropositive** and thus the most reactive metals occur in the bottom left-hand corner of the Periodic Table.
12. The **most electronegative** elements occur in the top right-hand corner of the Periodic Table.
13. s-Block elements tend to have **valencies** equal to their group number.
14. p-Block elements have common valencies equal to their group number and also equal to eight minus their group number.
15. d-Block elements exhibit **multiple valencies** and oxidation states.
16. The **reducing properties** of elements decrease across a group.
17. Reactivity with oxygen decreases across a group.
18. The ionic character of **oxides** decreases across a period whereas their covalent character increases.
19. Oxides, hydrides, hydroxides and halides all exhibit similar periodicity.
20. Lithium and magnesium have similar chemical properties and thus exhibit a **diagonal relationship**.
21. **Head elements** of the main groups have **anomalous properties**.

Examination Questions

1. This question concerns the Periodic Table and electronic structure.

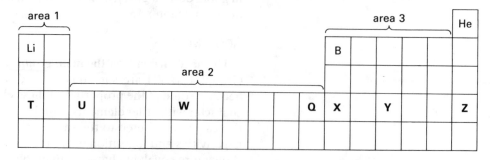

The diagram represents part of the Periodic Table of the elements; the positions of helium, lithium and boron are shown. (The letters in heavy type are not chemical symbols but indicate the positions of some other elements.)

(a) State in which areas the following kinds of elements are to be found:
  (i) s-block elements;
  (ii) p-block elements;
  (iii) d-block elements.

(b) (i) Write the electronic structure of the element **T** in terms of the occupancy of all its s and p levels.
  (ii) Is the first ionisation energy of **T** greater or less than that of lithium?
  (iii) Give a brief explanation of your answer to (ii).

(c) Elements **U** and **X** each form compounds in which they have an oxidation state of +3. Writing your answer in terms of numbers of s, p, or d electrons, as appropriate, state which electrons are lost, and from which level they are lost, in forming this oxidation state from: (i) **U** and (ii) **X**.

(d) Which one of the elements **U**, **W**, or **Q** is likely to show the greatest range of oxidation states?

(e) (i) Which one of the elements **X**, **Y**, or **Z** will have the greatest first ionisation energy?
  (ii) Give a brief explanation of your answer to (i).

(f) Elements **X** and **Y** each show a covalency of three.
  (i) Draw the expected shapes of the gaseous monomeric trichlorides of these elements.
  (ii) Why are the shapes different?

(O & C)

2. Give an account of the principal factors on which the trends and gradations in physical and chemical properties of elements in the same group of the Periodic Table depend.

From your knowledge of the elements of lower atomic number, make predictions about

(a) the physical nature of caesium, the fifth element in Group I, and its reaction with water,

(b) the solubility in water of barium salts; barium is the fifth element in Group II.

(c) the metallic character of bismuth, the fifth element in Group V.

(L)

3. (a) Describe *briefly* the basis for the arrangement of the elements in the Periodic Table. How do (i) the atomic radii and (ii) the first ionisation energies of the atoms vary in the first two periods?

Is the position of an element in the Periodic Table a guide to whether it should be classified as a metal or a non-metal? Explain your answer.

(b) Illustrate the transition from non-metallic to metallic properties exhibited by Group IV elements C, Si, Sn, and Pb by considering the properties of their oxides and chlorides.

(OLE)

4. (a) The first six ionisation energies of an element A are given below:

1st	2nd	3rd	4th	5th	6th	
787	1577	3230	4355	16090	19795	kJ mol^{-1}

   (i) To what group of the Periodic Table does A belong and why?
   (ii) What is the formula of the highest chloride formed by A?
   (iii) Given that the atomic number of A is greater than 10, predict the reaction of the highest chloride of A with water and state the reasons underlying your prediction.

(b) For an element B, the first ionisation energy of gaseous B$^-$ ions is 333 kJ mol^{-1}.
   (i) What is the electron affinity of gaseous B atoms?
   (ii) To what group of the Periodic Table is the element B likely to belong and why?

(OLE)

5. By referring to structure, bonding and reaction with water, describe the patterns in the chemistry of the chlorides of the elements in (i) the Period sodium to sulphur and (ii) Group V (N–Bi) of the Periodic Table.

(AEB, 1985)

6. This question is about trends in properties and the reactions of elements in that part of the Periodic Table shown.

	Groups						
	I	II	III	IV	V	VI	VII
2	Li	Be	B	C	N	O	F
3	Na	Mg	Al	Si	P	S	Cl
4	K	Ca	Ga	Ge	As	Se	Br
5	Rb	Sr	In	Sn	Sb	Te	I

Periods: 2, 3, 4, 5

(a) List, as in the printed table, the symbols for those elements which react readily with cold water to form alkaline solutions; and give details, with a balanced equation, for the reaction which you consider to be the most vigorous.

(b) List, as in the printed table, the symbols for those elements which form hydrides characterised by their having comparatively low boiling points. For the most stable hydride of each group, draw the electron bond diagram and sketch the shape of its molecule.

(c) For the elements of Period 3, in their highest oxidation states where there are several, state the formulae of the oxides and the corresponding acids or hydroxides which they form. Specify the oxidation number in each case.

(d) List, as in the printed table, the symbols for those elements of Period 3 and Group IV which form nitrates. Explain what can be deduced about the nature of an element from the results of its reactions (if any) with concentrated nitric acid and with dilute nitric acid.

(e) Write a concise comparative account of the chemistry of fluorine and iodine to illustrate the trends in properties of the elements in Group VII.

(SUJB)

7.

												C	N	O	F	Ne
Na	Mg											Al	Si	P	S	Cl	Ar
K	Ca	Sc	Ti	V	Cr	Mn	Fe	Co	Ni	Cu	Zn	Ga	Ge	As	Se	Br	Kr

Choose *only* from the *elements whose symbols appear above*:

(a) (i) *two* metals which react violently with cold water;

(ii) *two* metals inert enough at room temperature to be commonly used as protective coatings;

(iii) *one* element with both metallic and non-metallic physical and chemical properties;

(iv) elements which exist under ordinary conditions as molecules containing: 1 atom, 2 atoms, 4 atoms, 8 atoms (give one example of each);

(v) *two* elements that form tetrachlorides which are liquid at room temperature and pressure;

(vi) *two* elements which, with each other, form a compound with the *most ionic* crystal lattice (give the formula of compound so formed);

(vii) a metal that forms an ion with charge 2A which is coloured and easily oxidised in aqueous solution (give the formula of ion and its colour);

(viii) *one* metal which gives rise to *two* coloured ions in different oxidation states (both higher than A2) in aqueous solution (give the formulae of ions, its oxidation states, and colours).

(b) (i) Describe concisely how the electronic structures of the atoms are built up traversing the long period from potassium to krypton.

(ii) What feature of the electronic structures justifies the elements from gallium to bromine being placed in the same periodic groups as aluminium to chlorine respectively?

(NISEC)

8. Lithium is not a typical member of Group I (alkali metals) and fluorine is not a typical member of Group VII (halogens).

Support this statement by briefly discussing for *each* element, *two* appropriate features of the chemistry of the element or its compounds.

Contrast the chlorides of the elements magnesium to sulphur in the third period of the Periodic Table. Your answer should include (i) the types of bondings in the chlorides, (ii) their molecular shapes if covalent, (iii) their reactions with, or solubility in, water.

(AEB, 1984)

9. (a) Listed below are elements A to H and characteristics 1 to 8. Write down the letters and append to each letter the number which best fits the characteristics of the element. (*Each number must be used once only*.)

Element	Characteristic
A. Calcium	1. Forms an oxide which is freely soluble in water to give a solution of pH greater than 14.
B. Nitrogen	2. Forms a chloride which is a white solid, m.p. 875°C, freely dissolving in water to give a solution with good electrical conductance.
C. Chlorine	3. Forms a red oxide dissolving in sodium hydroxide to give a yellow solution.
D. Sodium	4. Forms a gaseous hydride freely dissolving in water to give an alkaline solution.
E. Tin	5. Forms only one white oxide which is insoluble in water but soluble in both acids and alkalis.

F. Sulphur   6. Forms at least two gaseous oxides which are explosive.

G. Chromium   7. Forms two chlorides, the higher one being a volatile fuming liquid.

H. Aluminium   8. Forms a hydride dissolving in water to form a weak acidic solution.

(b) Contrast the chemistry of calcium with that of sulphur. It will be sufficient to comment briefly on:
   (i) oxidation states;
   (ii) type of bonding in oxides, chlorides, hydrides.
   Illustrate your answer throughout with ionic and covalent bond diagrams.

(c) For calcium and sulphur, comment on:
   (i) the basic or acidic nature of the oxide and hydride;
   (ii) the extent to which their chlorides might hydrolyse.

(SUJB)

12 HYDROGEN AND WATER

What Did Bloom See on the Range?

What did Bloom see on the range?

On the right (smaller) hob a blue enamelled saucepan: on the left (larger) hob a black iron kettle.

What did Bloom do at the range?

He removed the saucepan to the left hob, rose and carried the iron kettle to the sink in order to tap the current by turning the faucet to let it flow.

Did it flow?

Yes. From Roundwood reservoir in county Wicklow of a cubic capacity of 2400 million gallons, percolating through a subterranean aqueduct of filter mains of single and double pipeage constructed at an initial plant cost of £5 per linear yard by way of the Dargle, Rathdown, Glen of the Downs and Callowhill to the 26 acre reservoir at Stillorgan, a distance of 22 statute miles, and thence, through a system of relieving tanks, by a gradient of 250 feet to the city boundary at Eustace bridge, upper Leeson street, though from prolonged summer drouth and daily supply of 12½ million gallons the water had fallen below the sill of the overflow weir for which reason the borough surveyor and waterworks engineer, Mr Spencer Harty, C. E., on the instructions of the waterworks committee, had prohibited the use of municipal water for purposes other than those of consumption (envisaging the possibility of recourse being had to the impotable water of the Grand and Royal canals as in 1893) particularly as the South Dublin Guardians, notwithstanding their ration of 15 gallons per day per pauper supplied through a 6 inch meter, had been convicted of a wastage of 20 000 gallons per night by a reading of their meter on the affirmation of the law agent of the corporation, Mr Ignatius Rice, solicitor, thereby acting to the detriment of another section of the public, selfsupporting taxpayers, solvent, sound.

What in water did Bloom, waterlover, drawer of water, watercarrier returning to the range, admire?

Its universality: its democratic equality and constancy to its nature in seeking its own level: its vastness in the ocean of Mercator's projection: its unplumbed profundity in the Sundam trench of the Pacific exceeding 8000 fathoms: the restlessness of its waves and surface particles visiting in turn all points of its seaboard: the independence of its units: the variability of states of sea: its hydrostatic quiescence in calm: its hydrokinetic turgidity in neap and spring tides: its subsidence after devastation: its sterility in the circumpolar icecaps, arctic and antarctic: its climatic and commercial significance: its preponderance of 3 to 1 over the dry land of the globe: its indisputable hegemony extending in square leagues over all the region below the subequatorial tropic of Capricorn: the multisecular stability of its primeval basin: its luteofulvous bed: its capacity to dissolve and hold in solution all soluble substances including millions of tons of the most precious metals: its slow erosions of peninsulas and downwardtending promontories: its alluvial deposits: its weight and volume and density: its imperturbability in lagoons and highland tarns: its gradation of colours in the torrid and temperate and frigid zones: its vehicular ramifications in continental lake-contained streams and confluent oceanflowing rivers with their tributaries and transoceanic currents: gulfstream, north and south equatorial courses: its violence in seaquakes, waterspouts, artesian wells, eruptions, torrents, eddies, freshets, spates, groundswells, watersheds, waterpartings, geysers, cataracts, whirlpools, maelstroms, inundations, deluges, cloudbursts: its vast circumterrestrial ahorizontal curve: its secrecy in springs, and latent humidity, revealed by rhabdomantic or hygrometric instruments and exemplified by the hole in the wall at Ashtown gate, saturation of air, distillation of dew: the simplicity of its composition, two constituent parts of hydrogen with one constituent part of oxygen: its healing virtues: its buoyancy in the waters of the Dead Sea: its persevering penetrativeness in runnels, gullies, inadequate dams, leaks on shipboard: its properties for cleansing, quenching thirst and fire, nourishing vegetation: its infallibillity as paradigm and paragon: its metamorphoses as vapour, mist, cloud, rain, sleet, snow, hail: its strength in rigid hydrants: its variety of forms in loughs and bays and gulfs and bights and guts and lagoons and atolls and archipelagos and sounds and fjords and minches and tidal estuaries and arms of sea; its solidity in glaciers, icebergs, icefloes: its docility in working hydraulic millwheels, turbines, dynamos, electric power stations, bleachworks, tanneries, scutchmills: its utility in canals, rivers, if navigable, floating and graving docks: its potentiality derivable from harnessed tides or watercourses falling from level to level: its submarine fauna and flora (anacoustic, photophobe) numerically, if not literally, the inhabitants of the globe: its ubiquity as constituting 90% of the human body: the noxiousness of its effluvia in lacustrine marshes, pestilential fens, faded flowerwater, stagnant pools in the waning moon.

Water: 'Its universality: its democratic equality and constancy to its nature in seeking its own level'. Rice is the staple diet of more than two thirds of the world's population. This picture shows a family—including the women—planting rice in a paddy field in Thailand

Having set the halffilled kettle on the now burning coals, why did he return to the stillflowing tap?

To wash his soiled hands with a partially consumed tablet of Barrington's lemonflavoured soap, to which paper still adhered (bought thirteen hours previously for fourpence and still unpaid for), in fresh cold neverchanging everchanging water and dry them, face and hands, in a long redbordered holland cloth passed over a wooden revolving roller.

What reason did Stephen give for declining Bloom's offer?

That he was hydrophobe, hating partial contact by immersion or total by submersion in cold water (his last bath having taken place in the month of October of the preceding year), disliking the aqueous substances of glass and crystal, distrusting aquacities of thought and language.

What impeded Bloom from giving Stephen counsels of hygiene and prophylactic to which should be added suggestions concerning a preliminary wetting of the head and contraction of the muscles with rapid splashing of the face and neck and thoracic and epigastric region in case of sea or river bathing, the parts of the human anatomy most sensitive to cold being the nape, stomach, and thenar or sole of foot?

The incompatibility of aquacity with the erratic originality of genius.

What additional didactic counsels did he similarly repress?

Dietary: concerning the respective percentage of protein and caloric energy in bacon, salt ling and butter, the absence of the former in the lastnamed and the abundance of the latter in the firstnamed.

Which seemed to the host to be the predominant qualities of his guest?

Confidence in himself, an equal and opposite power of abandonment and recuperation.

What concomitant phenomenon took place in the vessel of liquid by the agency of fire?

The phenomenon of ebullition. Fanned by a constant updraught of ventilation between the kitchen and the chimneyflue, ignition was communicated from the faggots of precombustible fuel to polyhedral masses of bituminous coal, containing in compressed mineral form the foliated fossilised decidua of primeval forests which had in turn derived their vegetative existence from the sun, primal source of heat (radiant), transmitted through omnipresent luminiferous diathermanous ether. Heat (convected), a mode of motion developed by such combustion, was constantly and increasingly conveyed from the source of calorification to the liquid contained in the vessel, being radiated through the uneven unpolished dark surface of the metal iron, in part reflected, in part absorbed, in part transmitted, gradually raising the temperature of the water from normal to boiling point, a rise in temperature expressible as the result of an expenditure of 72 thermal units needed to raise 1 pound of water from 50° to 212° Fahrenheit.

What announced the accomplishment of this rise in temperature?

A double falciform ejection of water vapour from under the kettlelid at both sides simultaneously.

For what personal purpose could Bloom have applied the water so boiled?

To shave himself.

What advantages attended shaving by night?

A softer beard: a softer brush if intentionally allowed to remain from shave to shave in its agglutinated lather: a softer skin if unexpectedly encountering female acquaintances in remote places at incustomary hours: quiet reflections upon the course of the day: a cleaner sensation when awaking after a fresher sleep since matutinal noises, premonitions and perturbations, a clattered milkcan, a postman's double knock, a paper read, reread while lathering, relathering the same spot, a shock, a shoot, with thought of aught he sought though fraught with nought might cause a faster rate of shaving and a nick on which incision plaster with precision cut and humected and applied adhered which was to be done.

James Joyce

LEARNING OBJECTIVES

After you have studied this chapter you should be able to

1. Explain the **ubiquitous** and **unique** nature of **hydrogen.**
2. Justify the position of hydrogen in the **Periodic table.**
3. Outline the principal methods of preparing hydrogen in the laboratory.
4. Describe the principal forms of hydrogen.
5. Describe the most important chemical and physical properties of hydrogen.
6. Compare the chemical and physical properties of the different types of **hydride.**
7. Outline the principal methods of manufacturing hydrogen.
8. Give examples of the major uses of hydrogen.
9. Explain what is meant by 'the **hydrogen economy**' and compare its advantages and disadvantages.
10. Describe the structure of water.
11. Indicate the **anomalous physical properties** of water.
12. Explain why water is sometimes called the '**universal solvent**'.
13. Summarise the most important chemical reactions involving water.
14. Explain the term **hydrological cycle.**
15. Give examples of the uses of water.
16. List the most important parameters for determining **water quality.**
17. Describe the principal types of **water pollution.**
18. Briefly describe the various methods of **water treatment.**

12.1 Hydrogen ▌ UBIQUITOUS AND UNIQUE

Ubiquitous Hydrogen

Hydrogen is the most abundant element in the Universe. It has been estimated that hydrogen makes up over 90% of the atoms and about 75% of the mass of the Universe. On Earth, hydrogen is the ninth most abundant element. It accounts for 0.76% of the Earth's mass and occurs in about as many different compounds as does carbon. The most important naturally occurring compound containing hydrogen is water. Hydrogen also occurs in organic substances such as coal and oil.

Hydrogen is the simplest of all elements. It is also ubiquitous. Every student is familiar with the reactions of metals such as magnesium and zinc with dilute mineral acids to produce hydrogen. They are familiar with the famous 'pop' test for hydrogen. Among the first formulae chemistry students learn are those of compounds containing hydrogen: water, H_2O; methane, CH_4; sulphuric acid, H_2SO_4; ammonia, NH_3; and ethanol, C_2H_5OH, for example.

Hydrogen is not only ubiquitous—that is, everywhere—it is also unique to the extent that it has a range of chemical and physical properties completely different from all other elements. It also forms a unique range of compounds. In addition it is the only element that has a chemical bond named after it (*see* section 2.1). It even has a bomb (*see* section 1.3), bacteria and an economy named after it (*see* below).

Hydrogen is the only element that exists as an inflammable gas. Indeed, it was called inflammable gas (*gas pinque*) by the Flemish chemist Johannes Baptista van Helmont (1579–1644) who first identified it. Hydrogen was first prepared in the laboratory by the action of acid on iron by Turquet de Mayerne (who died in 1655) and Robert Boyle in 1672. It was carefully studied by the English chemist and physicist Henry Cavendish in 1766. He called it inflammable air. However, the name hydrogen was coined by Lavoisier.

Lavoisier Names Hydrogen

The word 'hydrogen' was coined by the French chemist Antoine Laurent Lavoisier (1743–1794). The word derives from two Greek words—*hydro* and *genes*, meaning 'water' and 'forming' respectively.

Lavoisier is known as the founder of modern chemistry. His major contribution to chemistry was to disprove the *phlogiston theory*. According to this theory, combustible substances contained two components—phlogiston and calx. On burning the substance, the phlogiston was released leaving the calx—or ash. Lavoisier showed experimentally that oxygen from the air is involved in combustion. He also determined oxygen's role in respiration and was the first to distinguish between elements and compounds.

Antoine Laurent Lavoisier from a painting by Thulstrup

Structure of the Hydrogen Atom

The structure of the hydrogen atom is the simplest of all atomic structures. It consists of a nucleus containing a single proton surrounded by a single electron in the 1s orbital (*see* section 1.2). This simple structure makes the hydrogen atom distinctive in a number of ways. First of all, its outer shell is the only shell. Its only electron is thus not shielded from the nuclear charge by any inner-shell electrons. Secondly, this outer shell can gain or lose a single electron to achieve a stable electron configuration. Finally, since the atom consists of only one electron and one proton, the atom is small. Indeed, its covalent radius (0.029 nm) and its van der Waals radius (0.12 nm) are the smallest of any element (*see* section 2.2). These features account for many of the distinctive properties of hydrogen and also for its unique position in the Periodic Table.

Position in the Periodic Table

Since the hydrogen atom can lose a single electron to form a singly charged positive ion, the element is placed at the top of Group I in the Periodic Table. However, although hydrogen can become metallic under certain conditions (*see* figure 2.15) it is normally non-metallic in its properties. A comparison of its ionisaton energy with that of lithium and sodium shows that it is very different from the other Group I metals—that is, the alkali metals (*see* table 12.1).

The hydrogen atom can also, with difficulty, gain an electron to form an H^- ion. This would place it at the top of Group VII along with the halogens. However, hydrogen is not a p-block element and a comparison of its electron affinity (*see* section 2.1) with that of fluorine and chlorine shows that it is out of place here (*see* table 12.2).

Furthermore, although hydrogen forms diatomic molecules like the halogens, it forms a much stronger bond than that of the fluorine or chlorine molecules. This is reflected in the bond enthalpies (*see* section 5.3) of these bonds (*see* table 12.3).

Table 12.1 Ionisation energies of hydrogen, lithium and sodium

Element	Ionisation energy/kJ mol^{-1}
hydrogen	1310
lithium	520
sodium	500

Table 12.2 Electron affinities of hydrogen, fluorine and chlorine

Element	Electron affinity/kJ mol^{-1}
hydrogen	−72
fluorine	−332.6
chlorine	−364

Table 12.3 Mean bond enthalpies of hydrogen, fluorine and chlorine molecules

Bond	Mean bond enthalpy/kJ mol^{-1}
H—H	435
F—F	155
Cl	243

LABORATORY PREPARATION OF HYDROGEN

Hydrogen can be prepared in the laboratory in a number of ways.

By the Action of Dilute Acid on Metals

Metals above hydrogen in the electrochemical series (*see* section 10.4) react with dilute sulphuric acid and dilute hydrochloric acid to form a salt and hydrogen. Zinc and dilute hydrochloric acid are normally used,

$$Zn(s) + 2HCl(aq) \rightarrow ZnCl_2(aq) + H_2(g)$$

By Electrolysis

Electrolysis of dilute aqueous solutions of alkalis and acids produces hydrogen at the cathode,

$$2H_3O^+(aq) + 2e^- \rightarrow H_2(g) + 2H_2O(l)$$

By the Action of Alkalis on Zinc and Aluminium

Both zinc and aluminium react with aqueous solutions of sodium hydroxide or potassium hydroxide forming hydrogen,

$$Zn(s) + 2OH^-(aq) + 2H_2O(l) \rightarrow [Zn(OH)_4]^{2-}(aq) + H_2(g)$$
$$\text{tetrahydroxozincate(II)   ion}$$

$$2Al(s) + 2OH^-(aq) + 6H_2O(l) \rightarrow [2Al(OH)_4]^-(aq) + H_2(g)$$
$$\text{tetrahydroxoaluminate (III)   ion}$$

By the Hydrolysis of Ionic Hydrides

Ionic hydrides such as calcium hydride, CaH_2, react with cold water to produce hydrogen,

$$CaH_2(s) + 2H_2O(l) \rightarrow Ca(OH)_2(aq) + 2H_2(g)$$

FORMS OF HYDROGEN

Isotopes

Hydrogen exists in three isotopic forms, **protium** $_1^1H$, **deuterium** $_1^2H$ and **tritium** $_1^3H$ (*see* sections 1.1 and 4.1). Of naturally occurring hydrogen, 99.985% is $_1^1H$. The remaining 0.015% is deuterium. Tritium is an unstable radioactive isotope and thus only occurs in trace amounts. It is a β-emitter with a half-life of 12.3 years (*see* section 1.3).

All the isotopic forms of hydrogen have virtually the same chemical properties. Their physical properties are different, however. Some of the physical properties of hydrogen and deuterium are shown in table 12.4.

Table 12.4   Physical properties of H_2 and D_2

	H_2	D_2
relative molecular mass	2.016	4.028
melting point/°C	−259.2	−254.5
boiling point/°C	−252.6	−249.4
standard molar enthalpy of fusion/kJ mol^{-1}	117	219
standard molar enthalpy of evaporation/kJ mol^{-1}	904	1227

There is a corresponding deuterium compound for every hydrogen compound. The most important of these is deuterium oxide, D_2O. This is also known as **heavy water**. It is used as a moderator in some types of nuclear reactor (*see* section 1.3).

Deuterium oxide is prepared by the electrolysis of water. The electrolysis produces H_2 at the cathode leaving the residual water richer in D_2O. On average 100 litres of water yields 7.5 cm^3 of 60% D_2O by this method.

Other deuterium compounds are normally prepared from deuterium oxide. For example

$$2D_2O(l) + 2Na(s) \rightarrow 2NaOD(aq) + D_2(g)$$

Atomic Hydrogen

The hydrogen prepared in the laboratory preparations described above is invariably diatomic molecular hydrogen, H_2. Molecular hydrogen can be dissociated into atoms by using a high-energy source such as a discharge tube containing hydrogen at low pressure. Hydrogen is also atomised by an electric arc struck between tungsten electrodes. The atoms recombine on a metal surface resulting in the liberation of sufficient energy to raise the temperature to about 3500°C. This effect is used in the atomic hydrogen blowlamp for welding metals.

Atomic hydrogen is a powerful reducing agent. It reduces metal oxides and chlorides to the metal.

Nascent Hydrogen

Gaseous hydrogen, that is molecular hydrogen, is a poor reducing agent. This is due to its high bond energy, 435 kJ mol^{-1}. For example, when the gas is bubbled through iron(III) solution, no reduction takes place. However, if the hydrogen is generated in the solution containing the iron(III) ions, the ions are immediately reduced to iron(II),

$$Fe^{3+}(aq) + e^- \rightarrow Fe^{2+}(aq)$$

The hydrogen can be generated in this way by adding dilute sulphuric acid and zinc to the solution containing the iron(III) ions. Hydrogen generated in this way is sometimes called **nascent hydrogen**. The word nascent means new-born.

Orthohydrogen and Parahydrogen

The two protons in a hydrogen molecule, H_2, are bound by two 1s electrons in a $\sigma(s)$ bonding orbital (*see* section 2.1). The two electrons in this molecular orbital must have opposite spins. However, unlike the electrons, the two protons can have either parallel spins or opposite spins. When the spins are parallel, the hydrogen is called orthohydrogen (*see* figure 12.1). When the protons spin in opposite directions, the hydrogen is called parahydrogen.

Ordinary hydrogen is a mixture of orthohydrogen and parahydrogen. At very low temperatures, parahydrogen predominates. As the temperature is raised, the proportion of orthohydrogen increases until at 25°C the mixture contains about 75% orthohydrogen and 25% parahydrogen.

Parahydrogen can be prepared by passing ordinary hydrogen through a tube packed with charcoal and then cooling it to liquid-air temperature. These two forms of hydrogen do not differ in their chemical properties. However, they do differ slightly in their melting and boiling points (*see* table 12.5).

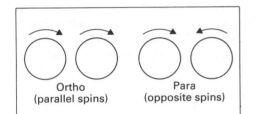

Figure 12.1 Orthohydrogen and parahydrogen

Table 12.5 Melting and boiling points of orthohydrogen and parahydrogen

	Ortho	Para
melting point/K	13.93	13.88
boiling point/K	20.41	20.29

Hydrogen Ions

The bare nucleus of a hydrogen atom is the proton, H^+. The isolated proton is unstable and rapidly combines with an electron to form a hydrogen atom. Isolated protons can be produced in hydrogen discharge tubes or by artificial nuclear transformation (*see* section 1.3), for example by bombarding nitrogen atoms with α-particles,

$$^{14}_{7}N + {}^{4}_{2}He \rightarrow {}^{17}_{8}O + {}^{1}_{1}H$$

In aqueous solution, protons are bound to water molecules by coordinate bonds, forming monohydrated protons, that is **oxonium ions**, H_3O^+ (*see* section 2.1).

Hydrogen also forms two other positive ions: the hydrogen molecule ion, H_2^+, and the hydrogen triatomic molecule ion, H_3^+. H_2^+ is very unstable. H_3^+ is more stable and can be detected in a discharge tube.

The hydrogen anion or hydride ion, H^-, is present in the hydrides of Group I metals and in the hydrides of Ca, Sr and Ba (*see* below). It consists of a proton associated with two electrons occupying the 1s orbital.

PHYSICAL PROPERTIES OF HYDROGEN

Hydrogen is a colourless, non-toxic gas with no taste or smell. Its density is lower than any other substance: at s.t.p. one litre ($1 \, dm^3$) of hydrogen has a mass of 0.0899 g. The two atoms of the hydrogen molecule, H_2, are bound by a single covalent bond. This gives each atom the stable helium electronic configuration.

The hydrogen molecule is non-polar. The intermolecular forces of attraction are thus weak. This is indicated by the low boiling and melting points of hydrogen (*see* table 12.4). The gas can be liquefied by compression and cooling in liquid nitrogen followed by sudden expansion.

Hydrogen is only slightly soluble in water. At s.t.p. less than 2 cm^3 of the gas dissolve in one litre of water.

CHEMICAL PROPERTIES OF HYDROGEN

Hydrogen can act as both a reducing agent and an oxidising agent (*see* section 10.2).

Reduction of Non-Metals

Hydrogen reduces the non-metals in Groups IV to VII, forming binary compounds. For example, when an arc is struck between carbon poles in an atmosphere of hydrogen, the hydrogen combines with the carbon to form mainly ethyne. Hydrogen combines with oxygen to form water. This reaction is described at the beginning of chapter 10. Hydrogen also reacts with boiling sulphur producing hydrogen sulphide,

$$H_2(g) \; + \; S(l) \; \rightarrow \; H_2S(g)$$

oxidation no.   0   0   +1 −2

The reactions of hydrogen with the halogens are described in chapter 16.

Reduction of Oxides and Unsaturated Compounds

At high temperatures, hydrogen reduces many metal oxides to the metals. For example

$$CuO(s) \; + \; H_2(g) \rightarrow Cu(s) \; + \; H_2O(g)$$

This type of reaction is used to extract tungsten and molybdenum from their oxides,

$$WO_3(s) \; + \; 3H_2(g) \rightarrow W(s) \; + \; 3H_2O(g)$$

In the presence of a catalyst such as finely divided nickel or platinum at a temperature of about 150°C, hydrogen reduces —C=C— and —C≡C— bonds in organic compounds. For example

$$H_2C{=}CH_2 + H_2 \rightarrow H_3C{-}CH_3$$

Oxidation of metals

At temperatures of up to 700°C, hydrogen combines directly with alkali metals and with calcium, strontium and barium in Group II to form ionic hydrides (*see* below). For example, hydrogen reacts with molten sodium above 300°C to form sodium hydride,

$$2Na(l) \; + \; H_2(g) \rightarrow 2NaH(s)$$

oxidation no.   0   0   +1   −1

TYPES OF HYDROGEN COMPOUND

Hydrogen, like carbon and oxygen, forms millions of compounds (*see* beginning of chapter 4). The vast majority of these compounds are organic compounds. The chemistry of organic compounds is discussed in chapters 17 to 20. Table 12.6 summarises the main types of compound formed by hydrogen. The chemistry of acids and alkalis is described in chapter 8. In this section we shall therefore confine our attention to the four types of hydride formed by hydrogen.

Table 12.6   Principal types of compound formed by hydrogen

Oxidation no. of hydrogen	Type of compound	Examples
+1	covalent hydrides	H_2O, CH_4, H_2S, NH_3, HCl
	acids	$H_2SO_4(aq)$
	alkalis	$NaOH(aq)$
	organic compounds	C_2H_5OH, $C_6H_{12}O_6$
0	interstitial hydrides	$ZrH_{1.92}$
−1	ionic hydrides	NaH
	complex hydrides	$Li[AlH_4]$

Ionic Hydrides

The alkali metals and some Group II metals are sufficiently electropositive to reduce hydrogen to the hydride ion, H^-. The hydrides formed are called the ionic hydrides. They are white crystalline solids formed by passing hydrogen over the heated metal (*see* above). The hydrides of Group I metals have crystalline structures similar to that of sodium chloride (*see* section 3.2).

The presence of the hydride ion in the ionic hydrides can be demonstrated by electrolysis of the hydride in fused alkali halides. Hydrogen gas is produced at the *anode*.

$$2H^-(l) \rightarrow H_2(g) + 2e^-$$

This reaction should be compared with the electrolysis of aqueous solutions (*see* section 10.3). In this case hydrogen gas is produced at the *cathode*.

Apart from lithium hydride, the ionic hydrides decompose below their melting points. They are unstable to moisture in air,

$$NaH(s) + H_2O(l) \rightarrow NaOH(aq) + H_2(g)$$

The reactions with water show that the hydride ion is a strong base,

$$H^- \quad + \quad H_2O \quad \rightarrow \quad H_2 + OH^-$$

base
(proton
acceptor)

acid
(proton
donor)

Complex Hydrides

These are compounds containing hydride ions coordinated to metal ions. The two most important are lithium tetrahydridoaluminate(III), $Li[AlH_4]$ (also known as lithium aluminium hydride) and sodium tetrahydridoborate(III), $Na[BH_4]$ (also known as sodium borohydride).

$Li[AlH_4]$ is prepared by the reaction of aluminium chloride with an excess of lithium hydride using dry ethoxyethane (ether) as a solvent,

$$4LiH + AlCl_3 \rightarrow Li[AlH_4] + 3LiCl$$

$Li[AlH_4]$ is a white crystalline solid with some covalent characteristics. It is hydrolysed by water,

$$Li[AlH_4](s) + 4H_2O(l) \rightarrow LiOH(aq) + Al(OH)_3(s) + 4H_2(g)$$

For this reason, reactions with Li[AlH$_4$] are carried out in dry ether solutions. It is used in organic chemistry to reduce aldehydes, ketones, esters and acids to alcohols (see section 19.2). It is also used to prepare the hydrides of other elements. For example

$$SiCl_4 + Li[AlH_4] \xrightarrow{\text{dry ethoxyethane}} SiH_4 + LiCl + AlCl_3$$
$$\text{silane}$$

Covalent Hydrides

Hydrogen forms hydrides with non-metals and weakly electropositive metals in Groups III to VII which are essentially covalent in character. The hydrides consist of molecules held together by weak van der Waals forces or by hydrogen bonds. They are thus usually gases, liquids, or solids with low melting and boiling points. The Group III hydrides are exceptional, since they are polymeric. For example, the formula of aluminium hydride is $(AlH_3)_n$.

The thermal stability of covalent hydrides decreases down a particular group of the Periodic Table. For example, hydrogen chloride is stable to heat whereas hydrogen iodide decomposes into its elements (see chapter 16).

The melting and boiling points and the enthalpies of vaporisation all tend to increase down a group. Exceptions are water, hydrogen fluoride and ammonia, all of which form hydrogen bonds (see section 2.2).

On crossing a period, the acid–base characteristics of the hydrides change from non-acidic and non-basic to basic to acidic. For example, in Period 3:

SiH_4	$\rightarrow$	PH_3	$\rightarrow$	H_2S	$\rightarrow$	HCl
silane		phosphine		hydrogen sulphide		hydrogen chloride
non-acidic and non-basic		weakly basic		weakly acidic		very acidic

Interstitial Hydrides

These are **non-stoichiometric** compounds (see section 4.1) formed by heating d-block metals in an atmosphere of hydrogen. In these compounds, hydrogen atoms are accommodated in the spaces between the metal atoms in the metallic lattice. The exact composition of the compound varies with temperature and pressure and so no definite formula can be ascribed to the compounds. Examples of interstitial hydrides are $TiH_{1.58}$, $ZrH_{1.92}$ and $VH_{0.6}$.

OCCURRENCE, MANUFACTURE AND USES OF HYDROGEN

Occurrence

We have already considered the ubiquitous nature of hydrogen. We saw that hydrogen is the most abundant element in the Universe. The element is found in the Sun and stars and, in particular, plays an important role in the proton–proton reactions that produce energy for the Sun and stars (see section 1.3). Free hydrogen is thought to be the major component of the planet Jupiter. In the interior of this planet, the pressure is so great that it is probable that liquid molecular hydrogen is converted to liquid metallic hydrogen (see section 2.1).

On Earth, free hydrogen accounts for less than one part per million of the atmosphere. It is found occluded in salt deposits, coal, rock and meteorites. It is also a component of the gases ejected by volcanoes. However, hydrogen continually diffuses from the Earth's atmosphere into space.

Occlusion
Occlusion is the retention of a small amount of a gas or liquid in a solid.

Manufacture

Several million tonnes of hydrogen are manufactured in the world each year. In the United Kingdom alone, about 500 000 tonnes are manufactured annually. The hydrogen is produced by a number of methods.

The Water Gas Process

This is also known as the Bosch process. Steam is passed over white-hot coke, producing **water gas**.

$$C(s) + H_2O(g) \rightleftharpoons CO(g) + H_2(g)$$
water gas

The water gas is mixed with more steam and passed over an iron(III) oxide catalyst at a temperature of 450°C.

$$CO(g) + H_2O(g) \rightleftharpoons CO_2(g) + H_2(g)$$

This is known as the **shift reaction**. The carbon dioxide is removed by scrubbing with water under pressure or by absorption in warm potassium carbonate solution.

From Natural Gas or Naphtha

Nowadays large quantities of hydrogen are produced by the **steam reforming** of natural gas or naphtha. Naphtha is a mixture of light hydrocarbons obtained from coal tar or petroleum. In the first stage of the process a mixture of natural gas or naphtha is passed over a nickel catalyst at a temperature of about 900°C,

$$CH_4(g) + H_2O(g) \rightleftharpoons CO(g) + 3H_2(g)$$
synthesis gas

The resultant mixture is called **synthesis gas** since it is the source of hydrogen in the Haber synthesis of ammonia (*see* section 7.2). The synthesis gas is then converted to carbon dioxide and water by the shift reaction and the carbon dioxide removed (*see* above).

Hydrogen is also produced from natural gas by the catalytic oxidation of methane by oxygen,

$$2CH_4(g) + O_2(g) \rightleftharpoons 2CO(g) + 4H_2(g)$$

Cracking and Reforming of Hydrocarbons

Hydrogen is produced as a by-product in oil refineries during the cracking and reforming of hydrocarbons (*see* chapter 18).

Electrolysis of Brine

Hydrogen is an important by-product of the electrolysis of brine (an aqueous solution of sodium chloride) in the manufacture of chlorine and sodium hydroxide (*see* chapter 13).

Uses

Manufacture of Ammonia

About 50% of hydrogen manufactured is used in the Haber synthesis of ammonia (*see* section 7.2). About one third of the ammonia produced is converted to nitric acid, which in turn is used to manufacture explosives, dyestuffs and nitrogenous fertilisers.

Manufacture of Inorganic Chemicals

Hydrogen is used in the manufacture of hydrogen chloride and hydrochloric acid (*see* chapter 16).

Manufacture of Organic Chemicals

Hydrogen is used in the synthetic production of methanol. The methanol is obtained by the reaction of hydrogen and carbon monoxide at a temperature of 400°C and a pressure of 300 atmospheres in the presence of zinc oxide and chromium(III) oxide catalyst,

$$2H_2(g) + CO_2(g) \xrightarrow{ZnO/Cr_2O_3} CH_3OH(g)$$

Figure 12.2 Hydrogen is used to make margarine

High in Polyunsaturates

Fats that contain a high proportion of fatty acids (*see* section 20.3) with several double bonds are called polyunsaturated fats. Semi-soft margarines 'high in polyunsaturates' are made by a process in which untreated oils rich in polyunsaturated fatty acids (such as sunflower, safflower, maize and soya oils) are blended with partially hardened fats. Polyunsaturated fats are thought to be less harmful to the heart and arteries than saturated fats and may even be beneficial.

Margarine is a butter substitute made by hardening vegetable and fish oils using hydrogen in the presence of a catalyst. During this process double bonds in the fatty acids in the oils are converted to single bonds. Semi-soft margarines are blends of untreated oils with partially hardened fats

---

The methanol is used as a solvent and also in the manufacture of other organic chemicals.

In Oil Refineries

About 12% of hydrogen produced is used in oil refineries to **hydrogenate** sulphur compounds in the oil. This enables the sulphur to be removed from the oil.

Manufacture of Margarine

Hydrogen is used in the hydrogenation of groundnut oil in the presence of a nickel catalyst to produce solid edible fats. During this process the carbon–carbon double bonds in the vegetable oil are converted to carbon–carbon single bonds. Fats containing carbon–carbon single bonds are said to be saturated whilst those with double bonds are said to be unsaturated.

Extraction of Metals

Hydrogen is used to extract metals such as molybdenum and tungsten from their oxides (*see* above).

Other Uses

Hydrogen is used in oxyhydrogen and atomic hydrogen torches (*see* above). Large electric alternators use hydrogen gas as a coolant and liquid hydrogen is used in low-temperature studies. This field of research is called **cryogenics**.

Large quantities of liquid hydrogen have also been used in the space industry as a rocket fuel and also to generate electric power in fuel cells (*see* chapters 5 and 10).

The Hydrogen Economy

As supplies of coal, oil and natural gas dwindle, there is an increasing search for economically viable alternative sources of energy. We have already considered these in chapter 5 where we saw that one possible future source of energy is hydrogen. A world economy based on hydrogen rather than oil as the primary source of energy is called the hydrogen economy. The use of hydrogen as a major source of energy has a number of advantages and disadvantages.

Advantages

(a) Hydrogen is the most abundant element in the Universe. It is also abundant on Earth in the form of water. There is therefore a virtually inexhaustible supply of the element. Furthermore, the only product of the combustion of hydrogen is water. Use of hydrogen as a fuel would thus produce water. This would immediately be returned to the atmosphere and thus enter the water cycle (*see* below).

(b) The ratio of energy output per unit mass of hydrogen is several times higher than for comparable fuels. The standard molar enthalpy of combustion is -285.8 kJ mol^{-1},

$$H_2(g) + \tfrac{1}{2}O_2(g) \rightarrow H_2O(g) \qquad \Delta H_m^{\ominus} = -285.8 \text{ kJ mol}^{-1}$$

Reference to table 5.2 shows that the molar enthalpy of combustion of hydrogen is lower than for comparable fuels such as methane, propane and ethanol. However, when the enthalpies of combustion per gram of fuel are compared, hydrogen is seen to have a high energy output (*see* table 12.7)

Table 12.7 Enthalpies of combustion

Substance	Formula	$\Delta H^{\ominus}/\text{kJ g}^{-1}$
hydrogen	$H_2(g)$	-142.9
carbon	$C(s)$	-32.8
methane	$CH_4(g)$	-55.6
propane	$C_3H_6(g)$	-52.9
butane	$C_4H_{10}(g)$	-49.6
benzene	$C_6H_6(l)$	-41.9
methanol	$CH_3OH(l)$	-22.7
ethanol	$C_2H_5OH(l)$	-30.4

(c) The use of hydrogen as a fuel does not cause pollution. As we have seen, the only product of the complete combustion of hydrogen is water. Any unburnt gas released into the atmosphere would be harmless since hydrogen is non-toxic.

Disadvantages

(a) Hydrogen is not as easy to store and use as petroleum. Hydrogen has to be stored in gas cylinders or liquid containers. Thus, although it might be possible to use hydrogen for industrial and domestic purposes in the same way that natural gas is used, it would be difficult to use hydrogen directly as a fuel for motor vehicles or jet planes. One possible solution of the problem is the use of interstitial hydrides (*see* above). Large quantities of hydrogen can be absorbed by metals such as palladium and later released on warming.

(b) The production of hydrogen is, at present, not economically viable. Hydrogen can be readily obtained from water by electrolysis (*see* chapters 5 and 10) but the process is capital-intensive and relatively inefficient. The use of thermochemical and photochemical methods of producing hydrogen from water and other materials is at present the subject of much research. However, they are unlikely to make a significant contribution to the world's energy needs for at least a decade or so.

SUMMARY

1. The **hydrogen atom** consists of one proton and one electron.
2. Hydrogen forms **diatomic molecules**.
3. Hydrogen is ubiquitous and has distinctive chemical and physical properties.
4. Hydrogen can be prepared in the laboratory by the action of water or dilute acid on metals or by electrolysis.
5. Hydrogen forms three isotopes: **protium**, **deuterium** and **tritium**.
6. **Atomic hydrogen** is a powerful reducing agent.
7. In water, **protons** form **oxonium** ions.
8. Hydrogen can act as a reducing or oxidising agent.
9. Hydrogen forms four types of **hydride**: ionic, complex, covalent and interstitial.
10. Hydrogen is the most abundant element in the Universe.
11. Hydrogen is manufactured from natural gas or naphtha by **steam reforming.**
12. Large quantities of hydrogen are used to manufacture ammonia.
13. A **hydrogen economy** would rely on hydrogen as the major source of energy.

12.2 Water | 1 000 000 000 000 000 000 Tonnes of It

Water is essential for life. About two-thirds of the human body is water

Water is the most common compound found on Earth. There are a million million million (10^{18}) tonnes of it and this covers some four fifths of the Earth's surface. It is the only chemical compound found naturally as a liquid, as a solid (ice) and as a gas (water vapour). Water plays a vital role in industry, in the home and in the laboratory, and it is essential for life. About two-thirds of the human body is water and many foods consist predominantly of water (*see* table 12.8).

Structure and Physical Properties

Ancient philosophers believed that water was one of the four elements—the others being earth, air and fire. This concept persisted into the Middle Ages. In 1781 Henry Cavendish showed that the combustion of hydrogen produced water. However, it was not until 1860 that Stanislao Cannizzaro finally established the formula of water as H_2O.

Water is a covalent molecular compound. In each molecule the oxygen atom has two lone pairs of electrons. This accounts for the bent tetrahedral shape of the molecule (*see* chapter 2).

Water is a clear, colourless liquid with a number of anomalous physical properties. It has, for example, abnormally high freezing and boiling points and surface tension (*see* section 2.2). Its enthalpies of vaporisation and fusion per gram are higher than almost any other substance. It is unique in that the density of the liquid at 4°C is higher than the density of ice. Thus ice floats on water.

These anomalous properties of water can be attributed to the hydrogen bonds that link the molecules together in both ice and liquid water (*see* section 2.1). Water is a poor conductor of electricity, although it becomes a good conductor when even small amounts of ionic substances are dissolved in it (*see* chapter 8).

The Universal Solvent

Water is used extensively as a solvent in industrial chemical processes and also in the laboratory. Water is also the universal solvent for biochemical reactions. The reason for this is that water is an excellent solvent for ionic compounds and also for many covalent compounds. The solvent properties of water are due to the polar nature of the water molecules. The water molecule has a relatively high dipole moment (*see* table 2.5). Thus, when ions dissolve in water the water molecules become oriented around the ions. This is known as solvation. Aqueous solutions of ionic compounds are electrolytes (*see* chapter 10).

The solubility of covalent compounds in water depends on their ability to form hydrogen bonds with the water molecules. The hydrogen bonds are the dipole–dipole interactions between the hydrogen atoms of the water molecules and the electronegative atoms of the solute molecules. Simple covalent compounds such as sulphur dioxide, ammonia and hydrogen chloride all dissolve in water. Oxygen, nitrogen and carbon dioxide are slightly soluble in water. Many organic compounds containing electronegative atoms such as oxygen or nitrogen are soluble in water. Examples are ethanol, C_2H_5OH, ethanoic acid, CH_3COOH, sucrose, $C_{12}H_{22}O_{11}$, and diethylamine, $(C_2H_5)_2NH$.

Finally, we should note that the presence of non-volatile solutes such as sodium chloride or sucrose in water lowers the vapour pressure and freezing point of water but elevates the boiling point (*see* section 6.3).

CHEMICAL REACTIONS INVOLVING WATER

Water is involved in numerous chemical reactions as a solvent, reactant or product. We have already looked at the solvent properties of water. Water is a

Table 12.8 Water content of some foods

Food	Percentage water by mass
lettuce	96
tomatoes	95
mushrooms	92
milk	87
oranges	86
apples	84
white fish	82
potatoes	76
eggs	75
beef	64

96% water!

product in many inorganic and organic chemical reactions. For example, it is the product of the neutralisation of an acid with a base. In organic chemistry, many condensation reactions involve the elimination of a water molecule. In this section we shall examine four of the most important types of chemical reaction in which water is involved as a reactant.

Acid–Base Reactions

Water is **amphoteric**. This means it can act either as an acid or as a base. Its amphoteric nature is due to auto-ionisation, that is self-ionisation,

$$2H_2O(l) \rightleftharpoons H_3O^+(aq) + OH^-(aq)$$

Water can thus act as a proton acceptor,

$$HCl + H_2O \rightleftharpoons H_3O^+ + Cl^-$$

or a proton donor,

$$NH_3 + H_2O \rightleftharpoons NH_4^+ + OH^-$$

These reactions are discussed in more detail in chapter 8.

Oxidation and Reduction

Water can act as an oxidising agent and a reducing agent. It oxidises metals above tin in the electrochemical series (*see* table 10.6). For example, in the reaction between sodium and water, the following oxidation process occurs:

$$Na(s) \rightarrow Na^+(aq) + e^-$$

The water itself is reduced during the reaction,

$$2H_2O(l) + 2e^- \rightarrow 2OH^-(aq) + H_2(g)$$

Another example is the reaction between magnesium and steam,

$$Mg(s) + H_2O(g) \rightarrow MgO(s) + H_2(g)$$

Water is also involved as an oxidising agent in corrosion processes (*see* section 10.4). For example, one of the processes involved in rusting is

$$2H_2O + O_2 + 4e^- \rightarrow 4OH^-$$

Water is an important reducing agent in certain biochemical processes. For instance, several stages of the citric acid cycle (*see* section 4.1) involve the oxidation of water,

$$2H_2O \rightarrow O_2 + 4H^+ + 4e^-$$

This electron transfer process is also important in the reduction of organic phosphate compounds during photosynthesis. Both the citric acid cycle and photosynthesis are complicated processes involving sequences of chemical reactions. In both cases the electron transfer processes are not yet fully understood.

Hydration

We have already seen that both cations and anions can be solvated by water molecules. This process is known as hydration. The water of hydration in salt crystals is known as water of crystallisation. The water molecules are normally bonded to the cation by coordinate bonds.

Hydrolysis

This is the reaction of an ion or molecule with water. The reaction between hydrogen chloride and water to form hydrochloric acid is an example of this type of reaction (*see* above). Another example is the hydrolysis of iron(III) chloride,

$$FeCl_3(aq) + 3H_2O(l) \rightleftharpoons Fe(OH)_3(s) + 3H^+(aq) + 3Cl^-(aq)$$

Ice melts under high pressure. This property enables us to skate on ice since the liquid water formed under the skate acts as a lubricant. The picture shows champion skaters Torvill and Dean who won an unprecedented number of perfect sixes in both the 1984 World Championships and the Los Angeles Olympics

Hydrolysis of organic compounds is also common. One of the most commonly cited examples is the hydrolysis of ethyl ethanoate.

$$CH_3COOC_2H_5 + H_2O \rightarrow CH_3COOH + C_2H_5OH$$

ethyl ethanoate   ethanoic acid   ethanol

WATER RESOURCES AND USES

The hydrological cycle was known in Old Testament times: 'All the rivers run into the sea; yet the sea is not full; unto the place from whence the rivers come, thither they return again.' (Ecclesiastes 1:7)

Of the 10^{18} tonnes of water on Earth, only 3% is fresh and 80% of this is unavailable to man. This is because it is locked in glacial ice on the ice caps. Water is available to man through the **hydrological cycle** (or water cycle). This is shown in figure 12.3. The cycle involves about 500 000 km³ of water annually through evaporation and precipitation as rain or snow. The theoretical maximum amount of water available to man is about 40 000 km³ per year. This is the water that *runs off* from the land into the sea (known as **run-off**).

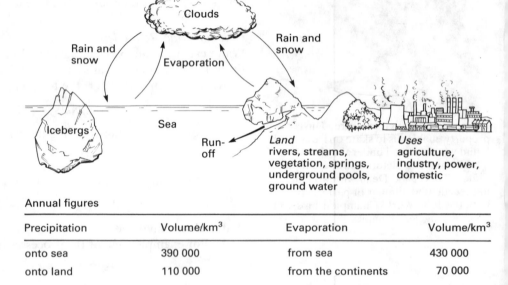

Annual figures

Precipitation	Volume/km³	Evaporation	Volume/km³
onto sea	390 000	from sea	430 000
onto land	110 000	from the continents	70 000
total	500 000	total	500 000

Figure 12.3 The hydrological cycle

The use of this water is often classified as '**in-stream**' use or '**withdrawal**' use. Alternatively, water is sometimes classified as non-consumed or consumed.

In-stream or non-consumed water can be re-used. Uses include

navigation,
generation of hydroelectric power,
fishing.

Withdrawal or consumed water, once used, is not available for re-use. It includes water that is consumed and later lost through evaporation or transpiration; water that is incorporated into products; and water that is discharged into the sea. About 2500 to 3000 km³ are withdrawn in the world each year, of which about 10% is used for domestic purposes, 8% for industry and a vast 82% for agricultural irrigation.

There are predictions that, by the twenty-first century, the amount of water used in the world will exceed the amount available from run-off. In order to meet this problem, various schemes for tapping more water are being developed. These include the following.

Transpiration is the loss of water vapour from plants.

Water Uses

Domestic Uses

Water is used in the home for drinking, for cooking, for washing clothes, the body and the car, in toilets and for watering the garden. About 10% is wasted.

In Europe, the average domestic consumption of water is about 230 litres per person per day. This is about the same as in the days of Imperial Rome. About 10% of the water withdrawn worldwide is used for domestic consumption.

Industrial Uses

Over 85% of water used in industry is for cooling purposes. The rest is used for washing, scrubbing, hydraulic transport and as a solvent. About half a million litres of water are used in the manufacture of a single four-door saloon car. This includes both in-stream and consumed water.

About 8% of the water consumed in the world is consumed by industry.

Agricultural Uses

Agriculture consumes 82% of the water withdrawn in the world. This is used for irrigation. About 11 000 million litres of water are needed to grow one tonne of cotton. A single cabbage needs 0.15 tonnes of water to grow to maturity.

A farmer firms up the muddy banks of a small irrigation canal in a sorghum field in the Sudan. 82% of water withdrawn in the world is used for irrigation

Power

Over 50% of the water supply in the UK is used by electricity power stations. The water is needed for hydroelectric power, for steam raising to turn turbines and as a coolant. Although power stations use vast quantities of water, there is virtually no loss since it is used 'in-stream'.

Increasing the Amount of Water Available from Run-off

Much of the water that runs off from the land into the sea is not harnessed for use. The siting of new reservoirs and the digging of new wells and boreholes to obtain ground water will increase the usage of the water before it is lost to the sea.

In hot weather, large quantities of water are lost from lakes and reservoirs through evaporation. This can be inhibited by covering the surface of the water with a thin film of the alcohol hexadecan-1-ol.

Using Sea-water and Brackish Water

Pure water can be obtained from sea-water by a technique known as flash distillation. This is distillation at low pressure. The method is energy-intensive and only economic in countries such as Kuwait where energy is relatively inexpensive and there is low rainfall.

Pure water can also be obtained by the electrodialysis (*see* section 10.3) of brackish water. Brackish water occurs in river estuaries. It is partially salty and partially fresh.

There are now around 2000 desalination plants in the world. Techniques used to desalinate water include not only flash distillation and electrodialysis but also freezing, ion exchange and reverse osmosis.

This installation pumps sea-water for Kuwait City Water Distillation and Power Plant

As far as the eye can see! A desalination plant in Saudi Arabia

WATER QUALITY

Since water is such a good solvent, it rarely occurs absolutely pure. The suitability of water for both aquatic and human use depends on its quality. Some of the parameters that determine the quality of water are shown in table 12.9. It is beyond the scope of this book to provide a detailed systematic treatment of the significance of each parameter and how it is determined experimentally. The determination of pH and conductivity are described in chapter 10. The hardness of water and its treatment are described in chapter 8.

The presence of micro-organisms in water is determined by measuring the **biological oxygen demand** (BOD) of the water. The oxygen content of the water is determined before and after incubation in the dark for 5 days at 20°C. It is measured in mg dm^{-3}. BOD is normally considered to be a measure of the pollution of water. When polluting organic matter is discharged into water, a natural purifying action tends to set in. This involves certain micro-organisms that use the oxygen dissolved in the water to oxidise the polluting substances. It is generally accepted that BOD values have the following significance:

BOD/mg dm^{-3}	Significance
less than 30	no pollution
30 to 80	mild pollution
over 80	severe pollution

WATER POLLUTION

Water pollution is the degradation of the quality of water by the introduction into rivers, streams, lakes and oceans of chemical, physical or biological materials. Pollution can take many forms.

Waste

Industrial effluents containing inorganic and organic wastes are often discharged into rivers and the sea. Each year thousands of environmentally untested chemicals are routinely discharged into waterways. Hundreds of these are new compounds. Although the industrial effluents are often treated, they still contain toxic substances that are hard to detect.

Domestic sewage containing, for example, synthetic detergents ends up in rivers and the sea. In addition, fertilisers drain from the land into waterways leading to lakes and the sea. All this poses a number of severe problems particularly with respect to enclosed bodies of water such as lakes, bays and fjords.

Solid Wastes

Suspended solids, when in excess, cut out light from the Sun and thus inhibit the process of photosynthesis in the aquatic environment. This in turn disturbs the food chains in these environments. In addition, solid wastes silt up rivers and navigational channels, with the result that frequent dredging is necessary.

Eutrophication

The effluents, sewage and fertilisers that are discharged into waterways contain high levels of nitrates and phosphates. This results in the over-fertilisation of enclosed bodies of waters, leading to an immediate increase in primary productivity. Primary productivity is the productivity of micro-organisms such as algae. Species such as the blue-green algae thus flourish. Unfortunately, blue-green algae are generally unsuitable as food for the fish. These algae also consume more oxygen than is produced in the water. The BOD of the water thus increases. Biological wastes such as wood pulp and untreated raw sewage also lead to a high BOD. Other plants and animal species cannot compete in this environment. In addition,

Tap water in urban areas is often recycled water. This means it might well have passed through several pairs of kidneys before being consumed again. Although this recycled water is perfectly safe to drink, some people find the prospect distasteful. They prefer to drink natural or sparkling water from a bottle.

Table 12.9 Some parameters that determine the quality of water

turbidity
colour
temperature
taste
odour
pH
conductivity
hardness
dissolved inorganic substances, e.g. nitrates, chloride, iron
dissolved organic substances, e.g. phenols
presence of microbiological organisms, e.g. bacteria
presence of algae and animals

24 tons of waste water are discharged for every ton of paper produced in a paper mill.

the micro-organisms responsible for decomposing dead plant and animal tissue thrive in this environment, consuming more oxygen and producing more nitrates and phosphates. Gradually, the number of species in the lake is considerably reduced. The most important victims are the fish. Eventually, the lake ages as the oxygen content is depleted by these algae and plant decomposers. It becomes like a swamp. This process is known as eutrophication.

A classic example of where this process has occurred is Lake Erie in the United States. Over 25 years, the nitrogen content of the lake has increased by 50% and the phosphorus by 500%. Most of this is due to the disposal into the lake of domestic effluent containing synthetic detergents. Synthetic detergents contain high levels of phosphates.

Sewage treatment is inadequate to cope with these problems since it removes solid material and only a small percentage of the dissolved nutrients.

Toxicity of Inorganic Waste

The discharge of industrial effluents into fresh waters and the sea has led to increased levels of toxic heavy metal ions such as cadmium, mercury and lead. A substantial proportion of these are absorbed or precipitated onto particulate matter in what is sometimes called a self-purification process. However, in enclosed bodies of water heavy metals can reach dangerously high levels.

The most famous case of this occurred in Minamata Bay in Japan. Industrial waste containing methyl mercury ethanoate was discharged into the bay. The mercury became incorporated into a food chain by algae in the bay. The algae were consumed by shellfish, which in turn were consumed by fish, which were finally consumed by the local population. The mercury levels in the fish were sufficient to cause birth defects and death. The disease was called Minamata disease.

There is also growing concern about the increasing levels of nitrates found in drinking water. High levels, it is claimed, may well cause stomach cancer and can cause the death of babies.

Disease

According to the International Labour Organisation (ILO), an estimated 70% of the world's population lack safe water. The problem is particularly acute in

In 1982, in Syria, a woman went to a doctor complaining of stomach pains. She was admitted to hospital where she had an operation on her stomach. A six-foot snake was found inside her. The doctors concluded that she must have drunk water polluted with snakes' eggs.

These flies can cause trachoma—a disease which causes blindness and afflicts hundreds of millions of people in the developing world

Many people in the developing world drink from, and wash in, contaminated water

developing countries. About 90% of all rural dwellers always drink from, and wash in, water that is contaminated. The World Health Organisation has estimated that 80% of the sickness and disease in the world is caused by inadequate water and sanitation. Water-borne diseases include cholera, typhoid, malaria, bilharzia and leprosy. In all, nearly 500 million people suffer from water-borne diseases.

For example, hundreds of millions suffer from a disease called trachoma, which often causes blindness. It is particularly prevalent amongst babies and children in developing countries such as Ethiopia. When a baby cries, the tears, which are made up of salt and water, run out of the eye and down the cheek. The water evaporates leaving the salt. Insects, such as the Ethiopian horsefly, feed on the salt and lay eggs. Because the babies are malnourished, they do not have enough energy to brush away the flies. The flies lay eggs, which hatch and eat the babies eyes. If they had enough food to give them strength to brush away the flies, or if they had access to clean water to wash away the salt and eggs, they would not catch the disease. The problem could be solved by spending no more than a few pence on each child.

However, the problem of contaminated water and disease is not just confined to developing countries. One-quarter of all Mediterranean beaches are considered dangerously filthy. Nearly all its mussels and oysters are unsafe to eat according to a pollution survey of the Mediterranean published in 1983 by the United Nations Environmental Programme (UNEP). Typhoid, paratyphoid, dysentery, polio, viral hepatitis and food poisoning are endemic in the area and there are periodic outbreaks of cholera. Most of the disease is caused by sewage. An estimated 85% of the waste from the sea's 120 coastal cities is flushed into the Mediterranean where holidaymakers and residents bathe and fish. Between Barcelona and Genoa about 200 tonnes of sewage is flushed out each year for every mile of coastline.

> Every year about 200 tonnes of sewage is flushed out for every mile of coastline in the Mediterranean. In 1980, Algeria, Egypt, France, Monaco, Tunisia and Turkey ratified an international agreement to tackle the most serious sources of pollution in the Mediterranean—at a cost of £6000 million. The agreement came into force in 1983.

A benthic (bottom) chamber is lowered into Buzzards Bay, Massachusetts, USA for chemical studies of PCBs and other toxic materials

Pesticides

The most toxic pesticides are the halogenated hydrocarbons such as DDT and polychlorinated biphenyls (PCBs). The chemistry of these is discussed further in chapter 16. Although DDT is banned in many parts of the world, it is still used and about 25% of it ends up in the sea. Unfortunately, these halogenated hydrocarbons are chemically stable and non-biodegradable. They thus accumulate in food chains. DDT can result in the sterilisation of complete stretches of waterways and is known to cause infertility in birds.

Oil Spillages

In the United States alone there are an estimated 13 000 oil spillages of various magnitude each year. Up to 12 million tonnes of oil entered the seas in one year alone. In the UK, over one million tonnes of used engine oil are poured down drains.

Oil spillages have several adverse effects on the marine environment. First of all, oil can smother birds, which soon die from drowning, exposure or starvation. Animals such as seals have been shown to be blinded by oil. Oil can also reduce the light entering enclosed waters, such as tide pools and coral reefs, and can raise temperature. This is particularly harmful to organisms that can only exist in a narrow temperature band. Oil also contains toxic components, such as aromatic hydrocarbons, which can kill certain forms of aquatic life—even at levels in the order of a few parts per million.

Stranded in the sludge. This conger eel was only one victim of the *Amoco Cadiz* oil slick which took a heavy toll on the local wildlife of Brittany

An aerial view of the stricken supertanker *Amoco Cadiz* which ran aground off the village of Portsall in Brittany, France, in March 1978, shedding its entire 220 000 tonnes of crude oil into the sea

Acid Rain

All rain is naturally acidic. Acid rain is a term that is used for rain with a pH of less than 5.6. Rain with a pH as low as 4.0 has been recorded in Germany, Scandinavia and North America. Acid rain is produced when the gases emitted by smelters, fossil-fuelled power stations, oil refineries, factories and car exhausts enter the atmosphere. These gaseous emissions contain sulphur and nitrogen oxides, which combine with the moisture and oxygen in the atmosphere to form sulphuric and nitric acids. These acids then fall to the ground—sometimes many hundreds of miles away from the source of pollution. In countries such as Canada, USA, East and West Germany and Czechoslovakia, thousands of lakes and rivers are no longer able to support plants and fish. The problem is compounded by the fact that water of low pH is able to leach, that is dissolve, toxic minerals including those containing aluminium and heavy metals such as cadmium and mercury from the ground. These would not normally be sufficiently soluble in water to be harmful.

Both sulphur dioxide and nitrogen dioxide can be removed from industrial emissions by scrubbing—although reducing it to a minimum is very expensive. Many industrialised countries are now introducing or tightening up legislation to ensure that sulphur dioxide and nitrogen dioxide emissions are reduced to more acceptable levels. In 1984, for example, a European Economic Community (EEC) commission recommended a 60% reduction in sulphur dioxide emissions and a 40% reduction in nitrogen oxides to be achieved by 1995.

Other Forms of Pollution

These include radioactive and thermal pollution. The main source of radioactive pollution of the sea is low-level radioactive waste discharged from nuclear power plants (*see* section 1.3). One of the major concerns here is that marine organisms such as algae accumulate or concentrate radioactive isotopes.

Thermal pollution can also be caused by fossil-fuel and nuclear-powered electricity generating stations. Waste heat from generating stations is discharged in the cooling water, which is then discharged into nearby waterways. The increased temperature accelerates certain biochemical processes and also reduces the level of dissolved oxygen in the water (*see* section 6.2). This results in rapid and often dramatic changes in the biological communities in the vicinities of the power stations. Delicately balanced reproductive cycles of organisms are disturbed. Algae tend to flourish in these conditions, whilst other aquatic organisms cannot survive.

WATER TREATMENT

Our water supply originates from rivers, streams, lakes and other sources. Before we can consume this water, it has to be treated to improve its quality. Treatment of water to meet drinking water standards involves a number of both physical and chemical processes.

Physical Processes Involved in Water Treatment

Screening
The first step in water treatment is to remove large floating objects and suspended debris. Later on in the treatment of the water, micro-screens are used to remove fine suspended material.

Aeration
The water may be aerated in a number of ways, for example, by waterfall cascades. This removes carbon dioxide, hydrogen sulphide and volatile oils, which can give the water a taste or odour. It also oxidises dissolved iron and manganese ions.

Flocculation
This involves the gentle agitation of water so that smaller particles will conglomerate to form large clumps which will settle to the bottom.

Sedimentation
In this process, particles are removed from the water by allowing them to settle on the bottom.

Filtration
This removes finely suspended material from the water by using a layer of sand—or sand with crushed coal—supported by a gravel bed.

A filter bed removes finely suspended material from water

Chemical Treatment

The chemical treatment of water varies depending on the quality of the water taken off from the river or reservoir. The most common forms of chemical treatment are as follows.

Coagulation

Coagulants are added to coagulate fine and colloidal suspended material. The coagulants form 'flocs'. These have sufficient particle size and density to be removed by sedimentation. Sodium aluminate and aluminium sulphate are commonly used to coagulate alkaline substances in water.

Disinfection

Micro-organisms in the water are destroyed by the addition of disinfectants—usually chlorine. This is normally the final stage in water treatment.

Water Softening

This is the removal of hardness from water caused by dissolved calcium and magnesium salts. In waterworks it is often achieved by the addition of calcium hydroxide or sodium carbonate to the water. Water can also be softened by the use of ion exchange resins (see chapter 8).

Adsorption

Adsorption is the taking up of one substance on the surface of another. In the water industry, activated charcoal is used to adsorb organic chemicals. Some of these cannot be removed by conventional water treatment processes.

Oxidation

Undesirable substances dissolved in water can be oxidised to less harmful forms by oxidation. For example, cyanides can be converted to cyanates by oxidation with ozone.

Desalination

We have already referred to this process above.

Treatment of Sewage and Industrial Effluent

The treatment of sewage and industrial effluent can be divided into three stages:

Primary Treatment

This involves screening the water for large objects and the removal of suspended material.

Secondary Treatment

In this stage, organic matter in the sewage and waste water is broken down by micro-organisms. This biodegradation of organic matter is encouraged by blowing air into the tanks.

The sludge formed in the primary and secondary treatment is dumped into the sea, used to fill quarries and reclaim land or, since it is rich in nitrogen and phosphorus, sold as solids to farmers to use as grassland fertiliser. The sludge can also be processed in methane digesters (see chapter 5) to produce usable methane gas. Conversion to methane cuts the waste volume by half and, by providing an inexpensive fuel, reduces the cost of running the sewage works.

Advanced Treatment

This is the biological, chemical and physical treatment of waste water to remove

- plant nutrients such as phosphates that promote excessive growth of algae in water,

- industrial inorganic pollutants such as dissolved heavy metal ions,
- non-biodegradable organic chemicals such as the halogenated hydrocarbons used in the manufacture of pesticides.

Advanced treatment can produce effluent that meets drinking water standards. Effluent from sewage works is normally returned to the rivers or allowed to flow into the sea. In Britain the effluent from sewage works must have a BOD of less than 20 mg dm^{-3} and contain no more than 30 mg dm^{-3} of suspended solids.

THE OCEANS

The oceans of the world not only provide a rich source of food for mankind but also a vast variety and in some cases an almost unlimited supply of raw materials.

Inorganic minerals of economic importance constantly flow into the oceans from a number of sources. Much of this accumulates on the ocean bottom. Rivers, for example, dump millions of tons of particulate mineral deposits into the oceans each year. Volcanic eruptions both on land and on the ocean floor also introduce many metals into the ocean. Some of these dissolve and some are deposited onto the ocean floor.

The total volume of the oceans of the world is about 1370 million cubic kilometres. Each cubic kilometre contains millions of tonnes of dissolved salts. The seven most abundant compounds found in sea-water are shown in table 12.10. Sea-water contains enormous reserves of most elements dissolved as ions or molecules. Table 12.11 shows the composition of the sea in terms of its top 15 elements.

Table 12.10 Concentration of the seven most abundant compounds in sea-water

Compound	Formula	Concentration/tonnes km^{-3}
sodium chloride	NaCl	27 500 000
magnesium chloride	MgCl$_2$	6 750 000
magnesium sulphate	MgSO$_4$	5 625 000
calcium sulphate	CaSO$_4$	1 800 000
potassium chloride	KCl	750 000
calcium carbonate	CaCO$_3$	111 250
potassium bromide	KBr	102 500

1 tonne = 1000 kg

Table 12.11 Percentage by mass of elements in sea-water

Element	Percentage by mass
oxygen	85.4
hydrogen	10.7
chlorine	1.85
sodium	1.03
magnesium	0.127
sulphur	0.087
calcium	0.040
potassium	0.038
bromine	0.0065
carbon	0.0027
nitrogen	0.0016
strontium	0.00079
boron	0.00043
silicon	0.00028
fluorine	0.00013

Sea-water is the world's principal source of bromine. Minerals commercially recovered from the shelf areas of the seas include not only oil, coal, natural gas and sulphur but also diamonds, gold and tin. The deep ocean floor, particularly of the Pacific Ocean, is strewn with ferromanganese nodules, which are richer in valuable metals such as copper, nickel and zinc than the ores currently being mined on the land. One estimate has put the total mass of deposits in the Pacific Ocean alone at more than 17×10^{11} tons.

Sea-weeds also provide a source of both inorganic and organic chemicals. Each year millions of tonnes of sea-weed are harvested for the production of chemicals such as iodine, alginic acid, laminarin and agar-agar (see chapter 16).

SUMMARY

1. Water is a **covalent** molecular compound.
2. Water has a number of **anomalous** physical properties due to **hydrogen bonding**.
3. Water is an excellent **solvent** for ionic compounds and many covalent compounds.
4. Water is **amphoteric**.
5. Water can act as an **oxidising agent** or **reducing agent**.

Cont'd. overleaf

6. **Hydrolysis** is the breaking up of ions or molecules by water.
7. Water has many domestic, industrial and agricultural uses.
8. **Biological oxygen demand** (BOD) is a measure of water pollution.
9. **Water pollution** is the degradation of the quality of water by chemical, physical and biological materials.
10. Pollution can be caused by domestic and industrial effluents and waste, pesticides and oil spillages.
11. The depletion of oxygen in a body of water due to the decay of organic matter is known as **eutrophication**.
12. **Acid rain** has a pH of less than 5.6.
13. **Water treatment** involves a number of physical and chemical processes.
14. The treatment of sewage and industrial effluent may be divided into three stages: primary, secondary and advanced.
15. **Oceans** provide not only food for mankind but also an almost unlimited source of raw materials.

Examination Questions

1. (a) Briefly describe the chemical properties of hydrogen and explain why the chemistry of hydrogen is unique among the elements in some respects.
    (b) Discuss the properties of the hydrides of **three** typical elements which are selected one from each of the s-, p- and d-blocks of the Periodic Table.
    (c) Comment on the observation that the mass spectrum of CD_3OH, made by the reaction between CD_3ONa and HCl, shows only **one** major peak (at mass number 35) whereas the mass spectrum of a sample of HDO, made by the interaction of NaOD and HCl, shows **three** major peaks (at mass numbers 18, 19 and 20).

(O & C)

2. (a) What enthalpy changes are involved in the formation of 1 mol of solid sodium hydride from its elements in the standard state?
    (b) The enthalpy of formation of solid sodium hydride is given as -57.3 kJ mol^{-1} whereas that for calcium hydride is -189.0 kJ mol^{-1}. What reasons can you suggest to account for this large difference?
    (c) **Two** other types of bonding shown by hydrogen are exemplified by the formation of hydrogen chloride from its elements and its subsequent solution in water.
        Describe these two types of bonding and account for the (i) pH values and (ii) electrical conductivity in both hydrogen chloride and its subsequent solution.
    (d) Lithium tetrahydridolaluminate(III) (lithium aluminium hydride) is a useful reagent in organic chemistry.
        (i) Give a balanced equation for the preparation of this compound.
        (ii) Draw a suitable structure for the compound.
        (iii) Give **two** reactions in which it exhibits reducing properties.

(AEB, 1984)

3. (a) The ionisation energy of hydrogen atoms is 1310 kJ mol^{-1} and the electron affinity of chlorine atoms is -347 kJ mol^{-1}.
        (i) Write the equations for the two processes.

(ii) Given the following additional information

$$H(g) + Cl(g) \rightleftharpoons HCl(g) \qquad \Delta H^{\ominus} = -432 \text{ kJ mol}^{-1}$$

$$HCl(g) \rightleftharpoons H^+(aq) + Cl^-(aq) \quad \Delta H^{\ominus} = -75 \text{ kJ mol}^{-1}$$

calculate the standard enthalpy change for the process

$$H^+(g) + Cl^-(g) \rightleftharpoons H^+(aq) + Cl^-(aq)$$

(b) The table below gives the enthalpies of hydration of some ions.

Ion	Cl^-	Br^-	I^-	Li^+	Na^+	K^+
Enthalpy of hydration/Kj mol^{-1}	-380	-350	-310	-520	-400	-320

(i) Using the result from (a), obtain the enthalpy of hydration of the proton.

(ii) Compare this value with those of the ions in the table and explain briefly why the proton has the value it does.

(JMB)

4. This question concerns the hydrides of a range of elements.
   (a) In each of the following cases, give the name of a hydride and write an equation for the reaction described.
      (i) A hydride which hydrolyses rapidly and extensively in cold water.
      (ii) A hydride which is spontaneously flammable in air.
      (iii) A hydride which, on electrolysis in the molten state, gives hydrogen at the anode.
      (iv) A hydride which, in water, gives a dibasic acid.
      (v) A hydride which, in water, gives a weak alkali.
      (vi) A hydride which can act as a ligand with d-block metal ions.
   (b) Give the names of *two* hydrides which combine with each other by addition and write an equation for the reaction.
   (c) Suggest an explanation for the ability of carbon to form long chain hydrocarbons.
   (d) The enthalpy of combustion of methane is highly exothermic. To what do you attribute the stability of methane in air?
   (e) Suggest a reason for the large difference in boiling points of methane and water.

(L)

5. The diagram below illustrates some of the reactions of hydrogen:

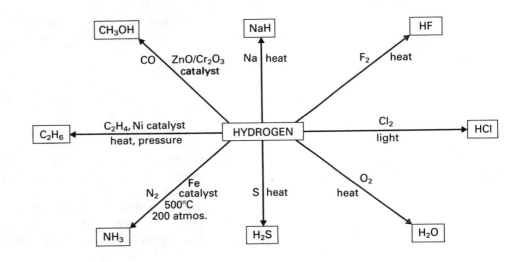

The following questions are based on the above diagram:

(a) Name the type of chemical reaction producing ethane.

(b) Which two products are basic in character?

(c) In which of the above reactions is hydrogen acting as an oxidising agent? Write a pair of ion–electron equations for this reaction.

(d) Hydrogen fluoride exhibits a high degree of hydrogen bonding, yet there is no such bonding in hydrogen. Explain why this is so.

(e) (i) Write a balanced equation for the formation of methanol.

    (ii) Assuming that the reaction goes to completion, what mass of methanol could be made from 2 litres of hydrogen, measured at s.t.p.?

(f) What role does light play in the formation of hydrogen chloride?

(g) In the industrial manufacture of ammonia, some plants operate at pressures greater than 200 atmospheres. State one advantage and one disadvantage of increasing the pressure beyond 200 atmospheres.

(h) Draw a diagram to show clearly the shape of a hydrogen sulphide molecule.

(i) The combustion of hydrogen is exothermic. Why, then, does energy have to be supplied?

(SEB)

6. (a) Describe one method of preparing hydrogen on a large scale, including, where appropriate, its separation from other products of the preparative reaction.

(b) How would you prepare and collect a sample of hydrogen, from which the major impurities have been removed, in the laboratory? (Note that concentrated sulphuric acid is not suitable for drying pure hydrogen because it is liable to form SO_2; another drying agent must be used.)

(c) Discuss the structure and bonding of the hydrides of

    (i) sodium,

    (ii) nitrogen (one hydride of your choice),

    (iii) iodine.

What reactions occur between these hydrides and water?

(O & C)

7. (a) Describe the structure of, and the bonding in, the water molecule.

(b) Describe what you would see when anhydrous copper(II) sulphate is added to an excess of water. Account for your observations by reference to the structure of, and the bonding in, the copper-containing species present in the resulting solution.

(c) Describe the reactions of each metal ion in solutions of the sulphates of chromium(III), iron(II), cobalt(II) and copper(II) with, in each case, an excess of (i) aqueous potassium hydroxide, (ii) aqueous ammonia.

(JMB)

8. (a) Explain the meaning of the term *hydrolysis*.

(b) When aluminium(III) chloride is hydrolysed, all the chlorine in the compound reacts. Write the equation for the hydrolysis of aluminium chloride.

(c) When magnesium(II) chloride is hydrolysed, only half the chlorine in the compound reacts. Write the equation for the hydrolysis of magnesium chloride.

(d) Sodium, magnesium, and aluminium have atomic numbers of 11, 12 and 13 respectively. These elements are in the same period of the Periodic Table. Sodium chloride does not hydrolyse at all.

(i) Say what is meant by atomic number.

(ii) Explain why these elements are in the same period of the Periodic Table.

(iii) Suggest reasons for the differences in the extent of hydrolysis of these three chlorides.

(e) Carbon and silicon are in the same group of the Periodic Table, the latter having the larger atomic number. Tetrachloromethane (carbon tetrachloride) is not hydrolysed at all but silicon tetrachloride is hydrolysed completely.

(i) Explain why these two elements are in the same group of the Periodic Table.

(ii) Suggest reasons for the differences in the extent of the hydrolysis of these two chlorides.

(f) A molten binary compound A is electrolysed under special conditions. At the cathode, a vapour is formed which can be condensed to a greyish solid that shines when freshly cut but is rapidly tarnished in damp air. At the anode, a gas can be collected which burns in chlorine forming 'misty' fumes.

(i) Write the name of the product at the cathode.

(ii) Write the name of the product at the anode.

(iii) Write an ionic formula for the binary compound A.

(iv) Special practical conditions would be necessary in this experiment if the two products are to be isolated. Suggest two such special conditions.

(g) If compound A from (f) above is put into water, the same anodic gas is evolved as in the electrolysis, and the final solution turns pink with phenolphthalein.

(i) Write an equation for the hydrolysis of A.

(ii) As a result of reaction (i), to what kind of use can A be put as a chemical agent?

(iii) Suggest a simple practical method by which A might be prepared.

(SUJB)

9. Under the appropriate chemical conditions water can behave as
(a) a Brønsted acid,
(b) A Brønsted base,
(c) a ligand,
(d) a nucleophile,
(e) an oxidising agent.

Select substances from the following list to illustrate each of these types of behaviour: $Al_2(SO_4)_3$(aq), CaO, CH_3Cl, CH_3CHO, $CuSO_4$(aq), Mg, Na, NH_2-NH_2, $SiCl_4$.

Your answer should refer to *two* substances for each of (a) to (e). It should also include equations wherever possible, names of products, and brief statements of the physical conditions under which reactions occur.

(NISEC)

13 THE s-BLOCK METALS

s-Block Metals and E numbers

If you look at the label on a food container you will often see E numbers listed with the ingredients (*see* figure 13.1). Each E number corresponds to a specific food additive. These additives include elements such as carbon, chlorine, gold(!) and silver. Most, however, are compounds—both inorganic and organic. Many of these compounds contain s-block metals. Calcium, magnesium, potassium and sodium compounds are all commonly used as food additives.

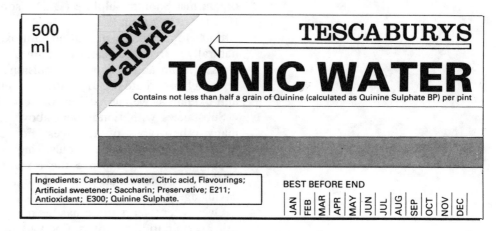

Figure 13.1 This low-calorie tonic water contains sodium benzoate (E211) and vitamin C (E300). The ham contains $NaNO_2$ (E250) and $NaNO_3$ (E251)

Food additives are substances added in small amounts to food. They fall into two main groups: (i) preservatives and anti-oxidants are added to prevent food deteriorating; (ii) the other group includes all additives used to enhance the texture, flavour or appearance of food.

The most common food additives are sugar and salt (NaCl). Salt has been used by man for thousands of years for flavouring and preserving food. Its importance in our life is reflected in phrases such as 'salt of the earth' and 'worth his salt'. Two thousand years ago Roman soldiers were paid in salt money. Indeed, the word 'salary' derives from this. Until the advent of refrigeration, bottling and canning, 'salting down' was one of the only methods of preserving vegetables, meat and fish.

Since foods are often imported from and exported to other countries, the European Economic Community (EEC) has introduced a list of additives that are generally recognised as safe. Each of the additives on the list has an E number. The aim of the list is to harmonise the labelling of food ingredients within the EEC countries. As part of the EEC consumer protection programme, the UK regulations on food labelling now require that the packaging of all foods shows a list of either E numbers or the actual names of the ingredients. The ingredients are usually listed in descending order of amount used.

Every so often new E numbers and thus additives are added to the EEC list and others are removed. Deletions may occur if the additive is found to present problems or if more suitable alternatives become available.

E numbers are divided into various categories. E100 to E180 are permitted colours. Many of these are organic substances and some are natural products (*see* chapter 20). Chlorophyll (E140) is extracted from nettles and grass and used to colour foods green. Beetroot red or betanin (E162) is a natural extract of beetroot. It gives a deep purple-red colour to food. One of the few inorganic compounds in this group is calcium carbonate (E170). This is used as a surface food colourant.

Additives E200 to E290 are preservatives. These prevent the growth of micro-organisms. Sodium sulphite (E221), sodium nitrite (E250) and sodium nitrate (E251) are all examples.

E300 to E321 are permitted anti-oxidants. L-Ascorbic acid (E300) occurs naturally in fresh fruits and vegetables. It is commonly known as vitamin C and is used as an anti-oxidant in emulsions of fats and oils and in iron mixtures. Sodium L-ascorbate (E301), which is prepared from vitamin C, is also used as an anti-oxidant as well as a colour preservative.

Substances with E numbers above 321 include emulsifiers, stabilisers and many other types of additives. For example, potassium dihydrogencitrate (E332) is used as an emulsifying salt, sodium dihydrogenphosphate(v) (E339) is used to improve the texture of foods, and magnesium carbonate (E504) is used as an alkali to reduce acidity in foods. It is also used as an anti-caking agent.

Over recent years there has been a growing debate about the desirability of using certain food additives. Food additives are often tested using high doses on animals such as rats and mice which only have short lifespans. Many consider that a high proportion of food additives have not been tested adequately under the conditions in which they are consumed by humans—that is, with small doses over a long period. It has been calculated that every person in the West consumes the equivalent of between 12 and 36 aspirins each day in the form of food additives. Some food additives such as E221, E250 and E251 (*see* above) are considered by some to be associated with hyperactivity in children and others such as potassium benzoate (E212) and calcium benzoate (E213) are known to be dangerous to asthmatics and aspirin-sensitive people. One of the most infamous food additives is sodium hydrogen L-glutamate (E621), commonly called monosodium glutamate. This compound is used in a variety of fast foods. It occurs naturally in a Japanese seaweed called Seatango and can be prepared commercially from sugarbeet pulp and wheat gluten. It is used as a flavour enhancer in high-protein foods. It is dangerous to asthmatics and can give rise to 'Chinese restaurant syndrome'. The symptoms of this syndrome include heart palpitations, dizziness, muscle tightening, nausea, headaches and weakness of the upper arms. Its use in baby foods is prohibited.

After you have studied this chapter you should be able to

1. Write down the **electronic configurations** of metals in Groups I and II.
2. Outline the main trends in the properties of these metals on descending Groups I and II.
3. Describe the general **physical properties** of metals in these two groups.
4. Describe and account for the **chemical reactivity** of the **s-block** metals.
5. Give typical examples of the reactions of s-block metals with
   (a) non-metals,
   (b) water and acids,
   (c) ammonia.
6. For the more important s-block compounds, briefly describe their
   (a) **structure** and **bonding**,
   (b) preparation,
   (c) chemical and physical properties.
7. Describe and account for the **patterns** in the **thermal stabilities** and **solubilities** of s-block oxo-compounds.
8. Give examples of the **anomalous properties** of lithium and beryllium.
9. Give examples to illustrate the **diagonal relationships** between
   (a) lithium and magnesium,
   (b) beryllium and aluminium.
10. Give examples of some common **minerals** in which the s-block metals occur.
11. Outline the most important commercial methods of **extracting**
    (a) sodium,
    (b) magnesium.
12. Briefly indicate how the following compounds are **manufactured**:
    (a) sodium hydroxide,
    (b) potassium hydroxide,
    (c) sodium carbonate.
13. Give examples of the more important **uses** of the s-block metals and their compounds.

13.1
Structure and Properties of the Metals

The word *alkali* derives from the Arabic word meaning 'the ashes of a plant'. Plants and their ashes contain mixtures of salts. Some of these salts, such as sodium carbonate and potassium carbonate, are alkaline.

WHAT ARE THE s-BLOCK ELEMENTS?

Elements in Groups I and II and helium in Group 0 of the Periodic Table (*see* figure 13.2) are known as the s-block elements. Apart from hydrogen and helium, they are all metals. Metals in Group I are called **alkali metals** since they react with water to form alkalis. With the exception of beryllium, metals in Group II are often loosely termed **alkaline earth metals**. The term *alkaline earth* refers to the oxides of these metals. These oxides react with water to form alkalis (*see* below).

Francium, at the bottom of Group I, and radium, at the bottom of Group II, are both radioactive. The isotopes of francium have very short half-lives and little is known about the element. It is formed during the radioactive decay of actinium

$$^{227}_{89}\text{Ac} \rightarrow {}^{223}_{87}\text{Fr} + {}^{4}_{2}\text{He}$$

What's in a Name?

Name	Symbol	Year discovered	Origin of name
Group I			
lithium	Li	1817	from *lithos*, meaning stone
sodium	Na	1807	from soda; the symbol derives from the element's Latin name, *natrium*
potassium	K	1807	from potash; the symbol derives from the element's Latin name, *kalium*
rubidium	Rb	1861	from *rubidus*, meaning red
caesium	Cs	1860	from *caesius*, meaning sky-blue
francium	Fr	1939	from France
Group II			
beryllium	Be	1798	from the mineral beryl
magnesium	Mg	1775	from Magnesium, an ancient city in Asia Minor
calcium	Ca	1808	from calx (calcium oxide)
strontium	Sr	1790	from Strontian, Scotland
barium	Ba	1808	from *barys*, meaning heavy or dense
radium	Ra	1898	from *radius*, meaning ray

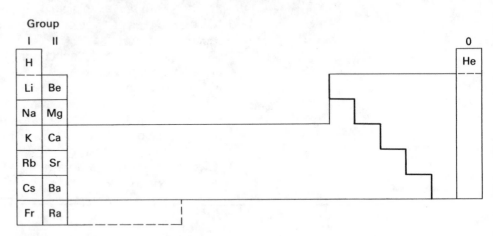

Figure 13.2 The s-block elements

STRUCTURE

The s-block metals all have one or two s electrons in their outer shells (*see* table 13.1). They tend to lose these electrons readily to form stable ions with the same electronic configurations as the noble gases (*see* section 2.1).

Since s-block metals are highly electropositive and thus tend to lose electrons readily, they do not occur naturally as metals. Their ions do occur widely, however. Their highly electropositive nature is demonstrated by relatively low first ionisation energies and low electronegativity values. The values in table 13.2 should be compared with those in table 1.3 and table 2.2. Metals in Group I form ions much more easily than do those in Group II. This is shown by comparing the first ionisation energies of metals in Group I with the sum of the first and second ionisation energies of metals in Group II. For example

$$Na \rightarrow Na^+ + e^- \qquad \Delta H_1^\ominus \qquad = 500 \text{ kJ mol}^{-1}$$

$$Mg \rightarrow Mg^{2+} + 2e^- \qquad \Delta H_1^\ominus + \Delta H_2^\ominus = 2190 \text{ kJ mol}^{-1}$$

In both groups the ionisation energies decrease down the group as the size of

Table 13.1 Electronic configurations of the s-block metals

Element	Atomic number	Electronic configuration	Configuration in outer shell
Group I			
lithium	3	2.1	$2s^1$
sodium	11	2.8.1	$3s^1$
potassium	19	2.8.8.1	$4s^1$
rubidium	37	2.8.18.8.1	$5s^1$
caesium	55	2.8.18.18.8.1	$6s^1$
francium	87	2.8.18.32.18.8.1	$7s^1$
Group II			
beryllium	4	2.2	$2s^2$
magnesium	12	2.8.2	$3s^2$
calcium	20	2.8.8.2	$4s^2$
strontium	38	2.8.18.8.2	$5s^2$
barium	56	2.8.18.18.8.2	$6s^2$
radium	88	2.8.18.32.18.8.2	$7s^2$

Table 13.2 Some properties of the s-block metals

	Metallic radius/nm	Ionic radius/nm	Atomic volume/cm^3 mol^{-1}	Pauling electro-negativity value	First ionisation energy/kJ mol^{-1}	Second ionisation energy/kJ mol^{-1}
Group I						
Li	0.155	0.060	13.1	1.0	520	7300
Na	0.190	0.095	23.7	0.9	495	4600
K	0.235	0.133	45.5	0.8	420	3100
Rb	0.248	0.148	55.8	0.8	400	2700
Cs	0.267	0.169	71	0.7	380	2400
Group II						
Be	0.112	0.031	4.9	1.5	900	1800
Mg	0.160	0.065	14.0	1.2	740	1450
Ca	0.197	0.099	26	1.0	590	1150
Sr	0.215	0.113	33.7	1.0	550	1060
Ba	0.222	0.135	39.3	0.9	500	970

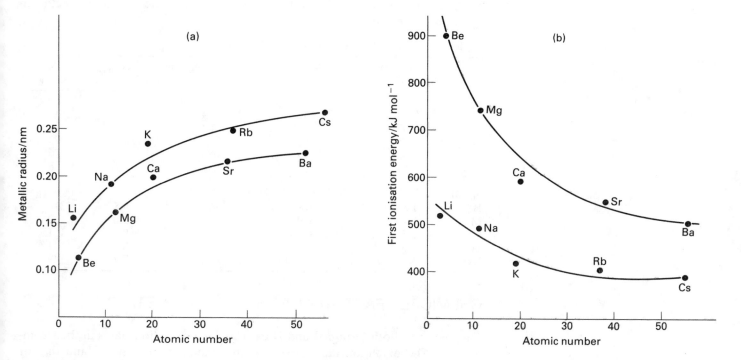

Figure 13.3 Trends in properties of s-block metals: (a) metallic radius; (b) first ionisation energy

Table 13.3 Crystal structures of Group II metals

Element	Crystal structure
beryllium	hexagonal close-packed
magnesium	hexagonal close-packed
calcium	hexagonal close-packed and face-centred cubic
strontium	face-centred cubic
barium	body-centred cubic

the atoms increases and the outer s electrons become increasingly shielded from the nucleus (*see* table 13.2 and figure 13.3a).

The s-block metals all exist in the solid state at room temperature and pressure. The solids have metal lattice structures. All Group I metals have body-centred cubic structures (*see* figure 3.20 and table 3.7). Group II metals have the crystal structures shown in table 13.3. None of the s-block metals forms allotropes.

PHYSICAL PROPERTIES

The metals in Group I are soft and have low densities compared with other metals. Indeed, lithium, sodium and potassium are unique in that they float on water—as they react! The hardness of the Group II metals is greater than that of Group I metals. Their densities are also higher, although low compared with those of the d-block metals.

Values for some of the physical properties of these elements are shown in table 13.4. The low values for the melting and boiling points and the enthalpies of fusion and vaporisation are due to the relatively weak metallic bonds in these metals. The metallic bonds are formed by the delocalised outer-shell s electrons, which form a 'glue' to hold the positive metal ions together. In a particular group, the larger the metallic radius the more thinly the delocalised electrons are spread around and thus the weaker the bond. Generally speaking

$$\text{Strength of metallic bond} \propto \frac{\text{Number of delocalised electrons per atom}}{\text{Metallic radius}}$$

This ratio is low for metals in Group I and even lower in Group II. This accounts for the low values of the melting and boiling points. On descending both groups the metallic radius becomes larger (*see* figure 13.3b) whilst the number of delocalised electrons remains the same in each group. The metallic bonds thus become weaker. This accounts for the decrease in hardness, melting and boiling points in Group I and the decrease in enthalpies of fusion and vaporisation down both groups. The values for melting and boiling points in Group II are irregular. This is largely due to the different crystal structures of the metals in this group (*see* above).

Table 13.4 Physical properties of the s-block metals

	Density/g cm^{-3}	Melting point/°C	Boiling point/°C	$\Delta H_{\text{fus,m}}^{\ominus}$/kJ mol^{-1}	$\Delta H_{\text{vap,m}}^{\ominus}$/kJ mol^{-1}
Group I					
Li	0.53	181	134	2.9	135
Na	0.97	98	883	2.6	98
K	0.86	64	774	2.3	79
Rb	1.53	39	688	2.3	76
Cs	1.87	28	678	2.1	68
Group II					
Be	1.85	1278	2970	9.8	310
Mg	1.74	649	1090	9.0	132
Ca	1.54	839	1484	8.7	161
Sr	2.6	769	1384	8.7	141
Ba	3.5	725	1640	7.7	149

CHEMICAL REACTIVITY OF THE s-BLOCK METALS

The metals in both Groups I and II exist in fixed oxidation states in their compounds. The oxidation number of Group I metals is usually +1 and that of Group II metals usually +2.

The metals in Group I are readily oxidised to monovalent ions, Na$^+$ for example. High second ionisation energies (*see* table 13.2) prohibit the formation of other ions such as Na^{2+}. Group II metals are readily oxidised to divalent ions, Mg^{2+} for example. The magnitudes of the third ionisation energies of Group II metals are comparable to the second ionisation energies of the Group I metals. Trivalent ions such as Mg^{3+} are thus not formed.

The ease of oxidation of Group I and II metals is reflected by the relatively high redox potentials of these metals (*see* tables 10.5 and 13.5). The negative values show that the metals are electropositive and tend to be oxidised. Thus, the reverse reactions in the half-reactions shown in table 13.5 tend to take place. Since they are readily oxidised and thus readily produce electrons, all the metals in Groups I and II are strong reducing agents. However, the metals cannot be used as reducing agents in aqueous solutions since they reduce the water instead in vigorous and violent reactions (*see* below). In non-aqueous conditions they reduce non-metals and some compounds such as ammonia.

The high values of the Group I and II redox potentials also suggest that the metals should be highly reactive. They are high in the electrochemical series (*see* chapter 10). In some types of reaction the reactivity tends to increase down the group, although this is not always the case. The reactivities of Group I metals with nitrogen and carbon (charcoal) are examples. In these cases metals at the top of the groups are generally more reactive.

Table 13.5   Standard redox potentials of the s-block metals

	Half-reaction	$E^{\ominus}/V$
Group I	$Li^+(aq) + e^- \rightleftharpoons Li(s)$	-3.04
	$Na^+(aq) + e^- \rightleftharpoons Na(s)$	-2.71
	$K^+(aq) + e^- \rightleftharpoons K(s)$	-2.93
	$Rb^+(aq) + e^- \rightleftharpoons Rb(s)$	-2.99
	$Cs^+(aq) + e^- \rightleftharpoons Cs(s)$	-3.02
Group II	$Be^{2+}(aq) + 2e^- \rightleftharpoons Be(s)$	-1.85
	$Mg^{2+}(aq) + 2e^- \rightleftharpoons Mg(s)$	-2.37
	$Ca^{2+}(aq) + 2e^- \rightleftharpoons Ca(s)$	-2.87
	$Sr^{2+}(aq) + 2e^- \rightleftharpoons Sr(s)$	-2.89
	$Ba^{2+}(aq) + 2e^- \rightleftharpoons Ba(s)$	-2.90

Reactions with Non-Metals

With Oxygen

The s-block metals, when freshly cut, have a bright lustre. However, they readily tarnish when they come into contact with the oxygen of the atmosphere. Thus, with the exception of beryllium and magnesium, they are stored under liquid paraffin to prevent contact with the atmosphere. Beryllium and magnesium form a protective oxide layer and thus tarnish comparatively slowly.

All the s-block metals burn in air to form one or more of three types of oxide (*see* table 13.6). For example, lithium burns in air to form lithium oxide, Li_2O, whilst sodium produces a mixture of sodium oxide, Na_2O, and sodium peroxide, Na_2O_2.

Table 13.6   Oxides of the s-block metals

	Normal oxide	Peroxide	Superoxide
Formula	O^{2-}	O_2^{2-}	O_2^-
Formed by burning in air	Li and Group II elements	Na and Ba	K, Rb, Cs
Example	$2Mg(s) + O_2(g) \rightarrow 2MgO(s)$	$2Na(s) + O_2(g) \rightarrow Na_2O_2(s)$	$K(s) + O_2(g) \rightarrow KO_2(g)$
Formed by other methods (*see* below)	all Group I and II metals	all Group I and II metals except Be	all Group I and II metals except Be, Mg, Li

With Hydrogen

All Group I and II metals except beryllium combine directly with hydrogen at temperatures between 300 and 700°C to form hydrides (*see* section 12.1). For example

$$Ca(s) + H_2(g) \rightarrow CaH_2(s)$$

With Halogens

All the s-block metals reduce halogens on heating (*see* chapter 16). For example

$$Na(s) \rightarrow Na^+(s) + e^-$$
$$\tfrac{1}{2}Cl_2(g) + e^- \rightarrow Cl^-(s)$$

The reactivity increases down the group.

With Nitrogen

The Group II metals and lithium in Group I all burn in nitrogen to form nitrides. For example

$$6Li(s) \ + N_2(g) \rightarrow 2Li_3N(s)$$
$$3Mg(s) + N_2(g) \rightarrow Mg_3N_2(s)$$

With Sulphur

All Group I and II metals react with sulphur to form sulphides. The reactions between the Group I metals and sulphur are vigorous. For example

$$2Na(s) + S(s) \rightarrow Na_2S(s)$$
$$Mg(s) + S(s) \rightarrow MgS(s)$$

With Carbon

Lithium and sodium and the Group II metals combine directly with charcoal to form ionic dicarbides (also known as acetylides):

$$2Na(s) + 2C(s) \rightarrow Na_2C_2(s)$$

$$Mg(s) + 2C(s) \rightarrow MgC_2(s)$$

The dicarbide ion has the structure $[C{\equiv}C]^{2-}$.

Reaction with Water and Acids

All Group I metals reduce cold water to hydroxides and hydrogen. For example

$$Na(s) \rightarrow Na(aq) + e^-$$
$$H_2O(l) + e^- \rightarrow OH^-(aq) \ + \tfrac{1}{2}H_2(g)$$

The reactivity increases down the group. For example, lithium reacts relatively slowly whereas potassium reacts explosively with water, igniting spontaneously and burning with a purple flame.

    Group II metals also reduce water to produce hydroxides and hydrogen

$$Ca(s) + 2H_2O(l) \rightarrow Ca(OH)_2(aq) + H_2(g)$$

However, the reactivity of Group II metals is less than that of the corresponding Group I metals. Like the Group I metals the reactivity increases down the group. Beryllium reacts hardly at all with either cold water or steam. Magnesium reacts only very slowly with cold water but vigorously with steam

$$Mg(s) + H_2O(g) \rightarrow MgO(s) + H_2(g)$$

    The reactivity of Group II metals with acids also decreases as the group is descended

$$Mg(s) + 2H^+(aq) \rightarrow Mg^{2+}(aq) + H_2(g)$$

Beryllium reacts only slowly with acids.

    All the alkali metals react explosively with acids.

Crystalline Alkalides

With the exception of lithium, solutions of the alkali metals in amines and ethers contain three species

    the alkali metal cation M^+
    the alkali metal anion M^-
    the **solvated electron** e^-_{solv}

The two ions are formed by disproportionation (*see* section 10.2)

$$2M(s) \overset{RNH_2}{\rightleftharpoons} M^+ + M^-$$

This shows that an alkali metal atom can either lose its single outer s electron to form the cation with a noble gas configuration or add an electron to form the anion with a filled outer s orbital.

    The cation in the solution can be stabilised by complexing it with a **crown ether**

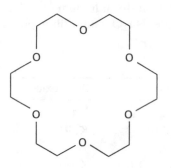

The salts formed in this way can be crystallised. They have the general formula $M^+L{\cdot}M^-$(s) and are known as **alkalides**.

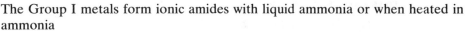

Reaction with Ammonia

The Group I metals form ionic amides with liquid ammonia or when heated in ammonia

$$Na(s) + NH_3(l) \rightarrow Na^+NH_2^-(s) + \tfrac{1}{2}H_2(g)$$

Calcium, strontium and barium all form metal amides and hydrogen with liquid ammonia. When heated in ammonia they form either the nitride or the hydride

$$3Mg(s) + 2NH_3(g) \rightarrow Mg_3N_2(s) + 3H_2(g)$$
$$3Ca(s) + 2NH_3(g) \rightarrow 3CaH_2(s) + N_2(g)$$

Flame Tests

The Group I metals and some of the Group II metals give distinctive flame tests. To perform the flame test, a platinum wire is moistened with concentrated hydrochloric acid, dipped into the substance under test, and then held in a non-luminous flame. The colours shown in table 13.7 are obtained.

Table 13.7 Flame colours of s-block metals

Metal	Flame colour	Metal	Flame colour
Group I		*Group II*	
lithium	scarlet	calcium	brick red
sodium	golden yellow	strontium	crimson
potassium	lilac	barium	apple green
rubidium	red		
caesium	blue		

Flame test

SUMMARY

1. The **s-block** metals form **stable ions** that have the same **electronic configurations** as the noble gases.
2. The s-block metals are highly **electropositive**.
3. The s-block metals have low **first ionisation energies** and low **electronegativity values**.
4. Ionisation energies decrease down Groups I and II.
5. The s-block metals are solids with **metal lattice structures**.
6. The low values of the **physical properties** of s-block metals are due to relatively weak **metallic bonds**.
7. Group I and II metals have relatively high **redox potentials** and thus
   (a) are easily oxidised,
   (b) act as strong **reducing agents**.
8. The s-block metals are highly reactive.
9. The s-block metals react with **non-metals** including
   (a) oxygen,
   (b) the halogens,
   (c) sulphur.
10. Group I metals reduce cold **water** to hydroxides and hydrogen.
11. Group II metals are less reactive with water than are the Group I metals.
12. Reactivity with water increases down both groups.
13. Many s-block metals give distinctive **flame tests**.

13.2
Compounds Formed by the s-Block Metals

The principal types of compound formed by the s-block metals are shown in table 13.8. In this section we shall review the structure and bonding, preparation and the chemical and physical properties of some of the more important of these compounds. We shall, in particular, look at the thermal stability and solubility of the oxo-compounds formed by the s-block metals.

Table 13.8 Principal types of compound formed by s-block metals

Type	Examples	
	Group I	Group II
binary compounds		
oxide	Li_2O, Na_2O_2, KO_2	MgO, BaO
hydride	NaH	CaH_2
halide	NaCl	$MgCl_2$
nitride	Li_3N	Mg_3N_2
sulphide	Na_2S	MgS
carbide	Na_2C_2	MgC_2
hydroxides	NaOH	$Ca(OH)_2$
salts		
carbonate	Na_2CO_3	$CaCO_3$
hydrogencarbonate	$NaHCO_3$	$Ca(HCO_3)_2$
nitrate	$NaNO_3$	$Mg(NO_3)_2$
sulphate	Na_2SO_4	$MgSO_4$
complexes	$K^+[Na(NH_2)_2]^-$	chlorophyll
organic compounds	C_2H_5Li	C_6H_5MgBr

THE OXIDES

We saw above that the s-block metals form three types of oxide and that these are produced by burning the elements in oxygen or air. The products of burning Group I metals in air tend to be mixtures of these oxides. The oxides can be prepared separately by temperature-controlled oxidation in air. For example

$$4Na(s) + O_2(g) \xrightarrow{180°C} 2Na_2O(s)$$
sodium monoxide

$$2Na(s) + O_2(g) \xrightarrow{300°C} Na_2O_2(s)$$
sodium peroxide

$$\underset{\text{(in excess)}}{Na(s)} + O_2(g) \xrightarrow{300°C} NaO_2(s)$$
sodium superoxide

The normal oxides can also be prepared by indirect methods; for example, by reducing the nitrate by its metal

$$10K(s) + 2KNO_3(s) \rightarrow 6K_2O(s) + N_2(g)$$

The normal oxides of Group II metals can be prepared by the thermal decomposition of the carbonate or nitrate. For example

$$MgCO_3(s) \rightarrow MgO(s) + CO_2(g)$$

The normal oxides are white, the peroxides are white or slightly coloured and the superoxides are highly coloured.

All three types of oxide are basic, the basic character increasing on descending the group. With the exception of magnesium oxide, the oxides of the s-block metals readily react with water to form strongly alkaline solutions. For example

$$Na_2O(s) + H_2O(l) \rightarrow 2NaOH(aq)$$

Magnesium oxide only reacts slowly with water. In general, the Group II oxides are less basic than the Group I oxides.

Both the peroxides and the superoxides react with acids at 0°C to form hydrogen peroxide

$$Na_2O_2(s) + 2HCl(aq) \rightarrow 2NaCl(aq) + H_2O_2(aq)$$
$$2KO_2(s) + 2HCl(aq) \rightarrow 2KCl(aq) + H_2O_2(aq) + O_2(g)$$

The peroxides and superoxides are powerful oxidising agents. For example, chromium(III) compounds can be oxidised to chromium(VI) by fusing the compound with sodium peroxide.

$$2Cr(OH)_3(s) + 3Na_2O_2(s) \rightarrow 2Na_2CrO_4(aq) + 2NaOH(aq) + 2H_2O(l)$$

OTHER BINARY COMPOUNDS

Hydrides

We have already discussed the chemistry of the hydrides in section 12.1. The hydrides of the s-block metals can be prepared by direct combination with hydrogen. An exception is beryllium hydride, BeH_2, which is prepared by reducing beryllium compounds such as dimethylberyllium, $(CH_3)_2Be$, with lithium tetrahydridoaluminate(III).

All the hydrides of the s-block metals are ionic, with the exception of BeH_2 and MgH_2, which are both covalent.

The s-block hydrides are hydrolysed by water, producing hydroxides and hydrogen. For example

$$NaH(s) + H_2O(l) \rightarrow NaOH(aq) + H_2(g)$$
$$CaH_2(s) + 2H_2O(l) \rightarrow Ca(OH)_2(aq) + 2H_2(g)$$

The general ionic equation for these reactions is

$$H^-(s) + H_2O(l) \rightarrow H_2(g) + OH^-(aq)$$

Since the hydride ion is readily oxidised to hydrogen, it is a powerful reducing agent.

Sodium and lithium also form the important hydrides—$NaBH_4$ and $LiAlH_4$ (see section 12.1).

Halides

The Group I halides are all essentially ionic in character, although in each case the degree of ionic character is less than 100% (see section 2.1). The physical properties of the Group I halides are those associated with ionic compounds. They have high melting and boiling points. These decrease on descending the group. They dissolve in water but are relatively insoluble in organic solvents. When molten and when in aqueous solution, they are electrolytes.

The degree of ionic character of the Group II halides increases on descending the group. Beryllium halides are covalent and the magnesium halides intermediate in character. The halides of metals lower in the group are essentially ionic.

The structure, preparation and properties of the halides in general are described in chapter 16.

Nitrides and Dicarbides

The nitrides and dicarbides of the s-block metals are formed by direct combination of the metal with nitrogen and carbon respectively. The most notable feature of their chemistry is their reaction with water. Lithium nitride, for example, is hydrolysed by water, forming the hydroxide and ammonia

$$Li_3N(s) + 3H_2O(l) \rightarrow 3LiOH(aq) + NH_3(g)$$

The dicarbides are hydrolysed to form ethyne. For example

$$Na_2C_2(s) + 2H_2O(l) \rightarrow 2NaOH(aq) + C_2H_2(g)$$
sodium dicarbide   ethyne

HYDROXIDES

The Group I hydroxides are white crystalline solids with the sodium chloride structure (*see* section 3.2 and figure 3.25).

With the exception of lithium hydroxide, they are deliquescent and dissolve in water readily with the evolution of heat.

Group II hydroxides are less soluble in water. The solubility of both Group I and II hydroxides increases on descending the group.

The soluble hydroxides are prepared in the laboratory by dissolving the metals in cold water. The less-soluble hydroxides of Group II may be prepared by precipitation from solutions of their salts by an alkali metal hydroxide. For example

$$MgCl_2(aq) + 2NaOH(aq) \rightarrow Mg(OH)_2(s) + 2NaCl(aq)$$

The Group I hydroxides are strong bases. When dissolved in water they form alkalis. The Group II hydroxides are less basic than the corresponding Group I hydroxides. In both groups the base strength increases on descending the group.

The hydroxides, particularly those of sodium and potassium, are important in both qualitative and quantitative chemical analysis. They also have a number of other important applications in both inorganic and organic chemistry.

Most Important Reactions

With Acids

Hydroxides neutralise acids to form salts and water. The general equation is

$$H_3O^+(aq) + OH^-(aq) \rightarrow 2H_2O(l)$$

They also react with acidic gases to form salts and water

$$2OH^-(aq) + CO_2(g) \rightarrow CO_3^{2-}(aq) + H_2O(l)$$
$$2OH^-(aq) + SO_2(g) \rightarrow SO_3^{2-}(aq) + H_2O(l)$$

With Metal Ions

Insoluble metal hydroxides can be prepared by adding hydroxide solution to an aqueous solution of a salt. For example

$$Cu^{2+}(aq) + 2OH^-(aq) \rightarrow Cu(OH)_2(s)$$
$$Fe^{3+}(aq) + 3OH^-(aq) \rightarrow Fe(OH)_3(s)$$

These reactions are particularly important in qualitative inorganic analysis.

The hydroxides of some metals are unstable and precipitate their oxides in

---

Deliquescence

Deliquescence is the absorption of water from the atmosphere to form a solution. Sodium and potassium hydroxides and calcium chloride are all deliquescent. They are used as drying agents in desiccators. The term *deliquescence* is sometimes confused with the term *hygroscopic*.

A **hygroscopic** substance can absorb water from the air without changing its physical state. Anhydrous cobalt chloride is an example of such a substance. Water changes solid anhydrous cobalt chloride from blue to pink (*see* chapter 14).

Efflorescence is the loss of water of crystallisation to the atmosphere. For example, crystals of sodium carbonate-10-water become powdery when exposed to air

$$Na_2CO_3 \cdot 10H_2O(s)$$
$$\rightarrow Na_2CO_3 \cdot H_2O(s) + 9H_2O(g)$$

alkaline solutions. For example

$$2Ag^+(aq) + 2OH^-(aq) \rightarrow Ag_2O(s) + H_2O(l)$$

Aqueous solutions of salts of metals such as aluminium, tin and lead produce amphoteric hydroxides (*see* chapter 15) on reaction with hydroxide solution. These amphoteric hydroxides are first precipitated

$$Al^{3+}(aq) + 3OH^-(aq) \rightarrow Al(OH)_3(s)$$

and then redissolve in excess hydroxide solution

$$Al(OH)_3(s) + OH^-(aq) \rightarrow Al(OH)_4^-(aq)$$
tetraaquaaluminium(III)ion

With Salts of Weak Bases

A hydroxide will displace a weak base from its salt

$$NH_4Cl(aq) + OH^-(aq) \rightarrow NaCl(aq) + H_2O(l) + NH_3(g)$$

This type of reaction is used in the laboratory preparation of ammonia. Solid reactants are used

$$2NH_4Cl(s) + Ca(OH)_2(s) \overset{heat}{\rightarrow} CaCl_2(s) + 2H_2O(g) + 2NH_3(g)$$

With Halogens

Chlorine, bromine and iodine all disproportionate in hydroxide solutions (*see* section 16.1).

Hydrolysis of Organic Compounds

Hydroxide solutions hydrolyse organic compounds such as esters (*see* chapter 19). For example

$$CH_3COOC_2H_5(aq) + OH^-(aq) \rightarrow CH_3COO^-(aq) + C_2H_5OH(aq)$$

SALTS

The salts of the Group I metals are the most ionic salts known although, as we have seen, the degree of ionic character is invariably less than 100%. The degree of covalent character depends on the ability of the cation to polarise the anion (*see* figure 2.11). We saw in chapter 2 that Fajans' rules predicted that the degree of covalency is high if the ionic charges are high, the cation small and the anion large. Since the Group I cations have a constant charge of 1+ and Group II cations a constant charge of 2+ the size of the cation is critical in determining the covalent character of a particular type of salt. Cations are smallest at the top of each group. They thus have the highest **polarising power** in the group and their compounds thus show the highest degree of covalent character. On descending each group, the cationic size increases. The degree of ionic character thus increases and the degree of covalent character decreases. Lithium and beryllium compounds such as LiI and BeCl$_2$ thus show more covalent character than the corresponding salts of metals lower in the group.

The same arguments concerning ion size and ionic and covalent character can also be applied to the hydroxides and oxides of s-block metals.

The salts of s-block metals have a number of notable properties. For example, sodium and potassium salts are all colourless—unless they contain a coloured anion such as manganate(VII), MnO$_4^-$. One of the most interesting features of their chemistry is the variation in thermal stability and solubility of the various groups of salts. These two features are discussed separately below.

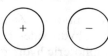

Polarisation

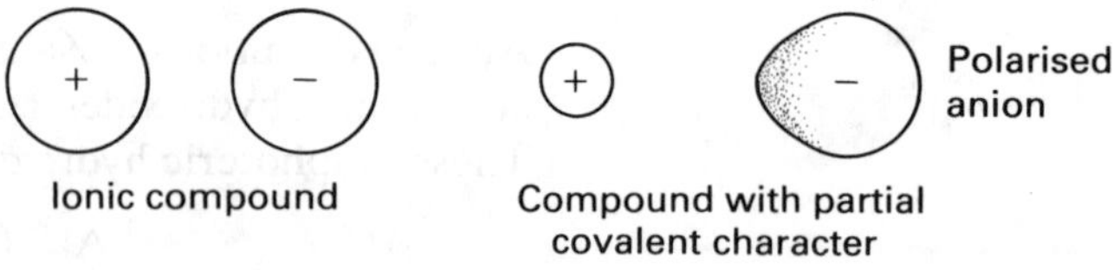

Polarised anion

Ionic compound

Compound with partial covalent character

Figure 13.4

Polarisation is the distortion of an anion by a positive charge (*see* figure 13.4). The ability of a cation to polarise an anion is called its polarising power. According to Fajans' rules, polarisation and thus degree of covalent character (*see* section 2.1) increases if

the cation is small,
the anion is large,
the charges on the ions are high.

An approximate measure of the polarising power of a cation is its **charge density**. This is given approximately by the ratio

charge : ionic radius

The charge density of Group I cations and thus their polarising power decreases on descending the group.

Ion	Charge	Ionic radius/nm	Charge density (= ratio)	
Li$^+$	+1	0.060	16.7	
Na$^+$	+1	0.095	10.5	
K$^+$	+1	0.133	7.5	polarising power decreases
Rb$^+$	+1	0.148	6.8	
Cs$^+$	+1	0.169	5.9	

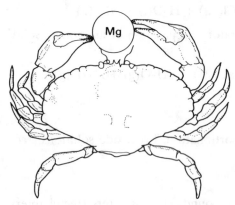

Figure 13.5   The word 'chelate' comes from the Greek word for 'crab's claw'

edta

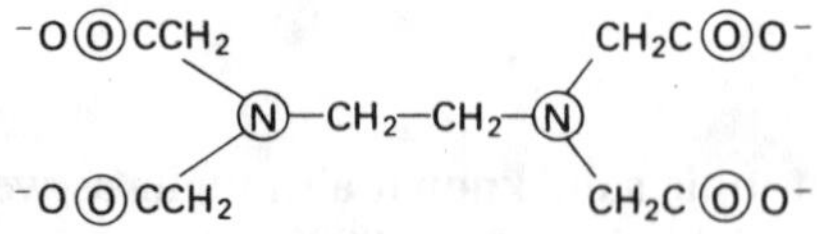

$^-$O O CCH$_2$ CH$_2$C O O$^-$
          N—CH$_2$—CH$_2$—N
$^-$O O CCH$_2$ CH$_2$C O O$^-$

The edta hexadentate ligand

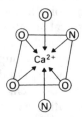

Part of the [Ca(edta)]$^{2-}$ complex ion

edta is the ligand formed from its parent acid H$_4$edta by loss of its protons. H$_4$edta is 1,2-bis[bis-(carboxymethyl)amino]ethane. The traditional name for this compound is ethylenediaminetetraacetic acid. Hence the acronym edta.

COMPLEXES

Since the Group I metal ions are large and do not have d electrons, they do not form complexes readily. They form fewer complex compounds than any other group of metal ions. The ability to form complexes decreases down the group. The complexes that are formed are mostly with **chelating** oxygen **ligands** (*see* chapter 14, and figure 13.5).

The Group II metal ions form more complexes than the alkali metal ions. They generally form them with strong complexing agents with oxygen and nitrogen donors. The chelating ligand edta is an example. This forms complexes with Mg^{2+} and Ca^{2+}. Titrations are performed using the disodium salt of edta in buffered solutions to estimate the concentration of Ca^{2+} and Mg^{2+} in water and thus the hardness of water.

ORGANIC COMPOUNDS

The Group I metals replace hydrogen in organic acids, forming salts such as sodium ethanoate, CH$_3$COONa. The sodium salts of long-chain fatty acids such as stearic acid are important in their use as soap. The alkali metals also form a number of alkyl and aryl compounds such as butyllithium, C$_4$H$_9$Li, which have important applications in organic preparative chemistry.

Magnesium forms an important group of organometallic compounds known as the Grignard reagents (*see* section 19.1). An example of a Grignard reagent is CH_3MgBr. It is formed by refluxing magnesium with bromomethane, CH_3Br, in dry ethoxyethane.

THERMAL STABILITY OF THE OXO-COMPOUNDS OF THE s-BLOCK METALS

Thermal stability of metallic compounds depends on the polarising power of the metal ions. Cations with low polarising power tend to form compounds that are relatively stable to heat. As polarising power increases, compounds become more covalent in character and thus less stable to heat. Since the metal ions in Group I have low charges but relatively large sizes, they have low polarising power. They are thus relatively stable to heat. Salts of the Group II metals are less stable (*see* table 13.9).

Table 13.9 Melting or decomposition temperatures of s-block compounds

	Melting point/°C				
	Sulphate	Carbonate	Nitrate	Hydroxide	Oxide
Group I					
Li	845	d723	264	450	>1700(Li_2O)
Na	884	851	307	318	s1275 (Na_2O)
K	1069	891	334	360	490 (K_2O_2)
Rb	1060	837	310	301	570 (Rb_2O_2)
Cs	1010	d610	414	272	400 (Cr_2O_2)
Group II					
Be	d550	d100	d60	d138	2530 (BeO)
Mg	d1124	d350	129	d350	2800 (MgO)
Ca	1450	d900	560	d580	2580 (CaO)
Sr	d1605	1290*	570	375	2430 (SrO)
Ba	1350	d1350	592	408	1923 (BaO)

s   sublimes
d   decomposes
*   melts under pressure

The oxides tend to be more stable than the carbonates or nitrates of a particular metal. This is because the oxide ion, O^{2-}, is relatively small compared with the carbonate or nitrate ion. Its polarisability is thus less than the other two types of ion. For this reason, some of the carbonates and nitrates in Groups I and II decompose on heating to the more stable oxide.

Carbonates and Hydrogencarbonates

The carbonates of sodium, potassium, rubidium and strontium are all stable to heat. The remaining carbonates of the s-block metals decompose on heating, forming an oxide and carbon dioxide

$$Li_2CO_3(s) \rightarrow Li_2O(s) + CO_2(g)$$
$$MgCO_3(s) \rightarrow MgO(s) + CO_2(g)$$

Note from table 13.9 that on descending Group II the carbonates are more stable to heat. This is because the cations become larger and thus their polarising power decreases.

Lithium and the Group II metals do not form solid hydrogencarbonates, although they exist in solution. On heating these solutions, the hydrogencarbonates decompose to form carbonates. The solid hydrogencarbonates all decompose at temperatures between 100 and 300°C, forming carbonates

$$2NaHCO_3(s) \rightarrow Na_2CO_3(s) + H_2O(g) + CO_2(g)$$

Nitrates

With the exception of lithium nitrate, all the Group I metal nitrates decompose on strong heating, forming nitrites and oxygen. For example

$$2NaNO_3(s) \rightarrow 2NaNO_2(s) + O_2(g)$$

Lithium nitrate and the Group II metal nitrates decompose on heating to form nitrogen dioxide, oxygen and the thermally stable oxides. For example

$$4LiNO_3(s) \rightarrow 2Li_2O(s) + 4NO_2(g) + O_2(g)$$
$$2Ca(NO_3)_2(s) \rightarrow 2CaO(s) + 4NO_2(g) + O_2(g)$$

Hydroxides and Oxides

The thermal stability of the Group I and II hydroxides follows a similar pattern to that of the carbonates and nitrates. Lithium hydroxide and the Group II hydroxides all decompose on heating to form stable oxides and water

$$2LiOH(s) \rightarrow Li_2O(s) + H_2O(l)$$
$$Mg(OH)_2(s) \rightarrow MgO(s) + H_2O(g)$$

Apart from lithium hydroxide, the hydroxides of the Group I metals are thermally stable.

The normal oxides of lithium and sodium and the Group II metals are particularly stable and only melt at high temperatures (*see* table 13.9). This is due to the relatively small size of the oxide ion, O^{2-}. The peroxides have lower melting points.

Sulphates and Halides

The sulphates and halides of the Group I and II metals are all thermally stable.

SOLUBILITIES OF GROUP I AND II SALTS AND HYDROXIDES IN WATER

There are a number of general patterns in the solubilities of the salts and hydroxides of the Group I and II metals in water. These are:

1. All Group I metal salts are soluble.
   Exception: lithium fluoride.
2. The solubilities of Group II salts that have an anion of charge 1− tend to be soluble. Examples include the chlorides, bromides, iodides and nitrates.
   Exceptions: the fluorides and hydroxides.
3. The solubilities of Group II salts that have an anion of charge 2− tend to be insoluble. These include the sulphates, carbonates, chromates (CrO_4^{2-}) and the ethanedioates ($C_2O_4^{2-}$) (also known as oxalates).
   Exceptions: some magnesium and calcium salts.
4. Solubilities of the Group I nitrates and most Group II salts tend to increase on descending the group.
   Exceptions: the fluorides in both groups and also the carbonates in Group I and the hydroxides in Group II. The pattern of the Group I chloride and sulphate solubilities is irregular.

Accounting for These Patterns

This is notoriously difficult since a number of factors are involved. For a compound to dissolve, the free energy change $\Delta G_{\text{solution}}$ must be negative. Applying equation (23) from section 5.3 we obtain

$$\Delta G_{\text{solution}} = \Delta H_{\text{solution}} - T\Delta S_{\text{solution}}$$

On dissolving, disorder increases and thus ΔS is positive. Thus if $\Delta G_{solution}$ is to be negative, the term $\Delta H_{solution}$ must either be negative or be positive but smaller than the term $T\Delta S_{solution}$.

The enthalpy of solution depends on two factors:

- **The hydration energy**: this is the energy evolved on dissolving in water.
- **The lattice energy**: this is the energy evolved on forming the lattice.

Before the compound can dissolve, its lattice has to be broken down. An amount of energy equal to the lattice energy must be absorbed to do this. This energy is the energy supplied by the hydration process. If it is insufficient to break down the lattice, the salt remains largely insoluble. However, if the hydration energy exceeds the lattice energy, the salt dissolves. Thus

$$\Delta H_{solution} = \Delta H_{hydration} - \Delta H_{lattice}$$

If

$$\Delta H_{hydration} > \Delta H_{lattice}$$

then $\Delta H_{solution}$ is negative and the salt dissolves.

However if

$$\Delta H_{hydration} < \Delta H_{lattice}$$

then $\Delta H_{solution}$ is positive and the salt is relatively insoluble.

In general we can say that the more negative $\Delta H_{solution}$ the more soluble the salt.

To account for the trends in the solubilities in a specific group of salts, we now need to consider two other factors: ionic size and ionic charge.

Ionic Size

As ionic size increases, the lattice and hydration energies both decrease. However, the lattice energy is influenced more by ionic size than by hydration energy. Lattice energy is, in fact, dependent on a number of factors. These include the reciprocal of interionic distance, $(r_+ + r_-)^{-1}$, where r_+ is the cationic radius and r_- the anionic radius. So

$$\text{Lattice energy} \propto \frac{1}{r_+ + r_-}$$

We should note that in any particular group of salts—such as the sulphates—the radius of the anion, r_-, remains constant. In this context there are two cases:

CASE 1

The anion is large compared to the cation. In this case r_+ can be regarded as insignificant and ignored. Thus,

$$\text{Lattice energy} \propto 1/r_-$$

Since r_- is constant, the lattice energy is also constant within a particular group of salts. The cationic radius does not influence it.

CASE 2

The sizes of the anion and cation are both of the same order. In this case, since r_- remains constant in a particular group of salts, the cationic radius becomes an important factor in determining the lattice energy.

Ionic Charge

Ions with a large size but a low charge have a low charge density. Smaller ions have higher charge densities. Smaller ions are thus more strongly attracted to ions of opposite charge in the lattice. The lattice energy is thus higher. However,

---

Example

Predict whether lithium iodide, LiI, is soluble in water or not, given that

hydration energy = $-824\ kJ\ mol^{-1}$
lattice energy   = $-763\ kJ\ mol^{-1}$

Solution

From these values we can see that

$\Delta H_{solution,m}$
$= (-824\ kJmol^{-1}) - (-763\ kJmol^{-1})$
$= -61\ kJ\ mol^{-1}$

Since $\Delta H_{solution}$ is negative, lithium iodide is soluble in water. (The solubility is 162 g per 100 g of water.)

a higher charge density on a cation also means that there is a greater attraction for the negatively charged oxygen atoms in water molecules. A higher charge density thus favours a higher hydration energy. However, it has less influence on the hydration energy than on the lattice energy.

We are now, finally, in a position to account for some of the patterns and exceptions noted at the beginning of this section.

1. The Group I metal salts and the Group II salts with anions of charge 1− are soluble because their hydration and lattice energies produce enthalpies of solution that are either negative or have a small positive value (*see* table 13.10).

Table 13.10 Enthalpies of solution

	$\Delta H^\ominus_{hyd,m}/$ kJ mol^{-1}	$\Delta H^\ominus_{latt,m}/$ kJ mol^{-1}	$\Delta H^\ominus_{soln,m}/$ kJ mol^{-1}	Solubility/ g per 100 g H_2O
LiI	−824	−763	−61	162
NaI	−711	−703	− 8	184
KI	−627	−647	+20	148
RbI	−598	−624	+26	163
CsI	−569	−601	+32	87

2. Lithium fluoride is insoluble because its enthalpy of solution has a relatively high positive value.

Table 13.11 Solubilities

	Solubility/g per 100 g H_2O				
	Fluoride	Sulphate	Carbonate	Nitrate	Hydroxide
Group I					
Li	0.13	35	1.29	85	12.9
Na	4.1	28	29.4	92	114
K	102	12.0	112	38	119
Rb	131	51.0	450	65	178
Cs	370	182	vs	27	320
Group II					
Be	550	41	–	108	i
Mg	0.013	36.4	0.06	73	0.0012
Ca	0.0016	0.21	0.0013	138	0.12
Sr	0.012	0.01	0.001	80	1.0
Ba	0.16	0.00025	0.002	10.1	4.7

vs = very soluble   i = insoluble

Table 13.12 Hydration energies

	Ionic radius/ nm	$\Delta H^\ominus_{hyd,m}/$ kJ mol^{-1}
Group I		
Li$^+$	0.060	−519
Na$^+$	0.095	−406
K$^+$	0.133	−322
Rb$^+$	0.148	−301
Cs$^+$	0.169	−276
Group II		
Be^{2+}	0.031	−2450
Mg^{2+}	0.065	−1920
Ca^{2+}	0.099	−1650
Sr^{2+}	0.113	−1480
Be^{2+}	0.135	−1380

3. Solubilities of the Group I nitrates and most Group II salts tend to decrease down a group. This is more noticeable in some of the Group II salts (*see* table 13.11). The reason can be attributed to two factors. First of all, the lattice energy in a particular group of salts remains constant owing to the large anions. Secondly, since the lattice energy is constant and since the hydration energy of the anion remains constant in any group of salts, the solubility depends entirely on the hydration energy of the cation. The hydration energy of the cation decreases as the cationic size increases (*see* table 13.12). The enthalpies of solution and thus the solubilities decrease on descending the group.

4. The fact that the solubilities of the fluorides in both groups and the solubilities of the hydroxides in Group II all increase on descending the groups can be attributed to the small size of the anion. The lattice energy can thus not be regarded as constant. The increased size of the cation on descending the group thus influences not only the hydration energy but also the lattice energy. The lattice energy decreases much more than the hydration energy and so the solubility increases.

These generalisations are only intended to serve as approximate guidelines to interpret or account for experimentally determined solubilities. The difficulty of relating solubilities to ionic size and thermodynamic properties, such as enthalpies of solution, is shown by close examination of the values of the enthalpies of solution and solubilities in table 13.10 and also the irregular patterns of the solubilities of Group I sulphates and chlorides (*see* table 13.11).

Hydrolysis of Group I and II Compounds

We have already seen that Group I and II oxides react with water to form alkalis. We also saw that the hydrides, nitrides and dicarbides are all hydrolysed by water, forming hydroxides and hydrogen, ammonia and ethyne respectively.

The alkali metal halides, nitrates and sulphates are all neutral in water. The carbonates and hydrogencarbonates, however, are hydrolysed to form alkaline solutions

$$CO_3^{2-}(aq) + H_2O(l) \rightleftharpoons HCO_3^-(aq) + OH^-(aq)$$
$$HCO_3^-(aq) + H_2O(l) \rightleftharpoons H_2CO_3(aq) + OH^-(aq)$$

In Group II the soluble salts of the strong acids are neutral or slightly acidic. For example, aqueous solutions of the chlorides and nitrates of strontium and barium are neutral, whereas those of magnesium and calcium are very slightly acidic.

Solid Hydrates

Table 13.13 Solid hydrates of s-block salts

Group I	Group II
$Na_2CO_3 \cdot 10H_2O$	$MgSO_4 \cdot 7H_2O$
$K_2CO_3 \cdot 2H_2O$	$CaSO_4 \cdot 2H_2O$

Since the hydration energies decrease down both Group I and II, the salts tend to have fewer molecules of water of crystallisation on descending the group (*see* table 13.13).

Lithium is exceptional in that almost all its salts are hydrated in the solid state.

ANOMALOUS PROPERTIES OF LITHIUM AND BERYLLIUM AND THEIR COMPOUNDS

We have already noted, in passing, certain anomalous properties of lithium and beryllium. For example, lithium is the only alkali metal to form a normal oxide on combustion. Beryllium, unlike the rest of the Group II metals, does not react noticeably with cold water or steam.

These and the other anomalous properties described below can be attributed to a number of factors related to the position of both elements at the top of their respective groups:

1. Both elements have high ionisation energies compared with other members of their groups (*see* table 13.2). This accounts for the covalent character of some of the compounds of lithium and beryllium.
2. The radii of the ions formed by these two elements are smaller than any other in the group. This results in a higher charge : radius ratio and thus polarising power for each element. This favours covalent character. It also results in high lattice energies for their compounds. The high lattice energies result in low solubilities.
3. Both elements are less electropositive and thus more electronegative than other members of their groups. This is reflected by their Pauling electronegativity values (*see* table 13.2). The high redox potential of lithium (*see* table 13.5) is out of character, as it would indicate that lithium is more highly electropositive than the other alkali metals. This is not the case. The high value is due to its high hydration energy (*see* table 13.12).

Lithium

Anomalous Properties of Lithium

1. Lithium is the only alkali metal to burn in air to form a normal oxide.
2. Lithium is the only alkali metal to combine directly with nitrogen to form a nitride.

Other properties of lithium, although not anomalous, are extreme in relation to the other alkali metals.

3. Many of the physical properties of lithium are higher than those of the other alkali metals. These include density, melting and boiling points and enthalpies of fusion and vaporisation. Lithium is also harder than the other alkali metals.
4. Lithium reacts more slowly with cold water than do the other alkali metals.

Anomalous Properties of Lithium Compounds

1. Lithium compounds are more covalent than compounds of the other alkali metals. For example, the lithium halides are more covalent. They are also more soluble in organic solvents. Alkyl and aryl lithium compounds such as CH_3Li are less reactive than the corresponding compounds of other alkali metals.
2. The hydroxide, carbonate and nitrate of lithium all decompose on heating to form the normal oxide. The hydroxides and carbonates of the other alkali metals are all thermally stable. The nitrates of the other alkali metals decompose on heating to form nitrites. Unlike the hydrogencarbonates of the other alkali metals, lithium hydrogencarbonate cannot be isolated as a solid. It only exists in aqueous solution.
3. Lithium salts of anions with a high charge density are less reactive and less soluble in water than are the corresponding salts of other alkali metals. Lithium hydroxide and lithium fluoride are examples. These properties are due to the relatively high lattice energies of these compounds.
4. Lithium hydroxide, unlike the other alkali metal hydroxides, is not a strong base.
5. Almost all the lithium salts are hydrated in the solid state. Comparatively few hydrates are found amongst the other alkali metal salts.
6. Lithium forms more stable complex compounds than do the other alkali metals.

Diagonal Relationship of Lithium with Magnesium

There are many similarities between the properties of lithium and its compounds and those of magnesium and its compounds. This is called a diagonal relationship (*see* chapter 11). For example, the following properties of magnesium are all analogous to the anomalous properties of lithium detailed above:

1. On combustion, magnesium forms a normal oxide.
2. Magnesium combines directly with nitrogen to form a nitride.
3. The magnesium halides are partially covalent in character. For example, they dissolve in many organic solvents.
4. The hydroxide, carbonate and nitrate of magnesium all decompose on heating to form a normal oxide. Magnesium hydrogencarbonate is thermally unstable and cannot be isolated as a solid.
5. Magnesium compounds such as the hydroxide, fluoride, carbonate and ethanedioate all show similar solubilities to those of the corresponding lithium compounds.

Beryllium

Anomalous Properties of Beryllium

1. Unlike the other Group II metals, beryllium does not combine with hydrogen to form a hydride.

2. Beryllium reacts with sodium hydroxide solution, forming hydrogen and the complex ion $[Be(OH)_4]^{2-}$.

Other properties of beryllium, although not anomalous, are extreme in relation to other metals in the group.

3. Many of the physical properties of beryllium are higher than those of other Group II metals. These include density, melting and boiling points and enthalpies of fusion and vaporisation. Beryllium is also harder than other Group II metals.
4. Beryllium reacts slowly with acids compared to the other Group II metals.
5. Beryllium salts are very soluble in water.

Anomalous Properties of Beryllium Compounds

1. Beryllium compounds are more covalent than the corresponding compounds of the other Group II metals. The oxide and hydroxide are both covalent. The halides of beryllium are polymeric covalent compounds. They dissolve in organic solvents.
2. The oxide and hydroxide of beryllium are both amphoteric.
3. The salts formed by beryllium and large anions are unstable. For example, beryllium is the only Group II metal not to form a stable carbonate. It does not form a hydrogencarbonate.
4. Beryllium has a strong tendency to form tetrahedral complexes, for example, $[Be(H_2O)_4]^{2+}$, $[Be(OH)_4]^{2-}$ and $[BeF_4]^{2-}$.

Diagonal Relationship of Beryllium with Aluminium

There are many similarities between the properties of beryllium and its compounds and those of aluminium and its compounds. It thus has a diagonal relationship with aluminium. The following properties of aluminium are all analogous to those of beryllium:

1. Aluminium reacts with alkalis to form hydrogen and the complex ion $[Al(OH)_4]^{-}$.
2. Aluminium is resistant to attack by acid due to the formation of a protective oxide coating.
3. All the halides except fluoride are covalent in the solid form.
4. The oxide and hydroxide of aluminium are both amphoteric.
5. Aluminium carbonate does not exist under normal conditions.
6. Aluminium forms complexes such as $[Al(H_2O)_6]^{3+}$, $[Al(OH)_4]^{-}$ and $[AlF_6]^{3-}$.

SUMMARY

1. Group I metals tend to burn in air forming mixtures of oxides.
2. The basic character of the Group I oxides increases on descending the group.
3. All the hydrides of the s-block metals are ionic, except BeH_2 and MgH_2.
4. The **degree of ionic character** of the Group II halides increases on descending the group.
5. Group I hydroxides are **strong bases.** They dissolve in water to form **alkalis.**
6. The **normal oxides**, sulphates and halides of the s-block metals tend to be **stable to heat.**
7. Carbonates of the s-block metals decompose on heating, with the exception of Na_2CO_3, K_2CO_3, Rb_2CO_3 and $SrCO_3$.
8. All nitrates of the s-block metals decompose on heating.
9. $\Delta H_{solution} = \Delta H_{hydration} - \Delta H_{lattice}$.
10. If $\Delta H_{hydration} < \Delta H_{lattice}$ then $\Delta H_{solution}$ is negative and the salt dissolves.

11. A higher **charge density** favours a higher **hydration energy**.
12. The size of a cation in a salt influences both hydration and **lattice energies**.
13. Lithium and beryllium exhibit **anomalous properties**. These include
    (a) high **ionisation energies**,
    (b) high **polarising power** of their ions and thus a higher degree of **covalent character** of their compounds,
    (c) high **electronegativity values**.
14. Lithium has a **diagonal relationship** with magnesium. For example, the compounds of both elements have similar solubilities.
15. Beryllium has a diagonal relationship with aluminium. For example, the oxides and hydroxides of both elements are **amphoteric**.

13.3
Occurrence, Manufacture and Uses

OCCURRENCE

Abundance

All the s-block metals are too reactive to occur as free elements. They occur in combination with other elements in mineral deposits and sea-water. Calcium, sodium, potassium and magnesium occur in considerable quantities in both the Earth's crust and ocean waters (*see* table 13.14). They are respectively the fifth,

River Dee in Grampian, Scotland. River water contains most s-block metals of which the four most abundant are sodium, magnesium, potassium and calcium

Table 13.14 Abundance of the s-block metals on Earth

	Abundance in Earth's crust/μg (element) per g (Earth's crust)	Concentration in typical river water/μg dm^{-3}	Concentration in ocean water/ μg dm^{-3}
Group 1			
Li	20	3	180
Na	2.4×10^4	9000	11.05×10^6
K	2.4×10^4	2300	4.16×10^5
Rb	90	1	120
Cs	3	0.05	0.5
Group II			
Be	2.8	< 0.1	0.6×10^{-3}
Mg	2.0×10^4	4100	1.326×10^6
Ca	4.2×10^4	1500	4.22×10^5
Sr	375	50	8.5×10^3
Ba	425	10	30
Ra	–	4×10^{-7}	1.0×10^{-7}

sixth, seventh and eighth most abundant elements on Earth. Strontium is only moderately abundant. The other metals in Groups I and II only constitute a minute proportion of the Earth's crust and ocean water.

Isotopes

Sodium, caesium and beryllium all occur almost entirely as single isotopes. The isotopic abundances of the s-block metals are shown in table 13.15.

Table 13.15   Abundances of the principal naturally occurring s-block metal isotopes

Element	Relative atomic mass	Isotopic abundance
Group I		
lithium	6.941	$^{6}_{3}Li$ 7.42%; $^{7}_{3}Li$ 92.58%
sodium	22.9898	$^{23}_{11}Na$ 100%
potassium	39.0983	$^{39}_{19}K$ 93.1%; $^{41}_{19}K$ 6.88%
rubidium	85.468	$^{85}_{37}Rb$ 72.15%; $^{87}_{37}Rb$ 27.85%
caesium	132.90122	$^{133}_{55}Cs$ 100%
Group II		
beryllium	9.0122	$^{9}_{4}Be$ 100%
magnesium	24.305	$^{24}_{12}Mg$ 78.70%; $^{25}_{12}Mg$ 10.13%; $^{26}_{12}Mg$ 11.17%
calcium	40.08	$^{40}_{20}Ca$ 96.947%; $^{44}_{20}Ca$ 2.083%
strontium	87.62	$^{86}_{38}Sr$ 9.86%; $^{87}_{38}Sr$ 7.02%; $^{88}_{38}Sr$ 82.56%
barium	137.33	$^{134}_{56}Ba$ 2.42%; $^{135}_{56}Ba$ 6.59%; $^{136}_{56}Ba$ 7.81%; $^{137}_{56}Ba$ 11.32%; $^{138}_{56}Ba$ 71.66%

Minerals

Sodium, potassium, magnesium and calcium all occur extensively in the Earth's crust in a variety of minerals (*see* table 13.16). Lithium, rubidium and caesium only occur in a few rare aluminosilicate minerals or in small quantities in the presence of the more common alkali metal minerals. Beryllium occurs principally as the mineral beryl. Strontium and barium minerals are also relatively rare.

Pollucite

One of the world's richest sources of caesium is at Bernic Lake, Manitoba. The deposits contain an estimated 300 000 tonnes of the mineral pollucite. This mineral is a hydrated aluminosilicate of caesium. It contains an average 20% by mass of caesium.

Table 13.16   Some important minerals containing s-block metals

Element	Mineral	Formula
Group I		
lithium	spodumene	$LiAl(SiO_3)_2$
sodium	rock salt	$NaCl$
	Chile saltpetre	$NaNO_3$
	cryolite	Na_3AlF_6
potassium	carnallite	$KCl \cdot MgCl_2 \cdot 6H_2O$
	saltpetre	KNO_3
	sylvite	KCl
Group II		
beryllium	beryl	$Be_3Al_2(SiO_3)_6$
magnesium	magnesite	$MgCO_3$
	dolomite	$MgCO_3 \cdot CaCO_3$
calcium	limestone	$CaCO_3$
	gypsum	$CaSO_4 \cdot 2H_2O$
	fluorspar	CaF_2
	fluorapatite	$CaF_2 \cdot 3Ca_3(PO_4)_2$
strontium	celestite	$SrSO_4$
	strontianite	$SrCO_3$
barium	barite	$BaSO_4$
	witherite	$BaCO_3$

What are Minerals?

Minerals are the natural crystalline materials that make up the Earth and most of its rock. They can be divided into three types depending on how they are formed.

Igneous Rocks

These are formed from **magma.** Magma is the molten material that exists below the Earth's crust. It is rich in silicon, oxygen, aluminium, sodium, potassium, calcium, iron and magnesium. These elements combine to form the various types of silicates. When the magma rises to the Earth's surface—through a volcano for example—it cools and hardens, forming igneous rocks. Micas, feldspars and quartz are all igneous rocks.

Metamorphic Rocks

Metamorphism means change in form. When certain rocks are exposed to high temperature and pressures, they undergo structural and chemical changes—without melting—to form different types of rocks. These are the metamorphic rocks. Marble is an example of a metamorphic rock. It is formed from limestone.

Sedimentary Rocks

All rocks slowly disintegrate as a result of weathering. This results in the formation of small rock particles in the form of clay, silt, sand and gravel. These eventually become cemented together, forming what is known as **clastic sedimentary rocks**. Sandstone is an example.

Weathering also causes some rocks to dissolve in water. These may precipitate and accumulate in layers on the beds of oceans or lakes. When cemented together they are called **chemical sedimentary rocks.** Rock salt, gypsum, limestone and dolomite are all examples of chemical sedimentary rocks.

Ores

These are minerals from which metals may be profitably extracted. The term is also extended to any mineral of value and non-metallic rocks such as coal.

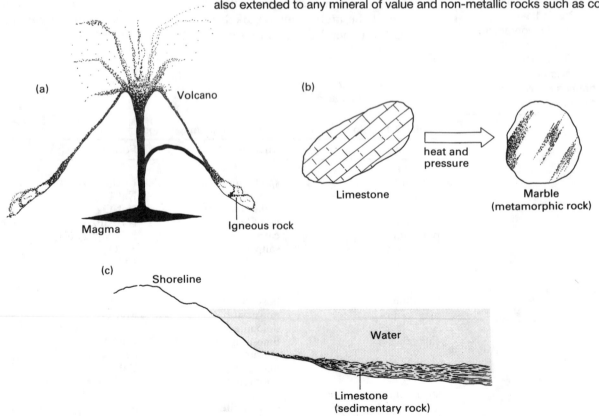

Figure 13.6   Types of rock: (a) igneous, (b) metamorphic, and (c) sedimentary

A volcano in White Island, New Zealand. Volcanoes produce a molten material known as magma. This solidifies to form igneous rocks. Quartz is an igneous rock

EXTRACTION OF s-BLOCK METALS

The Group I metals have high electrode potentials and are the strongest reducing agents known. They therefore cannot be extracted by reduction of their oxides. Since they are so electropositive, they also cannot be displaced from aqueous solutions of their salts by other metals. Furthermore, electrolysis of an aqueous solution of their salts leads to the formation of their hydroxides and not the metal itself. An amalgam of the metal and mercury is formed if a mercury cathode is used. However, recovery of the pure metal from the amalgam is difficult.

The alkali metals are extracted by electrolysis of their fused halides. A second salt, such as calcium chloride, is added to lower the melting point.

Group II metals are also strong reducing agents. They therefore cannot be reduced directly by carbon since they form dicarbides. The Group II metals can be extracted by electrolysis of their fused chlorides with sodium chloride added to lower the melting point. However, barium and strontium can only be extracted with difficulty by this method since they form colloidal suspensions. They are obtained from their oxides by thermal reduction using aluminium

$$3BaO(s) + 2Al(s) \rightarrow 3Ba(l) + Al_2O_3(s)$$

The oxide used is a **calcined** mineral. This is a carbonate mineral from which carbon dioxide has been removed by heating. For example

$$BaCO_3(s) \overset{heat}{\rightarrow} BaO(s) + CO_2(g)$$

witherite   calcined
                         witherite

> An **amalgam** is a solution of a metal in mercury.

Extraction of Sodium

Both sodium and chlorine are extracted from fused sodium chloride by electrolysis. Sodium chloride is obtained as a raw material by the following methods:

- Conventional rock salt mining.
- Solar evaporation. The Sun's heat is used to evaporate water from the ocean or inland salt lakes. The salt crystals settle out and the liquid is drained off. This produces a coarse form of salt.

- Solution mining. This is used to extract salt from underground salt deposits. Water is pumped into the deposit and the resulting **brine** (sodium chloride solution) is pumped to the surface. Salt crystals are obtained by vacuum evaporation. This method is also used to obtain salt from brines that occur naturally underground.

The Downs Cell

Sodium is extracted commercially from sodium chloride using the Downs cell (*see* figure 13.7). This is a cylindrical steel cell in which molten sodium chloride is electrolysed between a central graphite anode and a circular steel cathode. Sodium chloride melts at about 800°C. At this temperature, the steel cell corrodes rapidly. In addition, some of the sodium produced by the electrolysis dissolves in the molten electrolyte. This causes the electrolyte to become a metallic conductor and thus electrolytic decomposition stops. The problem is overcome by using an electrolyte containing 60% calcium chloride and 40% sodium chloride. The calcium chloride lowers the melting point to about 600°C.

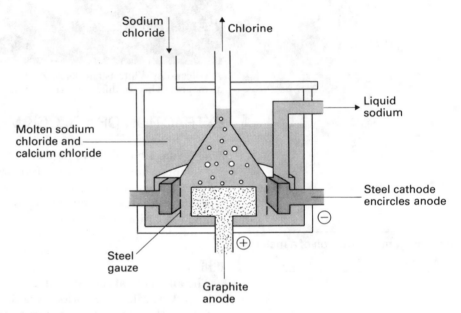

Figure 13.7   The Downs cell

A typical Downs cell operates with a current of 30 000 A and a voltage of about 7.0 V. The electrode reactions are

$$at\ the\ cathode \qquad Na^+(l) + e^- \rightarrow Na(l)$$
$$at\ the\ anode \qquad 2Cl_2(l) \rightarrow Cl_2(g) + 2e^-$$

Both the liquid sodium and the chlorine rise to the surface of the electrolyte. Since they react, they are kept apart by a steel gauze.

Manufacture of Magnesium by Thermal Reduction

Magnesium can be produced from magnesium oxide in a thermal reduction process known as the **Pidgeon process**. The process uses calcined dolomite as a raw material

$$MgCO_3 \cdot CaCO_3(s) \rightarrow MgO \cdot CaO(s) + 2CO_2(g)$$
$$\text{dolomite} \qquad\qquad \text{calcined dolomite}$$

The calcined dolomite is heated with an alloy of iron and silicon. The silicon reduces the magnesium oxide forming magnesium vapour

$$2MgO(s) + Si(s) \rightarrow SiO_2(s) + 2Mg(g)$$

The silicon(IV) oxide reacts with the calcium oxide to form calcium silicate, which is removed as slag

$$CaO(s) + SiO_2(s) \rightarrow CaSiO_3(l)$$
$$\text{slag}$$

Calcined magnesite is also used as a raw material in this process.

Commercial Extraction of Magnesium from Sea-water

Magnesium is also extracted commercially from sea-water or a slurry of calcined dolomite in sea-water. The procedure used varies from country to country, although there are essentially three stages:

1. Precipitation of magnesium hydroxide. The magnesium in the sea-water or slurry is precipitated as magnesium hydroxide. If sea-water alone is used, the precipitation is achieved by adding hydrated lime, $Ca(OH)_2$. If a slurry is used, the calcium hydroxide is already present due to the reaction between water and the calcium oxide in the calcined dolomite. The reaction for this stage is

$$Mg^{2+}(aq) + Ca(OH)_2(s) \rightarrow Mg(OH)_2(s) + Ca^{2+}(aq)$$

2. Formation of magnesium chloride. The magnesium hydroxide is converted to magnesium chloride using hydrochloric acid

$$Mg(OH)_2(s) + 2HCl(aq) \rightarrow MgCl_2(aq) + 2H_2O(l)$$

3. Electrolysis of molten magnesium chloride. Hydrated magnesium chloride, $MgCl_2 \cdot 6H_2O$, is recovered from the solution by evaporation in a furnace. This is then dehydrated to produce anhydrous magnesium chloride. A fused mixture containing 25% $MgCl_2$, 15% $CaCl_2$ and 60% $NaCl$ is then electrolysed in a cylindrical cell using carbon anodes and steel cathodes. The $NaCl$ and $CaCl_2$ are used to lower the melting point and also to increase the density of the electrolyte so that the molten magnesium floats to the surface of the electrolyte. The electrode reactions are

$$\text{at the cathode} \quad Mg^{2+}(l) + 2e^- \rightarrow Mg(l)$$
$$\text{at the anode} \quad 2Cl^-(l) \rightarrow Cl_2(g) + 2e^-$$

MANUFACTURE OF s-BLOCK COMPOUNDS

Group I and II compounds such as $NaOH$, KOH, Na_2CO_3, KCl, K_2CO_3, $CaCO_3$ and $Ca(OH)_2$ are all of immense industrial importance and are used in considerable quantities. We shall consider some of the uses of these and other s-block compounds below. In this section we shall outline the manufacture of four of these compounds.

Sodium Hydroxide

This is manufactured by the electrolysis of brine in either a mercury cathode cell or a diaphragm cell. Hydrogen and chlorine are important by-products of both methods.

The Mercury Cathode Cell

The mercury cathode or Castner–Kellner cell consists of a graphite or titanium anode and a flowing mercury cathode (*see* figure 13.8). The cell operates at about 4.3 V and 300 000 A. The chemistry of the manufacture of sodium hydroxide, hydrogen and chlorine using this type of cell falls into three parts:

1. Electrolysis. The electrolyte is concentrated **brine** (sodium chloride solution). The use of a mercury cathode enables sodium ions to discharge in prefer-

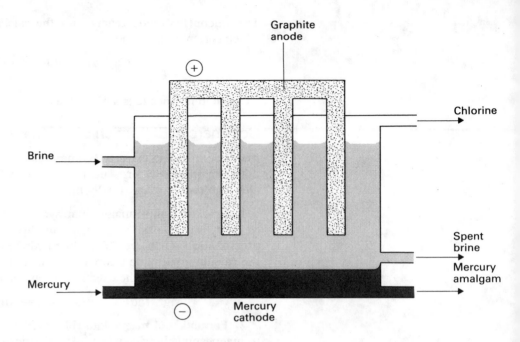

Figure 13.8 The mercury cathode cell

ence to hydrogen ions. Chloride ions are discharged at the anode. The electrode reactions are the same as for the Downs cell.

2. Formation of amalgam.   The sodium formed at the cathode dissolves in the mercury, forming a sodium amalgam

$$Na(s) + Hg(l) \rightarrow Na/Hg(l)$$

sodium amalgam

3. Reaction with water.   The amalgam flows into a separate chamber called a soda cell (not shown in figure 13.8.) In this cell the amalgam is mixed with water, producing sodium hydroxide and hydrogen. Pure mercury is regenerated during the process

$$2Na/Hg(l) + 2H_2O(l) \rightarrow 2NaOH(aq) + H_2(g) + 2Hg(l)$$

The regenerated mercury is recirculated.

The Diaphragm Cell

A diaphragm cell consists essentially of a graphite or titanium anode and a steel cathode separated by an asbestos diaphragm. There are several types of diaphragm cell. Figure 13.9 shows the main features of a Hooker diaphragm cell. This is the only type of diaphragm cell used in the UK.

The electrolyte is brine, which is run into the top of the cell. Chloride ions are discharged at the anode and the chlorine is liberated at the top of the cell. The remaining solution seeps through the asbestos diaphragm into the cathode department. Hydrogen and sodium hydroxide are produced at the cathode

$$2H_2O(l) + 2e^- \rightarrow 2OH^-(aq) + H_2(g)$$

The solution produced in the cathode compartment contains approximately equal amounts of sodium hydroxide and sodium chloride. This is run out of the cell. The solution is concentrated by evaporation so that the sodium chloride crystallizes out. This produces a solution containing about 50% by mass of sodium hydroxide and about 1% sodium chloride.

Potassium Hydroxide

Potassium hydroxide is also manufactured by both types of cell. A concentrated solution of potassium chloride instead of brine is used.

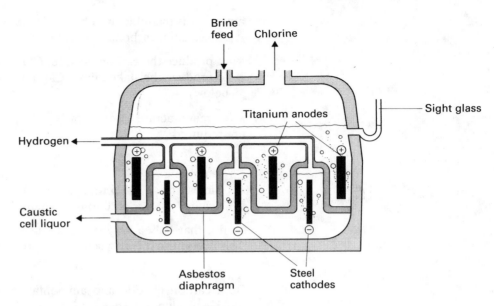

Figure 13.9 The diaphragm cell

Sodium Carbonate

Most sodium carbonate used in the UK is manufactured by the **Solvay ammonia–soda process**. This process consists of two stages:

1. Precipitation of sodium hydrogencarbonate. Carbon dioxide is pumped into the bottom of a Solvay tower (*see* figure 13.10). Brine, saturated with ammonia, is fed into the top of the tower. As it descends it meets with ascending carbon dioxide. The carbon dioxide dissolves in the water

$$CO_2(g) + H_2O(l) \rightarrow HCO_3^-(aq) + H_3O^+(aq)$$

The ammonia in the solution neutralises the oxonium ions, pulling the above equilibrium to the right

$$NH_3(aq) + H_3O^+(aq) \rightarrow NH_4^+(aq) + H_2O(l)$$

Sodium hydrogencarbonate is relatively insoluble and thus precipitates from the solution

$$Na^+(aq) + HCO_3^-(aq) \rightarrow NaHCO_3(s)$$

This leaves ammonium chloride in solution.

The overall process may be represented by the equation

$$H_2O(l) + NaCl(aq) + CO_2(g) + NH_3(g) \rightarrow NaHCO_3(s) + NH_4Cl(aq)$$

Since the overall process is exothermic, water is used to cool the tower.

A white sludge containing sodium hydrogencarbonate suspended in ammonium chloride solution is removed from the bottom of the tower. The sodium hydrogencarbonate is separated by vacuum filtration.

2. Production of sodium carbonate. The sodium hydrogencarbonate produced in stage 1 is heated in long iron tubes

$$2NaHCO_3(s) \overset{\text{heat}}{\rightarrow} Na_2CO_3(s) + H_2O(l) + CO_2(g)$$

Residual ammonium chloride sublimes off during this stage. The product is known as soda ash. It contains a small amount of sodium chloride.

Raw Materials

The raw materials for the Solvay process are brine and calcium carbonate (limestone), both of which are relatively inexpensive, and ammonia. The

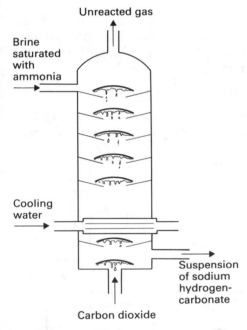

Figure 13.10 The Solvay tower

Potassium Carbonate

This cannot be manufactured by the Solvay process since potassium hydrogencarbonate is too soluble to be precipitated.

---

ammonia is manufactured by the Haber process (*see* section 7.2).
The calcium carbonate is required

- to produce the carbon dioxide, CO_2, used in stage 1;
- to produce the slaked lime, $Ca(OH)_2$, needed for recycling the ammonia (*see* below).

To manufacture 1 tonne of Na_2CO_3 the following amounts of raw materials are required:

brine	1.5 tonnes
ammonia	2 kg
limestone	1.5 tonnes
water	50.0 tonnes (mainly for cooling)

Notice that only a very small amount of ammonia is consumed in the process. This is because it is recycled in the process.

Recycling

Both carbon dioxide and ammonia are recycled in the process. The carbon dioxide produced in stage 2 in the production of soda ash is reintroduced into the Solvay tower. Ammonia is regenerated from the ammonium chloride solution produced in stage 1 by the addition of slaked lime.

$$Ca(OH)_2(s) + 2NH_4Cl(aq) \rightarrow CaCl_2(aq) + 2H_2O(l) + NH_3(g)$$

Products

The major product of the process is soda ash. Washing soda crystals, $Na_2CO_3\cdot10H_2O$, are made by dissolving the soda ash in hot water and crystallising. Pure sodium hydrogencarbonate is made by blowing carbon dioxide through sodium carbonate solution until it is saturated

$$Na_2CO_3(aq) + CO_2(g) \rightarrow H_2O(l) + 2NaHCO_3(s)$$

The only other product of the Solvay process is calcium chloride, which is produced during the recycling of ammonia. Calcium chloride has no large-scale uses and is thus largely a waste product.

The Solvay process is summarised schematically in figure 13.11.

Figure 13.11   The Solvay process

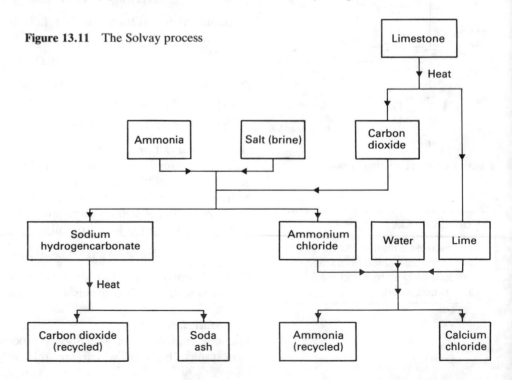

USES

Sodium and its Compounds

The annual world consumption of sodium and its compounds exceeds 100 million tonnes. The uses of sodium and its compounds are numerous. The treatment below only touches on some of the more important uses.

Sodium Metal

Owing to the high reactivity of sodium and also to the relatively high cost of manufacture, the metal is not manufactured in large quantities. However, significant quantities are used in

- Street lighting.
- Fast breeder reactors. An alloy of sodium and potassium is used as a coolant.
- Metallurgical processes—for example, in removing antimony from lead.
- The production of tetraethyllead(IV) which is used as an anti-knock additive in petrol.
- Organic processes as a catalyst. The sodium is usually dispersed on carbon.
- The manufacture of inorganic chemicals such as sodium cyanide.

Sodium Hydroxide

Well over 30 million tonnes of sodium hydroxide are manufactured each year. This is used to manufacture soaps, detergents and other cleaning agents, dyes, cosmetics and pharmaceuticals. About 15% of the sodium hydroxide manufactured in the UK is used to make rayon. Sodium hydroxide is also used to make organic compounds such as phenol and naphthol.

Sodium Carbonate

About 26 million tonnes of sodium carbonate are manufactured each year. The glass industry consumes about 7 million tonnes of soda ash each year. Large quantities are also used in sewage treatment and water softening. Sodium carbonate is also used in metal refining, textile dyeing and leather tanning. It is also used to produce compounds such as sodium arsenate and sodium silicate.

Other Sodium Compounds

Sodium silicate, Na_2SiO_3, commonly known as **water glass**, is used as a 'builder' in soaps and detergents. It is used as a lubricant, as a desiccant and as an adhesive in the manufacture of fibre-board and corrugated cardboard boxes.

Sodium arsenate(V) and sodium arsenate(III) are both used as pesticides. Sodium aluminium fluoride, in the naturally occurring form of cryolite, is used in the production of aluminium. Since the supply of cryolite is insufficient for this purpose, large quantities of sodium aluminium fluoride are also manufactured each year.

The Other Alkali Metals and their Compounds

Potassium

Potassium is an essential plant nutrient. Large quantities of potassium in the form of potassium nitrate are thus used as a fertiliser. Potassium nitrate is also used in the manufacture of explosives and glass.

Potash, K_2CO_3, is used in the manufacture of glass and soft soaps. Potassium phosphate(V) is used as a 'builder' to enhance the surfactant performance of detergents.

Lithium

Lithium is used as a metal in various alloys. Lithium carbonate is a white powder, and is used as an antidepressant drug. It is given by mouth to prevent

Sodium is used in street lighting

Sodium hydroxide is used to produce a variety of cosmetics

and treat schizophrenia and similar psychiatric disorders. Lithium hydroxide is used in the ventilation systems of submarines and spacecraft to absorb carbon dioxide. The uses of rubidium and caesium are limited by their high cost and highly reactive nature. However, both are used as components in photocells.

Calcium Compounds

By far the most important calcium compound is calcium carbonate. It is the major constituent of limestone, marble and chalk and also occurs in dolomite along with magnesium carbonate. Over 100 million tonnes of limestone are quarried in the UK each year. It is used in the building industry, for road making and in the manufacture of steel, sodium carbonate and glass.

Calcium carbonate is also used to manufacture calcium oxide. This is known as quicklime or, simply, lime. The calcium carbonate is heated to about 1200°C in a limekiln

$$CaCO_3(s) \rightarrow CaO(s) + CO_2(g)$$
$$\text{quicklime}$$

Calcium hydroxide is produced from quicklime by adding water to it. The process is known as slaking

$$CaO(s) + H_2O(l) \rightarrow Ca(OH)_2(s)$$
$$\text{slaked lime}$$

The Use of Calcium Compounds in the Building Industry

Mortar is formed by adding water to a mixture of sand and slaked lime to form a thick paste. As this dries out the slaked lime reacts with carbon dioxide from the atmosphere to form calcium carbonate

$$Ca(OH)_2(s) + CO_2(g) \rightarrow CaCO_3(s) + H_2O(g)$$

The formation of calcium carbonate causes the mortar to harden.

Cement is a mixture of calcium silicate and calcium aluminate. It is formed by heating clay with either limestone or calcium oxide obtained by heating calcium sulphate with coke; for example

$$4CaCO_3(s) + Al_2Si_2O_7(s) \rightarrow \underbrace{2CaSiO_3(s) + Ca_2Al_2O_5(s)}_{\text{cement}} + 4CO_2(g)$$

Plaster used in the building industry is a hydrated form of calcium sulphate with the formula $(CaSO_4)_2 \cdot H_2O(s)$. It is made by partially dehydrating gypsum, $CaSO_4 \cdot 2H_2O(s)$.

Other Uses of Calcium Compounds

Common glass is a mixture of calcium silicate and sodium silicate. It is also known as **soda glass**. It is made by blending a mixture of sand, sodium carbonate and limestone and heating it in a furnace.

Slaked lime, $Ca(OH)_2$, is used to make bleach (*see* section 16.3) and in the manufacture of calcium hydrogensulphate, $Ca(HSO_4)_2$. This is used in the paper industry. It is made by passing sulphur dioxide through slaked lime. Slaked lime is also used to neutralise acids in industrial effluents, to soften water and to treat sewage.

Calcium dicarbide is manufactured by heating quicklime with coke in an electric furnace

$$CaO(s) + 3C(s) \rightarrow CaC_2(s) + CO(g)$$

Large quantities of calcium dicarbide are used to manufacture ethyne (acetylene)

$$CaC_2(s) + 2H_2O(l) \rightarrow Ca(OH)_2(aq) + C_2H_2(g)$$

Limestone quarry in Derbyshire, England

Workers carrying loads of cement on their heads. The cement is being used to build the Emerald Dam as part of the Kundah Hydro Electric Scheme, India. Cement is a mixture of calcium silicate and calcium aluminate

The Other Group II Metals and their Compounds

Magnesium

Magnesium is used in the manufacture of light alloys, especially with aluminium. These alloys are used in automobile and aircraft construction. Magnesium is also used as a sacrificial anode (*see* chapter 10). Magnesium oxide is a constituent of some toilet preparations and toothpaste. 'Milk of magnesia' is a suspension of magnesium hydroxide in water. It is used to neutralise excess acidity in the stomach. Epsom salts contain magnesium sulphate. They are used as a mild laxative.

Barium

The 'barium meal' used in X-ray diagnostic work contains the highly insoluble barium sulphate. This compound is also used to make certain paints and pigments. Barium carbonate is used as a rat poison. Concrete and plaster containing barium salts are used as building materials to protect against radiation.

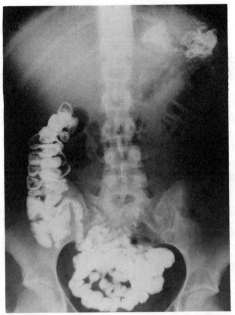

A 'barium' X-ray showing parts of the ileum, colon and stomach

Materials Used in Firework Manufacture	
Oxidising agents	Chlorates(V) of potassium and barium Chlorates(VII) of potassium and ammonium Nitrates of potassium, barium and strontium Peroxides of barium and strontium
Colouring agents	Barium nitrate and chlorate(V) for green Strontium carbonate, ethanedioate, nitrate for red Sodium ethanedioate, cryolite for yellow Copper(II) oxide, carbonate and chloride for blue
Fuels	Accaroid resin, gum copal, shellac for colours Charcoal and sulphur for non-colours Antimony sulphide
Colour intensifier	p.v.c.
Other materials	Titanium, aluminium, magnesium, antimony, lampblack, dextrin, gum arabic, starch, barium carbonate, calcium phosphate(V)

Barium carbonate is used to poison rats

October lst celebrations in Beijing, China Compounds of Group II metals such as strontium are used as colouring agents in fireworks

Beryllium and Strontium

Beryllium is used in alloys. Strontium and its compounds have no major uses although strontium salts are used to give a crimson colour in fireworks.

Biological and Agricultural Importance of the s-Block Elements and their Compounds

There are four s-block metals that play crucial roles in the structure of living organisms and in the biochemical processes that take place in these organisms. These metals are potassium, sodium, calcium and magnesium. All plants and animals contain these elements. They are thus of great importance in the biological and agricultural sciences.

Papyrus on the Zambesi River, near Victoria Falls, Zimbabwe. Papyrus ash contains a relatively high percentage of sodium chloride

---

Salt from Plants

One of the oldest ways of getting salt for cooking is by extracting it from plants. This method is still used today in parts of Africa where there are no salt mines or where the coast is a long distance away.

Different plants for salt making are favoured by different tribes throughout Africa. Papyrus, certain grasses, banana leaves and banana skins are a few examples of the type of plant material used. In East Africa, some tribes produce salt from ashes by burning dry leaves on the ground. This produces a mixture of black ash and partly burned leaves. The salts are extracted by adding water to dissolve the salts and then filtering (*see* figure 13.12).

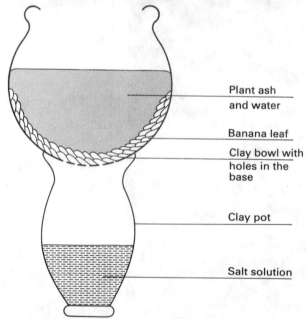

Figure 13.12   One method of producing salt in an African village

A clay bowl with holes in the bottom is used. The holes are covered inside the bowl with strips of banana leaf. The leaf has slits cut in it and this acts as a simple filter. The ash is put into the bowl and cold water is added. The water slowly soaks through the slits in the banana leaves into the pot below. More water is added when needed and the ash is pressed down from time to time. The solution of salt produced in this way is used immediately for cooking vegetables.

Plant ash is not common salt (sodium chloride) but a mixture of salts. Papyrus ash has a relatively high percentage of sodium chloride, up to 40% of all the salts present. Potassium chloride, sodium carbonate and potassium carbonate are also found. In the north of Ghana, West Africa, people still prefer to use plant ash salt, though common salt is readily available, because it neutralises the bitter taste of some vegetable soups, softens beans and helps to preserve stews a little longer. Common salt is also added to give the food more flavour.

Body Fluids and Homeostasis

Ions such as Na^+, K^+, Ca^{2+}, Mg^{2+}, Cl^- and HCO_3^- are found in the body fluids of all animals. These body fluids include blood plasma, sweat and cell fluids. For example blood plasma is a body fluid that contains relatively high concentrations of Na^+ and Cl^- ions, or in another word 'salt'. The salt causes the blood to have an osmotic pressure. This osmotic pressure must be regulated if animal cells are to function properly. The organ primarily responsible for regulating the osmotic pressure and thus the water and electrolyte levels in the blood is the kidney. The process is called **osmoregulation**. It is an example of **homeostasis**. Homeostasis is a process by which living organisms maintain their internal systems at equilibrium despite variations in the external conditions.

Sodium chloride is also one of the main constituents of sweat—as we all know from its taste. Sweating is another example of homeostasis. We sweat when we are hot in order to reduce our body temperature.

Homeostasis is essential to all living organisms. Without it they and we could not survive. Homeostasis involves all our bodily systems including the nervous system. And central to the central nervous system are those s-block ions again, in this case Na^+ and K^+.

The nervous system of an animal is composed of millions of nerve cells. The messages that a nervous system transmits to and from the brain of an animal are due to tiny electrical pulses which travel through the nerve cells. The electrical pulses result from sudden changes in the concentrations of Na^+ and K^+ ions in the nerve cells. The fluid *inside* animal cells, including nerve cells, has a relatively high concentration of K^+ ions whereas fluids *outside* the cells have a higher concentration of Na^+ ions. When K^+ diffuses out of a nerve cell, a potential difference is set up across the cell wall. This is due to the net negative charge inside the cell due to excess anions such as Cl^-. Upon stimulation the nerve cell allows Na^+ ions to enter the cell. This reverses the sign of the potential difference. Consequently an electrical impulse is sent from the cell to the other cells of the nerve.

Calcium

The adult human body needs to consume about 1 g of calcium each day. The principal sources of the element in our diet are dairy products such as milk and cheese. The element is essential for the normal growth and functioning of our bodies. On average the adult human body contains about 1 kg of calcium. About 99% of this occurs as calcium phosphate in our bones and teeth.

Calcium is also essential for the proper functioning of many of our bodily processes. Blood clotting, muscle contractions and our nervous system all depend on calcium. The concentration of calcium in our blood is about 0.1 g dm^{-3}. This level is maintained by hormones. The absorption of calcium from the food in our stomach is facilitated by vitamin D. A deficiency of vitamin D in our diets can result in conditions such as rickets. Liver and fish oils are good sources of vitamin D.

Magnesium

Magnesium is also essential to life. It is needed for the proper functioning of our muscles and nervous system. On average the adult human body contains about 25 g of magnesium. Most of this is concentrated in the bones. We consume magnesium when we eat green leafy vegetables. All green plants contain chlorophyll. Chlorophyll is essential for photosynthesis—the process by which green plants synthesise simple carbohydrates from water and carbon dioxide using light energy. Chlorophyll is a mixture of four pigments. Two of these—chlorophyll a and chlorophyll b—are organic complexes containing magnesium bonded by four heterocyclic nitrogens at the centre of a flat organic ring system known as porphyrin (*see* figure 13.13).

British heavyweight boxer Frank Bruno demonstrates an example of homeostasis ... sweating! Sodium and potassium ions play key roles in homeostasis

These young Kampuchean children are waiting for milk during the 1980 emergency. This milk contains important nutrients such as calcium, which is essential for the normal growth and functioning of their bodies

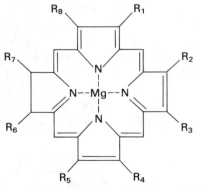

Figure 13.13 Skeleton of a chlorophyll molecule. R_1 to R_8 are organic side groups such as methyl and ethyl

Soil

The healthy growth and yields of agricultural crops depend on a number of factors. These include the availability of plant nutrients in the soil and the condition of the soil.

Plant nutrients may be divided into two groups. Nitrogen, phosphorus and potassium are often referred to as major or **macro-nutrients** because plants require them in relatively large quantities. Plants also require **micro-nutrients** or **trace elements**. These include boron, copper, iron, manganese, molybdenum and zinc. Calcium, magnesium and sulphur are also required to sustain healthy crop growth. These three elements may be regarded as intermediate between macro- and micro-nutrients.

Two of the four elements most commonly added to soils to make up for deficiencies are s-block elements. The four elements are nitrogen, phosphorus, potassium and calcium. All of these, except calcium, are added to fertilise the soil, that is to add plant nutrients to the soil to make up for deficiencies.

Potassium fertilisers are all derived from natural mineral deposits such as sylvite and carnallite (*see* table 13.16). The potassium is normally added in the form of potassium chloride. However, the quality of some crops, such as potatoes, tomatoes and soft fruit, suffers if the chloride concentration in the soil is too high. With such crops, potassium sulphate is used as a fertiliser.

The most common nitrogen fertiliser is ammonium sulphate. However, the application of ammonium sulphate has two particular disadvantages, especially for humid soils. First of all, application leads to the loss of calcium from the soil. Secondly, it lowers the pH of the soil. The reason is that the clay in the soil acts as an ion exchange resin (*see* section 8.2). The clay contains calcium ions, which are exchanged for ammonium ions when the fertiliser is applied (*see* figure 13.14a). The calcium ions are leached from the soil and carried away in drainage water. Furthermore, the ammonium ions on the clay slowly oxidise to form acid (*see* figure 13.14b).

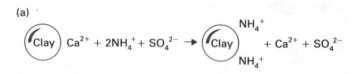

Figure 13.14

This young boy from Senegal almost certainly does not know that the optimum pH range for a good crop of potatoes is 4.8 to 5.7!

Ammonium fertilisers and related fertilisers such as urea, $CO(NH_2)_2$ (also known as carbamide), are thus often known as 'acid' fertilisers. The problem of acidity can be overcome by adding calcium carbonate to the fertiliser. Alternatively, nitrate fertilisers such as potassium nitrate or sodium nitrate are used.

Phosphorus fertilisers are derived from rock phosphates such as apatite. However, rock phosphate is insoluble in water and therefore inaccessible to plants when added to the soil. Rock phosphate is thus treated with concentrated sulphuric acid to convert it to calcium dihydrogenphosphate(v), which is soluble. The product is known as **superphosphate**:

$$Ca_3(PO_4)_2(s) + 2H_2SO_4(aq) + 4H_2O(l) \rightarrow \underbrace{Ca(H_2PO_4)_2(s) + 2CaSO_4 \cdot 2H_2O(s)}_{superphosphate}$$

rock phosphate

Triple superphosphate is a concentrated form of $Ca(H_2PO_4)_2$ obtained by treating rock phosphate with phosphoric acid.

Table 13.17 Optimum pH ranges of crops

Crop	Optimum pH range
potatoes	4.8 – 5.7
rye	5.0 – 6.0
turnips	5.8 – 6.8
barley	7.0 – 7.5
sugarbeet	7.0 – 7.5
wheat	6.5 – 7.5

Calcium is added to soils to increase the pH in a process known as liming. This is often necessary because the uptake of both macro- and micro-nutrients by particular crops depends on pH. The optimum pH of soils for most crops lies in the range 6.0 to 7.0, although some crops prefer a more acidic soil whilst others prefer a very weakly alkaline soil. Table 13.17 shows the optimum pH range for six different crops.

Calcium is added to the soil in one of the following forms:

- Calcium oxide, CaO. This is known as quicklime, burnt lime or, simply, lime.
- Calcium hydroxide, $Ca(OH)_2$. This is known as slaked lime or hydrated lime.
- Calcium carbonate, $CaCO_3$. This is normally added as finely crushed limestone or chalk.

The term **lime** is often loosely applied to any of these calcium compounds.

SUMMARY

1. The s-block metals **do not occur naturally** as free elements but **occur** widely in mineral deposits and sea-water.
2. Most s-block metals can be **extracted** by electrolysis of their fused halides. An impurity is added to lower the melting point.
3. Sodium is extracted commercially from molten sodium chloride using the **Downs cell**.
4. Magnesium is produced either by the reduction of **calcined dolomite** in the Pidgeon process or from sea-water.
5. Sodium hydroxide is manufactured by the electrolysis of brine using either a **mercury cathode cell** or a **diaphragm cell**.
6. Sodium carbonate is manufactured by the **Solvay ammonia–soda process**. The **raw materials** are brine, limestone and ammonia. The process involves two principal stages: (1) precipitation of sodium hydrogencarbonate; (2) conversion of sodium hydrogencarbonate to sodium carbonate (**soda ash**). Calcium chloride is a **waste product**.
7. Sodium hydroxide is used to make **cleaning agents, dyes** and **rayon**.
8. Potassium nitrate is used as a **fertiliser**.
9. Calcium compounds are used in the **building industry**.
10. Potassium, sodium, calcium and magnesium all play crucial roles in the structure and **biochemical processes** of living organisms.
11. The fertiliser **superphosphate** is produced from rock phosphate.
12. The pH of soils can be increased by a process known as **liming**. **Lime** is calcium oxide or a related compound.

Examination Questions

1. (a) Describe *three* characteristic features in the chemistry of the s-block elements (alkali and alkaline earth elements) and relate the features which you describe to the electronic structures of the atoms.
    (b) Standard electrode potentials $(E^\ominus/V)$ are listed below for certain s-block elements.

Li	−3.04	Be	−1.85
Cs	−2.92	Ba	−2.90

    Why are the values for Li and Cs approximately equal whereas the value for Ba is considerably more negative than that for Be?
    (c) Suggest a reason why barium fluoride and sulphate are much less soluble in water than potassium salts with the same anions. (Ionic radii/nm: K 0.133, Ba 0.135).

(d) Calcium carbonate is an important raw material in many chemical manufacturing processes. Give *two* examples of its large scale uses and explain its chemical function in *one* of these applications.

(O & C)

2. (a) By reference to the metals sodium and magnesium, discuss and explain the periodic trends associated with
   (i) atomic and ionic radii,
   (ii) ionisation energy,
   (iii) melting point,
   (iv) the basic nature of their oxides and hydroxides.

(b) Caesium is a member of the alkali metals. From a knowledge of *either* the chemistry of potassium *or* sodium, predict the chemical behaviour of caesium in the following situations.
   (i) The reation of the metal towards water.
   (ii) The reaction of the metal towards hydrogen.
   (iii) The stability of the metal nitrate towards heat.
   (iv) The use of the carbonate of the metal for softening permanently hard water.

(c) Describe the crystalline structure shown by sodium chloride and caesium chloride respectively and account for the difference.

(d) (i) What volume of hydrochloric acid, of concentration 0.2 mol dm^{-3}, would be required to convert 8.15 g of caesium carbonate into caesium hydrogencarbonate?
   (ii) What indicator would you use if it were necessary to check this volume by means of titration?

(AEB 1982)

3. (a) Listed below are substances A → H and properties 1 → 8. Write down the letters and append to each letter the number which best fits the properties of the substance.

Substance	Property
A. Barium chloride	1. Aqueous solution gives yellow precipitate with $HNO_3(aq)$
B. Sodium bromide	2. Aqueous solution gives brown precipitate with $NaOH(aq)$
C. Lead nitrate	3. Solid gives green flame test
D. Iron(III) chloride	4. Aqueous solution gives oxygen when $MnO_2(s)$ is added
E. Magnesium nitride	5. Brown gas formed when solid is heated alone.
F. Sodium thiosulphate	6. Brown gas formed when solid is heated with $MnO_2(s)/H_2SO_4(l)$
G. Hydrogen peroxide	7. Aqueous solution decolourises acidified $KMnO_4(aq)$, but does not give oxygen with $MnO_2(s)$
H. Sulphur dioxide	8. Alkaline gas is formed when hot water is added to solid

(b) Describe what happens when aqueous solutions of the following pairs are mixed, and write equations:
   (i) A and H ⎫
   (ii) B and C ⎬ in the presence of dilute acid
   (iii) B and G ⎪
   (iv) D and H ⎭

(c) Write an equation to show how E might be prepared, giving simple conditions.

(d) In what way is F useful in chemical analysis?

(SUJB)

4. The elements in Group II of the Periodic Table, in order of increasing atomic number, are: beryllium, magnesium, calcium, strontium, barium, and radium.
   (a) Predict the following, by comparison with the chemistry of the elements higher in the group:
      (i) the reaction of radium with water;
      (ii) the solubilities of radium nitrate, sulphate, and chloride;
      (iii) the acidic/basic character of radium oxide;
      (iv) the thermal stability of radium carbonate;
      (v) the type of bonding present in radium compounds and their general physical nature;
      (vi) the best method by which radium might be isolated from a suitable compound.
   (*Make sure you give an adequate explanation, giving conditions and equations.*)
   (b) Beryllium(II) chloride hydrolyses with complete removal of chlorine, but magnesium(II) chloride forms an oxychloride on hydrolysis. Explain.
   (c) Sodium chloride and magnesium(I) chloride have nearly the same lattice energy, but the latter is very difficult to prepare. Explain.
   (d) Radium has the largest atomic radius. Explain.

(SUJB)

5. (a) Explain why lithium differs from the other alkali metals in some respects and give *two* examples of the differences.
   (b) What type of chemical compounds constitutes the major source of alkali metals? Suggest a reason why the alkali metals occur in these forms.
   (c) Indicate by means of diagrams and equations how sodium carbonate is manufactured from a naturally occurring sodium compound. State *two* large scale uses of sodium carbonate.

(O & C)

6. 'The properties of the first member of a group of elements in the Periodic Table are not typical of those of the group as a whole.' Discuss this statement by referring to the chemistry of the elements of Group I (Li–Cs) and Group II (Be–Ba).
   In your answer you should give reasons for the differentiation between lithium and beryllium and the other members of their respective groups, illustrating your answer by reference to specific properties.

(JMB)

7. This question is concerned with the chemistry of sodium.
   (a) (i) What are the principles of a process by which metallic sodium may be obtained from sodium chloride? You may draw a labelled sketch if you wish.
      (ii) Why is sodium manufactured in this way?
      (iii) Give one use of sodium metal, either in industry or in the laboratory.
      (iv) Explain why a particular property of sodium makes it useful in this application.

(b) (i) Outline the principles of a process for the manufacture of sodium hydroxide from sodium chloride and draw a labelled sketch to illustrate the process.

(ii) Give one major industrial use of sodium hydroxide.

(iii) Explain why a particular chemical property of sodium hydroxide makes it useful in this application.

(O & C)

8. (a) What is meant by the term 'metallic bonding'?

(b) Why is the melting point of sodium higher than that of potassium?

(c) Answer the following about the chemistry of potassium and calcium.

(i) Give the formula of the major product of burning each of these metals in an excess of oxygen.

(ii) Which of the two metals forms the more basic hydroxide?

(iii) Which of these two metals forms the thermally more stable carbonate?

(iv) Give an equation for the thermal decomposition of the less stable carbonate.

(d) Why does ammonium chloride have a similar crystal structure and solubility in water to potassium chloride?

(JMB)

9. Sodium hydroxide is an important material in the production of several sodium compounds. The following flow scheme includes a range of compounds that can be made using reagents A–D.

$$\text{NaCl} \longrightarrow \text{Na}$$

$$\text{NaCl} + \text{NaOCl} \xleftarrow{\ A\ } \text{NaOH} \xrightarrow{\ B\ } \text{RCOO}^-\text{Na}^+ \ (R = \text{alkyl})$$

$$\downarrow \qquad\qquad\qquad \downarrow C$$

$$\text{NaClO}_3 \qquad \text{Na}_2\text{SO}_3 \xrightarrow{\ D\ } \text{Na}_2\text{S}_2\text{O}_3$$

(a) Name reagents A–D

(b) Briefly describe the essential chemistry of the extraction of sodium from salt.

(c) Sodium thiosulphate reacts with iodine according to the equation:

$$2\text{Na}_2\text{S}_2\text{O}_3 + \text{I}_2 \rightarrow 2\text{NaI} + \text{Na}_2\text{S}_4\text{O}_6$$

(i) Explain whether the iodine molecules are being oxidised or reduced.

(ii) What mass of iodine would be required to react with 30 cm^3 of 0.10 M sodium thiosulphate solution?

(iii) Explain how you would determine the end-point when iodine is titrated with sodium thiosulphate.

(d) Sodium thiosulphate, sodium sulphate and sodium sulphite are all white crystalline solids. Describe chemical tests that will distinguish between these three compounds.

(e) Sodium chlorate(I)* is converted by heat to sodium chlorate(V)† and sodium chloride. Write a balanced equation for this reaction and, using oxidation numbers, explain the redox changes taking place. (*sodium hypochlorite; †sodium chlorate).

(NISEC)

10. Calcium and zinc atoms have the electronic structures represented below.

$$Ca \quad 1s^2 2s^2 2p^6 3s^2 3p^6 4s^2$$
$$Zn \quad 1s^2 2s^2 2p^6 3s^2 3p^6 3d^{10} 4s^2$$

(a) Explain why the values of metallic and ionic radii are smaller for zinc than for calcium although the atom of zinc contains ten more electrons.

(b) Describe *two* chemical reactions of calcium or its compounds. How do the reactions of zinc or its compounds differ from those which you have described for calcium?

   Explain the reasons for the differences between the reactions of these two elements or their compounds.

(c) What are the principles of the methods used to obtain a typical electropositive metal like calcium and a less electropositive metal like zinc? Why are the methods different?

(d) Suggest a method which could be used to prepare a sample of any zinc compound from a mixture of calcium carbonate and zinc carbonate.

(O & C)

11. The following is a simple account of the production of magnesium from sea-water:

   'Sea-water is concentrated and calcium removed by the controlled addition of carbonate ions (A). The mixture is filtered and the clear solution treated by controlled addition of hydroxide ions (B). The precipitate is thermally decomposed (C) and the residue converted into anhydrous magnesium chloride (D). Magnesium is then obtained by electrolysis (E).'

(a) Write equations, as simply as possible, for reactions A, B, C and D.

(b) What is meant by 'controlled addition'?

(c) Outline the process E, emphasising what you consider to be the important points (details of plant are *not* required).

(d) Suggest why it is difficult to produce magnesium by heating the residue from the thermal decomposition with carbon. Mention a metallic ore reducible by carbon.

(e) Write *two* commercial uses for magnesium or its compounds.

(f) Group II metals can only form one oxidation state. Using a suitable electronic configuration diagram (outer shells only), show how they achieve this.

(g) Arrange the atoms of the Group II metals in order of decreasing atomic size (i.e. put the *smallest* last). Comment on the stability of the metal cation as the group is ascended.

(h) Suggest why beryllium chloride hydrolyses completely, magnesium chloride hydrolyses partially, but barium chloride does not hydrolyse at all

(i) State the 'flame' colours for calcium, strontium and barium. Suggest why magnesium compounds fail to produce a visible 'flame' colour.

(SUJB)

12. Barium sulphide can be formed by strongly heating barium sulphate with carbon:

$$BaSO_4(s) + 4C(s) = BaS(s) + 4CO(g) \qquad (i)$$

The sulphide can then react with sodium hydroxide according to the equation:

$$BaS(s) + 2NaOH(aq) = Ba(OH)_2(s) + Na_2S(aq) \qquad (ii)$$

The sodium sulphide can be removed and barium oxide obtained by thermal decomposition of barium hydroxide.

This series of reactions can be used to make barium oxide from the sulphate.

(a) Suggest a safety precaution which should be used if barium sulphate was heated with carbon in the laboratory. Give a reason.

(b) Outline a simple practical procedure you would use to remove the sodium sulphide in reaction (ii) and then obtain pure, dry barium oxide.

(c) (i) Write an equation for the thermal decomposition of barium hydroxide.

(ii) Calculate the percentage yield if 1.53 g of barium oxide is obtained from 4.66 g of barium sulphate. (O = 16; S = 32; Ba = 137).

(d) Write out the electronic configuration of barium (atomic number = 56).

(e) Explain what type of chemical bonding you would expect in barium oxide.

(f) Draw a diagram to show the electronic structure of barium oxide but showing the outer shells (orbitals) of electrons only.

(g) Barium oxide is often used because of its refractory properties. Explain what this means and how this property is related to your answer in (e) above.

(h) What is the flame colour of barium?

(i) Using a sample of barium hydroxide, what practical procedure would you adopt to obtain the flame colour of barium?

(j) Explain, in terms of electron transitions, why the flame colour of barium is obtained by using the procedure in (i).

(SUJB)

13. A sample of magnesium sulphate is contaminated with about 1% of barium sulphate.

(a) Outline the practical procedure you would use to remove the barium sulphate and to obtain dry crystals of hydrated magnesium sulphate.

(b) (i) Give the electron configuration of barium (atomic number 56).

(ii) How would you obtain the flame colour of barium from your sample of barium sulphate?

(iii) What is the colour of the barium flame?

(iv) What explanation can you give for the origin of the flame colour of barium?

(UCLES)

14. (a) Solutions containing beryllium and aluminium ions each produce a white precipitate on treatment with aqueous sodium hydroxide. With an excess of alkali these precipitates dissolve to give a beryllium-containing ion (relative formula mass = 77) and $[Al(OH)_6]^{3-}$ respectively.

(i) What is the formula of this beryllium-containing ion?

(ii) State the shape of each of these ions.

(b) The decomposition temperatures of the Group II carbonates are given below.

$BeCO_3$	$MgCO_3$	$CaCO_3$	$SrCO_3$	$BaCO_3$
370 K	470 K	1170 K	1550 K	1630 K

(i) Write the equation for the thermal decomposition of the carbonate ion, giving state symbols.

    (ii) Suggest, with reasons, what causes the variation in decomposition temperature.

(c) Large quantities of magnesium are used in the extraction of the element titanium, Ti. In this process, titanium(IV) chloride is reacted with magnesium to produce magnesium chloride and titanium.

    (i) Write the equation for this reaction.

    (ii) What mass of magnesium is required to react with 3.8 tonnes of titanium(IV) chloride?

(UCLES)

15. Name **four** macronutrient elements essential for the healthy growth of plants and in each case describe the chemical form in which the elements are normally absorbed by the plant root from the soil, the natural primary source of these elements and the typical deficiency symptoms observed in plants growing in soils with less than adequate supplies of the nutrient.

(JMB)

14 THE d-BLOCK ELEMENTS

Understanding the Electron

Henry Taube's 30 years of research on the mobility of electrons in metal ions forms not only the basis of modern inorganic chemistry, but has led to a fuller understanding of the biochemical reactions that maintain life. His reward is this year's Nobel prize for chemistry

Henry Taube, 1983 winner of the Nobel Prize in Chemistry

Until recently, inorganic chemistry was in the academic doldrums—living in the shadow of its more illustrious sister, organic chemistry. This is due in part to the vast number of carbon compounds that make up organic chemistry, and the importance of these to biochemistry and industry.

Meanwhile, poor inorganic chemistry was left standing at the bench. In the last couple of decades, however, it too has benefited from theoretical unification due in no small part to this year's Nobel prizewinner for chemistry, Professor Henry Taube of Stanford University in California, and his work on electron transfer reactions in metal complexes. Taube's work on the mechanisms by which transition metal complexes transfer electrons forms not only the basis of modern inorganic chemistry, but has led to deeper insights into the chemical processes that maintain life.

It was in the early 1950s, when an assistant professor at the University of Chicago, that Taube began work that was to lead to last week's prize.

He is the first to admit that his work is difficult to appreciate at first glance; indeed, a TV journalist asked Taube at Stanford University's press conference last week 'Doctor, can you explain in terms that we will understand what you won the award for? What your work is?' 'Nope' was Taube's reply. But he continued 'earlier this year, when I won another award, I was asked to write out my work in words the layman could understand. I found myself writing the lectures for the first year in general chemistry'. The anecdote illustrates how seminal his work has been to basic science.

But to begin at the beginning, the electrons of chemistry are passed between atoms and molecules, rather like money is passed between people. Each atom possesses a number of electrons but, in its chemical reactions, an atom attempts to lose or gain more so that it can attain a stable electron configuration. The number of electrons that are lost or gained, to some extent, determines what kind of bonding relationship each atom has with another. The electronic ties between atoms allows them to form stable structures ranging from simple crystals to complex biomolecules and, ultimately, life.

Where living things differ from crystals is that the former undergo continuous chemical reactions. Many of these involve electron transfer mechanisms vital to energy processes in the organism.

Precursor complex

Electron transfer
through bridge

Successor complex

Taube and his students recognised that the formation of ligand bridges between two interacting complexes was one of the fundamental mechanisms of electron transfer in such complexes. An example is given above.

One of the groups of enzymes that make this possible are those containing transition metals, such as iron, copper, cobalt and molybdenum. Transition metals have the ability of changing their full complement of stabilising electrons. In other words, they have variable valency. This makes them ideal candidates for transferring electrons and it is also the reason why they make such excellent conductors of electricity.

Transition metals are well known for forming coordination complexes. These are chemical compounds made up of a metal ion with non-metallic ions or molecules, called ligands, bonded to it. Ligands can donate at least one pair of electrons to the metal ion. Typical ligands include chloride ion (Cl^-), water and ammonia. Alfred Werner developed the basic theory of coordination compounds which revolutionised our approach to the structure of inorganic compounds and for which he was awarded the Nobel chemistry prize in 1913.

Typical Werner complexes are hexaaquairon(II), $Fe(H_2O)_6^{2+}$, which is light green in colour, and tetraamminecopper(II), $Cu(NH_3)_4^{2+}$, which is deep blue. In the iron complex, the iron ion sits in the middle of an octahedron defined by the six water ligands. The copper complex has the copper ion at the centre of a tetrahedron defined by the four ammonia ligands. Later theoretical work tried to classify how these complex ions react with each other in solution. Henry Taube's great contribution has been to do the careful experimental work that backs up the theory.

Taube was the first to point out the similarities between the well known substitution mechanisms of carbon chemistry with inorganic compounds. And he further showed that the speed at which ligands were substituted on a metal ion correlated with the electronic configuration of that metal ion. This classic work, which began in the 1950s, is now to be found in energy undergraduate textbooks. It was in fact the sparsity of knowledge in this area that led Taube to take up inorganic chemistry research in the first place.

Taube recalled that he was 'bored silly' by the task of preparing an undergraduate course in coordination chemistry. But he realised in the process how little was known about the differing rates of chemical reactions. 'It shows how important the connection between teaching and research can be' he commented.

Taube's name would be remembered in chemistry even without the Nobel prize because he has a compound named after him. The Creutz–Taube complex, discovered in 1972, consists of two pentammineruthenium complexes one in the II and the other in III oxidation state, joined by a special bridging ligand that should allow electrons to pass from one metal centre to the other. In other words, electrons should be delocalised over the two metal centres making the molecule symmetrical. If this were to occur, then it would mean that the metal centres would be both of fractional oxidation state (2.5). If on the other hand the ruthenium valencies are trapped one on each end, then it would mean that the ends of the molecule are not equivalent. This should be observable. It turns out that the answer you get depends on the spectroscopic technique used. This is because each technique operates within its own timescale. For relatively slow techniques like X-ray crystallography, the complex seems to have untrapped valence states delocalised over the molecule. For fast techniques like Mossbauer spectroscopy, the valence states appear trapped. The analogy is like trying to see the spokes of a rotating wheel.

Creutz–Taube complexes are important because they model the electron transfer events that go on between metal centres in coordination complexes. But being mixed valency compounds they also model the redox action of metals in the enzymes responsible for cellular energy transfer processes.

Taube's published work (over 200 research papers and a book) is not terribly easy to read but is experimentally solid and dependable.

Taube is no stranger to prestigious awards, only last August he received $150 000 as part of the Robert A. Welch award in chemistry founded in 1972 as part of the estate of a Houston multimillionaire. With the $190 000 attached to the Nobel prize, he now has $340 000 tax free. 'I got a screen door with the Welch Award'. He said ruefully when asked what he would do with the money: 'I will have a redwood frame and the Texans will inscribe it with a plaque reading "Robert A. Welch Memorial Door".'

The wit and humour that his colleagues praised so highly last week was well in evidence at the press conference. When asked by the TV journalist what his wife's reaction was, Taube replied: 'I guess she said "really"!'

Taube becomes the 10th living Nobel Laureate at Stanford (but he is probably the only one with a collection of 8000 78 rpm records, many of them from German opera). Remarkably, he is the fourth Nobel recipient from the same department. The other chemistry winners were Paul Ber (1980), Paul Flory (1974) and Linus Pauling (1954).

**Lionel Milgrom and
Ian Anderson**

LEARNING OBJECTIVES

After you have studied this chapter you should be able to

1. Name the elements in the **first series** of d-block elements and write their electronic configurations.
2. Describe the typical physical properties of the d-block elements.
3. Describe and give specific examples of the typical chemical properties of the d-block elements.
4. Explain the following terms:
   (a) **interstitial compound,**
   (b) **paramagnetism,**
   (c) **ligand,**
   (d) **coordination number,**
   (e) **chelation,**
   (f) **stability constant,**
   (g) **absorption spectrum.**
5. Outline the chemistry of each of the elements from chromium to zinc in the first series of d-block elements.
6. Discuss the **occurrence** of elements in the first series of d-block elements.

7. Outline the **manufacture** of
   (a) iron,
   (b) **steel** using the **basic oxygen process,**
   (c) titanium,
   (d) nickel,
   (e) copper.
8. Distinguish between the different types of steel and their uses.
9. Compare different types of **alloys** and their uses.
10. Give examples of the importance of d-block elements and their compounds
    (a) in **industry,**
    (b) in **biological systems**.

14.1
Structure and Properties of d-Block Elements

THE TRANSITION ELEMENTS

A **transition element** is often simply and broadly defined as an element that occurs in either the d- or f-blocks of elements. These are shown in figure 11.3. The elements in these blocks are effectively in transition from the electropositive elements of the s-block to the electronegative elements of the p-block.

Transition elements are strictly defined as elements with partially filled d- or f-sub-shells. This definition excludes both copper and zinc from the first transition series. Zinc is not normally regarded as a transition metal. Copper *is*, however. This is because it exhibits variable oxidation number—copper(I) and copper(II). The copper(II) ion has a partially filled 3d sub-shell. The electronic configuration of its outer shell is $3s^23p^63d^9$. The copper(II) ion is thus transitional in character. Most copper(II) compounds are coloured, for example.

In order to include copper but exclude zinc from the term, a transition element is sometimes defined as an element that forms at least one ion with a partially filled d- or f-sub-shell. However, this definition excludes not only zinc from the first transition series but also scandium. Both of these metals form only one common ion each and both contain filled sub-shells:

$$Sc^{3+} \qquad 3s^23p^6$$

$$Zn^{2+} \qquad 3s^23p^63d^{10}$$

These two metals are non-transitional in character. They only have one oxidation state in their compounds. Both show little catalytic activity and both generally form white compounds.

THE d-BLOCK ELEMENTS

The d-block elements are also known as the main transition elements. Their atoms are characterised by '**inner building**' of d-sub-shells. Thus the s-orbital in the outer shell is usually filled before the d-orbitals in the shell next to the outermost shell (*see* figure 14.1). This means that the 'last' electron added by the aufbau principle (*see* section 1.2) is an inner-shell electron rather than an outer-shell electron. The chemistry of these elements is characterised by the involvement of electrons from both these shells.

The d-block elements consist of three series—in Periods 4, 5 and 6 respectively.

Figure 14.1 Energy levels of orbitals in the M and N shells of elements in the first series of d-block elements

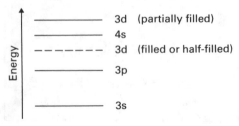

_____ 3d (partially filled)

_____ 4s

- - - - - - 3d (filled or half-filled)

_____ 3p

_____ 3s

Energy

The first series consists of the 10 elements scandium to zinc (*see* figure 14.2). It is characterised by 'inner building' of the 3d-orbitals (*see* table 14.1). The reason that the 4s-orbital is filled in preference to the five 3d-orbitals is that the 4s-orbital has a lower energy than the 3d-orbitals (*see* figure 14.1).

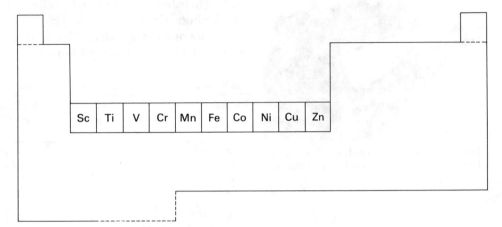

Figure 14.2   Elements in the first d-block series of elements

Table 14.1   Electronic configuration of elements in the series scandium to zinc

Element	Symbol	Atomic number	Electronic configuration			
			K	L	M	N
scandium	Sc	21	$1s^2$	$2s^2 2p^6$	$3s^2 3p^6 3d^1$	$4s^2$
titanium	Ti	22	$1s^2$	$2s^2 2p^6$	$3s^2 3p^6 3d^2$	$4s^2$
vanadium	V	23	$1s^2$	$2s^2 2p^6$	$3s^2 3p^6 3d^3$	$4s^2$
chromium	Cr	24	$1s^2$	$2s^2 2p^6$	$3s^2 3p^6 3d^5$	$4s^1$
manganese	Mn	25	$1s^2$	$2s^2 2p^6$	$3s^2 3p^6 3d^5$	$4s^2$
iron	Fe	26	$1s^2$	$2s^2 2p^6$	$3s^2 3p^6 3d^6$	$4s^2$
cobalt	Co	27	$1s^2$	$2s^2 2p^6$	$3s^2 3p^6 3d^7$	$4s^2$
nickel	Ni	28	$1s^2$	$2s^2 2p^6$	$3s^2 3p^6 3d^8$	$4s^2$
copper	Cu	29	$1s^2$	$2s^2 2p^6$	$3s^2 3p^6 3d^{10}$	$4s^1$
zinc	Zn	30	$1s^2$	$2s^2 2p^6$	$3s^2 3p^6 3d^{10}$	$4s^2$

          ↑   ↑ outer shell
'inner building' of d-sub-shell

It should be noted, however, that there are two anomalies. Both chromium and copper have single electrons in their 4s-orbitals. This is because half-filled or filled sub-shells are more stable than partially filled sub-shells. In the case of chromium, each of the five 3d-orbitals in the 3d-sub-shell contains a single electron. The sub-shell is thus half-filled. With copper, each 3d-orbital contains a pair of electrons.

PHYSICAL PROPERTIES

The d-block elements are all metals. Most have a bright lustre. They tend to be hard metals compared with the s-block metals. In particular:

- They have a **high tensile strength**. This means they can withstand high stresses and loads without fracturing.
- They are **ductile**. This means they can be drawn into wires.
- They are **malleable**. This means they can be beaten into sheets.

Most d-block elements crystallise in more than one form. Malleable and relatively soft metals such as copper crystallise in the face-centred cubic structure. Harder metals such as chromium crystallise in the body-centred cubic forms. Iron crystallises in both the face-centred cubic and body-centred cubic structures. Both cobalt and nickel crystallise in the face-centred cubic or hexagonal close-packed structures.

The d-block elements generally have high melting and boiling points (*see* table 14.2). Their hardness and high melting and boiling points result from the strong metallic bonds in the elements. This is due to their ability to release electrons from both outer and inner shells for bonding. For example, metals in the first series use both 3d- and 4s-electrons for bonding.

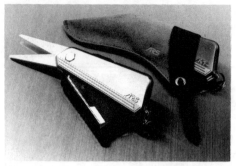

These scissors—used in industry—are made of zirconium, which is a d-block element

Table 14.2 Physical properties of elements in the series scandium to zinc

Element	Atomic (metallic) radius/nm	Density/ g cm^{-3}	Melting point/°C	Boiling point/°C	$\Delta H_{fus,m}^{\ominus}$/ kJ mol^{-1} at m.p.	$\Delta H_{vap,m}^{\ominus}$/ kJ mol^{-1} at b.p.
Sc	0.161	3.0	1540	2730	16	305
Ti	0.145	4.50	1670	3260	15.5	429
V	0.131	5.8	1900	3450	18	459
Cr	0.125	7.19	1900	2642	15	347
Mn	0.137	7.43	1250	2100	15	220
Fe	0.124	7.86	1540	3000	14	351
Co	0.125	8.90	1490	2900	15.5	381
Ni	0.125	8.90	1450	2730	18	372
Cu	0.128	8.96	1083	2600	13	305
Zn	0.133	7.14	419	906	7.4	115

The densities of d-block elements are also high compared to other metals. This is due to the relatively small radii of their atoms. The atomic radii of the metals change little across the first series (*see* table 14.2).

The d-block elements are good conductors of electricity. Those with a single outer s-electron above a half-filled or filled d-shell are particularly good conductors. Metals with a filled s-shell are less good. For example, the coinage metals copper, silver and gold, with their outer electronic structure $d^{10}s^1$, are better conductors than zinc, cadmium and mercury, which have the $d^{10}s^2$ structure. Chromium, molybdenum and tungsten, which all have d^5s^1 structure, have higher electrical conductivities than manganese, technetium and rhenium, which have the d^5s^2 structure.

CHEMICAL PROPERTIES

The electronegativities and ionisation energies of metals in the first series increases from scandium to zinc (*see* table 14.3). The elements thus become less metallic in character in crossing the series. This is also reflected in the increasingly positive redox potentials on crossing the series.

Table 14.3 Electronegativities, ionisation energies and standard redox potentials for elements in the series scandium to zinc

Element	Electronegativity	Ionisation energies/ kJ mol^{-1} at 298 K			Standard redox potentials/volts	
		1st	2nd	3rd	$M^{2+}+2e^- \rightarrow M$	$M^{3+}+3e^- \rightarrow M$
Sc	1.2	637	1241	2395	−2.08	
Ti	1.3	663	1315	2657	−1.63	−1.21
V	1.45	657	1420	2834	−1.18	−0.85
Cr	1.55	659	1598	2992	−0.91	−0.74
Mn	1.6	723	1515	3257	−1.18	−0.28
Fe	1.65	766	1567	2962	−0.44	−0.04
Co	1.7	765	1650	3234	−0.28	+0.40
Ni	1.75	743	1758	3400	−0.25	
Cu	1.75	751	1964	3558	+0.34	
Zn	1.6	912	1740	3837	−0.76	

The standard redox potentials of metals in the first series for the system M^{2+}/M are all negative with the exception of copper. These negative values would place the metals above hydrogen in the electrochemical series (*see* section 10.5). They thus might be expected to displace hydrogen from dilute mineral acids and form aqueous solutions of their metal ions. However, some of the metals react slowly with dilute acids, forming a protective oxide layer. Titanium, vanadium and chromium are all rendered passive in this way.

The d-block elements and their compounds have a number of characteristic chemical properties. These are summarised as follows:

- Formation of interstitial compounds
- Variable oxidation states
- Paramagnetism
- Formation of complex ions
- Formation of coloured compounds
- Ability to catalyse reactions

We shall examine the first five of these properties below and the ability to catalyse reactions in section 14.3.

Interstitial Compounds

d-Block elements tend to form **interstitial compounds** with non-metals with small atomic radii. These non-metals include hydrogen, carbon and nitrogen. The compounds were originally called 'interstitial' because it was thought that the non-metallic atoms found their way into the interstices in the metal lattice. **Interstices** are the spaces between atoms in the lattice. However, it is now known that the structure of an interstitial compound is different from that of the original metal. This indicates that strong bonding forces exist between the metallic and non-metallic atoms.

Interstitial compounds often have a non-stoichiometric composition and do not correspond to the normal oxidation states of the metal. Examples are, $TiH_{1.7}$, $PdH_{0.6}$ and $VH_{0.56}$. As we have already pointed out (*see* section 4.1) these substances are not strictly compounds at all. However, some interstitial compounds do have a stoichiometric composition. TiC, TiN and VN are examples.

Interstitial compounds have many of the properties of alloys containing d-block elements. For example, they are very hard, have high melting points and are good electrical conductors. Carbon steels are interstitial compounds. The interstitial carbon atoms prevent the iron atoms in the lattice from readily sliding over one another. This makes the iron harder and stronger but more brittle.

Variable Oxidation States

d-Block elements tend to exhibit more than one oxidation state in their compounds. The number of oxidation states increases across the series until a maximum is reached. It then decreases. In the first series, for example, manganese exhibits the greatest number of oxidation states.

Table 14.4 shows the stable oxides formed by metals in the first series. The table shows that all the metals in the series exhibit oxidation states of +2 or +3 or, in the case of iron and cobalt, both +2 and +3. The higher oxidation states occur in those compounds containing the more electronegative elements such as fluorine, chlorine and oxygen. These compounds contain the d-block element bound by a covalent bond to the non-metal or in the form of a complex ion. CrO_3 and TiO_2 are examples of covalent compounds and CrO_4^- and MnO_4^- examples of complex ions containing d-block elements in high oxidation states. Simple ions in high oxidation states such as Mn^{7+} are not formed since they would have too high a charge density.

Table 14.4 Stable oxides of elements in the first series of d-block elements

Element	Oxidation state						
	+1	+2	+3	+4	+5	+6	+7
Sc			Sc_2O_3				
Ti			Ti_2O_3	TiO_2			
V			V_2O_3		V_2O_5		
Cr			Cr_2O_3			CrO_3	
Mn		MnO		MnO_2			Mn_2O_7
Fe		FeO	Fe_2O_3				
Co		CoO	Co_2O_3				
Ni		NiO					
Cu	Cu_2O	CuO					
Zn		ZnO					

Only stable oxides are shown in table 14.4. d-Block elements also form unstable compounds in other oxidation states. For example, manganese can also form unstable compounds in which it exists in oxidation states +1, +3, +5 and +6. The stability of the compound relates to the ability of the inner core electrons in the atoms to screen the bonding electrons from the nuclear charge. As we have seen, the bonding electrons in metals in the first series are the 4s- and 3d-electrons.

Paramagnetism

Cations and compounds of d-block elements tend to be **paramagnetic**. The metal ions attract the magnetic lines of force in a magnetic field and move from the weakest part of the field to the strongest. The property arises because of the unpaired electrons in the ions. These unpaired electrons spin about their axes, generating a **magnetic moment**. The magnetic moment of an ion increases with the number of unpaired electrons. Thus, in the first series the ions manganese(II), Mn^{2+}, and iron(III), Fe^{3+}, both of which have five unpaired 3d-electrons, have the highest magnetic moments.

In the same series, cobalt, nickel and especially iron exhibit **ferromagnetism**. This is a form of paramagnetism in which the metals remain permanently magnetised when the magnetic field is withdrawn.

Some substances are repelled by a magnetic field. These substances are called **diamagnetic**. They repel the lines of force of a magnetic field and thus move from a strong part of the field to a weak part. Such substances contain no unpaired electrons. The ions of the s-block elements are diamagnetic.

> Two electrons in the same orbital have opposite spins. These give rise to a zero magnetic moment.

Complex Ions

One of the most important chemical properties of d-block elements is their ability to form complex ions. This ability is not confined to d-block elements. Aluminium and boron, for example, also form complex ions.

A **complex ion** is formed when one or more molecules or negatively charged ions become attached to a **central atom**. As we have seen, the molecule or negative ion that becomes attached is called a **ligand**. They are normally attached to the central atom by means of a **coordinate bond** (*see* section 2.1). The compounds formed are thus called **coordination compounds** and the number of ligands attached to the central atom is called the **coordination number**.

EXAMPLE

Coordination compound	hexaamminecobalt(III) chloride
Formula	$CoCl_3 \cdot 6NH_3$
Complex ion	hexaamminecobalt(III)
Formula	$[Co(NH_3)_6]^{3+}$
Central atom	cobalt
Ligands	six ammonia molecules
Coordination number	6

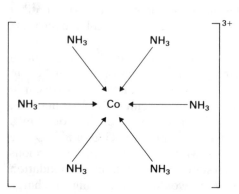

Structure of $[Co(NH_3)_6]^{3+}$ ion

The arrows in our example represent coordinate bonds or dative bonds as they are sometimes called (*see* section 2.1).

Simple ligands such as H_2O, NH_3, CN^- and Cl^- are called monodentate ligands since they can only form one coordinate bond. Monodentate means 'one tooth'. Bidentate ligands have 'two teeth'. In other words they form two coordinate bonds with the central atom. Ligands that form three coordinate bonds are called **tridentate**.

Complex ions formed by d-block elements may be neutral, positive or negative.

EXAMPLES

Neutral complex	tetracarbonylnickel(0)	$[Ni(CO)_4]$
Anionic complex	hexacyanoferrate(II)	$[Fe(CN)_6]^{4-}$
Cationic complex	hexaaquairon (III)	$[Fe(H_2O)_6]^{3+}$

Note that a segment of the Latin name is used for iron in its anionic complexes whereas in cationic complexes the name iron is used. The rules for naming complex ions and lists of the names and formulae of ligands and central atoms are given in chapter 4.

The charges on an ion are delocalised over the whole ion. The charge is the algebraic sum of the charge on the central ion and the charges on the ligands. For example, in the iron complexes shown above, the charges on the iron(II) and iron(III) central atoms and the cyano and water ligands are 2+, 3+, 1− and 0 respectively. The charges are calculated thus:

$$[Fe(CN)_6]^{4-} \qquad \text{charge} = (2+) + 6(1-) = 4-$$
$$[Fe(H_2O)_6]^{3+} \qquad \text{charge} = (3+) + 6(0) \quad = 3+$$

Some ligands are able to form ring structures with central atoms. This property is known as **chelation** and the resulting compounds known as **chelate compounds**. The ethane-1,2-diamine molecule is an example of a chelating ligand. It is bidentate (*see* figure 14.3). The ethane-1,2-diamine ligand is often represented as 'en', for example in the complex ion $[Co(en)_3]^{3+}$. It should be noted that since 'en' is bidentate, the coordination number in the complex is six.

The shape of a complex ion depends on its coordination number. Complexes with a coordination number of two are linear. Those with the coordination number of four are usually tetrahedral (*see* figure 14.4). However, some complex

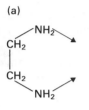
(a)

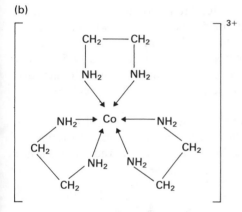
(b)

Figure 14.3 (a) Ethane-1,2-diamine bidentate chelating ligand. (b) Example of chelation: the complex ion $[Co(en)_3]^{3+}$

Coordination number 2

Linear

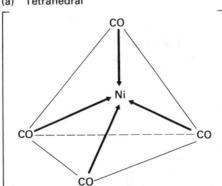

Coordination number 4
(a) Tetrahedral

(b) Square planar

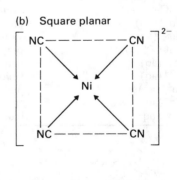

Figure 14.4 Shapes of complex ions

ions with coordination number of four are square planar. Complex ions with a coordination number of six often have octahedral structures (*see* figure 8.3, section 8.2).

1. The concentration units of stability constants are $(mol\ dm^{-1})^{-n}$ where n is the coordination number.
2. When considering the equilibrium concentrations of complex ions, the square brackets used for the formulae are often dropped. For example, $[Fe(H_2O)_6]^{2+}$ is thus written $Fe(H_2O)_6^{2+}$. The square bracket [ ] is used to indicate concentration.

Table 14.5 Logarithmic stability constants for octahedral complexes containing the ethane-1,2-diamine ligand

Formula of complex	lg K_{stab}
$[Mn(en)_3]^{2+}$	5.7
$[Fe(en)_3]^{2+}$	9.6
$[Co(en)_3]^{2+}$	13.8
$[Ni(en)_3]^{2+}$	18.1
$[Cu(en)_3]^{2+}$	18.7

In section 8.2 we considered the stability constants of complex ions. The stability constant of a complex ion is a measure of its stability. Thus, the greater its value the greater the stability of the complex. Some stability constants are extremely large. For example

$$[Fe(H_2O)_6]^{3+}(aq) + 6CN^-(aq) \rightarrow [Fe(CN)_6]^{4-}(aq) + 6H_2O(l)$$

$$K_{stab} = \left(\frac{[Fe(CN)_6^{4-}]}{[Fe(H_2O)_6^{2-}][CN^-]^6} \right)_{eq}$$

The value for this stability constant at 298 K is $10^{37}\ dm^{18}\ mol^{-6}$. This is extremely large and indicates that the formation of the hexacyanoferrate(II) complex goes virtually to completion. Since the values are often large, logarithmic stability constants are often used (*see* table 14.5).

Note that the stability of the complexes increases across the series from manganese to copper. This pattern is repeated for complexes of other ligands.

Colour

Compounds and ions of the d-block elements tend to be coloured. The colours of some of the more common ions of metals in the first series are shown in table 14.6.

Why are these ions coloured? When white light falls on a substance it may be totally reflected. In this case the substance appears white (*see* figure 14.5). If it is totally absorbed the substance appears black. However, if certain wavelengths are absorbed and others reflected, the substance appears coloured. The colour is due to the reflected wavelengths.

The absorption spectrum of the hydrated titanium(III) ion, $[Ti(H_2O)_6]^{3+}$, for example, shows that the blue, green, yellow and orange regions of the visible spectrum are absorbed (*see* figure 14.6). Light in the purple and red regions is not

Table 14.6 Colours of d-block ions

Ion	Colour
$Ti^{3+}(aq)$	purple
$V^{3+}(aq)$	green
$Cr^{3+}(aq)$	violet
$CrO_4^{2-}(aq)$	yellow
$Cr_2O_7^{2-}(aq)$	orange
$Mn^{2+}(aq)$	pink
$Mn^{3+}(aq)$	violet
$MnO_4^-(aq)$	purple
$Fe^{2+}(aq)$	pale green
$Fe^{3+}(aq)$	yellow
$Co^{2+}(aq)$	pink
$Ni^{2+}(aq)$	green
$Cu^{2+}(aq)$	blue

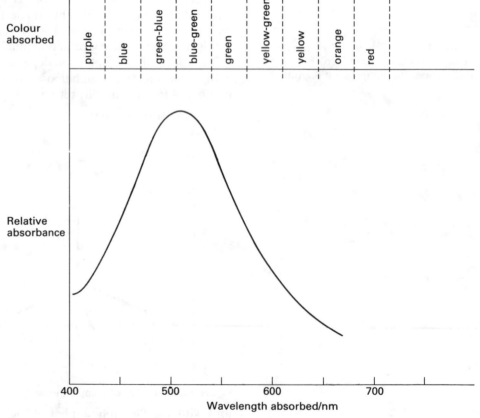

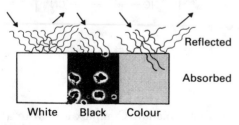

Figure 14.5 Schematic representation of light

Figure 14.6 Absorption spectrum of the $[Ti(H_2O)_6]^{3+}$ ion. The ion has a red-purple colour. All other colours are absorbed

Table 14.7 Colours of gemstones

Gemstone	Colour	Ion
blue sapphire	blue	V^{3+} or Co^{3+}
jade	green	Cr^{3+}
emerald	green	Cr^{3+}
ruby	red	Cr^{3+}
amethyst	purple	Mn^{3+}
peridot	light green	Fe^{2+}
garnet	red	Fe^{3+}
topaz	yellow	Fe^{3+}
aquamarine	pale blue	Fe^{3+}
turquoise	blue-green	Cu^{2+}

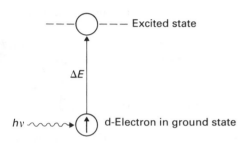

Figure 14.7 Excitation of a d-electron

absorbed to any significant extent. An aqueous solution of the ion thus appears purple.

The absorption of light is due to the presence of unpaired d-electrons in the ions. Aqueous solutions of Sc^{3+} and Zn^{2+} are colourless. These ions do not contain unpaired d-electrons. Unpaired electrons absorb light energy by becoming promoted from their ground-state energy levels to their excited-state energy levels (*see* section 1.2). The wavelength of the light absorbed depends on the energy difference ΔE between the ground state and excited state (*see* figure 14.7). The energy difference between the ground state and excited state in turn depends on the nature of the ligands in the coordination sphere of the d-block metal and the structure of the complex ion. The colour of a d-block metal ion thus also depends on the nature of the ligand and the structure of the ion. A common example of this is provided by cobalt complexes. The tetrachlorocobaltate(II) ion, $[CoCl_4]^{2-}$, is blue. When water is added to it, the pink octahedral hexaaquacobalt(II) ion is formed.

$$[CoCl_4]^{2-} + 6H_2O \rightleftharpoons [Co(H_2O)_6]^{2+} + 4Cl^-$$

blue   pink

(tetrahedral)   (octahedral)

It is this reaction that forms the basis of the 'cobalt chloride' test for water.

The colours of many gemstones are also due to the presence of traces of d-block ions (*see* table 14.7).

SUMMARY

1. d-block elements
   (a) are all metals,
   (b) are characterised by '**inner building**' of sub-shells,
   (c) have **variable oxidation states**,
   (d) have ions that tend to be **paramagnetic**,
   (e) form **complex ions**,
   (f) form **coloured compounds**,
   (g) act as **catalysts** for many chemical reactions.
2. The first, second and third series of the d-block elements occur in Periods 4, 5 and 6 respectively.
3. The d-block elements have high tensile strengths, high melting and boiling points and high densities. They are ductile and malleable.
4. Elements in the first series become less metallic on crossing the series.
5. The number of **ligands** attached to the **central atom** in a **coordination compound** is called the **coordination number**.
6. Coordination compounds in which the ligands form ring structures with the central atoms are known as **chelate compounds**.
7. The **stability constant** of a complex ion is a measure of its stability.
8. An **unpaired electron** of a d-block metal ion absorbs light of a specific wavelength and thus is promoted from its **ground state** to an **excited state**. The colour of the ion is due to unabsorbed light.

14.2
Chemistry of Seven d-Block Elements

In this section we shall outline the chemistry of seven elements in the first series of d-block elements. The occurrence, extraction and uses of these elements will be examined in the following section.

CHROMIUM

Chromium is a hard bluish white metal. It has the outer electronic configuration $3d^5 4s^1$.

As a metal, chromium is relatively unreactive. When red hot, it reacts with steam to form chromium(III) oxide, Cr_2O_3. The metal reacts slowly with dilute hydrochloric acid forming chromium(II) chloride

$$Cr(s) + 2HCl(aq) \rightarrow CrCl_2(aq) + H_2(g)$$

However, chromium(II) compounds are unstable and readily oxidise to chromium(III).

The two stable and thus most important oxidation states of chromium are +3 and +6.

Chromium(III)

This is the most common and most stable form of chromium. In solution, chromium(III) exists as the hexaaquachromium(III) ion, $[Cr(H_2O)_6]^{3+}$. This ion is violet when pure but usually appears green due to impurities. It undergoes hydrolysis, losing protons as follows:

$$[Cr(H_2O)_6]^{3+}(aq) + H_2O(l) \rightarrow [Cr(H_2O)_5(OH)]^{2+}(aq) + H_3O^+(aq)$$

The hydrated chromium(III) ion is thus stable in acid conditions. Acid shifts this equilibrium to the left. The addition of alkalis such as sodium hydroxide solution to hydrated chromium(III) ions produces a pale green precipitate of hydrated chromium(III) oxide, $Cr_2O_3 \cdot xH_2O$.

$$[Cr(H_2O)_6]^{3+}(aq) \underset{H_3O^+(aq)}{\overset{OH^-(aq)}{\rightleftharpoons}} Cr_2O_3 \cdot xH_2O(s)$$
$$\text{pale green}$$

The oxide dissolves in excess alkali forming a deep green solution

$$Cr_2O_3 \cdot xH_2O(s) \xrightarrow[\text{in excess}]{OH^-(aq)} [Cr(OH)_6]^{3-}(aq)$$
$$\text{deep green}$$

Chromium(III) compounds readily form complex ions. When ammonia solution is added in excess to a solution of a chromium(III) salt, the hexaamminechromium(III) ion, $[Cr(NH_3)_6]^{3+}$ is formed.

Chromium(III) salts can be oxidised to chromium(VI) by fusing them with sodium peroxide or by warming a solution of the salt with hydrogen peroxide under alkaline conditions. The reaction produces a mixture of chromate(VI), CrO_4^{2-}, and dichromate(VI), $Cr_2O_7^{2-}$ ions.

Chromium(VI)

Three of the most important chromium(VI) compounds are the following:

		Appearance
Chromium(VI) oxide	CrO_3	bright red needle-shaped crystals
Potassium chromate(VI)	K_2CrO_4	yellow
Potassium dichromate(VI)	$K_2Cr_2O_7$	orange

The structures of the chromate(VI) and dichromate(VI) ions are shown in figure 14.8.

Chromium(VI) oxide is an acidic oxide. It reacts with alkalis forming chromate(VI) ions

$$CrO_3(s) + 2OH^-(aq) \rightarrow CrO_4^{2-}(aq) + H_2O(l)$$
$$\text{bright red} \qquad\qquad \text{yellow}$$

Under acidic conditions the chromate(VI) ion is converted to the dichromate(VI) ion. The reaction is reversed under alkaline conditions

$$2CrO_4^{2-}(aq) + 2H_3O^+(aq) \underset{\text{alkali}}{\overset{\text{acid}}{\rightleftharpoons}} Cr_2O_7^{2-}(aq) + 3H_2O(l)$$
$$\text{yellow} \qquad\qquad\qquad \text{orange}$$

Dichromate(VI) is reduced to chromium(III) under acidic conditions

$$Cr_2O_7^{2-}(aq) + 14H^+(aq) + 6e^- \rightarrow 2Cr^{3+}(aq) + 7H_2O \qquad E^\ominus = +1.33V$$

The positive value of the **standard redox potential**, $E^\ominus$, shows that dichromate(VI) is an oxidising agent (*see* section 10.4). It is used as such in **volumetric analysis** to determine the concentration of iron(II) in acidic solution

$$Cr_2O_7^{2-}(aq) + 14H^+(aq) + 6Fe^{2+}(aq) \rightarrow 2Cr^{3+}(aq) + 6Fe^{3+}(aq) + 7H_2O(l)$$

It should be noted that the standard redox potential for Fe^{3+}/Fe^{2+} is less positive than that for $Cr^2O_7^{2-}/Cr^{3+}$

$$Fe^{3+}(aq) + e^- \rightarrow Fe^{2+} \qquad E^\ominus = +0.76 \text{ V}$$

The acidified iron(II) solution is pipetted into a conical flask and the dichromate(VI) solution added from a burette. A **redox indicator** such as barium diphenylamine sulphonate is used. The first drop of excess dichromate(VI) solution converts this colourless substance to a deep blue colour.

Potassium chromate (VI) can be obtained in a high degree of purity and can thus be used as a **primary standard**.

The standard redox potential for the $Cr_2O_7^{2-}/Cr^{3+}$ is less positive than that for the Cl_2/Cl^- system

$$\tfrac{1}{2}Cl_2(g) + e^- \rightarrow Cl^-(aq) \qquad E^\ominus = +1.36 \text{ V}$$

Dichromate(VI), unlike manganate(VII), does not, therefore, oxidise chloride ions.

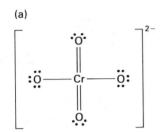

(a)

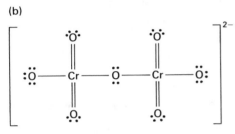

(b)

Figure 14.8 (a) Chromate(VI). (b) Dichromate(VI)

Primary Standards

A solution of accurately known concentration is a **standard solution**. A standard solution is prepared by dissolving a known mass of a primary standard in water and making it up to a known volume. A primary standard must be either absolutely pure or of known purity. Many different primary standards are used in **titrations**. The following are examples.

Type of titration	Primary standard
acid–base	Na_2CO_3
complex formation	$AgNO_3$
precipitation	$AgNO_3$
oxidation–reduction	$K_2Cr_2O_7$

MANGANESE

Manganese is a hard grey metal. It has the electronic configuration $3d^5 4s^2$.

The metal is attacked by water and reacts with acids forming manganese(II) ions

$$Mn(s) + 2HCl(aq) \rightarrow MnCl_2(aq) + H_2(g)$$

Manganese exhibits oxidation states +2, +3, +4, +6 and +7 in various compounds. The covalent character of manganese compounds increases as the oxidation state increases. The acidity of the oxides also increases with increasing oxidation state.

Manganese(II)

This is the most stable form of manganese. It has the outer electron configuration $3d^5$. Each of the five 3d-orbitals thus contains one electron.

Manganese(II) ions are hydrated in aqueous solution forming the pale pink hexaaquamanganese(II) complex ion, $[Mn(H_2O)_6]^{2+}$. This ion is stable in acidic conditions but produces a white precipitate of manganese(II) hydroxide, $Mn(OH)_2$, in alkaline solutions. Manganese(II) oxide is a basic oxide.

Manganese(III)

Manganese(III) only occurs in complexes. It is unstable. Under acid conditions manganese(III) **disproportionates** to manganese(II) and manganese(IV).

Manganese(IV)

The most important manganese(IV) compound is manganese(IV) oxide, MnO_2. This is a black compound, which is insoluble in water. It is thought to have an ionic structure. Its stability is due to its high lattice enthalpy.

Manganese(IV) oxide is weakly amphoteric. It is a powerful oxidising agent. For example, it releases chlorine from concentrated hydrochloric acid

$$MnO_2(s) + 4HCl(aq) \rightarrow MnCl_2(aq) + 2H_2O(l) + Cl_2(g)$$

This is sometimes used as a laboratory preparation for chlorine (*see* section 16.1).

Manganese(VI)

This is an unstable oxidation state. Potassium manganate(VI) can be prepared by fusing manganese(IV) oxide with a strong oxidising agent such as potassium chlorate(V) or potassium nitrate

$$3MnO_2(s) + KClO_3(l) + 6KOH(l) \rightarrow 3K_2MnO_4(l) + KCl(l) + 3H_2O(g)$$

Potassium manganate(VI) is green. It is only stable in alkaline solution. In acidic solution it disproportionates to manganese(IV) and manganese(VII)

$$\underset{\text{green}}{3MnO_4^{2-}(aq)} + 4H^+(aq) \rightarrow \underset{\text{black}}{MnO_2(s)} + \underset{\text{black}}{2MnO_4^-(aq)} + 2H_2O(l)$$

Manganese(VII)

Manganese exists in this oxidation state in the strongly acid oxide Mn_2O_7. However, by far the most important compound containing manganese(VII) is potassium manganate(VII), $KMnO_4$. The solid is highly soluble in water, forming a deep purpose solution. The ion manganate(VI), MnO_4^-, has a tetrahedral structure. Under slightly acidic conditions it gradually decomposes forming manganese(IV) oxide

$$4MnO_4^-(aq) + 4H^+(aq) \rightarrow 4MnO_2(s) + 2H_2O(l) + 3O_2(g)$$

Under alkaline conditions potassium manganate(VII) is reduced to the green potassium manganate(VI) and then to manganese(IV) oxide.

Potassium manganate(VII) is a powerful oxidising agent. Under acidic conditions it is reduced to manganese(II) ions. The standard redox potential for this system is $+1.52$ V. This is more than that for the Cl_2/Cl^- system. Manganate(VII) thus oxidises chloride to chlorine gas

$$\tfrac{1}{2}Cl_2(g) + e^- \rightarrow Cl^-(aq) \qquad E^\ominus = +1.36 \text{ V}$$

$$MnO_4^-(aq) + 8H^+(aq) + 5e^- \rightarrow Mn^{2+}(aq) + 4H_2O(l) \qquad E^\ominus = +1.52 \text{ V}$$

Thus

$$2MnO_4^-(aq) + 16H^+(aq) + 10Cl^-(aq) \rightarrow 2Mn^{2+}(aq) + 8H_2O(l) + 5Cl_2(g)$$

Potassium manganate(VII) is used widely as an odixising agent in the laboratory, for example

- in the preparation of oxygen and chlorine (*see* chapters 15 and 16),
- as a test for sulphur dioxide and hydrogen sulphide (*see* chapter 15),
- as an oxidant in preparative organic chemistry (*see* chapter 19),
- as a volumetric reagent in redox titrations.

It is used, for example, to estimate iron(II) and ethanedioates

$$5Fe^{2+}(aq) + MnO_4^-(aq) + 8H^+(aq) \rightarrow 5Fe^{3+}(aq) + Mn^{2+}(aq) + 4H_2O(l)$$

$$5C_2O_4^{2-}(aq) + 2MnO_4^-(aq) + 16H^+(aq) \rightarrow 10CO_2(g) + 2Mn^{2+}(aq) + 8H_2O(l)$$

However, since potassium manganate(VII) cannot be obtained in a high degree of purity, it cannot be used as a primary standard.

IRON

Iron is a grey metal with the outer electronic configuration $3d^6 4s^2$. When pure it is relatively soft, malleable and ductile. It is slowly attacked by moist air to form hydrated iron(III) oxide, $Fe_2O_3 \cdot xH_2O$, or **rust** as it is usually known (*see* section 10.5).

The metal reacts with steam, forming a black solid, Fe_3O_4 (iron(II) diiron(III) oxide)

$$3Fe(s) + 4H_2O(g) \rightleftharpoons 3Fe_3O_4(s) + 4H_2(g)$$

$Fe_3O_4(s)$ is an example of a mixed oxide (*see* section 15.4).

Iron is above hydrogen in the electrochemical series. It liberates hydrogen from dilute hydrochloric acid or dilute sulphuric acid

$$Fe(s) + 2HCl(aq) \rightarrow FeCl_2(aq) + H_2(g)$$

Iron exhibits several oxidation states in its compounds, of which the two most important are

iron(II)   electronic configuration $3d^6$

iron(III)   electronic configuration $3d^5$

Iron(III) is more stable than iron(II) since each of its five 3d-orbitals is singly occupied by an electron.

Iron forms two oxides, iron(II) oxide, FeO, and iron(III) oxide, Fe_2O_3, as well as the mixed oxide, Fe_3O_4 (*see* above). The oxides are basic, although Fe_2O_3 is slightly amphoteric in character.

The iron(II) ion, Fe^{2+}, is hydrated in aqueous solution. It exists as the pale green hexaaquairon(II) complex ion, $[Fe(H_2O)_6]^{2+}$ (*see* figure 14.9). This complex ion is formed, for example, when the metal reacts with dilute hydrochloric acid (*see* above). The ion is stable in acidic solutions. In neutral or alkaline solutions it is unstable and forms the hydrated iron(III) ion, $[Fe(H_2O)_6]^{3+}$ (*see* figure 14.9). This ion is pale purple when in solution. However, it readily hydrolyses, forming yellow aquahydroxo complexes. For example

$$[Fe(H_2O)_6]^{3+}(aq) \rightleftharpoons [Fe(H_2O)_5OH]^{2+}(aq) + H^+(aq)$$

pale purple   yellow

Iron(II) and iron(III) salts can be distinguished in the following ways.

Addition of an Alkali

Addition of an alkali such as sodium hydroxide solution to an aqueous solution of an iron(II) salt produces a dirty green precipitate of iron(II) hydroxide

$$[Fe(H_2O)_6]^{2+}(aq) + 2OH^-(aq) \rightarrow Fe(OH)_2(s) + 6H_2O(l)$$

dirty green

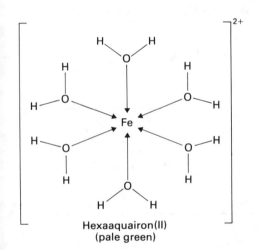

Hexaaquairon(II)
(pale green)

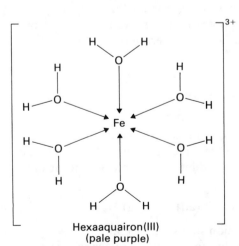

Hexaaquairon(III)
(pale purple)

Figure 14.9 Hexaaquairon complexes

The precipitate is slowly oxidised by air, forming hydrated iron(III) oxide, $Fe_2O_3 \cdot xH_2O$.

Addition of an alkali to a solution containing iron(III) produces a reddish brown precipitate of iron(III) hydroxide

$$[Fe(H_2O)_6]^{3+}(aq) + 3OH^-(aq) \rightarrow Fe(OH)_3(s) + 6H_2O(l)$$
$$\text{reddish brown}$$

Addition of Potassium Thiocyanate Solution

When a solution of potassium thiocyanate, KSCN, is added to a solution containing hydrated iron(III) ions a deep red colour is produced due to the formation of the hydrated thiocyanate complex, $[Fe(SCN)(H_2O)_5]^{2+}$. No colouration is produced with iron(II) ions. This test is very sensitive and will produce a red colour even in the presence of trace amounts of iron(III) ions.

Addition of Potassium Hexacyanoferrate(II) Solution and Potassium Hexacyanoferrate(III) Solutions

The addition of hexacyanoferrate(II) solution to solutions containing iron(III) ions produces a deep blue precipitate known as Prussian blue. The same precipitate is produced when a hexacyanoferrate(III) solution is added to a solution containing iron(II) ions. In this case the precipitate is known as Turnbull's blue. The reactions are summarised in table 14.8. For clarification the oxidation number of the iron atoms are included in the formulae. Prussian blue and Turnbull's blue are identical. They have the formula $K^IFe^{III}[Fe^{II}(CN)_6]$.

Both iron(II) and iron(III) exist in a number of important salts such as the sulphates and halides. They both also form **double salts** known as **alums** (*see* section 15.1)

ammonium iron(II) sulphate-7-water	$FeSO_4 \cdot (NH_4)_2SO_4 \cdot 7H_2O$
ammonium iron(III) sulphate-12-water	$NH_4Fe(SO_4)_2 \cdot 12H_2O$

Table 14.8 Hexacyanoferrate reactions

	Potassium hexacyanoferrate(II) $K_4[Fe^{II}(CN)_6](aq)$	Potassium hexacyanoferrate(III) $K_3[Fe^{III}(CN)_6]$
$Fe^{2+}(aq)$	white precipitate	deep blue precipitate (Turnbull's blue)
$Fe^{3+}(aq)$	deep blue precipitate (Prussian blue)	brown solution

These double salts are not coordination compounds. For example, in solution, ammonium iron(II) sulphate-7-water gives positive tests for three separate ions: NH_4^+, Fe^{2+} and SO_4^{2-}. In comparison, potassium hexacyanoferrate(II) only gives positive tests for two ions: K^+ and hexacyanoferrate(II), $[Fe(CN)_6]^{4-}$ It does not give a positive test for Fe^{2+}.

COBALT

Cobalt is a hard bluish white metal with the outer electronic configuration $3d^74s^2$. It is fairly unreactive compared with iron. At room temperature, it does not react with air or water and only reacts slowly with dilute hydrochloric acid or dilute sulphuric acid

$$Co(s) + 2HCl(aq) \rightarrow CoCl_2(aq) + H_2(g)$$

In compounds, cobalt can exist in oxidation states +2 or +3.

Cobalt(II) ions exist in aqueous solution as hexaaquacobalt(II) ions, $[Co(H_2O)_6]^{2+}$. These are stable and, as we saw above, pink. Addition of sodium hydroxide solution to a solution containing these ions produces a blue precipitate of cobalt(II) hydroxide. Addition of ammonia solution also produces this precipitate. The precipitate dissolves in excess ammonia, forming the hexaamminecobalt(II) ion. This is pale yellow.

$$[Co(H_2O)_6]^{2+}(aq) + 6NH_3(aq) \rightarrow [Co(NH_3)_6]^{2+}(aq) + 6H_2O(l)$$

Cobalt(II) occurs in a number of tetrahedral anionic complexes. An example is tetrachlorocobaltate(II), $[CoCl_4]^{2-}$. This can be produced by adding concentrated hydrochloric acid to an aqueous cobalt(II) salt

$$\underset{\text{pink}}{[Co(H_2O)_6]^{2+}(aq)} + 4Cl^-(aq) \rightleftharpoons \underset{\text{blue}}{[CoCl_4]^{2-}(aq)} + 6H_2O(l)$$

This reaction is reversed by adding water.

The hydrated cobalt(III) ion is a powerful oxidising agent. It is not stable in water, owing to the following reaction

$$4[Co(H_2O)_6]^{3+}(aq) + 2H_2O(l) \rightarrow [Co(H_2O)_6]^{2+}(aq) + 4H^+(aq) + O_2(g)$$

However, the cobalt can be stabilised in the +3 state by the presence of strong complexing ligands such as NH_3 and NO_2^- Hexaamminecobalt(III), for example, can be formed by the reaction of hydrated cobalt(II) ions with ammonia solution in the presence of air or hydrogen peroxide

$$\underset{\text{pink}}{[Co(H_2O)_6]^{2+}(aq)} + 6NH_3(aq) \rightarrow [Co(NH_3)_6]^{3+}(aq) + 6H_2O(l) + e^-$$

Solutions of cobalt(II) salts are also oxidised to the +3 state by excess sodium nitrite in the presence of ethanoic acid

$$[Co(H_2O)_6]^{2+}(aq) + 7NO_2^-(aq) + 2H^+ \rightarrow [Co(NO_2)_6]^{3+} + NO(g) + 7H_2O(l)$$

Sodium hexanitrocobaltate(III), $Na_3[Co(NO_2)_6]$ is a yellow octahedral complex and is soluble in water. It is used as a qualitative test for potassium ions. The reaction of the sodium compound with potassium ions produces $K_2[Co(NO_2)_6]$. This compound is yellow but unlike the sodium compound is insoluble in water. A yellow precipitate is thus obtained if the test is positive.

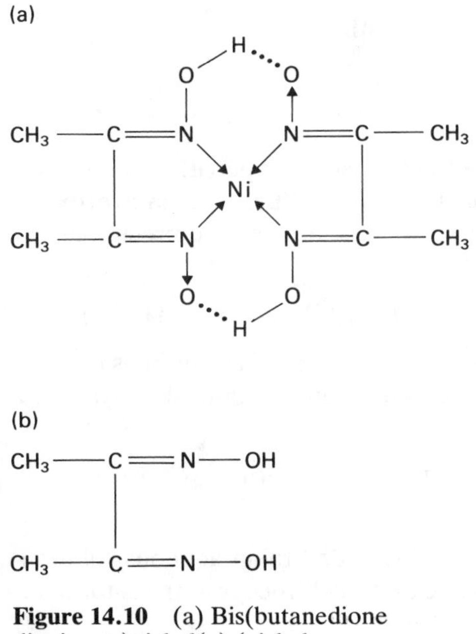

NICKEL

Nickel has the outer electronic configuration $3d^8 4s^2$. Its most stable oxidation state is +2 and, unlike other d-block metals, tends not to exist in other oxidation states. However, it does form complexes.

Aqueous solutions of nickel(II) salts contain the hexaaquanickel(II) ion, $[Ni(H_2O)_6]^{2+}$. When ammonia solution is added to a solution containing these ions, nickel(II) hydroxide is precipitated as a green gelatinous solid. This dissolves in excess ammonia due to the formation of hexaamminenickel(II) ions, $[Ni(NH_3)_6]^{2+}$.

Nickel forms both tetrahedral and square planar complexes. For example, tetrachloronickelate(II), $[NiCl_4]^{2-}$, is tetrahedral and tetracyanonickelate(II), $[Ni(CN)_4]^{2-}$, is square planar (see figure 14.4).

In both qualitative and quantitative analysis, nickel(II) ions are detected using an alkaline solution of butanedione dioxime—also known as dimethylglyoxime. The red coordination compound called bis(butanedione dioximato)nickel(II) is formed (see figure 14.10). Note that this is a chelate compound and that the butanedione dioximato ligand is bidentate.

(a)

(b)

Figure 14.10 (a) Bis(butanedione dioximato)nickel(II) (nickel dimethylglyoxime). (b) Butanedione dioxime

COPPER

Copper is a bright golden coloured metal. It has the outer electronic configuration $3d^{10}4s^1$. It is the least reactive of the metals in the first series of d-block elements.

It is not attacked by water and does not react with dilute hydrochloric or dilute sulphuric acids. However, it does react with concentrated sulphuric acid and both dilute and concentrated nitric acid (*see* chapter 15).

Copper exhibits two oxidation states in its compounds: +1 and +2. The +2 state is more stable.

Copper(I)

Many copper(I) compounds are white or colourless. This is because the five 3d-orbitals are completely filled with electrons. However, the oxide Cu_2O is reddish brown.

Copper(I) ions are unstable in aqueous solution since they readily disproportionate

$$2Cu^+(aq) \rightarrow Cu^{2+}(aq) + Cu(s)$$

Copper(I) does exist, however, in compounds that are insoluble in water or in complexes. For example, the dichlorocuprate(I) ion, $[CuCl_2]^-$, is stable. This can be made by adding concentrated hydrochloric acid to copper(I) chloride

$$CuCl(s) + Cl^-(aq) \rightarrow [CuCl_2]^-(aq)$$

Copper(I) chloride is a white insoluble solid. Like other copper(I) halides it is covalent in character and more stable than the copper(II) halide. Copper(I) chloride can be prepared by strongly heating copper(II) chloride

$$2CuCl_2(s) \rightarrow 2CuCl(s) + Cl_2(g)$$

Alternatively, it can be prepared by boiling a mixture of copper(II) chloride with copper in concentrated hydrochloric acid. This involves the formation of the complex ion $[CuCl_2]^-$ as an intermediate. When a solution containing this ion is poured into water, copper(I) chloride is precipitated.

Copper(I) chloride reacts with a concentrated solution of ammonia to form the diamminecopper(I) complex $[Cu(NH_3)_2]^+$. The complex is colourless in the absence of oxygen but turns blue when it reacts with oxygen.

Copper(II)

Copper(II) ions exist in aqueous solution as the hexaaquacopper(II) complex ion, $[Cu(H_2O)_6]^{2+}$. These have a characteristic blue colour. When sodium hydroxide solution is added to a solution containing these ions, a pale blue precipitate of hydrated copper(II) hydroxide is produced

$$[Cu(H_2O)_6]^{2+}(aq) + 2OH^-(aq) \rightarrow [Cu(H_2O)_4(OH)_2](s) + 2H_2O(l)$$

Hydrated copper(II) hydroxide is also produced if ammonia solution is used as an alkali. The hydroxide then dissolves in excess ammonia forming the royal blue diaquatetraammine complex

$$[Cu(H_2O)_4(OH)_2](s) + 4NH_3(aq) \rightarrow [Cu(NH_3)_4(H_2O)_2]^{2+}(aq) + 2OH^-(aq) + 2H_2O(l)$$
$$\text{pale blue solid} \qquad\qquad\qquad \text{royal blue solution}$$

If an excess of concentrated hydrochloric acid is added to an aqueous solution containing copper(II) ions, the anionic complex tetrachlorocuprate(II) is formed. This is yellow.

$$[Cu(H_2O)_6]^{2+}(aq) + 4Cl^-(aq) \underset{H_2O}{\overset{HCl}{\rightleftharpoons}} [CuCl_4]^{2-}(aq) + 6H_2O(l)$$
$$\text{blue} \qquad\qquad\qquad\qquad\qquad\qquad \text{yellow}$$

On dilution the solution turns green and finally blue due to the formation of hydrated copper(II) ions. The green colour results from the presence of both the yellow anionic complex and the blue hydrated copper(II) ions in solution.

Iodide ions can be used to reduce copper(II) to copper(I). This reaction is used in volumetric analysis to estimate the concentration of copper(II) in a solution. Potassium iodide is added to the solution containing the copper(II) salt. This results in a precipitate of white copper(I) iodide

$$2Cu^{2+}(aq) + 4I^-(aq) \rightarrow 2CuI(s) + I_2(aq)$$
$$\text{white}$$

The iodine formed in the reaction is then titrated against standard sodium thiosulphate solution.

Copper(II) is also reduced to copper(I) by aldehydes and reducing sugars such as glucose. This is the basis of Fehling's test. **Fehling's solution** is an alkaline solution containing copper(II) sulphate and potassium sodium 2,3-dihydroxylbutanedioate (otherwise known as potassium sodium tartrate or Rochelle salt). If Fehling's solution is warmed with a solution containing an aldehyde or reducing sugar, an orange precipitate of copper(I) oxide is produced

$$2Cu^{2+}(aq) + 2OH^-(aq) \rightarrow Cu_2O(s) + H_2O(l)$$

The potassium sodium salt forms a complex with the copper(II) ions. This prevents the formation of copper(II) hydroxide in the alkaline conditions.

ZINC

Zinc has the outer electronic structure $3d^{10}4s^2$. The metal displays few of the properties of the other d-block metals in the series. This is because all five of its 3d-orbitals are completely filled with electrons. Zinc exhibits only one oxidation state in its compounds. This is +2 and is due to the loss of the two 4s-electrons.

Zinc is a relatively reactive metal. It is used in the laboratory to prepare hydrogen from dilute hydrochloric acid

$$Zn(s) + 2H^+(aq) \rightarrow Zn^{2+}(aq) + H_2(g)$$

One of the most interesting compounds of zinc is zinc oxide, ZnO. This is a white powder which can be prepared by heating the metal in air or by heating zinc carbonate or zinc hydroxide. When the oxide is heated it turns yellow due to the loss of a small amount of oxygen. However, on cooling it regains the oxygen and becomes white again.

Zinc oxide is **amphoteric**. It reacts with acids forming zinc ions

$$ZnO(s) + 2H^+(aq) \rightarrow Zn^{2+}(aq) + 2H_2O(l)$$

With alkalis, zinc oxide forms the tetrahydroxozincate(II) anionic complex

$$ZnO(s) + 2OH^-(aq) + H_2O(l) \rightarrow [Zn(OH)_4]^{2-}(aq)$$
$$\text{tetrahydroxozincate(II)}$$

Zinc hydroxide is also amphoteric and like the oxide is insoluble in water. It is produced as a white gelatinous precipitate when an alkali is added to an aqueous solution of a zinc salt

$$Zn^{2+}(aq) + 2OH^-(aq) \rightarrow Zn(OH)_2(s)$$

When the hydroxide is added to an aqueous ammonia solution it dissolves forming tetraamminezinc, $[Zn(NH_3)_4]^{2+}$.

SUMMARY

1. In aqueous solution chromium(III) ions undergo hydrolysis

$$[Cr(H_2O)_6]^{3+}(aq) \xrightarrow{H_2O} [Cr(H_2O)_5(OH)]^{2+}(aq)$$

2. In excess ammonia, an aqueous solution of chromium(III) forms hexaamminechromium(III), $[Cr(NH_3)_6]^{3+}$.

3.
$$2CrO_4{}^{2-}(aq) \underset{\text{alkali}}{\overset{\text{acid}}{\rightleftharpoons}} Cr_2O_7{}^{2-}(aq)$$
$$\text{chromate(VI)} \qquad \text{dichromate(VI)}$$

4. Dichromate(VI) is an oxidising agent. It is used to determine iron(II).

5. MnO_2 and $KMnO_4$ are used in the laboratory preparation of chlorine.

6. Manganate(VII) is a powerful oxidising agent. It is used to determine iron(II) and ethanedioates.

7. Iron forms two oxides and a mixed oxide.

8. A solution of iron(III) ions undergoes **hydrolysis**

$$[Fe(H_2O)_6]^{3+}(aq) \rightarrow [Fe(H_2O)_5(OH)]^{2+}(aq)$$

9. Iron(II) can be distinguished from iron(III) by the addition of an alkali or by addition of potassium thiocyanate solution.

10.
$$[Co(H_2O)_6]^{2+}(aq) \xrightarrow{\text{excess } NH_3} [Co(NH_3)_6]^{2+}(aq)$$
$$\text{hexaaquacobalt(II)} \qquad\qquad \text{hexaamminecobalt(II)}$$

11.
$$[Co(H_2O)_6]^{2+}(aq) + 4Cl^-(aq) \rightarrow [CoCl_4]^{2-}(aq) + 6H_2O(l)$$
$$\text{(pink)} \qquad\qquad\qquad \text{tetrachlorocobaltate}$$
$$\text{(blue)}$$

12. When ammonia is added in excess to a solution of nickel(II) ions, hexaamminenickel(II) ions, $[Ni(NH_3)_6]^{2+}$, are formed.

13. Copper(I) ions are unstable in aqueous solution. They **disproportionate**.

14. Copper(I) halides are stable as solids and covalent in character.

15. On addition of alkali to aqueous solutions of copper(II) ions

$$[Cu(H_2O)_6]^{2+}(aq) \rightarrow [Cu(H_2O)_4(OH)_2](s)$$
$$\text{pale blue precipitate}$$

On addition of excess ammonia the precipitate dissolves

$$[Cu(H_2O)_4(OH)_2](s) \xrightarrow{\text{excess } NH_3(aq)} [Cu(NH_3)_4(H_2O)_2]^{2+}(aq)$$
$$\text{diaquatetraamminecopper(II)}$$
$$\text{(royal blue solution)}$$

16. Copper(II) is used to test for aldehydes and reducing sugars.

17. Zinc displays few properties characteristic of transition metals.

18. Zinc oxide and zinc hydroxide are **amphoteric**.

14.3
Occurrence, Extraction and Uses

In this section we shall confine our attention principally to the occurrence, extraction and uses of the more common and more important metals in the first series of d-block elements.

OCCURRENCE

The abundance, important sources and occurrence of metals in the first series of d-block elements are shown in table 14.9. The list is by no means exclusive. Many of the metals occur in many different **minerals** and these are distributed widely

Table 14.9 Occurrence of metals in the first series of d-block elements

Element	Abundance/grams of element per tonne of Earth's crust	Important sources		Occurrence
		Name	Formula	
scandium	22	thortveitite	$(Sc,Y)_2Si_2O_7$	Scandinavia, Malagasy
titanium	5.7×10^3	ilmenite	$FeO \cdot TiO_2$	Scandinavia, India, UK
		rutile	TiO_2	occurs widely: Brazil, Australia, USSR, Europe
vanadium	135	vanadite	$(PbO)_9(V_2O_5)_3PbCl_2$	USA, Mexico, USSR, Argentina, Scotland, Africa
chromium	100	chromite	$FeCr_2O_4$	USSR, South Africa, Turkey, Philippines, Zimbabwe
		crocoite	$PbCrO_4$	Brazil, Philippines, Tasmania, Zimbabwe
manganese	950	pyrolusite	MnO_2	West Germany, USSR, India, Brazil, Cuba, USA
		nodules	—	ocean floor
iron	5.6×10^4	haematite	Fe_2O_3	USSR, Switzerland, Italy, UK, Brazil
		magnetite	Fe_3O_4	Sweden, USA, South Africa, Canada, Italy, Norway
cobalt	25	cobaltite	$(Co,Fe)AsS$	USSR, Poland, UK, East Germany, India, Australia
nickel	75	pentlandite	$NiS \cdot 2FeS$	USA, Norway, Cuba, Japan
copper	55	native chalcopyrite cuprite malachite	Cu $CuFeS_2$ Cu_2O $CuCO_3 \cdot Cu(OH)_2$	USA, USSR, UK, Australia, Bolivia, Chile, Peru, Mexico, Zambia, Zaire, Zimbabwe and other countries
zinc	70	sphalerite (zinc blende)	ZnS	USA, Mexico, Britain and Europe
		smithsonite	$ZnCO_3$	

Transport of minerals at a manganese mine in Gabon

throughout the earth's crust. Note that the most abundant of the d-block elements on Earth is iron, followed by titanium. Indeed iron is the second most abundant metal on Earth—aluminium being the most abundant. Note also that most of the metals occur as oxides, sulphides or both. Only copper occurs in its native form. d-Block elements also occur on the Moon.

Lunar d-Block Elements

The mineralogy of the Moon is less complicated than that of the Earth. Two of the three major lunar minerals contain d-block elements. The three major minerals are:

ilmenite	$FeTiO_3$
pyroxene	$(Ca,Mg,Fe)SiO_3$
calcic plagioclase	$CaAl_2Si_2O_8$

The following minerals are also common on the Moon:

olivine	$(Mg,Fe)_2SiO_4$
pyroxferroite	$CaFe_6(SiO_3)_7$
cristobalite and tridymite	SiO_2

Some less common lunar minerals also contain d-block elements, for example:

ulvospinel	Fe_2TiO_4
armacolite	$(Fe,Mg)Ti_2O_5$
troilite	FeS
chromite	$FeCr_2O_4$

Metallic iron and nickel have also been found on the Moon.

Radioactive dating of lunar rocks shows that they crystallised 3.2 to 4.2×10^9 years ago. This is about the age of the oldest rocks found on Earth (*see* section 1.3).

Astronaut Edwin Aldrin (lunar module pilot) carries two components of the Early Apollo Scientific Experiments Package during the Apollo II lunar surface extravehicular activity in 1969

Open cut mining of copper ore in Papua New Guinea

(a)

(b)

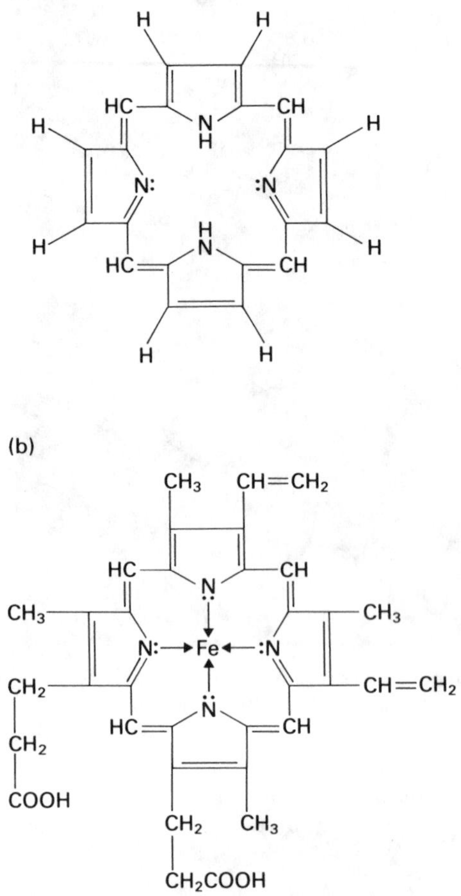

Figure 14.11   (a) The porphin molecule.
(b) Haem

d-Block elements are also found as **trace elements** in plants, animals and gemstones (*see* table 14.6). Iron, manganese, copper, nickel and zinc are all required by plants and animals and to a lesser extent so are cobalt and chromium. These trace elements are known as micronutrients (*see* also chapter 13). For example, iron occurs in blood as part of the protein haemoglobin. Haemoglobin has a large structure. It has a relative molecular mass of about 68 000 and the empirical formula

$$(C_{759}H_{1208}N_{210}S_2O_{204}Fe)_4$$

The basic unit that contains iron in haemoglobin is called haem. This contains iron(II) with a coordination number of four in a chelate structure.

Haem

The porphin molecule (*see* figure 14.11a) with its eight outermost hydrogens substituted by other groups is called a porphyrin. A porphyrin molecule can act as a tetradentate chelating ligand for metals such as magnesium, iron, copper and zinc. Chlorophyll (*see* figure 13.13) is a magnesium porphyrin complex. Haem is an iron porphyrin complex (*see* figure 14.11b).

We shall examine further the importance of some of these metals as trace elements in biological systems later in this section.

MANUFACTURE OF IRON AND STEEL

Iron comprises over 5% of the Earth's crust. The principal ores used for the extraction of iron are haematite, Fe_2O_3, and magnetite, Fe_3O_4. These ores may contain between 20 and 70% iron by mass. The main impurities present in the ores are sand (silicon(IV) oxide, SiO_2) and alumina (aluminium oxide, Al_2O_3).

The Earth's Core

Indirect evidence suggests that the Earth's core consists mainly of an iron alloy. The core has a radius of about 3470 kilometres (2150 miles). The radius of the Earth itself is 6370 kilometres (3960 miles). The inner core of the Earth is thought to be a solid with a radius of about 1200 kilometres (750 miles). Around this is a liquid outer core. The turbulent flow of the liquid in this part of the core generates the magnetic field of the Earth. Pressures inside the core range from 1.3 to 3.5 million atmospheres and the temperatures from 4000 to 5000°C.

The core is known to be predominantly iron although there is no consensus on its exact composition. About 8 to 10% of the core probably consists of other elements such as nickel, sulphur (as iron sulphide), oxygen (as iron oxide) and silicon (as iron silicide).

At least 12 countries in the world have proven reserves of over 10^9 tonnes of iron ores. These include Australia, Canada, the United States, South Africa, India, the USSR and France. Each year of the order of 700 million tonnes of steel are produced. The major producers are the Soviet Union, the USA and Japan—each of these producing well over 100 million tonnes annually. Production in the United Kingdom is of the order of 20 million tonnes per annum.

Manufacture of Iron

The manufacture of iron from iron ore takes place in two stages. First of all, the ore is prepared by crushing and heating. The crushing produces manageable lumps of up to about 10 cm in diameter. The lumps are then heated to remove water and volatile impurities.

In the second stage the iron ore is reduced to iron by carbon monoxide in a **blast furnace** (*see* figure 14.12). The **reduction** is carried out at temperatures of about 700°C

$$Fe_2O_3(s) + 3CO(g) \rightleftharpoons 2Fe(l) + 3CO_2(g)$$

An excess of carbon dioxide is used to increase the yield of iron.

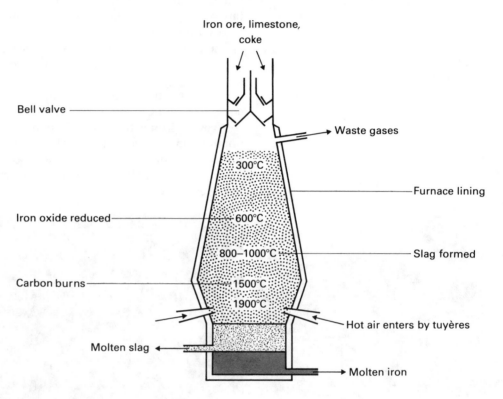

Figure 14.12   The blast furnace

The carbon monoxide is produced in the blast furnace from coke and air. The air is preheated to about 600°C and injected into the furnace through pipes called tuyères. The coke burns in the hot compressed air, forming carbon dioxide. The reaction is exothermic and the temperature rises to over 1700°C

$$C(s) + O_2(g) \rightarrow CO_2(g) \qquad \Delta H^{\ominus}_m = -406 \text{ kJ mol}^{-1}$$

The carbon dioxide rises up the furnace and reacts with more coke, forming carbon monoxide. This reaction is endothermic

$$CO_2(g) + C(s) \rightarrow 2CO(g) \qquad \Delta H^{\ominus}_m = +173 \text{ kJ mol}^{-1}$$

The iron produced by the reduction of the ore is impure due to the presence of sand and alumina (*see* above). These are removed by adding limestone to the furnace. At the temperatures inside the furnace the limestone undergoes thermal decomposition forming calcium oxide and carbon dioxide

$$CaCO_3(s) \xrightarrow{800°C} CaO(s) + CO_2(g)$$

The calcium oxide combines with the impurities forming **slag**. The slag contains calcium silicate and calcium aluminate

$$CaO(s) + SiO_2(s) \rightarrow CaSiO_3(l)$$
$$CaO(s) + Al_2O_3(s) \rightarrow CaAl_2O_4(l)$$

Iron melts at 1540°C (*see* table 14.2). The molten iron together with the molten slag sink to the base of the furnace. The molten slag floats on top of the molten iron. The two layers are tapped off periodically at different levels.

The blast furnace operates 24 hours a day, seven days a week. The raw materials are iron ore, coke and limestone. They are constantly charged into the top of the furnace. The iron is tapped from the furnace about four times a day, at five or six hour intervals. The iron leaves the furnace in a fiery stream with a temperature of about 1500°C. Furnaces vary in size and may produce between 1000 and 3000 tonnes of iron each day. In the United States some new furnaces have four tapholes and tap off molten iron continuously. These furnaces can produce up to 10 000 tonnes per day.

The iron produced by a blast furnace is run into moulds of sand. This iron is known as **pig-iron**. It is about 95% pure. Pig-iron is hard but brittle and melts at about 1200°C.

Cast iron is produced by melting a mixture of pig-iron, scrap iron and steel and coke. The molten iron is then cast into moulds and cooled.

Wrought iron is the purest form of commercial iron. It is produced by heating impure iron with haematite and limestone in a furnace. This increases the purity of the iron to about 99.5%. The melting point is increased to about 1400°C. Wrought iron is a strong, malleable and ductile form of iron. However, for many uses, it has been replaced by mild steel (*see* below).

Thirty thousand workers including several thousand women are employed at this steel plant in Bihar State, India. It is the largest steel plant in India

Wrought iron is the purest form of commercial iron

Manufacture of Steel

There are two types of steels. **Carbon steels** contain up to 1.5% carbon. **Alloy steels** contain not only small amounts of carbon but also other metals in small amounts. We shall examine the different types of steel and their properties and uses in detail below.

The Basic Oxygen Process

In recent decades the production of steel has been revolutionised by the development of the basic oxygen process (also known as the Linz–Donawitz or LD process). The process first became operational in 1953 in plants in two Austrian steel towns: Linz and Donawitz.

In this process the **basic oxygen converter** (*see* figure 14.13) is first tilted so that molten furnace iron and scrap iron can be charged. The converter is then returned to a vertical position. A water-cooled copper lance is lowered into the converter

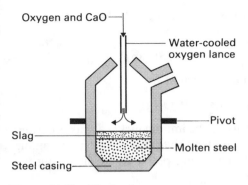

Oxygen and CaO

Water-cooled oxygen lance

Pivot

Slag

Molten steel

Steel casing

Figure 14.13   The basic oxygen converter

and oxygen and powdered lime (CaO) blown onto the surface of the molten iron. This is called the 'oxygen blow'. It lasts for just over 20 minutes and results in a violent reaction in which the impurities are oxidised. The charge in the converter is maintained in a molten state by the energy released in these oxidation reactions. The oxides combine with the lime, forming slag. The lance is removed and the converter tilted once again to remove the slag. After a second blow, the converter is tilted and the molten steel poured into a ladle.

The basic oxygen process is used mainly to produce carbon steels. Its major advantage is its efficiency. Between 300 and 350 tonnes of steel can be made in a converter in 40 to 45 minutes.

Nowadays all steel made in the United Kingdom is made by this process. The process also accounts for the major portion of current world steel production.

Electric Furnace

The electric furnace is used principally to convert scrap iron and steel into high-quality alloy steels such as stainless steel. The furnace is a round deep receptacle lined with refractory brick. The scrap metal is charged to the furnace while the roof is off. The roof is then replaced and carbon electrodes lowered through holes in the roof until they come into contact with the scrap. The power is switched on. Temperatures of well over 3000°C are generated by the arc which strikes between the electrodes. At these temperatures the scrap melts, forming the new steel. The furnaces generally produce between 25 and 50 tonnes of steel at a time.

EXTRACTION OF OTHER METALS IN THE FIRST TRANSITION SERIES

We shall now consider briefly the extraction of three other metals in the first transition series. The general principles of metal extraction were described in sections 5.3 and 10.5.

Titanium

The main source of metallic titanium is ilmenite. This is a mixed oxide of titanium and iron (see table 14.9). The ore has magnetic properties and so the two oxides can be separated magnetically.

The metal is most commonly extracted from its oxide by the Kroll process. The oxide is converted to titanium(IV) chloride by heating it with coke in a stream of chlorine at temperatures between 800 and 1000°C

$$TiO_2(s) + C(s) + 2Cl_2(g) \rightarrow TiCl_4(g) + CO_2(g)$$

Titanium(IV) chloride is a covalent liquid with a boiling point of 136°C. It is purified by fractional distillation and then reduced with molten magnesium in an atmosphere of argon

$$TiCl_4(l) + 2Mg(l) \rightarrow Ti(s) + 2MgCl_2(l)$$

The molten magnesium chloride is tapped off. The titanium is melted with an electric arc and formed into ingots.

In an alternative process, sodium is used in place of magnesium as the reducing agent.

Between 50 000 and 60 000 tonnes of titanium are produced in the world each year. The USSR produces about 50% of this, the USA 30% and Japan 17%. Only 3% is produced in the UK.

Nickel

The main source of nickel is the ore pentlandite. This is a sulphide of nickel and iron, NiS·2FeS. The sulphide is first roasted with limestone, sand and coke. The iron impurities are removed as slag. The nickel(II) sulphide is then converted to nickel(II) oxide in a furnace

$$2NiS(s) + 3O_2(g) \rightarrow 2NiO(s) + 2SO_2(g)$$

The nickel oxide is reduced to the metal using carbon

$$2NiO(s) + C(s) \rightarrow 2Ni(s) + CO_2(g)$$

The nickel is then purified **electrolytically** using the impure nickel as an anode and a sheet of pure nickel as the cathode. The electrolyte is a solution containing nickel(II) sulphate or nickel(II) chloride. These are prepared by dissolving the oxide or sulphide in the appropriate acid.

Copper

Over 360 ores are known to contain copper. Copper also occurs native. Copper is extracted from only a few of these ores. One of the most important is copper pyrites or chalcopyrites, $CuFeS_2$. This ore contains copper(II) sulphide and iron(II) sulphide.

The ore is first crushed and then concentrated by ore flotation (*see* chapter 10). The iron(II) sulphide is then converted to iron(II) oxide by roasting the ore in a limited supply of air

$$2CuFeS_2(s) + 4O_2(g) \rightarrow Cu_2S(s) + 3SO_2(g) + 2FeO(s)$$

Sand is added to convert the iron(II) oxide to a slag of iron(II) silicate, $FeSiO_3$.

The copper(I) sulphide is then reduced to copper by heating in a limited amount of air

$$Cu_2S(s) + O_2(g) \rightarrow 2Cu(l) + SO_2(g)$$

Stacking copper pigs at a copper smelting plant in Zaire

The molten copper is run off into moulds. On cooling, bubbles of nitrogen, oxygen and sulphur dioxide are released, giving the copper a blistered appearance. The impure '**blister copper**' is refined electrolytically. Pure copper is used as a cathode and the impure metal as an anode. Copper(II) sulphate is used as an electrolyte.

USES OF THE d-BLOCK ELEMENTS AND THEIR COMPOUNDS

The d-block elements and their compounds have a wide variety of uses in the laboratory, industrially, commercially and also biologically. We have already seen in the previous section and also in Section 10.2 how ions of d-block elements such as iron, chromium and manganese play an important part in redox titrations and laboratory preparations. In this section we shall confine our attention to the industrial, commercial and biological importance of these metals.

Structural Uses: Iron Alloys

A number of d-block elements are used extensively for structural purposes, principally in the form of alloys. An **alloy** is a mixture or solution of a metal with one or more other elements.

An alloy containing iron as the principal constituent is known as **a steel**. As we saw above, there are two types of steel: carbon steels and alloy steels.

Carbon Steels

These can be classified as mild steel, medium steel or high-carbon steel depending on their carbon content. The hardness of carbon steel increases with increasing carbon content. Mild steel, for example, is malleable and ductile. It is used when load-bearing is not a prime consideration. Some of the uses of carbon steels are shown in table 14.10. Carbon steels account for about 90% of all steel production.

Table 14.10   Carbon steels

Type of steel	Percentage of carbon	Uses
mild steel	0.2	general engineering: motor car bodies, wire, piping, nuts and bolts
medium steel	0.3–0.6	beams and girders, springs
high-carbon steel	0.6–1.5	drill bits, knives, hammers, chisels

Sparks fly in the manufacture of large steel cylinders in Bombay, India

Alloy steels

These steels contain up to 50% of one or more other metals such as aluminium, chromium, cobalt, molybdenum, nickel, titanium, tungsten and vanadium.

Stainless steels contain chromium and nickel. These metals improve the hardness of the metal and make it corrosion-resistant. The latter property is due to the formation of a fine film of chromium(III) oxide on the surface of the steel.

Tool steels include tungsten steel and manganese steel. The addition of these metals improves the hardness, toughness and heat resistance of the steel. They are used in rock drills, cutting edges and parts of machinery that are subject to heavy wear.

Silicon steels are used in the construction of electrical equipment such as electric motors, generators and transformers.

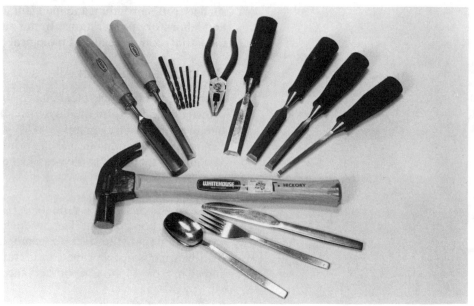

Some uses of high-carbon steels

Other Alloys

d-Block metals other than iron are also used as the main constituents of a number of alloys.

Titanium Alloys

Titanium readily alloys with metals such as tin, aluminium, nickel and cobalt. Titanium alloys are light, resistant to corrosion and strong at high temperatures. They are used in the aircraft industry to make turbine blades in jet engines.

A Boeing 747 has a mass of about 166 tonnes. Of this, 4 tonnes consists of titanium alloys and another 5 tonnes nickel alloys.

Titanium alloys are also used to make pacemakers. A pacemaker is an electronic device implanted in the chest wall to regulate abnormal heart rhythms.

Nickel Alloys

One of the most important nickel alloys is Monel metal. This contains 65% nickel, 32% copper and small amounts of iron and manganese. It is used to make condenser tubes, propeller shafts and in chemical, food and drug production plants. Another important nickel alloy is Nichrome. This contains 60% nickel, 15% chromium and 25% iron. An alloy of aluminium, cobalt and nickel called Alnico is used to make very strong permanent magnets.

Copper Alloys

Copper is used in a wide variety of alloys. The most important of these are shown in table 14.11.

Table 14.11   Copper alloys

Alloy	Composition
brasses	copper and zinc
bronze	copper and tin
phosphor bronze	copper, tin and phosphorus
copper coinage	copper, tin and zinc
silver coinage	copper and nickel

Copper and silver coinage

Industrial Catalysts

d-Block elements and their compounds find a wide variety of uses as industrial catalysts. The following examples are all taken from the first series of d-block elements.

Titanium(IV) Chloride, $TiCl_4$

This is used as a catalyst in the Ziegler method for polymerising alkenes (*see* chapter 20).

$$n(C_2H_4) \rightarrow (C_2H_4)_n$$

Vanadium(v) Oxide, V_2O_5

This is used in the following stage of the Contact process for the manufacture of sulphuric acid (*see* chapter 7).

$$2SO_2(g) + O_2(g) \rightarrow 2SO_3(g)$$

Iron or Iron(III) Oxide, Fe_2O_3

This is used in the Haber process for the synthesis of ammonia (*see* chapter 7)

$$N_2(g) + 3H_2(g) \rightarrow 2NH_3(g)$$

Nickel

This is used to harden vegetable oils, for example, in the manufacture of margarine.

Copper or Copper(II) Oxide

These are used in the dehydrogenation of ethanol to produce ethanal

$$CH_3CH_2OH(g) \rightarrow CH_3CHO(g) + H_2(g)$$

Other d-block metals used as industrial catalysts include rhodium from the second series of d-block elements and platinum from the third d-block series. Both are used in the Ostwald process for the manufacture of nitric acid (*see* chapter 15).

Pigments

We have already seen that one of the major characteristics of d-block elements is the tendency to form coloured compounds. Many gemstones, for example, depend on the presence of trace amounts of d-block metals for their colour (*see* table 14.6). The oxides of d-block elements are used to colour glasses. Cobalt(II) oxide, for example, imparts a deep blue colour to glass. A number of d-block metal compounds are used as pigments in a variety of industries.

Titanium(IV) Oxide, TiO_2

Over two million tonnes of TiO_2 are produced each year. It is used mainly as a white pigment in the manufacture of paints. It is also used in the manufacture of paper, plastics and textiles.

Chromium Compounds

Chrome alum (chromium(III) potassium sulphate-12-water), $KCr(SO_4)_2 \cdot 12H_2O$, is purple. It is used in dyeing and calico printing. Chromium(III) oxide, Cr_2O_3, is used as a green pigment. Chrome green, chrome oranges, chrome yellows and chrome reds are all pigments based on lead(IV) chromate.

Potassium Hexacyanoferrate(III), $K_3[Fe(CN)_6]$

This is used in dyeing, etching and in blueprint paper.

Cobalt Compounds

Cobalt blue is a pigment consisting of an aluminate of cobalt. Cobalt violets are purple and violet pigments produced by precipitating cobalt salts with alkaline earth phosphates.

Other Industrial Uses

So far we have considered the uses of d-block elements as structural alloys, industrial catalysts and pigments. The elements have a wide variety of other uses which do not conveniently fall into these categories.

Chromium. This is used to plate steel articles, for example, motor car fittings.

Cast iron. This is impure iron rather than an alloy. It is used to make a variety of objects such as fire grates, man-hole covers and gas stoves.

Cobalt. Cobalt-60 is used as a source of γ-radiation in the treatment of cancers.

Copper. The metal is used extensively in the electrical industry to make wires and cables and other conductors. It is also used to make copper waste pipes.

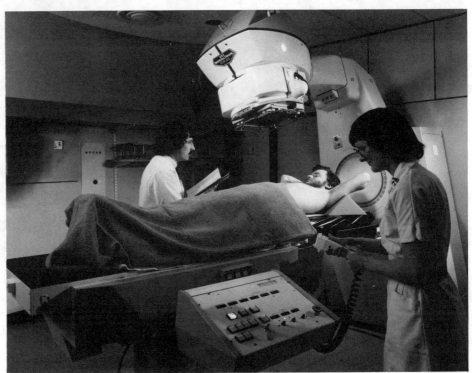

Cobalt-60 radiation therapy controlled by computer at the Royal Free Hospital, London

d-Block Elements in Biological Systems

d-Block elements play an important part in many biological systems. A healthy adult, for example, contains about four grams of iron. About two-thirds of this occurs in the form of haemoglobin. This is the red pigment of blood (*see* figure 14.11). Iron is also found in the muscle protein myoglobin and some iron is stored in organs such as the liver.

Table 14.12 Total body content of minerals for an average male adult

Major minerals	content/g	Trace elements	content/g
calcium	1000	fluorine	2.6
phosphorus	780	zinc	2.3
sulphur	140	copper	0.072
potassium	140	iodine	0.013
sodium	100	manganese	0.012
chlorine	95	chromium	< 0.002
magnesium	19	cobalt	0.0015
iron	4.2		

Elements that are only found in minute quantities are called trace elements. Table 14.12 shows the total mass of both major mineral elements and some trace elements in an average adult. Note that five of these elements are metals in the first d-block series. These and other d-block trace elements have various essential functions in biological systems.

Chromium is involved in the human body in the utilisation of glucose.

Manganese is associated with a number of enzyme systems. It is essential in the nutrition of birds and plants but less important for sheep and cattle. Although manganese is found in the human body, it has not been proved essential. Tea is rich in manganese. Plant products such as nuts, spices and whole cereals are also good sources of the element.

Manganese is essential in the nutrition of birds

A number of d-block metal trace elements have essential functions for the healthy growth of plants

Cobalt is also essential for cattle, sheep and man. Vitamin B$_{12}$, for example, is found to contain cobalt. The vitamin is used to treat pernicious anaemia and is required for the formation of DNA and RNA (*see* chapter 20).

Nickel is also found in human tissues, although once again there is no evidence that it is essential.

Copper is an important constituent of a number of enzymes and is required for the synthesis of haemoglobin. It is also required by plants. Sheep and other cattle are particularly susceptible to copper deficiency. Ewes deficient in copper can give birth to lambs afflicted by 'swayback'. This disease involves paralysis of their hindquarters. The only food that contains significant amounts of copper is liver. Sea foods, pulses, dried fruits and breakfast cereals all contain very small amounts of copper.

Zinc is associated with a number of enzymes. It is essential for the production of insulin and is a constituent of the enzyme anhydrase. This is required in respiration.

Zinc Deficiency Disease

Zinc deficiency disease was first recognised in the early 1960s in Iran and Egypt by Dr Ananda S. Prasad. He treated cases showing stunted growth and anaemia. Zinc deficiency has since been recognised as a complicating factor in children suffering from severe malnutrition. The element is necessary for the action of T-lymphocytes, without which the body's immune system cannot deal with infection.

Zinc supplements can help to treat heavy metal poisoning and some inherited conditions such as sickle cell anaemia (SCA). The latter is a hereditary disease of red blood cells found in some people of black African racial origin. In SCA, red cells curl up and are unable to carry oxygen. Too much calcium leads to changes in the charge distribution on the cell. However, added zinc competes with the calcium and makes the cell membrane less 'sticky' so that it does not curl up.

Zinc supplements have also been added to diets to help treat the slimming disease known as anorexia nervosa.

Starving children in northern Uganda. Zinc deficiency is one of the complicating factors in children suffering from severe malnutrition

SUMMARY

1. The most abundant d-block element on Earth is iron, followed by titanium.
2. d-Block elements occur as trace elements in plants, animals and gemstones.
3. Iron is commercially extracted from the ores haematite, Fe_2O_3, and magnetite, Fe_3O_4.
4. Iron is manufactured in a **blast furnace** by the **reduction** of iron ore with carbon monoxide. Limestone is added to remove impurities as **slag**.
5. **Carbon steels** are produced mainly by the **basic oxygen process** (also known as the Linz–Donawitz process).
6. An **electric furnace** is used to produce high-quality **alloy steels**.
7. Titanium is produced from ilmenite by the **Kroll process**. The titanium(IV) oxide from the ore is first converted to $TiCl_4$.
8. Nickel is extracted from the ore pentlandite. The nickel(II) sulphide from the ore is first converted to nickel(II) oxide. This is reduced to nickel using carbon.
9. Copper is extracted from chalcopyrites. The copper(I) sulphide from the ore is reduced by heating in a limited amount of air.
10. An **alloy** is a mixture or solution of a metal with one or more other elements.
11. **Steels** are alloys containing iron as the principal constituent.
12. The hardness of carbon steel increases with increasing carbon content.
13. Stainless steels, tool steels and silicon steels are all alloy steels.
14. Titanium and nickel alloys are used widely in industry. Copper alloys are used in coinage.
15. Titanium(IV) chloride, vanadium(V) oxide, iron(III) oxide, nickel and copper(II) oxide are all used as **catalysts** in industry.
16. d-Block metal oxides are used to colour glass, and other compounds are also used as **pigments**.
17. d-Block metals play important roles in biological systems. For example, haemoglobin, which is a red pigment of blood, contains iron.

Examination Questions

1. (a) Calcium, atomic number 20, has the electronic configuration Ar $\underset{4s}{\downarrow\uparrow}$. Similarly, by showing the direction of electron spin, write the configurations of the elements Sc, Ti, V, Cr, Mn, Fe, Co, Ni, Cu and Zn whose atomic numbers are 21 to 30 respectively.

   (b) From the series of elements 21 to 30, give:
   - (i) the formula of the tripositive ion (3+) which contains five *un*paired electrons, and write its configuration;
   - (ii) the formula of the dipositive ion (2+) with the smallest radius and write its configuration;
   - (iii) the configuration of the Mn^{2+} ion;
   - (iv) the formula of the stable monopositive ion (1+) which can be complexed with ammonia;
   - (v) the symbol of the element M which undergoes the reaction:

   $$MO_3^-(aq) + 2H^+(aq) \rightarrow MO_2^+(aq) + H_2O(l)$$

   and say whether or not this is a redox reaction, giving reasons.

   (c) Chromium(III) hydroxide dissolves in a hot mixture of sodium hydroxide and hydrogen peroxide solutions. Explain the role of the hydrogen peroxide and give the formula of the dinegative ion (2−) responsible for the colour of the final yellow solution.

(d) Giving reagents and conditions, mention industrial processes in which any *two* of the elements in (a) (or their compounds) act as catalysts. Name the catalysts.

(SUJB)

2. (a) Giving *two* illustrative examples of *each* property, describe *five* typical properties of the d-block elements.
   (b) The atomic number of iron is 26. Give the detailed electronic configurations of iron(III) and iron(II) ions. Describe how you would convert, quantitatively, iron(III) ions into iron(II) ions and how you would show, by suitable testing, that the conversion had been achieved.
   (c) 0.200 g of an alloy containing iron was dissolved in nitric acid and the iron was reduced to iron(II) ions. On titration, this solution required 20.0 cm^3 of a solution containing 0.0166 mol dm^{-3} of potassium dichromate(VI) for complete oxidation of the iron(II) ions. Calculate the percentage by mass of iron in the alloy. [A_r(Fe) = 56.0]

(WJEC)

3. Certain chemical characteristics are noted particularly in the chemistry of the d-block elements and their compounds but are nevertheless also to be seen in the chemistry of other elements.
   Illustrate this statement fully with respect to the properties mentioned below. As appropriate, your answer should include descriptions of compounds and their structures and equations to represent the reactions you discuss.
   (a) Variable oxidation state.
   (b) Catalytic behaviour.
   (c) Complex ion formation.
   (d) Colour in compounds.

(L)

4. (a) Write down the oxidation states of chromium, cobalt and nickel which commonly occur in aqueous solution.
   (b) Give *one* example of a reaction in which a first-row transition metal ion, M^{2+}(aq), acts as a reducing agent and *one* example in which an M^{2+}(aq) ion acts as an oxidising agent (balanced equations are not required).
   (c) Give the formula of the complex species formed in each of the following reactions carried out in aqueous solution:
      (i) $FeSO_4$ + excess KCN,
      (ii) $NiSO_4$ + excess KCN,
      (iii) $CuSO_4$ + excess concentrated HCl.
   (d) Indicate how, using simple chemical tests, you could detect:
      (i) an impurity of $NiSO_4$ in $FeSO_4$,
      (ii) an impurity of $Fe_2(SO_4)_3$ in $FeSO_4$.

(JMB)

5. (a) Show, by means of equations, how each of the following compounds can be prepared in the laboratory:
      (i) Iron(II) sulphate-7-water from iron.
      (ii) Tin(IV) chloride from tin.
      (iii) Potassium chromate(VI) from a chromium(III) salt.
   (b) (i) Write a balanced equation for the reaction between Fe^{2+} and $Cr_2O_7^{2-}$ in acidic, aqueous solution.
      (ii) A medicinal tablet for patients suffering from iron deficiency contains $FeSO_4 \cdot 7H_2O$ as the iron-containing ingredient. A tablet

weighing 0.2000 g was dissolved in an excess of dilute sulphuric acid and the resulting solution titrated with 0.0100 M potassium dichromate(VI); 11.51 cm^3 of the potassium dichromate(VI) solution were required for complete oxidation of the iron(II).

Calculate the percentage of $FeSO_4 \cdot 7H_2O$ in the tablet.

(JMB)

6. This question concerns the properties of d-block elements.
   (a) Give two properties of the atoms or ions of d-block elements which enable them to form strong bonds with ligands.
   (b) Which one of the halide ions (F$^-$, Cl$^-$, Br$^-$, I$^-$) would you expect to form the strongest bonds with an ion such as Fe^{3+}?
   (c) Draw the structures for one example of each of the following types of complexes formed by d-block elements, showing in each case the overall charge of the ion and the oxidation state of the d-block element:
      (i) a positively charged complex,
      (ii) a negatively charged complex.
   (d) Give two factors which enable d-block elements to have more than one oxidation state.
   (e) The ability of d-block elements to change oxidation state readily according to conditions forms the basis of many analytical and preparative methods in chemistry.
      (i) Give an example of the use of a compound of a d-block element in volumetric analysis. Show clearly the change in oxidation state of the d-block element.
      (ii) Give an example of the use of a compound of a d-block element as an oxidising agent in organic chemistry. Show clearly the change in oxidation state of the d-block element.

(O & C)

7. (a) Give *three* examples of ways in which chromium exhibits the properties associated with transition metals.
   (b) Comment on the fact that it is possible to isolate three different forms of the hexahydrate of chromium(III) chloride. How would you distinguish them chemically?
   (c) Describe briefly, but including all essential reagents and conditions, how you would prepare aqueous solutions of Cr^{2+}(aq) ions and of Cr$_2$O$_7^{2-}$(aq) ions, starting from hydrated chromium(III) chloride.
   (d) Equal volumes of acidic solutions containing, respectively, 1.00 mol dm^{-3} of Cr^{2+}(aq) and 1.00 mol dm^{-3} of Cr$_2$O$_7^{2-}$(aq) were mixed. Calculate the concentration(s) of any chromium-containing ions present in the solution after reaction.

(O & C)

8. (a) Describe and explain a reduction process involving the industrial extraction of *either* zinc, *or* iron *or* copper from a named source.
   (b) By referring to the electronic configuration of iron, explain the formation of the Fe^{2+} and Fe^{3+} ion. (Atomic number of iron is 26.)
      Compare the stability of these ions giving reasons for your answer.
   (c) The conversion of iron(II) ions to iron(III) ions can be carried out using nitric acid, hydrogen peroxide or potassium manganate(VII).
      For each reaction give:
      (i) the conditions of the reaction,
      (ii) an ionic equation to represent this change,
      (iii) the name or formula of the reduced species.

(d) Indicate *two* tests, which could be used to show that the oxidation from iron(II) to iron(III) had taken place.

(e) The hydrated iron(III) ion is considered to be an acid.
   (i) What is the formula of the hydrated ion?
   (ii) Explain its acid properties.

(f) (i) Give the structure of a complex anion involving iron.
   (ii) Describe briefly how this ion could be prepared in the laboratory.
   (iii) Give an ionic equation to represent this reaction.

(AEB, 1984)

9. (a) The oxidation state of copper can be raised from 0 to +2 by means of (i) an electrolytic method or (ii) the action of sulphuric acid.
   Explain both of these reactions.

(b) Identify the compounds A to F in the following reactions involving the chemistry of copper compounds. In each case write equations where possible to represent the reactions referred to in the descriptions.
   (i) When aqueous copper(II) sulphate was added to an alkaline solution of sodium potassium 2,3-dihydroxybutanedioate (tartrate) a deep blue solution A was obtained. On addition of a hot aqueous solution of an organic compound of empirical formula CH_2O, a red deposit B was obtained.
   (ii) On preparing concentrated aqueous copper(II) chloride, a brown solution C was obtained. On dilution with water a green solution was first obtained and on adding more water, a blue solution D was formed.
   (iii) Addition of aqueous copper(II) sulphate to aqueous potassium iodide gave a white precipitate E suspended in a brown solution F.

(c) Copper(I) chloride has *two* important uses in organic chemistry.
   (i) As a means of distinguishing between various alkynes.
   (ii) As a means of preparing chlorobenzene starting with phenylamine. What part is played by the copper salt in both these examples?

(AEB, 1982)

10. The percentage of zinc in an ore was estimated as follows:
   '3.20 g of the ore, P, was boiled with excess hydrochloric acid and filtered. To the filtrate, F, *excess* sodium hydrogencarbonate solution was added and the white precipitate formed, Q, filtered. Q was washed and then heated strongly to constant mass. The final residue, R, weighed 1.25 g.'

(a) Write:
   (i) the formula of the zinc ion in the filtrate, F;
   (ii) the *ionic* equation for the formation of the white precipitate, Q;
   (iii) the equation for the thermolysis of Q, and then calculate the percentage of zinc in the sample of the ore, P. (O = 16; Zn = 65)

(b) Explain:
   (i) why sodium hydroxide solution would be *un*suitable for precipitating the zinc ions;
   (ii) which soluble salt is removed in the washing of Q;
   (iii) why the residue, R, does not thermally decompose into its constituent elements;
   (iv) the principal and *unavoidable* sources of error in the estimation.

(c) Suggest:
   (i) why the electronic configuration of the zinc ion precludes it from being considered a transition element;
   (ii) *two* properties of zinc (or its compounds) which confirm the conclusion from (i) above in practice;

(iii) the formula of zinc uranyl ethanoate, a double salt containing the dioxouranium(VI) ion, by considering the charge on each ion involved.

(SUJB)

11. (a) Give an account of the industrial manufacture of iron. Your answer should be confined to (i) the chemical and physico-chemical principles and (ii) the economic considerations involved in the extraction process.
   (b) Briefly describe the changes in composition needed to produce steels from pig iron.
   (c) Name the products, and write a balanced equation, for the reaction of iron with steam.
   (d) Explain what happens when aqueous KCN is slowly added to aqueous $FeSO_4$ until the KCN is present in excess. What happens when Cl_2 is passed through the final solution?

(OLE)

12. Titanium occurs as the oxide, rutile, TiO_2. Extraction of the metal is by heating rutile with coke in a stream of chlorine. Under these conditions titanium(IV) chloride is formed as a volatile liquid, which is subsequently reduced to the metal with molten magnesium in an atmosphere of argon.
   (a) (i) Write an equation for the conversion of rutile to titanium(IV) chloride.
       (ii) Why is an argon atmosphere used during the reduction of the chloride?
   (b) Titanium has similar chemical properties to tin but rather different physical properties.
       (i) How would you expect titanium to react with hydrochloric acid?
       (ii) Describe *four* features of titanium and its compounds that are typical of transition element chemistry.
       (iii) Give one major use of titanium.
   (c) Titanium(IV) chloride dissolves in concentrated hydrochloric acid to give the ion $[TiCl_6]^{2-}$.
       (i) What is the coordination number of titanium in the ion?
       (ii) What is the oxidation number of titanium in the ion?
       (iii) Suggest a name for $[TiCl_6]^{2-}$.
       (iv) Draw the likely structure of $[TiCl_6]^{2-}$.
   (d) At room temperature titanium metal has a close-packed hexagonal structure which changes above 880°C to a body-centred cubic structure.
       Draw diagrams of the two structures.

(NISEC)

15 GROUPS III TO VI

Two British Women Chemists

In the early nineteenth century, two British women played a prominent role in chemistry. Elizabeth Fulhame performed original research on reduction which led her to challenge the contemporary theories. She deserves recognition as probably the first woman chemist of modern time. Jane Marcet wrote a classic introductory chemistry text which made a major impact on both sides of the Atlantic for nearly half a century.

In the history of modern chemistry, the earliest woman chemist to be noted is usually Marie Curie. Yet there are two British women whose individual contributions to chemistry in the late eighteenth and early nineteenth centuries deserve recognition. These forgotten women are Elizabeth Fulhame and Jane Marcet.

Elizabeth Fulhame

The only biographical information available on Elizabeth Fulhame is that she was married to a doctor. In 1780 she started a research project on the deposition of metals—particularly gold and silver—on silk fabrics so as to imitate gold and silver cloth. To accomplish the deposition, she systematically studied the reduction of metal salts of gold, silver, platinum, mercury, copper and tin. As reducing agents she used hydrogen gas, phosphorus, potassium sulphide, hydrogen sulphide, phosphine, charcoal and light.

It was the result of these thorough experiments which led her to challenge both the phlogiston and anti-phlogiston theories of the time. Her particular concern was that many reductions only occurred in the presence of water. This point was proved by some elegant experiments using ether and alcohol as solvents. Her explanation involved intermediate steps to the reactions (i.e. water as a catalyst), which was a novel idea at the time.

She noted that the presence of water also appeared to be necessary for the oxidation by air of carbon and of (white) phosphorus. Thus she considered the combustions to be a two-part process. In the carbon case, the carbon reacted with the oxygen of water to give carbon dioxide and hydrogen. The hydrogen gas combined with oxygen from the air to re-form the water. Although we now know that water is not essential for combustion, water has been shown to act as an oxidation catalyst for both carbon and hydrogen. Thus her findings have been at least partially vindicated.

She published this work in a book entitled *An essay on combustion* though she was obviously expecting a negative reception to her activities:

But censure is perhaps inevitable, for some are so ignorant, that they grow sullen and silent, and are chilled with horror at the sight of any thing that bears the semblance of learning, in whatever shape it may appear; and should the spectre appear in the shape of woman, the pangs which they suffer are truly dismal.

In spite of her forebodings, the work was well received and widely reviewed. Such prominent scientists as Priestley and Count Rumford discussed her findings, though both disagreed with her conclusions.

The book was published in the US and was also translated into German. It was in the US that she received the greatest acclaim. In an oration before the Chemical Society of Philadelphia it was stated that:

> Mrs Fulhame has now laid such bold claims to chemistry that we can no longer deny the sex the privilege of participating in this science also.

She was subsequently elected a corresponding member of the society. Her work was then forgotten until 1903 when it was briefly rediscovered by the famous J. W. Mellor who devoted almost a whole paper to an appreciation of her discoveries.

Jane Marcet

Jane Marcet

Substantially more is known about Jane Marcet (*née* Haldimand). Born in 1769, her interest in chemistry started with her marriage to Alexander Marcet, a physician who later became a prominent biochemist. Having difficulty understanding the chemical lectures of the time, she conversed with a friend on the topics and performed her own experiments. She found this experience gave her a much deeper understanding of the subject. As a result, she decided to write an introductory chemistry text in a conversational style accompanied by experimental work. She remarked in the preface:

> Hence it was natural to infer, that familiar conversation was, in studies of this kind, a most useful auxiliary source of information; and more especially for the female sex, whose education is seldom calculated to prepare their minds for abstract ideas, or scientific language.

Jane Marcet, too, was unsure of the reception of her work, commenting that:

> In writing these pages, the author was more than once checked in her progress, by the apprehension that such an attempt might be considered by some, either as unsuited to the ordinary pursuits of her sex, or ill-justified by her own imperfect knowledge of the subject.

In fact the work entitled *Conversations on chemistry* was a resounding success, becoming the most popular chemistry textbook in the first half of the nineteenth century. The text ran to at least 16 English editions, 15 American editions and three French editions over the period 1806 to 1853. In the US alone more than 160 000 copies had been sold by 1853.

The appeal of the book was twofold. Firstly, it was written as a conversation between a Mrs B (believed to be a Mrs Bryant) and two students, Emily, who is serious and hard working, and Caroline, whose lack of enthusiasm (except during exciting experiments) is summed up by her first statement in the book:

> To confess the truth, Mrs B, I am not disposed to form a very favourable idea of chemistry, nor do I expect to derive much entertainment from it.

It is the interplay of these three very believable characters which makes the material so much more interesting than a regular chemistry text.

The other factor in the success of the text was the inclusion of new discoveries in each successive edition. She introduced Sir Humphry Davy's isolation of the alkali metals before his findings were generally accepted. Even at 71 years old she was writing to Michael Faraday requesting permission to include his latest findings on electricity in the next edition of her text. It was of particular note that she should write to Faraday, for it was his reading of an early edition of *Conversations of chemistry* while he was a book binder's apprentice which developed his interest in the subject.

Jane Marcet subsequently authored many other texts in the *Conversations on . . .* series as well as several children's books. *Conversations on natural philosophy, Conversations on vegetable physiology* and *Conversations on political economy* were particularly popular, each one running to many editions.

Why have these two scientists been forgotten? In the case of Jane Marcet, it is because little interest is shown in the early history of chemical education. Any such history, though, should be sure to give a most prominent coverage of her text and its significance to chemical education in the first half of the nineteenth century. Elizabeth Fulhame was certainly recognised among her contemporaries as a pioneering woman chemist yet in more recent times only Mellor and Partington have noted her work. It is difficult to find a good explanation as to why she has been forgotten. Even though her chemical discoveries were not of the importance of other chemists of the era such as Priestley and Lavoisier, she should certainly be remembered both as an excellent experimentalist and as the first woman chemist of modern times.

G. W. Rayner-Canham

LEARNING OBJECTIVES

After you have studied this chapter you should be able to

1. Explain why boron is a semi-metal.
2. Give examples of some important compounds formed by boron.
3. Outline the uses of boron and its compounds.
4. Summarise the principal chemical properties of
   (a) aluminium,
   (b) aluminium oxide,
   (c) aluminium halides,
   (d) aluminium sulphate and the alums.
5. Outline the occurrence and extraction of aluminium.
6. List the more important uses of aluminium, its alloys and its compounds.
7. Compare the structures and chemical and physical properties of elements in Group IV.
8. Explain the term **allotropy** and give examples of this from Group IV.
9. Describe and give specific examples of the trend in chemical reactivity of the elements on descending Group IV.
10. Explain why carbon is unique.
11. Discuss the chemistry of the following Group IV compounds:
    (a) the carbon oxides,
    (b) the chlorides,
    (c) the hydrides,
    (d) the carbonates.
12. Sketch the atmospheric **carbon cycle**.
13. Outline the occurrence and extraction of the Group IV elements and indicate the principal uses of these elements and their compounds.
14. Discuss
    (a) the '**microchip revolution**',
    (b) lead pollution.
15. Compare the electronic configurations and chemical bonding of nitrogen and phosphorus.
16. Describe the allotropy of phosphorus.

(continued)

17. Summarise the occurrence, extraction and uses of nitrogen and phosphorus.
18. Outline the chemistry of
    (a) ammonia and its salts,
    (b) the nitrogen oxides,
    (c) nitrous acid and nitrites,
    (d) nitric acid and nitrates.
19. Discuss the pollution caused by nitrogen oxides.
20. Outline the manufacture of nitric acid and the nitrates.
21. Describe briefly the chemistry of
    (a) phosphine,
    (b) phosphorus chlorides,
    (c) phosphorus oxides,
    (d) phosphorus acids and their salts.
22. Outline the physical and chemical properties of oxygen.
23. Compare the different types of oxides.
24. Describe the various forms of sulphur.
25. Outline the occurrence, extraction and uses of sulphur.
26. Briefly describe the chemical properties of
    (a) hydrogen sulphide,
    (b) the oxides of sulphur,
    (c) sulphuric acid,
    (d) sulphates, sulphites and thio-compounds.

15.1 Aluminium and the Group III Elements

The five elements of Group III are shown in Figure 15.1. They form part of the p-block elements (*see* chapter 11). Each element in Group III has a characteristic s^2p^1 electronic configuration in its outer shell (*see* table 15.1).

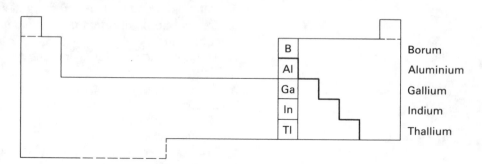

Figure 15.1   The Group III elements

Table 15.1   Atomic structure and ionisation energies of the Group III elements

Element	Atomic number	Configuration in outer shell	Atomic radius/nm	Ionic radius/nm	Ionisation energies/kJ mol^{-1}		
					First	Second	Third
boron	5	$2s^2 2p^1$	0.080	0.020	800	2427	3659
aluminium	13	$3s^2 3p^1$	0.125	0.050	578	1816	2744
gallium	31	$4s^2 4p^1$	0.125	0.062	579	1979	2952
indium	49	$5s^2 5p^1$	0.150	0.081	558	1817	2693
thallium	81	$6s^2 6p^1$	0.155	0.095	589	1970	2866

Boron is a semi-metal whereas aluminium is metallic. The other three elements in the group are weakly metallic. All Group III elements have a valency of 3.

BORON

Boron comprises about 0.001% of the Earth's crust. It occurs in a number of ores of which the most important is borax. This has the composition $Na_2B_4O_7 \cdot 10H_2O$.

Boron is a semi-metal since it has both metallic and non-metallic characteristics. It exhibits a **diagonal relationship** (*see* chapter 11) with silicon—an element which we shall discuss in detail later in this chapter. The non-metallic properties of boron are due to the small size of its atoms and its high ionisation energies. Indeed boron never forms the ion B^{3+}.

Boron forms a number of important compounds. These include the following.

The Borohydrides
Lithium tetrahydridoborate(III), $LiBH_4$, also known as lithium borohydride, is an example.

The Boranes
These are the boron hydrides. The simplest is diborane, B_2H_6. This is formed by the reaction of boron trichloride and lithium tetrahydridoaluminate(III).

Boron Trichloride
This is a colourless liquid. It is a covalent compound and exists as BCl_3 molecules. BCl_3 is hydrolysed by water, forming boric acid. BCl_3 forms a coordination compound with ammonia.

In this reaction the ammonia acts as a Lewis base since it is an electron donor. The BCl_3 acts as a Lewis acid since it is an electron acceptor (*see* section 3.1).

Boron Oxide B_2O_3
Like other soluble non-metallic oxides, this oxide is acidic. It reacts slowly with water, forming boric acid.

Boric Acid, H_3BO_3
This is also known as orthoboric acid. It is formed in the laboratory by the action of hydrochloric acid on borax solution

$$B_4O_7{}^{2-}(aq) + 2H^+(aq) + 5H_2O(l) \rightarrow 4H_3BO_3(s)$$

On cooling, the boric acid separates as flaky white crystals.

The Borates
Like silicon, boron has a high affinity for oxygen, forming a number of ring and chain structures with oxygen. For example, sodium metaborate, $Na_3B_3O_6$, and potassium metaborate, $K_3B_3O_6$, contain the cyclic $B_3O_6{}^{3-}$ anion.

Boron has a low electrical conductivity at room temperature. However, this increases greatly as temperature is raised. Boron is sometimes added to germanium and silicon to increase electrical conductivity of these semiconductor elements.

Calcium metaborate, which has the empirical formula CaB_2O_4, consists of long chains of borate anions $(B_2O_4{}^{2-})_n$ linked by Ca^{2+} ions.

Uses of Boron and its Compounds

Boron is important in the calcium cycle of plants. Borax or boric acid is thus often added as a fertiliser to soils deficient in boron. Boric acid is a mild disinfectant. It has been used as a food preservative although its use in this respect is prohibited in many countries.

New Ceramic Materials

Over recent years a number of new ceramic materials based on compounds of the Group III and IV elements have been developed and are becoming increasingly important in a variety of industries. The compounds include boron carbide, silicon carbide and silicon nitride. A ceramic alloy such as 'Sialon' is almost as hard as diamond, as strong as steel and as light as aluminium. Such ceramics are used for metal and wear-resistant machinery, for example. They can be used at temperatures of up to 1300°C and require no lubrication. Ceramic engines are already being tested and developed by the aerospace and automobile industries.

Borax is the most important commercial boron compound. Deposits of borax were first found in Tibet and later brought to Europe by the Arabs. Ever since the Middle Ages it has been used in pottery glazes. It combines with the oxides of d-block metals such as copper and cobalt to form beautifully coloured compounds. For example

$$CoO(s) + B_2O_3(s) \xrightarrow{\text{heat}} \underset{\text{deep blue}}{Co(BO_2)_2(s)}$$

These characteristic colours are used to identify d-block metals in the **borax bead test**.

Borax is also used in the manufacture of borosilicate glass. This type of glass has a high refractive index and is used in the manufacture of lenses. Since borax is a weak base, it is also used to prepare buffer solutions (*see* section 8.2) and photographic developers.

PHYSICAL AND CHEMICAL PROPERTIES OF ALUMINIUM

Physical Properties

Aluminium is 1.8 times as thermally conductive as copper and 9 times as thermally conductive as stainless steel.

Aluminium is a silvery white metal. It is an excellent conductor of both electricity and heat. It has a low density—about one-third of that of iron, copper and zinc. Even so, it is a very strong metal.

The three outer-shell electrons of an aluminium atom are delocalised through the aluminium lattice. The lattice has a face-centred cubic structure similar to that of tin and gold (*see* section 3.2). Aluminium is thus very malleable and ductile.

Chemical Properties

Aluminium forms both ionic and covalent compounds. However, its ionisation energy is high (*see* table 15.1). Furthermore, the charge-to-radius ratio of the Al^{3+} ion is very high compared with other metal cations in the period (*see* table 15.2).

Table 15.2 Charge-to-radius ratios of Na, Mg and Al

Cation	Charge	Ionic radius/nm	Charge to radius ratio/charge nm^{-1}
Na	+1	0.095	10.5
Mg	+2	0.065	30.8
Al	+3	0.050	60.0

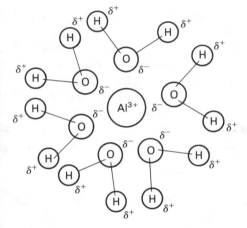

Figure 15.2 The hydrated aluminium ion

Anodising

Aluminium and light aluminium alloys can be protected even more by thickening the natural oxide layer in a process known as anodising. In this process the aluminium object is made the anode in an electrolytic cell in which the electrolyte is either chromic(VI) acid or sulphuric acid.

Since the Al^{3+} ion is small but has a high charge, it is highly polarising. The isolated Al^{3+} ion is thus found in only a few compounds such as anhydrous aluminium fluoride and aluminium oxide and even these compounds exhibit some covalent character. In aqueous solutions the Al^{3+} ion polarises water molecules, which then hydrate the cation (*see* figure 15.2). This hydration is very exothermic:

$$Al^{3+}(g) + 6H_2O(l) \rightarrow [Al(H_2O)_6]^{3+}(aq) \qquad \Delta H^{\ominus}_{hyd,m} = -4690 \text{ kJ mol}^{-1}$$

The standard redox potential of aluminium is -1.66 volts.

$$Al^{3+}(aq) + 3e^- \rightleftharpoons Al(s) \qquad E^{\ominus} = -1.66 \text{ V}$$

Aluminium is thus fairly high in the electrochemical series of elements (*see* section 10.5). Aluminium might thus be expected to react readily with oxygen and dilute mineral acids. However, when aluminium reacts with oxygen it forms a thin oxide layer which is non-porous. This makes the aluminium impervious to further attacks. The oxide layer can be removed by rubbing with mercury. The aluminium will then combine directly with oxygen and other non-metals such as sulphur and nitrogen. For example

$$4Al(s) + 3O_2(g) \rightarrow 2Al_2O_3(s)$$

Aluminium reacts with hot dilute hydrochloric and sulphuric acids, forming hydrogen:

$$2Al(s) + 6H^+(aq) \rightarrow 2Al^{3+}(aq) + 3H_2(g)$$

The reaction is slow at first due to the formation of the oxide layer. However, this is removed as the reaction progresses and the reaction becomes increasingly vigorous.

Aluminium is rendered '**passive**' by concentrated and dilute nitric acid and by concentrated sulphuric acid. In other words, aluminium does not react with these acids. This is caused by the formation of the thin oxide layer.

The metal is attacked by sodium hydroxide solution and other alkalis, forming tetrahydroxoaluminate(III) ions and hydrogen:

$$2Al(s) + 2OH^-(aq) + 6H_2O(l) \rightarrow 2[Al(OH)_4]^-(aq) + 3H_2(g)$$

Once the oxide layer is removed aluminium will act as a reducing agent in redox reactions (*see* section 10.2). It displaces metals low in the electrochemical series from their solutions. For example

$$2Al(s) + 3Cu^{2+}(aq) \rightarrow 2Al^{3+}(aq) + 3Cu(s)$$

A spectacular example of the reducing power of aluminium is provided by the **Thermit** (or thermite) reaction. This is the reaction between aluminium powder and iron(III) oxide. In the laboratory, it is usually initiated using magnesium ribbon as a fuse. The reaction is violent and generates sufficient energy to melt the iron which is formed:

$$2Al(s) + Fe_2O_3(s) \rightarrow Al_2O_3(s) + 2Fe(l)$$

The Thermit reaction is used in Thermit welding—to join rails, for example.

Aluminium Oxide, Al₂O₃

Aluminium oxide, or alumina as it is also called, exhibits both ionic and covalent characteristics. It has a melting point of 2045°C and when molten is an electrolyte. For this reason it is usually regarded as an ionic compound. However, in the solid state it has a macromolecular structure.

Corundum

Anhydrous forms of aluminium oxide occur naturally in the corundum group of minerals. Corundum is a very hard crystalline form of aluminium oxide. It is used as an abrasive and as a gemstone. Corundum gemstones are second only to diamond in hardness. Pure corundum is colourless but the presence of small amounts of the oxides of d-block metals as impurities gives the gemstones distinctive colours. The colour of ruby, for example, is due to the presence of Cr^{3+} and the colour of sapphire due to cobalt(II), iron(II) and titanium(IV) ions. The purple colour of amethyst is due to the presence of manganese. Artificial gemstones can be manufactured by fusing alumina with the appropriate d-block metal oxide. (*See* also tables 14.6 and 14.7.)

Aluminium oxide is insoluble in water and is amphoteric—reacting with both dilute acids and dilute alkalis. In acid the reaction is

$$Al_2O_3(s) + 6H^+(aq) \rightarrow 2Al^{3+}(aq) + 3H_2O(l)$$

In alkalis the tetrahydroxoaluminate(III) ion is formed

$$Al_2O_3(s) + 2OH^-(aq) + 3H_2O(l) \rightarrow 2[Al(OH)_4]^-(aq) \tag{1}$$

Aluminium Halides

The structure and bonding of the aluminium halides are described in section 16.2.

Aluminium chloride can be prepared by passing dry chlorine or dry hydrogen chloride over heated aluminium. For example

$$2Al(s) + 3Cl_2(g) \rightarrow 2AlCl_3(s)$$

With the exception of aluminium fluoride, the aluminium halides are hydrolysed by water:

$$AlCl_3(s) + 3H_2O(l) \rightarrow Al(OH)_3(s) + 3HCl(g)$$

For this reason the aluminium halides fume when in contact with moist air.

Aluminium Ions

We have already seen that Al^{3+} is hydrated in water. When aluminium salts are dissolved in water, the following equilibrium is established:

$$[Al(H_2O)_6]^{3+}(aq) + H_2O(l) \rightleftharpoons [Al(H_2O)_5OH]^{2+}(aq) + H_3O^+(aq)$$

The water acts as a base since it accepts a proton and the hydrated aluminium ion acts as an acid since it donates one. Aluminium salts are thus acidic. When an alkali such as NaOH(aq) or NH₄OH(aq) is added to a solution of an aluminium salt, hydrated aluminium oxide is precipitated:

$$[Al(H_2O)_6]^{3+}(aq) + 3OH^-(aq) \rightarrow [Al(H_2O)_3(OH)_3](s) + 3H_2O(l)$$
$$\text{white precipitate}$$

On addition of excess alkali the precipitate dissolves due to the formation of a soluble anion:

$$[Al(H_2O)_3(OH)_3](s) + OH^-(aq) \rightarrow [Al(H_2O)_2(OH)_4]^-(aq) + H_2O(l)$$

In this reaction the hydrated aluminium hydroxide acts as an acid. It also reacts with acids:

$$[Al(H_2O)_3(OH)_3](s) + H_3O^+(aq) \rightarrow [Al(H_2O)_4(OH)_2]^+ + H_2O(l)$$

Aluminium hydroxide is thus **amphoteric**.

Aluminium Sulphate and the Alums

The **alums** are **double salts** with the general formula $M(I)M(III)(SO_4)_2 \cdot 12H_2O$. $M(I)$ is a monovalent ion such as Na^+, K^+ or NH_4^+ and $M(III)$ a trivalent ion such as Al^{3+}, Cr^{3+} and Fe^{3+}. Typical alums are given in table 15.3.

Table 15.3 Typical alums

Name of alum	Formula	Common name
aluminium potassium sulphate-12-water	$KAl(SO_4)_2 \cdot 12H_2O$	alum or potash alum
chromium(III) potassium sulphate-12-water	$KCr(SO_4)_2 \cdot 12H_2O$	chrome alum
aluminium ammonium sulphate-12-water	$NH_4Al(SO_4)_2 \cdot 12H_2O$	ammonium alum
ammonium iron(III) sulphate-12-water	$NH_4Fe(SO_4)_2 \cdot 12H_2O$	ferric alum

In solution the alums give simple ions. For example, potash alum solution contains the ions K^+, Al^{3+} and SO_4^{2-}. The alums crystallise as octahedra.

Mordants

Potash alum and aluminium sulphate are used as mordants in dyeing cloths. The cloth is soaked with a solution of potash alum or aluminium sulphate. Alkali is then added. The salt solution acts as an acid, forming aluminium hydroxide (*see* above). The aluminium hydroxide deposits on the fibres of the cloth and 'bites' into the cloth. It also absorbs the dye and thus binds the dye to the cloth. The word 'mordant' comes from the Latin word *mordere* meaning to bite.

OCCURRENCE AND EXTRACTION OF ALUMINIUM

Aluminium is the most abundant metal and the third most abundant element on Earth. It makes up about 8% of the Earth's crust.

Owing to its high affinity for oxygen, aluminium does not occur naturally as a metal. It is found widely in the form of hydrated aluminium silicates in rocks such as clays, micas and feldspars. It is also found in ores such as bauxite. Bauxite contains hydrated aluminium oxide, $Al_2O_3 \cdot xH_2O$ where x ranges from 1 to 3. It is formed by the weathering of clays. During this process the silicates are leached out, leaving a residue rich in alumina.

Bauxite is the principal source of aluminium in the world today. Between 80 and 90 million tonnes of bauxite are produced each year. Almost 30% of this is produced in Australia and a further 15% in Jamaica. About 5 tonnes of bauxite are required to produce 1 tonne of aluminium. At current rates of production, there are sufficient bauxite reserves to supply the world with aluminium for several hundred years. The extraction of aluminium for bauxite takes place in two stages.

A precious resource—the Earth. Aluminium makes up about 8% of the Earth's crust. In this photograph taken from Apollo 17, the entire coastline of Africa is clearly visible. The Arabian Peninsula can be seen at the northeastern edge of Africa. The large island off the coast of Africa is the Malagasy Republic.

Bauxite

Bauxite is named after the town Les Baux in southern France, where, in 1821, a sample of the red clay-like sediment was obtained. Originally bauxite was thought to be a new mineral. However, subsequent analyses revealed a wide variation in the mineralogical composition, physical appearance and mode of occurrence of the ore. The term bauxite ore is applied to all deposits containing not less than 45% of one or more of the hydrated aluminium oxides and not more than 20% iron(III) oxide and 5% silica. The most important minerals in bauxite deposits are gibbsite, diaspore and boehmite.

Stage 1: Purification of Bauxite by the Bayer Process to Produce Alumina

Crude bauxite contains iron(III) oxide, silica and other impurities. If a bauxite ore is to be an economically attractive source of aluminium, it must contain at least 45% aluminium and less than 5% silica.

In the Bayer process the bauxite ore is first crushed and then mixed with sodium hydroxide (also known as caustic soda) solution. The mixture is pumped into a 'digester' where it is heated under high pressure. Sodium tetrahydroxoaluminate(III) is formed (*see* equation (1), p.540). Iron(III) oxide is removed from the mixture as 'red mud' by allowing it to settle out. The solution is then filtered and transferred to a precipitation tank where it is seeded with crystals of aluminium hydroxide. On seeding, the aqueous sodium tetrahydroxoaluminate(III) solution decomposes,

Bauxite storage, drying and shipping facilities at Port Kaiser on the southern coast of Jamaica

forming aluminium hydroxide. This grows into large crystals on the seed crystals. The equation for this process is

$$[Al(OH)_4]^-(aq) \xrightarrow{\text{seed}} Al(OH)_3(s) + OH^-(aq)$$

The sodium hydroxide formed in this process is recycled. The aluminium hydroxide crystals are filtered, washed and then 'roasted' in a rotary kiln at about 1000°C. Alumina is formed:

$$2Al(OH)_3(s) \xrightarrow{1000°C} Al_2O_3(s) + 3H_2O(g)$$

The melting point of alumina is 2045°C. The use of pure molten alumina as an electrolyte is thus neither practicable nor economic. The temperature is reduced by using a **eutectic mixture** (*see* section 6.2) consisting of a 5% solution of alumina in molten cryolite. This has a melting point of 970°C. Cryolite is an aluminium ore with the formula Na_3AlF_6. Cryolite is found naturally in Greenland, although large quantities are also produced synthetically.

Stage 2: Reduction of Alumina by Electrolysis

The electrolysis is carried out in a **Hall–Héroult cell** (*see* figure 15.3). Liquid aluminium is produced at the graphite cathode where it is tapped off. It is over 99% pure. The molten cryolite remains unchanged and so more alumina can be added as required. Oxygen is produced at the graphite anode. This burns away the anode, producing carbon monoxide and a smaller amount of carbon dioxide. An anode lasts about 20 days before it has to be replaced.

The electrode reactions are

at the cathode $\quad Al^{3+}(l) + 3e^- \rightarrow Al(s)$

at the anode $\quad \begin{cases} 2O^{2-} \rightarrow O_2(g) + 4e^- \\ 2C(s) + O_2 \rightarrow 2CO(g) \end{cases}$

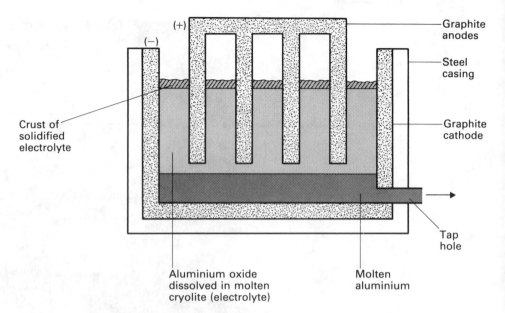

Figure 15.3 The Hall–Héroult cell

An operator looks into an electrolytic cell used for aluminium production

The process consumes large quantities of electricity. A direct current of over 100 000 A at a potential difference of about 5 V is used. Between 13 000 and 17 000 kW h are required to produce 1 tonne of aluminium. The process is thus only economic where electricity is inexpensive—for example, where hydroelectric power is available. A further problem is fluoride pollution in the vicinity of the works due to loss of cryolite.

USES OF ALUMINIUM AND ITS COMPOUNDS

Aluminium is one of the most versatile of all metals. It is used extensively in the transport industries—in the construction of aircraft, ships and cars, for example. It is used in the chemical industry as a reducing agent and in the food industry for packaging. In homes it is found in the form of kitchen foil, cooking utensils and window frames, for example.

Its versatility and widespread use is due to a number of physical and chemical properties which make it particularly attractive compared with other metals:

1. It is light yet strong. The strength-to-weight ratios of aluminium alloys are higher than any other commercial metals.
2. It is highly workable. Aluminium is both malleable and ductile. It can thus be rolled, pressed or extruded to any shape.
3. It resists corrosion. When exposed to air, aluminium and its alloys form a thin film of aluminium oxide, Al_2O_3, on the surface. The film is colourless, strong and does not flake. It thus seals off the metal from oxygen and prevents further oxidation.
4. It is an excellent conductor of both electricity and heat. Since it is such a good conductor of heat, it is used to make heat exchangers in the chemical, oil and other industries.
5. It is highly reflective. Aluminium is an excellent reflector of radiant energy. For this reason, it is commonly used in roofing to insulate buildings. Aluminium foil is also used to jam radar.
6. It is non-magnetic and is thus used in navigational equipment.
7. It is a good reducing agent and used for this purpose in the chemical and steel industries.
8. It is non-toxic and can thus be used for making food and brewing equipment and in packaging.

Alloys

Aluminium readily forms alloys with other metals. Metals used include copper, magnesium, nickel and zinc. Copper, nickel and zinc harden and increase the strength of the metal whilst magnesium improves its corrosion resistance. Silicon is also sometimes added to casting alloys to improve their fluidity and castability.

Two common alloys of aluminium are:

Duralumin	Al 95%; Cu 4%; Mg, Fe, Si 1%
Magnalium	Al 83%; Mg 15%; Ca 2%

Aluminium Compounds

We have already seen how alumina occurs naturally in the corundum group of minerals and how these are used as gemstones and abrasives. Alumina also has a number of other uses. Together with silica it is used as a refractory material for furnace lining. Alumina is also used in the chemical industry as a dehydrating agent and catalyst and in the laboratory as a packing material for chromatographic columns.

Potash alum and aluminium sulphate are, as we saw above, used as mordants for dyeing. Aluminium sulphate is also used as a coagulant in various ways. For example, it is used to precipitate colloidal matter from sewage.

Aluminium—a most versatile metal

Refractories

These are materials used for lining furnaces. They must withstand high temperatures and also molten metals, slags and hot gases. China clay and other clays, silica, magnesite, dolomite and alumina are all used as refractories.

SUMMARY

1. Elements in Group III have the **electronic configuration** s^2p^1 in their outer shells.
2. Boron exhibits a **diagonal relationship** with silicon.
3. Although aluminium is high in the electrochemical series, it does not react readily with oxygen and dilute acids due to the formation of a surface **oxide layer**.
4. Aluminium can act as a **reducing agent** in redox reactions.
5. Aluminium oxide exhibits both **ionic** and **covalent** characteristics.
6. Aluminium oxide is **amphoteric**.

(continued)

7. The **alums** have the general formula $M(I)M(III)(SO_4)_2 \cdot 12H_2O$.
8. **Bauxite** contains hydrated aluminium oxide.
9. Aluminium is extracted from bauxite in two stages:
   Stage 1: Conversion of bauxite to **alumina**.
   Stage 2: Electrolytic **reduction** of alumina to form aluminium.
10. Since aluminium and its **alloys** have a wide range of advantageous properties, they are used extensively in industry and in the home.

15.2
Group IV   THE ELEMENTS

The five elements of Group IV are shown in figure 15.4. Like the Group III elements they are p-block elements. Each element in Group IV has a characteristic s^2p^2 electronic configuration in its outer shell. Table 15.4 shows the electronic configurations and some of the properties of the Group IV elements. These and other chemical and physical properties are related to the structures of the elements:

- Carbon (diamond) has a **giant molecular structure** (*see* section 3.2).
- Silicon has a giant molecular structure similar to that of diamond.
- Germanium also has a giant molecular structure similar to that of diamond.
- Tin has a **giant metallic structure** (*see* also section 3.2).
- Lead has a giant metallic structure.

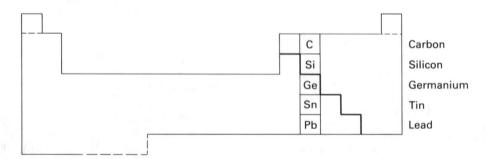

Figure 15.4   The Group IV elements

Both tin and lead have face-centred cubic structures.

Table 15.4   Electronic configurations and physical properties of the Group IV elements

Element	Atomic number	Configuration in outer shell	Atomic radius/nm	Density/ g cm^{-3}	Melting point/°C	Boiling point/°C
C	6	$2s^2 2p^2$	0.077	3.51(d) 2.26(gr)	3730(d)	4830(d)
Si	14	$3s^2 3p^2$	0.117	2.33	1410	2680
Ge	32	$4s^2 4p^2$	0.122	5.32	937	2830
Sn	50	$5s^2 5p^2$	0.142	7.3	232	2270
Pb	80	$6s^2 6p^2$	0.154	11.4	327	1730

d = diamond
gr = graphite

On descending the group, the atomic radius increases and the bonding between the atoms becomes weaker. The outer-shell electrons become increasingly delocalised and electrical conductivity therefore increases. The elements therefore become more metallic in nature:

- Carbon is a non-metal and in the form of diamond a non-conductor.
- Silicon and germanium are **semiconductors**.
- Tin and lead are metals and good conductors.

Furthermore, as the size of the atoms increases on descending the group, the bonds weaken. The density of the elements thus increases whilst the melting point, boiling point and hardness of the elements all decrease.

Allotropy

Silicon, germanium and lead each exist in one structural form only. However, both carbon and tin exist in more than one structural form. These are known as **allotropes** (*see* section 3.2).

Two allotropes of carbon exist. These are diamond and graphite. Their structures were described in section 3.2. The type of allotropy exhibited by carbon is known as **monotropy**. Monotropy is characterised by a number of features:

- The allotropes can all exist under a particular set of conditions; for example, both diamond and graphite exist at room temperature and pressure.
- There is no transition temperature at which one allotrope changes into another.
- One allotrope is more stable than the other. Graphite, for example, is more stable than diamond. The less stable forms are said to be **metastable**. Diamond is thus the metastable allotrope (or monotrope) of carbon.

Carbon also commonly exists in other elemental forms, such as charcoal, coke and carbon black. These are all impure forms of carbon. They are sometimes known as **amorphous** and were once thought to be a third allotropic form of carbon. Amorphous means 'without shape'. Amorphous carbon is now known to be microcrystalline graphite.

Tin exists in three allotropic forms. These are grey tin (α-tin), white tin (β-tin) and rhombic tin (γ-tin). The type of allotropy exhibited by tin is known as **enantiotropy**. This is characterised by the following two features:

- One allotrope changes to another at a definite temperature known as the transition temperature

$$\underset{\substack{\text{diamond}\\\text{structure}\\\text{(semiconductor)}}}{\alpha\text{-tin}} \underset{13.2°C}{\rightleftharpoons} \underset{\substack{\\ \underbrace{\qquad\qquad\qquad}\\\text{metallic structure}}}{\beta\text{-tin} \underset{161°C}{\rightleftharpoons} \gamma\text{-tin}}$$

- Each allotrope is only stable over a definite temperature range.

Reactivity of the Elements

The chemical reactivity of the elements generally increases down the group from carbon to lead. Only tin and lead are above hydrogen in the electrochemical series (*see* section 10.3). Lead reacts very slowly with dilute acids to release hydrogen. The reaction between tin and dilute acids is moderate.

Carbon is oxidised by hot concentrated acids such as concentrated nitric acid and concentrated sulphuric acid:

$$C(s) + 2H_2SO_4(l) \rightarrow CO_2(g) + 2H_2O(l) + 2SO_2(g)$$

Silicon does not react with any acid except hydrofluoric acid.

All the Group IV elements except carbon react with alkalis. For example, silicon reacts with dilute sodium hydroxide solution, forming sodium silicate:

$$Si(s) + 2OH^-(aq) + H_2O(l) \rightarrow SiO_3^{2-}(aq) + 2H_2(g)$$

Tin and lead react with hot concentrated sodium hydroxide solution, forming the stannate(IV) ion, SnO_3^{2-}, and plumbate(II) ion, PbO_3^{2-}, respectively.

All the Group IV elements except lead react with oxygen on heating to form dioxides. For example

$$C(s) + O_2(g) \rightarrow CO_2(g)$$

Lead combines with oxygen to form lead(II) oxide, PbO.

Carbon reduces steam (*see* section 12.1) and many metal oxides (*see* section 5.3).

THE COMPOUNDS

We have already noted that the metallic nature of the elements increases down the groups. Carbon and silicon form covalent compounds—either molecular or macromolecular. Only lead and tin form compounds with ionic character. These trends are reflected in the values for the first ionisation energy, electronegativity and standard redox potential of the elements (*see* table 15.5).

Table 15.5 Properties of Group IV elements

Element	First ionisation energy/kJ mol^{-1}	Electronegativity	Electrode potential, $E^{\ominus}$/V $M^{2+}(aq) + 2e^- \rightleftharpoons M(s)$
carbon	1086	2.5	–
silicon	787	1.8	–
germanium	760	1.8	0.23
tin	707	1.8	−0.14
lead	715	1.8	−0.13

Oxidation States

All the elements in Group IV form compounds in which they exist in an oxidation state of +4. The stability of compounds with elements in this oxidation state decreases on descending the group. For example, both carbon and silicon form four strong covalent bonds. These bonds are due to the promotion of an s electron into a p orbital and the formation of four sp^3 hybrid orbitals.

All the elements except silicon also form compounds in which they exist in an oxidation state of +2. This oxidation state arises from what is known as the **inert pair effect**. In large atoms such as those of tin and lead, the d and f electrons do not shield the outer-shell electrons as well as the inner core s and p electrons. The outer-shell s electrons in their spherical orbitals are, as a result, much more strongly attracted by the nucleus compared with the outer-shell p electrons. The outer-shell electrons effectively become sucked into the inner core of electrons and thus become 'inert'.

The stability of compounds with Group IV elements in oxidation state +2 increases down the group (*see* table 15.6). Thus, carbon monoxide, CO, is readily oxidised to the more stable carbon dioxide, CO$_2$. Compounds containing silicon in the +2 oxidation state are not common. Germanium exists in both oxidation states, the +4 state being more stable. For example, GeO$_2$ is more stable than GeO. The +4 state in tin compounds is slightly more stable than the +2 state. Consequently, Sn^{2+} tends to oxidise to Sn^{4+} and thus acts as a reducing agent. For example

$$Sn^{2+}(aq) + Hg^{2+}(aq) \rightarrow Sn^{4+}(aq) + Hg(l)$$

Table 15.6 Stability of Group IV elements

	Oxidation state	
	II	IV
carbon	low	high
silicon		↑
germanium	↓ stability	
tin		↑
lead	high	low

The +2 oxidation state in lead compounds is more stable than the +4 state. Lead(IV) thus tends to reduce to lead(II) and therefore can act as an oxidising agent. For example, lead(IV) oxide oxidises hydrochloric acid to chlorine:

$$PbO_2(s) + 4HCl(aq) \rightarrow PbCl_2(aq) + Cl_2(g) + H_2O(l)$$

Uniqueness of Carbon

Carbon is unique in its ability to bond with itself and form stable long-chain and ring compounds. This property is known as **catenation**. It is this property which allows carbon to form millions of compounds and have a whole branch of chemistry devoted to it.

Carbon readily catenates because the strength of the C–C bond is comparable to that of the C–Cl and C–O bonds. In comparison, the Si–O bond is much stronger than the Si–Si bond as the bond enthalpies in table 15.7 show.

Table 15.7 Carbon and silicon bond enthalpies

Bond	Average bond enthalpy/kJ mol^{-1}	Bond	Average bond enthalpy/kJ mol^{-1}
C–O	360	Si–O	464
C–Cl	338	Si–Cl	380
C–C	348	Si–Si	226

As we saw in section 2.1, carbon is also unique in its ability to hybridise. Its four bonding electrons can undergo hybridisation to form:

- *four sp^3 orbitals*—it is these orbitals which form single covalent bonds in a tetrahedral arrangement;
- *three sp^2 orbitals*—these have a planar shape and allow carbon to form double bonds;
- *two sp orbitals*—these have a linear shape and permit carbon to form triple bonds.

The Oxides

The Group IV elements form monoxides in which the elements have an oxidation state of +2 and dioxides in which their oxidation state is +4. The ability to form stable dioxides decreases down the group whereas the ability to form stable monoxides increases.

Carbon Dioxide, CO$_2$

Carbon dioxide normally exists in a gaseous state as linear molecules (*see* chapter 2). It is a colourless, odourless gas which is heavier than air. Solid CO$_2$ has a face-centred cubic structure similar to that of iodine (*see* section 3.2). It sublimes at −78°C.

Carbon dioxide is one of the normal products of combustion of carbon or any compound containing carbon when burned in a plentiful supply of oxygen or air. In the laboratory, it is normally prepared by the addition of dilute hydrochloric acid to calcium carbonate in the form of marble chips:

$$CaCO_3(s) + 2HCl(aq) \rightarrow CaCl_2(aq) + CO_2(g) + H_2O(l)$$

The gas is acidic, dissolving in water to form carbonic acid as part of the following sequence of equilibria:

$$CO_2(g) + H_2O(l) \rightleftharpoons \underset{\text{carbonic acid}}{H_2CO_3(aq)} \rightleftharpoons H^+(aq) + HCO_3{}^-(aq) \rightleftharpoons 2H^+(aq) + CO_3{}^{2-}(aq)$$

The test for carbon dioxide is to bubble it through dilute calcium hydroxide solution (limewater). This results in the formation of the insoluble calcium carbonate, which makes the solution cloudy:

$$Ca(OH)_2(aq) + CO_2(g) \rightarrow CaCO_3(s) + H_2O(l)$$

However, when CO_2 is added in excess the solution clears due to the conversion of the insoluble carbonate to the soluble hydrogencarbonate:

$$CaCO_3(s) + H_2O(l) + CO_2(g) \rightarrow Ca(HCO_3)_2(aq)$$

Carbon dioxide also forms carbonates and (when in excess) hydrogencarbonates when bubbled through other alkalis such as sodium hydroxide.

When heated with metals at the top of the electrochemical series, carbon dioxide is reduced to carbon. For example

$$2Mg(s) + CO_2(g) \rightarrow 2MgO(s) + C(s)$$

Silicon(iv) Oxide, SiO₂

Silicon(iv) oxide is also known as silica or silicon dioxide. It has a giant structure and exists in three allotropic forms: quartz, tridymite and cristobalite (*see* section 3.2).

Silicon(iv) oxide is an acidic oxide, reacting with alkalis to form silicates:

$$SiO_2(s) + 2OH^-(aq) \rightarrow SiO_3^{2-}(aq) + H_2O(l)$$

Other Group IV Dioxides

The remaining dioxides in Group IV, that is GeO_2, SnO_2 and PbO_2, all exhibit ionic character and are amphoteric. Lead(iv) oxide loses oxygen on heating:

$$2PbO_2(s) \rightarrow 2PbO(s) + O_2(g)$$

Carbon Monoxide, CO

Carbon monoxide is a colourless, odourless and very poisonous gas. It exists as a resonance hybrid:

$$:c \rightarrow \ddot{o}: \longleftrightarrow :c = \ddot{o}: \longleftrightarrow :c \equiv o:$$

It is only very slightly soluble in water, forming a neutral solution. At 150°C and a pressure of 10 atmospheres it reacts with sodium hydroxide solution, forming sodium methanoate:

$$NaOH(aq) + CO(g) \rightarrow HCOONa(aq)$$

It is readily oxidised and burns in air with a blue flame, forming carbon dioxide:

$$2CO(g) + O_2(g) \rightarrow 2CO_2(g)$$

It is a good reducing agent and is used industrially to extract iron and nickel, for example

$$Fe_2O_3(s) + 3CO(g) \rightleftharpoons 2Fe(l) + 3CO_2(g)$$

Carbon monoxide reacts with several d-block elements forming volatile carbonyl compounds. For example

$$Ni(s) + 4CO(g) \xrightarrow{60°C} [Ni(CO)_4](l)$$
$$\text{tetracarbonylnickel(0)}$$

These carbonyl compounds are covalent and have a tetrahedral structure.

Other Oxides

Both silicon(ii) oxide and germanium(ii) oxide are unstable. Germanium(ii) oxide tends to disproportionate (*see* section 10.2) on heating:

$$2GeO(s) \rightarrow GeO_2(s) + Ge(s)$$

oxidation number	+2	+4	0

Tin(II) oxide, SnO, and lead(II) oxide are both amphoteric.

Dilead(II) lead(IV) oxide, or red lead as it is usually known, is an example of a **mixed oxide**. It has the empirical formula Pb_3O_4 and consists of two parts PbO and one part PbO_2. On heating it yields oxygen:

$$2Pb_3O_4(s) \rightarrow 6PbO(s) + O_2(g)$$

The Chlorides

The tetrachlorides of the Group IV elements are covalent compounds existing as simple molecules with a tetrahedral structure (*see* figure 15.5). At room temperature they are all volatile liquids. Their thermal stability decreases down the group. Indeed lead(IV) chloride decomposes explosively at its boiling point.

Tetrachloromethane, CCl_4, can be prepared by bubbling chlorine through carbon disulphide:

$$CS_2(l) + 3Cl_2(g) \rightarrow CCl_4(l) + S_2Cl_2(l)$$

It can also be prepared by the reaction of chlorine and methane in the presence of ultra-violet light (*see* section 18.1):

$$CH_4(g) + 4Cl_2(g) \xrightarrow{\text{UV light}} CCl_4(l) + 4HCl(g)$$

The tetrachlorides of silicon, germanium and tin can all be prepared by heating the elements in chlorine. For example

$$Si(s) + 2Cl_2(g) \rightarrow SiCl_4(l)$$

When lead is heated in chlorine, lead(II) chloride is formed. Lead(IV) chloride is prepared by reacting lead(IV) oxide with excess concentrated hydrochloric acid in the presence of chlorine at 0°C. The overall reaction is:

$$PbO_2(s) + 4HCl(aq) \rightarrow PbCl_4(l) + 2H_2O(l)$$

All the tetrachlorides, except CCl_4, are hydrolysed by water. For example

$$SiCl_4(l) + 2H_2O(l) \rightarrow SiO_2(s) + 4HCl(g)$$

The ease of hydrolysis decreases from silicon to tin as the metallic nature of the elements increases.

The thermal instability of lead(IV) chloride results in it being rapidly hydrolysed in the presence of water. Lead(IV) oxide and some lead(II) chloride are formed.

Only germanium, tin and lead in Group IV form stable dichlorides. They are all solids and exhibit ionic characteristics.

Tin(II) chloride can be prepared by the reaction of tin with concentrated hydrochloric acid:

$$Sn(s) + 2HCl(aq) \rightarrow SnCl_2(aq) + H_2(g)$$

Lead(II) chloride is insoluble in water. It can thus be prepared by ionic precipitation, for example, by adding dilute hydrochloric acid or sodium chloride solution to an aqueous solution of lead(II) nitrate:

$$Pb^{2+}(aq) + 2Cl^-(aq) \rightarrow PbCl_2(s)$$

The golden yellow lead(II) iodide is prepared in a similar way (*see* section 4.2).

The Hydrides

The ability of Group IV elements to catenate (*see* above) decreases markedly down the group. This is demonstrated by the ability of the elements to form hydrides. Carbon forms numerous binary hydrides. Some are long-chain compounds whilst others are ring compounds. Collectively they are known as the **hydrocarbons**. They are discussed in detail in chapter 18.

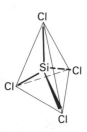

Figure 15.5 Silicon tetrachloride has a tetrahedral structure

Silicon forms a number of binary hydrides called **silanes**. Silanes of up to 11 silicon atoms have been synthesised. Germanium binary hydrides are known as **germanes**. Only three are known. Tin only forms two binary hydrides, SnH_4 and Sn_3H_8. Lead forms only one binary hydride, PbH_4.

From the point of view of Group IV chemistry, the most important hydrides are the tetrahydrides. These are all covalent, existing as simple molecules with a tetrahedral shape. They all have low melting and boiling points and exist as gases at room temperature. The thermal stability of the tetrahydrides decreases on descending the group. Indeed, lead(IV) hydride, PbH_4, is so unstable it cannot be isolated at room temperature.

The Carbonates

The carbonate ion, like other oxoanions, is a resonance hybrid of several limiting forms (*see* section 2.1 and figure 2.10).

Ammonium carbonate and all the Group I metal carbonates except lithium carbonate are soluble in water. All other carbonates are insoluble in water.

The carbonates decompose on heating to give the oxide and carbon dioxide. The most thermally stable carbonates are those of the Group II elements. Calcium and barium carbonates, for example, decompose at about 1000°C.

All carbonates react with dilute mineral acids to yield a salt, carbon dioxide and water. For example

$$MgCO_3(s) + 2HCl(aq) \rightarrow MgCl_2(aq) + CO_2(g) + H_2O(l)$$

If the salt formed is insoluble, it coats the carbonate and thus slows down the reaction.

OCCURRENCE AND EXTRACTION

Occurrence

Carbon is the sixteenth most abundant of elements, accounting for about 0.027% of the Earth's crust. It occurs naturally in its elementary form as diamonds in South Africa and Brazil and as graphite in Germany, Sri Lanka and the USSR. Coal contains up to 90% carbon (*see* chapter 5). Carbon occurs also in various forms in the fossil fuels, in the carbonate minerals such as calcite and dolomite and in all living matter. As carbon dioxide, carbon also accounts for about 0.046% by weight of the Earth's atmosphere. The carbon dioxide in the air forms the key part of the carbon cycle (*see* figure 15.6) in which carbon dioxide is consumed by plants in photosynthesis and regenerated by animals in respiration and in the combustion process.

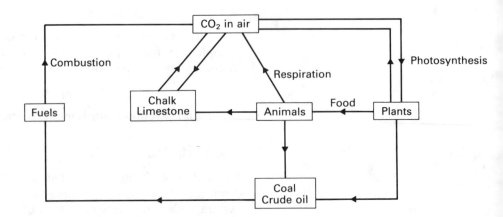

Figure 15.6   The atmospheric carbon cycle

The planet Venus—swathed in clouds of CO_2

The name *silicon* derives from the Latin *silex* or *silicis* which means flint.

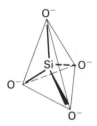

Figure 15.7   The silicate anion

(a)

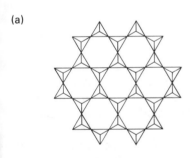

(b)

Figure 15.8   (a) Two-dimensional representation of the crystal structure of quartz. (b) Each silicon atom is joined to four oxygen atoms. In the two-dimensional representation, the fourth oxygen atom lies above the plane. It is not shown

The Planets

Venus is swathed in dense clouds which consist mainly of carbon dioxide. Mars has a thin atmosphere consisting mainly of carbon dioxide. Mercury has no atmosphere. It is now thought that the Earth began with an atmosphere of carbon dioxide like its neighbours.

Carbon dioxide is a product of respiration in animals. It combines with the haemoglobin in red blood cells and is carried to the lungs where it is exchanged for oxygen. Haemoglobin also combines with carbon monoxide to give a pink complex called carboxyhaemoglobin. This prevents normal oxygenation of the blood and can lead to death. For this reason, carbon monoxide is extremely poisonous.

Silicon is the second most abundant element on Earth, next to oxygen. It does not occur as a free element but occurs widely in the form of silica, SiO_2, and silicates. Sand, sandstone, quartz and flint are all forms of silica.

The Silicates

More than a 1000 silicate minerals are known. They make up about 75% of the Earth's crust—that is, the Earth's outer shell. The basic building block of all silicates is the silicon tetrahedron, which is bonded to four oxygen atoms to form the silicate anion, SiO_4^{4-} (*see* figure 15.7).

Common silicates may consist of these isolated tetrahedra or they may be linked together to form polymeric rings, chains, sheets and giant three-dimensional structures known as **framework silicates**.

The garnet family of minerals, of which the most common is almandine, $Fe_3Al_2(SiO_4)_3$, contain isolated tetrahedra. Many igneous and metamorphic rocks (*see* section 13.3) contain chain silicates. The most important group of chain silicates are the pyroxenes. Asbestos also has a polymeric chain structure (*see* section 3.2). Sheet silicates include kaolin, $Al_2Si_2O_5(OH)_4$, and other clays and also the micas and talc (*see* also section 3.2).

Framework silicates include the silica group, SiO_2, the feldspars and the zeolites. Quartz is a framework silicate. A two-dimensional representation is shown in figure 15.8 (*see* also section 3.2).

Silicate minerals include granite, feldspar, asbestos, clay and mica. Silicon is also present in the Sun and stars. It is a major constituent of a class of meteorites known as aerolites.

Germanium also occurs widely in the Earth's crust. The average abundance is about 7 grams per tonne. The element is found in bituminous coal and in small amounts in the ore germanite.

Tin is relatively rare, accounting for about 0.001% of the Earth's crust. Its most important ore is cassiterite, SnO_2. This is mined in Nigeria, Indonesia and several other countries.

Lead occurs in the Earth's crust to the extent of about 15 parts per million. Its most important ore is galena, lead(II) sulphide, PbS. This is mined in Canada and Australia.

Extraction

The demand for graphite and diamonds in the world exceeds the amount supplied by mining. About 70% of the **graphite** produced in the world is manufactured by the **Acheson process**. In this process a mixture of coke and sand (SiO_2) is heated at about 2000°C in an electric furnace for about 24 hours:

$$3C(s) + SiO_2(s) \rightarrow \underset{\text{graphite}}{C(s)} + Si(s) + 2CO(s)$$
$$\quad\underset{\text{coke}}{}\quad\underset{\text{sand}}{}$$

Industrial diamonds are made by heating graphite in the presence of a rhodium catalyst to about 3000°C using pressures in the region of 100 000 atmospheres.

Silicon is produced by the reduction of silica with carbon in an electric furnace:

$$SiO_2(s) + 2C(s) \rightarrow Si(s) + 2CO(s)$$

Very pure silicon is obtained by **zone refining**. A heating coil or furnace is used to melt a small zone at the end of a rod of silicon. The heater is slowly moved so that the molten zone moves along the rod from one end to the other. As it does, pure silicon crystallises out, leaving the impurities to collect in the molten zone. The impurities are thus swept to one end of the rod leaving the pure crystals in the rest of the rod.

Germanium is obtained from germanite by heating with hydrochloric acid and then by hydrolysing the germanium(IV) chloride which is formed. Germanium(IV) oxide is produced. The oxide is reduced with carbon or hydrogen. Zone refining is used to obtain very pure germanium from the product.

Tin is extracted from cassiterite by reduction with coke:

$$SnO_2(s) + 2C(s) \rightarrow Sn(l) + 2CO(g)$$

Lead is extracted from galena by roasting in air to produce lead(II) oxide which is then reduced with coke:

$$2PbS(s) + 3O_2 \rightarrow 2PbO(s) + 2SO_2(g)$$
$$PbO(s) + C(s) \rightarrow Pb(l) + CO(g)$$

USES OF CARBON

The various forms of carbon and its compounds have a wide variety of uses.

Diamonds are not only used as gemstones, they are also used for cutting and grinding wheels, in drills—for oil wells, for example—and in dies for making the tungsten filaments for light bulbs.

Graphite has numerous uses. It is used to make inert electrodes for electrolytic processes and as brushes for electric motors. It is used as a furnace lining, as a lubricant and as a moderator in nuclear reactors (*see* chapter 1). It is also baked with clay to make the 'lead' for lead pencils.

As we have seen, both **coke** and **carbon monoxide** are used industrially to extract metals from their ores.

The various forms of **charcoal** have a number of uses. Wood charcoal, which is obtained by burning wood in a limited supply of air, is used to absorb gases. This is because it is very porous and thus has a vast surface area. Animal charcoal, which is obtained by burning the bones of animals, is used to decolourise the juice of sugar cane in the manufacture of sugar. Carbon black is used to make ink, carbon paper and black shoe polish and is incorporated into rubber to make tyres.

Hydrocarbons and other organic chemicals have numerous uses. These are discussed in chapter 18.

Carbon dioxide is produced in the manufacture of quicklime (*see* chapter 13) and in fermentation processes (*see* chapter 9). It is used in fire extinguishers and in its solid form as 'dry ice' to produce low temperatures.

Black shoe polish contains carbon black

Carbon black is added to rubber to increase its strength and stiffness. This is why tyres are usually black

The body of this prototype ocean cruiser is made entirely from carbon-fibre-reinforced plastics (CFRP). These composite polymers are stronger than steel but lighter than aluminium

Table 15.8 Some uses of the silicates

detergents	roofing granules
pigments	building materials
cracking catalysts	ore flotation
silica gels and sols	water treatment
zeolites	textile bleaching
adhesives	foundry binders
cements	welding rods

Table 15.9 Silicon is a semiconductor

	Electrical resistance/ Ω cm^{-3}	
silver	0.00005	good conductor
silicon	50 000	semiconductor
mica	10^{12}	good insulator

USES OF SILICON

Silicon and its compounds, like carbon and its compounds, have extensive uses. Silicon is used to make microelectronic devices. Silica is used in glass manufacture and cement manufacture. Silicates have a wide range of uses (*see* table 15.8). In all these examples, sodium silicates are used. Finally, silicones are used to make synthetic rubbers, polishes and protective materials. We shall now outline in more detail three of these uses.

The Microchip Revolution

Over the last two decades or so, silicon has become exceedingly important as a semiconductor in the preparation of microelectronic devices known as 'microchips'.

A **semiconductor** has an electrical resistance somewhere between that of an electrical insulator and that of an electrical conductor (*see* table 15.9).

Semiconductors are often contaminated or 'doped' with controlled amounts of impurities. Doping reduces the gap between the conduction and valence bands (*see* section 2.1). It thus lowers the resistance of the semiconductor.

An **n-type** or negative-type semiconductor is a semiconductor which has been doped with a Group V element such as phosphorus. Since a phosphorus atom has five outer-shell electrons, the presence of phosphorus atoms in the silicon lattice results in a surplus of electrons and thus a net negative charge (*see* figure 15.9). This is why it is called an n-type semiconductor.

A **p-type** or positive-type semiconductor has a net positive charge due to the presence in the lattice of Group III atoms such as aluminium. Each aluminium atom creates an electron hole in the lattice and thus a positive charge.

A **junction diode** can be formed by placing n-type and p-type semiconductor electrodes together (*see* figure 15.10). Electrons flowing through the p-type electrode stop at the junction between the two electrodes. This is known as the pn junction. Electrons flowing in the other direction pass through the junction since they are passing through a lattice with surplus electrons into a lattice deficient in electrons. This same passage of electricity may also be regarded as the movement

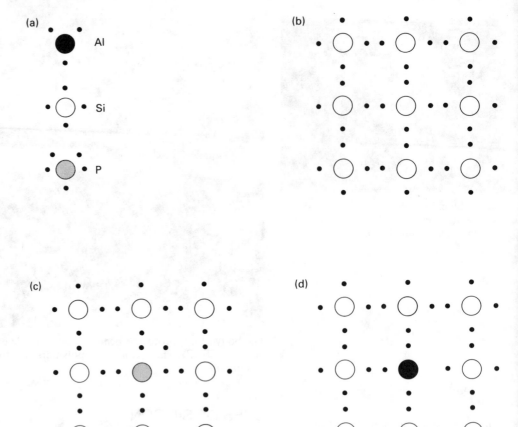

Figure 15.9 Doped silicon. (a) Atoms of Al, Si and P showing outer-shell electrons. (b) *Semiconductor:* each pair of electrons is a covalent bond. (c) *n-Type semiconductor:* the presence of a Group V atom such as phosphorus in the lattice adds an extra electron, and this decreases the electrical resistance. (d) *p-Type semiconductor:* the presence of a Group III atom in the lattice results in an electron 'hole' in the lattice

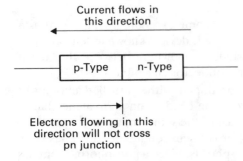

Figure 15.10 *Junction diode:* surplus electrons in the n-type semiconductor flow through the pn junction to fill 'holes' in the p-type semiconductor

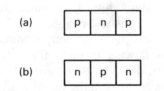

Figure 15.11 Transistors: (a) pnp transistor, (b) npn transistor

of electron holes or positive charge from the p-type electrode to the n-type electrode.

Junction diodes made from silicon are used as rectifiers to transform alternating current into direct current. A silicon-controlled rectifier (SCR) consists of n-type and p-type electrodes and also a third electrode known as the gate. An SCR permits the passage of direct current only when a small electrical signal is applied to the gate.

A **transistor** is a three-electrode semiconductor device with a thin layer of n-type (or p-type) semiconductor sandwiched between two regions of p-type (or n-type) semiconductor (*see* figure 15.11). This arrangement allows a large electric current flowing through the device to be controlled by a very small current or voltage. A pnp transistor works by 'hole conduction' whereas an npn transistor works by electron conduction.

Until the 1960s almost all transistors were housed in individual metal or plastic packages. Since then, **integrated circuits** have been developed. Nowadays, a single minute silicon chip in a program pocket calculator might contain over 30 000 transistors connected in a single integrated circuit.

Glass

Silicate glasses are formed by allowing molten silicates to cool. Soda glass consists of a mixture of calcium silicate, $CaSiO_3$, and sodium silicate, Na_2SiO_3. Reference

Borosilicate glass is widely used in chemical laboratories

Borosilicate glass can withstand temperatures up to 800°C

Lead crystal glass contains about 8% PbO

to its manufacture was made in the previous chapter. **Soda glass** is used for making windows and all kinds of flat glass.

Borosilicate glass contains about 81% SiO_2, 13% boron oxide, B_2O_3, and small amounts of sodium and aluminium oxides. Borosilicate glass can withstand temperatures up to 800°C and is much more resistant to attack by chemicals such as alkalis. The most common brand of borosilicate glass is Pyrex. Borosilicate glass is used in cooking and laboratory ware.

Lead glasses have a high refractive index and are used for lead crystal glassware. A typical lead glass contains about 8% lead(II) oxide, although fine crystal glass contains more.

Fibre glass is produced by dropping molten glass onto a refractory rotating disc. The glass flies off the disc forming the fibres. Fibre glass is used to make panels in cars and also to make furniture and aircraft components.

Glasses can be coloured by adding d-block metal oxides to the glass during manufacture. Cobalt imparts a blue or pink colour to glass depending on the amount of basic oxides such as CaO and Na_2O present. The colours of inexpensive brown and green glasses used in bottling wines and beers are due to iron compounds in the sand used to make the glass.

Optical fibres are made from silica glass. Silica glass is made by fusing silica. It is sometimes known as quartz glass. Silica glass has excellent optical transparency. However, the silica glass used for optical fibres has to be extremely pure. Impurities such as copper and iron have to be reduced to a level of less than one part in 10^{10}. For this reason the silica glass used for optical fibres is made directly by the vapour-phase reaction of oxygen and silicon(IV) chloride. Silicon(IV) chloride can be obtained in extremely high purity. It is known as 'electronic grade'.

Optical fibres consist of a central core region in which light is transmitted. The core is encased in a cladding of smaller refractive index so that light cannot escape. The fibre, which has the thickness of a human hair, is protected by a thin film of silicone or organic polymer.

Optical fibres are used to transmit television programmes, telephone conversations, computer outputs and other data. Some forecasts suggest that optical fibres will eventually replace the copper wire cables which are conventionally used for these purposes.

Water glass is an aqueous solution of sodium silicate, Na_4SiO_4. It is obtained by fusing silica with an alkali such as sodium hydroxide or sodium carbonate. Sodium silicate is a strong base. When it is acidified a gel is obtained. This is a polymeric acid thought to have the following structure:

$$\begin{array}{ccccc}
& H & & H & & H \\
& | & & | & & | \\
& O & & O & & O \\
& | & & | & & | \\
-O-&Si&-O-&Si&-O-&Si&-O- \\
& | & & | & & | \\
& O & & O & & O \\
& | & & | & & | \\
& H & & H & & H
\end{array}$$

On heating, this material dehydrates forming what is known as **silica gel**. Silica gel has a large surface area. It is used as a drying agent and an inert supporting material for some finely divided catalysts.

Silicones

These are polymeric chain compounds consisting of a backbone of alternately linked silicon and oxygen atoms. Alkyl or aryl groups (*see* chapter 17) are attached to the silicon atoms. For example:

$$H-O-\underset{\underset{CH_3}{|}}{\overset{\overset{CH_3}{|}}{Si}}-O-\left[\underset{\underset{CH_3}{|}}{\overset{\overset{CH_3}{|}}{Si}}-O\right]_n\underset{\underset{CH_3}{|}}{\overset{\overset{CH_3}{|}}{Si}}-O-H$$

Silicones are oils, greases or resinous or rubbery materials. They are made by the hydrolysis of chlorosilanes such as dimethyl chlorosilane, $(CH_3)_2SiCl_2$. The alkyl or aryl chlorosilanes are prepared by using Grignard reagents (*see* section 19.1) or by passing the alkyl or aryl halide vapour over pellets of silicon containing copper as a catalyst at a temperature of about 300°C:

$$Si + 2CH_3Cl \xrightarrow[300°C]{Cu} (CH_3)_2SiCl_2$$

The silicones are thermally stable and resist oxidation and attack by most chemical reagents. They are good water repellants and are used to waterproof material. They are also used as oils, lubricants and insulators and in varnishes, paints and polishes.

USES OF GERMANIUM, TIN AND LEAD

Germanium
Germanium, like silicon, is used as a semiconductor to manufacture microelectronic devices such as transistors.

Tin
Tin, because of its resistance to corrosion, is used as a protective coating for objects. Tin cans are steel cans coated with a thin layer of tin. However, once the tin coating is removed, the steel becomes exposed to the atmosphere and moisture and rapid corrosion occurs (*see* section 10.5).

Tin plating is accomplished by electrolysis or by dipping the article in molten tin. In the electrolytic method, an acidified solution of tin(II) sulphate, $SnSO_4$, is the electrolyte and the article to be plated is the cathode.

Many alloys contain tin. These include bronze, bearings and type metals. Pewter is an alloy of tin (or antimony) and lead, hardened with copper. Solder is an alloy of tin and lead.

> The use of tin dates back to the Bronze Age (2500 to 2000 BC). Bronze is an alloy of copper and tin. Bronze is easier to cast than copper and has superior mechanical properties.

Pewter tankards. Pewter is an alloy of tin (or antimony) and lead, hardened with copper

Lead

Lead is used as a roofing material, as cable casing, in accumulators and has been used since ancient times in coins, weights, ornaments, solder and in cooking utensils. It has also been used since Roman times for plumbing. However, because of its toxicity, lead water pipes are no longer used. Lead salts are used in paint manufacture and to make tetraethyllead(IV), $Pb(C_2H_5)_4$, which is used as an anti-knock additive in petrol.

Lead Pollution

Over recent years there has been much publicity over lead pollution. Lead and lead compounds are highly toxic when eaten or inhaled. The lead accumulates in the body since the rate of absorption is faster than the rate of excretion. Lead can cause lesions in the central nervous system. Symptoms include loss of appetite, vomiting, convulsions and brain damage. The pollution originates from a number of sources.

Paint

Paints containing lead are dangerous because some children chew the paintwork. In addition, paint can flake off and become mixed with dust. This can get onto children's hands and into their mouths as they eat and suck their fingers. This is particularly a problem in houses where the paint is old.

Plumbing and Drinking Utensils

Lead has been used in plumbing and drinking utensils since the Roman times. Until recently water pipes were commonly made of lead. Pewter, which, until the last century was used to make plates and beer mugs for example, contains 10 to 20% lead.

In soft waters, lead slowly combines with water and dissolved oxygen to form lead(II) hydroxide:

$$2Pb(s) + 2H_2O(l) + O_2(aq) \rightarrow 2Pb(OH)_2(aq)$$

The solubility of lead(II) hydroxide is 0.016 g per 100 g of water at 25°C. In acidic waters, increased amounts of lead dissolve. In hard water, lead(II) sulphate and lead(II) carbonate are formed. These have solubilities of 0.0045 and 0.00017 g per 100 g of water respectively. They form protective layers on the inner surfaces of the pipes.

Romans Killed by Lead

The symbol of lead, Pb, derives from the Latin word *plumbum*. Plumbum was well known in Roman times. It was widely used both in cooking implements and in water pipes. Bad wine was improved by adding lead. Pliny seemed to be aware of the problems of lead poisoning. He warned of the dangers of breathing lead fumes yet strangely advised that wine should be stored in lead vats. It appears that few Romans took his advice. Skeletons dug up from a Roman cemetery in Cirencester have been found to contain up to 10 times the concentration of lead found in contemporary man. It is thought that many of these died from lead poisoning. Indeed, some scholars have suggested that the Roman Empire declined and fell because lead poisoning sapped the abilities of its people.

Petrol

Lead is added to petrol in the form of tetraethyllead(IV) to prevent 'knocking'. These are uncontrolled explosions which can damage the high-compression-ratio engines fitted to cars. It has been estimated that every year between 7500 and

10 000 tonnes of lead are emitted into the atmosphere from car exhausts. Much of this is concentrated in urban areas where lead concentrations in the air, rain water and locally grown produce have been found to be much higher and in some areas up to twice the recommended limit. The lead levels in children's blood is high in these areas and this, it is alleged, has reduced their IQs.

In recent years there has been much pressure on Parliament to introduce measures to reduce lead in petrol from the current British level of 0.40 grams per litre to 0.15 grams per litre. In addition, it is recommended that the lead content in new household paints and the emissions from lead processing works should be reduced and that naturally acidic drinking water should be treated to reduce the amount of dissolved lead.

SUMMARY

1. Carbon, silicon and germanium all have **giant molecular structures**.
2. Tin and lead have **giant metallic structures**.
3. On descending Group IV, electrical conductivity increases and bonding becomes less covalent.
4. Carbon exhibits **monotropy**.
5. Tin exhibits **enantiotropy**.
6. Chemical reactivity increases down Group IV.
7. All Group IV elements except carbon react with alkalis.
8. Group IV elements except lead react with oxygen to form dioxides.
9. Carbon and silicon form covalent compounds.
10. Lead and tin form compounds with **ionic character**.
11. All Group IV elements form compounds in which they exist in oxidation state +4.
12. **Catenation** is the ability of an element to bond with itself. Carbon displays this property to a greater degree than any other element.
13. The ability of Group IV elements to form stable dioxides decreases down the group. The reverse is true for the monoxides.
14. Carbon dioxide is a gas and acidic.
15. Silicon(IV) oxide has a giant structure. It is also acidic.
16. Carbon monoxide is a poisonous gas. It is neutral.
17. The Group IV tetrachlorides are covalent compounds.
18. All the Group IV tetrachlorides except CCl_4 are **hydrolysed** by water.
19. The ability of Group IV elements to form hydrides decreases down the group.
20. Carbon dioxide forms a key part of the **carbon cycle**.
21. Silicates make up about 75% of the Earth's crust.
22. Graphite is manufactured from coke and sand by the **Acheson process**.
23. **Zone refining** is used to manufacture pure silicon.
24. An **n-type semiconductor** is doped with a Group V element.
25. A **p-type semiconductor** is doped with a Group III element.
26. Silicate glasses are formed by allowing molten silicates to cool.
27. **Optical fibres** are made from extremely pure silica glass.
28. **Silicones** are polymeric chain compounds.
29. Tin is corrosion-resistant.
30. The element lead and its compounds are highly toxic.
31. **Lead pollution** can originate from paint, plumbing, drinking utensils and petrol.
32. Lead is added to petrol to prevent **'knocking'**.

15.3
Group V: Nitrogen
and Phosphorus

The five elements of Group V are shown in figure 15.12. They are all p-block elements characterised by the s^2p^3 electronic configuration in their outer shells. Some of the properties of the Group V elements are shown in table 15.10. There is a transition from non-metallic to metallic properties down the group. Nitrogen and phosphorus are non-metals. Arsenic and antimony show some metallic character and bismuth is a metal.

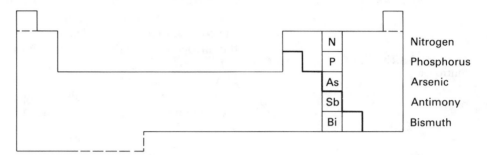

Figure 15.12   The Group V elements

Table 15.10   Properties of the Group V elements

Element	Atomic number	Configuration in outer shell	Atomic radius/nm	Density/ g cm^{-3}	Melting point/°C	Boiling point/°C
nitrogen	7	$2s^2 2p^3$	0.074	0.81	−210	−196
phosphorus	15	$3s^2 3p^3$	0.110	1.82(w)	44(w)	280(w)
arsenic	33	$3d^{10} 4s^2 4p^3$	0.121	5.73	sublimes at 610	
antimony	51	$4d^{10} 5s^2 5p^3$	0.141	6.7	630.5	1380
bismuth	83	$4f^{14} 5d^{10} 6s^2 6p^3$	0.152	9.8	271	1451

w = white

NITROGEN AND PHOSPHORUS

These are the two most important elements in the group.

Under ordinary conditions, nitrogen exists as simple, gaseous diatomic molecules, N_2. It is colourless, odourless, tasteless and particularly unreactive.

Phosphorus exists in three allotropic forms of which the two most common are red phosphorus and the less stable white phosphorus. The structure of these allotropes was described in section 3.2. Table 15.11 compares some of the properties of red and white phosphorus.

Table 15.11   Properties of white and red phosphorus

White phosphorus	red phosphorus
yellowy white waxy solid	deep red powder
consists of P_4 molecules	macromolecular structure
density = 1.82 g cm^{-3}	density = 2.34 g cm^{-3}
emits green phosphorescence in air	does not oxidise in air
stored under water	stored dry
melts at 44°C (under water)	sublimes at 416°C
soluble in organic solvents	insoluble in organic solvents
reacts with hot concentrated NaOH to form phosphine	does not react with NaOH

Although nitrogen is one of the most electronegative elements (*see* table 2.2) it is relatively unreactive. This is because its diatomic molecules are non-polar and also because the triple bond between them is very strong. The bond enthalpy is $+944$ kJ mol^{-1}.

Both nitrogen and phosphorus can exist in oxidation states ranging from -3 to $+5$. For example, in ammonia, NH_3, nitrogen has an oxidation number of -3 whereas in the nitrate ion, NO_3^- its oxidation number is $+5$.

Both elements can accept three electrons to complete their outermost p orbitals, thus forming nitride ions, N^{3-}, and phosphide ions, P^{3-}, respectively. However, both elements tend to form covalent compounds. They do so by forming four sp^3 hybrid orbitals arranged tetrahedrally (*see* figure 15.13). Three of these contain a single electron each. It is these which are shared to form the covalent bonds. The remaining sp^3 orbital contains a lone-pair of electrons. This can be used to form a coordinate bond. For example, in the ammonium ion, NH_4^+, the nitrogen lone-pair overlaps with the vacant s orbital of the hydrogen ion.

Phosphorus does not form ionic or coordinate bonds as readily as nitrogen. However, it can form five covalent bonds by promoting an electron from its 3s orbital to one of its 3d orbitals (*see* figure 15.14). Nitrogen cannot do this because there are no 2d orbitals (*see* section 1.2).

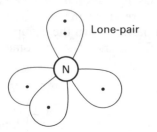

Figure 15.13 The four sp^3 hybrid orbitals of nitrogen

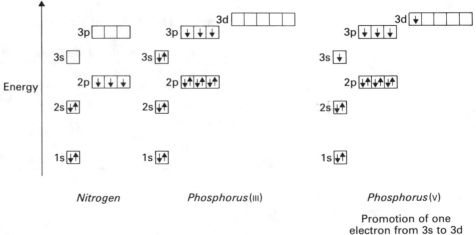

Figure 15.14 Electronic structure of nitrogen and phosphorus

Lightning strikes the Earth 100 times a second on average. Here we see it striking the top of Moscow University. During lightning flashes, nitrogen oxides are formed

OCCURRENCE, EXTRACTION AND USES OF NITROGEN AND PHOSPHORUS

Nitrogen occurs free in the atmosphere as diatomic molecules. About 78% by volume of the atmosphere is nitrogen. Nitrogen also occurs in plants and animals in the form of proteins. Animals obtain protein by consuming plants and other animals. Plants produce protein from the nitrates in the soil. The nitrates in the soil are formed from atmospheric nitrogen and ammonium compounds in the soil. The process of converting atmospheric nitrogen to a form usable by plants and animals is called **nitrogen fixation**.

Nitrogen fixation can occur in two ways:

- During lightning flashes some nitrogen and oxygen in the atmosphere combine to form nitrogen oxides. These dissolve in water, forming dilute nitric acid, which in turn produces nitrates in the soil.
- Nitrogen in the atmosphere is converted to ammonia, which is then converted to nitrate by bacteria by a process known as **nitrification**. Some bacteria exist freely in the soil whilst others occur in the root nodules of leguminous plants such as clover.

Leucaena leaves are used as an organic fertiliser in Thailand. *Leucaena* is a legume and, like all legumes, has a very high nitrogen content. It can thus be used in place of chemical fertiliser. The use of *Leucaena* is part of a watershed management and forest land use project in Thailand. The great forests of Thailand are rapidly being destroyed by people desperately in need of land for cultivation. This problem of deforestation is extremely serious for the entire country, as Thailand's main water supply is created in the forests of the north. With the forests rapidly shrinking, Thailand is having to face severe floods and droughts in the lowlands. Rapid deforestation is one of the world's major problems today

---

Nitrosamine

The level of nitrates in drinking water has been rising over recent years mainly because of the increased use of artificial nitrate fertilisers in agriculture. Although nitrate itself is reasonably harmless to adults, the body can convert it to nitrite. Both nitrates and nitrites (E250, E251 and E252 for example—*see* beginning of chapter 13) are also used to cure and preserve many foods, including ham, bacon, corned beef and some cheeses and fish. Some scientists now believe that the body may convert the nitrites to nitrosamines.

$$\text{Nitrosamine} \quad \begin{array}{c} R \\ \diagdown \\ N - N = O \\ \diagup \\ R' \end{array}$$

Nitrosamines are known to cause cancer in animals. Most of us are already exposed to low levels of nitrosamines from polluted air, cigarette smoke and some pesticides. Some experts suspect that nitrosamines may account for some 70 to 90% of the cancers believed to be caused by environmental factors.

Nitrates are also supplied to the soil in the form of fertilisers. Calcium nitrate, $Ca(NO_3)_2$, ammonium nitrate, NH_4NO_3, sodium nitrate, $NaNO_3$, and potassium nitrate, KNO_3, are all used as fertilisers (*see* chapter 13).

Plants obtain nitrates through their roots.

When animals and plants die, their proteins decompose, forming ammonium compounds. These eventually are converted by bacteria known as saprophytes, producing nitrates, which remain in the soil, and nitrogen, which is returned to the atmosphere.

All these processes can be summarised into a **nitrogen cycle** (*see* figure 15.15).

Over 50 million tonnes of nitrogen are produced in the world each year. Pure nitrogen along with oxygen and other gases such as argon are produced industrially by the **fractional distillation** of air. There are three stages in the process. First dust particles, water vapour and carbon dioxide are removed from the air. The air is then liquefied by subjecting it to low temperatures and high pressures. Finally the nitrogen, oxygen and argon are separated by fractional distillation.

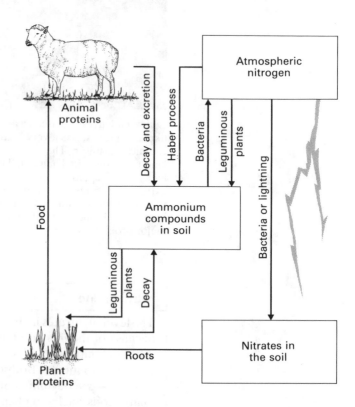

Figure 15.15   The nitrogen cycle

About three-quarters of the nitrogen produced in the UK each year is converted to ammonia (*see* section 7.2) and about one-third of this is converted to nitric acid (*see* below).

Nitric acid has a number of important **uses**:

- about 80% of that produced is used to make the fertiliser ammonium nitrate;
- it is used to manufacture synthetic fibres such as nylon;
- it is used to make explosives such as TNT and triglyceryl nitrate;
- it is used to nitrate aromatic amines for use in the dyestuffs industry.

Nitrates are used both in fertilisers and in explosives in various forms. For example, gunpowder is a mixture of sulphur, charcoal and sodium nitrate. Strontium nitrate and barium nitrate are used in flares and fireworks to produce red and pale-green colours, respectively.

TNT and Dynamite

TNT stands for trinitrotoluene. Dynamite contains glyceryl trinitrate absorbed on Kieselguhr. Nitric acid is used to make these and other explosive materials.

Trinitrotoluene (TNT)

$$CH_3$$

$$NO_2 \quad NO_2$$

$$NO_2$$

Glyceryl trinitrate (dynamite)

$$CH_2ONO_2$$
$$|$$
$$CHONO_2$$
$$|$$
$$CH_2ONO_2$$

Phosphate fertiliser plant in Bombay, India

Silver nitrate is also used to make the silver halides used in photography.

Nitrogen in its elemental form is used as an inert atmosphere in the manufacture of float glass, semiconductors, vitamin A, nylon and the lead–sodium alloy used to make tetraethyllead(IV). Liquid nitrogen is used as a refrigerant for preserving blood, bulls' semen and some foods.

Like nitrogen, phosphorus is also essential to life and occurs in all living organisms. It is found in animal bones and is needed by animals for respiration. This converts carbohydrates to energy.

Phosphorus also occurs in minerals such as apatite (rock phosphate). This contains calcium phosphate, $Ca_3(PO_4)_2$. About 125 million tonnes of rock phosphate are mined each year. Most of this is used to make phosphate fertilisers (*see* chapter 13).

White phosphorus is obtained from phosphate rock by heating the rock with coke and silica in an electric furnace to about 1500°C. This produces phosphorus(V) oxide, P_4O_{10}, which is then reduced to white phosphorus by heating with coke. Red phosphorus is obtained by heating white phosphorus in the absence of air at about 270°C for several days.

Red phosphorus is used to make matches. The side of the box is coated with the element. The head of a match is made of potassium chlorate(V), manganese(IV) oxide and sulphur. On striking the match, the phosphorus is oxidised. Most of the

Chemical polishing of car components. Many metal-finishing products are based on phosphate chemicals

white phosphorus produced nowadays is used to make phosphoric acid. Phosphoric acid is used to rustproof steel and for electrolytic and chemical polishing of aluminium and copper alloys. Dilute phosphoric acid is also used in the food and drink industry to adjust the acidity of jellies and soft drinks.

Pure calcium phosphate is also used in the food industry—in baking powder, for example. One of the most important phosphate compounds is sodium tripoly-phosphate, $Na_5P_3O_{10}$. This is used to make detergent powders and other types of water softener. Polyphosphates are also used to increase the water content of meats.

NITROGEN COMPOUNDS

Ammonia

Ammonia is a covalent compound consisting of molecules with a pyramidal shape (*see* section 2.2). It has a pungent smell and under ordinary conditions it is a colourless gas and is lighter than air.

Ammonia can be prepared in the laboratory by heating a mixture of ammonium chloride and calcium hydroxide:

$$2NH_4Cl(s) + Ca(OH)_2(s) \rightarrow CaCl_2(s) + 2H_2O(g) + 2NH_3(g)$$

Ammonia prepared in this way is first dried by passing it through calcium oxide and then collected by upward delivery in air.

An ammonia molecule can form a coordinate bond by donating the lone-pair of electrons on the nitrogen atom. Ammonia thus acts as a **Lewis base**. For example:

This reaction forms the basis for one of the tests for ammonia. When ammonia and hydrogen chloride gases are brought together, white clouds of ammonium chloride are formed:

$$NH_3(g) + HCl(g) \rightarrow NH_4Cl(s)$$

Ammonia is the most soluble of all gases in water. In aqueous solutions the following equilibrium exists:

$$NH_3(g) + H_2O(l) \rightleftharpoons NH_4^+(aq) + OH^-(aq)$$

This solution is sometimes called ammonium hydroxide solution. However, solid ammonium hydroxide cannot be obtained. The solution contains all four species shown in the equilibrium. The ammonia molecules are hydrogen-bonded to the water molecules. The ammonia also acts as a Lewis base by removing a proton from water molecules to form the ammonium ion, NH_4^+. The solution is a weak alkali. It has a pK_b of 4.76.

Ammonia solution precipitates insoluble metal hydroxides from solutions of their salts. For example

$$Cu^{2+}(aq) + 2OH^-(aq) \rightarrow \underset{\substack{\text{copper(II) hydroxide} \\ \text{(pale blue precipitate)}}}{Cu(OH)_2(s)}$$

Some of these metal hydroxides dissolve in excess ammonia solution to form complex anions. $[Cu(NH_3)_4(H_2O)_2]^{2+}$ and $[Ag(NH_3)_2]^+$ are two examples.

Ammonia is a reducing agent and will reduce chlorine and heated metal oxides:

$$2NH_3(g) + 3Cl_2(g) \rightarrow N_2(g) + 6HCl(g)$$

$$3CuO(s) + 2NH_3(g) \rightarrow 3Cu(s) + N_2(g) + 3H_2O(g)$$

Ammonia does not burn in air but will burn in oxygen with a pale yellow-green flame:

$$4NH_3(g) + 3O_2(g) \rightarrow 2N_2(g) + 6H_2O(g)$$

When a hot platinum catalyst is used, the following reaction occurs:

$$4NH_3(g) + 5O_2(g) \rightleftharpoons 4NO(g) + 6H_2O(g)$$

The reaction is involved in the manufacture of nitric acid by the Ostwald process (*see* below).

Ammonia can be liquefied easily by cooling and compression. Liquid ammonia has many properties which are similar to water. The molecules in liquid ammonia are hydrogen-bonded and thus its boiling point is higher than expected (*see* chapter 2). Both ammonia and water are poor conductors of electricity but excellent ionising solvents.

Sodium, potassium, barium and calcium all dissolve in liquid ammonia, forming blue solutions. The metals can be recovered by evaporating the ammonia. However, if the solutions are allowed to stand the solutions gradually become colourless due to the formation of metal amides, $NaNH_2$ for example. The metal amides are ionic: $Na^+ \, NH_2^-$.

Ammonium Salts

Ammonia and its aqueous solution readily react with acids to form ammonium salts. These are all ionic and contain the ammonium ion, NH_4^+. Ammonium salts are generally soluble in water, undergoing hydrolysis to form slightly acidic solutions:

$$NH_4^+(aq) + H_2O(l) \rightleftharpoons NH_3(aq) + H_3O^+(aq) \qquad pK_b = 4.76$$

All the ammonium salts are thermally unstable. The ammonium halides sublime on heating:

$$NH_4Cl(s) \rightleftharpoons NH_3(g) + HCl(g)$$

Ammonium salts of the oxoacids decompose on heating to form nitrogen or dinitrogen oxide:

$$NH_4NO_2(s) \rightarrow N_2(g) + 2H_2O(g)$$
$$NH_4NO_3(s) \rightarrow N_2O(g) + 2H_2O(g)$$
$$(NH_4)_2Cr_2O_7(s) \rightarrow N_2(g) + 4H_2O(g) + Cr_2O_3(s)$$

The last of these reactions is known as the 'volcano reaction'. All three reactions can be explosive.

The Oxides of Nitrogen

Nitrogen forms six oxides (*see* table 15.12) in which nitrogen exhibits oxidation numbers from +1 to +5. N_2O_4 is a dimer of NO_2 (*see* below). The other oxides are all reasonably stable except N_2O_3 which readily decomposes to NO and NO_2.

Table 15.12   The oxides of nitrogen

Name	Formula	Oxidation state of nitrogen	Appearance at 20°C
dinitrogen oxide	N_2O	+1	colourless gas
nitrogen monoxide	NO	+2	colourless gas
dinitrogen trioxide	N_2O_3	+3	blue liquid
nitrogen dioxide	NO_2	+4	brown gas
dinitrogen tetraoxide	N_2O_4	+4	yellow liquid
dinitrogen pentoxide	N_2O_5	+5	colourless gas

All the nitrogen oxides are **endothermic compounds** (*see* chapter 5).

Dinitrogen Oxide, N_2O

This was also known as nitrous oxide or laughing gas. The latter name arises from its use as an anaesthetic in dentistry since it can cause intoxication. N_2O can be prepared in the laboratory and is manufactured by the carefully controlled thermal decomposition of ammonium nitrate:

$$NH_4NO_3(s) \rightarrow N_2O(g) + 2H_2O(g)$$

Since this reaction can be explosive, the ammonium nitrate is best prepared *in situ*. A mixture of sodium nitrate and ammonium sulphate is heated. This produces the ammonium nitrate which decomposes as fast as it is produced.

The N_2O molecule exists as a resonance hybrid of two unsymmetrical forms with linear structures:

$$\overset{-}{:}\overset{..}{N}=\overset{+}{N}=\overset{..}{O}: \qquad \longleftrightarrow \qquad :N\equiv\overset{+}{N}-\overset{..}{\underset{..}{O}}:^{-}$$

N_2O has a sweetish smell and is very slightly soluble in water, producing a neutral solution. It is an oxidising agent and can support combustion—of carbon, sulphur and phosphorus, for example:

$$P_4(s) + 10N_2O(g) \rightarrow P_4O_{10}(s) + 10N_2(g)$$

Nitrogen Monoxide, NO

This is also known as nitric oxide or nitrogen oxide. It is prepared in the laboratory and industrially by the action of 50% nitric acid on copper:

$$3Cu(s) + 8HNO_3(aq) \rightarrow 3Cu(NO_3)_2(aq) + 4H_2O(l) + 2NO(g)$$

NO is also formed during lightning flashes in the atmosphere and by the passage of an electric discharge through a gaseous mixture of nitrogen and oxygen:

$$N_2(g) + O_2(g) \rightarrow 2NO(g)$$

It is also produced as an intermediate by the catalytic oxidation of ammonia in the Ostwald manufacture of nitric acid.

The NO molecule is a resonance hybrid of the following two forms:

Note that both these forms contain an unpaired electron. Nitrogen monoxide is thus **paramagnetic** (*see* previous chapter).

Nitrogen monoxide is a colourless gas which is virtually insoluble in water. In the solid and liquid states NO tends to dimerise, forming N_2O_2. It is a reducing agent and turns brown in air due to the formation of nitrogen dioxide:

$$2NO(g) + O_2(g) \rightarrow 2NO_2(g)$$

With iron(II) sulphate, NO forms the brown complex, $[Fe(H_2O)_5NO]^{2+}$. It is this complex which is formed in the brown ring test for nitrates (*see* below).

Nitrogen Dioxide, NO₂

Nitrogen dioxide is prepared in the laboratory by heating lead(II) nitrate (*see* chapter 6):

$$2Pb(NO_3)_2(s) \rightarrow 2PbO(s) + 4NO_2(g) + O_2(g)$$

NO_2 normally exists in equilibrium with its dimer, N_2O_4 (*see* also section 7.1):

$$N_2O_4(g) \rightleftharpoons 2NO_2(g)$$

N_2O_4 is pale yellow whereas NO_2 is brown. On cooling the gas, N_2O_4 condenses as a green liquid.

The NO_2 molecule is a resonance hybrid of the following two angular forms:

When two NO_2 molecules combine, the two odd electrons pair up to form a weak N–N bond. The resulting N_2O_4 is a resonance hybrid of the following two planar forms:

Nitrogen dioxide is a very poisonous gas. It dissolves in water forming nitrous and nitric acids:

$$2NO_2(g) + H_2O(l) \rightarrow \underset{\substack{\text{nitrous} \\ \text{acid}}}{HNO_2(aq)} + \underset{\substack{\text{nitric} \\ \text{acid}}}{HNO_3(aq)}$$

The gas itself turns blue litmus red although, unlike bromine (which also exists as a brown gas), it does not bleach litmus paper.

NO_2 decomposes on heating to form nitrogen monoxide:

$$2NO_2(g) \rightarrow 2NO(g) + O_2(g)$$

Pollution by the Nitrogen Oxides

The oxides of nitrogen are known as **primary pollutants** of the atmosphere. The oxides are emitted into the air during the combustion of fossil fuels. Power stations, oil refineries, factories and car exhausts have all contributed to this type of pollution. The oxides of nitrogen (collectively known as N_xO_y) can pollute in two ways.

First of all they can dissolve in water to form nitrous and nitric acids. These are known as **secondary pollutants** and together with sulphurous and sulphuric acid cause acid rain (*see* section 12.2).

An 'ecological protest' during the opening of a New York Automobile Show

Nitrogen oxides are emitted into the air during the combustion of fossil fuels

Secondly, the nitrogen oxides can combine with hydrocarbons to produce **photochemical smog**. The hydrocarbons are also emitted during fossil fuel combustion and thus are primary pollutants. The photochemical smog is produced in a complicated series of reactions involving **radicals** (*see* chapter 17). The first stage requires the presence of ultra-violet light from the Sun. This produces the following type of **photochemical reaction**:

$$NO_2(g) \rightarrow NO(g) + O^{\bullet}(g)$$
$$\text{radical}$$

The oxygen radical then reacts with oxygen molecules to form ozone molecules:

$$O^{\bullet}(g) + O_2(g) \rightarrow O_3(g)$$

Ozone is toxic to both animal and plant life. It is a secondary pollutant. If hydrocarbons are not present in the atmosphere, it combines with the nitrogen monoxide to regenerate nitrogen dioxide:

$$NO(g) + O_3(g) \rightarrow NO_2(g) + O_2(g)$$

The nitrogen dioxide is thus maintained in what is called a 'closed cycle'.

When hydrocarbons are present in the atmosphere, this cycle is disrupted. The

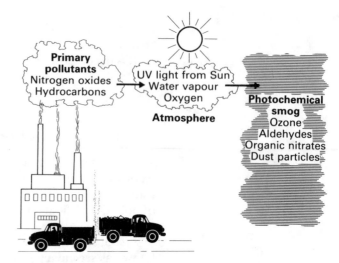

Figure 15.16 Photochemical smog

ozone, which is part of the cycle, reacts with unsaturated hydrocarbons to produce organic radicals such as:

$$CH_3O^{\bullet} \qquad HC{=}O^{\bullet}$$

These combine with the nitrogen oxides to produce aldehydes and organic nitrates of the following type:

These, together with the ozone, are the secondary pollutants which form the photochemical smog (*see* figure 15.16).

In many cities of the world, the problem is compounded by the formation of an inversion layer in the atmosphere (*see* figure 15.17). This is a layer of warm air which lies above the city and traps a layer of cooler air. This warm layer is usually

(a)

(b)

Figure 15.17 (a) No inversion; (b) inversion layer

dry and allows the maximum amount of sunlight to be transmitted. The secondary pollutants trapped in the lower layer thus accumulate. This photochemical smog can often be seen as a haze above cities during warm weather. The haze is due to particles in the smog.

Nitrous Acid and Nitrites

A pale blue aqueous solution of nitrous acid can be prepared in the laboratory by adding dilute hydrochloric acid to a cold dilute solution of sodium nitrite:

$$H^+(aq) + NO_2^-(aq) \rightarrow HNO_2(aq)$$

Sodium nitrite can be prepared by heating sodium nitrate strongly either by itself or with a reducing agent such as lead:

$$NaNO_3(s) + Pb(s) \rightarrow NaNO_2(s) + PbO(s)$$

The sodium nitrite is extracted with water.

Nitrous acid is a weak and unstable acid. It has a pK_a of 3.34. At room temperature it disproportionates, forming nitric acid and nitrogen monoxide:

$$3HNO_2(aq) \rightarrow HNO_3(aq) + 2NO(g) + H_2O(l)$$

oxidation state
of nitrogen $+3$ $+5$ $+2$

Nitrous acid and acidic solutions of nitrites are oxidising agents, although they do behave as reducing agents in the presence of strong oxidising agents such as acidified potassium manganate(VII). Acidified solutions of sodium nitrite are particularly important in organic chemistry where they are used to prepare diazonium salts (*see* chapter 19).

Nitric Acid and Nitrates

Pure nitric acid is a colourless fuming liquid. It is prepared in the laboratory by heating either sodium nitrate or potassium nitrate with concentrated sulphuric acid:

$$NaNO_3(s) + H_2SO_4(l) \rightarrow NaHSO_4(s) + HNO_3(g)$$

The product is normally yellow due to the presence of dissolved nitrogen dioxide formed by the thermal decomposition of the acid:

$$4HNO_3(l) \rightarrow 4NO_2(g) + 2H_2O(g) + O_2(g)$$

An aqueous solution of nitric acid is a typical strong acid. For example, it reacts with bases to form nitrates and with carbonates to give carbon dioxide.

In 1984, the longest load ever transported on British roads was moved some 5 miles to Billingham, Cleveland. The load was a 67.4 m long stainless-steel absorption column for a nitric acid plant at Billingham. Considerable modifications were required along the route. These included lifting of overhead cables and the removal of one complete roundabout. Investment in the new 330 000 tonne per year nitric acid plant was in the region of £30 million

Both dilute and concentrated nitric acid are oxidising agents. Concentrated nitric acid oxidises non-metals such as carbon and sulphur:

$$C(s) + 4HNO_3(l) \rightarrow 2H_2O(l) + 4NO_2(g) + CO_2(g)$$
$$S(s) + 6HNO_3(aq) \rightarrow 2H_2O(l) + 6NO_2(g) + H_2SO_4(aq)$$

The reactions of nitric acid with metals are varied. Calcium and magnesium react with very dilute nitric acid liberating hydrogen. Zinc reduces dilute nitric acid forming dinitrogen oxide. Generally, however, metals tend to react with dilute nitric acid to give nitrogen monoxide and with the concentrated acid to give nitrogen dioxide. Copper is an example of such a metal:

with 50% HNO₃ $3Cu(s) + 8HNO_3(aq) \rightarrow 3Cu(NO_3)_2(aq) + 4H_2O(l) + 2NO(g)$

with conc. HNO₃ $Cu(s) + 4HNO_3(aq) \rightarrow Cu(NO_3)_2(aq) + 2H_2O(l) + 2NO_2(g)$

It should be noted that these are the stoichiometric equations. These processes take place in a complicated sequence of steps.

Concentrated nitric acid reacts with iron, chromium and aluminium forming a thin protective film of oxide on the metal which renders the metals **passive**. Precious metals such as gold and platinum do not react with nitric acid.

Nitric acid also oxidises certain cations and anions. For example, iron(II) ions are oxidised to iron(III) ions:

$$3Fe^{2+}(aq) + 4H^+(aq) + NO_3^-(aq) \rightarrow 3Fe^{3+}(aq) + 2H_2O(l) + NO(g)$$

Iodide ions are oxidised to iodine:

$$6I^-(aq) + 8H^+(aq) + 2NO_3^-(aq) \rightarrow 3I_2(s) + 4H_2O(l) + 2NO(g)$$

Hydrogen sulphide and other inorganic covalent compounds are also oxidised by nitric acid:

$$3H_2S(g) + 2HNO_3(aq) \rightarrow 4H_2O(l) + 2NO(g) + 3S(s)$$

In organic chemistry, nitric acid is employed as a nitrating agent. A mixture of concentrated nitric and sulphuric acids is employed.

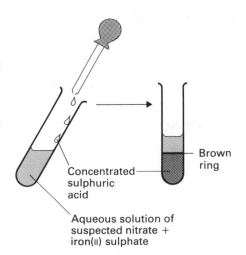

Concentrated sulphuric acid

Aqueous solution of suspected nitrate + iron(II) sulphate

Brown ring

Figure 15.18 The brown ring test for nitrates

Nitrates

Metallic nitrates can be made by the reaction of nitric acid with metals, oxides, hydroxides or carbonates. The nitrates are detected by the brown ring test (*see* figure 15.18).

The Brown Ring Test for Nitrates

The suspected nitrate is dissolved in water and added to iron(II) sulphate solution. Concentrated sulphuric acid is slowly run down the side of the test tube so that two layers form (*see* figure 15.18). If nitrate is present, it reacts with the sulphuric acid, forming nitric acid. The nitric acid reacts with the iron(II) sulphate forming the complex $[Fe(H_2O)_5NO]^{2+}$. This forms a brown ring between the two layers in the test tube and thus indicates the presence of nitrate.

Inorganic nitrates are all soluble in water and they are all thermally unstable. The metallic nitrates decompose to give the nitrite, oxide or metal itself depending on the position of the metal in the electrochemical series:

$$2KNO_3(s) \rightarrow 2KNO_2(s) + O_2(g)$$
$$2Cu(NO_3)_2(s) \rightarrow 2CuO(s) + 4NO_2(g) + O_2(g)$$
$$2AgNO_3(s) \rightarrow 2Ag(s) + 2NO_2(g) + O_2(g)$$

Ammonium nitrate forms dinitrogen oxide and water when heated:

$$NH_4NO_3(s) \rightarrow N_2O(g) + 2H_2O(l)$$

Manufacture of Nitric Acid and the Nitrates

Nitric acid is manufactured by the **Ostwald process**. There are three stages in this process.

Stage 1: The Catalytic Oxidation of Ammonia

Ammonia, obtained by the Haber process (*see* section 7.2) is mixed with air and rapidly passed over a platinum–rhodium catalyst at 900°C:

$$4NH_3(g) + 5O_2(g) \rightarrow 4NO(g) + 6H_2O(g) \qquad \Delta H_{r,m}^{\ominus} = -905.6 \text{ kJ mol}^{-1}$$

The reaction is sufficiently exothermic to maintain the temperature at 900°C.

Stage 2: Production of Nitrogen Dioxide

The gases from stage 1 are cooled and mixed with air. The nitrogen monoxide is oxidised to nitrogen dioxide:

$$2NO(g) + O_2(g) \rightarrow 2NO_2(g)$$

Stage 3: Production of Dilute Nitric Acid

The nitrogen dioxide from stage 2 is passed through water sprays in a steel absorption tower:

$$3NO_2(g) + H_2O(l) \rightarrow 2HNO_3(aq) + NO(g)$$

The nitrogen monoxide is recycled back to stage 2. The nitric acid produced has a concentration of about 50%. Distillation yields a constant-boiling mixture of 68%.

About 80% of the nitric acid manufactured in this way is neutralised with excess aqueous ammonia to produce ammonium nitrate:

$$NH_3(aq) + HNO_3(aq) \rightarrow NH_4NO_3(aq)$$

The ammonium nitrate is used as a fertiliser (*see* above).

PHOSPHORUS COMPOUNDS

Phosphine, PH_3

Phosphine is a colourless gas with a fishy smell. It is very poisonous. It can be prepared in the laboratory by boiling white phosphorus with a concentrated aqueous solution of sodium hydroxide:

$$P_4(s) + 3NaOH(aq) + 3H_2O(l) \rightarrow 3NaH_2PO_2(aq) + PH_3(g)$$
sodium
phosphinate

Phosphine prepared in this way is spontaneously inflammable due to the presence of unstable diphosphine, P_2H_4.

The phosphine molecule has a pyramidal shape like the ammonia molecule although the angles are different.

Phosphine is less basic than ammonia and unlike ammonia only slightly soluble in water. This is because phosphorus is less electronegative than nitrogen and thus does not form hydrogen bonds in water. The absence of hydrogen bonding in phosphine causes phosphine to have a lower boiling point than ammonia even though it has a higher relative molecular mass.

Phosphorus Chlorides

Phosphorus forms two chlorides: phosphorus trichloride, PCl_3, and phosphorus pentachloride, PCl_5.

Phosphorus trichloride is formed by passing chlorine over white phosphorus. The phosphorus burns with a pale green flame. The trichloride condenses as a colourless liquid.

The trichloride is hydrolysed by water, forming phosphonic acid and hydrogen chloride:

$$PCl_3(l) + 3H_2O(l) \rightarrow H_3PO_3(l) + 3HCl(g)$$

Phosphorus trichloride is used in organic chemistry as a chlorinating agent.

Phosphorus pentachloride can be prepared in the laboratory by the reaction of chlorine and phosphorus trichloride at a temperature of about 0°C. The reaction is reversible:

$$PCl_3(l) + Cl_2(g) \rightleftharpoons PCl_5(s)$$

The pentachloride is a pale yellow crystalline solid made up of tetrahedral $[PCl_4]^+$ and octahedral $[PCl_6]^-$ ions. In the vapour state PCl_5 exists as covalent molecules with a bipyramidal shape.

On heating, the pentachloride dissociates, forming the trichloride and chlorine.

Phosphorus pentachloride reacts violently with water, forming phosphoric(v) acid. The overall reaction is:

$$PCl_5(s) + 4H_2O(l) \rightarrow H_3PO_4(aq) + 5HCl(aq)$$

The pentachloride is also used as a chlorinating agent.

Phosphorus Oxides

Phosphorus(III) oxide is a white solid obtained by burning white phosphorus in a limited amount of air. When white phosphorus is burnt in a plentiful supply of air, **phosphorus(v) oxide** is formed as a white solid:

$$P_4(s) + 3O_2(g) \rightarrow P_4O_6(s)$$
$$P_4(s) + 5O_2(g) \rightarrow P_4O_{10}(s)$$

Owing to these reactions, white phosphorus is stored under water. Phosphorus(III) oxide reacts with oxygen when heated, forming phosphorus(v) oxide:

$$P_4O_6(s) + 2O_2(g) \rightarrow P_4O_{10}(s)$$

Both phosphorus(III) and phosphorus(v) oxides are acidic, reacting readily with water to form phosphonic and phosphoric(v) acids respectively:

$$P_4O_6(s) + 6H_2O(l) \rightarrow \quad \underset{\text{phosphonic acid}}{4H_3PO_3(aq)}$$

$$P_4O_{10}(s) + 6H_2O(l) \rightarrow \quad \underset{\text{phosphoric(v) acid}}{4H_3PO_4(aq)}$$

Both oxides must therefore be stored in airtight containers. Because of its high affinity for water, phosphorus(v) oxide is sometimes used as a drying agent.

Phosphorus Acids and their Salts

Phosphorus forms a number of oxoacids. Some of these are monomeric—phosphinic, phosphonic and phosphoric(v) acids for example. These can be monobasic (or monoprotic) or polybasic (or polyprotic). Phosphorus also forms polymeric oxoacids. These may have chain or ring structures. Diphosphoric(v) acid is a dimeric oxoacid of phosphorus.

The most important of all these acids is phosphoric(v) acid—also known as orthophosphoric(v) acid. This is a white deliquescent crystalline solid. An 85%

aqueous solution is known as 'syrupy phosphoric acid'. Phosphoric(v) acid is a weak triprotic acid:

$$H_3PO_4(aq) + H_2O(l) \rightleftharpoons H_3O^+(aq) + H_2PO_4^-(aq) \qquad pK_1 = 2.12$$

$$H_2PO_4^-(aq) + H_2O(l) \rightleftharpoons H_3O^+(aq) + HPO_4^{2-}(aq) \qquad pK_2 = 7.21$$

$$HPO_4^{2-}(aq) + H_2O(l) \rightleftharpoons H_3O^+(aq) + PO_4^{3-}(aq) \qquad pK_3 = 12.67$$

The trisodium and disodium salts of phosphoric(v) acid tend to be insoluble—although those of the alkali metals and ammonium are exceptions. Salts containing the dihydrogenphosphate(v) ion are more soluble. For example, calcium phosphate, $Ca_3(PO_4)_2$, which is found in rock phosphate, is insoluble, whereas calcium dihydrogenphosphate(v) is soluble. The latter is used as a phosphate fertiliser in superphosphate (*see* chapter 13).

Phosphorus Acids and their Salts

Phosphinic Acid, HPH$_2$O$_2$
(hypophosphorous acid)

Salt: sodium phosphinate, $Na^+PH_2O_2^-$

Phosphonic Acid, H$_2$PHO$_3$
(phosphorous acid or orthophosphorous acid)

Salts: disodium phosphonate, $(Na^+)_2PHO_3^{2-}$
sodium hydrogenphosphonate, $Na^+HPHO_3^-$

Phosphoric(v) Acid, H$_3$PO$_4$
(orthophosphoric acid)

Salts: trisodium phosphate(v), $(Na^+)_3PO_4^{3-}$
disodium hydrogenphosphate, $(Na^+)_2HPO_4^{2-}$
sodium dihydrogenphosphate, $Na^+H_2PO_4^-$

Diphosphoric(v) Acid, H$_4$P$_2$O$_7$
(pyrophosphoric acid)

Salts: tetrasodium diphosphate(v), $(Na^+)_4P_2O_7^{4-}$
disodium dihydrogendiphosphate(v), $(Na^+)_2H_2P_2O_7^{2-}$

This plant at Whitehaven, Cumbria, has a capacity to produce 160 000 tonnes per year of 'merchant-grade' phosphoric acid. The major part of the output goes to make sodium tripolyphosphate, a vital ingredient of washing powder detergents

This huge rotary filter forms part of the Whitehaven phosphoric acid plant

Phosphoric(v) acid is manufactured by heating rock phosphate with concentrated sulphuric acid or by dissolving phosphorus(v) oxide in water (*see* above).

SUMMARY

1. There is a transition from non-metallic to metallic properties on descending Group V.
2. Phosphorus exists in three **allotropic forms**.
3. Both nitrogen and phosphorus can exist in compounds in oxidation states ranging from -3 to $+5$.
4. The process of converting atmospheric nitrogen to a form usable by plants and animals is called **nitrogen fixation**. This process is part of the **nitrogen cycle**.
5. Nitrogen is produced industrially by the liquefaction and **fractional distillation** of air.
6. Phosphorus is needed by animals for their bones and for **respiration**.
7. Phosphorus occurs in minerals such as **apatite**.
8. Apatite is used to make phosphate fertilisers.
9. Ammonia is a covalent compound. Its molecules have a pyramidal shape.
10. Ammonia acts as a **Lewis base**.
11. Ammonia solution precipitates insoluble metal hydroxides from solutions of their salts.
12. Ammonia is a **reducing agent**.
13. Ammonium salts are thermally unstable.
14. All six nitrogen oxides are **endothermic compounds**.
15. The three most important nitrogen oxides are
    (a) dinitrogen oxide, N_2O,
    (b) nitrogen monoxide, NO,
    (c) nitrogen dioxide, NO_2.
    These are all simple molecular compounds existing as **resonance hybrids**. For example

$$:\!N\!=\!\ddot{O}\!: \quad\longleftrightarrow\quad {}^-\!:\!\ddot{N}\!=\!\ddot{O}\!:^+$$

16. Nitrogen oxides are **primary pollutants**. They undergo reactions in the atmosphere to form **secondary pollutants**. These give rise to **photochemical smog**.
17. Nitric acid is a typical **strong acid**.
18. Nitric acid is used as a **nitrating agent** in organic chemistry.
19. Inorganic nitrates are thermally unstable.
20. Nitric acid is manufactured by the **Ostwald process**. There are three stages:
    Stage 1: catalytic oxidation of ammonia.
    Stage 2: production of nitrogen dioxide.
    Stage 3: conversion of nitrogen dioxide to nitric acid.
21. Both phosphorus chlorides are hydrolysed by water.
22. Both phosphorus oxides are acidic.
23. Phosphorus forms a number of **oxoacids**. The most important of these is phosphoric(v) acid, H_3PO_4.

15.4
Group VI: Oxygen and Sulphur

The Chalcogens

Elements in Group VI are sometimes known as the chalcogens. The word 'chalcogen' derives from two Greek words meaning 'copper' and 'born'. This is because most copper ores consist of compounds containing oxygen or sulphur and many also contain small amounts of selenium and tellurium. Chalcocite, for example, is a major copper ore. It consists of copper(I) sulphide, Cu_2S. The most widespread copper mineral and an important copper ore is chalcopyrite, $CuFeS_2$. Elements which have a strong affinity for sulphur are known by geologists as chalcophiles. Lead is an example of a chalcophile.

The five elements of Group VI are shown in figure 15.19. They are all p-block elements characterised by the s^2p^4 electronic configuration in their outer shells. Some of the properties of the Group VI elements are shown in table 15.13. Like other groups of p-block elements there is a transition from non-metallic to metallic properties down the group. Oxygen and sulphur are non-metallic elements and selenium and tellurium are semiconductors. Polonium is metallic in character but highly radioactive.

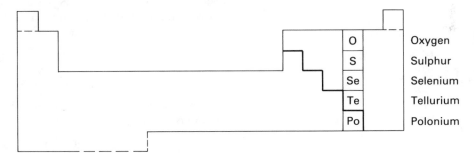

Figure 15.19   The Group VI elements

Table 15.13   Properties of the Group VI elements

Element	Atomic number	Configuration in outer shell	Atomic radius/nm	Melting point/°C	Boiling point/°C
oxygen	8	$2s^2 2p^4$	0.074	−219	−183
sulphur	16	$3s^2 3p^4$	0.104	119	445
selenium	34	$3d^{10} 4s^2 4p^4$	0.117	217	685
tellurium	52	$4d^{10} 5s^2 5p^4$	0.137	450	1390
polonium	84	$4f^{14} 5d^{10} 6s^2 6p^4$	0.152	–	–

The Group VI elements have a tendency to form compounds—particularly with hydrogen and the reactive metals—in which they exist in the −2 oxidation state.

OXYGEN

Oxygen is a colourless, odourless gas consisting of diatomic molecules. It is slightly more dense than air and slightly soluble in water.

It can be produced in the laboratory by a number of methods.

- By the catalytic decomposition of hydrogen peroxide:

$$2H_2O_2(l) \rightarrow 2H_2O(l) + O_2(g)$$

Manganese(IV) oxide is often used as a catalyst for this reaction.

- By the thermal decomposition of oxygen-rich compounds such as potassium manganate(VII), potassium nitrate, potassium chlorate(VII) and dilead(II) lead(IV) oxide:

$$2KClO_3(s) \rightarrow 2KCl(s) + 3O_2(g)$$

$$2Pb_3O_4(s) \rightarrow 6PbO(s) + O_2(g)$$

- By the action of sodium peroxide on water:

$$2Na_2O_2(s) + 2H_2O(l) \rightarrow 4NaOH(aq) + O_2(g)$$

- By the electrolysis of aqueous solutions of acids and alkalis:

$$4OH^-(aq) \rightarrow 2H_2O(l) + O_2(g) + 4e^-$$

The oxygen is discharged at the anode.

Oxygen has an electronegativity value of 3.5 and is thus highly electronegative and a strong oxidising agent. It combines with many elements to form oxides. The reactions are highly exothermic and in many cases this results in ignition of the element or compound.

Trioxygen (Ozone), O_3

Trioxygen, commonly known as ozone, is an allotrope of oxygen. It is a pale blue gas which is slightly soluble in water. In small concentrations it is non-toxic but in concentrations above 100 parts per million it is toxic.

Trioxygen consists of triatomic molecules which are resonance hybrids of the following two non-linear structures:

It occurs in the upper atmosphere as a result of the action of the Sun's ultra-violet radiation on atmospheric oxygen:

$$3O_2(g) \rightarrow 2O_3(g) \qquad \Delta H^{\ominus}_{r,m} = +284 \text{ kJ mol}^{-1}$$

Trioxygen can be produced by subjecting oxygen to a silent electrical discharge. It is endothermic and very unstable. High concentrations can be explosive.

Trioxygen is a stronger oxidising agent than oxygen, as the following electrode potentials show:

$$O_3(g) + 6H_3O^+(aq) + 6e^- \rightleftharpoons 9H_2O(l) \qquad E^{\ominus} = +2.07 \text{ V}$$

$$O_2(g) + 4H_3O^+(aq) + 4e^- \rightleftharpoons 6H_2O(l) \qquad E^{\ominus} = +1.23 \text{ V}$$

It oxidises lead(II) sulphide to lead(II) sulphate and acidified iron(II) and iodide ions to iron(III) ions and iodine, respectively:

$$PbS(s) + 4O_3(g) \rightarrow PbSO_4(s) + 4O_2(g)$$

$$2Fe^{2+}(aq) + 2H^+(aq) + O_3(g) \rightarrow 2Fe^{3+}(aq) + H_2O(l) + O_2(g)$$

$$2I^-(aq) + 2H^+(aq) + O_3(g) \rightarrow I_2(aq) + H_2O(l) + O_2(g)$$

Ozone also reacts with alkenes to split their double bonds in a process known as ozonolysis. Organic compounds known as ozonides are formed.

Occurrence and Uses of Oxygen

Oxygen accounts for about half of the mass of the Earth's crust. Most of this is in the form of silicates. Oxygen also accounts for 89% of the mass of the oceans of the world. The atmosphere contains 23% by mass of oxygen. This is equivalent to 21% by volume. This percentage is kept constant by the process of photosynthesis. In this process, green plants use sunlight to convert carbon dioxide and water to carbohydrates and oxygen.

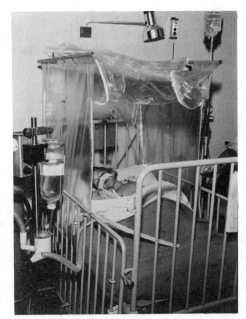

An oxygen tent

These two astronauts carry the oxygen they need on their backs. They are constructing an experimental assembly of structures outside their space vehicle

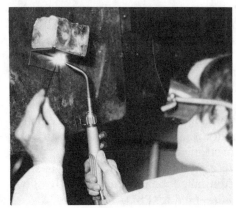

Oxygen is used for welding

Oxygen and its compounds are essential to life. They play important roles in metabolism and respiration. **Respiration** is the process by which animals generate the energy needed to sustain life. Oxygen is also required for combustion of gaseous fuels such as methane, liquid fuels such as oil and solid fuels such as coal.

The element is produced industrially by the liquefaction of air and its subsequent fractional distillation. Over 80 million tonnes of oxygen are produced each year. Most of this is used in the iron and steel industries where it is used to convert impure iron into steel (*see* previous chapter). Oxygen is used for welding and cutting. It is used extensively in the chemical industry to produce a variety of compounds. It is also used in spacecraft for the combustion of hydrogen and other fuels. Finally, oxygen is used during high-altitude flying, in surgery and is given to patients with breathing difficulties.

OXYGEN COMPOUNDS

The oxygen atom has six outer-shell electrons, two of which are unpaired (*see* figure 15.20). It can gain two electrons to fill its p orbitals and form the oxide ion, O^{2-}. In this state oxygen has an oxidation number of -2. Oxygen can also share its two unpaired 2p electrons to form two covalent bonds—as in water for example. Owing to the small size and high electronegativity of its atoms, oxygen can stabilise the higher oxidation states of other elements, chlorine(VII) in $Cl_2O_7^{2-}$ and chromium(VII) in $Cr_2O_7^{2-}$ for example.

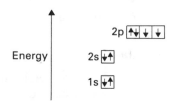

Figure 15.20 The electronic structure of oxygen

Oxygen forms numerous compounds of importance. We dealt with the structure, properties and importance of water in chapter 12. Organic compounds containing oxygen are considered in chapters 17–20. In this chapter we shall confine our attention to the oxides and to hydrogen peroxide.

The Oxides

Oxygen forms a great variety of oxides. These can be classified in various ways.

Metallic/Non-Metallic Oxides

Oxides may be simply classified as metallic or non-metallic. We shall see below that metallic oxides tend to be basic whereas non-metallic oxides tend to be acidic. For this reason, there is a strong tendency for metallic oxides to combine with non-metallic oxides to form salts. For example

$$Na_2O(s) + CO_2(g) \rightarrow Na_2CO_3(s)$$

Ionic/Covalent Oxides

Oxides may be ionic or covalent. Calcium oxide, CaO, is an example of an ionic oxide and carbon dioxide an example of a covalent oxide.

Classification According to Composition

Oxides may also be classified according to their composition. This classification does not distinguish between metallic and non-metallic oxides nor between ionic and covalent oxides. The following are the three most important types in this classification.

Normal oxides are oxides in which there are only bonds between the element and oxygen. They include magnesium oxide, MgO, sulphur trioxide, SO_3, and the macromolecular silicon(IV) oxide, SiO_2.

Peroxides are oxides in which there is bonding not only between the element and oxygen but also between two oxygen atoms. Sodium peroxide, Na_2O_2, and hydrogen peroxide are examples. They are strong oxidising agents.

Mixed oxides behave as though they are a mixture of two oxides. Dilead(II) lead(IV) oxide, Pb_3O_4, or red lead as it is often called, behaves as a mixture of two parts lead(II) oxide, PbO, and one part lead(IV) oxide, PbO_2.

Classification According to Acidic/Basic Properties

Finally oxides may be classified according to their acidic and basic properties.

Basic oxides are oxides of metals in low oxidation states. They contain the oxide ion, O^{2-}. These oxides react with acids to form a salt and water. For example

$$MgO(s) + 2HCl(aq) \rightarrow MgCl_2(aq) + H_2O(l)$$

Basic oxides of metals high in the electrochemical series dissolve in water to form alkalis:

$$O^{2-}(s) + H_2O(l) \rightarrow 2OH^-(aq)$$

For example

$$CaO(s) + H_2O(l) \rightarrow Ca(OH)_2(aq)$$

Acidic oxides are usually simple molecular oxides of non-metals or d-block elements in high oxidation states. They dissolve in water forming acids. For this reason they are called **acid anhydrides**. For example

$$SO_3(g) + H_2O(l) \rightarrow H_2SO_4(aq)$$

Amphoteric oxides are usually oxides of less electropositive metals such as aluminium, lead and zinc. They can behave as either acidic or basic oxides. For example, zinc oxide reacts as a base in the presence of a strong acid:

$$ZnO(s) + 2HCl(aq) \rightarrow ZnCl_2(aq) + H_2O(l)$$

In the presence of an alkali it reacts as an acid:

$$ZnO(s) + 2NaOH(aq) + H_2O(l) \rightarrow Na_2[Zn(OH)_4](aq)$$
sodium
tetrahydroxozincate(II)

Water is an amphoteric oxide (*see* chapter 12).

Neutral oxides react with neither acids nor bases to form salts. Examples are nitrogen monoxide, NO, and dinitrogen oxide, N_2O.

Oxides can be prepared by a variety of methods. Metallic oxides can be prepared by thermal decomposition of a salt. For example

$$CaCO_3(s) \rightarrow CaO(s) + CO_2(g)$$

Non-metallic oxides can be prepared by direct combination of the element with oxygen or by the oxidation of the element with nitric acid:

$$C(s) + O_2(g) \rightarrow CO_2(g)$$
$$C(s) + 4HNO_3(aq) \rightarrow CO_2(g) + 4NO_2(g) + 2H_2O(l)$$

Hydrogen Peroxide

Hydrogen peroxide is a pale blue liquid. It is weakly acidic. Molecules of hydrogen peroxide have the structure shown in figure 15.21. Like water molecules, hydrogen peroxide molecules are hydrogen-bonded. However, the –O–O– bond in the molecule is weak and thus hydrogen peroxide is unstable. Solutions of hydrogen peroxide spontaneously decompose into water and oxygen at room temperature:

$$2H_2O_2(g) \rightarrow 2H_2O(g) + O_2(g)$$

This reaction is accelerated by the presence of a catalyst such as manganese(IV) oxide (*see* above).

A dilute solution of hydrogen peroxide can be prepared in the laboratory by the action of dilute sulphuric acid on barium peroxide at 0°C:

$$BaO_2(s) + H_2SO_4(aq) \rightarrow BaSO_4(s) + H_2O_2(aq)$$

The product is filtered to remove the barium sulphate.

Hydrogen peroxide is generally used as a '20 volume' solution. One volume of this solution produces 20 volumes of oxygen gas. A 20 volume solution contains about 6% hydrogen peroxide.

Hydrogen peroxide is a strong oxidising agent. For example, a dilute solution will oxidise lead(II) sulphide to lead(II) sulphate. In the presence of a strong oxidising agent, however, hydrogen peroxide acts as a reducing agent (*see* section 10.2).

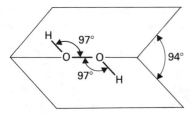

Figure 15.21 Structure of hydrogen peroxide

Air pollution causes the surfaces of paints containing 'white lead' to blacken. White lead is lead(II) carbonate. It reacts with hydrogen sulphide in the atmosphere to give lead(II) sulphide—a black compound. By treating this with hydrogen peroxide, lead(II) sulphate is formed. This is a white compound:

$$PbS(s) + 4H_2O_2(l) \rightarrow PbSO_4(s) + 4H_2O(l)$$
black   white

Hydrogen peroxide is used for a variety of purposes:

- to make bleaching agents for detergents and as a bleaching agent for textiles and paper;
- to prepare peroxides, particularly organic peroxides;

- in polymerisation reactions;
- to prepare antiseptics;
- to restore the original hues of paints containing lead;
- to combat odours in sewers.

SULPHUR

Sulphur is a yellow solid which is insoluble in water. It is non-toxic.

The element has several allotropes of which two of the most important are α-sulphur and β-sulphur (*see* also section 3.2 and figure 6.8). α-Sulphur (or rhombic sulphur) is stable up to 95.5°C. β-Sulphur (or monoclinic sulphur) is stable above 95.5°C. This type of allotropy is known as **enantiotropy** (*see* section 15.2).

Crystals of α-sulphur have a lemon yellow appearance whereas the needle-like crystals of β-sulphur have a deeper yellow colour. Both allotropes are composed of the S_8 puckered molecular rings.

Sulphur is also known in a number of other forms. **Roll sulphur** consists mainly of α-sulphur. **Flowers of sulphur** is a form of **amorphous sulphur** obtained when sulphur vapour is allowed to fall on a cold surface. **Colloidal sulphur** can be prepared by adding concentrated hydrochloric acid to sodium thiosulphate solution:

$$S_2O_3{}^{2-}(aq) + 2H^+(aq) \rightarrow H_2O(l) + SO_2(g) + S(s)$$

When α- or β-sulphur is heated the crystals melt to form a yellow liquid. On further heating, the liquid becomes red and almost black. The **viscosity** (thickness) of the liquid rises until about 200°C. Between 200°C and the boiling point of sulphur at 444°C the viscosity decreases. When boiling sulphur is poured into water, **plastic sulphur** is obtained. This is a soft brown elastic solid consisting of a random arrangement of chains of sulphur atoms. Plastic sulphur is unstable. The chains slowly break up and the sulphur atoms rearrange to re-form the S_8 molecules of rhombic sulphur.

Sulphur reacts with most metals to form sulphides. Some of these are non-stoichiometric 'compounds' (*see* section 4.1). Sulphur also combines with non-metals such as chlorine and oxygen:

$$2S(s) + Cl_2(g) \rightarrow S_2Cl_2(l)$$
$$\text{disulphur}$$
$$\text{dichloride}$$
$$S(s) + O_2(g) \rightarrow SO_2(g)$$

Sulphur dioxide is also formed by the reaction of concentrated sulphuric acid on sulphur:

$$S(s) + 2H_2SO_4(l) \rightarrow 2H_2O(l) + 3SO_2(g)$$

Occurrence, Extraction and Uses of Sulphur

Sulphur occurs naturally in an almost pure elemental form in several countries including Italy, Mexico, Japan, Poland and USA. Sulphur is also found in sulphide ores such as iron pyrites (FeS_2), zinc blende (ZnS) and galena (PbS) and in sulphate ores such as gypsum ($CaSO_4$) and barite ($BaSO_4$). Fossil fuels such as oil also contain sulphur compounds. Natural gas contains hydrogen sulphide. Sulphur also occurs in sea-water as sulphate. It is found in the tissues of all living plants and animals in the form of the amino acids cysteine, cystine and methionine.

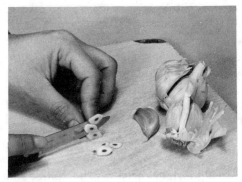

The odour of garlic is due to an
organosulphur compound

Sulphur Stinks

The odours of garlic, mustard, onions and cabbage are all due to
organosulphur compounds. The source of garlic odour, for example, is a
compound known as allicin. It has the following structure:

The substance that makes people cry when slicing onions is also a sulphur
compound. It is known as the **lacrimatory factor**. The compound has two
forms, known as *syn* and *anti*, of which the latter predominates:

syn form   *anti* form

The Frasch Process

Sulphur is extracted from underground deposits by the Frasch process (*see* figure
15.22). Hot compressed air and superheated steam are piped underground. This
forces water and molten sulphur to the surface. The solid is then cooled and
solidified. The sulphur is 99.5% pure and can be used directly. Frasch mining is
carried out in the USA, Iraq and Mexico. However, diminishing reserves together
with increased energy costs have resulted in a sharp cut-back in this type of mining
over recent years.

At one time sulphur used to be
collected by lowering men in baskets
into dormant volcanos where they
used to scrape sulphur from the inner
walls of the volcanos.

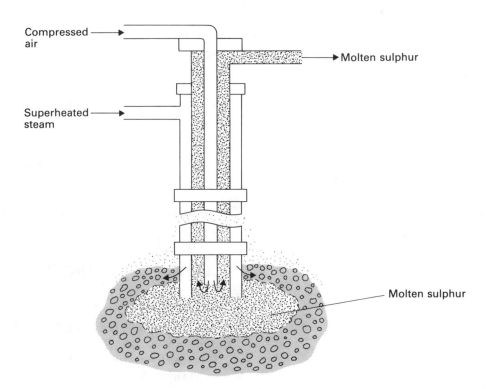

Figure 15.22 The Frasch pump

A sulphur store in a sulphuric acid plant

Elemental sulphur is also produced from the sulphides extracted from oil and natural gas. About 75% of elemental sulphur used in non-communist countries is produced in this way. The main sources of sulphur derived from natural gas are Canada, France, West Germany, Mexico, Saudi Arabia and the USA. In communist countries, sulphur is produced by underground melting and conventional mining as well as from natural gas and oil. Poland, which is a major sulphur exporter, operates the world's largest mine. Other sulphur sources are pyrites and other sulphide ores. These are particularly important in China, the USSR and some European countries.

Sulphur has a number of important uses:

- it is used to manufacture sulphur dioxide, sulphuric acid, hydrogen sulphide and carbon disulphide;
- vulcanisation of rubber;
- production of chemicals used in detergents;
- manufacture of gunpowder;
- in the preparation of skin ointments and drugs;
- manufacture of insecticides and fungicides.

Sulphur atoms have six electrons in their outer shells (*see* figure 15.23). The atoms can accept two electrons into their two singly occupied 3p orbitals to form the sulphide ion, S^{2-}. Sulphur atoms can also form two, four or six covalent bonds, that is, it can also exist in oxidation states +2, +4 and +6 as well as the −2 state in the sulphide ion. Sulphur atoms form two covalent bonds by sharing the two electrons in the singly occupied 3p orbitals. To form four covalent bonds, a sulphur atom promotes one of its paired 3p electrons to a 3d orbital—all of which are empty. Six covalent bonds can be formed by also promoting a 3s electron into another 3d orbital.

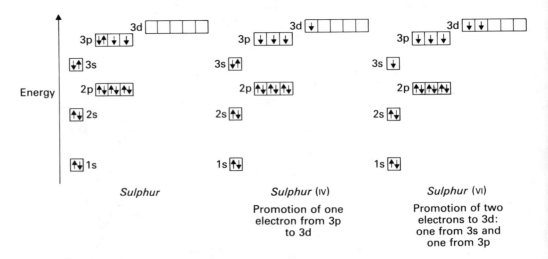

Figure 15.23 Electronic structure and valency of sulphur

SULPHUR COMPOUNDS

Hydrogen Sulphide

Hydrogen sulphide is a colourless and very toxic gas with a characteristic smell of rotten eggs.

It is usually prepared in the laboratory by adding dilute hydrochloric acid to iron(II) sulphide:

$$FeS(s) + 2HCl(aq) \rightarrow FeCl_2(aq) + H_2S(g)$$

It can also be prepared by adding cold water to aluminium sulphide:

$$Al_2S(s) + 6H_2O(l) \rightarrow 2Al(OH)_3(s) + 3H_2S(g)$$

The hydrogen sulphide prepared in this way is more pure.

Hydrogen sulphide is a covalent compound. Its molecules have a bent tetrahedral structure like that of water (*see* section 2.2). However, unlike water, hydrogen sulphide molecules do not form hydrogen bonds. This is because the sulphur atom is less electronegative but bigger than oxygen. Sulphur thus has a much lower charge density than oxygen. Owing to the absence of hydrogen bonding, hydrogen sulphide has a low boiling point compared with water. The absence of hydrogen bonding also accounts for its low solubility in water.

In solution, hydrogen sulphide is a weak dibasic acid:

$$H_2S(g) + H_2O(l) \rightleftharpoons H_3O^+(aq) + \underset{\text{hydrogensulphide ion}}{HS^-(aq)}$$

$$HS^-(aq) + H_2O(l) \rightleftharpoons H_3O^+(aq) + \underset{\text{sulphide ion}}{S^{2-}(aq)}$$

Hydrogen sulphide burns in air with a blue flame. In a limited supply of air, a deposit of sulphur is formed:

$$2H_2S(g) + O_2(g) \rightarrow 2H_2O(g) \rightarrow 2S(s)$$

In a plentiful supply of air sulphur dioxide is formed:

$$2H_2S(g) + 3O_2(g) \rightarrow 2H_2O(l) + 2SO_2(g)$$

Hydrogen sulphide is a reducing agent. For example, it decolourises bromine water and reduces iron(III) ions to iron(II) ions:

$$Br_2(aq) + H_2S(s) \rightarrow 2HBr(aq) + S(s)$$
$$2Fe^{3+}(aq) + H_2S(g) \rightarrow 2Fe^{2+}(aq) + 2H^+(aq) + S(s)$$

Hydrogen sulphide forms both sulphides and hydrogensulphides (*see* above). The sulphides of alkali metals dissolve in water to give alkaline solutions. This is due to hydrolysis:

$$S^{2-}(s) + H_2O(l) \rightleftharpoons HS^-(aq) + OH^-(aq)$$

Insoluble metal sulphides can be precipitated by passing hydrogen sulphide through an aqueous solution of a salt of the metal. For example, when hydrogen sulphide is passed through a solution containing lead(II) ions, a black precipitate of lead(II) sulphide is formed:

$$Pb^{2+}(aq) + H_2S(g) \rightarrow PbS(s) + 2H^+(aq)$$

This reaction is used as a test for hydrogen sulphide.

The Oxides of Sulphur

Sulphur forms a number of oxides, the two most important of which are sulphur dioxide, SO_2, and sulphur trioxide, SO_3. The latter is also known as sulphur(VI) oxide.

Sulphur dioxide is a dense colourless gas with a choking pungent smell. It can be prepared in the laboratory by burning sulphur in air or oxygen, by adding warm dilute acid to a sulphite or by heating concentrated sulphuric acid with copper:

$$S(s) + O_2(g) \rightarrow SO_2(g)$$
$$SO_3^{2-}(s) + 2H^+(aq) \rightarrow H_2O(l) + SO_2(g)$$
$$Cu(s) + 2H_2SO_4(l) \rightarrow CuSO_4(aq) + 2H_2O(l) + SO_2(g)$$

Sulphur dioxide is an acidic oxide. It readily dissolves in water to form sulphurous acid, H_2SO_3. This is partially ionised and thus a weak acid:

$$SO_2(g) + H_2O(l) \rightleftharpoons H_2SO_3(aq)$$

$$H_2SO_3(aq) + H_2O(l) \rightleftharpoons \underset{\text{hydrogensulphide ion}}{HSO_3^-(aq)} + H_3O^+(aq)$$

$$HSO_3^-(aq) + H_2O(l) \rightleftharpoons \underset{\text{sulphite}}{SO_3^{2-}(aq)} \; H_3O^+(aq)$$

When sulphur dioxide is bubbled through sodium hydroxide solution, sodium sulphite, Na_2SO_3, is formed. This reacts with excess sulphur dioxide, forming sodium hydrogensulphite, $NaHSO_3$. The reactions are analagous to the reaction of carbon dioxide and limewater.

Both gaseous sulphur dioxide and its aqueous solution are reducing agents. For example, the gas reduces lead(IV) oxide to lead(II) sulphate and sulphurous acid reduces manganate(VII) to manganese(II):

$$PbO_2(s) + SO_2(g) \rightarrow PbSO_4(s)$$

$$2MnO_4^-(aq) + 5SO_3^{2-}(aq) + 6H^+ \rightarrow 2Mn^{2+}(aq) + 5SO_4^{2-}(aq) + 3H_2O(l)$$

Sulphur dioxide is manufactured as a by-product in the manufacture of sulphuric acid during the roasting of sulphur or sulphide ore in air (*see* chapter 7). Sulphur dioxide is used in the production of wood pulp, the bleaching of textiles and to preserve fruit and vegetables.

Sulphur dioxide plays an infamous part in air pollution and particularly the formation of **acid rain**. In the atmosphere it is oxidised to sulphur trioxide:

$$2SO_2(g) + O_2(g) \rightarrow 2SO_3(g)$$

Under normal conditions this process is slow but in the presence of a catalyst it can be rapid—as in the Contact process. Small amounts of iron and manganese compounds in the atmosphere are thought to catalyse the oxidation. The sulphur trioxide then reacts with moisture to form clouds of acid rain (*see* also chapter 12).

Solid sulphur trioxide is polymorphic. The two most important forms are α-SO_3 and β-SO_3. The former is a trimer with a cyclic structure (*see* figure 15.24a). Its crystals are transparent. β-SO_3 has a linear polymerised structure (*see* figure 15.24b). This form exists as needle-like crystals. The α- and β-forms melt at 17°C and 30°C respectively. In the presence of moisture the α-form gradually changes into the β-form. Both forms boil at 45°C. Sulphur trioxide vapour consists of discrete, symmetrical planar molecules.

Sulphur trioxide can be prepared in the laboratory by heating iron(III) sulphate:

$$Fe_2(SO_4)_3(s) \rightarrow Fe_2O_3(s) + 3SO_3(g)$$

Sulphur trioxide is strongly acidic. It reacts exothermically with water to produce sulphuric acid:

$$SO_3(g) + H_2O(l) \rightarrow H_2SO_4(aq)$$

It is also a strong oxidising agent. For example, it oxidises hydrogen bromide to bromine.

(a)

(b)

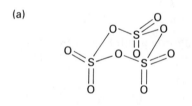

Figure 15.24 Sulphur trioxide: (a) cyclic structure, (b) linear structure

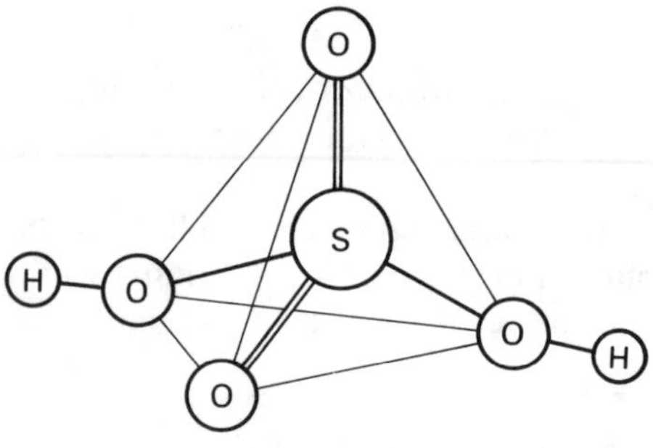

Sulphuric Acid

Sulphuric acid is a corrosive, colourless, oily liquid. It is one of the most important of all industrial chemicals. Its manufacture by the Contact process and its uses are described in chapter 7.

Pure sulphuric acid is a covalent compound. Its molecules have a tetrahedral structure (*see* figure 15.25).

Figure 15.25 Structure of sulphuric acid

Sulphuric acid boils and decomposes at 340°C, forming sulphur trioxide and steam:

$$H_2SO_4(l) \rightarrow SO_3(g) + H_2O(l)$$

The high boiling point and viscosity are thought to be due to hydrogen bonding between the hydrogen and oxygen atoms of neighbouring molecules.

Sulphuric acid is a strong dibasic acid. In water, it completely ionises, forming hydrogensulphate and sulphate ions:

$$H_2SO_4(l) + H_2O(l) \rightleftharpoons H_3O^+(aq) + HSO_4^-(aq)$$
$$HSO_4^-(aq) + H_2O(l) \rightleftharpoons H_3O^+(aq) + SO_4^{2-}(aq)$$

Sulphuric acid undergoes all the typical reactions of an acid. It reacts with metals to form sulphates and hydrogen, with carbonates to form sulphates, carbon dioxide and water, and with bases and alkalis to form sulphates or hydrogen-sulphates.

Concentrated sulphuric acid can react violently with water. For this reason the acid must always be added to water and not *vice versa*. The acid is also **hygroscopic**. This means it absorbs moisture from the air. Gases which do not react with sulphuric acid can be dried by passing them through the acid. Most important of all in this context, concentrated sulphuric acid acts as a **dehydrating agent**. For example, concentrated sulphuric acid will remove water from:

- blue copper(II) sulphate-5-water crystals to give anhydrous copper(II) sulphate;
- carbohydrates such as sucrose to give sugar charcoal

$$C_{12}H_{22}O_{11}(s) \rightarrow 12C(s) + 11H_2O(l)$$

paper is also charred by concentrated sulphuric acid—this is due to the action of the acid on the cellulose fibres;
- ethanol to give ethene

$$C_2H_5OH(l) \rightarrow CH_2{=}CH_2(g) + H_2O(l)$$

under controlled conditions, ethoxyethane (diethyl ether), $C_2H_5OC_2H_5$, is also formed by the dehydration of ethanol using concentrated sulphuric acid—ethoxyethane is manufactured in this way;
- a mixture of chlorobenzene and trichloroethanal (chloral) (*see* section 16.3).

Concentrated sulphuric acid is also a strong oxidising agent. It oxidises metals and non-metals. For example

$$Cu(s) + 2H_2SO_4(l) \rightarrow CuSO_4(aq) + 2H_2O(l) + SO_2(g)$$
$$C(s) + 2H_2SO_4(l) \rightarrow 2H_2O(l) + CO_2(g) + 2SO_2(g)$$

Other examples of the oxidising properties of concentrated sulphuric acid include the oxidation of iron(II) to iron(III) ions and bromide to bromine in the following reactions:

$$2FeSO_4(s) + 2H_2SO_4(l) \rightarrow Fe_2(SO_4)_3(aq) + 2H_2O(l) + SO_2(g)$$
$$2HBr(g) + H_2SO_4(l) \rightarrow 2H_2O(l) + SO_2(g) + Br_2(aq)$$

Finally, sulphuric acid is used as a sulphonating agent in organic chemistry (*see* chapter 17).

Sulphates, Sulphites and Thio-Compounds

The sulphate ion has a tetrahedral structure similar to that of the sulphuric acid molecule.

The salts are formed by the reaction of dilute sulphuric acid with a metal, its oxide, hydroxide or carbonate. Except for barium and lead(II) sulphates, they tend to be soluble in water. Calcium sulphate is only slightly soluble in water. The sulphates are generally stable to heat. The iron sulphates are exceptions. Iron(II) sulphate, for example, decomposes to form iron(III) oxide:

$$2FeSO_4(s) \rightarrow Fe_2O_3(s) + SO_3(g) + SO_2(g)$$

Sulphate ions in solution can be detected by the addition of aqueous barium chloride. If sulphate is present, a white precipitate of barium sulphate is formed:

$$Ba^{2+}(aq) + SO_4^{2-}(aq) \rightarrow BaSO_4(s)$$

Sulphates have several important uses. For example, magnesium sulphate, $MgSO_4 \cdot 7H_2O$ (known as Epsom salts), is used to make explosives and matches and also to fireproof fabrics. A saturated solution of the salt is used medicinally as an anti-inflammatory agent.

Group I metals also form hydrogensulphates. Sodium hydrogensulphate, $NaHSO_4$, is an example. This is soluble in water and decomposes on heating to give sodium pyrosulphate(V), $Na_2S_2O_7$, which in turn decomposes to the sulphate:

$$2NaHSO_4(s) \rightarrow Na_2S_2O_7(s) + H_2O(g)$$
$$Na_2S_2O_7(s) \rightarrow Na_2SO_4(s) + SO_3(g)$$

The most important sulphite is sodium sulphite, Na_2SO_3. This reacts with acid to give sulphur dioxide:

$$SO_3^{2-}(s) + 2H^+(aq) \rightarrow H_2O(l) + SO_2(g)$$

On boiling a solution of sodium sulphite with sulphur, a solution of sodium thiosulphate is formed:

$$Na_2SO_3(aq) + S(s) \rightarrow Na_2S_2O_3(aq)$$

Sodium thiosulphate is known as photographers' hypo. It is used as a fixing agent in the photographic process. It is also used in volumetric analysis to estimate iodine (*see* next chapter).

Sodium thiosulphate is a sulphur analogue of sodium sulphate. Another important thio-compound is potassium thiocyanate, KSCN. This is the sulphur analogue of potassium cyanate, KCNO. Potassium thiocyanate can be prepared by fusing potassium cyanide with sulphur:

$$8KCN(s) + S_8(s) \rightarrow 8KSCN(s)$$

Potassium thiocyanate is used in qualitative analysis to detect iron(III) ions. If these ions are present, a blood red colouration is produced due to the formation of the complex ion $[Fe(SCN)(H_2O)_5]^{2+}$.

SUMMARY

1. There is a transition from non-metallic to metallic properties on descending group VI.
2. Group VI elements tend to combine with hydrogen and reactive metals to form compounds in which the elements exist in the −2 oxidation state.
3. Oxygen is highly electronegative and a strong oxidising agent.
4. **Trioxygen** (also known as ozone) is a stronger oxidising agent than oxygen.
5. Oxygen is essential to life. It plays key roles in respiration and **metabolic processes.**

6. Oxygen forms a variety of oxides:

Type	Example	
	Formula	Name
metallic	CuO	copper(II) oxide
non-metallic	CO_2	carbon dioxide
ionic	CaO	calcium oxide
covalent	SO_2	sulphur dioxide
normal	MgO	magnesium oxide
peroxide	H_2O_2	hydrogen peroxide
mixed	Pb_3O_4	dilead(II) lead(IV) oxide
basic	CaO	calcium oxide
acidic	SO_3	sulphur trioxide
amphoteric	ZnO	zinc oxide
neutral	NO	nitrogen monoxide

7. Hydrogen peroxide is a strong oxidising agent.
8. Sulphur exhibits **enantiotropy**.
9. In the **Frasch process** sulphur is extracted from underground deposits by the use of compressed air and superheated steam.
10. Sulphur forms compounds in which it exists in oxidation states +2, +4, +6 and −2.
11. Hydrogen sulphide is a weak dibasic acid and a reducing agent.
12. Gaseous sulphur dioxide and its aqueous solution are reducing agents.
13. Sulphur trioxide is strongly acidic and a strong oxidising agent.
14. Sulphuric acid is a strong dibasic acid, a **dehydrating agent** and a strong oxidising agent.
15. Both sulphuric acid molecules and sulphate ions have tetrahedral structures.

Examination Questions

1. (a) Write equations to show how aluminium reacts with the following reagents:
   (i) concentrated aqueous sodium hydroxide;
   (ii) iron(III) oxide (on heating);
   (iii) concentrated hydrofluoric acid.
   (b) What features of the chemistry of aluminium are illustrated by each of these reactions?
   (c) Aluminium chloride is used in the Friedel–Crafts reaction and as a halogen carrier in organic chemistry. Given an example of one of these uses and explain why an aluminium compound should be particularly useful in such reactions.
   (d) Thallium is the heaviest element in Group III and is in the same period as lead. Suggest one way in which the chemistry of thallium might be expected to differ from that of aluminium, giving your reasons.

   (O & C)

2. Describe carefully the production of aluminium from pure aluminium oxide by an electrolytic process. Comment on the economics of this process and the steps taken to reduce the production costs. (No account of the purification of crude bauxite is required.)
   Discuss the reasons why (a) a thermal reduction process, and (b) electrolysis of the chloride, are unsuitable industrial methods for refining aluminium.

Explain the corrosion resistance of aluminium in atmospheric conditions and how this resistance may be increased commercially.

Mention *three* uses of aluminium, stating clearly in each case the property of the metal which makes it particularly appropriate material.

(L)

3. Discuss trends in Group IV of the Periodic Table, illustrating your answer with examples from the chemistry of carbon and lead only.

You should consider particularly;
(a) the hardness and melting points of the elements, (melting points: C(graphite), 3730°C; Pb, 327°C)
(b) the relative stabilities of the divalent and tetravalent compounds,
(c) the stabilities of the tetrachlorides to hydrolysis,
(d) the existence of a large number of hydrides of carbon,
(e) the acid–base character of the monoxides and dioxides.

(L)

4. (a) State two *chemical* properties of carbon which illustrate its special character among the elements of Group IV.
(b) Choose from the species tin(II), tin(IV), lead(II) and lead(IV), (i) an oxidising agent, (ii) a reducing agent and give an example of each chosen species reacting accordingly. Balanced equations are not required.
(c) Explain why CCl_4 does not react readily with water.
(d) Give the formulae of the tin-containing products when $SnCl_4$ reacts with
(i) water,
(ii) concentrated hydrochloric acid.

(JMB)

5. The following table lists some information concerning Group IV elements.

Element	Carbon	Silicon	Germanium	Tin	Lead
Atomic number	6	14	32	50	82
First ionisation energy/kJ mol^{-1}	1086	787	760	707	715
Melting point/°C	3730	1410	937	232	327

(a) (i) Write the full electronic structure of silicon.
(ii) Give *two* reasons why the first ionisation energy generally decreases as the group is descended.
(iii) Describe how the trend in melting points is related to the structure of the elements.
(b) The oxides of carbon and silicon differ greatly. Carbon dioxide is a gas and silicon dioxide is a solid.
(i) Compare the reactions of the dioxides with alkali, indicating the conditions required.
(ii) Carbon dioxide is a molecular compound. Using the outer electrons of both elements, draw 'dot and cross' diagrams to illustrate the electronic structure and shape of the carbon dioxide molecule.
(iii) Silicon dioxide can be reacted with magnesium to produce silicon. During this process magnesium silicide, Mg_2Si, is produced. The silicide reacts with hydrochloric acid to form silicon(IV) hydride which ignites immediately in air.
Write balanced equations for all the reactions taking place.

(c) When 0.143 g of tin was heated in a current of dry chlorine a colourless liquid chloride was formed which hydrolysed in water to give a solution that required 48.00 cm^3 of 0.1 M sodium hydroxide for neutralisation. (Relative atomic masses: Cl = 35.5, Sn = 119.0).
  (i) Use these data to calculate the formula of the tin chloride.
  (ii) Write an equation for the hydrolysis of the chloride.

(NISEC)

6.   Group V of the Periodic Table contains the elements N, P, As, Sb, Bi, in increasing order of atomic number. The most common oxidation states of this group are −3, 0, +3, +5.
(a) Write the formula of
  (i) the hydride of nitrogen in the −3 oxidation state;
  (ii) an oxide of As in the +3 oxidation state;
  (iii) an oxoacid of P in the +5 oxidation state.
(b) Write an equation showing the thermal dissociation of the compound in (a) (i), naming the required catalyst.
(c) Draw a bond diagram to show the molecular structure of the compound in (a) (i), commenting on the bond angles and the geometrical shape of the structure.
(d) Put the hydrides of this group in order of *increasing* stability to heat (put the *most* stable *last*).
(e) Red phosphorus and concentrated nitric acid react according to the equation:

$$P_4(s) + 20HNO_3(l) \rightarrow P_4O_{10} + 20NO_2(g) + 10H_2O(l).$$

  (i) Name the *type* of reaction involved.
  (ii) 1.42 g P_4O_{10}, in water, produces an acidic solution which neutralises 20.0 cm^3 1.00 M sodium hydroxide solution. Deduce the formula of the phosphorus acid formed. (P = 31, O = 16.)
  (iii) Say whether the N-containing acid of comparable formula will be more (or less) acidic than the acid in (ii). Explain your answer.
(f) Describe the action of heat on Group I and II nitrates, emphasising patterns of behaviour. Indicate, very briefly, how ionic radii can be used to interpret these.

(SUJB)

7. (a) From a consideration of the elements themselves, their hydrides and chlorides, explain why the inclusion of nitrogen and phosphorus in the same group of the Periodic Table is justified.
(b) What is meant by the term allotropy?
    Name *two* allotropes of phosphorus and explain briefly how each can be prepared. Give *two* physical and *two* chemical differences between the named allotropes.
(c) Nitric and phosphonic (phosphorous) acids can be prepared from oxides of nitrogen and phosphorus respectively. Give equations to illustrate these preparations.
(d) Explain why aqueous sodium nitrate and aqueous sodium phosphate(v) show different pH values.

(AEB, 1982)

8. (a) For the elements nitrogen and phosphorus compare the trihydrides and pentoxides as follows:
   Trihydrides
   (i) Shapes of the molecules including the bond angles;
   (ii) boiling points (account for the difference);
   (iii) stabilities to heat, giving a reason for the difference;
   (iv) basicities, illustrated by their reactions with water and acids.
   Pentoxides
   (i) Physical states at ordinary temperature and pressure;
   (ii) structures of the molecules;
   (iii) action of water (give equations) on the oxides;
   (iv) strengths of the oxoacids formed and the reactions of these with aqueous sodium hydroxide.
   (b) In terms of electronic structure and atomic size, account for the similarities and gradual changes in properties of the Group V elements on passing down the group.

   (AEB, 1982)

9. (a) (i) Give the formulae of **four** compounds in which the oxidation state of sulphur is -2, $+2$, $+4$ and $+6$ respectively.
      (ii) Describe briefly how each of these compounds could be prepared from the element sulphur.
   (b) Sulphuric acid is used in the following types of reactions: dehydration, oxidation, sulphonation and hydrolysis.
      For each of these types of reaction involving sulphuric acid give
      (i) a suitable balanced equation,
      (ii) an explanation of the part played by the acid.
   (c) There are two acid chlorides of sulphuric acid.
      (i) Give their formulae.
      (ii) Describe briefly how they are prepared.
      (iii) Describe the use of either **one** of these compounds in organic synthesis.
   (d) (i) What changes in appearance are noted when heating sulphur from its solid state at room temperature to its boiling point?
      (ii) How are these observations explained in terms of changes in the molecular structure of sulphur?

   (AEB, 1984)

10. Write an essay on the properties and uses of sulphuric acid. Include in your answer examples, where possible, from both inorganic and organic chemistry, of H_2SO_4 acting as (a) an acid, (b) a catalyst, (c) an oxidising agent, (d) a dehydrating agent, (e) an electrophile and (f) a sulphonating agent.

   (JMB)

16 THE HALOGENS AND THE NOBLE GASES

Dulce et Decorum Est

Wilfred Owen is one of the most admired poets of World War I. The following poem is one of his finest and most famous. He wrote it during the last year of his life—he was killed in action on 4 November 1918 at the age of 25 years. The poem includes a description of the death of a soldier from chlorine poisoning. The title comes from the line 'Dulce et decorum est pro patria mori' by the Latin poet Horace. The line means 'It is sweet and fitting to die for one's country.'

Dulce et Decorum Est

Bent double, like old beggars under sacks,
Knock-kneed, coughing like hags, we cursed through sludge,
Till on the haunting flares we turned our backs,
And towards our distant rest began to trudge.
Men marched asleep. Many had lost their boots,
But limped on, blood-shod. All went lame, all blind;
Drunk with fatigue; deaf even to the hoots
Of gas-shells dropping softly behind.

Gas! GAS! Quick, boys!—An ecstasy of fumbling,
Fitting the clumsy helmets just in time,
But someone still was yelling out and stumbling
And floundering like a man in fire or lime.—
Dim through the misty panes and thick green light,
As under a green sea, I saw him drowning.

In all my dreams before my helpless sight
He plunges at me, guttering, choking, drowning.

If in some smothering dreams, you too could pace
Behind the wagon that we flung him in,
And watch the white eyes writhing in his face,
His hanging face, like a devil's sick of sin;
If you could hear at every jolt, the blood
Come gargling from the froth-corrupted lungs, Bitter as the cud
Of vile, incurable sores on innocent tongues,—
My friend, you would not tell with such high zest
To children ardent for some desperate glory,
The old Lie: *Dulce et decorum est
Pro patria mori*.

Wilfred Owen

Wilfred Owen

1914–1918 War. German anti-aircraft gun crew in gas masks

'Behind the wagon that we flung him in'

Chemical Warfare

Modern chemical warfare started in World War I with the use of chlorine. Both the Allies and the Germans used the gas. It was released directly from cylinders. However, its effectiveness was dependent on the weather. Later in the war, phosgene and mustard gas were used.

$$\begin{array}{c} Cl \\ \diagdown \\ C=O \\ \diagup \\ Cl \end{array} \quad \begin{array}{l} \text{Phosgene} \\ \text{(carbonyl chloride)} \end{array} \qquad \begin{array}{c} Cl-CH_2-CH_2 \\ \diagdown \\ S \\ \diagup \\ Cl-CH_2-CH_2 \end{array} \quad \begin{array}{l} \text{Mustard gas} \\ \text{(2,2'-dichloro-} \\ \text{diethyl sulphide)} \end{array}$$

These gases were contained in artillery shells. They could be delivered on target and resulted in hundreds of thousands of casualties. Large stockpiles of mustard gas are still held in several countries, including the United States.

In World War II, organophosphorus nerve gases were stockpiled by the Germans and seized by the Soviet army at the end of the war. However, the major powers did not enter into chemical warfare during World War II.

Since World War II, several thousand chemical warfare agents have been tested and evaluated. During the Vietnam war, the United States used a variety of chemical defoliating and riot-control agents. These included the now infamous herbicide 2,4,5-T, otherwise known as Agent Orange. This chemical invariably contains small amounts of the highly toxic chemical dioxin.

TCDD or 'dioxin'
2,3,7,8-tetrachloro-
dibenzo-*p*-dioxin)

2,4,5-T or Agent Orange
(2,4,5- trichlorophenoxy-
ethanoic acid and esters)

The use and misuse of pesticides such as 2,4,5-T are described in section 16.3.

LEARNING OBJECTIVES

After you have studied this chapter you should be able to

1. Describe the atomic and molecular structures of the **halogens**.
2. Compare the physical properties of the halogens.
3. Describe the preparations of chlorine, bromine and iodine in the laboratory.
4. Outline the principal chemical characteristics and trends of elements in Group VII of the Periodic Table.
5. Give specific examples of how chlorine, bromine and iodine act as oxidising agents.
6. Describe and give specific examples of the reactions of chlorine, bromine and iodine with
   (a) metals,
   (b) non-metals including hydrogen,
   (c) water and alkalis,
   (d) organic compounds.
7. Define and give examples of **disproportionation**.
8. Indicate the anomalous nature of fluorine.
9. Compare the bonding and structure of the chlorides across a period and down Group VII.
10. Indicate how chlorides may be prepared in the laboratory.
11. Discuss the solubility in water and hydrolysis of halides.
12. Give examples of the most important reactions of halide ions.
13. Outline the preparations and trends in stability and properties of the hydrogen halides.
14. Give an example of an anomalous property of hydrofluoric acid.
15. Discuss and account for the various oxidation states of halogens in their compounds.
16. Outline the trends in properties and reactivity of the chloric acids and their salts.
17. Give examples of metal halide complex ions.
18. State the principal sources of the halogens and their compounds.
19. Outline the methods of manufacture of chlorine, hydrogen chloride and the chlorates.
20. Summarise the methods of extraction for fluorine, bromine and iodine.
21. Discuss the benefits and dangers of using organochlorine compounds.
22. Compare the atomic structures and physical properties of the noble gases.
23. Outline the chemical properties, preparation and structures of noble-gas compounds—with particular reference to xenon compounds.
24. Summarise the occurrence, extraction and uses of the noble gases and their compounds.

16.1
The Halogens

We have all heard of fluoride in toothpaste and chlorine in swimming pools. Bromide is well known as a tranquilliser and tincture of iodine as an antiseptic agent. Fluorine, chlorine, bromine and iodine are all halogens. The halogens are non-metals. They occur in Group VII of the Periodic Table (*see* figure 16.1).

What's in a Name?

Gen and the Halogens
With the exception of fluorine, the names of the halogens derive from Greek words:

Name	Origin	Meaning
halogen	*hals* (Gk)	salt
	gen (Gk)	producing
fluorine	*fluere* (L and F)	flow or flux (until AD 1500 the mineral fluorspar, CaF_2, was used as a flux in metallurgy)
chlorine	*chloros* (Gk)	greenish yellow
bromine	*bromos* (Gk)	stench
iodine	*iodes* (Gk)	violet
astatine	*astatos* (Gk)	unstable

(F = French; Gk = Greek; L = Latin)

The syllable *gen* occurs as a prefix or suffix in numerous scientific terms, for example, hydrogen, oxygen, nitrogen, gene, generator and antigen. It is normally associated with growth or production. For example halogen (= *hals* + *gen*) means salt producing. Oxygen (= *oxy* + *gen*) means acid producing. *Oxy* derives from the Greek word *oxus* meaning sharp. The Greek word *genes* means born. Similar words also occur in Old French (*genre/ gendre*) and Latin (*genus*). These words mean kind, type or class.

The *gen* in geneva has a completely different origin, however. Geneva is a spirit distilled from and flavoured with juniper berries. The word derives from the Old French word *genevre* and Latin word *juniperus*. The alcoholic beverage gin also has the same origin.

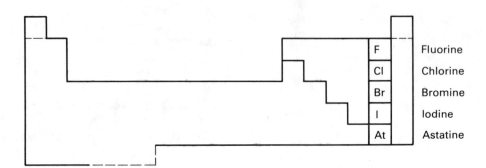

Figure 16.1 The halogens

STRUCTURE

Halogen atoms all have seven electrons in their outer shells. A halogen atom readily accepts another electron to form a halide ion. The halide ion has a stable octet of electrons. Halogens are highly electronegative and reactive and thus do not exist in their elemental forms in nature. However, halide ions do occur widely.

Halogens exist as diatomic molecules. The bonds in these molecules are single covalent bonds. They are formed by sharing two electrons—one from each atom.

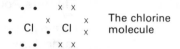

Table 16.1 shows some data on the structure of the halogens. Note how the atomic and ionic radii and also the bond lengths all increase down the group. On the other hand, the bond dissociation energies and thus the bond strengths decrease. Fluorine is an exception. Its low bond strength is probably due to the closeness of the atoms in the molecules. This results in repulsion between the non-bonding electrons and thus weakening of the bond.

Table 16.1   Structure and properties of the halogens

Element	Atomic number	Electron configuration	Configuration in outer shell	Atomic radius/nm	Ionic radius/nm	Bond length/nm	Bond dissociation energy/kJ mol^{-1}
fluorine	9	2.7	$2s^2 2p^5$	0.072	0.136	0.142	158
chlorine	17	2.8.7	$3s^2 3p^5$	0.099	0.181	0.200	242
bromine	35	2.8.18.7	$4s^2 4p^5$	0.114	0.195	0.229	193
iodine	53	2.8.18.18.7	$5s^2 5p^5$	0.133	0.216	0.266	151

LABORATORY PREPARATION

Chlorine

Chlorine can be prepared in the laboratory by the oxidation of concentrated hydrochloric acid with potassium manganate(VII):

$$2MnO_4^-(s) + 16H^+(aq) + 10Cl^-(aq) \rightarrow 2Mn^{2+}(aq) + 8H_2O(l) + 5Cl_2(g)$$

The chlorine is passed through water to remove hydrogen chloride and then through concentrated sulphuric acid to dry it. It can be collected by downward delivery—that is, upward displacement of air.

Chlorine may also be prepared in the laboratory from bleaching powder using dilute hydrochloric acid:

$$Ca(OCl)_2(s) + 4H^+(aq) + 2Cl^-(aq) \rightarrow Ca^{2+}(aq) + 2H_2O(l) + 2Cl_2(g)$$

Bromine

Bromine can be prepared in the laboratory by adding manganese(IV) oxide to concentrated sulphuric acid and potassium bromide. The hydrobromic acid formed by the reaction between the acid and bromide is oxidised by the manganese(IV) oxide:

$$KBr(s) + H_2SO_4(l) \rightarrow KHSO_4(s) + HBr(g)$$

then

$$MnO_2(s) + 4H^+(aq) + 2Br^-(aq) \rightarrow Mn^{2+}(aq) + 2H_2O(l) + Br_2(g)$$

The bromine is separated from the reaction mixture by distillation.

Iodine

Iodine can be prepared in a similar manner to bromine, but using potassium iodide. The iodine is separated from the reaction mixture by sublimation.

PHYSICAL PROPERTIES

- The halogens are toxic.

Chlorine Poisoning

Liquid chlorine can cause severe skin burns. The gas is a severe irritant, particularly to the eyes and respiratory system. It reacts with the water in the eyes, lungs and moist mucous membranes to form hydrochloric acid and chloric(I) acid (*see* 'Reactions with water and alkalis'). Symptoms of chlorine poisoning include burns to the eyes and respiratory system, irritative cough and, in severe cases, bloody sputum. Bronchial pneumonia can develop in lungs damaged by chlorine. The maximum toleration of chlorine in the air is 0.005 mg dm^{-3}. A level of 0.2 mg dm^{-3} is regarded as dangerous and 2 mg dm^{-3} can result in rapid death. However, fatalities are rare since people who have been exposed to the gas are usually removed in time.

- They have characteristic 'swimming pool' odours.
- They are volatile. This is because the van der Waals forces holding the molecules together are weak. Although van der Waals forces increase down the group, they are still sufficiently weak in iodine crystals to allow it to sublime on heating. A purple vapour is produced. The melting points and boiling points of the halogens are shown in table 16.2.

Table 16.2 Physical properties of the halogens

Element	Melting point/°C	Boiling point/°C	State and appearance at 20°C
fluorine	-220	-188	pale yellow gas
chlorine	-101	-34	pale yellow-green gas
bromine	-7	58	brown liquid with heavy brown vapour
iodine	114	183	shiny grey-black crytals

- They are coloured. The colour intensity increases down the group.
- They are slightly soluble in water. **Chlorine water** is a solution of chlorine in water. The halogens dissolve in organic solvents to form coloured solutions. For example, chlorine dissolves in tetrachloromethane to form a yellow solution, bromine to form a red solution and iodine a purple one.

CHEMICAL REACTIVITY

The halogens are the most reactive group of elements in the Periodic Table. They have low bond dissociation energies (*see* table 16.1) and, because they all have seven electrons in their outer shells, they are highly electronegative. Fluorine is the most electronegative and most reactive non-metal in the Periodic Table. Reactivity then decreases down the group. In the next section we shall look at the ability of halogens to oxidise both metals and non-metals and see how this ability decreases down the group from fluorine to iodine.

Halogens are Oxidising Agents

When hydrogen sulphide is bubbled through chlorine water, sulphur is precipitated. The equation is

$$8H_2S(g) + 8Cl_2(aq) \rightarrow 16HCl(aq) + S_8(s)$$

The chlorine oxidises the hydrogen sulphide by removing the hydrogen. Chlorine also oxidises iron(II) to iron(III). For example, when chlorine is shaken with an aqueous solution of iron(II) sulphate, iron(III) sulphate is formed:

$$2Fe^{2+}(aq) + Cl_2(g) \rightarrow 2Fe^{3+}(aq) + 2Cl^-(aq)$$

The half equation is

$$2Fe^{2+} \rightarrow 2Fe^{3+} + 2e^-$$

Another example is the synthesis of sodium chloride by burning sodium in chlorine:

$$2Na(s) + Cl_2(g) \rightarrow 2NaCl(s)$$

The sodium is oxidised since each sodium atom loses an electron to form a sodium ion:

$$2Na \rightarrow 2Na^+ + 2e^-$$

The chlorine accepts these electrons, forming chloride ions:

$$Cl_2 + 2e^- \rightarrow 2Cl^-$$

All halogens are oxidising agents. The most powerful is fluorine. The standard electrode potentials of the halogens are shown in table 16.3. The table shows that the oxidising strength of the halogens decrease down the group. This can be demonstrated by adding potassium bromide solution to a gas jar of chlorine. The chlorine oxidises the bromide to form bromine. The solution thus turns from colourless to brown:

$$Cl_2(g) + 2Br^-(aq) \rightarrow 2Cl^-(aq) + Br_2(aq)$$

Chlorine is thus more strongly oxidising than bromine. Similarly, if potassium iodide solution is mixed with bromine, a black precipitate of iodine is formed. The bromine oxidises the iodide:

$$Br_2(g) + 2I^-(aq) \rightarrow 2Br^-(aq) + I_2(s)$$

Both these reactions are examples of **displacement reactions**. In each case the more reactive, that is the more strongly oxidising, halogen displaces the less reactive halogen from solution.

Table 16.3 The standard electrode potentials of the halogens

Electrode reaction	Standard electrode potential, $E^{\ominus}$/V
$F_2(g) + 2e^- \rightleftharpoons 2F^-(aq)$	+2.87
$Cl_2(g) + 2e^- \rightleftharpoons 2Cl^-(aq)$	+1.36
$Br_2(l) + 2e^- \rightleftharpoons 2Br^-(aq)$	+1.09
$I_2(s) + 2e^- \rightleftharpoons 2I^-(aq)$	+0.54

Oxidation of Metals

Halogens readily oxidise metals. Fluorine oxidises all metals, including gold and silver, with ease. We have already seen how chlorine oxidises sodium to form sodium chloride. As another example, when a stream of chlorine is passed over heated iron filings, iron(III) chloride, a brown solid, is formed:

$$2Fe(s) + 3Cl_2(g) \rightarrow 2FeCl_3(s)$$

Even iodine slowly oxidises metals low in the electrochemical series. The ease with which halogens oxidise metals decreases down the group. This can be seen by comparing the energy evolved when the halides are formed from their elements. Table 16.4 shows how the standard enthalpies of formation of the sodium halides decrease down the group.

Table 16.4 The standard enthalpies of formation of the sodium halides

Halide	Standard enthalpy of formation, $\Delta H^{\ominus}_{f,m}$/kJ mol^{-1}
NaF	−573
NaCl	−414
NaBr	−361
NaI	−288

Oxidation of Non-Metals

Apart from nitrogen and most of the rare gases, fluorine oxidises all non-metals. Chlorine reacts with phosphorus and sulphur. Carbon, nitrogen and oxygen do not react directly with chlorine, bromine or iodine. The reactivities of the halogens with non-metals can be compared by examining their reactions with hydrogen (*see* table 16.5).

Oxidation of Hydrocarbons

Under certain conditions halogens oxidise hydrocarbons. Chlorine, for example, completely removes the hydrogen from turpentine:

$$C_{10}H_{16}(l) + 8Cl_2(g) \rightarrow 10C(s) + 16HCl(g)$$

Table 16.5 Reactions of the halogens with hydrogen

Reaction	Observations
$H_2(g) + F_2(g) \rightarrow 2HF(g)$	explosive
$H_2(g) + Cl_2(g) \rightarrow 2HCl(g)$	explosive in sunlight but slow in the dark
$H_2(g) + Br_2(g) \rightarrow 2HBr(g)$	requires heat and a catalyst
$H_2(g) + I_2(g) \rightarrow 2HI(g)$	slow, even on heating

The oxidation of ethyne can be explosive:

$$H\!-\!C\!\equiv\!C\!-\!H(g) + Cl_2(g) \rightarrow 2C(s) + 2HCl(g)$$

Reactions with Water and Alkalis

Fluorine reacts with cold water, forming hydrogen fluoride and oxygen:

$$2F_2(g) + 2H_2O(g) \rightleftharpoons 4HF(g) + O_2(g)$$

Chlorine dissolves slowly in water forming **chlorine water**. Chlorine water is slightly acidic due to the **disproportionation** (*see* section 10.2) of chlorine to form hydrochloric acid and chloric(ı) acid.

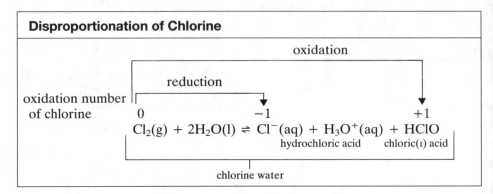

Bromine and iodine disproportionate in water in a similar way. The extent of disproportionation in water decreases from chlorine to iodine.

Chlorine, bromine and iodine also disproportionate in alkali. For example, in cold dilute alkali, bromine disproportionates to bromide and bromate(ı):

$$Br_2(g) + 2OH^-(aq) \rightarrow Br^-(aq) + BrO^-(aq) + H_2O(l)$$

When hot concentrated alkali is used, further disproportionation occurs:

$$\underset{\text{bromate(ı)}}{3BrO^-(aq)} \rightarrow \underset{\text{bromide}}{2Br^-(aq)} + \underset{\text{bromate(v)}}{BrO_3^-(aq)}$$

Iodate(ı) is unstable even in cold dilute alkali. It spontaneously disproportionates to iodide and iodate(v).

The reaction of fluorine with alkali, like its reaction with water, is exceptional. In cold dilute alkali, the following reaction occurs:

$$2OH^-(aq) + 2F_2(g) \rightarrow 2F^-(aq) + \underset{\substack{\text{oxygen} \\ \text{difluoride}}}{F_2O(g)} + H_2O(l)$$

In hot concentrated alkali, the reaction is

$$4OH^-(aq) + 2F_2(g) \rightarrow 4F^-(aq) + O_2(g) + 2H_2O(l)$$

Halogen Analyses

Halogens may be analysed both qualitatively and quantitatively using silver nitrate solution. For example

$$Ag^+(aq) + Br^-(aq) \rightarrow Ag\,Br(s)$$

Iodine can be determined both qualitatively and quantitatively using starch soluton. Since iodine is only sparingly soluble in water, it is usually estimated in the presence of potassium iodide. This is because iodine forms a soluble triiodide ion, I_3^-, with iodide:

$$I_2(aq) + I^-(aq) \rightleftharpoons I_3^-(aq)$$

Reducing agents, such as thiosulphate(VI), are determined with iodine/iodide solutions. Oxidising agents such as manganate(VII) can also be determined using iodine/iodide solutions. Oxidising agents displace the above equilibrium to the left, liberating iodine. The iodine is then titrated with thiosulphate(VI).

SUMMARY

1. All halogen atoms have seven electrons in their outer shells.
2. Halogens may be prepared by oxidation of the hydrohalic acid.
3. Halogens oxidise metals, non-metals and hydrocarbons.
4. Halogens **disproportionate** in water and alkalis to form halide ions, halate(I) and halate(V) ions.
5. The trends in the physical and chemical properties of the halogens down the group are shown in table 16.6

Table 16.6   Trends down Group VII

Property	F	Cl	Br	I
atomic radius	○		increases →	
ionic radius	○		increases →	
bond length in halogen molecule	○		increases →	
bond dissociation energy	×	○	decreases →	
colour intensity	○		increases →	
m.p. and b.p.	○		increases →	
chemical reactivity	○		decreases →	
electronegativity	○		decreases →	
oxidising power	○		decreases →	
standard electrode potential	○		decreases →	
extent of disproportionation in water	×	○	decreases →	

6. Fluorine is exceptional for the following reasons:
   (a) It has a low bond dissociation energy.
   (b) It exists in only one oxidation state in its compounds.
   (c) It is the most electronegative and most chemically reactive of all non-metals.
   (d) Its reactions with water and alkalis are different from those of the other halogens.

16.2 Halogen Compounds

The principal types of compounds formed by the halogens are shown in table 16.7. In this section we shall consider the

- structure and bonding,
- laboratory preparation,
- physical and chemical properties

of some of the more important of these compounds. The manufacture and uses of halogen compounds are dealt with in section 16.3.

Table 16.7   Types of halogen compound

Type	Example	
halides		
metal halides	sodium chloride	NaCl
hydrogen halides	hydrogen chloride	HCl(g)
other non-metal halides	phosphorus(III) chloride	PCl_3
hydrohalic acids	hydrochloric acid	HCl(aq)
oxohalogen compounds		
halic acids	chloric(v) acid	$HClO_3$
halates	sodium chlorate(v)	$NaClO_3$
oxides	chlorine dioxide	ClO_2
interhalogen compounds	iodine monochloride	ICl
organohalogen compounds	trichloromethane	$CHCl_3$

THE HALIDES

Structure and Bonding

Alkali metal halides are ionic compounds. With the exception of caesium halides, they crystallise with the structure of rock salt (*see* section 3.2). Group II metal halides are principally ionic, although some are partially covalent. Many crystallise as layer lattices (*see* section 3.2). Across a period, the halides become increasingly covalent (*see* table 16.8).

Table 16.8   The s- and p-block chlorides of period 3

	Group	Compound	m.p./°C	Type of bonding
s block	I	NaCl	801	ionic
	II	$MgCl_2$	708	ionic
p block	III	$AlCl_3$	190	partially covalent
	IV	$SiCl_4$	58	covalent
	V	PCl_3	−112	covalent
	VI	SCl_2	−78	covalent

Table 16.9   Ionic and covalent characteristics of iron and lead chlorides

Halide	Character	m.p./°C
$FeCl_2$	ionic	672
$FeCl_3$	covalent	306
$PbCl_2$	ionic	501
$PbCl_4$	covalent	−15

Many metal halides exhibit both ionic and covalent characteristics. Where a metal exhibits more than one oxidation state, the halide of the metal in the highest oxidation state is covalent. The one with the lowest oxidation state is ionic (*see* table 16.9).

Both ionic and covalent halides of metals in oxidation state(II) tend to crystallise in layer lattices. Copper(II) chloride is an interesting exception. It has a chain structure of indefinite length:

The covalent characteristics of the halides increase not only across a period but also down a group. The halides of aluminium provide an example:

- aluminium fluoride—ionic
- aluminium chloride—mainly covalent, crystallises as a layer lattice
- aluminium bromide—covalent, crystals consist of Al_2Br_6 molecules
- aluminium iodide—covalent, crystals consist of Al_2I_6 molecules

Preparation of Halides

Halides can be made by passing the halogen or hydrogen halide over the heated metal. For metals with two oxidation states, the hydrogen halide produces the halide of the metal in the lower oxidation state. The halide of the metal in the higher oxidation state is formed with the halogen. For example

$$Fe(s) + 2HCl(g) \rightarrow FeCl_2(s) + H_2(g)$$
$$2Fe(s) + 3Cl_2(g) \rightarrow 2FeCl_3(s)$$

Non-metal halides such as phosphorus trichloride can also be formed by direct combination with the halogen:

$$2P(s) + 3Cl_2(g) \rightarrow 2PCl_3(l)$$

Phosphorus pentachloride can be prepared by the action of chlorine on phosphorus trichloride:

$$PCl_3(l) + Cl_2(g) \rightleftharpoons PCl_5(s)$$

Metal halides may also be made by the action of the hydrohalic acid on the metal or its oxide, hydroxide or carbonate. For example

$$MgCO_3(s) + 2HCl(aq) \rightarrow MgCl_2(aq) + CO_2(g) + 2H_2O(l)$$

Upon evaporation and crystallisation, crystals of the hexahydrate $MgCl_2 \cdot 6H_2O$ are formed.

Insoluble halides can be obtained by precipitation from solutions of their ions. For instance, lead(II) iodide, an insoluble yellow compound, is made by mixing solutions of lead(II) nitrate and potassium iodide:

$$Pb^{2+}(aq) + 2I^-(aq) \rightarrow PbI_2(s)$$

Solubility in Water and Hydrolysis

Most metal halides are soluble in water. Notable exceptions are copper(I) and lead(II) halides. Silver chloride, bromide and iodide are also insoluble. Silver fluoride is soluble however. This is because silver fluoride is more ionic than the others. On the other hand, all the calcium halides are soluble in water except calcium fluoride. Calcium fluoride is insoluble because it has a high lattice energy.

Alkali metal halides dissolve in water to form neutral solutions. Then, along each period, halide solutions become increasingly acidic. This is due to acid–base reactions with water. The reactions of d-block metal halides with water can be complicated. They can take place in several steps and involve several species of hydrated metal ion. For example, iron(III) chloride reacts with water to form the following species:

$$[Fe(H_2O)_6]^{3+}$$
$$[Fe(OH)(H_2O)_5]^{2+}$$
$$Fe(OH)_2$$
$$[Fe_2(OH)_2(H_2O)_8]^{4+}$$
$$Fe_2O_3$$

The pH of the resultant solution is about 3.

The halides of p-block elements react with water to form solutions with pH values as low as 1. Once again the reactions are not simple. For example, phosphorus(III) chloride undergoes hydrolysis in a number of steps. The overall reaction is

$$PCl_3(l) + 6H_2O(l) \rightarrow 3H_3O^+(aq) + 3Cl^-(aq) + H_3PO_3(aq)$$

The hydrolysis of phosphorus(V) chloride is similar.

Silicon tetrachloride undergoes hydrolysis to form silicon dioxide. The overall reaction is

$$SiCl_4(l) + 6H_2O(l) \rightarrow 4H_3O^+(aq) + 4Cl^-(aq) + SiO_2(s)$$

Reactions of Halide Ions

Examples of the four most important reactions of halide ions are shown in table 16.10. Displacement reactions and the reactions of halide ions with concentrated sulphuric acid have already been described in the previous section. Halide ions react with lead(II) nitrate or lead ethanoate solutions forming white or yellow precipitates of the lead halide. The reaction between halides and silver nitrate solution forms the basis of both the qualitative and quantitative analysis of halides.

Table 16.10 Four important reactions of the halide ions

Reaction	Example	
1. displacement	$Cl_2(g) + 2I^-(aq)$	$\rightarrow 2Cl^-(aq) + I_2(s)$
2. with concentrated acid	$Cl^-(s) + H_2SO_4(l)$	$\rightarrow HCl(g) + HSO_4^-(s)$
3. with Pb^{2+}	$Pb^{2+}(aq) + 2Cl^-(aq)$	$\rightarrow PbCl_2(s)$
4. with Ag^+	$Ag^+(aq) + Cl^-(aq)$	$\rightarrow AgCl(s)$

The silver halides are unstable in the presence of sunlight. They decompose, forming silver and the halogen. For example

$$2AgI(s) \rightarrow 2Ag(s) + I_2(s)$$

It was this reaction which led to the development of the first photographic plate known as the daguerreotype.

In modern black-and-white photography, silver bromide is used since it is more sensitive to light than is silver iodide

In modern black-and-white photography, silver bromide is preferred since it is more sensitive to light than the iodide. Photographic plates and films are developed using thiosulphate solution. The thiosulphate reacts with the unreacted silver bromide forming the complex ion $[Ag(S_2O_3)_3]^{3+}$.

Silver bromide and silver chloride react with ammonia solution to form the diammine silver(I) ion, $[Ag(NH_3)_2]^+$. This reaction is used to help distinguish between the halide ions.

Halide ions are also ligands in many complex ions. For example, chloride ions remove water from the pink hexaaqua ion of cobalt forming the blue tetrachloro ion:

$$[Co(H_2O)_6]^{2+} + 4Cl^- \rightleftharpoons [CoCl_4]^{2-} + 6H_2O$$

pink hexaaqua   blue tetrachloro
(octahedral)   (tetrahedral)

This reaction is used as a test for the presence of water.

HYDROGEN HALIDES

Bonding and Acidity

All the hydrogen halides are colourless gases at 25°C. Gaseous HCl, HBr and HI all consist of covalent molecules. Hydrogen fluoride molecules associate with one another in both the gaseous and liquid states due to hydrogen bonding. This accounts for its abnormally high melting and boiling points (*see* table 16.11).

Table 16.11 The hydrogen halides

Hydrogen halide	Melting point/ °C	Boiling point/ °C	Bond dissociation energy/kJ mol^{-1}	Acid dissociation constant, K_a
hydrogen fluoride HF	−80	20	562	5.6×10^{-4}
hydrogen chloride HCl	−115	−85	431	1×10^7
hydrogen bromide HBr	−89	−67	366	1×10^9
hydrogen iodide HI	−51	−35	299	1×10^{11}

The hydrogen halides dissolve in organic solvents such as methylbenzene (toluene). The solutions are non-electrolytes and do not turn blue litmus red. All the hydrogen halides fume in moist air and dissolve in water to form hydrohalic acids. The H–F bond is stronger than those of the other hydrogen halides and thus it is only weakly acidic when dissolved in water. The others completely dissociate:

$$HCl(g) + H_2O(l) \rightarrow H_3O^+(aq) + Cl^-(aq)$$

With the exception of hydrogen fluoride, the hydrohalic acids are thus strong electrolytes. The bond strength of the hydrogen halides decreases down the group. The acidity thus increases. This is reflected by the increasing values of the acid dissociation constants, K_a (*see* table 16.11).

Preparation of the Hydrogen Halides

There are three principal methods of preparing the hydrogen halides in the laboratory.

By Direct Combination
Under suitable conditions the hydrogen halides can be prepared by direct synthesis (*see* table 16.5).

By Displacement by a Less Volatile Acid
Hydrogen chloride is more volatile than sulphuric acid (hydrogen sulphate). Thus, concentrated sulphuric acid displaces hydrogen chloride when added to sodium chloride:

$$NaCl(s) + H_2SO_4(l) \rightarrow NaHSO_4(s) + HCl(g)$$

Concentrated sulphuric acid cannot be used for the preparation of hydrogen bromide or hydrogen iodide since it oxidises these to the halogens (*see* below). However, concentrated phosphoric(v) acid can be used since it is relatively non-volatile and a poor oxidising agent:

$$NaBr(s) + H_3PO_4(l) \rightarrow HBr(g) + NaH_2PO_4(s)$$

Hydrogen iodide is prepared in a similar way.

By Hydrolysis of Phosphorus Trihalides
Hydrogen bromide can be prepared in the laboratory by adding bromine to red phosphorus and water. Phosphorus(III) bromide is formed *in situ* and immediately hydrolysed:

$$2P(s) + 3Br_2(l) \rightarrow 2PBr_3(l)$$
$$PBr_3(l) + 3H_2O(l) \rightarrow H_3PO_3(l) + 3HBr(g)$$

Hydrogen iodide is prepared in the laboratory by the parallel synthesis and hydrolysis of phosphorus(III) iodide. In this case water is added to a mixture of red phosphorus and iodine.

Hydrogen chloride may be prepared in a similar manner. However, the reaction between sulphuric acid and sodium chloride is more convenient and almost invariably used for its preparation in the laboratory.

Reduction of the Halide with Hydrogen Sulphide

Aqueous solutions of HBr and HI can be made by passing hydrogen sulphide through a mixture of the halogen and water:

$$Br_2(aq) + H_2S(g) \rightarrow 2HBr(aq) + S(s)$$

Reactions of the Hydrogen Halides

As Acids

Aqueous solutions of the hydrogen halides exhibit typical properties of strong acids:

- They turn blue litmus red.
- They neutralise bases.

$$HCl(aq) + NaOH(aq) \rightarrow NaCl(aq) + H_2O(l)$$

- They react with metals high in the electrochemical series.

$$Mg(s) + 2HCl(aq) \rightarrow MgCl_2(aq) + H_2(g)$$

- They react wth carbonates.

$$CaCO_3(s) + 2HCl(aq) \rightarrow CaCl_2(aq) + CO_2(g) + H_2O(l)$$

Oxidation

The ease of oxidation of the hydrogen halides increases down the group:

$$HCl < HBr < HI$$

Hydrogen chloride is only oxidised by the strongest oxidising agents such as potassium manganate(VII). Hydrogen bromide reduces (and is thus oxidised by) moderately strong oxidising agents such as concentrated sulphuric acid:

$$H_2SO_4(l) + 2HBr(g) \rightarrow SO_2(g) + 2H_2O(l) + Br_2(l)$$

Both hydrobromic acid and hydroiodic acid solutions are unstable in the presence of air. Hydrobromic acid liberates bromine and turns yellow. Hydroiodic acid turns brown due to the formation of iodine.

$$4I^-(aq) + 4H_3O^+(aq) + O_2(g) \rightarrow 2I_2(aq) + 6H_2O(l)$$

Aqua Regia

Aqua regia is a mixture consisting of

  3 parts concentrated hydrochloric acid
  1 part concentrated nitric acid

It was well known to the alchemists for its power to dissolve gold. Hence its name, *aqua regia*, which is the Latin term for 'royal water'. When gold is added to aqua regia, yellow crystals of the acid tetrachloroaurate(III), $HAuCl_4$, are produced:

$$Au(s) + 4H^+(aq) + NO_3^-(aq) + 4Cl^-(aq) \rightarrow [AuCl_4]^-(s) + NO(g) + 2H_2O(l)$$

Hydroiodic acid is a strong reducing agent. It reduces iron(III) ions to iron(II) ions:

$$2Fe^{3+} + 2e^- \rightarrow 2Fe^{2+}$$

With Ammonia

The hydrogen halides form a white smoke with concentrated ammonia.

$$HCl(g) + NH_3(g) \rightarrow NH_4Cl(s)$$

OXOHALOGEN COMPOUNDS

Oxidation States

All the halogens exist in the -1 oxidation state as halides. This is the only stable oxidation state of fluorine. The other halogens exhibit a number of positive oxidation states. These are due to the promotion of electrons from p orbitals into vacant d orbitals (*see* figure 16.2). These d orbitals have relatively low energies and are therefore easily accessible.

Chlorine and bromine both have the following stable oxidation states:

$$+1, +3, +4, +5, +6, +7$$

Iodine exhibits all these except $+4$ and $+6$.

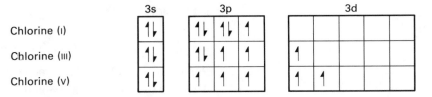

Figure 16.2   The promotion of chlorine's p electrons

Oxoacids of Chlorine

Table 16.2 shows the new and traditional names of the four oxoacids of chlorine and their salts. As the oxidation number of chlorine increases, both the thermal stability and the acid strength of the oxoacids increase:

$$HOCl < HClO_2 < HClO_3 < HClO_4$$

$HClO_3$ and $HClO_4$ are both strong acids. The other two are only partially dissociated in water and exist principally in molecular form. Only $HClO_4$ can be isolated in a pure state. The other oxoacids of chlorine only exist in solution. $HClO_4$ is one of the strongest acids known.

Table 16.12   Oxoacids of chlorine and their salts

Oxidation state	Acid			Anion		
	Formula	Systematic name	Traditional name	Formula	Systematic name	Traditional name
+1	HOCl	chloric(I)	hypochlorous	OCl$^-$	chlorate(I)	hypochlorite
+3	HClO$_2$	chloric(III)	chlorous	ClO$_2^-$	chlorate(III)	chlorite
+5	HClO$_3$	chloric(V)	chloric	ClO$_3^-$	chlorate(V)	hypochlorate
+7	HClO$_4$	chloric(VII)	perchloric	ClO$_4^-$	chlorate(VII)	perchlorate

The oxidising strength of the acids decreases with increasing oxidation number:

$$HOCl > HClO_2 > HClO_3 > HClO_4$$

$HOCl$ and $HClO_2$ are particularly good oxidising agents. For example, in acid solution $HOCl$ oxidises iron(II) ions to iron(III) ions:

$$2Fe^{2+}(aq) + H_3O^+(aq) + HOCl(aq) \rightarrow 2Fe^{3+}(aq) + 2H_2O(l) + Cl^-(aq)$$

In the presence of sunlight it decomposes to form oxygen:

$$2OCl^-(aq) \rightarrow 2Cl^-(aq) + O_2(g)$$

When warmed to about 75°C it disproportionates to chloride and chlorate(v):

$$3OCl^-(aq) \rightarrow 2Cl^-(aq) + ClO_3^-(aq)$$

Oxoacid Salts

The salts of the oxoacids are generally more stable than the acids, although the solid chlorate(III) salts can be dangerous. They detonate when heated and explode when they come into contact with combustible material. In solution, the oxidising strength of oxoacid salts increases with oxidation number. However, they are not as good oxidising agents as the acids.

The sodium and potassium salts of $HOCl$, $HClO_2$ and $HClO_3$ are important commercially. Their manufacture and uses are described in the next section. Potassium chlorate(v) is commonly used in the laboratory preparation of oxygen using manganese(IV) oxide as catalyst:

$$2KClO_3(s) \xrightarrow[\text{MnO}_2]{370°C} 2KCl(s) + 3O_2(g)$$

On heating the salt at lower temperature without a catalyst, potassium chlorate(VII) is formed:

$$4KClO_3(s) \xrightarrow{370°C} 3KClO_4(s) + KCl(s)$$

Potassium iodate(v), KIO_3, and potassium bromate(v) $KBrO_3$, are strong oxidising agents. They are used as oxidants in quantitative analysis.

SUMMARY

1. Across a period:
   (a) the halides become increasingly covalent and less ionic;
   (b) aqueous solutions of the halides become increasingly acidic due to hydrolysis.
2. Down Group VII:
   (a) the halides become increasingly covalent;
   (b) bond strength of the hydrogen halides decreases;
   (c) acidity of the hydrohalic acids decreases;
   (d) ease of oxidation of the hydrogen halides increases.
3. As **oxidation number** increases,
   (a) thermal stability of the oxoacids increases;
   (b) acidity of the oxoacids increases;
   (c) oxidising strength of the oxoacids decreases;
   (d) oxidising strength of the salts of the oxoacids increases.
4. Halides can be prepared by synthesis from their elements.
5. Hydrogen halides can be prepared by displacement by a less volatile acid.
6. Exceptional properties of fluorine compounds:
   (a) silver fluoride is soluble in water whilst calcium fluoride is not;
   (b) hydrogen fluoride has abnormally high melting and boiling points;
   (c) hydrogen fluoride is only weakly acidic when dissolved in water;
   (d) fluorine only exhibits one stable oxidation state.
      Other halogens exhibit a variety of oxidation states due to promotion of p electrons into easily accessible, low energy d orbitals.

16.3
Occurrence, Manufacture and Uses of Halogens and their Compounds

OCCURRENCE

The halogens are too reactive to occur as free elements. They occur as salts both in the Earth's crust and in sea-water. The salts are very soluble, particularly at higher temperatures. They are thus mainly found in those parts of the Earth's crust which have formed at low temperatures. Since the formation of the Earth, vast quantities of these salts have dissolved in ocean water. The average **salinity** of ocean water is about 35‰ (*see* below). Where seas or lakes have become enclosed, evaporation has resulted in high concentrations of salt. This has led to layers of salt depositing on the bottom. Even fresh-water lakes have eventually become salty due to this process. The Great Salt Lake in the USA is an example.

The Great Salt Lake of Utah, USA. A conveyor machine system stockpiles raw salt on the shores of the lake. The principal salts found in the lake are NaCl and Na_2SO_4. Thousands of tonnes of salt have been harvested from the lake without perceptibly reducing the supply. Virtually nothing grows in such a salt-saturated region. Brine flies and shrimps, together with some invertebrates and algae, have the lake to themselves

Salinity
Salinity (‰) is the mass in grams of dissolved inorganic matter in 1 kg of sea-water. Surface salinity of the oceans varies from about 34.0 to 36.0‰. It is highest in tropical regions. In these regions evaporation is relatively high and rainfall relatively low. Lower values are found in equatorial regions due to lighter winds and higher rainfall. Low values are also found near the coasts of continental shelves due to dilution by river water. Land-locked seas have higher salinities. The salinity of the Mediterranean Sea is 37–40‰ and the Red Sea 40–41‰, for example. The Dead Sea, which is the lowest point on the Earth's surface, has a salinity of about 240‰!

Table 16.13 shows the amounts of the halogens in the Earth's crust and in river and ocean water. In sea-water, underground brines and rivers the halogens exist as halide ions. Iodine also exists in water as iodate(v).

Table 16.13   Abundance of halogens on Earth

Number of grams of each halogen per tonne (1000 kg) of	F	Cl	Br	I
Earth's crust	0.625	0.130	0.0025	0.0005
typical river water	0.1	8	0.02	0.005
ocean water	1.4	20 000	68	0.06

Fluorine occurs in bones, teeth and blood

Fluorine is found in all natural waters, bones, teeth, blood and all plants. In the Earth's crust it occurs primarily as fluorspar, CaF_2 (*see* table 16.14). This mineral has a cubic crystal structure known as fluorite (*see* section 3.2). Fluorine also occurs as cryolite. Chlorine occurs mainly as rock salt. This has a cubic crystal structure known as halite. The principal source of bromine is sea-water. Iodine occurs as iodate in Chilean saltpetre and caliche. All four halogens also occur in a variety of other minerals, some of which only exist in relatively small quantities.

Table 16.14   Where do minerals containing halogens occur?

Halogen	Mineral	Formula	Occurrence
fluorine	fluorspar	CaF_2	USA, Europe—including UK—and other parts of the world
	fluorapatite	$CaF_2 \cdot 3Ca_3(PO_4)_2$	USA, Jordan
	cryolite	Na_3AlF_6	mainly Greenland, although deposits occur in the USSR, Spain and the USA
	villiaumite	NaF	minute quantities occur in West Africa and the USA
chlorine	rock salt	$NaCl$	Africa, Australia, Europe—including UK—the USA; in fact almost everywhere!
	carnallite	$KMgCl_3 \cdot 6H_2O$	Germany, Spain and the USA
	sal ammoniac	NH_4Cl	Europe—including UK—and South America
	sylvite	KCl	USA and several countries in Europe and South America
	nantokite	$CuCl$	Australia and Chile
	calomel	$HgCl$	USA and several countries in Europe
	cerargyrite	$AgCl$	Australia, South America and USA
bromine	bromyrite	$AgBr$	Europe, USSR, Australia, USA and Chile
iodine	(iodate(v))	IO_3^-	Chile, in caliche (impure $NaNO_3$) and saltpetre (KNO_3)
	miersite	AgI	Australia
	marshite	CuI	Australia, Chile

Fluorine and bromine exist almost entirely as single isotopes. The isotopic abundances of the halogens are shown in table 16.15.

Table 16.15   Abundances of the naturally occurring halogen isotopes

Halogen	Relative atomic mass	Isotopic abundance		
fluorine	18.99	^{19}F	approx. 100%	
chlorine	35.453	^{35}Cl 75.4%;	^{37}Cl 24.6%	
bromine	79.909	^{79}Br 50.5%	^{81}Br 49.4%	
iodine	126.90	^{127}I	approx. 100%	

Astatine

Twenty isotopes of this radioactive isotope are now known. Traces of the four isotopes ^{215}At, ^{217}At, ^{218}At and ^{219}At exist with naturally occurring isotopes of uranium, thorium and neptunium. ^{210}At is the longest-lived isotope. It has a half-life of 8.3 h. The total amount of astatine in the Earth's crust is thought to be no more than 30 grams. The isotopes $^{209-211}$At have been prepared by bombarding bismuth with α-particles (*see* section 1.3):

$$^{209}_{83}\text{Bi} + {}^{4}_{2}\text{He} \rightarrow {}^{211}_{85}\text{At} + 2{}^{1}_{0}\text{n}$$

Astatine, like iodine, is thought to accumulate in the thyroid gland.

MANUFACTURE OF CHLORINE

Electrolytic Methods

Chlorine is manufactured principally by the electrolysis of sodium chloride using graphite anodes. The reaction at the anode is

$$2\text{Cl}^- \rightarrow \text{Cl}_2(\text{g}) + 2\text{e}^-$$

The principal electrolytic methods are shown in table 16.16. The world production of chlorine is around 20 million tonnes per year. In the UK about 1 million tonnes of chlorine are produced annually. About 75% of this is produced using flowing mercury cells and about 20% using the Down's cell. The USA accounts for about 50% of the world's production of chlorine. Approximately 20% of this is made by the flowing mercury cell and about 80% by the other two methods. The cells are described in detail in section 13.3.

Table 16.16 Electrolytic methods of manufacturing chlorine

Cell	Electrolyte	Cathode	Anode	Other products
Down's	molten NaCl	iron	graphite	sodium
flowing mercury and soda cells (Solvay or Castner–Kellner cells)	concentrated brine	mercury	graphite	hydrogen and sodium hydroxide
diaphragm	concentrated brine	steel	graphite	hydrogen and sodium hydroxide

Deacon Process

This is a method for producing chlorine from waste hydrogen chloride. The hydrogen chloride is oxidised by atmospheric oxygen at 400°C in the presence of copper(II) chloride catalyst:

$$4\text{HCl}(\text{g}) + \text{O}_2(\text{g}) \xrightarrow[\text{CuCl}_2]{400°\text{C}} 2\text{H}_2\text{O}(\text{g}) + 2\text{Cl}_2(\text{g})$$

Storage

After manufacture, chlorine is cooled externally by water and dried using concentrated sulphuric acid. Finally, it is compressed into steel storage tanks.

MANUFACTURE OF THE OTHER HALOGENS

Fluorine

This is manufactured by the electrolysis of a molten mixture of hydrogen fluoride and potassium fluoride. The hydrogen fluoride is obtained from a mineral such as fluorspar, CaF_2, by treatment with concentrated sulphuric acid:

$$2\text{F}^- + \text{H}_2\text{SO}_4(\text{l}) \rightarrow \text{SO}_4^{2-} + 2\text{HF}(\text{g})$$

Bromine

Bromine is extracted from sea-water by treatment with chlorine. There are five basic steps in the process:

Hydrogen fluoride plant control room

1. Sea-water is treated with chlorine. This oxidises the bromide ions to bromine:

$$Cl_2(g) + 2Br^-(aq) \rightarrow Br_2(aq) + 2Cl^-(aq)$$

2. The bromine is expelled by blasting with air.
3. The air is mixed with sulphur dioxide and steam. Hydrogen bromide and sulphuric acid condense out:

$$SO_2(g) + Br_2(g) + 2H_2O(g) \rightarrow 2HBr(aq) + H_2SO_4(aq)$$

4. The acids are scrubbed (removed) from the vapours using fresh water.
5. The acid mixture is then treated with chlorine once again and the bromine expelled by steam:

$$2HBr(aq) + Cl_2(g) \rightarrow 2HCl(aq) + Br_2(aq)$$

How to Choose a Site for a Bromine Plant

These are some of the factors to be considered in choosing a site for a bromine plant:

1. The sea-water should be as rich in bromide as possible.
2. The sea-water should be as warm as possible to assist the 'blowing out' of bromine.
3. The sea-water should be as clean as possible. Sewage, for example, increases chlorine consumption.
4. A low tidal range is required to avoid extreme variation in pumping height.
5. The site should be as close to the sea as possible with good rocky foundations for buildings. A promontory would be ideal since it separates the inlet and outlet sea-water.
6. A fresh water supply is needed. A river is suitable.
7. Good access to raw materials such as chlorine and sulphur is required as well as access to users.
8. Labour should be available.

During the process the addition of acid is necessary to prevent the disproportionation of bromine. The maximum yield is obtained when the pH is below

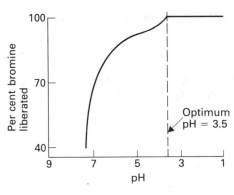

Figure 16.3 The yield of bromine as a function of pH

3.5 (*see* figure 16.3). About 22 000 tonnes of sea-water are needed to yield 1 tonne of bromine.

Iodine

About 50% of the world's iodine is extracted from Chilean nitrates such as caliche, $NaNO_3$. These nitrates contain about 0.2% sodium iodate. To extract the iodine, the sodium nitrate is first crystallised out. The iodate which remains is then reduced by adding sodium hydrogensulphate:

$$IO_3^-(aq) + 3HSO_3^-(aq) \rightarrow I^-(aq) + 3HSO_4^-(aq)$$

The iodine is precipitated by adding more iodate and acidifying:

$$5I^-(aq) + IO_3^-(aq) + 6H^+(aq) \rightarrow 3I_2(s) + 3H_2O(l)$$

The precipitate is filtered, washed and purified by sublimation.

Iodine is also obtained from some sea-weeds which are able to concentrate the element. The sea-weed is dried in the sun and then burnt to ash. This is called **kelp**. The kelp is treated with water and then concentrated until chlorides, carbonates and sulphates crystallise out. The solution is then treated with chlorine to precipitate the iodine.

These aged farmers in Bhutan are harvesting rice. They are affected by goitre. This is a malfunction of the thyroid gland mainly caused by lack of iodine in the diet. Bhutan is one of the least developed countries. It is situated in the northeastern Himalayas between the Tibetan plateau and the Assam–Bengal plains of northeastern India

MANUFACTURE OF CHLORINE COMPOUNDS

Several groups of chlorine compounds are manufactured in large quantities. These include organochlorine compounds (*see* section 19.1), polymers containing chlorine (*see* section 18.3 and 20.4), hydrogen chloride and chlorates.

Hydrogen Chloride

Over 5 million tonnes of hydrogen chloride are manufactured in the world each year. It is manufactured by burning chlorine in hydrogen using charcoal as a catalyst. In the UK the chlorine is obtained mainly by the electrolysis of brine (*see* section 13.3) and the hydrogen by the steam re-forming of natural gas (section 12.1).

Sodium Chlorate(I), NaOCl

This is manufactured by the electrolysis of *cold* brine. The chlorine evolved at the anode is allowed to mix with the sodium hydroxide produced at the cathode:

$$Cl_2(g) + 2OH^-(aq) \rightarrow OCl^-(aq) + Cl^-(aq) + H_2O(l)$$

Sodium Chlorate(v), NaClO$_3$

This is manufactured by the electrolysis of *hot* brine:

$$3Cl_2(g) + 6OH^-(aq) \rightarrow ClO_3^-(aq) + 5Cl^-(aq) + 3H_2O(l)$$

Hydrogen is liberated at the cathode. The overall reaction is

$$NaCl(aq) + 3H_2O(l) \rightarrow NaClO_3(aq) + 3H_2(g)$$

Sodium Chlorate(vII), NaClO$_4$

This is obtained commercially by the electrolysis of sodium chlorate(v):

$$ClO_3^-(aq) + H_2O(l) \rightarrow ClO_4^-(aq) + 2H^+(aq) + 2e^-$$

A line of electrolytic cells at a plant at Thunder Bay, Ontario, Canada, used for the production of sodium chlorates

USES OF THE HALOGENS AND THEIR COMPOUNDS

The halogens and their compounds have many uses. Some of these are listed in table 16.17. One of the most important uses of chlorine is as a bleaching agent. Bleaches are chemical substances which remove unwanted colours by either oxidation or reduction.

Colour in organic materials is due to double covalent bonds called **chromophores**. Oxidising bleaches break up these double bonds:

Chromophore

Table 16.17   Uses of halogens and their compounds

Fluorine	fluorides are used in toothpastes
	uranium hexafluorides are used for
	   separating uranium isotopes
	fluorocarbons such as Freon 12
	   (difluorodichloromethane) are
	   used as refrigerant gases and gas
	   propellants for aerosols
	Teflon (polytetrafluoroethene) is a
	   non-toxic, non-stick plastic
		
		Freon 12
		
		Teflon

Chlorine	manufacture of inorganic chlorides
	   such as $AlCl_3$, $FeCl_3$, PCl_3 and
	   PCl_5
	manufacture of hydrogen chloride
	manufacture of organochlorine
	   compounds (see section 18.3)
	manufacture of chlorates: chlorate(I)
	   is used for bleaching; chlorate(V) is
	   used in explosives, matches and
	   fireworks, weed killers, making
	   oxygen and in throat lozenges!
	organic solvents, e.g. trichloroethene
	   is used as a dry-cleaning agent
	bleaching agents
	sterilisation of water (e.g. in
	   swimming pools)
	manufacture of bromine (see above)
	manufacture of medicinal
	   compounds such as chloral
	   hydrate, a 'sleeping drug', and
	   hexachlorophene, a bacteriocide
	   used in mouthwashes
	recovery of tin and aluminium from
	   scrap
		
		Trichloroethene
		
		Chloral hydrate
		
		Hexachlorophene

Bromine	manufacture of petrol additives photography flame retardants medicine agriculture	

Iodine	dyes and in colour photography
	medicine, e.g. iodoform which is an
	   antiseptic agent; tincture of iodine
	   consists of iodine dissolved in
	   potassium iodide solution with
	   90% ethanol
	electric lighting (iodine vapour bulbs)
		
		Iodoform

Domestic bleach contains sodium chlorate(I)

Reducing bleaches convert the double bonds to single bonds:

Most domestic liquid bleaches contain sodium chlorate(I), NaOCl, whereas bleaching powders more often contain calcium chlorate(I), $Ca(OCl)_2$. These are examples of oxidising bleaches. Hydrogen peroxide and oxygen itself are also oxidising bleaches. Reducing bleaches include gaseous sulphur dioxide and sodium sulphite, Na_2SO_3.

Yellowish fabrics can be whitened by the use of detergents containing brighteners. These are also known as optical bleaches or fluorescent whitening agents. They work by producing blue light. Cotton fabrics but not synthetic fabrics can be bleached in this way. For this reason, optical bleaches are often incorporated into synthetic fibre before it is spun in yarn.

THE OTHER SIDE OF CHLORINE

The towns Seveso (Italy), Love Canal (Niagara Falls, USA) and Times Beach (Missouri, USA) are all famous for having been contaminated by a toxic substance called dioxin (*see* introductory article to this chapter). Dioxin is insoluble in water and clings to the soil where it can remain for up to a century.

Seveso
In Seveso in 1976 an explosion in a chemical plant sent a cloud of dioxin over nearly 4000 acres of the surrounding area. As a result, children were poisoned, babies were born deformed and thousands of animals had to be slaughtered. The land around Seveso was barren for several years.

Times Beach
In Times Beach in the early 1970s salvage oils were sprayed on the roads before they were paved. The oils were the waste from a chemical plant in Missouri which

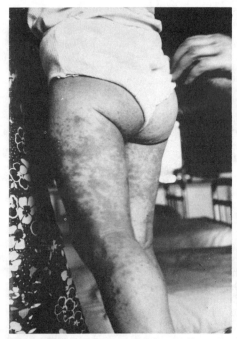

The legs of this Seveso child are affected by a skin rash brought on by exposure to dioxin

Rabbits killed by the leak of dioxin at Seveso

manufactured Agent Orange (*see* introductory article to this chapter). About 2500 residents were evacuated. In 1983 the US Government were still negotiating a plan to compensate the residents with $33 million for the loss of their homes. There are still thought to be at least 100 dioxin-contaminated sites in Missouri alone.

Love Canal

In Love Canal in 1978, after two years of arguing, residents had their homes declared uninhabitable—the reason: dioxin contamination.

Pesticides and Herbicides

There can be no doubt of the social and economic value of pesticides and herbicides. They have been responsible for eliminating pests which carry disease and also for vastly improving food production. The United Nations Food and Agriculture Organisation (FAO) has estimated that about 35 million tonnes of food or fodder products are destroyed each year by rats, insects, predators and fungi. In some of the Developing World, food lost in storage can run as high as 40%.

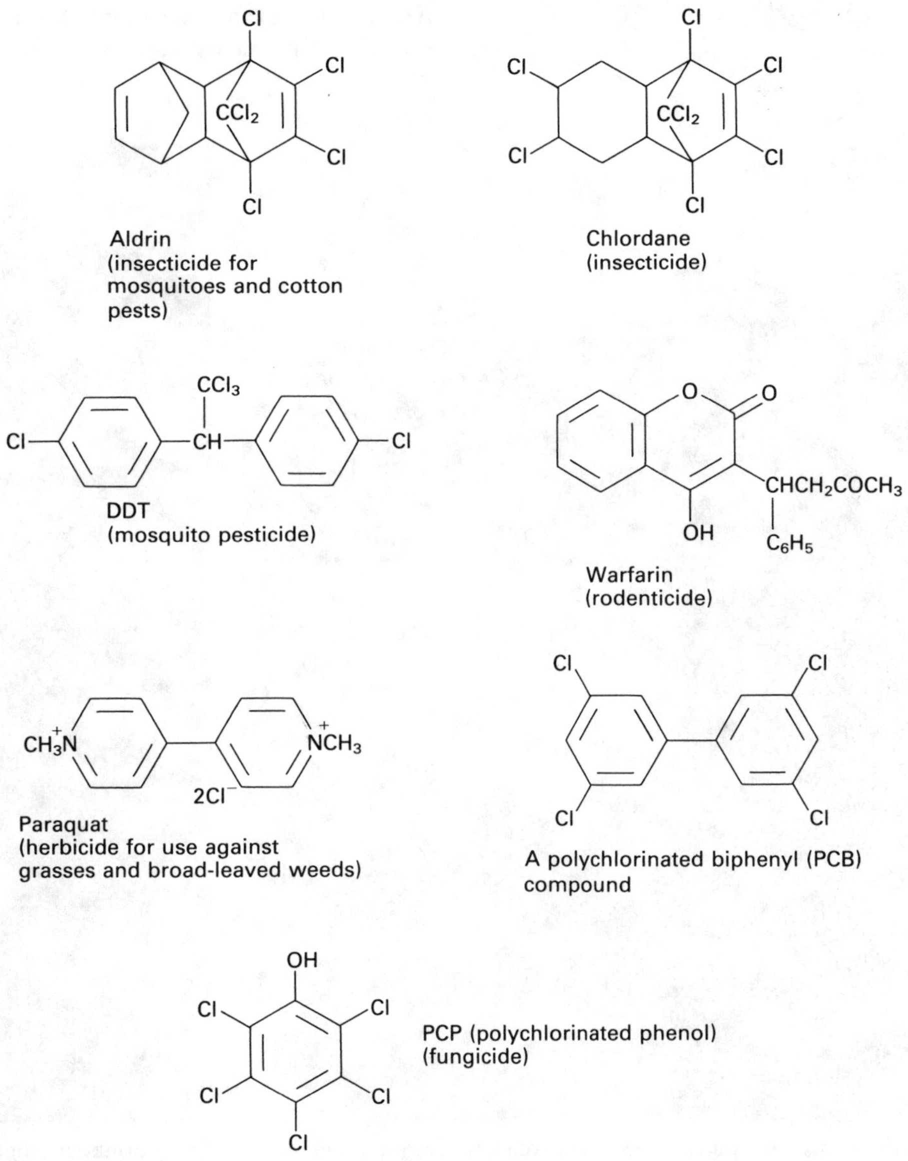

Aldrin
(insecticide for
mosquitoes and cotton
pests)

Chlordane
(insecticide)

DDT
(mosquito pesticide)

Warfarin
(rodenticide)

Paraquat
(herbicide for use against
grasses and broad-leaved weeds)

A polychlorinated biphenyl (PCB)
compound

PCP (polychlorinated phenol)
(fungicide)

Figure 16.4   (In)famous pesticides and herbicides

Chemical defences against such pestilence include attacks on the breeding grounds of mosquitoes and tsetse fly, the locust patrols of the Middle East and Africa and the use of selective weed killers. However, one of the problems associated with the use of organochlorine compounds is the impact on the ecological balance. The destruction of insect pests can lead to breaks in food chains, with the result that one pest may be replaced by another. Organochlorine compounds such as DDT, which is used to attack malaria-carrying mosquitoes, leave a residual toxicity in birds and fish. Even compounds which do not leave a residual toxicity can be dangerous. Paraquat is an example. Paraquat is not an organochlorine compound but a dimethylbipyridinium salt (*see* figure 16.4). In 1981, Thailand imported 4 million kilograms of paraquat to kill weeds in rubber estates and rice fields. High levels of paraquat and other pesticides and herbicides have been found in rivers, streams and canals in Thailand. The result has been called the country's worst man-made ecological disaster. Millions of fresh-water fish have been killed—and fish is the main source of protein for rural Thais.

There is another problem associated with the use of pesticides and herbicides. It is mishandling. Unfortunately—but as usual—it is the people of the Developing World who suffer most. In 1972, for example, the nations of the Developing World suffered half of the reported 250 000 cases of pesticide poisoning, including 6700 deaths. Yet they accounted for less than 15% of the world's consumption of pesticides. Chemicals responsible for such poisonings include paraquat, DDT,

Open drains and ponds suspected of containing mosquito larvae are sprayed with insecticide. The number of cases of malaria rose to over eight million by the mid 1970s in India

Clean drinking water is essential for good health. Drinking water from contaminated containers has resulted in many pesticide poisonings in the Developing World

PCP and aldrin. Chemicals like DDT and aldrin are still widely used in the Developing World although their use in Europe and North America has been severely restricted. In the Developing World, farm workers who use these toxic chemicals are often illiterate and have no training in their use and storage. They cannot read the instructions on the labels and quite often they or their wives and children use pesticide and herbicide containers as drinking or cooking utensils or for carrying water.

In 1976 about 30% of the pesticides exported by the United States were prohibited for use in the United States. In 1982 there were reports that British chemical companies were supplying African countries such as Kenya with aldrin. Aldrin's suspected health hazards include cancer, damage to the foetus and nervous disorders.

Even so, there are people who argue that such dangers are a small price to pay for the improved levels of food production in a world where over 400 million people are actually starving and another 400 million barely surviving.

SUMMARY

1. The main **sources**, methods of **manufacture** and **uses** are summarised in table 16.18.

   Table 16.18 Main sources, methods of manufacture and uses of the halogens and their compounds

	Main source	Principal method of manufacture	Main uses
fluorine	fluorspar, CaF_2	electrolysis of molten HF/KF	refrigerants, aerosols, extraction of uranium isotopes
chlorine	rock salt, NaCl	electrolysis of NaCl	manufacture of HCl, organic solvents and disinfectants
bromine	sea-water	extraction from sea-water	petrol additives and photography
iodine	Chilean nitrates	extraction from nitrates	medicines, dyes and photography
HCl		burning chlorine in hydrogen $H_2(g) + Cl_2(g) \rightarrow 2HCl(g)$	treating metal surfaces, making plastics and chlorine
chlorates		electrolysis of brine	bleaches, explosives

2. The use of **pesticides** and **herbicides** has both advantages and dangers.
   Advantages:
   reduction of disease;
   reduction of food wastage;
   improved levels of food production.
   Dangers:
   accidental explosions during manufacture;
   contamination from toxic waste from manufacture;
   pollution of environment, including long-term residual toxicity in animal food chains;
   disturbance of ecological balance leading to the replacement of one pest by another;
   mishandling often due to the lack of knowledge and training of users.

16.4
The Noble Gases

The noble gases occur in group 0 of the Periodic Table (*see* figure 16.5). Apart from helium, the noble gases have eight electrons in their outer electron shells. For this reason, the group number is often written as VIII. The noble gases are chemically unreactive and are thus often called inert gases. They are also known as rare gases.

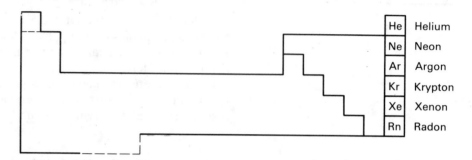

He	Helium
Ne	Neon
Ar	Argon
Kr	Krypton
Xe	Xenon
Rn	Radon

Figure 16.5 The noble gases

The elements are all monatomic gases at room temperature. There are no interactions between the atoms apart from weak van der Waals forces. The atomic structures and physical properties of the gases are shown in table 16.19. Melting points, boiling points and density all increase down the group. Indeed, helium has the lowest boiling point of any known substance.

Table 16.19 The atomic structures and physical properties of the noble gases

Element	Atomic number	Electron configuration in outer shell	Atomic radius/nm	First ionisation energy/kJ mol^{-1}	M.p./°C	B.p./°C	Density/g dm^{-3}
helium	2	$1s^2$	0.093	2370	−272	−269	0.18
neon	10	$2s^22p^6$	0.112	2080	−285	−246	0.90
argon	18	$3s^23p^6$	0.154	1520	−189	−186	1.78
krypton	36	$4s^24p^6$	0.169	1350	−157	−153	3.74
xenon	54	$5s^25p^6$	0.190	1170	−112	−108	5.89
radon	86	$6s^26p^6$	–	–	−72	−62	9.73

NOBLE-GAS COMPOUNDS

The most common noble-gas compounds are the fluorides of xenon.

Xenon(II) Fluoride

This compound can be prepared by direct combination of xenon and fluorine using light from a mercury arc. It can also be prepared by treating xenon with oxygen monofluoride at −120°C:

$$Xe(s) + F_2O_2(g) \rightarrow XeF_2(s) + O_2(g)$$

This compound is a linear molecule. It hydrolyses in the presence of alkali to form oxygen:

$$2XeF_2(s) + 4OH^-(aq) \rightarrow 2Xe(g) + 4F^-(aq) + 2H_2O(l) + O_2(g)$$

Krypton(II) fluoride, KrF_2, and radon(II) fluoride, RnF_2, are also known.

Xenon(IV) Fluoride

This compound is prepared by direct combination of xenon and fluorine at 400°C:

$$Xe(g) + 2F_2(g) \rightarrow XeF_4(s)$$

The Remarkable Superfluid Properties of Liquid Helium-4

Helium exists as two isotopes, helium-4 and the rare helium-3. Below 2.2 K, liquid helium-4 exists as a mixture of two fluids. One is an ordinary fluid component and the other a superfluid component. This fluid has such a low viscosity that it is almost frictionless. It ignores the rotation of the Earth and will even flow uphill. It will escape from a non-porous vessel by climbing out over the edge. This is called the creeping film effect (*see* figure 16.6). Helium is a superconductor of heat to such an extent that it is very difficult to set up temperature differences in the liquid. It does not boil and is the only liquid which cannot be solidified by reducing the temperature. A pressure of 26 atmospheres is needed to solidify it. Superfluid helium will pass through the fine pores of an unglazed earthenware jug (*see* figure 16.6).

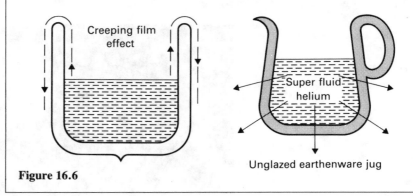

Figure 16.6

Xenon(IV) fluoride has a square planar structure:

The compound disproportionates in water to form xenon(VI) oxide and xenon:

$$6XeF_4(s) + 12H_2O(l) \rightarrow 2XeO_3(aq) + 24HF(aq) + 4Xe(g) + 3O_2(g)$$

Xenon(VI) Fluoride

This compound is prepared by direct combination at 300°C under pressure. It has a distorted octahedral structure. The compound reacts with silica to form tetra-fluoroxenon(VI) oxide:

$$SiO_2(s) + 2XeF_6(l) \rightarrow SiF_4(g) + 2XeOF_4(l)$$

Argon, krypton and xenon all form **clathrate compounds**. These are also known as cage or enclosure compounds. Xenon hydrate, $Xe \cdot 6H_2O$, for example, consists of xenon atoms trapped in a network of water molecules. The water molecules are held together by hydrogen bonds. When quinol is crystallised from water in xenon, krypton or argon atmospheres under pressure, noble-gas clathrate compounds with quinol are formed.

Table 16.20 Occurrence of noble gases in the atmosphere

	Percentage in atmosphere
helium	0.00052
neon	0.0018
argon	0.93
krypton	0.0011
xenon	0.0000087

OCCURRENCE, EXTRACTION AND USES

Neon, argon, krypton and xenon occur only in the atmosphere (*see* table 16.20).

Helium occurs not only in the atmosphere but also in natural gas deposits in the United States and elsewhere. It is second only to hydrogen in abundance in the Universe. Radon exists only in trace quantities in the atmosphere. It is radioactive.

Radon-222, its most common isotope, has a half-life of 3.823 days. This isotope is formed by the α-decay of radium:

$$^{226}_{88}\text{Ra} \rightarrow {}^{222}_{86}\text{Rn} + {}^{4}_{2}\text{He}$$

It is thought that every square mile of soil on Earth to a depth of six inches contains about one gram of radium.

The noble gases are extracted from liquid air by fractional distillation followed by selective absorption on charcoal.

USES

Helium has a density twice that of hydrogen. Even so, because it is far safer than hydrogen, it is used to inflate aircraft tyres and balloons and also used in space rockets.

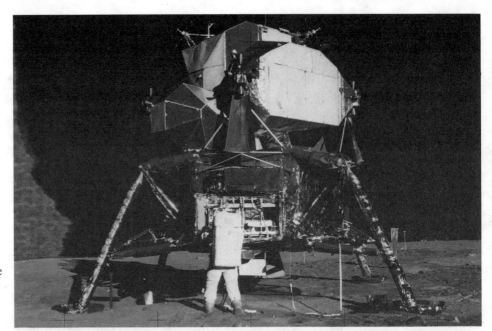

Helium is used on a large scale to pressurise liquid fuel rockets. About 13 million cubic feet of helium were required to fire each Saturn booster for the Apollo lunar missions

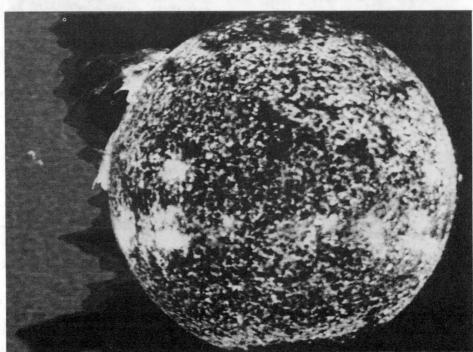

This huge solar eruption of helium spans more than 588 000 kilometres (365 000 miles)

A mixture consisting of 80% helium and 20% oxygen is used as an artificial atmosphere by divers. It is used in preference to nitrogen because it is less soluble than nitrogen in blood. Nitrogen can cause decompression sickness known as the 'bends'. Both helium and argon are used to provide inert gas shields for welding. Helium is also used as a protective atmosphere for growing crystals of germanium and silicon.

Argon is used to fill electric light bulbs and various types of fluorescent and photo tubes.

Neon is used extensively in advertising signs and also in television tubes. It is also increasingly being used as a refrigerant.

Krypton is used to fill fluorescent lights and photographic flash lamps.

Radon is used in the treatment of cancer.

Xenon is used for making electron tubes and stroboscopic lamps. It is also used for bubble chambers in atomic energy reactors.

SUMMARY

1. **Noble gases** are monatomic and unreactive.
2. Helium has exceptional properties. It has a very low boiling point and virtually no viscosity in its liquid state.
3. The most common noble-gas compounds are the fluorides of xenon, XeF_2, XeF_4 and XeF_6. XeF_2 is linear and XeF_4 has a square planar structure.
4. The most abundant noble gas in the atmosphere is argon. Helium also occurs in underground deposits.
5. The noble gases are extracted from air by fractional distillation.
6. Helium is used by the aircraft and space industries. It is also used by divers. Helium and argon are used for arc welding. The other noble gases are principally used in various forms of lighting.

Examination Questions

1. Write an essay on *either* A *or* B.

    A *The halogens and their compounds.* Reference should be made to their natural occurrence, electronic structure and the preparation, properties, reactions and uses of the elements and their important compounds. In the cases of preparations and reactions, equations should be given.

    B *The noble gases.* Reference should be made to their natural occurrence, history of discovery, isolation, uses, compounds and electronic structure and importance in relation to the Periodic Table and the theory of chemical bonding.

    (NISEC)

2. (a) Discuss trends in the chemical properties of the halogens F, Cl, Br, I by considering their reactions with metals, hydrogen and aqueous sodium hydroxide.
    (b) Give two characteristic properties of (i) metallic oxides, (ii) non-metallic oxides, (iii) amphoteric oxides, (iv) non-metallic chlorides.

    (OLE)

3. (a) Which of the halogens is the most powerful oxidising agent?
    (b) Indicate, by stating the necessary reagents and observations, how you

could distinguish between sodium bromide and sodium iodide using a simple chemical test.

(c) State the bond type in SF_6. State, and account for, the shape of this molecule.

(JMB)

4. On mixing a solution of silver nitrate with dilute hydrochloric acid a white precipitate is immediately formed.
   (a) Write an ionic equation for the reaction.
   (b) Why is this a fast reaction?
   (c) The precipitate darkens rapidly. Why?

(SEB)

5. Compare the chemical and physical properties of the two compounds in any **three** of the following pairs of chlorides, explaining the role of the electronic structures of the elements involved, where appropriate:
   (a) CCl_4 and $SnCl_4$,
   (b) $AlCl_3$ and PCl_3,
   (c) $CaCl_2$ and $FeCl_2$,
   (d) HCl and NaCl.

(O & C)

6. (a) State what you would observe, name the gaseous product(s) and write equations for the reactions which occur between hot concentrated sulphuric acid and (i) sodium fluoride and (ii) sodium bromide.
   (b) State what you would observe when a dilute aqueous solution of silver nitrate is added to dilute aqueous solutions of (i) sodium fluoride and (ii) sodium bromide.
   (c) Explain why in dilute aqueous solution hydrogen fluoride is a weaker acid than hydrogen chloride.

(JMB)

7. (a) Outline an industrial method for the manufacture of chlorine.
   (b) The oxidising power of chlorine can be illustrated by its behaviour towards iron, ethane, potassium bromide and tin(II) chloride.
      For **each** example:
      (i) Write a suitable balanced equation for the redox reaction.
      (ii) Give the conditions required for each reaction.
      (iii) Explain the different interpretations of the term oxidation as applied to the above reactions.
   (c) Explain why:
      (i) silver chloride is insoluble in water whilst silver fluoride is soluble.
      (ii) aqueous iron(III) chloride liberates carbon dioxide from sodium carbonate.
      (iii) tetrachloromethane (carbon tetrachloride) behaves differently to silicon tetrachloride in its behaviour towards water.

(AEB, 1984)

8. The following is a student's account of the extraction of bromine from magnesium bromide in sea-water.
   'The sea-water is concentrated and the calculated quantity of chlorine (as chlorine water) is added. The chlorine is obtained by electrolysing salt. An organic solvent is added to the aqueous mixture and the bromine extracted.

Finally, the bromine is separated from the organic solvent and then stored.'

(a) Briefly decribe, without details of plant, how chlorine is obtained by electrolysing salt. Write the relevant electrode equation.

(b) Write an *ionic* equation for the reaction of chlorine with magnesium bromide in aqueous solution.

(c) What is meant by 'the calculated quantity of chlorine'? Why must excess of chlorine be avoided?

(d) Name an organic solvent which could have been used.

(e) Describe, in some detail, the *practical* procedure for the extraction of bromine from the aqueous mixture.

(f) Explain why bromine dissolves preferentially in the organic layer.

(g) How would bromine be removed from the solvent in the final stage?

(h) Compare and contrast the reactions of solid sodium chloride, bromide, and iodide with concentrated sulphuric acid.

In what way would these reactions be different if phosphoric acid were used? Explain.

(SUJB)

9. *Laminaria* sea-weed concentrates iodide ions derived from sea-water. The iodine can be estimated:

'The sea-weed is burned in a limited supply of air. The ash is boiled with water, filtered, and the filtrate treated with an excess of acidified hydrogen peroxide solution. The liberated iodine is extracted in trichloromethane, the appropriate layer separated and titrated against standard sodium thio-sulphate solution ($Na_2S_2O_3$).'

(a) Why is the sea-weed burned to ash?

(b) In what form is the iodine liberated by boiling with water?

(c) Explain the *type* of change undergone by iodide ions in the reaction with hydrogen peroxide. Write the reaction equation.

(d) Why is it necessary to use an organic extraction rather than titrate the liberated iodine directly?

(e) In which layer will the iodine concentrate during the extraction?

(f) What will be the colour of the iodine in this layer?

(g) Where are the principal sources of iodine loss in the entire method of extraction?

(h) 1 kg of sea-weed produced sufficient iodine to react with 100.0 cm³ of 0.100 mol dm⁻³ sodium thiosulphate solution.

Calculate the mass of iodine extracted from 1 kg of sea-weed.

(i) $^{131}_{53}I$ is used as a radioactive tracer in the diagnosis of thyroid gland malfunction. Explain the significance of the numbers quoted on the iodine symbol.

(SUJB)

10. This question concerns compounds of the element xenon.

(a) Xenon and fluorine react to form at least three different fluorides. After separation and purification, a sample of one of these compounds was analysed:

Mass of xenon fluoride = 0.490 g

Volume of xenon obtained on removing the fluorine from the sample (measured at 25°C and 1 atm) = 48 cm³

Calculate the empirical formula for this fluoride of xenon.

(Relative atomic masses: Xe = 131, F = 19. 1 mole of gas at 25°C and 1 atm occupies 24 dm³)

(b) Calculate the standard heat of formation of the fluoride $XeF_4(g)$ from the following bond energy terms, and comment on its likely stability.

$$E(Xe\text{–}F) = 130 \text{ kJ mol}^{-1} \qquad E(F\text{–}F) = 158 \text{ kJ mol}^{-1}$$

(Note: You may wish to construct an energy cycle diagram and use it to answer this question.)

(c) The oxide XeO_3 is produced by the hydrolysis of the fluoride XeF_6. Write an equation for this reaction.

(L)

17 FUNDAMENTALS OF ORGANIC CHEMISTRY

The Poor World Needs Chemists

Developing countries will remain poor unless they can exploit natural resources to their own advantage. To do this requires indigenous skills in chemistry.

Four-fifths of the world's scientists and technologists work in Europe, the USSR and North America. Roughly four-fifths of the world's people live elsewhere. According to the Science Policy Research Unit at Sussex University, less than 1% of research carried out in the developed countries has any significance for the developing world, and half that research effort is devoted to military and related activities. In order to redress this global imbalance in scientific expertise, Third World countries need to build up their own capabilities in science, technology and especially in chemistry. The chemist plays three essential roles in the Third World: the productive role, the protective role and the problem-solving role.

Third World countries are poor by many criteria. They are poor in terms of their provision of health and education services, the average calorific intake of their people, their average per capita incomes, and their gross national products. They are also poor in terms of their production and consumption of chemicals. The United Nations Industrial Development Organisation estimates that from 1960 to 1970 developing countries increased their share of the world's chemical production from 4.7% to 5.2%. In the same period, developing countries increased their share of world consumption of chemicals from 7.2% to 7.6%. The average per capita consumption for all developing countries was about an eighth of the world average. Yet many of these countries are rich in organic and mineral resources. For example, 80% of our tin and 75% of our bauxite come from the Third World.

Much of the chemical knowledge, expertise and resources needed to solve problems in the Third World already exists. The problem is to bring resources to focus on these problems, and to adapt knowledge to local conditions, constraints and opportunities. Chemistry that works in cold climates does not always work in hot climates.

One of the greatest challenges is medicine. 'Tropical diseases' affect more than 800 million people, almost one-quarter of the world's population. More than 300 million people suffer from the parasitic infection filariasis. Cholera, yellow fever, sleeping sickness leishmaniasis and leprosy affects tens of millions of people in tropical regions. Yet there is often no effective therapy for these diseases.

Schistosomiasis, a debilitating parasitic disease, affects some 400 million people in Africa, Asia and Latin America. The cause is a blood fluke, the *Schistosoma mansoni*, which depends on an aquatic snail as an intermediate host. The parasite multiplies inside the snail, eventually emerging as fork-tailed larvae. These invade the human body through the skin. Scientists have attacked the problem from

Research on insects in a laboratory in Nigeria. The aim is to assess the ability of the insects to spread diseases

every angle, by education, improving hygiene and eradicating the snail. Now Otto Gottlieb, Professor of Chemistry at the University of Sao Paulo, and Walter Mors, a Brazilian specialist in the chemistry of natural products, have drawn attention to a possible new approach: chemicals to stop larvae penetrating the skin. One candidate, which occurs in the heartwood of a number of tropical trees, is lapachol and its derivatives.

These compounds not only prevent schistosomiasis, but also prevent abnormal cell division. The Brazilians are producing lapachol for oral administration as part of a drug therapy for cancer, and the authorities have approved it for clinical trials in humans. Recent laboratory tests have shown that lapachol might also inhibit *Trypanosoma cruzi*, the protozoan that causes Chagas' disease. This form of sleeping sickness affects millions of Brazilians.

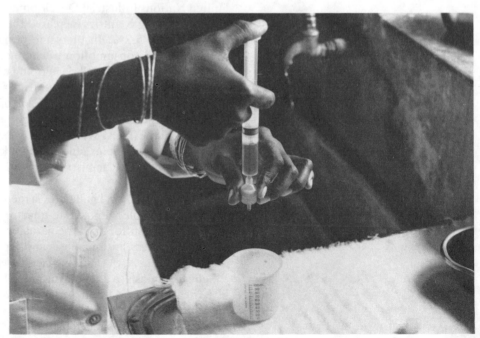

Filtration of a urine sample in a test for schistosomiasis at a school in Zanzibar

Malaria surveillance workers take blood samples from suspected cases in a village in India

Chinese scientists are doing considerable work on medicinal plants, particularly for treating malaria. They already claim a breakthrough comparable with the development of chloroquine, the standard drug for treating malaria. The new drug is *Ching Hao Su*. It is extracted from wormwood, which according to ancient Chinese records was first used to treat malaria more than 1000 years ago. Scientists at the Institute of Chinese Materia Medica, attached to the Academy of Traditional Chinese Medicine, first extracted the drug in 1972. Since then, the Chinese have done an immense amount of work, discovering that its crystalline chemical structure differs from those of other antimalarial drugs.

African researchers have also been investigating extracts from medicinal plants. In Nigeria, Donald Ekong, Professor of Chemistry at the University of Ibadan, has investigated the fruit of *Xylopia aethiopica*. The dried fruits of this plant are used in many parts of Africa in traditional obstetrics and folk medicine. Ekong analysed the fruits and found a new compound, xylopic acid. The publication of the work prompted researchers in Ghana to study the substance's antimicrobial properties. They found it was very active against two bacteria, *Staphylococcus aureus* and *Bacillus subtilis*, and the yeast *Candida albicans* which is a common agent of vaginal infections.

The world's food problem is as big as its medical one. The United Nations Food and Agriculture Organisation estimates that around 450 million people are chronically undernourished. The World Bank, using different criteria, puts the figure at over 1000 million. The very size of the problem effectively excludes imported solutions, so the only hope must lie within the borders of the countries that suffer the most.

The developing world has no choice but to build a self-reliant system of food production. But the improvements that have so far been effective—the so-called Green Revolution programmes—have been highly energy-intensive because of their dependence on irrigation, fertilisers and mechanisation. Such solutions are too expensive for many poor countries. Research must now turn towards farming strategies that conserve energy.

New areas of research include improving plants' photosynthetic efficiency, increasing their uptake of nutrients and water, and reducing farmers' dependence on synthetic fertilisers and chemical pesticides.

One target for research is the cowpea *Vigna unguiculat* L. *Walp*, an important source of plant protein in the tropics. The yield of the plant is low because the

Modern industry in Morocco through the eye of a camel!

Swine fever isolation unit of the East African Community veterinary research laboratories in Nairobi, Kenya

Liquid organic fertiliser being applied to a vegetable plot in China

flower buds and immature fruit frequently drop off. At the University of Ibadan, chemists have analysed the growth hormones in cowpeas to determine the role they play in this abscission (separation). They have shown that immature fruits possess only growth inhibitors—abscissic acid and its metabolites—whereas more mature fruits contain significant amounts of growth agents called gibberellins.

At the University of Ghana, Dr Samuel Sefa-Dedeh has been looking at ways of increasing the productivity of grain legumes in the tropics. One problem with these vegetables is that they need a lot of cooking to make them digestible and palatable. Sefa-Dedeh is looking at the hydration properties of legumes with the idea of developing a variety that takes less time. Examinations with a scanning electron microscope show that cowpeas that have been stored for a long time readily lose protein when soaked, whereas samples stored without soaking do not lose protein. Sefa-Dedeh is now developing a simple process to prepare a high protein food from cereals and legumes.

One interesting potential source of energy in the Third World is the water hyacinth and other aquatic weeds. The weeds are a problem in many countries, including Bangladesh, India and Malaysia. They block waterways and hinder fishing and irrigation. The main culprit is *Eichhorniacrassipes*, otherwise known as the water hyacinth. In India, up to a quarter of total cultivable waters are infested with the weed. In some states, infestation reaches 40%. The increasing use of manures and fertilisers in farmland means that the problem is getting worse.

Past attempts to control the weed have involved mechanical harvesters, weed-eating fish and chemical herbicides. None has been a success. Now researchers are starting to look on the hyacinth as a resource. The plant's high ash content has led to its use for manure, and its conversion into compost. It has also been examined as a potential animal feed. However, the Central Food Technological Research Institute in Mysore has shown that, although the hyacinth is relatively rich in crude protein, protein extracted from it contains large quantities of unfavourable minerals. Apart from protein, chemists have extracted carotenes, vitamin A and hormones from the hyacinth. They have also found growth-regulating substances, which could accelerate the growth of crops while inhibiting the roots. Scientists at the Central Mechanical Engineering Research Institute in Durgapur, India, have found that hyacinth containing 60–70% moisture ferments quickly. Under optimum conditions fermentation starts within two to three days, and gives off usable quantities of gas after 15 to 20 days. The compost left has a high manure value.

In the final analysis, the Third World will never solve its problems without developing its greatest resource: people. Any country that wants to develop its industrial and agricultural production and the services it provides to its people must develop education in science and technology. Generally speaking, self-sufficiency in scientific and technological expertise is a characteristic of all industrial countries, large and small. Long-term and self-sustaining development thus requires a country to become the master of its own science and technology, and not the servant of other people's.

New Scientist

LEARNING OBJECTIVES

After you have studied this chapter you should be able to

1. Draw the structures of the name typical
   - (a) **aliphatic, alicyclic, aromatic** and **heterocyclic** compounds,
   - (b) **alkanes, alkenes** and **alkynes**,
   - (c) **saturated** and **unsaturated hydrocarbons**,
   - (d) **primary, secondary** and **tertiary amines** and **alcohols**.
2. Write the **systematic names** of simple organic compounds given their **structural formulae**.
3. Draw the structural formulae of simple organic compounds given their systematic names.
4. Define and give an example of **structural isomerism**.
5. Name and write the structures of the more important organic **functional groups** and simple compounds that contain these groups.
6. Explain the term **homologous series**.
7. Write the names, molecular formulae and structural formulae of the first six members of the homologous series of
   - (a) alkanes,
   - (b) alcohols,
   - (c) amines.
8. Describe the general physical properties of organic compounds, and explain these in terms of bonding.
9. Outline the main methods of preparing, identifying and determining the structures of organic compounds.
10. Describe typical bonding and molecular structure of carbon compounds.
11. Explain the term **conformation** and give an example.
12. Distinguish between **structural isomers** and **stereoisomers**.
13. Give examples of **geometrical isomerism** and **optical isomerism**.
14. Explain the term **optical activity**.
15. Outline the main features of a **polarimeter** and its operation.
16. Distinguish between **homolytic** and heterolytic fission.
17. Distinguish between the terms **carbanion** and **carbocation**.
18. Summarise the factors affecting the reactivity of organic compounds.
19. Write a short account on each of the following:
    - (a) **inductive effect**,
    - (b) **resonance effects**,
    - (c) **conjugated system**,
    - (d) **steric hindrance**.
20. Define and give examples of
    - (a) **nucleophiles**,
    - (b) **electrophiles**.
21. Give an example of *nucleophilic substitution*.
22. Explain the terms S_N1 **reaction** and S_N2 **reaction**.
23. Give an example of
    - (a) **electrophilic substitution**,
    - (b) **electrophilic addition**,
    - (c) an **elimination** reaction.

17.1
Organic Chemistry

WHAT IS ORGANIC CHEMISTRY?

Organic chemistry is the chemistry of compounds which contain the element carbon. Most of these compounds also contain hydrogen and many also contain oxygen or other elements. There are a few compounds containing carbon, however, which are not normally classified as **organic compounds**. Carbon monoxide, carbon dioxide and the metal carbonates are examples.

Although organic substances such as sugars, alcohol and vinegar had been known for thousands of years, it was not until the eighteenth century that organic compounds were first isolated. One of the first scientists to study organic compounds was the self-taught chemist, Carl Wilhelm Scheele (1742–1786). He obtained and purified a number of organic acids and other organic compounds from plant and animal sources. He isolated 2-hydroxypropanoic acid (commonly known as lactic acid) from milk, for instance, and showed that this acid was the cause of turning milk sour. Scheele was also the first person to prepare chlorine and a number of other inorganic elements and compounds.

During the eighteenth century chemists believed that organic compounds could only be synthesised by means of a 'life force' in living cells. This was called the vitalistic theory of organic chemistry. However, in 1828, the German chemist Friedrich Wohler (1800–1882) prepared urea (also known as carbamide) by heating an aqueous solution of ammonium cyanate:

$$NH_4CNO \xrightarrow{\text{heat}} CO(NH_2)_2$$

ammonium   urea
cyanate   (carbamide)

This was the first synthesis of an organic compound. It heralded the decline of the vitalistic theory. The term 'biochemistry' is now used for the chemistry of living things and life processes.

Why is Organic Chemistry Important?

In this and the following chapters we shall discover numerous instances of the importance of organic chemistry. All living things contain organic compounds.

> Note that state symbols are usually omitted in chemical equations representing organic reactions. State symbols are more important in physical and inorganic chemistry where reactions often involve ions and aqueous solutions. For example, it is important to distinguish between the electrolyte NaCl(aq) and the non-electrolyte NaCl(s). Organic reactions, on the other hand, often take place in non-aqueous solvents.

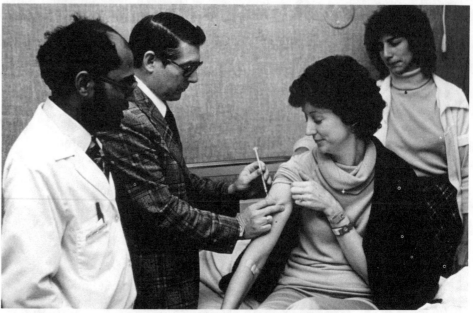

Many modern medicines and drugs are synthetic organic compounds

Aerial spraying of insecticide onto cotton in Southern Turkey

Table 17.1 Some organic materials

Naturally occurring	Synthetic
carbohydrates	plastics
proteins	many medicines and drugs
fats and oils	insecticides
vitamins	many dyes

Table 17.2 Growth in known organic compounds

Year	Number of known compounds
1880	12 000
1910	150 000
1940	500 000
1960	1 000 000
1970	2 000 000
1980	5 500 000

Furthermore, many of the modern products and materials upon which we depend are organic (*see* table 17.1).

A measure of the importance of organic chemistry nowadays can be gauged from the almost exponential growth in the number of known organic compounds over the last century (*see* table 17.2).

Why is it that carbon can form such a vast array of naturally occurring and synthetic compounds? The answer lies in its unique ability to catenate and thus form chain and ring structures. It also has the ability to form single, double and triple covalent bonds not only with itself but also with other elements. This unique nature of carbon is discussed in section 15.2.

NOMENCLATURE

We pointed out at the beginning of chapter 4 that, with the discovery of so many new chemical compounds, an unambiguous system of nomenclature is absolutely essential. The full description of such a system (or systems) of nomenclature requires several volumes. In this section we shall thus only consider the basic and bare essentials of organic nomenclature.

Classes of Organic Compounds

Organic compounds are sometimes classified by the structure of their carbon skeleton. There are four classes:

Aliphatic compounds. These are compounds with open chains of carbon atoms. The open chains may be unbranched or branched and may contain single, double or triple bonds or combinations of these.

Alicyclic compounds. These are organic compounds with closed rings of carbon atoms. The rings may contain single or multiple bonds.

Aromatic compounds. These are compounds containing at least one benzene ring. This is a ring of six carbon atoms in which electrons in p orbitals are delocalised, forming a π-electron cloud (*see* figure 2.8).

Heterocyclic compounds. These are compounds based on a closed ring made up of atoms of carbon and one or more other elements. Heterocyclic compounds typically contain nitrogen, oxygen or sulphur atoms in their rings. They may form single or multiple bonds with the other atoms in the ring.

Examples of each class of compound are shown in figure 17.1. Note that the use of fully displayed structural formulae can be cumbersome, particularly for ring

Authoritative sources of chemical nomenclature include:

'IUPAC Glossary of Terms Used in Physical Chemistry' (Recommendations 1982). *Pure and Applied Chemistry*, 1983, **55**, 1281–371.

'IUPAC Nomenclature of Inorganic Chemistry' (the 'Red Book'), and 'How to Name an Inorganic Substance', Pergamon Press, 1977.

'IUPAC Nomenclature of Organic Chemistry' (the 'Blue Book'), Pergamon Press, 1979.

'IUPAC Compendium of Macromolecular Nomenclature' (the 'Purple Book'), Blackwell Scientific Publications, 1987.

'IUPAC Compendium of Analytical Nomenclature' (the 'Orange Book'), Pergamon Press, 1978.

'Manual of Symbols and Terminology for Physicochemical Quantities and Units' (the 'Green Book'), Pergamon Press, 1979.

A research chemist involved in organic synthesis might typically prepare several hundred and possibly over a thousand new compounds during his or her lifetime.

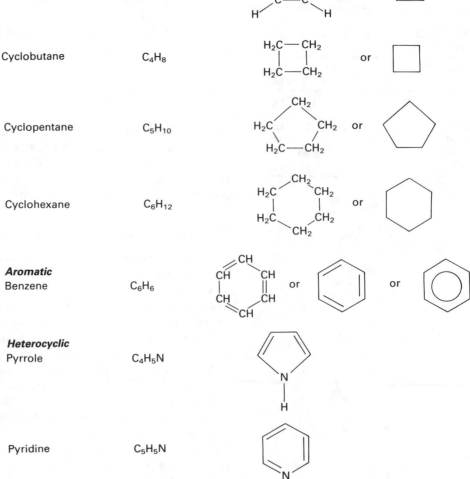

Figure 17.1 Classes and examples or organic compounds

structures. It is therefore customary to omit the —CH_2— in alicyclic or hetero-cyclic structures. Similarly, in open chain structures it is often convenient to display only the multiple bonds or, in the case of isomers (*see* below), only the branching.

Alicyclic chemistry and heterocyclic chemistry are rather specialised branches of chemistry. In this book, therefore, we shall principally be concerned with the chemistry of aliphatic and aromatic compounds.

HYDROCARBONS

Compounds which contain carbon and hydrogen only are called **hydrocarbons**. Hydrocarbons may be aliphatic, alicyclic or aromatic. They cannot be hetero-cyclic of course! Aromatic hydrocarbons are sometimes called **arenes**.

Aliphatic Hydrocarbons

The three most important groups of aliphatic hydrocarbons are alkanes, alkenes and alkynes.

Alkanes

These are aliphatic hydrocarbons in which the carbon atoms are joined by single covalent bonds only. Compounds in which the atoms are joined by single covalent bonds only are called **saturated** compounds.

Examples of alkanes are propane (*see* figure 17.1), methane and ethane:

Methane   Ethane

The alkanes have the general formula C_nH_{2n+2} (*see* table 17.3). The term *alkane* can be split into two parts:

- *alk-*   This is the general term for meth-, eth-, prop- and so on. This indicates the number of carbon atoms in the alkane or alkyl grooup.
- *-ane*   This indicates that the hydrocarbon is saturated.

The traditional name for alkenes is *paraffins*. The word 'paraffin' is derived from the Latin words *parum affinis* which mean lack of affinity. Alkanes are stable and inert towards many other substances.

Table 17.3   Alkanes and alkyl groups

Alkane	Formula	Alkyl group	Formula
methane	CH_4	methyl	CH_3—
ethane	C_2H_6	ethyl	C_2H_5—
propane	C_3H_8	propyl	C_3H_7—
butane	C_4H_{10}	butyl	C_4H_9—
pentane	C_5H_{12}	pentyl	C_5H_{11}—
hexane	C_6H_{14}	hexyl	C_6H_{13}—

Alkenes

These are aliphatic hydrocarbons which contain a double bond:

Propene is an example (*see* figure 17.1). Alkenes have the general formula C_nH_{2n} (*see* table 17.4).

Table 17.4 Alkenes

Name		Formula	
IUPAC	Traditional	Molecular	Structural
ethene	ethylene	C_2H_4	$CH_2{=}CH_2$
propene	propylene	C_3H_6	$CH_3{-}CH{=}CH_2$

The letters *-en-* in the middle of the IUPAC name are used for compounds containing carbon–carbon double bonds.

Alkynes

These aliphatic hydrocarbons contain a triple bond $-C{\equiv}C-$. The simplest alkyne ethyne (also known as acetylene). Alkynes have the general formula C_nH_{2n-2} (*see* Table 17.5).

Table 17.5 Alkynes

Name		Formula	
IUPAC	Traditional	Molecular	Structural
ethyne	acetylene	C_2H_2	$CH{\equiv}CH$
propyne	methylacetylene	C_3H_4	$CH_3C{\equiv}CH$

The letters *-yn-* in the IUPAC name indicates a triple bond.

Alkenes and alkynes are **unsaturated hydrocarbons**. Unsaturated hydrocarbons contain multiple carbon–carbon bonds.

Structural Isomers

The aliphatic hydrocarbon C_4H_{10} has two possible structures. The carbon atoms can either be linked together to form an unbranched chain (sometimes called a straight chain) or they can be linked together to form a branched chain. The displayed formulae of the two types of structures are shown in figure 17.2. The unbranched chain structure of butane can be displayed in a number of ways. Three possible displayed formulae are shown in figure 17.2. All three are identical since they all have the same carbon skeleton and each respective carbon atom has the same number of hydrogen atoms attached to it. Conventionally butane is displayed as structure (a) in figure 17.2.

The branched chain structure of C_4H_{10} is also shown in figure 17.2. Notice that the central carbon atom has three other carbon atoms attached to it but only one hydrogen atom.

The unbranched chain structure and branched chain structure of C_4H_{10} are called structural isomers. **Structural isomers** have the same molecular formulae but differ in structure.

C_4H_{10} is the simplest aliphatic hydrocarbon to have structural isomers. The alkanes methane, ethane and propane each exist in one structural form only. The number of possible structural isomers of an alkane increases with the number of carbon atoms. Whereas C_4H_{10} can only have two structural isomers, C_5H_{12} has three structural isomers. These are shown in figure 17.3. Notice that the structure of the carbon skeleton is different for each isomer.

In the IUPAC rules of nomenclature, isomers are named as derivatives of the longest unbranched chain of carbon atoms. This is called the **main chain** or **root**. It is named after the unbranched alkane with the same number of carbon atoms. If, for example, a main chain has three carbon atoms, it is named *prop-*. If it has six carbon atoms it is named *hex-* and so on. The side chains are named after the alkyl groups which replace the hydrogen atoms in the main chain. The six simplest alkyl groups are shown in table 17.3. Finally, in the IUPAC system of nomenclature, the position at which the side chain or group is attached to the main chain is indicated where necessary by a number. This is the lowest possible number of carbon atoms from the end of the main chain. The number is called a **locant**.

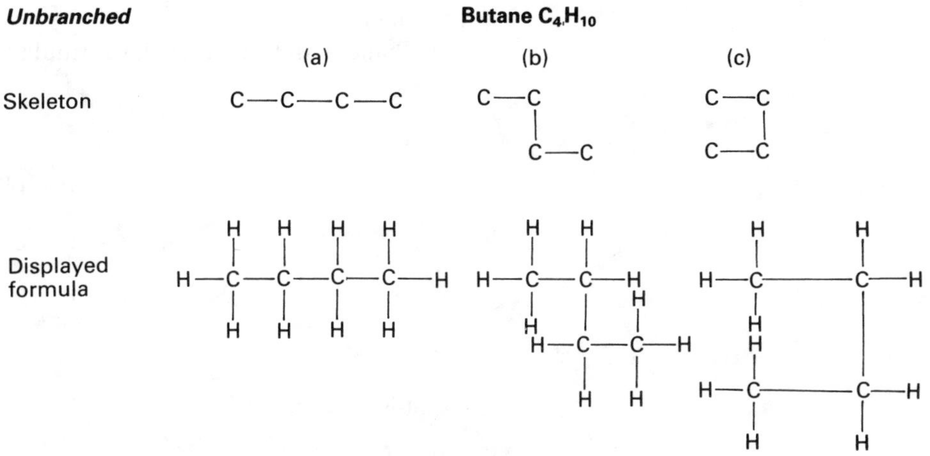

Unbranched   **Butane C$_4$H$_{10}$**

These are all equivalent. They can
be written CH$_3$—CH$_2$—CH$_2$—CH$_3$ or CH$_3$(CH$_2$)$_2$CH$_3$

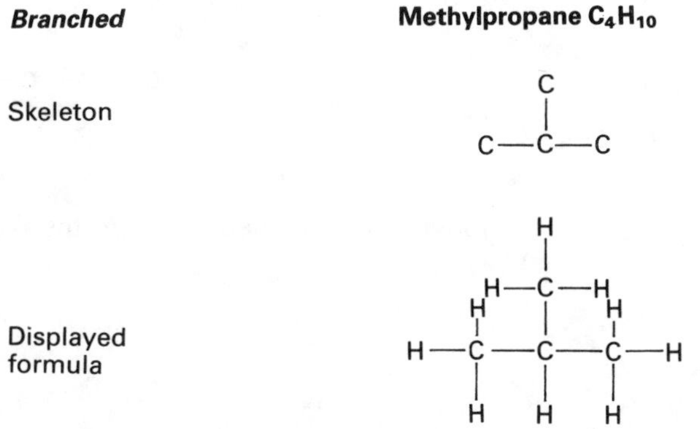

Branched   **Methylpropane C$_4$H$_{10}$**

The structure may also be written as CH$_3$—CH—CH$_3$ or just (CH$_3$)$_3$CH.
Methylpropane is also known as isobutane or 2-methylpropane

Figure 17.2   Unbranched chain and branched chain hydrocarbons

Name	Skeleton	Displayed formula
Pentane	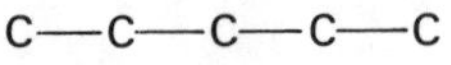	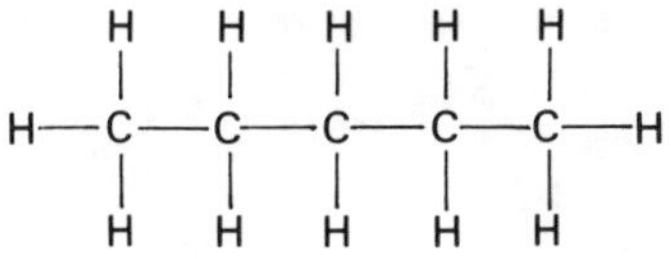
2-methylbutane	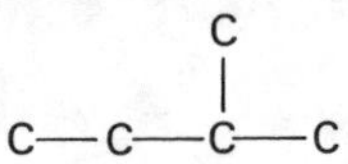	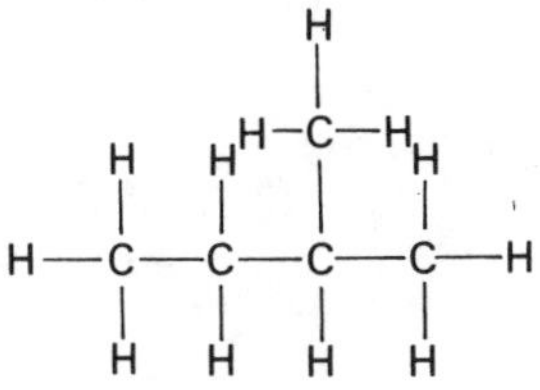
2,2-dimethylpropane	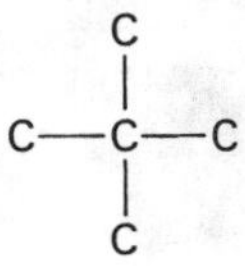	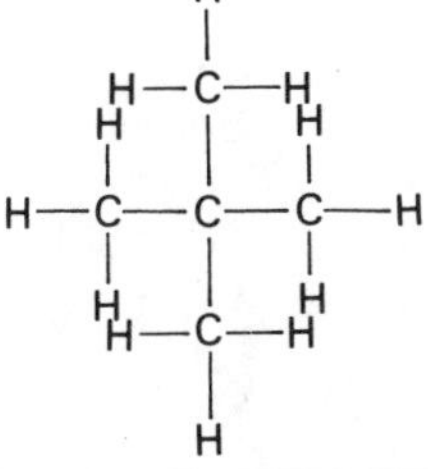

Figure 17.3   C$_5$H$_{12}$ isomers

EXAMPLE

An alkane with the molecular formula C_8H_{18} has the following structure:

What is its systematic name?

SOLUTION

The carbon skeleton for this structural isomer is:

The longest unbranched chain in this skeleton is

C—C—C—C—C   or   C—C—C—C—C—C

Both of these are equivalent. They represent the hexane skeleton (see table 17.3). Hydrogen atoms on both the second and third carbon atoms from the end of the hexane chain have been replaced by methyl groups. The isomer is thus the 2,3-dimethyl derivative of hexane. The full name is 2,3-dimethylhexane.

The alcohol with the formula C_3H_7OH has two structural isomers. These are propan-1-ol and propan-2-ol—also known as n-propyl alcohol and isopropyl alcohol respectively. In the latter nomenclature, the prefix n- stands for normal and refers to unbranched structures. The prefix iso refers to branched chain structures

Table 17.6 Unsaturated structural isomers

Molecular formula	Structural formula		
C_4H_8	$CH_3CH_2CH{=}CH_2$ or $CH_2{=}CHCH_2CH_3$		$CH_3CH{=}CHCH_3$
	but-1-ene		but-2-ene
C_4H_6	$CH_3CH_2C{\equiv}CH$ or $CH{\equiv}CCH_2CH_3$		$CH_3C{\equiv}CCH_3$
	but-1-yne		but-2-yne

All classes of organic compounds and their derivatives exhibit structural isomerism. Examples of structural isomers of alkenes and alkynes based on the root *but-* are shown in table 17.6. Another example is dimethylbenzene. Its three structural isomers are shown in figure 17.4. In this case the root of the name is not a chain of carbon atoms but the benzene ring. Numbers are assigned to the carbon atoms in the ring to indicate the position of the groups or side chains.

IUPAC name	Traditional name	Structural formula
1,2-dimethylbenzene	*ortho*-xylene (*o*-xylene)	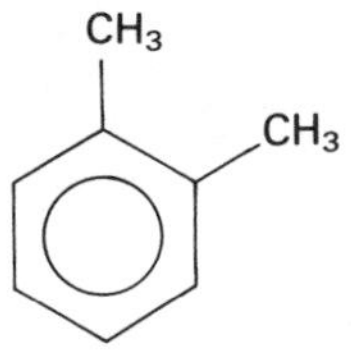
1,3-dimethylbenzene	*meta*-xylene (*m*-xylene)	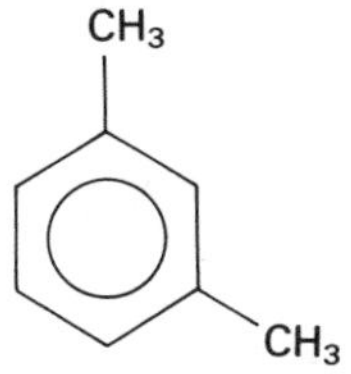
1,4-dimethylbenzene	*para*-xylene (*p*-xylene)	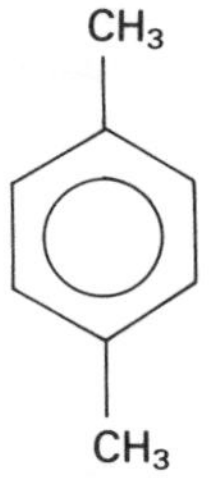

Figure 17.4 The three dimethylbenzene structural isomers

FUNCTIONAL GROUPS

Many of the physical and chemical properties of hydrocarbon derivatives depend on the group attached to the hydrocarbon root rather than on the root itself. Consider ethanoic acid, for example:

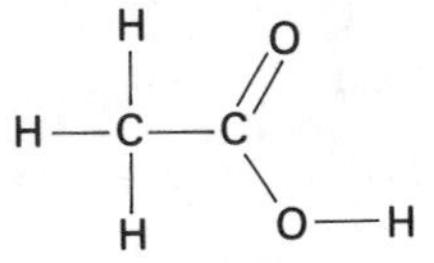

Ethanoic acid
(traditional name: acetic acid)

The root of this name is *ethan-*. This, as we have seen, corresponds to the alkane with two carbon atoms, that is ethane. However, both the physical and chemical properties of ethanoic acid derive not so much from the root but from the *-oic* group:

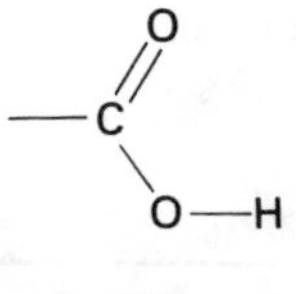

$$-\text{oic}$$

For example, ethanoic acid is a colourless liquid, soluble in water and a weak acid (*see* chapter 8), whereas ethane is a gas which is insoluble in water. The properties of ethanoic acid are thus very different from ethane although both compounds have the same hydrocarbon root. The distinctive properties of ethanoic acid are due to the *-oic* part rather than the *ethan*-part. This part of the compound is thus called the functional group.

A list of functional groups and their associated types of compounds is shown in figure 17.5. This list is not comprehensive. Note that a functional group is sometimes included in the name of the compound as a **suffix** (it is put *after* the root) whereas at other times it is included as a **prefix** (it is put *before* the root). This is not arbitrary. There are rules, based on convention, which determine when a functional group should be used as a prefix and when as a suffix.

Type of compound	Structure of functional group	Prefix— or —suffix	Example (common name in brackets)
Alcohol	—OH	—ol	Ethanol C_2H_5—OH
Carboxylic acid	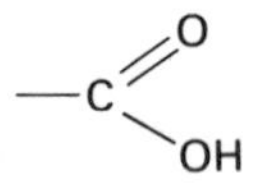	—oic acid	Ethanoic acid (acetic acid) 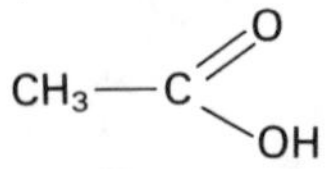
Carboxylate ion	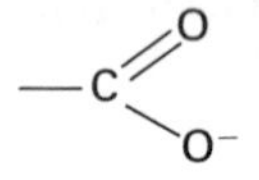	—oate	Sodium ethanoate (sodium acetate) 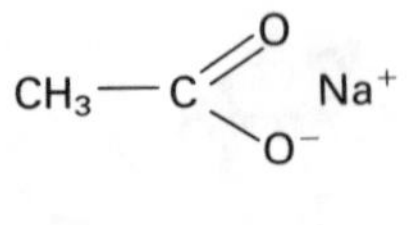
Ester	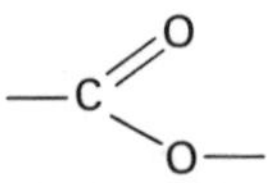	—oate	Ethyl ethanoate (ethyl acetate) 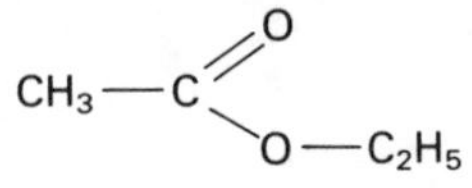
Acid anhydride	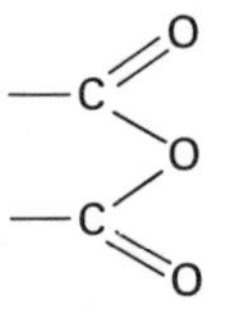	—oic anhydride	Ethanoic anhydride (acetic anhydride) 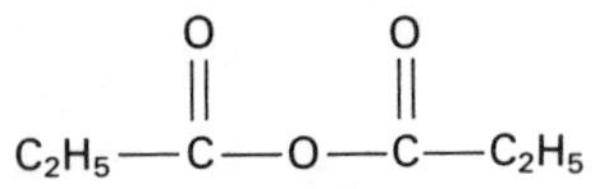

Figure 17.5   Functional groups

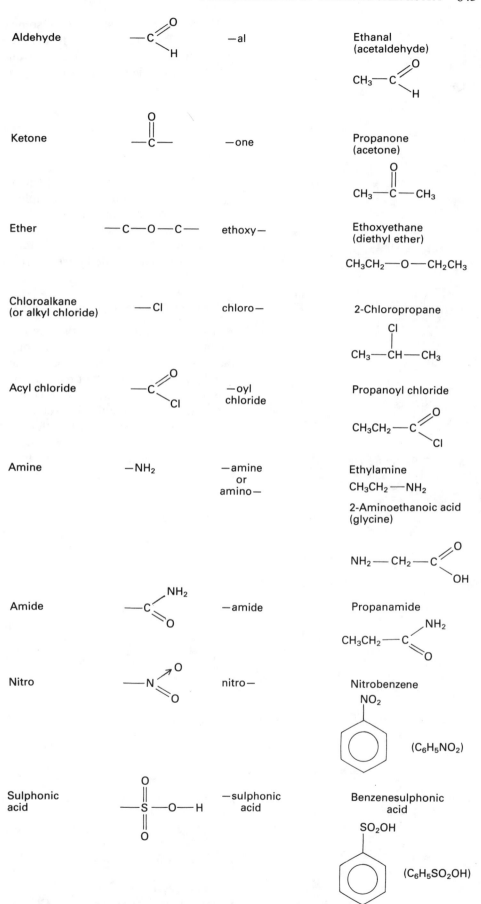

Figure 17.5 (*continued*)

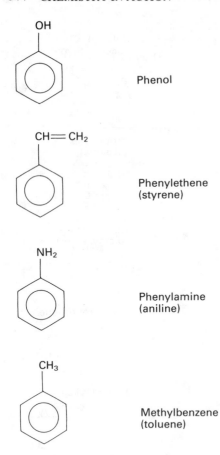

Figure 17.6 Nomenclature of simple aromatic compounds

Note also that the structure of some of the functional groups include a carbon atom. This carbon atom may form part of the root as well as part of the functional group. We have already met one example of this in the case of ethanoic acid. Another example is propanone:

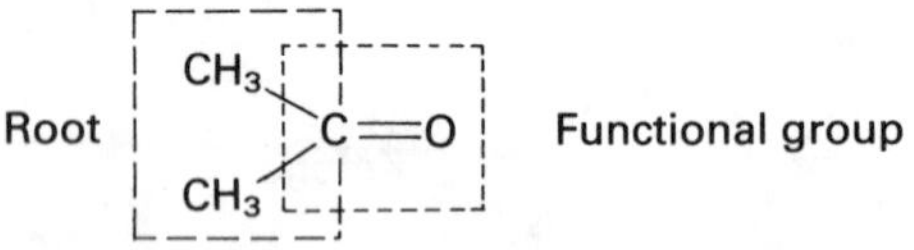

$$\text{Root} \quad \boxed{\begin{matrix} CH_3 \\ \quad \quad C=O \\ CH_3 \end{matrix}} \quad \text{Functional group}$$

Finally, we should note that the nomenclature shown in figure 17.5 is not fully systematic. Ethylamine, for example, consists of the prefix *ethyl-* and the suffix *-amine*. It does not have a root.

The nomenclature of some of the simpler aromatic compounds is also not fully systematic. For example, the systematic name for the root C_6H_5— is benzene. However, it is also called *phenyl-* after phenol (*see* figure 17.6).

Shorthand for Alkyl- and Aryl- Groups

As a matter of convenience the symbol R is often used as shorthand in the formula when the root is *alkyl-*, *aryl-* or even hydrogen (*see* table 17.7). This enables us to focus attention on the functional group which, as we saw above, is usually the most important part of a molecule from the chemist's point of view.

Sometimes the symbol Ar or Ph is used for *aryl-* to distinguish it from *alkyl-*. If it is necessary to distinguish between two—as in the case of esters—then R and R' are used (*see* figure 17.7). If a third symbol is required then R″ is used. These symbols are useful in distinguishing between primary, secondary and tertiary compounds. These have respectively one, two and three alkyl and/or aryl groups attached to a specific atom related to the functional group (*see* figure 17.8).

An ester

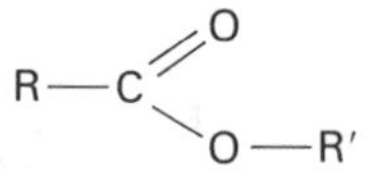

$$R-C{\overset{\displaystyle O}{\underset{\displaystyle O-R'}{\big\langle}}}$$

Example

R is CH_3—

R' is C_5H_{11}—

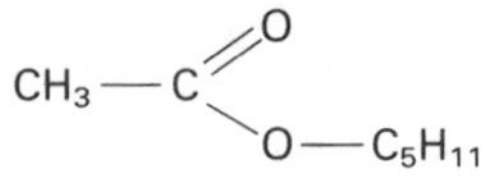

$$CH_3-C{\overset{\displaystyle O}{\underset{\displaystyle O-C_5H_{11}}{\big\langle}}}$$

Pentyl ethanoate

Figure 17.7 R and R'

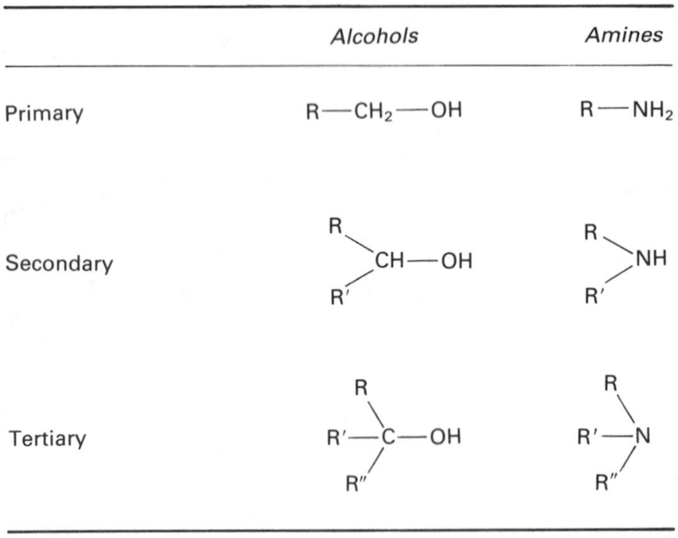

	Alcohols	Amines
Primary	$R-CH_2-OH$	$R-NH_2$
Secondary	$\begin{matrix} R \\ \quad CH-OH \\ R' \end{matrix}$	$\begin{matrix} R \\ \quad NH \\ R' \end{matrix}$
Tertiary	$\begin{matrix} R \\ R'-C-OH \\ R'' \end{matrix}$	$\begin{matrix} R \\ R'-N \\ R'' \end{matrix}$

Figure 17.8 Primary, secondary and tertiary amines and alcohols. *Note:* Any two or all three of R, R' and R″ may be identical

Homologous Series

A homologous series of organic compounds is a series in which each successive member increases by the unit —CH_2—. The first six members of the homologous series of alkanes are shown in table 17.3. The simplest member of this series is

Table 17.7 Use of R in the formula of alkyl chlorides of general formula R—Cl

R	Formula	Name
H—	H—Cl	hydrogen chloride
CH₃—	CH₃—Cl	chloromethane
CH₃CH₂—	CH₃CH₂—Cl	chloroethane
C₆H₅—	C₆H₅—Cl	chlorobenzene

methane. The next is ethane and so on. The difference between methane and ethane is —CH_2—. Table 17.8 shows the first six members of the homologous series of alcohols with the general formula $C_nH_{2n+1}OH$ and the first six members of the homologous series of amines with the general formula $C_nH_{2n+1}NH_2$.

In general, the methods of preparation and chemical properties of each member of a particular series are similar.

Table 17.8 Homologous series

(a) Alcohols $C_n H_{2n+1}OH$

Name	Molecular formula	Structural formula
methanol	CH_3OH	CH_3OH
ethanol	C_2H_5OH	CH_3CH_2OH
propan-1-ol	C_3H_7OH	$CH_3CH_2CH_2OH$
butan-1-ol	C_4H_9OH	$CH_3CH_2CH_2CH_2OH$
pentan-1-ol	$C_5H_{11}OH$	$CH_3CH_2CH_2CH_2CH_2OH$
hexan-1-ol	$C_6H_{13}OH$	$CH_3CH_2CH_2CH_2CH_2CH_2OH$

(b) Amines $C_nH_{2n+1}NH_2$

Name	Molecular formula	Structural formula
methylamine	CH_3NH_2	CH_3NH_2
ethylamine	$C_2H_5NH_2$	$CH_3CH_2NH_2$
propylamine	$C_3H_7NH_2$	$CH_3CH_2CH_2NH_2$
butylamine	$C_4H_9NH_2$	$CH_3CH_2CH_2CH_2NH_2$
pentylamine	$C_5H_{11}NH_2$	$CH_3CH_2CH_2CH_2CH_2NH_2$
hexylamine	$C_6H_{13}NH_2$	$CH_2CH_2CH_2CH_2CH_2CH_2NH_2$

PHYSICAL PROPERTIES OF ORGANIC COMPOUNDS

Almost all organic compounds are **covalent**. They thus exist as molecules. These molecules can range from the simple to the macromolecular. The physical properties of organic compounds thus depend to a great extent on the size, shape and structure of these molecules. They are also influenced by the polarisation of covalent bonds. This is common when a molecule contains an atom which is more electronegative than carbon. Chlorine or oxygen are examples:

$$-\overset{|}{\underset{|}{C}}{}^{\delta+}-\overset{\delta-}{Cl} \qquad \overset{|}{\underset{|}{C}}{}^{\delta+}=\overset{\delta-}{O}$$

These are called **dipoles**.

Melting Points and Boiling Points

In general, organic compounds are gases, liquids or relatively low-melting solids. This is because the organic molecules are only held together by weak molecular forces known as van der Waals forces. In contrast, salts are held together in their crystals by powerful ionic forces and thus have very high melting points (*see* chapter 2).

Both melting point and boiling point increase with increasing number of carbon atoms in a homologous series. We shall meet an example of this when we come to consider the alkanes in the following chapter. In general, branched chain isomers are more volatile (that is they have lower boiling points) than the corresponding unbranched chain isomers.

Alcohols and carboxylic acids, both of which contain the —OH group, have abnormally high boiling points. This is due to intermolecular hydrogen bonding (*see* chapter 2). Intramolecular hydrogen bonding can result in low boiling points for certain organic liquids. A classic example is the low boiling point of 2-nitrophenol (*see* section 2.2).

Solubility

Non-polar compounds such as the hydrocarbons are generally insoluble in—that is immiscible with—water. However, they do dissolve in non-polar solvents such as trichloromethane (chloroform) and methylbenzene (toluene).

The more polar organic compounds such as the alcohols and carboxylic acids tend to be more soluble in polar solvents such as water. This is due to the formation of hydrogen bonds between the solute and solvent molecules. However, the solubility decreases rapidly as the number of carbon atoms increases.

Densities and Viscosities

As the number of carbon atoms and thus the relative molecular masses of compounds increase along a homologous series so do the densities and viscosities of the compounds. The trend is thus similar to that for melting and boiling points. The branched chain isomers also have lower densities and viscosities than the unbranched chain isomers.

PREPARATION AND IDENTIFICATION OF ORGANIC COMPOUNDS

The preparation and identification of an organic compound can involve a number of important procedures. These may be grouped under the following headings:

- Preparation
- Separation
- Purification
- Analysis
- Structure determination

We have already considered some of these procedures in preceding chapters. For example, the determination of relative molecular masses and the empirical and molecular formulae of compounds are described in chapters 3, 4 and 6. Mass spectrometry is considered in chapter 1 and various methods of chromatography in chapter 6. We shall not describe classical methods of analysis since, over recent decades, these have largely been superseded by mass spectrometry, chromatography and spectroscopy.

Preparation

The preparations of specific organic compounds are described in the following chapters. Many of these reactions occur slowly at room temperature. It is therefore quite common to carry out the reaction by **heating under reflux**. The higher temperature speeds up the reaction.

A typical apparatus used for heating under reflux is shown in figure 17.9. A **water condenser** is attached vertically to the reaction flask. The flask contains the reaction mixture. This may be dissolved in an organic solvent such as methylbenzene. The mixture is heated to its boiling point and maintained at this temperature for the duration of the reaction. The vapour from the mixture condenses in the condenser and continuously drips back into the flask.

Separation and Purification

On completion of a reaction the reaction flask will often contain not only the desired product but also other products, by-products of side reactions, unreacted reactants and the solvent. If the product is required pure, then it is necessary to isolate it from the other substances in the flask. Various separation techniques are used for this purpose depending on the nature of the required product and the other substances in the flask.

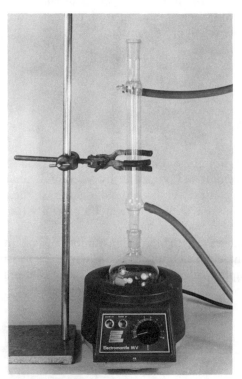

Heating under reflux

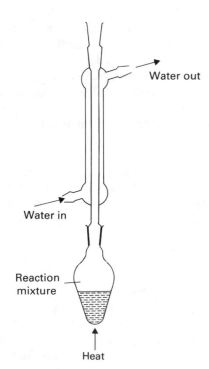

Water out

Water in

Reaction mixture

Heat

Figure 17.9 Reflux apparatus

Distillation

If the required product is the only volatile substance in the reaction flask then it can be separated from the other substances by distillation. A typical distillation apparatus is shown in figure 17.10. For liquids which boil at temperatures lower than about 140°C at atmospheric pressure, a water condenser is used. The liquid is heated to its boiling point. Its vapour distils over, condenses and is collected. Porcelain chips are often used in the reaction flask to ensure steady boiling and prevent bumping.

If the boiling point of the liquid is higher than 140°C an **air condenser** is used. An air condenser is a straight glass tube without a jacket.

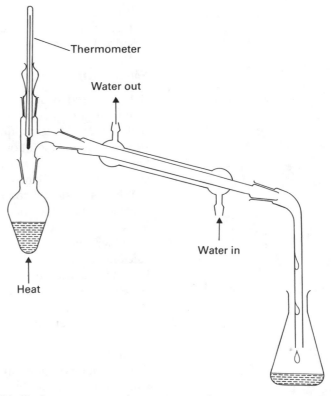

Thermometer

Water out

Water in

Heat

Figure 17.10 Distillation apparatus using a water condenser

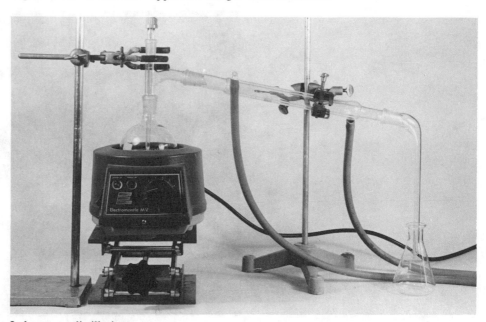

Laboratory distillation

Since many organic liquids are flammable, care has to be taken when heating the reaction flask. For low-boiling liquids a water bath is often used. For larger-scale preparations an electric heating mantle is commonly used.

Some organic liquids thermally decompose at temperatures below their boiling points at atmospheric pressure. In such cases it is necessary to distil the liquid at reduced pressure in order to lower the boiling point. To do this a vacuum pump is connected to a side arm in the collection flask. This method of distillation is called **vacuum distillation**.

If the required product is one component of a zeotropic mixture of two or more volatile liquids then **fractional distillation** can be used to separate the components. The theory and practice of this method of distillation is described in section 6.2.

If the product is a liquid with a high boiling point and if it is immiscible with water, then it can be separated by steam distillation. This technique is also described in section 6.2.

Solvent Extraction

If the desired product is the only component in a reaction mixture which is soluble in a particular solvent then it can be isolated by solvent extraction (*see* section 6.2). The technique is particularly useful for separating an organic product from an aqueous solution containing inorganic impurities. The solution is shaken with an organic solvent which is immiscible with water but in which the product is soluble. Ethoxyethane (diethyl ether) is often used for this purpose.

When an organic compound is extracted from an aqueous solution using an organic solvent the solvent often becomes wet. Before it can be distilled off to leave the pure product the solvent must therefore be dried. Sodium wire is used to dry ethers and some liquid hydrocarbons. Anhydrous sodium, magnesium and calcium sulphates are all used for drying organic solvents and liquids. Potassium hydroxide is used to dry amines but cannot be used for acids, esters or phenols since it reacts with them. Similarly, the drying agent calcium chloride is often used in organic preparations although it cannot be used for acidic solutions, alcohols, amines and phenols since it reacts with them.

Crystallisation

This method is commonly used to separate and purify solids. A solvent is used in which solubility of the solid increases with temperature. The crude solid is

Bottom discharge batch centrifuge used in industry for separating solids from liquids

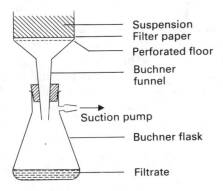

Figure 17.11 Crystallisation using a Buchner flask and Buchner funnel

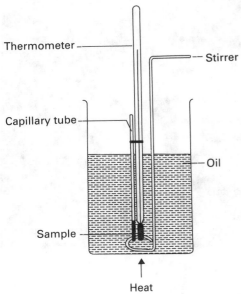

Figure 17.12 Determination of melting point

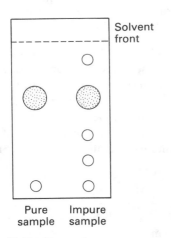

Figure 17.13 Paper chromatogram

dissolved in the minimum amount of hot solvent in order to give an almost saturated solution. The solution is then filtered to remove insoluble impurities and then allowed to cool. The solid crystallises out. The crystals are filtered under reduced pressure using a Buchner funnel and flask (*see* figure 17.11). They are then either dried or recrystallised to obtain a more pure product.

Chromatography

Column chromatography is often used to separate a mixture containing two or more solid products. The solid mixture is dissolved in a solvent and allowed to pass down a column. Various fractions containing the separated components are collected. The separate components can be recovered from solution by solvent evaporation and crystallisation. The method is described in more detail in section 6.3.

Gas chromatography can also be used to separate mixtures of gases, liquids and volatile solids. This method is also described in section 6.3.

How Do We Know if a Product is Pure?

We saw in section 6.2 that most pure solids have characteristic sharp melting points. A pure organic solid melts over a degree or two whereas an impure solid may melt over a range of five or more degrees Celsius. Melting point can be determined using the apparatus shown in figure 17.12. A small amount of the solid is gently tapped down into the closed end of a glass capillary tube. This is attached to a thermometer by a rubber band. The thermometer is then lowered into a beaker containing a high-boiling oil. The oil is heated slowly with stirring. The temperatures at which the solid begins to melt and is completely melted are recorded. The method of mixed melting points for determining the identity of solids is described in chapter 6.

A pure solid will also produce a single spot on a TLC plate or paper chromatogram (*see* figure 17.13 and also section 6.3). For volatile substances—that is gases, liquids and solids with high vapour pressures—gas chromatography can be used to determine purity. A pure compound will produce a single peak whereas impurities will show up as additional peaks (*see* figure 17.14).

Thin-layer chromatography, paper chromatograpy and gas chromatography can all be used to identify either pure compounds or the components in a mixture.

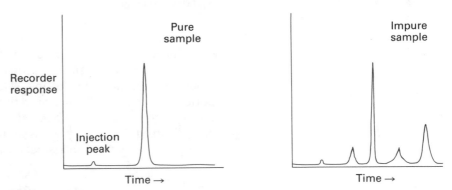

Figure 17.14 Gas chromatogram

Spectroscopy

Spectroscopic methods are used extensively for research and analysis in chemistry and also in other sciences such as astronomy. We shall describe below three of the most frequently used techniques: infra-red (IR) spectroscopy, ultra-violet–visible (UV/VIS) spectroscopy and nuclear magnetic resonance (NMR) spectroscopy.

In the spectroscopic method the sample substance is exposed to radiation. Different chemical species absorb radiation to different extents depending on their atomic and molecular structures. It is thus possible to deduce certain features of the chemical structure of an unknown substance by examining its spectrum.

Infra-red Spectroscopy

IR spectroscopy depends on the **vibrations** of atoms with respect to one another. The most important are the mutual vibrations of two atoms bonded to each other. The vibrations of larger parts of the molecule are not so important.

The two vibrating atoms are regarded as two point masses connected by a spring. The spring represents the bond between the two atoms. The frequency of the vibration corresponds to the frequency of the absorbed radiation. This depends on the masses of the atoms and the bond strength. The frequency is characteristic of the two bonded atoms. The IR spectrum of an organic compound, for example, immediately reveals whether the molecule contains a carbonyl group, C=O. Other functional groups can also be immediately recognised from the IR spectrum.

An infra-red spectrometer

In molecules consisting of many atoms, vibrations are very complex. Three different modes of vibration occur in the water molecule, H_2O. In a relatively complicated molecule such as propanone, CH_3COCH_3, which has 10 atoms, 24 modes of vibration are possible. However, not all vibrational modes are important. The most distinct vibration in the propane molecule is that of the carbonyl group, C=O. The remainder of the molecule can be regarded as being partially fixed. Most compounds containing this functional group exhibit a similar absorption of IR frequencies between 5.0×10^{13} and 5.6×10^{13} Hz.

The **hertz**, Hz, is the SI unit of frequency. The hertz is equivalent to second^{-1}. In IR spectroscopy it is more common to express frequencies as **wavenumbers**. A wavenumber is the number of waves per centimetre. It can be found by dividing the frequency in hertz units by the speed of light. The **speed of light** is 2.998×10^{10} cm s^{-1}. Thus, for example, the wavenumber of a frequency with the value 5.0×10^{13} Hz can be calculated as follows: Since

$$5.0 \times 10^{13} \text{ Hz} \equiv 5.0 \times 10^{13} \text{ s}^{-1}$$

$$\text{Wavenumber} = \frac{5.0 \times 10^{13} \text{ s}^{-1}}{2.998 \times 10^{10} \text{ cm s}^{-1}}$$

$$= 1678 \text{ cm}^{-1}$$

IR spectroscopy has now developed into a common and routine technique. Organic chemists use it to identify organic compounds. The spectra are obtained by measuring a substance's absorption of IR radiation at different frequencies. The frequencies absorbed correspond to the frequencies with which the atoms in the molecules vibrate with respect to one another. Groups such as C—C, C=O and C—H absorb radiation with wavenumbers in the range of 1300 cm^{-1} to 3000 cm^{-1} (see figure 17.15). Specific absorption peaks indicate the presence of these groups in a molecule. This part of the spectrum is known as the **band region**. Vibrational frequencies in the 600 cm^{-1} to 1300 cm^{-1} region characterise the whole molecule. This is called the **fingerprint region** since it can be used to distinguish between molecules with the same functional groups. For example, propanone, CH_3COCH_3, and butanone, $CH_3COCH_2CH_3$, have the same spectra in the band region. However, in the fingerprint region the spectra differ sufficiently to enable the two compounds to be distinguished from one another (see figure 17.15).

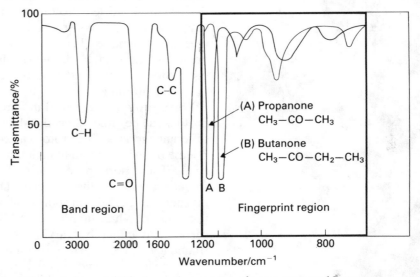

Figure 17.15 Infra-red (IR) absorption spectra of propanone and butanone

Ultra-violet–Visible Spectroscopy

Ultra-violet–visible spectroscopy depends on the electron transitions which occur from one energy level to another due to the absorption of radiation. Colourless compounds absorb in the UV region of the spectrum whereas coloured compounds absorb in the visible part of the spectrum.

The light source usually has wavelengths in the range 0.2×10^{-6} to 0.8×10^{-6} m. Two separate sources are normally used—one for the UV region and another for the visible region. Glass lenses and cells cannot be used for UV measurements since they absorb in this region. Quartz is thus used.

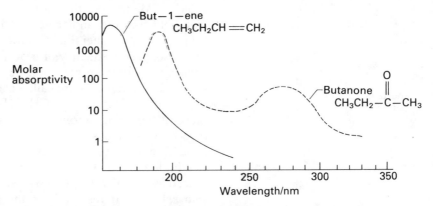

Figure 17.16 Ultra-violet (UV) absorption spectra of but-1-ene and butanone

UV spectroscopy is commonly used to measure concentrations of solutions. Such experiments rely on the **Beer–Lambert law** which states that the amount of light absorbed by a solution is proportional to the concentration of the solute and the path length of the light. Normally the absorption at a single wavelength is measured. This is usually in the region of the spectrum where absorption is strongest. The extent of UV absorption is known as the **molar absorptivity**.

The wavelength at which absorption occurs depends on the types and number of double bonds present. The absorption of UV light raises a molecule's outer (or valence) electrons from the lowest energy states to excited states. For example, the UV spectrum of but-1-ene exhibits a strong absorption peak due to excitation of electrons on the C=C group (*see* figure 17.16). Butanone has two peaks resulting from the transitions of different C=O electrons.

Nuclear Magnetic Resonance Spectroscopy

NMR spectroscopy is most frequently used to obtain information about the structures of organic compounds. The technique depends on the magnetic properties of atomic nuclei. Protons and neutrons in the atomic nuclei behave as tiny magnets. This magnetism cancels out in some nuclei but in others a residual magnetism remains. This is called **nuclear spin**. Carbon-12 and oxygen-16 have no net nuclear spin. Hydrogen-1 (also known as the proton) does show magnetic properties however and may be studied by NMR spectroscopy. This is often known as proton nuclear magnetic resonance (proton NMR) spectroscopy.

When protons are placed in a magnetic field an interaction takes place. The protons can be either aligned with the field or opposed to it. These two states have different energies. The proton can convert from one to the other by absorption or release of energy. Even at high magnetic fields these energy differences are very small. The absorption of energy takes place at low frequencies—in the radio-frequency region of the spectrum. The frequencies absorbed by protons and other magnetic nuclei are proportional to the strength of the magnet used in the instrument. When running an NMR spectrum, therefore, the scanning can be carried out either by varying the magnetic field strength or by varying the frequency source. The resulting spectrum is the same. An older convention was to quote the low-field and high-field parts of the spectrum, but the modern norm is to quote frequency instead..

Different nuclei have different resonance frequencies. Thus an instrument with a given magnet strength needs to have one frequency source for recording proton spectra and different frequency sources for other magnetic nuclei. Consequently a scan for protons does not show absorptions for other elements such as carbon or oxygen.

Absorption occurs at the **resonance frequency**. The absorption peak of the protons of tetramethylsilane (TMS) is used for reference. At a magnetic field strength of 1.4092 tesla (this is the SI equivalent of 14 092 gauss) the TMS protons come to resonance at 60 MHz. The protons in TMS are very **shielded** from the magnetic field by electrons surrounding them in the molecule. The resonance peak of TMS protons therefore occurs at a low frequency. This is equivalent to a high field. In other molecules protons are less shielded and so their resonance frequencies are higher than that for TMS. These protons are said to be **deshielded**. The resonance frequency of a proton or group of protons thus depends on the environment of the proton or group in the molecule. The resulting differences in resonance frequencies are called **chemical shifts**.

The ethanol molecule, CH_3CH_2OH, has three types of hydrogen atoms: those of the methyl group CH_3—, those of —CH_2— and those of —OH. Each type of proton has different resonance frequencies. The NMR spectrum of ethanol thus has three separate peaks.

With high-resolution instruments, peaks can often be seen to split into closely spaced finer peaks. This is known variously as spin coupling, spin–spin coupling or **spin splitting**. It results from the shielding effect on protons by the field of neighbouring protons. Splitting of peaks in the NMR spectrum provides infor-

NMR spectroscopists at work

$$CH_3 - Si - CH_3$$

with CH_3 above and CH_3 below the Si.

Tetramethylsilane (TMS)

MHz is the symbol for megahertz. One megahertz is 10^6 hertz.

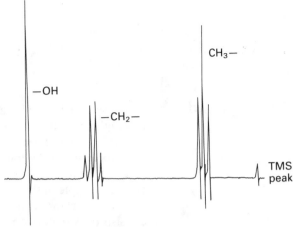

Figure 17.17 Nuclear magnetic resonance (NMR) spectrum of ethanol

mation on how groups are arranged in the molecule. For example, protons on neighbouring atoms in ethanol interact with one another. Each peak is thus split into multiple peaks (*see* figure 17.17).

SUMMARY

1. **Organic chemistry** is the chemistry of carbon compounds.
2. All living things contain **organic compounds**.
3. **Aliphatic** compounds consist of either **unbranched** or **branched** chains of carbon atoms.
4. **Alicyclic** compounds have closed rings of carbon atoms.
5. The **aromatic** ring of benzene consists of six carbon atoms with **delocalised** π electrons.
6. **Heterocyclic** compounds have a closed ring of carbon atoms and one or more atoms of other elements.
7. **Saturated** compounds contain single **covalent bonds** only.
8. **Unsaturated** compounds contain **multiple carbon–carbon bonds**.
9. **Structural isomers** have the same **molecular formulae** but differ in structure.
10. The physical and chemical properties of compounds depend, to a large extent, on their **functional groups**.
11. Each successive member in a **homologous series** increases by $-CH_2-$.
12. Organic compounds are mainly covalent.
13. Organic compounds are usually gases, liquids or relatively low-melting solids.
14. **Hydrocarbons** are **non-polar** and thus insoluble in water but soluble in non-polar solvents.
15. Many organic compounds can be prepared by heating the reaction mixture under **reflux**.
16. Methods of separating organic compounds include **vacuum distillation, fractional distillation, solvent extraction, crystallisation** and **chromatography**.
17. The **purity** of an organic compound can be determined from its **melting point** or by chromatography.
18. **Infra-red spectroscopy** is used to identify organic compounds and specific groups of atoms in these compounds.
19. **Ultra-violet–visible spectroscopy** is used to determine the concentrations of organic compounds in solution and also the types and numbers of multiple bonds present in a molecule.
20. **Nuclear magnetic resonance spectroscopy** is used to determine the structure of organic compounds.

17.2 Molecular Geometry of Organic Compounds

In this section we shall start by reviewing and developing the topic of bonding and molecular structure of carbon compounds. Carbon has the electron configuration $1s^2 2s^2 2p^2$. In section 2.1 we saw that the four electrons in the 2s and 2p orbitals can **hybridise** to form four sp^3 **orbitals**, all of which are equivalent and indistinguishable. These four orbitals form a **tetrahedral** structure. The methane molecule is a classic example of this type of structure (*see* figure 17.18). Each of the four hybrid orbitals overlaps with a hydrogen 1s orbital forming a σ bond. Each σ bond consists of two electrons—one from the carbon atom and one from the hydrogen atom.

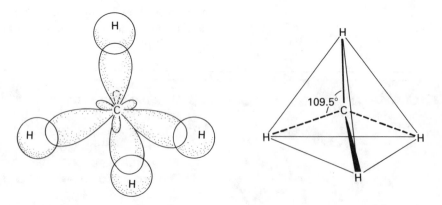

Figure 17.18   The methane molecule

The 2s orbital and two of the three 2p orbitals can also hybridise to form three sp^2 orbitals. These have a **planar** structure. In this case the carbon is left with a spare electron in a p orbital. This can join with a p electron on an adjacent carbon atom to form a π orbital. This is the case in the ethene molecule (*see* figure 17.19). The double bond in this molecule consists of one σ bond and one π bond. The π bond is represented in figure 17.19 as two clouds.

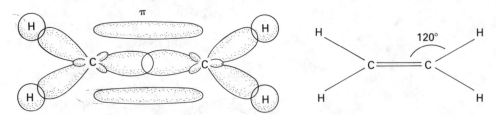

Figure 17.19   The ethene molecule

In the case of ethyne (*see* figure 17.20) the 2s orbital and one of the 2p orbitals hybridise to form two sp orbitals. These have a **linear** structure. Each carbon atom is left with two electrons in two separate p orbitals. These form two π bonds in planes intersecting each other at right angles. The triple bond in the ethyne molecule thus consists of one σ bond and two π bonds.

In aromatic ring structures the p electrons of each of the six carbon atoms become **delocalised**, forming a **π-electron cloud** (*see* figure 2.8).

All **saturated** organic compounds contain σ covalent bonds only. Figures 17.21 and 17.22 show the bonding in the propane and methanol molecules. In both figures each overlapping orbital represents a σ bond. Figure 17.22 also shows the

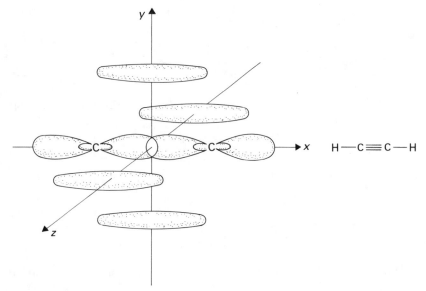

Figure 17.20 The ethyne molecule

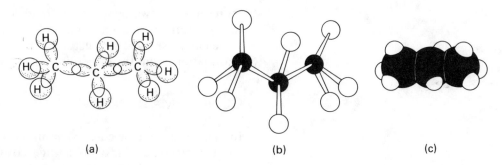

Figure 17.21 Models of propane: (a) the propane molecule; (b) ball-and-stick model; (c) space-filling model

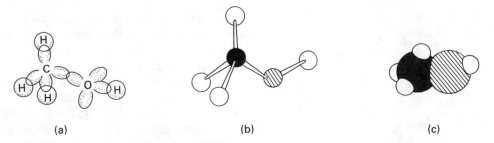

Figure 17.22 Models of methanol: (a) the methanol molecule; (b) ball-and-stick model; (c) space-filling model

two non-bonding orbitals of the oxygen atom. These each contain two electrons. In **Lewis formulae** the **non-bonding electrons** on the oxygen atom are represented by pairs of dots thus

$$- \overset{..}{\underset{..}{O}} -$$

The three-dimensional arrangement of atoms in an organic molecule is often represented by a model. Two types of model are commonly in use. These are the ball-and-stick models and the space-filling models. Both **ball-and-stick** and **space-filling** models of the propane and methanol molecules are shown in figures 17.21 and 17.22 respectively.

In writing, the structures of organic molecules can be displayed to show the three-dimensional arrangement of the atoms (*see* figure 17.23a) or they can be

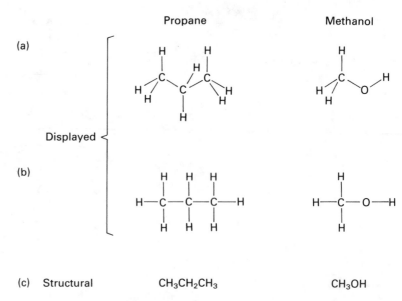

Figure 17.23 Displayed and structural formulae

represented in two dimensions (figure 17.23b). The latter is usually used unless the geometry of the molecule is being considered. More often than not it is sufficient to show the structural formulae only (figure 17.23c). This gives no information about the three-dimensional arrangement of atoms in the molecule.

CONFORMATIONS

In many organic molecules **rotation** of one part of the molecule with respect to another part is possible about a single covalent —C—C— bond. To illustrate this, let us consider a long unbranched chain alkane with the formula RCH_2—CH_3. R could be any alkyl or substituted alkyl group for example. In this type of molecule the methyl group CH_3— is continuously and freely rotating about the —C—C— bond (*see* figure 17.24a). This free rotation leads to two extreme arrangements of the methyl group in relation to the rest of the molecule. These arrangements are called conformations. **Conformations** are the various spatial arrangements of atoms in a molecule.

In our example, the two extreme conformations are called the **staggered** and **eclipsed** conformations. We can visualise these if we imagine looking along the —C—C bond from the direction of the methyl group. The two conformations are shown in figure 17.24b. These are known as Newman projections. The staggered arrangement is the more stable of the two since all three hydrogen atoms on the methyl group are at the maximum distance from the rest of the molecule. In the eclipsed conformation on the other hand, all three hydrogen atoms are closer to the remainder of the molecule. They will thus normally experience repulsive forces which force them into the staggered conformation. Conformations in between staggered and eclipsed are known as **skew** conformations.

Free rotation is only normally possible about σ bonds and not π bonds. Rotation is thus not possible about the —C=C— bond or the —C≡C— bond. In the ethene and ethyne molecules, for example, the spatial arrangement of all the carbon and hydrogen atoms is fixed with respect to one another.

Movement is possible, however, in ring structures containing single covalent bonds. The most stable conformation of the cyclohexane ring structure, for example, is a puckered arrangement known as the **chair** conformation. The carbon skeleton of this ring structure is shown in figure 17.25a. Another possibility is the **boat** conformation (figure 17.25b) although this is less stable than the chair conformation. Note that in both the chair and boat conformations the bonds of the carbon atoms have a tetrahedral arrangement.

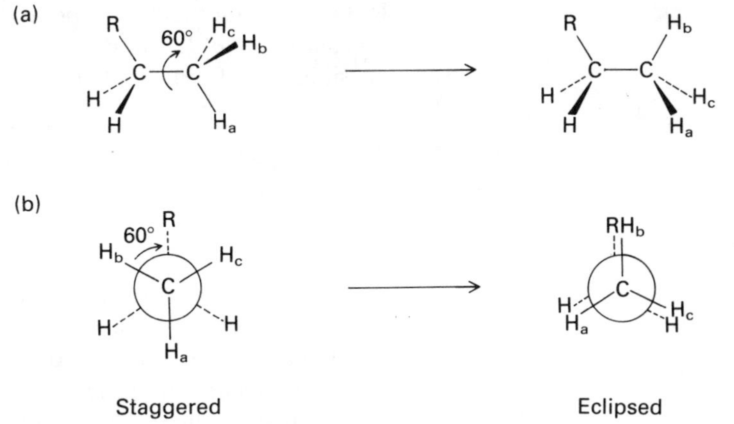

(a)

(b)

Staggered   Eclipsed

Figure 17.24   Conformations

(a)   (b)

Figure 17.25   Six-membered ring conformations: (a) chair; (b) boat

STEREOISOMERISM

Both inorganic and organic compounds can exhibit stereoisomerism. For example, the compound N_2F_2 exhibits geometrical isomerism and the octahedral anion complex ion $[Cr(C_2O_4)_3]^{3-}$ exhibits optical isomerism.

Isomers are groups of compounds which have the same molecular formulae but different arrangements of atoms.

There are two general types of isomer. **Structural isomers**, as we saw above, have the same molecular formulae but their atoms are linked together in different sequences. **Stereoisomers** also have the same molecular formulae but in this case they have different spatial arrangements of their atoms. Stereoisomerism can be divided into two categories: geometrical isomerism and optical isomerism.

Before we go on to consider each type of stereoisomerism we should note that isomers—whether structural or stereo—are compounds which can be isolated and have distinct chemical and/or physical properties.

Geometrical Isomerism

Geometrical isomers differ in the geometrical arrangements of their atoms. 1,2-Dibromoethene is an example of a compound exhibiting geometrical isomerism. Its molecular formula is $C_2H_2Br_2$ and its structural formula CHBrCHBr. A molecule with the structural formula CHBrCHBr can have two distinct geometrical arrangements. These are called *cis* and *trans* isomers. These isomers have very distinctive physical properties. For example, their melting and boiling points are quite different:

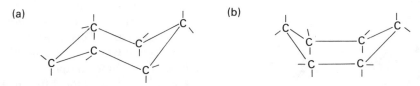

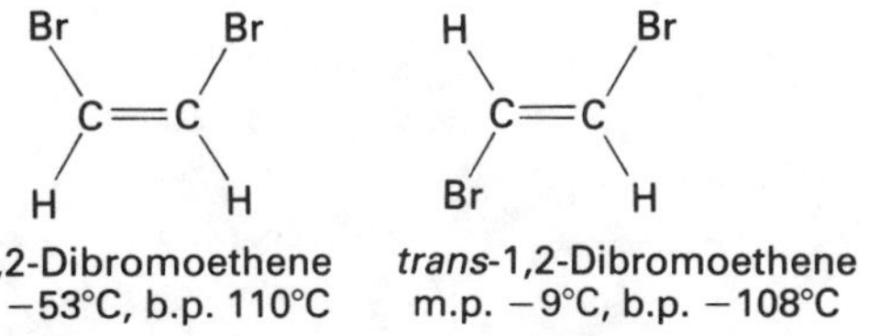

cis-1,2-Dibromoethene   *trans*-1,2-Dibromoethene
m.p. −53°C, b.p. 110°C   m.p. −9°C, b.p. −108°C

The word *cis* means 'on the same side' whereas *trans* means 'on opposite sides'. This type of isomerism is sometimes known as *cis–trans* isomerism. The double bond in the molecule fixes the atoms in either of these two arrangements.

Rotation about the double bond is not possible. We should compare this with 1,2-dibromoethane. In this compound free rotation about the carbon–carbon bond is possible. The following two arrangements are thus not isomers:

These two arrangements are equivalent to the staggered and eclipsed conformations we referred to above. A molecule might be in the staggered conformation at one instant and in the eclipsed conformation the next. It is thus not possible to isolate molecules in either conformation. Since the spatial arrangement is not fixed and since the separate molecules cannot be isolated we cannot regard these as isomers.

Finally, we should note that a compound with the molecular formula $C_2H_2Br_2$ can exist as either of two structural isomers. These are CHBrCHBr which, as we have just seen, can have two geometrical isomers. The other has the structural formula CH_2CBr_2. This structure does not have geometrical isomers. The formulae CBr_2CH_2 and CH_2CBr_2 are simply the same structures written differently. They represent the same compound.

A good example of the difference in chemical properties of geometrical isomers is provided by the geometrical isomers of butenedioic acid (*see* figure 17.26). The *cis* isomer is commonly known as maleic acid and melts at 139–140°C. When heated to 160°C or to 100°C under reduced pressure it loses water to give the

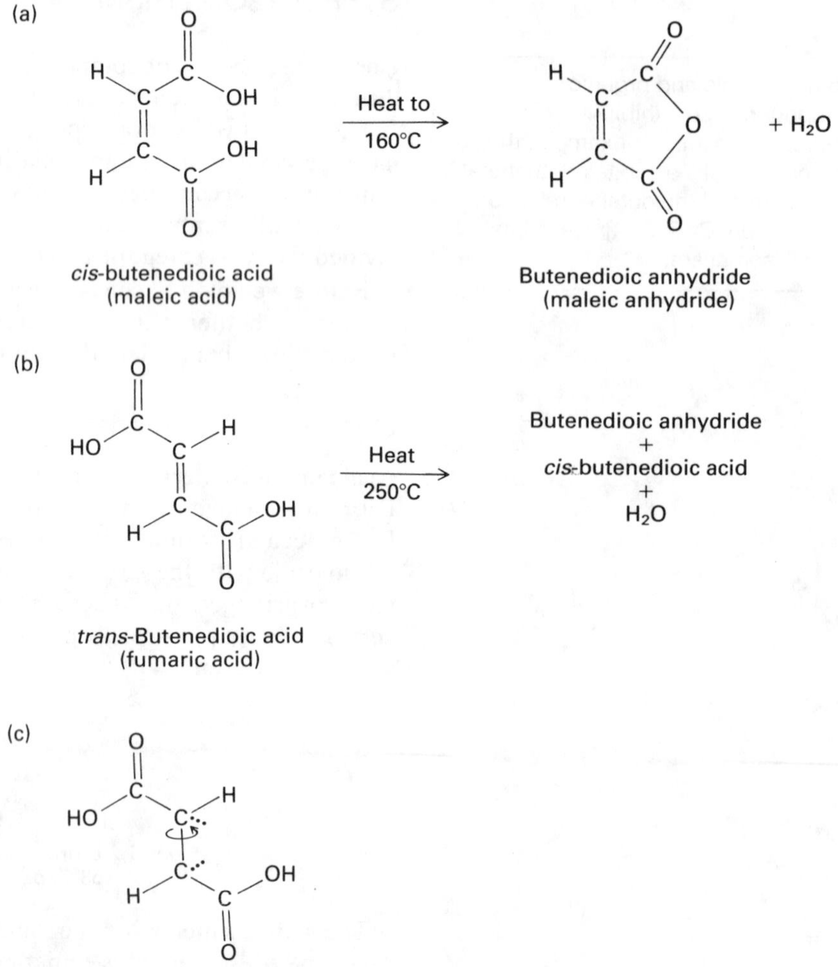

(a)

cis-butenedioic acid
(maleic acid)

Heat to 160°C

Butenedioic anhydride
(maleic anhydride)

+ H_2O

(b)

trans-Butenedioic acid
(fumaric acid)

Heat 250°C

Butenedioic anhydride
+
cis-butenedioic acid
+
H_2O

(c)

Figure 17.26

anhydride in low yield (figure 17.26a). The *trans* isomer, commonly known as fumaric acid, melts at 300°C. When heated to 200°C it sublimes. On further heating to between 250 and 300°C it rearranges to give the *cis* isomer and a small amount of the anhydride (figure 17.26b). The reaction mechanism involves the breaking of the π bond between the two carbon atoms. This is followed by rotation about the σ bond (*see* figure 17.26c) before the π bond is reformed.

Optical Isomerism

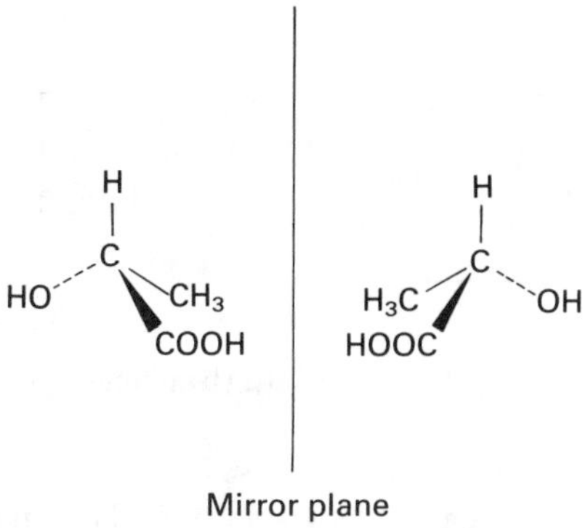

Any carbon atom with four different atoms or groups attached to it is **asymmetric**. A molecule containing one or more asymmetric carbon atoms is usually—although not always—also asymmetric. Asymmetric atoms and asymmetric molecules are known as chiral atoms and chiral molecules respectively. An example of a compound with a single asymmetric carbon atom is 2-hydroxypropanoic acid (otherwise known as lactic acid). The central carbon is asymmetric since it is bonded to four different atoms or groups. Thus, no matter how the molecule is rotated, twisted or turned, it cannot be superimposed on its mirror image (*see* figure 17.27). The two molecules, which are mirror images of each other, are thus isomers. They are called **enantiomers**.

Enantiomers may exist separately or as mixtures. A mixture containing equal numbers of moles of each enantiomer is known as a **racemic mixture**. Separation of a racemic mixture into the two pure enantiomers is known as **resolution**. The crystals of enantiomers are mirror images of one another.

A pair of enantiomers are chemically and physically identical with each other in almost every respect except for one vital difference. This difference is their **optical activity**.

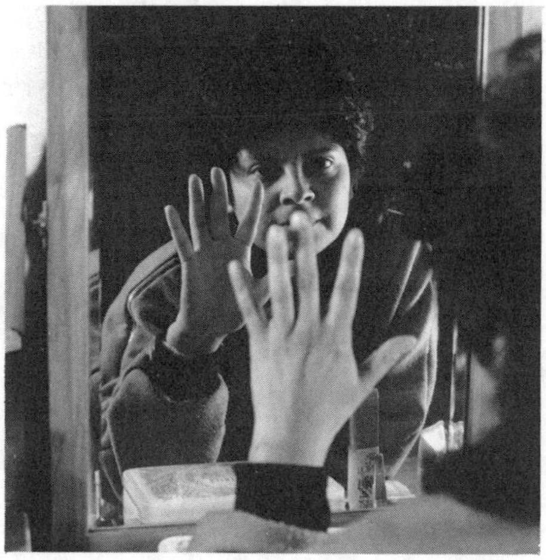

Mirror plane

Figure 17.27 2-Hydroxypropanoic acid (lactic acid)

Chirality

If we hold our left hand up into a mirror it appears in the mirror as our right hand. The **mirror image** of our left hand is thus our right hand. Now imagine—God forbid—that both our hands are cut off at the wrists in an accident and by mistake the right hand is sewn back onto the left wrist. No matter how it is sewn back—whether palm up or palm down—it would not be the same as the left hand. Thus, although the right hand is the mirror image of the left hand it is not superimposable. This property is known as **chirality**. It extends to anything which can be right-handed or left-handed —even the feet! The word 'chiral' comes from the Greek word for hand.

Chirality arises from lack of symmetry. Lack of symmetry is called **asymmetry**. Any pair of objects which are mirror images of each other are asymmetric.

What is Optical Activity?

The waves of a ray of normal light vibrate in all directions at right angles to the direction in which the ray is travelling. Figure 17.28a shows a ray of light vibrating in four different directions, that is in four different planes. A cross-section of these planes in also shown in the figure. Light which vibrates in one plane only is called plane-polarised light. This is represented in figure 17.28b.

A compound which can rotate plane-polarised light so that the light vibrates in a different plane is said to be **optically active**. In order that a compound be optically active, its molecules (or ions) must be asymmetric. All compounds which contain a single asymmetric carbon atom exhibit optical activity.

All enantiomers exhibit this property. They are thus sometimes known as **optical isomers**. If one enantiomer rotates a plane of light clockwise, then the other will rotate it anticlockwise. Clockwise rotation is called **dextrorotation** (*see* figure 17.29). The enantiomer which exhibits dextrorotation is given the sign (+).

Side view

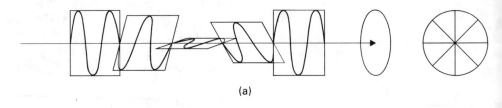

(a)

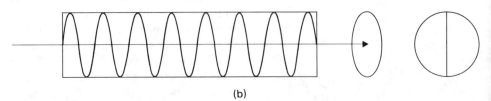

(b)

Figure 17.28   Unpolarised and plane-polarised light: (a) unpolarised; (b) plane-polarised

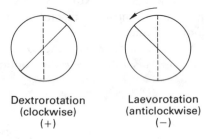

Dextrorotation   Laevorotation
(clockwise)   (anticlockwise)
(+)   (−)

Figure 17.29   Rotation

Anticlockwise rotation of plane-polarised light is called **laevorotation.** The enantiomer which does this is given the sign (−).

Optical isomerism is of immense importance in biological systems. For example, the amino acids used in the construction of proteins are all optically active—except the simplest member, which does not have an asymmetric carbon. The (+) and (−) isomers of the amino acid with the IUPAC name 2-aminopropanoic acid but commonly known as alanine are shown in figure 17.30. Only the (+) isomer (on the left) occurs in nature. If we replace the CH_3 group in alanine with other groups R, then we find that all the other naturally occurring amino acids have the same configuration as that of (+)-alanine. However, the sign of rotation may be (+) or (−) depending on the group R. Many carbohydrates are optically active. Glucose is an example.

We have already noted that a pair of enantiomers are chemically and physically identical except in respect to optical activity. The chemical activity of each of a pair of enantiomers may be completely different in reactions with other optically active compounds. Many biochemical reactions are very stereospecific. This is particularly true for enzymes.

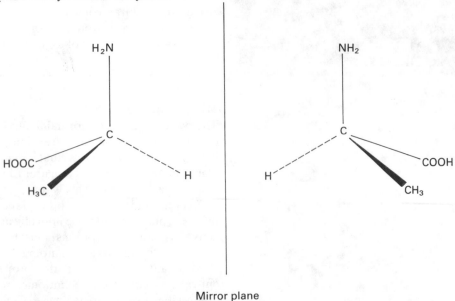

Mirror plane

Figure 17.30   2-Aminopropanoic acid (alanine)

The Polarimeter

The angle through which an enantiomer rotates plane-polarised light is specific for that enantiomer. The angle can be measured using a polarimeter (*see* figure 17.31). This normally uses a monochromatic source of light such as a sodium lamp. **Monochromatic light** consists of a single wavelength only whereas ordinary white light consists of light with all wavelengths over the visible range. The monochromatic light is unpolarised. It is therefore passed through a polariser which converts it to **plane-polarised light**. The plane-polarised light then passes through a tube containing a solution of the sample whose angle of rotation is to be measured. On emerging from the sample tube the plane-polarised light has been rotated either clockwise or anticlockwise through the angle to be measured. The direction of rotation is defined with respect to the observer and measured in a polarimeter. The angle is measured using an analyser. This is a device which only allows plane-polarised light to pass through it. It is initially set in line with the plane-polarised light transmitted by the polariser before it is rotated by the sample. Plane-polarised light rotated by the sample cannot pass through the analyser when it has this initial setting. The analyser is then slowly rotated until the rotated light can pass through. At this point its transmission plane is in line with that of the rotated plane-polarised light (*see* figure 17.31b). The angle at this point is measured.

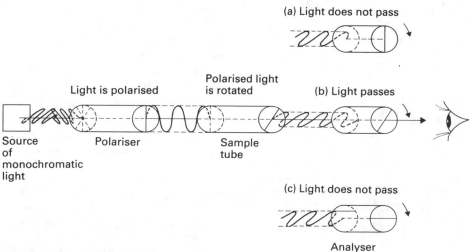

Figure 17.31   A polarimeter

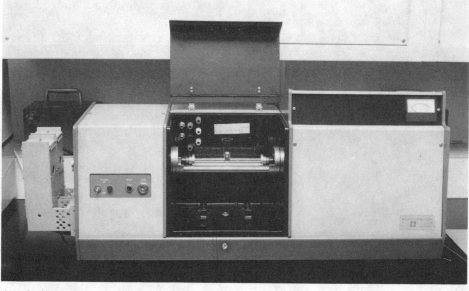

A polarimeter

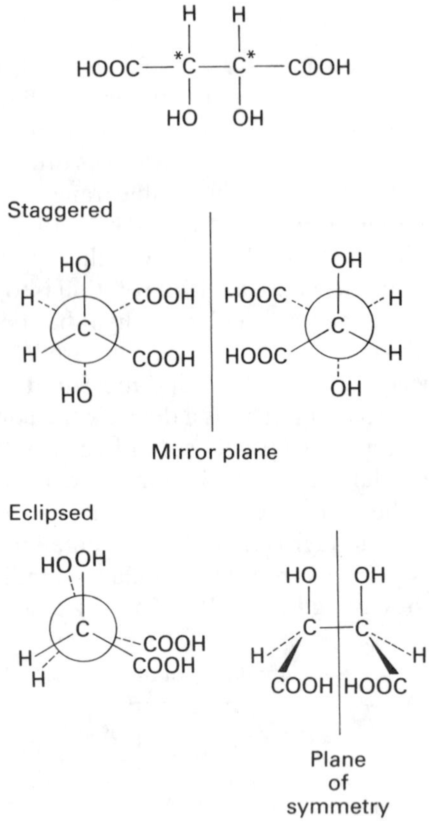

Figure 17.32 2,3-Dihydroxybutanedioic acid (tartaric acid)

Compounds Containing Two or More Asymmetric Carbon Atoms

A compound containing two or more asymmetric carbon atoms can exist as three or more stereoisomers. An example is 2,3-dihydroxybutanedioic acid (commonly known as tartaric acid). Its two asymmetric carbon atoms are labelled with asterisks in figure 17.32. This acid has three stereoisomers. Newman projections of these are shown in figure 17.32. All three isomers may exist in staggered, eclipsed or skew conformations. Two of the three isomers are completely asymmetric. They possess neither a **plane of symmetry** nor a **centre of symmetry** no matter how one half of the molecule is rotated with respect to the other. These two isomers form non-superimposable mirror images of one another. They are therefore enantiomers. In figure 17.32 they are shown in their staggered conformations. The two asymmetric carbons of one of the enantiomers rotates plane-polarised light to the right. The enantiomer is thus dextrorotatory and given the symbol (+). The asymmetric carbon atoms in the other enantiomer rotate the plane-polarised light to the left. The enantiomer is thus laevorotatory and given the symbol (−).

The third stereoisomer also has two asymmetric carbon atoms but overall the molecule is symmetric. This molecule has a plane of symmetry at right angles to the bond joining the two central carbon atoms. This stereoisomer is shown in the eclipsed conformation in figure 17.32. Since the molecule is symmetric it is not optically active. One of the asymmetric carbon atoms rotates plane-polarised light to the right and the other rotates it equally to the left. The overall effect is zero.

When a stereoisomer containing two or more asymmetric carbon atoms is **optically inactive** due to the symmetry of the molecule it is said to be **internally compensated**. Any stereoisomer which is not optically active (and thus not an enantiomer) is called a **diastereoisomer**. The isomer with the eclipsed conformation in figure 17.32 is thus a diastereoisomer.

A racemic mixture of two enantiomers is also optically inactive. This is because the dextrorotation of one enantiomer compensates for the laevorotation of the other. The net rotation is thus zero. A racemic mixture is given the symbol (±) and said to be **externally compensated**.

SUMMARY

1. Methane has a **tetrahedral** structure.
2. Ethene has a **planar** structure.
3. Ethyne has a **linear** structure.
4. The p electrons in an aromatic ring structure are **delocalised**, forming a π**-electron cloud**.
5. **Conformations** are the various spatial arrangements of atoms in a molecule.
6. **Staggered** and **eclipsed** are two extreme types of conformation. **Skew** conformations are in between these two extremes.
7. The cyclohexane ring structure may exist as **chair** or **boat** conformations.
8. **Stereoisomers** have different spatial arrangements of their atoms.
9. **Geometrical isomers** differ in the geometrical arrangements of their atoms. **Cis–trans isomerism** is an example of this type of isomerism.
10. **Optical isomers** are molecules which are non-superimposable **mirror images** of one another. They are known as **enantiomers**.
11. A **racemic mixture** consists of equal amounts of each of a pair of enantiomers. The mixture is **optically inactive**.
12. A compound which rotates **plane-polarised light** is said to be **optically active**.
13. The molecules of an optically active compound are **asymmetric**.
14. A **polarimeter** is used to measure the angle through which an enantiomer rotates plane-polarised light.

17.3 Reaction Mechanisms in Organic Chemistry

Since organic compounds are predominantly covalent compounds, their reactions inevitably involve the breaking and formation of covalent bonds. In this section we shall first examine how covalent bonds are broken and then see how this applies to the mechanisms of various types of organic reaction.

BREAKING COVALENT BONDS

We have already seen that a covalent bond consists of a shared pair of electrons. Single covalent bonds can be represented by short straight lines. For example:

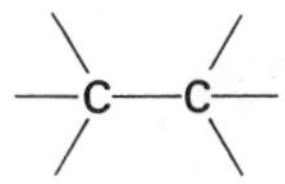

To emphasise that the bond between the two carbon atoms consists of two electrons we could represent this bond by two dots—as in the Lewis formulae (*see* section 2.1):

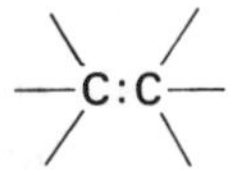

If we take this as an example we can see that the bond between the two carbon atoms can break in one of two ways. These are known as homolytic fission and heterolytic fission.

Homolytic Fission
In this type of fission the two shared electrons in the bond are split equally between the two atoms:

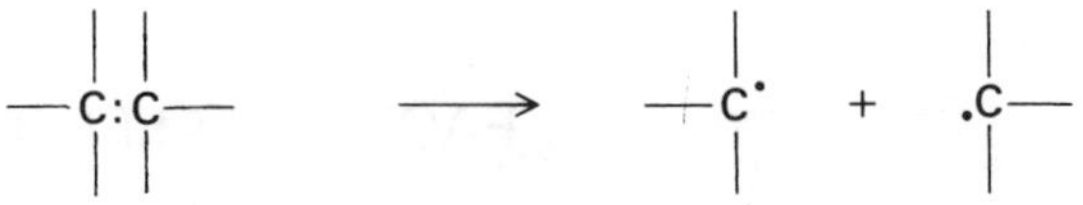

The resulting species are known as **radicals**. They are reaction intermediates and many only exist for a split second. Homolytic fission does not only occur between single carbon–carbon bonds. It can occur with any type of single covalent bond. For example:

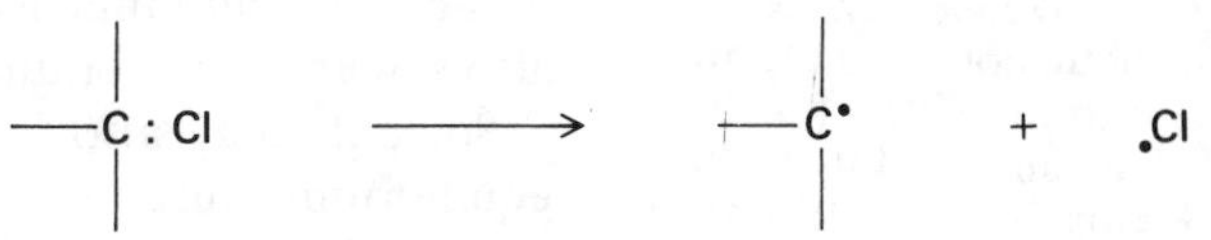

The movement of each single electron in homolytic fission is often depicted by a barbed or half-arrow thus:

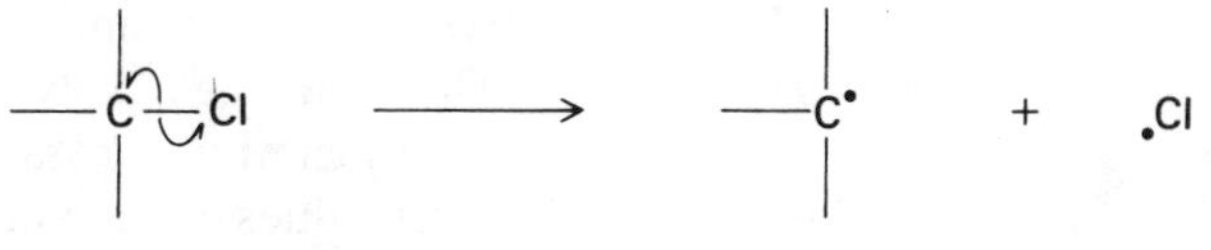

Radicals are also known as 'free radicals'. However radicals can be trapped in a way that they are no longer free to move. The term 'radical' is thus preferred to the term 'free radical' nowadays.

Homolytic fission normally occurs in reactions which take place in the gas phase or in non-polar solvents. The reactions are often catalysed by light or by other radicals. We shall meet some examples of this in the following chapter.

Heterolytic Fission

In this type of fission the two shared electrons in the bond are split unequally between the two atoms. One of the atoms keeps both electrons. As a result it acquires a negative charge. The other atom is deficient in one electron and thus has a positive charge. A species which contains a carbon atom with a negative charge is known as a **carbanion**. One with a positive charge on a carbon atom is called a **carbocation**. This was formerly known as a carbonium ion.

The heterolytic fission of a single carbon–carbon bond can occur in two ways. For example:

The movement of a pair of electrons during the heterolytic fission of a covalent bond is represented by a curly arrow thus:

The ions formed by heterolytic fission are often reaction intermediates. Thus, like radicals, these have only the shortest lifespans. Since organic reactions involving heterolytic fission of bonds produce ions, these reactions tend to take place in polar solvents.

FACTORS AFFECTING REACTIVITY OF ORGANIC COMPOUNDS

A number of factors influence the breaking and formation of bonds and thus the reactivity of organic compounds. We have already seen that the chemical properties of groups of compounds are often dominated by the functional group. However, even in a specific reaction with another compound, reactivity will vary widely. We saw in chapter 9 that temperature, pressure, state of subdivision and the presence of a catalyst all influence the rate at which a reaction proceeds. In chapter 7 we also saw that temperature can influence the equilibrium constant.

Solvents may also have a pronounced effect on a chemical process and its equilibrium constant. A classic example is provided by hydrogen chloride. When dissolved in water it ionises to form a strong acid (*see* section 8.1). The solution is thus a strong electrolyte. However, a solution of hydrogen chloride in methylbenzene does not conduct electricity. This type of effect is known as a **solvent effect**. Solvent effects include hydrogen bonding and dipole–dipole interactions between solvents and solute molecules (*see* chapter 2).

The influence of solvent on a weak carboxylic acid is shown by the values for the acid dissociation constant of ethanoic acid in various solvents (*see* table 17.9). The highest values occur when the solvent contains the highest proportion of water—a

Table 17.9 Effect of solvent on the acid dissociation constant of ethanoic acid at 25°C

Solvent	K_a/mol dm^{-3}
water	1.74×10^{-5}
90% water–10% methanol	1.25×10^{-5}
80% water–20% methanol	8.34×10^{-6}
80% water–20% dioxan	5.11×10^{-6}
55% water–45% dioxan	4.93×10^{-7}
30% water–70% dioxan	4.78×10^{-9}
benzene	0

polar solvent. When the proportion of a less polar solvent such as methanol or dioxan is increased, the acid dissociation constant increases.

Dioxan

$$\begin{array}{c} H_2C \overset{\displaystyle O}{<} CH_2 \\ H_2C \underset{\displaystyle O}{<} CH_2 \end{array}$$

Solvent effects may be classed as an environmental effect on the reactivity of a molecule. The environment of a functional group within a molecule can exert an influence on the reactivity of the functional group. This type of effect is known as a **structural effect**.

Structural effects within a molecule can be divided into two categories: electronic effects and steric effects.

Electronic Effects

Electronic effects may themselves be divided into several categories. The two most important are the inductive effect and resonance effect.

The Inductive Effect

This applies only to single covalent bonds between unlike atoms. Such bonds are polarised due to the different electronegativities of the two atoms. We have already seen that the covalent bond between a carbon atom and a chlorine atom is polarised. This can be represented as:

$$ -\overset{|}{\underset{|}{C}}{}^{\delta+}\!\!-Cl^{\delta-} $$

This polarisation can equally be represented by an arrow:

$$ -\overset{|}{\underset{|}{C}}\!\rightarrow Cl $$

The arrow shows that the carbon atom repels electrons and the chlorine atom, due to its higher electronegativity, attracts electrons. The chlorine atom thus acquires a greater electron density and the carbon atom a lower electron density. This shift in electron density from one atom to another and the resultant polarisation of the bond is known as the **inductive effect**.

Most atoms and groups are more electronegative than carbon and thus withdraw electrons from carbon. This withdrawal of electrons is called the negative inductive effect or $-I$ effect. However, some atoms and groups are less electronegative than the carbon atom and thus donate electrons to the carbon atom. This is known as the positive inductive effect or $+I$ effect. Alkyl groups are known to have a positive inductive effect. The effect increases with the number of alkyl groups substituted on the carbon atom thus:

$$ CH_3 \rightarrow CH_2 - \qquad \begin{array}{c} CH_3 \searrow \\ \qquad CH - \\ CH_3 \nearrow \end{array} \qquad \begin{array}{c} CH_3 \\ \downarrow \\ CH_3 \rightarrow C - \\ \uparrow \\ CH_3 \end{array} $$

Primary   Secondary   Tertiary

Positive inductive effect increases →

Resonance Effects

These occur in molecules with multiple bonds and operate through π orbitals. The effect results in the redistribution of π electrons and leads to the stabilisation of molecules and some radicals and carbon ions. An example of this occurs in the hydrolysis of 3-bromoprop-1-ene (allyl bromide) using aqueous alcoholic sodium hydroxide. The product is prop-2-en-1-ol (allyl alcohol):

$$CH_2 = CH - CH_2Br \longrightarrow CH_2 = CH_2OH$$

The reaction is thought to proceed via the propene carbocation intermediate. This exists as a **resonance hybrid** of two **limiting forms** (*see* chapter 2):

$$CH_2 = CH - CH_2^+ \longleftrightarrow {}^+CH_2 - CH = CH_2$$

The arrow $\longleftrightarrow$ indicates that the carbocation exists somewhere between the two forms.

The stabilisation of this resonance hybrid is due to the delocalisation of π electrons. This delocalisation of electrons can be represented as follows. First, the two π electrons of the unsaturated bond of one of the limiting forms is attracted to the positively charged carbon atom:

$$CH_2 = CH - CH_2^+ \longleftrightarrow {}^+CH_2 - CH = CH_2$$

The two π electrons in the other canonical form are similarly attracted to the positively charged carbon atom:

$$^+CH_2 - CH = CH_2 \longleftrightarrow CH_2 = CH - CH_2^+$$

However, it should be stressed that a resonance hybrid is *not* an equilibrium mixture of the two canonical forms but rather an average of the two.

The resonance effect is also known as the **mesomeric effect** or **conjugative effect**. The latter is used because the effect is particularly important in conjugated systems. Buta-1,3-diene is an example of a compound with such a system. The resonance effect is transmitted along the conjugated chain of carbon atoms. The two limiting forms are:

$$H_2C = CH - CH = CH_2 \longleftrightarrow {}^+H_2C - CH = CH - CH_2^-$$

Compounds containing the carbonyl group also exhibit the resonance effect:

$$\diagdown C = O \longleftrightarrow \diagdown \overset{+}{C} - \overset{-}{O}$$

> A **conjugated system** is a system of alternate single and double bonds in a molecule.

Steric Effects

The most important example of the effect of steric factors on chemical reactivity is **steric hindrance**. Steric hindrance can occur when large groups on a molecule 'get in the way' and thus hinder the reaction. An example is provided by the 2,6-disubstituted benzoic acids:

Because of the presence of the adjacent methyl groups, the carboxylic group is not free to rotate about its bond between the carbon atom and the benzene ring. It is thus fixed in a position at right angles to the benzene ring. As a result of these steric factors, the 2,6-disubstituted benzoic acids are resistant to normal methods of esterification.

Steric hindrance inhibits the *completely* free rotation about most single carbon–carbon bonds.

NUCLEOPHILES AND ELECTROPHILES

Table 17.10 Nucleophiles and electrophiles

Nucleophiles	Electrophiles
OH^-	H_3O^+
Cl^-	$>C^+{-}OH$
Br^-	$\geqslant C^+$
CN^-	
CH_3O^-	BF_3
NH_3	
H_2O	$AlCl_3$
ROH	

A chemical reaction between two reactants may be regarded as the attack of one species on the other. In this case the attacking species is called the **reagent** and the species which is under attack is called the **substrate**. The substrate is the reactant containing atoms to which bonds are made or whose bonds are broken.

Reagents in organic reactions are classified as nucleophiles or electrophiles. The word nucleophile means 'nucleus-loving' whereas the word electrophile means 'electron-loving'.

A **nucleophile** is a species which attacks a carbon atom by **donating an electron pair**. It is thus a **Lewis base** (*see* section 8.1).

An **electrophile** is a species which attacks a carbon atom by **accepting an electron pair**. It is thus a **Lewis acid**.

Examples of nucleophiles and electrophiles are shown in table 17.10.

TYPES OF ORGANIC REACTION

Organic reactions may be divided into two broad categories:

Homolysis reactions. These are radical reactions. We shall examine this type of reaction further in the following chapter. The kinetics and mechanism of this type of reaction were discussed in chapter 9.

Heterolysis reactions. These are essentially ionic reactions. They can be divided into three groups:

- substitution,
- addition,
- elimination.

Substitution Reactions

In these reactions an atom or group of atoms is displaced by another atom or group. For this reason, these reactions are sometimes known as displacement reactions. An example of this type of reaction is provided by the hydrolysis of chloromethane, to form methanol:

$$HO^- + CH_3{-}Cl \longrightarrow HO{-}CH_3 + Cl^-$$

The hydroxide ion is a nucleophile. This type of substitution is thus known as nucleophilic substitution. It is given the symbol S_N. The chloride ion is called the **leaving group**.

We can depict the nucleophile as Nu^- and the leaving group as Le. Using these symbols we can write the following general reaction for nucleophilic substitution at a saturated carbon atom in the alkyl group R:

$$Nu^- + R{-}Le \longrightarrow Nu{-}R + Le^-$$

Rate studies of this type of reaction show that there are two possible types of reaction for S_N reactions.

S_N1 Reactions

For some S_N reactions the rate equation (*see* section 9.1) is

$$\text{Rate} = k[\text{R—Le}]$$

The reaction is thus **first order** with respect to R—Le but **zero order** with respect to Nu⁻. First-order kinetics is good evidence that the rate-determining step is unimolecular. The reaction is thus given the symbol S_N1.

Since the nucleophile is not involved in the rate-determining step, the mechanism must involve at least two steps. The following mechanism has been proposed for such reactions:

> The reaction is zero order with respect to Nu⁻ since the rate of reaction is independent of the concentration of Nu⁻. Thus
>
> $$\text{Rate} = K[\text{Nu}^-]^0[\text{R—Le}]^1$$
> $$= K[\text{R—Le}]$$
>
> (since $[\text{Nu}^-]^0 = 1$)

$$(1) \quad \text{R—Le} \xrightarrow{\text{Slow}} \text{R}^+ + \text{Le}^-$$

$$(2) \quad \text{Nu}^- + \text{R}^+ \xrightarrow{\text{Fast}} \text{Nu—R}$$

The first step is the ionisation of R—Le to form a carbocation, R^+. This is the rate-determining step.

An example of an S_N1 type reaction is the alkaline hydrolysis of tertiary alkyl halides. For example

$$\text{HO}^- + (\text{CH}_3)_3\text{CBr} \longrightarrow (\text{CH}_3)_3\text{C}^+ + \text{Br}^-$$

In this case the rate equation is given by

$$\text{Rate} = k[(\text{CH}_3)_3\text{CBr}]$$

S_N2 Reactions

The rate equation of some S_N reactions is

$$\text{Rate} = k[\text{Nu}^-][\text{R—Le}]$$

In this case the reaction is first order with respect to the nucleophile and first order with respect to R—Le. It is thus second order overall. This provides good evidence that the rate-determining step is usually bimolecular. The reaction is thus given the symbol S_N2. Since both the nucleophile and the substrate R—Le are involved in the rate-determining step, it is thought to proceed in a single step involving a transition state (*see* section 9.2):

$$\text{Nu}^- + \text{R—Le} \longrightarrow [\text{Nu}\cdots\text{R}\cdots\text{Le}]^- \longrightarrow \text{Nu—R} + \text{Le}^-$$
$$\text{Transition state}$$

The hydrolyses of primary alkyl halides under alkaline conditions are S_N2 reactions. For example:

The rate equation for this reaction is

$$\text{Rate} = k[\text{OH}^-][\text{CH}_3\text{CH}_2\text{Br}]$$

So far we have only considered nucleophilic substitution at a saturated carbon atom. Nucleophilic substitution is also possible at an unsaturated carbon atom:

This type of reaction is called **nucleophilic acyl substitution**.

Electrophilic Substitution

Electrophilic substitution is also possible on benzene rings. In this type of substitution two of the delocalised π electrons on the benzene ring are donated to the electrophile. An unstable π complex containing both an electrophile and a leaving group is formed as an intermediate. These complexes can be illustrated with an incomplete circle to depict the loss of the two π eletrons:

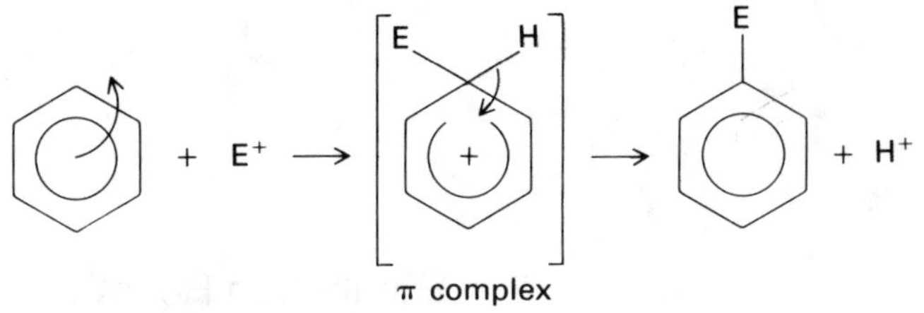

π complex

The nitration of benzene is an example of electrophilic substitution:

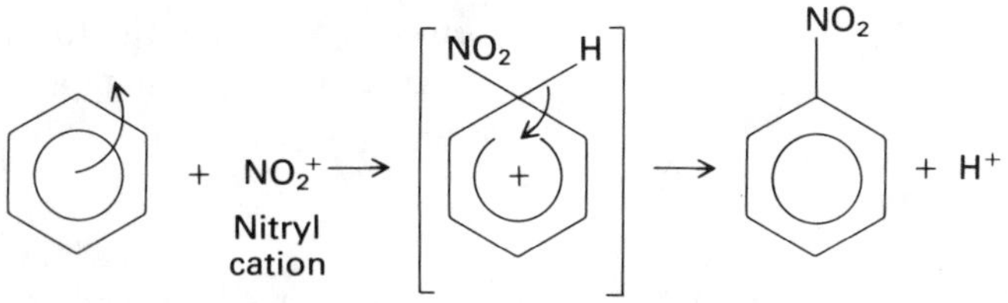

The nitration is carried out under reflux at 55 to 60°C using a nitrating mixture. This contains equal amounts of concentrated nitric acid and sulphuric acid. The two acids react to generate the nitryl cation, NO_2^+:

$$HNO_3 + 2H_2SO_4 \rightarrow NO_2^+ + H_3O^+ + 2HSO_4^-$$

Addition Reactions

In this type of reaction an electrophile or nucleophile is added to an unsaturated carbon atom. We shall consider one example each of electrophilic addition and nucleophilic addition.

An example of **electrophilic addition** is the reaction between hydrogen bromide and an alkene. The hydrogen bromide can be generated by adding concentrated sulphuric acid to sodium bromide (*see* section 16.2). The hydrogen bromide molecule is polar. This is because the bromine atom exerts a $-I$ effect on hydrogen. The molecule thus acts as a strong acid. The reaction is thought to take place in two stages. During the first stage the positively charged hydrogen atom attacks the double bond which acts as a source of electrons. An activated complex and a bromide ion are formed:

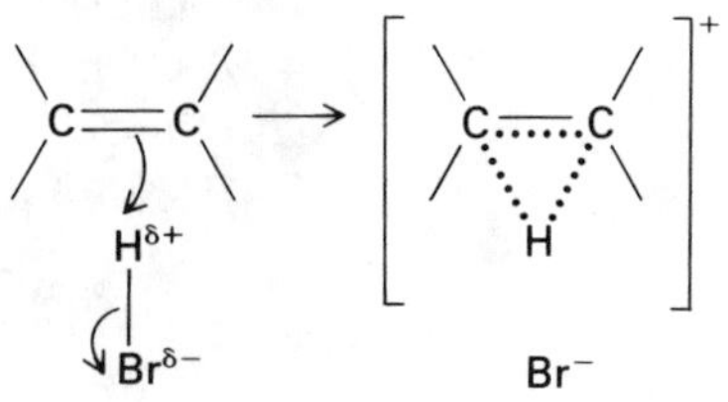

The bromide ion then attacks the complex, forming the alkyl bromide:

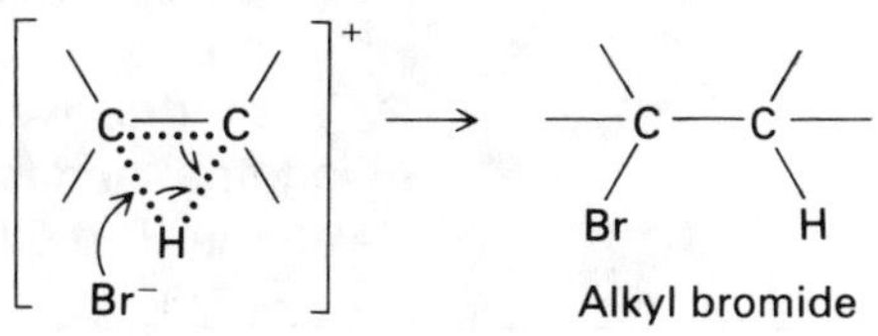

Alkyl bromide

The addition of hydrogen cyanide to an aldehyde or ketone is an example of nucleophilic addition. The aldehyde or ketone is first treated with an aqueous solution of sodium cyanide, NaCN. Excess mineral acid is then added, generating the hydrogen cyanide, HCN. The nucleophile is the cyanide ion, CN^-. This attacks the positively charged carbon atom on the carbonyl group. The positive charge and polarity of the carbonyl group arise from the resonance effect described above.

The reaction may be represented as follows:

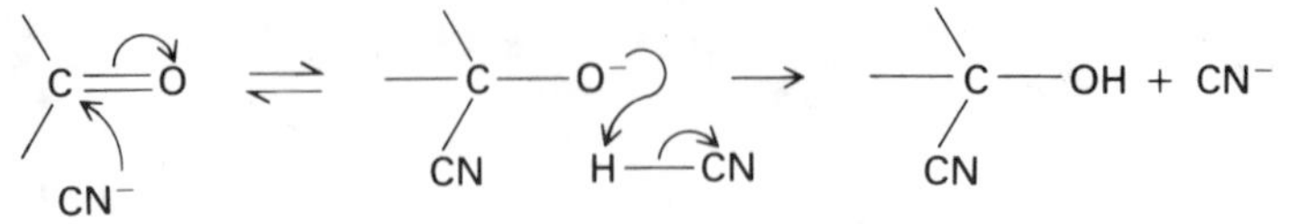

Elimination Reactions

These are the reverse of addition reactions. They involve the removal of atoms or groups of atoms from two carbon atoms joined by a single covalent bond. The removal results in the formation of a multiple bond.

An example of such a reaction is the base-catalysed elimination of hydrogen and a halogen from an alkyl halide. For example:

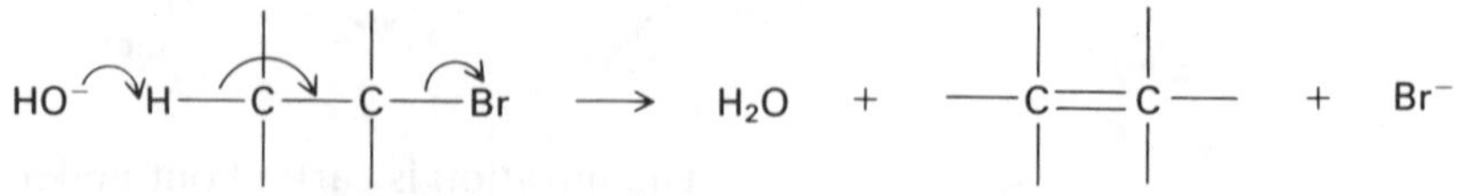

The reaction can be carried out by treating the alkyl halide with potassium hydroxide in ethanol at 60°C.

It should be noted that treatment of an alkyl halide with a hydroxide also results in nucleophilic substitution (*see* above). As a result both substitution and elimination reactions run concurrently, producing a mixture of substitution and elimination products. Which reaction predominates depends on a number of factors including the medium in which the reaction is carried out. Nucleophilic substitution of alkyl halides is carried out in the presence of water. Elimination reactions, on the other hand, are carried out in the absence of water and at a higher temperature.

SUMMARY

1. In **homolytic fission** of a bond the two shared electrons are split equally between the two atoms.
2. In **heterolytic fission** of a bond the two shared electrons are split unequally between the two atoms.
3. A **carbanion** is an ion containing a carbon atom with a negative charge.
4. A **carbocation** is an ion containing a carbon atom with a positive charge.
5. **Solvent effects** can have a pronounced effect on chemical processes and their equilibrium constants.
6. The influence of the environment of a functional group within a molecule on the reactivity of the functional group is known as a **structural effect**.
7. Electronic effects and steric effects are structural effects.
8. Two of the most important **electronic effects** are inductive effects and resonance effects.
9. The **inductive effect** is the shift in electron density from one atom to another and the resultant **polarisation** of the bond between the two atoms. The effect can be positive or negative.

10. Species with multiple bonds can exist as **resonance hybrids** of two or more **limiting forms**.
11. The **resonance effect** (or conjugative effect) is the stabilisation of resonance hybrids due to the **delocalisation** of electrons.
12. **Steric hindrance** can occur when large groups on a molecule 'get in the way' and hinder the reaction.
13. A **nucleophile** is a species which attacks a carbon atom by donating an electron pair. It is a **Lewis base**.
14. An **electrophile** is a species which attacks a carbon atom by accepting an electron pair. It is a **Lewis acid**.
15. **Homolysis** reactions are **radical** reactions.
16. **Heterolysis** reactions are essentially **ionic** reactions.
17. The displacement of a group in a molecule by a nucleophile is known as **nucleophilic substitution**. The displaced group is called the **leaving group**.
18. **Electrophilic substitution** on a benzene ring involves the donation of two delocalised electrons to an electrophile.
19. In **electrophilic addition** an electrophile is added to an unsaturated carbon atom.
20. The addition of hydrogen cyanide to an aldehyde or ketone is an example of **nucleophilic addition**.
21. In **elimination reactions** atoms or groups of atoms are removed from two carbon atoms joined by a single covalent bond. A multiple bond is formed as a result.

Examination Questions

1. (a) Give the structures of: A, 1,2-dibromopropane; B, 2-methyl-2-chloro-propane; C, 1-bromo-2-chloroethane.
   (b) (i) How would you obtain A from propene?
       (ii) Give a mechanism for the reaction.
   (c) (i) How would you convert B into the corresponding alcohol?
       (ii) Give a mechanism for the reaction.
   (d) Give a reaction scheme for converting C into butanedioic acid. You need only give reagents, reaction conditions and the structures of intermediate compounds.

   (L)

2. (a) State which of the following are polar compounds. Draw the structural formula of *one* of the polar compounds, indicating the bond polarity: A, $CH_3CH_2CH_3$; B, CH_3COCH_3; C, CH_3CH_2Cl; D, $CH_2{=}CH_2$.
   (b) Give the name of *two* classes of compounds which conform to each of the general formula of (i) $C_nH_{2n+2}O$ and (ii) $C_nH_{2n}O$.
   (c) Give and name the mechanism for the reaction of propene with concentrated sulphuric acid.

   (JMB)

3. (a) A non-cyclic organic compound X has a molecular formula $C_4H_{10}O$. Give the seven possible structures of X, labelling them to A to G.
   (b) Indicate, by using the appropriate letters, the isomer(s) which will
       (i) form hydrogen on adding sodium,
       (ii) when oxidised, form a ketone having the same number of carbon atoms,

(iii) give a mixture of two alkenes on dehydration.

(c) Give reagents and conditions to show how you would carry out (i) the oxidation and (ii) the dehydration reactions in (b).

(d) In fact X gives hydrogen when treated with sodium, forms an aldehyde on oxidation, and one alkene on dehydration. Using the appropriate letters, indicate possible structures of X.

(e) Select any **one** of the compounds listed in (a) and indicate how you would prepare it from any other compound containing four carbon atoms per molecule.

(L)

4.   What are the chief characteristics of a homologous series?
Give an account of the chemistry of
(a) aliphatic aldehydes, and
(b) aliphatic carboxylic acids.

The first member of a homologous series often exhibits some properties which are not typical of the series as a whole. Illustrate this statement by reference to methanoic acid.

(L)

5.   Compound A is a monoacid base and has the composition, 78.5% carbon, 8.4% hydrogen, and 13.08% nitrogen. 2.14 g of A react with 20.0 cm^3 of 1.0 mol dm^{-3} hydrochloric acid. With nitric(III) acid (nitrous acid), A gives B, C_7H_8O. B, on oxidation, forms C, $C_7H_6O_2$.

(H = 1; C = 12; N = 14)

(a) Deduce the molecular and structural formula of A, explaining the reaction with hydrochloric acid.

(b) Write the structural formulae of B and C, explaining the reactions involved.

(c) By giving reagents and reaction conditions, show the steps by which A can be obtained from phenylethanoic acid.

(d) Write the structural formula of D which is an isomer of A. Mention **two** reactions in which D will behave differently from A.

(e) How would you distinguish chemically between A and D? Describe the observations which will enable you to decide.

(SUJB)

6.   What do you understand by the terms (a) *structural isomerism*, (b) *cis-trans isomerism*?

A gaseous hydrocarbon X contains 85.7% of carbon by mass. When 0.140 g of X was introduced into a gas syringe, its volume (after correction to s.t.p.) was found to be 56.0 cm^3. When X was shaken with aqueous bromine the latter was decolourised.

Three structural isomers A, B and C of the hydrocarbon X were found to have the following properties.

(i) A exists as a pair of *cis–trans* isomers.

(ii) B underwent oxidation under certain conditions to produce methanal and a compound Y (empirical formula C_3H_6O) which gave an orange precipitate with 2,4-dinitrophenylhydrazine reagent and a red-brown precipitate on boiling with Fehling's solution.

(iii) C can be obtained by the dehydration of 2-methylpropan-2-ol.

Deduce the **full** structural formula of **each** of the isomers A, B and C, explaining your reasoning and giving balanced equations where possible.

(UCLES)

7.   Insects use certain small organic molecules to communicate with their fellows by releasing these molecules in small amounts. The structural formula of one such molecule is shown below. It is used to signal 'alarm' in ants.

$$CH_3CH_2CHCOCH_2CH_3$$
$$|$$
$$CH_3$$

(a) (i) Give the systematic name for the structure above.

     (ii) The molecule contains a chiral centre. Write down the formula and mark this centre with an asterisk [star] and explain what is meant by a chiral centre.

    (iii) Draw the *two* stereoisomers (optical isomers) of this molecule.

(b) Another alarm chemical found in ants is hex-2-enal:

$$CH_3CH_2CH_2CH{=}CHCHO$$

   (i) How would you show, by a chemical test, that hex-2-enal is unsaturated? Give an equation for the reaction involved.

  (ii) Hex-2-enal can exist as *two* geometric isomers. Draw the structures of the isomers and label the *cis* structure.

(c) Both of these alarm pheromones, as they are known, have relatively small molecules.

   (i) What advantage, to the insect, is there in the small molecular mass of the alarm pheromones?

  (ii) The small molecular size is increased when the compounds are reacted with 2,4-dinitrophenylhydrazine.
    Write an equation for the reaction of this reagent with one of the pheromones.

 (iii) What is the usual purpose of forming 2,4-dinitrophenylhydrazones from aldehydes or ketones?

(NISEC)

8. (a) What is meant by the term *optical activity*?

(b) What conditions must be fulfilled in order that a compound may exhibit optical activity?

(c) What are the essential requirements for a structure to exhibit geometrical (*cis–trans*) isomerism?

(d) Indicate which of the following structures may exhibit stereoisomerism, and, where they do, draw a diagram of each stereoisomer. You should indicate clearly what type(s) of stereoisomerism is(are) involved.

$$CH_3CH_2CH_2CH(OH)CH_3$$
$$CH_3CH_2CH(OH)CH_2CH_3$$
$$CH_3CH{=}CHCOOCH(CH_3)C_6H_5$$
$$CHCOOH$$
$$\diagup\diagdown$$
$$CH_2{-}CHCOOH$$

(O & C)

9. (a) Discuss optical isomerism with reference to 2-hydroxypropanoic (lactic) acid, including in your discussion an explanation of the terms *enantiomer* and *racemate*.

(b) Outline, by means of equations and by stating essential experimental conditions, how 2-hydroxypropanoic acid can be prepared (i) from ethanal and (ii) from 2-bromopropanoic acid. Include reaction mechanisms where appropriate.

(c) Account for the fact that 2-hydroxypropanoic acid prepared according to method (i) above is optically inactive whereas that obtained from natural sources is optically active.

(JMB)

10. How are the procedures of
(a) refluxing,
(b) distillation, and
(c) recrystallisation
carried out in chemistry laboratories?
  Explain the purpose of carrying out the procedures by reference to a *different* chemical reaction in each part of your answer, giving formulae and equations where appropriate.

(L, Nuffield)

11. Spectroscopic techniques suggest that an organic compound, W, has a relative molecular mass of 74.0 and contains a carbonyl group. W contains only the elements carbon, hydrogen, and oxygen. On complete combustion, 0.0444 g of W gave 0.0792 g of carbon dioxide and 0.0324 g of water.

$$(H = 1; C = 12; O = 16)$$

(a) Use the above data to find the molecular formula of W. It is essential that you explain each step of your working clearly.
(b) Suggest *four* possible structural formulae for W and give systematic (IUPAC) names for each isomer.
(c) Describe chemical tests which could be used to distinguish between *any two* of your isomers in (b) and write balanced equations for the reactions.
(d) Suggest how the mass of carbon dioxide may have been obtained in the quantitative analysis of W.

(SUJB)

12. (a) Explain briefly what is meant by the term *homolytic fission*.
(b) State the reaction conditions under which methylbenzene (toluene) is converted into (chloromethyl)benzene. Write an equation to indicate the formation of the reactive inorganic species and outline a mechanism for the reaction with methylbenzene.
(c) Write an equation for the reaction of (chloromethyl)benzene with potassium cyanide. State the type of reaction taking place and outline a mechanism.

(JMB)

13. This question is concerned with the nature of organic reactions, and their mechanisms. For parts (a), (b) and (c), consider the two reactions A and B below.

A $\quad\quad\quad$ $CH_3I + OH^- \rightarrow CH_3OH + I^-$
B $\quad\quad\quad$ $CH_2{=}CH_2 + Br_2 \rightarrow BrCH_2CH_2Br$

(a) From the compounds/ions which take part in these two reactions, give
  (i) one which is an addition product,
  (ii) two which are behaving as nucleophiles,
  (iii) two which are behaving as electrophiles,
  (iv) one which is behaving as a leaving group.
(b) Which, if either, of these reactions is an elimination reaction? (Write A, B, both or neither.)

(c) (i) What is a nucleophile?

    (ii) For each of the nucleophiles which you pick out in (a) (ii), indicate what the essential feature is in that species which enables it to be a nucleophile.

(d) Give an example of an electrophilic substitution reaction, carefully specifying which reactant is the electrophile.

(e) (i) Give an example of a free radical.

    (ii) What feature of this species determines that it is a free radical?

    (iii) Give the overall equation for a reaction which involves free radicals (you are not expected to show the mechanism).

(O & C)

14. Nucleophilic substitution takes place when bromoalkanes are hydrolysed with alkali in boiling aqueous ethanol.

(a) (i) What do you understand by the term 'nucleophilic substitution'?

    (ii) Write a balanced equation for the reaction of 1-bromopropane with hydroxide ion.

    (iii) How would you determine the rate of this reaction?

(b) The rate expression for the reaction is:

$$\text{Rate} = k[C_3H_7Br][OH^-]$$

    (i) What is the overall order of the reaction?

    (ii) What are the units for the rate constant k?

    (iii) What would be the effect of doubling the concentration of hydroxide ion?

(c) (i) Suggest a mechanism for the hydrolysis of 1-bromopropane which is based upon the kinetics of the reaction.

    (ii) Draw an energy profile for the reaction.

(d) (i) Compare the reactivities of bromobenzene and 1-bromopropane with hydroxide ions.

    (ii) How do the conditions for the reaction of bromine with benzene differ from those for the reaction with propane?

(NISEC)

15. The nitration and chlorination of benzene are examples of *electrophilic substitution* reactions.

(a) Explain what is meant by the terms:

    (i) a *substitution* reaction;

    (ii) an *electrophile*.

(b) (i) Under what conditions does the nitration of benzene to nitrobenzene take place?

    (ii) Under what conditions does the chlorination of benzene to chlorobenzene take place?

(c) Write the formulae for the electrophiles in both the *nitration* and *chlorination* of benzene, and, by means of equations, indicate how they are formed under the conditions stated in (b).

(AEB, 1982)

18 HYDROCARBONS

Petroleum Geochemistry and Fossil Fuel Exploration

Geochemistry and Exploration

Early theories about the controlling principle of petroleum occurrences were often limited in concept in that they mainly addressed the question of 'where' accumulations were located. It has become clear during the past 20 years, that to be able to answer the question *where*? it is also necessary to evaluate *why, when* and *how much* petroleum is present in a basin, and to understand and establish the generation, migration and accumulation processes. This understanding is essential if we are to improve our petroleum exploration success ratio.

The formation of hydrocarbons is currently understood as a complex series of geochemical processes (*see* figure 18.1) within a source rock by which the original components of biological systems (natural products) are converted to hydro-carbons, and to a lesser extent, into polar compounds of varying degree of thermodynamic stability during sedimentation and burial at elevated temperature nd pressure in the subsurface. The primary migration of liquid and gaseous

Offshore oil-well

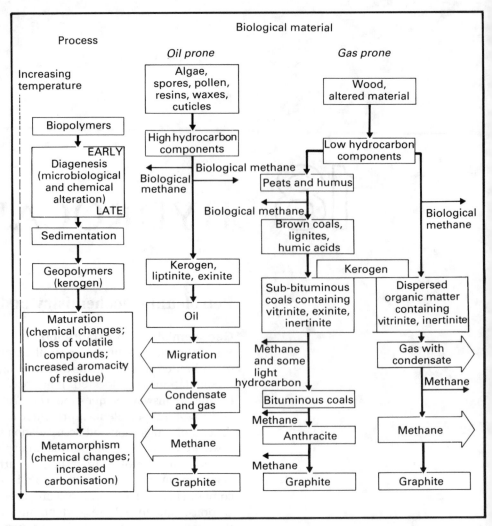

Figure 18.1   Geochemical processes leading to the formation of hydrocarbons

products out from the source rock and the subsequent secondary migration (via carrier horizons, faults, etc.) into porous reservoir rocks leads to the formation of hydrocarbon accumulations where further migration is halted by a trap.

Compounds with similar chemical structures are found in the organic extracts from sediments and also in petroleum. The compounds of major significance to the petroleum geochemist are those possessing 'biological marker' ('chemical fossil') characteristics. Such hydrocarbons are closely related to the compounds occurring in biological systems (e.g. lipids, pigments and metabolites) from which petroleum is formed. These compounds not only demonstrate the biogenic origin of the hydrocarbons, but also are capable of yielding very specific information regarding the hydrocarbon source rocks, maturation and generation, migration and the biodegradation of petroleum.

Hydrocarbon Source Rocks

The term hydrocarbon source rock is probably best defined as 'a fine-grained sediment that in its natural setting has or could generate and release significant amounts of oil and/or gas'. The classification of source rocks is defined in terms of amount and type of organic matter, its state of organic maturation (changes caused by temperatures of approximately 50–180°C), and the nature and amount of hydrocarbons capable of being produced. Organic matter ('kerogen')* in

---

* Kerogen (*keros* = wax; *gen* = that which produces) is the disseminated organic matter of rocks that is insoluble in organic solvents, non-oxidising mineral acids and bases.

sediments occurs in many different forms, but can be classified into four main types:

- *Liptinites* have a very high hydrogen, but low oxygen, content due to the presence of aliphatic carbon chains. They are considered to have been derived mainly from algal material (often bacterially degraded) and have high potential for petroleum.
- *Exinites* contain a high hydrogen content (but lower than liptinites), with aliphatic chains and some saturated naphthene and aromatic rings and oxygen-containing functional groups. This organic matter is derived from membranous plant materials such as spores, pollen, cuticle and other structured portion of plants. Exinites have a good potential for oil, can generate condensate† and have a good potential for gas at higher maturation levels.
- *Vitrinites* have a low hydrogen content, high oxygen content and consist mainly of aromatic structures with short aliphatic chains connected by oxygen-containing functions. They are mostly derived from structured woody(ligno-cellulose) materials and have a limited potential for oil, but a high potential for gas.
- *Inertinites* are the black opaque debris (high carbon, low hydrogen) that are derived from highly altered woody precursors. They have no potential for oil or gas.

The main factors for recognition of a hydrocarbon source rock are its content of kerogen, its type of organic matter and stage of organic maturation. Good source rocks ideally require about 2–4% organic matter content of a suitable type to generate and release their hydrocarbons. Under favourable geochemical conditions, oil can be generated from sediments containing *liptinite* and *exinite* organic matter. Gas is usually generated from *vitrinite*-rich source rocks or by thermal cracking of previously generated oil.

The burial of a sedimentary basin results in the organic matter contained being subjected to an increasingly higher temperature, which causes the thermal degradation of kerogen to form petroleum-like products. There are limits to the time and temperature (depth) at which petroleum can be formed in commercial amounts. Temperature can be traded for time (assuming a first-order reaction and applying Arrhenius equation). For example, if a quantity of petroleum formed in approximately 20 million years at 100°C, it would take about 40 million years to form at 90°C or 80 million years at 80°C. The rate of hydrocarbon generation from kerogen appears to double for each 10°C increment. However, the chemical composition of kerogen is very variable and this time–temperature relationship is only a useful working approximation.

Current geochemical evaluations show a typical North Sea basin thermal gradient of about 1.6–1.9°F per 100 ft. Such gradients suggest that the organic-rich source rocks generated their liquid hydrocarbons at a burial depth of about 8000–13 000 ft, which were attained 50–80 million years ago. Lighter oils and condensates† were probably generated in the zone from 13 000 to 16 000 ft, and the methane (dry gas) generation zone is estimated to be at depths in excess of 16 000 ft.

Jim Brooks

---

† Condensate = hydrocarbon mixture that is gaseous in the reservoir, but condenses into liquid when produced.

LEARNING OBJECTIVES
After you have studied this chapter you should be able to

1. Indicate how the molecular formulae of gaseous hydrocarbons can be determined by **eudiometry**.
2. Describe the general physical properties of **alkanes, alkenes, alkynes** and **arenes**.
3. Compare and give specific examples of the chemical properties of alkanes, alkenes, alkynes and arenes—particularly with respect to
   (a) **combustion** and other oxidation reactions,
   (b) reactions with halogens and hydrogen halides,
   (c) **hydration**,
   (d) **cracking**.
4. Describe the laboratory preparation of **ethene** and **ethyne**.
5. Explain the term **aromaticity**.
6. Discuss the structure of **benzene**.
7. Describe briefly the following chemical reactions of benzene:
   (a) **nitration**,
   (b) **halogenation**,
   (c) **sulphonation**,
   (d) **Friedel–Crafts reactions**,
   (e) **addition** reactions.
8. Outline the chemical properties of **methylbenzene**.
9. Give examples of the **directing power of substituents** on the benzene ring.
10. Write a short account of the occurrence of hydrocarbons.
11. Briefly indicate how hydrocarbons are manufactured from **coal**.
12. Outline the main stages in **petroleum refining**.
13. Compare the properties and uses of the main **fractions** produced by the **distillation** of **crude oil**.
14. List the main uses of alkanes, alkenes, alkynes and arenes.
15. Discuss the various types of **pollution** resulting from the use of crude oil and its refined products.

18.1 Aliphatic Hydrocarbons

We saw in the previous chapter that **hydrocarbons** are covalent compounds containing hydrogen and carbon only. In this section we are concerned with aliphatic hydrocarbons only. These are unbranched chain or branched hydrocarbons. Three of the most important groups of aliphatic hydrocarbons are shown in table 18.1.

The alkanes are **saturated** compounds since they contain single covalent bonds only. The alkenes and alkynes are **unsaturated** compounds since they contain double and triple covalent bonds respectively.

EUDIOMETRY

When any hydrocarbon burns in excess oxygen the following general equation applies:

$$C_xH_y + \left(x + \frac{y}{4}\right)O_2 \rightarrow xCO_2 + \frac{y}{4}H_2O \qquad (1)$$

x and y are whole numbers.

Table 18.1 Aliphatic hydrocarbons

	General formula of homologous series	Traditional name	Example
alkanes	C_nH_{2n+2}	paraffins	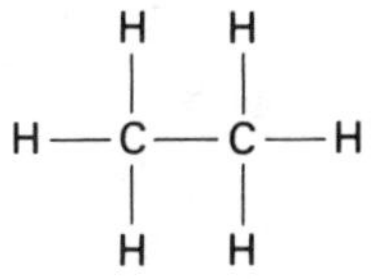 Ethane
alkenes	C_nH_{2n}	olefins	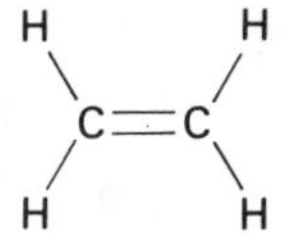 Ethene
alkynes	C_nH_{2n-2}	acetylenes	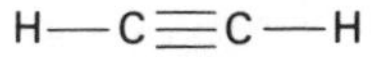 $H-C\equiv C-H$ Ethyne

This general reaction forms the basis of a technique used to determine the molecular formulae of gaseous hydrocarbons. The technique is known as **eudiometry**. A known volume of the gaseous hydrocarbon is mixed with a known volume of oxygen in a graduated combustion tube. The oxygen is in excess. The mixture is ignited. On cooling the water vapour formed in the reaction condenses leaving carbon dioxide and excess oxygen as the only gases in the combustion tube. This volume is measured. The gases are then shaken with potassium hydroxide solution. This absorbs the carbon dioxide, leaving the excess oxygen.

A eudiometric determination thus involves four measurements—all made at the same temperature and pressure:

V_1 = volume of hydrocarbon
V_2 = total volume of oxygen
V_3 = volume of excess oxygen + volume of carbon dioxide
V_4 = volume of excess oxygen

Thus,

Volume of oxygen consumed = $V_2 - V_4$
Volume of carbon dioxide = $V_3 - V_4$

From equation (1) and according to Avogadro's hypothesis we see that one volume of C_xH_y produces x volumes of CO_2. V_1 volumes of C_xH_y therefore produces xV_1 volumes of CO_2. Thus

$$xV_1 = V_3 - V_4$$
$$x = \frac{V_3 - V_4}{V_1} \tag{2}$$

By similar arguments it is possible to show that

$$x + \frac{y}{4} = \frac{V_2 - V_4}{V_1}$$

Thus,

$$y = 4\left[\left(\frac{V_2 - V_4}{V_1}\right) - x\right] \tag{3}$$

EXAMPLE

10 cm^3 of a gaseous hydrocarbon were bubbled into the combustion tube of an eudiometer. Oxygen was then bubbled in until the total volume was 110 cm^3. After ignition and then cooling the volume was 80 cm^3. The gases in the tube were then shaken with potassium hydroxide solution. This caused the volume to be reduced to 50 cm^3. What was the hydrocarbon?

SOLUTION

Values for x and y in the general formula C_xH_y can be calculated from equations (2) and (3). The volumes needed to solve these equations are obtained from the experimental data:

$$V_1 = 10 \text{ cm}^3$$
$$V_2 = 110 \text{ cm}^3 - 10 \text{ cm}^3 = 100 \text{ cm}^3$$
$$V_3 = 80 \text{ cm}^3$$
$$V_4 = 50 \text{ cm}^3$$

Substituting values for V_1, V_3 and V_4 into equation (2) we obtain

$$x = \frac{80 \text{ cm}^3 - 50 \text{ cm}^3}{10 \text{ cm}^3} = 3$$

Substituting values for V_1, V_2 and V_4 into equation (3) we obtain

$$y = 4\left[\left(\frac{100 \text{ cm}^3 - 50 \text{ cm}^3}{10 \text{ cm}^3}\right) - 3\right] = 8$$

The hydrocarbon thus has the molecular formula C_3H_8. This is propane.

ALKANES

The names and formulae of the first six alkanes in the homologous series are given in table 17.3. The IUPAC rules for nomenclature of alkanes is also described in the same section.

Physical Properties

Both the boiling and melting points of the alkanes increase with the number of carbon atoms. Figure 18.2 shows how the boiling points of the first eight alkanes in the homologous series increase with increasing number of carbon atoms. The alkanes with one to four carbon atoms—that is from methane to butane—are all colourless gases at room temperature and pressure. Alkanes with five to 16 carbon atoms are liquids and those with more carbon atoms are solids.

All alkanes are insoluble in water but soluble in organic solvents such as ethanol and ethoxyethane.

Laboratory Preparation

Alkanes are readily available from natural sources (*see* section 18.3 below) and so it is not usually necessary to prepare them either in the laboratory or industrially. Furthermore, alkanes are normally used as mixtures and so it is rarely necessary to separate and purify a specific alkane. However, pure alkanes can be prepared in the laboratory by a number of routes. For example, alkanes can be prepared by reduction of the corresponding alkenes using a finely divided palladium or platinum catalyst (*see* below). Alkanes can also be prepared in the laboratory by the hydrolysis of alkylmagnesium halides (Grignard reagents, *see* chapter 19).

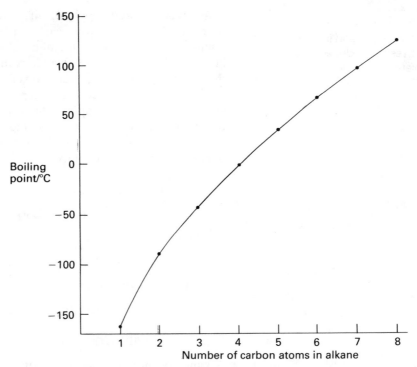

Figure 18.2 Boiling points of the first eight unbranched chain alkanes in the homologous series

Chemical Properties

Alkanes are relatively inert compounds. For example, they do *not* react with:

- strong acids such as hydrochloric acid;
- strong bases such as sodium hydroxide;
- strong oxidising agents such as potassium manganate(VII);
- strong reducing agents such as sodium.

Their lack of reactivity is due to the lack of electron-deficient sites on the alkane molecules. This is because hydrogen and carbon have similar electronegativity values (*see* table 2.2).

However, alkanes do undergo **substitution**—with chlorine for example—and, of course, they undergo **combustion**. In both these types of reactions the covalent bonds in the alkane molecules undergo homolytic fission (or **homolysis**). This means that they are **radical** reactions.

Combustion

Alkanes burn in a plentiful supply of air or oxygen to produce carbon dioxide and water. In a limited supply of oxygen, alkanes burn to form carbon monoxide and water.

Cigarette smoke contains relatively high concentrations of carbon monoxide

Carbon monoxide is produced when car engines are run in enclosed spaces. High levels are often found in busy traffic areas where the streets are enclosed by tall buildings. Levels of over 60 ppm have been detected in inner cities whereas the World Health Organisation (WHO) recommends a limit of 9 ppm over an eight-hour period and no more than 36 ppm over an hour period. These limits are recommended because of the toxicity of carbon monoxide. It competes with oxygen for the haemoglobin in blood and thus interferes with respiration. Exposure to carbon monoxide causes tiredness and headaches. Carbon monoxide is also produced in relatively high concentrations in cigarette smoke.

Carbon black, which is used as a pigment for paper and inks and in rubber tyres, is produced by the incomplete combustion of alkanes.

In a very limited supply of air, alkanes burn to form carbon as one of the products. It is the glowing carbon particles in flames which cause them to be luminous. For example, methane burns in a Bunsen burner with a hot non-luminous flame when the air-hole is fully open. If it is closed and the air supply thus severely restricted, a luminous flame is produced.

The mechanisms of combustion processes are complicated and not fully understood. They involve complex sequences of radical reactions. The overall reactions can, however, be represented as simple stoichiometric equations. Combustion of methane, for example, will involve one or more of the following reactions depending on the availability of air or oxygen:

in a plentiful supply of oxygen

$$CH_4(g) + 2O_2(g) \rightarrow CO_2(g) + 2H_2O(g)$$

in a limited supply of oxygen

$$CH_4(g) + \tfrac{3}{2}O_2(g) \rightarrow CO(g) + 2H_2O(g)$$

in a low supply of oxygen

$$CH_4(g) + O_2(g) \rightarrow C(s) + 2H_2O(g)$$

Reactions with Chlorine

Alkanes do not react with chlorine in the dark. However, in the presence of sunlight they undergo a series of substitution reactions. For example, methane and chlorine react explosively in sunlight, producing a mixture of chloromethane, CH_3Cl, dichloromethane, CH_2Cl, trichloromethane, $CHCl_3$, and tetrachloromethane, CCl_4.

This reaction is an example of a photochemical chain reaction (*see* section 9.2). It takes place in three stages:

Initiation. This is the homolytic fission of chlorine molecules producing chlorine radicals:

$$Cl_2 \xrightarrow{h\nu} 2Cl^\bullet$$

Propagation. This involves a number of reactions resulting in the formation of the products CH_3Cl, CH_2Cl_2, $CHCl_3$ and CCl_4. For example:

(i)	$CH_4 + Cl^\bullet \rightarrow CH_3^\bullet + HCl$
(ii)	$CH_3^\bullet + Cl_2 \rightarrow CH_3Cl + Cl^\bullet$
(iii)	$CH_3Cl + Cl^\bullet \rightarrow CH_2Cl^\bullet + HCl$
(iv)	$CH_2Cl^\bullet + Cl_2 \rightarrow CH_2Cl_2 + Cl^\bullet$

Termination. This occurs when two radicals combine. The energy evolved in the process is dissipated to a third body. This may be any molecule present in the reaction system or the walls of the reaction vessel. Termination steps include:

$$Cl^\bullet + Cl^\bullet \rightarrow Cl_2$$
$$CH_3^\bullet + Cl^\bullet \rightarrow CH_3Cl$$

Alkanes react violently with fluorine. Their reactions with bromine are much slower than those with chlorine. They do not normally react with iodine.

The reactions of alkanes with bromine are far more selective than with chlorine. For example, in the presence of sunlight chlorine reacts with propane to give an almost 50:50 mixture of 1-chloropropane, CH_3—CH_2—CH_2Cl, and 2-chloropropane, CH_3—$CHCl$—CH_3. With bromine, however, the corresponding reaction produces about 97% 2-bromopropane, CH_3—$CHBr$—CH_3, and only 3% 1-bromopropane, CH_3—CH_2—CH_2Br.

Cracking

This is the process in which C—C bonds in long chain alkane molecules are broken, producing smaller molecules of both alkanes and alkenes (*see* also section 18.3 later). The composition of the products depends on the conditions under which the cracking takes place.

Thermal Cracking

This type of cracking is also known as **pyrolysis**. The alkane is heated to a temperature between 450 and 700°C. At these temperatures carbon–carbon bonds in the alkane molecules undergo homolytic fission. The reactions proceed by chain mechanisms involving a variety of radicals. Let us take a relatively simple example—that of butane.

Initiation. In this stage methyl, ethyl and propyl radicals are produced:

$$CH_3—CH_2—CH_2—CH_3 \rightarrow CH_3—CH_2^{\bullet} + CH_3—CH_2^{\bullet}$$
ethyl radicals

or

$$CH_3—CH_2—CH_2—CH_3 \rightarrow CH_3^{\bullet} + CH_3—CH_2—CH_2^{\bullet}$$
methyl   propyl
radical   radical

Propagation. The alkyl radicals react with butane molecules to form butyl radicals. For example

The C—C bonds in the butyl radicals then undergo homolytic fission to produce alkene molecules and more radicals:

A variety of other propagation steps are also possible.

Termination. As usual, this occurs when two radicals combine to form a molecule. For example

$$CH_3^{\bullet} + CH_3^{\bullet} \rightarrow CH_3—CH_3$$
ethane

In practice the thermal cracking of butane produces a mixture of methane, ethane, ethene, propene, 1-butene, 2-butene and hydrogen.

Catalytic Cracking

In this method of cracking the alkane is passed over a catalyst at a temperature between 400°C and 500°C. The catalyst used is normally aluminium oxide mixed with either silica or chromium(VI) oxide. Like thermal cracking, catalytic cracking produces a mixture of shorter chain alkanes and short chain alkenes. However, unlike thermal cracking, catalytic cracking has an ionic mechanism. The acidic oxides used as catalysts promote the formation of carbon ions.

Catalytic cracking can be demonstrated in the laboratory using the apparatus shown in figure 18.3. The presence of alkenes can be established by shaking with bromine water. Unsaturated compounds decolourise bromine water (*see* below).

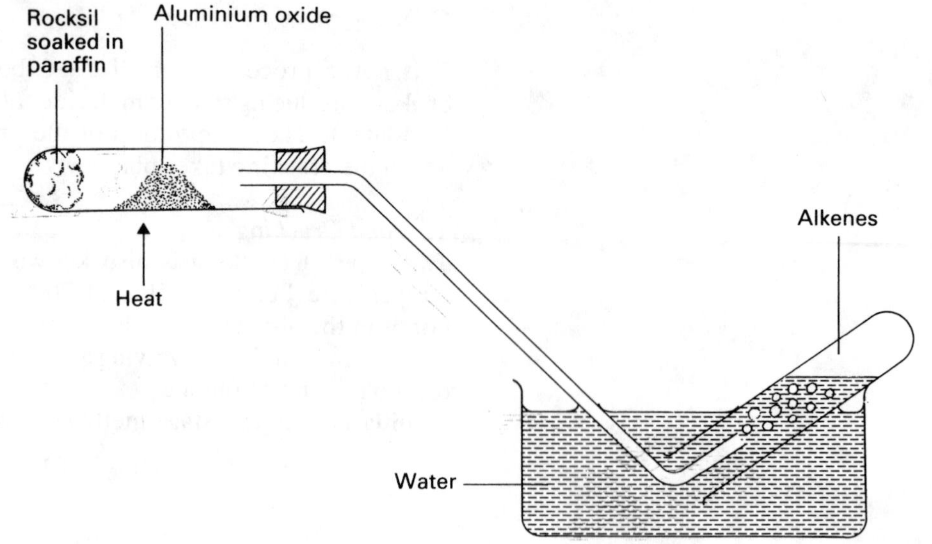

Figure 18.3 Catalytic cracking of alkanes

Care must be taken to remove the delivery tube from the water when heating is stopped or when the rate of formation of gas bubbles slows down. This is to avoid the water sucking back up the delivery tube.

Nitration

Alkanes can be nitrated by nitric acid. The reaction occurs in the vapour phase at temperatures of about 300°C:

$$C_2H_6 + HNO_3 \rightarrow \underset{\text{nitroethane}}{C_2H_5NO_2} + H_2O$$

Higher alkanes form mixtures of products. These are separated by fractional distillation. Nitroalkanes are used as solvents.

ALKENES

Alkenes are unsaturated aliphatic hydrocarbons with one or more carbon–carbon double bonds. The double bond locks the two carbon atoms together in a **planar structure** with bond angles of 120°:

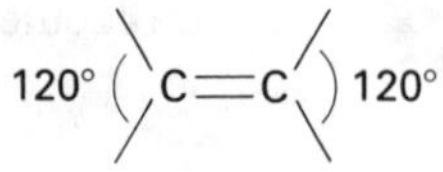

$$120° \left(\diagdown \right) C = C \left(\diagup \right) 120°$$

The two simplest members of the homologous series of alkenes with the general formula C_nH_{2n} are ethene and propene:

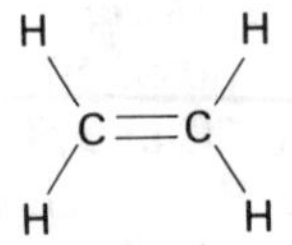

Ethene

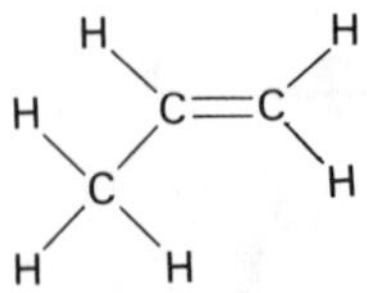

Propene

Members of the series with four or more carbon atoms exhibit positional isomerism. For example, the alkene with the formula C_4H_8 has three isomers, two of which are positional isomers:

$$\overset{4}{CH_3}-\overset{3}{CH_2}-\overset{2}{CH}=\overset{1}{CH_2} \qquad \overset{4}{CH_3}-\overset{3}{CH}=\overset{2}{CH}-\overset{1}{CH_3}$$

But-1-ene   But-2-ene

Note that the chain is numbered from the end nearer to the double bond. The position of the double bond is identified by the lower of the two numbers describing the positions of the two carbon atoms forming the double bond. The third isomer is branched:

$$\begin{array}{cc} CH_3 & H \\ \diagdown & \diagup \\ C & =C \\ \diagup & \diagdown \\ CH_3 & H \end{array}$$

2-methylpropene

The number of isomers of an alkene increases with the number of carbon atoms. For example, hexene has three positional isomers:

$$CH_3-CH_2-CH_2-CH_2-CH_2=CH_2 \qquad CH_3-CH_2-CH_2-CH=CH-CH_3$$

Hex-1-ene   Hex-2-ene

$$CH_3-CH_2-CH=CH-CH_2-CH_3$$

Hex-3-ene

Alkenes which contain two double bonds are known as **dienes**. Compounds with the system $C=C=C$ are known as **allenes**. Those with two double bonds separated by a single bond, thus, $C=C-C=C$, are known as **conjugated dienes**. One of the most important conjugated dienes is buta-1,3-diene, also known simply as butadiene:

$$CH_2=CH-CH=CH_2$$

Compounds containing three double bonds are known as trienes. Hydrocarbons with multiple double bonds are called polyenes.

Physical Properties

Alkenes have slightly lower melting and boiling points than their corresponding alkanes. For example, the boiling point of pentane is 36°C whereas that of pent-1-ene is 30°C. Ethene, propene and the three butene isomers are all gases at room temperature and pressure. Alkenes with between five and 15 carbon atoms are liquids. Like the alkanes their volatility increases with branching. Alkenes with 15 or more carbon atoms are solids.

Alkenes dissolve in non-polar organic solvents such as tetrachloromethane and benzene. They are almost completely insoluble in water.

Laboratory Preparation

The two principal methods of preparing alkenes in the laboratory are the dehydration of alcohols and the dehydrohalogenation of haloalkanes. Ethene, for example, can be prepared by the dehydration of ethanol using concentrated sulphuric acid in excess at 170°C (*see* section 19.2):

$$CH_3-CH_2-OH \rightarrow CH_2=CH_2 + H_2O$$

Ethene can also be prepared from ethanol by passing the ethanol vapour over heated aluminium oxide. The apparatus shown in figure 18.3 can be used for this purpose.

The second common method of preparing alkenes is by the base-catalysed dehydrohalogenation of haloalkanes:

$$CH_3CH_2Br + OH^- \rightarrow CH_2=CH_2 + Br^- + H_2O$$

bromoethane   ethene

The mechanism for this type of elimination reaction is described in section 17.3.

Reactions of the Alkenes

Alkenes are much more reactive than alkanes. This is because the π electrons of a double bond attract electrophiles (*see* section 17.3). The reactions of the alkenes are thus predominantly electrophilic **addition** reactions to the double bond:

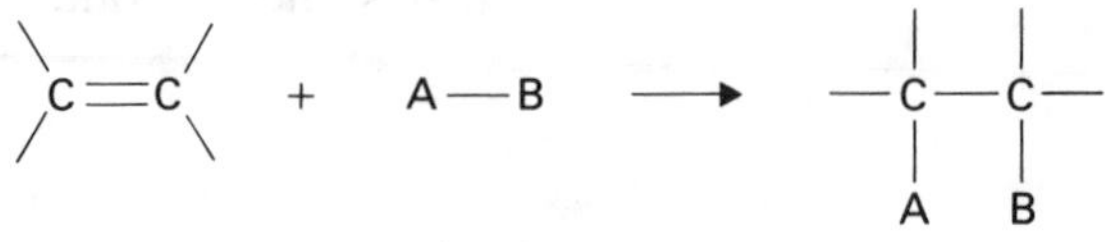

Many of these reactions have ionic mechanisms (*see* section 17.3).

Hydrogenation

When an alkene such as ethene is mixed with hydrogen and passed over a platinum catalyst at room temperature or a nickel catalyst at about 150°C, addition of hydrogen occurs at the double bond. The corresponding alkane is formed:

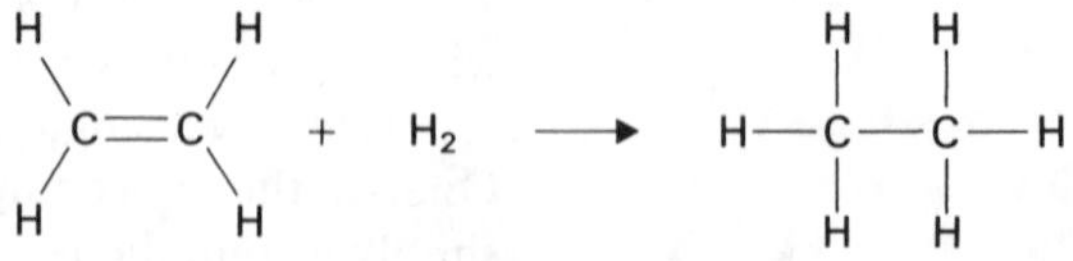

This type of reaction is an example of **heterogeneous catalysis**. The mechanism is described in section 9.2 and shown in figure 9.20.

Addition of Halogens

Chlorine or bromine readily add across the double bond of an alkene in a non-polar solvent such as tetrachloromethane or hexane. The reaction proceeds by means of an ionic mechanism involving a carbocation. The double bond induces a dipole in the halogen molecule:

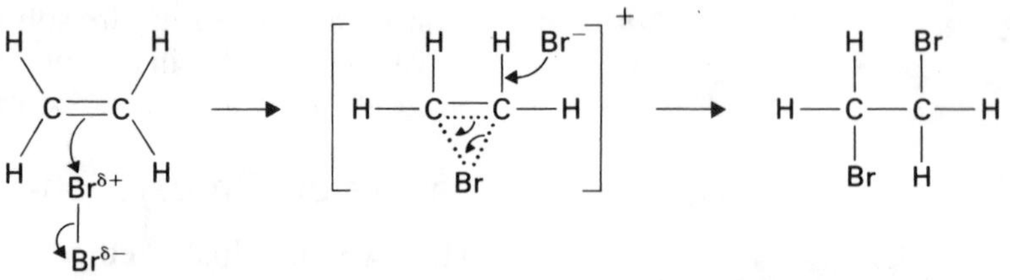

A solution of bromine in hexane or tetrachloromethane is thus decolourised when shaken with an alkene. This also occurs when an alkene is shaken with bromine water. Bromine water is a solution of bromine in water. It contains bromic(I) acid, HOBr. This adds across the double bond forming a bromoalcohol. For example:

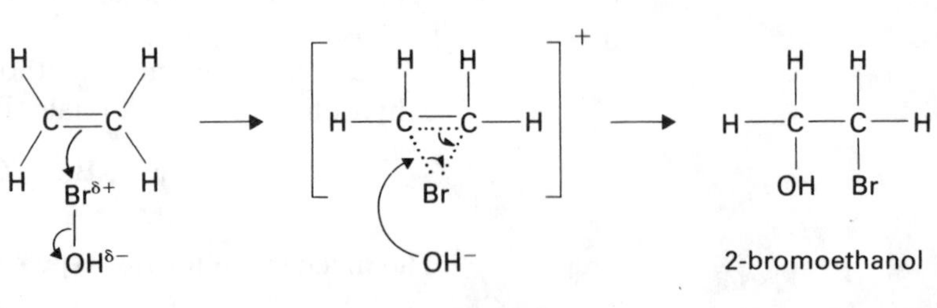

2-bromoethanol

Addition of Hydrogen Halides

The mechanism for this type of reaction is described in section 18.3. An example is provided by the addition of hydrogen chloride to propene:

$$CH_3—CH{=}CH_2 + HCl \rightarrow CH_3—CHCl—CH_3$$

propene 2-chloropropane

Note that the product is 2-chloropropane and not 1-chloropropane:

$$CH_3—CH_2—CH_2Cl$$

1-chloropropane

In such additions, the more electronegative atom or group always adds to the carbon atom bound to the least number of hydrogen atoms. This is a form of **Markownikoff's rule**.

The preference of the electronegative atom or group for the carbon atom with the least number of hydrogen atoms arises from the increasing stability of the carbocation as the number of alkyl substituents on the carbon atom increases. This increasing stability is due to the inductive effect which results in the alkyl groups being electron donating:

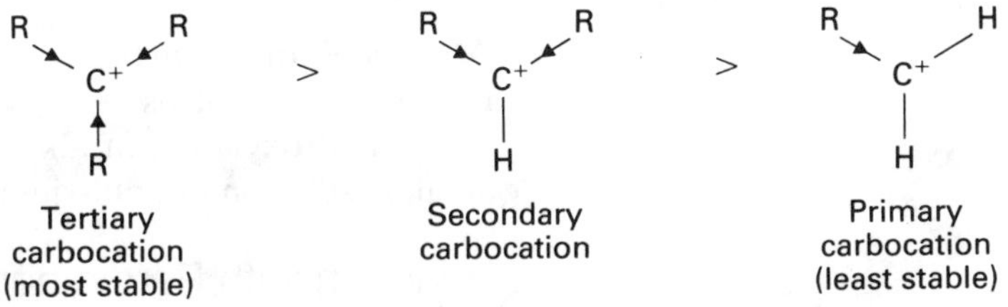

Tertiary carbocation (most stable) > Secondary carbocation > Primary carbocation (least stable)

In the presence of an organic peroxide, propene reacts with hydrogen bromide to form 1-bromopropane, $CH_3—CH_2—CH_2—Br$. This is known as an **anti-Markownikoff product**. It is formed as the result of a radical mechanism rather than an ionic mechanism.

Hydration

Alkenes react with cold concentrated sulphuric acid to form alkyl hydrogen-sulphates. For example

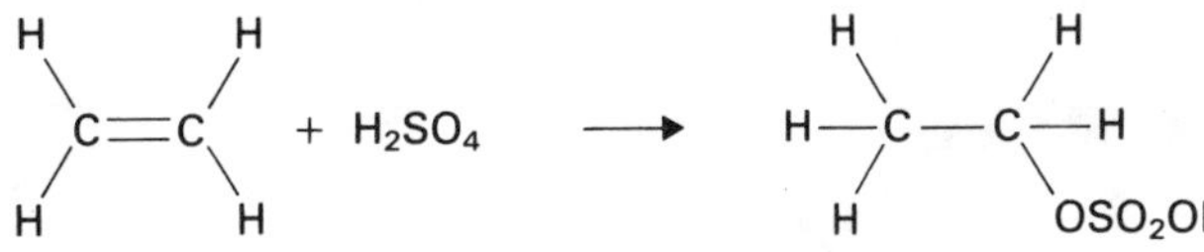

This is an addition reaction since the acid adds across a double bond. It is the reverse of the dehydration of ethanol to form ethene. The mechanism is similar to the addition of hydrogen halides across a double bond. It involves an intermediate carbocation. When the product of the reaction is diluted and warmed, it hydrolyses, forming ethanol:

$$CH_3—CH_2—O—SO_2—OH + H_2O \rightarrow CH_3CH_2OH + H_2SO_4$$

The addition of sulphuric acid to an alkene obeys Markownikoff's rule:

Reaction with Acidified Potassium Manganate(VII)

The purple colour of an acidified solution of potassium manganate(VII) disappears when the solution is shaken with an alkene. This alkene is hydroxylated, resulting in the formation of diols. For example, when excess ethene is shaken with acidified $KMnO_4$, ethane-1,2-diol (also known as ethylene glycol) is formed:

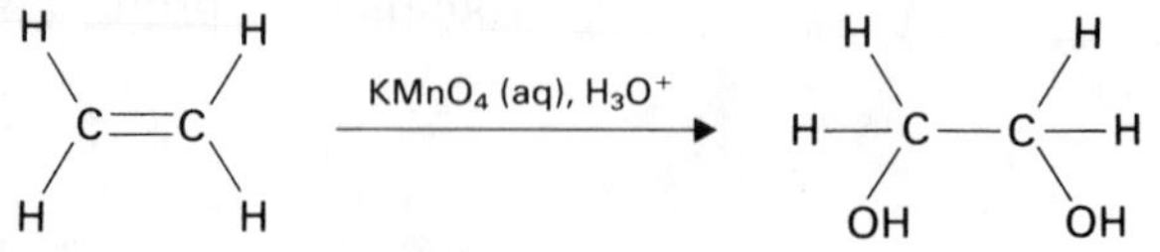

When an alkene is shaken with excess manganate(VII) solution **oxidative cleavage** occurs, yielding aldehydes and ketones:

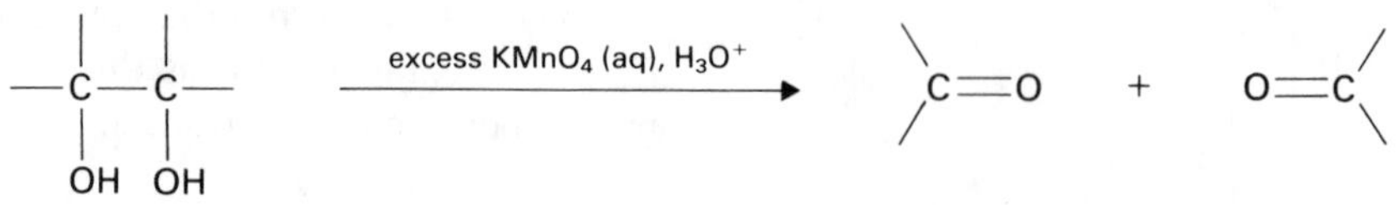

Aldehydes formed in this way undergo further oxidation by the manganate(VII) to form carboxylic acids.

The hydroxylation of alkenes to form diols may also be carried out using an alkaline solution of potassium manganate(VII).

Reaction with Peroxybenzoic Acid

Alkenes react with peroxyacids such as peroxybenzoic acid to form alkene oxides. For example

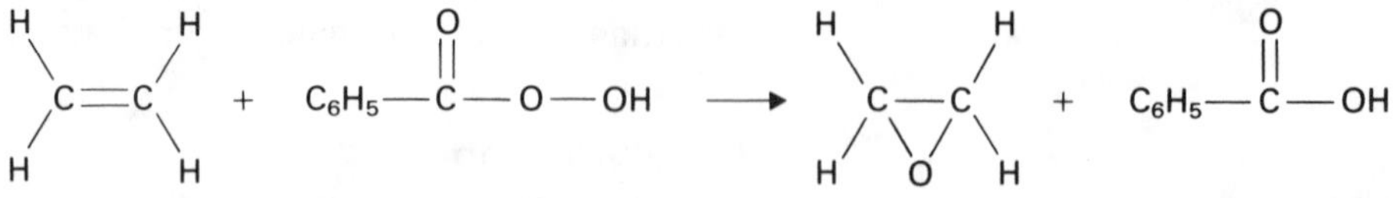

peroxybenzoic acid — epoxyethane — benzoic acid

On warming the epoxyethane with dilute acid, ethane-1,2-diol is formed:

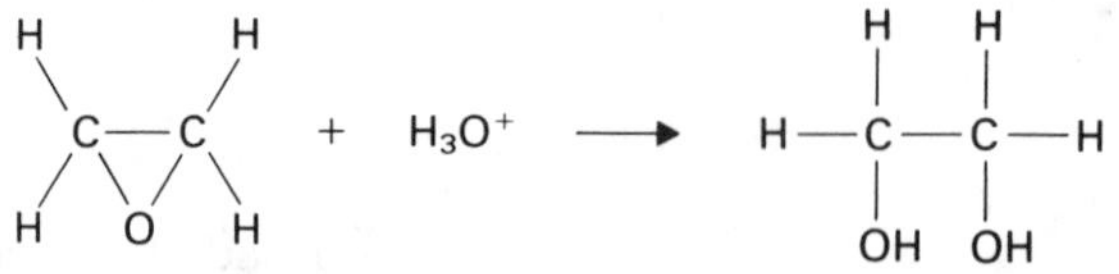

Reactions with Oxygen

In common with all hydrocarbons, the alkenes burn in a plentiful supply of air, forming carbon dioxide and water:

$$C_2H_4(g) + 3O_2(g) \rightarrow 2CO_2(g) + 2H_2O(g)$$

In a limited supply of air, carbon monoxide and water are formed:

$$C_2H_4(g) + 2O_2(g) \rightarrow 2CO(g) + 2H_2O(g)$$

Since alkenes have a relatively higher percentage of carbon than the corresponding alkanes, they burn with more smoky flames. This is due to the formation of carbon particles:

$$C_2H_4(g) + O_2(g) \rightarrow 2C(s) + 2H_2O(g)$$

When an alkene is mixed with oxygen and passed over a silver catalyst at a temperature of about 200°C, the alkene oxide is formed:

$$CH_2{=}CH_2 + \tfrac{1}{2}O_2 \xrightarrow[200°C]{Ag} \underset{\underset{O}{\diagdown\diagup}}{H_2C{-}CH_2}$$

Ozonolysis

When trioxygen (commonly known as ozone) is bubbled through a solution of an alkene in trichloromethane or tetrachloromethane at a temperature below 20°C, an alkene ozonide is formed:

ethene ozonide

The ozonides are unstable and can be explosive. They undergo hydrolysis to form aldehydes or ketones. For example

$$+ H_2O \longrightarrow 2HCHO + H_2O_2$$
methanal

In this case some of the methanal reacts with the hydrogen peroxide to form methanoic acid:

$$HCHO + H_2O_2 \rightarrow HCO_2H + H_2O$$
methanoic
acid

Polymerisation

The simple alkenes can be polymerised to form macromolecular compounds with the same empirical formula as the original alkene:

$$n(CH_2{=}CH_2) \rightarrow [{-}CH_2{-}CH_2{-}]_n$$

This reaction occurs when ethene is subjected to high pressures at 120°C in the presence of oxygen as a catalyst. The polymerisation is also catalysed at lower pressures by the presence of a 'Ziegler' catalyst. One of the most common 'Ziegler' catalysts is a mixture of triethylaluminium and titanium(IV) chloride.

We shall consider polymerisation in more depth in section 18.3 below.

ALKYNES

Alkynes are unsaturated aliphatic hydrocarbons with one or more carbon–carbon triple bonds. The triple bonds have a linear structure (*see* section 2.1). Those alkynes with a single triple bond form a homologous series with the general formula C_nH_{2n-2}. The simplest member of this series is ethyne (also known as acetylene). This has the formula $H—C\equiv C—H$.

The alkynes are named in the same way as alkanes except that the suffix -yne is used. Thus

$$\overset{5}{CH_3}—\overset{4}{CH_2}—\overset{3}{C}\equiv\overset{2}{C}—\overset{1}{CH_3}$$
pent-2-yne

The alkynes have similar melting and boiling points to those of the alkanes and alkenes. They increase with increasing number of carbon atoms. Ethyne, propyne and but-1-yne are all gases at room temperature and pressure. But-2-yne has a boiling point of 27°C. Higher alkynes are liquids. Like the alkenes and alkanes the alkynes are insoluble in water but soluble in non-polar organic solvents.

Laboratory Preparation

Ethyne can be prepared by the hydrolysis of calcium dicarbide using cold water:

$$CaC_2 + 2H_2O \rightarrow Ca(OH)_2 + C_2H_2$$
calcium   ethyne
dicarbide

Higher alkynes can be made by the dehydrohalogenation of the dihaloalkanes. This involves the elimination of two hydrogen halide molecules. The preparation is carried out by refluxing the dihaloalkanes with a solution of potassium hydroxide in ethanol. For example

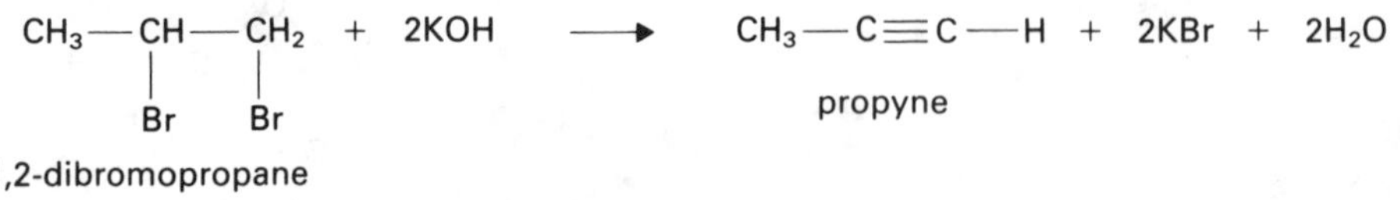

$$CH_3—\underset{\underset{Br}{|}}{CH}—\underset{\underset{Br}{|}}{CH_2} + 2KOH \longrightarrow CH_3—C\equiv C—H + 2KBr + 2H_2O$$
1,2-dibromopropane   propyne

The higher alkynes can also be prepared by the reaction of sodium dicarbide (sodium acetylide) with primary alkyl halides. For example

$$CH\equiv CNa + CH_3I \rightarrow CH\equiv CCH_3 + NaI$$
sodium   iodomethane   propyne
dicarbide

This reaction is an example of nucleophilic substitution. The dicarbide carbanion is the nucleophile:

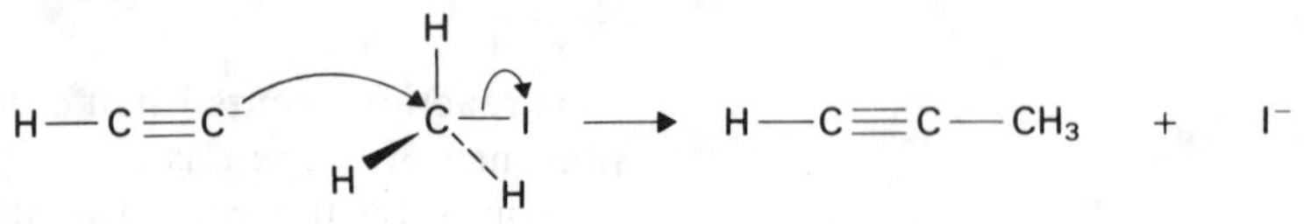

Reactions of the Alkynes

In many reactions the alkynes are more reactive than the corresponding alkenes. Alkynes can undergo electrophilic addition due to the availability of the π electrons in their triple bonds. For unsymmetrical alkynes and unsymmetrical reagents the Markownikoff rule applies. However, for addition reactions catalysed by a peroxide the anti-Markownikoff product is obtained due to a radical mechanism. Alkynes can also undergo double addition. In this type of addition two molecules are added across the triple bond:

$$-C\equiv C- \ + \ 2AB \ \longrightarrow \ \begin{matrix} A & B \\ | & | \\ -C-C- \\ | & | \\ A & B \end{matrix}$$

Alkynes also undergo homolytic fission with electrophilic reagents such as chlorine.

Reactions with Halogens

In the presence of a catalyst such as aluminium chloride or iron(III) chloride, ethyne undergoes electrophilic addition with both chlorine and bromine. For example

$$H-C\equiv C-H \ \xrightarrow{Cl_2} \ \underset{\text{1,2-dichloroethene}}{ClCH\!=\!CHCl} \ \xrightarrow{Cl_2} \ \underset{\text{1,1,2,2-tetrachloroethane}}{H-\overset{\displaystyle Cl}{\underset{\displaystyle Cl}{C}}-\overset{\displaystyle Cl}{\underset{\displaystyle Cl}{C}}-H}$$

In the absence of a catalyst, ethyne reacts explosively with chlorine producing red flames and clouds of black soot:

$$CH\equiv CH + Cl_2 \rightarrow 2C + 2HCl$$

The reaction can be demonstrated in a spectacular fashion by generating both the ethyne and chlorine *in situ*. This can be done by adding a mixture of calcium dicarbide and potassium manganate to 50% concentrated hydrochloric acid.

Like alkenes, when an alkyne is shaken with a solution of bromine dissolved in tetrachloromethane the solution is decolourised:

$$-C\equiv C- \ \xrightarrow{Br_2} \ BrCH\!=\!CHBr \ \xrightarrow{Br_2} \ Br_2CH-CHBr_2$$

In this type of addition reaction the intermediate dihaloalkene can be isolated.

Addition of Hydrogen Halides

Alkynes undergo electrophilic addition with the hydrogen halides, although the reactions are slower than those of the corresponding alkenes:

$$CH\equiv CH \ \xrightarrow{HCl} \ \underset{\text{chloroethene}}{CH_2\!=\!CHCl} \ \xrightarrow{HCl} \ \underset{\text{1,1-dichloroethane}}{CH_3-CHCl_2}$$

Note the Markownikoff addition of the second molecule of HCl. This reaction is catalysed by the presence of mercury(II) ions. The intermediate chloroethene (vinyl chloride) can be isolated and polymerised (*see* section 18.3).

The reactions of alkynes with hydrogen bromide are faster than those with hydrogen chloride but slower than those with hydrogen iodide.

Addition of Hydrogen

Ethyne is reduced by hydrogen at room temperature in the presence of metal catalysts such as platinum or palladium. A nickel catalyst at 150°C can also be used:

$$CH{\equiv}CH \xrightarrow[\substack{\text{metal} \\ \text{catalyst}}]{H_2} CH_2{=}CH_2 \xrightarrow[\substack{\text{metal} \\ \text{catalyst}}]{H_2} CH_3{-}CH_3$$

By using modified catalysts, this type of alkyne reaction can be stopped at the alkene.

Addition of Water

When ethyne is bubbled into a solution containing sulphuric acid and mercury(II) sulphate at about 60°C, ethanal is formed:

$$H{-}C{\equiv}C{-}H \;+\; H_2O \longrightarrow CH_3{-}\underset{\underset{H}{\big|}}{\overset{\overset{O}{\|}}{C}}$$

ethanal

Reactions with Metals and Metal Ions

A hydrogen atom bonded to an alkynic carbon atom shows weak acidic properties. Sodium, for example, displaces one of the hydrogen atoms in ethyne to form sodium dicarbide:

$$2CH{\equiv}CH + 2Na \xrightarrow{\text{liquid } NH_3} 2CH{\equiv}CNa + H_2$$

sodium dicarbide

This is a substitution reaction. It is carried out in liquid ammonia.

Substitution also occurs when ethyne is bubbled into aqueous ammoniacal solutions of copper(I) chloride or silver nitrate at room temperature. In the case of copper(I) chloride a red precipitate of copper(I) dicarbide is formed:

$$CH{\equiv}CH + 2Cu^+ \rightarrow CuC{\equiv}CCu + 2H^+$$

copper(I) dicarbide

With silver nitrate a white precipitate of silver dicarbide is formed:

$$CH{\equiv}CH + 2Ag^+ \rightarrow AgC{\equiv}CAg + 2H^+$$

silver dicarbide

Combustion

Alkynes are endothermic compounds (*see* chapter 5). This means that they have positive enthalpies of formation. For example

$$\Delta H_{f,m}^{\ominus} \;(C_2H_2, 298\text{ K}) = +227\text{ kJ mol}^{-1}$$

The combustion of ethyne (commonly known as acetylene) in oxygen is thus strongly exothermic:

$$2C_2H_2 + 5O_2 \rightarrow 4CO_2 + 2H_2O \qquad \Delta H_{c,m}^{\ominus} (298\text{ K}) = -1257\text{ kJ mol}^{-1}$$

Oxy-acetylene welding relies on the high temperatures produced from this reaction.

In air, the combustion of ethyne is incomplete. Since ethyne has such a high proportion of carbon, it burns with a highly luminous flame due to the presence of carbon particles.

Polymerisation

Ethyne is polymerised when passed through a copper tube at about 300°C. Benzene is formed:

$$3CH \equiv CH \xrightarrow{\text{Cu, 300°C}}$$

The copper acts as a catalyst.

SUMMARY

1. The molecular formulae of gaseous hydrocarbons can be determined by **eudiometry**. In this technique a measured volume of hydrocarbon is burnt in excess oxygen.
2. The melting and boiling points and thus **volatility** of **aliphatic hydrocarbons** increase with increasing number of carbon atoms.
3. **Ethene** can be prepared in the laboratory from ethanol or bromoethane.
4. **Ethyne** can be prepared in the laboratory from calcium dicarbide.
5. The aliphatic hydrocarbons all burn in excess oxygen, forming carbon dioxide and water.
6. **Unsaturated** aliphatic hydrocarbons are oxidised by acidified potassium manganate(VII) solution.
7. Unsaturated hydrocarbons undergo **addition reactions** with hydrogen, halogens and hydrogen halides.
8. In electrophilic addition to a double bond, the more electronegative atom or group of atoms always adds to the carbon atom bound to the least number of hydrogen atoms. This statement is a form of **Markownikoff's rule**.
9. Both alkenes and alkynes can be
   (a) **hydrated**,
   (b) **polymerised**.
10. Alkanes undergo **substitution** reactions with chlorine. These are **chain reactions** involving **homolytic fission** of the covalent bonds. There are three stages in the reaction:
   (a) **initiation**,
   (b) **propagation**,
   (c) **termination**.
11. The **thermal cracking** of alkanes also proceeds by chain mechanisms involving homolytic fission of covalent bonds.
12. The **catalytic cracking** of alkanes has an **ionic mechanism**.
13. Alkenes undergo **ozonolysis** to form unstable alkene ozonides.
14. Alkynes react with metals and thus show acidic properties.

18.2 Arenes

We saw in section 17.1 that hydrocarbons may be classified as aliphatic, alicyclic or aromatic. Aromatic hydrocarbons are known as **arenes**. Some of the more common arenes are shown in figure 18.4. Note how the systematic names are often completely different from the traditional names. Note also how the benzene ring is numbered for systematic nomenclature.

The term **aromatic** was originally used for naturally occurring products which were sweet smelling, that is compounds which had an aroma. Since this group included many compounds with benzene rings, the term aromatic became associated with compounds with benzene rings.

	Systematic name	Traditional name
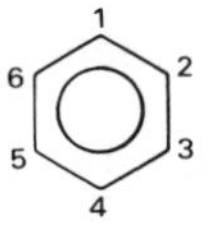 benzene structure (positions 1–6)	Benzene	Benzene
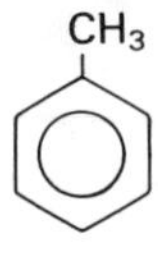 CH₃ structure	Methylbenzene	Toluene
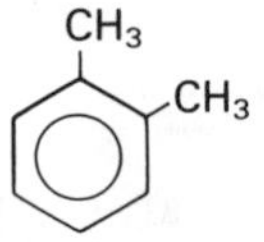 1,2-dimethyl structure	1,2-Dimethylbenzene	*ortho*-Xylene (*o*-xylene)
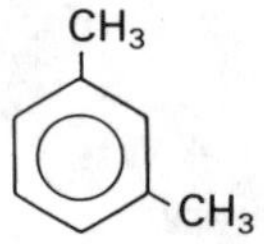 1,3-dimethyl structure	1,3-Dimethylbenzene	*meta*-Xylene (*m*-xylene)
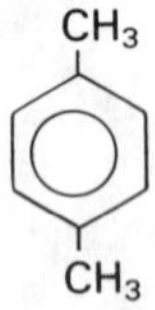 1,4-dimethyl structure	1,4-Dimethylbenzene	*para*-Xylene (*p*-xylene)
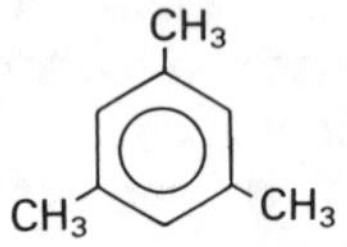 trimethyl structure	1,3,5-Trimethylbenzene	Mesitylene
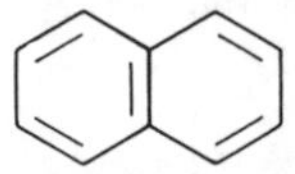 naphthalene structure	Naphthalene	Naphthalene
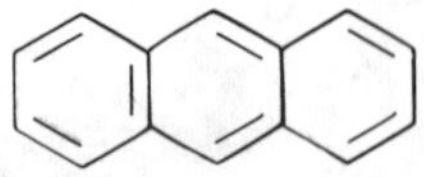 anthracene structure	Anthracene	Anthracene

Figure 18.4 Arenes

Bitter Almonds

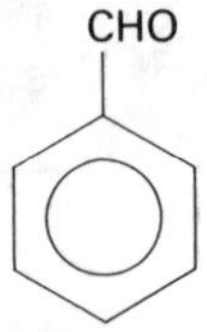

CHO

Benzaldehyde

Benzaldehyde has a pleasant aromatic smell like that of bitter almonds. It is sometimes known as oil of almonds or oil of bitter almonds since it can be extracted from bitter almonds. It is used to flavour foods, in perfumery as well as in the manufacture of dyestuffs and pharmaceuticals.

Nowadays the term aromatic is used for any compound with one or more benzene rings—whether they smell sweetly or not. The term **aromaticity** or **aromatic character** is also used to describe certain properties of benzene and related compounds. For example:

1. Aromatic compounds burn with smoky flames. This is due to the high proportion of carbon in such compounds.
2. Aromatic compounds do not readily undergo addition reactions. This is due to the stability of the benzene ring (*see* below). For example, aromatic compounds do not decolourise bromine water.
3. Aromatic compounds undergo substitution reactions.

Table 18.2 shows how the characteristic chemical properties of the arenes compare with those of the aliphatic hydrocarbons.

The physical properties of the monocyclic arenes have more in common with aliphatic hydrocarbons with a similar number of carbon atoms. For example, the

Table 18.2 Comparison of the chemical properties of the arenes with the aliphatic hydrocabons

Characteristic reactions	Alkanes	Alkenes	Alkynes	Arenes
combustion in air	yes	yes	yes	yes
type of flame	luminous	smoky	very smoky	very smoky
oxidation with $KMnO_4$	no	yes	yes	benzene ring not oxidised
addition reactions, e.g. with bromine water	no	yes	yes	no
substitution reactions	yes	no	yes	yes

monocyclic arenes are colourless volatile liquids. Since they are non-polar they are miscible with other non-polar solvents but immiscible with water. The arenes and their related compounds can also be highly toxic.

Toxicity of Benzene

Inhalation of benzene vapour causes dizziness and headaches. High concentrations can lead to unconsciousness. The vapour irritates the eyes and mucous membranes. Liquid benzene is absorbed through the skin and this can result in poisoning. It is extremely poisonous if taken by mouth. Repeated inhalation of low concentrations of benzene over long periods can have chronic effects including severe or even fatal blood disease. Containers of benzene and all other toxic chemicals must bear the hazard warning symbol showing the skull and crossbones.

Figure 18.5 Benzene is highly toxic. (The full list of hazard symbols is given in appendix B)

Friedrich August Kekulé von Stradonitz was a nineteenth-century German chemist. In 1854 he found the first known organic acid containing sulphur—thioacetic acid (thioethanoic acid). He also determined the structure of diazo compounds. However, his best known discovery (in 1866) concerned the structure of benzene. He showed that the double bonds alternated around the ring. The idea first came to him in a dream. Later he showed that the two possible arrangements are identical and that the benzene ring oscillated between the two structures. He thus anticipated the concept of resonance which was first introduced in the early 1930s

BENZENE

The first cyclic structure for benzene was proposed by Kekulé in 1865:

If benzene had this structure the 1,2-disubstituted benzenes would have two isomers. For example

However, it has never been possible to separate two isomers of any 1,2-disubstituted benzenes.

Note that ring hydrogen atoms are commonly omitted in these representations of benzene molecules and their derivatives.

Kekulé later suggested that a benzene molecule existed as two rapidly alternating structures:

The benzene molecule is now known to exist as a resonance hybrid of these two limiting forms (*see* section 2.1). The resonance hybrid can also be viewed in terms of its molecular orbitals. We saw in section 3.1 that the 2p electrons in the π bond orbitals of the carbon atoms are delocalised. They form a π electron cloud. The benzene molecule can then be conveniently represented as:

Experimental evidence confirms that benzene exists with this structure. If it existed as the original Kekulé structure with three conjugated double bonds, then benzene would undergo addition reactions like the alkenes. As we have seen, it does not. Furthermore, benzene is more stable than it would be if it contained three distinct double bonds. In section 5.3 we saw how the enthalpy of hydrogenation of benzene to form cyclohexane is more negative than three times the enthalpy of hydrogenation of cyclohexene. The difference is variously known as the delocalisation enthalpy, resonance energy or stabilisation energy of benzene.

Finally, the lengths of the carbon–carbon bonds in the benzene ring are intermediate between those in alkanes and alkenes. We would expect this if the bonds are a hybrid between single and double bonds. The actual bond lengths are shown in table 18.3.

The benzene molecule is planar and has the geometry shown in figure 18.6.

Table 18.3 Carbon–carbon bond lengths

		Bond length/nm
alkane	C—C	0·154
benzene ring	C⋯C	0·139
alkene	C=C	0·134

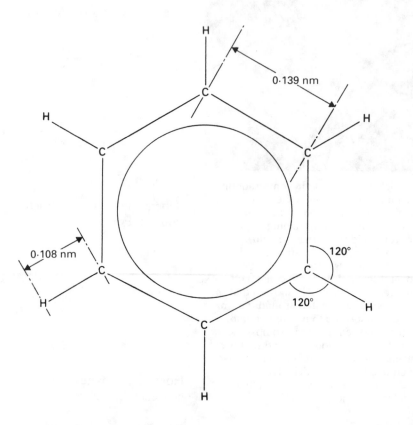

Figure 18.6 The geometry of the benzene molecule

Physical Properties

Benzene is a colourless liquid which freezes at 5.5°C and boils at 80°C. It has a distinctive aromatic odour but, as we saw above, it is highly toxic. Benzene is immiscible with water and forms the upper of the two layers. It is miscible with non-polar organic solvents, however, and is itself a good solvent for organic compounds.

Chemical Properties

Although benzene does undergo certain addition reactions (*see* below), it does not exhibit typical alkene reactivity. For example, it does not decolourise bromine water or manganate(VII) solution. Furthermore, it does not undergo addition reactions with strong acids such as hydrogen chloride or sulphuric acid.

Benzene does, however, undergo a range of electrophilic substitution reactions. The products of this type of reaction are aromatic since they retain the delocalised π-electron system. The general mechanism for the substitution of a hydrogen atom on the benzene ring by an electrophile is described in section 17.3. The nitration, halogenation, sulphonation and Friedel–Crafts reactions of benzene are all examples of electrophilic substitution.

Nitration

Benzene can be nitrated using a mixture of concentrated nitric and sulphuric acids:

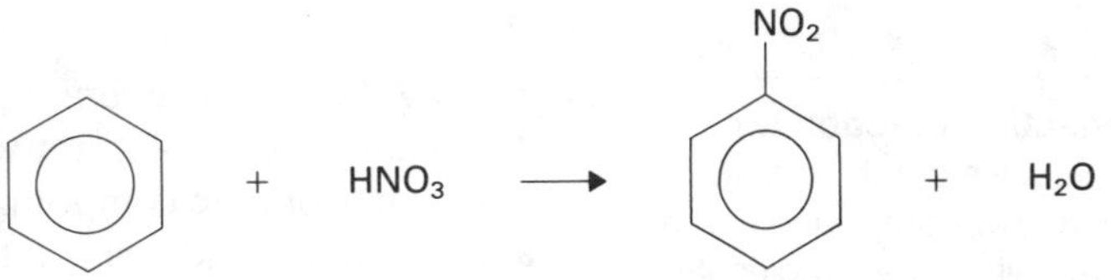

The conditions and mechanism for this reaction are described in section 17.3.

Nitrobenzene is a pale yellow liquid with a characteristic smell of almonds. During the nitration of the benzene, crystals of 1,3-dinitrobenzene are also formed in the following side reaction:

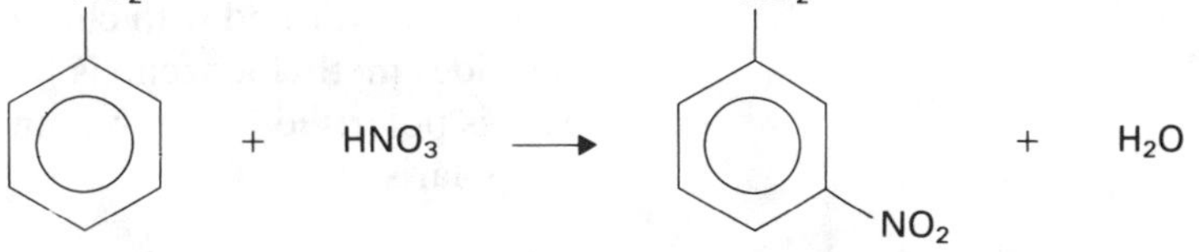

Halogenation

When benzene is mixed with chlorine or bromine in the dark, no reaction occurs. However, in the presence of Lewis acid catalysts, electrophilic substitution occurs. Typical catalysts are iron(III) bromide and aluminium chloride. They work by inducing polarisation in the halogen molecule, which then forms a complex with the catalyst:

$$Br \!-\! Br \quad FeBr_3 \quad \longrightarrow \quad Br^+[FeBr_4]^-$$

It is doubtful whether Br^+ exists as free ions. Even so, it is useful to formulate the reaction in this way. The mechanism for the bromination of benzene using an iron(III) bromide carrier can be represented as follows:

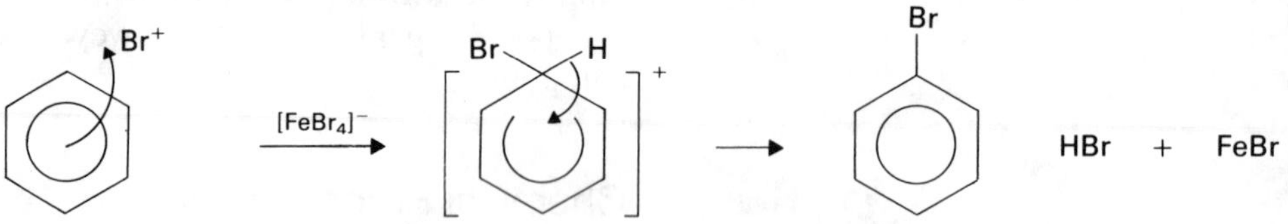

Sulphonation

Benzene can be sulphonated by refluxing it with concentrated sulphuric acid for several hours. Alternatively the benzene can be warmed with fuming sulphuric acid. This contains sulphur trioxide. The mechanisms for this reaction can be represented as follows:

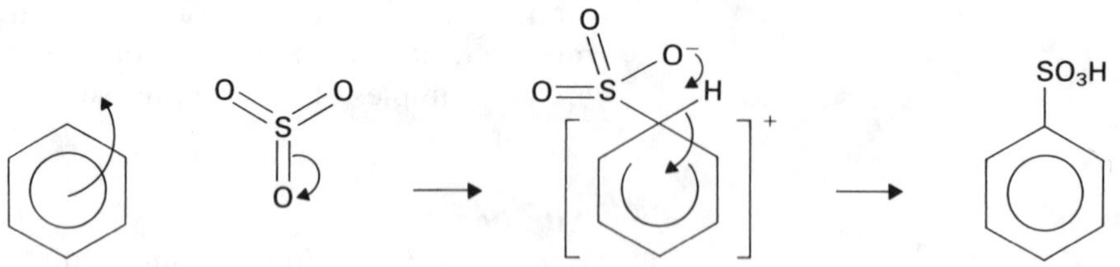

benzenesulphonic acid

In a **condensation reaction** two reactant molecules combine to form a molecule of a new compound and at the same time eliminate a molecule of a relatively simple compound such as water or hydrogen chloride.

Friedel–Crafts Reactions

The name Friedel–Crafts was originally applied to the condensation reactions between aromatic compounds and alkyl halides in the presence of an anhydrous aluminium chloride catalyst.

Nowadays, the name Friedel–Crafts is applied to any electrophilic substitution reaction of an aromatic compound in which the electrophile is a carbocation or a highly polarised complex with a positively charged carbon atom. The electrophilic reagent is typically an alkyl halide or acid chloride although it might also be an alkene or alcohol, for example. The catalyst used is more often than not anhydrous aluminium chloride. Friedel–Crafts reactions are normally divided into two categories, alkylation and acylation.

Alkylation. In this type of Friedel–Crafts reaction, one or more hydrogen atoms on a benzene ring are substituted by alkyl groups. For example, when benzene is warmed with chloromethane in the presence of anhydrous aluminium chloride, methylbenzene is formed. The electrophilic reagent is chloromethane. This is polarised by aluminium chloride in the same way that halogen molecules are polarised:

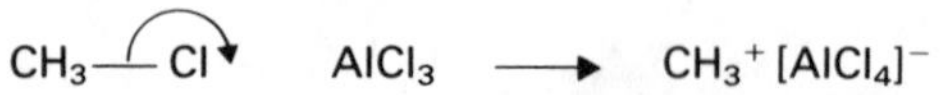

The mechanism for the reaction can then be represented as follows:

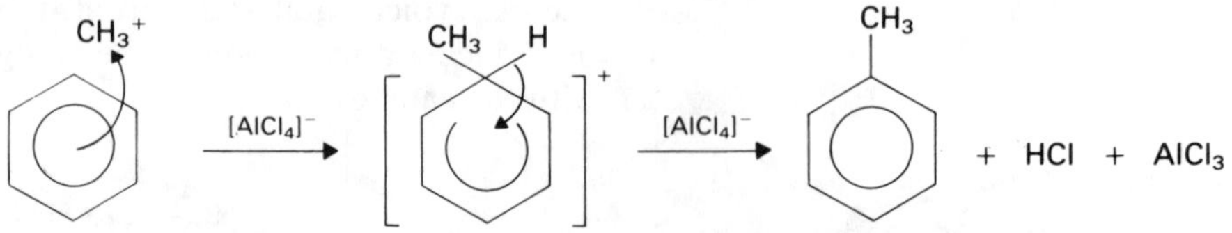

The acyl group has the general formula

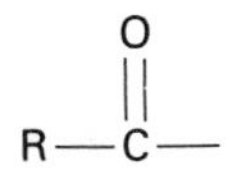

$$R-\overset{\overset{\displaystyle O}{\|}}{C}-$$

Acyl compounds are named by replacing the -ic of the carboxylic acid from which they derive with -yl. For example:

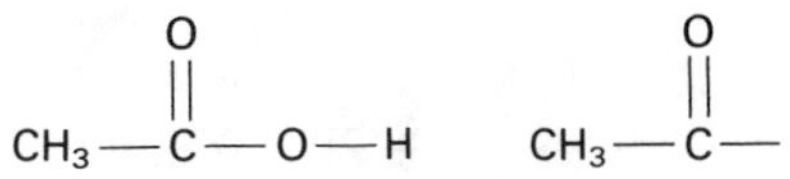

$$CH_3-\overset{\overset{\displaystyle O}{\|}}{C}-O-H \qquad CH_3-\overset{\overset{\displaystyle O}{\|}}{C}-$$

Ethanoic acid   Ethanoyl

We should note that in this condensation reaction between benzene and chloromethane, a molecule of hydrogen chloride is eliminated. We should also note that it is doubtful whether the methyl carbocation, CH_3^+, exists as a free ion.

The alkylation of benzene using chloromethane with an anhydrous aluminium chloride catalyst does not stop at methylbenzene. Further alkylation of the benzene ring occurs, resulting in the formation of 1,2-dimethylbenzene:

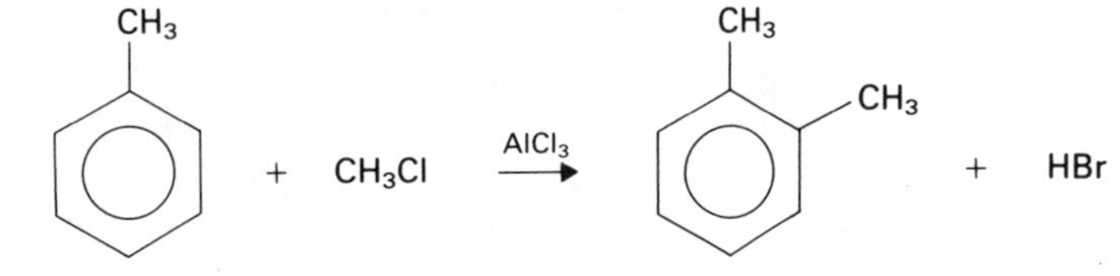

Methylbenzene $+ \quad CH_3Cl \quad \xrightarrow{AlCl_3}$ 1,2-Dimethylbenzene $+ \quad HBr$

Acylation. In a Friedel–Crafts acylation reaction, a hydrogen atom on a benzene ring is substituted by an acyl group. The product is thus an aromatic ketone. The acylation is carried out using an acid chloride or acid anhydride in the presence of an anhydrous aluminium chloride catalyst. For example:

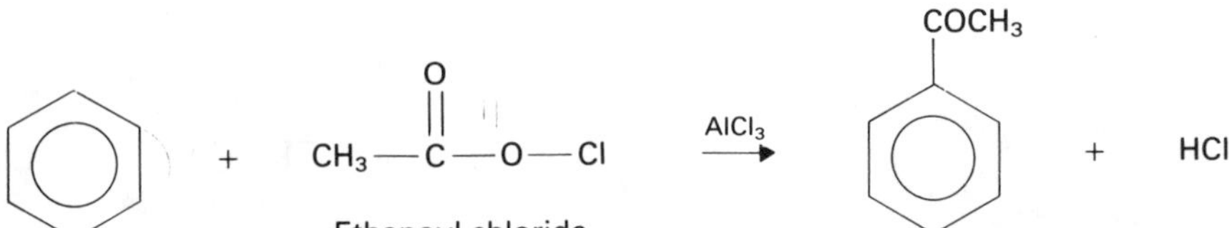

     $+ \qquad CH_3-\overset{\overset{\displaystyle O}{\|}}{C}-O-Cl \qquad \xrightarrow{AlCl_3}$ $+ \quad HCl$

Ethanoyl chloride   Phenylethanone

Note that this is a condensation reaction in which a molecule of hydrogen chloride is eliminated. Note also that the name 'phenyl' is often used for the benzene ring in compounds where benzene is not the principal group:

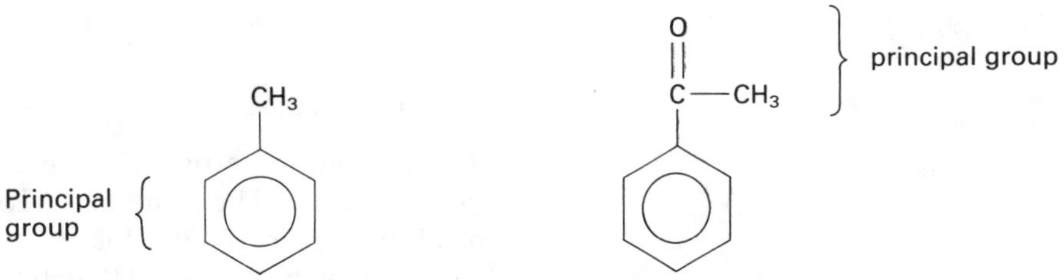

Principal group $\left\{ \vphantom{\begin{array}{c}a\\a\end{array}} \right.$ Methylbenzene   Phenylethanone $\left. \vphantom{\begin{array}{c}a\\a\end{array}} \right\}$ principal group

Addition Reactions of Benzene

Although the characteristic reactions of benzene involve electrophilic substitution, benzene also undergoes certain addition reactions. One of these we have already met. It is the hydrogenation of benzene (see section 5.3). When a mixture of benzene and hydrogen is passed over a finely divided nickel catalyst at 150 to 160°C, a sequence of reactions occurs resulting in the formation of cyclohexane. The overall stoichiometric equation can be written as:

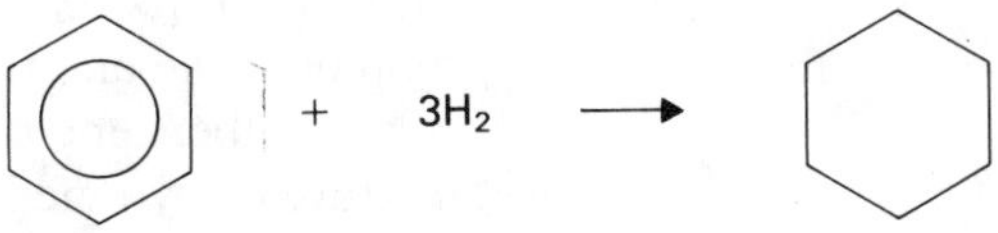

$+ \quad 3H_2 \quad \longrightarrow$

Benzene can also undergo addition with chlorine in the presence of ultra-violet light or direct sunlight. The reaction involves a complex radical mechanism. The final product is 1,2,3,4,5,6-hexachlorocyclohexane:

A similar reaction occurs between benzene and bromine in the presence of ultra-violet light or sunlight.

Oxidation

Benzene and the benzene ring are generally resistant to oxidation even by strong oxidising agents such as acidified or alkaline potassium manganate(VII). However, benzene and other aromatic hydrocarbons do burn in air or oxygen with a very smoky flame characteristic of hydrocarbons with a high carbon content.

METHYLBENZENE

Methylbenzene
(toluene)

Physical Properties

The physical properties of methylbenzene (or toluene as it is also known) are similar to those of benzene. It is a colourless liquid which is insoluble in water but soluble in organic solvents. Like benzene it is a good solvent for organic compounds. It is now more widely used than benzene as a solvent since it is far less toxic than benzene.

Chemical Properties

The reactions of methylbenzene can be divided into two categories: (a) those involving the benzene ring and (b) those involving the methyl group.

The Aromatic Ring

Methylbenzene undergoes all the electrophilic substitution reactions we described for benzene above. It thus undergoes nitration, halogenation, sulphonation and Friedel–Crafts reactions. In all these reactions methylbenzene is more reactive than benzene and its reactions are faster.

Methylbenzene can be nitrated in the same way as benzene. The product is a mixture of two methylnitrobenzene isomers:

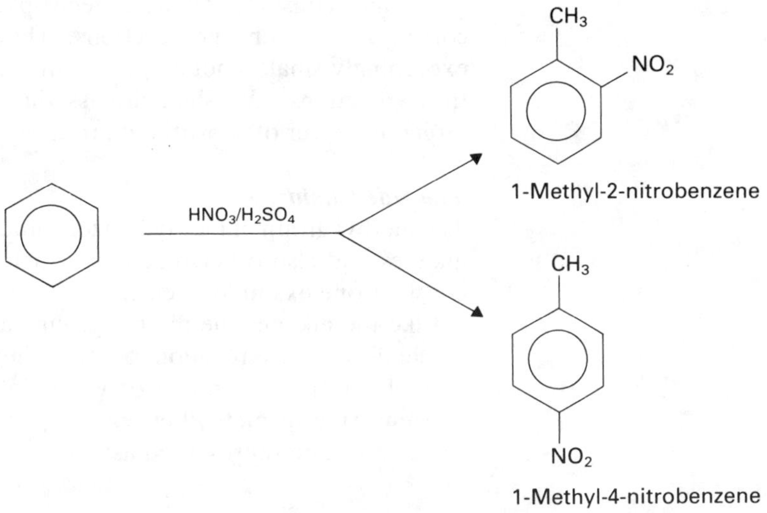

1-Methyl-2-nitrobenzene

1-Methyl-4-nitrobenzene

The benzene ring of methylbenzene can be chlorinated by bubbling chlorine through the liquid in the presence of aluminium chloride in the dark. The aluminium chloride acts as a catalyst. Once again, the 2- and 4-substituted isomers are formed:

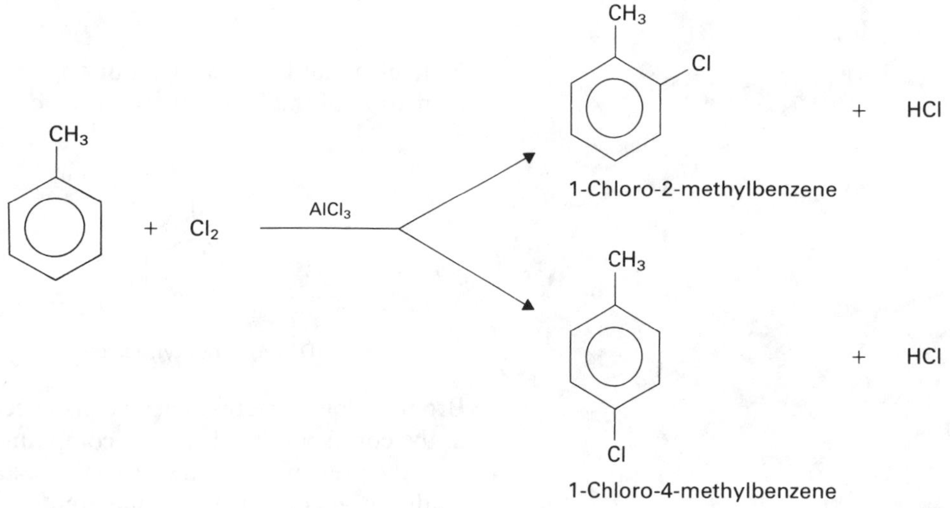

1-Chloro-2-methylbenzene

1-Chloro-4-methylbenzene

The sulphonation of methylbenzene with concentrated sulphuric acid also leads to a mixture of the 2- and 4-substituted isomers:

2-Methylbenzenesulphonic acid

4-Methylbenzenesulphonic acid

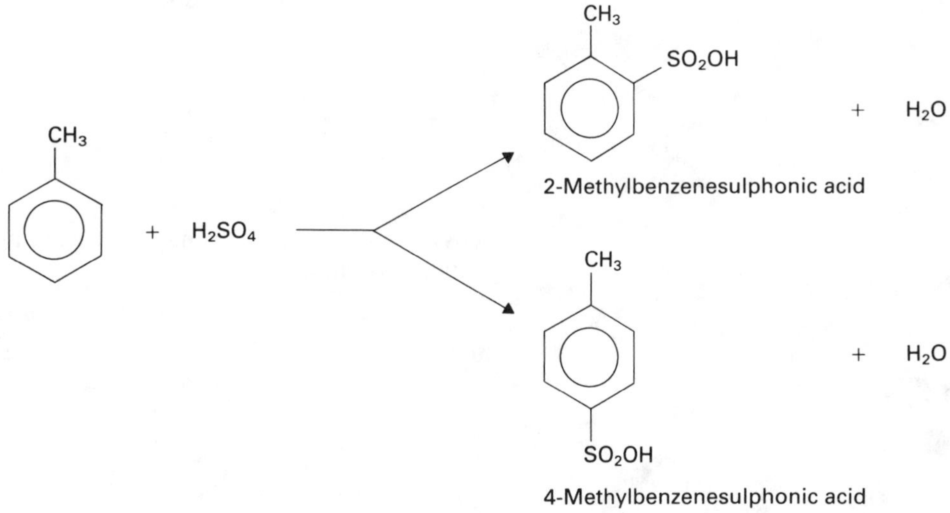

The mechanism for all these electrophilic substitution reactions is similar to the corresponding benzene reactions. The 3-substituted isomers are produced in exceedingly small amounts, and can for all intents and purposes be ignored in these reactions. We shall discuss this below when we come to consider the directing power of substituent groups.

The Side Chain

The methyl group of methylbenzene undergoes certain reactions characteristic of alkanes and also other reactions which alkanes do not undergo. We shall now consider one example of each.

Like the alkanes the methyl group can be halogenated by means of a radical mechanism. This reaction occurs when chlorine is bubbled through boiling methylbenzene in sunlight or ultra-violet light. Note that halogenation of the benzene ring in methylbenzene requires completely different conditions. The reaction can be represented as:

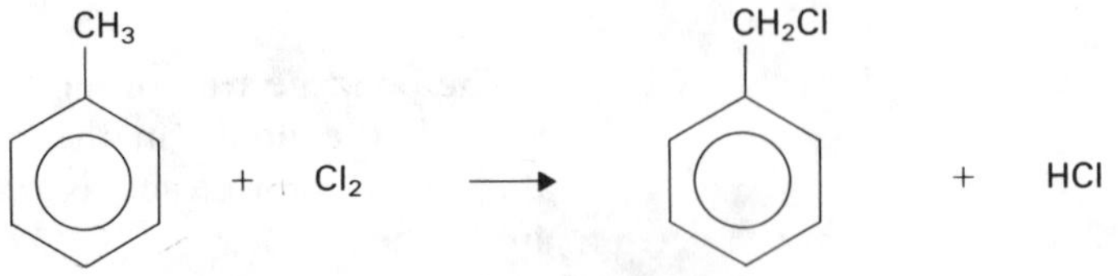

$$CH_3 \text{-benzene} + Cl_2 \longrightarrow CH_2Cl \text{-benzene} + HCl$$

Chloromethylbenzene

Note also that this is a substitution reaction. Further halogenation results in the formation of the following compounds:

$CHCl_2$-benzene

(Dichloromethyl)benzene

CCl_3-benzene

(Trichloromethyl)benzene

Bromination of methylbenzene also occurs under the same conditions, resulting in the corresponding bromine compounds.

In the previous section we saw that alkanes are inert to oxidation even by strong oxidising agents such as potassium manganate(VII). The methyl side chain in methylbenzene on the other hand does undergo oxidation even by relatively mild oxidising agents such as manganese(IV) oxide:

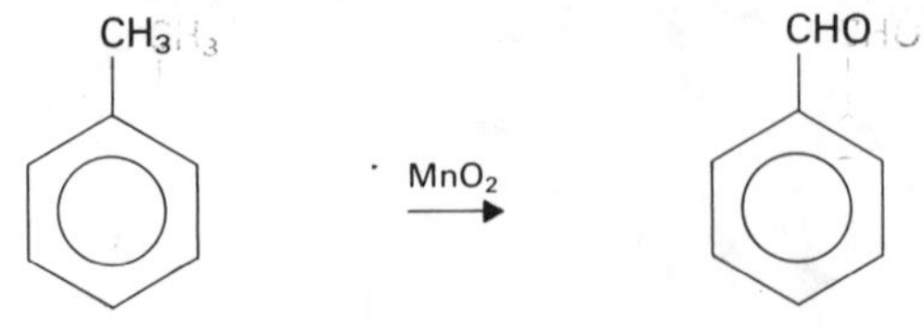

$$CH_3 \text{-benzene} \xrightarrow{MnO_2} CHO \text{-benzene}$$

Benzaldehyde

With a stronger oxidising agent such as potassium manganate(VII) further oxidation occurs:

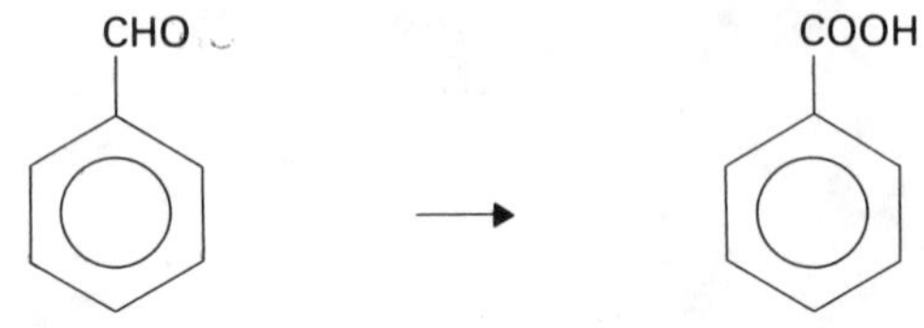

$$CHO \text{-benzene} \longrightarrow COOH \text{-benzene}$$

Benzoic acid

Directing Power of Substituents on the Benzene Ring

We saw above that electrophilic substitution of methylbenzene results in the formation of 2- and 4-substituted isomers of methylbenzene. Because of this the methyl group is said to be **2,4-directing**. A number of other substituents on the benzene ring are also 2,4-directing with respect to electrophilic substitutions. For such reactions we can write the general equation:

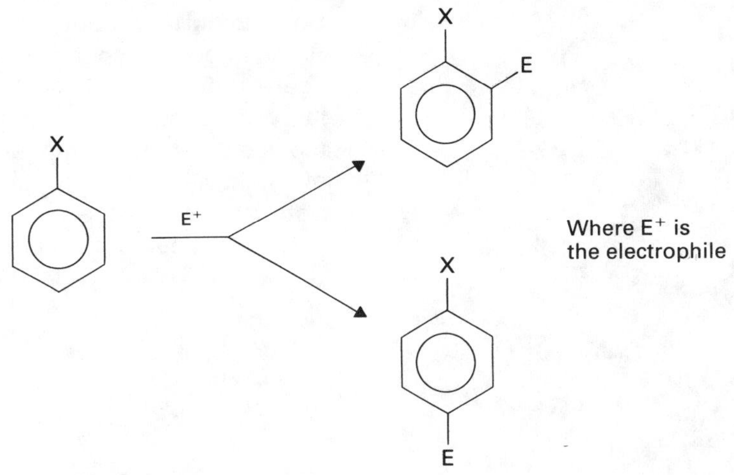

Where E$^+$ is the electrophile

E$^+$ is the electrophile and X is the 2,4-directing substituent. These substituents are usually saturated groups. They include —CH$_3$, —Cl, —OH, —NH$_2$ and —OCH$_3$. Under certain conditions, 2,4-directing substituents are also 6-directing. For example

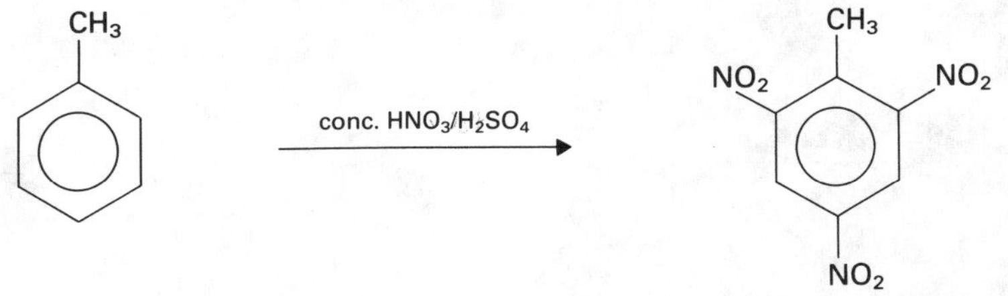

Methyl-2,4,6-trinitrobenzene
(trinitrotoluene or TNT)

The reaction rates of electrophilic substitution at the 2,4-positions are generally faster than the corresponding reactions with benzene.

Unsaturated substituents tend to be **3-directing**:

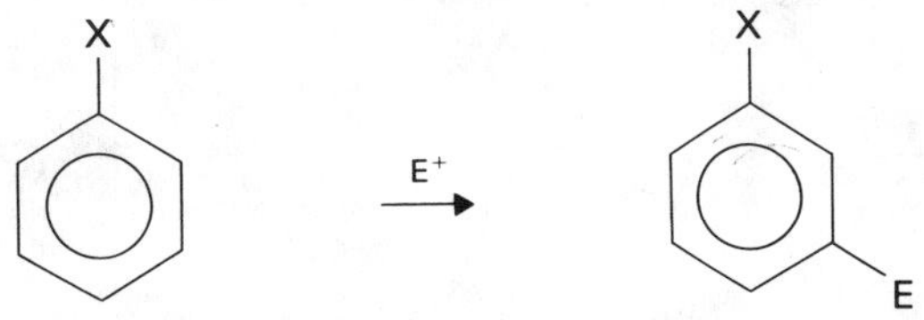

Groups which are 3-directing include —NO$_2$, —SO$_2$OH, —COOH and —CN. The reaction rates of electrophilic substitutions at the 3-position are slower than the corresponding reactions with benzene.

The directing power of substituents depends on whether the substituent donates electrons to the benzene ring or withdraws electrons from it. 2,4-Substitution results from either the positive inductive ($+I$) effect or positive mesomeric ($+M$) effect. 2,4-Directing groups donate electrons to the benzene ring by means of these effects and thus activate the ring. They are thus called **activating groups**. 3-Directing groups withdraw electrons from the ring by means of the $-I$ and $-M$ effects. They are called **deactivating groups**.

SUMMARY

1. **Aromatic** compounds
   (a) burn with a smoky flame,
   (b) undergo **substitution** reactions,
   (c) do not readily undergo **addition** reactions.
2. The benzene molecule exists as a **resonance hybrid** of two **limiting** forms.
3. Some important chemical reactions of **benzene** are shown in figure 18.7.
4. Some important chemical reactions of **methylbenzene** are shown in figure 18.8.
5. In a **condensation reaction** two reactant molecules combine to form a molecule of a new compound and at the same time eliminate a molecule of a relatively simple compound such as water or hydrogen chloride.
6. **Saturated substituents** on the benzene ring are **2,4-directing**.
7. **Unsaturated substituents** on the benzene ring are **3-directing**.

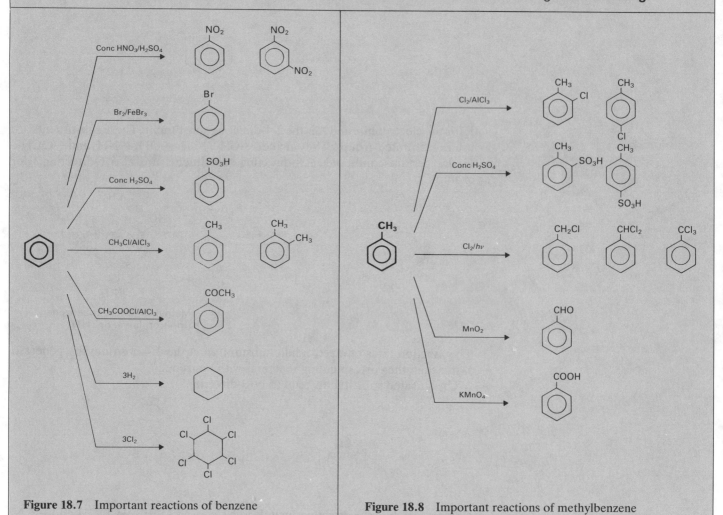

Figure 18.7   Important reactions of benzene

Figure 18.8   Important reactions of methylbenzene

18.3 Occurrence, Production and Uses

OCCURRENCE

Hydrocarbons occur naturally in fossil fuels: petroleum, coal and peat. The term 'petroleum' which means 'oil from rock', includes both crude oil and natural gases containing hydrocarbons. Petroleum was formed some 100 to 200 million years ago from microscopic marine plants and animals that became incorporated in the sediments and rocks formed at the bottom of the sea. Coal and peat, on the other hand, began to form up to 340 million years ago from terrestrial vegetation (*see* reading at beginning of this chapter).

Natural gas and crude oil normally occur together with water in traps or reservoirs in between rock layers (*see* figure 18.9). The term 'natural gas' also applies to the gases formed naturally from the decomposition of coal. Both natural gas and crude oil are produced on all continents except Antarctica. The four largest producers of natural gas in the world are the USSR, Algeria, Iran and the United States. The four largest producers of crude oil are the USSR, Saudi Arabia, Kuwait and Iran.

Natural gas consists mainly of methane. It has the composition shown in table 18.4.

Table 18.4 Composition of natural gas

		Percentage
methane	CH_4	88–95
ethane	C_2H_6	3–8
propane	C_3H_8	0.7–2.0
butane	C_4H_{10}	0.2–0.7
pentane	C_5H_{12}	0.03–0.5
carbon dioxide	CO_2	0.6–2.0
nitrogen	N_2	0.3–3.0
helium	He	0.01–0.5

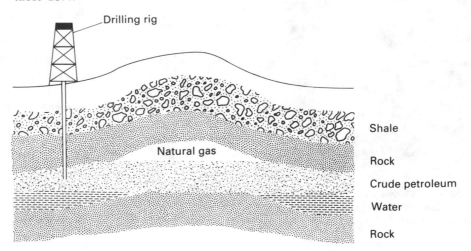

Figure 18.9 Natural gas and crude petroleum occur in traps between rock layers

Crude oil is a liquid varying in colour from dark brown or green to almost colourless. It contains a large number of alkanes. These are unbranched chain alkanes, branched alkanes and cycloalkanes with between five and 40 carbon atoms. The cycloalkanes are known in industry as **naphthenes**. Crude oil also contains about 10% aromatic hydrocarbons and small amounts of other compounds containing sulphur, oxygen and nitrogen.

Coal, as we shall see below, is an important source of aromatic compounds. We have already discussed the composition of coal in section 5.4.

Hydrocarbons occur naturally not only in fossil fuels but also in some living tissues. **Natural rubber** is an example of a naturally occurring hydrocarbon polymer. The rubber molecule consists of thousands of methylbuta-1,3-diene (also known as isoprene) units (*see* figure 18.10). Methylbuta-1,3-diene has the following structure:

$$CH_2{=}\overset{\displaystyle CH_3}{\underset{\displaystyle |}{C}}{-}CH{=}CH_2$$

Methylbuta-1,3-diene
(isoprene)

Pouring latex into reception tanks at a factory in Nigeria

Planting rubber bushes in Brazil

Limonene is found in the peels of citrus fruits. The compound is a member of a group of hydrocarbons known as terpenes

Natural Rubber

About 90% of the natural rubber produced in the world today is produced in Asia from the Brazilian Para rubber tree, *Hevea brasiliensis*. The tree's sap, which is a water latex, is tapped from the tree by scoring the bark with a knife. The latex contains about 30% rubber. This is suspended as small particles in water. The sap is strained into aluminium tanks and acid added to coagulate the rubber.

Many other naturally occurring compounds contain isoprene units. Limonene, for example, contains two **isoprene** units. Limonene is a major constituent of the oils extracted from the peels of citrus fruits such as lemon and orange. The compound is a member of a group known as the **terpenes**. Terpenes are C_{10} compounds containing two isoprene units joined head to tail. Compounds with four isoprene units (C_{20} compounds) are known as diterpenes and those with six units are called triterpenes (C_{30} compounds). Squalene, which is found in shark liver oil, is a triterpene. The tetraterpenes (C_{40} compounds) contain eight isoprene units. Tetraterpenes are found in pigments in vegetables and animal fats. Their colour is due to the long **conjugated** systems of double bonds. β-Carotene, for example, provides the distinctive orange colour in carrots.

(a) Rubber

(b) Limonene

(c) Squalene

(d) β-Carotene

Figure 18.10 Hydrocarbons in living tissues

MANUFACTURE OF HYDROCARBONS

Alkanes, alkenes, alkynes and arenes are all obtained by refining oil (*see* below). Coal is also an important source of hydrocarbons. Bituminous coal is heated in the absence of air in a retort. This produces coke, coal tar, ammonia, hydrogen sulphide and coal gas. This process is called the **destructive distillation** of coal. Distillation of the coal tar produces arenes (*see* table 18.5).

Table 18.5 Some aromatic compounds produced by fractional distillation of coal tar

Fraction	Boiling point range/°C	Major components
light oil	80–170	benzene, methylbenzene
middle oil (carbolic)	170–230	phenol, naphthalene
heavy oil (creosote)	230–270	phenol, naphthalene, anthracene
green oil	270–400	anthracene
residue	over 400	pitch

> In less industrialised countries, hydrocarbons such as methane and ethene are increasingly being produced from biomass (*see* chapter 5). Biogas is principally methane. Ethene can be manufactured by dehydrating ethanol obtained by fermentation.

The coke can be converted to **water gas** by the reaction with steam:

$$C(s) + H_2O(g) \rightarrow CO(g) + H_2(g)$$

Alkanes and alkenes can be produced from water gas by the **Fischer–Tropsch** process. The water gas is mixed with hydrogen and passed over an iron, cobalt or nickel catalyst at an elevated temperature and under a pressure of 200 to 300 atmospheres. For example:

$$CO(g) + 3H_2(g) \rightarrow CH_4(g) + H_2O(g)$$
$$2CO(g) + 4H_2(g) \rightarrow 2H_2O(g) + C_2H_4(g)$$

Methanol and other compounds containing oxygen can also be obtained from water gas by the Fischer–Tropsch process:

$$CO(g) + 2H_2(g) \xrightarrow{\text{catalyst}} CH_3OH(g)$$

A zinc and chromium(III) oxide catalyst is used at 300°C and a pressure of 300 atmospheres.

Calcium dicarbide is also manufactured from coke by heating it with calcium oxide to above 2000°C in an electric furnace:

$$CaO + 3C \rightarrow CaC_2 + CO$$

On treating calcium dicarbide with water, ethyne is produced (*see* section 18.1). This thus provides an alternative route for the manufacture of unsaturated hydrocarbons from coke.

PETROLEUM REFINING

Crude oil is a complex mixture of hydrocarbons and other compounds. As such it has little use. It thus has to be refined into products which are useful. Crude oil is thus transported either by tanker or by overland pipelines to refineries.

Refining oil involves a number of physical and chemical processes: fractional distillation, cracking, re-forming and sulphur removal.

Fractional Distillation

Crude oil is separated into a number of components by means of simple distillation, fractional distillation and vacuum distillation columns. The nature of the plant in

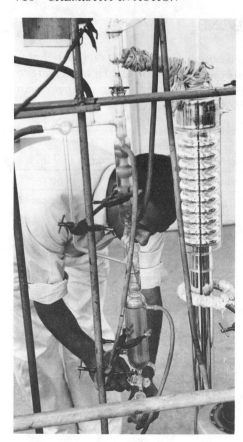

Extraction and distillation laboratory at the Indian Institute of Petroleum

Aerial view of the Isle of Grain refinery in Kent

an oil refinery and also the number and nature of the fractions produced varies depending on the composition of the crude oil and on the demand for the various fractions.

The crude oil is first stripped of dissolved gaseous components by simple distillation. **Primary distillation** then separates the oil into a gaseous fraction, a light distillate, a middle distillate and a residue. Further fractional distillation of the light and middle distillates and **vacuum distillation** of the residue produces more fractions. The boiling point ranges and composition of various fractions are shown in table 18.6. A schematic diagram of a primary distillation column is shown in figure 18.11.

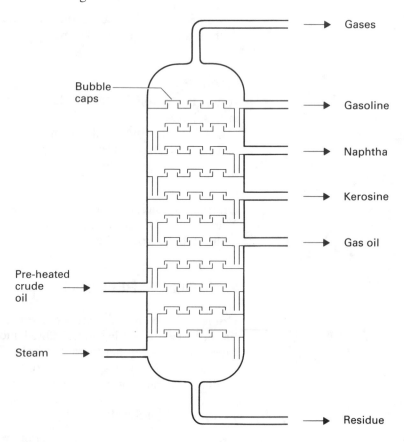

Figure 18.11   Primary distillation of crude oil

Table 18.6 Typical fractions of crude oil

Fraction	Boiling point range/°C	No. of carbon atoms in molecule	Percentage of crude oil by mass
gases	less than 40	1–4	3
gasoline	40–100	4–8	7
naphtha	80–180	5–12	7
kerosine	160–250	10–16	13
residue:			
lubricating oil and wax	350–500	20–35	25
bitumen	over 500	over 35	25

Gases

These are simple unbranched chain alkanes such as ethane, propane and the butanes. This fraction is sometimes known as **refinery gas**. The gases are stripped from the crude oil before primary distillation or separated from the gasoline fraction after primary distillation. They are used as gaseous fuels or liquefied under pressure to form liquefied petroleum gas (LPG). This is sold as heating fuel or used as cracker feedstock for the production of ethene.

Gasoline

This is the petrol fraction. It consists of a mixture of hydrocarbons including unbranched chain and branched alkanes. The combustion characteristics of unbranched chain alkanes are not ideally suited to internal combustion engines. This fraction is thus sometimes thermally re-formed (*see* below) to convert the unbranched chain molecules to branched molecules. Before use as petrol the fraction is normally blended with branched chain alkanes, cycloalkanes and aromatic compounds obtained from other fractions by either catalytic cracking or re-forming.

The quality of petrol is indicated by its **octane number**. This is the percentage by volume of 2,2,4-trimethylpentane (also known as iso-octane) in a mixture of 2,2,4-trimethylpentane and heptane (an unbranched chain alkane) which gives the same knocking characteristics as the petrol under test.

A poor fuel has a zero octane number whereas a good fuel has an octane number of 100. The octane number of the gasoline fraction obtained from crude oil is normally less than 60. The combustion characteristics of petrol are improved by adding the **anti-knock additive** tetraethyllead(IV), $Pb(C_2H_5)_4$ (*see* section 15.2). This is a colourless liquid obtained by heating chloroethane with an alloy of sodium and lead:

$$4C_2H_5Cl + 4Na/Pb \rightarrow Pb(C_2H_5)_4 + 4NaCl$$

On combustion of petrol containing this additive, particles of lead and lead(II) oxide are formed. These slow down certain steps in the combustion of the fuel and thus prevent knocking. 1,2-Dibromoethane is also added to the petrol. This reacts with the lead and lead(II) to form lead(II) bromide. Since lead(II) bromide is volatile it is removed in the exhaust gases of the car (*see* section 15.2).

Naphtha

Petroleum naphtha is a 'cut' taken between gasoline and kerosine during the fractional distillation of crude oil. It consists predominantly of alkanes (*see* table 18.7). Most petroleum naphtha is reformed to produce extra gasoline. However, a significant proportion is used as **feedstock** to produce other chemicals (*see* below).

> Naphtha is also obtained by fractional distillation of the light oil fraction obtained from coal tar (*see* table 18.5). Coal tar naphtha has a high proportion of aromatic hydrocarbons.

Jet aircraft use kerosine as a fuel

Table 18.7   Percentage composition of typical Middle Eastern naphtha

	Number of carbon atoms					Total
	5	6	7	8	9	
unbranched chain alkanes	13	7	7	8	5	40
branched chain alkanes	7	6	6	9	10	38
cycloalkanes	1	2	4	5	3	15
aromatics	–	–	2	4	1	7
total						100

Kerosine

Kerosine consists of aliphatic alkanes, naphthenes (*see* above) and aromatic hydrocarbons. Some is refined for use as paraffin and some is cracked to produce gasoline. However, it is predominantly used as a fuel for jet aircraft.

Gas Oil

This is also known as diesel. Some of this is cracked to form refinery gas and gasoline. However, it is predominantly used as fuel for diesel engines. In a diesel engine, fuel is ignited by compression. Diesel engines thus do not have sparking plugs. The diesel fuel is known as DERV (Diesel Engine Road Vehicle). Gas oil is also used to heat furnaces in industry.

Residue

This is the fraction left when the other fractions have been removed from the oil. Most of it is used as **fuel oil** to heat boilers and raise steam in factories, power stations and ships. Some, however, is vacuum distilled to produce lubricating oils and paraffin wax. Lubricating oils are purified by solvent extraction. The dark viscous material left after the vacuum distillation of the residue is known as bitumen or asphalt. This is used to surface roads.

We have seen how fractional distillation and vacuum distillation together with solvent extraction separate crude oil into various useful fractions. These are all physical processes. Chemical processes are also used in refining oil. These processes can be divided into two categories: cracking and re-forming.

Cracking

In this process larger molecules in the higher-boiling fractions of crude oil are broken down to lower-boiling fractions with smaller molecules. Cracking is necessary since the demand for lower-boiling fractions—particularly gasoline—often outstrips the supply from the fractional distillation of crude oil.

Cracking also produces alkenes needed as chemical feedstock. There are three important types of cracking: hydrocracking, catalytic cracking and thermal cracking.

Hydrocracking

High-boiling fractions such as waxes and heavy oils can be converted to lower-boiling fractions by hydrocracking. In this process the fraction to be cracked is heated at a very high pressure in hydrogen. As the large molecules break up hydrogen is added. Saturated smaller molecules are thus produced. The method is used to produce gas oil and gasolines from heavier fractions.

Catalytic Cracking

This method produces a mixture of saturated and unsaturated products. It operates at a relatively low temperature and uses a catalyst such as a mixture of silica and alumina. Catalytic cracking is used to produce high-grade petrol and unsaturated hydrocarbons from heavy fractions.

Thermal Cracking

The large hydrocarbon molecules in a heavy fraction can be broken down to smaller molecules by heating the fraction to a temperature above its boiling point. Like catalytic cracking, a mixture of saturated and unsaturated products is obtained. For example:

$$\underset{\text{decane}}{C_{10}H_{22}} \rightarrow \underset{\text{octane}}{C_8H_{18}} + \underset{\text{ethene}}{CH_2{=}CH_2}$$

Thermal cracking is particularly important in the production of unsaturated hydrocarbons such as ethene and propene. Plants known as **steam crackers** are used for thermal cracking. The hydrocarbon feedstock is first heated in a furnace to about 800°C. Steam is then added to dilute the feedstock. This increases the yield of alkenes. The larger molecules are cracked into smaller molecules. The hot

Catalytic cracking plant in Venezuela

Table 18.8 Products of steam cracking different feedstocks

Product	Feedstock	
	Ethane percentage, by mass	Naphtha percentage by mass
hydrogen	10	1
methane	6	15
ethene	76	30
propene	3	16
butene	1	5
buta-1,3-diene	2	5
petrol	2	23
fuel oil	–	4

gases are cooled to about 400°C by water in a quench boiler. The boiler produces high-pressure steam. The gases then pass to a fractionator where they are cooled to 40°C. The larger molecules condense to form petrol and fuel oil. The remaining gases are compressed in a compressor driven by the high-pressure steam from the quench boiler. The products are finally separated in fractional distillation columns.

In Europe, naphtha is the main cracker feedstock for the production of unsaturated hydrocarbons. In the USA, the principal feedstock is ethane. This is readily available from refineries as a component of liquid petroleum gas (LPG) or from natural gas and oil wells as a component of natural gas liquid NGL. Propane, butane and gas oil are also used as feedstock for steam cracking. The products of cracking ethane and naphtha are shown in table 18.8.

Cracking takes place by means of a radical mechanism (*see* section 18.1 above).

Re-forming

Whereas cracking processes break down larger molecules into smaller molecules, re-forming processes restructure molecules or build them up to form larger molecules. Re-forming is used in refining crude oil to convert low-grade gasoline fractions to higher-grade fractions. It is also used to produce petrochemical feedstock. Re-forming processes may be divided into three categories: isomerisation, alkylation, and cyclisation and aromatisation.

Isomerisation

During this process the molecules of one isomer undergo rearrangement to form another isomer. The process is important in upgrading the gasoline fraction obtained from the primary distillation of crude oil. We have already seen that this fraction contains an undesirable proportion of unbranched chain alkanes. This can be converted to branched alkanes by heating the fraction to 500–600°C under a pressure of 20 to 70 atmospheres. This process is known as **thermal re-forming**.

Unbranched chain alkanes can also be isomerised by catalytic re-forming. For example, butane can be isomerised to form 2-methylpropane by using an aluminium chloride catalyst at a temperature of 100°C or higher:

$$CH_3CH_2CH_2CH_3 \xrightarrow{AlCl_3} (CH_3)_3CH$$

This reaction has an ionic mechanism involving carbocations (*see* section 17.3).

Alkylation

During this process alkanes and alkenes which have been formed by cracking are put back together to form higher-grade gasolines. The alkanes and alkenes generally have two to four carbon atoms. The process is carried out at a low temperature using a strongly acidic catalyst such as sulphuric acid:

$$(CH_3)_3CH + (CH_3)_2C{=}CH_2 \xrightarrow[25°C]{conc. H_2SO_4} (CH_3)_3CCH_2CH(CH_3)_2$$

2,2,4-trimethylpentane
(iso-octane)

The reaction takes place by means of an ionic mechanism involving the carbocation $(CH_3)_3C^+$.

Cyclisation and Aromatisation

When the gasoline and naphtha fractions of the primary distillation of crude oil are passed over a catalyst such as platinum or molybdenum(VI) oxide on aluminium oxide at 500°C under 10 to 20 atmospheres pressure, cyclisation and then aromatisation of hexane and larger unbranched chain alkanes occur:

$$CH_3CH_2CH_2CH_2CH_2CH_3 \xrightarrow{-H_2} \bigcirc \xrightarrow{-3H_2} \bigcirc$$

Hexane   Cyclohexane   Benzene

The removal of hydrogen from the hexane and then the cyclohexane is known as **dehydrogenation**. This type of re-forming is a type of cracking process. It is variously known as platforming, catalytic re-forming or simple re-forming. Sometimes hydrogen is introduced into the system to prevent the alkane decomposing to carbon and to maintain the activity of the catalyst. In this case the process is known as hydroforming.

Removal of Sulphur

Crude oil contains hydrogen sulphide and other compounds containing sulphur. The amount of sulphur in the oil varies depending on the source of the oil. North Sea oil has a low sulphur content. On distillation of crude oil the organosulphur compounds break down to form additional hydrogen sulphide. The hydrogen sulphide ends up in the refinery gas or the fraction known as LPG (*see* above). Since hydrogen sulphide is a weak acid, it can be removed by treatment with a weak base. Alternatively the sulphur can be recovered by burning the hydrogen sulphide in air and passing the combustion products over an aluminium oxide catalyst at 400°C. The overall reaction can be represented as:

$$2H_2S(g) + O_2(g) \rightarrow 3S(s) + 2H_2O(g)$$

About 75% of all elemental sulphur used nowadays by industry in non-Communist countries is extracted from crude oil and natural gas (*see* section 15.4).

USES OF THE HYDROCARBONS

About 90% of all crude oil produced is used as fuel. Even though the percentage used for petrochemicals is small it is of great significance. Many thousands of organic compounds are made from the refined products of oil. These in turn are used to manufacture thousands of products which satisfy the basic needs of modern society and also provide its luxury and leisure consumer items (*see* figure 18.12).

A North Sea oil-rig. North Sea oil has a low sulphur content

Sulphur recovery plant for the extraction of sulphur from refinery gases at Pembroke, Wales

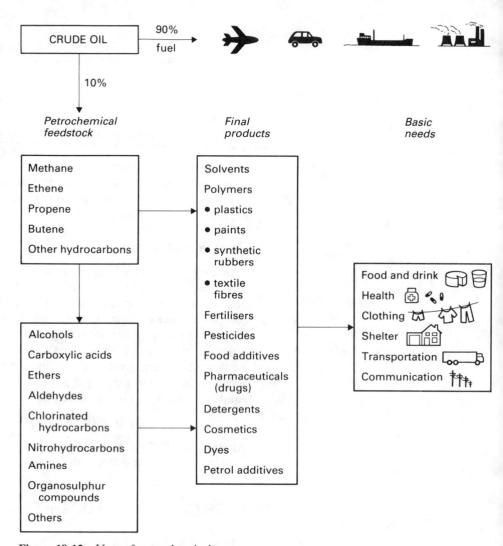

Figure 18.12 Uses of petrochemicals

Although the groups of chemicals shown in figure 18.12 are broadly classified as petrochemicals since they are produced from petroleum, it should be noted that many organic chemicals—particularly aromatic compounds—are manufactured from coal tar and other sources. However, about 90% of the total amount of all organic chemical feedstock is obtained from petroleum.

We shall now examine some specific examples of how the hydrocarbons are used as chemical feedstock.

A petroleum refinery in Iraq. Many thousands of organic compounds are made from the refined products of oil

The Alkanes

Methane has a number of uses as well as being an important fuel. It is used to make **synthesis gas** or syn-gas as it is also known. Like water gas—which is produced from coke and steam—synthesis gas is a mixture of carbon monoxide and hydrogen. It is made by heating methane or naphtha to about 750°C under a pressure of about 30 atmospheres in the presence of a nickel catalyst:

$$CH_4 + H_2O \rightarrow CO + 3H_2$$

In section 7.2 we saw how synthesis gas is used to manufacture hydrogen for the Haber synthesis of ammonia.

Synthesis gas is also used to manufacture methanol and other organic compounds. In the manufacture of methanol the synthesis gas is passed over a solid zinc oxide–copper catalyst at 250°C and 50 to 100 atmospheres pressure:

$$CO + 2H_2 \rightarrow CH_3OH$$

The synthesis gas required for this process has to be very pure.

Methanol can be broken down catalytically to form synthesis gas. It thus provides a convenient means of transporting the gas. Methanol is an important petrochemical feedstock. It is used, for example, in the manufacture of ethanoic acid:

$$CH_3OH + CO \rightarrow CH_3COOH$$

The catalyst for this process is the soluble rhodium anionic complex $[Rh(CO)_2I_2]$. Ethanoic acid is manufactured in this way since the industrial demand for it outstrips the amount available from the fermentation processes.

Soluble rhodium compounds may also be used one day as homogeneous catalysts in the manufacture of ethane-1,2-diol from synthesis gas:

$$2CO + 3H_2 \rightarrow HOCH_2CH_2OH$$

The reaction occurs at 300°C and 500–1000 atmospheres pressure. At present the process is not economically viable. The product—which is also known as ethylene glycol—is used as antifreeze and to manufacture polyesters such as Terylene.

Methane is also used to manufacture chloromethanes such as trichloromethane (commonly known as chloroform). The chloromethanes have a variety of uses. Chloromethane itself is used to manufacture silicones, for example.

Finally, methane is increasingly being used in the manufacture of ethyne:

$$2CH_4 \rightarrow CH{\equiv}CH + 3H_2$$

This reaction occurs at about 1500°C. The heat required to raise the gas to this temperature is produced by burning methane in a limited supply of air.

Ethane also has a number of important uses. It is used to manufacture ethyl chloride. As we saw above, this is used to manufacture tetraethyllead(IV). In the United States, ethane is an important source for the manufacture of ethene (*see* table 18.8).

Propane is important in the manufacture of aldehydes such as methanal and ethanal. These are particularly important in the manufacture of plastics (*see* chapter 20). Butane is used to make buta-1,3-diene which, as we shall see below, is used to make synthetic rubber.

The Alkenes

Ethene

By far the most important alkene and also the most important petrochemical is ethene. It is the raw material for a number of plastics:

Poly(ethene). This is also known as polyethylene or polythene.

Poly(chloroethene). This is also known as polyvinylchloride or pvc. It is made from chloroethene (also known as vinyl chloride) which in turn is made from ethene as follows:

$$CH_2{=}CH_2 + Cl_2 \rightarrow ClCH_2{-}CH_2Cl$$
1,2-dichloroethane

The 1,2-dichloroethane is manufactured in either the liquid or vapour phase using a zinc chloride or iron(III) chloride catalyst. When 1,2-dichloroethane is heated to 500°C in the presence of pumice under 3 atmospheres pressure, chloroethene is formed:

$$ClCH_2{-}CH_2Cl \rightarrow CH_2{=}CHCl + HCl$$
chloroethene
(vinyl chloride)

Chloroethene is also manufactured by heating ethene, hydrogen chloride and oxygen to 250°C using a copper(II) chloride catalyst:

$$2CH_2{=}CH_2 + 2HCl + O_2 \rightarrow 2CH_2{=}CHCl + H_2O$$

Polyester fibres. Terylene is an example. It is made from ethane-1,2-diol, which in turn is made from epoxyethane as follows:

A safety flaring process being tested at the £500 million ethene cracker plant in Fife, Scotland. It was one of the most expensive safety test 'burns' ever. The ethene produced by the plant is piped to Grangemouth, a grid system and a tanker terminal on the River Forth. The ethene plant adjoins a £400 million natural gas liquid (NGL) fractionation plant opened in 1986

Ethane-1,2-diol is also used as antifreeze and for making detergents.

Ethanol is manufactured by the hydration of ethene using phosphoric acid catalyst on silica:

$$CH_2{=}CH_2 + H_2O \xrightarrow[\text{300°C, 60 atm}]{H_3PO_4} \underset{\text{ethanol}}{C_2H_5OH}$$

Ethanol is used to make ethanal (*see* section 19.2). It is also used as a solvent for lacquers and varnishes and also in cosmetic and toilet preparations.

Finally, ethene is used to make chloroethane, which, as we saw above, is used to make the anti-knock additive, tetraethyllead(IV).

Propene

Propene, like ethene, is used to manufacture a wide range of organic chemicals. Many of these are used to make plastics and resins:

Poly(propene). Propene is the monomer of poly(propene):

$$n(CH_3CH{=}CH_2) \rightarrow [{-}CH_3CHCH_2{-}]_n$$

Propanone and propenal. Propanone (also known as acetone) is widely used as a solvent and also for the manufacture of the plastic known as Perspex. It is manufactured from (1-methylethyl)benzene (*see* below) or from propan-2-ol. The latter is made from propene as follows:

Oxidation of propene over a copper(II) oxide catalyst at 350°C yields the aldehyde propenal:

$$CH_3CH{=}CH_2 + O_2 \rightarrow \underset{\text{propenal}}{CH_2{=}CHCHO}$$

Propane-1,2,3-triol. The propan-2-ol, hydrogen peroxide and propenal manufactured in the above process can be used to manufacture propane-1,2,3-triol (otherwise known as glycerol):

Glycerol is used to make cellophane film.

Propenenitrile. This is also known as acrylonitrile. This is used to make synthetic fibres, rubbers and plastics. It is made by passing propene, ammonia and air over a molybdate catalyst at 450°C:

$$2CH_3CH=CH_2 + 2NH_3 + 3O_2 \rightarrow 2CH_2=CH-CN + 6H_2O$$
propenenitrile

Methylbuta-1,3-diene. This is also known as isoprene. This is polymerised to make synthetic rubbers. It is made in the following stages:

$$C_3H_6 \xrightarrow[\text{Ni, 150°C, 200 atm}]{\text{tripropylaluminium, } (C_3H_7)_3Al} CH_2=CCH_2CH_2CH_3$$
propene

with CH_3 group on the carbon.

$$CH_2=CCH_2CH_2CH_3 \xrightarrow[\text{150–300°C}]{H_3PO_4} (CH_3)_2C=CHCH_2CH_3$$

with CH_3 group.

$$(CH_3)_2C=CHCH_2CH_3 \xrightarrow[\text{700°C}]{\text{HBr, steam}} CH_2=C-CH=CH_2$$
isoprene

with CH_3 group on the central carbon.

Epoxypropane. This is used to manufacture polyurethane foams, polyesters and detergents. It is made as follows:

$$C_3H_6 \xrightarrow[\text{35°C}]{Cl_2 + H_2O} CH_3CHOHCH_2Cl$$
l-chloropropan-2-ol

$$CH_3CHOHCH_2Cl \xrightarrow[\text{100°C}]{Ca(OH)_2} CH_3CH-CH_2$$
with O bridging.
epoxypropane

Butenes and Buta-1,3-diene

But-1-ene, but-2-ene and buta-1,2-diene are used to make synthetic rubbers. If the butenes are used as starting materials they are first converted to buta-1,3-diene by dehydrogenation using a chromium(III) oxide catalyst mixed with aluminium oxide. For example

$$CH_3-CH=CHCH_3 \xrightarrow[\text{600°C}]{Al_2O_3/Cr_2O_3} CH_2=CHCH=CH_2 + H_2$$
but-2-ene   buta-1,3-diene (butadiene)

The Alkynes

The most important alkyne is ethyne (acetylene). This has a number of uses:

● It is used as a fuel in oxy-acetylene torches for cutting and welding metals. When ethyne burns in pure oxygen, a flame temperature of about 3000°C can be achieved.
● It is used to make chloroethene, although the primary starting material for the manufacture of chloroethene is now ethene (*see* above).
● Ethyne is used to manufacture the solvent 1,1,2,2-tetrachloroethane.

The Arenes

Both benzene and methylbenzene are produced in large quantities during the refining of crude oil. Since more methylbenzene is produced than needed, some of it is converted to benzene. The methylbenzene is mixed with hydrogen and passed over a platinum catalyst on aluminium oxide at 600°C under pressure:

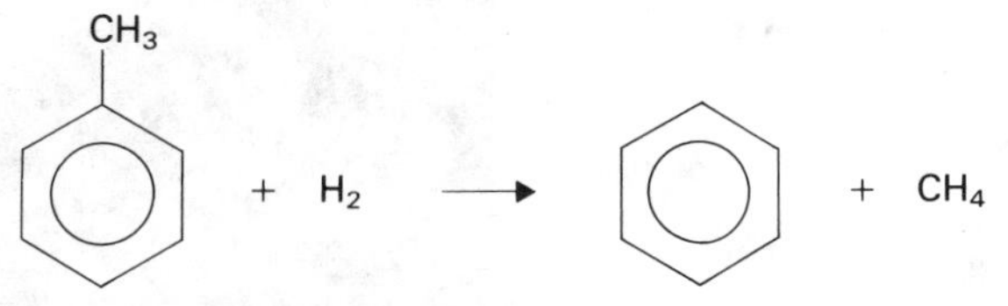

This process is known as **hydroalkylation**.

Benzene is used as a starting material for the manufacture of plastics:

(1-Methylethyl)benzene. This is also known as cumene or 2-phenylpropane. This is used to make phenol and propanone. Phenol is used to make a variety of resins and plastics. The three stages in the manufacture are:

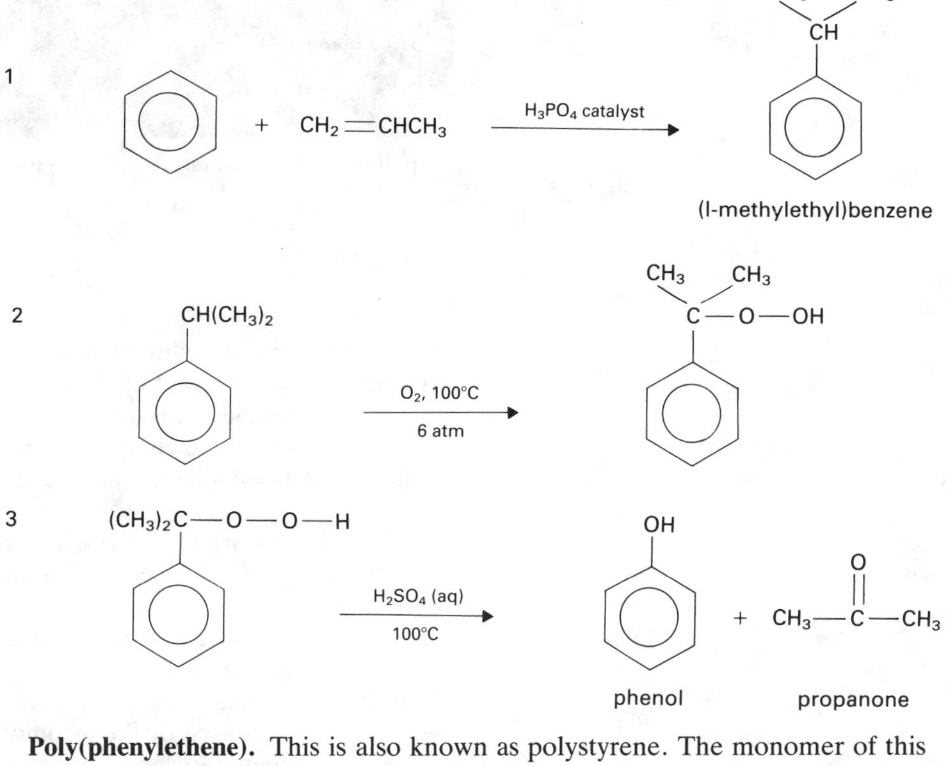

1

(l-methylethyl)benzene

2

3

phenol   propanone

Poly(phenylethene). This is also known as polystyrene. The monomer of this polymer is phenylethene (styrene). This is made from benzene as follows:

Friedel–crafts

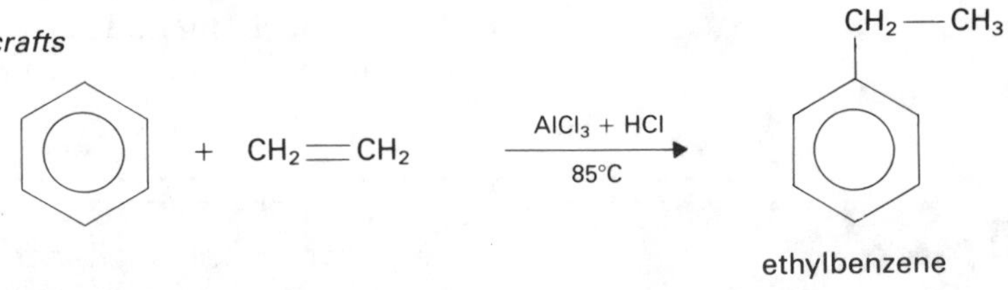

ethylbenzene

Catalytic dehydrogenation

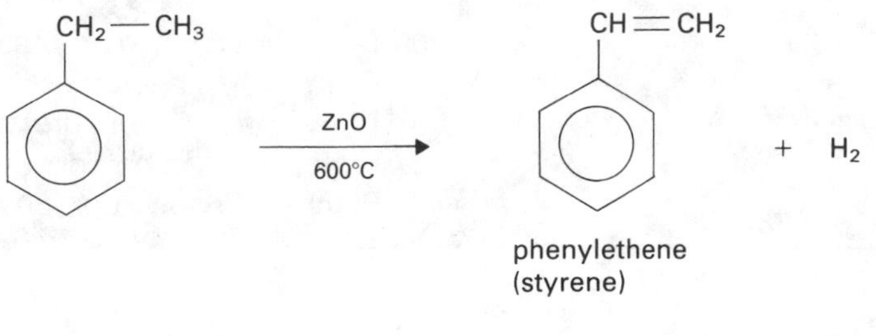

phenylethene
(styrene)

Oil pollution on a beach in Antigua, Leeward Islands, West Indies

POLLUTION

Crude oil and its refined products are responsible for various types of pollution. The most important pollutants are:

Oil spillages.   (*See* section 11.2.)

Carbon monoxide. This is produced by the incomplete combustion of fuels in air. It combines with haemoglobin and is thus highly toxic. It may affect mental alertness and in a confined space a concentration of 10% in air can be fatal within 2 minutes.

Unburnt hydrocarbons. These are also produced by incomplete combustion of fuels. In strong sunlight these can contribute to photochemical smog (*see* section 15.3).

Lead compounds. These result from the use of tetraethyllead(IV) as a petrol additive (*see* section 15.2).

Particulates. These are solid particles. They include carbon and unburnt hydrocarbons produced by the incomplete combustion of fuels. They can contribute to the formation of smog.

Nitrogen and sulphur oxides. Nitrogen and sulphur compounds are present in trace amounts in fuels. They react with oxygen in the air to form acidic oxides. These can cause acid rain (*see* section 11.2).

SUMMARY

1. **Hydrocarbons** occur naturally in **fossil fuels**.
2. Coke and **coal tar** are produced by the **destructive distillation** of coal.
3. **Coal tar** is rich in aromatic compounds.
4. **Coke**, when heated with steam, produces water gas.
5. **Water gas** is a mixture of carbon monoxide and hydrogen.
6. Water gas can be converted to alkanes and alkenes by the **Fischer–Tropsch** process.

7. **Petroleum refining** involves a number of chemical and physical processes:
   (a) simple, fractional and vacuum **distillation**,
   (b) hydro-, catalytic and thermal **cracking**,
   (c) **re-forming**,
   (d) sulphur removal.
8. The main fractions produced by distillation of crude oil are:
   (a) gases,
   (b) gasoline (petrol),
   (c) naphtha,
   (d) kerosine,
   (e) gas oil (diesel oil),
   (f) residue, including lubricating oils, waxes and bitumen.
9. Cracking proceeds by a **radical mechanism**.
10. The three main categories of re-forming are:
    (a) isomerisation (thermal and catalytic reforming),
    (b) alkylation,
    (c) cyclisation and aromatisation.
11. About 90% of the refined products of crude oil are used as **fuel**.
12. The remaining 10% is used as chemical feedstock to produce a variety of organic compounds (*see* table 18.9). These are used to make solvents, plastics, pharmaceuticals and a variety of other products.

Table 18.9 Hydrocarbons as chemical feedstock

Feedstock	Chemical products
alkanes	
methane	methanol, ethanoic acid, chloromethane, ethene
ethane	ethyl chloride, tetraethyllead(IV)
propane	methanal, ethanal
alkenes	
ethene	poly(ethene), poly(chloroethene), polyesters, ethanol, ethanal
propene	poly(propene), propanone, propenal, propane-1,2,3-triol, propenenitrile, epoxypropane
butenes	synthetic rubber
alkynes	
ethyne	chloroethene, 1,1,2,2,-tetrachloroethane
arenes	
benzene	(1-methylethyl)benzene, phenol, poly(phenylethene)

Examination Questions

1. 10.0 cm^3 of a gaseous hydrocarbon was exploded with 100.0 cm^3 of oxygen. The total volume after the explosion was 75.0 cm^3 which decreased to 25.0 cm^3 on shaking the gaseous mixture with aqueous sodium hydroxide. (All volumes measured at room temperature and pressure.)
    (a) Say why there was a decrease in volume on shaking with alkali. Write an equation.
    (b) Deduce the volume of carbon dioxide formed in the explosion.
    (c) Deduce the volume of oxygen that *actually reacted* with 10.0 cm^3 of the hydrocarbon.
    (d) Why was the hydrocarbon mixed with *excess* of oxygen before the explosion?
    (e) Calculate the molecular formula of the hydrocarbon setting out your working clearly.

(f) Write the structural formula of four possible compounds the hydrocarbon could be and give them their IUPAC names.

(g) One of the isomers from part (f) is dissolved in tetrachloromethane and ozonised oxygen is bubbled through the solution. The product of this reaction is reduced with hydrogen and platinum catalyst. The final reaction yields a mixture of ethanal and propanone. Identify the isomer and explain both reactions using appropriate equations to illustrate your answer.

(SUJB)

2. Bromoalkanes may react with alcoholic potassium hydroxide solution to form alkenes.

(a) Write an equation for the reaction of 1-bromobutane with alcoholic potassium hydroxide.

(b) (i) What is the maximum volume of the alkene (gaseous) measured at s.t.p. which could be obtained from 6.85 g of 1-bromobutane? (RAM: H 1, C 12, Br 80. Molar volume of an ideal gas at s.t.p. = 22.4 dm^3)

   (ii) The volume of alkene actually obtained is less than the value calculated. What other transformation is occurring in the reaction mixture?

(c) Give the structural formulae of the isomeric alkenes obtained by treating 2-bromobutane with alcoholic potassium hydroxide.

(d) Draw the structural formula of the alkene obtained by reacting 2-bromo-2-methylpropane with alcoholic potassium hydroxide.

(e) Equal quantities of 1-bromobutane A and 2-bromo-2-methylpropane B were separately allowed to react with alcoholic potassium hydroxide under identical conditions. Which haloalkane, A or B, gave the larger quantity of alkene?

(f) Give a simple chemical test which would allow you to distinguish between 1-bromobutane and butanoyl bromide.

(O & C)

3. (a) The primary process in the thermal cracking of hydrocarbons is the rupture of the weakest C—C bond to form two radicals, as illustrated by

$$C_5H_{12} \rightarrow C_2H_5^{\bullet} + (CH_3)_2CH^{\bullet}$$

   (i) What type of chemical reaction is involved in thermal cracking?
   (ii) Name an alternative type of cracking.
   (iii) State *one* advantage which this alternative has over thermal cracking.
   (iv) Write an equation for the formation of propene from the radical $(CH_3)_2CH^{\bullet}$.

(b) (i) Outline the stages in the formation of chloromethane from methane and chlorine at 450°C.
   (ii) Explain briefly why the chloromethane obtained in this way is impure.

(JMB)

4. The following scheme shows some synthetic pathways starting from ethene.

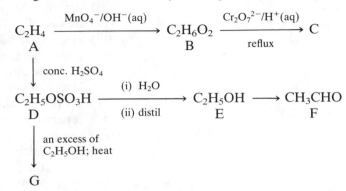

(a) Give the full structural formulae of compounds B, C, D and G.
(b) What reagents and conditions are used to convert E into F?
(c) Write equations showing how C reacts with
   (i) concentrated sulphuric acid,
   (ii) acidified aqueous potassium manganate(VII).
(d) Draw the full structural formula of the repeat unit of each polymer which can be obtained from
   (i) A alone,
   (ii) B and benzene-1,4-dicarboxylic acid.

(UCLES)

5. (a) Explain the theoretical basis of the Markownikoff rule by considering the addition of hydrogen bromide to but-l-ene.
   (b) Addition of bromine water to an alkene A, C_4H_8, gave a compound B. Acid-catalysed dehydration of B yielded two isomeric bromoalkenes C and D, C_4H_7Br. Cleavage of compound C with trioxygen (ozone), followed by reductive hydrolysis, resulted in ethanal being isolated as one of the reaction products. Similar treatment of compound D yielded propanal rather than ethanal.
   Explain the observations described above and identify compounds A, B, C and D.

(JMB)

6. Give the structural formulae of benzene, cyclohexene and cyclohexane, and describe the bonding in benzene.
   Benzene is said to contain a delocalised system of electrons. What do you understand by the term 'delocalised'? Give two pieces of evidence that benzene does in fact contain such a system.
   Compare and contrast the reactions of benzene, cyclohexene and cyclohexane with bromine, in the mole ratio 1:1.
   Give mechanisms for the reactions of bromine with (i) cyclohexene, (ii) cyclohexane.

(L)

7. A method of preparing phenol from benzene is outlined below.

$$C_6H_6 \xrightarrow{\text{Step 1}} C_6H_5NO_2 \xrightarrow{\text{Step 2}} C_6H_5NH_2 \xrightarrow{\text{Step 3}} C_6H_5N_2^+Cl^- \xrightarrow{\text{Step 4}} C_6H_5OH$$

Suggest suitable reagents for carrying out each of the above steps. For each step state the type of reaction taking place and write a balanced equation.

(JMB)

8. (a) For the *primary* distillation of crude petroleum
    (i) describe the process briefly with a fully labelled diagram;
    (ii) explain the principle involved;
    (iii) give (on the diagram, if you wish) commercial names for the separated fractions at this stage;
    (iv) say how different separated fractions vary as to molecular mass, boiling point, and viscosity;
    (v) say (on the diagram, if you wish) to which final commercial use each separate fraction is put.
  (b) Explain the meaning of the following types of reaction, giving **one** example in each case from the *petrochemical* industry:
    (i) cyclisation;
    (ii) hydrogenation;
    (iii) isomerisation;
    (iv) catalytic cracking.
  (c) Comment on the relationship between the demand for petrol *vis-à-vis* diesel fuel and the usefulness of (b) (iv) above.
  (d) How may hydrocarbons, of petrochemical origin, be converted industrially into:
    (i) polyvinyl chloride, $(CH_2{=}CHCl)_n$;
    (ii) dichloroethane, CH_2ClCH_2Cl?

(SUJB)

9. Starting from ethene, propene, ethyne or a petroleum fraction, outline the production of **five** of the following industrial chemicals:
  (a) ethanol,
  (b) vinyl acetate,
  (c) acrylonitrile,
  (d) acetone,
  (e) trichloroethylene,
  (f) acetic acid.

(NISEC)

10. Describe the *chemistry* of **three** environmental problems that have arisen from the introduction of three products or processes by industry. Suitable examples include detergents, insecticides, aerosols, waste gases from factories and motor cars.

(L, Nuffield)

19 FUNCTIONAL GROUPS

Bhopal's Poison Gas Tragedy—Could it Happen in Britain?

The [recent] tragic accident at Union Carbide India Ltd's (UCIL) pesticide plant at Bhopal, and the gas explosion in Mexico City, highlight the necessity of employing stringent safety precautions in the chemical and allied industries. These include early warning devices, emergency plans, risk analysis, etc.—but measures can only be taken to *reduce* the probability of such accidents occurring: some element of risk will always exist.

It was the explosion at Flixborough in 1974 and the accident in Seveso, Italy, that focused European attention on the importance of safety at industrial plants and the need for statutory controls to reduce the likelihood of such disasters happening. In Britain, the result was the setting up of the Major Hazards Advisory Committee, established to advise the Health and Safety Commission (HSC) on which installations would present a major hazard to both employees and the public. Its three reports have been acted upon. In 1982 the Notification of Installations Handling Hazardous Substances Regulations were introduced, covering some 1500—2000 sites. More recently, new regulations—the Control of Industrial Major Accident Hazards (CIMAH) Regulations 1984—have been introduced from the beginning of this year [1985]. The new regulations implement an EEC directive on major hazards—known colloquially as the Seveso Directive. These new statutory controls

Firemen spray water on a temporary sacking barrier at the plant in Bhopal soon after the leak of methyl isocyanate. Three tons of the gas killed more than 2500 people

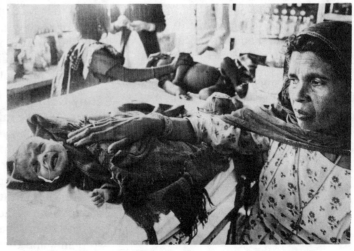

A mother sits beside a child in hospital. The child is suffering from methyl isocyanate poisoning

Bhopal Chronology

1977 Bhopal plant commissioned.

December 1981 A man was killed by phosgene gas at the UCIL plant in Bhopal.

1982 (approx). Last US technician left the plant.

1982 Another phosgene gas leak affected some 30 workers at the plant. A journalist in India reported in 1982 that there had been operating problems at the plant since it opened.

2 December 1984 Leak at UCIL plant occurred at about midnight.

3 December By the end of the day 410 people had died and 20 000 needed hospital treatment

3 December Five of the plant managers were placed under house arrest.

6 December UC chairman, Warren Anderson, arrived in India from the US.

7 December Anderson reached Bhopal and was arrested. He was released after six hours.

7 December By the end of the week, it was reported that over 2000 had died and more than 50 000 needed hospital treatment.

9 December Anderson left India secretly. UC announced that it would give $1m aid for victims of the leak and a similar amount for the construction of an orphanage at Bhopal. The offer was rejected.

14 December 150 000 fled Bhopal in fear at the proposed restarting of the plant.

16 December 'Operation Faith' got under way—the plant was restarted to use up the remaining stock of MIC; estimated to be about 15 tonnes.

21 December Claimed that 19 tonnes of MIC had been processed and that three tonnes still remained at the plant.

5 January 1985 First statement from the head of the investigation into the accident. Suggested that water got into the system and that safety devices were inefficient.

January 1985 Estimates put dead at more than 2500, with some 90 000 undergoing hospital treatment.

require 250 or so sites with large inventories of dangerous substances, specified in the regulations by both type and quantity, to submit on- and off-site emergency plans and to provide information to the local populace.

The disaster at the UCIL pesticide plant at Bhopal in India, a subsidiary of Union Carbide (UC), USA, has again drawn attention to the devastating consequences that can follow industrial accidents, which the legislation is designed to avert. An estimated 2500 people were killed in the surrounding area following an escape of the intermediate methyl isocyanate (MIC)—used by UC in the manufacture of its pesticide called Sevin.

The leak occurred from one of three underground tanks where MIC is stored under pressure. One of the tanks was empty and each of the other two contained an estimated 15 tons of MIC. It appears that a build up of temperature and pressure ruptured a safety valve. Under such conditions, the escaping MIC vapour, on entering a scrubbing tower, should have been neutralised by a sodium hydroxide solution and then flared off. However, the vapour was released in its toxic form without being neutralised by the scrubbing tower.

What Went Wrong?

It has been suggested in a number of press reports that the scrubber was undergoing maintenance and was not in use at the time of the accident. UC, however claims that the scrubber was operational but was unable to cope with the unprecedented rapid flow of gas forced upon it. In addition it has been reported that the flare was turned off.

But why was there a temperature and pressure build up? The failure of the refrigeration system might have allowed a rise in temperature but it is thought more likely that a runaway reaction occurred. According to Professor T. A. Kletz of Loughborough University, contamination could have triggered a self-reaction in the MIC—a form of polymerisation—rapidly generating a heat and pressure build up in the storage vessel. It is reported that Dr S. Varadarajan, the top scientist for the Indian Government heading the inquiry, has suggested that a water leak caused the incident. Half a kilogram of water entering the storage tanks, he says, could lead to a runaway reaction.

Dr P. Merriman, of the Chemical Industries Association (CIA) told *Chemistry in Britain* at the time of the disaster that there were a number of possibilities that could have caused the escape of MIC, one of which was the entry of water, though it would only react slowly. However, if, for example, some sodium hydroxide from the scrubbing device leaked into the storage tank as well, it would act as a catalyst, causing a violent reaction. The MIC 'polymerisation' reaction can also occur if the compound lies without being used. This may be substantiated by the reports that MIC production at Bhopal had been halted in October. Another source of contamination might have been rust which could build up in the tanks while they were not in use, particularly since one was empty at the time of the accident. Varadarajan suggested that following water contamination, phosgene—used in the production process—entered the tank, causing the violent reaction.

Safety Controls

In retrospect the safety devices at the Bhopal plant were inadequate. Varadarajan is quoted as saying that the cooling system on the MIC storage tank was inadequate and liquid level controls did not work at the time of the accident. He was also unconvinced that the scrubber had been turned on.

Bhopal is one of two plants owned by UC that manufacture MIC. Its other plant is located at Institute, West Virginia, USA, and is both older and larger. After the accident at the Bhopal plant MIC production was halted at Institute. UC in the USA has been distancing itself from the UCIL plant; originally UC stated that the two plants and their process safety standards were identical and that the equipment at Bhopal had been made by a US manufacturer to its specifications. However, in a carefully worded statement, Jackson Browning, director of UC's health safety and environmental affairs, later said that standards employed at

UCIL were set by UC, USA, but that the design to meet these standards originated in India.

Unlike the Institute plant Bhopal did not have what was originally reported as a 'computerised safety system', which would have warned of temperature and pressure rises. But, according to a spokesman for UC in the UK, Browning, at a conference in the USA, said that the safety devices, including flares, scrubbers and cooling systems, are part of both the Bhopal and Institute plants but that '. . . the computer at the Institute, West Virginia, plant is for plant production operations and to log in data and it does not operate safety equipment . . . '. However, even if it did not operate safety devices, such a process control computer at Bhopal would have given an early warning of heat and pressure build up.

Could it Happen Here?

The question now being asked is could such an event occur in Britain? In the case of MIC the answer must be no, since MIC is no longer produced in this country. Under the new CIMAH regulations, any company storing more than 1 ton of the material would come under the control of these regulations. Ciba-Geigy in the UK, although it does not manufacture MIC, uses it in the production of herbicides at Grimsby. MIC processing ceased at the plant at the time of the accident but was resumed five weeks later, following discussions with the HSE.

According to the Department of Employment (DE), 'Britain has one of the tightest and most sophisticated systems for controlling the risk of major industrial accidents of any country in the world'. When the CIMAH regulations were introduced DE said that 'these new regulations . . . reduce the likelihood of such accidents and minimise the consequences'. Where a potential hazard exists, companies must show that hazards have been assessed and measures taken to control any safety risks.

LEARNING OBJECTIVES

After you have studied this chapter you should be able to

1. Write the names and **structural formulae** of typical **examples** of compounds in each class shown in table 19.1.

   Table 19.1 Classes of compounds

halogenoalkanes	carboxylic acids
halogenoalkenes	esters
halogenoarenes	acid chlorides
Grignard reagents	acid anhydrides
alcohols	amides
phenols	nitriles
ethers	amines
aldehydes	diazonium salts
ketones	

2. Describe typical **physical properties** of these compounds.
3. Outline their methods of **preparation** in the laboratory.
4. Give examples of their important chemical properties.
5. Indicate how they are **manufactured**.
6. List their important **uses**.
7. Indicate which types of compounds **occur naturally** and give examples.
8. Show how it is possible to
   (a) **ascend** the **homologous series** in certain groups, for example from ethanoic acid to propanoic acid;
   (b) **descend** the **homologous series** in certain groups, for example from ethanol to methanol.
9. Show how it is possible to **convert** a compound in one group to a compound in another group; for example, an aldehyde to a ketone.

19.1
Organic Halogen
Compounds

We saw in section 17.1 that organic compounds are often classified according to their functional groups rather than their carbon skeletons. A functional group is a reactive site on a molecule. It gives rise to many of the characteristic chemical and physical properties of the compound.

Formulae and names of some important organic halogen compounds are shown in table 19.2. For our purposes organic halogen compounds may conveniently be divided into three categories: halogenoalkanes, halogenoalkenes and halogeno-arenes. However, it should be noted that other types of organic compound also contain halogens. One example is the acid chlorides. We shall consider these in section 19.3.

Table 19.2 Organic halogen compounds

Halogenoalkanes	Name	Other names
CH_3Cl	chloromethane	methyl chloride
CH_3Br	bromomethane	methyl bromide
$CHCl_3$	trichloromethane	chloroform
CHI_3	triiodomethane	iodoform
CH_3CH_2Br	bromoethane	ethyl bromide
CH_2BrCH_2Br	1,2-dibromoethane	
Halogenoalkenes		
$CH_2{=}CHCl$	chloroethene	vinyl chloride
$CHCl{=}CCl_2$	trichloroethane	

Halogenoarenes

bromobenzene

(bromomethyl) benzene

HALOGENOALKANES

Under the IUPAC system of nomenclature, monohalogen derivatives of alkanes are known as halogenoalkanes. They are alternatively known as alkyl halides. For example, the IUPAC name for the compound with the formula CH_3CH_2Cl is chloroethane. It is alternatively known as ethyl chloride.

Halogenoalkanes are produced by the reaction between the alkane and the halogen in the presence of sunlight. These are radical reactions and, as we saw in section 18.1, produce not only monohalogen derivatives but also the dihalogen and trihalogen derivatives.

Halogenoalkanes can be prepared by various methods in the laboratory. They can, for example, be prepared from alcohols or from alkenes by the addition of hydrogen halides (*see* sections 17.3 and 18.1).

Nucleophilic Substitution

Much of the chemistry of the halogenoalkanes results from the polar nature of the carbon–halogen bond:

$$\underset{}{\rangle} C^{\delta+}{-}Hal^{\delta-}$$

The polarity is due to the inductive effect (*see* section 17.3). In the presence of a nucleophile the bond can undergo heterolytic fission under suitable conditions:

$$Nu \overset{\curvearrowleft}{} R \overset{\curvearrowright}{-} \text{Hal} \longrightarrow Nu-R + \text{Hal}^-$$

This type of reaction is known as nucleophilic substitution. The mechanisms for such types of reaction are described in section 17.3.

The hydrolysis of halogenoalkanes, Williamson's synthesis, formation of esters and nitriles and the reactions of halogenoalkanes with ammonia are all examples of nucleophilic substitution.

Hydrolysis

Halogenoalkanes are hydrolysed to form alcohols when refluxed with aqueous solutions of alkalis. For example

$$\underset{\text{1-bromobutane}}{CH_3CH_2CH_2CH_2Br} \xrightarrow[\text{reflux}]{OH^-(aq)} CH_3CH_2CH_2CH_2OH$$

An aqueous solution of potassium hydroxide is usually used. We should note that when an ethanolic solution of potassium hydroxide is used, alkenes are formed by elimination of the hydrogen bromide (*see* below).

A halogenoalkane is also hydrolysed when boiled with a suspension of silver oxide in water. The latter is known as moist silver oxide:

$$2R-\text{Hal} + Ag_2O + H_2O \rightarrow 2Ag\text{Hal} + 2R-OH$$

Conversions

Hydrolysis is a typical reaction of the halogenoalkanes. The general equation can be represented as:

$$R-\text{Hal} \xrightarrow[\text{reflux}]{OH^-(aq)} R-OH$$

Lists of typical reactions for various functional groups are given in the summaries in this chapter.

It is not always possible to convert one compound into another in a single step. For example, to convert a halogenoalkane into a carboxylic acid requires at least two steps:

$$\underset{\text{bromoethane}}{CH_3CH_2Br} \xrightarrow[\text{reflux}]{OH^-(aq)} \underset{\text{ethanol}}{CH_3CH_2OH}$$

$$CH_3CH_2OH \xrightarrow[\substack{\text{excess acidified} \\ K_2Cr_2O_7}]{[O]} \underset{\text{ethanoic acid}}{CH_3COOH} \text{ (\textit{see} section 19.2)}$$

The second step is a typical reaction of an alcohol. Conversions of one compound to another may thus often be accomplished by several routes. These routes may require one, two, three or more steps.

Williamson's Synthesis

When a halogenoalkane is refluxed with an alcoholic solution of sodium alkoxide, an ether is produced. For example

$$\underset{\text{bromoethane}}{CH_3CH_2Br} + \underset{\text{sodium ethoxide}}{CH_3CH_2O^-Na^+} \rightarrow \underset{\text{ethoxyethane}}{CH_3CH_2-O-CH_2CH_3} + NaBr$$

This is known as Williamson's synthesis. The nucleophile in this reaction is $CH_3CH_2O^-$. The sodium alkoxide is produced by the reaction of sodium with the alcohol:

$$2Na + 2CH_3CH_2OH \rightarrow 2CH_3CH_2O^-Na^+ + H_2$$

Formation of Esters

An ester is produced when a halogenoalkane is refluxed with a sodium or silver salt of a carboxylic acid:

$$CH_3CH_2Br + \underset{\text{sodium ethanoate}}{CH_3COO^-Na^+} \rightarrow \underset{\text{ethylethanoate}}{CH_3COOCH_2CH_3} + NaBr$$

In this reaction the nucleophile is ethanoate, CH_3COO^-.

Formation of Nitriles

When a halogenoalkane is refluxed with an ethanolic solution of potassium or sodium cyanide, a nitrile is formed:

$$CH_3CH_2Br + CN^- \rightarrow \underset{\substack{\text{propanenitrile}\\\text{(ethyl cyanide)}}}{CH_3CH_2CN} + Br^-$$

Since an extra carbon atom is added to the carbon skeleton, this reaction provides a means of **ascending the homologous series** (see bottom of page).

Reaction with Ammonia

When a halogenoalkane is heated with an ethanolic solution of ammonia in a sealed tube, a mixture of primary, secondary and tertiary amines and the quaternary ammonium salt is produced. For example, when bromoethane is heated with an ethanolic solution of ammonia, the following amines are formed:

$CH_3CH_2NH_2$	ethylamine	(primary amine)
$(CH_3CH_2)_2NH$	diethylamine	(secondary amine)
$(CH_3CH_2)_3N$	triethylamine	(tertiary amine)
$(CH_3CH_2)_4N^+Br^-$	tetraethylammonium bromide	(quaternary ammonium salt)

If ammonia is used in excess, the yield of primary amine improves:

$$CH_3CH_2Br + NH_3 \rightarrow CH_3CH_2NH_2 + HBr$$

Example of ascending the homologous series

How would you convert ethanoic acid to propanoic acid?

Solution

One possible route involves the following four steps:

1. Ethanoic acid is first converted to ethanol by treatment with tetrahydridoaluminate(III) in dry ethoxyethane (diethyl ether):

$$CH_3CO_2H \xrightarrow[\text{ethoxyethane}]{LiAlH_4} CH_3CH_2OH \quad \textit{(see section 19.3)}$$

2. The ethanol is converted to iodoethane by the reaction with concentrated hydroiodic acid. This can be generated *in situ* by the reaction of concentrated sulphuric acid and potassium iodide:

$$CH_3CH_2OH \xrightarrow[\text{reflux}]{HI} CH_3CH_2I \quad \textit{(see section 19.2)}$$

3. The iodoethane is refluxed with potassium cyanide yielding propanenitrile:

$$CH_3CH_2I \xrightarrow[\text{reflux}]{KCN} CH_3CH_2CN$$

4. Finally, the nitrile is hydrolysed to form propanoic acid by boiling with dilute acid or alkali:

$$CH_3CH_2CN \xrightarrow{\text{hydrolysis}} CH_3CH_2CO_2H \quad (see \text{ section 19.3})$$

Elimination Reactions

If a halogenoalkane is refluxed with an ethanolic solution of potassium hydroxide solution rather than an aqueous solution, elimination of hydrogen and the halogen occurs, resulting in the formation of an alkene:

$$CH_3CH_2I + OH^- \rightarrow CH_2{=}CH_2 + H_2O + I^-$$
$$\text{iodoethane} \qquad\qquad \text{ethene}$$

$$CH_3CHBrCH_3 + OH^- \rightarrow CH_3CH{=}CH_2 + H_2O + Br^-$$
$$\text{2-bromopropane} \qquad\qquad \text{propene}$$

The mechanism of this type of reaction was described in section 17.3. As we saw in that section, elimination and nucleophilic substitution reactions tend to occur concurrently. Conditions determine which one predominates.

When a dihalogenoalkane is refluxed with an ethanolic solution of potassium hydroxide, an alkyne is produced:

$$CH_3CHCl_2 + 2OH^- \rightarrow CH{\equiv}CH + 2Cl^- + 2H_2O$$
$$\text{1,1-dichloroethane} \qquad\qquad \text{ethyne}$$

$$CH_2BrCH_2Br + 2OH^- \rightarrow CH{\equiv}CH + 2Br^- + 2H_2O$$
$$\text{1,2-dibromoethane}$$

Grignard Reagents

Francois Auguste Victor Grignard (1871–1935) published his first paper on the reactions of organomagnesium compounds in anhydrous ether in 1901. In 1912 he won the Nobel Prize for Chemistry for the discovery of Grignard reagents. Grignard became Professor of Chemistry in Lyon, France, in 1919.

Victor Grignard

These are organomagnesium compounds with the general formula RMgHal. They are formed by the reaction of magnesium with a halogenoalkane in dry ethoxyethane (commonly known as ether):

$$CH_3CH_2Br + Mg \rightarrow CH_3CH_2MgBr$$
$$\text{bromoethane} \qquad\qquad \text{ethylmagnesium}$$
$$\text{bromide}$$

Tetrahydrofuran is also used as a solvent for this reaction:

Tetrahydrofuran

Grignard reagents contain polar carbon–metal bonds and provide a source of nucleophilic carbon. Although not fully ionised they can be regarded as carbanion donors:

$$RMg^+Hal^- \longleftrightarrow R^- Mg^{2+}Hal^-$$

Grignard reagents are widely used in the synthesis of alkanes, alcohols, ketones, carboxylic acids and other types of organic compound.

Alkanes are produced by hydrolysis of the Grignard reagent using dilute acid:

$$CH_3CH_2MgBr + H_2O \rightarrow CH_3CH_3 + Mg^{2+} + Br^- + OH^-$$
$$\text{ethylmagnesium} \qquad\qquad \text{ethane}$$
$$\text{bromide}$$

Primary, secondary and tertiary **alcohols** can all be prepared by the reaction of Grignard reagents with aldehydes and ketones:

$$CH_3CH_2MgBr + HCHO \longrightarrow CH_3CH_2CH_2OH$$

methanal   propan-1-ol

$$CH_3CH_2MgBr + CH_3CHO \longrightarrow CH_3CH_2-\overset{\overset{\displaystyle OH}{|}}{\underset{\underset{\displaystyle H}{|}}{C}}-CH_3$$

ethanal   butan-2-ol

$$CH_3CH_2MgBr + CH_3-\overset{\overset{\displaystyle O}{\|}}{C}-CH_3 \longrightarrow CH_3-\overset{\overset{\displaystyle OH}{|}}{\underset{\underset{\displaystyle CH_3}{|}}{C}}-CH_2CH_3$$

propanone   2-methylbutan-2-ol

Ketones can be prepared by addition of a Grignard reagent to a nitrile group followed by hydrolysis of the product with dilute acid:

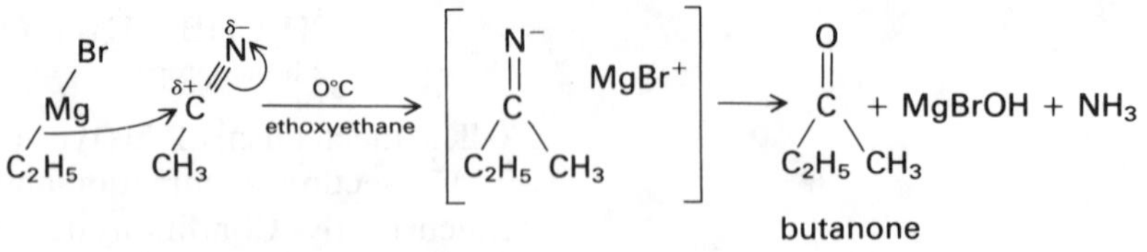

Carboxylic acids can be prepared by the reaction of carbon dioxide with a Grignard reagent dissolved in ethoxyethane followed by hydrolysis of the product with dilute acid:

$$RMgHal + CO_2 \longrightarrow R-\overset{\overset{\displaystyle O}{\|}}{C}-O-MgHal \xrightarrow{H_3O^+} R-\overset{\overset{\displaystyle O}{\|}}{C}-OH + MgHalOH$$

carboxylic acid

Other **organometallic compounds** have similar properties to Grignard reagents. Organolithium compounds, for example, also provide a source of nucleophilic carbon:

$$R-Li \longleftrightarrow R^-Li^+$$

alkyllithium

Like the Grignard reagents, the alkyllithium compounds can also be prepared from halogenoalkanes. For example

$$CH_3CH_2CH_2CH_2Cl + 2Li \xrightarrow{epoxyethane} CH_3CH_2CH_2CH_2Li + LiCl$$

1-butyllithium

HALOGENOALKENES

By far the most important halogenoalkene from a commercial point of view is chloroethene (also known as vinyl chloride), $CH_2\!=\!CHCl$.

Chloroethene can be prepared from 1,2-dichloroethane or directly from ethene (*see* section 18.3). It can also be prepared by the addition of hydrogen chloride to ethyne (*see* section 18.1).

Another important halogenoalkene is trichloroethene. This is produced by passing 1,1,2-tetrachloroethane vapour over barium hydroxide:

$$CHCl_2CHCl_2 \xrightarrow{Ba(OH)_2} CHCl\!=\!CCl_2 + HCl$$

1,1,2,2-Tetrachloroethane is prepared by the addition of chlorine to ethyne (*see* section 18.1).

Unlike the halogenoalkanes, neither chloroethene nor trichloroethene undergo nucleophilic substitution. For example, neither are hydrolysed by an aqueous solution of potassium hydroxide.

HALOGENOARENES

Bromobenzene and chlorobenzene can be prepared by electrophilic substitution of benzene using a catalyst. This type of halogenation is discussed in section 18.2.

Halogenoarenes can also be prepared from the corresponding diazonium salt using a solution of the copper(I) halide in the concentrated halogen acid. For example

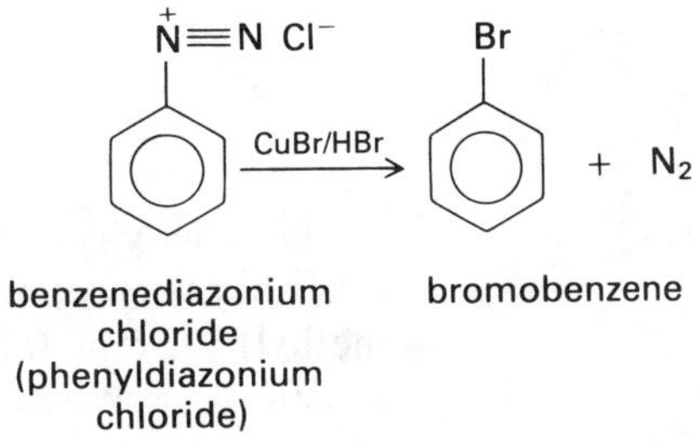

benzenediazonium
chloride
(phenyldiazonium
chloride)

bromobenzene

This is known as the **Sandmeyer reaction**. Chlorobenzene can also be prepared in this way.

The benzene ring in halogenoarenes can undergo further halogenation. For example, when chlorine is bubbled through chlorobenzene in the presence of iron(III) chloride catalyst, a mixture of 1,2-dichlorobenzene and 1,4-dichlorobenzene is formed:

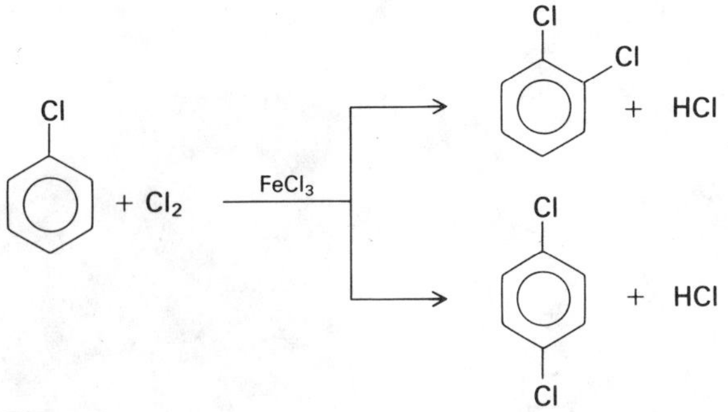

Nitration of chlorobenzene with a mixture of concentrated nitric acid and sulphuric acid gives both 1-chloro-2-nitrobenzene and 1-chloro-4-nitrobenzene:

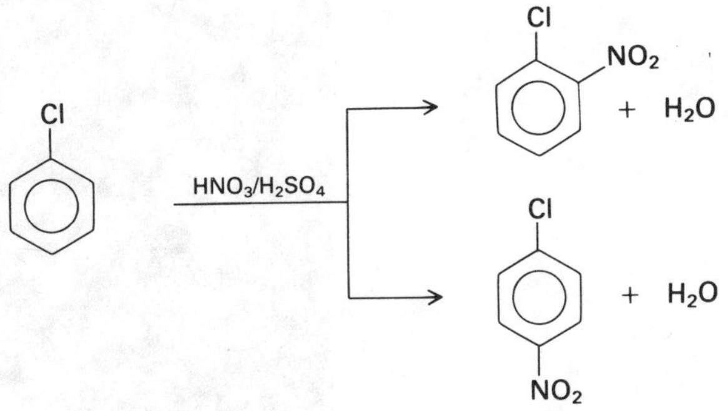

The reactivity of the halogen atom in halogenoarenes is lower than that in halogenoalkanes. This is because the benzene ring lowers the polarity of the carbon–halogen bond. As a result, halogenoarenes are unreactive with nucleophilic reagents. They do not undergo hydrolysis to form phenols, for example, except under vigorous conditions. For instance, 300°C and a pressure of about 200 atmospheres are required for the following reaction to occur:

$$\text{Cl} \quad + 2\text{NaOH} \longrightarrow \text{O}^-\text{Na}^+ \quad + \text{H}_2\text{O} + \text{NaCl}$$

On addition of dilute hydrochloric acid to the product, phenol is formed:

$$\text{O}^-\text{Na}^+ \quad + \text{HCl (aq)} \longrightarrow \text{OH} \quad + \text{NaCl}$$

The reactivity of halogenoarenes with the halogen atom in a side chain, on the other hand, is comparable to that of the halogenoalkanes. For example, (bromomethyl) benzene (also known as benzyl bromide) readily undergoes nucleophilic substitution. When it is refluxed with an aqueous solution of sodium hydroxide, phenylmethanol (also known as benzyl alcohol) is formed:

$$\text{CH}_2\text{Br} \quad \xrightarrow{\text{NaOH (aq)}} \quad \text{CH}_2\text{OH} \quad + \text{Br}^-$$

(bromomethyl) benzene   phenylmethanol

Fluoroaromatic plant control room

MANUFACTURE AND USES OF ORGANIC HALOGEN COMPOUNDS

Manufacture

Many organic halogen compounds are manufactured from the hydrocarbons produced in petroleum refining. **Halogenoalkanes** are produced either by the direct halogenation of alkanes or addition of a hydrogen halide to an alkene (*see* above and also section 18.1). They are manufactured from alcohol and the hydrogen halide using a suitable catalyst. For example

$$CH_3OH + HBr \xrightarrow[\text{heat}]{\text{catalyst}} CH_3Br + H_2O$$

Chloroethene and **trichloroethene** are manufactured by the methods outlined above and in sections 18.1 and 18.3. **Chlorobenzene** is manufactured by passing a mixture of benzene, hydrogen chloride and oxygen over a copper(II) chloride catalyst at 250°C. The overall reaction can be represented as:

This process is known as the **Raschig process**.

Uses of Organic Halogen Compounds

Organic Synthesis

Organic halogen compounds are used widely in organic synthesis. Halogenoalkanes, for example, are used to prepare alcohols, ethers, esters, nitriles and amines. This is due to their ability to undergo nucleophilic substitution (*see* above). Grignard reagents are also used widely in organic synthesis. These can be prepared by the reaction of magnesium with halogenated alkanes, alkenes and arenes.

Petrol Additives

1,2-Dibromoethane and tetraethyllead(IV) (which is made from chloroethane) are added to petrol to improve its combustion characteristics (*see* section 18.3 and 15.2).

Polymers

Chloroethene (also known as vinyl chloride) is used to manufacture poly(chloroethene)—also known as polyvinyl chloride (*see* section 18.3). Chloromethane, CH_3Cl, and Grignard reagents such as CH_3MgCl are used in the manufacture of silicone polymers (*see* section 15.2).

Solvents

The halogenoalkanes in particular are widely used as solvents in the laboratory, in industry and commercially. For example, trichloromethane (also known as chloroform) is used as a solvent for organic syntheses and for solvent extraction (*see* chapter 6).

Dichloromethane is used as a paint stripper. Trichloroethene is commonly used as a dry cleaning agent in laundering and for removing oils and greases from metal surfaces.

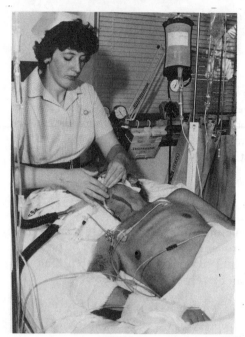

A patient recovers from an anaesthetic after an operation. Halothane is widely used as an anaesthetic in hospital. It is an organohalogen compound

Chloroform Extraction of Caffeine from Tea Leaves

Caffeine occurs in tea leaves. It is soluble in hot water but more soluble in trichloromethane (chloroform). Solvent extraction can be used to separate caffeine from other materials present in tea leaves. The tea leaves are boiled with water. This extracts caffeine and other substances from tea leaves. The solution is cooled and then shaken with chloroform in a separating funnel. The caffeine passes from the water into the chloroform. The chloroform layer is separated from the water layer. Finally, the chloroform is carefully evaporated off leaving the pure solute, caffeine.

Caffeine is also found in cola, cocoa and coffee beans. It is a heterocyclic compound with the structure:

This type of structure is known as an xanthine.

Caffeine is a respiratory stimulant. When consumed in large amounts it can cause sleeplessness and hallucinations. It is also a weak diuretic. A diuretic increases the excretion of urine.

Tea picking in India. Caffeine occurs in tea leaves and also cola, cocoa and coffee beans

Anaesthetics

Trichloromethane (chloroform) was once used extensively as an anaesthetic in surgery. However, it is now considered too toxic for use. Repeated small doses can cause cirrhosis of the liver. Halothane is now widely used in hospitals. It has the structure:

Halothane (2-bromo-2-chloro-1,1,1-trifluoroethane)

Chloroethane, CH_3CH_2Cl, is used to induce general anaesthesia. It is also used as a local anaesthetic by spraying on the skin. Since it is volatile (its boiling point is 12.5°C), it readily evaporates. The cooling results in temporary localised anaesthesia.

Halogenated hydrocarbons are widely used as pesticides

Pesticides

Halogenated hydrocarbons are widely used as pesticides. For example, bromo-ethene is used as a rodenticide and also as a fumigating agent. The pesticides aldrin, chlordane, DDT and PCBs are all chlorinated hydrocarbons. Their use and abuse is discussed further in chapter 16.

Freons

These are chlorofluorocarbons. One of the most important is dichlorodifluoro-methane. This is used as a refrigerant and as an aerosol propellant (see section 16.3). Freons are also used as solvents since they are inert, stable and non-flammable. They are odourless and their toxicity is low.

Pollution

Organic halogen compounds also give rise to various types of pollution. This is discussed in section 16.3.

SUMMARY

1. Typical reactions of the halogenoalkanes.

R—Hal

$\xrightarrow{\text{OH}^-(aq)}$		R—OH
$\xrightarrow{\text{R'ONa}}$		R—O—R'
$\xrightarrow{\text{R'CO}_2\text{Na}}$		R'—CO$_2$R
$\xrightarrow{\text{CN}^-}$		R—CN
$\xrightarrow{\text{excess NH}_3}$		R—NH$_2$
$\xrightarrow[\text{ethoxyethane}]{\text{Mg}}$		R—MgHal

$$-\overset{|}{\underset{|}{C}}-\overset{|}{\underset{|}{C}}-\text{Hal} \xrightarrow[\text{C}_2\text{H}_5\text{OH}]{\text{OH}^-} \overset{/}{\underset{\backslash}{C}}=\overset{\backslash}{\underset{/}{C}} + H_2O + Hal^-$$

continued overleaf

2. Typical reactions of Grignard reagents.

$$R-MgHal \xrightarrow{\text{dil } H_3O^+} R-H + MgHal_2$$

$$R-MgHal \xrightarrow{\text{HCHO}} R-CH_2OH$$

with $CH_3\overset{\displaystyle O}{\overset{\|}{C}}CH_3$ gives $R-\overset{\displaystyle OH}{\underset{\displaystyle CH_3}{\overset{|}{\underset{|}{C}}}}-CH_3$

with (1) CO_2 (2) H_3O^+ gives $R-\overset{\displaystyle O}{\overset{\|}{C}}-OH$

3. Typical reactions of halogenoarenes.

19.2
Alcohols, Phenols and Ethers

ALCOHOLS

Alcohols containing one hydroxyl group only are called **monohydric alcohols**. These may be saturated or unsaturated aliphatic alcohols and aromatic alcohols (*see* table 19.3). Saturated monohydric alcohols form a homologous series with

Table 19.3 Alcohols

Monohydric alcohols	Name	Other names
primary		
CH_3OH	methanol	methyl alcohol
CH_3CH_2OH	ethanol	ethyl alcohol or, simply, alcohol
$CH_3CH_2CH_2OH$	propan-1-ol	n-propyl alcohol
CH_2OH attached to benzene ring	phenylmethanol	benzyl alcohol
secondary		
OH attached to CH_3CHCH_3	propan-2-ol	isopropyl alcohol
tertiary		
$CH_3-C(CH_3)(CH_3)-OH$	2-methylpropan-2-ol	tert-butyl alcohol

Polyhydric alcohols

dihydric   (also known as diols or glycols)

$HO-CH_2CH_2-OH$	ethane-1,2-diol	ethylene glycol

trihydric   (also known as triols)

$CH_2-CH_2-CH_2$ each with OH	propane-1,2,3-triol	glycerol or glycerine

the general formula $C_nH_{2n+1}OH$. The two simplest alcohols in the series are methanol and ethanol. These are both **primary alcohols** since the hydroxyl group is attached to a carbon atom which is attached to two or three hydrogen atoms.

Alcohols in the series with three or more carbon atoms exist in two or more structural isomeric forms. For example, the alcohol with the empirical formula C_3H_7OH can exist as two structural isomers—propan-1-ol or propan-2-ol. Propan-1-ol is a primary alcohol whereas propan-2-ol is a **secondary alcohol** since the hydroxyl group is attached to a carbon atom which is attached to two alkyl groups. **Tertiary alcohols** such as 2-methylpropan-2-ol have three alkyl groups attached to the carbon atom to which the hydroxyl group is attached.

Alcohols containing two or more hydroxyl groups are called **polyhydric alcohols**. Examples are ethane-1,2-diol which is a **dihydric alcohol** and propane-1,2,3-triol which is a **trihydric alcohol** (*see* table 19.3).

Physical Properties

All the alcohols shown in table 19.3 are liquids at room temperature and pressure. The boiling points and densities of the alcohols increase with increasing relative molecular mass. The boiling points of the alcohols are much higher than alkanes with similar relative molecular mass. This is due to intermolecular hydrogen bonding resulting from the highly polar nature of the hydroxyl group:

$$R-O^{\delta-} \cdots H^{\delta+}-O_{\delta-}-R$$

Methanol, ethanol and propanol are all miscible with water in all proportions due to this ability to form hydrogen bonds. The solubility in water decreases as the number of carbon atoms increases. All the alcohols are soluble in most organic solvents.

Preparations of Monohydric Alcohols

The alcohols can be prepared in the laboratory by a variety of methods.

From alkenes. Aliphatic monohydric alcohols can be prepared by the hydration of alkenes using an acid (*see* section 18.1).

From organic halogen compounds. When an aliphatic or aromatic halide is refluxed with aqueous alkali, the corresponding alcohol is produced (*see* previous section).

From carbonyl compounds. Both aliphatic and aromatic alcohols can be produced by reduction of the corresponding aldehydes, ketones or carboxylic acids. These reactions are described in the following section.

From amines. Primary aliphatic alcohols can be produced by the reaction of alkylamines with nitrous acid (*see* section 19.4).

From Grignard reagents. Primary, secondary and tertiary alcohols can all be prepared by the reaction of Grignard reagents with aldehydes or ketones (*see* previous section).

By fermentation. Ethanol can be prepared by the fermentation of a warm sugar solution containing yeast. The ethanol is separated by distillation (*see* chapter 6 and section 9.5).

Chemical Properties of the Alcohols

Basicity and Acidity of Alcohols

Alcohols possess both basic and acidic properties. In acidic media alcohols act as Lewis bases. They do so by accepting protons from the acid to form substituted oxonium ions:

The ROH_2^+ ion then undergoes nucleophilic attack, resulting in loss of a water molecule and substitution by the nucleophile. This type of mechanism is thought to occur in the halogenation of alcohols using hydrogen halides and in the dehydration of alcohols to form alkenes and ethers (*see* below). In both types of reaction fission of the R—OH bond occurs.

Alcohols act as acids by fission of the RO—H bond. This occurs in the reactions of alcohols with reactive metals such as sodium:

$$2RO\text{—}H + 2Na \rightarrow \underset{\text{sodium alkoxide}}{2RO^-Na^+} + H_2$$

This reaction is far less vigorous than the reaction of sodium with water. Indeed, alcohols are much weaker acids than water. The acid dissociation constant of ethanol, for example, shows that the equilibrium lies predominantly to the left:

$$C_2H_5OH + H_2O \rightleftharpoons C_2H_5O^- + H_3O^+$$

$$K_a = \frac{[C_2H_5O^-][H_3O^+]}{[C_2H_5OH]} = 10^{-16} \text{ mol dm}^{-3}$$

The pK_a values for water, ethanol and methanol are compared in table 19.4.

The acidic strength of alcohols decreases in the order:

Primary > Secondary > Tertiary

Table 19.4 pK_a values of water and alcohols

	pK_a
water	14
methanol	15.5
ethanol	16

Halogenation of Alcohols

Alcohols react with hydrogen halides forming halogenoalkanes:

$$R\text{—}OH + HX \rightarrow R\text{—}X + H_2O$$

Dry hydrogen halide gas is bubbled through the alcohol under reflux. Zinc chloride is used as a catalyst for the reaction with hydrogen chloride. Alternatively, the concentrated hydrohalic acid is used. This can be generated *in situ* by the reaction of concentrated sulphuric acid on a halide salt; for example

$$2KBr + H_2SO_4 \rightarrow KHSO_4 + HBr$$

Halogenation of alcohols by this method occurs in three steps:

1. Protonation of the alcohol.

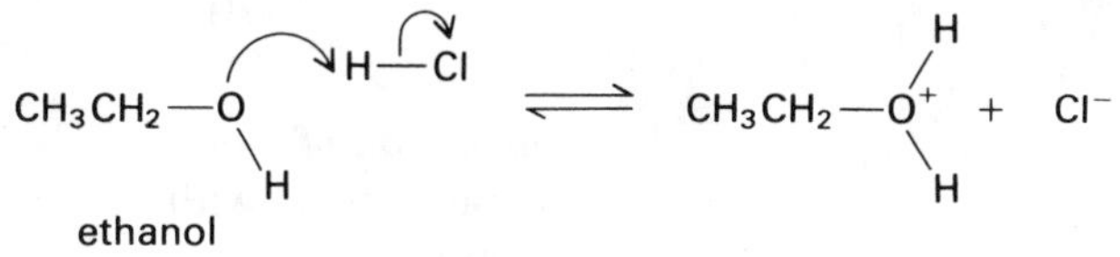

2. Loss of water.

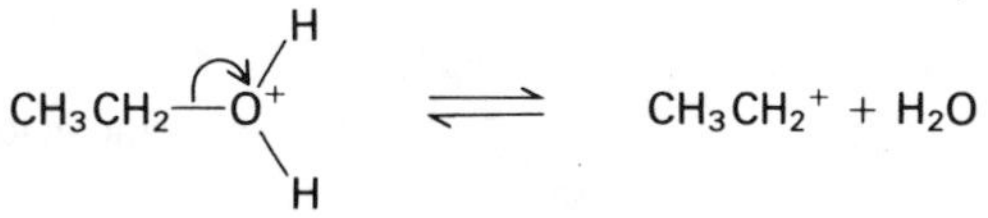

3. Nucleophilic attack by the halide ion.

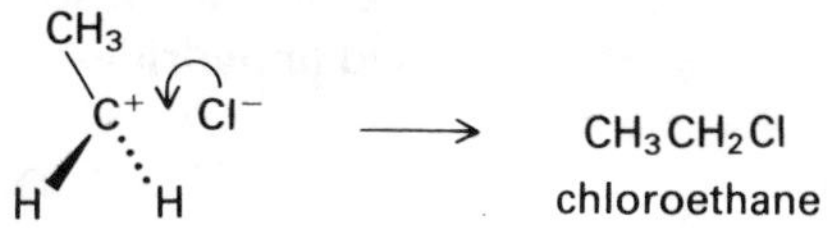

Steps 2 and 3 are an S_N1 mechanism. The rate-determining step is step 2.

Alcohols can also be halogenated using a phosphorus halide such as phosphorus tribromide or phosphorus pentachloride:

$$3CH_3CH_2OH + PBr_3 \rightarrow 3CH_3CH_2Br + H_3PO_3$$

The phosphorus tribromide is generated *in situ* by adding bromine to a mixture of the alcohol and red phosphorus and then refluxing. The reaction of phosphorus pentachloride with anhydrous ethanol is carried out at room temperature:

$$CH_3CH_2OH + PCl_5 \rightarrow CH_3CH_2Cl + PCl_3O + HCl$$

Sulphur dichloride oxide (also known as thionyl chloride) can also be used to chlorinate alcohols:

$$CH_3CH_2OH \quad + \quad \underset{\substack{\text{sulphur dichloride}\\\text{oxide}}}{SCl_2O} \quad \rightarrow CH_3CH_2Cl + SO_2 + HCl$$

Dehydration

Addition of concentrated sulphuric acid to a primary alcohol such as ethanol at 180°C results in the formation of ethene as the main product:

$$CH_3CH_2OH \xrightarrow[180°C]{\text{conc. } H_2SO_4} CH_2{=}CH_2 + H_2O$$

Alternatively, the alcohol vapour can be passed over aluminium oxide at 350°C (*see* section 18.1). The first step in the process is protonation of the alcohol:

$$CH_3CH_2OH + H^+ \rightleftharpoons CH_2CH_2\text{—}\overset{+}{O}H_2$$

The next step is the loss of water to form the ethyl carbocation:

$$CH_3CH_2 \!-\! \overset{+}{O}H_2 \rightleftharpoons CH_3\overset{+}{C}H_2 + H_2O$$

The hydrogensulphate ion, HSO_4^-, then removes a proton from the carbocation to give ethene:

$$CH_3\overset{+}{C}H_2 + HSO_4^- \rightleftharpoons CH_2\!=\!CH_2 + H_2SO_4$$

When an excess of ethanol is used at the lower temperature of 140°C, the carbocation attacks another molecule of ethanol forming ethoxyethane:

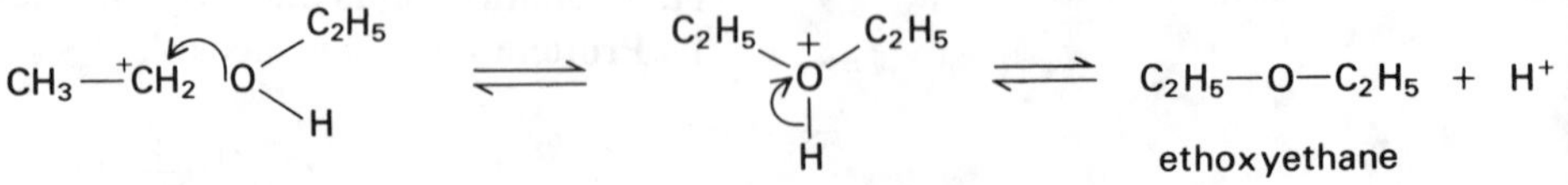

Formation of Esters

Alcohols react with both carboxylic acids and inorganic acids to form esters. For example

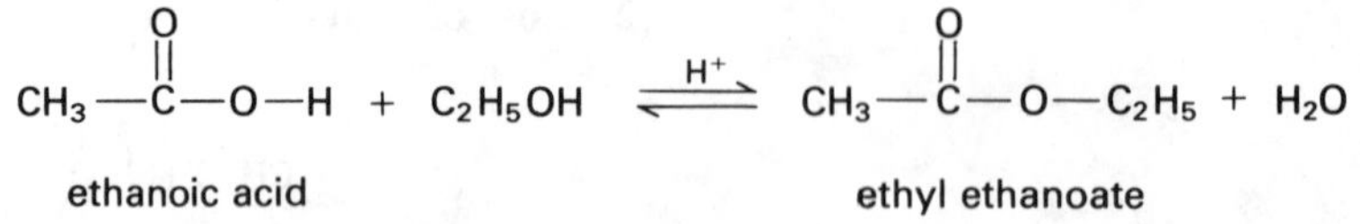

This reaction is slow. Concentrated sulphuric acid is used as a catalyst and also as a dehydrating agent. Removal of the water pulls the equilibrium to the right.

Esters are also formed by the reactions between alcohols and acid chlorides or acid anhydrides:

This reaction takes place vigorously in the cold. For the reaction with ethanoic anhydride, however, refluxing is necessary:

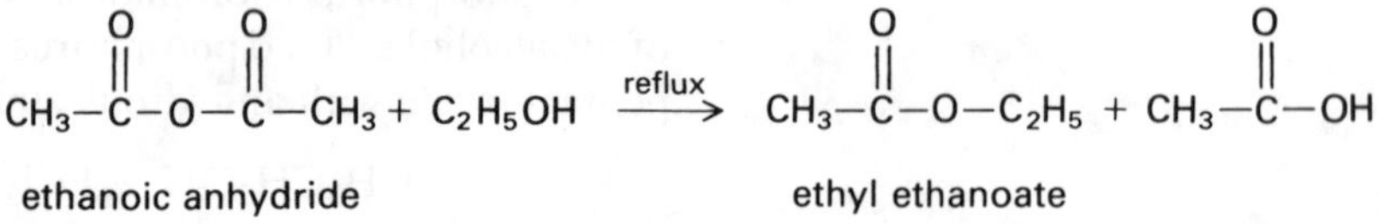

Ethanol reacts with concentrated sulphuric acid at 0°C to form ethyl hydrogensulphate:

$$C_2H_5OH + H_2SO_4 \rightarrow C_2H_5\!-\!O\!-\!SO_2\!-\!OH + H_2O$$
ethyl hydrogensulphate

At higher temperatures dehydration of the alcohol to form an alkene or an ether occurs in the presence of concentrated sulphuric acid (*see* above).

Oxidation of Alcohols

Primary and secondary alcohols can be oxidised using oxidising agents such as potassium manganate(VII) in dilute acid or potassium dichromate(VI) in dilute acid. Primary alcohols are oxidised to form aldehydes:

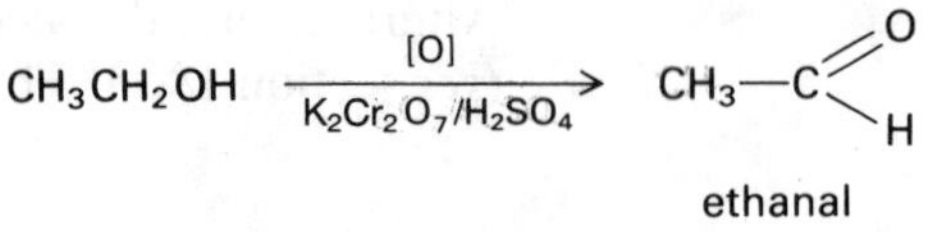

If the oxidising agent is used in excess, further oxidation to the carboxylic acid occurs:

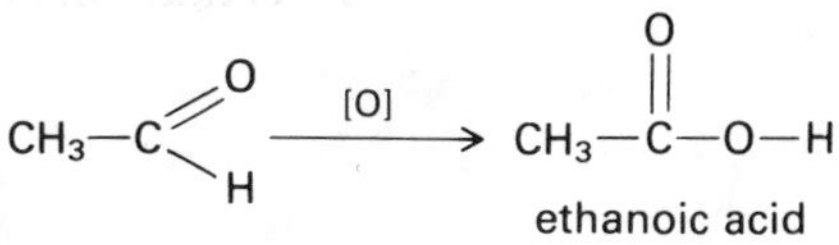

$$CH_3-C\underset{H}{\overset{O}{\diagdown}} \xrightarrow{[O]} CH_3-\overset{O}{\overset{\|}{C}}-O-H$$

ethanoic acid

Primary alcohols can also be oxidised to aldehydes by passing the alcohol vapour over a copper catalyst at 500°C or by passing a mixture of alcohol vapour and air over a silver catalyst at the same temperature:

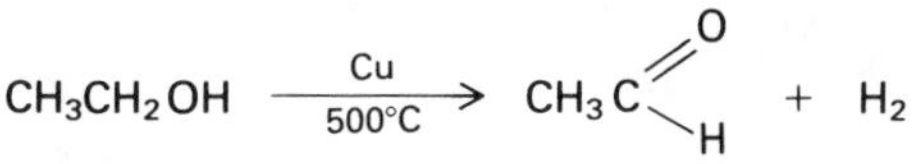

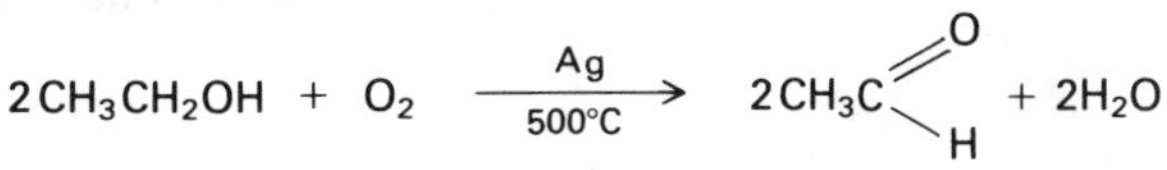

$$CH_3CH_2OH \xrightarrow[500°C]{Cu} CH_3C\underset{H}{\overset{O}{\diagdown}} + H_2$$

$$2CH_3CH_2OH + O_2 \xrightarrow[500°C]{Ag} 2CH_3C\underset{H}{\overset{O}{\diagdown}} + 2H_2O$$

Secondary alcohols are oxidised to ketones when oxidised with an oxidising agent such as potassium dichromate(VI) in acid:

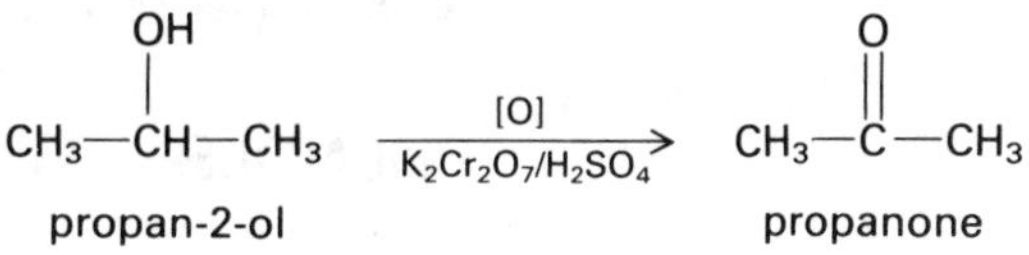

$$CH_3-\overset{OH}{\overset{|}{CH}}-CH_3 \xrightarrow[K_2Cr_2O_7/H_2SO_4]{[O]} CH_3-\overset{O}{\overset{\|}{C}}-CH_3$$

propan-2-ol   propanone

Further oxidation to a carboxylic acid does not occur except under severe conditions.

Tertiary alcohols are normally resistant to oxidation.

Alcohols burn in air or oxygen with clear, non-sooty flames, forming carbon dioxide and water:

$$C_2H_5OH + 3O_2 \rightarrow 2CO_2 + 3H_2O$$

The Haloform Reaction

Ethanol and secondary alcohols containing

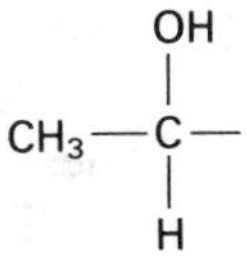

$$CH_3-\overset{OH}{\underset{H}{\overset{|}{\underset{|}{C}}}}-$$

are oxidised to triiodomethane when warmed with an alkaline solution of iodine or a solution of potassium iodide and sodium chlorate(I). This is known as the **iodoform reaction**. It is an example of a haloform reaction. If potassium bromide or potassium chloride are used instead of potassium iodide, then tribromomethane, $CHBr_3$ (bromoform), or trichloromethane, $CHCl_3$ (chloroform), are formed, respectively.

Ethanol and methyl ketones also undergo this reaction (*see* section 19.3).

In the iodoform reaction the triiodomethane precipitates as yellow crystals. The stoichiometric equation for the overall reaction using sodium hydroxide as an alkali is:

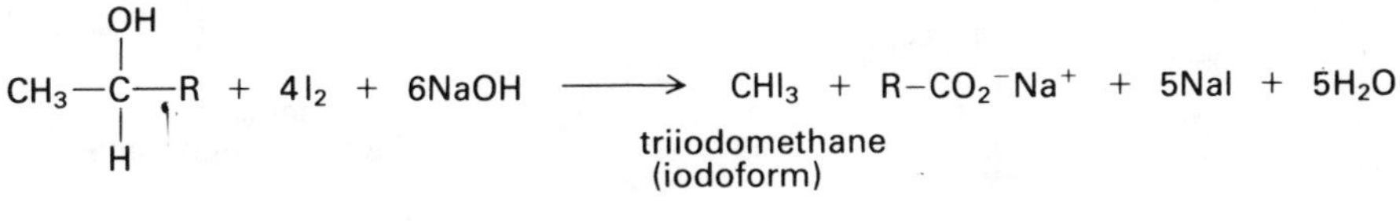

$$CH_3-\overset{OH}{\underset{H}{\overset{|}{\underset{|}{C}}}}-R + 4I_2 + 6NaOH \longrightarrow CHI_3 + R-CO_2^-Na^+ + 5NaI + 5H_2O$$

triiodomethane (iodoform)

R may be H, as in the case of ethanol, or an alkyl group. The reaction takes place in the following three steps.

1. Oxidation of the alcohol to an aldehyde or ketone.

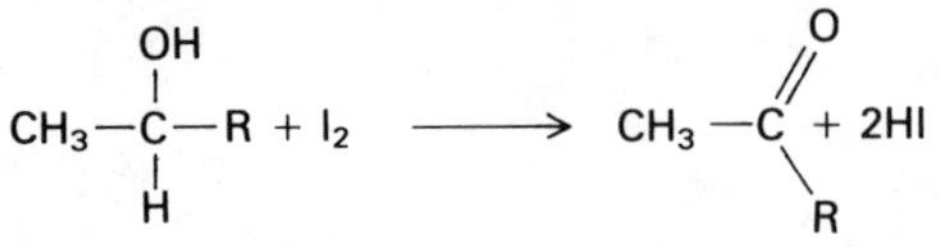

2. Substitution of the iodine atoms in the methyl group.

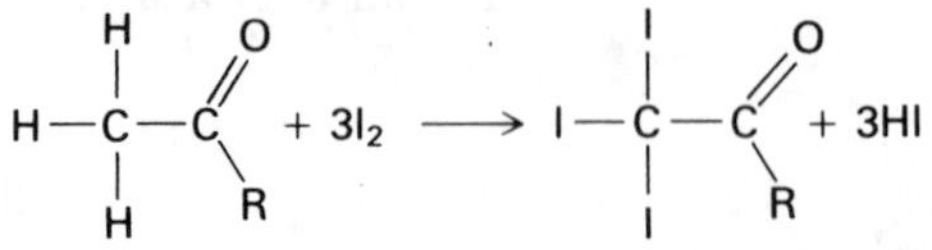

3. Hydrolysis under alkaline conditions to form the triiodomethane.

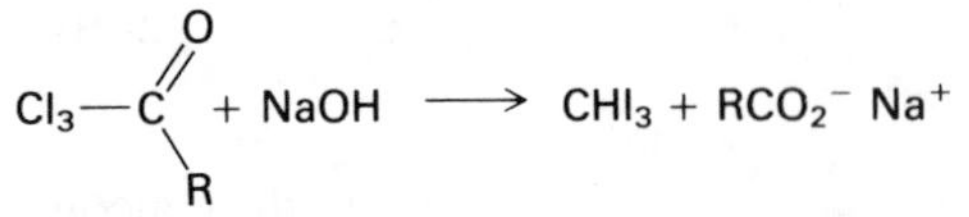

$$CI_3-C(=O)-R + NaOH \longrightarrow CHI_3 + RCO_2^- Na^+$$

Only steps 2 and 3 apply to ethanal and methyl ketones which undergo this type of reaction (*see* section 19.3).

PHENOLS

Compounds with one or more hydroxyl groups attached to the benzene ring are known as phenols. The most important of these is phenol itself:

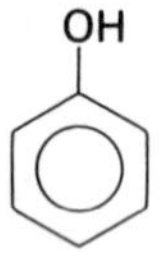

The simple phenols are all solids with low melting points. Phenol is a colourless crystalline solid with a melting point of 43°C. It has a characteristic smell. Like the alcohols, the phenols have higher boiling points than might be expected. This is due to intermolecular hydrogen bonding. We have already noted how 2-nitrophenol has a lower boiling point than 4-nitrophenol. The former exhibits intramolecular bonding whereas the latter exhibits intermolecular bonding and is thus less volatile (*see* section 2.2).

The phenols are only sparingly soluble in water but soluble in organic solvents including alcohols and ethers. Phenol is only partially miscible with water at temperatures less than 66°C. Above 66°C it is miscible in all proportions (*see* figure 6.22 and section 6.2).

Laboratory Preparation

Phenol can be prepared in the laboratory by fusing the anhydrous sodium salt of benzenesulphonic acid with solid sodium hydroxide at 300 to 350°C followed by addition of dilute hydrochloric acid:

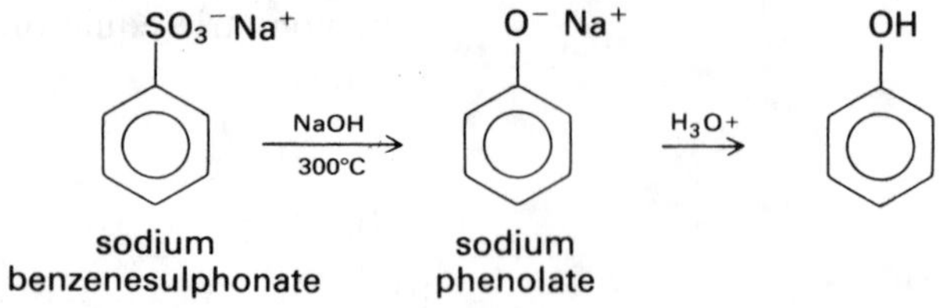

Benzenesulphonic acid is prepared by the sulphonation of benzene (*see* section 18.2). Neutralisation of the acid with sodium hydroxide produces the sodium salt.

Phenol is also produced when an aqueous solution of benzenediazonium chloride is warmed above 10°C:

Benzenediazonium chloride is prepared by the diazotisation of phenylamine (*see* section 19.4).

Chemical Properties of Phenols

The Hydroxyl Group

Acidity. The pK_a value of phenol is 9.95. It is thus slightly acidic although a stronger acid than methanol, ethanol and water (*see* table 19.4). The phenolate ion produced by loss of an H^+ ion is stabilised by delocalisation of the negative charge:

These are known as limiting forms (*see* sections 2.1 and 18.2).

Like the alcohols, phenol reacts with strongly electropositive metals such as sodium to produce hydrogen:

However, unlike the alcohols, phenols react with sodium hydroxide. For example

Phenol is not as strongly acidic as the carboxylic acids. Carboxylic acids such as ethanoic acid or benzoic acid liberate carbon dioxide from sodium hydrogen-carbonate or sodium carbonate whereas phenol does not. This reaction is used to distinguish carboxylic acids from phenols.

Ester formation. Although phenol does not react with carboxylic acids to form esters, it does react with acid chlorides in alkaline solutions:

This type of reaction is called **acylation**.

Ether formation. Phenol reacts with halogenoalkanes under alkaline conditions to form ethers:

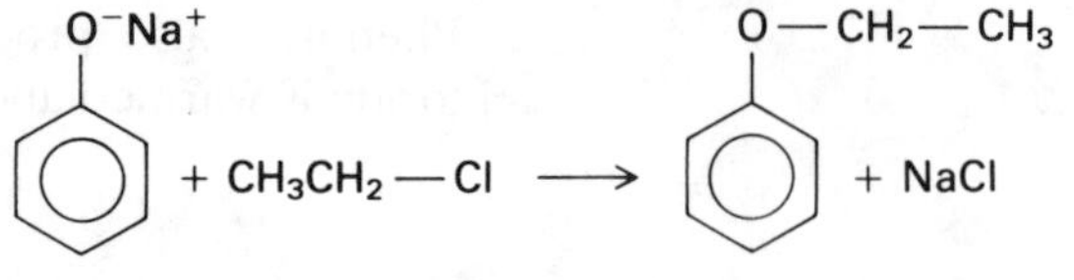

This is an example of Williamson's synthesis (*see* previous section).

Reaction with phosphorus pentachloride. Unlike the alcohols, phenol does not react with hydrogen halides or phosphorus trihalides. However, it does react slowly with phosphorus pentachloride to give a poor yield of chlorobenzene:

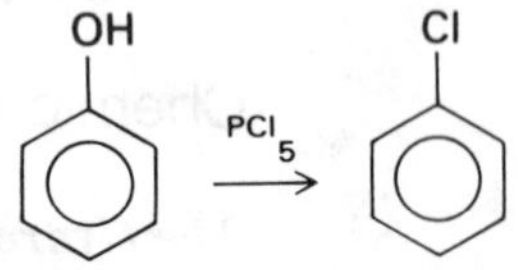

Reaction with iron(III) chloride. When a neutral solution of iron(III) chloride is added to phenol a complex is formed with a violet colour. This is used as a test for phenol. The reaction is characteristic for compounds containing the **enol** group:

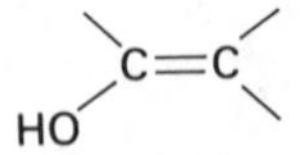

The Benzene Ring

The benzene ring in the phenol molecule undergoes electrophilic substitution more readily than benzene alone. This is because the non-bonding electrons on the oxygen atom are drawn into the ring, thereby activating it. The hydroxyl group is 2,4-directing with respect to electrophilic substituents (*see* section 18.2).

Halogenation. Halogenation of phenol occurs under much milder conditions than with benzene. For example, when bromine water is added to an aqueous solution of phenol, a white precipitate of 2,4,6-tribromophenol is formed:

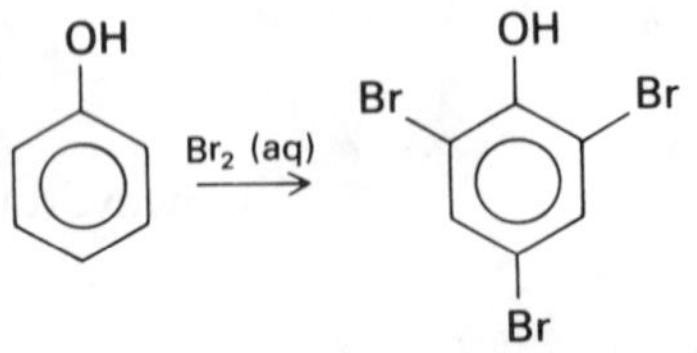

We saw in section 18.2 that bromination of benzene required the presence of a catalyst.

Nitration. Phenol can be nitrated using dilute nitric acid. A mixture of 2-nitrophenol and 4-nitrophenol is produced:

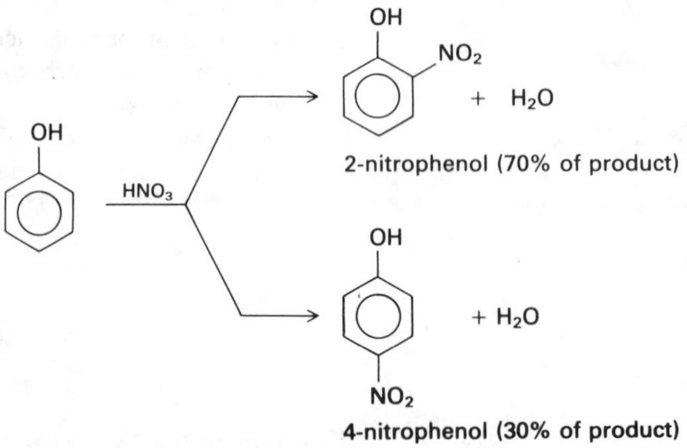

2-nitrophenol (70% of product)

4-nitrophenol (30% of product)

Once again we should compare these mild conditions with the conditions for the corresponding reaction of benzene. Nitration of benzene requires a mixture of concentrated nitric acid and sulphuric acid (*see* section 18.2).

2-Nitrophenol and 4-nitrophenol are both stronger acids than phenol. They both have pK_a values of approximately 7.2. This is because the nitro group is electron withdrawing. As a result, the benzene ring withdraws more electrons from the oxygen atom in the hydroxyl group.

Sulphonation. Concentrated sulphuric acid reacts with phenol to produce a mixture of hydroxybenzenesulphonic acids:

2-hydroxybenzenesulphonic acid
(15% of product)

4-hydroxybenzenesulphonic acid
(85% of product)

The two products formed in this reaction react with concentrated nitric acid to form 2,4,6-trinitrophenol. This is a yellow crystalline solid commonly known as picric acid:

2,4,6-trinitrophenol
(picric acid)

Owing to the combined electron-withdrawing power of the three nitro groups, picric acid is a relatively strong acid. It has a pK_a value of 1 and will liberate carbon dioxide from sodium carbonate solution.

Coupling reactions. An alkaline solution of phenol reacts with benzenediazonium chloride solution to form an orange precipitate of 4-hydroxyphenylazobenzene:

The product is an azo dye. This type of reaction is known as a coupling reaction.

ETHERS

Ethers have the general formula R—O—R'. All the ethers shown in table 19.5 are either gases or volatile liquids with the exception of phenoxybenzene. Their boiling points are approximately the same as alkanes with similar relative molecular masses. However, since ether molecules do not associate by hydrogen bonding, they have much lower boiling points than isomeric alcohols (*see* table 19.6).

Table 19.5   Some ethers

Formula	Name	Other names
CH_3—O—C_2H_5	methoxyethane	methyl ethyl ether
C_2H_5—O—C_2H_5	ethoxyethane	diethyl ether or ether
⬡—O—CH_3	methoxybenzene	methyl phenyl ether or anisole
⬡—O—⬡	phenoxybenzene	diphenyl ether

Table 19.6

Compound	Empirical formula	Structural formula	Relative molecular mass	Boiling point/°C
pentane	C_5H_{12}	$CH_3(CH_2)_3CH_3$	72	36
ethoxyethane	$C_4H_{10}O$	C_2H_5—O—C_2H_5	74	35
butan-1-ol	$C_4H_{10}O$	$CH_3(CH_2)_3OH$	74	118

Laboratory preparation

Symmetrical ethers such as ethoxyethane, C_2H_5—O—C_2H_5, can be prepared by the partial dehydration of alcohols using concentrated sulphuric acid and an excess of the alcohol:

$$2C_2H_5OH \xrightarrow{\text{H}_2\text{SO}_4} \underset{\text{ethoxyethane}}{C_2H_5\text{—O—}C_2H_5} + H_2O$$

The dehydration of alcohol is discussed above.

Both symmetrical ethers such as ethoxyethane and **unsymmetrical ethers** such as methoxyethane, CH_3—O—C_2H_5, and ethoxybenzene, C_6H_5—O—C_2H_5, can be prepared from halogenoalkanes and alcohols or phenols by Williamson's synthesis (*see* above).

Chemical Properties of Ethers

Ethers are much less reactive than the alcohols. Since no hydrogen atom is attached to the oxygen atom, they do not possess the acidic properties of alcohol. They do not, for example, react with sodium. However, they do possess weak basic properties due to the lone-pairs of electrons on the oxygen atom.

Aliphatic ethers function as Lewis bases in acidic media. They dissolve in strong mineral acids, forming disubstituted oxonium salts:

When aliphatic ethers are heated with concentrated hydroiodic acid, iodo-alkanes are formed:

$$R\text{—O—}R' + 2HI \rightarrow RI + R'I + H_2O$$

For example, the reaction of ethoxyethane with hydriodic acid yields iodomethane:

$$C_2H_5\text{—O—}C_2H_5 + 2HI \rightarrow 2C_2H_5I + H_2O$$

Combustion. Ethers are highly inflammable and can be explosive when their vapours are mixed with air. The usual products of combustion are formed:

$$C_2H_5\text{—O—}C_2H_5 + 6O_2 \rightarrow 4CO_2 + 5H_2O$$

Cyclic Ethers

The most important cyclic ethers are epoxyethane and tetrahydrofuran:

$$H_2C\!-\!CH_2 \qquad H_2C\!-\!CH_2$$

epoxyethane

tetrahydrofuran

Epoxyethane and tetrahydrofuran have boiling points of 13°C and 65°C, respectively. Epoxyethane undergoes hydration to form ethane-1,2-diol (see section 18.3).

MANUFACTURE AND USES OF ALCOHOLS, PHENOLS AND ETHERS

Manufacture

The manufacture of various alcohols, phenols and ethers is described in section 18.3.

Uses

Organic syntheses.   Alcohols and phenols are used widely both in the laboratory and in industry to synthesise other organic compounds. For example, methanol is used to produce methanal and ethanol to produce ethanal (see following section).

Solvents. Methanol and ethanol are both used commercially, industrially and in the laboratory as solvents. Methanol is used to remove paints and varnishes. Ethoxyethane is used as a solvent for fats and oils. It is used extensively in 'ether' extractions to separate and purify organic compounds. It is also particularly valuable as a solvent for metal hydrides and Grignard reagents. Since these compounds readily undergo hydrolysis, anhydrous conditions are necessary for storage of the ethoxyethane. The ethoxyethane is thus usually stored over sodium wire for this purpose. Tetrahydrofuran is also used as a solvent for Grignard reagents. It is less volatile than ethoxyethane and thus safer and more convenient to use.

Polymers. The alcohols and phenols are widely used in the manufacture of plastics and epoxy resins (see chapter 20). For example, phenol is used to make Bakelite, nylon and epoxy resins.

Explosives. Propane-1,2,3-triyl trinitrate, more commonly known as nitroglycerine, is manufactured from propane-1,2,3-triol:

$$CH_2\!-\!OH \qquad\qquad CH_2\!-\!O\!-\!NO_2$$
$$CH\!-\!OH + 3HNO_3 \longrightarrow CH\!-\!O\!-\!NO_2 + 3H_2O$$
$$CH_2\!-\!OH \qquad\qquad CH_2\!-\!O\!-\!NO_2$$

propane-1,2,3-triyl trinitrate
(nitroglycerine)

Nitroglycerine is not only used as an explosive but also in medicine to treat heart diseases such as angina pectoris.

2,4,6-Trinitrophenol (picric acid) and its salts are also explosive. Picric acid was used in World War I in the form of lyddite. In the laboratory it is used to identify amines and polynuclear aromatic hydrocarbons. It is usually stored under water in a bottle with a cork.

Nitroglycerine

Nitroglycerine was first made in 1846 by the Italian chemist Ascario Sobrero. It is very sensitive and can be detonated by fairly slight shock. The Swedish chemist Alfred Nobel started manufacturing nitroglycerine in a factory near Stockholm in 1862. In 1864 the factory was wrecked by an explosion which killed five people including Nobel's young brother Emil. The same year Nobel found that nitroglycerine could be made safe and transportable by mixing it with kieselguhr. In 1867 Nobel patented the mixture under the name 'dynamite'.

Cordite, an explosive used in shells and bullets, contains about 30% nitroglycerine and 65% guncotton. Guncotton, which is cellulose trinitrate, was also first made by Nobel. Gelignite contains about 60% nitroglycerine.

Ethanol

Ethanol is consumed in the form of beers, wines and spirits (*see* chapters 6 and 9). The approximate percentages by mass of ethanol in these beverages are shown in table 19.7.

Table 19.7 Approximate percentages by mass of ethanol in alcoholic beverages

	Percentage ethanol by mass
beers	4–6
wines	9–12
fortified wines	20
spirits	40–50

Rectified spirit is a constant-boiling mixture containing 96% ethanol and 4% water. It is produced by the fractional distillation of aqueous solutions of ethanol. For industrial use, benzene is added to ethanol. The water in rectified spirit can be removed by refluxing over calcium oxide. The product is known as **absolute alcohol** and contains less than 0.5% water.

Industrial methylated spirits contain about 95% rectified spirit and 5% methanol. The latter is added to make it unfit for consumption. The methylated spirits which are used as a fuel for portable stoves contain methyl violet as a pink dye.

Beer contains between 4% and 6% ethanol

Plant for producing absolute alcohol

Phenol

Phenol is not only an important raw material for the manufacture of resins and plastics, it is also used to make azo dyes, herbicides such as 2,4-D and germicides such as 'Dettol' and 'TCP' (*see* table 19.8).

Table 19.8   Products manufactured from phenol

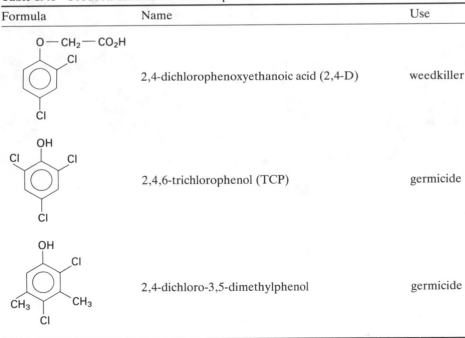

Formula	Name	Use
	2,4-dichlorophenoxyethanoic acid (2,4-D)	weedkiller
	2,4,6-trichlorophenol (TCP)	germicide
	2,4-dichloro-3,5-dimethylphenol	germicide

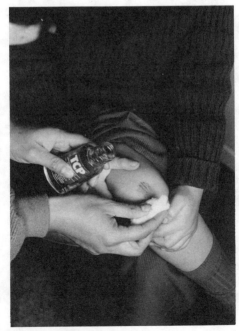

2,4,6-Trichlorophenol (TCP) is a germicide

Antifreeze mixtures contain
ethane-1,2-diol (ethylene glycol)

The term **germicide** includes both disinfectants and antiseptics. Strictly speaking **antiseptics** destroy micro-organisms on living tissue whereas **disinfectants** are used on inanimate surfaces. Ethanol, propan-2-ol and phenol are also germicides. Phenol, or carbolic acid (*see* above), is no longer used as a disinfectant since it is corrosive to the skin.

Ethoxyethane (Ether)
Ethoxyethane was once used extensively as an anaesthetic. However, it causes irritation of the respiratory tract. Its volatility can also cause problems in hot climates and it is an explosion hazard.

Ethane-1, 2-diol (Ethylene Glycol)
This is used as an antifreeze in car radiators and also as a de-icing fluid.

SUMMARY

1. Typical reactions of alcohols.

$$R-OH$$

$$\xrightarrow{Na} R-O^-Na^+$$

$$\xrightarrow[\substack{or\ PHal_3 \\ or\ PHal_5}]{H-Hal} R-Hal$$

$$\xrightarrow[\substack{or\ R'-COCl \\ or\ R'-CO-O-COR'}]{R'-CO_2H} R'-CO_2R$$

$$C_2H_5OH \xrightarrow{Conc.\ H_2SO_4}$$

$$\xrightarrow{180°C} \overset{H}{\underset{H}{}}C=C\overset{H}{\underset{H}{}}$$

$$\xrightarrow{140°C} C_2H_5-O-C_2H_5$$

$$\xrightarrow{0°C} C_2H_5-O-SO_3H$$

Oxidation

$$R-CH_2-OH \longrightarrow R-\overset{O}{\overset{\|}{C}}-H \longrightarrow R-\overset{O}{\overset{\|}{C}}-O-H$$

$$R-CHOH-R' \longrightarrow R-\overset{O}{\overset{\|}{C}}-R'$$

$$alcohols \xrightarrow[combustion]{complete} CO_2 + H_2O$$

Haloform

$$CH_3-\overset{OH}{\underset{H}{\overset{|}{C}}}-R \xrightarrow{I_2/NaOH} CHI_3 + R-CO_2^-\ Na^+$$

(cont'd)

2. Important reactions of phenol.

19.3 The Carbonyl Group

The carbonyl group $\diagdown C{=}O$

occurs in aldehydes, ketones, carboxylic acids, acid chlorides, acid anhydrides and amides. The carbon atom in this group is sp^2 hybridised. The group thus has a planar structure (*see* figure 2.7):

about 120° $\diagdown C{=}O$

The carbon is bound to the oxygen by a σ bond and a π bond. The carbonyl group is polarised due to the mesomeric effect. It thus has a higher electron density around the more electronegative oxygen atom:

$\diagdown C\overset{\delta+}{=}\overset{\delta-}{O}$

The fractional positive charge on the carbon atom makes it susceptible to nucleophilic attack.

ALDEHYDES AND KETONES

Aldehydes and ketones are known as **carbonyl compounds**. Aldehydes have a single alkyl or aryl group attached to the carbon atom of the carbonyl group, whereas ketones have two:

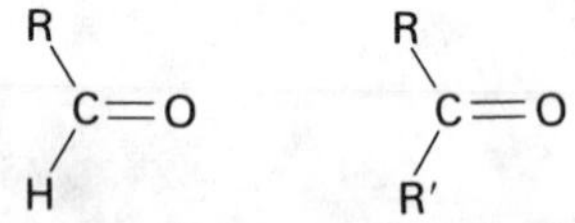

R and R' can be either alkyl or aryl. The homologous series of aldehydes and ketones both have the general formula $C_nH_{2n}O$. Some of the more important aldehydes and ketones are shown in table 19.9.

Table 19.9   Aldehydes and ketones

	Name	Other names
Aldehydes		
H—CHO	methanal	formaldehyde
CH₃—CHO	ethanal	acetaldehyde
CH₃CH₂—CHO	propanal	propionaldehyde
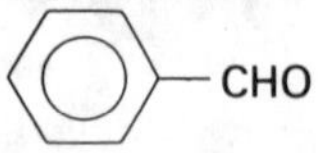	benzaldehyde	
Ketones		
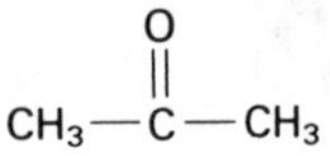	propanone	acetone
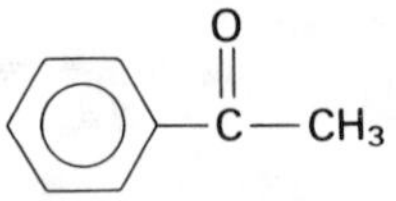	phenylethanone	acetophenone
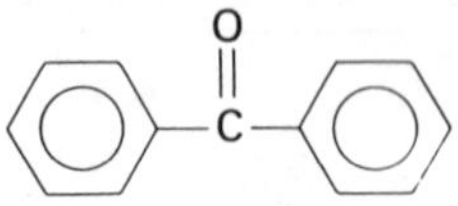	diphenylmethanone	benzophenone
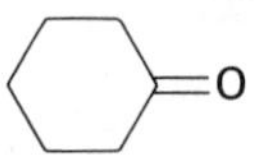	cyclohexanone	

Physical Properties

Apart from methanal, which is a gas, members of the homologous series of both aldehydes and ketones are liquids. Their boiling points tend to be higher than alkanes with similar relative molecular mass. This is due to intermolecular attractions resulting from the polar nature of the carbonyl group. The carbonyl group is not sufficiently polar to form hydrogen bonds with other molecules of the compound. In water, however, hydrogen bonds are formed with the carbonyl groups. Consequently, the simpler aldehydes and ketones, such as ethanal and propanone, are readily soluble in water:

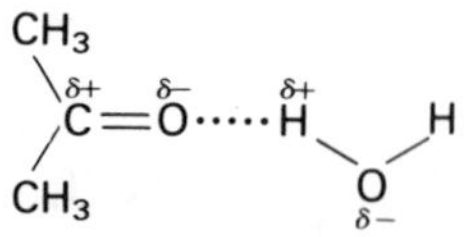

Preparation of Aldehydes and Ketones

Both aldehydes and ketones can be prepared in the laboratory by oxidation of alcohols using acidified potassium dichromate(VI) (*see* previous section). Aldehydes are prepared from primary alcohols:

$$R-CH_2OH \xrightarrow{[O]} R-CHO + H_2O$$

To prevent further oxidation to a carboxylic acid, the aldehyde is distilled off (*see* figure 19.1).

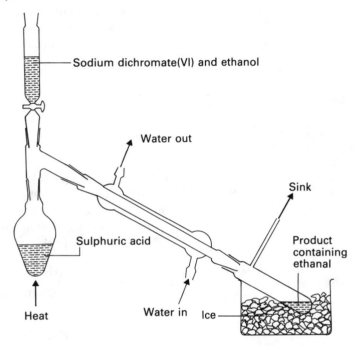

Figure 19.1 Preparation of ethanal. Ethanol and sodium dichromate(VI) are added to hot sulphuric acid. Immediately the ethanal is formed it distils off. It has a lower boiling point than ethanol. This prevents the ethanal from being oxidised to ethanoic acid. Ethanol stays in solution until it is oxidised. Pure ethanal is obtained by redistillation

Ketones are prepared from secondary alcohols:

$$\begin{matrix} R \\ \diagdown \\ CHOH \\ \diagup \\ R' \end{matrix} \xrightarrow{[O]} R-\overset{\displaystyle O}{\overset{\|}{C}}-R' + H_2O$$

Aldehydes are named so since they can be obtained from ALcohols by DEHYDrogenation.

Both aldehydes and ketones can be prepared by dehydrogenation of the primary or secondary alcohol respectively. A finely divided copper catalyst heated to about 300°C is used for this purpose:

$$RCH_2OH \xrightarrow[300°C]{Cu} R-CHO + H_2$$

$$\begin{matrix} R \\ \diagdown \\ CHOH \\ \diagup \\ R' \end{matrix} \xrightarrow[300°C]{Cu} R-\overset{\displaystyle O}{\overset{\|}{C}}-R' + H_2$$

Aldehydes and ketones can also be prepared from Grignard reagents (*see* section 19.1).

Chemical Properties of Aldehydes and Ketones

Addition Reactions

Both aldehydes and ketones undergo nucleophilic addition. The general mechanism for nucleophilic addition to the carbonyl group is described in section 17.3.

Ketones are generally less reactive than aldehydes towards nucleophiles. This is because alkyl groups are electron donating. Since ketones contain two such groups, the positive charge on the carbonyl carbon atom is smaller than that in an aldehyde.

Addition of hydrogen cyanide. Both aldehydes and ketones undergo nucleophilic addition with hydrogen cyanide to yield 2-hydroxynitriles (also known as cyanohydrins). For example, when ethanal is treated with sodium cyanide and then excess mineral acid, the following reaction occurs:

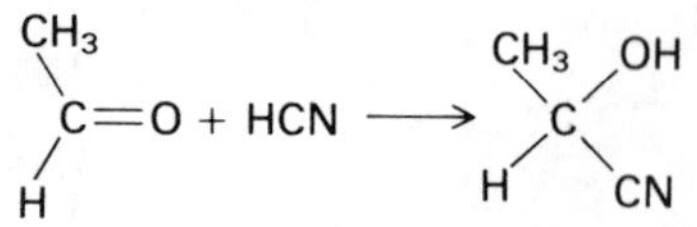

2-hydroxypropanenitrile
(acetaldehyde cyanohydrin)

The mechanism for this nucleophilic addition is described in section 17.3.

Hydrogensulphite addition. Aldehydes and methyl ketones undergo nucleophilic addition to form crystalline ionic compounds when treated at room temperature with a saturated aqueous solution of sodium hydrogensulphite, $NaHSO_3$. The nucleophile in this case is the hydrogensulphite ion, HSO_3^-:

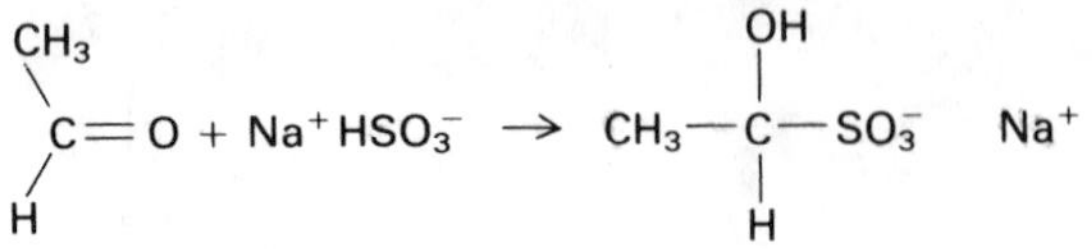

ethanal

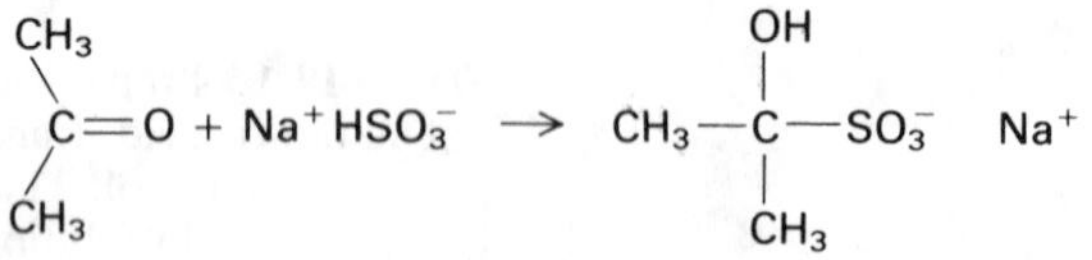

propanone

Ethanal undergoes this reaction more readily than propanone.

Addition of hydrogen. Aldehydes and ketones can add hydrogen across the carbonyl group to form primary alcohols and secondary alcohols respectively. A platinum or nickel catalyst is used:

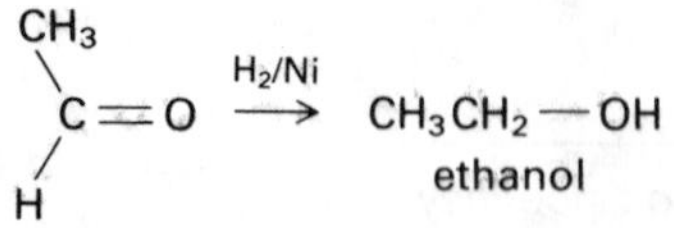

ethanol

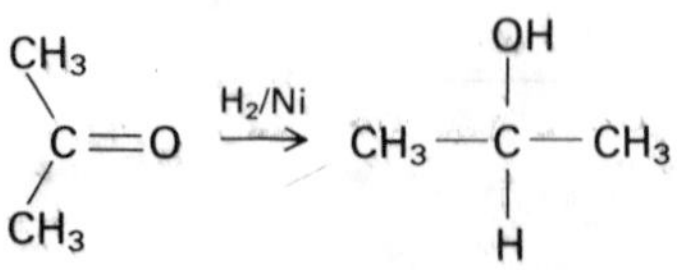

propan-2-ol

Alternatively, lithium tetrahydridoaluminate(III) $LiAlH_4$ (also known as lithium aluminium hydride), or sodium tetrahydridoborate(III), $NaBH_4$ (also known as sodium borohydride), can be used as reducing agents. In the case of lithium tetrahydridoaluminate(III), anhydrous ethoxyethane or tetrahydrofuran is used as a solvent. The tetrahydridoaluminate(III) ions provide a source of nucleophilic hydride ions:

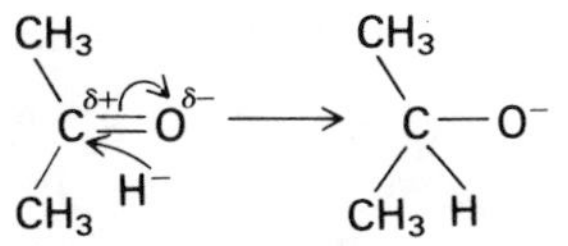

The reaction is completed by adding water to hydrolyse the product:

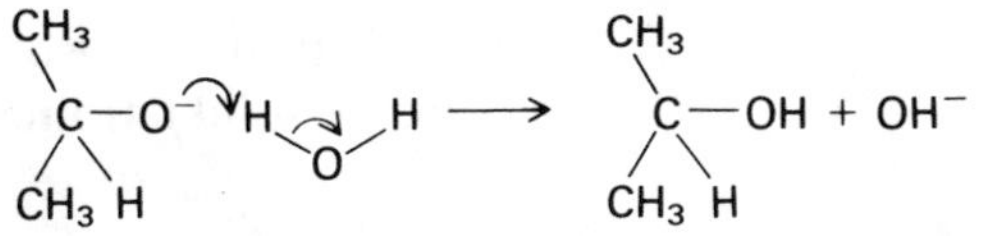

The water also destroys excess lithium tetrahydridoaluminate(III). Reduction using sodium tetrahydridoborate(III) is less vigorous. Water or aqueous methanol or ethanol is used as a solvent.

Addition of Grignard reagents. Methanal, aldehydes and ketones all react with Grignard reagents in dry ethoxyethane to give primary, secondary and tertiary alcohols respectively (*see* section 19.1).

Addition of alcohols. Aldehydes, but not ketones, combine with alcohols to form acetals in the presence of hydrogen chloride. The hydrogen chloride acts as a catalyst:

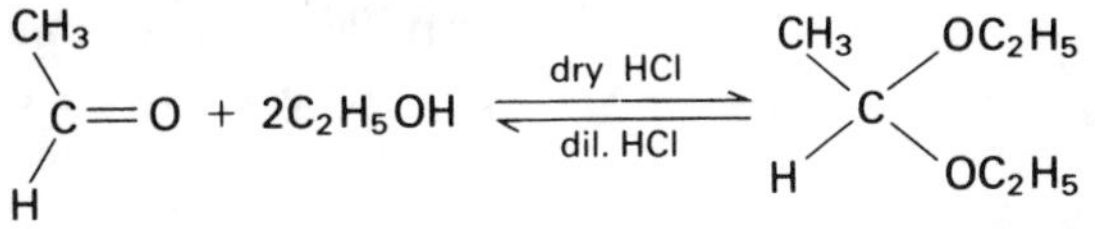

1,1-diethoxyethane
(an acetal)

Both the alcohol and the hydrogen chloride must be dry. The reaction is reversed when dilute hydrochloric acid is added. Without the catalyst, a hemiacetal is formed:

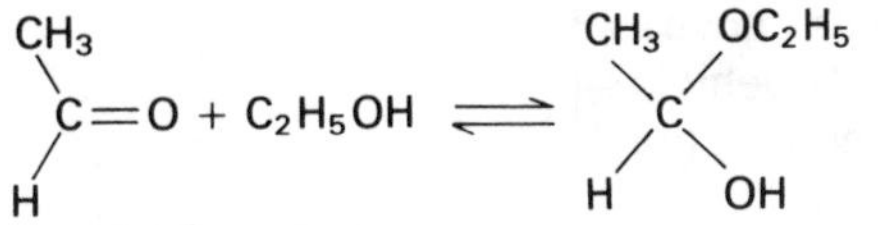

1-ethoxy-1-ethanol
(a hemiacetal)

Polymerisation

Aldehydes readily form a number of addition polymers. For example, evaporation of an aqueous solution of methanal (known as formalin) yields poly(methanal). This has the structure:

$$\cdots-O-CH_2-O-CH_2-O-CH_2-\cdots$$

Poly(methanal) is sometimes known as paraformaldehyde.

Distillation of an acidified solution of methanal produces a solid with the following cyclic structure:

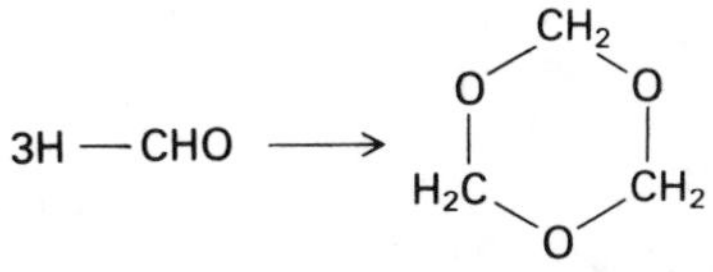

$$3H-CHO \longrightarrow$$

methanal trimer

The compound is also known as metaformaldehyde or trioxane.

When a few drops of concentrated sulphuric acid are added to ethanal, a vigorous reaction ensues, resulting in the formation of a liquid **trimer**:

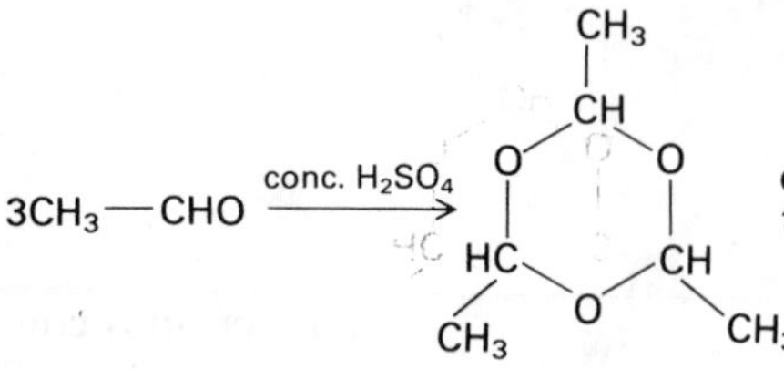

$$3CH_3{-}CHO \xrightarrow{\text{conc. }H_2SO_4} \text{ethanal trimer}$$

This trimer is also known as paraldehyde. If the reaction is carried out at 0°C, a white solid **tetramer** with the following structure is formed:

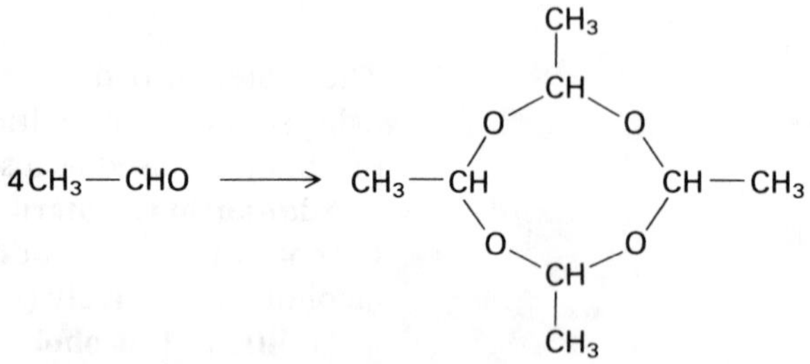

$$4CH_3{-}CHO \longrightarrow$$

The product is also known as metaldehyde.

Ethanal reacts with a cold dilute solution of an alkali such as sodium hydroxide or potassium carbonate to yield the **dimer** 3-hydroxybutanal. Dehydration of the product gives but-2-enal:

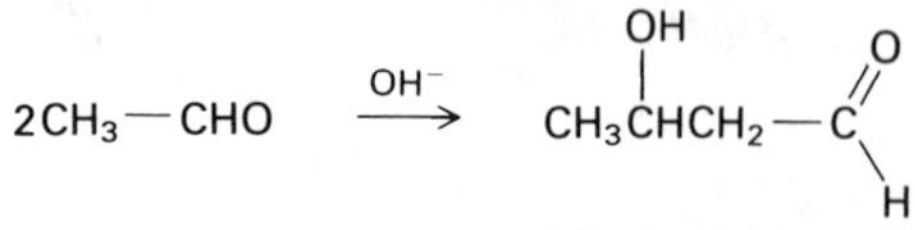

$$2CH_3{-}CHO \xrightarrow{OH^-}$$

3-hydroxybutanol
(aldol)

$$\longrightarrow \quad CH_3CH{=}CH{-}C + H_2O$$

but-2-enal

Aldol is the trivial name given to 3-hydroxybutanal. The name derives from the presence of both ALDehyde and alcohOL functional groups in the dimer.

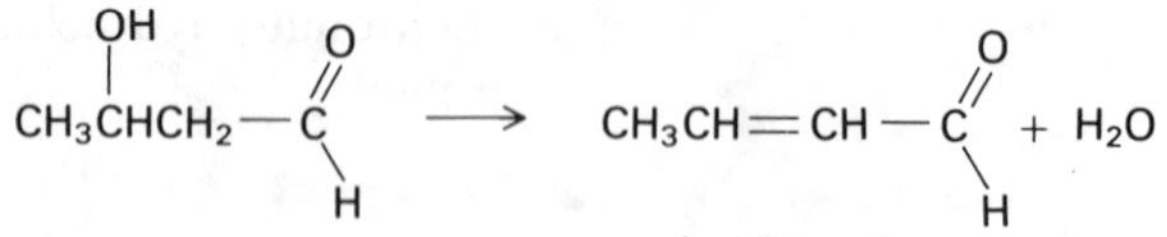

The overall reaction is known as the **aldol condensation**.

Condensation Reactions

Aldehydes and ketones undergo nucleophilic addition reactions with ammonium derivatives followed by the elimination of water. Such reactions are known as condensation reactions or addition–elimination reactions. The reactions can be represented in general form as follows:

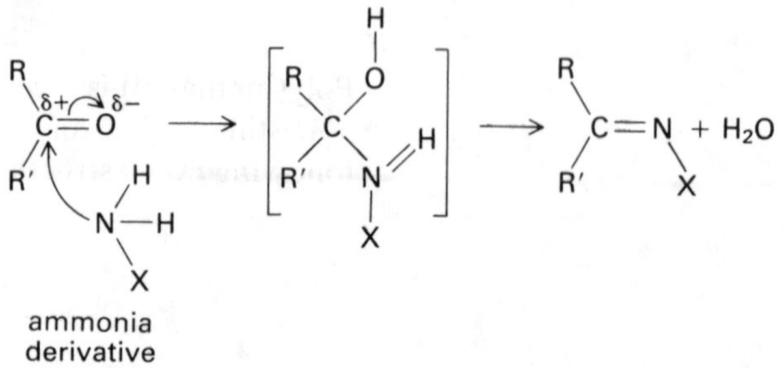

$$\longrightarrow \quad C{=}N + H_2O$$

ammonia
derivative

Examples of this type of reaction are the reactions between hydroxylamine and carbonyl compounds. The products are crystalline solids known as oximes:

$$
\begin{array}{c}
CH_3 \\
| \\
C{=}O + NH_2OH \longrightarrow \\
| \\
H
\end{array}
\qquad
\begin{array}{c}
CH_3 \\
| \\
C{=}N + H_2O \\
| \quad | \\
H \quad OH
\end{array}
$$

hydroxylamine   ethanal oxime

$$
\begin{array}{c}
CH_3 \\
| \\
C{=}O + NH_2OH \longrightarrow \\
| \\
CH_3
\end{array}
\qquad
\begin{array}{c}
CH_3 \\
| \\
C{=}N + H_2O \\
| \quad | \\
CH_3 \quad OH
\end{array}
$$

propanone oxime

Aldehydes and ketones undergo condensation reactions with 2,4-dinitrophenyl-hydrazines to form 2,4-dinitrophenylhydrazones. These are yellow or orange crystalline solids with sharp melting points.

phenylethanone   2,4-dinitrophenyl hydrazine

phenylethanone
2,4–dinitrophenylhydrazone

Oxidation of Carbonyl Compounds

Aldehydes are readily oxidised to carboxylic acids by oxidising agents such as acidified dichromate(VI) or manganate(VII) solutions:

$$
CH_3CHO \xrightarrow{Cr_2O_7^{2-}/H_2SO_4} CH_3COOH
$$

ethanoic acid

Aldehydes can be oxidised by **Tollen's reagent**. This is a solution of silver nitrate in aqueous ammonia. It contains the diamminesilver(I) complex cation, $Ag(NH_3)_2^+$. Aldehydes reduce this ion to metallic silver. This is deposited as a '**silver mirror**' on the inner wall of the test tube:

$$
CH_3CHO + 2Ag(NH_3)_2^+ + H_2O \rightarrow CH_3COOH + 2Ag + 2NH_4^+
$$

silver
mirror

Aliphatic aldehydes also reduce the copper(II) ions in Fehling's solution. This is an alkaline solution of copper(II) sulphate and potassium sodium 2,3-dihydroxy-butanedioate (also known as potassium sodium tartrate or Rochelle salt). A reddish precipitate of copper(I) oxide is formed:

$$
CH_3CHO + 2Cu^{2+} + NaOH + H_2O \rightarrow CH_3COONa + Cu_2O + 4H^+
$$

Aromatic aldehydes such as benzaldehyde do not react in this way.

Ketones give no reaction with either Tollens' reagent or Fehling's reagent.

Ketones are only oxidised by strong oxidising agents such as hot nitric acid and chromic(VI) acid:

$$
\begin{array}{c}
CH_3 \\
| \\
C{=}O \\
| \\
C_2H_5
\end{array}
\xrightarrow[\text{acid}]{\text{hot concentrated}}
CH_3COOH + C_2H_5COOH
$$

ethanoic   propanoic
acid   acid

Both ketones and aldehydes burn readily with clear flames to give carbon dioxide and water.

Halogenation

Aldehydes and ketones react with phosphorus pentachloride under anhydrous conditions to give dichloroalkanes:

$$CH_3CHO + PCl_5 \longrightarrow CH_3CHCl_2 + POCl_3$$

1,1-dichloroethane

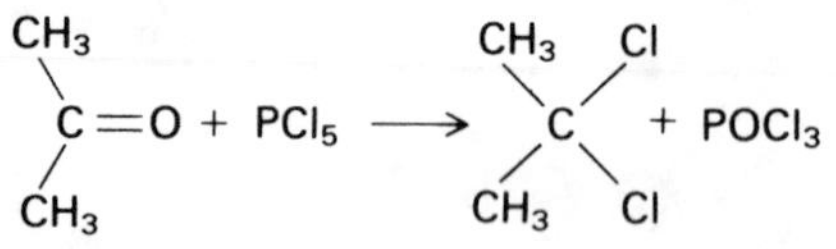

$$\begin{array}{c} CH_3 \\ \diagdown \\ C=O \\ \diagup \\ CH_3 \end{array} + PCl_5 \longrightarrow \begin{array}{c} CH_3 \quad Cl \\ \diagdown \quad \diagup \\ C \\ \diagup \quad \diagdown \\ CH_3 \quad Cl \end{array} + POCl_3$$

2,2-dichloropropane

When chlorine is bubbled through ethanal, a colourless oily liquid called trichloroethanal is formed:

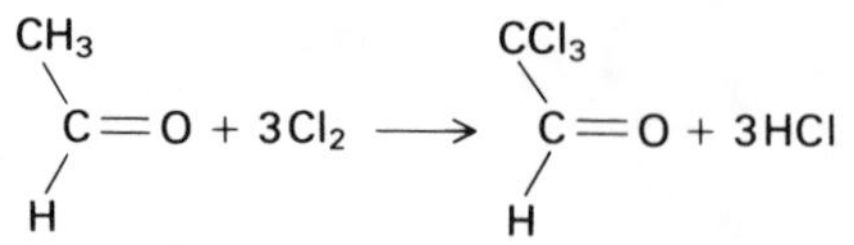

$$\begin{array}{c} CH_3 \\ \diagdown \\ C=O \\ \diagup \\ H \end{array} + 3Cl_2 \longrightarrow \begin{array}{c} CCl_3 \\ \diagdown \\ C=O \\ \diagup \\ H \end{array} + 3HCl$$

The product is also known as chloral.

Ethanal and methyl ketones undergo the haloform reactions (*see* previous section):

$$CH_3CHO \xrightarrow{3Br_2} CBr_3CHO \xrightarrow{OH^-} CHBr_3 \quad + \quad HCO_2^-$$

tribromomethane   methanoate
(bromoform)   ion

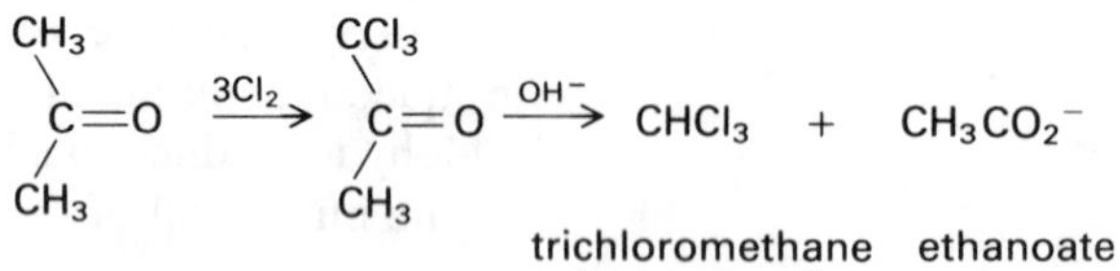

$$\begin{array}{c} CH_3 \\ \diagdown \\ C=O \\ \diagup \\ CH_3 \end{array} \xrightarrow{3Cl_2} \begin{array}{c} CCl_3 \\ \diagdown \\ C=O \\ \diagup \\ CH_3 \end{array} \xrightarrow{OH^-} CHCl_3 \quad + \quad CH_3CO_2^-$$

trichloromethane   ethanoate
(chloroform)   ion

MANUFACTURE AND USES OF CARBONYL COMPOUNDS

Manufacture

Methanal is manufactured by passing a mixture of methanol vapour and air over a heated copper or silver catalyst:

$$2CH_3OH + O_2 \xrightarrow[500°C]{Ag} 2HCHO + 2H_2O$$

Ethanal is manufactured from ethanol in a similar manner:

$$2C_2H_5OH + O_2 \xrightarrow[500°C]{Ag} 2CH_3CHO + 2H_2O$$

Both aldehydes and ketones are manufactured from alkenes by the **Wacker process**. A gaseous mixture of the alkene and air is passed into a solution of copper(II) chloride and palladium(II) chloride at 20 to 60°C:

$$CH_2{=}CH_2 + PdCl_2 + H_2O \rightarrow CH_3CHO + 2HCl + Pd$$
$$CH_3CH{=}CH_2 + PdCl_2 + H_2O \rightarrow CH_3COCH_3 + 2HCl + Pd$$

The palladium produced in these processes is converted back to palladium(II) chloride in the following reactions:

$$2Pd + O_2 + 4HCl \rightarrow 2PdCl_2 + 2H_2O$$
$$Pd + 2CuCl_2 \rightarrow PdCl_2 + 2CuCl$$

The copper(I) chloride is oxidised back to copper(II) chloride as follows:

$$4CuCl + O_2 + 4HCl \rightarrow 4CuCl_2 + 2H_2O + Cl^-$$

Propanal and propanone can both be made from propene by the methods described in section 18.3. Propanone is also produced as a by-product in the manufacture of phenol from benzene and propene (*see* section 18.3 also).

Uses

Organic syntheses. Carbonyl compounds are used in the laboratory and industrially to synthesise a range of chemical intermediates and final products. Ethanal, for example, is used to make ethanoic acid and its derivatives. It is also used to make chloral and ethanal trimer and tetramer.

Plastics and resins. Carbonyl compounds are also important in the manufacture of plastics. Methanal, for example, is involved in the manufacture of Bakelite and propanone in the manufacture of Perspex.

Solvents. Propanone is used widely as a solvent in both the laboratory and industry. In industry, for example, propanone is used as a solvent for cellulose nitrate and cellulose ethanoate.

Other uses. Methanal in the form of formalin is a powerful disinfectant. Formalin is also used to preserve animal specimens. Chloral hydrate, $CCl_3CH(OH)_2$, and ethanal trimer are both used as hypnotic drugs. A hypnotic drug induces sleep by depressing the central nervous system. Chloral is also a component of some analgesics. However, its main use is in the manufacture of the insecticide DDT. Ethanal tetramer (metaldehyde) is used as a slug poison. Benzaldehyde, which has a characteristic almond odour, is used to flavour foods. Aldehydes are also used in the manufacture of dyestuffs.

Acetone Breath

Propanone (commonly known as acetone) is present in small amounts in the blood and urine of humans. In diabetics, however, the concentration is higher than normal. Diabetics evacuate propanone through the lungs and exhale it. This causes an odour known as 'acetone breath'.

CARBOXYLIC ACIDS

Some common carboxylic acids are shown in table 19.10. They all contain the carboxyl group as the functional group. This consists of a carbonyl group linked to a hydroxyl group. Acids containing one such functional group are known as **monocarboxylic acids**. They have the suffix *-oic*. Those containing two carboxyl groups are called dicarboxylic acids. They have the suffix *-dioic*.

Saturated aliphatic monocarboxylic acids form a homologous series. They have the general formula $C_nH_{2n+1}CO_2H$. Unsaturated aliphatic dicarboxylic acids can exist as geometrical isomers (*see* section 17.2).

Physical Properties

The lower members of the homologous series of the saturated monocarboxylic acids are liquids with distinctive pungent odours. Ethanoic acid, for example, has a characteristic smell which is recognisable in vinegar. Anhydrous ethanoic acid is a liquid at room temperature. It freezes at 17°C to form a solid which looks like ice. This is known as glacial acetic acid.

The dicarboxylic acids shown in table 19.10 are all white crystalline solids at room temperature. The lower members of both the monocarboxylic and dicarboxylic acids are soluble in water. Solubility decreases as relative molecular mass increases.

Table 19.10 Carboxylic acids

Formula	Name	Other names
Monocarboxylic acids		
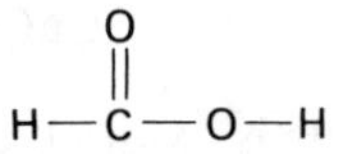 H—C(=O)—O—H	methanoic acid	formic acid
C_2H_5—C(=O)—O—H	ethanoic acid	acetic acid
CH_3—C(=O)—O—H	propanoic acid	propionic acid
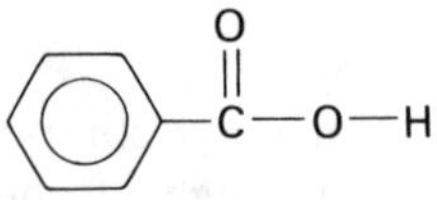 (benzene ring)—C(=O)—O—H	benzoic acid	
$CH_3(CH_2)_{16}CO_2H$	octadecanoic acid	stearic acid
Dicarboxylic acids		
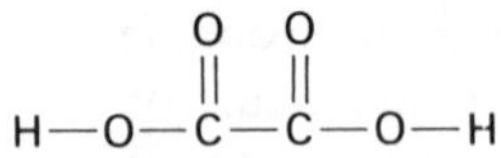 H—O—C(=O)—C(=O)—O—H	ethanedioic acid	oxalic acid
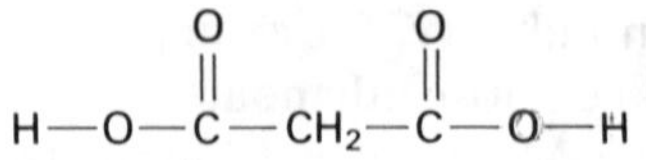 H—O—C(=O)—CH_2—C(=O)—O—H	propanedioic acid	malonic acid
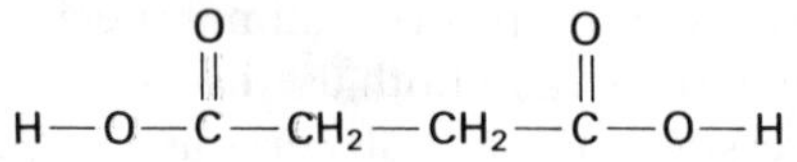 H—O—C(=O)—CH_2—CH_2—C(=O)—O—H	butanedioic acid	succinic acid
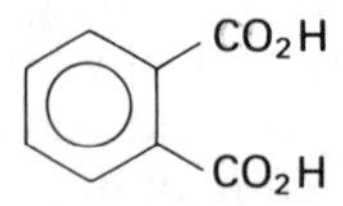 (benzene ring with CO_2H, CO_2H at 1,2)	benzene-1,2-dicarboxylic acid	phthalic acid
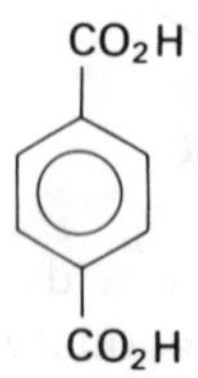 (benzene ring with CO_2H, CO_2H at 1,4)	benzene-1,4-dicarboxylic acid	terephthalic acid

In the liquid state and in non-aqueous solvents pairs of monocarboxylic acid molecules are joined together by hydrogen bonds to form dimers:

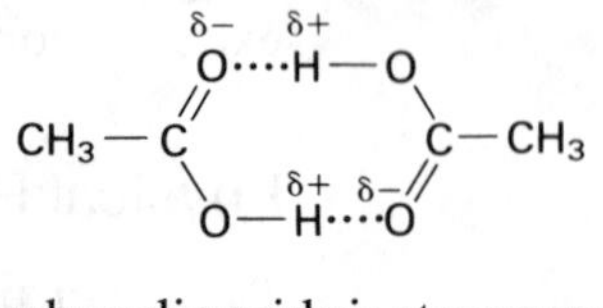

The hydrogen bonding in carboxylic acids is stronger than that in alcohols. This is due to the highly polar nature of the carboxyl group caused by the withdrawal of electrons away from the hydrogen atom towards the carbonyl oxygen atom:

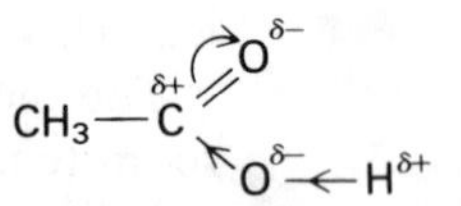

As a result, carboxylic acids have relatively high boiling points (*see* table 19.11).

Table 19.11

Compound	Empirical formula	Structural formula	Relative molecular mass	Boiling point/°C
ethanoic acid	$C_2H_4O_2$	CH_3COOH	60	118
ethanol	C_2H_6O	CH_3CH_2OH	46	78
propan-1-ol	C_3H_8O	$CH_3CH_2CH_2OH$	60	97

Preparation

Monocarboxylic acids can be prepared from primary alcohols and aldehydes by oxidation with excess acidified potassium dichromate(VI) solution:

$$RCH_2OH \xrightarrow[\text{acid}]{K_2Cr_2O_7} RCHO \xrightarrow[\text{acid}]{K_2Cr_2O_7} RCO_2H$$

Monocarboxylic acids and their **salts** can be prepared by hydrolysis of a nitrile or amide:

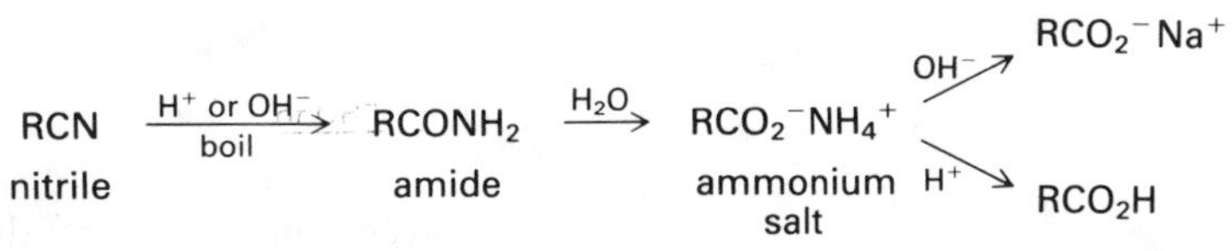

The preparation of **carboxylic acids** by the reaction of Grignard reagents and carbon dioxide is described in section 19.1.

Benzoic acid can be prepared by oxidation of the methyl side chain in methylbenzene (*see* section 18.2).

Benzoic acid can also be prepared from benzaldehyde by the **Cannizzaro reaction**. The benzaldehyde is treated with 40 to 60% sodium hydroxide solution at room temperature. Simultaneous oxidation and reduction results in the formation of benzoic acid and phenylmethanol respectively:

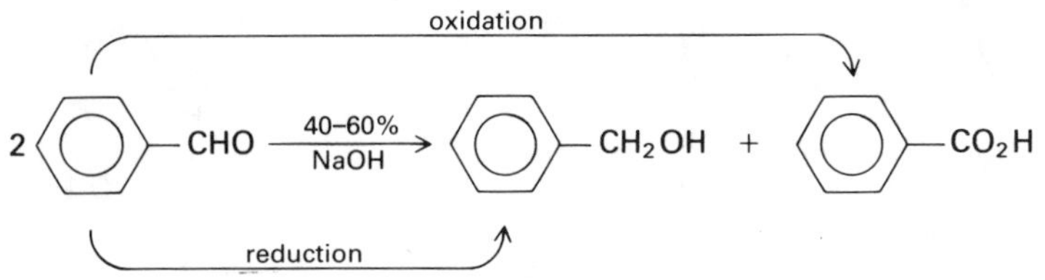

The Cannizzaro reaction is characteristic of aldehydes which have no α-hydrogen atoms. These are hydrogen atoms attached to the carbon atom adjacent to the aldehyde group:

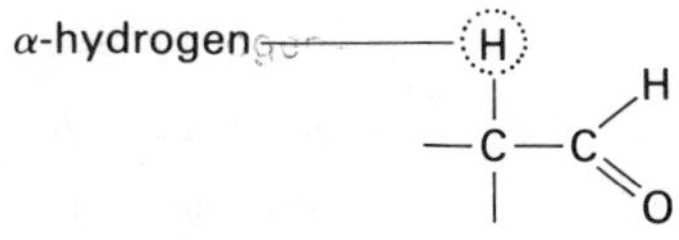

Since methanal has no α-hydrogen atoms, it also undergoes the Cannizzaro reaction. Aldehydes containing at least one α-hydrogen undergo base-catalysed aldol condensation in the presence of sodium hydroxide solution (*see* above).

Chemical Properties

Although the carboxyl group contains a carbonyl group, carboxylic acids do not exhibit some of the characteristic reactions of aldehydes and ketones. For example, they do not undergo addition or condensation reactions. This is because the carbon atom of the carboxyl group is less positive than that in the aldehyde or ketone groups.

Acidity

The withdrawal of electrons away from the carboxyl hydrogen atom weakens the O—H bond. As a result the carboxyl group can lose a proton. Monocarboxylic acids are thus monobasic acids. In aqueous solutions the following equilibrium is established:

$$R-C\underset{O-H}{\overset{O}{\diagup}} + H_2O \rightleftharpoons R-C\underset{O^-}{\overset{O}{\diagup}} + H_3O^+$$

The carboxylate ion exists as a resonance hybrid of two limiting forms:

$$R-C\underset{O^-}{\overset{O}{\diagup}} \longleftrightarrow R-C\underset{O}{\overset{O^-}{\diagup}}$$

This can be represented as

$$R-C\underset{O}{\overset{O}{\diagup}} {}^-$$

The resonance energy of the carboxylate ion stabilises the ion. Carboxylic acids are thus much stronger acids than alcohols. Even so, owing to the covalent nature of carboxylic acid molecules, the above equilibrium lies predominantly to the left. Carboxylic acids are thus weak acids. Ethanoic acid, for example, has a K_a value of 1.75×10^{-5}.

Substituents in a carboxylic acid molecule strongly influence the strength of the acid due to the inductive effect. Substituents such as chlorine are electron withdrawing $(-I)$. Withdrawals of electrons away from the carboxyl hydrogen strengthens the acid. On the other hand, substituents such as alkyl groups are electron donating $(+I)$. They weaken the acid:

Increasing acid strength →

$$CH_3 \to C \to CO_2H \qquad H-C-CO_2H \qquad Cl \leftarrow C \leftarrow CO_2H$$

propanoic acid   ethanoic acid   monochloroethanoic acid

The effect of substituents on acid strength is clearly shown by the trend of pK_a values of the acids (*see* table 19.12).

Table 19.12 pK_a values of carboxylic acids

Carboxylic acid		pK_a	
propanoic acid	CH_3CH_2COOH	4.87	
ethanoic acid	CH_3COOH	4.76	
monochloroethanoic acid	$CH_2ClCOOH$	2.86	increasing acid strength
dichloroethanoic acid	$CHCl_2COOH$	1.29	
trichloroethanoic acid	CCl_3COOH	0.65	

Salt Formation

Carboxylic acids exhibit the normal properties of acids. They react with reactive metals, bases, alkalis, carbonates and hydrogencarbonates to form salts (*see* table 19.13). Both soluble and insoluble carboxylic acids undergo the reactions shown in this table.

Table 19.13 Formation of salts from carboxylic acids

				Salt
$2R—CO_2H$	+	Mg metal	→	$2R—CO_2^-Mg^{2+} + H_2$
$2H—CO_2H$	+	CaO(s) base	→	$2H—CO_2^-Ca^{2+} + H_2O$
$CH_3—CO_2H$	+	NH_3(g) base	→	$CH_3—CO_2^-NH_4^+ + H_2O$
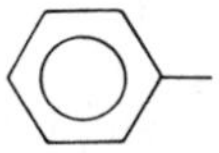$—CO_2H$	+	NaOH (aq) alkali	⟶	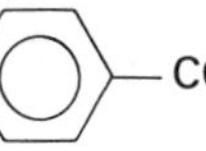$—CO_2^-Na^+ + H_2O$
$2CH_3—CO_2H$	+	Na_2CO_3 carbonate	→	$2CH_3—CO_2^-Na^+ + CO_2 + H_2O$
$CH_3—CO_2H$	+	$NaHCO_3$ hydrogencarbonate	→	$CH_3—CO_2^-Na^+ + CO_2 + H_2O$

In common with other salts of weak acids, the carboxylate salts react with excess mineral acids to form the original acid. For example, when sodium hydroxide solution is added to a suspension of the insoluble benzoic acid in water, the acid dissolves due to the formation of sodium benzoate. If sulphuric acid is then added, benzoic acid is precipitated:

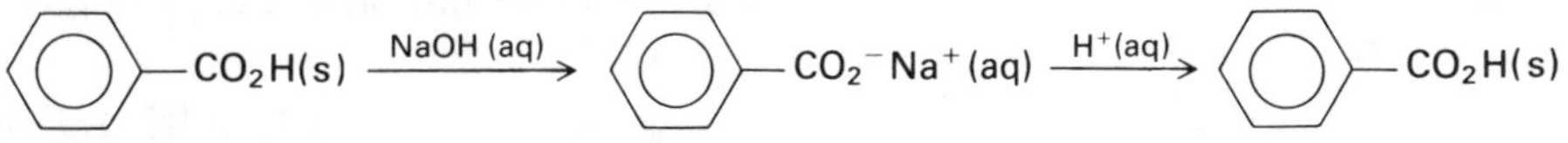

Esterification

When a carboxylic acid is heated with an alcohol in the presence of a concentrated mineral acid an ester is formed. Esterification requires fission of the alcohol molecules. There are two possibilities:

1. Alkoxy–hydrogen fission. In this case the alcohol oxygen atom ends up in the ester molecule:

$$R—\overset{\overset{O}{\|}}{C}\wr O—H + H\wr O^*—R' \longrightarrow R—\overset{\overset{O}{\|}}{C}—O^*—R' + H_2O$$

2. Alkyl–hydroxy fission. With this type of fission the alcohol oxygen atom ends up in the water molecule:

$$R—\overset{\overset{O}{\|}}{C}—O\wr H + H—O^*\wr R' \longrightarrow R—\overset{\overset{O}{\|}}{C}—O—R' + H_2O^*$$

Which applies can be determined experimentally by carrying out the esterification using alcohol containing the ^{18}O isotope (*see* section 1.3). This is called **isotopic labelling**. Determination of the relative molecular mass of the ester by mass spectrometry shows whether it is labelled with an isotope or not. For primary alcohols the ester is found to be labelled:

$$CH_3—\overset{\overset{O}{\|}}{C}—OH + \underset{\text{labelled ethanol}}{C_2H_5—{}^{18}O—H} \rightleftharpoons \underset{\text{labelled ethyl ethanoate}}{CH_3—\overset{\overset{O}{\|}}{C}—{}^{18}O—C_2H_5} + H_2O$$

This shows that methoxy–hydrogen fission occurs in the methanol molecule during this reaction.

Halogenation

Carboxylic acids react with phosphorus pentachloride and sulphur dichloride to form acid chlorides. For example

$$\langle\bigcirc\rangle\!-\!CO_2H + PCl_5 \longrightarrow \langle\bigcirc\rangle\!-\!\overset{\overset{\displaystyle O}{\|}}{C}\!-\!Cl + PCl_3O + HCl$$

benzoyl chloride   phosphorus trichloride oxide

Both benzyl chloride and phosphorus trichloride oxide are liquids and therefore have to be separated. The use of sulphur dichloride oxide is often preferred since the hydrogen chloride and sulphur dioxide can be driven off as gases leaving the acid chloride:

$$CH_3\!-\!CO_2H + SCl_2O \longrightarrow CH_3\!-\!\overset{\overset{\displaystyle O}{\|}}{C}\!-\!Cl + HCl + SO_2$$

When chlorine is bubbled through boiling ethanoic acid in the presence of a red phosphorus or iodine catalyst and sunlight, monochloroethanoic acid is formed:

$$CH_3COOH + Cl_2 \rightarrow CH_2ClCO_2H + HCl$$

Further chlorination of the products yields the di- and trisubstituted products:

$$CH_2ClCOOH + Cl_2 \rightarrow CHCl_2COOH + HCl$$
$$CHCl_2CO_2H + Cl_2 \rightarrow CCl_3COOH + HCl$$

Reduction

Carboxylic acids can be reduced to the corresponding alcohol using lithium tetrahydridoaluminate(III) in dry ethoxyethane. An alkoxide intermediate is formed first. Hydrolysis of this intermediate yields the alcohol:

$$CH_3\!-\!\overset{\overset{\displaystyle O}{\|}}{C}\!-\!OH \xrightarrow[\text{(2) } H_2O]{\text{(1) LiAlH}_4 \text{ in ethoxyethane}} CH_3CH_2OH$$

Carboxylic acids are resistant to reduction by many common reducing agents. The acids cannot be reduced to the corresponding aldehyde directly.

Oxidation

With the exception of methanoic acid and ethanedioic acid, carboxylic acids do not undergo oxidation readily. Methanoic acid and its salts are oxidised by potassium manganate(VII). Methanoic acid also reduces Fehling's reagent and gives a 'silver mirror' when warmed with ammoniacal silver nitrate solution. Carbon dioxide and water are formed:

$$HCOOH \xrightarrow{\text{[O]}} CO_2 + H_2O$$

Ethanedioic acid is also oxidised by potassium manganate(VII) to form carbon dioxide and water:

$$\overset{\displaystyle O}{\underset{\displaystyle O}{\|}}\!\!\overset{C}{\underset{C}{\big|}}\!\!\overset{OH}{\underset{OH}{}} \xrightarrow{\text{[O]}} 2CO_2 + H_2O$$

Dehydration

Distillation of a carboxylic acid with a dehydrating agent such as phosphorus(v) oxide, P_4O_{10}, results in the elimination of a molecule of water and the formation of the acid anhydride:

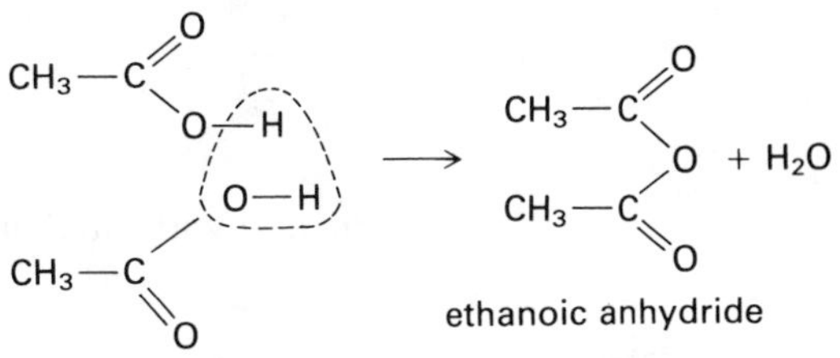

ethanoic anhydride

Methanoic acid and ethanedioic acid are exceptional once again. Dehydration of methanoic acid or its potassium or sodium salts with concentrated sulphuric acid yields carbon monoxide and water:

$$H\text{---}CO_2H \xrightarrow{\text{conc. } H_2SO_4} CO + H_2O$$

The dehydration of sodium methanoate with concentrated sulphuric acid is a standard laboratory preparation of carbon monoxide. Dehydration of ethanedioic, acid with hot concentrated sulphuric acid produces a mixture of carbon monoxide and carbon dioxide:

$$\begin{array}{c} CO_2H \\ | \\ CO_2H \end{array} \xrightarrow{\text{heat}} CO + CO_2 + H_2O$$

Carboxylates

The sodium and potassium salts of carboxylic acids are white crystalline solids. They readily dissolve in water, forming strong electrolytes.

Electrolysis of a sodium or potassium carboxylate salt dissolved in aqueous methanol yields alkanes and carbon dioxide at the anode and hydrogen at the cathode:

at the anode $2R\text{---}CO_2^- \rightarrow R\text{---}R + 2CO_2 + 2e^-$

at the cathode $2H^+ + 2e^- \rightarrow H_2$

This method of producing alkanes is known as the **Kolbé synthesis**.

Alkanes are also produced when a sodium or potassium carboxylate is heated with sodium hydroxide or soda lime. Methane can be prepared in the laboratory as follows:

$$\underset{\text{sodium ethanoate}}{CH_3CO_2^-Na^+} + NaOH \rightarrow Na_2CO_3 + CH_4$$

> **Soda lime** is a mixture of sodium hydroxide and calcium hydroxide.

Aromatic sodium or potassium carboxylates yield arenes:

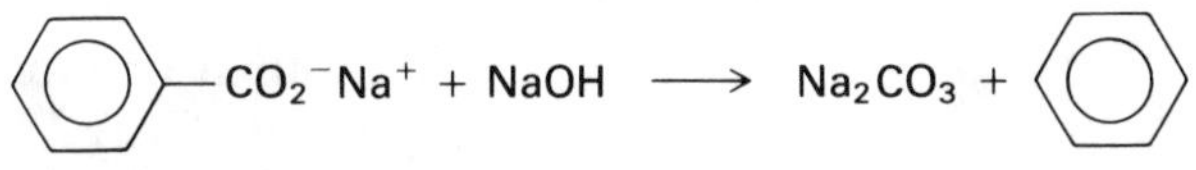

sodium benzoate

Sodium carboxylates form acid anhydrides when heated with acid chlorides:

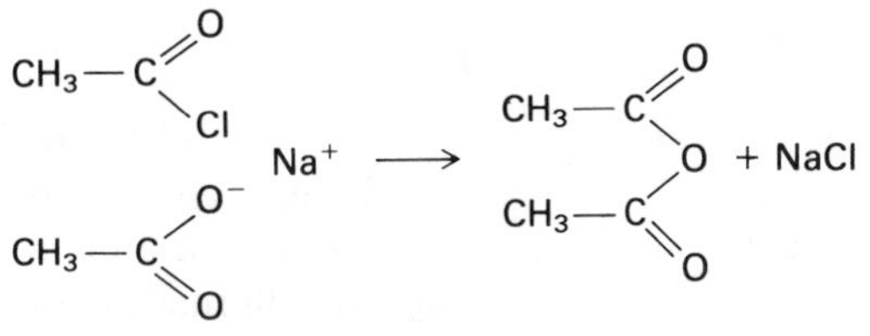

Calcium carboxylates are also white crystalline solids which are generally soluble in water. On heating they produce a poor yield of ketone:

$$2CH_3CO_2^- \, Ca^{2+} \xrightarrow{\text{heat}} \underset{\text{propanone}}{\begin{array}{c} CH_3 \\ \diagdown \\ C=O \\ \diagup \\ CH_3 \end{array}} + CaCO_3$$

If heated with calcium methanoate an aldehyde is formed:

$$2CH_3CO_2^- \, Ca^{2+} + 2HCO_2^- \, Ca^{2+} \longrightarrow \underset{\text{ethanal}}{\begin{array}{c} CH_3 \\ \diagdown \\ C=O \\ \diagup \\ H \end{array}} + 2CaCO_3$$

Ammonium salts are also white crystalline solids soluble in water. When heated strongly, an amide is formed:

$$CH_3-CO_2^- \, NH_4^+ \longrightarrow \underset{\text{ethanamide}}{CH_3-C\overset{\textstyle O}{\underset{\textstyle NH_2}{\diagup}}} + H_2O$$

$$\underset{\text{ammonium ethanoate}}{}$$

ACID DERIVATIVES

Acid derivatives are compounds in which the hydroxyl group of the carboxylic acid has been replaced by another functional group. However, they all have the **acyl group**

$$\begin{array}{c} R \\ \diagdown \\ C=O \end{array}$$

in common. R may be an alkyl or aryl group (*see* table 19.14). Although nitriles do not possess the acyl group, they are normally considered as acid derivatives since they can be obtained from carboxylic acids.

Acid Chlorides

Acid chlorides are colourless liquids with pungent smells. They are prepared by the reaction of the carboxylic acid and phosphorus(v) chloride or sulphur dichloride oxide (*see* above).

Acid chlorides are very reactive compounds. The carbonyl group readily undergoes nucleophilic attack by nucleophiles such as —OH, —OR′ and —NH₂. The reaction of acid chlorides with alcohols is an example of **nucleophilic acyl substitution** or more simply **acylation**. The mechanism of the reaction can be represented as:

$$R-\overset{\delta+}{C}\overset{\diagup O^{\delta-}}{\underset{\diagdown Cl}{}} \; \underset{\diagup}{\overset{O}{\underset{H}{R'}}} \; \rightleftharpoons \; R-C\overset{\diagup O^-}{\underset{\diagdown Cl}{}} \; \underset{R'\;\;\;H}{\overset{O_+}{}} \longrightarrow R-C\overset{\diagup O}{\underset{\diagdown OR'}{}} + HCl$$

Nucleophilic substitution of the ethanoyl group $CH_3CO—$ is known as **ethanoylation**. The mechanism for nucleophilic acyl substitution is often compared to condensation reactions since it involves the **addition** of a nucleophile followed by the **elimination** of chloride (*see* section 17.3).

Table 19.14 Acid derivatives

Derivative	Examples		

acid

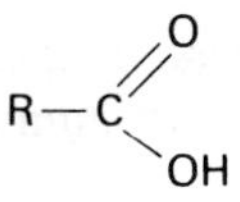

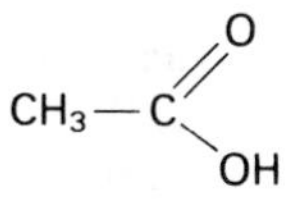

ethanoic acid   benzoic acid

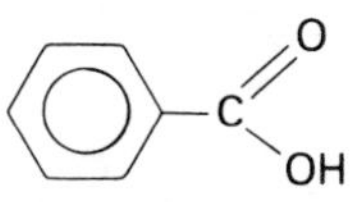

acid chloride

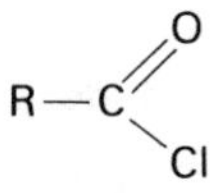

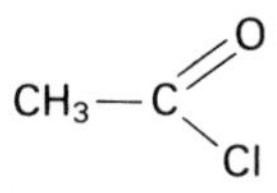

ethanoyl chloride   benzoyl chloride

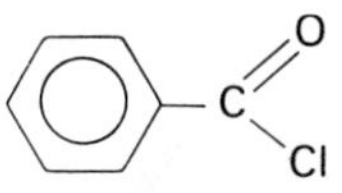

acid anhydride

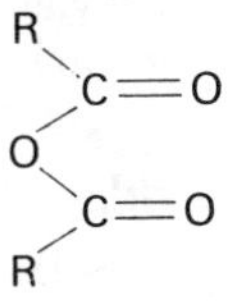

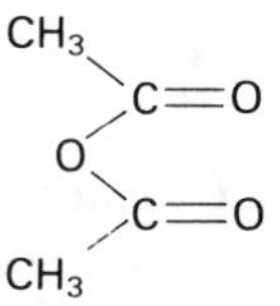

ethanoic anhydride   benzoic anhydride

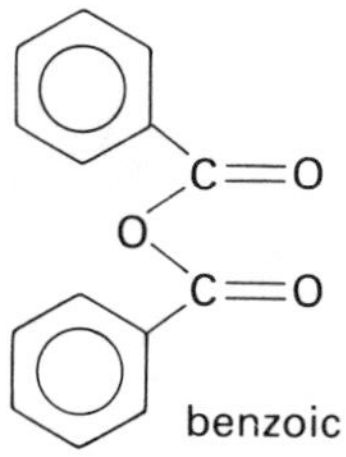

ester

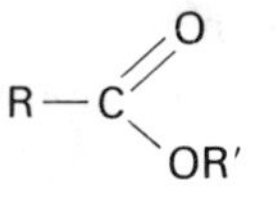

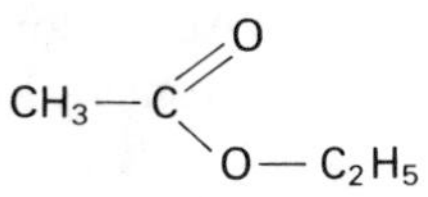

ethyl ethanoate   methyl benzoate

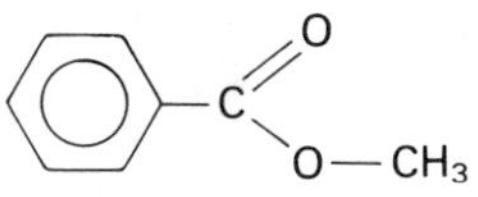

amide

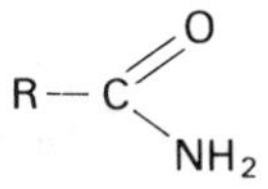

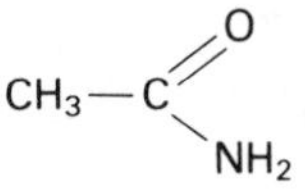

ethanamide   benzamide

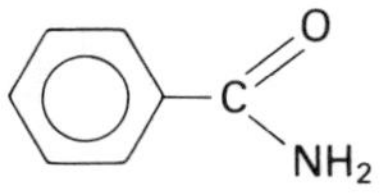

nitrile

$$R-C\equiv N \qquad CH_3-C\equiv N \qquad$$

ethanenitrile

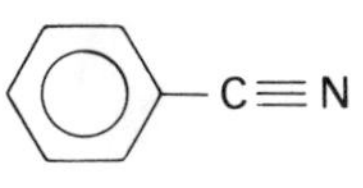

benzonitrile

Hydrolysis

Aliphatic acid chlorides are vigorously hydrolysed by cold water, forming the carboxylic acid and hydrogen chloride:

$$CH_3-C\underset{Cl}{\overset{O}{\big<}} + H_2O \longrightarrow CH_3-C\underset{OH}{\overset{O}{\big<}} + HCl$$

Acid chlorides fume in moist air due to the production of hydrogen chloride by this reaction. Aromatic acid chlorides are also hydrolysed but more slowly:

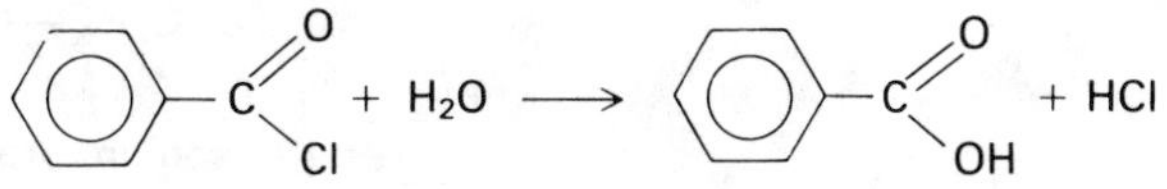

$$C_6H_5-C\underset{Cl}{\overset{O}{\big<}} + H_2O \longrightarrow C_6H_5-C\underset{OH}{\overset{O}{\big<}} + HCl$$

Ester Formation

Acid chlorides react with alcohols and phenol to produce esters and hydrogen chloride. These reactions are described in section 19.2.

Reaction with Ammonia and Amines

Acid chlorides react with ammonia and amines to form amides and substituted amides respectively:

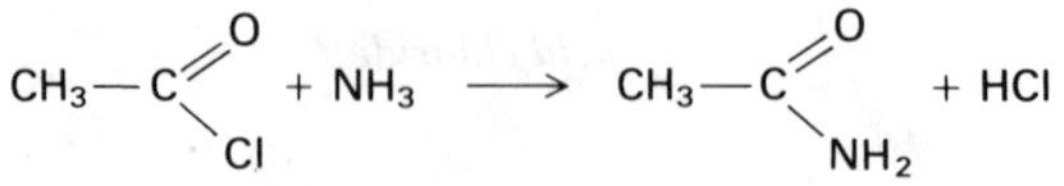

$$CH_3 - C \underset{Cl}{\overset{O}{<}} + NH_3 \longrightarrow CH_3 - C \underset{NH_2}{\overset{O}{<}} + HCl$$

ethanoyl chloride   ethanamide

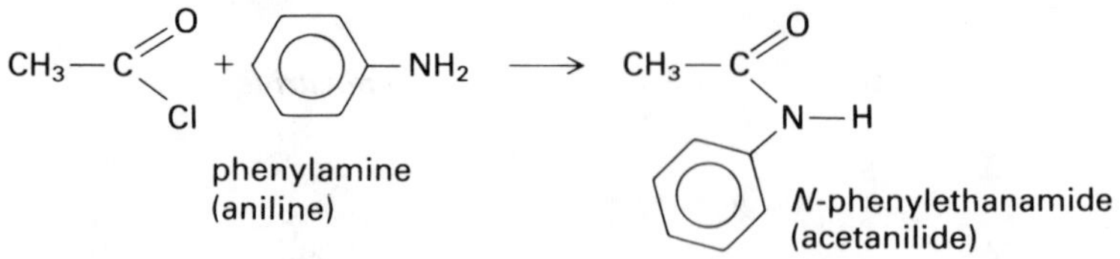

$$CH_3 - C \underset{Cl}{\overset{O}{<}} + \langle \bigcirc \rangle - NH_2 \longrightarrow$$

phenylamine
(aniline)

N-phenylethanamide
(acetanilide)

Formation of Acid Anhydrides

Acid chlorides form acid anhydrides when distilled with the sodium salt of a carboxylic acid (*see* above).

Friedel–Crafts Reactions

Acid chlorides react with benzene in the presence of aluminium chloride to produce aromatic ketones. These are examples of Friedel–Crafts reactions (*see* section 18.2).

Reduction

Acid chlorides are reduced to aldehydes by hydrogen in the presence of a poisoned palladium catalyst:

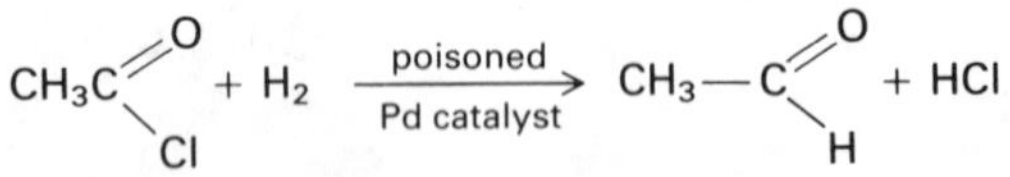

$$CH_3C \underset{Cl}{\overset{O}{<}} + H_2 \xrightarrow[\text{Pd catalyst}]{\text{poisoned}} CH_3 - C \underset{H}{\overset{O}{<}} + HCl$$

This is known as the **Rosenmund reaction**. With lithium tetrahydridoaluminate(III), acid chlorides are reduced to alcohols:

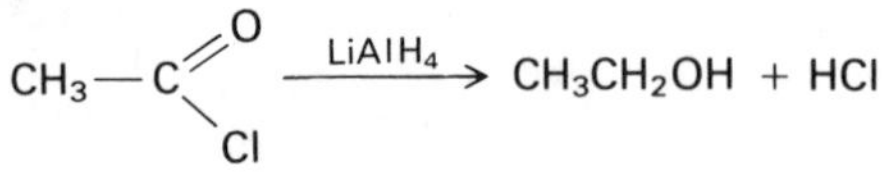

$$CH_3 - C \underset{Cl}{\overset{O}{<}} \xrightarrow{\text{LiAlH}_4} CH_3CH_2OH + HCl$$

Acid Anhydrides

Like the acid chlorides, these are colourless liquids with pungent smells. They are prepared by the reaction of an acid chloride with an anhydrous sodium carboxylate (*see* above).

    Mixed acid anhydrides can be prepared by the reaction of an acid chloride and the sodium salt of another carboxylic acid. For example

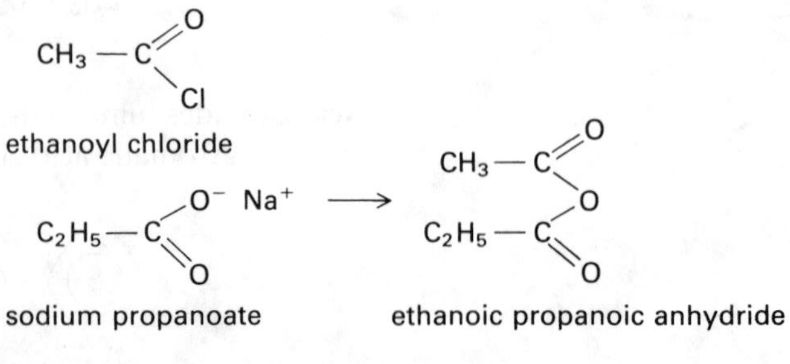

$$CH_3 - C \underset{Cl}{\overset{O}{<}}$$

ethanoyl chloride

$$C_2H_5 - C \overset{\diagup O^- \; Na^+}{\underset{\diagdown O}{}} \longrightarrow$$

sodium propanoate

$$CH_3 - C \overset{O}{\underset{}{<}}$$
$$C_2H_5 - C \overset{O}{\underset{O}{<}}$$

ethanoic propanoic anhydride

Acid anhydrides undergo nucleophilic substitution in the same way as acid chlorides although more slowly. They are hydrolysed to form carboxylic acids and react with alcohols and phenols to produce esters and carboxylic acids:

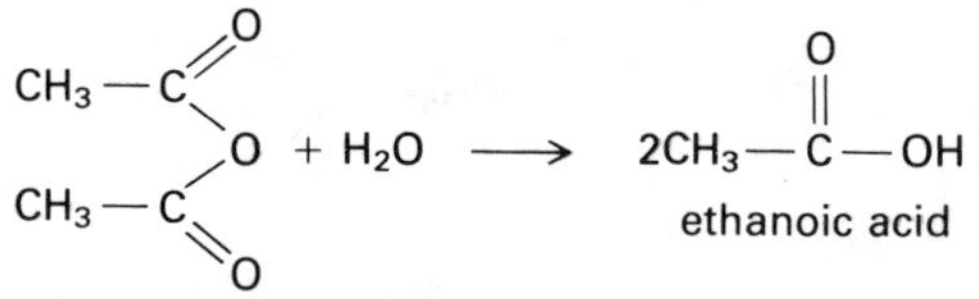

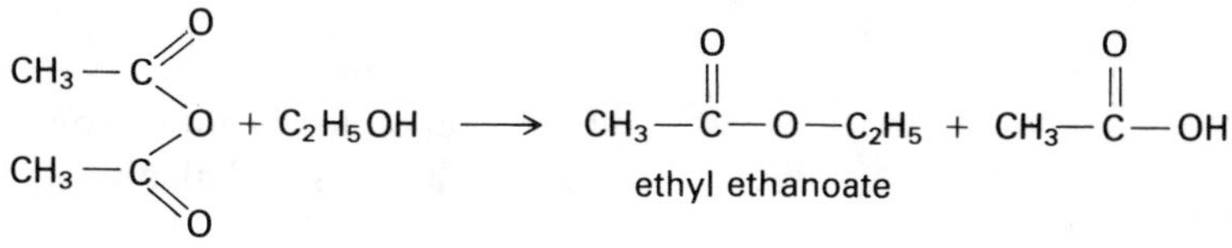

They also react with ammonia and amines to yield amides and substituted amides respectively:

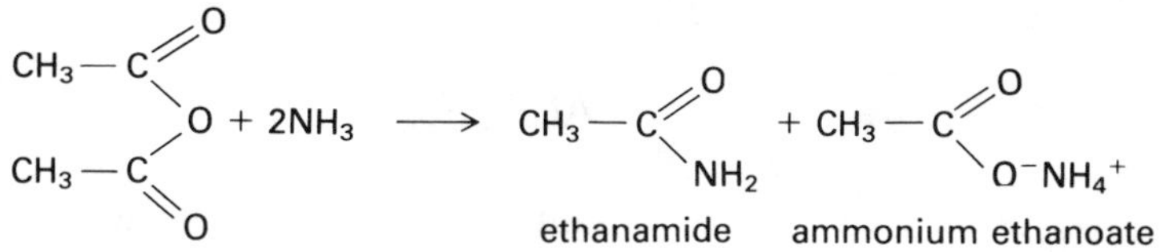

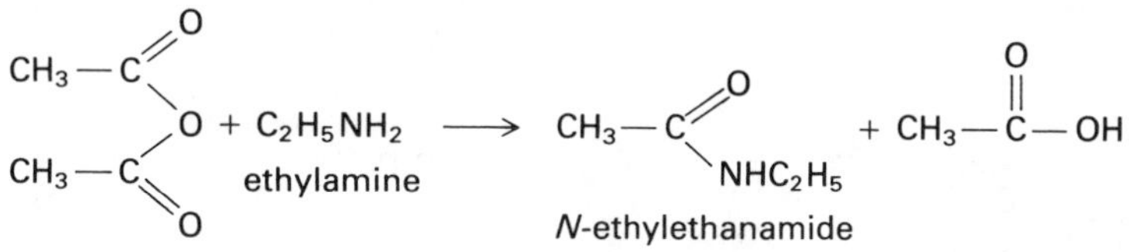

Esters

The lower esters are colourless liquids with fruity smells. Esters tend to be insoluble in water but soluble in organic solvents. They can be prepared by the reaction of a carboxylic acid and an alcohol in the presence of concentrated sulphuric acid. This type of reaction is known as **esterification**. They are also prepared by the reaction of either an acid chloride or acid anhydride with an alcohol. The latter method is preferred since it is rapid and goes to completion. Examples of both methods of preparation are described above.

Hydrolysis

Esters are hydrolysed when refluxed with a dilute mineral acid or dilute alkali. **Acid hydrolysis** yields the carboxylic acid and the alcohol:

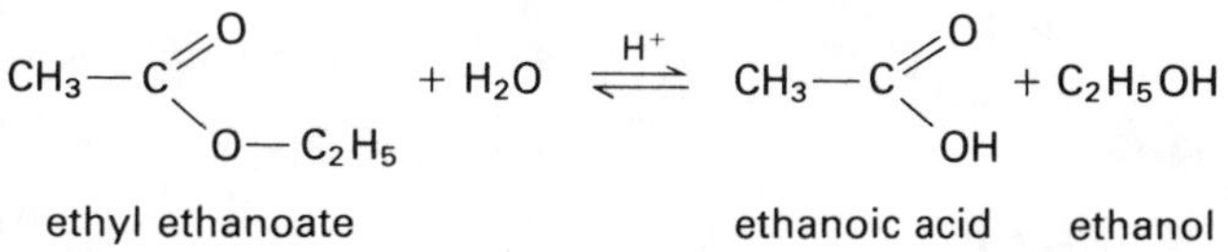

This reaction is the reverse of esterification. The kinetics of acid hydrolysis of ethyl ethanoate are described in chapter 9.

Alkaline hydrolysis of esters yields an alcohol and a salt of the carboxylic acid:

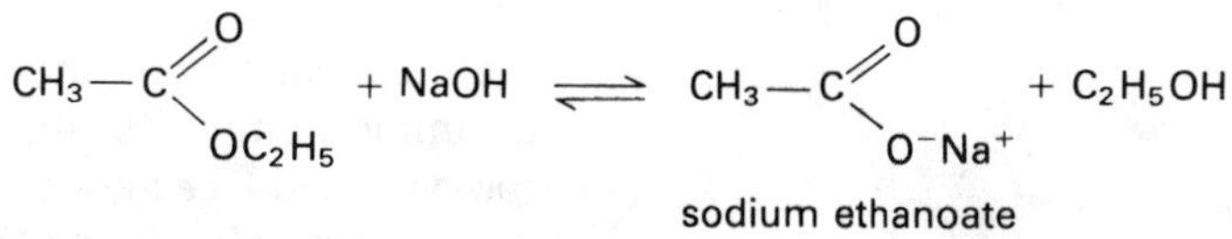

This type of reaction is known as **saponification** (*see* chapter 20).

Reduction

Esters are reduced by lithium tetrahydridoaluminate(III) in anhydrous ethoxy-ethane, yielding a mixture of two alcohols:

$$CH_3-C\overset{O}{\underset{O-CH_2-C_6H_5}{}} \xrightarrow{LiAlH_4} C_2H_5OH + C_6H_5-CH_2OH$$

phenylethylethanoate   ethanol   phenylmethylethanol

Reaction with Ammonia

Esters react slowly with a concentrated aqueous or alcoholic solution of ammonia to form amides and alcohols:

$$CH_3-\overset{O}{\overset{\|}{C}}-O-C_2H_5 + NH_3 \longrightarrow CH_3-\overset{O}{\overset{\|}{C}}-NH_2 + C_2H_5OH$$

ethanamide

AMIDES

Simple amides contain the functional group

$$-\overset{O}{\overset{\|}{C}}-NH_2$$

For example:

CH$_3$—C(=O)NH$_2$	C$_6$H$_5$—C(=O)NH$_2$	CH$_3$—C(=O)NHCH$_3$	NH$_2$—C(=O)NH$_2$
ethanamide (acetamide)	benzamide	N-methylethanamide (N-methylacetamide)	urea (carbamide)

Amides with one or both of the amide hydrogen atoms substituted by alkyl groups are known as **N-substituted amides**. Urea (also known as carbamide) is an example of a **diamide**.

Apart from methanamide, $HCONH_2$, which is a liquid at room temperature, the amides are white crystalline solids. The melting and boiling points are thus high compared to the other simple derivatives of the carboxylic acids. This is because of the association of the amide molecules due to hydrogen bonding:

$$\underset{H}{\overset{R}{>}}N-\overset{\delta+}{C}=O^{\delta-}\cdots\cdots\overset{\delta+}{H}-\underset{H}{N}-C\overset{O}{\underset{R}{}}$$

The simpler amides are all soluble in water and all the amides are soluble in common organic solvents.

Preparation

The amides can be prepared by heating the ammonium salt of the carboxylic acid. The ammonium salts are prepared by adding ammonium hydroxide to the carboxylic acid (*see* above).

Amides can also be prepared by the reaction of ammonia or an amine with either an acid chloride or acid anhydride (*see* above).

Chemical Properties

Basicity

The amides are only very weakly basic. The pK_b of ethanamide, for example, is 14.1. As a result the amides do not react with mineral acids to form salts. They are also neutral to litmus.

The amide group exists as a resonance hybrid of two limiting forms:

$$R-C\overset{O}{\underset{\underset{H}{N}}{}}H \longleftrightarrow R-C\overset{O^-}{\underset{\underset{H}{N^+}}{}}H$$

The lone-pair of electrons on the nitrogen atom is thus less available for donation than the pair on the nitrogen atom in an amine.

Hydrolysis

Amides are hydrolysed when heated with a dilute acid or dilute alkali:

$$CH_3-\overset{O}{\overset{\|}{C}}-NH_2 \ + \ HCl \xrightarrow{(aq)} CH_3CO_2H \ + \ NH_4Cl$$

$$CH_3-\overset{O}{\overset{\|}{C}}-NH_2 \ + \ NaOH \xrightarrow{(aq)} CH_3CO_2^-Na^+ \ + \ NH_3(g)$$

Dehydration

Amides are dehydrated when heated with the dehydrating agent phosphorus(v) oxide. A nitrile is formed:

$$CH_3-\overset{O}{\overset{\|}{C}}-NH_2 \xrightarrow{P_4O_{10}} CH_3-C\equiv N + H_2O$$
$$\text{ethanenitrile}$$

Reaction with Nitrous Acid

Amides react with nitrous acid with the evolution of nitrogen:

$$CH_3-\overset{O}{\overset{\|}{C}}-NH_2 + HNO_2 \longrightarrow CH_3-CO_2H + N_2 + H_2O$$

Hofmann Degradation

When simple amides are treated with bromine and sodium hydroxide solution, primary amines are formed:

$$R-\overset{O}{\overset{\|}{C}}-NH_2 \xrightarrow{Br_2/OH} R-NH_2$$

August Wilhelm von Hofmann (1818–1892) is famous for his work on organonitrogen compounds—especially dyes

This elimination of a carbonyl group provides a means of shortening the length of a carbon chain by one carbon atom and thus **descending the homologous series**. It is known as the Hofmann degradation.

Example of descending the homologous series

How could you convert ethanol to methanol?

Solution
Ethanol can be converted to methanol by means of the following four steps:

1. Ethanol is first converted to ethanoic acid by oxidation:

$$C_2H_5OH \xrightarrow{K_2Cr_2O_7/H_2SO_4} CH_3COOH \quad (\textit{see} \text{ section } 19.2)$$

2. Ethanoic acid is converted to ethanamide by addition of excess ammonia solution followed by heating:

$$CH_3COOH \xrightarrow[\text{(2) heat}]{\text{(1) } NH_3} CH_3CONH_2$$

3. The amide is heated with bromine and sodium hydroxide solution. This is the Hofmann degradation. It removes a carbon atom:

$$CH_3CONH_2 \xrightarrow[\text{(2) NaOH}]{\text{(1) } Br_2} CH_3NH_2 \quad \text{methylamine}$$

4. The methylamine is treated with a cold solution of sodium nitrite in dilute hydrochloric acid. The nitrous acid is produced in situ:

$$CH_3NH_2 \xrightarrow{HNO_2} CH_3OH \quad (\textit{see} \text{ section } 19.4)$$

Urea

Urea (also known as carbamide) is a white crystalline solid. It is soluble in water and ethanol but insoluble in ethoxyethane. It was first obtained by Wohler in 1828 by heating an aqueous solution of ammonium cyanate (*see* section 17.1).

Urea is hydrolysed by both dilute and mineral acids and alkaline solutions:

$$H_2N-\overset{\overset{\displaystyle O}{\|}}{C}-NH_2 + H_2O \longrightarrow CO_2 + 2NH_3$$

> This hydrolysis is catalysed by the enzyme *urease* which is present in some soil bacteria. The ammonia produced is converted to nitrates which are then absorbed by plants.

Like the amides, urea also reacts with nitrous acid with the evolution of nitrogen:

$$H_2N-\overset{\overset{\displaystyle O}{\|}}{C}-NH_2 + 2HNO_2 \longrightarrow CO_2 + N_2 + 3H_2O$$

On heating, urea forms a compound known as **biuret**:

$$H_2N-\overset{\overset{\displaystyle O}{\|}}{C}-NH_2 + H_2N-\overset{\overset{\displaystyle O}{\|}}{C}-NH_2 \rightarrow H_2N-\overset{\overset{\displaystyle O}{\|}}{C}-NH-\overset{\overset{\displaystyle O}{\|}}{C}-NH_2 + NH_3$$
$$\text{biuret}$$

When a drop of copper(II) sulphate is added to an alkaline solution of biuret, a purple colour is produced. Peptides and proteins give a similar reaction. This is known as the **biuret reaction** or biuret test. Biuret, peptides and proteins all contain the peptide linkage $-CO-NH_2-$.

Nitriles

Nitriles are also known as alkyl cyanides. They are colourless liquids. The simpler nitriles such as ethanenitrile are soluble in water. All nitriles are soluble in organic solvents.

Nitriles can be prepared by the dehydration of amides (*see* above). They are also prepared by refluxing a halogenoalkane with an ethanolic solution of potassium or sodium cyanide (*see* section 19.1).

On boiling with a dilute mineral acid or dilute alkali, nitriles are hydrolysed. The carboxylic acid is formed:

$$CH_3-C\equiv N + HCl + 2H_2O \rightarrow CH_3COOH + NH_4Cl$$

$$CH_3-C\equiv N + NaOH + H_2O \rightarrow CH_3COO^-Na^+ + NH_3$$

When treated with a reducing agent such as sodium and ethanol or lithium tetrahydridoaluminate(III) in ethoxyethane, nitriles are reduced to primary amines:

$$CH_3-C\equiv N \xrightarrow[\text{in ethoxyethane}]{\text{LiAlH}_4} CH_3CH_2NH_2$$

$$\text{ethylamine}$$

OCCURRENCE, MANUFACTURE AND USES OF CARBOXYLIC ACIDS AND THEIR DERIVATIVES

Occurrence

Carboxylic acids occur widely in nature. Some of these are shown in table 19.15. Many long chain saturated and unsaturated acids and their esters occur naturally in fats and oils. The chemistry of fats and oils is described in the following chapter. Amides also occur in plants and animals. For example, the B vitamin commonly known as nicotinic acid or niacin and its derivative nicotinamide are found in yeast, liver, leafy green foods and milk. *N*-Methylnicotinamide is an important urinary metabolite of the vitamin:

Carbamide is also involved in these metabolic processes.

Table 19.15   Naturally occurring carboxylic acids

Formula	Systematic name (common name)	Occurrence
HCO_2H	methanoic acid (formic acid)	ants and stinging nettles
CH_3CO_2H	ethanoic acid (acetic acid)	vinegar
$CH_3(CH_2)_2CO_2H$	butanoic acid (butyric acid)	rancid butter
$CH_3(CH_2)_4CO_2H$	hexanoic acid (caproic acid)	goats' milk
$\begin{matrix} CO_2H \\ | \\ CO_2H \end{matrix}$	ethanedioic acid (oxalic acid)	rhubarb leaves
$CH_3CHOHCO_2H$	2-hydroxypropanoic acid (lactic acid)	human breast milk, cow's milk
$\begin{matrix} CH_2-CO_2H \\ | \\ HO-C-CO_2H \\ | \\ CH_2-CO_2H \end{matrix}$	2-hydroxypropane-1,2,3-tricarboxylic acid (citric acid)	citrus fruits
	3-pyridinecarboxylic acid (nicotinic acid, niacin)	a B vitamin found in plants and animals

Breast feeding in Zimbabwe. Human breast milk contains lactic acid

Manufacture

Methanoic acid. This is made by the reaction of carbon monoxide with sodium hydroxide solution at 200°C under high pressure:

$$NaOH + CO \rightarrow HCO_2^-Na^+$$
$$\text{sodium methanoate}$$

The salt is then hydrolysed using dilute sulphuric acid and the methanoic acid is distilled off:

$$HCO_2^-Na^+ \xrightarrow{H^+} H-CO_2H + Na^+$$

Ethanoic acid. This is manufactured by the oxidation of butane in air at 200°C using a cobalt(II) ethanoate catalyst. Methanoic acid is also produced in this process.

Vinegar is produced by the atmospheric oxidation of wine or beer. The process is catalysed by the enzymes present in bacteria.

Benzoic acid. This is manufactured by the catalytic oxidation of methylbenzene with air.

Ethanedioic acid. This is manufactured from sodium methanoate. The salt is first heated and then the ethanedioate is precipitated as a calcium salt by adding calcium hydroxide. Hydrolysis of the dried product using dilute sulphuric acid yields ethanedioic acid:

$$2HCO_2Na \xrightarrow{heat} \begin{array}{c} CO_2^-\,Na^+ \\ | \\ CO_2^-\,Na^+ \end{array} \xrightarrow{Ca(OH)_2} \begin{array}{c} CO_2^- \\ | \\ CO_2^- \end{array} Ca^{2+} \xrightarrow{2H^+} \begin{array}{c} CO_2H \\ | \\ CO_2H \end{array}$$

Ethanoic anhydride. This is manufactured by heating propanone vapour over a nickel and chromium alloy catalyst at 750°C:

$$CH_3-\overset{\overset{\displaystyle O}{\|}}{C}-CH_3 \xrightarrow[750°C]{Ni/Cr} CH_2{=}C{=}O + CH_4$$
$$\text{ethanone}$$

Ethanone is a highly reactive gas. To produce the anhydride the ethanone is first condensed by cooling and then passed through ethanoic acid:

$$CH_2{=}C{=}O + CH_3CO_2H \longrightarrow \begin{array}{c} CH_3-C{\displaystyle\diagup^O}_{\diagdown O} \\ CH_3-C{\diagup^O}_{\diagdown O} \end{array}$$

Amides. These are manufactured by the reaction of ammonia with acid chlorides or anhydrides. Alternatively, they are produced by heating the ammonium salts of carboxylic acids (*see* above).

Urea. This is produced by heating carbon dioxide with excess ammonia at 200°C and 200 atmospheres pressure:

$$CO_2 + 2NH_3 \xrightarrow[200\ atm]{200°C} \begin{array}{c} NH_2 \\ \diagdown \\ C{=}O \\ \diagup \\ NH_2 \end{array}$$

Uses

Carboxylic acids, esters, acid anhydrides and urea are all used extensively to manufacture plastics or resins (*see* chapter 20). Large quantities of the acids and their esters are also used in soap manufacture (*see* chapter 20 also). Both ethanoic acid and ethyl ethanoate are used in industry and in the laboratory as solvents. The acids and esters are also used in the food industry. Esters, for example, are used as food flavourings. Benzoic acid is used as food preservative and calcium propanoate as an additive in bread manufacture. Urea is used as a fertiliser.

Calcium propanoate occurs naturally in some cheeses and is added to others

Carboxylic acids and their derivatives are also used widely in the pharmaceutical industry. Aspirin, for example, is manufactured by the ethanoylation of 2-hydroxybenzoic acid using ethanoic anhydride:

2-hydroxybenzoic acid (salicylic acid)   aspirin

2-Hydroxybenzoic acid is manufactured by heating sodium phenolate with carbon dioxide under pressure and then hydrolysing the salt which is formed:

phenol   sodium phenolate   sodium 2-hydroxybenzoate (sodium salicylate)

The **barbiturates** are alkyl-substituted derivatives of barbituric acid (*see* table 19.16).

Barbiturate

Barbituric acid is made by the reaction of urea and propanedioic acid in the presence of phosphorus trichloride oxide:

propanedioic acid (malonic acid)   barbituric acid

Table 19.16   Some barbiturates

Alkyl substituent R		Name
C_2H_5-	ethyl	barbitone (veronal)
CH_3CH_2-C- with CH_3 groups	3,3-dimethylbutyl (isoamyl)	amylobarbitone (amyltal)
(phenyl ring)-	phenyl	phenobarbitone (luminal)

Barbiturates used to be used as hypnotics and sedatives. They have now been replaced by the benzodiazepines. However, barbiturates are still used as anaesthetics and in the treatment of epilepsy.

SUMMARY

1. Typical reactions of aldehydes and ketones.

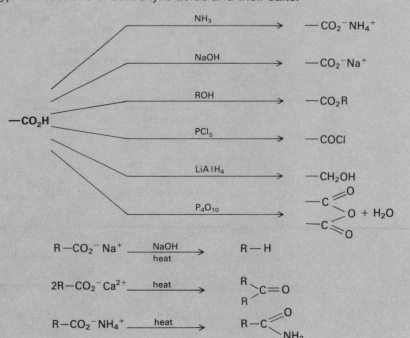

2. Typical reactions of carboxylic acids and their salts.

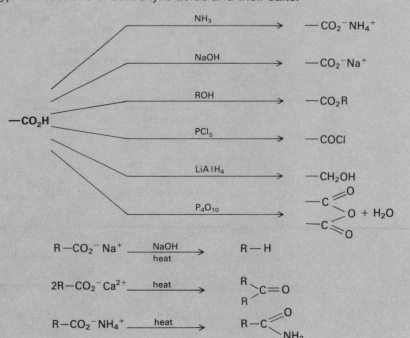

3. Typical reactions of acid chlorides.

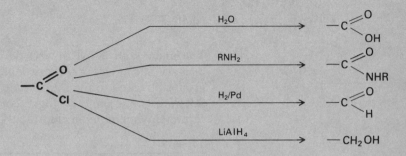

4. Typical reactions of acid anhydrides.

5. Typical reactions of esters.

6. Typical reactions of amides.

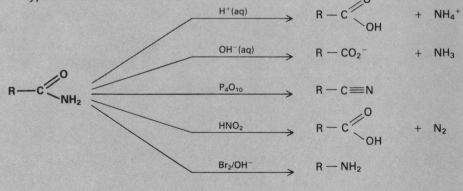

19.4
Amines and
Related
Compounds

AMINES

The amines are alkyl or aryl derivatives of ammonia. They are classified as primary, secondary or tertiary depending on whether they have one, two or three groups attached to the nitrogen atom (*see* table 19.17). **Quaternary ammonium salts** are the organic analogues of inorganic ammonium salts. An example is tetramethylammonium chloride, $(CH_3)_4N^+Cl^-$.

Table 19.17 Amines

	General formula	Alkylamine	Arylamine
Primary		methylamine	phenylamine (aniline)
		ethylamine	(phenylmethyl)amine (benzylamine)
Secondary		dimethylamine	N-methylphenylamine (N-methylaniline)
			diphenylamine
Tertiary		trimethylamine	N,N-dimethylphenylamine (N,N-dimethylaniline)
			triphenylamine

The structure of an aliphatic amine molecule is closely related to the structure of an ammonia molecule. The nitrogen atom is sp³ hybridised with a lone-pair of electrons in one orbital. The molecule has a trigonal pyramidal structure (see table 2.9).

Physical Properties

The simpler amines are gases or low-boiling liquids with characteristic fishy smells. The amines are polar compounds due to the lone-pairs of electrons on the nitrogen atoms. As a result, both primary and secondary amines associate by intermolecular hydrogen bonding:

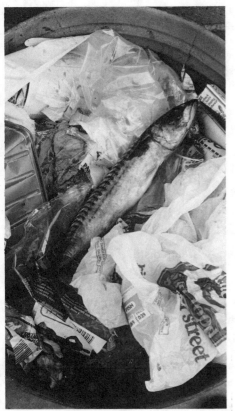

'What have we here?'

Fish Gas

> TRINCULO   What have we here? a man or a fish?
> dead or alive?
> A fish, he smells like a fish . . . a
> very ancient and fish-like smell . . .
> a kind of not-of-the-newest poor-john
>
> (*The Tempest*, William Shakespeare, Act 2, Scene 2)

Trimethylamine has a characteristic fish odour. Indeed, when it was first isolated from the distillation of herring-brine with lime in 1851–52 it was called 'fish gas'.

Trimethylamine is produced by **anaerobic respiration** of marine animals. This is respiration without oxygen. The amine is also produced by the breakdown of proteins by micro-organisms in the sediments of eutrophic lakes. It has also been isolated in human tears, gastric juices, liver bile, gall bladder bile and eye-socket fluid.

Trimethylamine can also cause an embarrassing condition known as fish odour syndrome. This is a condition where breath, sweat and urine smell strongly of rotting fish. It results from uncontrolled bacterial degradation of trimethylamine oxide. This releases trimethylamine in the gut and bowel. In healthy bodies, trimethylamine is taken up by the liver and reconverted to the odourless oxide. However, for patients with a metabolic liver malfunction this does not occur. The condition can be partially relieved by using drugs or by dietary control. In the latter, fish, liver, eggs, kidneys and other foods with a high choline content are excluded from the diet.

The boiling points of amines are thus higher than the alkanes of similar relative molecular masses although not as high as those of the corresponding alcohols.

Amines dissolve in water, forming hydrogen bonds. They are also soluble in organic solvents.

Preparation

From Ammonia

When halogenoalkanes are heated with ammonia under pressure, progressive alkylation of ammonia occurs. This yields a mixture of amine salts (*see* table 19.18).

Table 19.18   Progressive alkylation of ammonia

$$CH_3Cl + NH_3 \rightarrow CH_3NH_2 + HCl \rightleftharpoons CH_3NH_3{}^+Cl^-$$
$$CH_3Cl + CH_3NH_2 \rightarrow (CH_3)_2NH + HCl \rightleftharpoons (CH_3)_2NH_2{}^+Cl^-$$
$$CH_3Cl + (CH_3)_2NH \rightarrow (CH_3)_3N + HCl \rightleftharpoons (CH_3)_3NH^+Cl^-$$
$$CH_3Cl + (CH_3)_3N \rightarrow (CH_3)_4N^+Cl^-$$

The amines are separated by addition of an alkali followed by fractional distillation. The alkali frees the amines from the salts. For example

$$CH_3NH_3{}^+Cl^- + NaOH \rightarrow CH_3NH_2 + NaCl + H_2O$$

Since this method of preparing amines is cumbersome, it is not normally used in the laboratory.

Aryl halides do not react with ammonia and so cannot be prepared in this way.

From Nitro Compounds

Primary amines can be prepared by reduction of the nitro compound using hydrogen and a nickel catalyst or with lithium tetrahydridoaluminate(III):

$$R-NO_2 \xrightarrow{\text{LiAlH}_4} R-NH_2$$

Aryl amines are usually prepared by the reduction of nitro compounds. For example, phenylamine is prepared in the laboratory by reduction of nitrobenzene using tin and concentrated hydrochloric acid:

Phenylamine is freed from the complex salt by addition of excess sodium hydroxide:

Phenylamine is separated from the mixture by steam distillation.

From Nitriles

Primary amines can also be prepared by the reduction of nitriles using lithium tetrahydridoaluminate(III) in ethoxyethane (*see* previous section).

Hofmann Degradation

Primary amines can be prepared from amides by the reaction with bromine and sodium hydroxide solution (*see* previous section).

CHEMICAL PROPERTIES

Basicity

Amines are basic. This is due to their ability to donate the lone-pair of electrons on the nitrogen atom to an acid. For example

Methylammonium chloride and other similar quaternary ammonium salts are white crystalline solids with high melting points. They are soluble in water.

When an amine is dissolved in water, it forms an alkaline solution due to the following equilibrium:

$$CH_3\text{---}NH_2 + H_2O \rightleftharpoons CH_3\text{---}\overset{+}{N}H_3 + OH^-$$

The extent to which this equilibrium lies to the right is a measure of the strength of the base. This is given by the base dissociation constant (*see* chapter 8):

$$K_b = \frac{[CH_3\overset{+}{N}H_3][OH^-]}{[CH_3NH_2]}$$

The values of the base dissociation constants of ammonia and some amines at 25°C are shown in table 19.19. These show that the simple amines are similar in strength if not slightly stronger than ammonia. The arylamines on the other hand are far weaker bases. This is because the lone-pair of electrons on the nitrogen atom is delocalised round the benzene ring. It is thus less available for donation to an acid. The delocalisation of the lone-pair of electrons on the nitrogen atom in phenyl-amine can be represented by the following limiting structures:

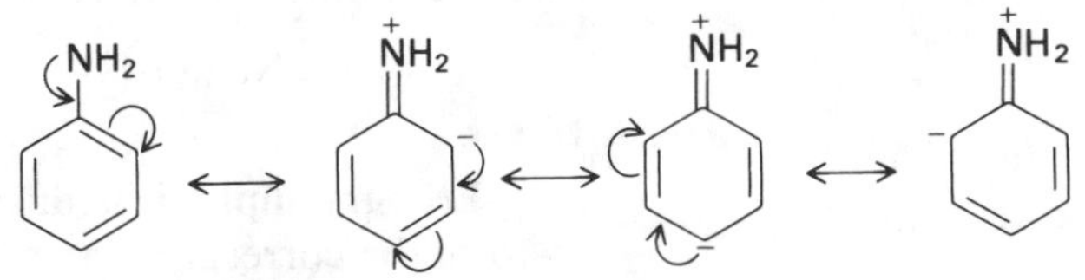

Table 19.19 Base dissociation constants of ammonia and some amines

Formula	Name	K_b/mol dm^{-3}	pK_b
NH_3	ammonia	1.8×10^{-5}	4.75
CH_3NH_2	methylamine	4.4×10^{-4}	3.36
$(CH_3)_2NH$	dimethylamine	5.4×10^{-4}	3.27
$(CH_3)_3N$	trimethylamine	6.5×10^{-5}	4.19
$C_6H_5NH_2$	phenylamine	4.2×10^{-10}	9.38

Salt Formation

Aliphatic amines and phenylamine react with mineral acids to form salts. Salts are formed, for example, in the preparation of amines from ammonia (*see* above). Phenylamine reacts with mineral acids to form phenylammonium chloride:

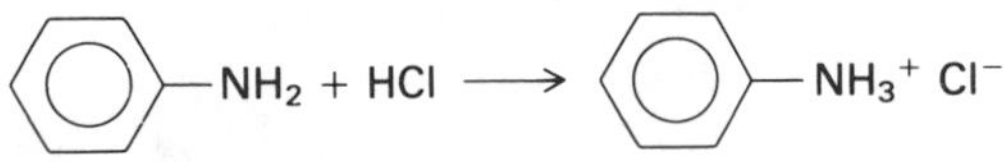

The phenylamine can be regenerated by treatment with excess sodium hydroxide solution:

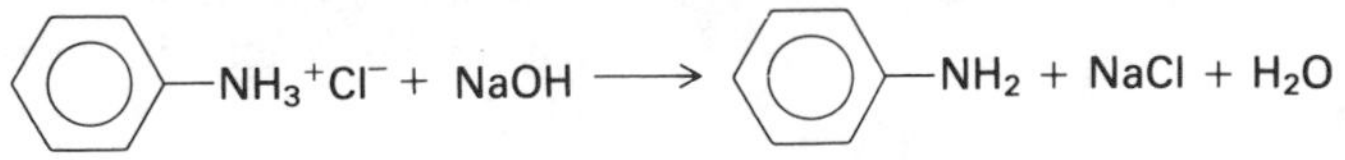

Acylation

Primary and secondary amines are acylated in the cold by acid chlorides and acid anhydrides. Substituted amides are formed:

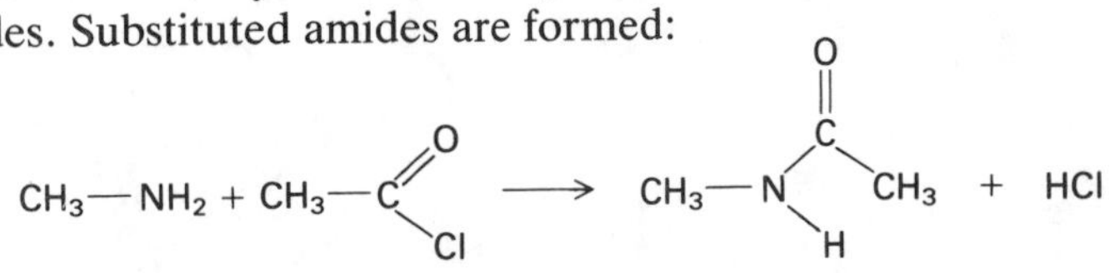

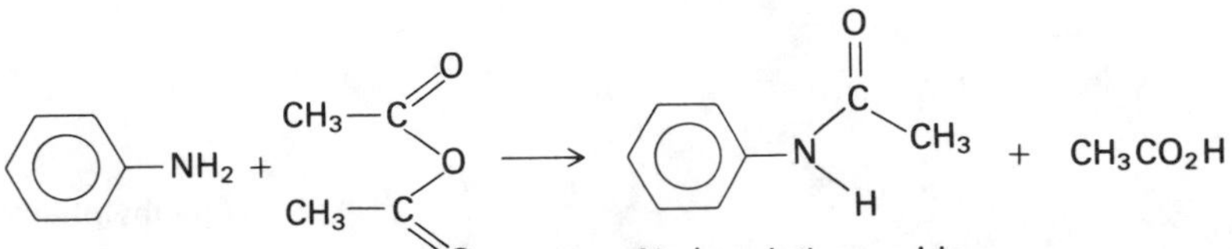

N-phenylethanamide

Tertiary amines are not acylated since they do not have a hydrogen atom attached to the nitrogen atom.

Primary and secondary amines can be benzoylated using benzoyl chloride in excess sodium hydroxide:

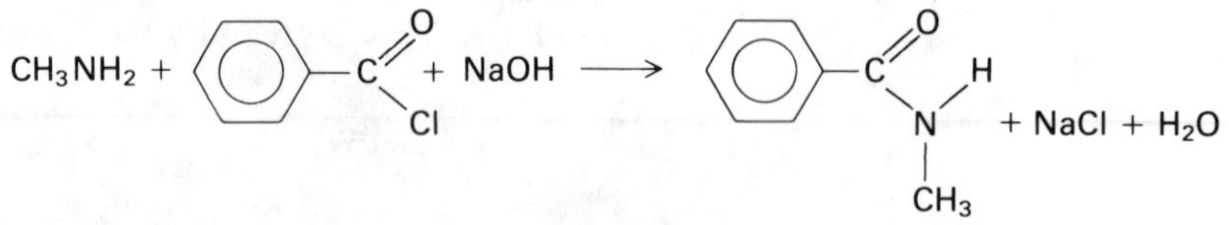

N-methylbenzamide

Alkylation with Halogenoalkanes

The alkylation of amines is described above in the section on the preparation of amines from ammonia.

Reaction with Nitrous Acid

Nitrous acid is prepared *in situ* by the reaction of dilute hydrochloric acid with sodium nitrite in the cold:

$$NaNO_2(s) + HCl(aq) \rightleftharpoons NaCl(aq) + H—O—N{=}O(aq)$$
nitrous acid

Primary aliphatic amines. Nitrous acid reacts with primary aliphatic amines to form the corresponding alcohol and nitrogen gas:

$$CH_3CH_2NH_2 + HONO \longrightarrow CH_3CH_2OH + H_2O + N_2$$
ethylamine   ethanol

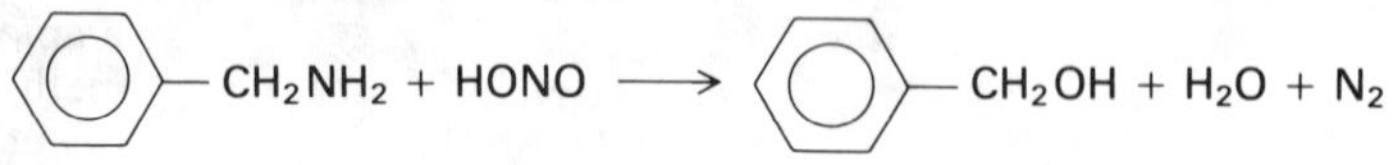

(phenylmethyl)amine   phenylmethanol
(an aliphatic amine)   (benzyl alcohol)

An unstable diazonium ion $R—\overset{+}{N}{\equiv}N$ is formed as an intermediate during these reactions.

Primary arylamines. Cold nitrous acid reacts with phenylamine and other primary arylamines to form relatively stable diazonium salts. This type of reaction is known as **diazotisation**:

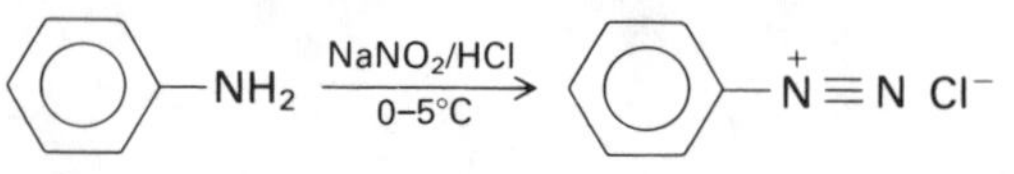

The chemistry of these salts is outlined below.

Secondary aliphatic and aromatic amines. Both aliphatic and aromatic secondary amines react with nitrous acid to form nitroso compounds. These are yellow oils:

dimethylnitrosoamine

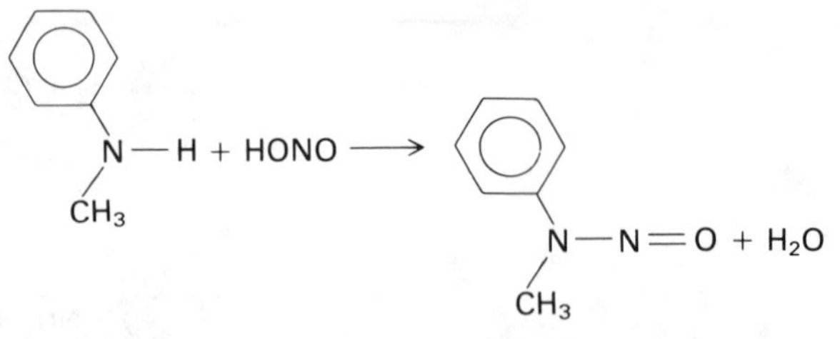

N-methylphenylamine   *N*-nitroso-*N*-methylphenylamine

These are known as **nitrosation** reactions.

Tertiary aliphatic amines. Since these contain no replaceable hydrogen atoms on the nitrogen atom, they do not undergo diazotisation or nitrosation. They react with nitrous acid to form nitrite salts. These are unstable and cannot be isolated:

$$R_3N + HONO \rightarrow R_3NH^+ + NO_2^-$$
unstable nitrite

Tertiary arylamines. These undergo nitrosation in the 4-position on the benzene ring. The compounds thus undergo *C*-substitution rather than *N*-substitution:

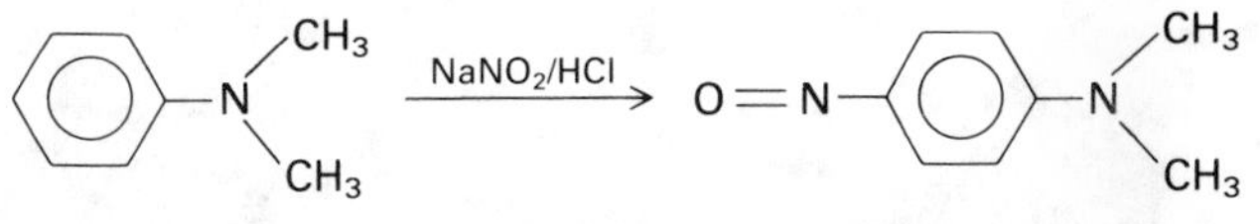

4-nitroso-*N*,*N*-dimethylphenylamine

Formation of Isocyano Compounds

Isocyano compounds are formed when primary aliphatic or aromatic amines are warmed with trichloromethane and an ethanolic solution of potassium hydroxide:

$$C_2H_5-NH_2 + CHCl_3 + 3KOH \rightarrow C_2H_5-\overset{+}{N}\equiv\overset{-}{C} + 3KCl + 3H_2O$$
isocyanoethane

These compounds are known as isocyanides or carbylamines. They have distinctive bad smells. The reaction is used as a test for primary amines.

Bromination

Phenylamine reacts readily with bromine to form 2,4,6-tribromophenylamine:

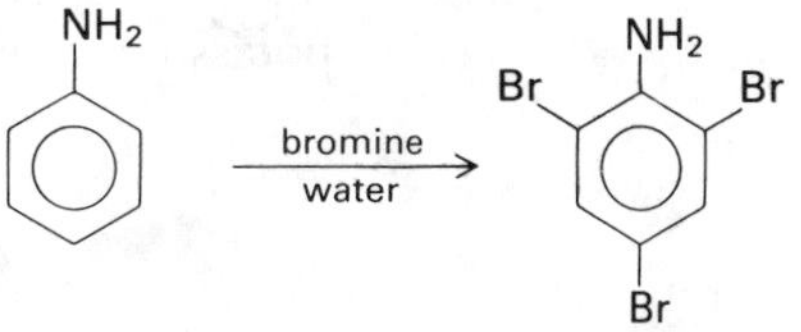

Formation of Complex Ions

As with ammonia, aqueous solutions of amines form coloured complexes with d-block metal ions (*see* chapter 14). For example

$$Cu^{2+} + 4C_2H_5NH_2 \longrightarrow [Cu(C_2H_5NH_2)_4]^{2+}$$
royal blue

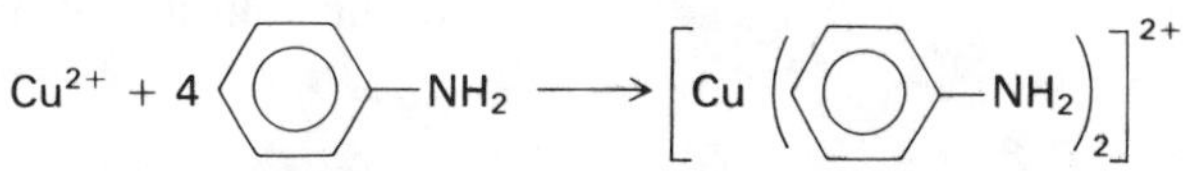

lime green

Combustion

Amines burn in air with a yellow flame to produce carbon dioxide, water and nitrogen:

$$2C_2H_5NH_2 + 7O_2 \rightarrow 4CO_2 + 6H_2O + N_2$$

DIAZONIUM SALTS

We saw above that arylamines react with nitrous acid in the cold to form diazonium salts. The most important of these is benzenediazonium chloride:

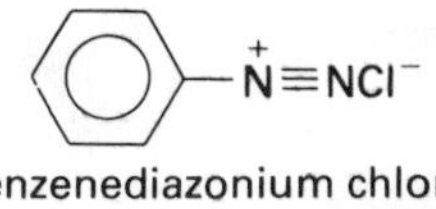

benzenediazonium chloride

Chemical Properties

Diazonium salts are stable in solution in the cold but are explosive when solid. They undergo two important types of reaction. First of all they react with nucleophilic reagents with the evolution of nitrogen. Secondly, they undergo coupling reactions.

Replacement of Nitrogen

With hydroxyl. When a diazonium salt solution is warmed to just above 10°C, it reacts with water, forming phenol:

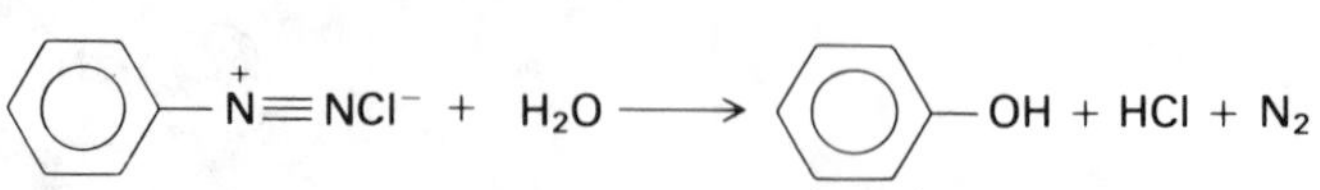

phenol

With halogens. Diazonium compounds react with copper(I) halides dissolved in the concentrated halogen acid to form halogenoarenes and nitrogen. This is the **Sandmeyer reaction** (*see* section 19.1). It is used to prepare chlorobenzene and bromobenzene. Iodobenzene is formed by warming the diazonium salt with an aqueous solution of potassium iodide:

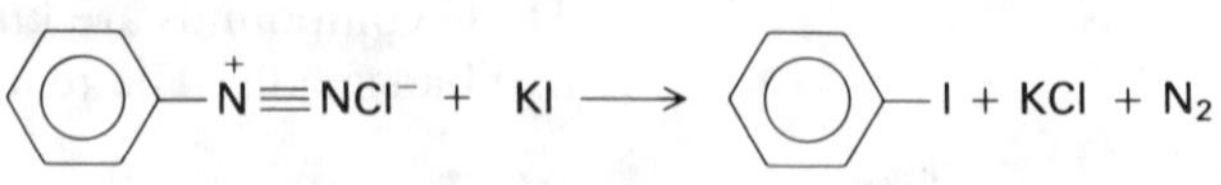

iodobenzene

With nitriles. The two nitrogen atoms in a diazonium compound can be replaced by a nitrile group by heating the compound with a solution containing potassium cyanide and copper(I) cyanide:

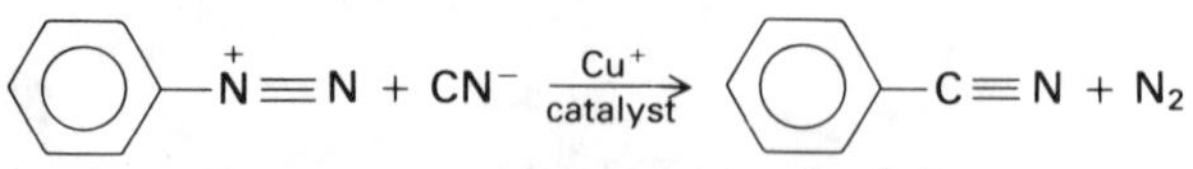

benzonitrile

Coupling Reactions

In these reactions—unlike the previous ones—the nitrogen atoms of the diazonium compound are retained.

Coupling with phenols. When a solution of diazonium salt is added to an alkaline solution of phenol, an orange azo compound is precipitated:

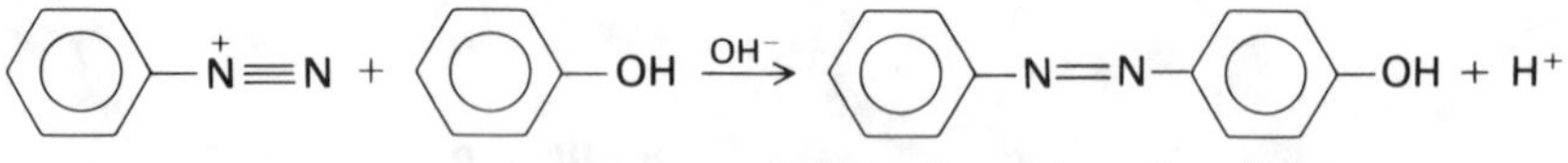

(4-hydroxyphenyl)azobenzene

Diazonium salts react with alkaline solutions of naphthalen-2-ol to give a red precipitate of 1-(phenylazo)naphthalen-2-ol:

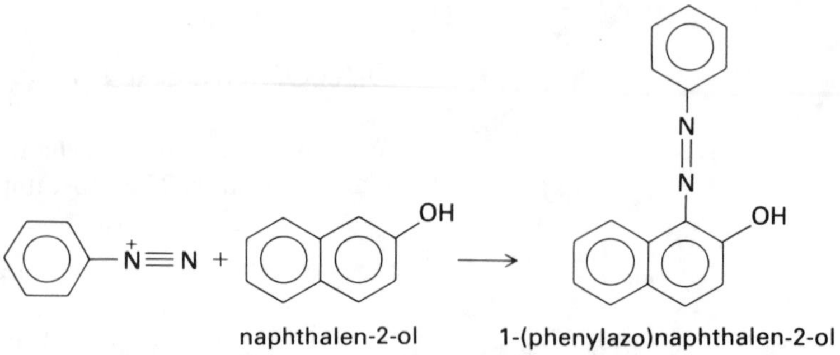

naphthalen-2-ol   1-(phenylazo)naphthalen-2-ol

Naphthalene is numbered as follows:

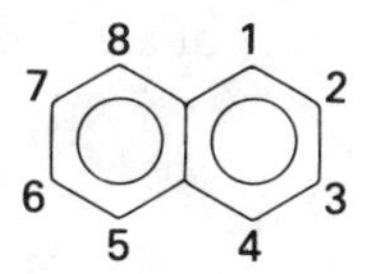

Naphthalen-2-ol is also known as β-naphthol.

Chromophores

The bright colours of these azo compounds are due to the absorption of light of particular wavelengths by the group —N≡N—. Such a group is known as a chromophore. Other chromophores include —C≡C—, C≡O, —NO₂ and —NO.

This reaction is used as a test for primary arylamines. The material to be tested is first treated with dilute hydrochloric acid and sodium nitrite at 0–5°C in order to convert it to the diazonium salt.

Coupling with amines. Diazonium salts couple with primary, secondary and tertiary arylamines. In acidic solution the coupling occurs at the 4-position of the benzene ring of the amine. The reactions are thus examples of C-substitution

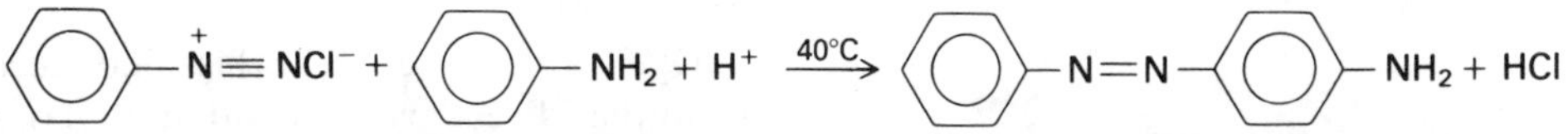

phenylamine

4-(phenylazo) phenylamine
(orange precipitate)

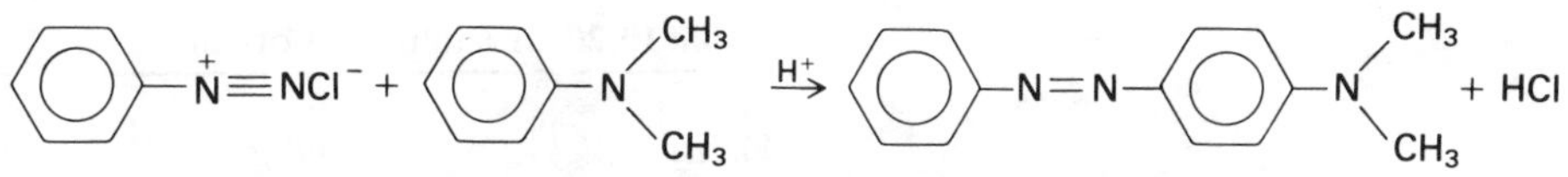

N,N-dimethylphenylamine   4-(phenylazo)-N,N-dimethylphenylamine
(green precipitate)

MANUFACTURE AND USES

Manufacture

Some **primary aliphatic amines** are produced by the reduction of nitroalkanes.

Methylamines and the **ethylamines** are manufactured by passing the alcohol and ammonia under pressure over a catalyst such as aluminium oxide:

$$CH_3OH + NH_3 \xrightarrow[400°C]{Al_2O_3} CH_3NH_2 + H_2O$$

$$CH_3NH_2 + CH_3OH \longrightarrow (CH_3)_2NH + H_2O$$

Aliphatic amines are also prepared by the reactions between halogenoalkanes and ammonia.

Phenylamine is prepared by the reduction of nitrobenzene.

Uses

Dyes and Pigments

The use of natural dyes such as indigo has been documented since 3000 BC. In Europe, the textile dyeing industry started in the sixteenth century with the use of indigo. In 1856 Sir William Henry Perkin discovered the dye aniline mauve.

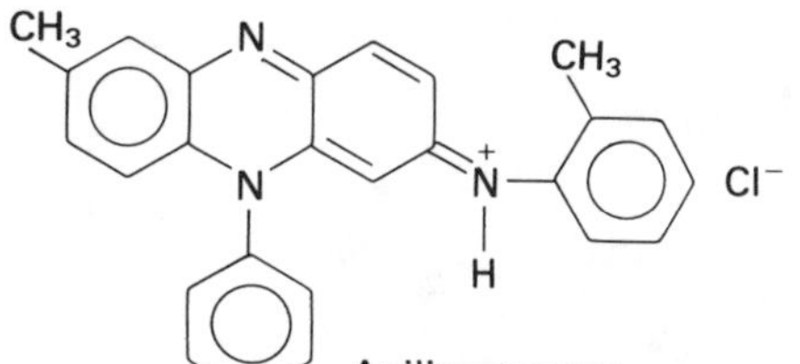

Aniline mauve

At the time he was studying phenylamine (also known as aniline). This was a coal tar derivative. He later opened a factory to produce the substance. The first natural dye to be produced synthetically was alizarin. This dye is found in cochineal and was first produced in 1868. In 1880 indigo was also synthesised.

Dyes chemically bind to the material which they colour. **Pigments**, on the other hand, do not chemically bind to the material which they colour. Many organic dyes and pigments contain amine groups or are derivatives of azobenzene:

Azobenzene

Dyes are sometimes classified according to their chemical structure. For example, direct green B and methyl orange (*see* table 19.20) are examples of azo dyes. Alizarin is an anthraquinone dye. Dyes with the indigo structure are known as indigoid dyes. Aniline mauve is an oxazine dye and crystal violet a triaryl-methane. There are other classes as well.

Table 19.20   Examples of organic dyes

Vat dye	indigo	
Mordant dye	alizarin	
Direct dye	direct green B	
Disperse dye	disperse red 9	
Acid dye	methyl orange	
Basic dye	crystal violet	

Acid dyes are used to dye nylon fabrics

Batch reactors used for making fine chemicals

Fine and Heavy Chemicals

Dyes and antioxidants are manufactured by the fine chemicals industry. **Fine chemicals** are those produced in relatively small quantities—usually no more than tens or hundreds of thousands of tonnes annually. Fine chemicals also include pesticides, pharmaceuticals and photographic chemicals. **Heavy chemicals** are those manufactured in large quantities—in millions of tonnes annually. Sulphuric acid and ammonia are heavy chemicals.

Dyes are more usually classified by their manner of application to the fabric.

Vat dyes. These are very fast. A dye is **fast** if it is not affected by conditions of use such as temperature, moisture and light. Vat dyes are insoluble in water. Before application to a fabric they are reduced in a vat to a water-soluble form. The fabric is then dyed and exposed to air or an oxidising agent. On oxidation, the dye becomes insoluble. Indigo is a vat dye. It is used to dye cotton. Indigo manufacture has undergone rapid growth in recent years to meet the demand for blue jeans.

Mordant dyes. These require a mordant such as alum in order to become attached to the fabric. Alizarin is an example of a mordant dye.

Direct dyes. These can be applied directly to the fabric without a mordant. An example is direct green B.

Disperse dyes. These are insoluble in water. They are applied as near-colloidal aqueous dispersions. Disperse red 9 is an example of such a dye. Disperse dyes are used to dye polyester fibres.

Acid or anionic dyes. These are generally sodium salts of sulphonic acids. They are used to dye nylon, wool and silk. An example is methyl orange.

Basic or cationic dyes. These generally contain a quaternary ammonium group. They are used to dye cotton, silk and polyacrylonitrile fibres. Crystal violet is an example.

Stabilisers

Amines are also used as stabilisers. Stabilisers prevent or retard deterioration of a substance. They are used extensively in petrol, food, cosmetics and synthetic polymers. Since deterioration is usually due to oxidation, stabilisers are normally **antioxidants**.

N-Phenylnaphthalen-1-amine is an important amine antioxidant. This is used in synthetic rubbers—in tyres, for example—at concentrations of 0.5 to 2%. It is inexpensive due to the simplicity of its manufacture.

naphthalen-2-ol   +   phenylamine $\xrightarrow{\text{heat}}$ N-phenyl-naphthalen-1-amine

Drugs

Amines are used widely as drugs. Antihistamines are examples.

$$H—C=C—CH_2—CH_2—NH_2$$

Histamine

Histamine is found naturally in nearly all body tissues. It is also released in allergic conditions such as hay fever. Antihistamines are used to relieve such allergic reactions. Some antihistamines are shown in table 19.21.

Table 19.21 Antihistamines

Type of amine	Example	
	Chemical name	Trade name
ethanolamine	diphenhydramine	Benadryl
ethenediamine	mepyramine	Anthisan
propylamine	chlorpheniramine	Piriton

Amines and their derivatives are also used as tranquillisers, analgesics and bactericides. They are also used as drugs to treat tropical diseases such as trypanosomiasis (also known as sleeping sickness) and malaria. Three examples are shown in table 19.22.

Table 19.22 Examples of amines used as drugs

chlorpromazine hydrochloride (tranquilliser)

pethidine hydrochloride (analgesic)

4–aminobenzenesulphonamide (sulphanilamide) (bactericide)

Other Uses

Pesticides. Amines are used as starting materials in the manufacture of certain pesticides. For example, the toxic methyl isocyanate which is used to make pesticides (*see* story at beginning of chapter) is made from methylamine and another very toxic compound known as phosgene:

$$CH_3—NH_2 \ + \ COCl_2 \ \rightarrow \ CH_3N=C=O$$

methylamine   carbonyl dichloride (phosgene)   methyl isocyanate (MIC)

Plastics. Amines are used to manufacture plastics such as nylon and polyurethane (*see* chapter 20).

Malaria-carrying mosquito (*Anopheles stephensi*) puncturing the skin of its victim

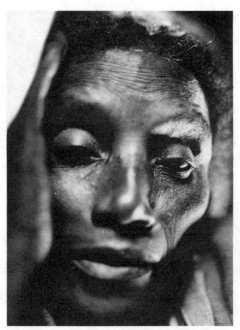

The face of malaria in Mexico

Victim of sleeping sickness in the Cameroons

SUMMARY

1. Typical reactions of the amines and diazonium salts.

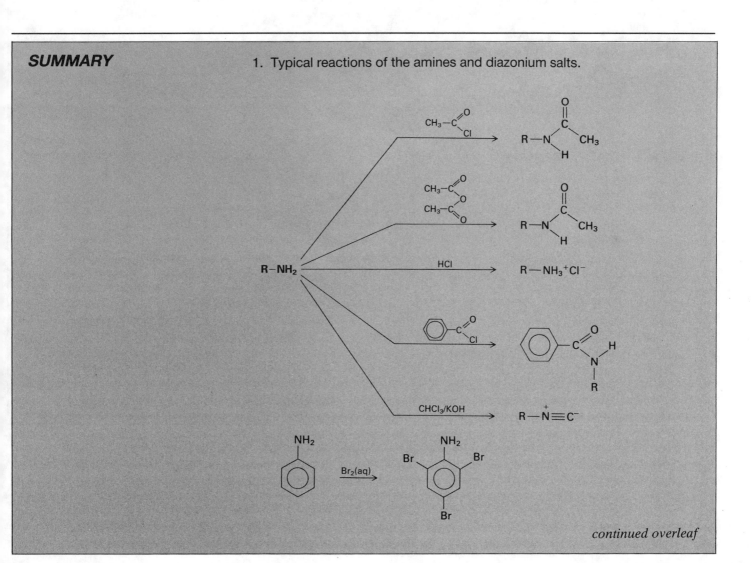

continued overleaf

Reactions with nitrous acid, HONO

Diazonium salts

Examination Questions

1. It may be said that 'organic chemistry is the chemistry of functional groups'. Discuss this statement by comparing and contrasting the reactions of
   (a) the OH groups in C_6H_5COOH and C_6H_5OH,
   (b) the unsaturated carbon–carbon bonds in $CH_2=CHCH_3$ and C_6H_6,
   (c) the Cl groups in CH_3COCl and CH_3CH_2Cl.

   (JMB)

2. The chemical properties of a functional group depend on whether it is attached to an alkyl group, an acyl group or an aryl group'. Discuss this statement with reference to the —OH, —Cl and —NH_2 groups.
   Explain any differences in properties as far as you can in terms of the electron distribution in the molecules concerned.

   (UCLES)

3. Write an account of the chemistry of halogenoalkanes.
   You should consider the reactions suitable for their preparations and their characteristic reactions. You should also refer to some industrial applications and to the mechanism of at least one of their reactions.

   (L, Nuffield)

4. State briefly, giving reaction conditions and writing equations, how you would replace the chlorine atom in 1-chlorobutane with the groups —OH, —NH_2 and —CN.
   Give a mechanism for one of the reactions, and indicate in what ways it differs from the mechanism for the corresponding reaction with the isomer, 2-chloro-2-methylpropane.
   How may the following conversions be carried out?
   (a) $CH_3CH_2CH_2CH_2Br$ into $CH_3CH_2CHBrCH_3$.
   (b) $CH_2=CHCl$ into $\{CH_2CHCl\}_n$.
   Mention any two uses of the product in (b).

   (L)

5. (a) Show, by giving equations, conditions and reaction types, how the following conversions may be effected:
   (i) iodoethane into ethanol,
   (ii) 2-bromopropane into propene,
   (iii) iodoethane into ethyl propanoate.
   (b) Give *one* chemical test which would distinguish between the members of each of the following pairs of compounds:
   (i) 1-chloropropane and ethanoyl (acetyl) chloride,
   (ii) 2-chloropropane and 2-bromopropane,
   (iii) 1-chloropropane and chlorobenzene.
   In each case give the reagents and conditions for the reactions you describe.

   (JMB)

6. This question concerns the chemistry of compounds containing hydroxyl groups.
   (a) Show by means of equations (with essential conditions and reagents) how the following transformations may be affected:
   (i) ethane to ethanol,
   (ii) benzene to phenol.

    (b) State the reagents you would use to convert:
       (i) butan-l-ol to 1-bromobutane,
       (ii) butanoic acid to butanoyl chloride.
    (c) If you were provided with a solution containing a mixture of phenol and benzoic acid in ethoxyethane (diethyl ether) how would you isolate the phenol and the acid?

(O & C)

7. Outlined below are two commercial routes for obtaining 'Aspirin' and 'Disprin' from phenol:

    (a) Of the alternative routes, which is the more economic, and why?
    (b) Explain why A, B and C are all sparingly soluble in water but soluble in aqueous sodium hydroxide. What would be observed if excess acid is added to the alkaline solutions? Why is benzaldehyde insoluble in cold alkali?
    (c) Describe specific tests to identify
       (i) the —OH group in A;
       (ii) the —CHO group in B;
       (iii) the —COOH group in C.
    (d) Name the reagent for carrying out the conversions
       (i) B → C;
       (ii) C → D;
       (iii) D → E.
    (e) 'Disprin' is said to be more effective for relieving pain than 'Aspirin' because it is more soluble in water. Why is this so?
    (f) Describe the reaction of phenol with chlorine. In what way is the product useful medicinally?

(SUJB)

8. This question is concerned with reactions of simple carbonyl compounds, principally aldehydes and ketones.
    (a) Give the structure of the products and state what type of reaction occurs between the following compounds (for example, the type of reaction between $CH_2=CH_2$ and Br_2 is an *addition* reaction; the mechanism is *not* required):
       (i) CH_3COCH_3 and HCN;
       (ii) CH_3COCl and NH_3.

(b) The iodoform reaction, in the case of propanone, can be described broadly as taking place in two stages:

$$CH_3COCH_3 \xrightarrow{A} CH_3COCI_3 \xrightarrow[(NaOH)]{B} CH_3COONa + HCI_3$$

In what way(s) might reaction B be said to be unusual?

(c) (i) Give the structure of the product obtained from ethanal (acetaldehyde) by the aldol reaction.

(ii) Indicate, with a brief statement of your reasoning, which of the following molecules would undergo the aldol reaction when treated with dilute aqueous alkali: CH_3COCH_3; CH_3COCI_3; $(CH_3)_3CCHO$.

(iii) What is the effect of *concentrated* aqueous alkali on ethanal (acetaldehyde)?

(iv) What is the importance of the aldol reaction?

(O & C)

9. This question is concerned with the properties of ethanal (acetaldehyde), a simple aldehyde.

(a) Give a reagent, or set of reagents, which will convert ethanal into:

(i) CH_3CO_2H,

(ii) $CHI_3 + HCO_2Na$,

(iii) $CH_3CH(OH)_2$.

(b) Give balanced equations for reactions (a) (ii) and (iii) above.

(c) Give the structure of the products obtained when ethanal reacts with:

(i) 2,4-dinitrophenylhydrazine,

(ii) hydrogen cyanide,

(iii) dilute, aqueous alkali.

(O & C)

10. Review the chemistry of organic acids and bases, considering their physical properties, preparation and reactions. Quote formulae and equations wherever appropriate. You should also consider their relative strengths and make a comparison with *appropriate* inorganic substances.

(L, Nuffield)

11. (a) For each of the following parts (i) to (x) *one or more* of the alternatives A, B, C, D listed below is (are) correct. Decide which of the alternatives is (are) correct and write the appropriate letter(s) in answer to the question.

A: $CH_3CH_2CH_2CHO$ B: $CH_3CH_2COCH_3$

C: $CH_3COOCH_2CH_3$ D: CH_3CH_2COOH

(i) Which will be completely miscible with water?

(ii) Which gives an acidic solution in water?

(iii) Which will react with hydroxylamine (NH_2OH)?

(iv) Which is obtained by the reaction between an acid and an alcohol?

(v) Which is obtained by oxidising a secondary alcohol?

(vi) Which, on refluxing with aqueous sodium hydroxide, will give two *different organic* products?

(vii) Which compounds are isomeric?

(viii) Which polymerises with alkali?

(ix) Which can react with an alcohol to form an ester?

(x) Which can form hydrogen bonds?

(b) Name the class of organic compounds to which A, B, C and D belong.

(c) For the sequence:

$$A \xrightarrow{X} CH_3CH_2CH_2 - \underset{\underset{CN}{|}}{\overset{\overset{H}{|}}{C}} - NH_2 + H_2O$$

$$\downarrow Y \text{ (Hydrolysis)}$$

$$E$$

(i) Name the reagents X and Y and complete the structure of E.
(ii) Name the class of compound to which E belongs and say why this class of compound is important biochemically.
(iii) By using two structural formulae of E show how they could enter into a condensation reaction and write the structural formula of the product.

(SUJB)

12. (a) What is the effect of each of the following reagents upon primary amines?
    (i) hydrogen chloride,
    (ii) ethanoyl (acetyl) chloride,
    (iii) iodomethane (methyl iodide),
    (iv) nitrous acid.
(b) On the basis of their reactions with reagents (ii) and (iii), to what mechanistic class of reagent do amines belong?
(c) To what classes of compound do the products from (i), (ii), and (iii) belong?
(d) Indicate in each case in (a) how, if at all, the original amine may be regenerated from the product in a single reaction.

(O & C)

13. A compound, A, C_7H_5N, reacts with lithium tetrahydridoaluminate ($LiAlH_4$) to give B, C_7H_9N, which is soluble in dilute hydrochloric acid, but not in water.

Treatment of B with cold, freshly prepared aqueous nitrous acid gives a gas (one mole for every mole of B), and when the resultant solution is added slowly to alkaline solutions of phenols, no coloured precipitates are produced.

From the solution obtained from B and nitrous acid, C, C_7H_8O, can be isolated which, on oxidation with aqueous sulphuric acid and sodium dichromate(VI), gives D, $C_7H_6O_2$.

Compound D is also obtained by acidifying the solution given by boiling A with aqueous sodium hydroxide.

Treatment of D with phosphorus pentachloride gives E, C_7H_5OCl, which reacts with C to give F, $C_{14}H_{12}O_2$, and with benzene in the presence of aluminium chloride to give G, $C_{13}H_{10}O$. G gives a crystalline derivative with hydroxylamine.

Identify compounds A to G inclusive, giving your reasoning and accounting for every reaction described.

(O & C)

14. (a) Listed below are *types* of organic reaction A to H and numbered conversions 1 to 8. Write down the letters A to H and append to each letter the numbered conversion which is most appropriate to the reaction type. (*Each number must be used only once.*)

Type	Conversion
A Disproportionation	1 CH_3OH to $CH_3OCH_2CH_3$
B Oxidation	2 $nCH_2{=}CH_2$ to $(CH_2CH_2)_n$
C Isomerisation	3 $CH_3CH{=}CHCH_3$ to $CH_3CH_2CH{=}CH_2$
D Polymerisation	4 $CH_3CH_2CH_2Br$ to $CH_3CH{=}CH_2$
E Elimination	5 C_6H_6 to $C_6H_5CH_3$
F Hydrolysis	6 $2C_6H_5CHO$ to $C_6H_5CH_2OH + C_6H_5COOH$
G Etherification	7 $CH_3CH_2CH_2OH$ to CH_3CH_2CHO
H Alkylation	8 $H_2NCH_2CONHCH_2COOH$ to $2H_2NCH_2COOH$

(b) Name the reagents in the following conversions:
   (i) C_6H_6 to $C_6H_5CH_3$,
   (ii) $CH_3CH_2CH_2Br$ to $CH_3CH{=}CH_2$,
   (iii) $2C_6H_5CHO$ to $C_6H_5CH_2OH + C_6H_5COOH$,
   (iv) $CH_3CH_2CH_2OH$ to CH_3CH_2CHO.

(c) How, by means of *one* chemical test in each case, would you distinguish between the following pairs? Quote the reagents and describe the *observations* which would enable you to decide firmly:
   (i) $CH_3CH_2CH(OH)CH_3$ and $(CH_3)_3COH$;
   (ii) CH_3COCH_3 and $CH_3CH_2COCH_2CH_3$.

(SUJB)

15. Describe and briefly explain what happens in **each** of the following experiments and write balanced equations for the reactions that occur.
   (a) Propan-2-ol is warmed with acidified aqueous potassium dichromate(vi).
   (b) Ethanedioic acid is heated with concentrated sulphuric acid.
   (c) Aqueous bromine is added to aqueous phenol.
   (d) Phenylethene (a liquid) is allowed to stand in the air for some time.
   (e) Cold, aqueous sodium nitrite is slowly added to a solution of phenylamine in hydrochloric acid at 5°C and the resulting mixture is poured into an alkaline solution of phenol.

(UCLES)

16. How, if at all, and under what conditions does aqueous sodium hydroxide react with
   (a) ammonium ethanoate,
   (b) ethanamide,
   (c) but-2-ene,
   (d) phenol,
   (e) a typical fat?
   What role does the hydroxide ion play in reactions (a) and (b)?

(O & C)

17. How would you distinguish *chemically* between the following compounds? For each pair, give the reagents, conditions, and observations for each compound. (Equations are *not* necessary in this question.)
   (a) C_2H_5Br and C_2H_5I;
   (b) Hydroxybenzene, C_6H_5OH, and hexan-1-ol, $C_6H_{13}OH$;
   (c) $CH_3CH_2CONH_2$ and $CH_3CONHCH_3$;
   (d) But-1-ene and but-2-ene.

(SUJB)

18. (a) (i) Give suitable reagent(s) for **each** of the conversions in A to E, Use a *different* reagent in each case.

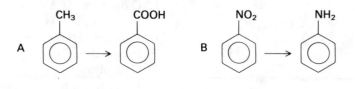

C CH₃COCH₃ ⟶ CH₃CHOHCH₃   D CH₃CH=CH₂ ⟶ CH₃CH—CH₂

E CH₃CH=CH₂ ⟶ CH₃CH₂CH₃

   (ii) Give the structure(s) of the intermediates formed in the transformation in A.

(b) What major difficulty can arise in the preparation of CH₃CHO by the oxidation of CH₃CH₂OH, and how is this difficulty overcome?

(JMB)

20 NATURAL PRODUCTS AND POLYMERS

'Superbugs' Start to Buzz

It's a field ranging from the outer limits of science fiction to the homely and mundane—from research into human cloning to making cheese. It involves both happy chance (the discovery of penicillin) and warnings from no less an agency than Porton Down's Microbiological Research Establishment that experiments could trigger a disaster on the scale of a twentieth-century Black Death. It's biotechnology.

This new 'buzz' word covers a science which the Sumerians used to make 19 different varieties of beer in 3000 BC, and which promises to be for the 1980s what

Biotechnology also involves cell culture

An industrial fermenter. Biotechnology includes fermentation processes

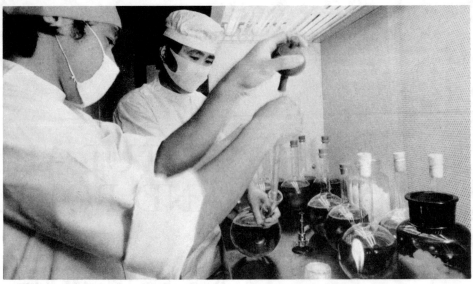

Japanese scientists working on interferon—the so-called 'wonder-drug' for cancer

the silicon chip was to the 1970s—the great white hope for our industrial future. This hope may or may not be fulfilled—we won't know until at least the 1990s—but it is a field where large oil and pharmaceutical companies are putting increasing sums of research investment because they dare no longer ignore the possibilities.

Biotechnology includes genetic engineering—manipulating the DNA, or genetic coding system, in cells to produce new cells with chosen properties.

It also covers the use of enzymes as catalysts to increase chemical reaction rates—recently the discovery of the techniques for using 'restriction enzymes' has made it possible for scientists to control the splicing of DNA with much greater precision. This cuts down the dangers of Porton Down's unplanned 'biohazards'.

Biotechnology includes, too, the kind of cell culture and fermentation processes which have been with us for years in making champagne, beer, or penicillin, but which now give hints of huge commercial possibilities.

Some claims for these possibilities—for example, transforming the world's deserts into fertile food producers within 20 years with solar bacterial biomass plants—may sound far-fetched. But in the field of health care at least, biotechnology is already breaking through with commercial possibilities.

Five British hospitals are testing insulin made by genetic engineering, and a multi-million-pound race is on between the Danish Novo Industri and the American Eli Lilly—the latter with a genetic engineering product—to market synthetic insulin in European and American markets worth at least £100 million a year.

Interferon, the so-called wonder-drug for cancer, is now made commercially through biotechnology, and genetic engineering has made the hormone that eliminates dwarfism. The cloning of antibodies promises an end to hepatitis.

This area of genetic engineering is a field where the Americans, with their close contacts between universities and commercial companies, are well ahead of the rest.

In other applications of biotechnology than health, the possibility of commercial benefits is finely balanced.

Three fields are capable of exploitation within 10–15 years: energy, chemicals, and plant-breeding for agriculture.

With chemicals, one problem to be overcome—apart from the fine balance between the comparative production costs of conventional and biotechnological methods—is that the chemical produced often kills the biocatalyst.

On the agricultural side, British biotechnology is ahead of the rest of the world through the Government-funded Agricultural Research Council. The effort is concentrated on increasing yield, and though there is no gene that codes for yield, there are some that code for resistance to disease: we may soon be able to create plants immune from debilitating diseases and thus increase crops.

The other area under research is the use of tissue culture to make plants grow from single identical cells. This plant cloning is not yet at the commercial stage, but Unilever is working on oil palm and hopes soon to be able to increase yields to commercially viable levels.

Biotechnology development in energy in Europe and America will centre on making gas and liquid fuel from coal, and possibly on improving the viscosity of water to give it greater pushing power when it is pumped into oil wells to force oil to the surface. This could have great significance in the 1990s for harvesting 'difficult' oil fields, including those off North-West Scotland, which are so far not viable.

Alcohol fuels from sugar cane or trees, however, will be developed mainly in the Third World, though Alcon Biotechnology—an alliance between Allied Breweries and the John Brown Group—is using biotechnology for this purpose and has taken a demonstration plant to the Philippines.

Brazil already runs significant numbers of its cars on ethanol made from sugar cane—but its government made it cheap to buy alcohol cars and then reduced the subsidy on the fuel, so the savings are so far questionable. Zimbabwe, when it was a siege economy, made tractor fuel from sunflower oil, and South Africa has a liquids-from-coal programme as a hedge against being cut off from other sources of energy.

Biotechnology in most countries is closely connected with government, universities and private sector industry in concert—and each country is proceeding where it sees its particular needs. The Australians have developed a protein to make the wool fall off sheep to cut shearing costs, for example.

Jane McLoughlin

LEARNING OBJECTIVES

After you have studied this chapter you should be able to

1. Give the names and draw the structures of simple **amino acids**.
2. Explain the terms
   (a) **essential amino acids**,
   (b) **zwitterion**,
   (c) **isoelectric point**.
3. Outline how amino acids may be separated and detected.
4. Draw a **peptide link** and give an example of a **polypeptide**.
5. Distinguish between the terms **globular protein** and **fibrous protein**.
6. Explain the term **prosthetic group** and give an example.
7. Outline how proteins may be separated and detected.
8. With reference to **proteins**, distinguish between the terms primary **structure**, secondary structure, tertiary structure and quaternary structure.
9. Indicate how peptides may be synthesised.
10. Distinguish between the terms **monosaccharide, disaccharide** and **polysaccharide** and give an example of each.
11. Draw the structure of a typical **aldose** and **ketose**.
12. Draw (a) the **Haworth projections** and (b) the **chair conformations** of both **anomers** of D-glucose.
13. Explain the term **mutarotation** and indicate how it can be measured.
14. Outline the chemical properties of **carbohydrates**.
15. Explain the term **reducing sugar** and give an example.
16. Outline how sugars may be detected and characterised.
17. Give an account of the importance of **nucleic acids** in living organisms.

(cont'd)

18. Distinguish between the terms **nucleotides** and **nucleosides**.
19. Write a brief account on **replication** and **protein biosynthesis**.
20. Explain the term **lipid**.
21. Give examples of naturally occurring **fats** and **oils**.
22. Distinguish between the terms **saturated fatty acid** and **unsaturated fatty acid**.
23. Write brief accounts on each of the following:
    (a) **soaps** and **detergents**,
    (b) **waxes**,
    (c) **phospholipids** and **glycolipids**,
    (d) **vitamins**,
    (e) **alkaloids**.
24. Give examples of naturally occurring **polymers**.
25. Distinguish between the terms polymer and **plastic**.
26. Distinguish between and give examples of **thermoplastics** and **thermosetting resins**.
27. Write brief accounts on
    (a) **addition polymerisation**,
    (b) **condensation polymerisation**,
    (c) **natural rubber** and **synthetic rubber**,
    (d) **cellulose fibres**.

20.1 Amino Acids, Peptides and Proteins

NATURAL PRODUCTS

The term **natural products** is usually applied to the organic compounds which naturally occur in living organisms. They fall into a number of important classes. The most important of these are shown in table 20.1.

Many naturally occurring compounds have very high relative molecular masses. Molecules with a relative molecular mass in excess of 10 000 are known as **macromolecules**. Macromolecules are essential to all life processes. Enzymes, haemoglobin, starch and DNA are all biological macromolecular compounds. Many of these macromolecules are built up from smaller molecules to form long chains. During this process water molecules are eliminated. The process thus involves **condensation** reactions.

If the smaller molecules are all the same or if several different smaller molecules are repeated in a regular pattern along the chain then the macromolecule is known as a **polymer**. The smaller molecules from which they are constructed are known as **monomers**. Both natural and synthetic polymers can have relative molecular masses of up to several million. Cotton, rubber, wood and wool are all naturally occurring polymeric materials. Poly(ethene) and poly(chloroethene) (also known as polythene and polyvinyl chloride (or pvc) respectively) are examples of synthetic polymers.

The proteins are a major group of natural products. They occur in all living plant and animal tissues. The flesh of all humans, animals, birds and fish is made of protein. So are yeasts, moulds and bacteria. Smaller amounts of protein also occur in leaves, seeds, stems and storage organs of plants. Proteins are essential for the maintenance of the structure and proper functioning of all living organisms. For example, enzymes are proteins which catalyse the chemical reactions that take place in living organisms (*see* section 9.2).

Table 20.1 Natural products

Class	Examples
proteins	enzymes and hormones
amino acids	glycine
carbohydrates	glucose, sucrose, cellulose, starch and glycogen
nucleic acids	DNA and RNA
lipids	oils, fats and waxes including the terpenes and steroids

Amino acids contain primary amine groups. Glycine is an example. Proline, which also occurs in proteins, has a secondary amine group as part of a heterocyclic ring. It is an amino acid.

Twelve of the 20 amino acids occurring in proteins can be synthesised by the human body. The remaining eight must be consumed in our diet. They are known as **essential amino acids**.

Proteins are polyamides. Their basic units are the **amino acids**. Proteins may be regarded as polymers of amino acids. The amino acids are joined together by **peptide links**. A peptide link is formed by the elimination of a water molecule.

AMINO ACIDS

The amino acids are bifunctional compounds containing both acidic carboxyl groups and basic amine groups. The amino acids in proteins are mostly α-**amino acids**. These are amino acids in which both the carboxyl and amino groups are attached to the same carbon atom. The simplest example is glycine:

$$H_2N-\underset{\underset{H}{|}}{\overset{\overset{H}{|}}{C}}-COOH$$

The α-carbon atom in all α-amino acids except glycine is asymmetric. All the α-amino acids except glycine thus exhibit **chirality** (*see* section 17.2) and thus **optical isomerism**.

Optical isomers of alanine

The amino acids in proteins are all L-isomers.

Twenty amino acids (18 α-amino acids and two imino acids) commonly occur in proteins. The amino acids differ from one another by their side chains. When writing out the structures of proteins or polypeptides, it is usual to use an abbreviaton of three letters for each amino acid (*see* table 20.2). The names given in table 20.2 are the original names and they are the ones commonly used. However, each amino acid also has a systematic name. For example, the systematic name of glycine is aminoethanoic acid and that of alanine is 2-aminopropanoic acid.

Table 20.2 Some amino acids: general formula

$$H_2N-\underset{\underset{|}{\overset{|}{CH}}}{}-COOH \quad (R \text{ above } CH)$$

Amino acid	Abbreviation	R—
glycine	Gly	H—
alanine	Ala	CH_3—
valine	Val	$(CH_3)_2CH$—
leucine	Leu	$(CH_3)_2CHCH_2$—
serine	Ser	$HO-CH_2$—
tyrosine	Tyr	$HO-\langle\bigcirc\rangle-CH_2$—
aspartic acid	Asp	$HOOC-CH_2$—
glutamic acid	Glu	$HOOC-CH_2CH_2$—
cysteine	Cys	$HS-CH_2$—
asparagine	Asn	$\underset{H_2N}{\overset{O}{\underset{\diagdown}{\overset{\parallel}{C}}}}-CH_2$—
lysine	Lys	$H_2NCH_2CH_2CH_2$—

Amino acids may be classified as **neutral**, **basic** or **acidic** depending on how many amine and carboxyl groups they have. For example, glycine and alanine are both neutral since they each have one amine group and one carboxyl group. Aspartic acid, which has two carboxyl groups, is acidic. On the other hand lysine, which has two amine groups, is basic.

Amino acids can exist in various ionic forms in aqueous solution depending on pH and on whether they are neutral, acidic or basic. For a neutral amino acid the following species may exist in aqueous solution depending on pH:

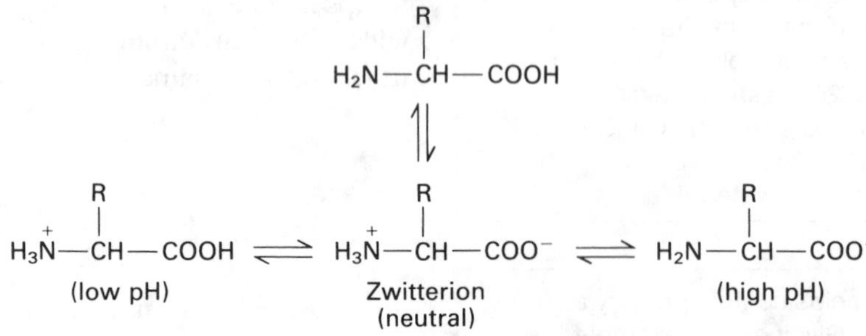

The form in which a positive and negative charge are both present is known as a **zwitterion**. This is the predominant form in neutral solution. The pH at which the amino acid has no net charge is known as the **isoelectric point**. Each amino acid has a characteristic isoelectric point. For many α-amino acids the value lies a little above or below 6.00. For example, the isoelectric points of glycine and alanine are 5.97 and 6.02 respectively.

Separation and Detection of Amino Acids

A mixture of amino acids can be separated by

- paper chromatography (*see* section 6.3),
- ion exchange chromatography (*see* section 8.2),
- electrophoresis (*see* section 9.1).

Amino acids can be detected using a compound known as **ninhydrin**:

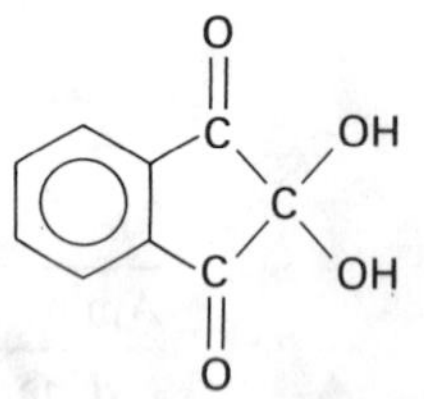

Ninhydrin oxidises amino acids forming coloured products. Most α-amino acids give a blue or purple colour. Proline gives a yellow colour.

PEPTIDES

Two amino acids can link together with the elimination of a water molecule to form a **dipeptide**. The group which links the two amino acids together is known as a peptide link. This is an amide group:

$$H_2N-CH-C\overset{\displaystyle O}{\underset{(O-H}{\overset{\diagup}{\diagdown}}} \quad\overset{+}{}\quad \overset{H}{\underset{H)}{N}}-\overset{CH_3}{\underset{}{CH}}-COOH$$

Glycine Alanine

↓

$$H_2N-CH-C-N-CH-COOH + H_2O$$

Peptide link

We can represent the formation of this link using the abbreviations for amino acids as follows:

NH$_2$—Gly—COOH + NH$_2$—Ala—COOH ⟶ NH$_2$—Gly—Ala—COOH + H$_2$O

Peptide link

Dipeptide

In this type of notation it is conventional to start with H$_2$N— and end with —COOH.

A **tripeptide** contains three amino acids linked together. For example

H$_2$N—Gly—Lys—Ala—COOH

A **polypeptide** contains several amino acids linked together. For example

H$_2$N—Tyr—Lys—Gly—Asn—Leu—Gly—COOH

Polypeptides are thus polyamides. The amino acids which form a polypeptide chain are often known as amino acid residues. A polypeptide chain containing repeating sequences of amino acid residues may be regarded as a polymer.

PROTEINS

Proteins are naturally occurring polypeptides with high relative molecular masses. These may range from 10 000 or so to several millions.

Proteins which consist solely of α-amino acids as their basic units are known as **simple proteins**. These are often classified on the basis of their solubility in water.

Globular proteins are simple proteins which are either soluble in water or form colloidal solutions with water. They are all soluble in dilute aqueous solutions of acids, bases and salts but precipitated by concentrated salts and acids and organic solvents. Albumin, which is a globular protein found in egg white, is soluble in water. Protein hormones such as insulin, blood proteins and enzymes are all globular proteins.

The solubility and biological activity of globular proteins are reduced when they are subjected to high temperatures or extremes of pH. This is known as **denaturation**. During this process the proteins change from a regular folded structure (*see* below) to a random irregular structure. This occurs on boiling an egg, for example. Although denaturation is normally irreversible, it does not involve breaking the polypeptide chain.

Fibrous proteins are simple proteins which are insoluble in water. Since they are found in skin, hair and muscle they are sometimes known as **structure proteins**. Examples include:

- the keratins—these are found in wool, hair and nails;
- the collagens—these occur in skin and connective tissue;
- fibroin—this is found in silk;
- elastins—these are found in lungs, arteries and some ligaments.

Fibrous proteins are found in skin, hair and muscle

Table 20.3   Prosthetic groups

Protein	Prosthetic groups
nucleoprotein	nucleic acid
lipoprotein	lipid
glycoprotein	carbohydrate
haemoglobin	porphyrin

Some proteins contain not only amino acids but also other groups. These non-amino acid groups are known as **prosthetic groups**. Some examples are given in table 20.3.

Separation, Detection and Physical Properties of Proteins

Proteins normally occur in mixtures. They may be separated by a technique known as **fractional precipitation**. This technique depends on the different solubilities in salt solutions or organic solvents such as propanone or ethanol. On the careful addition of one of these, certain proteins are precipitated whilst others remain in solution. Adjustment of the pH of an aqueous solution of proteins can also result in the selective precipitation of one or more of them.

Certain types of chromatography are unsuitable for the separation of proteins. For example, paper chromatography causes denaturation. However, both column chromatography and ion exchange chromatography can be used for their separation. Electrophoresis is also used on a small scale.

Proteins do not have sharp melting points since they decompose on heating. However, the purity, and also the relative molecular mass, of a protein can be determined by **ultracentrifugation**. Ultracentrifuges spin samples of the protein solution at a very high speed. The rate at which the proteins move to the end of the sample cell and form a sediment depends on their shape, density and relative molecular mass. The rate of sedimentation of pure proteins differs from that of impure proteins.

The presence of proteins in a solution can be detected by the **biuret test** (*see* section 19.3). In this test, an alkaline copper(II) solution is added to the solution. The presence of proteins is indicated by a purple colour. The biuret reaction can also be used to determine the concentration of proteins in solution. The intensity of the purple colour, which can be determined by UV spectroscopy (*see* section 17.1) is proportional to the concentration of the protein.

Since proteins contain large numbers of ionisable amino and carboxyl groups they are electrolytes. At low pH values the positively charged amino groups predominate and thus the proteins are cationic. At high pH values proteins are anionic. Like the amino acids proteins have characteristic isoelectric points (*see* table 20.4). Proteins are least soluble at their isoelectric point.

The pK_a values of amino acids may change once they become part of a peptide or protein chain. This is because the ionisation of an amino acid in a chain is influenced by the presence of neighbouring charged groups and other factors in its immediate environment.

Table 20.4   Isoelectric points of proteins

Proteins	Isoelectric point
pepsin	1.0
serum albumin	4.7
insulin	5.4
haemoglobin	7.2
ribonuclease	9.6
lysozyme	11.1

Structure of Proteins

The structures of proteins are very complex. It is therefore customary to consider four separate elements of protein structure. These are known as the primary, secondary, tertiary and quaternary structures.

Primary Structure
This is the sequence of covalently linked amino acid residues in the polypeptide. For example, ox insulin consists of two polypeptide chains consisting of 21 and 30 amino acid residues respectively (*see* figure 20.1).

The number and variety of amino acids in a protein or polypeptide can be determined by hydrolysis. This can be accomplished by heating proteins or polypeptides with a dilute acid such as hydrochloric acid. Partial hydrolysis yields a mixture of peptides and amino acids. Proteins yield a maximum of 20 amino acids. These can be separated by chromatography or electrophoresis. The amino acids are then identified by using a suitable colour reagent such as ninhydrin. Of

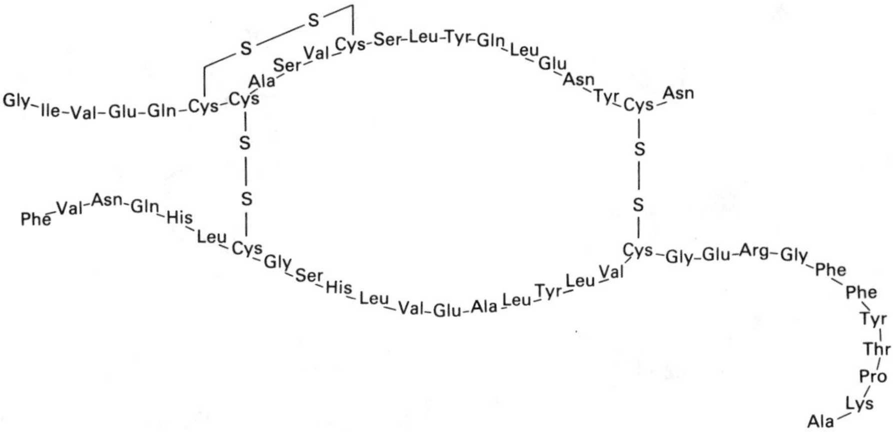

Figure 20.1 Ox insulin consists of two polypeptide chains with 21 and 30 amino acid residues respectively. The two chains are linked by disulphide bridges

Frederick Sanger

The technique of determining the amino acid sequence in proteins by partial breakdown of the proteins was pioneered by the English biochemist Frederick Sanger. In 1953 he determined the complete structure of the protein insulin. For this he was awarded the Nobel Prize for Chemistry in 1958. He also jointly won the 1980 Nobel Prize for Chemistry together with Paul Berg and Wally Gilbert for work on various aspects of DNA. Frederick Sanger was only the third person to win two Nobel Prizes in scientific disciplines.

course, some amino acids may occur repeatedly in the amino acid sequence of a polypeptide chain. The amounts of each amino acid present in the chain can be determined from the colour intensity. This can be measured by UV spectroscopy (*see* section 17.1).

The actual sequence of amino acids in a polypeptide chain can be determined by **end-group analysis**. The amino group at one end of the polypeptide chain is **labelled** using 2,4-dinitrofluorobenzene:

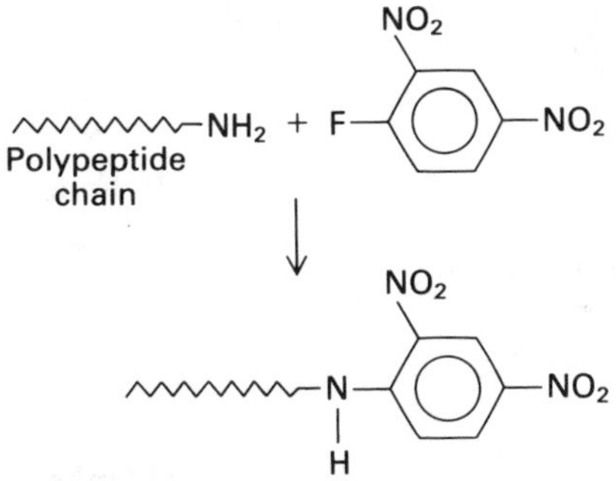

After hydrolysis of the protein, the labelled amino acid can be identified by means of paper chromatography. It appears as a yellow spot on the chromatogram. Certain selective chemical reagents and enzymes can remove one amino acid at a time from the end of the chain. However, this procedure becomes more difficult and more inaccurate the more the chain is shortened. It is more common to use a technique called **peptide mapping**. The polypeptide chain is broken down by a **cleaving reagent** to separate peptide fragments. These can be more easily 'sequenced'. By using two or more cleaving reagents a complete map of the polypeptide chain can be determined. The chain is cleaved by either 'chemical' hydrolysis or by use of an enzyme such as trypsin.

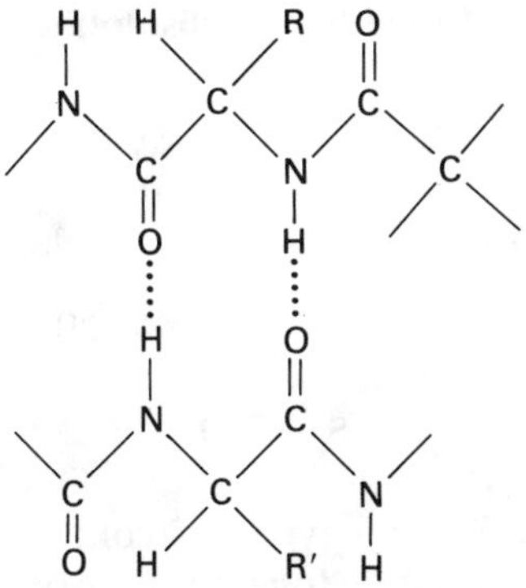

Secondary Structure
This is the regular folding of the 'backbone' of the polypeptide chain due to intramolecular hydrogen bonding between the carboxyl and amino groups (*see* figure 20.2). Intramolecular hydrogen bonding in globular proteins can result in a spiral structure known as the α-**helix** (*see* figure 20.3). In fibrous proteins this type of bonding can lead to the pleated sheet structure (*see* figure 20.4a). This type of structure is also obtained by intermolecular hydrogen bonding between different chains in a fibrous protein (*see* figure 20.4b).

Figure 20.2 Hydrogen bonding between carboxyl and amino groups

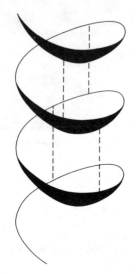

Figure 20.3 α-Helix

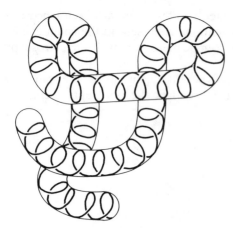

Figure 20.5 Folding of the α-helix

Hydrophobic means water-hating.
Hydrophilic means water-loving.

Cysteine and Cystine

Cysteine is an amino acid. Cystine consists of two cysteine units linked together by a disulphide bond.

(a)

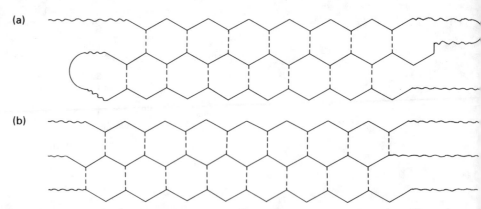

(b)

Figure 20.4 Pleated sheet structure: (a) hydrogen bonding within the same chain; (b) hydrogen bonding between different chains

Tertiary Structure

This is the three-dimensional folding of the polypeptide chain superimposed on the α-helical or pleated sheet regions of the chain. For example, the α-helix may fold back on itself (*see* figure 20.5). The tertiary structure is due to **crosslinking** between parts of the polypeptide chain. This cross linking is due to bonding and interactions between the side chains of the amino acid residues. This crosslinking can take a number of forms.

Hydrogen bonding between side chains.

$$\text{Ser}-CH_2-O-H\cdots O=\underset{\underset{OH}{|}}{C}-CH_2-\text{Asp}$$

Ionic bonding. Side chain crosslinking can occur as a result of ionic bonding between the anionic and cationic side chains. For example

$$\text{Glu}-CH_2-CH_2-\underset{\underset{O}{\|}}{C}-O^-\ ^+NH_2=\underset{\underset{NH_2}{|}}{C}-NH-(CH_2)_3-\text{Arg}$$

Hydrophobic bonding. In globular proteins about half the amino acid residues have hydrophobic side chains. As a result, proteins in aqueous solutions tend to fold so that the hydrophobic side chains become clustered inside the folds. The polar side chains, which of course are hydrophilic, are on the outside or surface of the molecule. In this way water is excluded from the interior of the molecule.

Covalent bonding. The most common form of interchain bonding is the disulphide bond formed between the sulphur atoms of two cysteine residues. Ox insulin consists of two polypeptide chains linked together by **disulphide bridges** (*see* figure 20.1).

$$\underset{\underset{COOH}{|}}{\overset{\overset{H}{|}}{H_2N-C}}-CH_2-SH \qquad \underset{\underset{COOH}{|}}{\overset{\overset{H}{|}}{H_2N-C}}-CH_2-S-S-CH_2-\underset{\underset{COOH}{|}}{\overset{\overset{H}{|}}{C}}-NH_2$$

Cysteine
(Cys)

Cystine
(Cys—S—S—Cys)

Covalent crosslinks also exist in collagen. Several types of collagen occur although all consist of three polypeptide chains twisting together to form a triple helix (*see* figure 20.6). Covalent crosslinks occur between lysine chains both within the triple helix and also between triple helices.

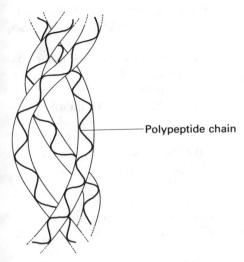

Figure 20.6 Collagen consists of three polypeptide chains twisting together in a triple helix. Up to 25% of the total protein in the bodies of mammals consists of collagen. It occurs under skin and in teeth, bone, tendon and cartilage

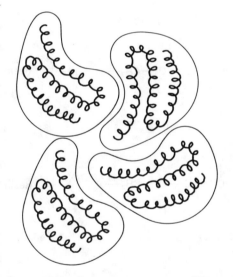

Figure 20.7 Quaternary structure. The identical or similar sub-units fit together in a symmetrical arrangement

Quaternary Structure

Some proteins, such as haemoglobin, have a quaternary structure. The quaternary structure is composed of several sub-units. Each sub-unit is a separate polypeptide chain. In some proteins, the sub-units are identical or similar whilst in others there may be several different types of chain. The sub-units fit together in a symmetrical arrangement (*see* figure 20.7). The quaternary structure is held together by the non-covalent bonds between side chains—as in the tertiary structure.

Synthesis of Peptides

The synthesis of peptides from amino acids has yielded much information about protein structure over recent decades. It is also possible nowadays to synthesise peptides with biological activity. Peptide synthesis has become of immense importance in **genetic engineering** and **biotechnology** (*see* reading at the start of this chapter).

As we saw above, amino acids have two functional groups—the carboxyl group and the amino group. These are known as the **C-terminal** and **N-terminal** respectively. In order to synthesise a dipeptide from two amino acids it is necessary to 'block' the C-terminal on one and the N-terminal on the other by use of a protecting group. This prevents reaction at the respective terminals. The peptide link between the two amino acids is then formed after which the terminals are unblocked. Peptide synthesis thus involves three stages.

1. Blocking.

$$X + H_2N-\underset{\underset{\text{R}}{|}}{CH}-COOH \longrightarrow \underset{\underset{\text{H}}{|}}{X}N-\underset{\underset{\text{R}}{|}}{CH}-COOH$$

$$H_2N-\underset{\underset{\text{R}'}{|}}{CH}-COOH + Y \longrightarrow H_2N-\underset{\underset{\text{R}'}{|}}{CH}-COOY$$

X and Y are two protecting groups.

2. Peptide formation.

$$\underset{\underset{\text{H}}{|}}{X}N-\underset{\underset{\text{R}}{|}}{CH}-COOH + H_2N-\underset{\underset{\text{R}'}{|}}{CH}-COOY \longrightarrow \underset{\underset{\text{H}}{|}}{X}N-\underset{\underset{\text{R}}{|}}{CH}-\underset{\overset{\text{O}}{||}}{C}-\underset{\underset{\text{H}}{|}}{N}-\underset{\underset{\text{R}'}{|}}{CH}-COOY + H_2O$$

3. Unblocking.

$$\underset{\underset{\text{H}}{|}}{X}N-\underset{\underset{\text{R}}{|}}{CH}-\underset{\overset{\text{O}}{||}}{C}-\underset{\underset{\text{H}}{|}}{N}-\underset{\underset{\text{R}'}{|}}{CH}-COOY \longrightarrow \underset{\underset{\text{H}}{|}}{X}N-\underset{\underset{\text{R}}{|}}{CH}-\underset{\overset{\text{O}}{||}}{C}-\underset{\underset{\text{H}}{|}}{N}-\underset{\underset{\text{R}'}{|}}{CH}-COOH + Y$$

The process can be repeated with further amino acids to form tripeptides and longer chain peptides:

$$\underset{\underset{\text{H}}{|}}{X}N-\underset{\underset{\text{R}}{|}}{CH}-\underset{\overset{\text{O}}{||}}{C}-\underset{\underset{\text{H}}{|}}{N}-\underset{\underset{\text{R}'}{|}}{CH}-COOH + H_2N-\underset{\underset{\text{R}''}{|}}{CH}-COOY$$

(1) Peptide formation
(2) Unblocking

$$\underset{\underset{\text{H}}{|}}{X}N-\underset{\underset{\text{R}}{|}}{CH}-\underset{\overset{\text{O}}{||}}{C}-\underset{\underset{\text{H}}{|}}{N}-\underset{\underset{\text{R}'}{|}}{CH}-\underset{\overset{\text{O}}{||}}{C}-\underset{\underset{\text{H}}{|}}{N}-\underset{\underset{\text{R}''}{|}}{CH}-COOH + Y$$

Bruce Merrifield, winner of the 1984 Nobel Prize for Chemistry

Numerous protecting groups are known for both C-terminals and N-terminals of the amino acids. Various methods of removing the protecting groups are also known.

Until the late 1950s and early 1960s peptide synthesis was a slow and laborious process. It was then revolutionised by the pioneering work of the American chemist Bruce Merrifield. In 1963 he published a method for the solid phase synthesis of peptides which has become known as the **Merrifield technique**. This involves attaching a blocked amino acid to a resin support. The resin was chloromethyl polystyrene. This is an insoluble gelatinous substance. The Merrifield technique for synthesising a dipeptide involves the steps shown in figure 20.8.

In 1964 Merrifield used this technique to synthesise a tissue hormone known as bradykinin. This contains nine amino acid residues. The Merrifield technique has since been used extensively in organic chemistry. It is also used in genetic engineering to synthesise short DNA sequences for structure determination.

In 1984, Merrifield won the Nobel Prize for Chemistry for his pioneering work on polymer supported peptide synthesis.

1. The C-terminal is attached to the resin. Merrifield used 'tertiary-butyloxycarbonyl' (or t-BOC) to block the N-terminal

2. The N-terminal is unblocked

3. The peptide link is formed with another amino acid with its N-terminal blocked

4. The dipeptide is detached from the resin and the N-terminal unblocked

Figure 20.8   The Merrifield technique for synthesising a dipeptide

SUMMARY

1. **Proteins** consist of **amino acids** joined together by **peptide links**.
2. Amino acids exhibit **optical isomerism**.
3. **Essential amino acids** cannot be synthesised in the human body. They must be consumed in our diet.
4. A **zwitterion** contains both a negative charge and a positive charge.
5. **Polypeptides** contain several amino acids linked together.
6. **Globular proteins** are either soluble in water or form colloidal solutions with water.
7. **Fibrous proteins** are insoluble in water.
8. **Fractional precipitation**, column chromatography, ion exchange chromatography and electrophoresis are all used to separate proteins.
9. The relative molecular masses of proteins can be determined by **ultracentrifugation**.
10. The sequence of amino acids in a protein is known as the **primary structure**.
11. The regular folding pattern of a protein or polypeptide chain is known as its **secondary structure**. It arises from hydrogen bonding between the carboxyl and amino groups. The spiral **helix** structure is an example.
12. The **tertiary structure** of a protein arises from crosslinking between side chains of the amino acid residues.
13. A **quaternary structure** consists of several sub-units each containing a separate polypeptide chain.
14. Peptides can be synthesised by the **Merrifield technique**. The N-terminal of an amino acid is blocked and the C-terminal attached to a resin support. The N-terminal is then unblocked and the peptide link formed.

20.2 Carbohydrates and Nucleic Acids

CARBOHYDRATES

Carbohydrates and their derivatives are one of the major classes of naturally occurring organic compounds. They are distributed widely in both plants and animals.

Carbohydrates are produced by plants in the process known as **photosynthesis**. The overall stoichiometric process can be represented as

$$6CO_2 + 6H_2O \xrightarrow[\text{chlorophyll}]{\substack{\text{energy from}\\\text{sunlight}}} \underset{\text{glucose}}{C_6H_{12}O_6} + 6O_2$$

Glucose and other simple carbohydrates are then converted to more complicated carbohydrates such as starch and cellulose. Between 60 and 90% of all solid plant material is made up of carbohydrates. Plants use carbohydrates as a chemical store of energy from sunlight. The plants use these carbohydrates to produce energy by an oxidation process known as respiration. This is effectively the reverse of photosynthesis

$$C_6H_{12}O_6 + 6O_2 \rightarrow 6CO_2 + 6H_2O + \text{energy}$$

The carbohydrates in plants also provide a source of energy for animals. Foods we consume such as bread, biscuits, cakes, potatoes and cereals are all high in carbohydrates (*see* section 5.4).

In one year about 200 000 000 000 tonnes of carbon dioxide are taken from the atmosphere during photosynthesis by green plants and bacteria. During the same time about 130 000 000 000 tonnes of oxygen are released into the atmosphere and 50 000 000 000 tonnes of organic carbon compounds are synthesised.

Photosynthesis on a massive scale: a rainforest in Uganda

One of the most important of these energy providing carbohydrates—at least as far as human beings are concerned—is sucrose. This occurs in sugar beet, sugar cane and other plants. Indeed it is known as cane sugar. Carbohydrates and especially cellulose are also important in providing the structural material of plants. They are also components of nucleic acids and thus play an important role in the biosynthesis of proteins.

The name carbohydrate derives from the French term *hydrates de carbone* or carbon hydrates. This is because most carbohydrates have the general formula $C_x(H_2O)_y$.

Carbohydrates may be classified as monosaccharides, disaccharides and polysaccharides.

Sugar factory in Colombia

Monosaccharides

These have between three and six carbon atoms. Monosaccharides with the general formula $C_5H_{10}O_5$ are known as **pentoses**. Ribose is an important example of a pentose. Monosaccharides with the general formula $C_6H_{12}O_6$ are known as **hexoses**. Glucose is a hexose.

Disaccharides

These have the general formula $C_{12}H_{22}O_{11}$. These are formed by the condensation of two monosaccharide units and the loss of a molecule of water. Lactose, sucrose and maltose are all disaccharides. Disaccharides formed from two hexose units have the general formula $C_{12}H_{22}O_{11}$.

Both the monosaccharides and disaccharides are white crystalline solids which are soluble in water and have a sweet taste. They are known as **sugars**. Table sugar is sucrose.

Polysaccharides

These are polymers of the monosaccharides. The general formula of polysaccharides formed from hexose units is $(C_6H_{10}O_5)_n$. Polysaccharides may be composed of one or more monosaccharide monomers. Some contain hundreds or thousands of monosaccharide units and thus have high relative molecular masses.

Cellulose and starch are the two most widely occurring polysaccharides in nature. Polysaccharides tend to be insoluble in water or form colloids with water. Cellulose, for example, is insoluble whereas starch forms a colloid.

Structure of Carbohydrates

Aldoses and Ketoses

Carbohydrates may be regarded as aldehydes or ketones containing a number of hydroxyl groups. Monosaccharides with an aldehyde group are known as **aldoses** whereas those with the ketone group are known as **ketoses**. Examples are ribose and fructose respectively:

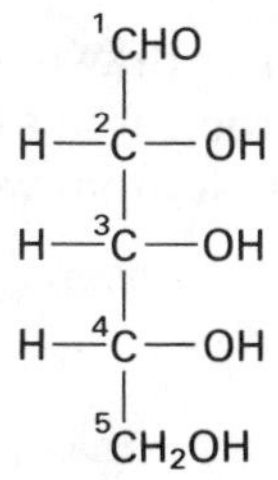

Ribose

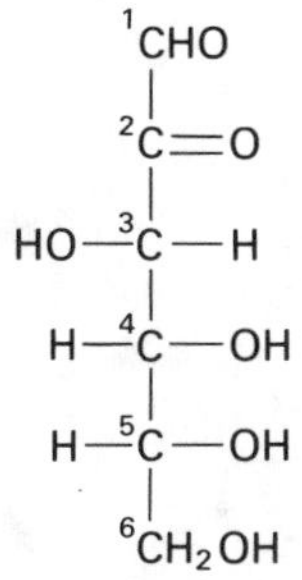

Fructose

Configurations of monosaccharides displayed in this way are known as Fischer conventional projections.

An aldose which is also a pentose is known as an **aldopentose**. Ribose is an example. A ketose which is also a hexose is known as a **ketohexose**. Fructose is an example. Glucose is an example of an **aldohexose**.

Both aldoses and ketoses contain asymmetric carbon atoms. The aldohexoses, for example, have four asymmetric carbon atoms. There are eight different aldohexoses. Each has two optical isomers—making 16 in all. All 16 isomers have been isolated. The most important of these are the two optical isomers of glucose:

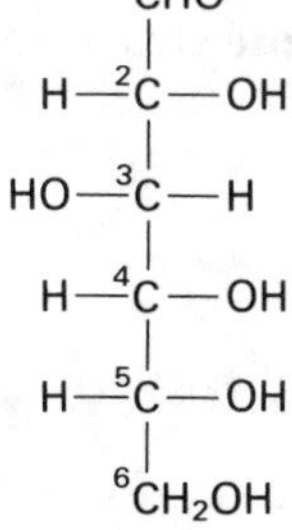

D-Glucose

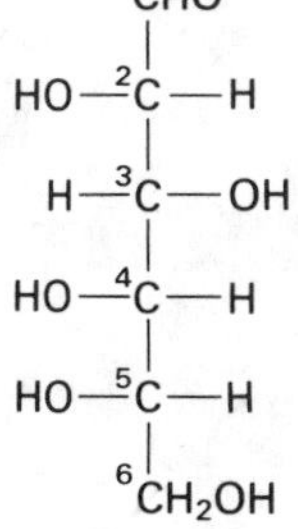

L-Glucose

For glucose, the prefixes D– and L– indicate the **configuration** of the hydroxyl group at the C–5 atom. The arrangement of hydroxyl groups at the four asymmetric carbon atoms, C–2, C–3, C–4 and C–5, shown in this structure is the configuration of glucose. Each of the eight hexoses has a specific D-configuration. Their L-configurations are always the mirror images of their D-configurations. D-Glucose rotates plane-polarised light to the right. It is thus dextrorotatory and given the sign (+). L-Glucose rotates plane-polarised light to the left. It is thus laevorotatory and given the sign (−). The D– stands for *dextro* (meaning right) since most D sugars rotate plane-polarised light to the right. However, this is not always the case. D-Fructose is laevorotatory (−). The L– stands for *laevo* (meaning left) since most (but not all) L-sugars rotate plane-polarised light to the left.

Ring Structures

Solid monosaccharides do not exist as the open chain structures shown above. They exist rather as ring structures. D-Glucose, for example, can exist in two separate crystalline forms known as α-D-glucose and β-D-glucose. These forms are known as **anomers**:

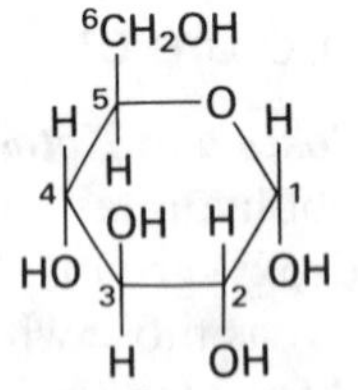

α-D-Glucose 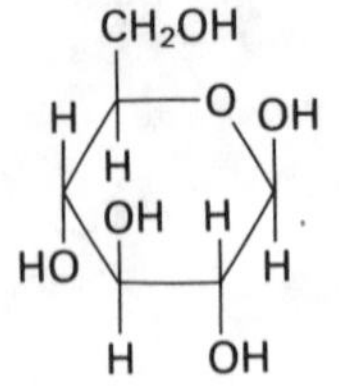   β-D-Glucose

Structures drawn in this form are known as **Haworth projections**. Notice that the ring carbon atoms have been omitted. Both forms of glucose are **hemiacetals** since at the C–1 atom we have

or

All pentoses and hexoses exist predominantly as six-membered rings containing an oxygen atom:

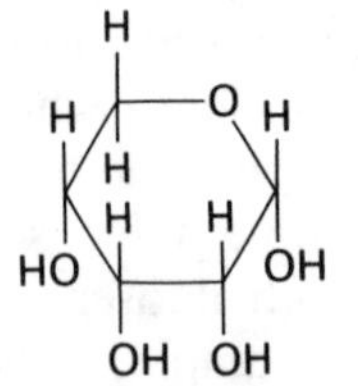

α-D-Ribose
(α-D-ribopyranose)

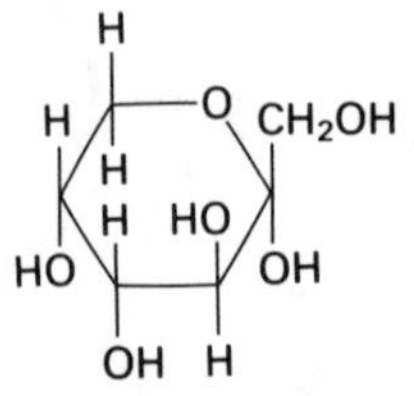

β-D-Fructose
(β-D-fructopyranose)

These six-membered rings are known as **pyranose** structures.

Monosaccharides, their derivatives and other carbohydrates often have **furanose** structures. These are five-membered rings containing an oxygen atom:

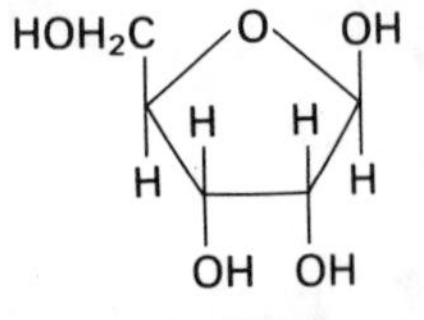

β-D-Ribose
(β-D-ribofuranose)

Pyranose ring systems normally exist in a **chair conformation**:

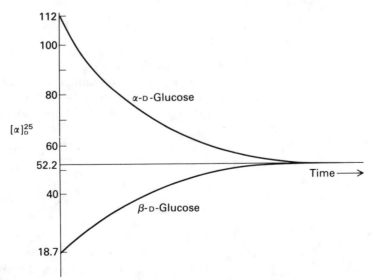

α-D-Glucose
(α-D-glucopyranose)

β-D-Glucose
(β-D-glucopyranose)

The substituents on the carbon atoms in this type of conformation are either **axial** or **equatorial**:

ax = axial
eq = equatorial

Mutarotation

α-D-Glucose and β-D-glucose have different melting points and also different specific rotations (*see* table 20.5). The specific rotation can be determined using a polarimeter. When either anomer of glucose is dissolved in water, it converts to the other until an equilibrium mixture of the two is formed together with a minute amount of the open chain form:

Table 20.5 Properties of D-Glucose

	α	β
$[\alpha]_D^{25}$/deg	+112	+18.7
melting point/°C	146	150

The **specific rotation**, $[\alpha]_D^{25}$, of a compound is the angle through which the plane of polarisation of a ray of sodium D light is rotated at 25°C by a column 100 cm long containing a solution of the compound of concentration 1 g cm^{-3}.

α-D-Glucose
36%

Open chain
form
0·02%

β-D-Glucose
64%

This change in optical rotation is known as **mutarotation** (*see* figure 20.9).

Figure 20.9   Mutarotation of α-D-glucose and β-D-glucose

Disaccharides

These can be represented conveniently as Haworth projections. For example, sucrose, which is formed by the condensation of α-D-glucose in the pyranose form and β-D-fructose in the furanose form, may be represented as

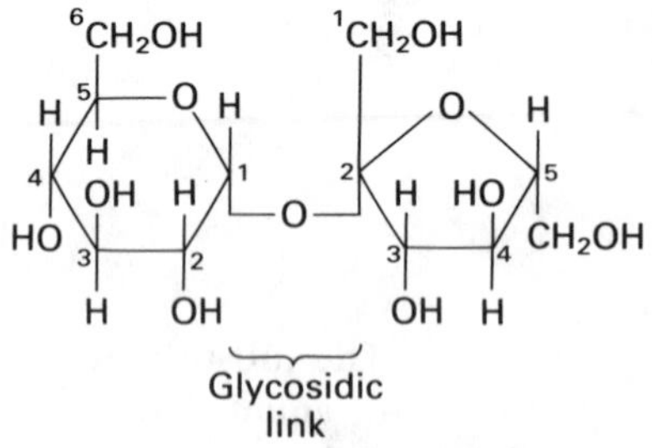

Sucrose

The link between the two sugars is known as a **glycosidic link**. In the case of sucrose, the link is between the C–1 atom of glucose in the α-configuration and the C–2 atom of fructose. The link is thus known as an α 1,2 bond.

Lactose is a disaccharide in which β-D-galactose is linked at the C–1 atom to the C–4 atom of β-D-glucose. This is called a β 1,4 bond. Both hexoses exist in the chair conformation:

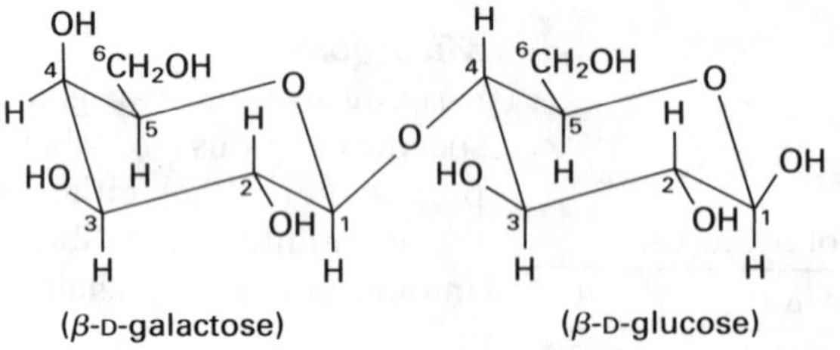

β-Lactose

(β-D-galactose)   (β-D-glucose)

Note that the C–1 atom of the glucose on the right is in the β-configuration.

Maltose consists of two glucose units with a α 1,4 bond:

Maltose is present in germinating cereal seeds. The name maltose derives from malt, which is the extract of germinated barley. Malt is used to make beer and whisky (*see* chapter 6, initial article and section 9.2).

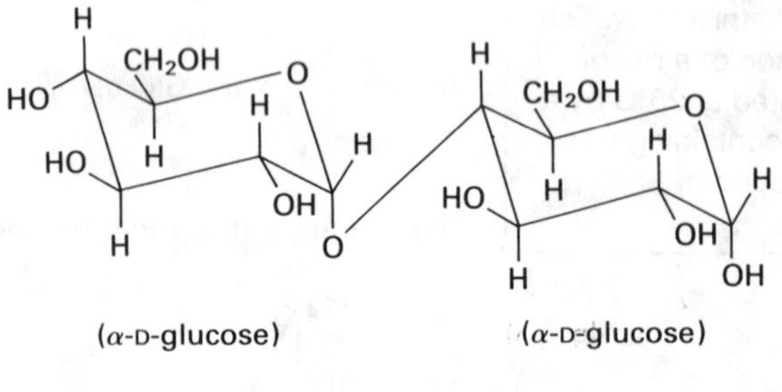

α-Maltose

(α-D-glucose)   (α-D-glucose)

Polysaccharides

One of the most important polysaccharides is cellulose. This is a major constituent of the walls of plant cells—hence its name. Cellulose is a polymer made up completely of β-D-glucose units linked by β 1,4 bonds:

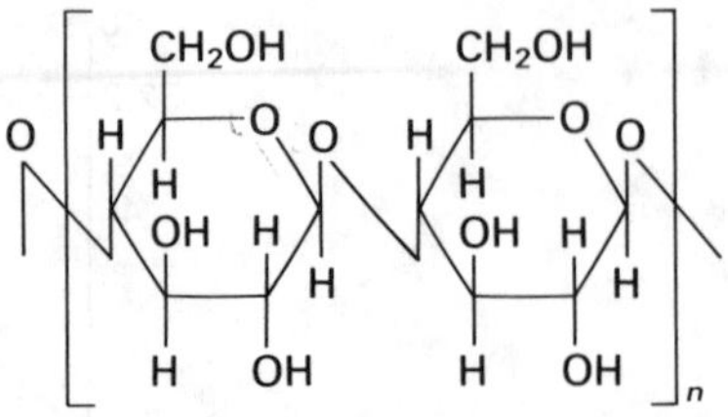

Cellulose may consist of a chain of more than 10 000 β-D-glucose units and have a relative molecular mass of more than one million.

Starch is also a naturally occurring polysaccharide. It exists in two forms: amylose, which is soluble in water, and amylopectin, which is insoluble in water.

Amylose is a straight chain polymer of α-D-glucose units with α 1,4 bonds:

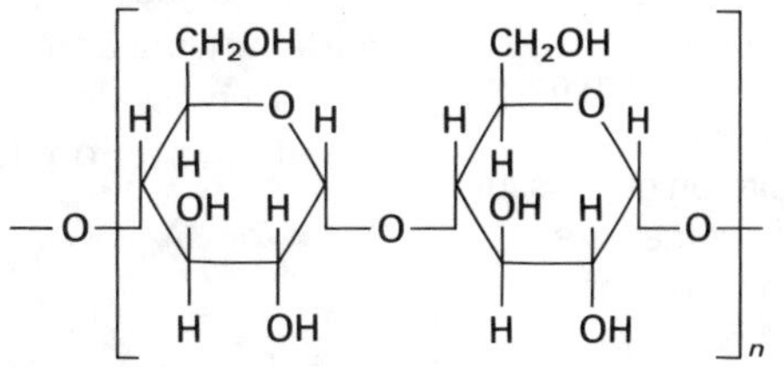

Amylopectin also consists of α-D-glucose units but it has a branched structure with both α 1,4 and α 1,6 bonds:

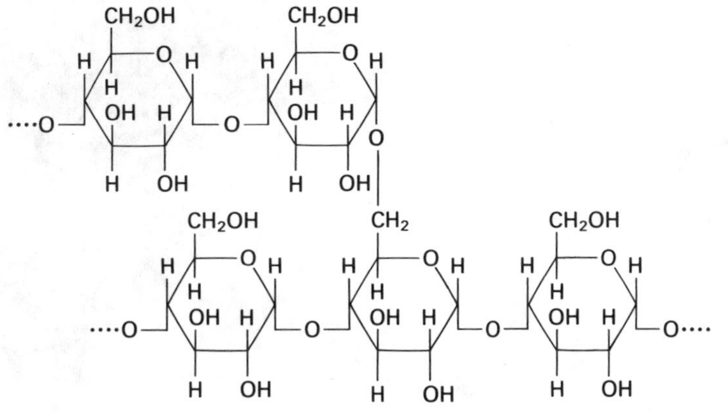

Most plants use starch as a store of carbohydrates and thus energy. Starch is also an essential part of the diet of many animals including humans. It is stored in the livers of animals in the form of glycogen—also known as animal starch. Glycogen has the same chemical structure as amylopectin.

Reactions of the Carbohydrates

Hydrolysis
Monosaccharides do not undergo acid hydrolysis. Disaccharides, on the other hand, are hydrolysed by dilute mineral acids yielding the two hexoses. Sucrose, for example, produces glucose and fructose:

$$C_{12}H_{22}O_{11} \xrightarrow{H^+} C_6H_{12}O_6 + C_6H_{12}O_6$$

D-sucrose   D-glucose   L-fructose

In this process the specific rotation of the solution changes from $(+)$ to $(-)$. The process is thus known as the **inversion of sucrose**. The product is known as invert sugar.

Polysaccharides like starch and glycogen are hydrolysed by acids to give monosaccharides:

$$(C_6H_{10}O_5)_n + nH_2O \xrightarrow{H^+} nC_6H_{12}O_6$$

Disaccharides such as sucrose and polysaccharides such as starch also undergo hydrolysis in the presence of enzymes in the digestive system. For example, when starch is consumed, amylase and other enzymes in saliva and the digestive tract break down the glycosidic links in starch in a complex process which results in the formation of maltose and then glucose. The glucose is absorbed into the bloodstream and then broken down in the respiration process to produce water, carbon dioxide and energy. Like plant starch, glycogen is also hydrolysed to glucose. Sucrose—or cane sugar as it is commonly known—is hydrolysed in the presence of

Lactose Intolerance

The disaccharide lactose is found only in milk. It is hydrolysed in the presence of the enzyme lactase to give its constituent monosaccharides: glucose and galactose. The enzyme is secreted in the small intestine. However, some population groups in Eastern and African countries are deficient in this enzyme. They are thus intolerant of milk. This is known as lactose intolerance. For this reason milk is not used in Chinese cooking (except in one small region).

enzymes to form glucose. Starch, glycogen and sucrose are all major sources of energy of the human body.

All human beings are unable to digest cellulose although it is a component of dietary fibre (roughage). This is because humans do not have the enzymes needed to hydrolyse the β 1,4 bonds in cellulose. However, cows and other ruminant or cud-chewing animals do have the required enzymes and can thus digest cellulose.

Starch is also hydrolysed during fermentation processes. This is described briefly in section 9.2.

Foods rich in dietary fibre

Glycoside Formation

The hydroxyl groups of sugars can behave in the same way as the hydroxyl groups of alcohols and form esters and ethers. These are known as glycosides. If the sugar is glucose, for example, the compound formed is called a glucoside. Methyl β-D-glucoside, which is an ether, is an example.

Methyl
α-D-glucoside

The glycosides are particularly important in the biochemical processes that take place in both plants and animals. For example, phosphate esters of ribose are particularly important in respiration and photosynthesis. One of the most important is adenosine triphosphate (ATP). This is a triphosphate ester of adenosine. Adenosine is the condensation product of adenine and ribose. It is a nucleoside (*see* below). The phosphate groups in ATP store the energy needed by cells for muscle contraction. The energy is released as the phosphate groups split off to form adenosine diphosphate (ADP) and adenosine monophosphate (AMP) (*see* figure 20.10).

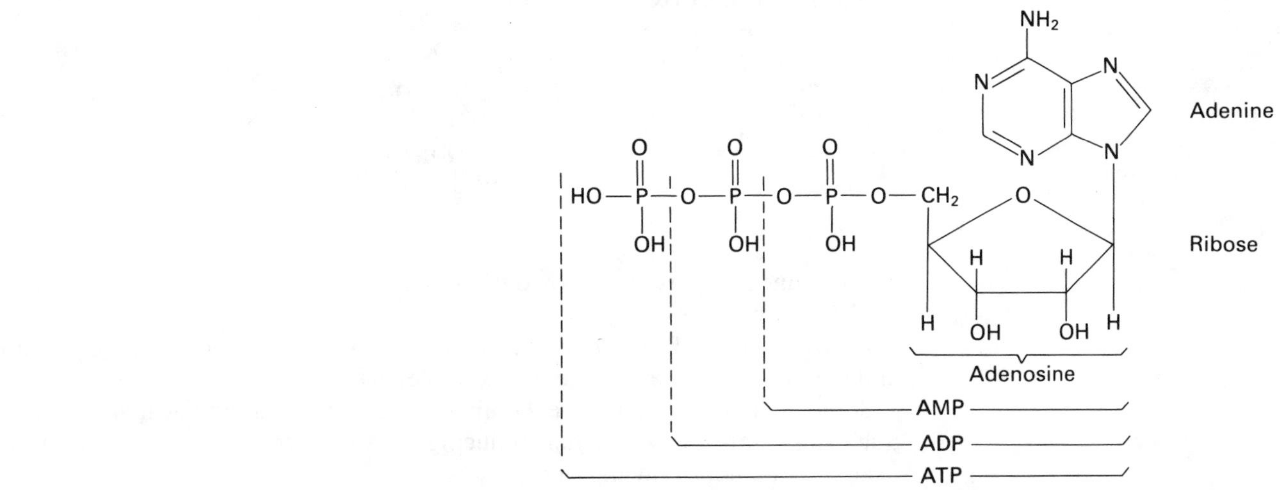

Figure 20.10   Structure of adenosine, adenosine monophosphate (AMP), adenosine diphosphate (ADP) and adenosine triphosphate (ATP)

Reduction and Oxidation of Sugars

Aldoses and ketoses can both be reduced chemically and biologically to polyhydric alcohols. For example, glucose can be reduced to sorbitol:

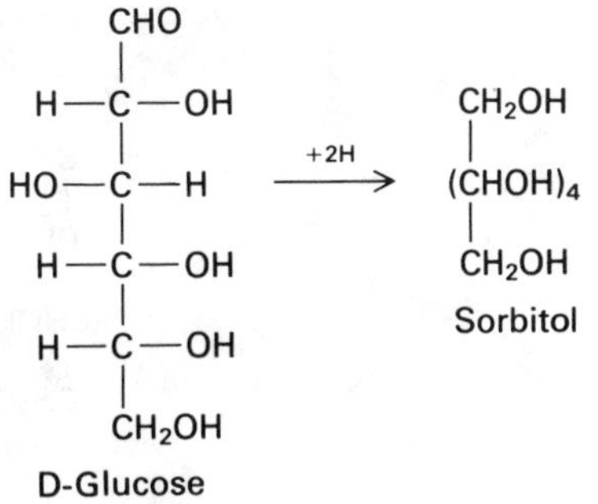

Sorbitol has a sweet taste and is used by diabetics as a substitute for cane sugar.

Certain sugars such as glucose, ribose, maltose and lactose are known as **reducing sugars**. Although these sugars may exist in the hemiacetal form they have a potentially free aldehyde group. For this reason they are also known as **free sugars**. The free aldehyde group can be oxidised to a carboxylic acid.

Reducing sugars reduce copper(II) in Fehling's solution to copper(I) oxide and ammoniacal silver nitrate in Tollens' reagent to silver. The aldehyde group can also be oxidised to bromine water.

When glucose is oxidised two acids are formed. Oxidation of the C–1 aldehyde results in the formation of gluconic acid:

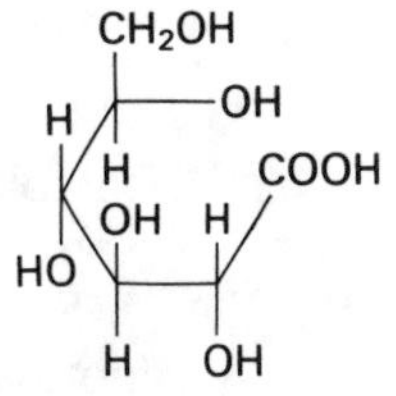

Ketoses do not undergo this type of oxidisation at C–1. Glucose also forms glucuronic acid by oxidation of C–6:

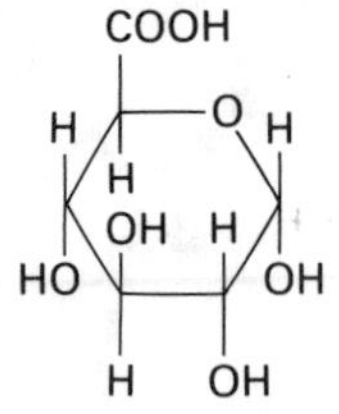

Ketoses undergo this type of oxidation at either the C–6 position (for hexoses) or at C–5 (for pentoses).

Ketoses such as fructose are reducing agents. They reduce Fehling's solution and ammoniacal silver nitrate for example.

Some sugars do not reduce Fehling's solution or ammoniacal silver nitrate. Such sugars are known as non-reducing sugars. Sucrose is the most important example of a non-reducing sugar.

Detection and Characterisation of Sugars

Reducing sugars such as glucose can be detected using reagents such as Fehling's solution, Benedict's solution or Tollens' reagent.

Carbohydrates containing an aldehyde or keto group—or if in the cyclic form, a potential aldehyde or keto group—react with phenylhydrazine in the cold forming the corresponding phenylhydrazones. These are usually soluble in water. If the sugar is heated to 100°C with excess phenylhydrazine, compounds known as osazones are formed (*see* figure 20.11). Glucose and mannose (both aldoses) and fructose (a ketose) all form the same osazones.

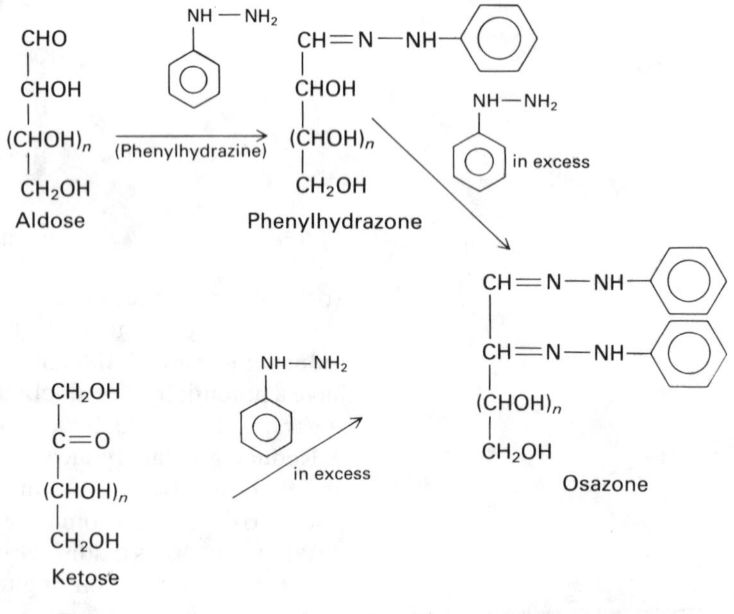

Figure 20.11 Formation of osazones from aldoses and ketoses

Osazones are usually yellow crystalline solids which are only sparingly soluble in cold water. The shapes of the crystals and the rates at which they form are sometimes used to identify sugars.

Starch and glycogen can both be identified using a dilute solution of iodine in potassium iodine solution. Starch gives a deep blue colour whereas glycogen gives a red wine colour. Cellulose does not give a colouration with iodine solution.

NUCLEIC ACIDS

Nucleic acids are organic acids of high relative molecular masses. They occur in all living organisms, playing an essential role in the biosynthesis of proteins and in the transmission of hereditary characteristics. The nucleic acids embody the **genetic code** which dictates the specific amino acid sequences in proteins. It is this genetic information which programmes the structure and metabolic activity of living organisms.

Nucleic acids are polymers. The monomeric unit is the nucleotide. They are thus polynucleotides.

Deoxyribonucleic acid (DNA) and **ribonucleic acid** (RNA) are both nucleic acids. The sugar component of DNA is 2-deoxyribose:

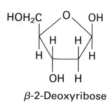

β-2-Deoxyribose

In RNA the sugar component is ribose:

Ribose

The structures of DNA and RNA are shown in figure 20.12.

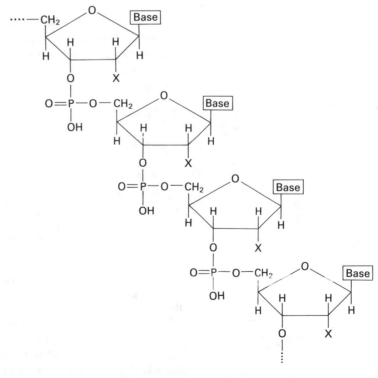

Figure 20.12 The structure of deoxyribonucleic acid (DNA) and ribonucleic acid (RNA), where X = H in DNA and X = OH in RNA

Nucleotides and Nucleosides

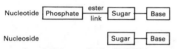

A nucleotide consists of three components—phosphate, sugar and base—whereas a nucleoside consists of only two components—sugar and base:

Nucleotide | Phosphate — ester link — Sugar — Base

Nucleoside | Sugar — Base

Adenosine triphosphate (ATP) is a nucleotide whereas adenosine is a nucleoside.

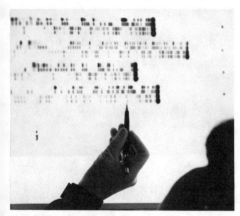

Analysis of nucleotide sequences

DNA sculpture at the Biomedical Centre, Uppsala, Sweden

Four different bases are present in DNA. Two of these are pyrimidine bases and two are purine bases (*see* figure 20.13). Note that each base is represented by a capital letter.

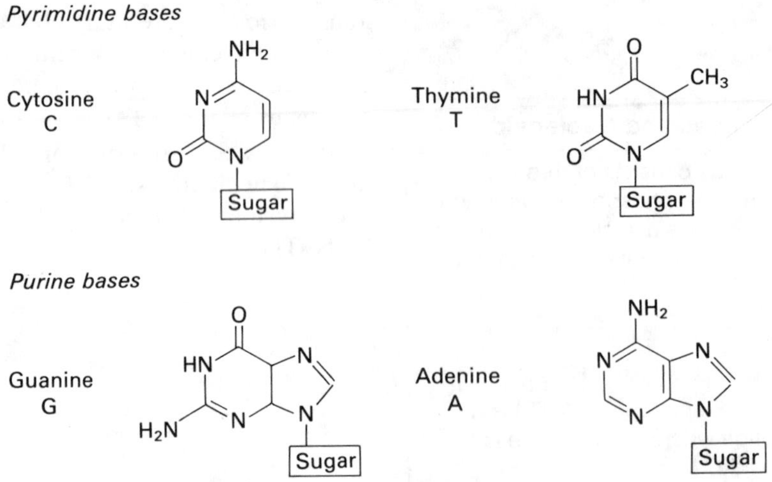

Pyrimidine bases

Cytosine
C

Thymine
T

Purine bases

Guanine
G

Adenine
A

Figure 20.13   The four bases present in DNA

RNA contains the same bases as DNA except that thymine is replaced by uracil:

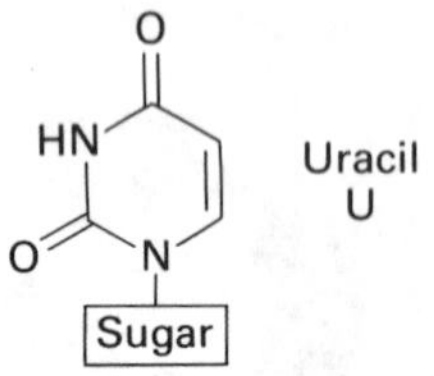

Uracil
U

Replication and Protein Biosynthesis

DNA is an essential constituent of almost all living organisms. It stores the genetic information that is transmitted from one generation to the next. DNA occurs in the chromosomes which are found in the nuclei of cells. A chromosome consists of a thread-like complex of DNA and protein coiled into a tightly packed structure (*see* figure 20.14). In the higher organisms up to 65% of chromosomes is protein.

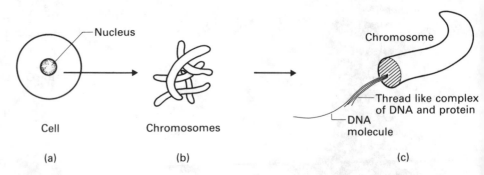

Cell

Chromosomes

Nucleus

Chromosome

Thread like complex of DNA and protein

DNA molecule

(a)

(b)

(c)

Figure 20.14   A cell nucleus (a) contains a set number of chromosomes (b). These contain the genes or genetic code. A chromosome consists of a thread-like complex of DNA and protein coiled into a tightly packed structure (c)

DNA is usually—but not always—**double stranded**. The two strands are held together by **hydrogen bonds** (*see* chapter 2) formed between pairs of bases. A **base pair** consists of a purine base and a pyrimidine base. In DNA only two types of base pair are possible (figure 20.15).

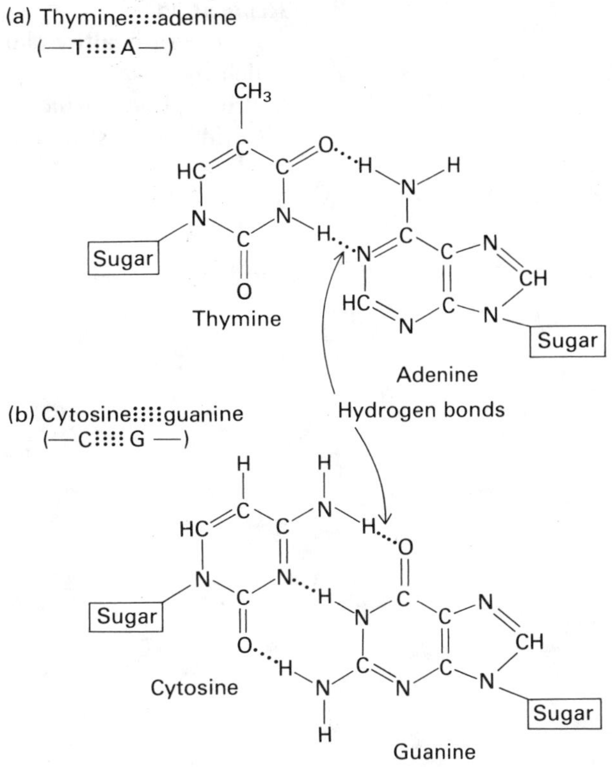

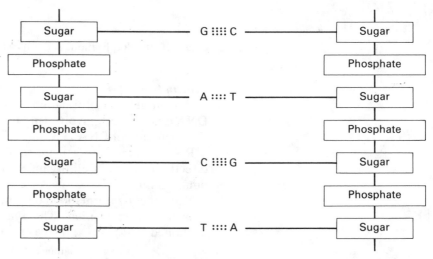

(a) Thymine::::adenine
(—T:::: A—)
Sugar
Thymine
Sugar
Adenine
Hydrogen bonds
(b) Cytosine::::guanine
(—C:::: G —)
Sugar
Cytosine
Sugar
Guanine

Figure 20.15 Base pairs in DNA: (a) thymine–adenine; (b) cytosine–guanine. The two bases in each pair are held together by hydrogen bonds

Each strand in DNA consists of a specific sequence of bases. The sequence must match exactly an antiparallel sequence of bases in the other strand. For example, the sequence G–A–C–T in one strand must match the sequence C–T–G–A in the other strand (*see* figure 20.16).

Sugar
G :::: C
Sugar
Phosphate
Phosphate
Sugar
A :::: T
Sugar
Phosphate
Phosphate
Sugar
C :::: G
Sugar
Phosphate
Phosphate
Sugar
T :::: A
Sugar

Figure 20.16 A sequence of bases in one strand of DNA match an antiparallel sequence of bases in the other strand

G::::C
A : : : : T
C : : : : G
T::::A

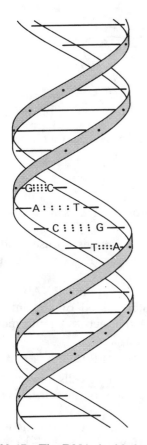

Figure 20.17 The DNA double helix

The three-dimensional structure of DNA was elucidated by James Watson and Francis Crick in Cambridge. Their results were based on X-ray diffraction patterns obtained from DNA by Maurice Wilkins and Rosalind Franklin in London. They found that the two strands of DNA are coiled in the form of a double helix (*see* figure 20.17). Watson, Crick and Wilkins were awarded the Nobel Prize for Physiology and Medicine in 1962.

Replication

The cells of healthy plants and animals are always dividing and replacing those which die away. This process is known as replication. During replication, the hydrogen bonds which link the two strands of the double helix are broken. The resulting single strands act as templates for new double helices (*see* figure 20.18).

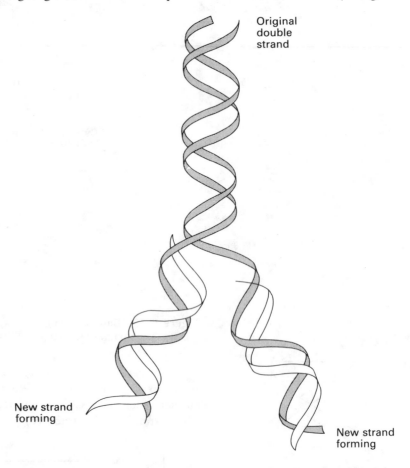

Original double strand

New strand forming

New strand forming

Figure 20.18   Replication. A double strand divides into two single strands

Protein Synthesis

Although genetic information is stored in the sequence of bases in DNA, the DNA is not directly involved in the synthesis of proteins. It acts indirectly through the formation of RNA.

RNA contains the genetic code of some of the simplest viruses and is also directly involved in all organisms in protein synthesis. RNA may be single or double stranded.

The genetic information contained in DNA is passed onto RNA in a process known as **transcription**. The synthesis of protein based on the information contained in RNA is known as **translation**. RNA thus acts as an intermediate

$$\text{DNA} \xrightarrow{\text{transcription}} \text{RNA} \xrightarrow{\text{translation}} \text{protein}$$

Both DNA and RNA are synthesised in the nuclei of cells whereas protein synthesis occurs in the material surrounding the nucleus of the cell. This material is known as **cytoplasm**.

The RNA involved in the synthesis of proteins is single stranded and is divided into three types: ribosomal, messenger and transfer.

Ribosomal RNA (rRNA). This forms the structural part of ribosomes. These consist of RNA and protein and occur in the cytoplasm. They are the sites of protein synthesis.

Messenger RNA (mRNA). This carries genetic information from the DNA in the chromosomes in the cell nuclei and takes it to the ribosomes in the cytoplasm. Messenger RNA effectively programmes the ribosomes to synthesise protein in a specific amino acid sequence. This information is contained in the specific sequences of bases in the mRNA. This sequence of bases is transcribed from the DNA.

The genetic information contained in mRNA is translated into a specific amino acid sequence by the **triplet code**. In this code any sequence of three bases corresponds to a specific amino acid. For example

GCU	translates to alanine
GGU	translates to glycine

'Code words' such as GCU are known as **codons**. An amino acid may have several codons. For example, leucine can have the codons UUA, UUG, CUU, CUC, CUA and CUG. The codons, UAA, UAG, and UGA are termination signs— or 'full stops'. They tell the ribosome to stop adding amino acids to the polypeptide. The ribosome then releases the polypeptide.

Transfer RNA (tRNA). This is also known as soluble RNA. It consists of a group of small molecules. Each type of tRNA carries a specific amino acid to the ribosomes in the cytoplasm for possible use in the protein synthesis.

Ribosomal RNA accounts for up to 80% of the RNA in a cell. About 5% is mRNA and 15% tRNA.

SUMMARY

1. **Carbohydrates** have the general formula $C_x(H_2O)_y$.
2. Carbohydrates may be classified as monosaccharides, disaccharides or polysaccharides.
3. **Aldoses** and **ketoses** contain **asymmetric** carbon atoms.
4. **Monosaccharides** exist mainly as ring structures.
5. Six-membered rings containing an oxygen atom are known as **pyranose** structures.
6. Pyranose structures normally exist in a **chair conformation**.
7. A **disaccharide** consists of two monosaccharides joined by a **glycosidic link**.
8. Cellulose is an example of a **polysaccharide**. It is made up of β-D-glucose units.
9. **Hydrolysis** of polysaccharides yields monosaccharides.
10. **Esterification** of sugars yields glycosides.
11. Sugars with a free aldehyde groups are known as free sugars or **reducing sugars**.
12. Reducing sugars can be detected using Fehling's solution.
13. Sugars can be characterised by osazone formation.
14. **Nucleic acids** are polynucleotides.
15. A **nucleotide** has the structure: phosphate–sugar–base.
16. **Deoxyribonucleic** acid (DNA) stores genetic information. It has a **double helix** structure.
17. DNA passes on genetic information to **ribonucleic acid** (RNA) in a process known as transcription.
18. RNA is directly involved in protein synthesis.

20.3
Fats, Oils and
Other Natural Products

Fats and oils belong to a group of compounds known as **lipids**. They are found in living tissues. Lipids are soluble in organic solvents such as propanone, ethanol and trichloromethane but insoluble in water. They include fatty acids, fats, oils, waxes, phospholipids, glycolipids, steroids and some vitamins. Lipids are defined on the basis of a specific physical property—namely solubility. They do not have any common structural feature.

FATS AND OILS

Fats and oils contain one or more triglycerides. Triglycerides are esters of fatty acids (*see* below). Fats and oils are distinguished on the basis of their melting points. Fats are solid at room temperature but become liquid on heating. Oils are fats which are liquid at room temperature. Fats and oils do not dissolve in water. When mixed vigorously with water they tend to emulsify.

In modern industrialised countries fats account for up to 45% of the total energy intake of humans. Such a high proportion of fat in the diet is almost certainly undesirable. Many modern diseases—especially heart diseases—are attributed to the high level of fat in the diet. In some Third World countries, on the other hand, fat may account for as little as 10% of the total energy intake.

Fats are important to the body (figure 20.19) since they provide an efficient means of storing energy. A limited amount of energy can be stored as carbohydrate in the form of glycogen (*see* previous section). However, surplus energy supplied in the form of proteins, carbohydrates or fat can be stored as fat.

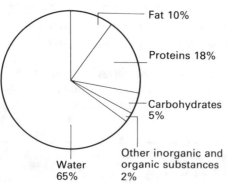

Figure 20.19 Composition of a typical human body

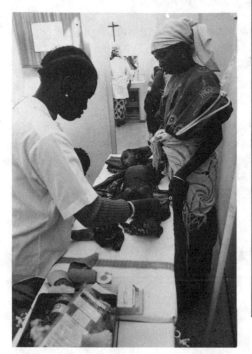

An infant suffering from malnutrition in Senegal. In some developing countries, fat may account for as little as 10% of the total energy intake.

Oils

Oils are neutral liquids. They may be divided into three main classes.

Fixed or fatty oils
These are derived from animal, vegetable and marine sources. They are esters of fatty acids.

Mineral Oils
These are derived from petroleum, coal and shale and consist of hydrocarbons. Paraffin oil is an example.

Essential oils
These are volatile substances with characteristic odours extracted from plants. They are hydrocarbons or their simple derivatives. For example, bitter almond oil is benzaldehyde and oil of wintergreen is the methyl ester of 2-hydroxybenzoic acid (salicylic acid). Oil of wintergreen is used as a rubefacient. Rubefacients are used in rubbing ointments to generate 'warmth' when treating muscle pains and rheumatism. Some essential oils are shown in table 20.6. They are called essential oils because they are extracted from the essence of plants.

Fat in the diet and in the body also has a number of other important functions. For a start, fats contribute significantly to the texture and palatability of foods. They also aid digestion by, for example, lubricating foods such as bread and help in swallowing. Fats also provide an important source of other nutrients—especially

Table 20.6 Some essential oils

Structure	Oil	Source
	limonene	peel of citrus fruits
	vanillin	vanilla orchid
	zingiberone	ginger
	eugenol	oil of cloves
	cinnamaldehyde	cinnamon oil
	menthol	peppermint
	methyl salicylate	oil of wintergreen (extracted from the bark of certain birch trees)

certain vitamins. These are known as fat-soluble substances. For example, milk fat and thus butter and cream and also fish liver oils all contain vitamin A and some vitamin D. Vegetable oils such as those found in peanuts and wheatgerm contain vitamin E. The average fat content of some foods is shown in table 20.7.

Table 20.7 Average fat content of some foods

Food	Grams of fat in 100 grams of food
milk	3.8
cheese (e.g. Cheddar)	33.0
eggs	10.9
chicken	17.7
butter	82.0
cooking oil	99.9
potatoes	0
roasted peanuts	49.0
white bread	1.7

The extraction of fat during food testing in a laboratory in Bangkok, Thailand

FATTY ACIDS AND THE TRIGLYCERIDES

Naturally occurring fats and oils are the esters formed by propane-1,2,3-triol (also known as glycerol or glycerine) and fatty acids. They are known as triglycerides. Fatty acid is a general name for a monobasic aliphatic carboxylic acid, RCOOH. They are often represented as

$$\sim\!\!\sim\!\!\sim\!\!\sim\!\!\sim\!\!\sim\text{COOH}$$

Hydrolysis of a triglyceride ester yields propane-1,2,3-triol and the fatty acids:

$$
\begin{array}{ccc}
\text{H}_2\text{COOR} & & \text{CH}_2\text{OH} \qquad\quad \text{R}\;\text{—COOH} \\
| & \xrightarrow{\text{hydrolysis}} & | \\
\text{HCOOR}' & & \text{CHOH} \quad + \quad \text{R}'\text{—COOH} \\
| & & | \\
\text{H}_2\text{COOR}'' & & \text{CH}_2\text{OH} \qquad\quad \text{R}''\text{—COOH} \\
& & \text{Propane-1,2,3-triol} \qquad \text{Fatty} \\
& & \text{(glycerol or glycerine)} \quad\;\; \text{acids}
\end{array}
$$

R, R′ and R″ may be the same or different.

Fatty acids may be divided into three categories: saturated, unsaturated and branched and cyclic.

Saturated Fatty Acids

These have the general formula $CH_3(CH_2)_nCOOH$ where n may range from 2 to over 20. When n is low the acid is known as a short chain fatty acid. If it is high then

The Greek word for butter is *butyrus*.

it is known as a long chain fatty acid. An example is butanoic acid (also known as butyric acid) obtained from milk fat including butter. Palmitic acid and stearic acid are also examples (*see* table 20.8). Both of these occur in the form of triglycerides in almost all animal and plant fats and oils.

Table 20.8   Saturated fatty acids

Common name	Systematic name	Formula
butyric acid	butanoic acid	$CH_3(CH_2)_2COOH$
palmitic acid	hexadecanoic acid	$CH_3(CH_2)_{14}COOH$
stearic acid	octadecanoic acid	$CH_3(CH_2)_{16}COOH$

Unsaturated Fatty Acids

These contain at least one double bond in the chain and may have short chains or long chains. One of the most abundant unsaturated fatty acids in all plants and land animals is oleic acid. This is named from olive oil. It is found in olive oil and pork fat. Its systematic name is *cis*-octadec-9-enoic acid and it has the formula

$$CH_3(CH_2)_7CH=CH(CH_2)_7COOH$$

The **degree of unsaturation** in any fat or oil is given by its **iodine number**. This is the mass of iodine in grams which reacts with 100 grams of the fat or oil. Oils tend to have iodine numbers above 70 whereas those of fats are generally below 70.

Triglycerides whose fatty acid components have either short chains or a high degree of unsaturation tend to have lower melting points than the triglycerides of long chain saturated acids. The former thus exist as oils at room temperature whereas the latter exist as solid fats. Triglycerides obtained from plants have a higher proportion of unsaturated fatty acids and thus exist as oils. Animal fats on the other hand, have a higher proportion of saturated fatty acids. This can be seen by comparing the distribution of fatty acids in olive oil (a plant oil) with that of butter (an animal fat) (*see* table 20.9). Note that both the saturated and unsaturated fatty acids have even numbers of carbon atoms. This is typical of the fatty acids of most fats and oils.

Table 20.9   Typical values of the distribution of fatty acids in olive oil and butter

Type of fatty acid	No.of C atoms	Percentage by mass of fatty acid in	
		olive oil	butter
saturated	4	–	4
	6–10	–	5
	12	–	5
	14	trace	12
	16	10	27
	18	2	10
	total	12	61
unsaturated	16	–	5
	18	84	28
	total	84	33

Margarine is a butter substitute containing vegetable oils and fats. The degree of unsaturation of these fats and oils is reduced by hydrogenation of the double bonds using a nickel catalyst. Soft and hard margarines differ in the extent to which the double bonds have been hydrogenated. Soft margarines have a higher degree of unsaturation. Vitamins A and D are added to margarines.

Branched and Cyclic Fatty Acids

Naturally occurring branched and cyclic fatty acids are uncommon compared to other fatty acids.

An example of a cyclic fatty acid is chaulmoogric acid:

$$\text{(CH}_2)_{12}\text{COOH}$$

Chaulmoogric
acid

This is one of three cyclic fatty acids which occur in chaulmoogra oil. This oil is found in the seeds of an East Indian plant. The oil was used to treat leprosy in China and India for hundreds of years.

Soaps and Detergents

Soaps are the metal salts—commonly sodium and potassium salts—of long chain fatty acids—mainly palmitic, stearic and oleic acids. They are made by boiling either tallow or a vegetable oil such as coconut oil with sodium or potassium hydroxide

$$
\begin{array}{l}
\text{H}_2\text{COOC(CH}_2)_{16}\text{CH}_3 \\
| \\
\text{HCOOC(CH}_2)_{16}\text{CH}_3 + 3\text{NaOH} \longrightarrow 3\text{CH}_3(\text{CH}_2)_{16}\text{COO}^- \text{Na}^+ + \\
| \\
\text{H}_2\text{COOC(CH}_2)_{16}\text{CH}_3
\end{array}
$$

Sodium octadecanoate
(sodium stearate)

$$
\begin{array}{l}
\text{CH}_2\text{OH} \\
| \\
\text{CHOH} \\
| \\
\text{CH}_2\text{OH}
\end{array}
$$
Propane-1,2,3-triol

When potassium hydroxide is used, a softer, milder soap is obtained. The process of making soap is known as **saponification**. On completion of saponification, sodium chloride is added to 'salt out' the soap (*see* section 8.2).

> Tallow is the fat obtained from cattle and sheep used for making soap and candles.

Soaps are metal salts of long chain fatty acids

Soapy lather turns this little boy in Upper Volta into a bizarre piece of sculpture

The cleaning action of soap is due to the affinity of soap anions for both grease and water. The anionic carboxylic group has an affinity for water. It is **hydrophilic**. The hydrocarbon chain of the fatty acid has an affinity for grease, oil or dirt. This is the **hydrophobic** or water-hating end. This end dissolves in the grease deposit and encapsulates it to form a **micelle**. The micelles are removed by rinsing with water (*see* figure 20.20).

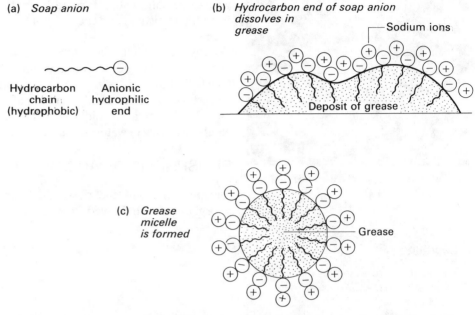

(a) *Soap anion*

Hydrocarbon   Anionic
chain   hydrophilic
(hydrophobic)   end

(b) *Hydrocarbon end of soap anion dissolves in grease*

Sodium ions

Deposit of grease

(c) *Grease micelle is formed*

Grease

Figure 20.20   How soap cleans. The hydrocarbon chains of soap anions (a) dissolve in grease (b). Grease micelles are formed (c) and carried away by rinsing

Soaps form insoluble calcium salts with the calcium ions in hard water. This is known as 'scum' and it results in a wastage of soap. With synthetic detergents a scum is not formed and so there is no wastage.

$$R-\underset{\substack{\| \\ O}}{\overset{O}{C}}-O^- \ Na^+ \qquad R-\underset{\substack{\| \\ O}}{\overset{O}{\underset{\|}{S}}}-O^- \ Na^+$$

A soap   A typical synthetic
                          detergent

A **detergent** is a cleansing agent although the term is commonly taken to mean a soapless or synthetic detergent which does not produce a scum in hard water. The alkylbenzene sulphonates are common examples of detergents:

$$CH_3-(CH_2)_n-CH_2-\underset{}{\bigcirc}-SO_3^- \ Na^+$$

Alkylbenzene sulphonate

Straight chain alkylbenzene sulphonates are slowly biodegradable. As the number of branches in the alkyl group increases the ability of bacteria to break down the alkyl group decreases. Highly branched alkylbenzene sulphonates do not undergo biodegradation and for this reason are no longer used in the United Kingdom.

A typical packet of detergent contains about 20% of the detergent and about 30% inorganic phosphates. The phosphates remove soluble calcium salts. Unfortunately, the phosphates end up in the sewage which is often released into streams, rivers, lakes or the oceans. The phosphates act as nutrients for certain algae. This results in a proliferation of green plants—especially in enclosed waters such as lakes and fjords. The green plants consume the oxygen in the water with the result that the aquatic plants and animals eventually die and decay. Such a process is known as **eutrophication**.

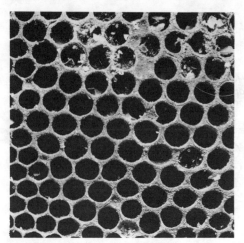

Honeycombs of bees are a source of $C_{15}H_{31}COOC_{30}H_{61}$!

Lecithins

The lecithins are examples of the type of phospholipid shown here. They are choline esters:

R = saturated fatty acid
R′ = unsaturated fatty acid
R″ = choline, $HOCH_2CH_2\overset{+}{N}(CH_3)_3$

Lecithins are important constituents of cell membranes—particularly in the brain. They are involved in the metabolism of fat by the liver.

WAXES

Waxes are plastic—that is pliable—materials which can be moulded when warm but become hard when cold. They are insoluble in water. Naturally occurring waxes are solid esters formed by fatty acids with alcohols other than glycerol. Both the fatty acid and the alcohol components may each contain over 30 carbon atoms.

Beeswax, which is a white or yellowish plastic substance obtained from the honeycombs of bees, contains an ester of palmitic acid with the formula $C_{15}H_{31}COOC_{30}H_{61}$. Wool wax contains a substance known as lanolin. This consists of the cholesterol esters of palmitic, oleic and stearic acids. Cholesterol is an example of a steroid (*see* below). It is resistant to acids and alkalis and emulsifies easily in water. It is used for making skin ointments and hair preparations.

PHOSPHOLIPIDS AND GLYCOLIPIDS

Phospholipids are the phosphate derivatives of lipids. One major group of phospholipids are derived from propane-1,2,3-triol. They have the general formula

$$H_2C-O-R$$
$$R'-O-C-H \qquad O$$
$$H_2C-O-P-OR''$$
$$OH$$

A phospholipid

Phospholipids are found in animal tissues and organs. In the human body they are produced by the liver and the small intestine and are involved in many metabolic processes.

Glycolipids are carbohydrate derivatives of the lipids. The carbohydrate is usually glucose or galactose. Some are derived from propane-1,2,3-triol and have the general formula

$$H_2COR$$
$$R'OCH$$
$$H_2COR''$$

A glycolipid

where R″ is the carbohydrate.

STEROIDS AND HORMONES

The **steroids** all contain the following ring system:

One of the most important steroids is **cholesterol**. It has the structure

Cholesterol is found in virtually all animal tissues particularly in the blood, brain and also in the spinal column and gallstones.

Cholesterol derivatives are important in the digestive systems—occurring in bile acids—and also as hormones. However, a diet which is rich in animal fats and thus cholesterol can—together with other factors—lead to a disease known as atherosclerosis. In this disease, the blood flow through arteries becomes restricted due to deposits of cholesterol derivatives. This can result in heart attacks, strokes and gangrene. Cigarette smoking, obesity, inactivity and a diet high in refined sugar also contribute to atherosclerosis.

Some of the most important cholesterol derivatives in the digestive system are the sodium salts of cholic acid and related acids:

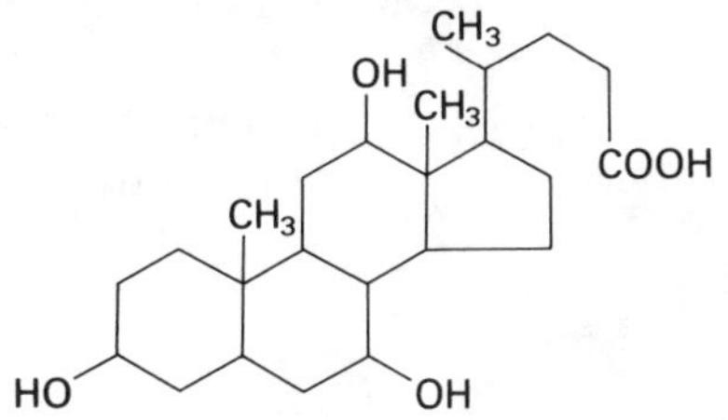

Cholic acid

These are found in bile. Bile is secreted by the liver and stored in the gall bladder. It emulsifies fats and thus helps their digestion by enzymes such as lipase. This enzyme breaks down the fat into fatty acids and propane-1,2,3-triol.

The **sex hormones** are also steroids. These control the sexual development and functioning of the body. An example is progesterone:

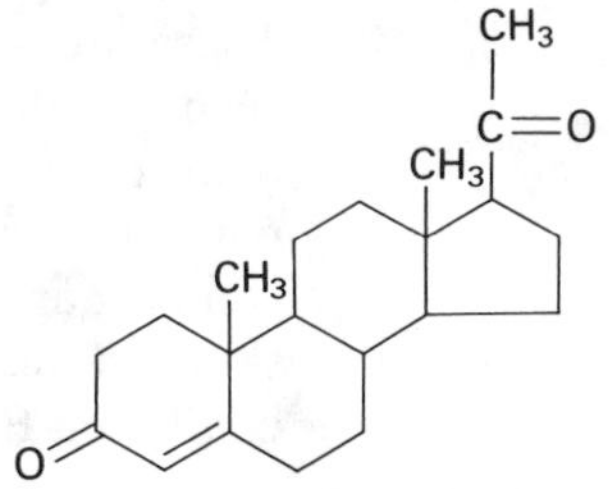

Progesterone

Synthetic female sex hormones are used as oral contraceptives.

Male sex hormones are known as **androgens**. Testosterone, for example, is excreted by the testes:

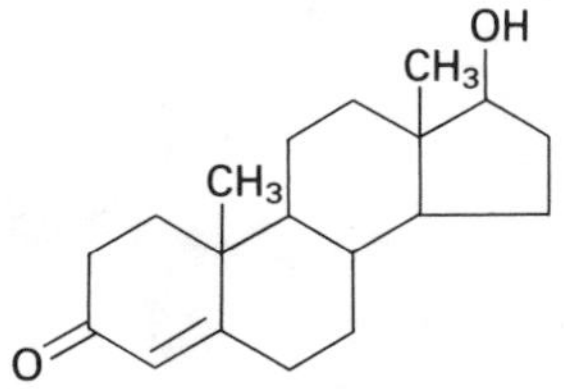

Testosterone

Steroids occur not only in animals but also in plants. Ergosterol, for example, occurs in yeast:

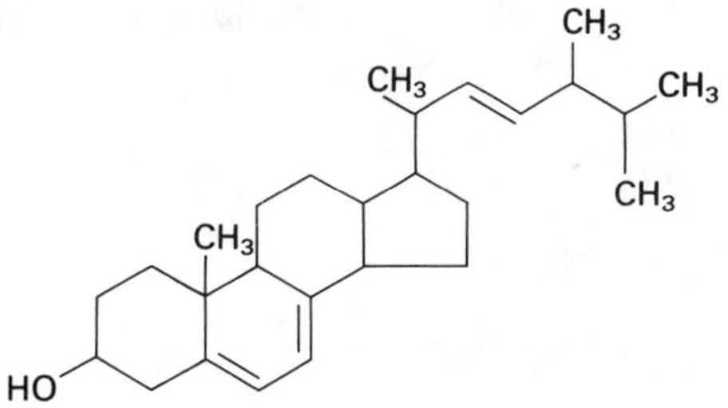

Ergosterol

Ergosterol is a plant sterol. Sterols are steroid alcohols. Cholesterol is also a sterol.

VITAMINS

These organic compounds were formerly thought to be amines. They were named 'vital amines', hence 'vitamins'.

Vitamins are a group of organic compounds which are required in very small amounts for the healthy growth and functioning of animal organisms. They cannot be made by organisms and thus have to be supplied in the diet. They do not have any chemical structural features in common and they are chemically different from the three main nutrients—fats, carbohydrates and proteins. Absence of a vitamin in a diet can cause a specific deficiency disease.

Vitamins A, D, E and K are all fat-soluble substances. The vitamin B complex and vitamin C are water-soluble.

Vitamin A

This is also known as retinol. It is a highly unsaturated alcohol with the structure

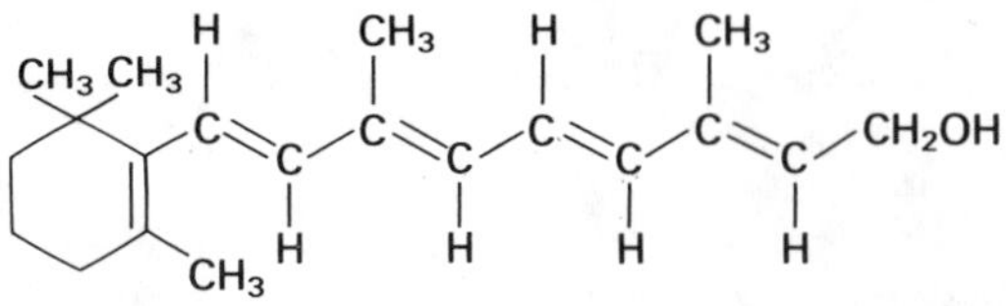

Vitamin A occurs free or as an ester in milk, butter, eggs, fish liver oils and vegetables such as cabbage and carrots. It is required for growth and vision. Deficiency can cause stunted growth and blindness.

Vitamin B Complex

These are a group of water-soluble vitamins which are found together in foods such as milk, liver and cereals. They do not have a common chemical structural feature. They include vitamins B_1, B_2, B_6, B_{12}, biotin, folic acid, nicotinic acid and pantothenic acid. They all function as coenzymes (see section 9.5). The vitamin B complex are required for the release of energy from food and to promote healthy skin and muscle. Deficiency can cause various diseases. For example, deficiency of vitamin B_1 (thiamin) can lead to beriberi. Deficiency of vitamin B_{12} can cause pernicious anaemia. B vitamins such as vitamin B_2 (riboflavin) and niacin (nicotinic acid) are commonly added to breakfast cereals.

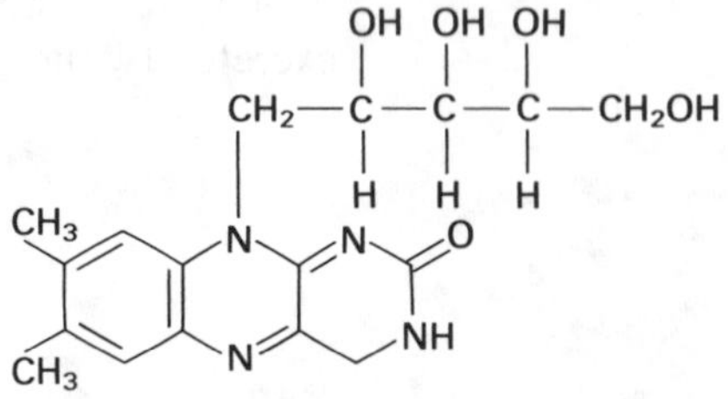

Vitamin B_2
(riboflavin)

Vitamin C

This is also known as ascorbic acid.

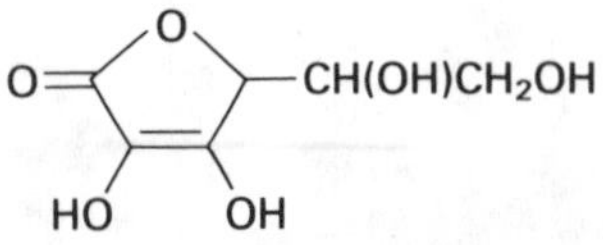

Vitamin C
(ascorbic acid)

It is found in citrus fruits, green vegetables, potatoes and other fruits and vegetables. Vitamin C is necessary for maintaining healthy skin and helps cuts and abrasions to heal properly. Deficiency leads to scurvy.

Vitamin D

This is a fat-soluble vitamin which regulates the absorption of calcium and phosphate from the intestine and promotes the formation of bone. Vitamin D occurs in two forms. Vitamin D_2 (also known as calciferol) is produced from ergosterol (*see* above) by plants when exposed to sunlight. The other, vitamin D_3 (also known as cholecalciferol), is produced in the skin when exposed to sunlight.

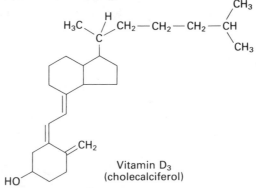

Vitamin D_3
(cholecalciferol)

Deficiency of vitamin D due to lack of sunlight or a poor diet can lead to the development of rickets.

ALKALOIDS

These are a wide group of nitrogen-containing bases which are found in the seeds, berries and roots of certain plants. They have potent effects on the body and can be very poisonous. Strychnine, for example, is an alkaloid. Alkaloids are used as drugs. Morphine, quinine, atropine and codeine are all alkaloids. Caffeine, which is a stimulant found in tea and coffee, is one of a group of compounds derived from the alkaloids known as purines.

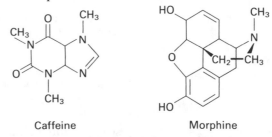

Caffeine   Morphine

The medicinal plant *Rauwolfia serpentina*. Its powdered whole root has been used in India for centuries as a tranquilliser for the mentally ill. In 1952 the alkaloid reserpine was isolated from the crude drug. Other *Rauwolfia* alkaloids have since been isolated. The powdered root was first used in Western medicine in 1953 and by the mid-1950s reserpine was in general use as a drug. Reserpine has now been replaced in psychiatric therapy by other drugs. However it is used in treating high blood pressure (hypertension). The drug is still obtained from plant sources although it has been synthesised. It has the formula $C_{33}H_{40}N_2O_9$

SUMMARY

1. **Lipids** are soluble in organic solvents but insoluble in water.
2. **Fats** and **oils** contain one or more **triglycerides**.
3. Triglycerides are esters of **fatty acids**.
4. Fats are solid and oils liquid at room temperature.
5. Triglyceride esters yield propane-1,2,3-triol and fatty acids on **hydrolysis**.
6. **Unsaturated fatty acids** contain at least one double bond.
7. The **degree of unsaturation** in a fat or oil is given by its **iodine number**.
8. Soaps are metal salts—commonly the sodium and potassium salts—of long chain fatty acids.
9. The anionic end of a soap anion is **hydrophilic**, that is, it has an affinity for water. The hydrocarbon chain of a soap anion has an affinity for grease but not water. It is **hydrophobic**.
10. Phosphates from detergents can cause **eutrophication**.
11. **Phospholipids** are phosphate derivatives of lipids.
12. Cholesterol and the sex hormones are **steroids**.
13. **Vitamins** are required in very small amounts for the healthy growth and functioning of animal organisms.

20.4
Polymers

A **polymer** is a compound consisting of large molecules built up by the repetition of small molecular units. Indeed, the term 'polymer' derives from the Greek word *poly* meaning 'many' and *meros* meaning 'parts'. The simple compound from which a polymer is made is known as a **monomer**. The monomer is usually the same as or similar to the repeat unit in the polymer. In the case of poly(ethene), for example, the repeat unit is $-CH_2-CH_2-$:

$$n \ CH_2\text{=}CH_2 \ \rightarrow \ \text{---}[CH_2\text{--}CH_2]_n$$

ethene poly(ethene)

monomer *polymer*

A copolymer is built up of more than one type of monomer.

Polymers may be linear, branched chain or have three-dimensional network structures (*see* figure 20.21). Network polymers are also known as crosslinked polymers.

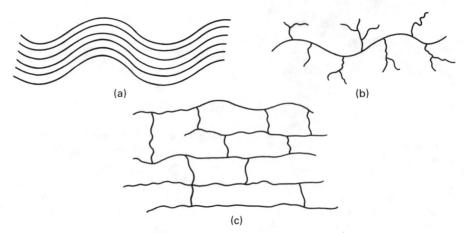

Figure 20.21 Polymer structures: (a) linear polymers; (b) branched chain polymers; (c) crosslinked polymers

- **Linear polymers** are well packed and thus have high densities, high tensile strength and high melting points. High-density polythene is an example.
- **Branched chain polymers** are irregularly packed and thus have lower tensile strength and melting points than linear polymers. Low-density polythene is an example.
- **Crosslinked** or **network polymers** are hard, rigid and brittle. Bakelite is an example.

The repeat units in **linear copolymers** may be arranged randomly, in blocks or alternately. For example, a copolymer built up from monomer A and monomer B might have the following arrangements:

linear alternating copolymer	–A–B–A–B–A–B–A–B–A–B–A–B–
linear block copolymer	–A–A–A–B–B–B–A–A–A–B–B–B–
linear random copolymer	–A–A–B–A–B–A–B–B–B–A–B–A–

The length of a polymer chain is the number of repeat units in the chain. This is called the **degree of polymerisation**. The degree of polymerisation of **high polymers** may range from several hundreds to tens of thousands. Relative molecular masses of these polymers vary from 10 000 to over 1 000 000.

Naturally occurring polymers include starch, cellulose and other polysaccharides, proteins, nucleic acids and natural rubber. Poly(ethene), pvc, Bakelite, nylon and silicones are examples of synthetic polymers.

The term **plastic** refers to any solid material which becomes mobile when heated and can thus be cast into moulds, extruded to form rods or tubes or used to form laminated produces (layers or sheets) and surface coatings. However, the term plastics is normally taken to include synthetic polymers only.

Plastics which become soft and melt on heating and as a result can be moulded or remoulded are known as thermosoftening plastics or simply thermoplastics. Poly(ethene) and Perspex are examples of synthetic thermoplastics. Natural thermoplastics include shellac and the cellulose plastics. Thermoplastics are generally linear polymers or linear copolymers. Their ability to soften easily is due to the absence of strong bonds between the chains.

Plastics consisting of three-dimensional networks cannot be softened easily on heating and thus cannot be remoulded. They are called **thermosetting resins** or thermosetting plastics. They are hard and infusible (this means they cannot be fused or melted). This is due to crosslinking. Examples of thermosetting resins include

- phenol resins,
- amino resins,
- epoxy resins,
- unsaturated polyester resins,
- urethane foams.

The term 'resin' is often loosely applied to synthetic polymers. At one time the term only applied to viscous sticky materials secreted as sap by trees, plants and some insects. Natural resins include shellac (the basis of lacquer), laudanum (tincture of opium) and myrrh (a gum resin used in medicine and perfumery).

Natural and synthetic polymers which can be drawn out as threads to be spun and woven are known as fibres. They include rayon and nylon. Natural and synthetic rubbers are also polymers. They are known as elastomers due to their property of elasticity. Elasticity is the tendency of an object or material to revert to its original shape or size after stretching, compression or deformation.

POLYMERISATION

The process of forming polymers from monomers is known as polymerisation. There are two modes of polymerisation—addition polymerisation and condensation polymerisation.

Addition Polymerisation

Addition polymerisation involves the repeated addition of monomers to the polymer chain. The monomers are unsaturated compounds and commonly derivatives of ethene.

Addition polymerisation is also known as chain reaction polymerisation since the process involves chain reactions (*see* chapter 9). The mechanism of the chain reaction can involve radicals, ions or coordination compounds.

As a typical example of a polymerisation mechanism, let us consider the radical polymerisation of ethene. This process can be initiated by the hydroxyl radical $HO\cdot$. It involves the usual three steps of a chain mechanism:

1. **Initiation**.
$$HO\cdot + H_2C{=}CH_2 \rightarrow HO{-}CH_2{-}CH_2\cdot$$
2. **Propagation**.

$$HO{-}CH_2{-}CH_2\cdot + H_2C{=}CH_2 \rightarrow HO{-}CH_2{-}CH_2{-}CH_2{-}CH_2\cdot$$
$$HO{-}CH_2{-}CH_2{-}CH_2{-}CH_2\cdot + H_2C{=}CH_2 \rightarrow$$
$$HO{-}CH_2{-}CH_2{-}CH_2{-}CH_2{-}CH_2{-}CH_2\cdot$$

The chain grows longer by further additions of ethene molecules.

3. **Termination**. This can occur when two radicals combine.

$$-CH_2{-}CH_2\cdot + \cdot CH_2{-}CH_2{-}CH_2{-} \rightarrow -CH_2{-}CH_2{-}CH_2{-}CH_2{-}CH_2{-}$$

Addition polymers have the same empirical formula as their monomers. They are linear or branched and are thus thermoplastic.

Poly(ethene)

This has the trade name polythene. It is manufactured by various techniques. **Low-density poly(ethene)**—which has a relative molecular mass of less than 300 000—is made by the **ICI high-pressure process**. In this process, oxygen is used as a radical initiator. Ethene, containing a trace of oxygen, is compressed under a pressure of over 1500 atmospheres at a temperature of about 200°C. Organic peroxides are used to initiate the reaction.

Over recent decades the application of science and technology to traditional agriculture in Third World countries has steadily increased production of food grains. However, it has been estimated that up to 20% of this produce never reaches the consumer. One of the problems is inadequate storage facilities. Stored grain becomes vulnerable to attack by insects, rodents, moisture, fungus, mites and birds. In this picture we see the use of polythene sheeting to cover stored grain bags prior to fumigation against insects at a village near Madras, India

High-density poly(ethene)—which can have a relative molecular mass of up to 3000 000—is made by the **Ziegler process**. The process is carried out at about 60°C and a pressure of 2 to 6 atmospheres. The ethene is passed through an inert aromatic hydrocarbon solvent containing a suspension of triethylaluminium and titanium(IV) chloride as catalyst. The polymerisation proceeds by an ionic mechanism. On completion of the polymerisation, dilute acid is added to decompose the catalyst. The polymer is then separated by filtration.

Low-density poly(ethene) is used to make squeeze bottles, washing-up bowls, bags for food and clothes and numerous other items. High-density polythene is used to make more rigid items such as refrigerator ice trays and milk bottle crates.

Poly(propene)

This is also known as polypropylene. It is made by the polymerisation of propene:

$$n\,CH_3-CH{=}CH_2 \longrightarrow \left[CH-CH_2\right]_n$$
$$\underset{\displaystyle CH_3}{|}$$

High-density poly(propene) is made by a process developed by the Italian chemist Giulio Natta. The propene is passed under pressure through heptane containing a suspension of an organometallic catalyst.

The German chemist Karl Ziegler (1898–1973) and the Italian chemist Giulio Natta (1903–1979) shared the Nobel Prize for Chemistry in 1963 for their work on polymerisation.

Poly(propene), in common with polymers formed from monosubstituted alkenes, can exist in three configurations. If all the methyl groups (or substituents) lie above the plane of the main chain of carbons then the polymer is said to be **isotactic** (*see* figure 20.22). If the methyl groups or substituents lie alternately above and below the plane the polymer is said to be **syndiotactic**. A random configuration is called **atactic**. By varying the catalyst and the conditions of the polymerisation, it is possible to control the percentage composition of the different configurations in the product.

Poly(propene) is used to make pipes, valves, wrapping films, ropes and other products.

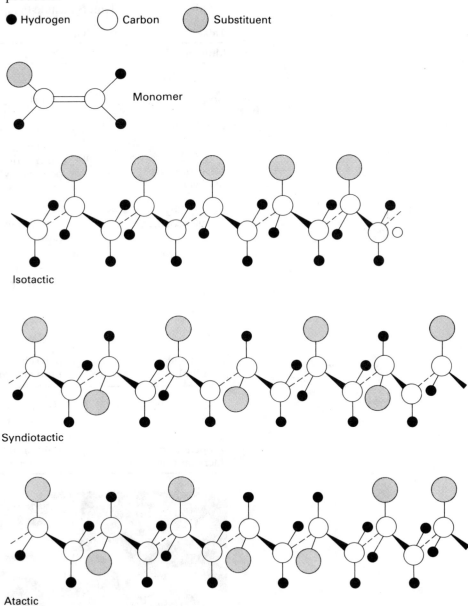

Figure 20.22 Configuration of polymers formed from monosubstituted alkenes

Poly(chloroethene)

This is also known as polyvinyl chloride or pvc. It is made by the polymerisation of chloroethene (or vinyl chloride):

$$n\text{CH}_2\!\!=\!\!\text{CHCl} \longrightarrow \ \ \text{---}\!\!\left[\text{CH}_2\!-\!\text{CH}\right]_n\!\!\text{---}$$

$$\underset{\text{Chloroethene}}{} \qquad \underset{\text{Poly(chloroethene)}}{\overset{|}{\text{Cl}}}$$

The manufacture of chloroethene is described in chapter 18. Poly(chloroethene) is manufactured by heating the chloroethene in an inert solvent containing a radical initiator.

Esters of benzene-1,2-dicarboxylic acid are added to pvc to soften it. Such additives are known as **plasticisers.** Materials which are added to plastics to increase their bulk are known as fillers. Car tyres contain about 80% of the filler carbon black. Other substances which are added to plastics are stabilisers—mainly antioxidants—and dyes and pigments to colour the plastics.

Poly(chloroethene) is used to make records, insulators, floor coverings, pipes and a variety of household goods.

Other addition polymers include Perspex, PTFE and polystyrene (*see* table 20.10).

Table 20.10   Addition polymers

Monomer	Polymer	Uses
$nCH_2{=}CH{-}COOCH_3$ (with CH_3 on the CH) methyl 2-methyl-propenoate (*also known as* methyl methacrylate)	$+CH_2{-}C+_n$ (with CH_3 above and $COOCH_3$ below the C) poly(methyl 2-methyl-propenoate) (*also known as* poly(methyl methacrylate) or Perspex)	windows, lenses, packaging
$nCF_2{=}CF_2$ → tetrafluoroethene	$+CF_2{-}CF_2+_n$ poly(tetrafluoroethene) (*also known as* PTFE *or* Teflon)	electrical insulators, non-stick coatings on frying pans
(phenyl ring)$CH{=}CH_2$ phenylethene (styrene)	$+CH{-}CH_2+_n$ (with phenyl ring) poly(phenylethene) (*also known as* polystyrene)	hot drink cups, toys, household articles
$nCH_2{=}CHCN$ → propenonitrile (acrylonitrile)	$+CH_2{-}CH+_n$ (with CN below) poly(propenonitrile) (*also known as* Acrilan)	fabrics

Polystyrene trays used for packaging tomatoes

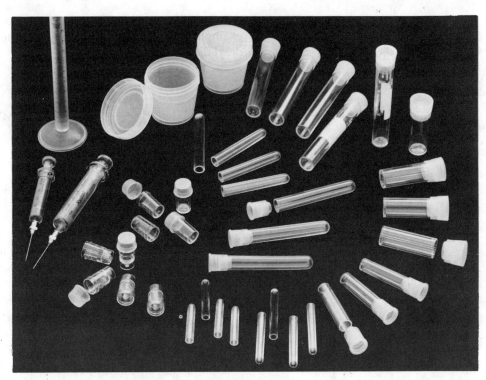

Disposable hospital ware. The tubes, phials and sample bottles are made of crystal polystyrene and the screw-top jars from low-density polythene

Condensation Polymerisation

Condensation polymerisation involves a series of condensation reactions involving two monomers. Each monomer normally contains two functional groups. The condensation reactions result in the loss of small molecules—usually water.

Polyesters

These are condensation polymers. They are used as synthetic fibres in place of cotton and wool. Polyesters are formed by the condensation of a dihydric alcohol and dibasic carboxylic acid. For example, poly(ethane-1,2-diyl benzene-1,4-dicarboxylate) which is marketed under the trade name Terylene or, in the USA, Dacron, is formed by heating ethane-1,2-diol with benzene-1,4-dicarboxylic acid. Note that both these monomers are bifunctional. The former is a dihydric alcohol and the latter a dibasic carboxylic acid. The name Terylene derives from terephthalic acid and ethylene glycol.

$$n\left\{\cdots HO - \overset{O}{\overset{\|}{C}} - \bigcirc - \overset{O}{\overset{\|}{C}} - OH + HO - CH_2CH_2OH \cdots \right\}$$

Benzene-1,4-dicarboxylic acid (terephthalic acid)   Ethane-1,2-diol (ethyleneglycol)

$$\cdots \left[\overset{O}{\overset{\|}{C}} - \bigcirc - \overset{O}{\overset{\|}{C}} - O - CH_2CH_2 - O \right]_n \cdots + 2nH_2O$$

Terylene

Nylon

This is also a condensation polymer. It is used as a synthetic fibre. There are various types of nylon. They are all polyamides. Some are formed by the condensation of diamines with chloride derivatives of dicarboxylic acids. For example, nylon 6,6 is formed by heating hexane-1,6-diamine with hexane-1,6-dioyl dichloride:

Hexane-1,6-diamine   Hexane-1, 6-dioyl dichloride

$$+HN-(CH_2)_6-NH-\overset{O}{\overset{\|}{C}}-(CH_2)_4-\overset{O}{\overset{\|}{C}}+_n \ + \ 2n\,HCl$$

Nylon 6,6

Nylon 6 is formed by the prolonged heating of caprolactam with a trace of water. Caprolactam is a heterocyclic compound. It is synthesised from cyclohexanone oxime, which in turn is synthesised in several stages from phenol:

Phenol   Cyclohexanone oxime   Caprolactam

Other Condensation Polymers

Polyurethanes, polyureas and the silicones are also condensation polymers. They have the linkages shown in table 20.11.

Table 20.11   Condensation polymer linkages

Polymer	Linkage
polyesters	$-O-\overset{O}{\overset{\|}{C}}-$
nylon	$-\overset{O}{\overset{\|}{C}}-NH-$
polyurethanes	$-O-\overset{O}{\overset{\|}{C}}-NH-$
polyureas	$-NH-\overset{O}{\overset{\|}{C}}-NH-$
silicones	$-\underset{R}{\overset{R}{Si}}-O-\underset{R}{\overset{R}{Si}}-$ or $-\underset{R}{\overset{R}{Si}}-O-\underset{\underset{R-Si-R}{O}}{\overset{R}{Si}}-$

NATURAL AND SYNTHETIC RUBBERS

We have already seen how **natural rubber** is obtained from the latex which is found in the bark of tropical and sub-tropical trees (*see* chapter 18). We also saw how rubber consists of methylbuta-1,3-diene units. The average chain length is about 5000 units. These can exist in two isomeric forms:

cis-isomer

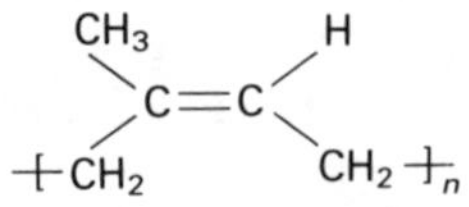

trans-isomer

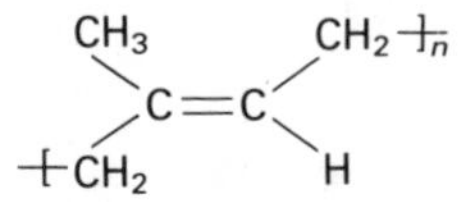

Natural rubber is mainly the *cis*-isomer, which is found in elastic. The *trans*-isomer is non-elastic. This is called gutta-percha—obtained from a species of tree called (Malay) percha.

Early rubber products such as tubing, elastic bands and waterproof products did not last long and became hot and sticky in hot weather. In common with other thermoplastics, natural rubber also became hard and brittle on cooling. This was due to a certain amount of crosslinking between the polymer chains.

In 1838 an American, Charles Goodyear, discovered that these defects could be overcome by heating rubber with sulphur. The process is known as **vulcanisation** and is due to the formation of disulphide crosslinks between the polymeric chains:

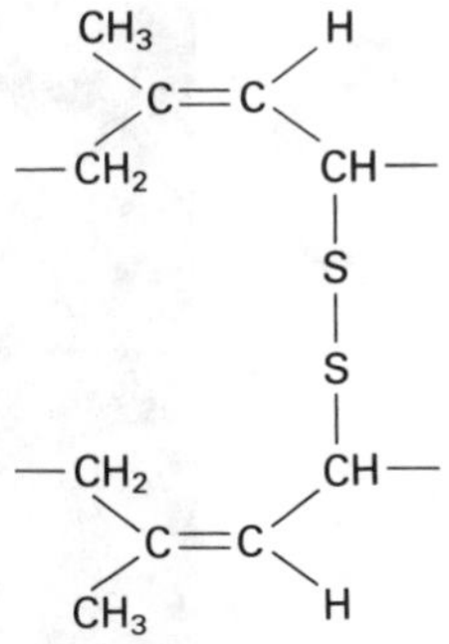

Vulcanisation is a slow process. The process is speeded up by accelerators. These are often sulphur-containing organic compounds. They form radicals in the process. Accelerators are often mixed with activators or promoters—typically metallic oxides such as zinc oxide.

Natural rubbers are sensitive to heat, light and especially oxygen. Ageing of the rubbers can be retarded by the use of anti-oxidants such as the phenyl naphthalen-1-amines (*see* section 19.4).

The first **synthetic rubber** was produced by the polymerisation of 2-chorobuta-1,3-diene:

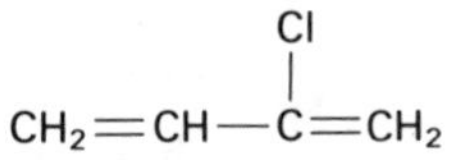

$$CH_2 = CH - \underset{\underset{Cl}{|}}{C} = CH_2$$

2-Chlorobuta-1,3-diene
(chloroprene)

The product is known as **neoprene**. This polymer is resistant to chemical attack and is still used to make hoses for petrol and oil and containers for corrosive chemicals.

Various types of rubber are made from buta-1,3-diene and its derivatives. The most important in terms of quantities produced and used is styrene–butadiene rubber (SBR). SBR is produced by the copolymerisation of phenylethene (styrene) and buta-1,3-diene (butadiene):

Neoprene-coated steel water-injection pipes are laid in the Central Cormorant oil-field in the North Sea

SBR can be vulcanised in the same way as for natural rubber. The above equation shows a 1:1 ratio between the two monomers. In practice, the ratio is usually three parts buta-1,3-diene to one part phenylethene. An increase in the phenylethene content increases the plasticity of the rubber. High phenylethene polymers are used in latex paints.

Between 60 and 70% of all rubber is used to make tyres. The next most important rubber product is footwear, although this accounts for only about 4% of total consumption. SBR is usually used for car tyres because of its good wear and grip. Larger tyres contain mixtures of natural and synthetic rubbers and aircraft tyres are 100% natural rubber. Natural rubber is far better than synthetic rubber in withstanding heat.

Both natural and synthetic rubber products contain only about 60% pure rubber. We have already seen how sulphur and additives are added to improve the properties of the rubber. A filler such as carbon black is also added to increase the strength and stiffness of the rubber. This is why tyres are usually black. Oil is also added during rubber manufacture to help processing and to reduce costs.

In the last few decades the production of natural rubber has failed to meet the demand for rubber and the production of synthetic rubber has now overtaken that of natural rubber (*see* table 20.12).

Table 20.12   Annual production of natural and synthetic rubbers

| Year | Annual production/tonnes $\times 10^3$ | |
	Natural	Synthetic
1900	45	–
1925	535	–
1950	1890	534
1975	3315	6855
2000	8000 (est)	18000 (est)

CELLULOSE FIBRES

Natural fibres such as cotton, flax and jute contain up to 90% cellulose. We have already seen that cellulose is a naturally occurring polymer—a polysaccharide. Cellulose is used to manufacture a number of semi-synthetic fibres. These are known as rayon. The cellulose required for this manufacture is normally obtained from wood.

Cotton picking near Selcuk, Turkey

Wood consists of about 50% cellulose and 30% lignin. Lignin is another naturally occurring polymer—although it is not a carbohydrate. The skeleton structure of the lignin unit is

To produce cellulose a softwood such as pine or fir is usually used. The lignin is removed by heating wood shavings with a solution of calcium hydrogensulphite, $Ca(HSO_3)_2$, containing excess sulphur dioxide. The lignin dissolves and the

Paper

Wood pulp is used to make paper. In the paper making process chlorine and chlorates are used for bleaching. Materials such as clay, talc, titanium(IV) oxide and barium sulphate are also used to improve the weight and quality of the paper.

cellulose fibres are separated by filtration. The product is beaten to produce wood pulp. Pure cellulose is obtained by treating the wood pulp with Schweitzer's reagent. This is an ammoniacal solution of copper(II) hydroxide. On addition of dilute mineral acid, the pure cellulose is precipitated.

Cellulose ethanoate (cellulose acetate) is made by the acetate process. The cellulose is treated with ethanoic acid or ethanoic anhydride. The hydroxyl groups in the cellulose are replaced by ethanoate groups. Cellulose ethanoate is non-flammable and has a shiny appearance. It is used to make lacquers and varnishes and cine films.

Nitrocellulose is obtained by treating cellulose with nitric and sulphuric acids. The OH groups in the cellulose are replaced in succession by ONO_2 ester groups. Nitrocellulose with low nitrogen content is used in lacquers. As the nitrogen content increases the nitrocellulose oxidises faster. Nitrocellulose with a high nitrogen content oxidises explosively. This type of nitrocellulose is called guncotton and used as a propellant for bullets and shells.

Celluloid—which was once used for making cine films—is obtained by treating cellulose with dilute nitric acid and mixing the product with camphor. When ignited celluloid burns furiously. It is known to have caused serious fires in cinemas and hospital X-ray areas. It has been replaced by cellulose ethanoate in the manufacture of films.

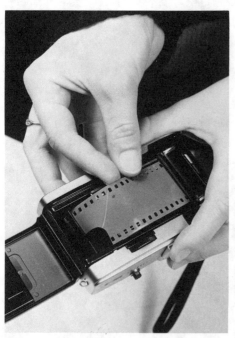

Celluloid has been replaced by cellulose ethanoate (cellulose acetate) in the manufacture of films

Nitrocellulose with a high nitrogen content oxidises explosively

Rayon is manufactured by adding cellulose to a mixture of sodium hydroxide and carbon disulphide. The product is an aqueous solution of sodium cellulose xanthate known as **viscose**:

$$\boxed{\ \ } \!-\! OH + NaOH + CS_2 \longrightarrow \boxed{\ \ } \!-\! O \!-\! C \!\underset{S^- \ Na^+}{\overset{O}{<}} + H_2O$$

Cellulose   Carbon   Sodium cellulose xanthate
        disulphide

When the viscose solution is forced through very fine holes into a bath of dilute sulphuric acid a fine thread of viscose rayon is precipitated. This thread can be spun and woven to make fabrics.

Cellophane is produced by forcing the viscose solution through fine slits into dilute sulphuric acid. Cellophane is a widely used transparent film used to package food, although increased amounts of poly(propene) are now being used for this purpose.

SUMMARY

1. A **polymer** consists of small repeating units known as **monomers**.
2. A **copolymer** consists of more than one type of monomer.
3. Polymers may be linear, branched chain or crosslinked.
4. A **plastic** is a solid material which becomes mobile when heated.
5. **Thermoplastics** soften upon heating and can be remoulded.
6. **Thermosetting resins** do not soften easily on heating and cannot be remoulded.
7. **Addition polymerisation** involves the repeated addition of monomers to the polymer chain. The process involves a chain reaction.
8. Poly(ethene) and poly(propene) are examples of addition polymers.
9. **Condensation polymerisation** involves a series of condensation reactions between two monomers.
10. Polyesters and nylon are examples of condensation polymers.
11. **Natural rubber** consists mainly of *cis*-methylbuta-1,3-diene units.
12. Styrene–butadiene rubber (SBR) is produced by the copolymerisation of phenylethene (styrene) and **buta-1,3-diene**.
13. **Cellulose** is used to manufacture rayon. This is a semi-synthetic fibre.

Examination Questions

1. The following is part of a protein chain (the bond angles are not correctly shown):

$$-N-CH-C-N-CH_2-C-N-CH-C-N-CH_2-C-$$

Draw the structure of two amino acids obtained on hydrolysis of this protein.

(SEB)

2. (a) 2,6-Diaminohexanoic acid (lysine) has the formula

$$H_2N(CH_2)_4CH(NH_2)COOH$$

Write down the formula of the predominant ionic organic species present in (i) strongly acidic solution, (ii) strongly alkaline solution, (iii) neutral solution, i.e. at pH 7.
(iv) Write down the structure of poly(lysine) showing the repeating unit of the polymer.
(v) If a specific poly(lysine) molecule contains 100 lysine residues, what are the sign and magnitude of its charge at neutral pH?

(b) Name and describe briefly *two* types of ordered structure which are found in many proteins but are usually absent in polypeptides.

(JMB)

3. Discuss the primary, secondary and tertiary structures of proteins, their hydrolysis to amino acids and the separation and analysis of mixtures of amino acids by paper chromatography.

(NISEC)

4. Describe the structures of *two* monosaccharides, *one* disaccharide, and the polysaccharides starch and cellulose.
Explain the importance of starch and cellulose in nature.

(NISEC)

5. (a) The principal carbohydrate component of milk is the disaccharide lactose.
    (i) One of the saccharide units in lactose is glucose. Sketch the structure of $(+)$-glucose pyranoside.
    (ii) Name the other saccharide unit in lactose.
    (iii) Explain, with reference to their molecular structures, why lactose will reduce Fehling's solution whereas the disaccharide sucrose, also containing a glucose sub-unit, will not.
  (b) The fat component of milk is present in the form of emulsified droplets.
    (i) What is the important structural requirement for a molecule to act as an emulsifying agent?
    (ii) Name *one* natural emulsifier in milk.
    (iii) How does the structure of the emulsion in milk differ from that in butter?
  (c) Dairy products are sources of vitamins A and B_2 and of minerals.
    (i) With which component of milk is each of the vitamins associated?
    (ii) What are the functional groups present in vitamin A?
    (iii) How does the structure of vitamin A reflect its stability?
    (iv) Name the polyvalent cation whose most important dietary source is dairy products.

(JMB)

6. (a) By means of general formulae (using R to denote alkyl groups) indicate the chemical structure of (i) proteins, (ii) fats and oils. Mark clearly, and name, the important linkages present in these food components.
  (b) Explain the functions of proteins and fats and oils in the diet and give one important example of each.
  (c) Give equations to show the products of hydrolysis of proteins and fats and oils. Indicate the conditions necessary to bring about the hydrolysis in the laboratory.
  (d) Glucose, maltose and starch are carbohydrates. By means of simple diagrams indicate the relationship between them and state the product of hydrolysis of these carbohydrates.
  (e) 'Cellulose is a carbohydrate similar in many respects to starch. It has no nutritional value but is essential in the human diet.' Explain this statement.

(UCLES)

7. (a) Copolymerisation of monomers A and B can produce a variety of polymers with a range of properties which are related to the proportions of A and B used. Condensation polymerisation of monomers C and D produces a single polymer with a fixed property.
      Give a specific example with equations, of the formation of
    (i) a copolymer.
    (ii) a condensation polymer, and account for the differences indicated above.
  (b) Explain how
    (i) ethene can produce both low-density and high-density polymers,
    (ii) propene can produce polymers with three different types of structure.
  (c) Explain the differences between thermosetting and thermoplastic materials and give an example of each.

(UCLES)

8. (a) The mechanism for an addition polymerisation can be divided into three stages, initiation, propagation and termination. Consider the polymerisation of an ethenyl (vinyl) monomer CH_2=CHX, using a suitable initiator, and answer the following questions.

    (i) Name a suitable initiator and state the appropriate experimental conditions.

    (ii) Indicate by means of an equation, or otherwise, the propagation step.

    (iii) Indicate by means of an equation, or otherwise, the termination step.

(b) Explain why an addition polymerisation reaction usually leads to a product having a wide distribution of chain lengths.

(c) Describe in molecular terms the change which takes place when an elastomer becomes a glass at the glass transition temperature.

(JMB)

9. (a) What is meant by condensation polymerisation?

(b) One possible route for the industrial production of nylon 6,6 (so called because each starting material contains six carbon atoms) is as follows:

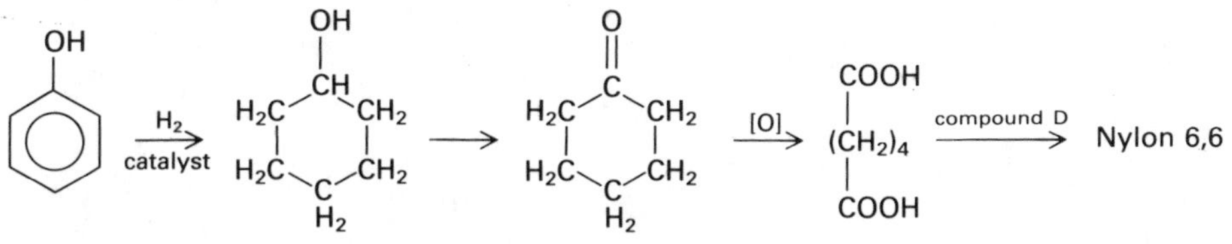

    (i) Write a balanced equation for the first step of the scheme, A to B.

    (ii) Name, or give the formula of, a reagent that could be used for converting B to C.

    (iii) Name, or give the formula for, compound D.

(c) Terylene was so named as it was prepared from '*tere*phthalic acid', i.e. benzene-1,4-dicarboxylic acid, and 'eth*ylene* glycol', i.e. ethane-1,2-diol.

    (i) Give the full structure of these two monomers.

    (ii) Write the full structural formula of a repeat unit of the polymer.

    (iii) What is the name of the functional group which links the monomers in the polymer chain?

    (iv) Benzene-1,4-dicarboxylic acid (molar mass 166 g) can be manufactured by oxidising 1,4-dimethylbenzene (molar mass 106 g). Write a balanced equation for this reaction using '[O]' as the oxidising agent.

    (v) What mass of 1,4-dimethylbenzene would be required to produce 5 kg of benzene-1,4-dicarboxylic acid if the yield of the diacid is only 90% of the theoretical yield?

(NISEC)

APPENDICES

**Appendix A
Physicochemical
Terms, Symbols
and Units**

The physicochemical terms, symbols and units used in this book are based on the International System of Units (commonly referred to as SI after the French *Système Internationale d'Unités*) and the recommendations of the International Union of Pure and Applied Chemistry (IUPAC). SI provides an unambiguous set of standard units for universal use.

There are seven **base SI units** from which units may be derived for all other physical quantities. SI is a **decimal system**. Derived units may thus be obtained by multiplication or division of the base units together with, where appropriate, multiplication by decimal multiples or sub-multiples. SI is also a **coherent system**. This means that if SI basic units are substituted into an equation the answer will automatically be in an SI unit.

The Seven SI Base Units

The names and symbols of the seven dimensionally independent physical quantities and their SI base units are shown in Table A.1.

Table A.1   The seven SI base units

Physical quantity	Symbol for quantity	Base unit	Symbol for base unit
length	l	metre	m
mass	m	kilogram	kg
time	t	second	s
electric current	I	ampere	A
thermodynamic temperature	T	kelvin	K
amount of substance	n	mole	mol
luminous intensity	I_v	candela	cd

SI Prefixes

The prefixes and symbols used to indicate multiples and sub-multiples of both SI base units and SI derived units are shown in Table A.2.

Table A.2   SI prefixes

Multiple	Prefix	Symbol	Sub-multiple	Prefix	Symbol
10	deca	da	10^{-1}	deci	d
10^2	hecto	h	10^{-2}	centi	c
10^3	kilo	k	10^{-3}	milli	m
10^6	mega	M	10^{-6}	micro	μ
10^9	giga	G	10^{-9}	nano	n
10^{12}	tera	T	10^{-12}	pico	p
10^{15}	peta	P	10^{-15}	femto	f
10^{18}	exa	E	10^{-18}	atto	a

Exception. Note that decimal multiples of the kilogram (kg) are formed by attaching an SI prefix to the gram (g) and not the kilogram even though the kilogram is the SI base unit of mass.

Derived SI Units

Table A.3 shows the recommended names and symbols of some SI derived units. Note that other SI derived units are commonly used for some of the physical quantities shown in this table. For example dm^3 and $mol\ dm^{-3}$ are commonly used for volume and concentration respectively. The litre (symbol l) is also acceptable in place of dm^3 for volume.

Table A.3   Some SI derived units

Physical quantity		Derived unit		
Name	Symbol	Name	Symbol	Definition
volume	V	cubic metre	m^3	
density	ρ	kilograms per cubic metre	$kg\ m^{-3}$	
speed of electromagnetic waves	c	metres per second	$m\ s^{-1}$	
frequency	ν	hertz	Hz	s^{-1}
force	F	newton	N	$kg\ m\ s^{-2}$
pressure	p	pascal	Pa	$kg\ m^{-1}\ s^{-2}$ $(= N\ m^{-2})$
energy enthalpy heat work	E H q w	joule	J	$kg\ m^2\ s^{-2}$
entropy	S	joules per kelvin	$J\ K^{-1}$	
concentration of solute B	$[B]$	moles per cubic metre	$mol\ m^{-3}$	
elementary charge (of a proton)	e	coulomb	C	A s
potential difference electromotive force standard electrode potential	V E $E^{\ominus}$	volt	V	$m^2\ kg\ s^{-3}\ A^{-1}$
electrical resistance	R	ohm	Ω	$m^2\ kg\ s^{-3}\ A^{-2}$
electrical conductance	G	siemens	S	$m^{-2}\ kg^{-1}\ s^3\ A^2$ $(= \Omega^{-1})$

Physicochemical Constants

The values of some useful physicochemical constants are shown in Table A.4.

Table A.4   Physicochemical constants

Quantity	Symbol	Value
speed of light in a vacuum	c	$2.998 \times 10^8\ m\ s^{-1}$
elementary charge (of proton)	e	$1.602 \times 10^{-19}\ C$
mass of electron	m_e	$9.110 \times 10^{-31}\ kg$
mass of proton	m_p	$1.673 \times 10^{-27}\ kg$
mass of neutron	m_n	$1.675 \times 10^{-27}\ kg$
atomic mass unit	m_u	$1.661 \times 10^{-27}\ kg$
Avogadro constant	L	$6.022 \times 10^{23}\ mol^{-1}$
gas constant	R	$8.314\ J\ K^{-1}\ mol^{-1}$
Boltzmann constant	$k = R/L$	$1.381 \times 10^{-23}\ J\ K^{-1}$
Faraday constant	$F = Le$	$9.648 \times 10^4\ C\ mol^{-1}$

Planck constant	h	6.626×10^{-34} J s
gravitational constant	$G^{\ominus}$	6.672×10^{-11} N m^2
molar volume of a gas at s.t.p.	V_{m}	2.24×10^{-2} m^3 mol^{-1}
Rydberg constant	R_{∞}	1.097×10^{7} m^{-1}

Tables and Graphs

A physical quantity is the product of a **numerical value** (also known as a **pure number**) and a unit

$$\text{physical quantity} = \text{numerical value} \times \text{unit}$$

example   5 kg   =   5 $\times$ kg

Entries in tables are numerical values. These express the ratio of the physical quantity to the unit

$$\text{numerical value} = \frac{\text{physical quantity}}{\text{unit}}$$

example   5 $=$ $\dfrac{5 \text{ kg}}{\text{kg}}$

The heading of each column or row of numerical values of a physical quantity in a table should thus be the physical quantity divided by units. Two examples of headings are shown in Table A.5.

The axes of graphs are number lines. They should be labelled with numerical values only. The ratio of the physical quantity to the unit represented by the numerical values on the axis is shown adjacent to the axis. A graph labelled in this way is shown in figure A.1. The graph is a plot of the values in table A.5.

The vertical axis of a graph (that is, the y axis) is used for the dependent variable and the horizontal axis (the x axis) is used for the independent variable.

Table A.5   Relation between time and mass of product

Time/s	Mass of product/kg
0	0
60	0.90
120	1.25
180	1.45
240	1.50

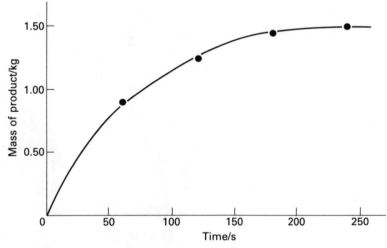

Figure A.1   Axes of graphs are labelled with numerical values

Appendix B
Safety Symbols

UK and EEC (European Economic Community) legislation requires all dangerous substances to be labelled and packaged to warn and protect the user. These regulations are known as **supply requirements**. Dangerous substances must also be labelled and packaged to warn and protect people engaged in road transport, the emergency services and the general public. These are known as **conveyance requirements**. All dangerous substances must be labelled for supply or conveyance or both.

CORROSIVE

OXIDISING

EXPLOSIVE

TOXIC

HARMFUL
or IRRITANT

HIGHLY
FLAMMABLE

Figure B.1   Supply symbols

Poster at an iron and steel plant in India

Supply symbols are shown in figure B.1. They must be shown with the designation of the product or substances, and the name and address of the manufacturer or other supplier. The label must also include appropriate 'risk phrases' and 'safety phrases'.

Conveyance labels must show one of the 13 diamond hazard warning signs (see figure B.2), the designation of the substance or product, the substance identification number and the name and address or telephone number of a person with the relevant expert knowledge of the dangers. If the volume in question is greater than 25 litres, then the nature of the dangers to which the substance may give rise and appropriate precautions that should be taken must also be known.

Safety signs give warning of a hazard and may be fixed to doors or cupboards in laboratories, for example. Examples are shown in figure B.3. Signs which indicate that certain behaviour is prohibited are known as **prohibition signs**. **Mandatory signs** indicate that a specific course of action must be taken. **Safe condition signs** provide information on safe conditions.

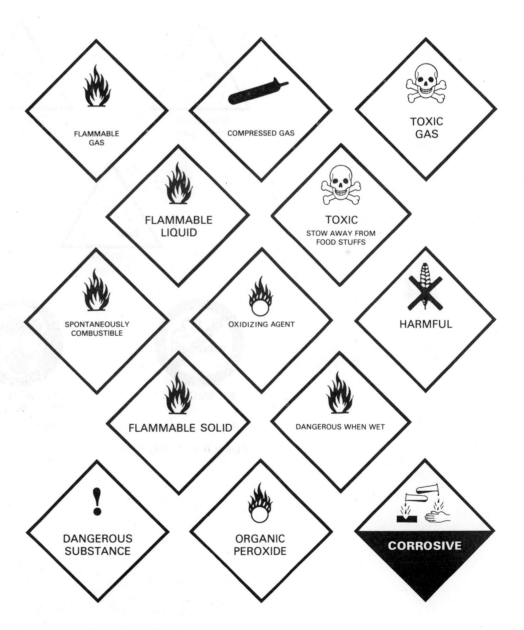

Figure B.2 Hazard warning signs

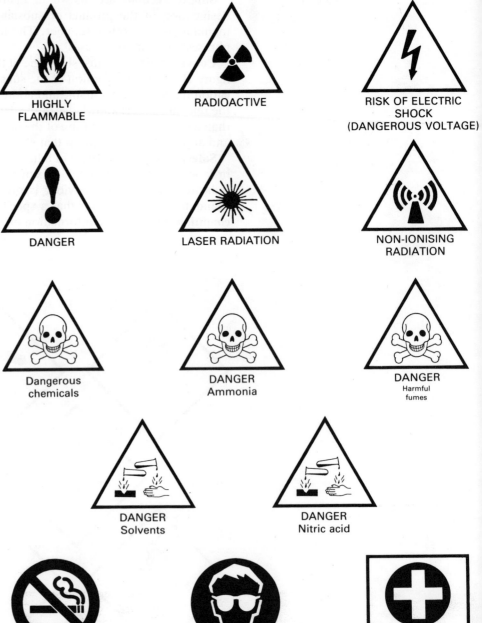

HIGHLY
FLAMMABLE

RADIOACTIVE

RISK OF ELECTRIC
SHOCK
(DANGEROUS VOLTAGE)

DANGER

LASER RADIATION

NON-IONISING
RADIATION

Dangerous
chemicals

DANGER
Ammonia

DANGER
Harmful
fumes

DANGER
Solvents

DANGER
Nitric acid

NO SMOKING

Prohibition sign

EYE PROTECTION
MUST BE WORN

Mandatory sign

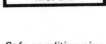

first-aid

Safe condition sign

Figure B.3   Safety signs

Appendix C
Relative Atomic Masses

Table C.1 lists elements with atomic numbers 1 to 103 in their alphabetical order together with their symbols, atomic numbers and relative atomic masses. Values for elements with neither a stable isotope nor a characteristic isotope composition are shown in brackets. They represent the most stable known isotopes. Relative atomic masses of many elements are not invariant but depend on the origin and treatment of the material. The relative atomic mass values in the fourth column are considered reliable to ± 1 in the last digit, unless otherwise stated. Relative atomic masses approximated to three significant figures are shown in the fifth column. These approximate values will be suitable for most calculations encountered by the student.

Table C.1 The elements

Element	Symbol	Atomic number	Relative atomic mass	Approximate relative atomic mass
actinium	Ac	89		(227)
aluminium	Al	13	26.98154	27.0
americium	Am	95		(243)
antimony	Sb	51	121.75±3	122
argon	Ar	18	39.948	39.9
arsenic	As	33	74.9216	74.9
astatine	At	85		(210)
barium	Ba	56	137.33	137
berkelium	Bk	97		(249)
beryllium	Be	4	9.01218	9.01
bismuth	Bi	83	208.9804	209
boron	B	5	10.811±5	10.8
bromine	Br	35	79.909	79.9
cadmium	Cd	48	112.41	112
caesium	Cs	55	132.9054	133
calcium	Ca	20	40.08	40.1
californium	Cf	98		(251)
carbon	C	6	12.011	12.0
cerium	Ce	58	140.115±4	140
chlorine	Cl	17	35.453	35.5
chromium	Cr	24	51.996	52.0
cobalt	Co	27	58.9332	58.9
copper	Cu	29	63.546±3	63.5
curium	Cm	96		(247)
dysprosium	Dy	66	162.50±3	162
einsteinium	Es	99		(254)
erbium	Er	68	167.26±3	167
europium	Eu	63	151.96	152
fermium	Fm	100		(253)
fluorine	F	9	18.998403	19.0
francium	Fr	87		(223)
gadolinium	Gd	64	157.25±3	157
gallium	Ga	31	69.723±4	69.7
germanium	Ge	32	72.61±2	72.6
gold	Au	79	196.9665	197
hafnium	Hf	72	178.49±2	178
helium	He	2	4.00260	4.00
holmium	Ho	67	164.9303	165
hydrogen	H	1	1.00794±7	1.01
indium	In	49	114.82	115
iodine	I	53	126.9045	127
iridium	Ir	77	192.22±3	192
iron	Fe	26	55.847±3	55.8
krypton	Kr	36	83.80	83.8
lanthanum	La	57	138.905	139
lawrencium	Lw	103		(257)
lead	Pb	82	207.2	207

Element	Symbol	Atomic number	Relative atomic mass	Approximate relative atomic mass
lithium	Li	3	6.941±2	6.94
lutetium	Lu	71	174.967	175
magnesium	Mg	12	24.305	24.3
manganese	Mn	25	54.9380	54.9
mendelevium	Md	101		(256)
mercury	Hg	80	200.59±3	201
molybdenum	Mo	42	95.94	95.9
neodymium	Nd	60	144.24±3	144
neon	Ne	10	20.1797±6	20.2
neptunium	Np	93		(237)
nickel	Ni	28	58.69	58.7
niobium	Nb	41	92.9064	92.9
nitrogen	N	7	14.0067	14.0
nobelium	No	102		(253)
osmium	Os	76	190.2	190
oxygen	O	8	15.9994±3	16.0
palladium	Pd	46	106.42	106
phosphorus	P	15	30.9376	31.0
platinum	Pt	78	195.08±3	195
plutonium	Pu	94		(242)
polonium	Po	84		(210)
potassium	K	19	39.0983	39.1
praseodymium	Pr	59	140.9077	141
promethium	Pm	61		(147)
protactinium	Pa	91		(231)
radium	Ra	88		(226)
radon	Rn	86		(222)
rhenium	Re	75	186.207	186
rhodium	Rh	45	102.9055	103
rubidium	Rb	37	85.468	85.5
ruthenium	Ru	44	101.07±2	101
samarium	Sm	62	150.36±3	150
scandium	Sc	21	44.95591	45.0
selenium	Se	34	78.96±3	79.0
silicon	Si	14	28.085	28.1
silver	Ag	47	107.868	108
sodium	Na	11	22.98977	23.0
strontium	Sr	38	87.62	87.6
sulphur	S	16	32.066±6	32.1
tantalum	Ta	73	180.9479	181
technetium	Tc	43		(98.9)
tellurium	Te	52	127.60±3	128
terbium	Tb	65	158.9253	159
thallium	Tl	81	204.383	204
thorium	Th	90	232.0381	232
thulium	Tm	69	168.9342	169
tin	Sn	50	118.71	119
titanium	Ti	22	47.88±3	47.9
tungsten	W	74	183.85±3	184
uranium	U	92	238.0289	238
vanadium	V	23	50.9415	50.9
xenon	Xe	54	131.29±2	131
ytterbium	Yb	70	173.04±3	173
yttrium	Y	39	88.9058	88.9
zinc	Zn	30	65.39±2	65.4
zirconium	Zr	40	91.22	91.2

ANSWERS TO NUMERICAL PROBLEMS

Chapter 1

4. (i) $(TCDD)^+$: 320, 322, 324, 326, 328. $(DDE)^+$: 316, 318, 320, 322, 324
5. (b) 72.5% ^{63}Cu, 27.5% ^{65}Cu
   (c) 220, 222, 224, 226. Relative heights: 3 : 4 : 4 : 1
6. (c) 17% (Californian), 19% (Italian)
9. (c) (iii) 24 days
10. (e) 37 625
12. (b) 63 years

Chapter 2

3. (c) $-159\ kJ\ mol^{-1}$

Chapter 3

1. (c) (ii) 209
2. 0.25 atm (O_2), 0.75 atm (N_2)
3. (c) (i) 102
4. (b) 1.027 : 1
6. (b) (i) $0.205\ J\ g^{-1}\ K^{-1}$
7. (a) (ii) 137.6
8. (c) 6.05×10^{23}

Chapter 4

2. +2; +4; −3; +5; +3
5. (b) $125\ cm^3$ (c) $75\ cm^3$
6. +4 (A), +2 (B)
7. (a) (i) 10^{-2} mol (ii) 0.18 g   (b) (i) 10^{-2} mol
   (ii) 0.96 g   (c) (i) 5×10^{-3} mol (ii) 0.32 g
   (d) (i) 0.54 g (ii) 3×10^{-2} mol

Chapter 5

1. (c) $-76\ kJ\ mol^{-1}$
2. (b) 39.3% (propane), 60.7% (butane)   (c) $+60\ kJ\ mol^{-1}$
3. (a) (ii) $-204\ kJ\ mol^{-1}$ (b) (ii) $+431\ kJ\ mol^{-1}$ (C−C),
   $+416\ kJ\ mol^{-1}$ (C−H) (iii) $+346\ kJ\ mol^{-1}$
5. (b) (ii) $-32.8\ kJ\ mol^{-1}$ (c) (ii) $+23.3\ kJ\ mol^{-1}$
   (d) $+39.7\ kJ\ mol^{-1}$

6. (c) -104.1 kJ mol^{-1}
8. (b) -124 kJ mol^{-1} (d) -248 kJ mol^{-1}

Chapter 7

1. (d) 54.7
2. (b) (ii) 62.3 kPa (iii) 17.8%
4. (a) (i) 0.995 kPa (ii) 0.233 kPa
5. (c) 0.43 mol (ethanoic acid)
6. (c) (ii) N_2: 1/10, 2×10^6 Pa; H_2: 3/10, 6×10^6 Pa; NH_3: 6/10, 1.2×10^7 Pa (iii) 3.3×10^{-13}
7. (b) (i) 43.9% (iii) 1.5×10^5 Pa

Chapter 8

3. (c) (i) 2 (ii) 3.4 (iii) 5.8
4. (c) (i) 2 (ii) 12 (iii) 7 (iv) 2.5 (d) 8.2 g
5. (a) 2 (c) 0.02 mol dm^{-3} (d) 4×10 mol^3 dm^{-9}
6. (b) (i) 5.6×10^{-10} mol dm^{-3} (ii) 3.02
7. (d) (i) 1 (ii) 13.3 (iii) 2.5 (iv) 2 (f) (ii) 9.2
8. (i) 0.05 mol dm^{-3} (ii) 75.0 cm^3
9. 2.4
10. (e) 3.9
11. (a) 1.3×10^{-5} (AgCl), 1.0×10^{-4} (Ag$_2$CrO$_4$)
(b) 4×10^{-3} mol dm^{-3}, 0.68 g (c) 9.6×10^{-3} g
12. (c) (i) 4.31×10^{-2} mol dm^{-3}
(ii) 8.62×10^{-2} mol dm^{-3} (iii) 3.2×10^{-4} mol^3 dm^{-9}
(iv) 2.8×10^{-2} mol dm^{-3} (v) 84.6 cm^3
13. (a) (i) 10 dm^3
(ii) $[Ba^{2+}][SO_4^{2-}] = 1.25 \times 10^{-5}$ mol^2 dm^{-6} ($> k_{sp}$)

Chapter 9

2. (b) (ii) 3.06×10^{-3} mol dm^{-3} s^{-1} (ii) $\times$ 1/9
(iii) $\times$ 4

Chapter 10

1. (c) before 3.11, after 4.95
2. (b) (i) 2.06×10^5 cm^{-1} (iii) 389.5 Ω^{-1} cm^2 mol^{-1}
3. (d) (ii) 2×10^{-3} mol (iii) 6×10^{-3} mol
6. (c) (i) 0.18 g (ii) M^{2+} (iii) 6.011×10^{23}
7. (b) 8.02 g
9. (d) 0.42 g excess of KMnO$_4$
14. (c) (i) 0.10 mol (ii) Fe(III)

Chapter 11

4. (b) (i) -333 kJ mol^{-1}

Chapter 12

3. (a) (ii) -1470 kJ mol^{-1} (b) (i) -1090 kJ mol^{-1}
5. (e) (i) 1.43 g

Chapter 13

2. (d) (i) 250 cm^3
9. (c) (ii) 0.762 g
12. (c) (ii) 50%
14. (c) (ii) 1.0 tonne

Chapter 14

2. (c) 55.8%
5. (b) 96.0%
7. (d) 0.83 mol dm^{-3} (Cr$_2$O$_7^{2-}$), 1.33 mol dm^{-3} (Cr^{+3})
10. (a) (iii) 31.3%

Chapter 16

10. (b) −204 kJ mol^{-1}

Chapter 18

1. (b) 25 cm^3 (c) 75 cm^3
2. (b) (i) 1.12 dm^3

Chapter 20

9. (c) (v) 3.55 kg

Chapter 14

Chapter 16

Chapter 19

Chapter 20

INDEX

transistor 556
transition, of electron 13
transition element 496
transition metal complexes 493
transition state 325, 668
transition state theory 325–326
transition temperature 111, 208
translation 826
transmutation 31
transpiration 434
transuranium elements 33
triads 398
  law of 398
tribromomethane 762
2,4,6-tribromophenol 748
2,4,6-tribromophenylamine 787
tricarboxylic acid (TCA) cycle 122
trichloroethanal 762
trichloroethane 617, 730
trichloroethene 734
trichloromethane 730, 762
(trichloromethyl)benzene 704
2,4,6-trichlorophenol (TCP) 753
tridymite 111
trienes 687
triglyceride ester, hydrolysis 830
triglycerides 828, 830–833
trigonal bipyramidal shapes 75
trigonal planar shapes 75–76
trihydric alcohols 741
triiodide 603
triiodomethane 730, 745
trimethylamine 782–783
trimethylbenzene 696
2,2,4-trimethylpentane 714
2,4,6-trinitrophenol 749, 751
trinitrotoluene, see TNT
trioxygen 580, 691
tripeptide 807
triphenylamine 782
triple point 207–208
triple superphosphate 484
triplet code 827
trisodium phosphate(V) 576
tritium 424
tritium bomb 47
tRNA, see transfer RNA
tropical diseases 629
Trouton's rule 158
trypanosomiasis 34
tsetse fly 34
tungsten 426, 498
Turnbull's blue 508
turpentine 601
Tyndall effect 235
tyrosine 805

U

ultra-violet light 15
ultra-violet visible spectroscopy 651–652
ultra–violet wavelength 13
ultracentrifugation 808
unit cell 105
universal indicator 284
universal solvent 432
Universe 422
  hydrogen in 428
unnilhexium 402
unnilpentium 402

unnilquadium 402
unsaturated compounds 680
  reduction 426
unsaturation, degree 831
uranium 25, 38
  hexafluorides 96
uranium–lead method 37
uranium minerals 30
uranium salts, separation 223
uranium-235 41
uranium-238 33
uranium-238 series 30
urea 634, 774, 776
  hydrolysis 335, 776
  manufacture 778
  uses 778–779
urease 776
useful work 182

V

vacuum distillation 710
valence band 68–69
valence electrons 57, 68
valence shell 57, 60
valence-shell electron-pair repulsion 74–76
valency 57, 130–131, 396
  electronic theory 57
  of metals 131
  multiple 131
  of non-metals 131
valine 805
van der Waals constants 98
van der Waals equation 98
van der Waals force 58, 70–71, 81
van der Waals radius 73–74
  of hydrogen 423
van't Hoff equation 233
vanadium 497
vanadium(V) oxide 523
  as catalyst 262, 322
vanillin 829
vaporisation methods 92
vapour pressure 100, 220
  lowering 228–229
  of solids 112
  of water 100
vapour pressure curve 208
  for ice 207
vegetable oils 191
velocity constant 314
Venus 553
Victor Meyer's method 93–94
vinyl chloride 718, 730, 841
viscosity 848
  of liquids 101
visible wavelength 13
vitamin A 836
vitamin B complex 836
vitamin C 836
vitamin D 837
vitamins 836–837
vitrinites 679
volatility 77
volcano 473, 585
volcano reaction 568
volt 348
volume 88
volumetric analysis 299, 505
von Laue, Max 104
vulcanisation 845

W

Wacker process 762
warfarin 619
washing soda 478
waste, solid 437
water 79–80
  acid–base reactions 433
  amphoteric nature 433
  brackish 436
  chemical reactions 432–434
  of crystallisation 433, 460, 467
  de-ionised 300
  drop 101
  hard 299
  and hydrogen 419–443
  ionic product 280–281
  oxidation 433
  phase diagram 206
  physical properties 432
  pK_a values 742
  pollution 437–441
  quality 437
  reduction 433
  self-ionisation 280
  softening 442
  structure 432
  uses 434–436
water condenser 646
water cycle 434
water gas 429, 709
water glass 479
water hyacinth 632
water softening 299–300
water treatment 441–443, 560
  adsorption 442
  aeration 441
  chemical 442–443
  coagulation 442
  desalination 442
  disinfection 442
  filtration 441
  flocculation 441
  oxidation 442
  screening 441
  sedimentation 441
Watson, James 825
wave equation 6
wave numbers 650
wave theory 20
wave–particle paradox 6
waxes 834
weedkiller 753
weight 124
Werner, Alfred 494
Wheatstone bridge circuit 349
whisky 201–202
Wilkins, Maurice 825
Williamson's synthesis 731–732, 750
Wilson cloud chamber 27
Wilson, C. T. R. 27
wine 338
witherite 473
  calcined 473
Wöhler, Friedrich 634
wood 148, 196, 847
work 153, 382
wrought iron 518

X

X-ray crystallography 103–105, 495
X-ray diffraction 74